Basic Control Engineering

B YOUSEFZADEH

MTech, BSc, CEng, MIEE,
MIERE, MBIM, FRSA

Formerly Lecturer in Control
and Computer Engineering
Barnet College England

Pitman

PITMAN PUBLISHING LIMITED
128 Long Acre London WC2E 9AN

Associated Companies
Pitman Publishing New Zealand Ltd, Wellington
Pitman Publishing Pty Ltd, Melbourne

First published in Great Britain 1979
Reprinted with corrections 1982
Reprinted 1983

Camera copy prepared by Morgan-Westley MW

ISBN 0 273 01035 2

Printed and Bound in Great Britain
at The Pitman Press, Bath

Preface

In recent years, the subject of control has become an integral part of almost every technical and scientific course of study and in particular of all branches of engineering. The applications of control systems cover a very wide scope, ranging from the design of simple appliances used in the kitchen to the study of the physiology of the human body.

Most books in the field of control engineering have, in the main, been either highly advanced, using high levels of mathematics, or largely descriptive, leaving out some essential details. An attempt has been made here to write a book which is rich in explanation and yet contains sufficient mathematics so that the essential details are easily analysed and understood.

This book is intended to fulfil the needs of students studying engineering courses in which basic control is an integral part. The contents have been carefully chosen to cover the syllabus of a wide range of existing courses. However, it is hoped that this is the first book of its kind which will be suitable for the new Higher Technician Certificates and Diplomas.

Although the text is written in an integrated form and should be studied from the first chapter to the last, a student with some background knowledge of control engineering may choose any chapter and follow it through without any need to refer to other chapters.

Since this is an elementary introduction to control engineering, all examples are considered to be linear with constant parameters.

In many places a basic knowledge of fundamental electrical and mechanical principles is assumed, although in some cases the necessary explanation is given. A basic knowledge of complex numbers and differential calculus is also assumed. The idea of the s-plane is explained so that time-domain analysis can be followed. Although differential equations are used in many examples in Chapter 5 and 6, no attempt is made to solve these equations and therefore one should not assume the use of advanced mathematics because of the appearance of some complicated-looking expressions.

I would like to thank the following examining bodies for granting me permission to make use of some of their past examination questions:

The Council of Engineering Institutions
The City and Guilds of London Institute
Middlesex Polytechnic
Polytechnic of Central London
Barnet College
Mander College

The accuracy of all solutions is, of course, solely my responsibility.

Finally, I wish to thank my wife Joyce for helping with the typing of the manuscript and for her patience while the book was being prepared.

Bijan Yousefzadeh

Dedicated to my dearest children,
Paul-Darius and Bettina Ziva

Contents

1 Introduction to Control

Feedback control is now a fundamental feature of modern industry and present-day technological society, simply because, in order to utilize natural resources and forces of nature efficiently, and to use energy purposefully and for the benefit of mankind, some form of control is needed. This chapter introduces the basic idea of feedback, together with the manipulation of feedback systems using transfer functions, block diagram algebra, and differential equations.

1.1 INTRODUCTION

Control engineering is, then, primarily concerned with understanding and controlling natural resources and forces of nature purposefully and for the benefit of mankind. That is, it is concerned with the design and development of machines and equipment by which man can utilize power.

Early man relied upon his strength and that of animals to supply energy for doing work. He then supplemented his energy and that of animals by utilizing power from natural sources, such as the wind for powering windmills for example, and waterfalls for turning water wheels. By using simple mechanical devices such as wheels and levers, he accomplished such major tasks as the building of the pyramids and the construction of the Roman cities.

Early machines and equipment used for control were primarily of a manually operated nature requiring frequent adjustments and resets in order to maintain and/or achieve the desired performance. In 1769 James Watt developed and used his flyball governer for controlling the speed of a steam engine. It is generally agreed that this is the first automatic feedback controller used in an industrial process.

Control systems, defined and discussed in detail in Chapter 9, can generally be classified into two groups depending on whether they use feedback or not. Feedback control systems themselves can be divided into two broad categories: regulator systems and follow-up systems.

The parameter of the system which it is required to control (temperature, velocity, position, etc.) is called the controlled variable. The basic element of the system (e.g. steam engine) is often referred to as the plant, and the input to the system is called the reference variable.

A REGULATOR SYSTEM is a system whose main function is to maintain the controlled variable constant. In these systems, the reference variable is not changed frequently, and under normal operating conditions a finite error exists. This is necessary to generate the power which is continuously required to produce the output.

A simple example of a regulating system is the central heating system of a house. Such a system employs feedback and, although it will respond to a change in the thermostat setting, its main function is to maintain the desired room temperature despite changes in outdoor temperature.

A FOLLOW-UP SYSTEM is a feedback control system whose prime function is to keep the controlled variable in close correspondence with a reference variable which is frequently, or continuously, changed. Servomechanisms are typical examples of follow-up systems and are discussed in depth in the following chapters.

1.2 FEEDBACK

The operation of a system may be controlled externally (by an operator) or automatically (by the system itself). When the control action of a system is independent of the output, the system is said to be an OPEN-LOOP control system. However, if the control action is somehow dependent on the output, the system is called a CLOSED-LOOP or FEEDBACK control system.

Figure 1.1 shows an open-loop and a closed-loop system. In a closed-loop system the output is fed back and compared with the desired input so that any necessary corrective action may be taken. The open-loop system, however, relies for its action on the settings of its components.

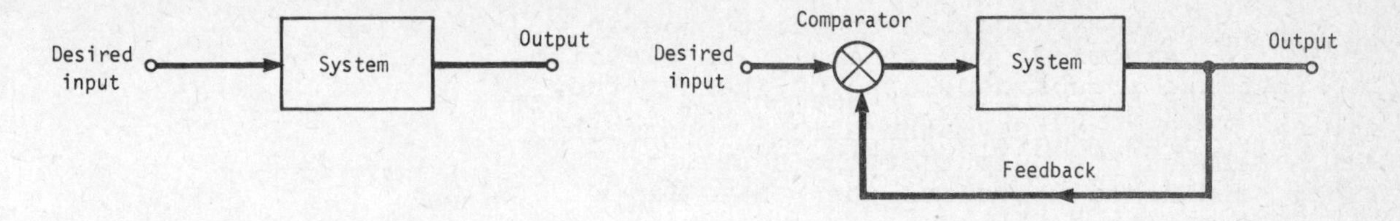

(a) OPEN-LOOP SYSTEM (b) CLOSED-LOOP SYSTEM

Fig.1.1

An example of the action of an open-loop system is a car-wash machine in which all cars receive the same amount of washing irrespective of how dirty they are. Here, the output is the cleanliness of the cars, which corresponds to a given setting of the machine (amount of water and washing time). A human being however, who washes cars, automatically makes sure that the dirtier cars receive more attention than others. He is therefore a closed-loop control system.

Other simple examples include the lighting of a room, a refrigerator, and a traffic-light system. The lighting of a room is an open-loop system. Once the light is turned on it will stay on (until it is switched off, of course) irrespective of whether the room is dark or light. So, as given by the definition, the action depends on the setting of the switch.

A refrigerator is a closed-loop system. Its temperature is measured by a thermostat which turns the motor ON when the temperature falls below the desired value and turns it OFF when it reaches the desired value.

Traffic-light systems vary in operation. Some systems are closed-loop and some are open-loop. In the open-loop type there is a timing mechanism which is set to switch the lights at regular intervals, irrespective of the volume of traffic. In the closed-loop type however, the amount of traffic passing through the junctions is monitored (electronically or otherwise) and the duration of the "red" and "green" conditions are adjusted for each part of the junction accordingly.

The features of open-loop and closed-loop systems may be summarized as follows:

OPEN-LOOP:	CLOSED-LOOP:
(1) They are simple.	(1) They are highly accurate.
(2) Their accuracy is determined by the calibration of their elements.	(2) They are more complex.
(3) They are not generally troubled with instability.	(3) Non-linearities and distortion are greatly reduced.
	(4) They have wide bandwidth.
	(5) They suffer from instability problems.

1.3 TRANSFER FUNCTION

The TRANSFER FUNCTION of an element or a system may be defined as the mathematical expression which relates the output of the system to its input, and hence describes the behaviour of the system. Thus,

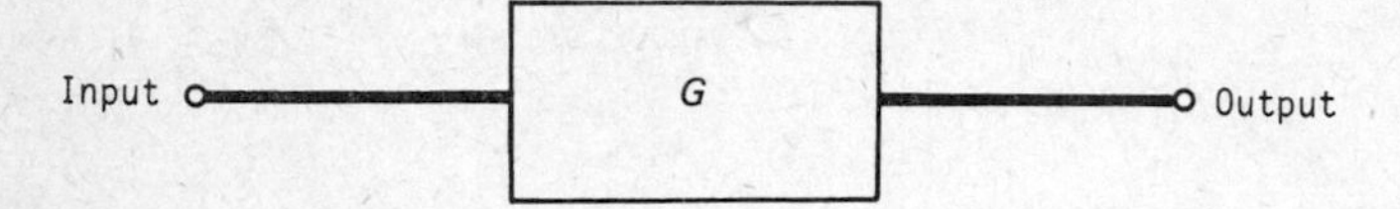

Output = G × input
where G is the transfer function

Consider the simple circuit shown in Figure 1.2. Using Kirchhoff's Law, the input voltage will be given by:

$$v_i = iR_1 + iR_2$$

i.e.

$$i = \frac{v_i}{R_1 + R_2}$$

also

$$v_o = iR_2$$

Substituting for i

$$v_o = \frac{R_2}{R_1 + R_2} v_i$$

Thus

$$\frac{v_o}{v_i} = \frac{R_2}{R_1 + R_2}$$

The transfer function of the circuit is therefore

$$\frac{R_2}{R_1 + R_2}$$

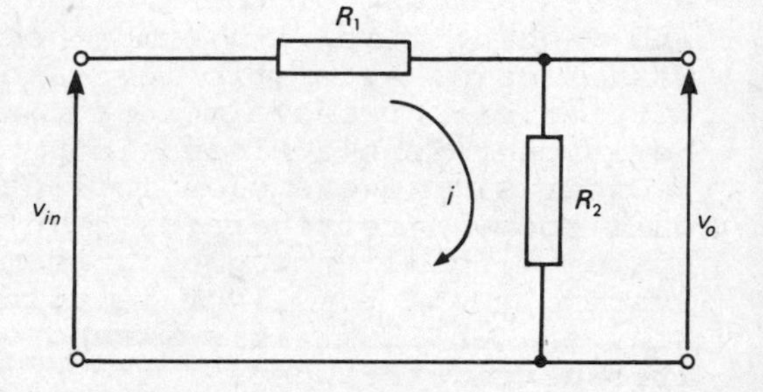

Fig.1.2

Example 1.1

Obtain the transfer function of the circuit shown in Figure 1.3.

Solution

$$v_i = iR + i\frac{1}{j\omega C} \quad \text{so} \quad i = \frac{v_i}{R + 1/j\omega C}$$

Also

$$v_o = i\frac{1}{j\omega C}$$

Substituting for i

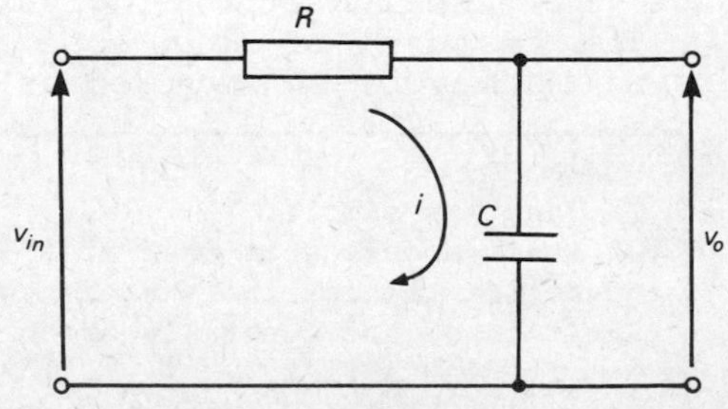

Fig.1.3

$$v_o = \frac{v_i}{R + 1/j\omega C} \times \frac{1}{j\omega C}$$

Thus

$$\frac{v_o}{v_i} = \frac{1}{j\omega RC + 1} \qquad (1.1)$$

Also

$$\frac{v_o}{v_i} = \frac{1/RC}{j\omega + 1/RC} \qquad (1.2)$$

Both equation (1.1) and (1.2) represent the transfer function of the circuit of Figure 1.3.

Example 1.2

Determine the transfer function of the network shown in Figure 1.4.

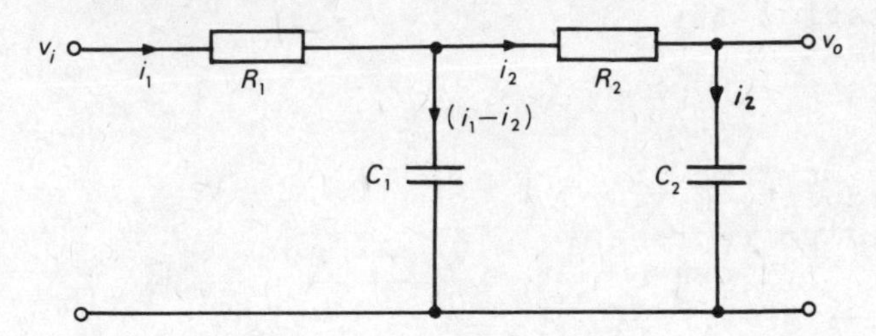

Fig.1.4

Solution

$$v_i = i_1R_1 + (i_1 - i_2) \times 1/j\omega C_1 \qquad \text{(i)}$$

For the second loop

$$i_2R_2 + (i_2 - i_1) \times 1/j\omega C_1$$
$$+ i_2 \times 1/j\omega C_2 = 0 \qquad \text{(ii)}$$

and $v_o = i_2 \times 1/j\omega C_2$ (iii)

or $i_2 = j\omega C_2 v_o$ (iv)

Rearranging (i)

$$v_i = (R_1 + 1/j\omega C_1)i_1 - (1/j\omega C_1)i_2 \qquad \text{(v)}$$

Rearranging (ii)

$$i_1 = j\omega C_1(R_2 + 1/j\omega C_1 + 1/j\omega C_2)i_2$$

Substituting for i_1 in (v)

$$v_i = j\omega C_1(R_2 + 1/j\omega C_1 + 1/j\omega C_2)$$
$$\times (R_1 + 1/j\omega C_1)i_2 - (1/j\omega C_1)i_2$$

Substituting for i_2 from (iv)

$$v_i = j\omega C_1(R_2 + 1/j\omega C_1 + 1/j\omega C_2)$$
$$\times\ (R_1 + 1/j\omega C_1)j\omega C_2 v_o - (1/j\omega C_1)j\omega C_2 v_o$$

Simplifying and rearranging

$$v_i = [(j\omega)^2 R_1R_2C_1C_2 + j\omega R_1C_1 + j\omega R_2C_2 + j\omega R_1C_2 + 1]v_o$$

The transfer function is therefore given by

$$\frac{v_o}{v_i} = \frac{1}{R_1R_2C_1C_2(j\omega)^2 + (R_1C_1 + R_2C_2 + R_1C_2)j\omega + 1}$$

1.4 BLOCK DIAGRAM ALGEBRA

In Chapters 5, 6 and 7, block diagrams are used as they are a convenient representation of system components. They can be used for analysis and often simplify the treatment of a system.

Once the transfer function of an element is determined, the details of its circuitry may be ignored and a block will be used to represent the element. The circuit of Figure 1.2 can therefore be represented in block diagram form as

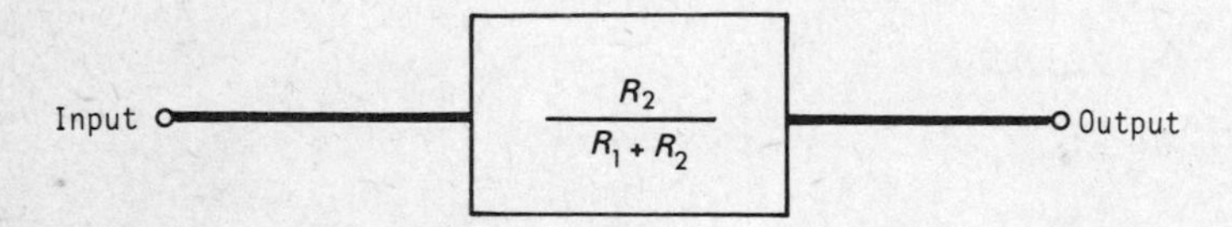

1.4.1 BLOCKS IN CASCADE

As shown in Figure 1.5, several blocks in cascade can be replaced by one block whose transfer function is the product of the individual blocks.

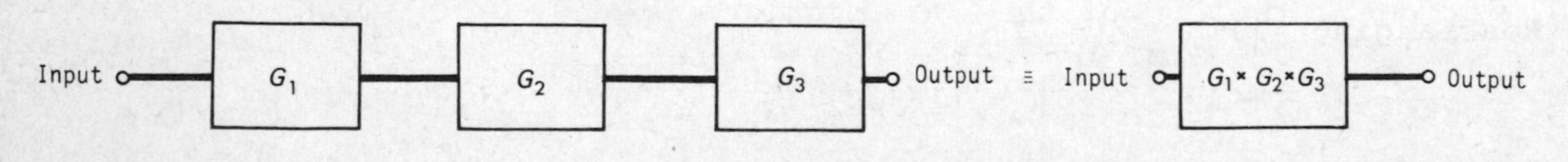

Fig.1.5 Blocks in cascade

1.4.2 SUMMING JUNCTION

A summing junction, shown in Figure 1.6, produces the algebraic sum of the quantities applied to it. A summing junction would also be called a comparator if its output is the difference between its two inputs.

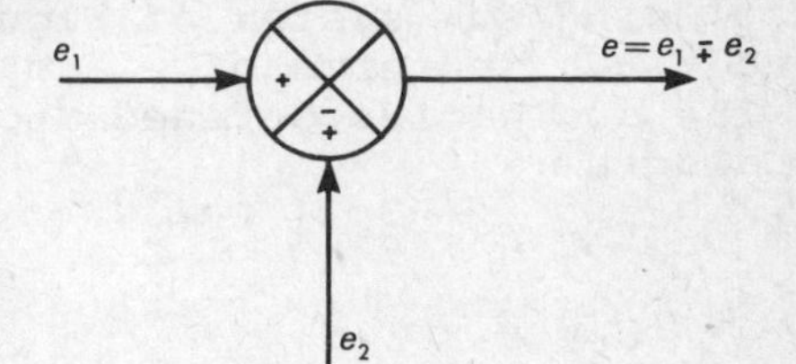

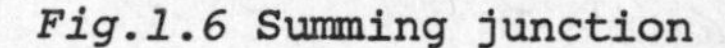

Fig.1.6 Summing junction

1.4.3 CANONICAL FORM

The simplest form of a closed-loop system is shown in Figure 1.7 in which the forward element has a transfer function *G* and the feedback element has a transfer function *H*.

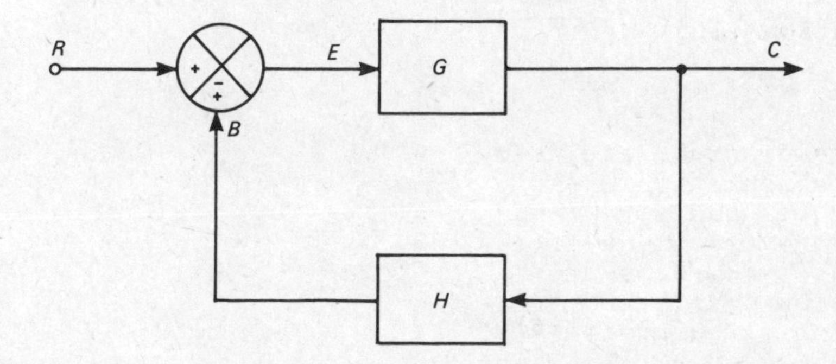

Fig.1.7 Canonical representation

Complicated systems may be simplified to this canonical form from which the following definitions are made:

Open-loop transfer function	$G \times H$
Closed-loop transfer function (control ratio)	C/R
Error ratio	E/R
Primary feedback ratio	B/R

A system in which $H = 1$ is called a unity feedback system.

A closed-loop system may employ positive or negative feedback. In the simple system of Figure 1.7, if the output of the summing junction is the sum of the two signals applied to it, then the system is said to use positive feedback. If however the signal fed back from the output of the system is subtracted from the input signal, the output of the summing junction will indicate the difference between the two quantities and negative feedback is said to operate.

Using the system of Figure 1.7, we will now derive, in terms of G and H, expressions for the quantities defined above. From the diagram:

$$E = R \mp B \tag{1.3}$$

$$C = G \times E \tag{1.4}$$

$$B = H \times C \tag{1.5}$$

Substituting for E in equation (1.4)

$$C = G(R \mp B) = GR \mp GB$$

Substituting for B from equation (1.5)

$$C = GR \mp GHC$$

$$C \pm GHC = GR$$

$$C(1 \pm GH) = GR$$

Thus

$$\frac{C}{R} = \frac{G}{1 \pm GH} \tag{1.6}$$

Substituting for B in equation (1.3)

$$E = R \mp HC$$

Substituting for C from equation (1.4)

$$E = R \mp HGE$$

$$E \pm HGE = R$$

$$E(1 \pm GH) = R$$

Thus

$$\frac{E}{R} = \frac{1}{1 \pm GH} \tag{1.7}$$

Substituting for C in equation (1.5)

$$B = HGE$$

Substituting for E from equation (1.3)

$$B = HG(R \mp B)$$

$$= GHR \mp BGH$$

$$B \pm BGH = GHR$$

$$B(1 \pm GH) = GHR$$

Thus

$$\frac{B}{R} = \frac{GH}{1 \pm GH} \tag{1.8}$$

Example 1.3

By using block diagram algebra reduce the block diagram of Figure 1.8 to its canonical form and hence determine:

(i) *the forward transfer function*
(ii) *the feedback transfer function*
(iii) *the open-loop transfer function*
(iv) *the closed-loop transfer function*
(v) *the error ratio, and*
(vi) *the primary feedback ratio.*

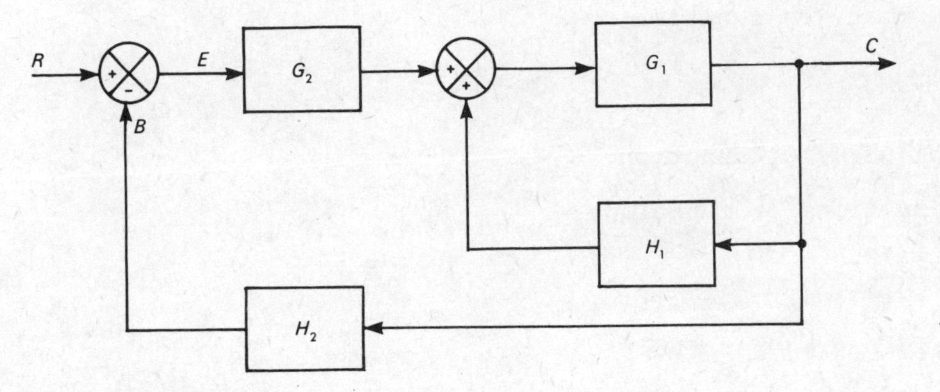

Fig.1.8

Solution

This system contains two feedback loops: one positive (the inner loop) and one negative (the outer loop). Using equation (1.6), the inner loop containing G_1 and H_1 can be replaced by one block whose transfer function is

$$\frac{G_1}{1 - G_1H_1}$$

The diagram reduces to

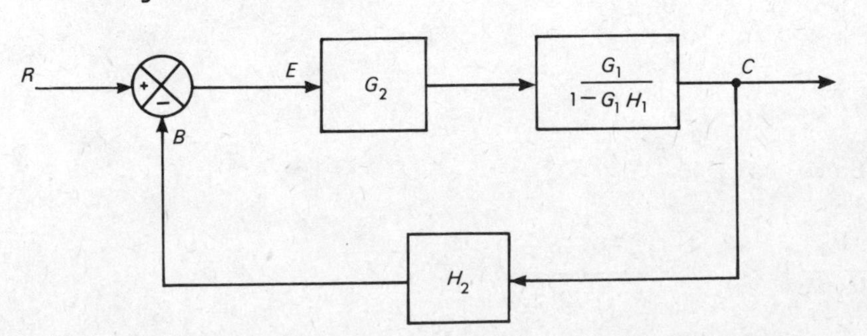

and in canonical form to

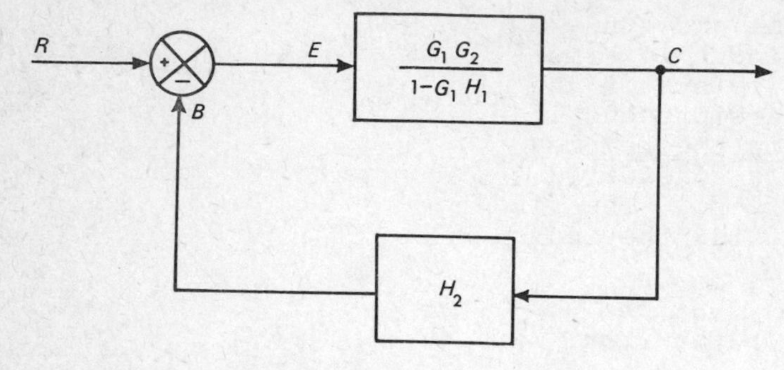

Comparing this with Figure 1.7:

(i) The forward transfer function:

$$G = \frac{G_1 G_2}{1 - G_1 H_1}$$

(ii) The feedback transfer function: $H = H_2$

(iii) The open-loop transfer function:

$$G \times H = \frac{G_1 G_2 H_2}{1 - G_1 H_1}$$

(iv) The closed-loop transfer function:
from equation (1.6),

$$\frac{C}{R} = \frac{G}{1 + GH} = \left(\frac{G_1 G_2}{1 - G_1 H_1}\right) \div \left(1 + \frac{G_1 G_2 H_2}{1 - G_1 H_1}\right)$$

$$= \frac{G_1 G_2}{1 - G_1 H_1 + G_1 G_2 H_2}$$

(v) The error ratio:
from equation (1.7),

$$\frac{E}{R} = \frac{1}{1 + GH} = \frac{1}{1 + G_1 G_2 H_2/(1 - G_1 H_1)}$$

$$= \frac{1 - G_1 H_1}{1 - G_1 H_1 + G_1 G_2 H_2}$$

(vi) The primary feedback ratio:
from equation (1.8),

$$\frac{B}{R} = \frac{GH}{1 + GH} = \left(\frac{G_1 G_2 H_2}{1 - G_1 H_1}\right) \div \left(1 + \frac{G_1 G_2 H_2}{1 - G_1 H_1}\right)$$

$$= \frac{G_1 G_2 H_2}{1 - G_1 H_1 + G_1 G_2 H_2}$$

1.4.4 MULTI-INPUT SYSTEMS

Often it is necessary to analyse and study a system in which more than one input is simultaneously applied at different points of the system. When multiple inputs are present in a linear system, the superposition theorem may be used. That is, each input is considered alone, setting others equal to zero. The final solution is then obtained by adding the individual answers together.

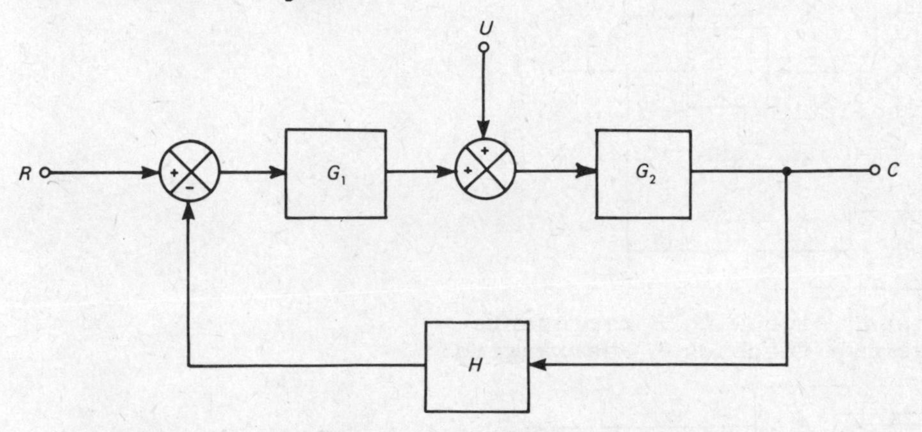

Fig.1.9

Consider the multi-input system of Figure 1.9 in which R is the input and U represents total noise and unwanted signals. Assuming that the system is linear, each input will be considered separately.

Thus, let $U = 0$, and C_R be the output due to R only:

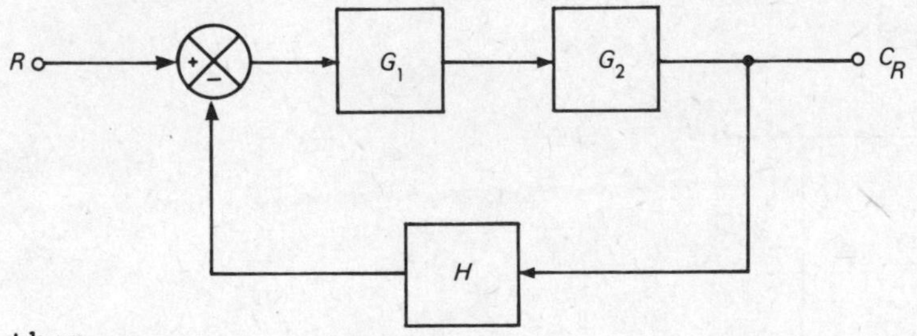

then

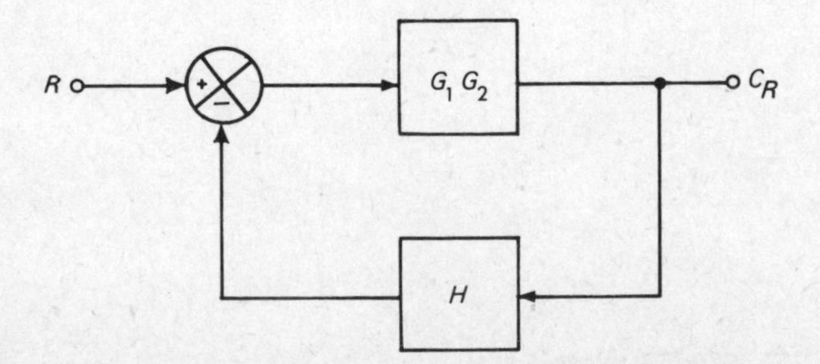

Now, from equation (1.6)

$$C_R = \frac{G_1 G_2}{1 + G_1 G_2 H} R$$

Now let $R = 0$, and C_U be the output due to U only:

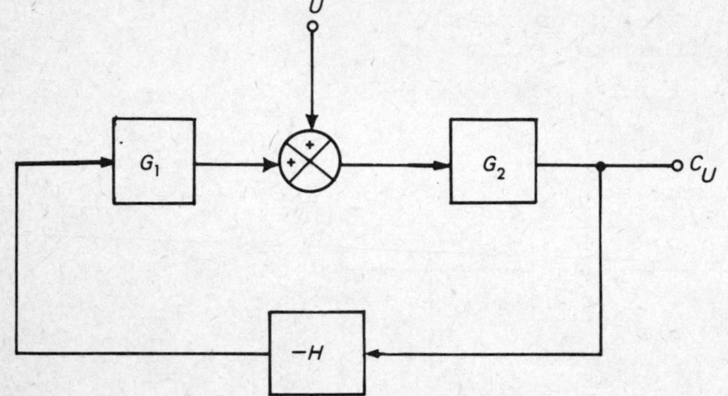

Note that the "minus" sign with H represents the effect of negative feedback. Rearranging:

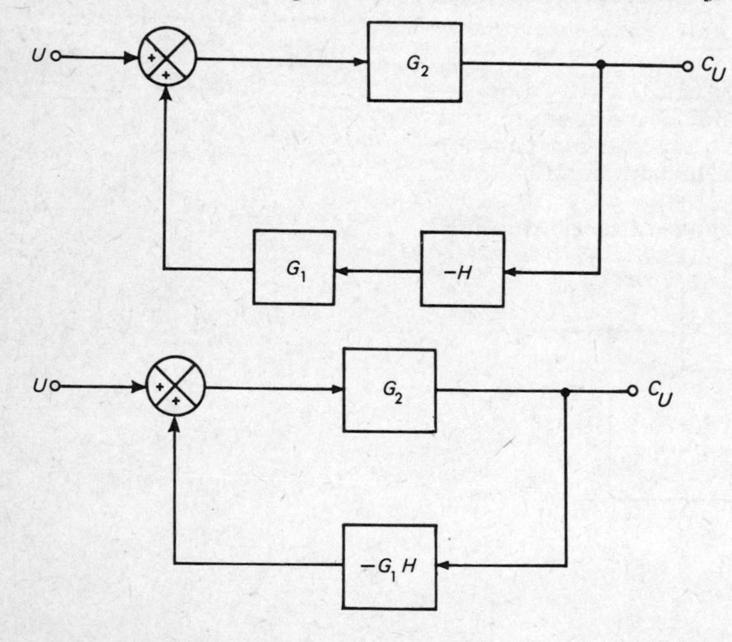

Now, from equation (1.6)

$$C_U = \frac{G_2}{1 + G_1 G_2 H} U$$

Adding the two outputs

$$C = C_R + C_U = \frac{G_1 G_2}{1 + G_1 G_2 H} R + \frac{G_2}{1 + G_1 G_2 H} U$$

Thus

$$C = \left(\frac{G_2}{1 + G_1G_2H}\right)(G_1R + U)$$

1.5 INPUT SIGNALS

Input to control systems may be simple time-varying quantities or complicated and random. In the testing of systems however, the following inputs are very useful:

(i) the step input (ii) ramp input
(iii) acceleration input (iv) sinusoidal input
and (v) the impulse input.

Fig.1.10 **Input signals**

STEP INPUT A step, shown in Figure 1.10(*a*), is a sudden change in the value of a physical quantity, ideally in zero time. Due to practical limitations it is impossible to achieve an *ideal* step input, but provided the rise is very fast compared with the response time of the system, the small rise time can be ignored. A step input is often used to obtain the transient response of a system.

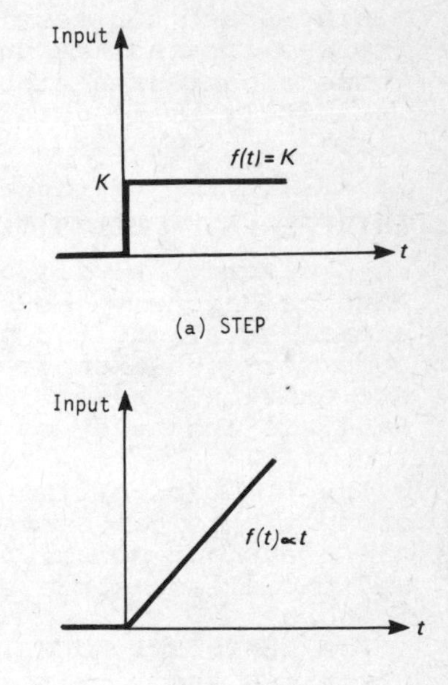

RAMP INPUT As shown in Figure 1.10(*b*), a ramp or constant velocity input is a function whose magnitude increases linearly with time. Rotation of the input shaft at a constant angular velocity, for example, represents a velocity input.

A major problem associated with the ramp input is that under steady-state condition, there is always a difference between the input and output quantities. This error known as velocity error or velocity lag (discussed in Section 5.5 on page 86) is actually a positional error and does not imply a difference between the input and output velocities. A ramp input is often used in testing the steady-state performance of a system.

ACCELERATION INPUT An acceleration or parabolic input is one in which the input control is rotated at constant acceleration. As shown in Figure 1.10(*c*), this function is proportional to t^2.

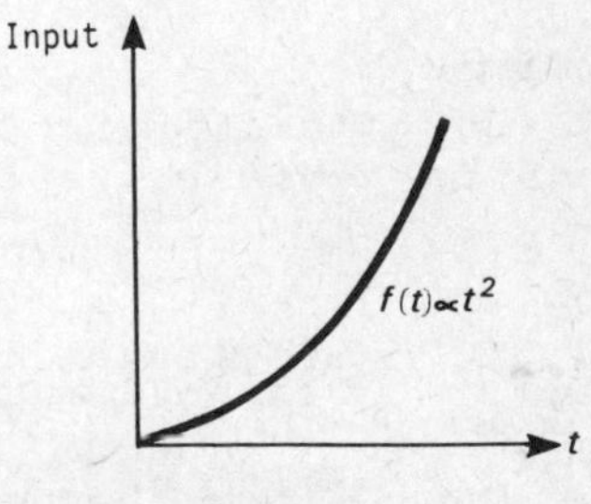

(c) ACCELERATION

SINUSOIDAL INPUT As shown in Figure 1.10(*d*), the input control is subjected to a sinusoidally varying signal. Oscillation of the input shaft at a constant frequency and magnitude represents a sinusoidal input. As described in Chapter 6 Section 6.5, sinusoidal inputs are used when performing frequency response tests.

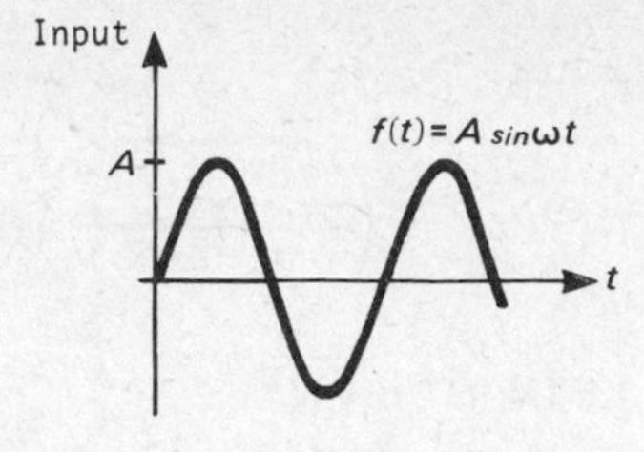

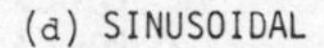
(d) SINUSOIDAL

UNIT-IMPULSE INPUT An *impulse* is a special type of pulse which ideally starts at $t = 0$, has an infinite height, and a duration of zero. In practice, a perfect impulse cannot be achieved and for testing purposes the unit-impulse is used. A unit-impulse is a pulse whose area represents one unit and has infinitely small duration, as shown in Figure 1.10(*e*).

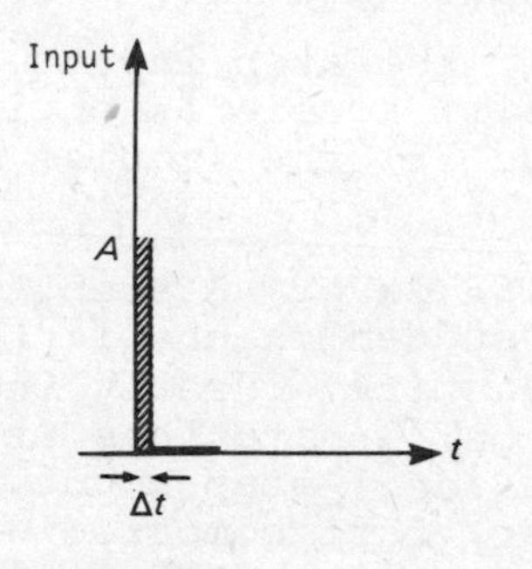

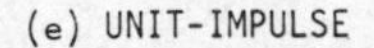
(e) UNIT-IMPULSE

1.6 DIFFERENTIAL EQUATIONS

In Chapter 7 it will be demonstrated how physical systems can be represented by differential equations and simulated on an analogue computer (see Example 7.9). These equations are extremely useful for the mathematical analysis and study of control systems' behaviour.

The solution of the differential equations of most physical systems contains two parts: the transient solution (complementary function) and the steady-state solution (particular integral).

The TRANSIENT SOLUTION is the part which describes the *transient response* of the system and its value approaches zero as time approaches infinity. The transient performance is normally tested with a *unit-step* input.

The STEADY-STATE SOLUTION describes the state of the system once the transients have died away and the system is under steady-state condition.

Example 1.4

Derive the differential equation for the remote position-control system of Figure 1.11.

Solution

Let J = total inertias of the motor and load referred to load side (kg-m^2)

F = viscous friction constant (N-m per rad/s)

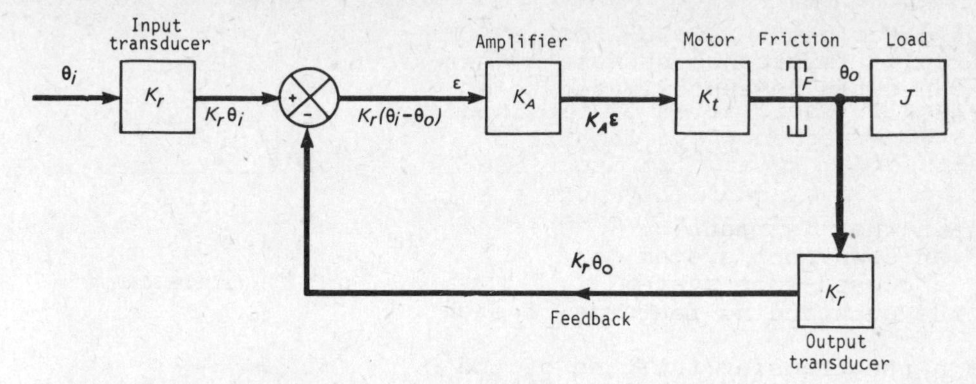

Fig.1.11 Remote position control system

T = driving torque produced by the motor (N-m)

K_A = amplifier transconductance (S)

K_t = motor torque constant (N-m/A)

K_r = transducers and error detector constant (V/rad)

ε = the error voltage (V).

The motor driving torque is proportional to the error voltage. Thus, from the diagram

$$T = K_A K_t \varepsilon \quad \text{and} \quad \varepsilon = K_r(\theta_i - \theta_o)$$

Neglecting any load torque, the driving torque produced by the motor must be sufficient to overcome both the inertia torque and viscous friction torque.

$$\text{Inertia torque} = J\frac{d^2\theta_o}{dt^2}$$

(Inertia × Acceleration)

$$\text{Viscous friction torque} = F\frac{d\theta_o}{dt}$$

(Friction × Angular velocity)

Thus,

$$J\frac{d^2\theta_o}{dt^2} + F\frac{d\theta_o}{dt} = K_A K_t \varepsilon \tag{1.9}$$

$$= K_A K_t K_r(\theta_i - \theta_o)$$

Let $K_A K_t K_r = K$ (N-m/rad); then, rearranging

$$J\frac{d^2\theta_o}{dt^2} + F\frac{d\theta_o}{dt} + K\theta_o = K\theta_i \tag{1.10}$$

Equation (1.10) describes a SECOND-ORDER SYSTEM and is of particular importance because higher-order systems can often be approximated to a 2nd order system.

EXERCISES

(1.1) (*a*) Explain what is meant by
(i) an open-loop system, and
(ii) a closed-loop system
listing the outstanding features of each system.
(*b*) Obtain the transfer function of the circuit shown in Figure 1.12. Draw block diagrams to represent (i) the open-loop condition, and (ii) the closed-loop condition.

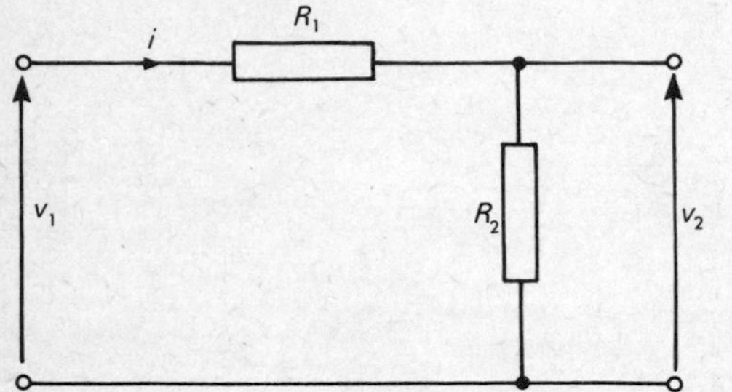

Fig.1.12

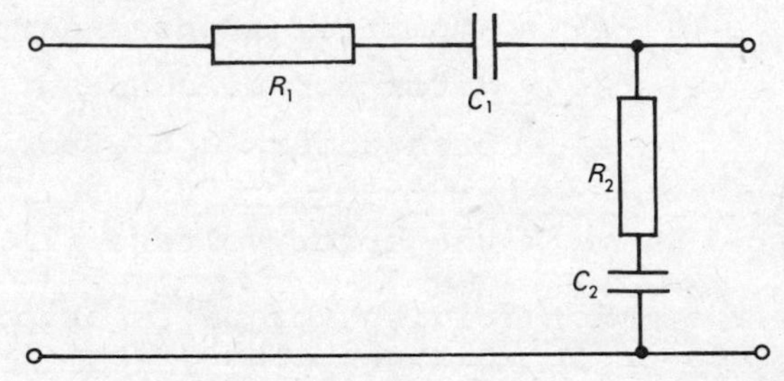

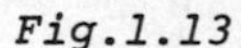

Fig.1.13

(1.2) Show that the transfer function of the phase-lead circuit of Figure 6.34 is given by

$$\frac{j\omega + 1/R_1C}{j\omega + (1/R_1C + 1/R_2C)}$$

(1.3) Show that the transfer function of the network shown in Figure 1.13 is given by

$$\frac{R_2C_2 + 1/j\omega}{(R_1 + R_2)C_2 + C_2(1/C_1 + 1/C_2)/j\omega}$$

(1.4) Determine the transfer function of the network shown in Figure 1.14.

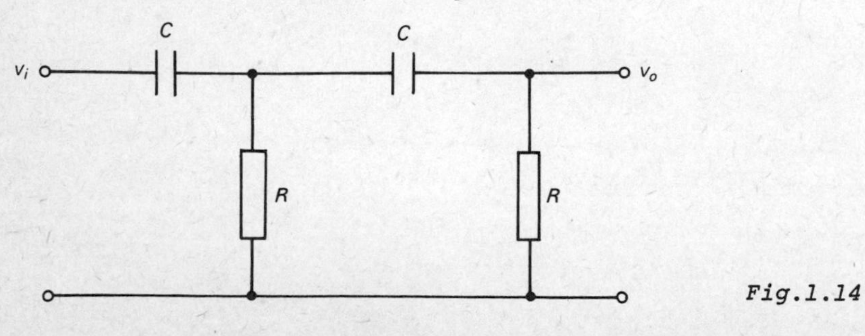

Fig.1.14

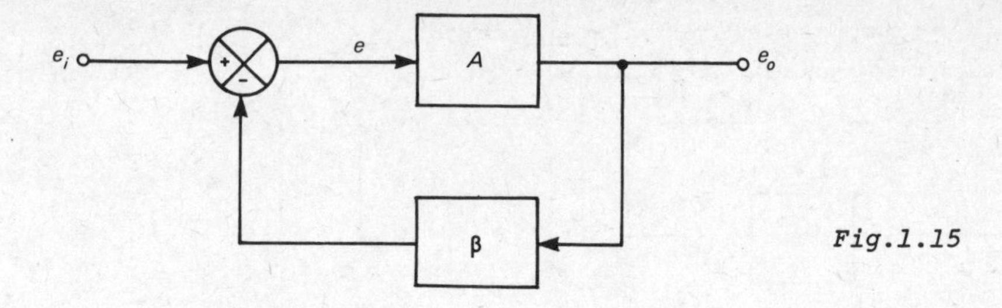

Fig.1.15

(1.5) In Figure 1.15, assuming negative feedback, derive the following functions:

(a) $\dfrac{e_o}{e_i} = \dfrac{A}{1 + A\beta}$ (b) $\dfrac{e}{e_i} = \dfrac{1}{1 + A\beta}$

The voltage gain of an amplifier, using negative feedback, is 25. Without feedback the amplifier gain can be considered infinite. Determine the feedback fraction β.

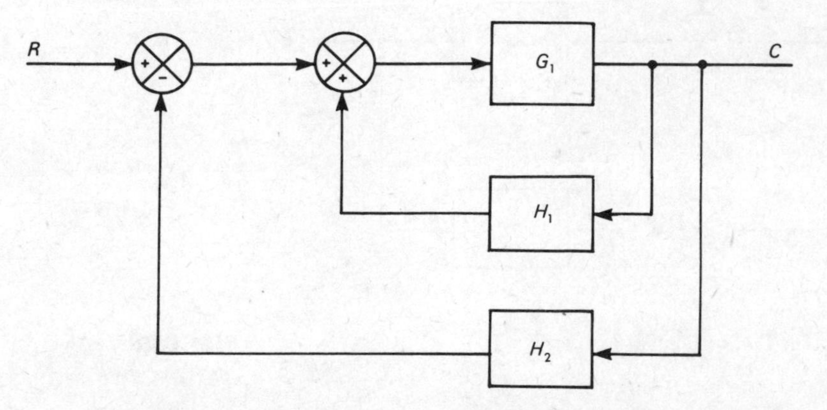

Fig.1.16

(1.6) Explain what is meant by the term "control ratio" as referred to control systems. Hence obtain an expression for the control ratio in terms of the forward transfer function G and the feedback transfer function H.

The block diagram of a system is given in Figure 1.16. Using block diagram algebra, show that the control ratio is given by

$$\frac{G_1}{1 + G_1(H_2 - H_1)}$$

(1.7) Reduce the block diagram of Figure 1.17 to canonical form and hence obtain an expression for the closed-loop transfer function.

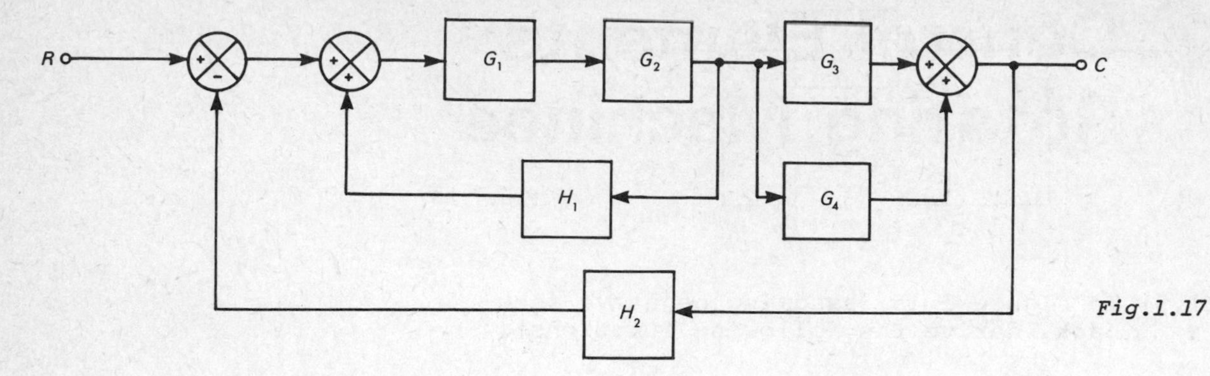

Fig.1.17

(1.8) Determine an expression for the output C of the linear system shown in Figure 1.18.

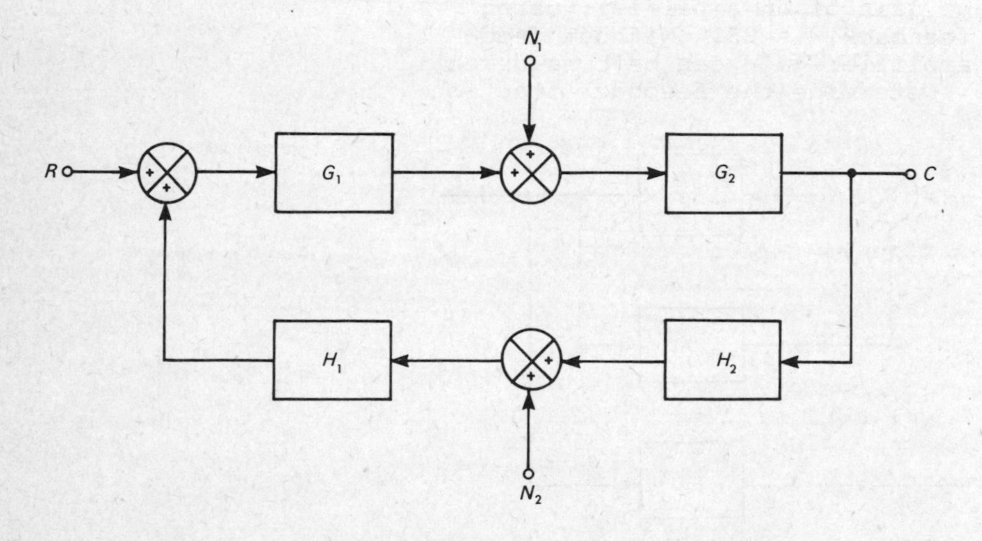

Fig.1.18

(1.9) Explain the difference between "open-loop" and "closed-loop" speed control of a d.c. motor. Draw a block diagram for a closed-loop speed-control system and state typical examples of apparatus that may be used in each section of the block diagram.

Explain what is meant by (*a*) response time, and (*b*) accuracy of control. Upon what factors does each depend?

2 Control Elements: Rotating Machines

The behaviour of a system depends on the characteristics of the components used and the manner in which they are interconnected. Electric motors and generators are commonly used in systems for position control and speed control and, in some cases, for power amplification. This chapter deals with the operation, the characteristics and the application of generators and motors as used in control systems.

2.1 GENERATORS

When a conductor is moved relative to a magnetic field, an e.m.f. will be induced in the conductor. This, illustrated in Figure 2.1, is the principle on which electric gener-

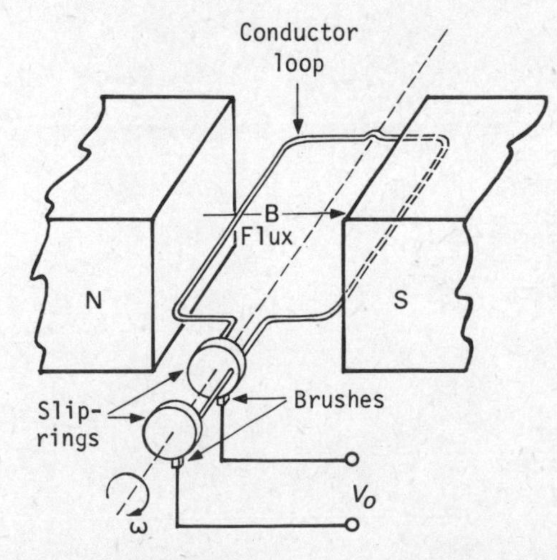

(a) A.C. TYPE

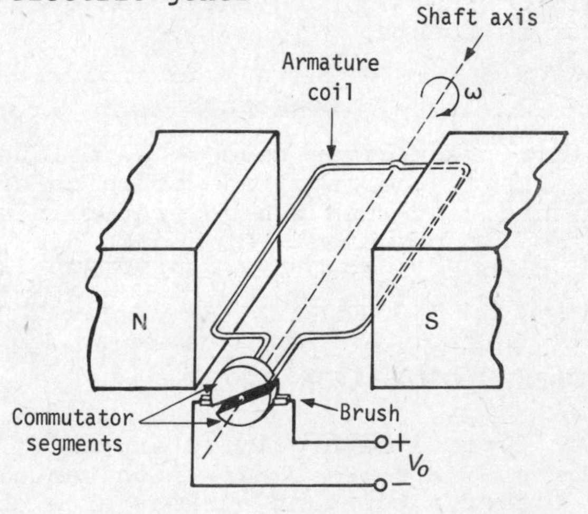

(b) D.C. TYPE

Fig.2.1 Simple generator

ators operate. A generator may be of the a.c. or d.c. variety. In an a.c. generator, the armature is rotated at constant speed in a d.c. magnetic field. The sinusoidal voltage induced in the conductors causes a current to flow, which is taken through a pair of slip rings and brushes.

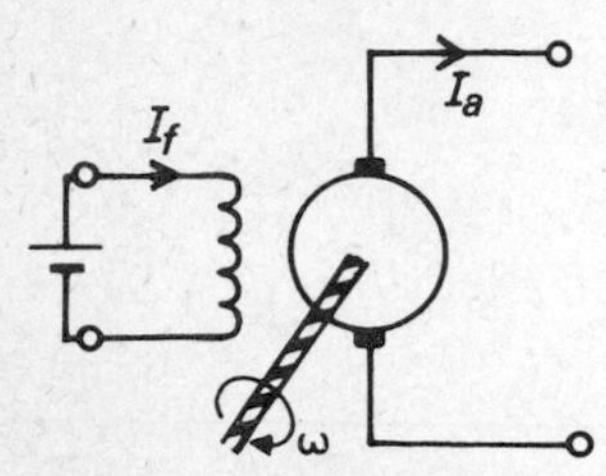

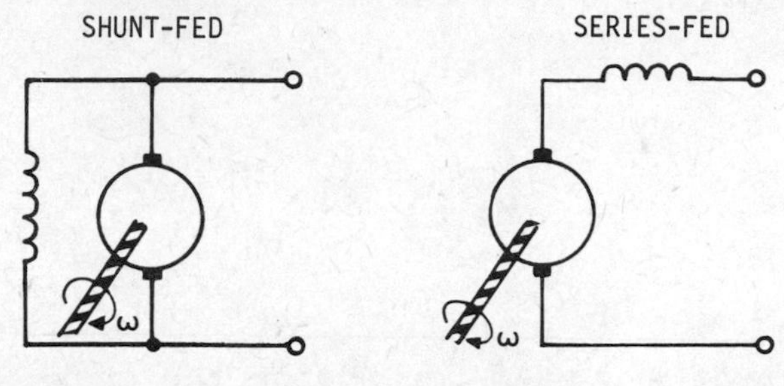

Fig.2.2

In a d.c. generator, however, a commutator is used instead of the slip rings. Its purpose is to rectify the induced voltage and produce a unidirectional output potential. As shown in Figure 2.2, a d.c. generator may be either self-excited or separately-excited. Most d.c. generators used in control systems are of the separately-excited type, in which the field current is supplied by a separate d.c. power source.

In these generators, the average induced e.m.f. in the armature is proportional to both the flux ϕ (which is itself proportional to the field current I_f) and the speed ω at which the shaft is rotated. Thus

$$E = K \times \phi \times \omega \quad \text{where } K \text{ is a physical constant}$$

The residual flux gives rise to a residual output voltage, the magnitude of which can be quite significant when the generator is used in a closed-loop system. High-gain d.c. generators, also called ROTARY AMPLIFIERS, are specially designed to overcome this problem.

2.1.1 CROSS-FIELD GENERATOR

A popular type of high-gain rotary amplifier is the *cross-field generator* which can also be used as either a constant current or a constant voltage source.

A cross-field generator is basically a separately-excited d.c. generator driven by an induction-type motor. As shown in Figure 2.3(*a*) there are four brushes placed at right angles, two of which are short-circuited. The output is taken from the other set of brushes.

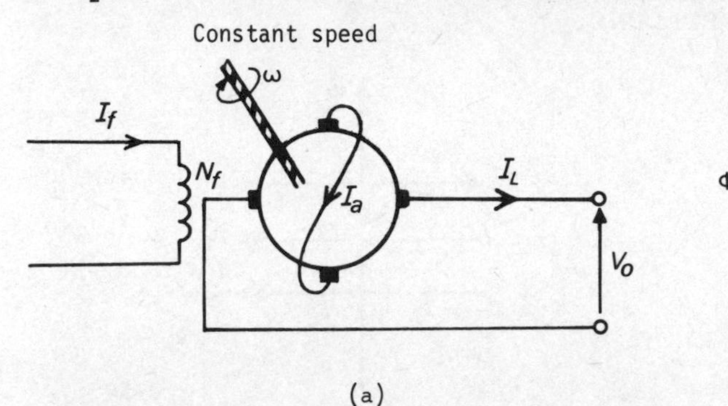

(a)

Φ_a Φ_L Φ_f

(b)

Fig.2.3 Cross-field generator

The field current I_f, which acts as the controlling current, sets up a flux ϕ_f. Since the two normal brushes are short-circuited, a large current I_a flows and sets up a large flux ϕ_a at right angles to the controlling flux. When the load current I_L flows, a reaction flux ϕ_L will be set up opposing ϕ_f (Lenz's Law), as shown in Figure 2.3(*b*).

Without any load connected, the output voltage V_o will be proportional to I_a and hence to I_f. As shown in Figure 2.4(*a*) this variation is similar to the magnetization curve of a d.c. machine but rotated through 90°.

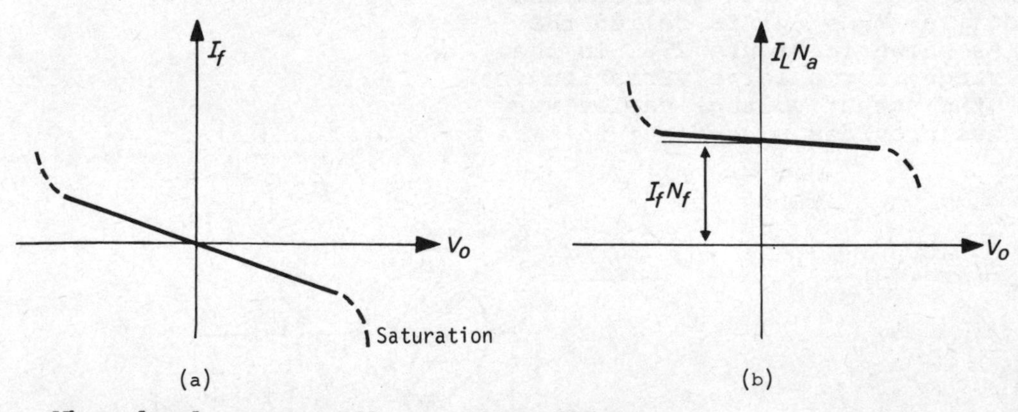

(a) (b)

Fig.2.4

When load current flows, the effective ampere-turns will be

$$I_f N_f - I_L N_a$$

where N_a is the effective armature turns. It can be seen that for a given field current, the load characteristic will be similar to the open circuit characteristic but effectively shifted vertically; the actual value of current being dependent on the ampere-turns as shown in Figure 2.4(*b*).

Provided the "effective" ampere-turns is small, in the unsaturated range, the load current remains almost constant. This *constant current* generator is known as the METADYNE.

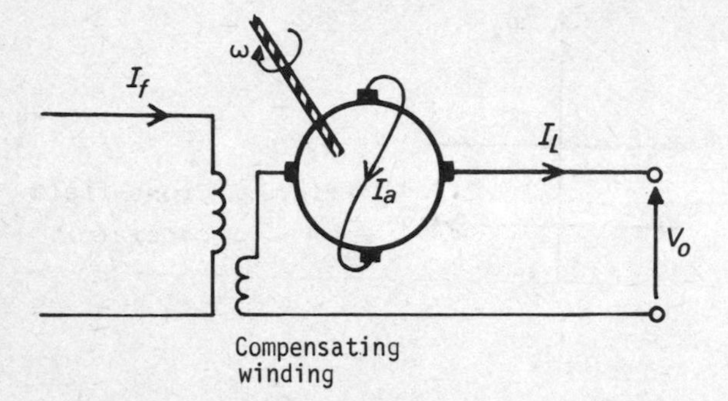

(a)

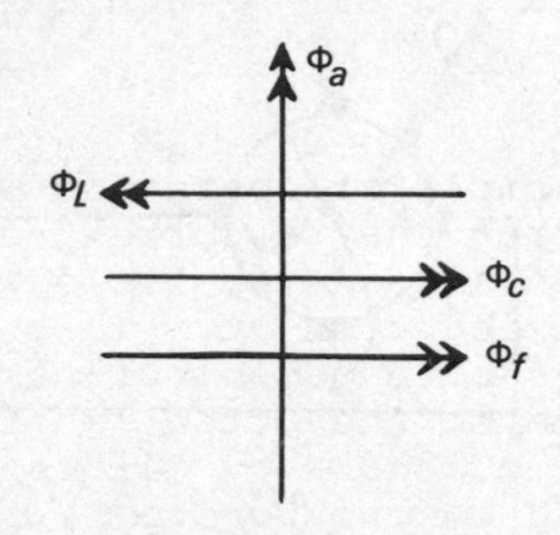

(b) *Fig.2.5*

The effect of reaction flux (acting as negative feedback), set up by the flow of load current, can be reduced by including a set of *compensating* windings in the path of the load current, placed so as to aid the field flux ϕ_f, as shown in Figure 2.5. This will have the effect of turning the characteristic clockwise and increasing the gain at the expense of an increase in the response time.

Depending on the percentage cancellation of the fluxes, different characteristics may be obtained. A machine in which 100% compensation is used (i.e. $\phi_L = \phi_c$) is called the AMPLIDYNE. As shown in Figure 2.6, in the unsaturated range, for a large variation in load current the output voltage varies very little and thus provides a *constant voltage* generator.

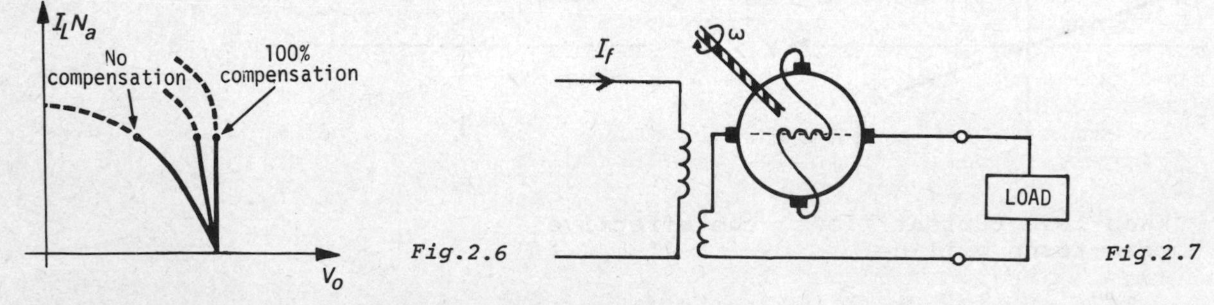

Fig.2.6 *Fig.2.7*

An additional winding at right angles to the main field winding, shown in Figure 2.7, will provide a flux assisting the armature field. The necessary ampere-turns can therefore be attained at a lower current, which reduces the overall size, the armature heating, and the wear of brushes and commutator.

The ratio of output power to input power is a measure of the amplification and, since a field current of say 50 mA can control output power in the range of kilowatts, high power amplification can be achieved.

2.2 MOTORS

The principle of operation of a motor is based on the fact that a current-carrying conductor placed in a magnetic field will experience a force acting upon it. This force is used to rotate an arrangement of conductors which are connected to the output shaft.

Both d.c. and a.c. motors are used extensively in control systems. The construction of d.c. motors is basically the same as that of d.c. generators. In a d.c. motor the force on each conductor is given by

$$F = B \times I \times \ell \times \sin\theta$$

where B is the flux density of the magnetic field, I is the current passing through the conductor, ℓ is the active length of the conductor, and θ is the angle between the conductor and the field. This force produces a unidirectional torque on the armature (the conductors are usually wound in slots in a laminated armature) and will make it rotate. The magnitude of the torque produced is given by

$$T = K \times \phi \times I_a$$

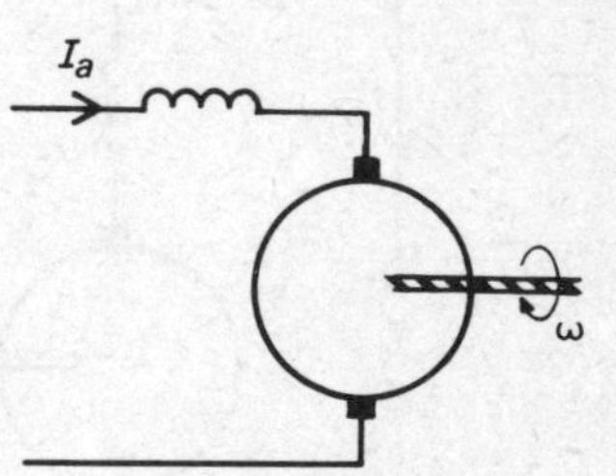

(a) SERIES MOTOR

where I_a is the armature current, ϕ is the flux per pole (which is proportional to the field current), and K is a constant.

The rotation of the armature in the magnetic field will cause an e.m.f. to be induced, in the conductors, by the generator action. This e.m.f., known as the BACK E.M.F., is such that it opposes the supply voltage to the motor (Lenz's Law).

Depending on the application, a motor can be of series, shunt, compound, or separately-excited type. Motors used in control systems are generally the separately-excited type.

When the field winding of a motor is in series with the armature, the motor is called a SERIES MOTOR. As shown in Figure 2.8(a), in this case separate control of the field is not possible. The output torque of such a machine varies in the manner shown in Figure 2.8(b).

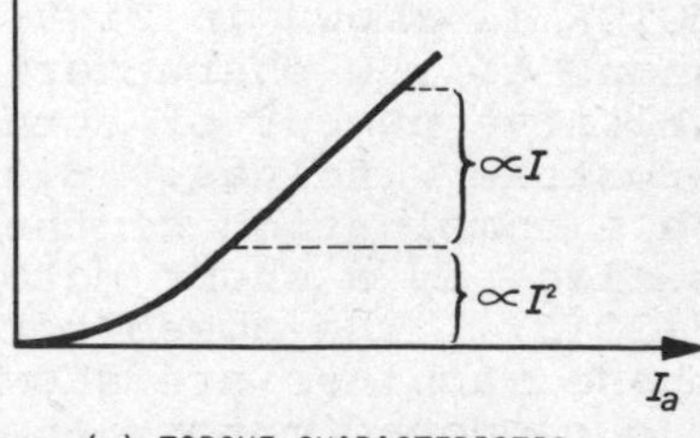

(b) TORQUE CHARACTERISTIC

Fig.2.8

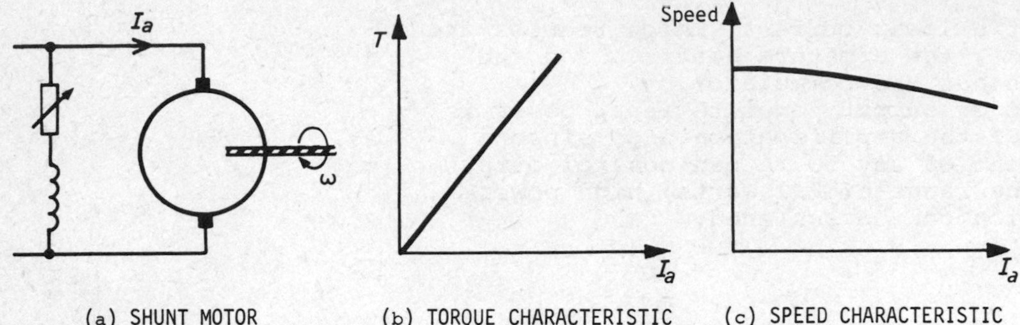

(a) SHUNT MOTOR (b) TORQUE CHARACTERISTIC (c) SPEED CHARACTERISTIC *Fig.2.9*

In a SHUNT MOTOR the field winding is in parallel with the armature winding as shown in Figure 2.9(*a*). For a constant value of field current, the torque is almost a linear function of armature current. As shown in Figure 2.9(*b*), if there is an increase in load, the armature current will increase, thus increasing the output torque. For a constant value of field current, the speed/armature current characteristic of Figure 2.9(*c*) shows that speed remains almost constant over a large variation of the armature current.

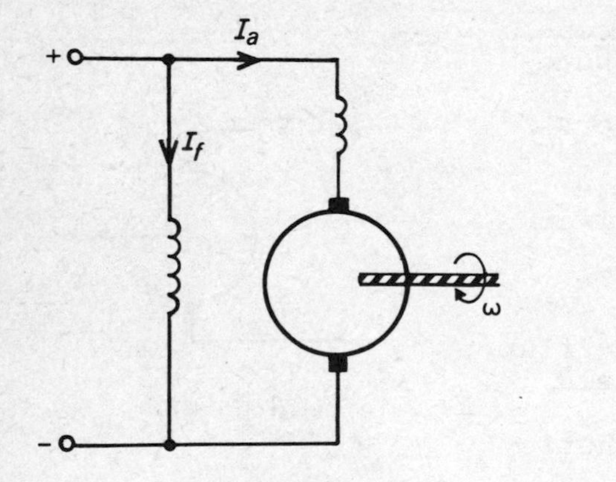

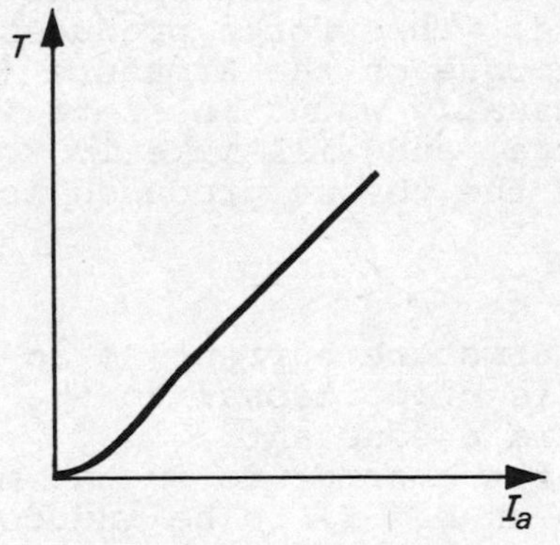

(a) COMPOUND MOTOR (b) TORQUE CHARACTERISTIC *Fig.2.10*

The circuit diagram of a simple COMPOUND MOTOR is shown in Figure 2.10(*a*). The exact shape of the characteristic will depend on the relative number of turns in the field and armature windings. But, in general, it will be a combination of the characteristic of a series and a shunt motor, as shown in Figure 2.10(*b*). The speed/torque characteristics of these machines are shown in Figure 2.11. For the compound motor, the characteristics depend on whether the magnetic fields are additive (cumulative compound) or opposing (differential compound).

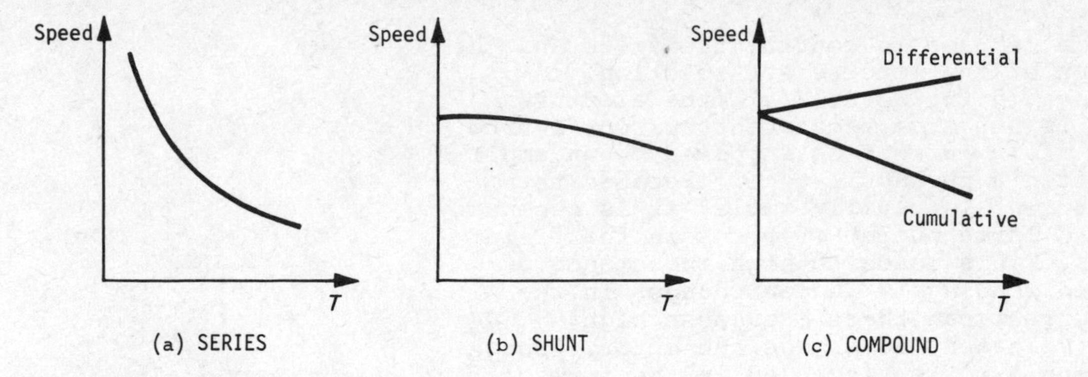

(a) SERIES (b) SHUNT (c) COMPOUND *Fig.2.11*

2.3 D.C. SERVOMOTORS

D.C. servomotors are specially designed d.c. motors with high starting torque and low inertia. They are used in systems where a high starting torque and a wide speed range may be required to operate in either direction of rotation.

There are two main reasons for using d.c. machines in control systems:

(*a*) the ease with which their speed can be controlled,

(*b*) the capability of providing power amplification.

In large d.c. servomotors, the field winding is supplied from a constant voltage source, while the armature circuit is used for controlling the speed. The torque developed will then be given by

$$T = KI_fI_a$$

where I_f and I_a are the field current and armature current respectively, and K is the motor constant.

Small d.c. servomotors use the field winding for speed control, the armature being supplied from a constant current source. The torque developed will then be proportional to the armature current and the "effective" flux set up in the field windings.

A widely used d.c. motor is the SPLIT-FIELD MOTOR which is, basically, a separately-excited d.c. motor in which the field winding is centre tapped, with the two halves wound in opposition.

The flux set up in the field winding is proportional to the effective current I_e which is the difference between the currents in each half of the field winding. When currents are equal, effective flux is zero and no torque will be developed. An unbalanced current can therefore be used to control the speed and the direction of the motor shaft rotation.

As shown in Figure 2.12(*a*), the armature circuit is fed from a constant current source and the field current is supplied by an amplifier, often a push-pull type. Because the field current is usually small, it is necessary to have a large number of turns in the field windings. This leads to high inductance of the field windings. Sudden changes in the field current can therefore cause high induced voltages and damage the motor. Surge protection is often included in the form of shunts across the field windings.

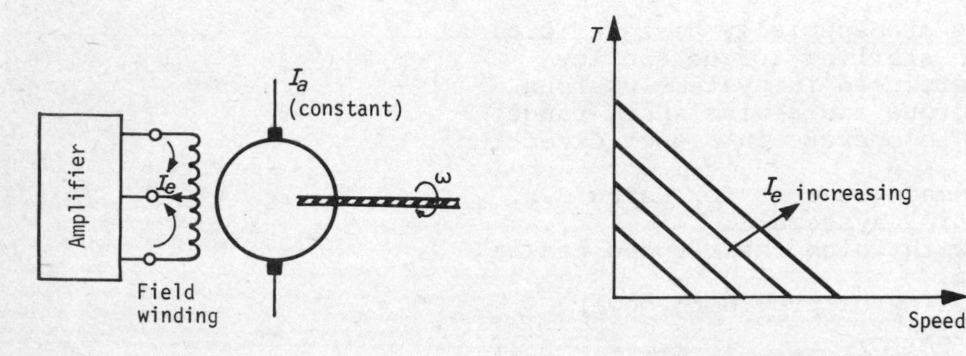

Fig.2.12 Split-field motor

This type of motor finds extensive application in low power servos. Since one of the requirements of servomotors is to have high starting torque and low inertia, the armature is designed to be long and thin.

Figure 2.12(*b*) shows typical torque/speed characteristics for various values of effective current.

2.4 A.C. SERVOMOTORS

Most a.c. motors used in control systems are of the two-phase induction type suitable for simple and low-power applications. As shown in Figure 2.13(*a*), the stator has two field windings placed at right angles to each other in order to produce the rotating field on which the motor action depends.

One phase is supplied from a constant a.c. reference voltage V_R, while the other phase acts as the controlling field and is supplied from the output of the servo amplifier. The speed of rotation is proportional to the control current I_c, the phase of which determines the direction of rotation. Typical torque/speed characteristics are shown in Figure 2.13(*b*). It is important to note that the curve for zero control current goes through the origin and that the slope is negative. This means that when the control current becomes zero, the motor develops a decelerating torque, causing it to stop. The curve also shows a large torque at zero speed.

The rotor core material is of low magnetic reluctance with high electrical resistance. This will prevent single-phasing, improve torque/load linearity, and give high starting torque which is essential for fast response in many systems such as positional servos.

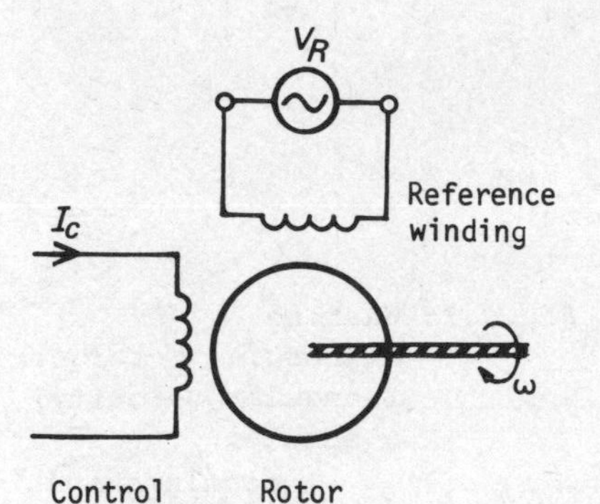

(a) TWO-PHASE MOTOR

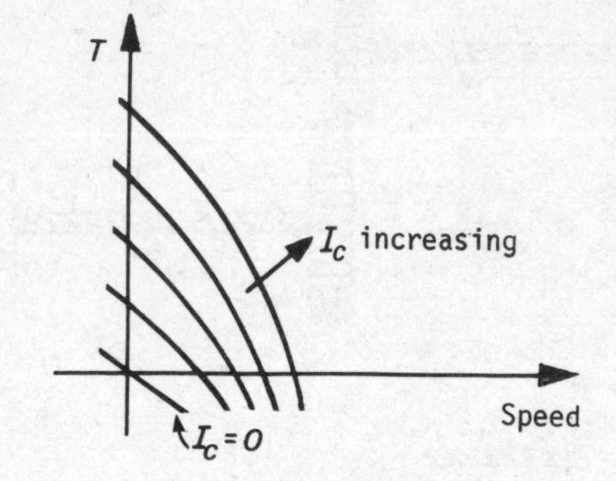

(b) TORQUE/SPEED CHARACTERISTICS

Fig.2.13

Figure 2.14 shows the torque/speed characteristic for an ordinary motor and a specially designed servomotor.

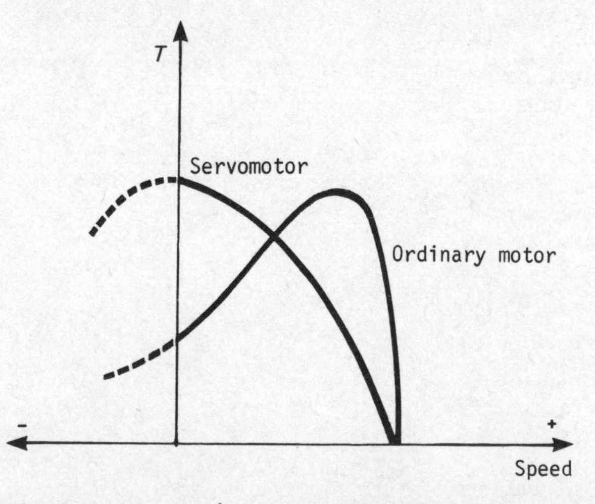

Fig.2.14 Torque/speed characteristics

2.5 GEARING

Control systems with rotating elements often have a system of gearing because:

(a) It is usually economic to design servo-motors which run at a much higher speed than that required by the load.

(b) It permits a higher load acceleration for a given motor.

A gearbox in a mechanical system serves much the same purpose as a transformer in an electrical circuit. Thus for ease of calculation, quantities such as inertia can be "referred" to the load side.

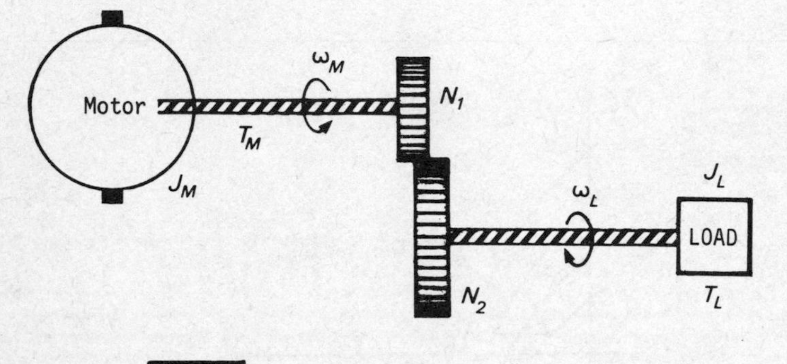

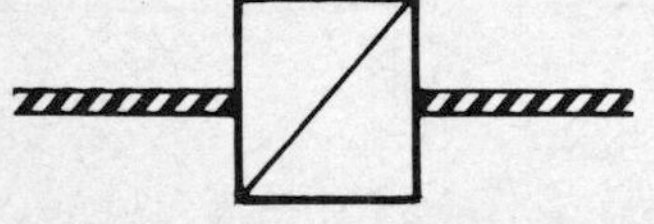

Block diagram

Fig.2.15 Gearing
J inertia; T torque;
ω angular velocity;
α angular acceleration;
N number of teeth
J_2 motor inertia "referred" to the load side

Consider the gear train of Figure 2.15. In the ideal case, mechanical power into the gear system equals the mechanical power out of the system. Thus,

$$T_M\omega_M = T_L\omega_L \tag{2.1}$$

The gear ratio is related to the angular acceleration and angular velocity by the equation:

$$\frac{N_1}{N_2} = \frac{\omega_L}{\omega_M} = \frac{\alpha_L}{\alpha_M} \tag{2.2}$$

But $T = \alpha J$, or

$$\alpha = T/J \tag{2.3}$$

For the load side

$$\alpha_L = T_L/J_2 \tag{2.4}$$

From equation (2.2)

$$\alpha_M = \frac{N_2}{N_1}\,\alpha_L$$

From equation (2.4)

$$\alpha_M = \frac{N_2}{N_1} \times \frac{T_L}{J_2} \tag{2.5}$$

From equations (2.1) and (2.2)

$$T_L = \frac{N_2}{N_1} T_M$$

Substituting in equation (2.5)

$$\alpha_M = \left(\frac{N_2}{N_1}\right)^2 \times \frac{T_M}{J_2}$$

i.e.

$$\alpha_M = \frac{T_M}{J_2 \div (N_2/N_1)^2} \tag{2.6}$$

Comparing equations (2.3) and (2.6), it can be seen that the motor inertia is given by

$$J_M = \frac{J_2}{(N_2/N_1)^2}$$

Thus, the motor inertia "referred" to the load side is given by

$$J_2 = J_M\left(\frac{N_2}{N_1}\right)^2 \tag{2.7}$$

Total inertia at the load side is therefore given by

$$J_T = J_L + J_M\left(\frac{N_2}{N_1}\right)^2 \tag{2.8}$$

Using equation (2.3), the load acceleration will be

$$\alpha_L = \frac{(N_2/N_1)T_M}{J_L + J_M(N_2/N_1)^2} \tag{2.9}$$

It can be shown (by differentiating α_L w.r.t. (N_2/N_1) and setting it equal to zero) that for maximum load acceleration, which is an essential requirement, the gear ratio is given by

$$\frac{N_2}{N_1} = \sqrt{\frac{J_L}{J_M}} \tag{2.10}$$

Differentiating

$$\frac{d\alpha_L}{dn} = \frac{n T_m \cdot 2n J_m - (n^2 J_m + J_L) T_m}{(n^2 J_m + J_L)^2}$$

$$n^2 J_m - J_L = 0$$

$$n = \sqrt{\frac{J_L}{J_m}}$$

However, because of the maximum speed requirements, it is often not possible to use the optimum gear ratio which would give the maximum load acceleration.

Example 2.1

(*a*) *Calculate which of the following motors gives the greater load acceleration:*
Motor A which has a maximum torque of 0.15 N-m *and an inertia of* 3 kg-m^2 *or*
Motor B which has a maximum torque of 0.2 N-m *and an inertia of* 4.5 kg-m^2.
(*b*) *Determine the gear ratio necessary to obtain greatest load acceleration, and hence calculate the maximum load acceleration, if the load inertia is* 450 kg-m^2, *using the motor with the greater load acceleration.*
(*c*) *If the load of part* (*b*) *is to be slewed through an angle of* 90° *in minimum time, calculate the maximum shaft speed of the selected motor.*

Solution:
(*a*) When matching of load to motor is made with a free choice of the gear ratio, the motor with the larger value of $T/\sqrt{J}$ will have the better load acceleration.

$$\text{For Motor A:} \quad \frac{T_A}{\sqrt{J_A}} = \frac{0.15}{\sqrt{3}} = 86.62 \times 10^{-3} \text{ N-kg}^{-\frac{1}{2}}$$

$$\text{For Motor B:} \quad \frac{T_B}{\sqrt{J_B}} = \frac{0.2}{\sqrt{4.5}} = 94.28 \times 10^{-3} \text{ N-kg}^{-\frac{1}{2}}$$

Thus Motor B gives the greater load acceleration.

(*b*) Using equation (2.10), the gear ratio should be

$$n = \frac{N_2}{N_1} = \sqrt{\frac{J_L}{J_M}} = \sqrt{\frac{450}{4.5}} = 10$$

For maximum load acceleration, using equation (2.9):

$$\alpha_L = \frac{nT_M}{J_L + n^2 J_M} = \frac{10 \times 0.2}{450 + 10^2 \times 45}$$

$$= 2.2 \times 10^{-3} \text{ rad/s}^2$$

(*c*) Maximum shaft speed is given by

$$\omega_M = \sqrt{\frac{n^3 \theta_L T_M}{2J_L}}$$

where θ_L is the angle through which the load is to be slewed.

$$\omega_M = \sqrt{\frac{10^3 \times \pi \times 0.2}{2 \times 2 \times 450}} = 0.599 \text{ rad/s}$$

EXERCISES

(2.1) Using suitable diagrams and typical characteristics, explain the operation of a metadyne generator. Suggest an application of such machines in servo systems.

(2.2) Describe, with the aid of sketches, the construction and operation of a split-field d.c. servomotor. Draw and explain the circuit normally used to obtain the armature supply for such a motor from an a.c. supply. With the aid of a block diagram show how such a motor would be used in a closed-loop control system.

(2.3) With the aid of sketches describe the construction and operation of a two-phase a.c. servomotor. Draw the block schematic diagram of a null balancing a.c. servo system and explain how velocity feedback may be used to improve the system transient response.

3 Control Elements: Transducers

A transducer is a device which converts energy from one form to another. Alternatively, a transducer may be defined as a sensing device that converts physical phenomena and chemical composition into electric, pneumatic, or hydraulic output signals. Depending upon the principle of operation, transducers may be classified under two main categories: those based on electric effects and those based on mechanical effects. Only the former will be considered here.

Electrical transducers will accept any one of the following energy inputs and produce an electrical output in analogue or digital form which is proportional to the input: mechanical, thermal, chemical, gravitational, and electromagnetic. In this chapter the principle of operation and the construction of some of the more commonly used transducers will be introduced.

3.1 CHOICE OF A TRANSDUCER

Present-day technology has produced numerous types of transducer. The choice of a transducer depends mainly on the application and circumstances. For example, a transducer used in a space-ship should be as small, light and reliable as possible, cost being of less importance.

The following factors should be considered in the selection of a transducer. The relative importance of each factor depends on the application.

(*a*) Accuracy
(*b*) Frequency response
(*c*) Range
(*d*) Sensitivity
(*e*) Resolution
(*f*) Reliability
(*g*) Ease of application
(*h*) Linearity
(*i*) Size and weight
(*j*) Environmental effects
(*k*) Noise level
(*l*) Apparatus requirements
(*m*) cost

3.2 RESISTANCE POTENTIOMETERS

A linear resistance potentiometer is the simplest type of resistance transducer in which, under no load, the voltage obtained from the slider is directly proportional to the position of the slider. The majority of resistive potentiometers used in control systems are specially designed and constructed for a long working life, good resolution, and low noise. In a wire-wound type, the voltage at the slider changes in discrete steps as the slider moves from turn to turn. This means that the accuracy with which the input can be measured is limited to these increments. The RESOLUTION of a potentiometer can thus be defined as the ratio of the voltage between adjacent turns of wire to the input voltage. This is a measure of the degree to which small changes of the input can be discriminated by the potentiometer. From the definition,

$$\text{Resolution} = \frac{\Delta V}{V} = \frac{\Delta V}{n \times \Delta V} = \frac{1}{n} \times 100\%$$

where ΔV is the voltage between adjacent turns, V is the total voltage across the potentiometer, and n is the number of turns (see Figure 3.1 (*a*)).

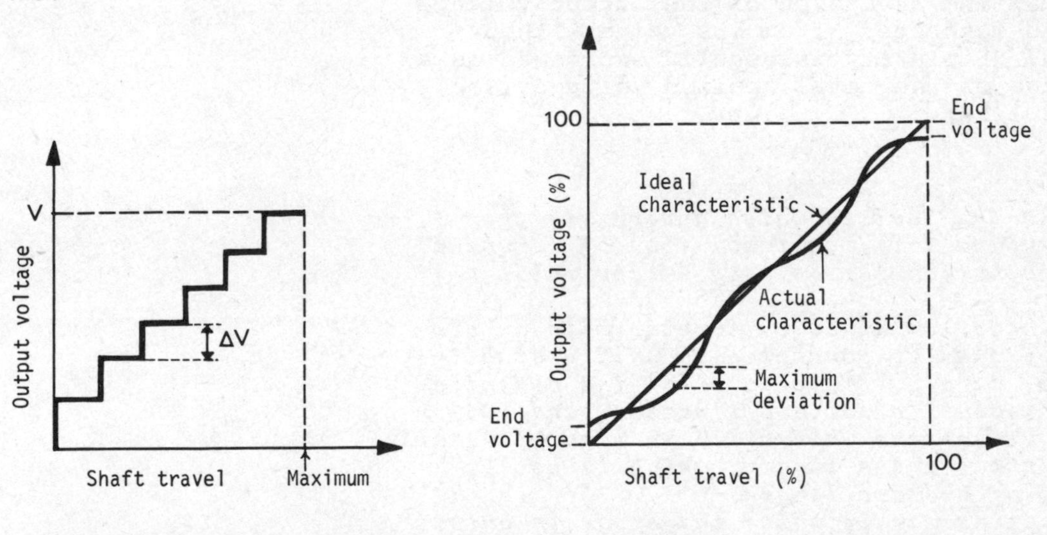

(a) POTENTIOMETER RESOLUTION (b) POTENTIOMETER LINEARITY *Fig.3.1*

As shown in Figure 3.2(*a*), a potentiometer can be calibrated and used for the measurement of linear displacement. For the measurement of angular displacement, a rotary potentiometer may be used. In the wire-wound type an insulated resistance wire is wound helically around an insulating card or rod and then bent into a circular shape as shown in Figure 3.2(*b*). The potentiometer is supplied from a constant voltage V_i. The wiper is connected to the rotating shaft and provides a voltage proportional to the angular position θ of the shaft. Rotary potentiometers can be either single turn (for one revolution) or multi-turn (for several revolutions). The main problems associated with this type of transducer are firstly the fact that a full 360° rotation is not always possible (usually between 300° and 330° rotation), and secondly, that the accuracy of the measurement depends on the resolution of the potentiometer.

One major problem associated with resistance potentiometers is that the linearity of the relation between the shaft position and output voltage is upset if appreciable current is taken from the slider (i.e. under heavy load). This problem is overcome in *cam corrected* potentiometers in which a cam mechanism is used to correct the intrinsic errors by either advancing or retarding the motion of the wiper arm. Good linearity of about 0.01 per cent is obtainable in these potentiometers.

The *linearity* of a potentiometer is defined as the maximum deviation of the output voltage with load connected, from its value without any load. Linearity is usually expressed as a percentage of the total applied voltage (see Figure 3.1(*b*)).

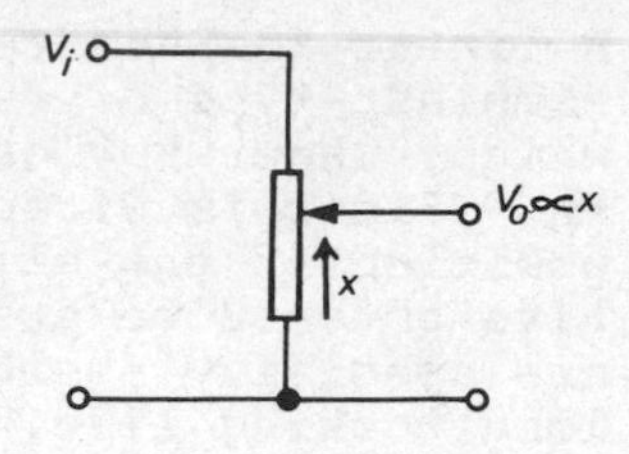

(a) SLIDER-TYPE

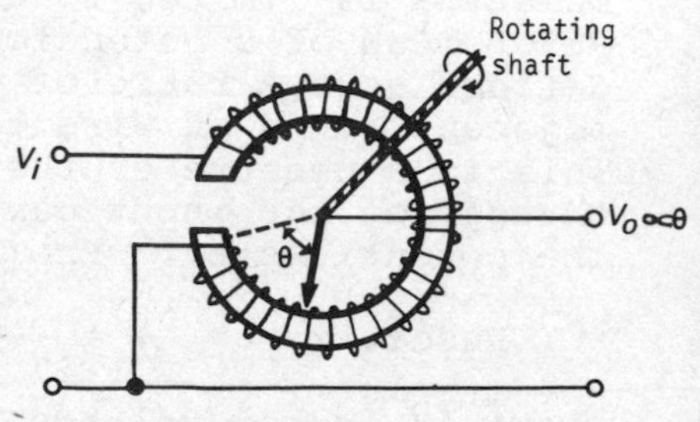

(b) ROTARY-TYPE

Fig.3.2

Example 3.1

(a) State the factors which govern (i) the linearity and (ii) the resolution of a typical rotary potentiometer as used in control systems.
(b) A linear 100 kΩ *potentiometer with an arc length of* 344° *is connected to a* 15 V *stabilised d.c. supply. Calculate (i) the potentiometer constant in volts per radian, and (ii) the output voltage when a* 100 kΩ *load resistor is connected to the slider, which is set at one half of the arc length.*

Discuss the way in which the error in output voltage will vary as the slider moves along the arc.

Solution:
(*a*)(i) The two major factors affecting the linearity of a potentiometer are the construction of the device and the resistance value of the applied load.

To achieve good linearity, the cross-section of the wire used, its resistivity and the spacing between the individual turns of the wire should be consistent. Also, the loading effect should be kept to a minimum by ensuring that the load resistance is much larger than the resistance of the potentiometer.

(ii) The resolution of a wire-wound potentiometer was defined as the reciprocal of the total number of turns. Thus, for a given wire, the larger the number of turns per unit length of the potentiometer, the smaller and hence the better the resolution.

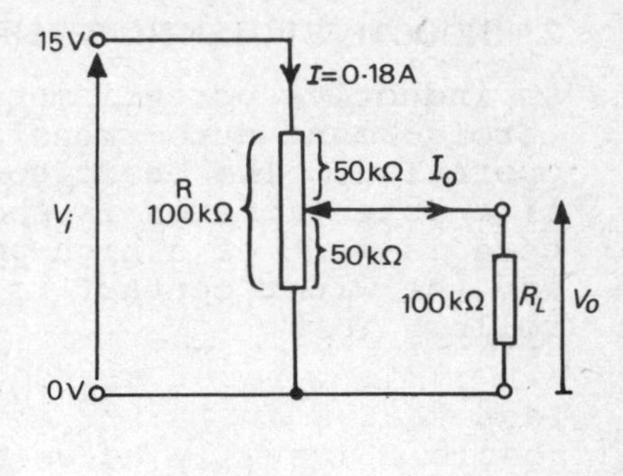

(*b*)(i) Arc length = $344° = \dfrac{344 \times 2\pi}{360}$ rad

$$\text{Potentiometer constant} = 15 \div \frac{344 \times 2\pi}{360}$$

$$= 2.5 \text{ V/rad}$$

(ii) With no load $V_o = \frac{1}{2}V_i = \frac{1}{2} \times 15 = 7.5$ V

The effect of the load is to shunt half of the potentiometer and draw current, as shown in Figure 3.3(*a*).

$$R_p = \frac{100 \times 50}{100 + 50} \text{ k}\Omega = \frac{100}{3} \text{ k}\Omega$$

$$R_T = 50 + \frac{100}{3} = \frac{250}{3} \text{ k}\Omega$$

$$I = 15 \div \left(\frac{250}{3} \times 10^3\right) = 0.18 \text{ A}$$

then

$$V_o = I \times R_p = 0.18 \times \frac{100}{3} = 6 \text{ V}$$

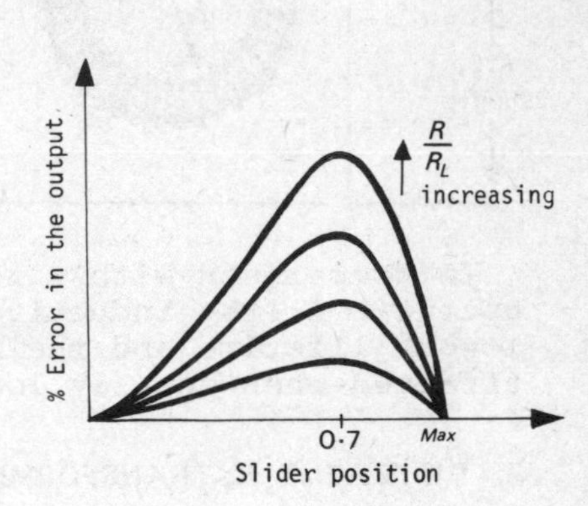

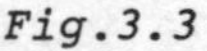

(b)

Fig.3.3

As the slider moves from one end of the potentiometer to the other, the percentage error in the output voltage increases to a maximum, at around 70% of the effective length of the arc, then reduces to zero. This is illustrated in Figure 3.3(*b*).

Potentiometers are used extensively in control systems as error detectors. In a servo system, for example, the difference between the input and output shaft positions (the error) can be obtained by using two similar potentiometers as shown in Figure 3.4.

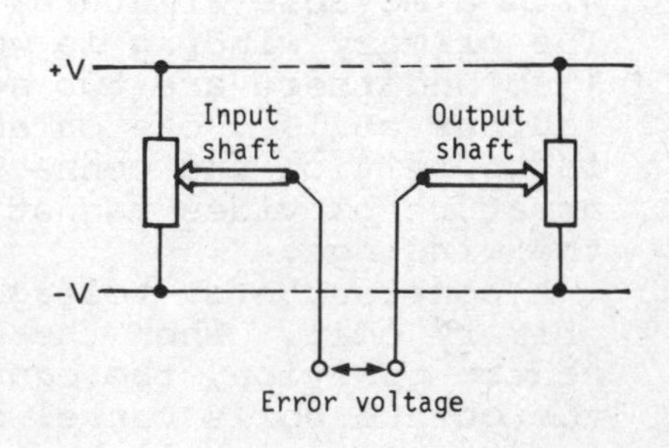

Fig.3.4

3.3 INDUCTIVE POTENTIOMETERS

An inductive potentiometer is a precision toroid-wound auto-transformer designed for a.c. operation. The basic construction of an inductive potentiometer is shown in Figure 3.5. The core is made of a high-permeability material and the wiper contact is a self-aligning carbon roller.

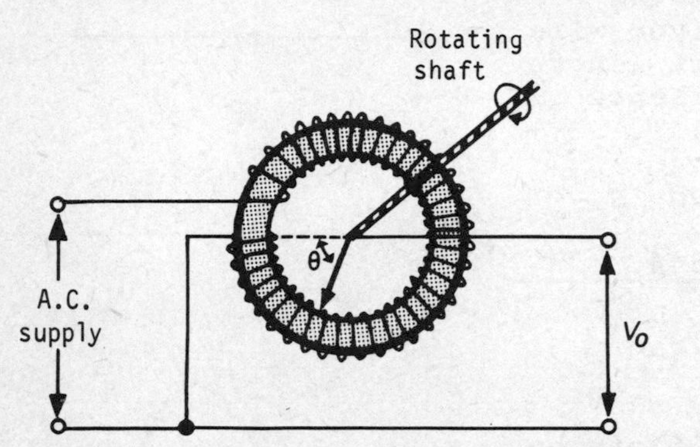

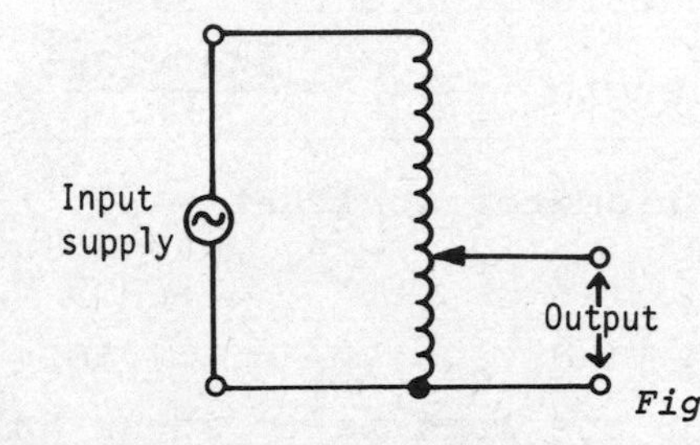

Fig.3.5 Inductive potentiometer

In comparison with wire-wound resistance potentiometers, inductive potentiometers have longer lifetime and their linearity is less affected when heavily loaded.

3.4 DIFFERENTIAL TRANSFORMER

Variation in dimensions of an airgap in a magnetic path changes the reluctance (magnetic resistance) of the circuit which will in turn produce a variation in the inductance. This is the principle of operation of a group of transducers which are used for measuring small linear and angular displacements.

A *differential transformer*, also called an E-I transformer, is basically an E-type core with a movable armature, shown in Figure 3.6. The primary winding is wound on the centre limb and there are two secondary windings (output coils), one on each outer limb, wound in opposition and connected in series. The armature provides magnetic coupling between the windings.

A constant a.c. voltage is applied to the primary coil. When the armature is in its centre position, the equal voltages induced in the output coils cancel each other and the output voltage will be zero. A displacement of the armature in either direction unbalances the flux paths and results in unequal voltages being induced in the secondary windings. The difference of the two voltages appears at the output and gives a linear measure of the position of the armature from its centre position. The phase of the output voltage depends upon the direction of the armature displacement from its centre position, shown in Figure 3.7(*a*).

For measurement of small angular displacements, an E-I transformer in which the armature is pivotted in the centre can be used. As shown in Figure 3.7(*b*), any angular displacement of the armature from its "level" position causes unequal e.m.f.s to be induced in the two output coils. As in the previous case, the magnitude and phase of the output voltage give a direct indication of the angle and direction of the displacement.

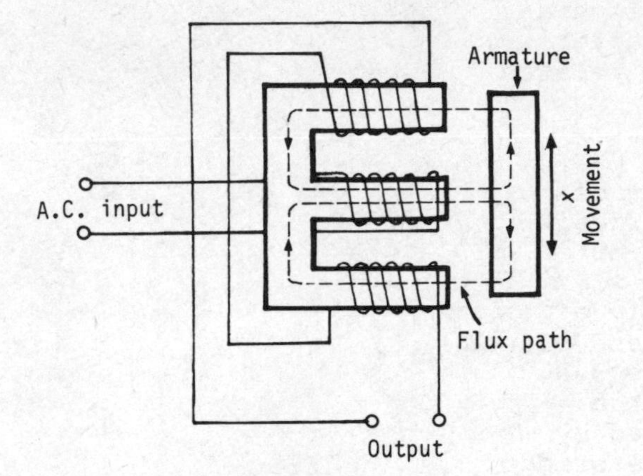

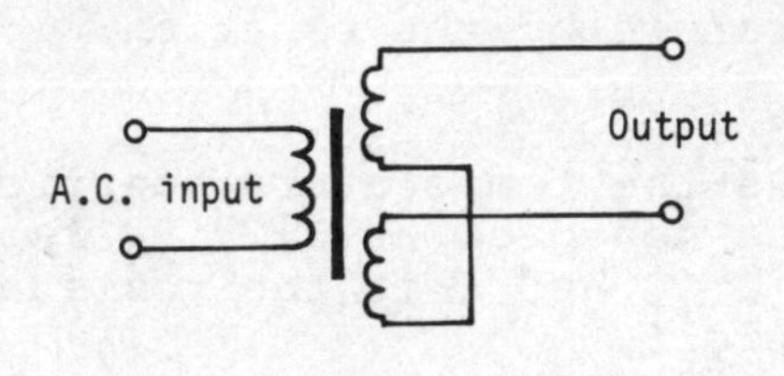

Fig.3.6 E-I transformer

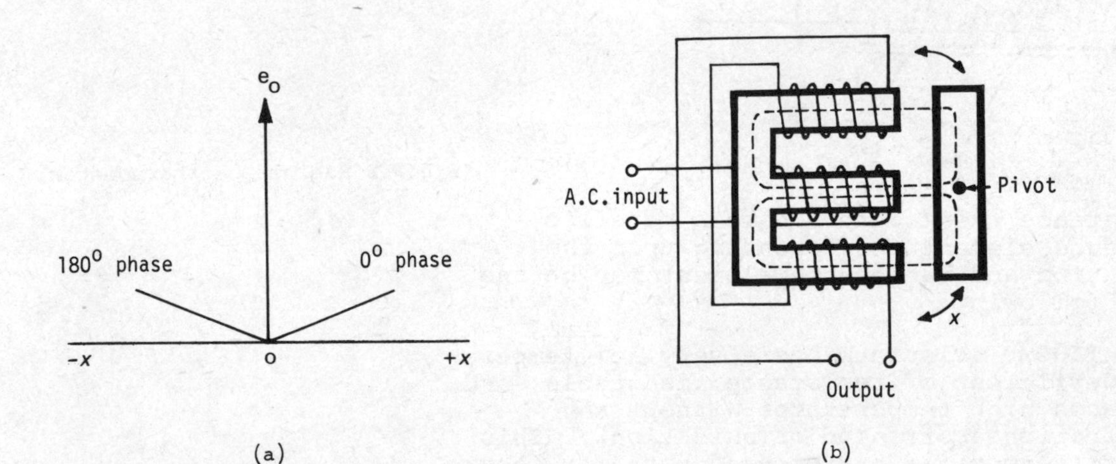

Fig.3.7

3.5 TEMPERATURE MEASUREMENT

Control of many industrial processes requires accurate measurement of temperatures ranging from about - 250°C to + 2000°C. Temperatures in this range can be accurately measured by various electrical methods based on the following effects:

(i) Variation of resistance with temperature from $R \simeq R_0(1 + \alpha t)$.
(ii) Thermoelectric effect.

3.5.1 RESISTANCE THERMOMETERS

The principle of operation of a resistance thermometer depends on the fact that the resistance of materials varies with temperature. In the linear part of the characteristic, the resistance of the element is related to temperature by the equation:

$$R \simeq R_0(1 + \alpha t)$$

where α is the temperature coefficient of resistance measured at 0°C, R_0 is the resistance at 0°C, and R is the resistance at temperature t.

A resistance thermometer is basically a resistor mounted on a suitable frame and placed in a sealed capsule as shown in Figure 3.8. Connecting leads are brought out and the variation of the resistance is measured by either a potentiometer or a Wheatstone bridge.

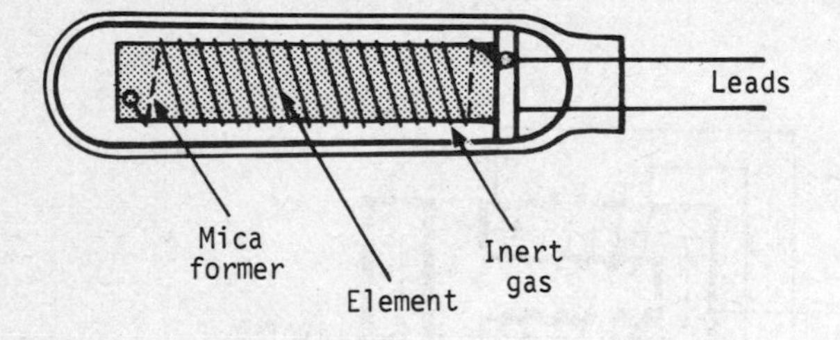

Fig.3.8 Resistance thermometer

Resistance thermometers have a metallic resistance element and depending upon the application and range, the element may be one of the following:

(1) PLATINUM. Platinum has a very low temperature coefficient of resistance, is stable, and withstands high temperatures without any deterioration (corrosion or oxidation). This type of thermometer can be used both for laboratory standards and for industrial applications over the range - 200°C to + 760°C.

(2) COPPER. The temperature coefficient of copper is slightly greater than that of platinum, but it does offer a linear characteristic over the range - 50°C to + 200°C. Compared with platinum it has a slower response to temperature changes and it also tends to oxidize at high temperatures. However, copper thermometers are very stable over a long period of time.

(3) NICKEL. Nickel thermometers have a very high temperature coefficient of resistance and are generally the cheapest of the three. They are less stable and less accurate than both platinum and copper thermometers, and are best suitable for temperature measurements from about - 75°C to + 150°C.

(4) THIN-FILM TYPE. Thin-film resistance thermometers are physically much smaller, have faster response time, and are a great deal more sensitive to temperature variation. These features make them ideal for use in spaceships.

For high accuracy measurement, the error introduced by variation of the resistance of the connecting leads must be eliminated.

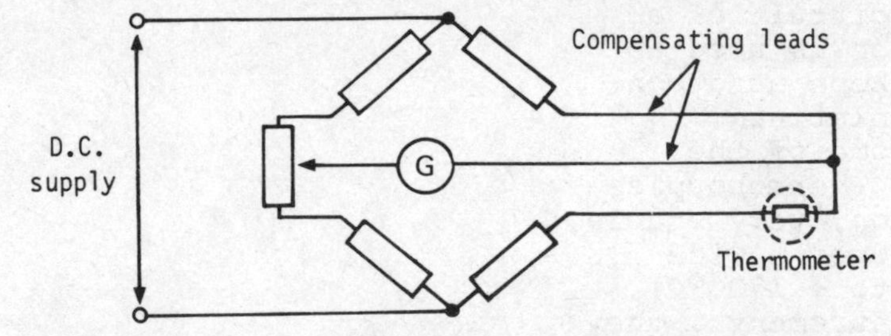

Fig.3.9

Figure 3.9 shows a 3-wire Wheatstone bridge in which compensating leads are included in each side of the bridge and one in series with the galvanometer. The bridge is balanced by using the rheostat.

3.5.2 THERMISTORS

Thermistors are special thermal resistors with high negative temperature coefficient of resistance. Some special-purpose ones have a positive temperature coefficient. Materials commonly used are metallic oxides of either cobalt, nickel, manganese, or copper.

Advantages of thermistors include their extremely high temperature sensitivity, good stability, and accuracy of better than ± 0.1°C (some thermistors, when used in conjunction with a Wheatstone bridge, can give an accuracy of up to 0.001°C). They can be operated over the range - 200°C to + 260°C with relatively simple circuits. Their main disadvantage is their non-linear resistance/temperature characteristic. Applications of thermistors include measurements of air and fuel temperatures.

3.5.3 THERMOCOUPLES

The principle of operation of thermocouples is based on the fact that when two dissimilar metals are joined together at both ends, an electric current will flow if the junctions are at different temperatures. The current flow is due to an induced e.m.f. which is proportional to the difference in temperature.

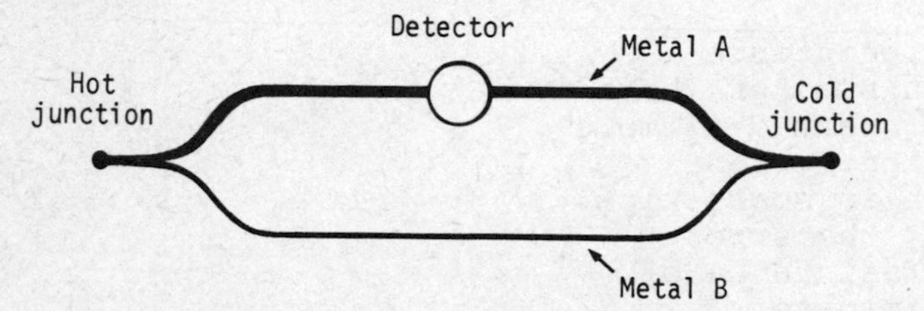

Fig.3.10 Thermocouple

Figure 3.10 shows the basic circuit of a thermocouple. The cold junction is used as the reference junction and is kept at a constant temperature. Tungsten, platinum, rhodium, copper and iron are some of the metals used in thermocouples. Thermocouples are quite simple in construction, very stable, small, cheap, and can be used over a wide range of temperature (- 225°C to + 1700°C). However, for high accuracy measurements, they need constant reference temperature and accurate measuring apparatus. Their applications include measurement of temperature in nuclear reactors.

Example 3.2

A platinum resistance of 100 Ω *at* 20°C *increases to* 273.6 Ω *when placed in an oven. If the temperature coefficient of resistance of platinum is* 0.0039/°C *and is assumed to be constant over the range of temperature, calculate the temperature of the oven.*

Solution

Using the equations for two different temperatures:

$$R_1 \simeq R_0(1 + \alpha t_1) \quad \text{and} \quad R_2 \simeq R_0(1 + \alpha t_2)$$

Dividing

$$\frac{R_1}{R_2} = \frac{1 + \alpha t_1}{1 + \alpha t_2}$$

Rewriting

$$R_1(1 + \alpha t_2) = R_2(1 + \alpha t_1)$$

or

$$t_2 = \frac{R_2(1 + \alpha t_1) - R_1}{R_1\alpha}$$

Substituting values:

$$t_2 = \frac{273.6(1 + 0.0039 \times 20) - 100}{100 \times 0.0039} = 499.8^{\circ}\text{C}$$

3.6 STRAIN GAUGES

In engineering, the term "strain" may be defined as the change in length of a body per unit length when the body is stressed or acted upon by a force. The principle of operation of a RESISTANCE STRAIN GAUGE is based on variations of resistance of an element when its dimensions are altered due to a stress or a force. Measurements of mechanical strain, force, loads and torque can thus be made using strain gauges of the variable resistance type. Although strain gauges are themselves transducers, they can also be employed as part of other transducers such as torque meters, diaphragm-type pressure gauges, or accelerometers as the mechanical/electrical conversion element.

Bonded resistance strain gauges can be of the wire type, metal foil type, or even semiconductor crystal type, all operating on the same principle.

Consider the expression for resistance of an element in terms of its dimensions:

$$R = \rho\ell/A$$

where ρ is the resistivity of the material, A is the cross-sectional area, and ℓ is the length of the element. A change in the resistance due to strain can be expressed as:

$$\frac{\Delta R}{R} = \frac{\Delta\rho}{\rho} + \frac{\Delta\ell}{\ell} - \frac{\Delta A}{A}$$

where $\Delta\ell/\ell = \varepsilon$ = strain,

and $\Delta A/A = -2\nu(\Delta\ell/\ell)$ where ν is Poisson's ratio

Thus

$$\frac{\Delta A}{A} = -2\nu\varepsilon$$

and

$$\frac{\Delta R}{R} = \frac{\Delta\rho}{\rho} + \varepsilon + 2\nu\varepsilon = \frac{\Delta\rho}{\rho} + \varepsilon(1 + 2\nu)$$

Assuming that the change in resistivity, $\Delta\rho/\rho$, is negligible, then

$$\frac{\Delta R}{R} \simeq \varepsilon(1 + 2\nu)$$

which shows that resistance change is proportional to strain ε.

The term

$$K = \frac{\Delta R/R}{\Delta\ell/\ell}$$

is called the *sensitivity factor* or *gauge factor* and can be positive or negative. The value of K varies between - 12 and + 6 depending on the metal used. In semiconductor type resistance gauges the gauge factor may vary between - 100 and + 200.

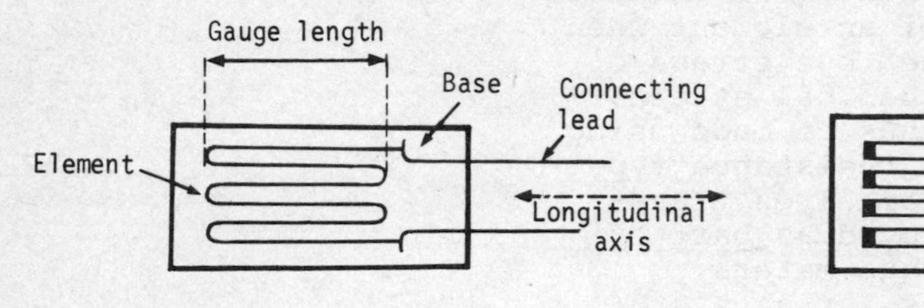

Fig.3.11 Bonded resistance strain gauge

A BONDED STRAIN GAUGE consists of a thin piece of material (normally paper, but glass fibres for special applications) on which is cemented a resistive element as shown in Figure 3.11. The gauge is treated with adhesive on the back or may be of the self-adhesive type so that it can be stuck on the surface of the member under test. When the gauge is bonded to the body under test and the body is subjected to a stress, the gauge base will be elongated and the bonded element will be lengthened. At the same time there will be a contraction along the axes at right-angles to the direction of the strain. The dimensional changes will alter the resistance of the element in proportion to the strain.

The change in resistance of the gauge element is very small (of the order of 1-2%) and often a Wheatstone bridge is used to measure its value. Since the strain gauge is often used under conditions of varying temperature, compensation is necessary. This is achieved by including a DUMMY GAUGE near the active one,

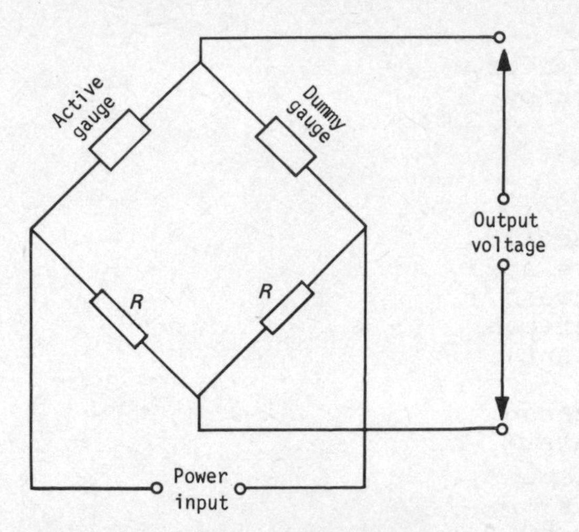

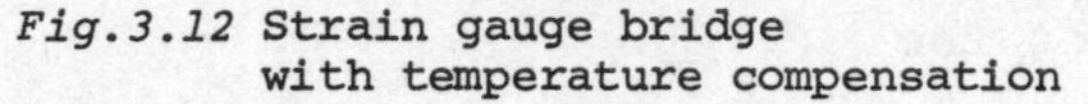
Fig.3.12 Strain gauge bridge with temperature compensation

on an unstressed part of the member under test, but in the adjacent arm of the Wheatstone bridge as shown in Figure 3.12.

Resistors R_1 and R_2 are used to balance the bridge for zero output. When the bridge is in operation, any change in the resistance of the active element causes a voltage, proportional to the change, to appear at the output. This voltage is of the order of 20 mV for wire type elements and about 200 mV for the semiconductor type.

It is often desirable to choose a gauge material whose coefficient of expansion is close to that of the body under test. This will reduce any variation in resistance of the element due to an *apparent* stress of the material.

Applications of resistance strain gauges are enormous. Although a strain gauge by itself is a transducer, it is also used in conjunction with other transducers. In a pressure-type transducer for example, four strain gauges are employed in the arms of a Wheatstone bridge which is bonded to the outside of a tube or diaphragm on to which the pressure is applied. The expansion or contraction of the surface (due to the change in pressure) is sensed by the strain gauge. Other applications of resistance strain gauges include measurements of static deformation, study of dynamic behaviour of systems (in accelerometers, torque meters, load cells, etc.). They are also used for measurement of strain in bone structures such as the human skull.

3.7 CAPACITIVE TRANSDUCERS

The capacitance of a parallel plate capacitor, in terms of physical quantities, is given by the expression

$$C = \frac{\varepsilon A}{d}$$

where ε is the permittivity of the dielectric (dielectric constant), A is the effective area of the plates, and d is the distance between the plates. A capacitive type transducer can be based on change in either of these quantities.

The most common types of capacitive transducers are based on variation of the plate separation and are used extensively for measurement of air, gas or fluid pressure. Displacements of several millimetres can be measured by varying the effective area of the plates.

The change in the capacitance can be measured in several ways. One method is by including the capacitance in the tuning circuit of an oscillator. Any variation in the resonant frequency of the oscillator (frequency modulated signal) gives a direct indication of pressure or displacement. Alternatively the capacitance can be measured using an a.c. bridge in which the transducer capacitance forms the unknown component.

3.8 SYNCHROS

A synchro is basically a single-phase a.c. machine with a stator and a rotor. The laminated stator is usually a slotted cylinder which carries three windings in the slots with their axes 120° apart. Although there are three windings, since the induced voltages are in phase, the synchro is not a three-phase machine. The rotor is of two-pole construction and often H-shaped, and carries a single winding. Connections to the rotor winding are made through slip rings which are mounted on the rotor shaft. The basic construction of a synchro element is shown in Figure 3.13.

Synchros are used extensively for error detection as well as data transmission. Depending on their exact function and construction, they can be divided into three groups: (i) synchro transmitters, (ii) synchro receivers, and (iii) synchro control transformers.

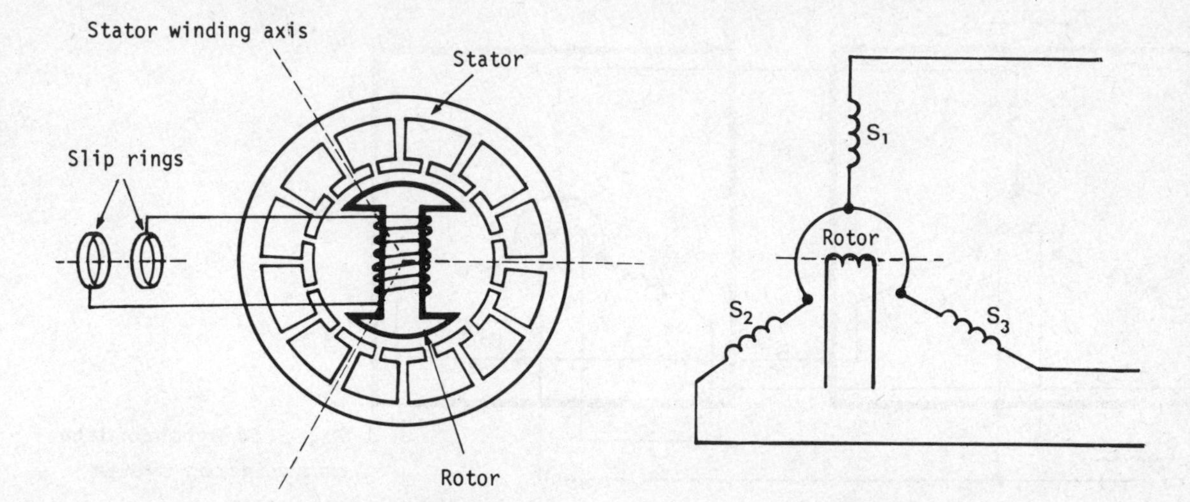

Fig.3.13 Synchro unit

SYNCHRO TRANSMITTER, X This unit generates and transmits an electrical signal which corresponds to the angular position of its rotor.

SYNCHRO RECEIVER, R The output of this unit is the angular position of its rotor which corresponds to an electrical signal input.

SYNCHRO CONTROL TRANSFORMER, CT This unit produces an electrical output signal whose magnitude is proportional to the angle of rotation of the unit rotor with respect to the magnetic field set up by its stator winding.

The main constructional differences between a transmitter and a receiver are: the provision of a mechanical damper (often a small flywheel) in the receiver to prevent its rotor from oscillating; and the fact that the rotor of the transmitter has a concentrated winding, while the receiver as well as the control transformer have a distributed rotor winding. The control transformer is very similar to the transmitter except that its stator windings have more turns and therefore higher impedance, drawing less current.

One of the applications of synchro units is for data transmission. A SYNCHRO DATA TRANSMISSION SYSTEM uses a transmitter unit and a receiver unit connected as shown in Figure 3.14. The rotor windings of both units are excited from the same single-phase a.c. source and the corresponding stator terminals are connected together.

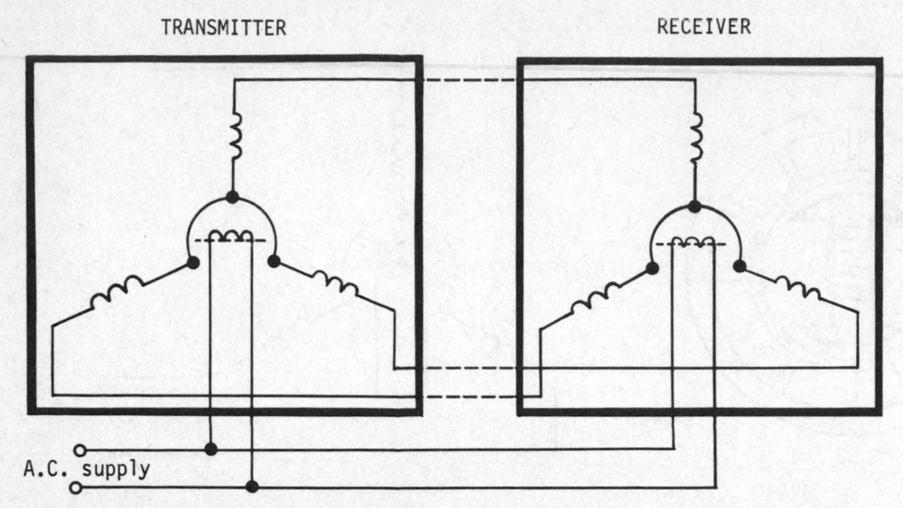

Fig.3.14 Synchro data transmission system

When the rotor windings are energized, alternating fields are set up and e.m.f.s are induced in the stator windings of both units. The magnitudes of these induced voltages depend on the relative positions of the rotor and stator windings. If the two rotors are in alignment, equal e.m.f.s induced in the stator windings will prevent any current flow through the stator windings. When the transmitter rotor moves through an angle, say θ, there will be a difference in the induced voltages and a current will flow through the stator windings. This current causes a torque to act upon the receiver rotor, moving it until its angular position is in alignment with the transmitter rotor.

An angular position error detecting system using two synchros is shown in Figure 3.15. The basic arrangement consists of a transmitter and a control transformer. The rotor of the transmitter is energized from an a.c. supply and e.m.f.s of different magnitudes are induced in the three stator windings. The magnitudes of these e.m.f.s will depend on the angular position of the rotor, but they are, of course, all in phase. Currents will flow in the stator windings of the control transformer and set up an alternating magnetic field whose direction will correspond to the direction of the field in the transmitter.

The input and output shafts are lined up by arranging the transformer rotor axis to be at right angles to the magnetic field of the stator. There is then, ideally, no coupling, and zero voltage is induced in the rotor winding. When the transmitter rotor (input shaft) is rotated, the transmitter field will

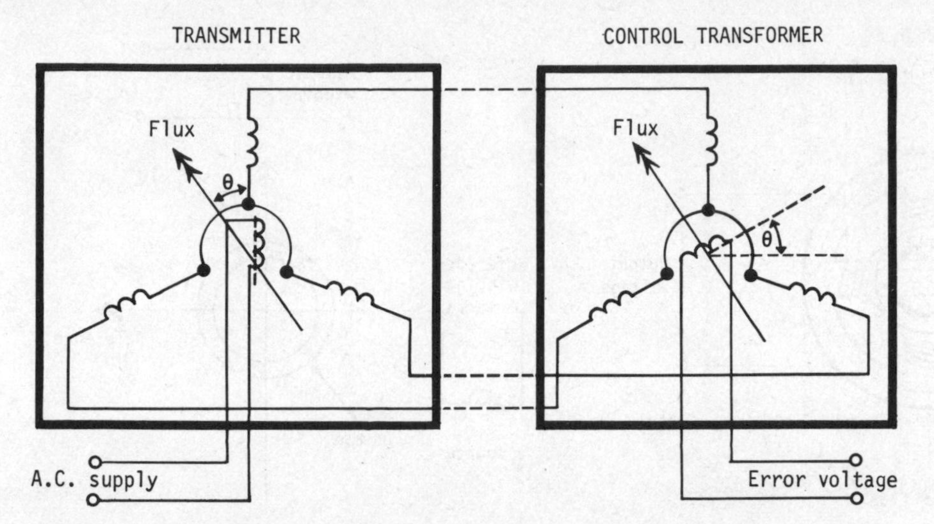

Fig.3.15 Synchro error detecting system

also rotate through an angle, say θ, causing the transformer field to be rotated through the same angle. This misalignment causes an error voltage to be induced in the windings of the transformer rotor. The magnitude of the error voltage is proportional to the degree of misalignment between the two shafts, and its phase depends on the direction of error.

3.9 TACHOGENERATORS

A tachogenerator is a generator which provides a voltage proportional to angular velocity and is used extensively in feedback control systems. Tachogenerators can be of permanent magnet d.c. type or a.c. type.

A D.C. TACHOGENERATOR is a small generator which has fixed excitation. The magnetic field is usually produced either by a small permanent magnet, or by windings supplied from a d.c. power source. The generated voltage in the output windings is proportional to the speed of rotation. The polarity of the output voltage depends on the direction of rotation and, in some applications, this is a great advantage. An output of 10-20 V per 1000 rev/min and a linearity of better than 0.1 per cent over their speed range are features of d.c. tachogenerators. The main problem associated with d.c. tachogenerators is the need for commutator and brushes which (i) require maintenance and (ii) produce high-frequency ripple superimposed on the output voltage.

These problems are overcome in a.c. tachogenerators, a popular form of which is the drag-cup generator. A DRAG-CUP TACHOGENERATOR, also called an *induction* tachogenerator, is

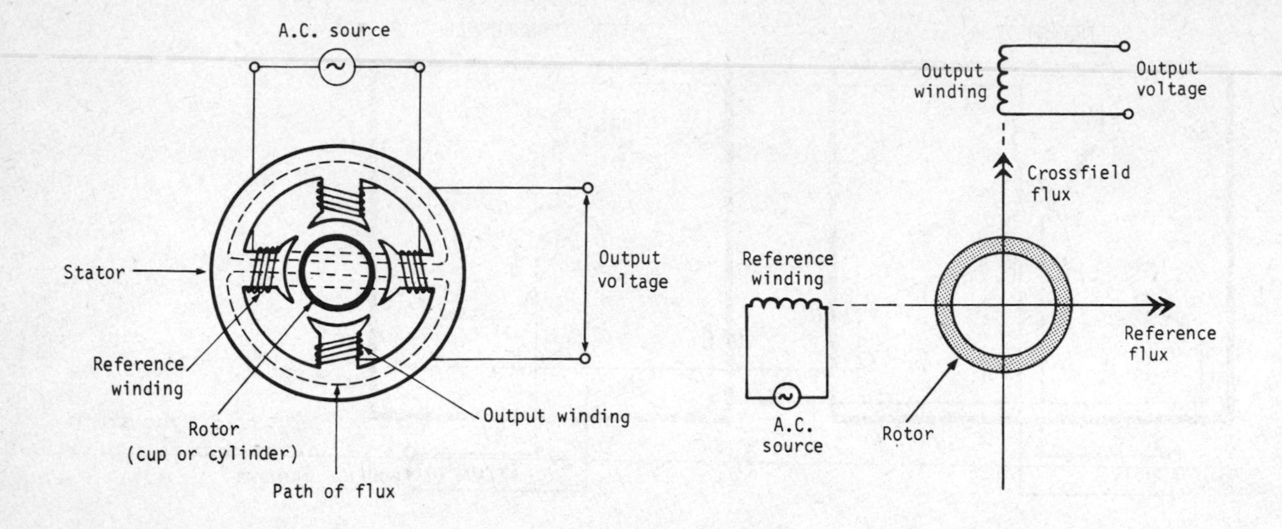

Fig.3.16 Drag-cup tachogenerator

basically an a.c. machine which has two phase windings with their axes at 90° to each other. One set of winding, the reference winding, is energized from an a.c. source with constant frequency and amplitude and provides the magnetic field. The other set of winding, the output winding, is used as voltage pick-up. The rotor is in the shape of a cup or cylinder, usually made of copper or aluminium. The laminated stator may carry both windings, as shown in Figure 3.16(*a*) or alternatively one winding may be housed inside the rotor (this winding is not connected to the rotor and therefore it does not rotate).

When the rotor is stationary, there is ideally no coupling between the input and output windings, and the output voltage will be zero. Rotation of the cup causes an e.m.f. to be induced in it which sets up eddy currents. These currents produce a magnetic cross-field which is inherently perpendicular to the field of the primary winding and hence along the axis of the secondary windings, as shown in Figure 3.16(*b*). The coupling of this flux causes a voltage, proportional to the speed of rotation, to be induced in the output winding.

3.10 MEASUREMENT OF ACCELERATION

Acceleration of a body can be defined in two ways:

(i) in terms of rate of change in velocity ($a = dv/dt$), or

(ii) in terms of force acting on a mass ($F = m \times a$).

Usually, the operation of angular accelerometers is based on the former, and linear accelerometers use the latter principle.

The DRAG-CUP tachogenerator described earlier can be used to generate a voltage proportional to the angular acceleration of the cup (i.e. a rotating shaft).

Suppose that the reference windings of the tachogenerator of Figure 3.16 are fed from a d.c. source. When the rotor has a constant speed, an e.m.f. is induced in it, causing a current to flow, which sets up a magnetic field at right angles to the main field. The flux linkage between the two coils depends on the speed of rotation. At constant speed, the flux linkage does not change and no voltage is induced in the output windings. If, however, the rotor accelerates or decelerates, the flux

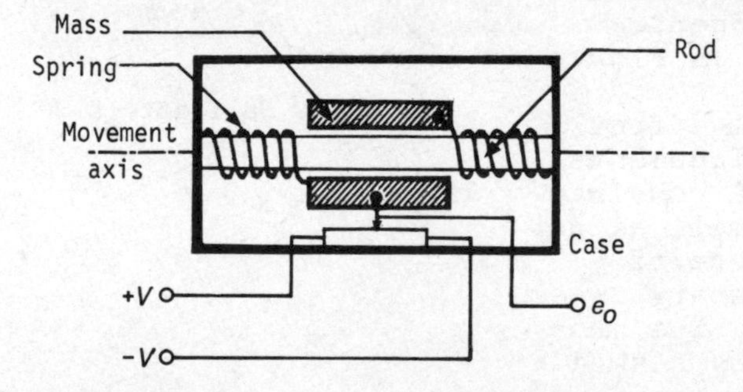

Fig.3.17 Spring-mass transducer

linkage will change at a rate proportional to the acceleration and induce a voltage in the output coils. This voltage is proportional to the angular acceleration of the rotating shaft.

The SPRING-MASS transducer provides a voltage proportional to linear acceleration. The basic construction is shown in Figure 3.17 and consists of a mass which is supported by a rod and is free to move along one axis only. The mass is restrained at both sides by slightly compressed rings. A potentiometer, fixed to the case, is supplied from equal but opposite polarity voltages (± V). The wiper of the potentiometer is fixed to the mass and provides the output voltage e.

Under steady state conditions, the wiper is in the centre of the potentiometer and the output voltage is zero. When the transducer is accelerated, the mass moves relative to the case by an amount proportional to the acceleration. Since the wiper is fixed to the mass, it will move with it and provide a voltage which is proportional to the acceleration.

This type of transducer is rather simple, cheap, and free from crosstalk. However, it suffers from poor threshold level (due to friction), and poor resolution (due to the use of the potentiometer). In a sophisticated version, the mass is lubricated so as to minimize friction. The potentiometer is replaced by an E-I transformer, and the effect of mechanical springs is produced by the use of a linear motor.

3.11 HALL EFFECT

When a current-carrying semiconductor is placed in a magnetic field which is perpendicular to the direction of current flow, an electric potential difference is set up across the semiconductor in the direction perpendicular to both current flow and the magnetic field. This phenomenon, illustrated in Figure 3.18, is known as the Hall effect.

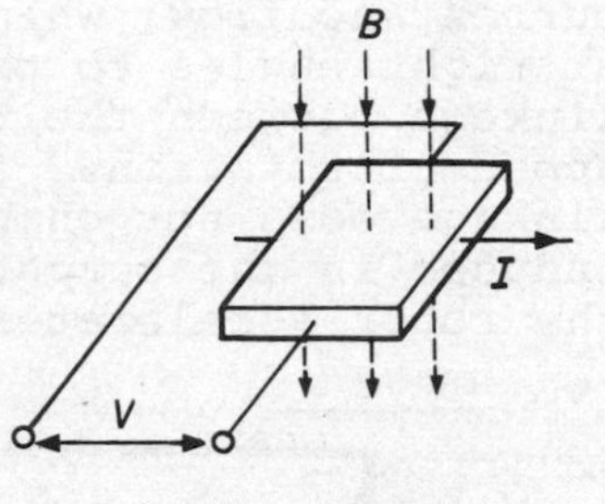

Fig.3.18 Hall effect
$V \propto I \times B$
i.e. $V = K \times I \times B$

One of the applications of the Hall effect is for the measurement of small displacements. A HALL-EFFECT DISPLACEMENT TRANSDUCER consists of a C-shaped magnetic core with a small airgap as shown in Figure 3.19. A small piece of semiconductor material with the necessary connections is placed in the airgap. The magnetic field for the gap is provided by a small magnet which is connected to the moving object to be monitored. If the current I is kept constant, the induced voltage will be directly proportional to the flux density in the airgap.

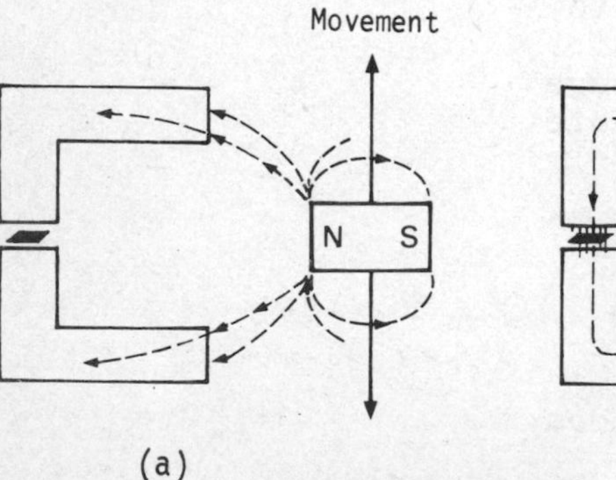

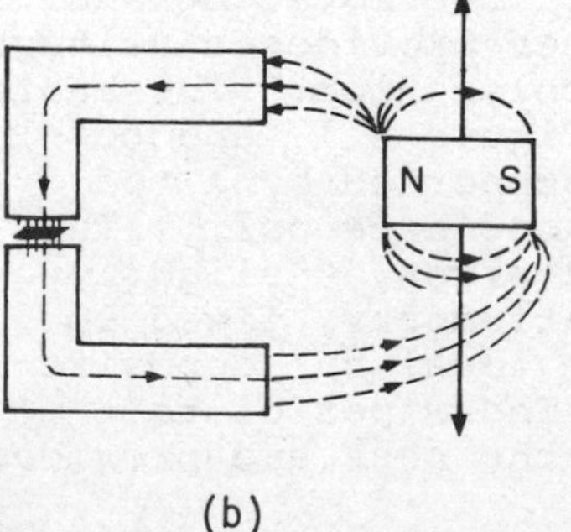

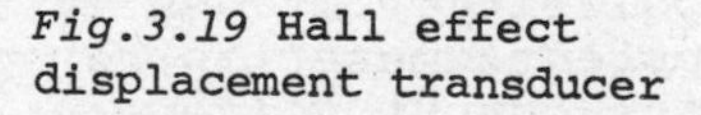
Fig.3.19 Hall effect displacement transducer

When the magnet is centred, as shown in Figure 3.19(*a*), flux distribution is symmetrical about both halves of the core and the effective field in the gap is zero; the output voltage is thus zero.

Small displacement of the magnet, as shown in Figure 3.19(*b*), alters the distribution of flux (increasing the flux in one side and reducing that of the other side). A voltage is then induced whose magnitude is linearly proportional to the displacement and whose polarity depends on the direction of the displacement. Transducers based on this principle are very sensitive and can detect very small displacements.

3.12 PHOTOELECTRIC EFFECT

Transducers based on the photoelectric effect are used for measurement of illumination and the detection of light. The basic principle of operation of cells used in such transducers may be one of the following:

PHOTOEMISSIVE CELL
The liberation of electrons from a surface subjected to light is known as the *photoemissive effect*. A photoemissive cell consists of a vacuum-filled or gas-filled tube in which electrons are emitted from the photosensitive cathode, accelerated and collected by the anode.

PHOTOCONDUCTIVE CELL
The change in conductivity of certain substances, when subjected to light is known as the *photoconductive effect*. Semiconductors such as selenium, the selenides, tellurides and germanium have this property. A photoconductive cell consists simply of a piece of semiconductor material such as cadmium sulphide or lead selenide whose resistance value depends upon the light intensity reaching it. A feature of photoconductive cells is their high sensitivity to infra red radiation.

PHOTOVOLTAIC CELL
Production of an e.m.f. by radiant light energy incident upon the junction of two dissimilar materials is known as the *photovoltaic effect*. A photovoltaic cell consists of two semiconductor materials forming a junction. Such a cell is self-generating and produces a voltage proportional to the light intensity reaching its junction.

Photovoltaic cells are used in digital computer peripheral equipments. Paper-tape readers and punched card readers, for example, employ photo-cells to detect the presence or the absence of a hole through which a beam of light passes. The punched tape is read by passing it between a light source and a reading head (photocells). When a hole passes under the head, the cell is energized and produces an electric impulse.

Other photocells commonly in use are *photodiodes*, *phototransistors* and *photothyristors* all of which operate on similar principles.

3.13 DIGITAL TRANSDUCERS

The output of digital transducers is in the form of electrical pulses, resulting in high degrees of accuracy often required in systems such as machine tool control.

A very popular and widely used digital transducer is the SHAFT ENCODER (a form of analogue-to-digital converter) which converts the angular position of a shaft into the corresponding digital code in the form of electrical pulses. The most usual form uses a disc and may be optical type, magnetic type, or brush type.

The disc of an *optical type* is usually glass with the code pattern represented by transparent and photoprinted opaque sections. A narrow-line light source is placed on one side of the disc and photoelectric sensors are mounted on the other side of the disc in the same radial line as shown in Figure 3.20(*a*). When the light source is energized, the sensors produce a coded digital output which represent the angular position of the shaft.

The *brush-type* disc is coded using a conducting material, shown black in Figure 3.20(*b*) and (*c*). The white areas are non-conducting and the pattern forms a code such as pure binary or Gray (see Chapter 8, sections 8.1.1 and 8.2.1). A common voltage is connected to all conducting areas and small carbon brushes, pressed against the disc by springs, are used to pick up the output voltages. Figure 3.20(*b*) shows a disc with pure binary code, and a disc with Gray code is shown in Figure 3.20(*c*). The decimal-binary relationship for these two codes is given in Table 8.6 of page 217.

As the shaft and disc rotate, brushes make contact with the conducting areas. The electrical pulses obtained give a code combination which is a direct indication of the angular position of the shaft.

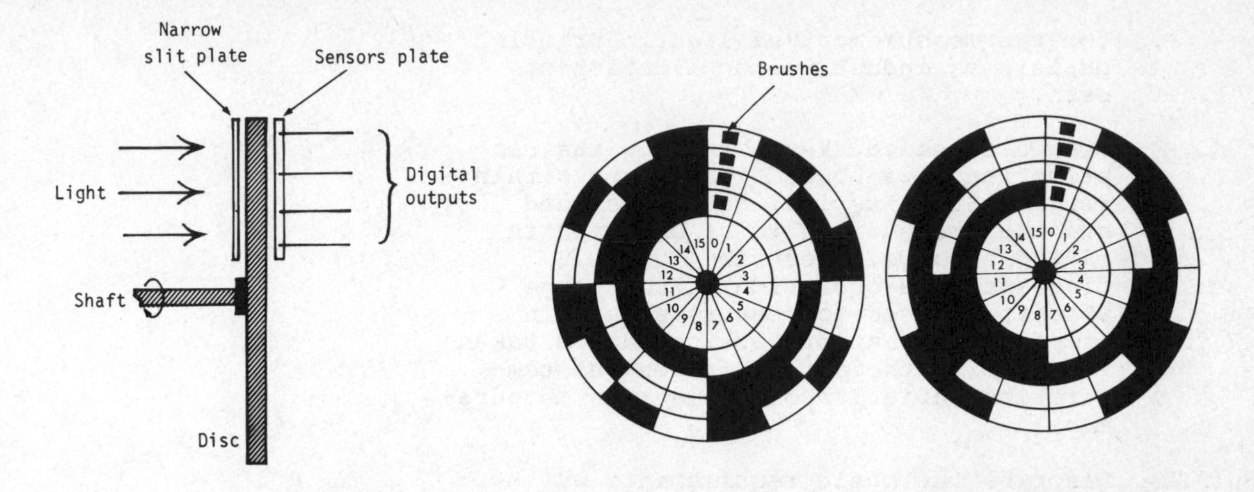

Fig.3.20

The use of pure binary code can lead to large errors. A deflection on the innermost track for example, will cause positions 0 and 16 to be indistinguishable, giving rise to a possible error of 180°. In order to avoid this, Gray code (see Chapter 8, section 8.2.1) is often used. As shown in Table 8.6, any two adjacent code combinations (including 15 and 0) differ by only one binary digit. On the disc, this gives a change of only one track when going from any one section to the next, thus reducing the possible error to 22.5°. Discs which employ codes with up to 19 digits have been produced and very high degrees of accuracy are obtained from them.

EXERCISES

(3.1) Write a short, concise account of the consideration which should be used in selecting a transducer for a given task. Explain the principle of operation of one type of electrical energy converting transducer.

(3.2) Explain what is meant by "variation of resistance with temperature" and illustrate, with the aid of a sketch, what practical application this effect might have.

(3.3) Explain, using labelled diagrams, the construction and operation of (*a*) a typical transducer for the measurement of temperature (within the range 200-1000°C), and (*b*) a typical transducer for the measurement of light. Briefly explain an industrial application of each.

(3.4) Draw a labelled sketch to show the construction of a bonded resistance strain gauge. State typical materials used and the precautions to be observed in fixing a gauge. Describe, using diagrams, the operation of the gauge when it is used to measure strain in a structure. What precaution can be taken to minimize the effect of ambient temperature variation on the strain measurement?

(3.5) Describe the basic requirements which have to be considered in the selection of a transducer for a particular application. Explain with the aid of suitable diagrams the principle of operation of a transducer which produces an electrical signal proportional to angular position. Support your exposition with the relevant theory and indicate the factors which limit the accuracy of the device.

(3.6) Sketch the construction of the following transducers: (*a*) synchro, (*b*) drag-cup tachogenerator, (*c*) E-I transformers. Describe briefly, using appropriate diagrams, an application of each one.

(3.7) Compare the advantages and disadvantages of the following transducers when used as error-forming devices in a.c. servo-systems: (*a*) a pair of driven rheostats, (*b*) a transmitter and control transformer (synchro pair). Give a circuit diagram showing how each would be embodied in an a.c. position control system. Explain how the error signal is formed for either (*a*) or (*b*).

(3.8) It is required to convert the following variables to electrical signals: (*a*) pressure, (*b*) linear displacement, (*c*) angular velocity, (*d*) angular acceleration. Choose *any three* and explain the construction and physical action of the transducer required in each case.

(3.9) Describe, with the aid of suitable sketches, the principles of operation

of *three* different types of transducer, as follows:
(*a*) one to trigger a fire alarm system by sensing temperature,
(*b*) one to measure liquid flow in a ventilation cooling system,
(*c*) one to read information in an optical paper tape reader.

(3.10) With the aid of sketches explain the construction and operation of a synchro.
Describe, with a circuit diagram, the operation of a pair of such devices to obtain a signal proportional to the angular positional difference between two shafts. Show how the error signal may be used to position the controlled shaft.

(3.11) Describe with the aid of sketches the principles of operation of the transducers listed below.
(*a*) A.C. tachogenerator
(*b*) D.C. tachogenerator
(*c*) Synchro transmitter and receiver
(*d*) Linear differential transformer with d.c. output.
Outline a suitable application and the function of each transducer when it is used in a control system.

(3.12) (*a*) What is a transducer? Why are transducers used?
(*b*) With the aid of sketches, describe the principle of operation of *two* types of transducers which may be used in digital computer peripheral equipment.

(3.13) An electromagnetic counter and a photocell with their associated circuitry are to be used to count the number of objects passing along a conveyor belt. Draw a simple circuit and explain its action. Explain briefly the operation of the type of photocell you choose.

(3.14) Describe the coded disc methods for the digital encoding of angular position. Indicate factors limiting accuracy, and describe the disc coding used to avoid ambiguity in reading.
There may be some difficulty in making calculations using data from coded discs. Comment on this and say how the difficulty may be overcome.

4 Control Elements: Controllers

A controller may be defined as a device or element which controls the transfer of power from an external source to its output in accordance with a command or input signal.

In many control systems the control signal is often too small to be of direct use and has to be amplified. Since energy cannot be created or destroyed but can only be transferred from one form (or source) to another, amplifiers are often classified under controllers. Thus, a controller may also be defined as a device which receives a low power control signal, together with power from an external source, and supplies a larger amount of controlled power.

This chapter deals with the basic principle of operation of controllers commonly used in control systems.

4.1 TYPES OF CONTROLLER

Broadly speaking, controllers may be divided into two groups:

(*a*) *Continuous*: the output varies continuously in accordance with the input.

(*b*) *Discontinuous*: the output power is controlled in discrete steps.

Depending on the principle of their operation, controllers may be one of the following types.

ELECTRICAL Electrical amplifiers cover a wide range. The continuous types can be

(*a*) *Electronic* These are normally used for low-power systems and include both a.c. and d.c. amplifiers.

(*b*) *Rotary* Strictly speaking, these are electromechanical amplifiers and include devices such as the metadyne and amplidyne (see page 22) in which output powers of the order of kilowatts are obtainable.

(*c*) *Magnetic* Magnetic amplifiers are current-operated devices and, although are probably at their best when delivering output powers of less than 1 kW, they are generally used for high power applications. They are robust and extremely reliable, but are limited in their operation within 50 Hz to 5 kHz frequency range.

The discontinuous types include thyristor control, ignitron and thyratron amplifiers which are suitable for high power applications.

MECHANICAL Most mechanical amplifiers incorporate electrical motors, and are therefore not strictly mechanical. In a mechanical torque amplifier for example, the torque transmitted from the input shaft to the output shaft is amplified by the action of a rotating drum driven, through a gear system, by a constant speed motor.

PNEUMATIC These controllers usually use compressed air as the control medium. They are very reliable and safe and for this reason find extensive use in process control systems. Pneumatic controllers are also used in applications such as aero-space projects where high pressures of the order of 20×10^6 N/m^2 are employed. One major advantage of pneumatic systems is the accessibility and convenience of using air.

HYDRAULIC The control medium in hydraulic systems is an incompressible fluid such as oil. In general, hydraulic elements are very rapid-acting and are more rugged than corresponding electrical components. These are major factors in applications such as aircraft, where vibration, shock, and noise pickup may cause damage to fine wires and adversely affect the normal operation of electrical equipment. In high pressure hydraulic systems, very large forces are obtained and can be used for rapid acceleration and accurate positioning of heavy loads. Hydraulic motors are generally much smaller than electric motors of equal power output. This results in considerable size and weight savings. Hydraulic systems, however, require elaborate power supplies and pressure-regulating devices. Machine-tool, speed control, and position control systems are some examples in which hydraulic controllers are used.

4.2 ELECTRONIC AMPLIFIERS

Electronic amplifiers may be of the thermionic or semiconductor type. Depending on the application, these in turn may be d.c. amplifiers or a.c. amplifiers. One type, particularly important and commonly used in electronic circuits and systems, is the *operational amplifier*. An operational amplifier is basically a high-gain direct-coupled amplifier with very high input impedance and very low output impedance. These amplifiers employ an odd number of stages so that, of the two inputs available, one is inverting. Their high gain, differential input, and low drift properties

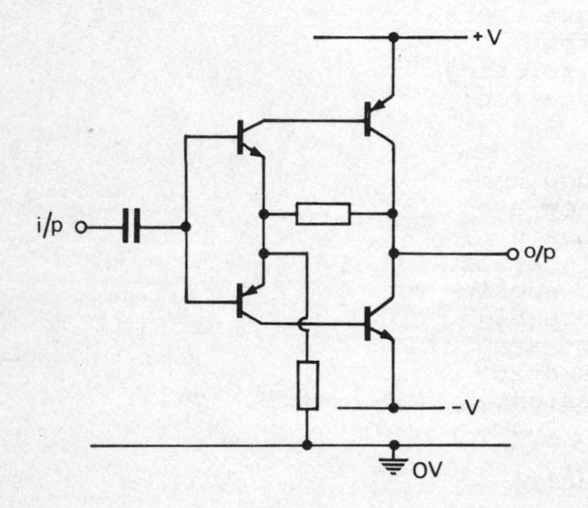

A.C. SERVO-AMPLIFIER

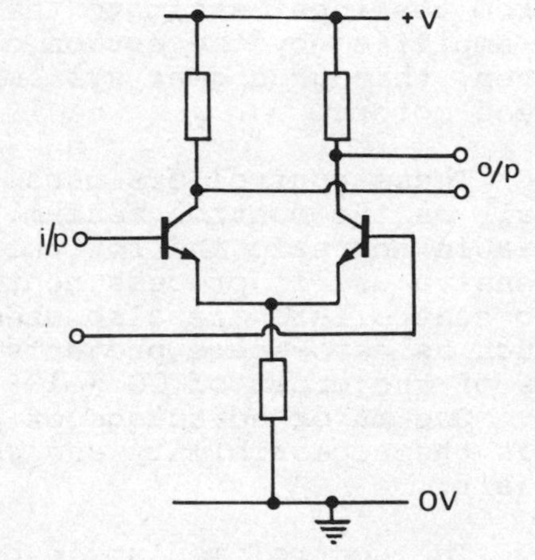

D.C. AMPLIFIER *Fig.4.1*

make them suitable for a wide range of applications, including analogue computers (see Chapter 7).

The principles of operation of electronic amplifiers are covered in detail in most electronic textbooks and will therefore not be discussed further. Figure 4.1 shows two simplified arrangements of commonly used transistorized amplifiers.

4.3 MAGNETIC AMPLIFIERS

The basic principle of operation of magnetic amplifiers relies on controlling the impedance and hence the current through one inductor by passing a direct current through another inductor. This is achieved as follows.

In the three-limbed core of Figure 4.2(*a*), the outer limb windings (a.c. windings) are

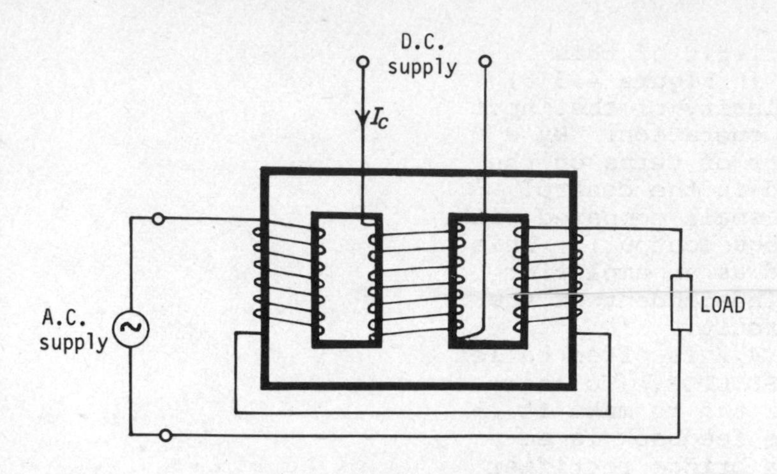

(a) BASIC ARRANGEMENT

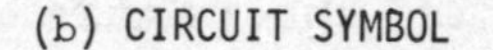

(b) CIRCUIT SYMBOL

Fig.4.2
Transductor

identical, wound in series opposition, and connected to an a.c. supply. When there is no direct current, the effective flux in the centre limb, due to the a.c. supply, is therefore zero and no voltage will be induced in its windings (control windings).

When a direct current is passed through the centre limb windings, a flux is set up in the core. This reduces the permeability μ of the core and hence reduces the inductance and impedance of the a.c. windings, thus allowing more alternating current to flow. The alternating current through a load connected in series with the a.c. windings can therefore be controlled by the direct current.

As shown in Figure 4.3(*a*), the effect of the control current is to shift the operation into the saturation region of the *B*-*H* characteristic and hence alter the permeability.

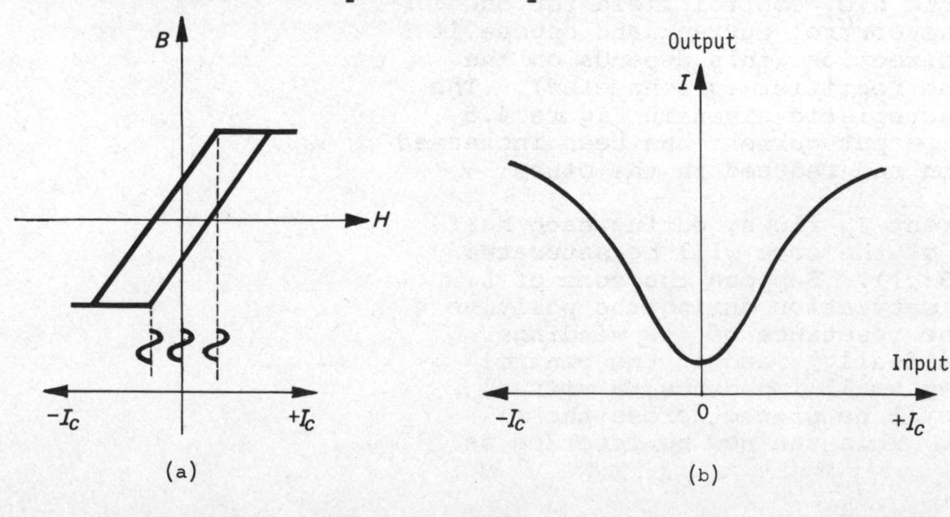

Fig.4.3

The input/output characteristic of this simple arrangement is shown in Figure 4.3(*b*) and it is clear that the polarity of the input current does not affect the operation. By a suitable choice of the number of turns on the windings, the power required in the control winding (input) can be made small compared with the power in the a.c. windings (output). This device can therefore be used as an amplifier with an output practically independent of the resistance of the output circuit.

The arrangement of Figure 4.2 is often called a saturable reactor or TRANSDUCTOR. To increase the gain of this transductor and to make it polarity sensitive, positive feedback is employed in conjunction with a bridge rectifier as shown in Figure 4.4. The feedback winding is wound on the centre limb and carries the

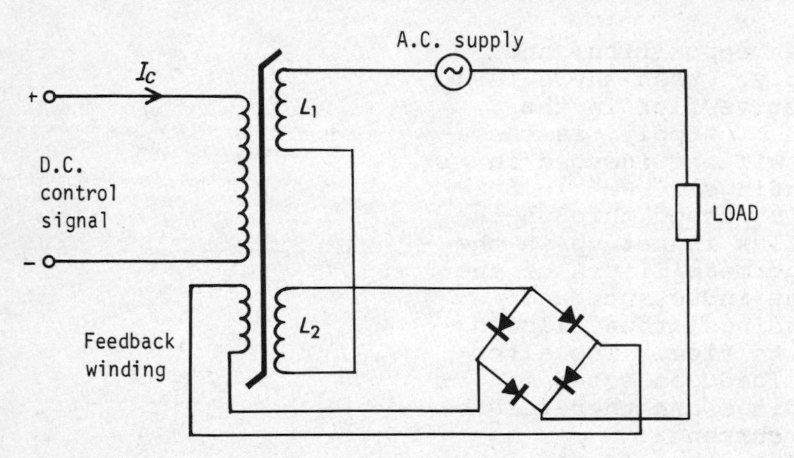

Fig.4.4 **Transducer with positive feedback**

unidirectional current supplied from the rectifier. The flux set up by the feedback winding will thus aid the d.c. control field for one direction of the control current and oppose it for the other direction (this depends on the way in which the rectifier is connected). The resultant characteristic given in Figure 4.5 shows that the output current has been increased in one direction and reduced in the other direction.

When the current I_C flows, during each half-cycle one side of the core will be saturated (see Figure 4.3(*a*)). Suppose the core of L_1 is driven into saturation during the positive half-cycle. The reactance of its windings falls to zero (ideally), and if the control circuit has very small impedance, a virtual short-circuit will be placed across the windings of L_2. This can now be regarded as

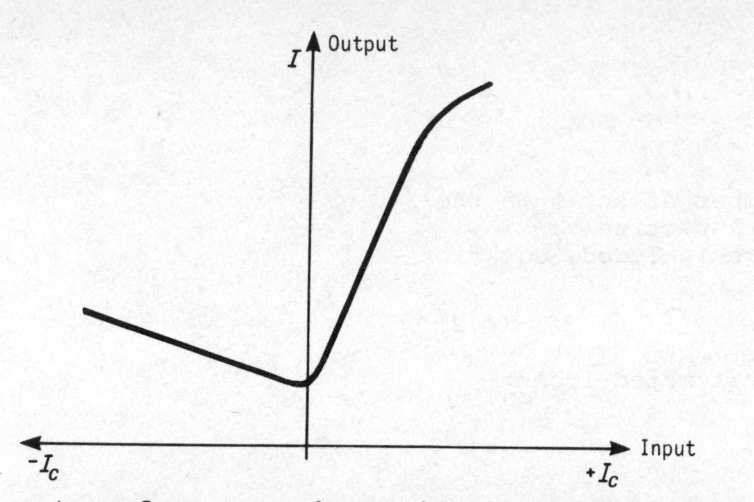

Fig.4.5

a transformer on short-circuit and the windings of L_2 will prevent any change of flux in its core. The flux in both cores will then stay constant (i.e. no voltage will be induced in them) and for the rest of the half-cycle the supply voltage will appear across the load. During the negative half-cycle the core of L_2 will be saturated and the above action will be repeated. Input/output waveforms are shown in Figure 4.6.

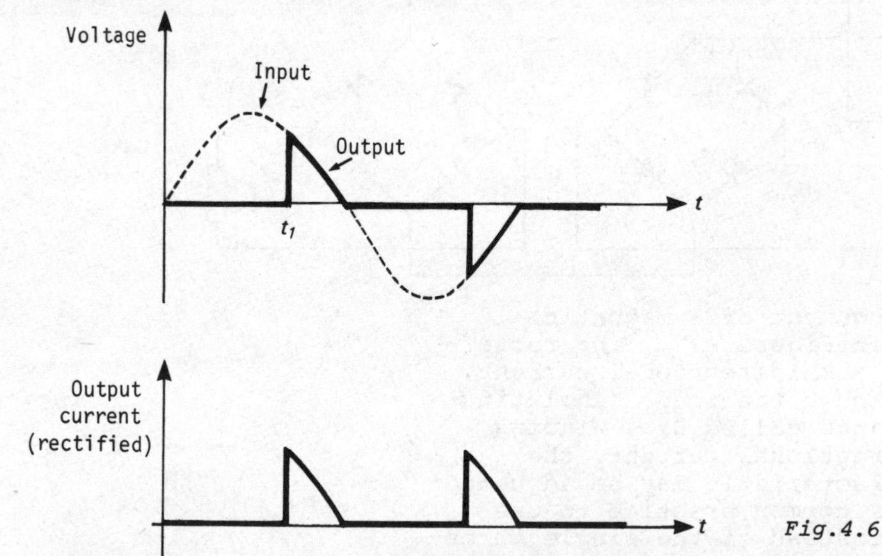

Fig.4.6

The *firing time* t_1 is inversely proportional to the frequency of the a.c. supply voltage and can be reduced by increasing the control current I_c.

Without the feedback windings, the current gain of the amplifier can be obtained from the approximate equation:

$$N_a I_a = N_c I_c \qquad (4.1)$$

thus

$$\text{Current gain} = \frac{I_c}{I_a} = \frac{N_a}{N_c}$$

where N_a and N_c are the number of turns in the a.c. and control windings respectively.

When feedback windings are included, equation (4.1) will be modified to

$$N_a I_a = N_c I_c + N_{fb} I_a \qquad (4.2)$$

and the power gain can be estimated from

$$\text{Power gain} \simeq \frac{I_a^2 R_L}{I_c^2 R_c}$$

where R_c and R_L are the resistances of the control circuit and the load respectively.

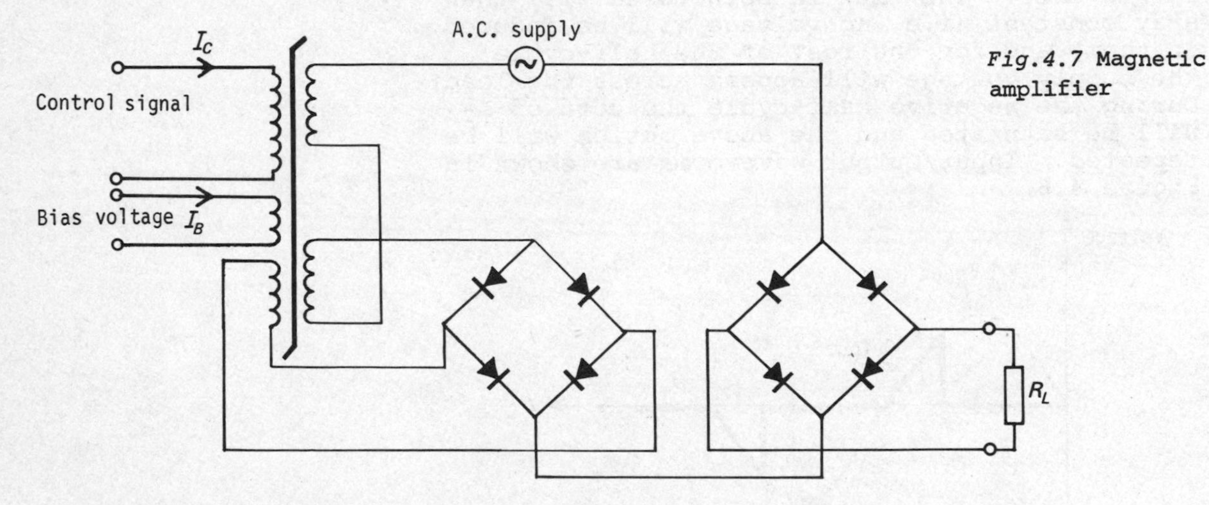

Fig.4.7 **Magnetic amplifier**

The complete arrangement of a magnetic amplifier is shown in Figure 4.7. The rectifier bridge provides a unidirectional current, often necessary, through the load. Including another set of windings called Bias Winding, which carry a unidirectional current, the position of the characteristic may be adjusted horizontally. It is common practice to use two separate cores instead of the single three-limbed core. The control windings are then wound over the a.c. windings so as to improve the magnetic coupling between the windings.

The magnetic properties of the core used in magnetic amplifiers include a narrow hysterisis loop, high maximum permeability, sharp knee at saturation, and for high power applications, high saturation flux density.

The main advantages of magnetic amplifiers are:

(1) Isolation between input and output circuits.
(2) Low power dissipation.
(3) Robust and extremely reliable.

The main disadvantages are:

(1) Limited frequency range, usually used within 50 Hz to 5 kHz.
(2) Time lag due to their inductive nature.

Example 4.1

During a test on a magnetic amplifier, the following results were obtained:

Control current mA	-4	-3	-2	-1	0	0.5	1	2	3	4
Output current mA	4	3	2	1	10	32	43	46.5	47.4	48

The amplifier control circuit had a resistance of 350 Ω and the load was a 100 Ω resistor. Estimate the best power gain.

Solution

The characteristic is plotted and shown in Figure 4.8.

In the linear region, the current gain is given by

$$\frac{\delta I_a}{\delta I_c} = \frac{40 - 10}{0.7} \simeq 42.9$$

and the power gain will be

$$\frac{P_a}{P_c} = \frac{I_a{}^2 R_L}{I_c{}^2 R_c} = \left(\frac{I_a}{I_c}\right)^2 \times \frac{R_L}{R_c}$$

$$= (42.9)^2 \times \frac{100}{350} \simeq 525$$

4.4 THYRISTORS

Thyristors are a group of controllable electronic switches with self-latching characteristics. Broadly speaking, they may be of reverse-blocking type, generally known as THYRISTOR, or of bi-directional type known by their trade name of TRIAC. The difference between the two will be discussed later.

The thyristor is a four-layer silicon semiconductor junction device as shown in Figure 4.9. The characteristics of the thyristor,

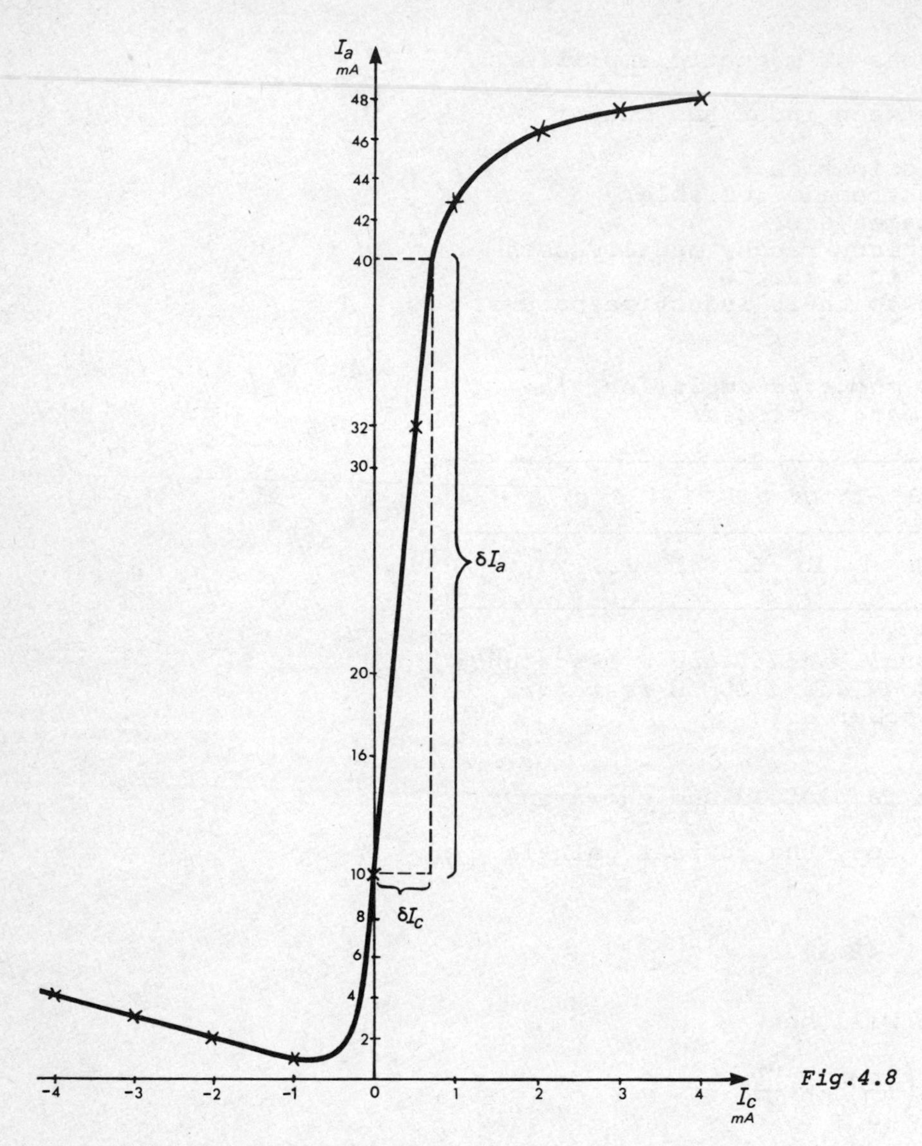

Fig.4.8

shown in Figure 4.10, are similar to those of a gas-filled triode in the forward direction, and the reverse characteristic follows the shape of a reverse biased diode.

The principle of operation of these devices is rather complex, but in brief may be explained as follows. Consider a thyristor as two transistors combined as shown in Figure 4.11.

When the anode A is negative with respect to the cathode K, emitter base junctions J_1 and J_3 are reverse biased and both transistors will be switched off. If the reverse voltage is increased, a small reverse-bias current will flow until reverse breakdown occurs.

When A is positive with respect to K, junctions J_1 and J_3 are both forward biased and J_2

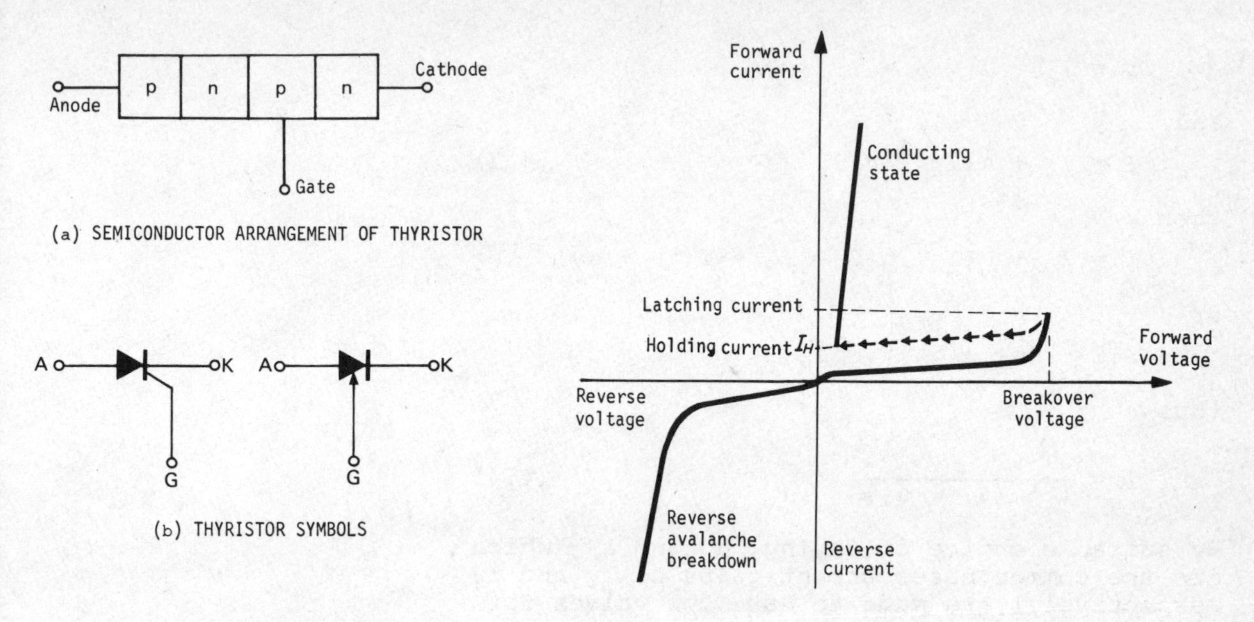

Fig.4.9

Fig.4.10 **Static characteristic of thyristor**

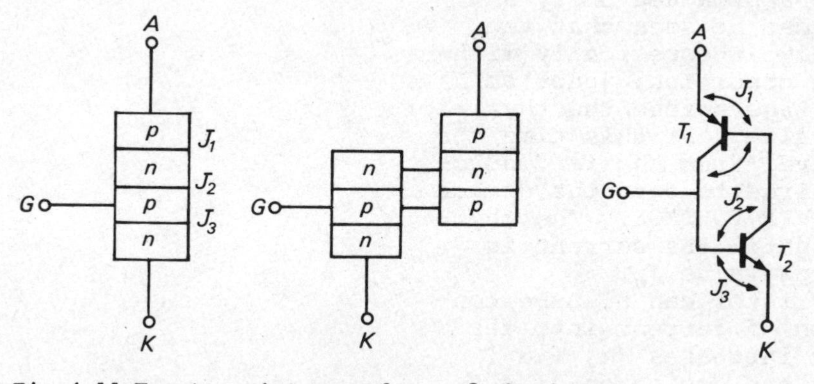

Fig.4.11 **Two-transistor analogy of thyristor**

will be reverse biased. Holes injected from the emitter of T_1 into its base will be collected across J_2. This contributes the flow of a hole-current $\alpha_1 I_h$. Similarly, the electrons injected from the emitter of T_2 into its base across J_3 will be collected across J_2. This contributes the flow of an electron-current $\alpha_2 I_e$. At the same time, in the reverse-biased diode forming junction J_2, the movement of holes from n to p and electrons from p to n causes a small current I_O to flow.

By Kirchhoff's law, currents at J_1 and J_3 must be equal. Also, the total external current flowing from A to K equals that passing across J_2. Thus,

$$I_h = I_e = I$$

and

$$I = I_o + \alpha_1 I_h + \alpha_2 I_e \qquad (4.3)$$

then

$$I = I_o + \alpha_1 I + \alpha_2 I = I_o + I(\alpha_1 + \alpha_2)$$

or

$$I[1 - (\alpha_1 + \alpha_2)] = I_o$$

thus

$$I = \frac{I_o}{1 - (\alpha_1 + \alpha_2)} \qquad (4.4)$$

By suitable choice of doping, α_1 and α_2 (which are the common base current gains of T_1 and T_2 respectively) are made to have low values for very small emitter currents (i.e. small applied voltage), rising towards unity as the current increases. As the voltage across the device is increased, $(\alpha_1 + \alpha_2)$ approaches unity and from equation (4.4) it can be seen that the current increases rapidly (theoretically without limit). Under this condition, junction J_2 breaks down and the voltage across the thyristor falls to a very small value. At this time the thyristor is said to have *fired* and is conducting. The voltage required to turn the device ON is called the BREAKOVER VOLTAGE. The thyristor will now stay ON until the current is reduced below the holding value I_H.

The firing of the thyristor can also be controlled by the injection of current into the gate. This effectively increases $(\alpha_1 + \alpha_2)$ towards unity and turns the device on at a lower applied voltage. Thus, if the voltage applied between A and K is kept below the breakover voltage, switching can only take place when there is sufficient gate current. Figure 4.12 shows a family of characteristics for different values of gate current.

A thyristor may be used as a control element in an a.c. circuit. The period for which the thyristor conducts is called the *conduction angle* and the time at which it is turned on is referred to as the *firing angle*.

Figure 4.13 shows two thyristors connected in an arrangement known as the INVERSE-PARALLEL configuration. The trigger pulses to the gates are arranged such that thyristor TH_1 is triggered at angle α^o and thyristor TH_2 is

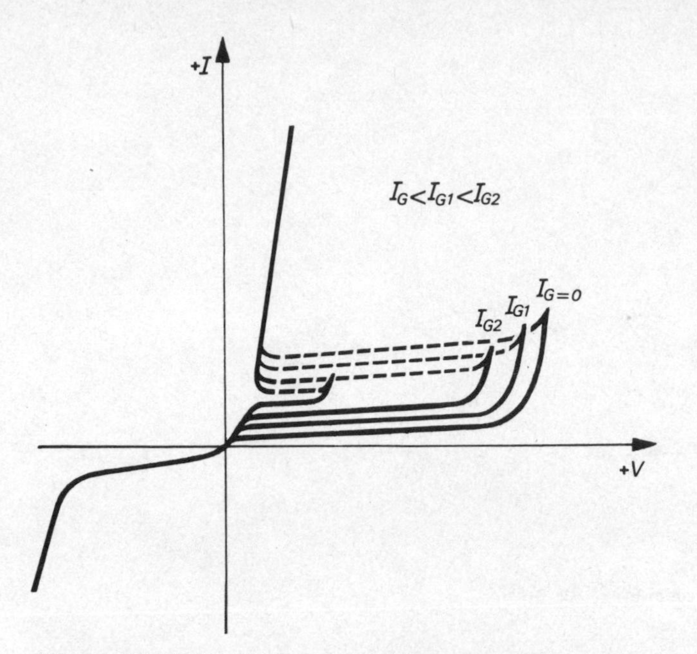

Fig.4.12 Effect of increasing gate current on thyristor characteristic

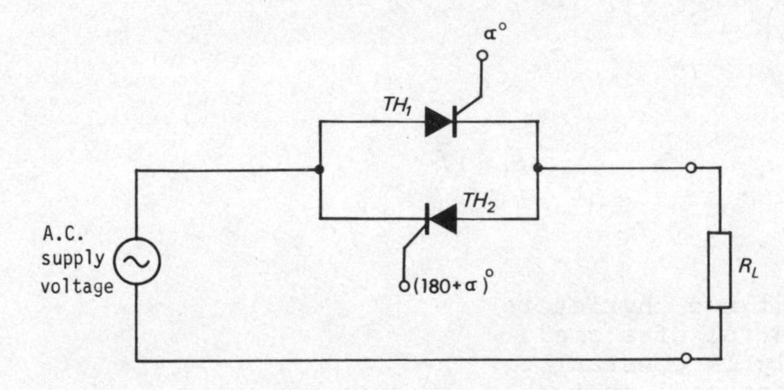

Fig.4.13 Inverse parallel configuration

triggered at $(180 + \alpha)^{\circ}$. When TH_1 is turned on, current flows through the load until the anode voltage becomes zero at 180°. Both thyristors will then stay non-conducting until TH_2 is triggered. Once again, current flows through the load until the end of the cycle (360°) at which time the anode voltage becomes zero and TH_2 turns off. The load current is therefore in the form of positive and negative pulses as shown in Figure 4.14.

It is seen that the load is supplied with current pulses whose duration determines the output power and can be controlled by varying the trigger angle α.

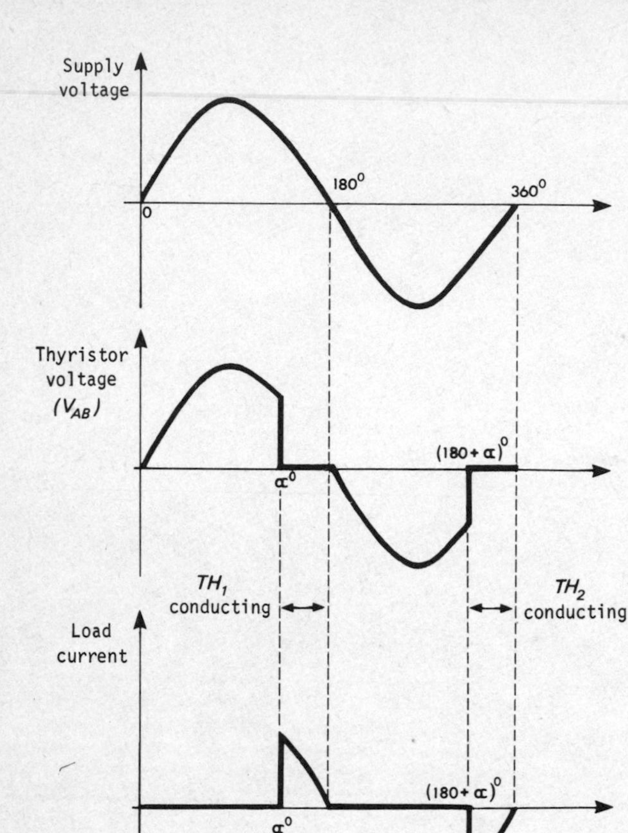

Fig.4.14

Figure 4.15 shows the use of two thyristors in a bridge for the speed control of a small motor. Since the field supply is constant, the motor speed will be proportional to the armature current which is controlled by the bridge. The desired speed is set by the reference voltage V_R and the actual speed is measured by the tachogenerator which provides voltage V_{fb} proportional to the speed of the motor. The difference between the two speeds is represented by the error signal $\varepsilon = V_R - V_{fb}$. The two thyristors are triggered from the pulse-generating network which is controlled by the error signal.

In practice the error signal is often too small and must be amplified. The block diagram of a more practical system is shown in Figure 4.16 in which the power supplied to the motor is under the control signals from both output current and output speed.

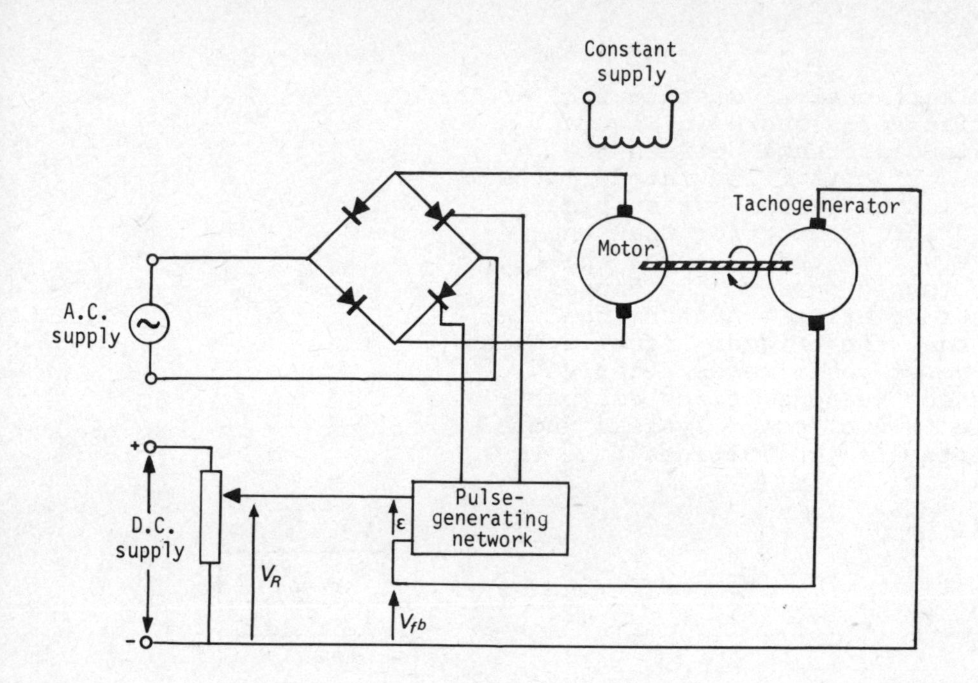

Fig.4.15

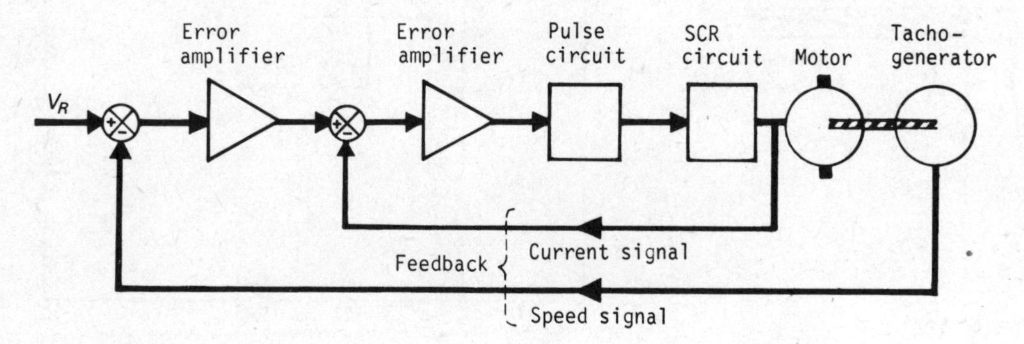

Fig.4.16

4.4.1 TRIAC

A triac is a bidirectional thyristor with static characteristice as shown in Figure 4.17(*a*). The main difference between the thyristor and the triac is illustrated in their reverse characteristics. The thyristor can only be triggered into conduction when the anode is positive w.r.t. the cathode, whilst the triac can be turned on with the anode either positive or negative. Another feature of the triac is that the trigger signal applied to the gate may be of positive or negative polarity. For these reasons triacs are particularly suitable in a.c. power systems where very large currents at high voltages have to be controlled.

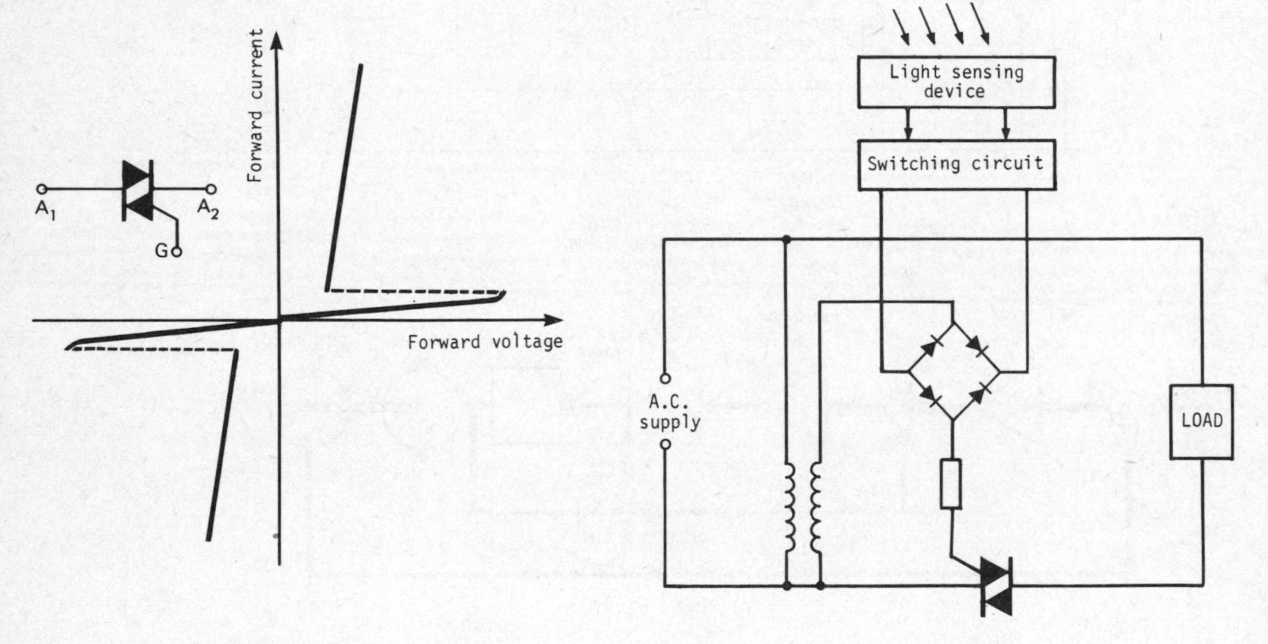

(a) STATIC CHARACTERISTIC AND CIRCUIT SYMBOL OF TRIAC

(b) SIMPLE APPLICATION OF TRIAC

Fig.4.17

Figure 4.17(*b*) shows a simple arrangement employing a triac in a light sensing circuit which could be used for applications such as burglar alarms, automatic opening of doors, switching motors, etc. When the light falls on the light sensor, a current is generated, turning the switching circuit on. This in turn shorts out the bridge and fires the triac into conduction allowing load current to flow.

4.5 PHASE-SENSITIVE RECTIFIERS

In a.c./d.c. control systems, the error signal (often produced by synchros) is an alternating signal and must be converted to a d.c. signal in order to operate the d.c. motor. A *phase-sensitive rectifier* converts an a.c. signal to a d.c. signal whose magnitude and polarity are proportional to the magnitude and phase of the a.c. signal.

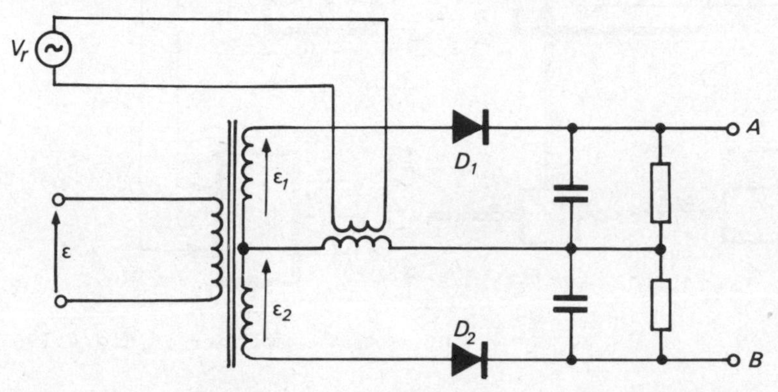

Fig.4.18 Phase-sensitive rectifier

A phase-sensitive rectifier (p.s.r.), shown in Figure 4.18, is basically a rectifier followed by an *R-C* smoothing network. The reference voltage V_r is chosen to be larger than the peak value of the error voltage ε so that during the negative half-cycles of the reference voltage both diodes will be reverse biased and non-conducting.

During the positive half-cycles of the reference voltage three conditions are possible:

(1) When the error voltage is zero, both diodes will be forward biased and conducting. Equal currents will pass through the diodes and the potential difference between A and B will be zero.

(2) When the error voltage is in phase with the reference voltage, the voltage applied to D_1 will be $V_r + \varepsilon_1$, and that applied to D_2 will be $V_r - \varepsilon_2$. D_1 will be forward biased and conducting and D_2 will be reverse biased. A will therefore be positive with respect to B.

(3) When the error voltage is antiphase with the reference voltage, the voltage applied to D_1 will be $V_r - \varepsilon_1$ and that applied to D_2 will be $V_r + \varepsilon_2$. D_1 will be reverse biased, while D_2 will be conducting. A will therefore be negative with respect to B.

The d.c. signal between A and B will therefore represent the magnitude and phase of the a.c. error signal.

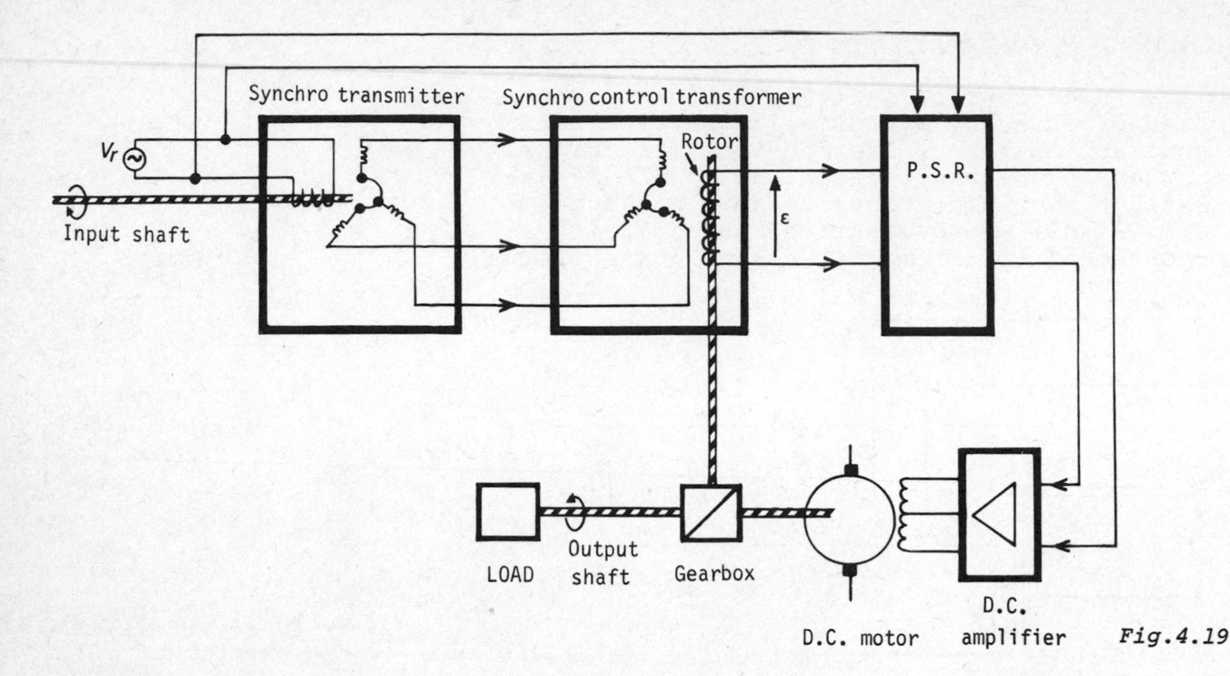

Fig.4.19

The block diagram of an a.c./d.c. position control system incorporating a phase-sensitive rectifier, synchros, a d.c. amplifier, and a split-field motor is shown in Figure 4.19. The operation of the system is as follows.

When the input and output shafts are in line, the transformer rotor axis is at right angles to the magnetic field of its stator and, hence, no voltage will be induced in the rotor windings. When the input shaft is rotated through an angle θ, the transmitter field will also rotate through the same angle, causing the transformer field to be rotated. This misalignment causes an error voltage to be induced in the windings of the transformer rotor. The magnitude of the error voltage is proportional to the degree of misalignment between the two shafts, and its phase depends on the direction of error.

The error signal is fed into the phase-sensitive rectifier which provides a d.c. signal whose magnitude and polarity are proportional to the magnitude and phase of the error voltage. The output signal from the p.s.r. is fed into the d.c. amplifier which drives the split-field d.c. motor in such a direction as to bring the output shaft into alignment with the input shaft. The motor will continue to rotate until the two shafts are lined up. At this time the transformer rotor axis will be at right angles to the magnetic field, the error voltage induced will be zero, and the motor will stop.

A gearbox is included so that a high speed motor can be used for low speed operations, thus providing high power-to-weight ratio.

4.6 HYDRAULIC CONTROLLERS

The two basic types of hydraulic controllers are the pump-controlled and the valve-controlled types. The pump-controlled system is reasonably simple and economic for controlling one motor. The system consists of a single pump, with variable stroke, which supplies a fixed-stroke motor. If two or more motors are to be controlled, the valve-controlled system is used. In this case each motor is controlled independently by a servo-valve.

A typical form of spool valve is shown in Figure 4.20. When the spool is in its central position (x = 0) both outlets A and B are cut off. Movement of the spool to the right allows the high-pressure input supply to be directed to channel A whilst channel B is connected to

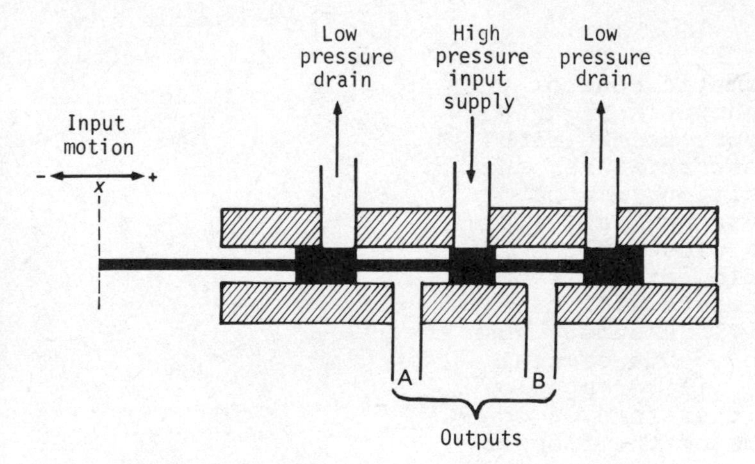

Fig.4.20 Simple form of spool valve

the low-pressure drain. When the spool is moved to the left, output A is connected to the drain and B is connected to the high-pressure input. The output pressures at A and B can therefore be controlled by the displacement of the spool.

The main problem associated with this valve is the existence of axial reaction forces. This can be reduced by using a flapper-nozzle valve (see section 4.7) operating a piston valve.

4.7 PNEUMATIC CONTROLLERS

Generally speaking, pneumatic controllers are cheaper, simpler and more reliable than comparable electrical and hydraulic controllers. Most pneumatic controllers consists of two stages: a nozzle-flapper valve which acts as a preamplifier (converting small deviation movements into an air pressure), followed by some form of pilot valve which acts as the main amplification stage.

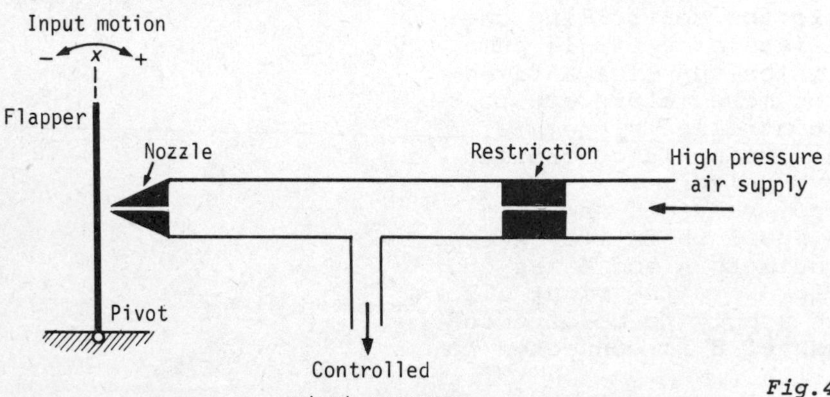

Fig.4.21 Flapper-nozzle valve

The simplest form of a pneumatic controller is the FLAPPER-NOZZLE VALVE shown in Figure 4.21. The motion of the flapper to the right or left alters the output pressure. When the flapper is in its central position ($x = 0$), the output pressure will have a nominal value, say P_0. As the flapper moves close to the nozzle it will restrict the flow of air from the nozzle and the output pressure will rise, approaching that of the supply pressure. When the flapper moves to the left of its central position, the output pressure will drop below P_0. The output pressure can therefore be controlled by the small movements of the flapper.

EXERCISES

(4.1) In the magnetic amplifier of Figure 4.4 the a.c. winding has 150 turns with negligible resistance and the control winding has 7500 turns with 250 Ω resistance. The feedback winding has 100 turns. Calculate the power gain of the amplifier when the load is a 100 Ω resistor.

(4.2) Discuss the principle of operation of each of the following amplifiers:
(*a*) hydraulic (*b*) magnetic (*c*) rotating

Give *one* example of the use of each, saying why it would be particularly suitable for that application.

(4.3) State, with reference to their characteristics, which type of amplifier could be used for the following applications
(*a*) linear positioning of a large machine tool bed
(*b*) amplification of the output of a thermocouple
(*c*) amplification of a.c. with large power output
(*d*) d.c. control of an a.c. supply
(*e*) profile tracing in machine tool control.

(4.4) Discuss, with the aid of appropriate diagrams, *two* of the following:
(*a*) a synchro pair with a phase-sensitive rectifier
(*b*) a d.c. servo amplifier
(*c*) a d.c. split-field motor with a d.c. tachogenerator
(*d*) a d.c. constant-current, armature supply for a servomotor.

(4.5) Sketch the circuit diagram of a shunt (COWAN) phase-sensitive rectifier, including a smoothing filter. With the aid of sinusoidal waveforms of signal and reference voltages having (*a*) in-phase and (*b*) anti-phase relationship, explain the operation of the rectifier. Discuss the advantages and disadvantages of including such a rectifier and filter in a control system.

(4.6) Explain briefly the basic principles of operation of a controlled semiconductor rectifier (thyristor). Describe *two* advantages in triggering the gate with voltage pulses and explain why a train of pulses may be required for each cycle of anode supply when the load is inductive.

(4.7) Describe any one type of solid state controllable rectifier. Show how this could be embodied in a system for the control of the voltage or current delivered to a variable load, and briefly describe the system operation. State what safety features should be incorporated in the system.

5 System Performance

In this chapter we study the responses of a system when it is subjected to various types of input signal. Damping, an essential part of any stable control system, and the effects of different methods of damping are explained.

5.1 SYSTEM RESPONSE

The UNIT-STEP is the most useful and convenient input for the study of the transient performance. Figure 5.1 shows a typical response to a unit-step input from which the following definitions describe the characteristics of a system.

(1) OVERSHOOT This is the maximum difference between the transient and steady-state response for a unit-step input. It is a measure of relative stability and is often expressed as a percentage of the final value of the output.

(2) DELAY TIME t_d This is a measure of the speed of response and is the time required for the response to reach 50% of its final value.

(3) RISE TIME t_r The time taken for the response to rise from 10 to 90 per cent of its final value.

(4) SETTLING TIME t_s This is defined as the time required for the response to reach and remain within a specific percentage of its final value (usually 2 to 5 per cent). The response falls and stays within ± 1% of its final value in approximately 5τ seconds, where τ is the time constant.

(5) TIME CONSTANT τ This is a measure of the rate of decay of the response:

$$\tau = 2J/F \qquad (5.1)$$

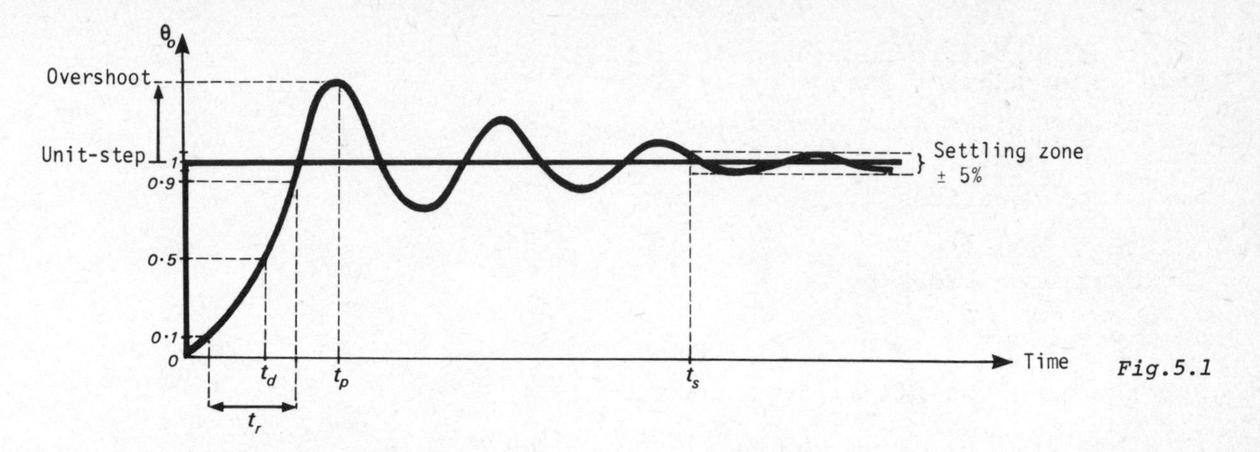

Fig.5.1

5.2 DIMENSIONLESS EQUATIONS

Depending on the amount of damping and frictional force in a system, the response to a unit-step input may be any one of the four shown in Figure 5.2.

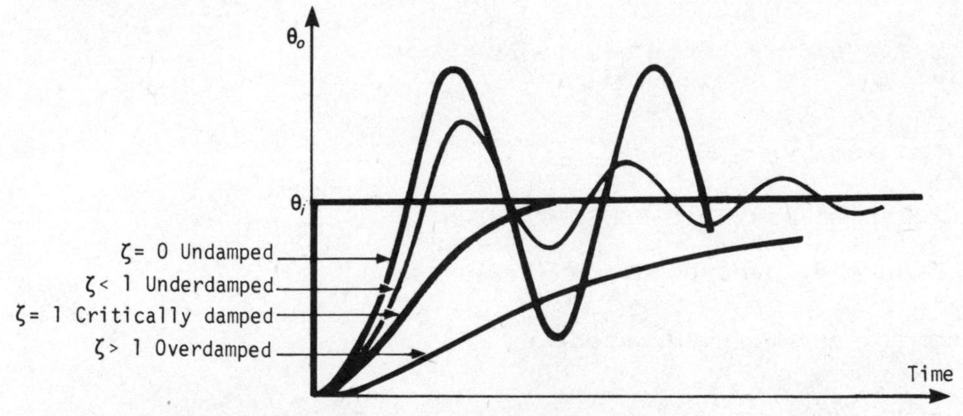

Fig.5.2 Various responses of a second-order system to a step input

Assuming that total frictional force and damping is F, and that needed for the system to be critically damped is F_c, then

$$\frac{F}{F_c} = \zeta \quad \text{(zeta)} \quad \textit{damping ratio} \qquad (5.2)$$

where

$$F_c = 2\sqrt{(KJ)} \qquad (5.3)$$

Without any damping the system would oscillate at an angular frequency ω_n given by

$$\omega_n = \sqrt{\frac{K}{J}} \quad \textit{undamped natural frequency} \qquad (5.4)$$

When the system is underdamped, it will oscillate at an angular frequency ω_d given by

$$\omega_d = \omega_n\sqrt{(1 - \zeta^2)} \quad \textit{damped natural frequency} \qquad (5.5)$$

Now, from equations (5.2) and (5.3)

$$F = 2\zeta\sqrt{(KJ)} \qquad (5.6)$$

Dividing both sides by J

$$\frac{F}{J} = 2\zeta\sqrt{\frac{K}{J}}$$

Substituting for $\sqrt{(K/J)}$ from equation (5.4)

$$\frac{F}{J} = 2\zeta\omega_n \qquad (5.7)$$

Now, dividing the differential equation of equation (1.10) by J

$$\frac{d^2\theta_o}{dt^2} + \frac{F}{J}\frac{d\theta_o}{dt} + \frac{K}{J}\theta_o = \frac{K}{J}\theta_i$$

Substituting for F/J and K/J from equations (5.4) and (5.6)

$$\frac{d^2\theta_o}{dt^2} + 2\zeta\omega_n\frac{d\theta_o}{dt} + \omega_n^2\,\theta_o = \omega_n^2\,\theta_i \qquad (5.8)$$

Equation (5.8) is known as the DIMENSIONLESS EQUATION.

Referring to Figure 5.1 and equation (5.2):

If $F < F_c$ then $\zeta < 1$,
system is *underdamped*; used in servos.

If $F = F_c$ then $\zeta = 1$,
system is *critically damped*; special applications only.

If $F > F_c$ then $\zeta > 1$,
system is *overdamped*; hardly ever used.

If $F = 0$ then $\zeta = 0$,
system is *undamped*; unstable.

Example 5.1

In a remote position-control system using synchros for error detection and viscous friction damping, the synchro output is 0.75 V *per degree of error and the coefficient of viscous friction is* 40 N-m per rad/s. *The motor torque constant referred to the load side is* 80×10^{-3} N-m/mA *and the total inertia of the system is*

0.85 kg-m^2. *If the system is to have a damping ratio of 0.7, determine*

(*a*) *the transconductance of the amplifier*
(*b*) *the undamped natural frequency of the system*
(*c*) *the steady-state velocity error when the input shaft is rotated at a constant speed of* 15 rev/min.

Solution

(*a*) From equation (5.6)

$$F = 2\zeta\sqrt{(KJ)}$$

then

$$K = \frac{F^2}{4\zeta^2 J} = \frac{40^2}{4 \times 0.7^2 \times 0.85} = 960.3 \text{ N-m/rad}$$

but

$$K = K_A K_t K_r$$

where K_r is the synchro constant in V/rad, K_t is the motor torque constant in N-m/mA, and K_A is the transconductance of the amplifier in mS. Thus,

$$K_r = 0.75 \times \frac{180}{\pi} = 43 \text{ V/rad}$$

Now

$$960.3 = K_A \times 80 \times 10^{-3} \times 43$$

$$K_A = \frac{960.3}{80 \times 10^{-3} \times 43} = 279.1 \text{ mS}$$

(*b*) From equation (5.4)

$$\omega_n = \sqrt{\frac{K}{J}} = \sqrt{\frac{960.3}{0.85}} = 33.6 \text{ rad/s}$$

(*c*) The differential equation of the system is

$$J\frac{d^2\theta_o}{dt^2} + F\frac{d\theta_o}{dt} + K\theta_o = K\theta_i$$

Under steady-state conditions the system (both the input and output) will rotate at constant velocity, but the output shaft will lag behind the input shaft position by a constant amount. This is called VELOCITY LAG or *velocity error*. Now, because the speed of rotation is constant, the acceleration term will be zero. Thus,

$$F\frac{d\theta_o}{dt} + K\theta_o = K\theta_i \qquad (5.9)$$

where $d\theta_o/dt = \omega_o$ is the angular velocity of the output shaft.

Then

$$F\omega_o = K(\theta_i - \theta_o) = K\theta_{ss}$$

where θ_{ss} is the steady-state velocity error.
Under steady-state condition $\omega_o = \omega_i$.

$$\theta_{ss} = \frac{F\omega_i}{K} \tag{5.10}$$

but $\omega_i = \dfrac{15 \times 2\pi}{60} = 1.57$ rad/s

$$\theta_{ss} = \frac{40 \times 1.57}{960.3} \times \frac{360}{2\pi} = 3.74^\circ$$

Example 5.2

In the system of Figure 1.11, (p.15) the motor is geared to the load through a 100:1 gearing. Given the following details for the system components, calculate the undamped natural frequency of oscillation.

Motor inertia $J_M = 3.2 \times 10^{-6}$ kg-m^2

Motor torque constant $K_t = 0.1 \times 10^{-3}$ N-m/mA

Amplifier transconductance $K_A = 40$ mA/V

Transducers constant $K_r = 1$ V per degree of error

Solution

Referring the quantities to the load shaft:

$$J' = n^2 J_M = 100^2 \times 3.2 \times 10^{-6}$$

$$K' = nK_A K_t K_r$$

but $K_r = 1$ V/degree of error $= 1 \times 360/2\pi$ V/radian of error

$$\omega_n = \sqrt{\frac{K'}{J'}} = \sqrt{\frac{100 \times 40 \times 0.1 \times 10^{-3} \times 1 \times 360}{100^2 \times 3.2 \times 10^{-6} \times 2\pi}}$$

$$= 26.8 \text{ rad/s}$$

5.3 TRANSIENT SOLUTION

Equation (5.8) on page 78 describes the behaviour of a 2nd order system in terms of its transient parameters and is therefore very useful. Solutions of this equation will now be considered for various inputs.

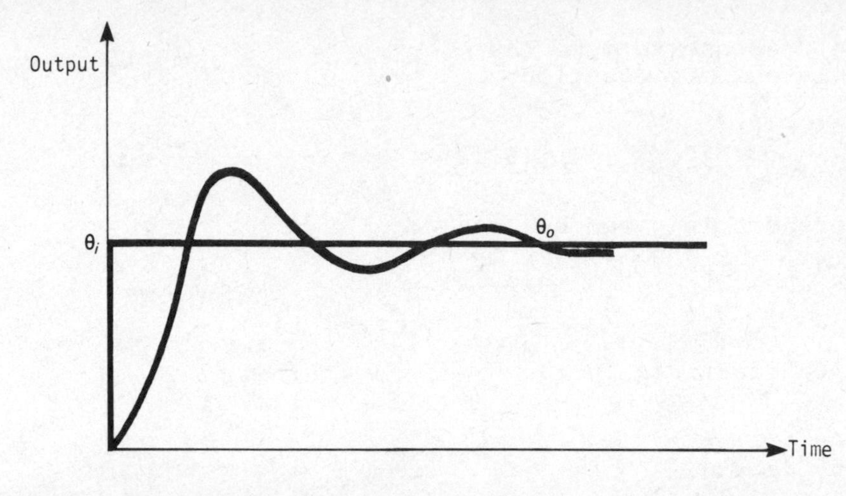

Fig.5.3 Response of an underdamped system to a step input

5.4 RESPONSE TO A STEP-INPUT

For a damping ratio of about 0.7 (typical of most servo systems), the response to a step-input is as shown in Figure 5.3. It is seen that eventually (i.e. under steady-state condition) the output equals the input and hence the steady-state error is zero.

Thus, for a step-input $\theta_{ss} = 0$.

The time solution of equation (5.8) in terms of ζ and ω_n is given by

$$\theta_o = \theta_i \left[1 - \frac{e^{-\zeta\omega_n t}}{\sqrt{(1 - \zeta^2)}} \sin\left\{\omega_n\sqrt{(1 - \zeta^2)}\,t + \phi\right\}\right] \tag{5.11}$$

where $\tan\phi = \dfrac{\sqrt{(1 - \zeta^2)}}{\zeta}$. (5.12)

Figure 5.4 shows a family of responses for various values of ζ.

If ζ is increased in value, it is evident from equation (5.11) and Figure 5.4 that

(*a*) the response becomes less oscillatory,
(*b*) the magnitude of the overshoot decreases,
(*c*) the response time increases and the system becomes sluggish.

The response time can be reduced by increasing the gain of the amplifier. However, this increases ω_n and reduces the value of ζ (see equations 5.4 and 5.7) and may cause instability. A damping ratio between 0.6 and 0.7 is found to give the fastest response with smallest overshoot. It should be noted that a system has a high speed of response if its time constant is small. From equation (5.1)

$$\text{Time constant } \tau = \frac{2J}{F} = \frac{1}{\zeta\omega_n} \tag{5.13}$$

The time taken for the response to make the overshoot can be calculated from equation (5.14):

$$t_p = \frac{\pi}{\omega_n\sqrt{(1 - \zeta^2)}} \qquad (5.14)$$

and the percentage overshoot is given by

$$100e^{-\zeta\pi/\sqrt{(1-\zeta^2)}} \qquad (5.15)$$

Example 5.3

A second-order system is described by the differential equation

$$1.5\frac{d^2\theta_o}{dt^2} + 3\frac{d\theta_o}{dt} + 6\theta_o = 6\theta_i$$

Determine the percentage overshoot of the response and the time taken to reach this overshoot when the system is subjected to a unit-step input

Solution
Dividing both sides of the equation by 1.5

$$\frac{d^2\theta_o}{dt^2} + 2\frac{d\theta_o}{dt} + 4\theta_o = 4\theta_i$$

Comparing the coefficients of this equation with those of equation (5.8):

$$2\zeta\omega_n = 2 \quad \text{and} \quad \omega_n^2 = 4$$

Thus, $\omega_n = 2$ and $\zeta = 0.5$.

Now, the percentage overshoot is given by

$$100e^{-\zeta\pi/\sqrt{(1-\zeta^2)}} = 100e^{-0.5\pi/\sqrt{(1-0.5^2)}}$$

$$= 100e^{-0.5\pi/0.866}$$

$$= 100e^{-1.814} \simeq 16.3\%$$

Also

$$t_p = \frac{\pi}{\omega_n\sqrt{(1 - \zeta^2)}} = \frac{\pi}{2 \times 0.866} \simeq 1.814 \text{ s}$$

Example 5.4

(a) *With the aid of diagrams, explain the meaning of the following terms when applied to the response of a control system:*

(i) *peak overshoot,*
(ii) *time to first zero,*
(iii) *settling time.*

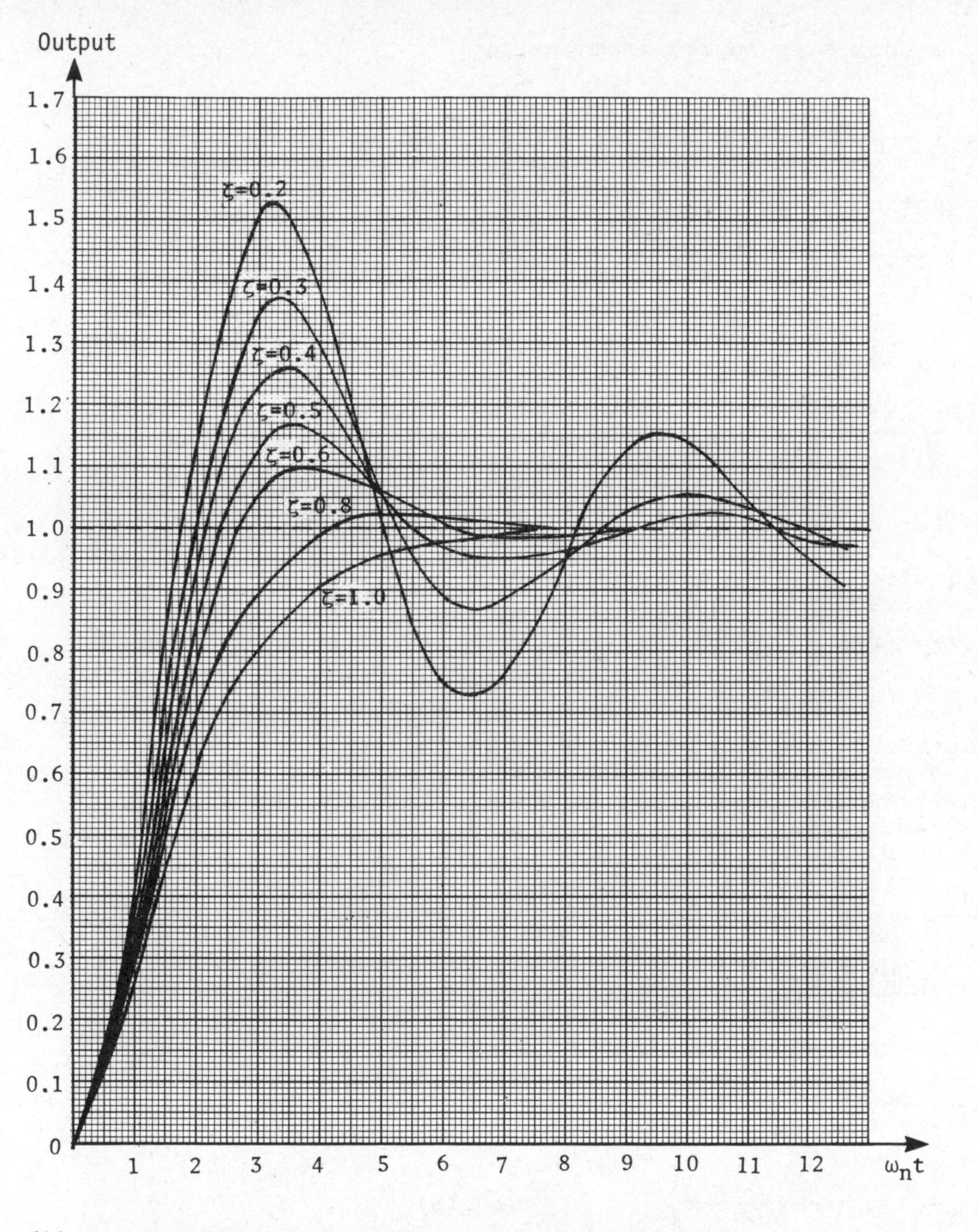

Fig.5.4

(*b*) *A second-order position control system has a damping ratio of* 0.5 *and an undamped natural frequency of* 6 rad/s. *Determine for a unit-step input:*

(i) *output response as a function of time,*
(ii) *value of the % peak overshoot,*
(iii) *settling time.*

Solution

(*a*) See section 5.1 and Figure 5.1.

(*b*) (i) For a unit-step input, from equation (5.11),

$$\theta_o(t) = 1 - \frac{e^{-\zeta\omega_n t}}{\sqrt{(1 - \zeta^2)}} \sin\left[\omega_n\sqrt{(1 - \zeta^2)}t + \phi\right]$$

but, from equation (5.12),

$$\tan\phi = \frac{\sqrt{(1 - \zeta^2)}}{\zeta} = \frac{\sqrt{(1 - 0.5^2)}}{0.5} = 1.732\,06$$

$$\phi = 60^\circ$$

Substituting values:

$$\theta_o(t) = 1 - \frac{e^{-0.5\times 6\times t}}{\sqrt{(1 - 0.5^2)}} \times \sin\left[6\sqrt{(1 - 0.5^2)}t + 60^\circ\right]$$

$$= 1 - 1.1547e^{-3t}\sin(5.196t + 60^\circ)$$

(ii) From equation (5.15), the percentage overshoot is given by

$$100e^{-\zeta\pi/\sqrt{(1-\zeta^2)}} = 100e^{-0.5\pi/\sqrt{(1-0.5^2)}}$$
$$= 16.3\%$$

(iii) The settling time can be estimated by putting the exponential term in the time-response equation of (5.11) equal to the percentage value specified.
Let the % be p. Then,

$$\frac{e^{-\zeta\omega_n t}}{\sqrt{(1 - \zeta^2)}} = \frac{p}{100}$$

$$e^{-\zeta\omega_n t} = \frac{p\sqrt{(1 - \zeta^2)}}{100}$$

Taking $\log_e$ of both sides

$$-\zeta\omega_n t = \log_e\left(\frac{p\sqrt{(1 - \zeta^2)}}{100}\right)$$

then

$$t_s \simeq -\frac{1}{\zeta\omega_n}\log_e\left(\frac{p\sqrt{(1 - \zeta^2)}}{100}\right) \qquad (5.16)$$

Since no value has been specified for p, 5% will be taken. Then, substituting values in equation (5.16):

$$t_s \simeq -\frac{1}{0.5 \times 6}\log_e\left(\frac{5\sqrt{(1 - 0.5^2)}}{100}\right) \simeq 1.046 \text{ s}$$

Example 5.5

A position-control system employs velocity feedback for damping and uses a gear system to match the load to the motor. Assuming that the system is subjected to a constant load torque of T_L N-m, obtain the differential equation of the system and hence derive an expression for the steady-state error under a constant-velocity input.

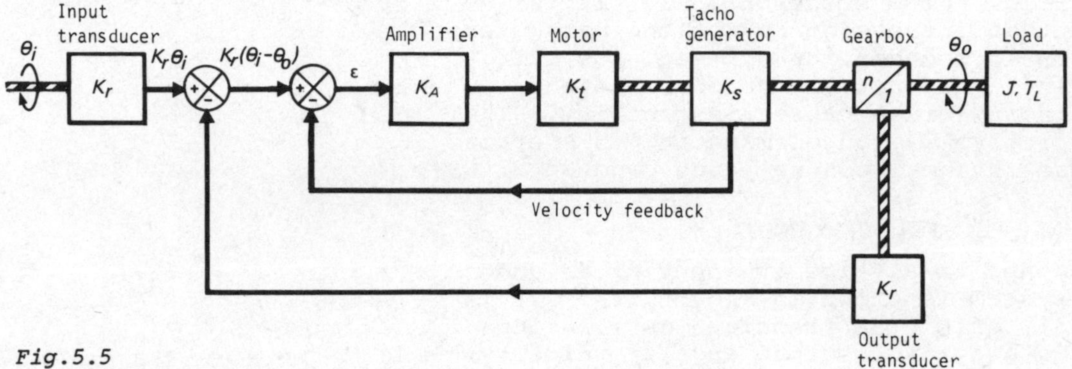

Fig.5.5

Solution

A block diagram for such a system is shown in Figure 5.5. If a system is subjected to a constant-load torque, the motor must have an input signal applied to it even under steady-state condition so as to produce the necessary torque. Thus

$$J\frac{d^2\theta_o}{dt^2} + T_L = nK_AK_t\varepsilon$$

$$J\frac{d^2\theta_o}{dt^2} + T_L = nK_AK_t\left\{K_r(\theta_i - \theta_o) - nK_s\frac{d\theta_o}{dt}\right\}$$

Rearranging, the differential equation will be

$$J\frac{d^2\theta_o}{dt^2} + n^2K_AK_tK_s\frac{d\theta_o}{dt} + nK_AK_tK_r\theta_o = nK_AK_tK_r\theta_i - T_L \tag{5.17}$$

Under steady-state condition

$$\frac{d^2\theta_o}{dt^2} = \frac{d\theta_o}{dt} = 0$$

then

$$nK_AK_tK_r\theta_o = nK_AK_tK_r\theta_i - T_L$$

but $\theta_i - \theta_o = \theta_{ss}$ thus

$$\theta_{ss} = \frac{T_L}{nK_AK_tK_r} \qquad (5.18)$$

The steady-state error due to load torque is often called OFF-SET and must be made as small as possible. From equation (5.18) it is evident that increasing any of the parameters K_A, K_t or K_r reduces the off-set. However, increasing K_t is not always possible and increasing K_A often causes instability. The most effective way of eliminating this error is to use integral control (see page 94).

5.5 RESPONSE TO A VELOCITY INPUT

The response to a velocity input of an underdamped system is shown in Figure 5.6. It is seen that, after the transient oscillations have died away, the output angular velocity equals the input angular velocity. However,

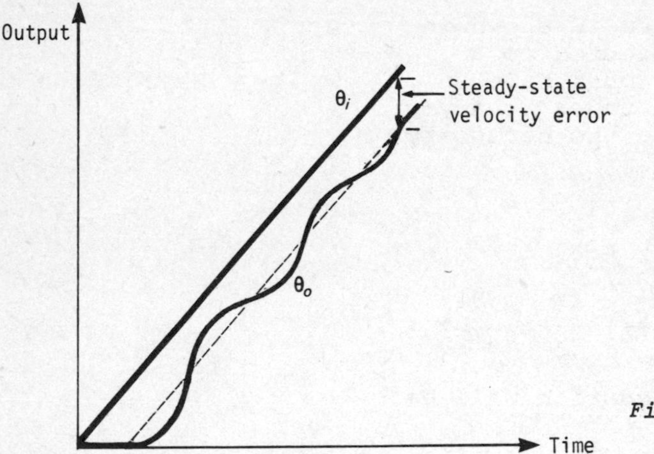

Fig.5.6 Response of an underdamped system to a velocity input

the output shaft will lag the input shaft position by a small amount known as the VELOCITY ERROR or VELOCITY LAG. This error must exist however small it might be, as the motor can only provide torque if it receives an input from the amplifier.

For an input of $\theta_i = \omega_i t$, the solution of equation (5.8) will be

$$\theta_o = \omega_i\left[t - \frac{2\zeta}{\omega_n} + \frac{e^{-\zeta\omega_n t}}{\omega_n\sqrt{(1 - \zeta^2)}} \times \sin\left(\omega_n\sqrt{(1 - \zeta^2)}t + \phi\right)\right]$$

(5.19)

where

$$\phi = 2\tan^{-1}\left(\frac{\sqrt{(1 - \zeta^2)}}{\zeta}\right) \qquad (5.20)$$

Under steady-state condition, the output equation becomes

$$\theta_{o(ss)} = \omega_i t - \frac{2\zeta\omega_i}{\omega_n}$$

Thus, the error will be

$$\theta_{ss} = \theta_i - \theta_o = \omega_i t - \omega_i t + \frac{2\zeta\omega_i}{\omega_n}$$

$$= \frac{2\zeta\omega_i}{\omega_n} \qquad (5.21)$$

It should be noted that this is in fact the same as equation (5.10).

In position-control systems such as machine tool or radar systems where the output should follow the input position precisely, velocity error is most undesirable. The magnitude of velocity error can be reduced by increasing the gain of the amplifier. However, as in the case of step-input, this makes the system more oscillatory.

5.6 RESPONSE TO A SINUSOIDAL INPUT

For a sinusoidal input of the form

$$\theta_i = A\sin\omega t$$

the steady-state output will be of the form

$$\theta_{o(ss)} = B\sin(\omega t + \phi)$$

It is seen that the angular frequency of the output is the same as the input but there is a difference in their amplitudes and the output lags the input by an angle ϕ. This type of input is very useful for study of the frequency response of a system. (See page 116.)

For a sinusoidal input of maximum value $\hat{\theta}_i$ and angular frequency ω, the steady-state solution of equation (1.10) is given by

$$\theta_{o(ss)} = \frac{\hat{\theta}_i \omega_n^2 \sin(\omega t + \phi)}{\sqrt{[(\omega_n^2 - \omega^2)^2 + (2\zeta\omega_n\omega)^2]}} \quad (5.22)$$

where

$$\tan\phi = \frac{2\zeta\omega_n\omega}{\omega_n^2 - \omega^2} \quad (5.23)$$

Example 5.6

The angular position of a radar aerial is controlled by a closed-loop control system to follow an input wheel. The input wheel is maintained in sinusoidal oscillation through ± 36° *with an angular frequency* ω = 2 rad/s. *The moving part of the aerial system, which is critically damped, has a moment of inertia of* 200 kg-m² *and a viscous frictional torque of* 1600 N-m per rad/s. *Calculate the amplitude of swing of the aerial and the time lag between the aerial and the input wheel.*

Solution

From equation (5.7)

$$\omega_n = \frac{F}{2\zeta J} = \frac{1600}{2 \times 1 \times 200} = 4 \text{ rad/s}$$

From equation (5.22), under steady-state condition, the amplitude of swing is given by

$$\theta_{o(ss)} = \frac{\hat{\theta}_i \omega_n^2}{\sqrt{[(\omega_n^2 - \omega^2)^2 + (2\zeta\omega_n\omega)^2]}}$$

$$= \frac{36 \times 4^2}{\sqrt{[(4^2 - 2^2)^2 + (2 \times 1 \times 4 \times 2)^2]}}$$

$$= \frac{576}{\sqrt{400}} = \pm 28.8^\circ$$

From equation (5.23), the phase lag is given by

$$\tan\phi = \frac{2\zeta\omega_n\omega}{\omega_n^2 - \omega^2} = \frac{2 \times 1 \times 4 \times 2}{4^2 - 2^2} = \frac{4}{3}$$

Thus $\phi = 53.2^\circ$

Now, since ω = 2 rad/s, then $f = 2/2\pi$ Hz and the period of one oscillation is

$$T = \frac{1}{f} = \pi \text{ seconds}$$

Then, a phase lag of 53.2° corresponds to

$$t = \frac{T \times \phi}{360} = \frac{\pi \times 53.2}{360} = 0.465 \text{ s}$$

5.7 DAMPING

It has been shown that, in order for a system to have fast response, the oscillations must die away quickly. Thus, some form of damping is necessary.

Viscous damping produces a frictional force which is proportional to relative motion of bodies. However, this kind of damping involves heat dissipation and power loss, and except in small position servos it is rarely used.

5.7.1 VELOCITY FEEDBACK

Another method of damping is the use of velocity feedback. That is, a voltage proportional to the output angular velocity is fed back and subtracted from the error voltage. The power loss due to friction is therefore minimized, leaving full motor torque to accelerate the

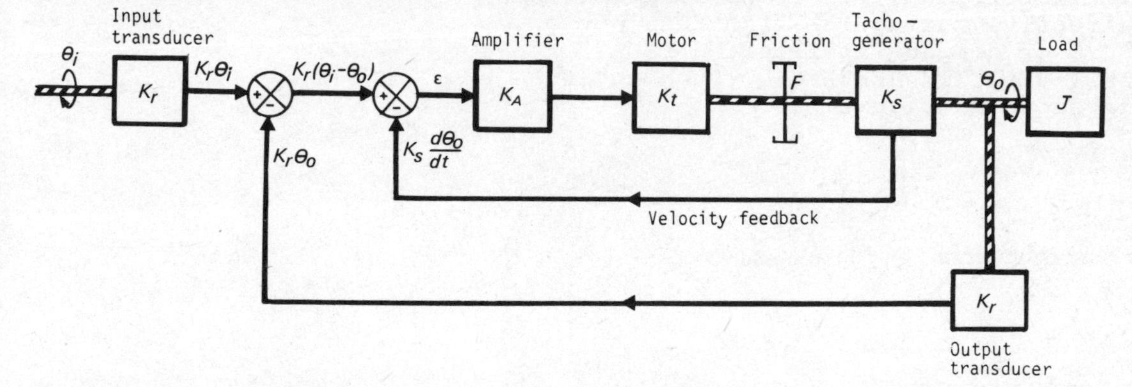

Fig.5.7 System using velocity feedback and viscous friction damping

load giving faster response. A position-control system using both viscous friction and velocity feedback for damping is shown in Figure 5.7. The differential equation of the system will be derived as follows. From equation (1.9)

$$J\frac{d^2\theta_o}{dt^2} + F\frac{d\theta_o}{dt} = K_A K_t \varepsilon$$

but

$$\varepsilon = K_r(\theta_i - \theta_o) - K_s\frac{d\theta_o}{dt}$$

therefore

$$J\frac{d^2\theta_o}{dt^2} + F\frac{d\theta_o}{dt} = K_A K_t\left[K_r(\theta_i - \theta_o) - K_s\frac{d\theta_o}{dt}\right] \tag{5.24}$$

Rearranging, the differential equation of the system will be

$$J\frac{d^2\theta_o}{dt^2} + (F + K_AK_tK_s)\frac{d\theta_o}{dt} + K_AK_tK_r\theta_o = K_AK_tK_r\theta_i \qquad (5.25)$$

Although velocity feedback overcomes the problem of heat dissipation due to friction, it does not reduce velocity error. This is because under steady-state condition the tachogenerator still feeds a constant voltage back.

A capacitor in series with the feedback signal will allow feedback during the transient period, but blocks the constant voltage under steady-state condition. In practice a simple *C-R* network is often used (which acts as a differentiator) and, since the feedback is only effective during the transient period, this form of damping is referred to as TRANSIENT VELOCITY FEEDBACK. Such an arrangement is shown in Figure 5.8.

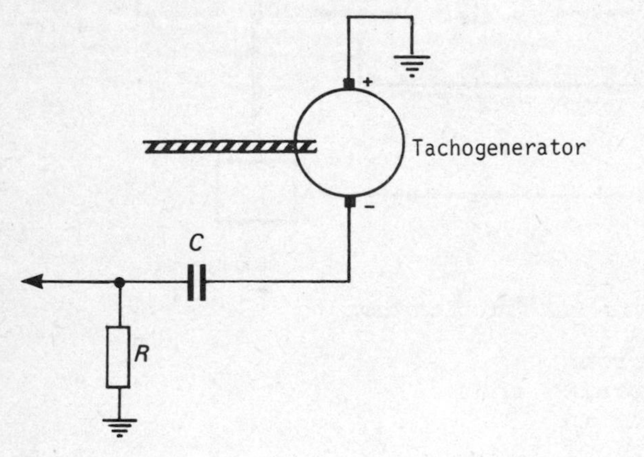

Fig.5.8

Example 5.7

The system of Figure 5.7 *is used for controlling the position of a rotatable mass. Total inertia of the system, referred to the mass, is* 100 kg-m^2 *and the system is critically damped with an undamped natural frequency of* 15 rad/s. *If the frictional torque is half that of the velocity feedback, determine the feedback torque per unit angular velocity.*

Solution

From equation (5.25), let

$$K_AK_tK_r = K \quad \text{and} \quad F + K_AK_tK_s = F'$$

then, comparing coefficients with those of equations (1.10) and (5.8), and from equations (5.4) and (5.6),

$$\omega_n^2 = K/J$$

thus

$$K = J\omega_n^2 = 100 \times 15^2 = 22\,500$$

and

$$F' = 2\zeta\sqrt{(KJ)} = 2 \times 1\sqrt{(22\,500 \times 100)} = 3\,000$$

But $F' = F + K_A K_t K_s$. Letting $K_A K_t K_s = K_V$, then since $F = \frac{1}{2}K_V$,

$$F' = \frac{1}{2}K_V + K_V$$

$$K_V = \frac{2}{3}F' = \frac{2}{3} \times 3000 = 2000 \text{ N-m per rad/s}$$

5.7.2 ERROR-RATE

Damping should be most effective when error is increasing most rapidly. This implies that the actuating signal feeding the motor should be a function of both the error and its rate of change.

A common method of damping a system and achieving the desired response is to use ERROR-RATE DAMPING. In such systems the error signal is passed through a network which gives an output proportional to the error plus its rate of change.

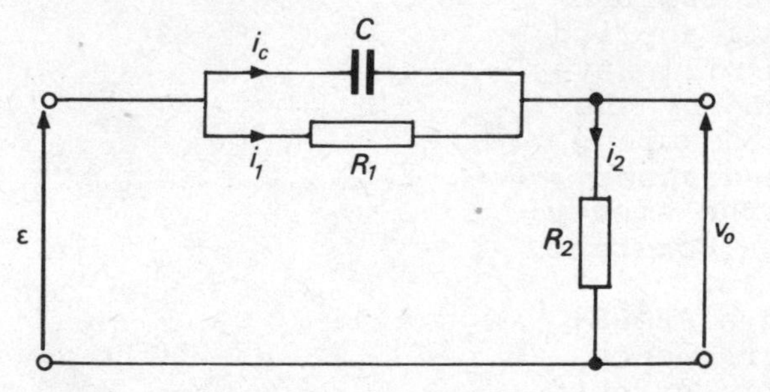

Fig.5.9 Phase-lead circuit

Consider the circuit shown in Figure 5.9 where R_2 is small compared with R_1. The operation of the circuit is basically such that the R_1R_2 path provides a signal proportional to the error voltage, whilst the C-R_2 path acts as a differentiator and provides a signal proportional to the rate of change of the error voltage.

Under steady-state condition the capacitor acts as an open circuit and

$$v_o \simeq \frac{R_2}{R_1 + R_2}\varepsilon$$

Since R_2 is very small compared with R_1, the output voltage will be a small fraction of the input error voltage. Thus, the voltage across R_1 and C will be approximately equal to the input voltage:

$$v_c = v_{R_1} = \varepsilon - v_o \simeq \varepsilon$$

then

$$i_c \simeq C\frac{d\varepsilon}{dt} \quad \text{and} \quad i_1 \simeq \frac{\varepsilon}{R_1}$$

but

$$v_o \simeq (i_1 + i_c)R_2 \simeq \left(\frac{\varepsilon}{R_1} + C\frac{d\varepsilon}{dt}\right)R_2$$

$$v_o \simeq \frac{R_2}{R_1}\varepsilon + R_2C\frac{d\varepsilon}{dt} \tag{5.26}$$

The circuit of Figure 5.9 thus provides an output which is proportional plus derivative of the input error voltage. This voltage is then fed into the amplifier to provide the actuating signal for the motor.

Consider a servo system incorporating error-rate damping. When a step input is applied the load moves towards its new desired position. The magnitude of error decreases and therefore the error-rate is negative. The signal applied to the amplifier is thus reducing and will have a damping effect on the torque produced by the motor. This effectively increases the damping ratio of the system and reduces the overshoot.

Error-rate damping does not affect the steady-state error because, when the error is constant, its rate of change is zero.

The circuit of Figure 5.9 is called a PHASE-LEAD or PHASE-ADVANCE CIRCUIT. The frequency characteristics of this circuit and its effect on systems' stability is dealt with in detail in chapter six.

Example 5.8

Draw the block diagram of a system incorporating error-rate damping and obtain the differential equation for the system.

Solution

In any practical system, there is always some

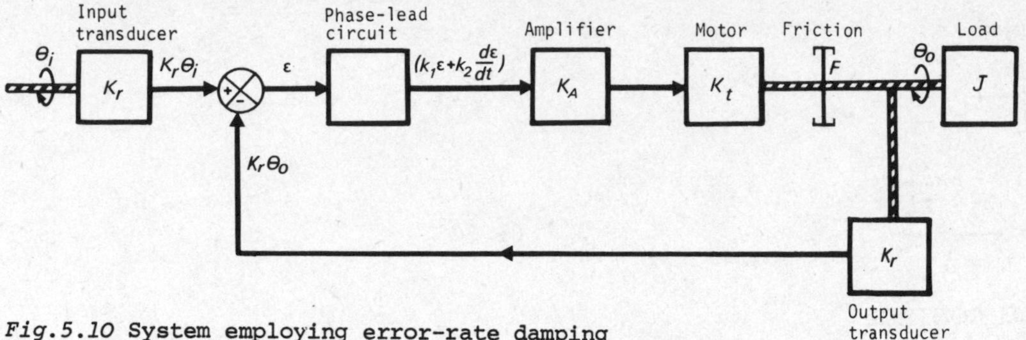

Fig.5.10 System employing error-rate damping

friction, and so, even when error-rate damping is employed, the effect of such friction must be taken into account. Figure 5.10 shows the block diagram of a position-control system using error-rate damping.

From equation (5.26), the input to the amplifier is given by

$$k_1\varepsilon + k_2\frac{d\varepsilon}{dt}$$

where $k_1 \simeq R_2/R_1$ and $k_2 = R_2C$

Then, using equation (1.9),

$$J\frac{d^2\theta_o}{dt^2} + F\frac{d\theta_o}{dt} = K_AK_t\varepsilon$$

$$J\frac{d^2\theta_o}{dt^2} + F\frac{d\theta_o}{dt} = K_AK_t\left(k_1\varepsilon + k_2\frac{d\varepsilon}{dt}\right) \qquad (5.27)$$

but

$$\varepsilon = K_r(\theta_i - \theta_o)$$

and

$$\frac{d\varepsilon}{dt} = K_r\frac{d\theta_i}{dt} - K_r\frac{d\theta_o}{dt}$$

then

$$J\frac{d^2\theta_o}{dt^2} + F\frac{d\theta_o}{dt} = K_AK_tK_rk_1\theta_i - K_AK_tK_rk_1\theta_o + K_AK_tK_rk_2\frac{d\theta_i}{dt} - K_AK_tK_rk_2\frac{d\theta_o}{dt}$$

Let $K_AK_tK_r = K$. Then the differential equation will be

$$J\frac{d^2\theta_o}{dt^2} + (F + Kk_2)\frac{d\theta_o}{dt} + Kk_1\theta_o = Kk_1\theta_i + Kk_2\frac{d\theta_i}{dt} \qquad (5.28)$$

For this system

$$\omega_n = \sqrt{\frac{Kk_1}{J}} \qquad (5.29)$$

and

$$2\zeta\omega_n = \frac{F + Kk_2}{J} \qquad (5.30)$$

5.7.3 INTEGRAL CONTROL

One of the most effective ways of eliminating steady-state error and off-set is to use integral control, in which the amplifier input is the sum of two quantities:

(i) the error voltage, and
(ii) the time integral of the error voltage.

As long as there is an error, the amplifier has an input, and its output will continue to increase (integral effect), energizing the motor until the input and output quantities are brought into line.

Consider the circuit of Figure 5.11 where R_2 is very small compared with R_1. Since

$$\varepsilon = iR_1 + v_o$$

then

$$i = \frac{\varepsilon - v_o}{R_1}$$

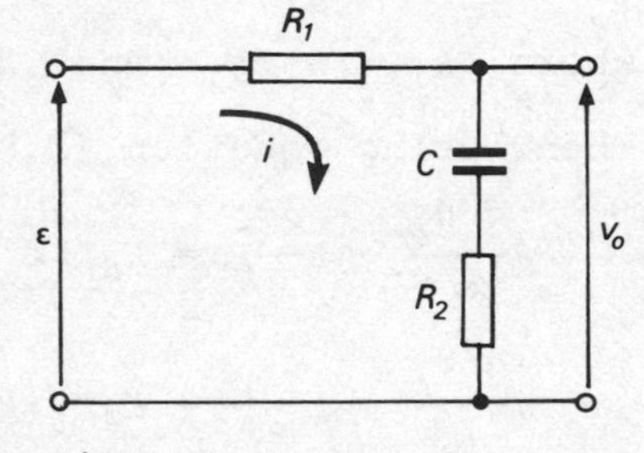

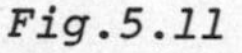
Fig.5.11

Since R_2 is very small compared with R_1, the output voltage will be a small fraction of the input voltage. Thus,

$$i \simeq \frac{\varepsilon}{R_1}$$

but

$$v_o = v_{R_2} + v_c = iR_2 + \frac{1}{C}\int i\,dt$$

Substituting for i

$$v_o \simeq \frac{R_2}{R_1}\varepsilon + \frac{1}{R_1 C}\int \varepsilon\,dt \qquad (5.31)$$

It is seen that the output voltage is the sum of two terms: one proportional to error, and one proportional to the integral of error. The circuit of Figure 5.11 is called a PHASE-LAG CIRCUIT and its frequency response characteristics are dealt with in Chapter 6.

Example 5.9

A small second-order position-control system incorporates viscous friction for damping and integral control in order to eliminate steady-state error. Draw the block diagram and explain the operation of the system. Show that the effect of integral control is to convert the system into a third-order system.

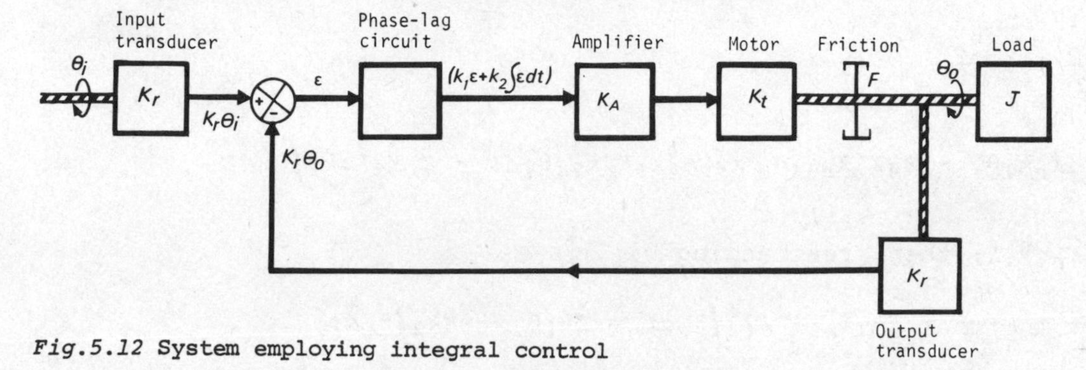

Fig.5.12 System employing integral control

Solution
The block diagram of such a system is shown in Figure 5.12. A phase-lag circuit is placed in front of the amplifier to provide the proportional plus integral error signal.

When the output shaft is not in line with the input shaft, an error voltage is detected by the comparator and is fed into the lag circuit. The capacitor charges and provides an increasing signal to the amplifier. The motor will be driven in such a way so as to bring the two shafts in line. When the error voltage is zero, the capacitor çannot discharge instantaneously so that there is still an input to the amplifier sufficient to drive the output shaft in line against any friction or load torque.

From the diagram:

$$J\frac{d^2\theta_o}{dt^2} + F\frac{d\theta_o}{dt} = K_A K_t \varepsilon'$$

where ε' is the output of the lag circuit and is given by

$$\varepsilon' \approx k_1\varepsilon + k_2\int\varepsilon dt$$

thus

$$J\frac{d^2\theta_o}{dt^2} + F\frac{d\theta_o}{dt} = K_A K_t\left(k_1\varepsilon + k_2\int\varepsilon dt\right)$$

but

$$\varepsilon = K_r(\theta_i - \theta_o)$$

and

$$\int \varepsilon dt = K_r\left(\int \theta_i dt - \int \theta_o dt\right)$$

then

$$J\frac{d^2\theta_o}{dt^2} + F\frac{d\theta_o}{dt}$$

$$= K_A K_t K_r k_1 \theta_i - K_A K_t K_r k_1 \theta_o + K_A K_t K_r k_2 \int \theta_i dt - K_A K_t K_r k_2 \int \theta_o dt$$

Let $K_A K_t K_r = K$; then, rearranging

$$J\frac{d^2\theta_o}{dt^2} + F\frac{d\theta_o}{dt} + Kk_1\theta_o + Kk_2\int \theta_o dt = Kk_1\theta_i + Kk_2\int \theta_i dt$$

Differentiating both sides

$$J\frac{d^3\theta_o}{dt^3} + F\frac{d^2\theta_o}{dt^2} + Kk_1\frac{d\theta_o}{dt} + Kk_2\theta_o = Kk_1\frac{d\theta_i}{dt} + Kk_2\theta_i \tag{5.32}$$

This differential equation shows that the application of integral control has changed the second-order system into a third-order type.

Example 5.10

In a second-order feedback control system it is required that the overshoot of the step-response should not exceed 15%. Determine the corresponding limiting values of the damping ratio and the resonance peak of the frequency response M_{pf}.

Find, also, the natural undamped frequency of the system if the peak overshoot is reached in 0.2 s, and the value of the resonant angular frequency ω_{rf} in the frequency domain.

Solution

Consider the differential equation of such a system:

$$\frac{d^2\theta_o}{dt^2} + 2\zeta\omega_n\frac{d\theta_o}{dt} + \omega_n^2\theta_o = \omega_n^2\theta_i$$

For a sinusoidal input $\theta_i(t) = \hat{\theta}_i \sin\omega t$, where $\hat{\theta}_i$ is the maximum value of the input signal and ω is its angular frequency, the instantaneous steady-state output is given by

$$\theta_{o(ss)}(t) = \frac{\hat{\theta}_i \omega_n^2 \sin(\omega t + \phi)}{\sqrt{\left[(\omega_n^2 - \omega^2)^2 + (2\zeta\omega_n\omega)^2\right]}} \tag{5.33}$$

Dividing through by ω_n^2 and letting $u = \omega/\omega_n$, the equation for the magnitude expressed as a function of ju will be

$$\left|\frac{\theta_o}{\theta_i}(ju)\right| = \frac{1}{\sqrt{\left[(1 - u^2)^2 + (2\zeta u)^2\right]}} \tag{5.34}$$

Let M be the magnitude; then the maximum value of M occurs when $dM/du = 0$. Thus,

$$\frac{dM}{du} = -\frac{4u^3 - 4u + 8\zeta^2 u}{2\times\sqrt{\left[u^4 + 1 - 2u^2(1 - 2\zeta^2)\right]^3}} = 0$$

or $4u^3 - 4u + 8\zeta^2 u = 0$

then

$$u = \sqrt{(1 - 2\zeta^2)} \tag{5.35}$$

Let the value of ω at which M is maximum be ω_{rf}, then

$$\frac{\omega_{rf}}{\omega_n} = \sqrt{(1 - 2\zeta^2)}$$

$$\omega_{rf} = \omega_n\sqrt{(1 - 2\zeta^2)} \tag{5.36}$$

The resonance peak of the frequency response M_{pf} may now be found by substituting the value of $u = \sqrt{(1 - 2\zeta^2)}$ in the equation for M:

$$M = \frac{1}{\sqrt{\left[(1 - u^2)^2 + (2\zeta u)^2\right]}}$$

then

$$M_{pf} = \frac{1}{2\zeta\sqrt{(1 - \zeta^2)}} \tag{5.37}$$

To find the *limiting values* of ζ:

From equation (5.35), it is evident that

$2\zeta^2 < 1$ thus $\zeta^2 < 0.5$ and $\zeta < 0.707$

Now, it is given that the percentage overshoot should not exceed 15%. From equation (5.15),

$$15 = 100e^{-\zeta\pi/\sqrt{(1-\zeta^2)}}$$

$$0.15 = e^{-\zeta\pi/\sqrt{(1-\zeta^2)}}$$

$$\log_e(0.15) = -\frac{\zeta\pi}{\sqrt{(1-\zeta^2)}} = -1.897$$

Squaring both sides and rearranging:

$$\zeta = \frac{1.897}{\sqrt{(\pi^2 + 1.897^2)}} = 0.588$$

The limiting values of ζ are therefore given by

$$0.707 > \zeta > 0.588$$

From equation (5.37),

$$M_{pf} = \frac{1}{2\zeta\sqrt{(1-\zeta^2)}} = \frac{1}{2 \times 0.588\sqrt{(1-0.588^2)}}$$

$$= 1.05$$

and, from equation (5.14),

$$t_p = \frac{\pi}{\omega_n\sqrt{(1-\zeta^2)}}$$

$$\omega_n = \frac{\pi}{t_p\sqrt{(1-\zeta^2)}} = \frac{\pi}{0.2\sqrt{(1-0.588^2)}}$$

$$\simeq 19.42 \text{ rad/s}$$

Also, from equation (5.36),

$$\omega_{rf} = \omega_n\sqrt{(1-2\zeta^2)} = 19.42\sqrt{(1-2\times 0.588^2)}$$

$$\simeq 15.72 \text{ rad/s}$$

Example 5.11

A closed-loop remote position-control system consists of an amplifier controlling a motor that drives an inertia and viscous friction load through a gearbox.

The input signal to the amplifier is equal to the difference between a reference voltage and a voltage proportional to the output load shaft position.

Details of the system are as follows:

Error detector	K_r = 1 V per degree of error
Amplifier	Gain K_A = 100 mA/V
Motor	Torque $K_t = 10^{-1}$ N-m/A
	Moment of inertia $J_M = 10^{-6}$ kg-m^2
	Viscous damping
	$F_M = 7 \times 10^{-6}$ N-m per rad/s
Gearbox	Ratio 100:1
	Transmission assumed ideal
Load	Moment of inertia $J_L = 10^{-2}$ kg-m^2
	Viscous damping
	F_L = 1.0 N-m per rad/s

Draw the block schematic diagram of the system and hence determine

(i) the undamped resonance frequency of the system
(ii) the damping ratio of the system.

If the damping ratio is to be doubled, by means of feedback applied to the amplifier from a tachogenerator connected to the output shaft, determine the tachogenerator constant required.

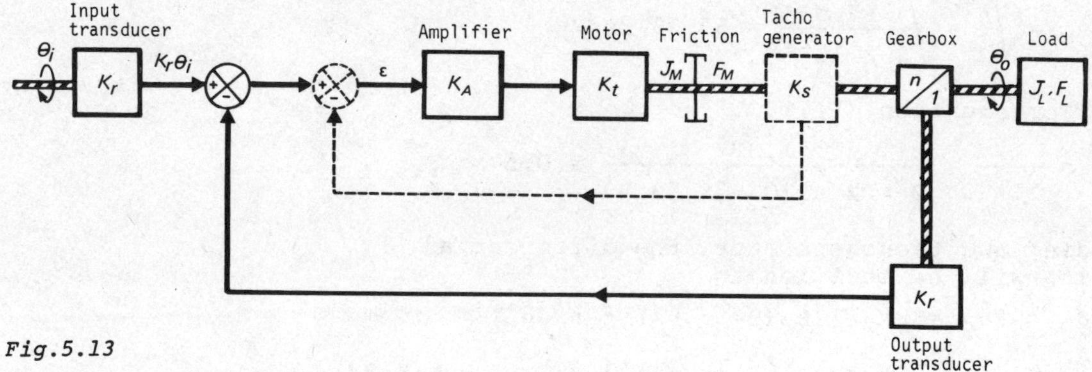

Fig.5.13

Solution
The block diagram of such a system is shown in Figure 5.13.
Referring quantities to the load shaft, total inertia is

$$J = n^2 J_M + J_L$$

$$= 100^2 \times 10^{-6} + 10^{-2} = 2 \times 10^{-2} \text{ kg-m}^2$$

and total viscous damping is

$$F = n^2 F_M + F_L$$

$$= 100^2 \times 7 \times 10^{-6} + 1.0 = 1.07 \text{ N-m/rad/s}$$

The differential equation for the system is given by

$$J\ddot{\theta}_o + F\dot{\theta}_o + K\theta_o = K\theta_i$$

where $K = nK_t K_A K_r$.

But $K_r = 1 \text{ V/}^\circ = 1 \times 360/2\pi = 57.3 \text{ V/rad}$

and $K_t = 10^{-1} \text{ N-m/A} = 10^{-4} \text{ N-m/mA}$

thus

$$K = 100 \times 10^{-4} \times 100 \times 57.3$$

$$= 57.3 \text{ N-m/rad of error}$$

(Note that nK_t "refers" the motor torque constant to the load shaft.)
Substituting values:

$$2 \times 10^{-2}\ddot{\theta}_o + 1.08\dot{\theta}_o + 57.3\theta_o = 57.3\theta_i$$

(i) $\omega_n = \sqrt{\frac{K}{J}} = \sqrt{\frac{57.3}{2 \times 10^{-2}}} = 53.53 \text{ rad/s}$

(ii) From equation (5.7),

$$\zeta = \frac{F}{2J\omega_n} = \frac{1.07}{2 \times 2 \times 10^{-2} \times 53.53} = 0.5$$

Including the tachogenerator, the differential equation will be modified to

$$J\ddot{\theta}_o + F\dot{\theta}_o = nK_t K_A \left[K_r(\theta_i - \theta_o) - nK_s\dot{\theta}_o\right]$$

$$J\ddot{\theta}_o + (F + n^2 K_A K_t K_s)\dot{\theta}_o + nK_t K_A K_r \theta_o = nK_t K_A K_r \theta_i$$

For the damping ratio to be doubled, without changing other parameters, the damping should be doubled. Thus

$$F + n^2 K_A K_t K_s = 2F$$

or

$$K_s = \frac{F}{n^2 K_A K_t} = \frac{1.07}{100^2 \times 100 \times 10^{-4}}$$

$$= 10.7 \times 10^{-3} \text{ V/rad/s}$$

EXERCISES

(5.1) Explain what is meant by the following terms as applied to control engineering:
(*a*) open-loop system gain
(*b*) undamped natural frequency
(*c*) velocity error.
A control system has an open-loop gain K(N-m/rad), a coefficient of viscous friction F(N-m-sec) and a load inertia J(kg-m^2). Discuss fully what effect each of the above factors has on (i) the undamped natural frequency and (ii) the steady-state velocity error, assuming a ramp input to the system.

(5.2) (*a*) Given that the equation for a second order d.c. position servomechanism is

$$J\ddot{\theta}_o + F\dot{\theta}_o + K\theta = 0 \quad \text{where } \theta = \theta_o - \theta_i$$

obtain the equation in terms of damping ratio and undamped natural frequency. Hence, or otherwise, derive an expression for the steady-state error when the system is subjected to a constant velocity input.
(*b*) In a servomechanism used for controlling the position of a rotatable mass the moment of inertia of the moving parts is 0.55 kg-m^2 and the motor torque is 800 N-m per radian of misalignment. The friction torque is 28 N-m/rad-s^{-1}. Determine by how much the friction torque must be increased in order for the system to be critically damped.

(5.3) (*a*) Show that a simple position-control system may be described by the following differential equation:

$$\frac{d^2\theta_o(t)}{dt^2} + 2\zeta\omega_n\frac{d\theta_o(t)}{dt} + \omega_n^2\,\theta_o(t)$$
$$= \omega_n^2\,\theta_i(t)$$

Explain the meaning of the coefficients of the equation in terms of the system parameters.
(*b*) If in such a system the total moment of inertia of all moving parts is 200 kg-m^2, the motor torque is 800 N-m per radian of error, and the torque due to viscous friction is 400 N-m per rad s^{-1}. Determine the output θ_o as a function of time for a unit step-input.

(5.4) A second-order position-control system has an inertia referred to the output shaft of 1 kg-m^2. The systems differential equation is given by

$$J\ddot{\theta}_o + F\dot{\theta}_o + K\theta_o = K\theta_i$$

Determine the values of F and K such that the maximum overshoot is 20% and the response reaches this peak in 1 second.

(5.5) Show that the operation of an error-actuated automatic control system for the position control of a rotating mass, with viscous friction damping, may be described by a differential equation of the form

$$\frac{d^2\theta_o}{dt^2} + 2\zeta\omega_n\frac{d\theta_o}{dt} + \omega_n^2\theta_o = \omega_n^2\theta_i$$

In such a system, the inclusive moment of inertia of the mass is 400 kg-m^2, the correcting torque is 3600 N-m per radian of misalignment, and the torque due to viscous friction is 800 N-m per rad/s. Show that the transient response is oscillatory.

(5.6) Explain, using a labelled block diagram, a procedure for obtaining the transient response of a torque-controlled servomechanism driving a radar antenna, emphasizing any special feature of the test equipment.
  Sketch the expected output shaft response assuming the system to be (*a*) undamped (*b*) critically damped (*c*) overdamped.
Discuss the factors which affect the undamped natural frequency of the control system.

(5.7) The angular position of a rotating mass, subject to viscous damping, is controlled by a simple position-control servomechanism. Prepare a diagram of a suitable scheme to accomplish this, using the hardware known to you from your laboratory.
  Set up the equation of motion of the system, clearing stating any assumptions you make.
  In a particular system, the moment of inertia of the rotating parts referred

to the controlled shaft is 500 kg-m^2 and the motor produces a torque of 4500 N-m per radian of misalignment. Calculate the required viscous friction constant which gives critical damping. What is the natural frequency of the system?

(5.8) Sketch circuit diagrams of resistance-capacitance networks which will provide an output which approximate to
(*a*) proportion plus derivative of input
(*b*) proportion plus integral of input.
Show how each of these networks could be incorporated in a closed-loop control system to improve its performance. Discuss for each case the effect on the transient and steady-state response of the control system obtained by including the network.

(5.9) Sketch the circuit diagram of a resistance-capacitance network which will provide an output voltage which is a proportion plus the integral of the input voltage.
The network is connected in cascade with the amplifier input circuit of a closed-loop position-control system. Assuming a constant velocity input, explain the setting-up procedure and compensation action on velocity misalignment error. Discuss briefly how over-compensation can result in instability.

(5.10) The position-control system of Figure 5.5 is used to drive a load through a gearbox of ratio 20:1. Total moment of inertia of the system is 25×10^{-3} kg-m^2, the amplifier transconductance is 200 mA/V, the transducers have a constant of 25 V/rad, and the motor torque constant is 10×10^{-3} N-m/mA. The system is critically damped by combined viscous friction and velocity feedback. If the tachogenerator constant is 0.03 V per rad/s, derive the differential equation of the system and hence calculate
(*a*) the undamped natural frequency
(*b*) the coefficient of viscous damping
(*c*) the steady-state velocity error when the input velocity is 0.2 rad/s.

(5.11) A simple position-control system is used to position a load having moment of inertia J and viscous friction F. The control system uses linear potentiometers as position transducers each with transfer function of K_p, an amplifier with a voltage-to-current gain of G, and a motor with a torque-current constant of K_m. Neglecting field time constant and motor inertia and friction:

(*a*) Sketch a diagram to show the electrical and mechanical interconnections to give a working arrangement of the above system.

(*b*) Draw a block diagram of the system to include the transfer function of each component and hence obtain the closed-loop transfer function of the system.

(*c*) Determine the undamped natural frequency and damping ratio in terms of the system components.

Sketch the output $\theta_o(t)$ for a unit step input.

(5.12) A servo system has an undamped natural frequency of 5 Hz. When a step input is applied, the output shaft reaches to and remains within 1% of the input position in 1 second. Determine the steady-state error when a velocity input of $\pi/12$ rad/s is applied.

(5.13) Discuss the factors influencing the choice of damping ratio of a simple position-control system making reference to overshoot, rise time, settling time, and steady-state velocity lag. Explain one method of increasing the damping ratio without increasing the velocity lag.

(5.14) In a d.c. position-control servomechanism, an inertia and viscous friction load is driven by a motor with a constant armature current. The field current of the motor is supplied from a d.c. amplifier, the input to which is the difference between the voltages obtain from the input and output potentiometers. Derive the differential equation of motion of the system and hence find the value of amplifier gain to give an undamped natural frequency of 5 rad s^{-1} if the values of the system constants are as follows:

Combined inertia of load and motor = 0.6 kg-m^2

Potentiometer constant = 5.0 V-rad^{-1}

Motor torque constant = 2 N-m A^{-1}

(5.15) In a second-order feedback control system, it is required that the peak overshoot of the step-response should not exceed 15%, reached in 0.2 s. Determine, proving formulae relating to frequency response, the limiting values of the damping ratio ζ, the resonance peak of the frequency response M_{pf} and the resonant angular frequency ω_{rf}.

6 Stability

The most essential requirement of any control system is that the output should at all times be under the control of the input. If the output of a system is bounded for every bounded input, the system is said to be stable.

In this chapter various techniques for determining system stability are described. Frequency response methods and their applications form the major part of this chapter.

6.1 CHARACTERISTIC EQUATION

Consider the nth order differential equation

$$\frac{d^n y}{dt^n} + a_{n-1}\frac{d^{n-1} y}{dt^{n-1}} + \ldots + a_1\frac{dy}{dt} + a_0 y = a_0 x \qquad (6.1)$$

where y is the input quantity and x is the output quantity.

Let D be the differential operator such that

$$D \equiv \frac{d}{dt} \quad \text{and} \quad D^n \equiv \frac{d^n}{dt^n} \qquad (6.2)$$

Then, rewriting equation (6.1),

$$D^n y + a_{n-1}D^{n-1}y + \ldots + a_1 Dy + a_0 y = a_0 x$$

$$\left(D^n + a_{n-1}D^{n-1} + \ldots + a_1 D + a_0\right)y = a_0 x$$

From the definition, the transfer function will be

$$\frac{y}{x} = \frac{a_0}{D^n + a_{n-1}D^{n-1} + \ldots + a_1 D + a_0}$$

The equation

$$D^n + a_{n-1}D^{n-1} + \ldots + a_1 D + a_0 = 0 \quad (6.3)$$

which is formed by equating the denominator of the transfer function to zero is called the CHARACTERISTIC EQUATION.

Example 6.1

Determine the characteristic equation of the system whose differential equation is given by

$$\frac{d^2\theta_o}{dt^2} + 4\frac{d\theta_o}{dt} + 3\theta_o = \theta_i$$

Solution

$$D^2\theta_o + 4D\theta_o + 3\theta_o = \theta_i$$

$$\frac{\theta_o}{\theta_i} = \frac{1}{D^2 + 4D + 3}$$

The characteristic equation will be

$$D^2 + 4D + 3 = 0$$

6.2 THE s-PLANE

The complex quantity $s = \sigma + j\omega$, where σ and ω are real variables, is of particular interest in control engineering. A plane in which the real axis is represented by σ and the imaginary axis is represented by ω, is referred to as the *complex plane* or the *s-plane*.

s is the Laplace operator and it transforms a function from the time domain to the frequency domain.

Figure 6.1 shows an s-plane with three complex variables s_1, s_2, and s_3.

6.2.1 POLES AND ZEROS

Most transfer functions are complex and may be expressed, in terms of s, as the ratio of two polynomials in the product form of equation (6.4):

$$F(s) = \frac{(s - Z_1)(s - Z_2) \ldots}{(s - P_1)(s - P_2) \ldots} \quad (6.4)$$

Any value of s making $F(s)$ zero is called a ZERO of $F(s)$.

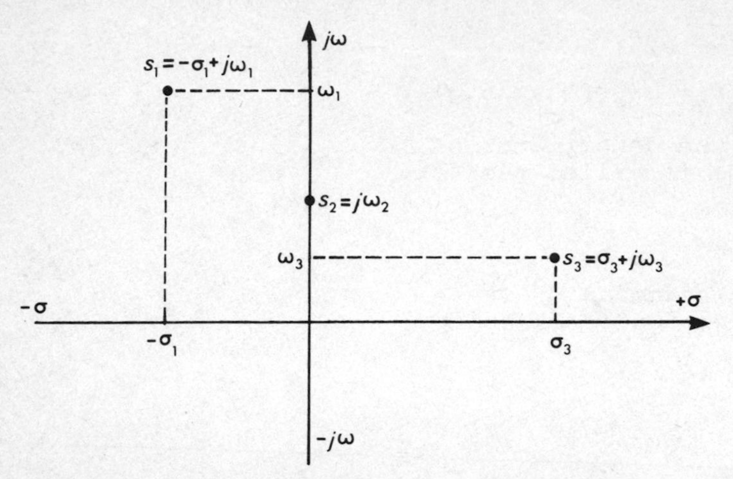

Fig.6.1 The s-plane

A value of s which makes $F(s)$ infinite is called a POLE of $F(s)$.

A closed path showing the variation of s or of $F(s)$ is known as a CONTOUR.

Example 6.2

Determine the poles and zeros of the function

$$F(s) = \frac{(s + 2)(s - 5)}{s(s + 2 + j)(s + 2 - j)}$$

Solution

For $s = -2$ and $s = +5$ the function becomes zero.

For $s = 0$, $s = -2 - j$ and $s = -2 + j$ the function becomes infinite.

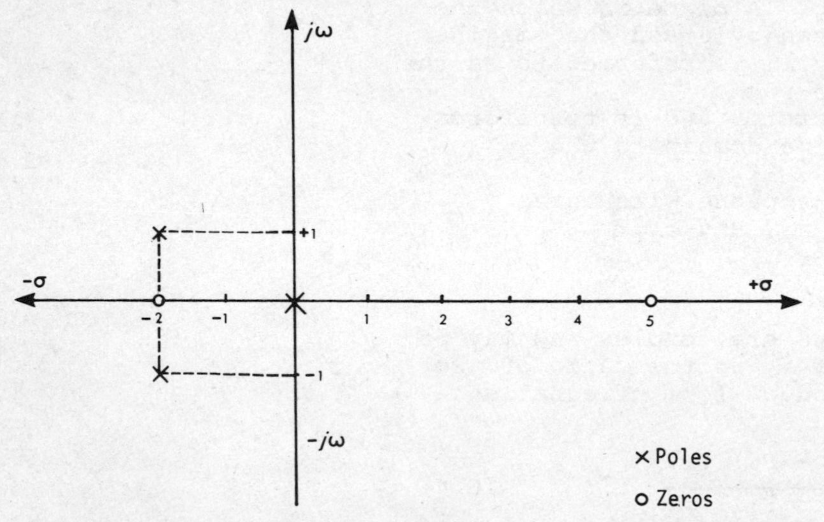

Fig.6.2 Poles and zeros

Thus, the function has two zeros at $s = -2$ and $s = +5$ and three poles at $s = 0$, $s = -2 - j$ and $s = -2 + j$, as shown on the s-plane of Figure 6.2.

Example 6.3

Determine, and plot on an s-plane, the poles and zeros of

$$F(s) = \frac{s^2 - 9}{s^3(s + 2)(s - 6)}$$

Solution

Zeros are at $s = -3$ and $s = +3$. Poles are at $s = -2$, $s = +6$, and a *triple pole* at $s = 0$. These are shown on the s-plane of Figure 6.3.

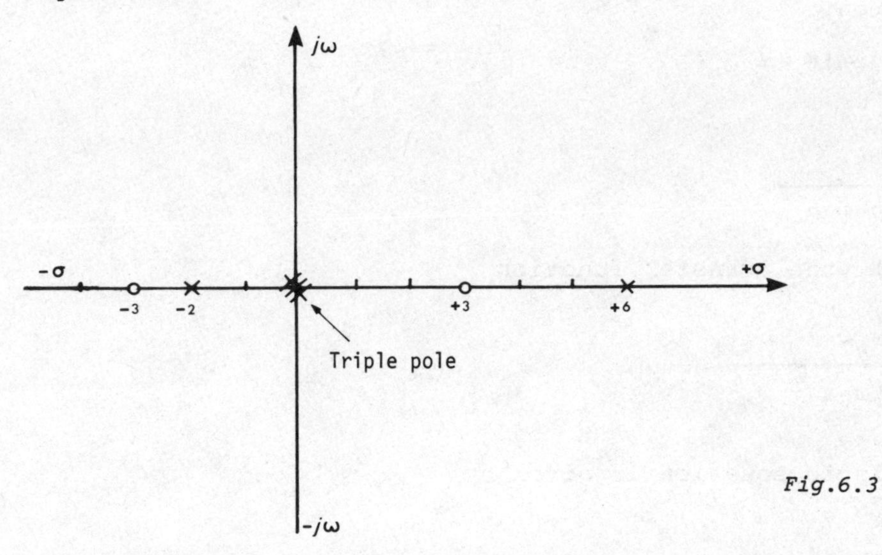

Fig.6.3

It should be noted that in the complex equation

$$s = \sigma + j\omega$$

if the real part $\sigma = 0$, then

$$s = j\omega \qquad (6.5)$$

It can be shown that the differential operator D may be replaced by s and leave the function mathematically unaltered. Thus,

$$D \equiv s \qquad (6.6)$$

The characteristic equation of Example 6.1 can therefore be written in terms of s as

$$s^2 + 4s + 3 = 0$$

Example 6.4

The differential equation of a second-order control system, in terms of damping ratio and undamped natural frequency, is given by

$$\frac{d^2\theta_o}{dt^2} + 2\zeta\omega_n\frac{d\theta_o}{dt} + \omega_n^2\,\theta_o = \omega_n^2\,\theta_i$$

(i) *Derive an expression, in terms of s, for the transfer function.*
(ii) *Write down the characteristic equation.*
(iii) *Determine the poles of the transfer function.*

Solution

(i) From equation (6.2), rewriting the differential equation,

$$(D^2 + 2\zeta\omega_n D + \omega_n^2)\,\theta_o = \omega_n^2\,\theta_i$$

then

$$\frac{\theta_o}{\theta_i} = \frac{\omega_n^2}{D^2 + 2\zeta\omega_n D + \omega_n^2}$$

From equation (6.6), the transfer function will be

$$\frac{\theta_o(s)}{\theta_i(s)} = \frac{\omega_n^2}{s^2 + 2\zeta\omega_n s + \omega_n^2}$$

(ii) The characteristic equation is given by

$$s^2 + 2\zeta\omega_n s + \omega_n^2 = 0 \qquad (6.7)$$

(iii) *The roots of the characteristic equation are in fact the poles of the transfer function.* Solving equation (6.7),

$$s = -\,\zeta\omega_n \pm j\omega_n\sqrt{(1 - \zeta^2)}$$

It is seen that there are two poles at

$$s_1 = -\,\alpha + j\omega_d \quad \text{and} \quad s_2 = -\,\alpha - j\omega_d$$

where $\omega_d = \omega_n\sqrt{(1 - \zeta^2)}$ is the damped natural frequency and $1/\alpha = 1/\zeta\omega_n = \tau$ is the time constant of the system. (See Chapter 5, page 76.)

Example 6.5

Figure 6.4 shows the positions of eight poles on an s-plane. Explain the significance of their locations.

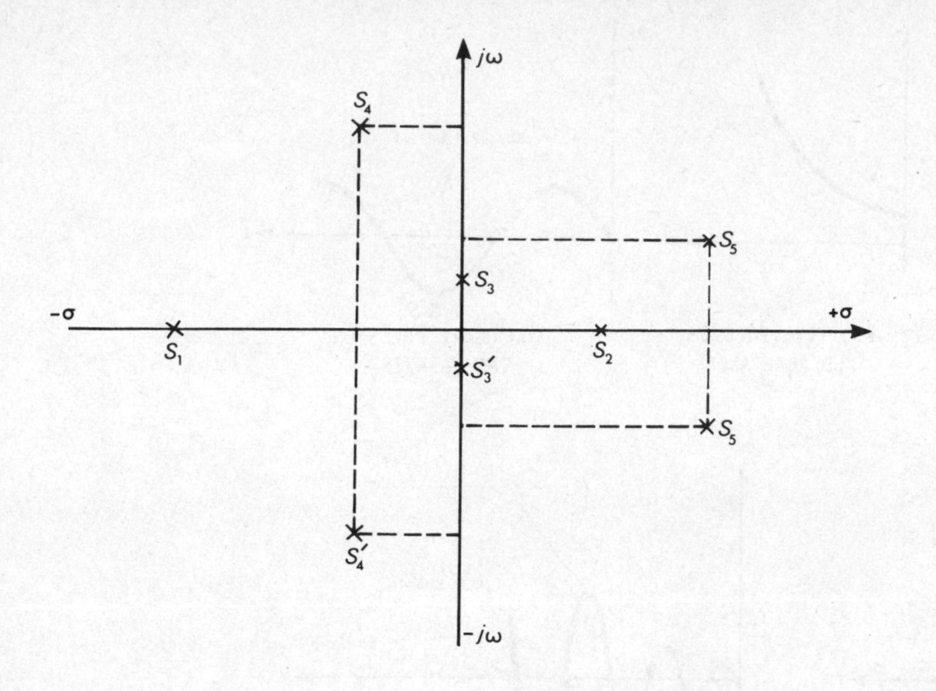

Fig.6.4

Solution

Consider the complex quantity $s = \sigma + j\omega$.

s_1 A pole *on* the real axis implies a simple exponential term in the solution (as $\omega = 0$, there will be no oscillation). Since s_1 is on the l.h.s., σ is negative and the exponential will be a decaying one.

s_2 This is also *on* the real axis but because it is on the r.h.s. it represents an exponential rise.

s_3 Poles *off* the real axis only occur in conjugate pairs, thus s_3 and s_3'. These poles imply an oscillatory term in the solution. Because they are *on* the $j\omega$ axis the real part is zero and thus there is no exponential term.

s_4 Conjugate pairs s_4 and s_4' are off both axes. Such poles on the l.h.s. imply an oscillatory term which decays exponentially.

s_5 Conjugate pairs s_5 and s_5' are also off both axes and since their real part is positive, they represent an oscillatory term which increases in magnitude exponentially.

These responses are sketched in Figure 6.5.

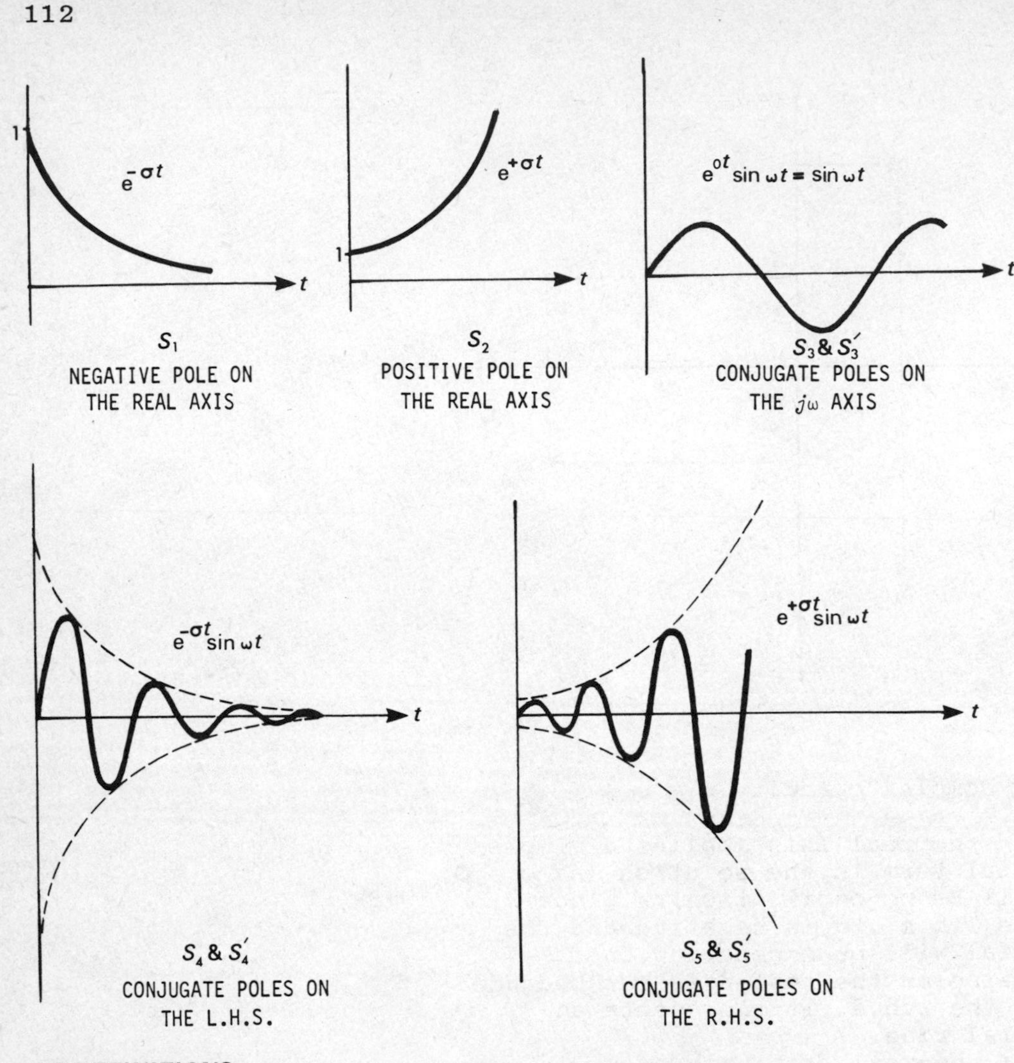

Fig.6.5

6.3 DEFINITIONS

Definitions of a stable system are numerous. *A system is said to be* STABLE *if for every bounded input the output remains bounded.* A bounded input is one whose magnitude is always less than a finite value.

Alternatively, *for a system to be stable, none of the poles of the closed-loop transfer function may lie in the right-hand half of the s-plane.*

There are various methods for investigating the stability of a system. These are dealt with in the following sections.

6.4 ROUTH STABILITY CRITERION

The ROUTH CRITERION is a method of determining whether or not a system is stable by examining its characteristic equation. A system having a characteristic equation of the form

$$a_n s^n + a_{n-1} s^{n-1} + \ldots + a_1 s + a_0 = 0 \quad (6.8)$$

will be stable if

(i) all coefficients of s from n to 0 are present and are positive, and
(ii) all the terms in the first column of the "array" formed from the characteristic equation have the same sign.

Any change of sign indicates a pole in the right-hand half of the s-plane.

The array is formed as follows:

$$\begin{array}{|llll}
a_n & a_{n-2} & a_{n-4} & \ldots \\
a_{n-1} & a_{n-3} & a_{n-5} & \ldots \\
b_1 & b_2 & b_3 & \ldots \\
c_1 & c_2 & \ldots & \\
d_1 & d_2 & \ldots & \\
e_1 & \ldots & & \\
\cdot & & & \\
\cdot & & & \\
\cdot & & &
\end{array}$$

where a_n, a_{n-1}, etc., are the coefficients of the characteristic equation and

$$b_1 = \frac{a_{n-1} \times a_{n-2} - a_n \times a_{n-3}}{a_{n-1}}$$

$$b_2 = \frac{a_{n-1} \times a_{n-4} - a_n \times a_{n-5}}{a_{n-1}} \quad (6.9)$$

$$c_1 = \frac{b_1 \times a_{n-3} - a_{n-1} \times b_2}{b_1}$$

$$c_2 = \frac{b_1 \times a_{n-5} - a_{n-1} \times b_3}{b_1} \quad (6.10)$$

$$d_1 = \frac{c_1 \times b_2 - b_1 \times c_2}{c_1}$$

$$d_2 = \frac{c_1 \times b_3 - b_1 \times c_3}{c_1} \quad (6.11)$$

$$e_1 = \frac{d_1 \times c_2 - c_1 \times d_2}{d_1} \qquad \text{etc.} \qquad (6.12)$$

Example 6.6

The characteristic equation of a system is given by

$$\tfrac{1}{2}s^4 + 4s^3 + 12s^2 + 16s + 8 = 0$$

Determine whether it represents a stable or an unstable system.

Solution

Forming the array:

½	12	8
4	16	0
b_1	b_2	
c_1	c_2	
d_1		
e_1		

From equations (6.9), (6.10), (6.11) and (6.12):

$$b_1 = \frac{4 \times 12 - \frac{1}{2} \times 16}{4} = 10$$

$$b_2 = \frac{4 \times 8 - \frac{1}{2} \times 0}{4} = 8$$

$$c_1 = \frac{10 \times 16 - 4 \times 8}{10} = 12.8$$

$$c_2 = \frac{10 \times 0 - 4 \times 0}{10} = 0$$

$$d_1 = \frac{12.8 \times 8 - 10 \times 0}{12.8} = 8$$

$$e_1 = \frac{8 \times 0 - 12.8 \times 0}{8} = 0$$

½	12	8
4	16	0
10	8	
12.8	0	
8		
0		

There are no sign changes in the terms of the first column; the system is stable.

Example 6.7

(*a*) *The transfer function of a control element is given by*

$$\frac{K}{s^3 + 4s^2 + 11s}$$

This element is connected in a unity feedback circuit. Calculate the minimum positive value of K at which the system becomes unstable.
(b) Given that K = 8 and s = - 1 is a root of the characteristic equation, determine the response of the feedback system to a unit step input.

Solution

(*a*) The characteristic equation is

$$s^3 + 4s^2 + 11s + K = 0$$

Applying Routh stability criterion, the array will be

$$\begin{array}{|ll} 1 & 11 \\ 4 & K \\ b_1 & b_2 \\ c_1 & c_2 \\ d_1 & \end{array}$$

$$b_1 = \frac{4 \times 11 - 1 \times K}{4} = \frac{44 - K}{4}$$

$$b_2 = \frac{4 \times 0 - 1 \times 0}{4} = 0$$

$$c_1 = \frac{[(44 - K)/4] \times K - 4 \times 0}{(44 - K)/4} = K$$

$$c_2 = 0$$

$$d_1 = 0$$

From the array, for stability

$$\frac{44 - K}{4} > 0 \quad \text{and} \quad K > 0$$

or $K < 44$ and $K > 0$

The minimum positive value of K at which the system becomes unstable is thus 44.

(*b*) The closed-loop transfer function will be

$$\frac{\theta_o(s)}{\theta_i(s)} = \frac{8}{s^3 + 4s^2 + 11s + 8}$$

$$= \frac{8}{(s + 1)(s^2 + 3s + 8)}$$

For a unit step input $\theta_i(s) = 1/s$. Hence

$$\theta_o(s) = \frac{8}{s(s + 1)(s^2 + 3s + 8)}$$

Using partial fractions,

$$\theta_o(s) = \frac{A}{s} + \frac{B}{s+1} + \frac{Cs + D}{s^2 + 3s + 8}$$

where $A = 1$, $B = -4/3$, $C = 1/3$ and $D = -1/3$.
Hence,

$$\theta_o(s) = \frac{1}{s} - \frac{4}{3}\left(\frac{1}{s+1}\right) + \frac{1}{3}\left(\frac{s-1}{s^2 + 3s + 8}\right)$$

$$= \frac{1}{s} - \frac{4}{3}\left(\frac{1}{s+1}\right) + \frac{1}{3}\left[\frac{(s+3/2)}{(s+3/2)^2 + (\sqrt{23}/2)^2} - \frac{5/2}{(s+3/2)^2 + (\sqrt{23}/2)^2}\right]$$

Using inverse Laplace transforms and converting to the time domain:

$$\theta_o(t) = 1 - \frac{4}{3}e^{-t} + \frac{1}{3}e^{-3t/2}\left\{\cos(\sqrt{23}/2)t - \frac{5}{\sqrt{23}}\sin(\sqrt{23}/2)t\right\}$$

6.5 FREQUENCY RESPONSE TESTING

The principle of frequency response testing is based on comparing the input and output of the element under test, over a wide range of frequencies, when subjected to a sinusoidal input signal.

The input and output amplitudes and phases are compared by using appropriate meters and/or an oscilloscope. A simple and convenient method uses Lissajous figure and is illustrated in Figure 6.6.

The time base of the OSCILLOSCOPE is switched off and the input and output signals are applied to the Y and X inputs respectively. If the two signals are in phase, thus $\phi = 0^\circ$, a single-line trace will appear on the screen along the XX axis. If they are antiphase, $\phi = 180^\circ$, the trace will be a straight line along the YY axis. When there is a phase angle ϕ, the trace will appear as an ellipse and the phase angle can be determined using

$$\sin\phi = \frac{AB}{CD}$$

The accuracy of the phase measurement can be improved by applying equal amplitude signals to the oscilloscope. This can be obtained by passing the input signal through an attenuator. The setting of the attenuator will then

Fig.6.6

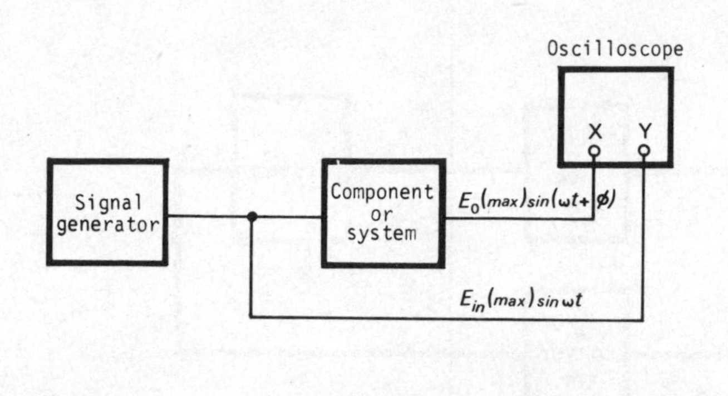

(a) FREQUENCY RESPONSE TESTING

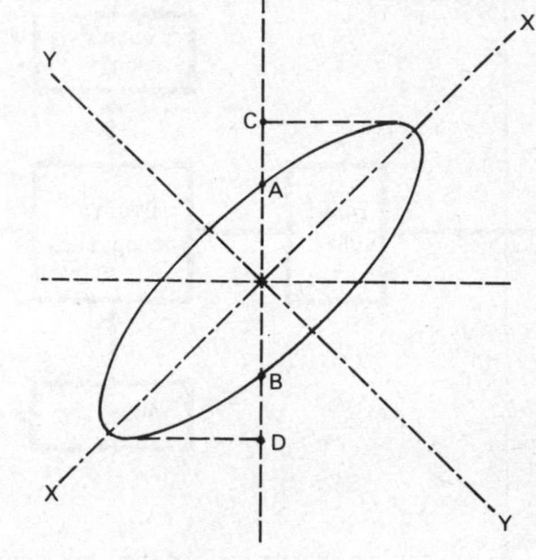

(b) LISSAJOUS FIGURE

indicate the attenuation between the input and output of the element under test.

The oscilloscope used for this purpose must have direct-coupled amplifiers, preferably calibrated, in the X and Y inputs. For measurements of very low frequencies the storage-type oscilloscope is preferable.

In modern and sophisticated test systems, small digital computers known as "minicomputers" are used in *digital transfer function analysers*. A simple block diagram of such a test system is shown in Figure 6.7.

The MINICOMPUTER is programmed to provide a digital output signal whose value follows a sinusoidal law. This signal is fed into the digital-to-analogue convertor, the output of which is a sinewave and is applied to the system or component under test. The output of the system which is an analogue signal is passed through the analogue-to-digital convertor and then fed into the analyser. The input is in the form of a computer program, fed into the input teletype, and the output may be a print-out on a printer or a plot on an X-Y plotter.

The program instructs the computer to test the system over a range of frequencies, compute the required information, and output it in a useful form. The information required may be a comparison between the input and output amplitudes, their phase difference, harmonic contents in the output, etc.

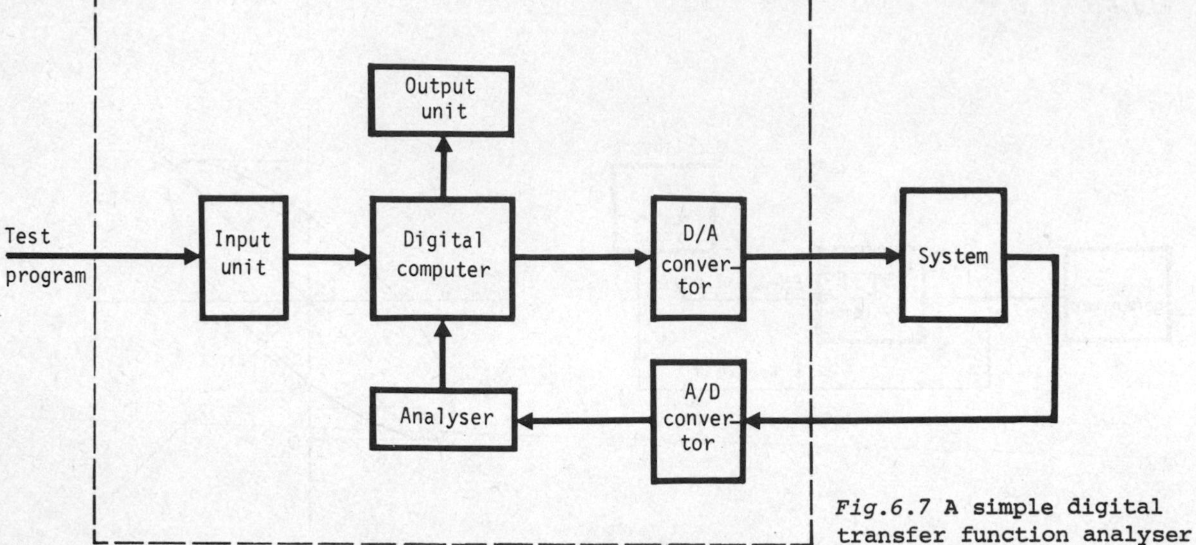

Fig.6.7 **A simple digital transfer function analyser**

Digital analysers of this kind are extremely useful for systems where a detailed study is necessary. An added feature is that any variation in the characteristics of components and devices may be compensated for by suitable programming.

6.6 NYQUIST DIAGRAMS

The open-loop frequency response of a system represented as a polar plot is called a NYQUIST DIAGRAM. The merit of Nyquist diagrams is that they are based on open-loop measurements and calculation, and, therefore, stability of a system can be determined without actually closing the feedback loop.

Example 6.8

Sketch the Nyquist diagram for a system whose open-loop transfer function is given by

$$G(j\omega) = \frac{50}{(1 + j2.5\omega)(1 + j0.05\omega)}$$

Solution

Using the complex number table given in Appendix 1, the gain and phase angle are calculated for various values of ω, as shown in the table. These results are plotted and shown on the Nyquist diagram of Figure 6.8.

ω (rad/s)	$(1 + j2.5\omega)$	$(1 + j0.05\omega)$	$\lvert G\rvert \angle\phi^\circ$
0.3	$1.25\angle 37^\circ$	$1\angle 0^\circ$	$40\angle -37$
0.4	$1.41\angle 45^\circ$	$1\angle 0^\circ$	$35.4\angle -45$
0.5	$1.62\angle 51^\circ$	$1\angle 0^\circ$	$30.8\angle -51$
0.7	$2.01\angle 60^\circ$	$1\angle 0^\circ$	$24.8\angle -60$
1	$2.7\angle 68^\circ$	$1\angle 0^\circ$	$18.5\angle -68$
2	$5.1\angle 78^\circ$	$1\angle 5^\circ$	$9.8\angle -83$
4	$10\angle 84^\circ$	$1.02\angle 11^\circ$	$4.9\angle -95$
8	$20\angle 87^\circ$	$1.07\angle 21^\circ$	$2.3\angle -108$
32	$80\angle 89^\circ$	$1.8\angle 58^\circ$	$0.33\angle -126$

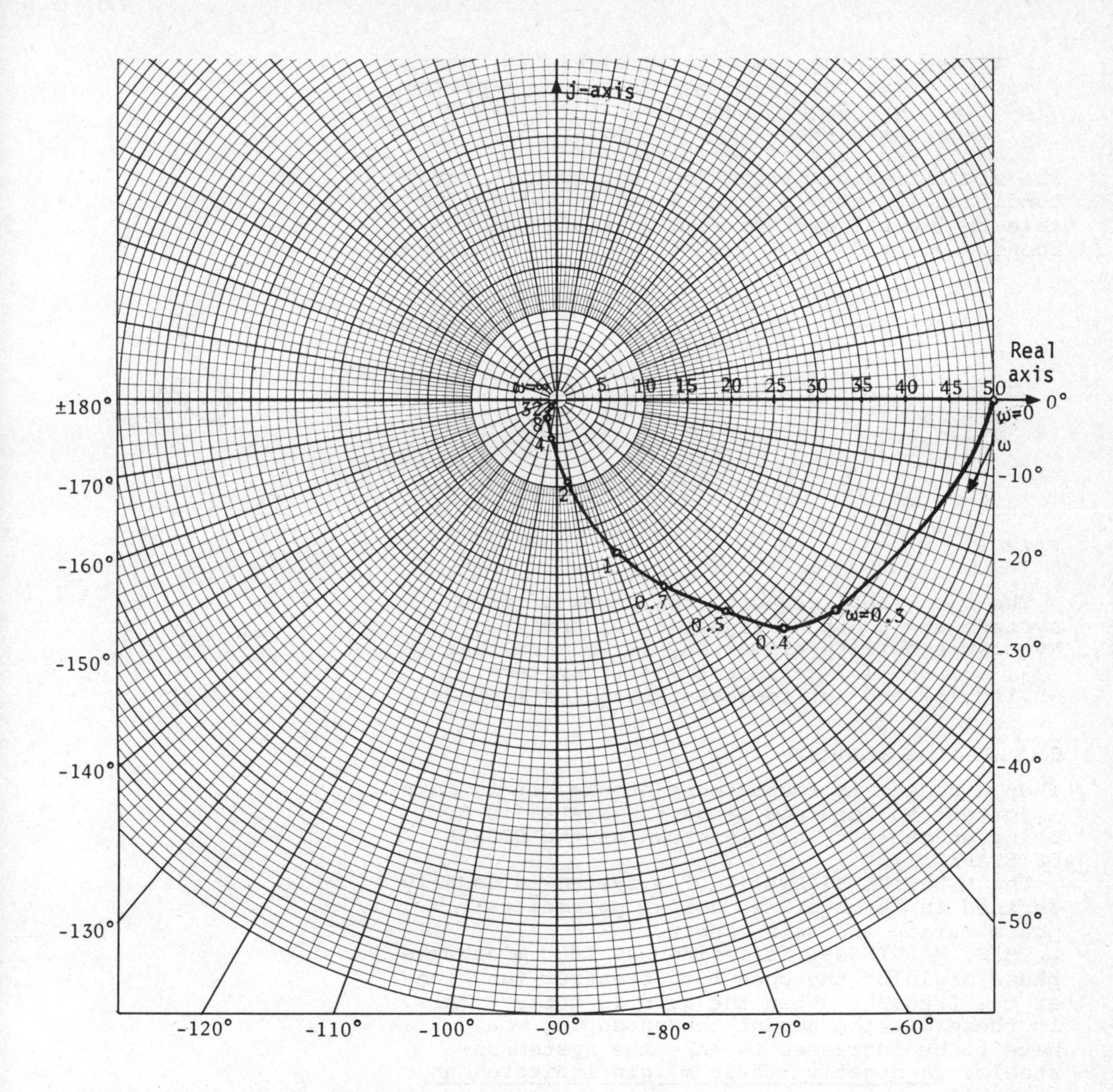

Fig.6.8 Nyquist plot

6.6.1 NYQUIST STABILITY CRITERION

The Nyquist stability criterion is based on the open-loop frequency response and is a graphical method of determining the stability of a system under closed-loop conditions. A detailed study of the subject is rather involved and is beyond the scope of this book. The simplified criterion may be stated as follows:

A closed-loop system will be stable if, when moving along the open-loop frequency response, plotted on a Nyquist diagram, in the direction of increasing frequency, the point (-1,j0) lies on the left of the locus.

The simplified criterion does not account for conditionally stable systems and is not applicable to unusual systems with unstable open-loops.

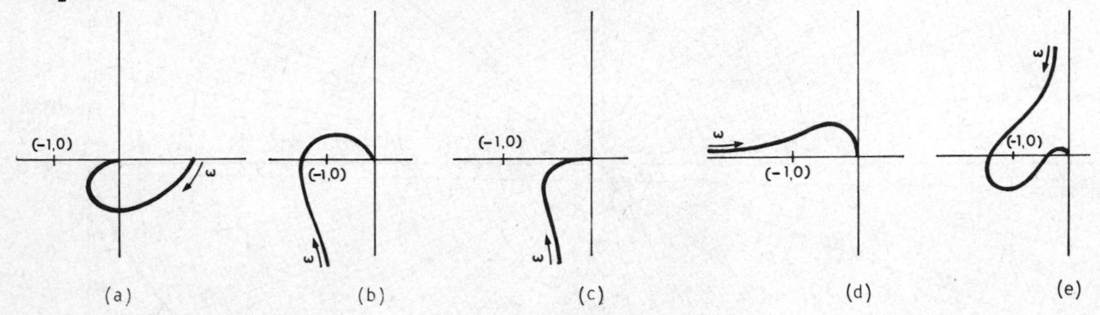

Fig.6.9

The open-loop frequency responses of five systems are shown in Figure 6.9. Applying Nyquist stability criterion, it is seen that (*a*), (*c*) and (*e*) represent *stable* systems, whilst (*b*) and (*d*) represent *unstable* systems.

6.6.2 RELATIVE STABILITY

When a system is found to be stable, it is most important to know how close the system is to being unstable. This is known as the DEGREE OF STABILITY.

The RELATIVE STABILITY of a system is usually defined in terms of two design parameters: phase margin and gain margin.

PHASE MARGIN ϕ_{PM} is defined as 180° minus the phase angle of the open-loop transfer function at the frequency when the gain is unity. This is therefore the amount the phase lag would have to be increased to make the system unstable. A negative phase margin indicates an unstable system.

GAIN MARGIN G_M is defined as the reciprocal of the magnitude of the open-loop transfer function (the gain) when the phase shift is 180°. This is therefore the factor by which the gain must be increased in order to produce instability. Gain margin is usually expressed in dB and if it is negative, it indicates an unstable system.

The frequency at which the gain is unity is called the GAIN CROSSOVER FREQUENCY ω_1, and the frequency at which the phase angle is 180° is called the PHASE CROSSOVER FREQUENCY ω_π.

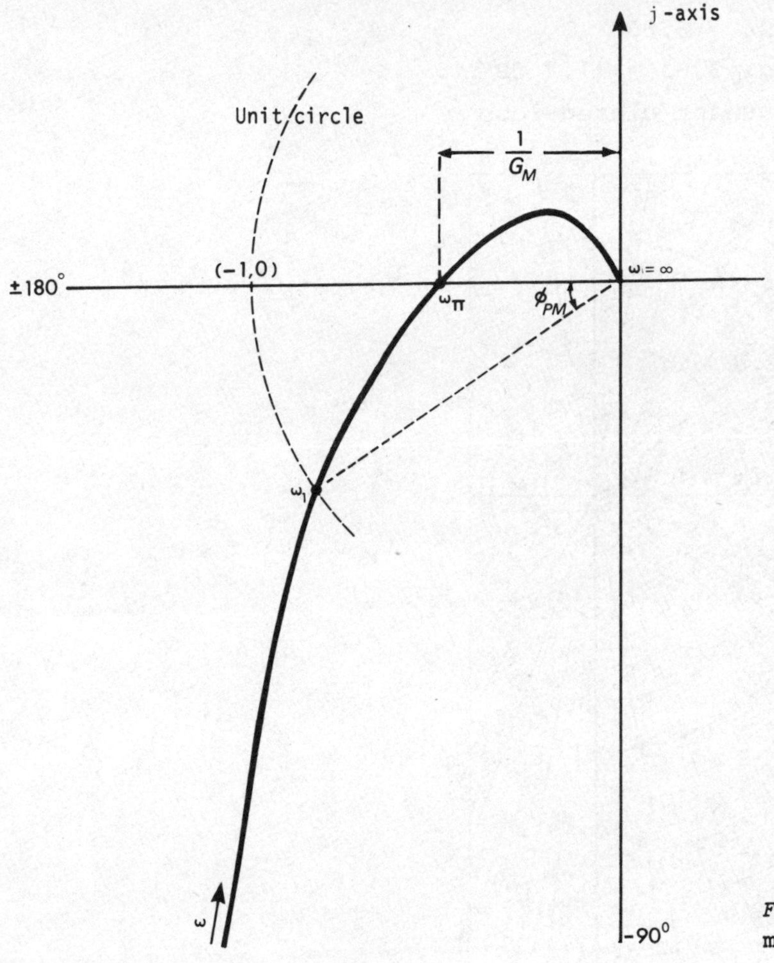

Fig.6.10 Phase margin and gain margin of a stable system

Example 6.9

A system has the following open-loop frequency response:

ω (rad/s)	2	3	4	5	6	8	10	30
Gain	2.8	1.9	1.3	0.9	0.68	0.4	0.26	0.12
φ (degree)	-120	-130	-140	-149	-157	-170	-180	-200

Plot the Nyquist diagram and determine the phase margin and the gain margin of the system. Would the system be stable under closed-loop condition?

Solution
The Nyquist plot is shown in Figure 6.11. From the plot:

Phase margin $\phi_{PM} = 180 - 146 = +34^\circ$

Gain margin $G_M = 1/0.26 = 3.85$

$= 20\log_{10} 3.85 = 11.7$ dB

The system will be stable under closed-loop condition.

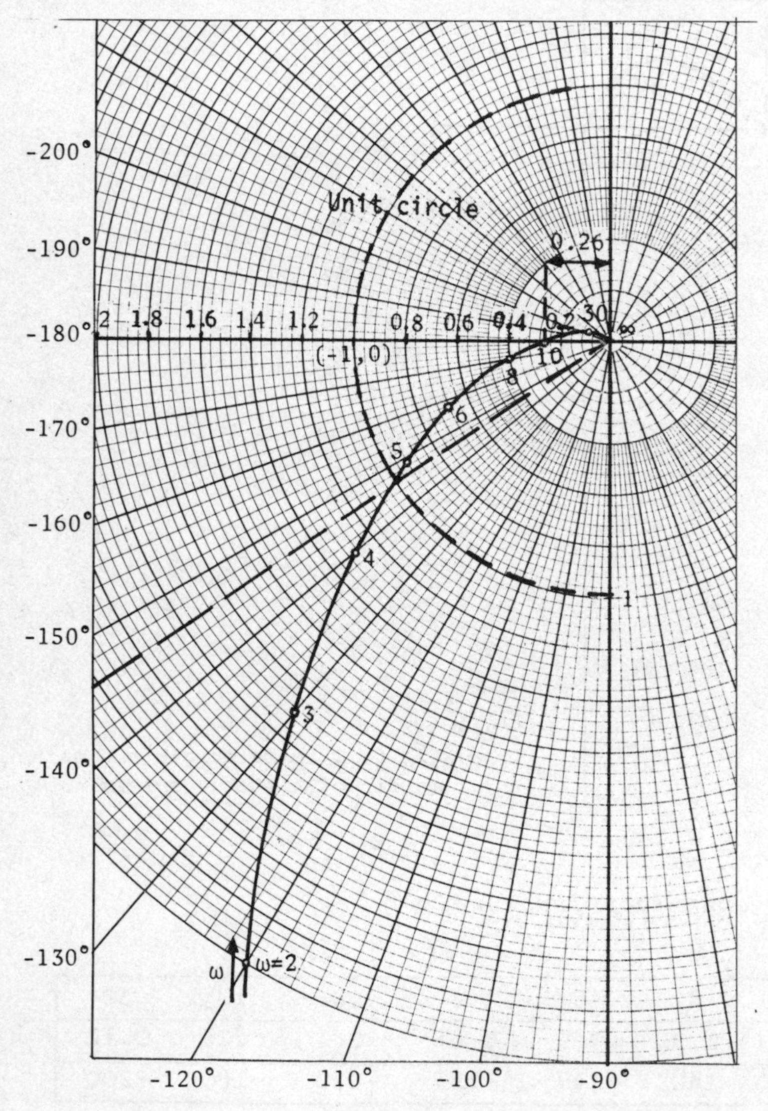

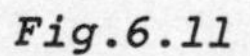
Fig.6.11

Example 6.10

A feedback control system has an open-loop transfer function:

$$G(s) = \frac{50}{s(1 + 0.1s)(1 + 0.5s)}$$

Prepare the Nyquist diagram and hence determine the phase and gain margins. Deduce from these values whether the system is stable, stating the reasons for your answer.

Solution

Rewriting the transfer function in terms of $j\omega$:

$$G(j\omega) = \frac{50}{j\omega(1 + j0.1\omega)(1 + j0.5\omega)}$$

Using the complex number table of Appendix 1, the magnitude and phase angle of the open-loop transfer function G are calculated for various values of ω and given in the table.

ω (rad/s)	$(j\omega)$	$(1+j0.1\omega)$	$(1+j0.5\omega)$	$\lvert G\rvert\angle\phi^\circ$
1.0	$1\angle 90^\circ$	$1\angle 6^\circ$	$1.12\angle 27^\circ$	$44.6\angle -123^\circ$
2.0	$2\angle 90^\circ$	$1.02\angle 11^\circ$	$1.41\angle 45^\circ$	$17.4\angle -146^\circ$
2.5	$2.5\angle 90^\circ$	$1.03\angle 14^\circ$	$1.62\angle 51^\circ$	$12\angle -155^\circ$
3.0	$3\angle 90^\circ$	$1.04\angle 17^\circ$	$1.8\angle 56^\circ$	$8.9\angle -163^\circ$
3.5	$3.5\angle 90^\circ$	$1.06\angle 19^\circ$	$2.01\angle 60^\circ$	$6.7\angle -169^\circ$
4.0	$4\angle 90^\circ$	$1.08\angle 22^\circ$	$2.23\angle 63^\circ$	$5.2\angle -175^\circ$
5.0	$5\angle 90^\circ$	$1.12\angle 27^\circ$	$2.69\angle 68^\circ$	$3.3\angle -185^\circ$
7.0	$7\angle 90^\circ$	$1.22\angle 35^\circ$	$3.64\angle 74^\circ$	$1.6\angle -199^\circ$
10	$10\angle 90^\circ$	$1.41\angle 45^\circ$	$5.1\angle 79^\circ$	$0.69\angle -214^\circ$
20	$20\angle 90^\circ$	$2.23\angle 63^\circ$	$10.05\angle 84^\circ$	$0.11\angle -237^\circ$

The Nyquist diagram is plotted and shown in Figure 6.12. From the diagram:

Phase margin $\phi_{PM} = 180 - 208 = -28^\circ$

Gain margin $G_M = 1/4.2 = 0.238$

$= 20\log_{10} 0.238 = -12.47$ dB

The phase margin is negative indicating that when the phase is 180° the gain is more than unity. The system is therefore unstable. Also, applying the Nyquist stability criterion, it is seen that moving along the response in the direction of increasing frequency, the $(-1, j0)$ point lies to the right.

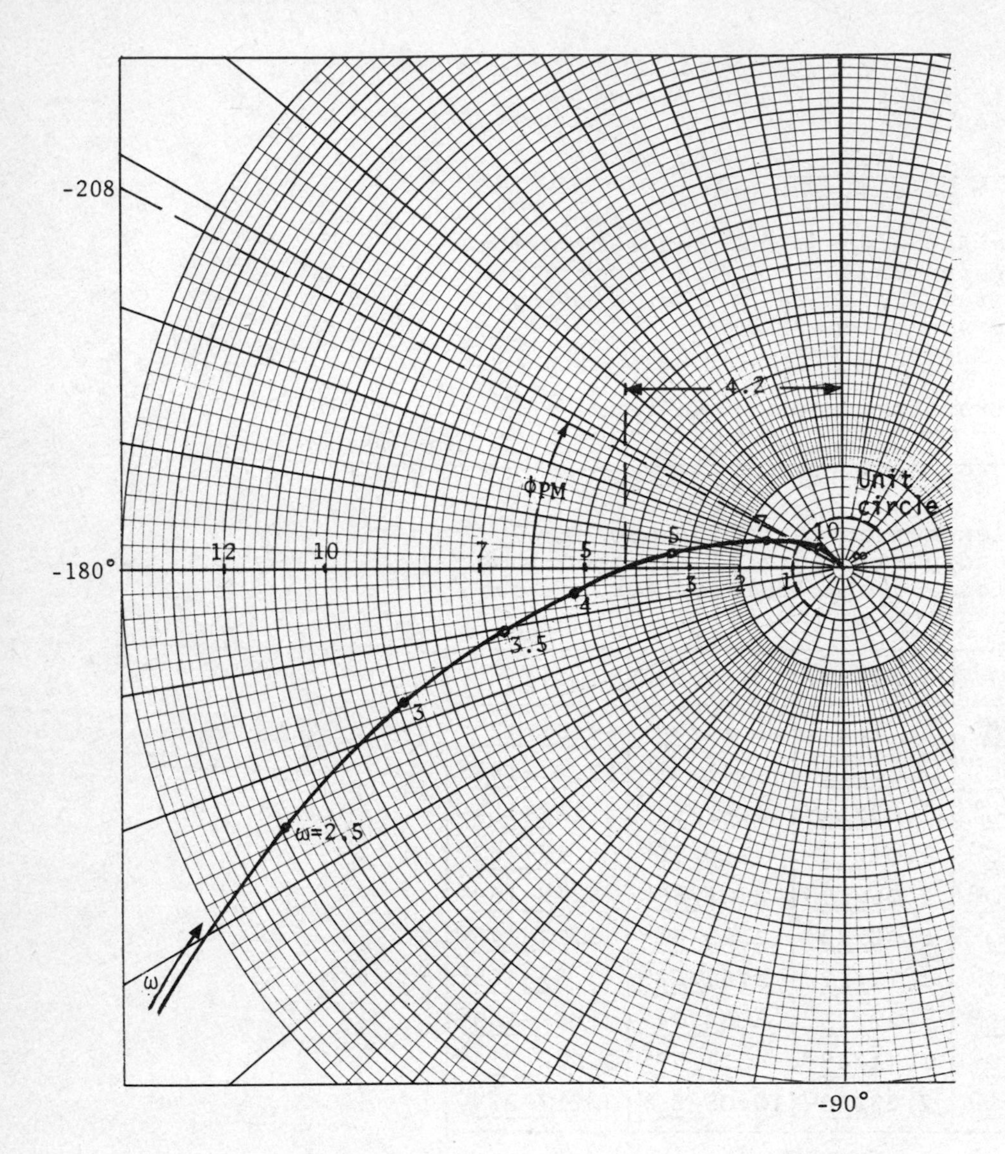

Fig.6.12

6.6.3 GAIN AND PHASE CIRCLES

From the open-loop frequency response, it is possible to determine the magnitude and phase of the closed-loop transfer function of the system. This is achieved by using lines of constant amplitude and lines of constant phase angle.

Consider a unity feedback system with forward transfer function $G(j\omega)$. From equation (1.6), the closed-loop transfer function will be

$$\frac{G(j\omega)}{1 + G(j\omega)}$$

Let the magnitude of this be M and the argument be N. Then, lines of constant magnitude drawn with

radius $\left|\frac{M}{M^2 - 1}\right|$ and centres at $[-M^2/(M^2-1),0]$

are called M-CIRCLES.

Similarly, lines of constant phase angle drawn with

radius $\frac{\sqrt{(N^2 + 1)}}{2N}$ and centres at $(-\frac{1}{2},1/2N)$

are called N-CIRCLES.

From the data given in Appendix 2, a family of M-circles and N-circles are drawn and shown in Figures 6.13 and 6.14 respectively (pages 126,127).

Example 6.11

The open-loop frequency response of a control system is as follows:

ω (rad/s)	1	2	3	4	5	10	20	50
$\lvert G\rvert$	5	3.8	2.7	2	1.3	0.88	0.4	0.2
ϕ (deg)	-94	-100	-108	-114	-122	-130	-150	-168

Plot the Nyquist diagram and determine the maximum value of M and the frequency at which it occurs.

Solution

The Nyquist plot is shown in Figure 6.15. In order to obtain the maximum value of M, it is necessary to find an M-circle that is tangential to the plot. By trial and error, the $M = 1.2$ circle fulfils this condition. From the diagram, the maximum value of M occurs at ω_{rf} = 8 rad/s.

6.6.4 INVERSE NYQUIST PLOT

It is often more convenient to design and investigate properties of a system from an INVERSE NYQUIST PLOT on which the quantities plotted are the reciprocals of those used in the Nyquist diagram. This is particularly useful for systems in which the feedback loop contains frequency-dependent components (feedback compensation for example). The following example illustrates the use of inverse Nyquist plots.

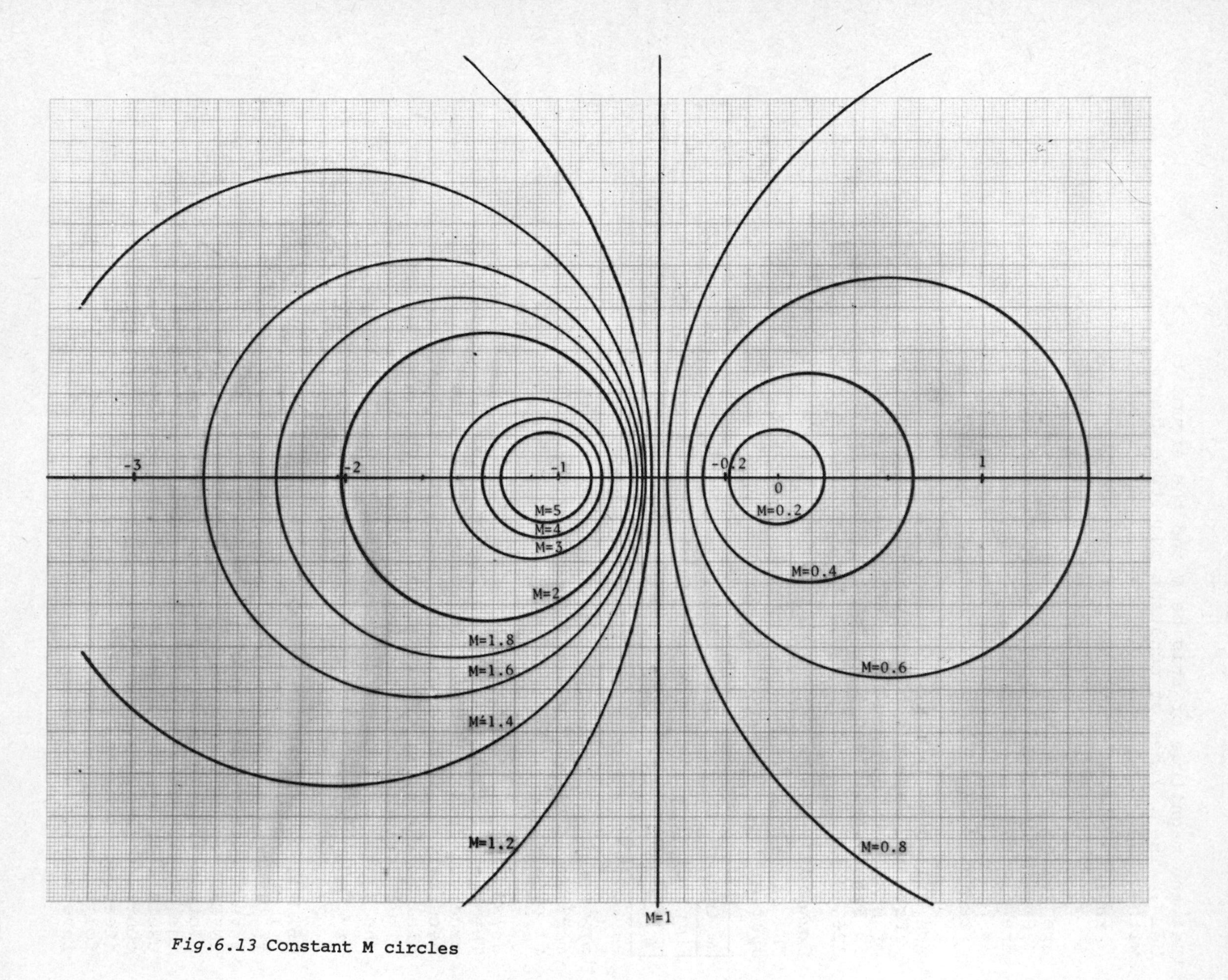

Fig.6.13 Constant M circles

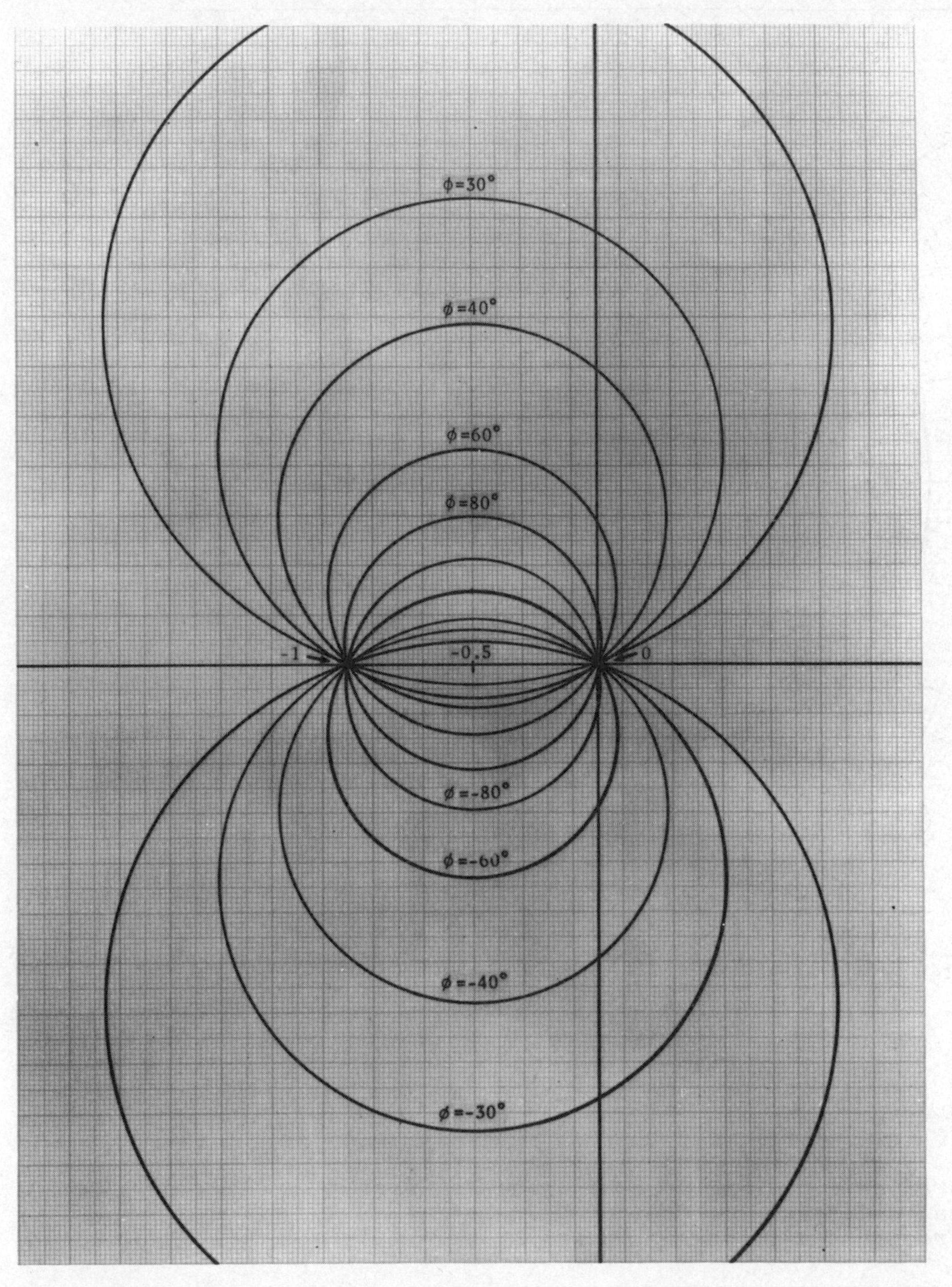

Fig.6.14 Constant N circles

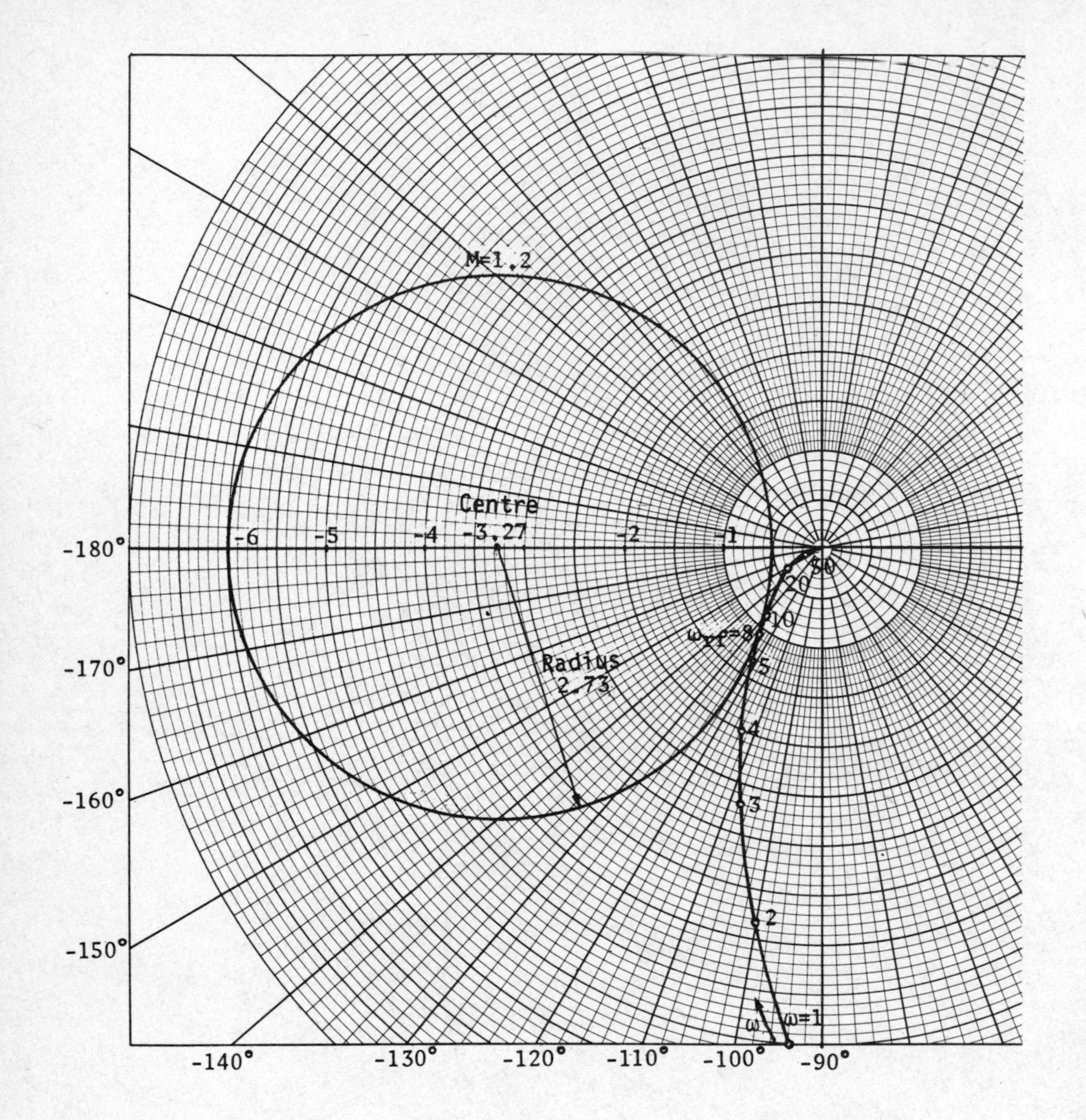

Fig.6.15

Example 6.12

The open-loop frequency response of a control system is given below. Plot an inverse Nyquist diagram and determine the maximum value of M and the frequency at which it occurs.

ω (rad/s)	1.0	1.5	2.0	3.0	5.0	10
$\|G\|$	5	2.5	1.72	0.96	0.67	0.51
φ (deg)	-110	-120	-125	-134	-140	-143

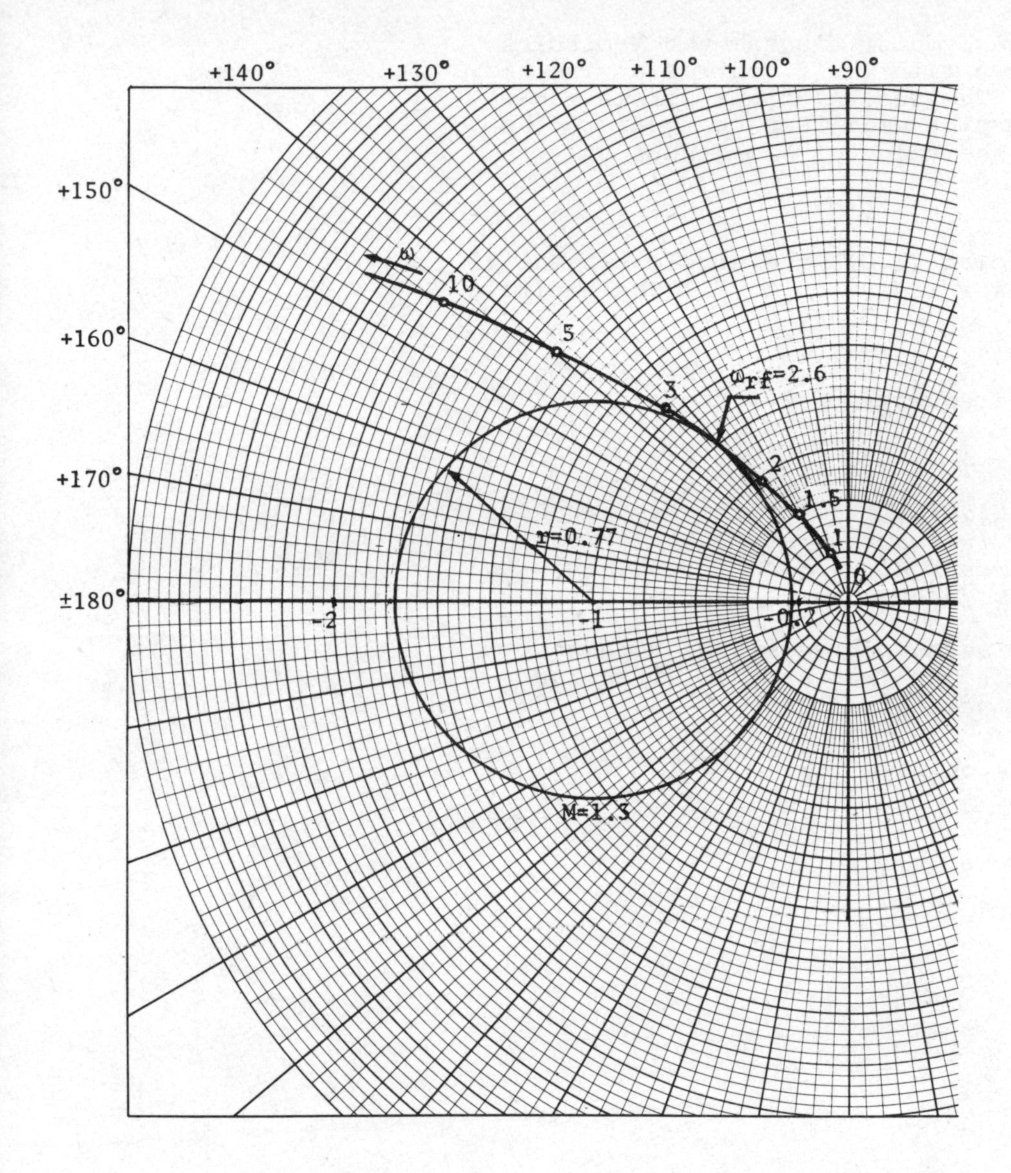

Fig.6.16
Inverse
Nyquist plot

Solution
The parameters required for the inverse Nyquist diagram are $1/|G|$ and $-\phi$ as follows:

ω(rad/s)	1.0	1.5	2.0	3.0	5.0	10
$1/\|G\|$	0.20	0.40	0.58	1.04	1.50	1.96
$-\phi$ (deg)	+110	+120	+125	+134	+140	+143

These figures are plotted and the inverse Nyquist diagram is shown in Figure 6.16.

In an inverse Nyquist diagram the M-circles are always drawn with their centres at the (-1,0) point. The radius of the circle tangential to the plot will then be equal to the reciprocal of the value of M_{pf}. Thus

$$M_{pf} = \frac{1}{r}$$

With its centre at (1,0), the circle tangential to the plot has a radius of 0.77. The M-circle gives

$$M_{pf} = 1/0.77 = 1.3$$

at an angular frequency of 2.6 rad/s.

Example 6.13

The open-loop Nyquist plot of a unity feedback control system is shown in Figure 6.17.
(*a*) *Derive a graphical method of obtaining the closed-loop frequency response of the system.*
(*b*) *Plot the closed-loop magnitude response, and estimate the frequency at which the peak magnitude would occur.*

Solution begins on page 132.

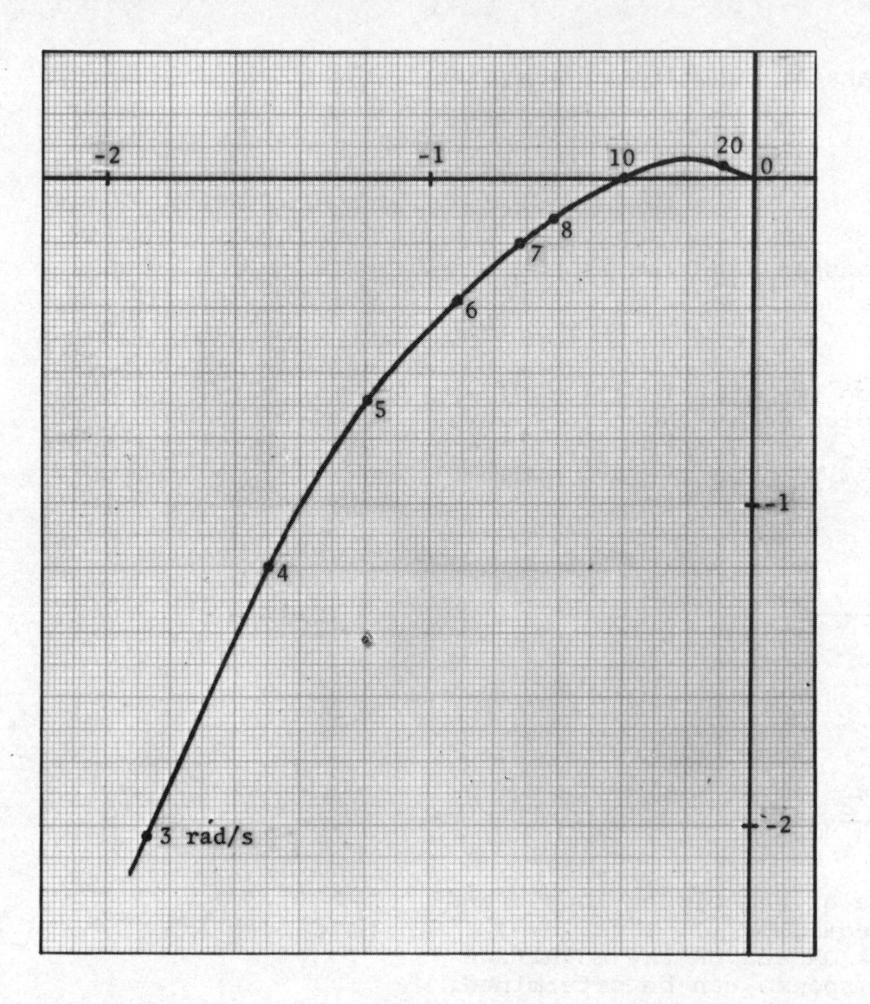

Fig.6.17

Solution
(*a*) The open-loop transfer function of a system is given by

$$G = \frac{\theta_o}{\theta} \qquad \text{(i)}$$

where $\theta = \theta_i - \theta_o$.

The closed-loop transfer function is

$$\frac{G}{1 + GH} = \frac{\theta_o}{\theta_i} \qquad \text{(ii)}$$

Consider the open-loop transfer function $G(j\omega)$ and let the Nyquist plot be as shown in Figure 6.18(a). The vector $\vec{OA}$ represents θ_o/θ at some particular frequency ω_1. Also,

$$\begin{aligned} \vec{BA} &= \vec{BO} + \vec{OA} \\ &= 1 + \vec{OA} \\ &= 1 + \frac{\theta_o}{\theta} = \frac{\theta + \theta_o}{\theta} \end{aligned}$$

Thus $\vec{BA} = \dfrac{\theta_i}{\theta}$

Then

$$M = \frac{\vec{OA}}{\vec{BA}} = \frac{\theta_o/\theta}{\theta_i/\theta} = \frac{\theta_o}{\theta_i}$$

This is the magnitude of the closed-loop transfer function at a frequency ω_1. Thus, by measuring the lengths OA and BA the magnitude of the closed-loop response can be determined.

By simple geometry it can be seen that θ_o lags θ_i by the angle α and as shown in the diagram $\phi = \alpha + \beta$.

(*b*) The closed-loop magnitude response is plotted using the relation $M = \vec{OA}/\vec{BA}$ for various frequencies and is shown in Figure 6.18(b). From the plot it is seen that the peak magnitude occurs at $\omega = 6$ rad/s.

ω	3	4	5	6	7	8	10	20
$\vec{OA}$	2.75	1.9	1.38	1.0	0.77	0.62	0.42	0.1
$\vec{BA}$	2.2	1.3	0.7	0.4	0.34	0.42	0.58	0.92
M	1.25	1.46	1.97	2.5	2.26	1.47	0.72	0.01

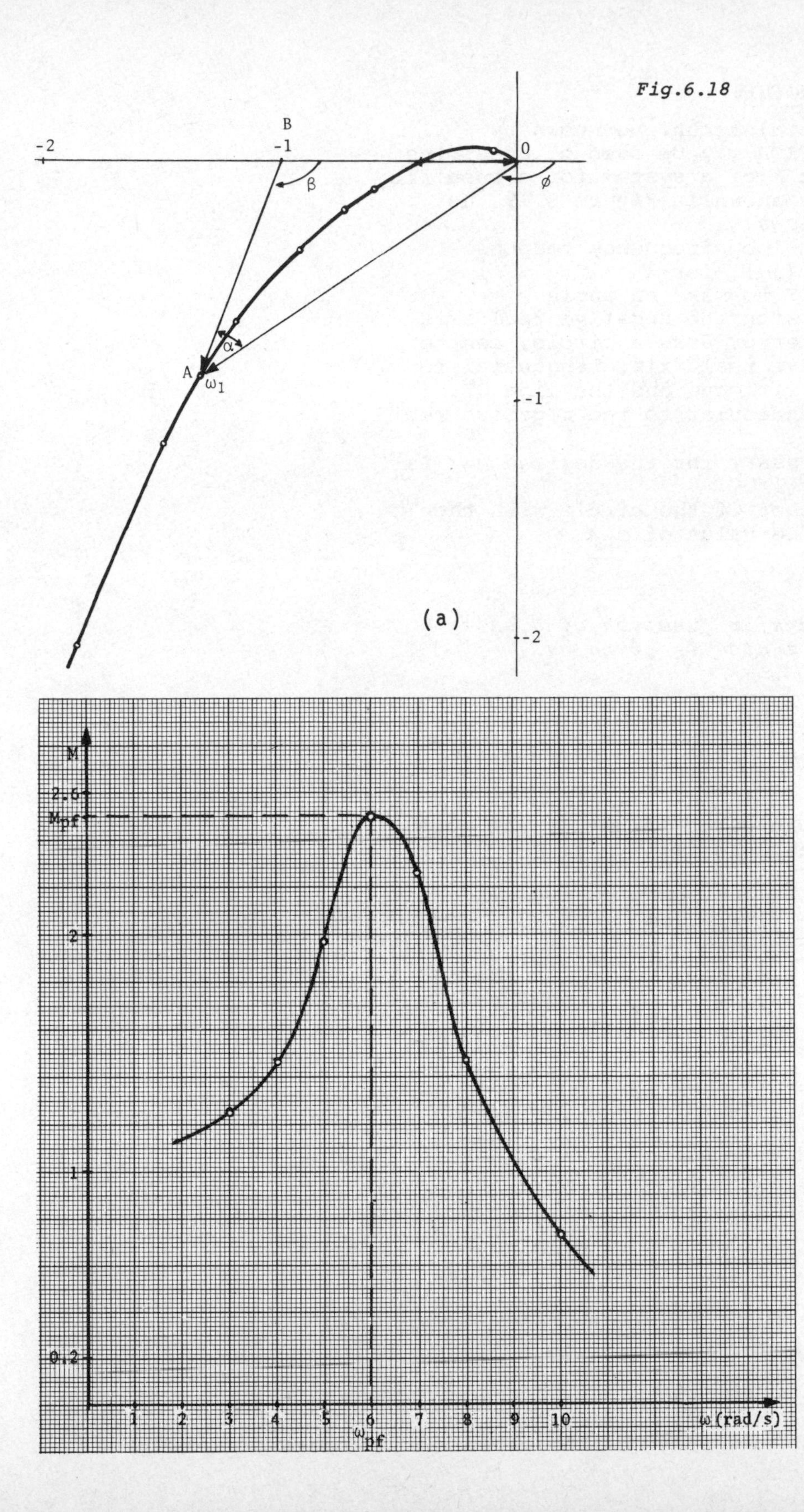
Fig.6.18
B
-2
-1
0
β
φ
α
A
ω1
-1
-2
(a)
(b)
M
2.6
Mpf
2
1
0.2
1
2
3
4
5
6
7
8
9
10
ωpf
ω(rad/s)

6.7 BROWN'S CONSTRUCTION

A simple geometrical technique known as BROWN'S CONSTRUCTION may be used to determine the gain constant K of a system for a specific value of M_{pf}. As shown in Figure 6.19 the steps are as follows:

(1) Plot the open-loop frequency response [locus of $KG\,(j\omega)$] for $K = 1$.
(2) Draw a line OT to make an angle $\psi = \sin^{-1}(1/M_{pf})$ with the negative real axis.
(3) By trial and error draw a circle, centre on the negative real axis, tangential to both the $KG(j\omega)$ locus and the line OT.
(4) Draw BC perpendicular to the negative real axis.
(5) The gain necessary for the desired M_{pf} is then given by 1/OC.
(6) Point of contact of the circle with the locus gives the value of ω_{rf}.

Example 6.14

The open-loop transfer function of a unity feedback control system is given by

$$G(j\omega) = \frac{K}{j\omega(1 + j0.1\omega)}$$

Determine the gain K such that $M_{pf} = 1.4$.

Solution
The following table gives values of the open loop frequency response for $K = 1$.

ω (rad/s)	$(j\omega)$	$(1 + j0.1\omega)$	$\lvert G\rvert \angle\phi^{\circ}$
2	$2\angle 90^{\circ}$	$1.02\angle 11^{\circ}$	$0.49\angle -101^{\circ}$
4	$4\angle 90^{\circ}$	$1.08\angle 22^{\circ}$	$0.23\angle -112^{\circ}$
6	$6\angle 90^{\circ}$	$1.17\angle 31^{\circ}$	$0.14\angle -121^{\circ}$
8	$8\angle 90^{\circ}$	$1.28\angle 39^{\circ}$	$0.098\angle -129^{\circ}$
10	$10\angle 90^{\circ}$	$1.41\angle 45^{\circ}$	$0.07\angle -135^{\circ}$
12	$12\angle 90^{\circ}$	$1.55\angle 50^{\circ}$	$0.054\angle -140^{\circ}$
15	$15\angle 90^{\circ}$	$1.8\angle 56^{\circ}$	$0.037\angle -146^{\circ}$
20	$20\angle 90^{\circ}$	$2.23\angle 63^{\circ}$	$0.022\angle -153^{\circ}$
50	$50\angle 90^{\circ}$	$5.1\angle 79^{\circ}$	$0.004\angle -169^{\circ}$

Evaluating ψ:

$$\psi = \sin^{-1}(1/M_{pf}) = \sin^{-1}(1/1.4) = 45.6^{\circ}$$

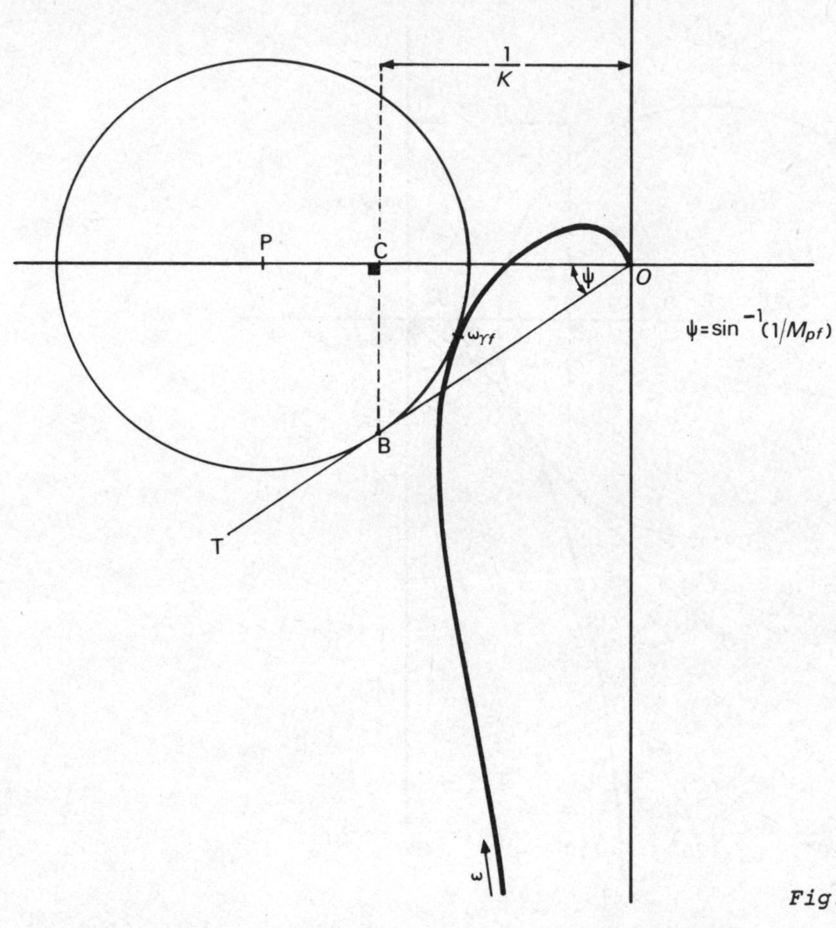

Fig.6.19 Brown's construction

The Nyquist diagram and necessary constructions are shown in Figure 6.20. From the diagram $1/K = 0.065$. Thus

$$K = 1/0.065 = 15.4$$

6.8 CLASSIFICATION OF SYSTEMS

In general, the open-loop transfer function of a control system may be written in the form

$$\frac{K(1 + s\tau_a)(1 + s\tau_b)\ \ldots}{s^n(1 + s\tau_1)(1 + s\tau_2)\ldots}$$

where K is the numerical value of the system gain at zero frequency, and τ_1, τ_2, ... τ_a, τ_b, etc. are the time constants of various sections of the system.

One way of identifying the nature of a system is by classifying it according to the value of n, which determines the number of poles at the origin of the s-plane and has a predominant effect upon the behaviour of the system. A

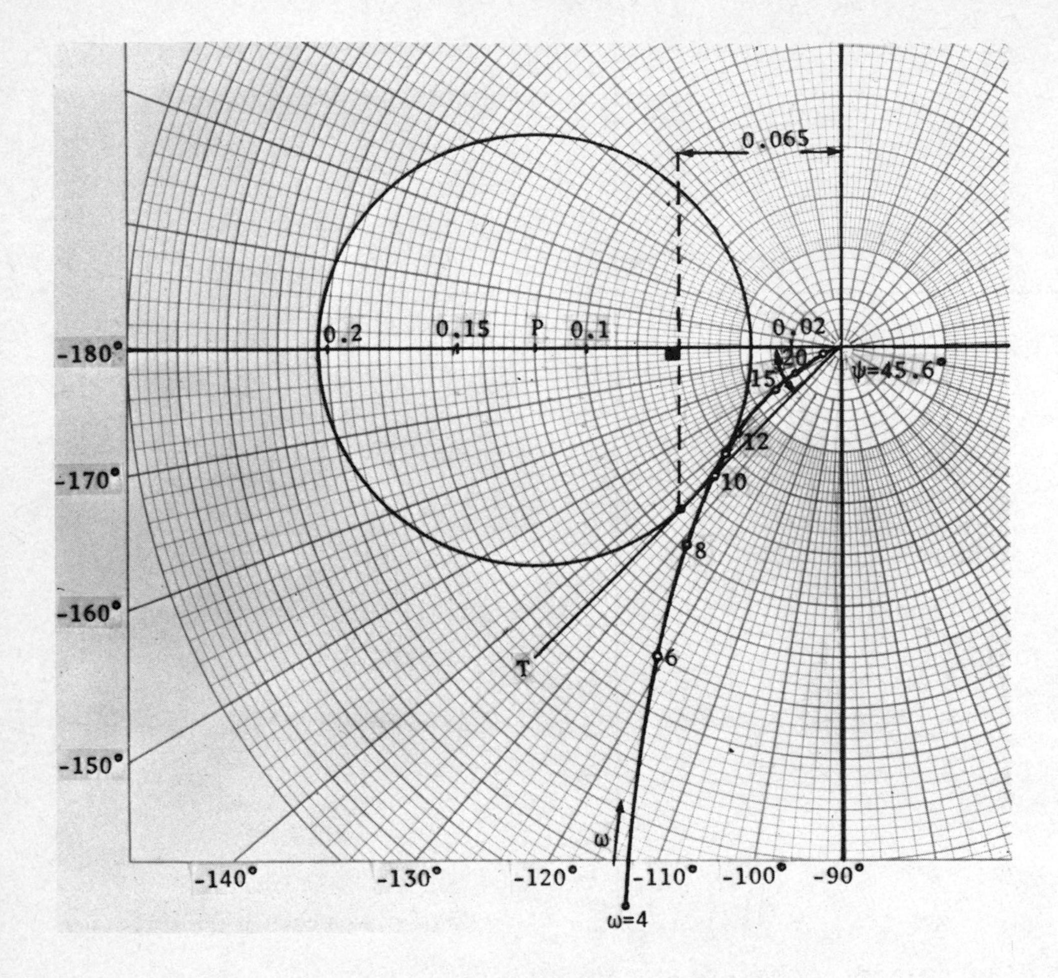

Fig.6.20

type 0 system results when there is no integration, as in a proportional control. In an integral control in which there is one integrator in the forward element, n would be 1 and the system is a type I. Thus,

n = 0 a type 0 system (such as velocity control)

n = 1 a type I system (such as position control)

n = 2 a type II system (such as integral compensated position control)

It should be noted that the higher the value of n, the more difficult it is to stabilize the system.

It is important to avoid confusion between the type and the order of a system. The ORDER OF A SYSTEM is the order of the denominator of the open-loop transfer function. Hence, a system with

$$G(s) = \frac{15(1 + 2s)}{s(1 + s)(1 + s + 2s^2)}$$

for example, is a type I system of *fourth* order.

Example 6.15

Using sketches, show how n determines the general shape of the open-loop frequency response.

Solution
Replacing s by $j\omega$, then as ω tends to zero, all brackets in the expression of the open-loop transfer function tend to unity and so the transfer function tends to $K/(j\omega)^n$.

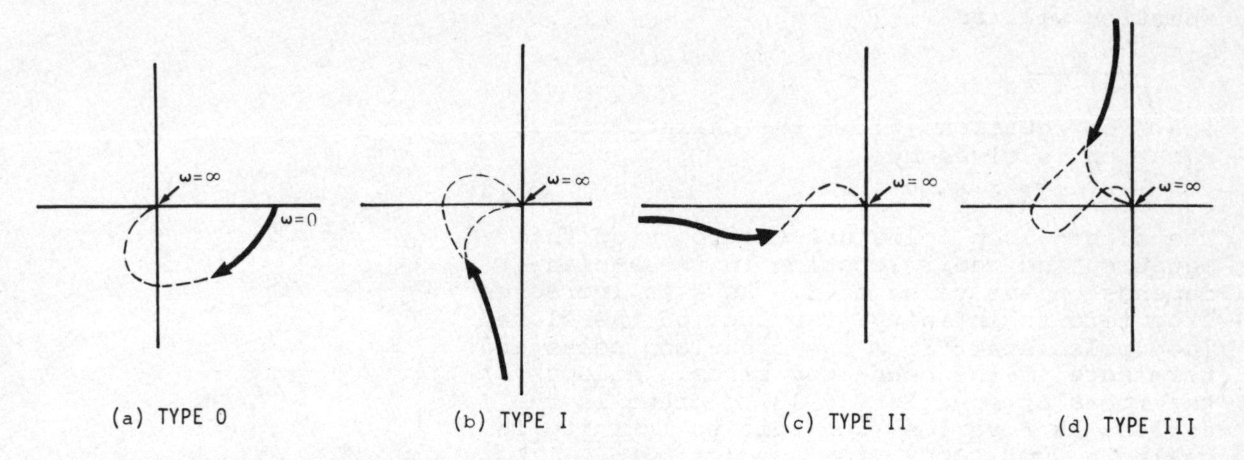

Fig.6.21 Typical open-loop frequency responses for type 0, I, II, III systems

For $n = 0$ $K/(j\omega)^n$ tends to K. The locus starts on the positive real axis, Figure 6.21(*a*).

For $n = 1$ $K/(j\omega)^n$ tends to $-j\infty$ as ω tends to zero. The locus starts at infinity close to the $-j$-axis, Figure 6.21(*b*).

For $n = 2$ $K/(j\omega)^n$ tends to $-\infty$ as ω tends to zero. The locus starts at infinity close to the negative real axis, Figure 6.21(*c*).

For $n = 3$ $K/(j\omega)^n$ tends to $+j\infty$ as ω tends to zero. The locus starts at infinity close to the $+j$-axis, Figure 6.21(*d*).

6.9 ROOT-LOCUS

The ROOT-LOCUS technique, based on the positions of poles and zeros in the s-plane, is a graphical means of determining the roots of the characteristic equation of a closed-loop control system.

From equation (1.6), the closed-loop transfer function of a unity feedback system is given by

$$\frac{G}{1 + G}$$

The open-loop transfer function G can be expressed in terms of a polynomial in s as $K/P(s)$, where K is the open-loop gain constant. Substituting for G, the closed-loop transfer function will be

$$\frac{K}{P(s) + K}$$

and from equation (6.3), the characteristic equation is given by

$$P(s) + K = 0 \qquad (6.13)$$

The closed-loop poles are the roots of this equation and their location in the s-plane depends on the value of K. As K is increased from zero to infinity, the loci of the closed-loop poles start from the open-loop poles and terminate at the open-loop zeros. A locus of the roots of equation (6.13) plotted in the s-plane as K varies from zero to infinity is called a ROOT-LOCUS PLOT.

The following rules are often helpful when plotting a root-locus diagram.

Rule 1 Each locus starts at an open-loop pole when $K = 0$ and either terminates at an open-loop zero or approaches infinity along an asymptote when $K = \infty$.

Rule 2 There are as many branches of the loci as there are open-loop poles.

Rule 3 The loci are always symmetrical about the real axis.

Rule 4 If there are m open-loop poles and n open-loop zeros and $n > m$, then $m - n$ branches of the loci approach infinity.

Rule 5 Any part of the real axis which lies to the *left* of an *odd* number of poles and zeros forms part of the loci.

Rule 6 The asymptotes cross the real axis at a point x given by

$$x = \left(\sum(\text{open-loop poles}) - \sum(\text{open-loop zeros})\right) \Big/ (m - n) \quad (6.14)$$

Rule 7 The angle which the asymptotes make with the real axis is given by

$$\alpha = \pm \frac{k \times 180}{m - n} \quad \text{where } k \text{ is an odd integer} \quad (6.15)$$

Rule 8 The value of K at the crossing points of the loci with the j-axis can be determined either using Routh's criterion or from *Rule 9*.

Rule 9 The crossing points of the loci with the j-axis can be determined as follows:

(i) Replace s by $j\omega$ in the characteristic equation.
(ii) Put the imaginary terms equal to zero.
(iii) Solve for ω.
(iv) Solve for K.

Rule 10 There is always a *breakaway point* between any two adjacent poles on the real axis which are connected by a branch of the loci.

Rule 11 The loci leaves the real axis (breakaway point) where K is maximum. Thus, from the characteristic equation $dK/ds = 0$.

Rule 12 *Angle of departure*, defined as the angle at which the loci leaves complex poles, can be determined from the relation

$$\sum\theta_Z - \sum\theta_p = \pm 180^\circ \quad (6.16)$$

where $\sum\theta_p$ is the sum of all the angles subtended by the poles and $\sum\theta_Z$ is the sum of the angles subtended by the zeros.

Example 6.16

Plot the root-locus diagram for the system whose open-loop transfer function is

$$GH(s) = \frac{K}{s(s^2 + 2s + 2)}$$

Solution
(i) Factorizing the denominator:

$$GH(s) = \frac{K}{s(s+1+j)(s+1-j)}$$

(ii) There are three open-loop poles at $s = 0$, $s = -1 - j$, and $s = -1 + j$, and no zeros.
(iii) There will be three branches of the loci (*Rule 2*).
(iv) The asymptotes will cross the real axis at x (*Rule 6*):

$$x = \frac{\sum P - \sum Z}{m - n} = \frac{[(0) + (-1+j) + (-1-j)] - [(0)]}{3 - 0} = -\frac{2}{3}$$

(v) The asymptotes will make an angle α (*Rule 7*):

$$\alpha = \pm \frac{k \times 180}{m - n} = \frac{1 \times 180}{3 - 0} = \pm 60^{\circ} \quad \text{for } k = 1$$

For $k = 3$, α will be 180° which indicates that one branch of the loci approaches infinity along the real axis.

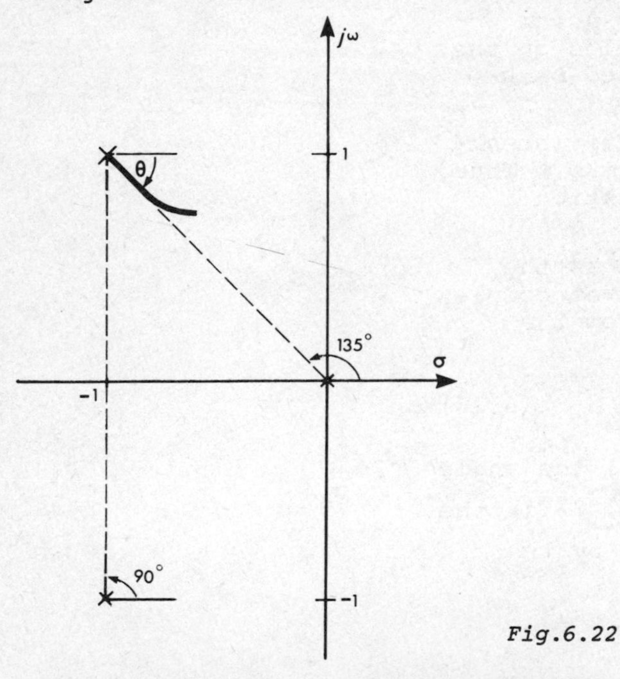

Fig.6.22

(vi) To find the angle at which the loci leaves the complex poles *Rule 12* is applied. Using equation (6.16) and from Figure 6.22,

$$\sum \theta_Z - \sum \theta_P = \pm 180^{\circ}$$

$$[(0)] - [(135) + (90) + (\theta)] = \pm 180^{\circ}$$

thus $\theta = \pm 45^{\circ}$

(vii) The characteristic equation is

$$s(s^2 + 2s + 2) + K = 0$$

or $s^3 + 2s^2 + 2s + K = 0$

Let $s = j\omega$ (*Rule 9*):

$$(j\omega)^3 + 2(j\omega)^2 + 2(j\omega) + K = 0$$

$$-j\omega^3 - 2\omega^2 + j2\omega + K = 0$$

$$(K - 2\omega^2) + j(2\omega - \omega^3) = 0$$

Putting the imaginary part equal to zero:

$$2\omega - \omega^3 = 0 \quad \text{i.e. } \omega = \pm\sqrt{2}$$

Putting the real part equal to zero:

$$K - 2\omega^2 = 0$$

Substituting for ω,

$$K = 2\omega^2 = 2 \times 2 = 4$$

The loci will cross the j-axis at $\omega = \pm\sqrt{2}$ and the value of K at these points is 4.
The root-locus plot is shown in Figure 6.23.

6.9.1 LINES OF CONSTANT VALUE

Figure 6.24 shows four lines representing constant values of ζ, ω_n, ω_d and $1/\tau$ in the s-plane. These lines, when used in conjunction with the root-locus, can be used to design and/or determine system parameters.

VALUE OF K AT A GIVEN POINT ON THE LOCI
(I) *If the point is on the real axis,* simply substitute the value of s for that point in the characteristic equation and calculate the absolute value of K.

Example 6.17
Determine the value of K at s = − 9 for the system with open-loop transfer function

$$KG(s)H(s) = \frac{K(s + 3)}{s(s + 5)}$$

Solution
The characteristic equation is

$$s(s + 5) + K(s + 3) = 0$$

then

$$K = \frac{-s(s + 5)}{(s + 3)}$$

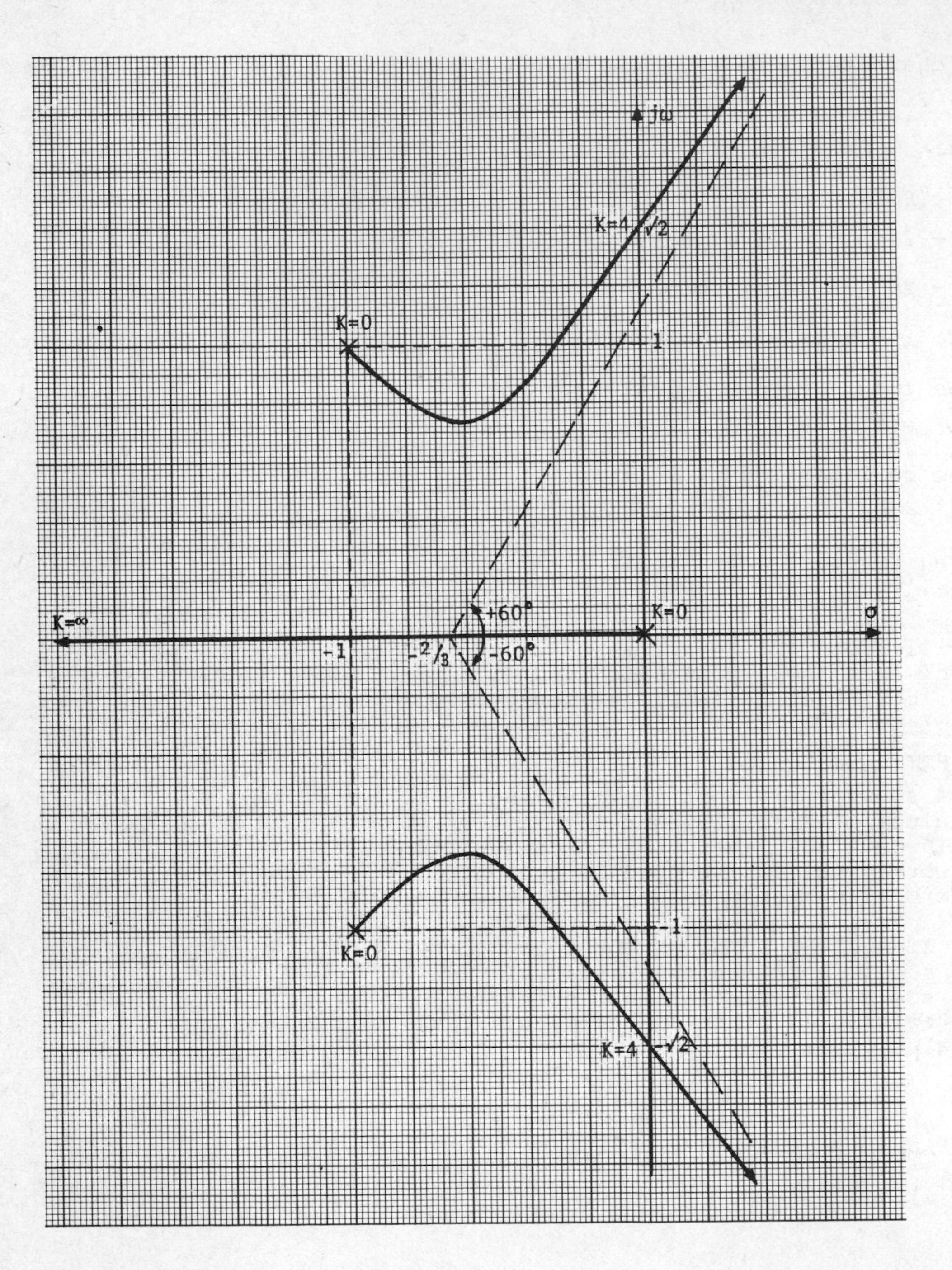

Fig.6.23

Substituting for *s*:

$$K = \left| \frac{-(-9)(-9 + 5)}{(-9 + 3)} \right| = 6$$

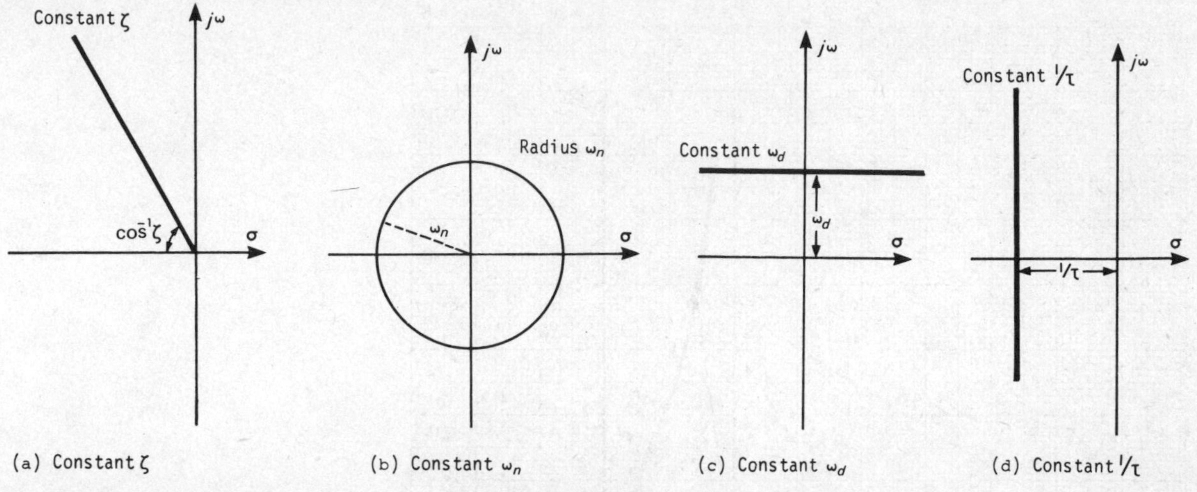

Fig.6.24

(II) *If the point is not on the real axis* a graphical method is often simpler. The procedure is as follows:
(1) Draw vectors from all the open-loop poles to the point.
(2) Measure the lengths of these vectors and multiply them.
(3) Draw vectors from all the open-loop zeros to the point.
(4) Measure their lengths and multiply them.
(5) To obtain K divide the result obtained in (2) by the result obtained in (4).

Example 6.18

Determine the value of K at point P *on the root-locus of Figure 6.25.*

Solution
From Figure 6.25,

$$K = \frac{2.15 \times 2.04 \times 3.78}{2.33} = 7.114$$

Example 6.19

Explain with the aid of sketches how the performance of a control system may be predicted from a root-locus diagram.
A control system with unity negative feedback has a forward path transfer function given by

$$\frac{k}{s(s + 1)(s + 3)}$$

Sketch the root-locus diagram for the system, and determine the value of k when the damping ratio is 0.5.

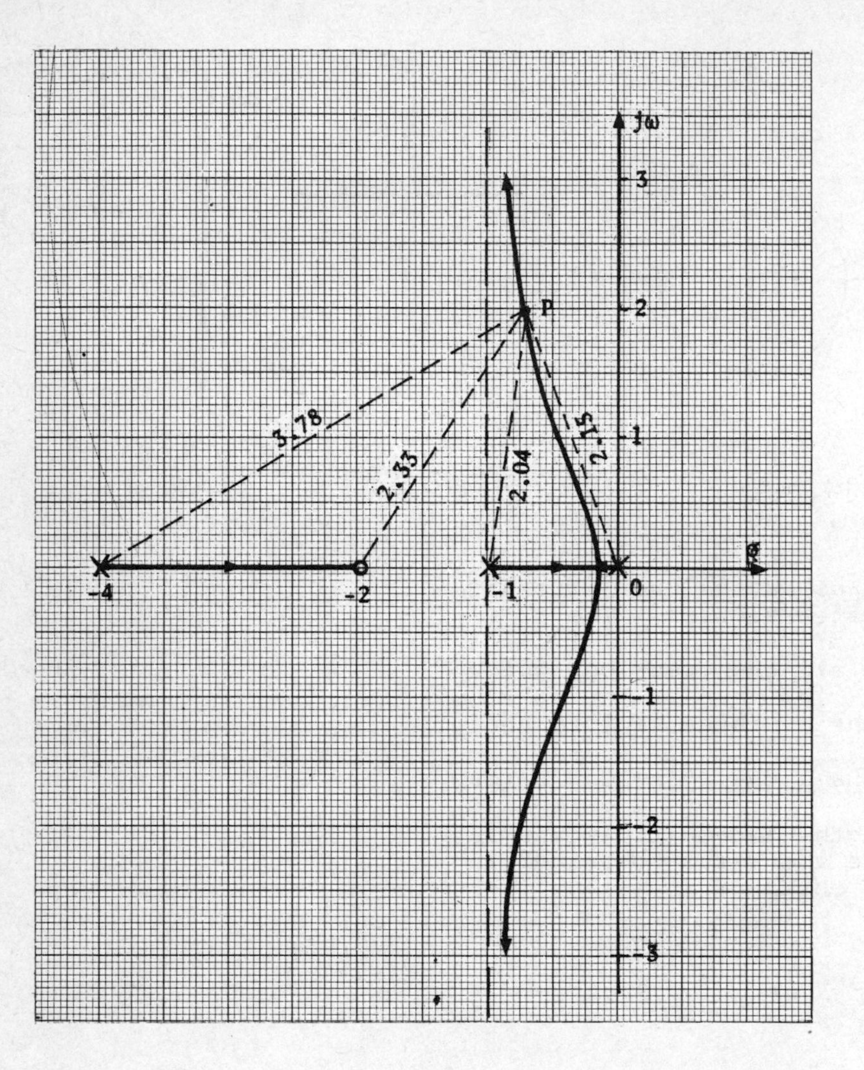

Fig.6.25

Solution
The major difficulty in the analysis of control systems is often with finding the roots of the high-order (higher than cubic) differential equation describing the performance of a system. Moreover, it is often necessary to observe the effects of parameter changes on the roots of the characteristic equation. As explained in Section 6.9 a root-locus diagram is based on the positions of the open-loop poles and zeros in the s-plane and shows the loci of the roots plotted as a function of some parameter, usually the loop gain. It was shown that a system is stable under closed-loop condition when all the roots of the characteristic equation (closed-loop poles) lie in the left-hand side of the s-plane. Also, in Example 6.5 and Figures 6.4 and 6.5 it was shown that the location of the

poles determines the system response. A root-locus diagram can therefore be used to determine the limits of stability as well as predicting the performance of a system under closed-loop condition.

In this example there are three open-loop poles at $s = 0$, $s = -1$ and $s = -3$, and no zeros. Hence, there will be three branches of the loci. From equation (6.14), the asymptotes cross the real axis at

$$x = \frac{[(0) + (-1) + (-3)] - [(0)]}{3 - 0} = -\frac{4}{3}$$

From equation (6.15), the angle the asymptotes make with the real axis is

$$\alpha = \pm \frac{1 \times 180}{3} = \pm 60^{\circ}$$

To determine the position of the breakaway point which lies between 0 and -1, $dK/ds = 0$. The characteristic equation is

$$s(s + 1)(s + 3) + K = 0$$

Rearranging,

$$K = -s^3 - 4s^2 - 3s$$

$$\frac{dK}{ds} = -3s^2 - 8s - 3 = 0$$

Hence, $s = -2.2$ or $s = -0.45$ of which only the -0.45 point lies between 0 and -1 and is therefore a breakaway point. To determine the crossing points of the loci with the j-axis, *Rule 9* is applied. Substituting $j\omega$ for s in the characteristic equation:

$$s^3 + 4s^2 + 3s + K = 0$$

$$(j\omega)^3 + 4(j\omega)^2 + 3(j\omega) + K = 0$$

or

$$(K - 4\omega^2) + j(3\omega - \omega^3) = 0$$

$$3\omega - \omega^3 = 0 \qquad \text{i.e. } \omega = \pm\sqrt{3}$$

Also

$$K - 4\omega^2 = 0 \qquad \text{i.e.} \quad K = 4\omega^2 = 4 \times 3 = 12$$

The loci will cross the j-axis at $\omega = \pm\sqrt{3}$ and the value of K at these points is 12. The root-locus diagram is shown in Figure 6.26.

For the damping ratio to be 0.5, a line is drawn from the origin at $\cos^{-1}0.5 = 60^{\circ}$ to the negative real axis. The intersection of this line with the loci determines the required value of K.

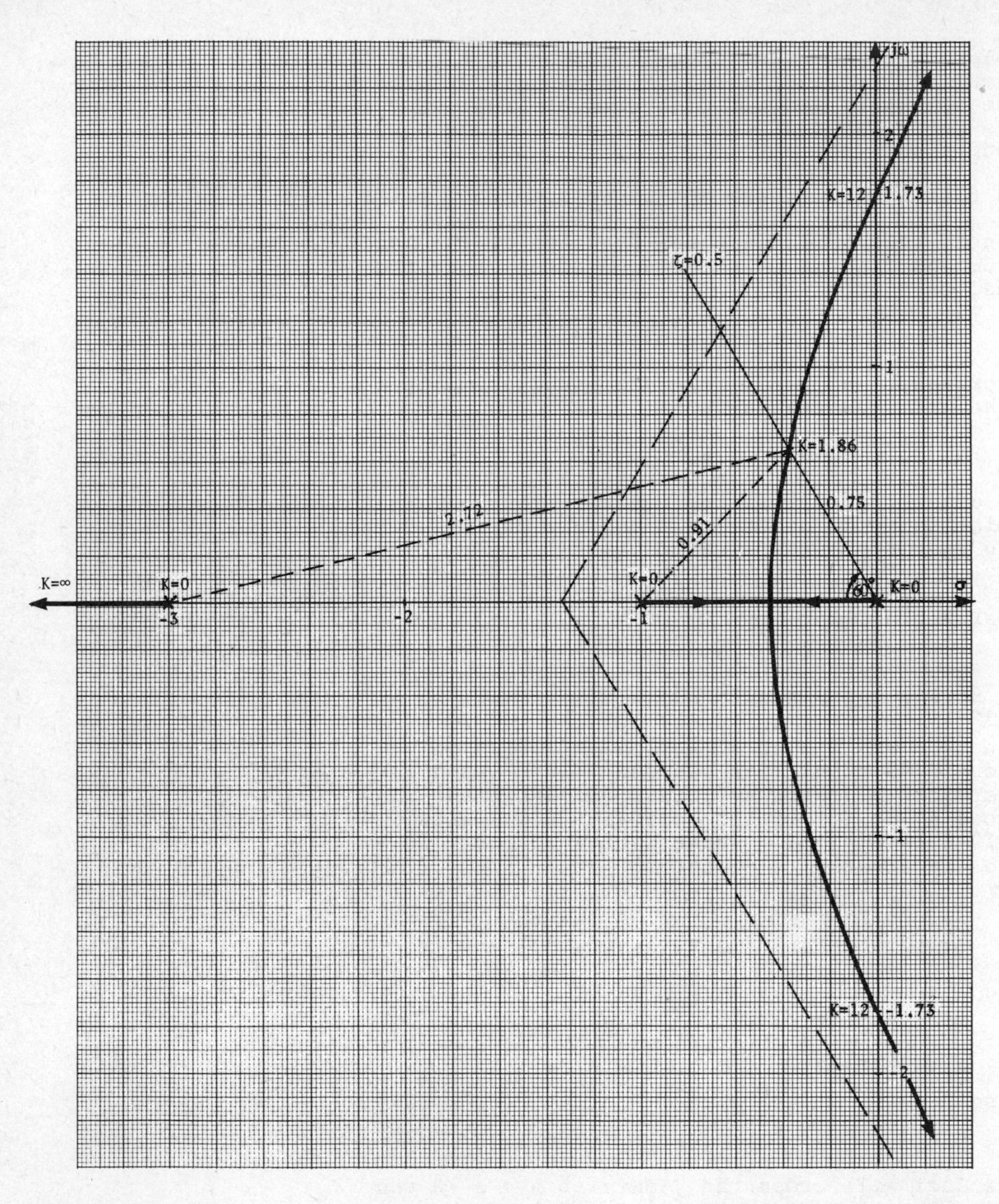

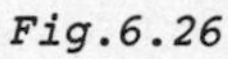
Fig.6.26

From the diagram,

$$K = (0.75) \times (0.91) \times (2.72) = 1.86$$

6.10 BODE PLOT

The BODE PLOT is a logarithmic diagram in which the modulus (amplitude) $|G(j\omega)|$ in decibels and the argument (phase) ϕ in degrees of a transfer function $G(j\omega)$ are plotted separately against the logarithm of frequency.

In general, the transfer function to be plotted will have the form

$$G(j\omega) = \frac{K(1 + j\omega\tau_a)(1 + j\omega\tau_b)\ \dots}{(j\omega)^n(1 + j\omega\tau_1)(1 + j\omega\tau_2)\ \dots} \qquad (6.17)$$

The log-modulus (LM) will be

$$20\log_{10}|G(j\omega)| = 20\log_{10}K + 20\log_{10}|1 + j\omega\tau_a| + 20\log_{10}|1 + j\omega\tau_b| + \dots - 20\log_{10}|(j\omega)^n| - 20\log_{10}|1 + j\omega\tau_1| - 20\log_{10}|1 + j\omega\tau_2| - \dots \qquad (6.18)$$

and the phase angle ϕ is given by

$$\angle G(j\omega) = \angle(1 + j\omega\tau_a) + \angle(1 + j\omega\tau_b) + \dots - n\angle 90^\circ - \angle(1 + j\omega\tau_1) - \angle(1 + j\omega\tau_2) - \dots \qquad (6.19)$$

Bode diagrams have the following advantages over Nyquist diagrams:

(1) The multiplication and division of transfer functions, which are long and tedious on the Nyquist diagram, become simply a matter of addition or subtraction on a Bode diagram.

(2) Common transfer functions have simple plots and are easily determined when straight-line approximations are used.

(3) The logarithmic scale used for frequency expands the low frequency region and this is often useful.

(4) The analysis and design of series compensation is greatly simplified.

Figure 6.27 shows a typical Bode diagram. It often proves easier to choose the 0 dB level of the LM axis to coincide with the -180° level of the phase angle axis. From the diagram it is seen that the gain margin is the gain measured when the phase lag is 180°, and the phase margin is the phase angle measured from the -180° line when the gain is 0 dB (unity). *It is clear that a system will be stable if the phase margin is positive.*

On a Bode diagram, an *octave* is a doubling of frequency and the change of frequency by a factor of ten is called a *decade*.

The process of plotting the Bode diagram may be simplified and speeded up by recognizing that factors commonly found in transfer functions are of the forms

K, $(j\omega)^{\pm n}$, $(1 + j\omega\tau)^{\pm n}$, and

$[1 + j\omega(2\zeta/\omega_n + (j\omega)^2/\omega_n^2]^{\pm n}$

where n is an integer.

Figures 6.28 and 6.29 illustrate some simple plots. For more complex factors the following procedure will be used.

The LM is

$$20\log_{10}|1/(1+j\omega\tau)| = 20\log_{10}[1/\sqrt{(1+\omega^2\tau^2)}]$$

$$= 20\log_{10}[1+(\omega\tau)^2]^{-\frac{1}{2}} = -10\log_{10}[1+(\omega\tau)^2]$$

for $\omega\tau \ll 1$ LM $\simeq -10\log_{10}(1) \simeq 0$ dB (low frequencies)

for $\omega\tau = 1$ LM $= -10\log_{10}(2) \simeq -3$ dB (at $\omega = 1/\tau$)

for $\omega\tau \gg 1$ LM $\simeq -20\log_{10}(\omega\tau)$ (high frequencies)

This means that at high frequencies the plot is a straight line with a negative slope of 20 dB/decade, as shown in Figure 6.29. The LM can therefore be approximated by two straight-line asymptotes, shown dotted in Figure 6.30, the actual curve passing through the -3 dB point at frequency $\omega = 1/\tau$. The asymptotes meet at this frequency which is known as the *corner frequency* or *break frequency*.

The phase angle is given by

$$\phi = -\tan^{-1}\omega\tau$$

for $\omega\tau \ll 1$ $\phi \simeq 0^\circ$ (low frequencies)

for $\omega\tau = 1$ $\phi = -45^\circ$ (at $\omega = 1/\tau$)

for $\omega\tau \gg 1$ $\phi \simeq -90^\circ$ (high frequencies)

A few other values can be given to $\omega\tau$ to complete the plot.

For most practical purposes asymptotic approximation is acceptable. The maximum error due to such approximation is 3 dB and occurs at the corner frequency. The Bode diagram is shown in Figure 6.30.

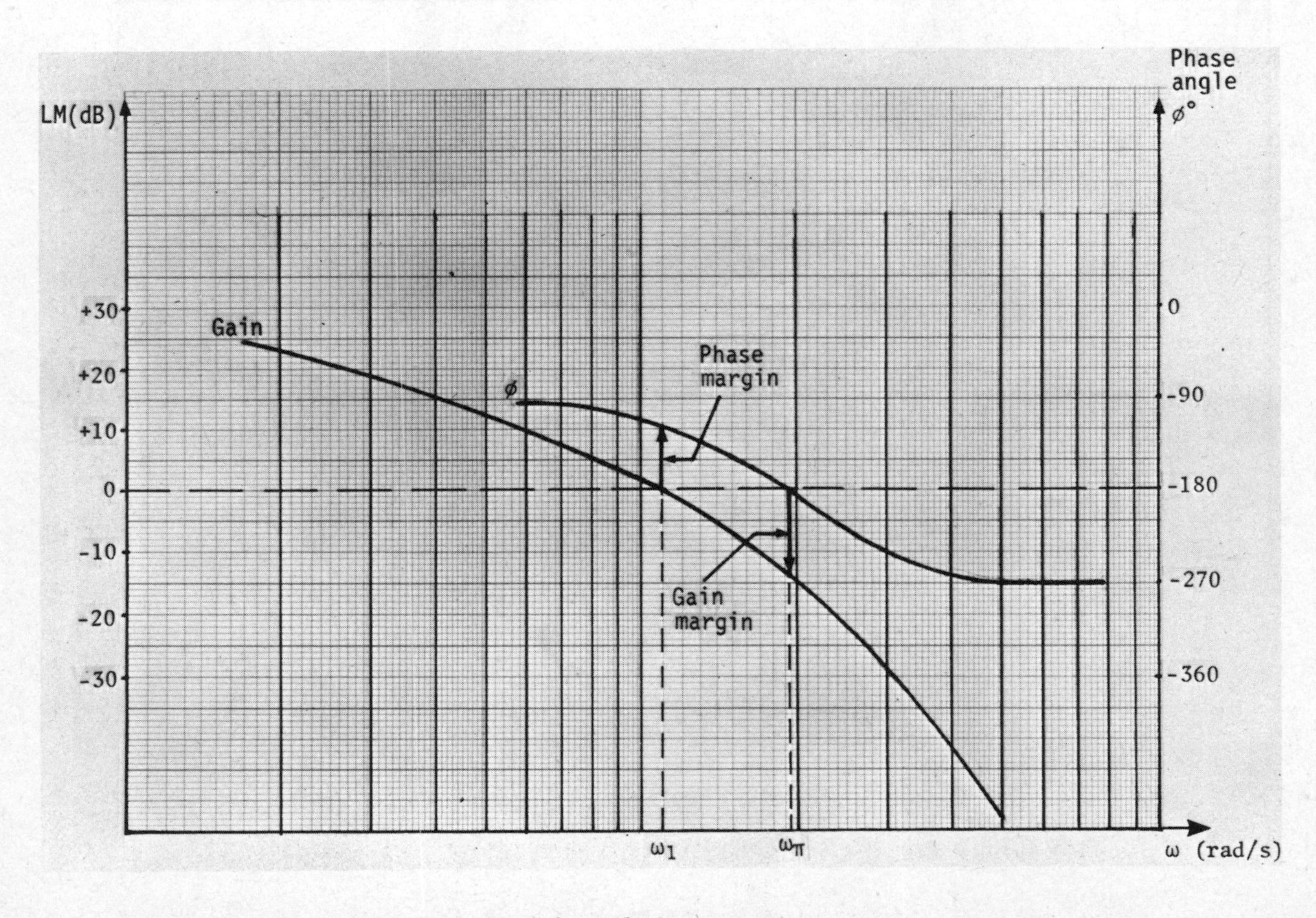

Fig.6.27 Bode diagram showing gain and phase margins

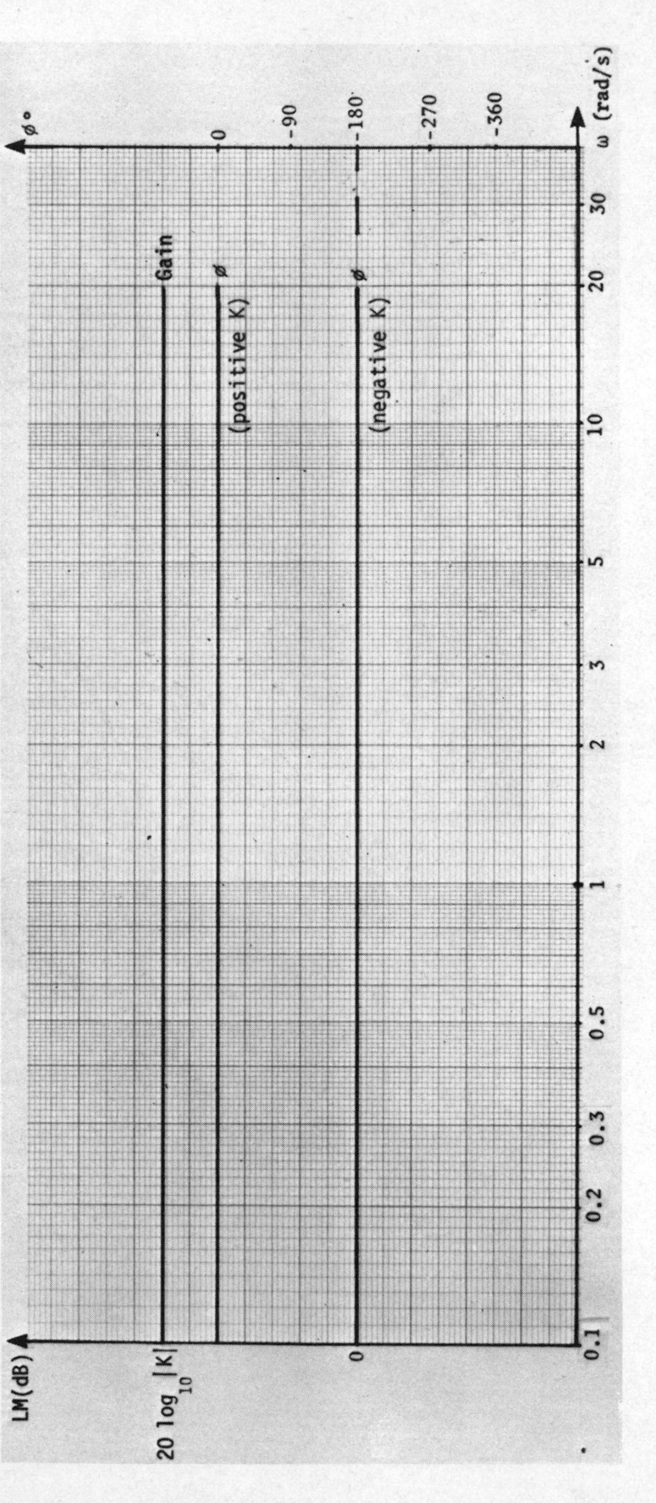

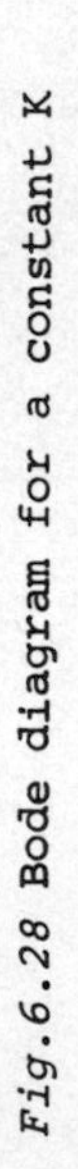

Fig.6.28 Bode diagram for a constant K

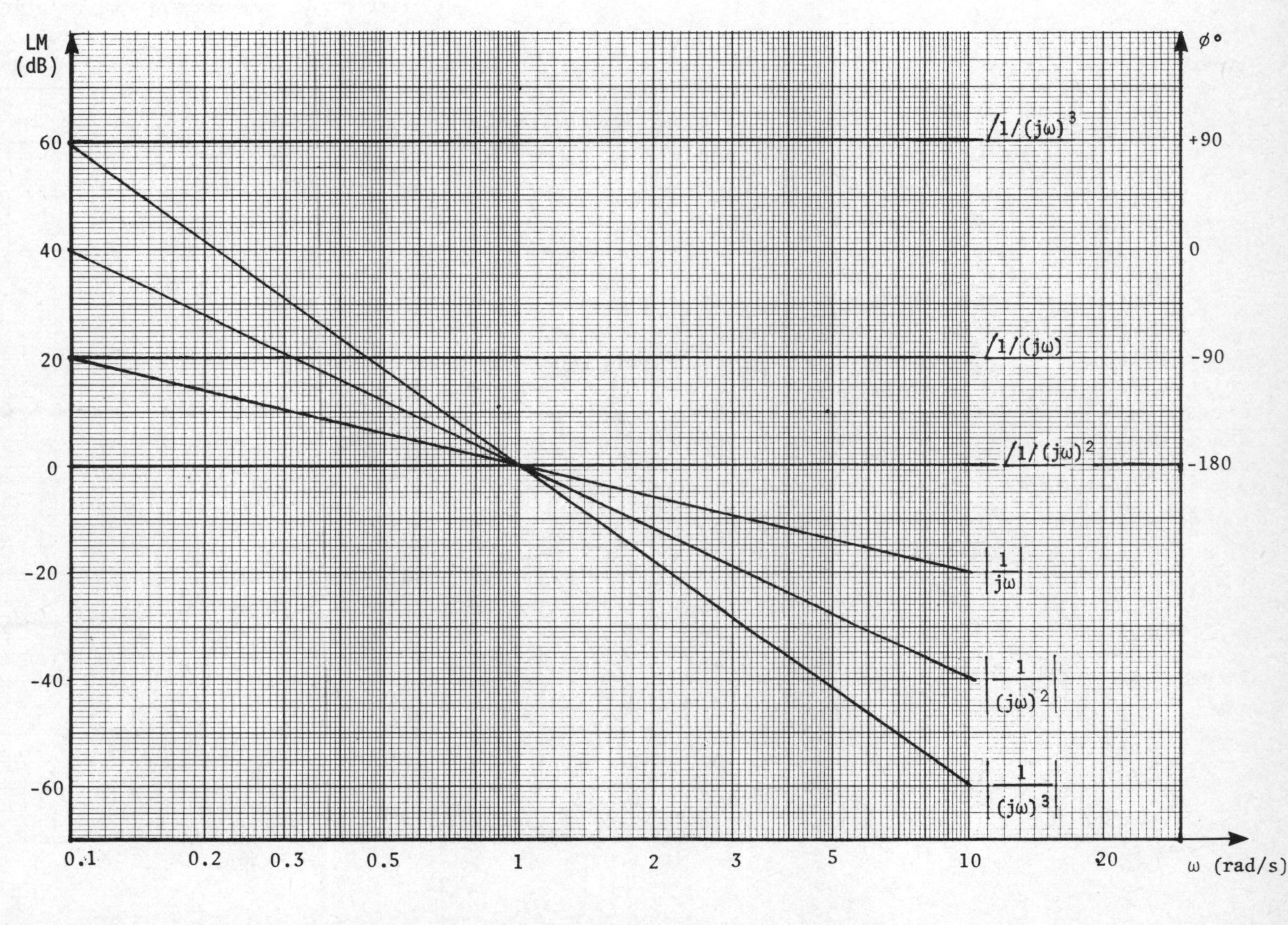

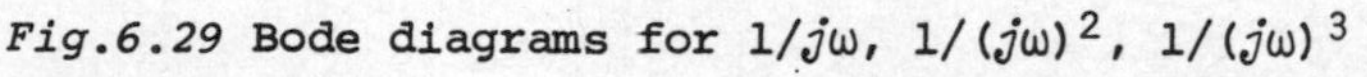

Fig.6.29 Bode diagrams for $1/j\omega$, $1/(j\omega)^2$, $1/(j\omega)^3$

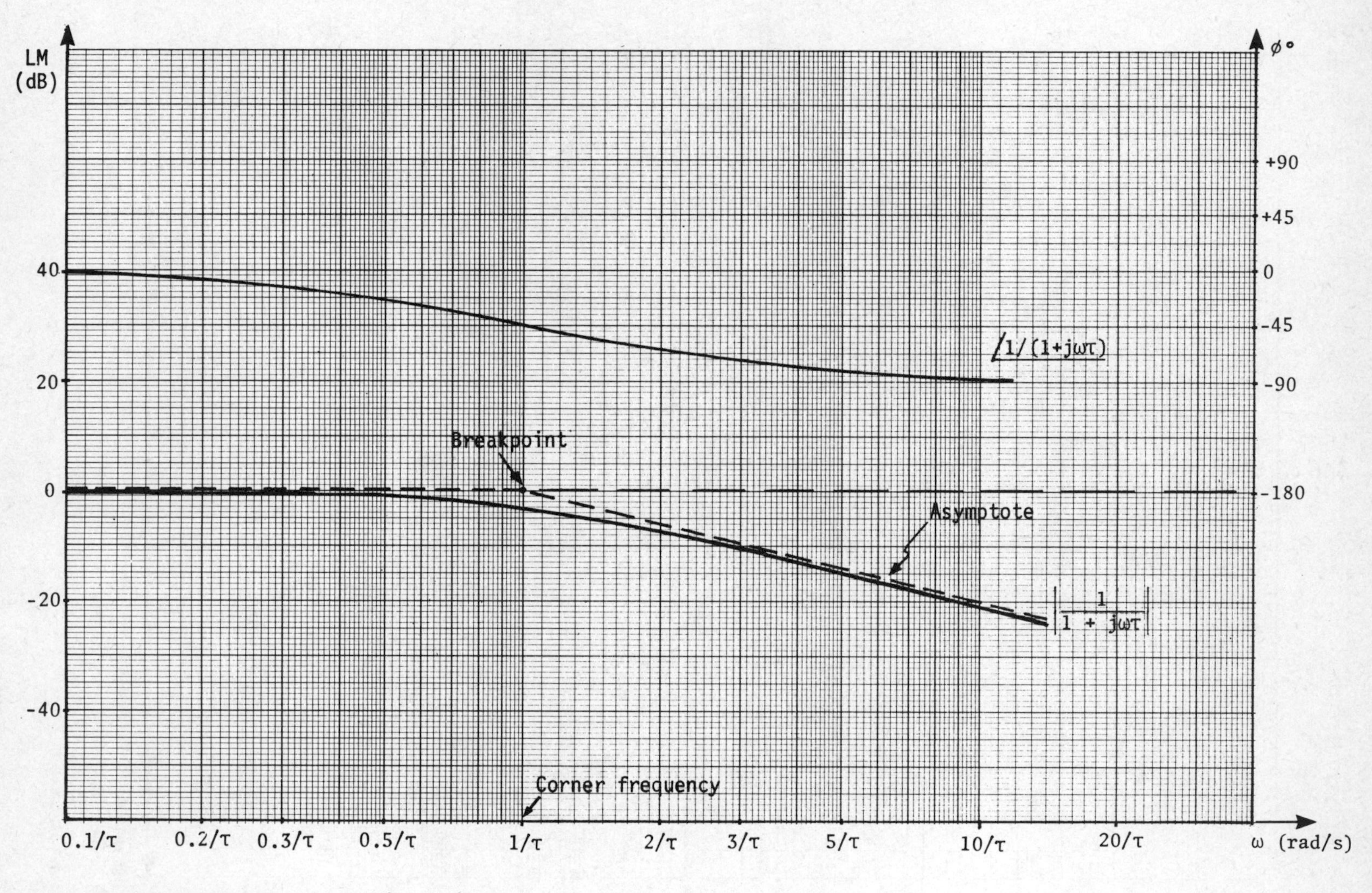

Fig.6.30 Bode diagram for $1/(1 + j\omega\tau)$

Example 6.20

The open-loop harmonic frequency response of a control system is

Frequency (rad/s)	0.1	0.4	1	4	10	40
Gain (dB)	43	31	23	11	0	-24
Phase lag (deg)	94	100	108	144	180	244

(*a*) *Plot the Bode diagram.*
(*b*) *Discuss briefly the closed-loop stability of the system.*
(*c*) *Determine the change in gain necessary to provide adequate stability and state the new gain and phase margins.*

Solution
(*a*) The Bode diagram is shown in Figure 6.31.
(*b*) From the diagram it is seen that when the phase angle is -180°, the gain is 0 dB (unity); both the phase margin and the gain margin are therefore zero. This system is *marginally stable* or strictly speaking unstable.
(*c*) As a rule of thumb, a phase margin between 30° and 60° and a gain margin between 8 dB and 20 dB ensure adequate stability. Hence, reducing the gain by 13 dB gives the system good stability margins.
As shown in the diagram the new phase margin is $+44^\circ$ and the new gain margin is -13 dB.
Note that although the reduction of gain in such cases ensures stability, it has the adverse effect of increasing the steady-state velocity error.

Example 6.21

The forward-path transfer function of a unity feedback control system is given below. Plot an approximate Bode diagram and hence determine the gain and phase margins of the system.

$$\frac{1.43}{s(1 + 0.5s)(1 + 0.01s)}$$

Solution
Substituting $j\omega$ for s

$$\frac{1.43}{j\omega(1 + j0.5\omega)(1 + j0.01\omega)}$$

Figure 6.32 shows the LM and phase plots for each term separately. Addition of the individual plots gives the overall Bode diagram of Figure 6.33. From the diagram,

Phase margin $\phi_{PM} = +\ 52^\circ$

Gain margin $G_M = +\ 37$ dB

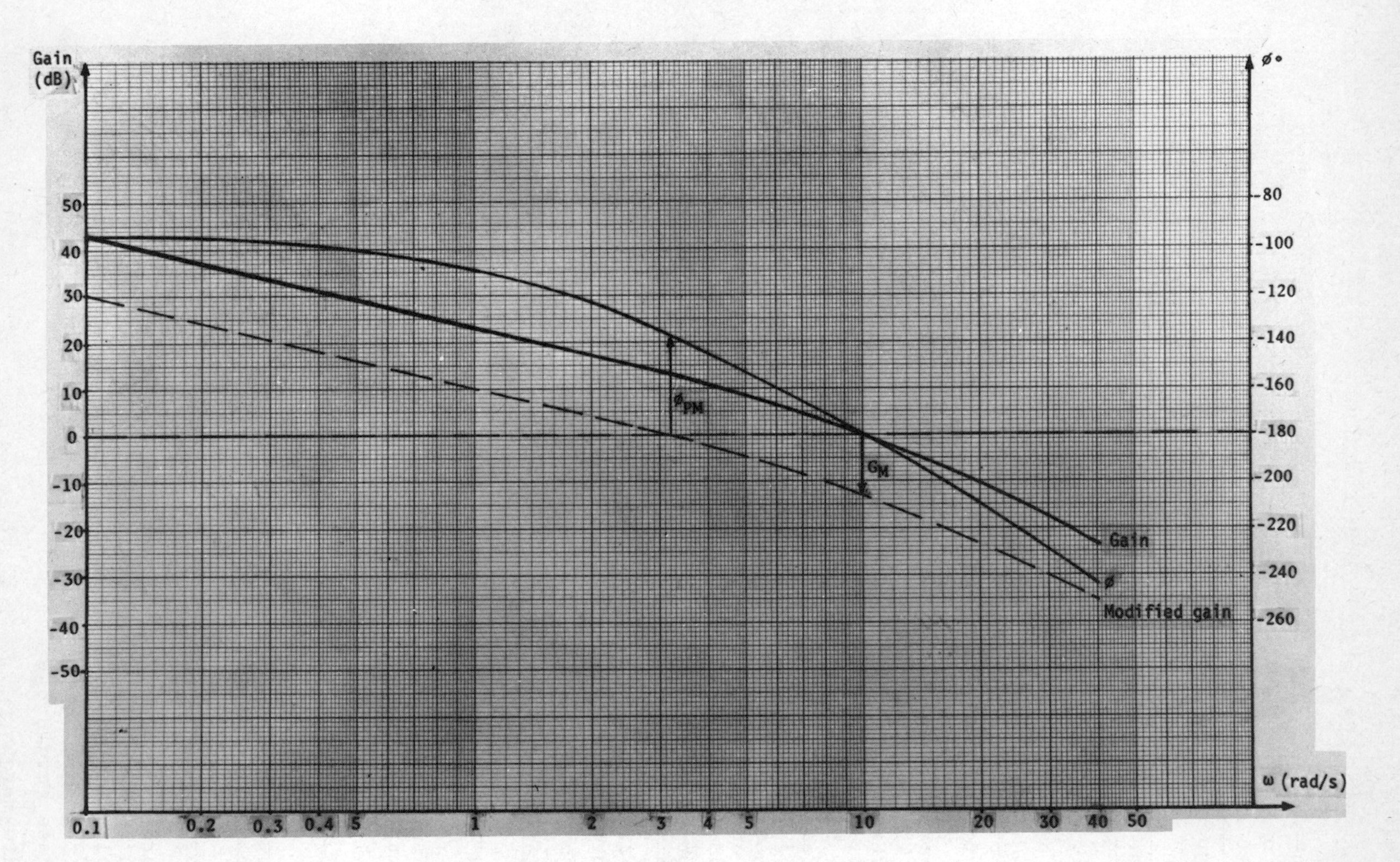

Gain (dB)
50
40
30
20
10
0
-10
-20
-30
-40
-50
φ°
-80
-100
-120
-140
-160
-180
-200
-220
-240
-260
ω (rad/s)
0.1
0.2
0.3
0.4
5
1
2
3
4
5
10
20
30
40
50
Gain
φ
Modified gain
G_M
φ_PM

Fig.6.31

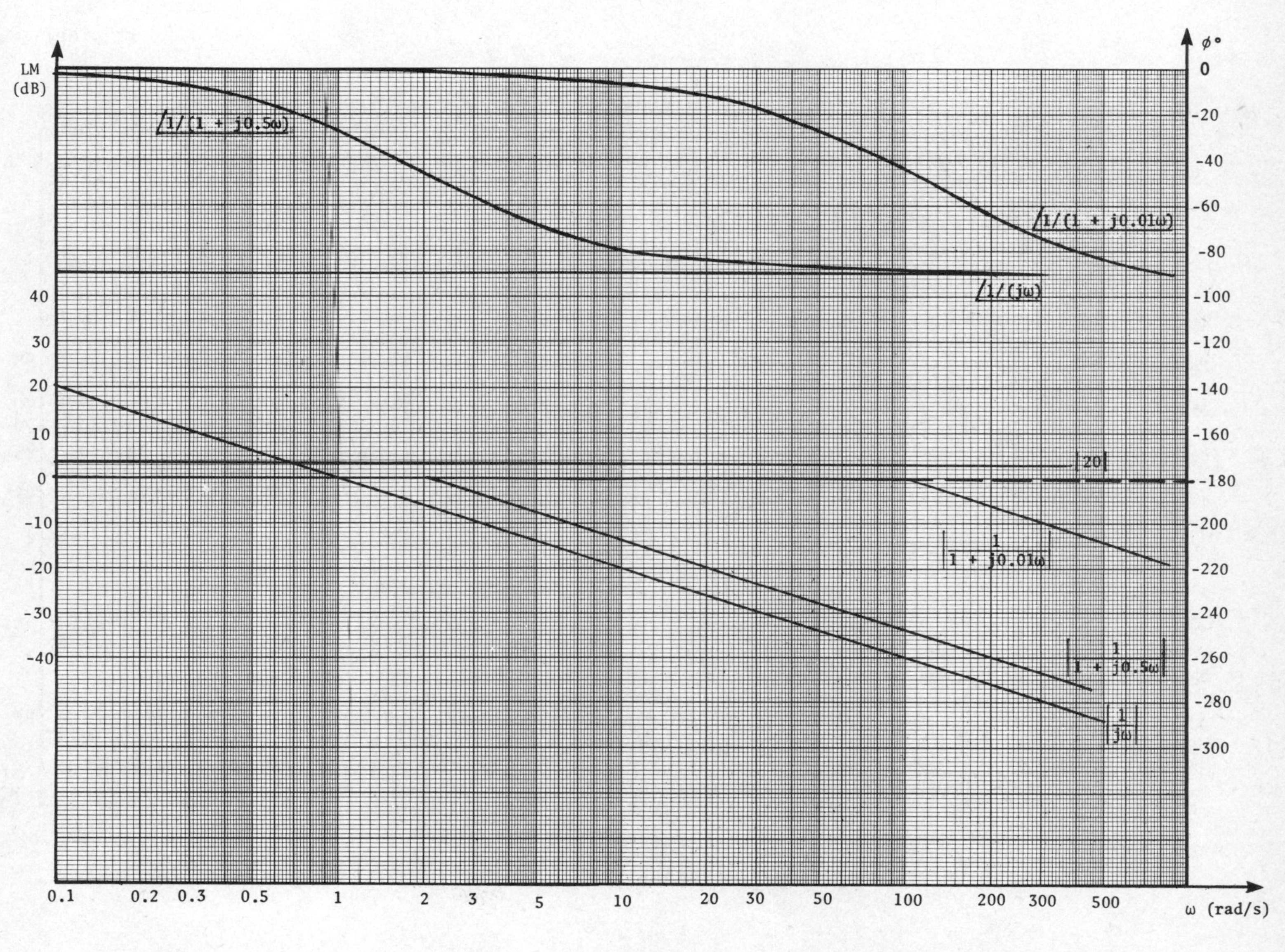

LM (dB)
40
30
20
10
0
-10
-20
-30
-40
φ°
0
-20
-40
-60
-80
-100
-120
-140
-160
-180
-200
-220
-240
-260
-280
-300
0.1
0.2
0.3
0.5
1
2
3
5
10
20
30
50
100
200
300
500
ω (rad/s)
/1/(1 + j0.5ω)
/1/(1 + j0.01ω)
/1/(jω)
|20|
|1/(1 + j0.01ω)|
|1/(1 + j0.5ω)|
|1/(jω)|

Fig.6.32

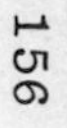

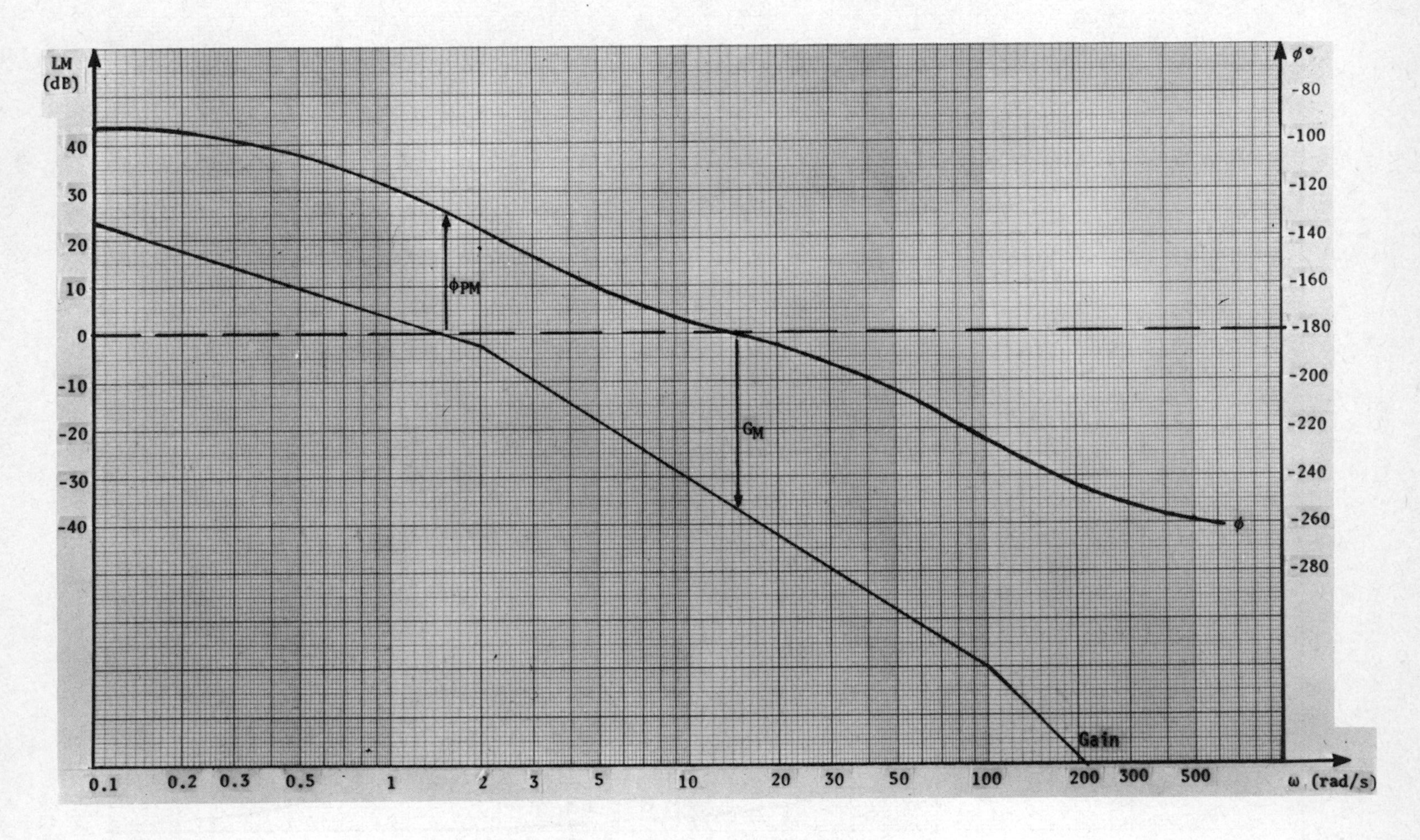
LM (dB)
40
30
20
10
0
-10
-20
-30
-40
φ°
-80
-100
-120
-140
-160
-180
-200
-220
-240
-260
-280
ω (rad/s)
0.1
0.2
0.3
0.5
1
2
3
5
10
20
30
50
100
200
300
500
φ
Gain
G_M
ϕ_{PM}

Fig.6.33

6.11 COMPENSATION

The term compensation is often used when an element is introduced in either the forward path (series compensation) or the feedback path (feedback compensation) in order to stabilise an unstable system or to increase the stability margins. The two most commonly used elements are the *phase-lead* and *phase-lag* circuits. The general effects of series compensation on a system may be summarized as follows:

(a) PHASE-LEAD NETWORK
(1) The relative stability is improved.
(2) The overall gain is reduced.
(3) The bandwidth of the system is increased.
(4) The overshoot is decreased.
(5) The rise time is usually shortened.

(b) PHASE-LAG NETWORK
(1) The relative stability is improved.
(2) The steady-state error may be reduced.
(3) The bandwidth of the system is reduced.
(4) The error constant is usually increased.
(5) The response time is increased.

6.11.1 PHASE-LEAD CIRCUIT

A simple phase-lead network is shown in Figure 6.34. The transfer function of this circuit is given by

$$\frac{v_o(s)}{v_i(s)} = \frac{\alpha(1 + s\tau)}{(1 + s\alpha\tau)} \tag{6.20}$$

where $\alpha = R_2/(R_1 + R_2) < 1$ and $\tau = R_1C$.
Substituting $j\omega$ for s

$$\frac{v_o(j\omega)}{v_i(j\omega)} = \frac{(1 + j\omega\tau)}{(1 + j\omega\alpha\tau)}$$

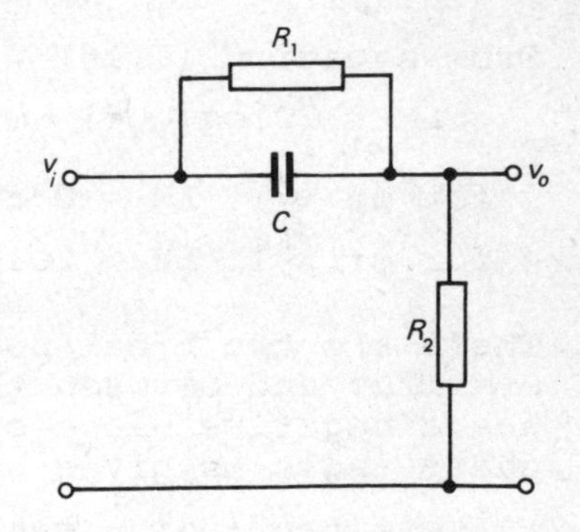

Fig.6.34 Phase-lead network

From equation (6.18),

$$LM = 20\log_{10}\alpha + 20\log_{10}[\sqrt{(1 + \omega^2\tau^2)}]$$
$$- 20\log_{10}[\sqrt{(1 + \omega^2\alpha^2\tau^2)}]$$

for $\omega\tau \ll 1$ $LM = 20\log_{10}\alpha$ dB (which is negative)

for $\omega\tau \gg 1$ $LM = 0$ dB

There are two break points at $\omega = 1/\tau$ and $\omega = 1/\alpha\tau$ and between these limits the asymptote has a slope of 20 dB/decade. The phase angle is given by

$$\phi = \tan^{-1}\omega\tau - \tan^{-1}\omega\alpha\tau$$

At very low and very high frequencies the phase angle is zero. The maximum phase angle (lead) is given by

$$\phi_{max} = \sin^{-1}\left(\frac{1 - \alpha}{1 + \alpha}\right) \qquad (6.21)$$

and this occurs at a frequency ω_{max} such that

$$\omega_{max} = \frac{1}{\tau\sqrt{\alpha}} \qquad (6.22)$$

The Bode diagram is shown in Figure 6.35(*a*) and the Nyquist diagram which is a semicircle is shown in Figure 6.35(*b*).

6.11.2 PHASE-LAG CIRCUIT

A simple phase-lag network is shown in Figure 6.36. The transfer function of this circuit is

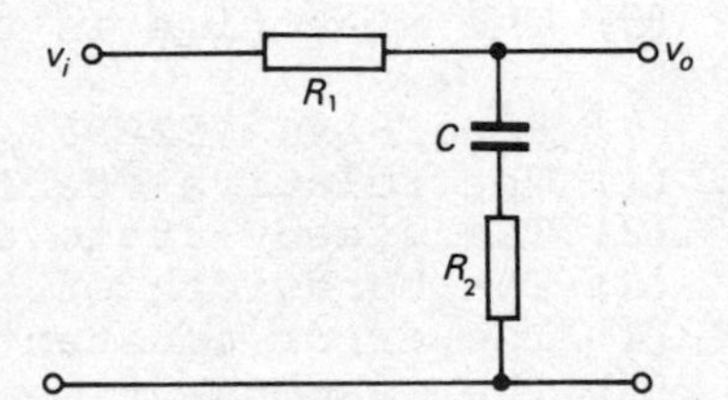

Fig.6.36 Phase-lag network

$$\frac{v_o(s)}{v_i(s)} = \frac{1 + s\alpha\tau}{1 + s\tau} \qquad (6.23)$$

where $\alpha = R_2/(R_1 + R_2) < 1$ and $\tau = C(R_1 + R_2)$.
Substituting $j\omega$ for s

$$\frac{v_o(j\omega)}{v_i(j\omega)} = \frac{1 + j\omega\alpha\tau}{1 + j\omega\tau}$$

From equation (6.18)

$$\text{LM} = 20\log_{10}\sqrt{(1 + \omega^2\alpha^2\tau^2)} - 20\log_{10}\sqrt{(1 + \omega^2\tau^2)}$$

for $\omega\tau \ll 1$ LM = 0 dB

for $\omega\tau \gg 1$ LM = $20\log_{10}\alpha$ dB (which is negative)

There are two break points at $\omega = 1/\tau$ and $\omega = 1/\alpha\tau$ and between these limits the asymptote has a negative slope of 20 dB/decade. The phase angle is given by

$$\phi = \tan^{-1}\omega\alpha\tau - \tan^{-1}\omega\tau$$

At very low and very high frequencies the phase angle is zero. The maximum phase angle (lag) is given by equation (6.21) which can also be written as

$$\phi_{max} = \tan^{-1}\left(\frac{1 - \alpha}{2\sqrt{\alpha}}\right) \qquad (6.24)$$

and this occurs at a frequency ω_{max} such that

$$\omega_{max} = \frac{1}{\tau\sqrt{\alpha}} \qquad (6.25)$$

The Bode diagram is shown in Figure 6.37(*a*) and the Nyquist diagram which is a semicircle is given in Figure 6.37(*b*).

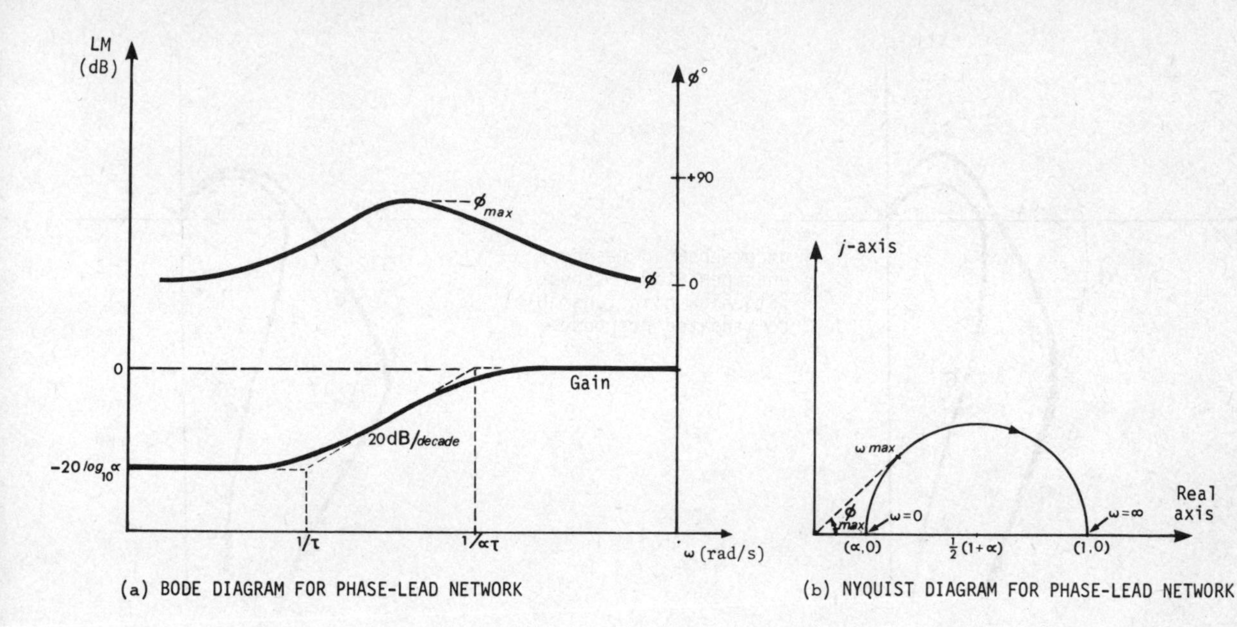

(a) BODE DIAGRAM FOR PHASE-LEAD NETWORK

(b) NYQUIST DIAGRAM FOR PHASE-LEAD NETWORK

Fig.6.35

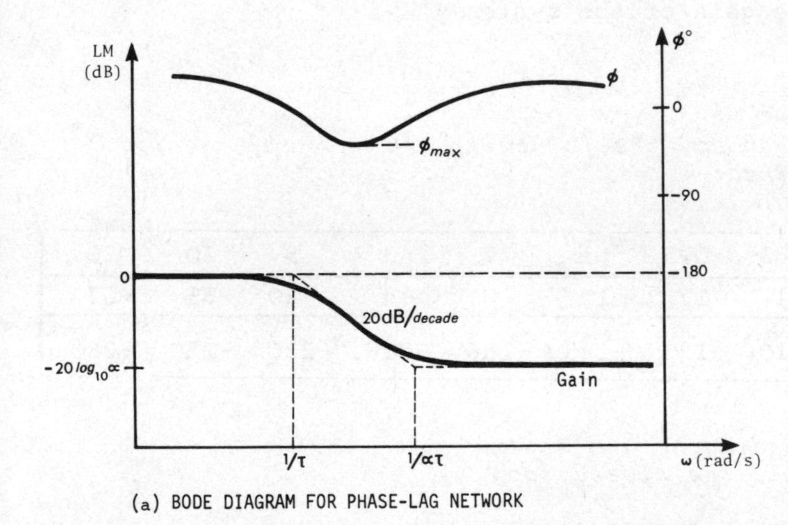

(a) BODE DIAGRAM FOR PHASE-LAG NETWORK

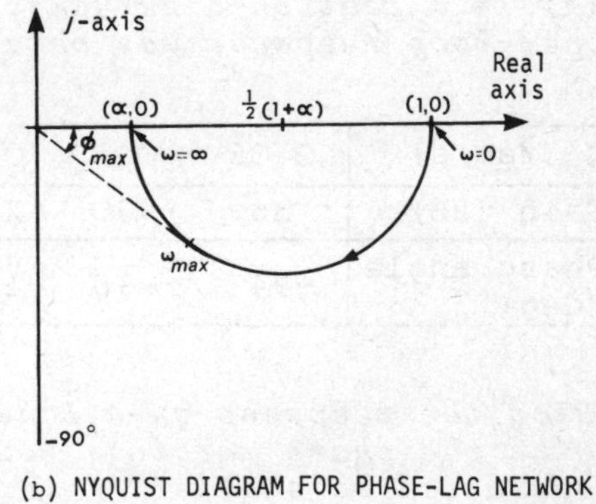

(b) NYQUIST DIAGRAM FOR PHASE-LAG NETWORK

Fig.6.37

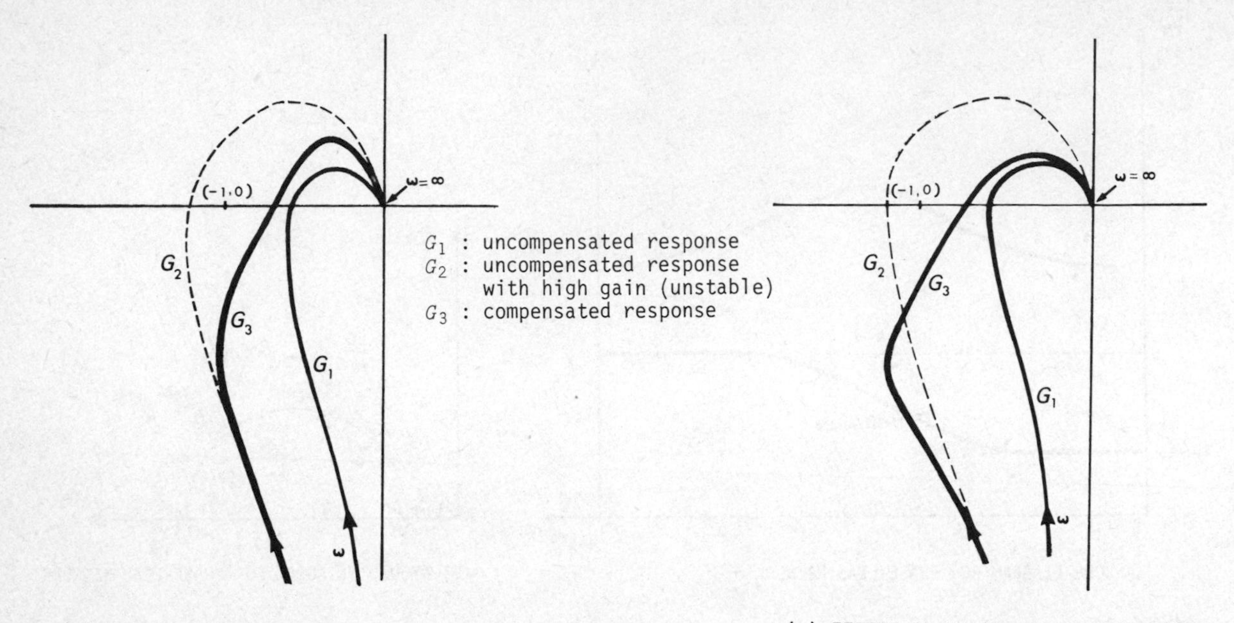

(a) EFFECT OF PHASE-LEAD COMPENSATION (b) EFFECT OF PHASE-LAG COMPENSATION

Figure 6.38 shows the effect of series compensation on the open-loop frequency response of a stable system. It should however be noted that the true effect of series compensation can only be optimised if the gain of the system is also varied.

Example 6.22

(a) In a test on a servomechanism, the following open-loop response was obtained:

ω (rad/s)	0.05	0.1	0.3	0.5	1	2	3	5	10	20
Gain (dB)	36	30	21	17	10	0	-8	-20	-39	-57
Phase angle (deg)	-94	-98	-110	-120	-145	-186	-208	-230	-250	-260

Plot the response on a Bode diagram and determine the phase margin. Would the closed loop operation be stable?
(b) The frequency response of a compensating network is given below. The circuit is connected in cascade with the servomechanism of part (a) of this question. Draw the frequency response of the combined system and hence determine
(i) the type of compensating network used,
(ii) the new phase margin
(iii) the gain margin.

ω (rad/s)	0.05	0.1	0.3	0.5	1	2	3	5	10	20
Gain (dB)	-22	-20	-13	-9	-3	0	0	0	0	0
Phase angle (deg)	+11	+24	+50	+56	+61	+56	+40	+35	+14	+5

Solution
(*a*) The response is plotted and shown in Figure 6.39. From the diagram, the phase margin is $\phi_{PM} = -6^{\circ}$.
Since the phase margin is negative, the system will be unstable under closed-loop operation.

(*b*) The compensated response is obtained by adding the two and is also shown in Figure 6.39 (dotted).
(i) The compensating network has a phase-lead characteristic with its maximum value at $\omega \doteq 1$ rad/s. The gain characteristic follows the general shape of Figure 6.35. The network is therefore a phase-lead network.
(ii) From the diagram the new phase margin is

$$\phi_{PM} = +\,50^{\circ}$$

(iii) The new gain margin is

$$G_M = +\,12 \text{ dB}$$

Example 6.23

A frequency response test on an active phase-advance compensator produced the following results.

Angular frequency ω (rad/s)	Output magnitude / Input magnitude	Phase shift (degrees advanced)
1.0	3.0	55
1.5	4.2	53
2.0	5.5	50
2.2	5.5	48
2.5	6.0	46
3.0	6.8	42

The compensator is inserted in the forward path of a system having an open-loop transfer function

$$\frac{\theta_o(s)}{\varepsilon(s)} = \frac{2}{s(1 + s)^2}$$

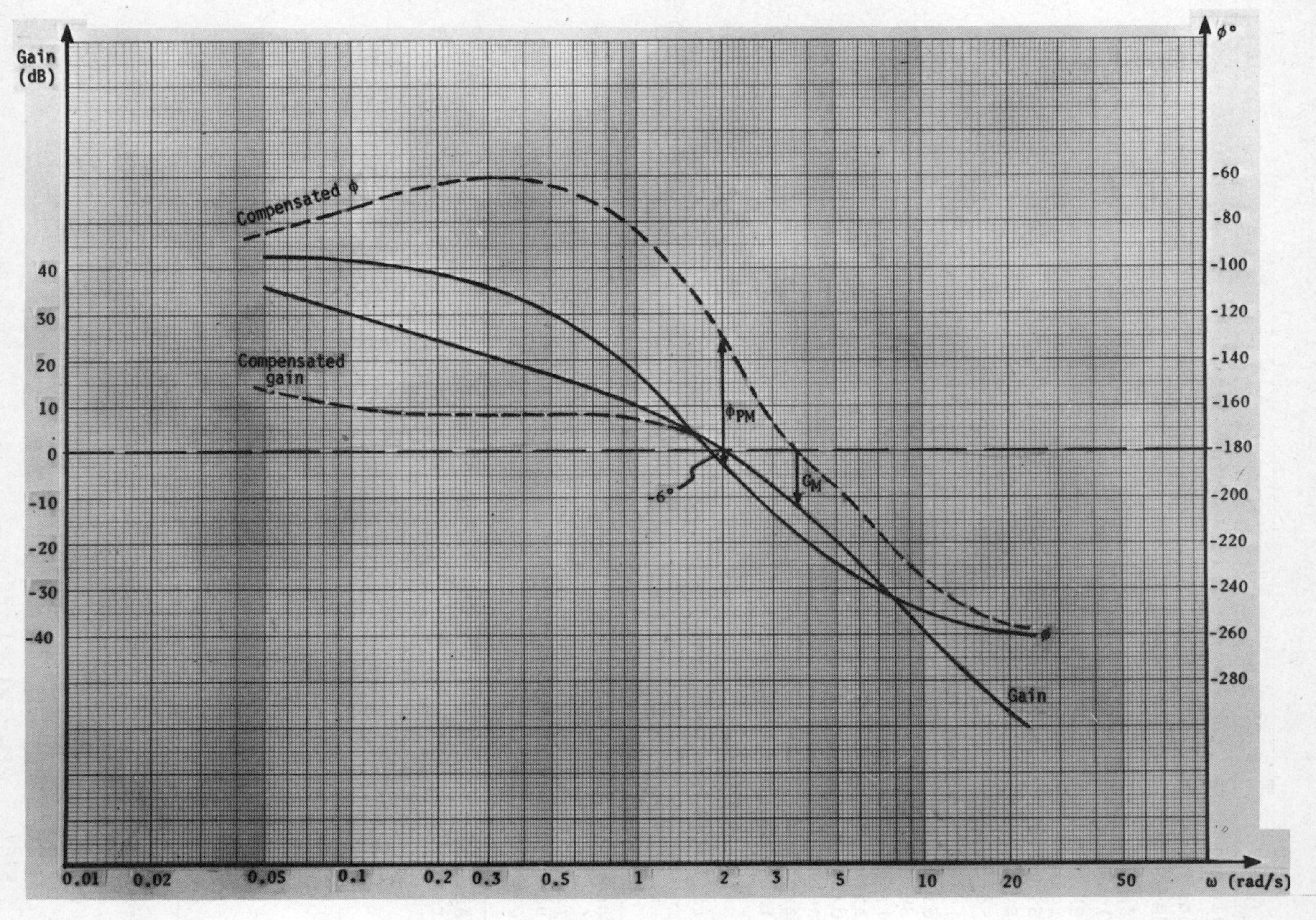

Gain (dB)
φ°
ω (rad/s)
40
30
20
10
0
-10
-20
-30
-40
-60
-80
-100
-120
-140
-160
-180
-200
-220
-240
-260
-280
0.01
0.02
0.05
0.1
0.2
0.3
0.5
1
2
3
5
10
20
50
Compensated φ
Compensated gain
Gain
φ
ϕ_{PM}
G_M
-6°

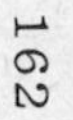

Fig.6.39

Plot the frequency response of the compensated system and hence determine (i) the phase margin, (ii) the gain margin.

Solution

Substituting $j\omega$ for s

$$\frac{\theta_o(j\omega)}{\varepsilon(j\omega)} = \frac{2}{j\omega(1 + j\omega)^2}$$

Using the table of Appendix 1, the magnitude and phase angle of the open-loop transfer function are calculated and are as follows:

ω (rad/s)	$j\omega$	$(1 + j\omega)^2$	$\left\lvert\frac{\theta_o}{\varepsilon}\right\rvert \angle\phi^\circ$
1	$1\angle 90^\circ$	$1.99\angle 90^\circ$	$1\angle -180^\circ$
1.5	$1.5\angle 90^\circ$	$3.24\angle 112^\circ$	$0.4\angle -202^\circ$
2	$2\angle 90^\circ$	$4.97\angle 126^\circ$	$0.2\angle -216^\circ$
2.2	$2.2\angle 90^\circ$	$5.86\angle 132^\circ$	$0.15\angle -222^\circ$
2.5	$2.5\angle 90^\circ$	$7.24\angle 136^\circ$	$0.11\angle -226^\circ$
3	$3\angle 90^\circ$	$9.99\angle 144^\circ$	$0.067\angle -234^\circ$

In order to obtain the overall response, the two magnitudes are multiplied whilst the phase angles are added. The magnitude and phase of the compensated system are as follows:

ω(rad/s)	1	1.5	2	2.2	2.5	3
Magnitude	3	1.68	1.1	0.83	0.66	0.46
Phase	-125°	-149°	-166°	-174°	-180°	-192°

The frequency response of the compensated system is shown in the Nyquist diagram of Figure 6.40.

From the diagram,

(i) Phase margin $\phi_{PM} = 180 - 168 = +12^\circ$

(ii) Gain margin $G_M = \frac{1}{0.66} = 1.5$

$= 3.52$ dB

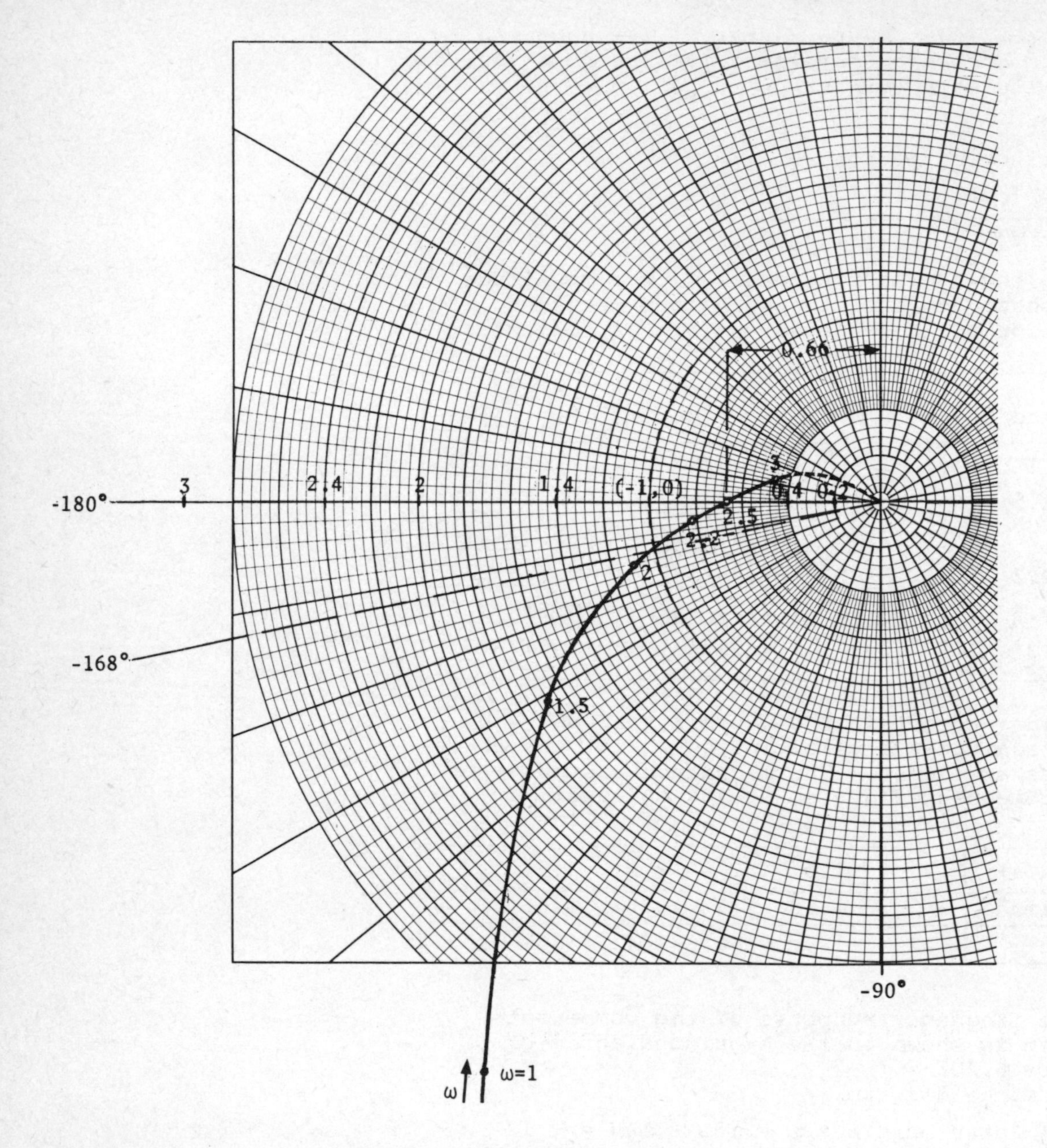

Fig.6.40

EXERCISES

(6.1) Determine the characteristic equations of systems with the following differential equations:

(*a*) $25\frac{d^2y}{dt^2} + 40\frac{dy}{dt} + 2y = x$

(*b*) $\frac{d^3\theta_o}{dt^3} + 8\frac{d^2\theta_o}{dt^2} + 25\frac{d\theta_o}{dt} + 20\theta_o = \theta_i$

(*c*) $\frac{d^3\theta_o}{dt^3} + 2\frac{d^2\theta_o}{dt^2} + 5\frac{d\theta_o}{dt} + 10\theta_o = \theta_i$

(*d*) $\ddot{\theta}_o + 2\zeta\omega_n\dot{\theta}_o + \omega_n^2\theta_o = \omega_n^2\theta_i$

(6.2) Using the Routh stability criterion, determine whether the following characteristic equations represent a stable or an unstable system:

(*a*) $s^2 + 4s + 20 = 0$

(*b*) $4s^4 + 8s^3 + 5s^2 + 5s + 2 = 0$

(*c*) $s^3 + 2s^2 + 2s + 6 = 0$

(*d*) $s^3 + 7s^2 + 7s + 46 = 0$

(6.3) Determine the range of values of K for which the system whose characteristic equation is $s^3 + 6s^2 + 16s + K = 0$ remains stable.

(6.4) A control system has the open-loop transfer function

$$\frac{\theta_o}{\theta_e} = \frac{1}{s^4 + 3s^3 + 14s^2 + 18s + 20}$$

Determine whether it represents a stable or an unstable system.

(6.5) (*a*) Describe with the aid of a diagram how a frequency response test is carried out. Explain how the data obtained from the test can be used to assess the stability of a control system. Mention particularly gain and phase margins.
(*b*) A control system has an open-loop transfer function:

$$\frac{\theta_o}{\theta_e}(s) = \frac{K}{s(1 + 2s)(1 + 4s)}$$

Show that the gain margin for unity feedback is $(1 - 4K/3)$. Hence find the values of K which make the system unstable.

Note:

$$\tan(\tan^{-1}x + \tan^{-1}y) = (x + y)/(1 - xy)$$

(6.6) Define the term "absolute stability". Explain why one is concerned with the "degree of stability" rather than "absolute stability" when considering system performance.

The forward-path transfer function of a unity feedback control system is given by

$$KG(s) = \frac{K}{s(1 + T_1 s)(1 + T_2 s)}$$

Derive an expression for the maximum possible value of K such that the system will remain stable.

(6.7) Plot the Nyquist diagram for each of the following and determine whether they represent a stable or an unstable system:

(a) $KGH(j\omega) = 5/j\omega(1 + j0.5\omega)$

(b) $KGH(j\omega) = 100/j\omega(1 + j0.8\omega)(1 + j0.1\omega)$

(c) $KGH(j\omega) = 10/j\omega(1 + j0.1\omega)(1 + j0.05\omega)$

(d) $KGH(j\omega) = \dfrac{1 + j4\omega}{4(j\omega)^2(1 + j0.25\omega)(1 + j0.1\omega)}$

(6.8) Explain the experimental procedure for making an open-loop harmonic response test on a servo system. State the merits of this test and the information derived.

Sketch typical Nyquist diagrams for systems which would be (a) stable, (b) unstable, in the closed-loop condition.

Indicate for case (a) how the gain and phase margins can be determined.

(6.9) (a) State Nyquist's stability criterion.
(b) Plot the following servomechanism frequency response on a Nyquist diagram and determine:
(i) the servomechanism order,
(ii) the gain constant, and
(iii) the gain constant necessary for $M_{max} = 1.58$.

Frequency (Hz)	0.5	0.7	0.85	1.0	1.5	2.0	2.5	3.0	4.0	5.0
Gain (ratio)	7	4.7	2.8	2.2	1.65	1.27	1.1	0.95	0.55	0.3
Phase angle (deg)	-100	-116	-137	-145	-158	-163	-166	-170	-180	-190

(6.10) Describe, using a block diagram, how the open-loop harmonic response of a control system can be obtained. The harmonic response of a control system is as follows:

Phase lag	-100°	-120°	-150°	-180°	-210°	-240°	-270°
θ_o/θ	1.5	0.8	0.7	0.75	0.65	0.45	0

(where θ_o is angular output and θ is angular error).

Plot the Nyquist diagram and state whether the control system will be stable in the closed-loop condition. Determine the gain and phase margins.

(6.11) State the Nyquist stability criterion.

The frequency response of a servomechanism is given below. Determine, graphically, the servomechanism's order and the gain constant of the system.

f (Hz)	0.5	0.75	0.9	1.3	1.6	2	2.5	3	3.5	4	4.5	5
G (ratio)	6.8	5.5	4.3	3.2	2.6	2.1	1.6	1.15	0.85	0.75	0.6	0.3
ϕ (deg)	-94	-96	-100	-110	-120	-130	-140	-155	-170	-180	-190	-220

(6.12) Give a short exposition of the Nyquist stability criterion.

A control system with unity negative feedback has an open-loop transfer function

$$G(s) = \frac{120}{s(1 + 0.01s)(1 + 0.02s)}$$

where s is the Laplace operator. Sketch the Nyquist plot for the system. Comment briefly on the stability of the system indicating how it could be improved and including the effects these improvements would have on the plot.

(6.13) A unity feedback control system has an open-loop transfer function

$$KH(j\omega) = \frac{20}{(1 + j2\omega)(1 + j0.5\omega)}$$

Determine the maximum value of the magnitude of the closed-loop frequency response M_{pf} and the frequency ω_{rf} at which it occurs, using (*a*) the Nyquist plot, (*b*) the inverse Nyquist plot.

(6.14) Using Brown's construction method, determine the value of K for the system whose open-loop transfer function is

$$KG(j\omega) = \frac{K}{j\omega(1 + j0.1\omega)}$$

so that $M_{pf} = 1.4$. Estimate the corresponding value of ω_{rf}.

(6.15) Explain how a frequency response test could be performed on either (*a*) an a.c. servo, or (*b*) a d.c. servo.
State clearly how amplitude and phase measurements are performed for the test described. Sketch a typical frequency response curve, suitably labelled, for a stable system.

(6.16) Sketch root-locus plots for the following:

(*a*) $GH(s) = K/s(s + 2)(s + 4)$

(*b*) $GH(s) = K(s + 4)/s(s + 2)$

(*c*) $GH(s) = 2K(s + 0.1)/(s + 1)(s + 0.2)$

(*d*) $GH(s) = K(s + 1)/s^2(s + 5)(s + 3)$

(*e*) $GH(s) = K(s+1)(s+2)/s(s+4)(s+3)$

(6.17) A unity feedback control system has an open-loop transfer function

$$KG(s) = \frac{K(s + 2)}{s(s + 20)(s + 10)}$$

Plot the root-locus diagram and determine the value of K to give a damped natural frequency of 2.5 rad/s. What would the damping ratio be for this value of K?

(6.18) (*a*) Define the terms (i) Stable, (ii) Gain margin, (iii) Phase margin, (iv) Gain crossover frequency, and (v) Phase crossover frequency, as referred to in control systems.

Using diagrams show how these terms can be illustrated on: (i) a Nyquist diagram and (ii) a Bode plot.

(*b*) Discuss briefly the advantages and disadvantages of a "frequency response" analysis of control systems.

(6.19) The following results were obtained for an open-loop frequency response test on a control system:

Frequency (rad/s)	0.1	0.5	1.0	5.0	10	50	100
Amplitude (dB)	37	24	17	-5	-19	-60	-78
Phase-lag (deg)	90	105	124	186	214	256	266

Plot the Bode diagram and (*a*) determine the gain and phase margins, and (*b*) comment fully on the practical implications of these results.

(6.20) Draw Bode diagrams, using asymptotic approximations, for systems having the open-loop transfer functions given below. Estimate gain-margin and phase-margin in each case.

(*a*) $$\frac{1.43}{j\omega(1 + j\omega 0.5)(1 + j\omega 0.01)}$$

(*b*) $$\frac{10(1 + j\omega)}{(1 + j\omega 0.5)(1 + j\omega 0.1)(1 + j\omega 0.01)}$$

(6.21) The open-loop transfer function of a control system may be expressed as

$$KG(j\omega) = \frac{K}{j\omega(1 + j\omega)(1 + 4j\omega)}$$

Make gain (dB) and phase-angle plots to a $\log\omega$ base for $K = 4$ and estimate the value to which K must be reduced to make the system marginally stable.

(6.22) Draw circuit diagrams of resistance-capacitance networks capable of providing (*a*) phase advance, (*b*) phase retard, (*c*) phase advance and retard.

Sketch separate attenuation/frequency and phase/frequency characteristics for each of the networks.

Describe an application of one of the above networks in a servo-system.

(6.23) Explain how the degree of stability of a system may be judged from the Nyquist and Bode diagrams, illustrating your answer with sketches.

Draw the circuit diagram of a phase-advance network and show how it may be used to improve the degree of stability of a system with particular reference to its effect on both Nyquist and Bode plots.

(6.24) (*a*) State briefly why "compensation" is often needed in control systems.
(*b*) Draw the circuit diagram and give a clear labelled diagram showing the frequency and phase characteristics of (i) a phase-lead compensator, and (ii) a phase-lag compensator.
(*c*) Compare the effects of phase-lead and phase-lag circuits on a system's performance.

(6.25) (*a*) The open-loop response of a servomechanism is

ω rad/s)	0.3	0.6	1.0	2.0	3.0	4.0	5.0	7.0	10	20	30
Gain (dB)	50	43	36	24	15	7	0	-9	-18	-36	-44
Phase angle (deg)	-100	-108	-117	-135	-149	-161	-175	-192	-210	-235	-249

By plotting the frequency response on a Bode diagram, determine the gain and phase margins of the servomechanism.
(*b*) A gain-compensated phase advance circuit has the following frequency response:

ω (rad/s)	1.0	2.0	5.0	10	20	30
Gain (dB)	0	+0.5	+3.0	+7.0	+12.0	+15.0
Phase angle (deg)	+3	+18	+45	+59	+63	+64

If the phase advance circuit were included in the servomechanism system of part (*a*), draw the modified frequency response. Thus determine the new gain and phase margins and the gain and phase crossover frequencies.

7 Analogue Computing Techniques

The behaviour of a system can often be described by a set of mathematical equations. Analogue computers are very useful for solving such equations and find extensive application in the study of what may be generally classified as *system dynamics*: that is, the study of the transient and steady-state behaviour of a wide variety of physical systems.

Analogue computers are also very useful in simulating a control system. Since variables are represented by voltages, the effect of variation in control elements and parameters can be simply obtained by adjusting the settings of potentiometers. Also, because of the nature of the output devices used in conjunction with analogue computers, the performance of the control system can be observed directly and almost immediately on an X-Y plotter or an oscilloscope.

This chapter deals with the basic principle and elements of analogue computers and shows how these elements are used in analogue computation. Methods of simulation and solution of problems are illustrated with particular emphasis on amplitude and time scaling.

In order that flow diagrams can easily be drawn from equations, potentiometer coefficients are shown in square brackets [ ] followed by the input gain of amplifiers, while scaled variables are given in round brackets ().

7.1 INTRODUCTION

The analogue computer is basically an electronic simulator in which continuously variable voltages are used to represent the system variables such as temperature, pressure, velocity, etc. The *independent* variable in these machines is always represented by *time*.

The basic mathematical operations which may be performed on an analogue computer may be generally classified under three main headings:

(1) Summation
(2) Integration
(3) Multiplication

The computer uses three types of elements:

LINEAR ELEMENTS These are part of the computer hardware and include the *summers*, *integrators*, and *coefficient potentiometers*.

NON-LINEAR ELEMENTS These are also part of the computer hardware and the main elements are the *multipliers*, *diode function generators*, and *comparators*.

CONTROL ELEMENTS These consist of the *control panel*, the *timer unit*, the *patch panel*, and the *switching and logic modules*.

7.2 LINEAR COMPUTING METHODS

The three main linear elements were listed above as the summer, the integrator, and the coefficient potentiometer (also called coefficient multiplier). A component which is used most extensively in many of these computer elements is the operational amplifier. An OPERATIONAL AMPLIFIER is basically a high-gain low-drift direct-coupled amplifier, generally with a differential input, and offers very high input impedance and very low output impedance. The block diagram of an operational amplifier is shown in Figure 7.1.

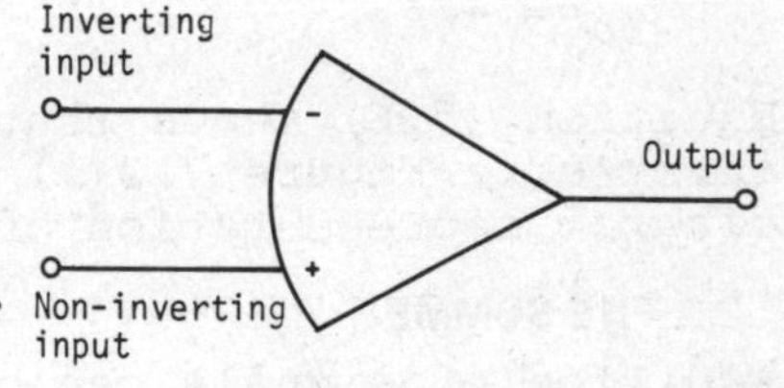

Fig.7.1 Operational amplifier

7.2.1 THE INVERTER

The operational amplifier can be used in conjunction with two suitable resistors, shown in Figure 7.2(*a*), to achieve the inversion of a signal. Assuming that the amplifier has a very high open-loop voltage gain *A*, and that it takes very little current, then the *summing point* S will be virtually at earth potential.

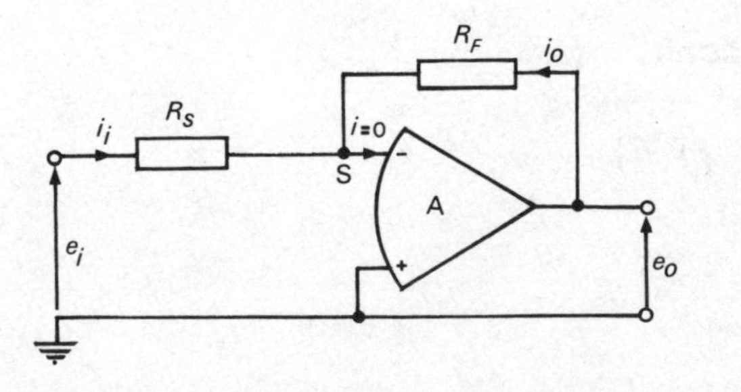

(a) CIRCUIT ARRANGEMENT

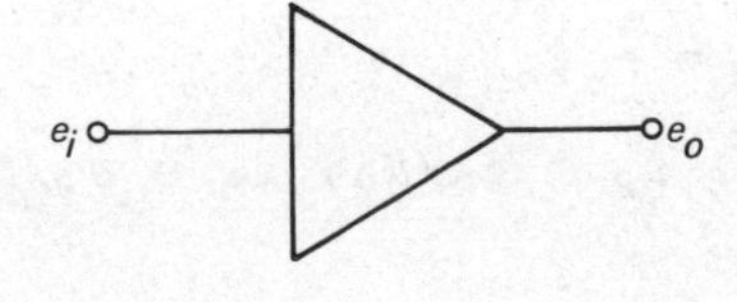

(b) BLOCK DIAGRAM

Fig.7.2 Inverter

Then,

$$e_i = i_i R_S \tag{7.1}$$

$$i_i = - i_o \tag{7.2}$$

$$e_o = i_o R_F \tag{7.3}$$

or $$e_o = - i_i R_F \tag{7.4}$$

Dividing equation (7.4) by equation (7.1),

$$\frac{e_o}{e_i} = \frac{-i_i R_F}{i_i R_S} = - \frac{R_F}{R_S}$$

i.e.

$$e_o = - \frac{R_F}{R_S} e_i \tag{7.5}$$

If $R_S = R_F$, then

$$e_o = - e_i \tag{7.6}$$

Equation (7.6) shows that the input has been inverted. Figure 7.2(*b*) shows the block diagram representation of an inverter.

7.2.2 THE SUMMER

Addition of signals can be achieved by the circuit of Figure 7.3 which has an inherent

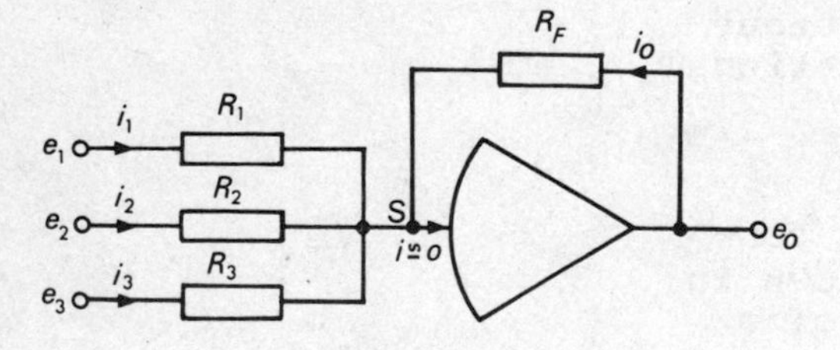

(a) CIRCUIT ARRANGEMENT

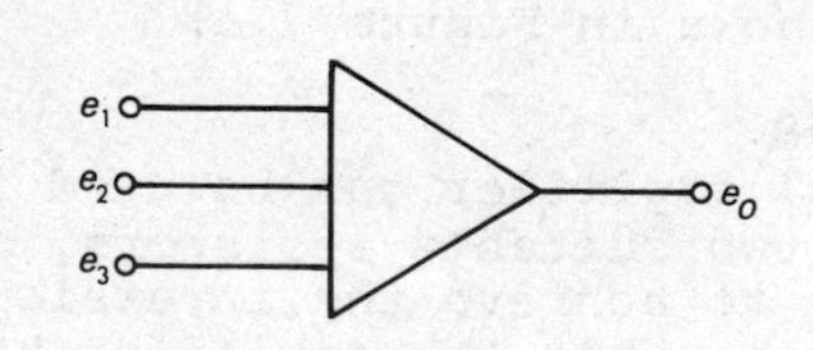

(b) BLOCK DIAGRAM

Fig.7.3 Summing amplifier

sign reversal. Making the usual assumptions, and applying Kirchhoff's law:

$$i_1 + i_2 + i_3 = - i_o \tag{7.7}$$

where $i_1 = e_1/R_1$, $i_2 = e_2/R_2$, $i_3 = e_3/R_3$, $i_o = e_o/R_F$.

Substituting in equation (7.7),

$$\frac{e_1}{R_1} + \frac{e_2}{R_2} + \frac{e_3}{R_3} = - \frac{e_o}{R_F}$$

$$e_o = -R_F\left(\frac{e_1}{R_1} + \frac{e_2}{R_2} + \frac{e_3}{R_3}\right)$$

or $$e_o = -\left(\frac{R_F}{R_1}e_1 + \frac{R_F}{R_2}e_2 + \frac{R_F}{R_3}e_3\right) \qquad (7.8)$$

The ratios R_F/R_1, R_F/R_2, etc. are the constants associated with the inputs e_1, e_2, etc. respectively.

In the special case when $R_F = R_1 = R_2 = R_3$, equation (7.8) becomes

$$e_o = -(e_1 + e_2 + e_3)$$

7.2.3 THE INTEGRATOR

The circuit arrangement of a simple analogue integrator and its block diagram representation is shown in Figure 7.4. The basic function of

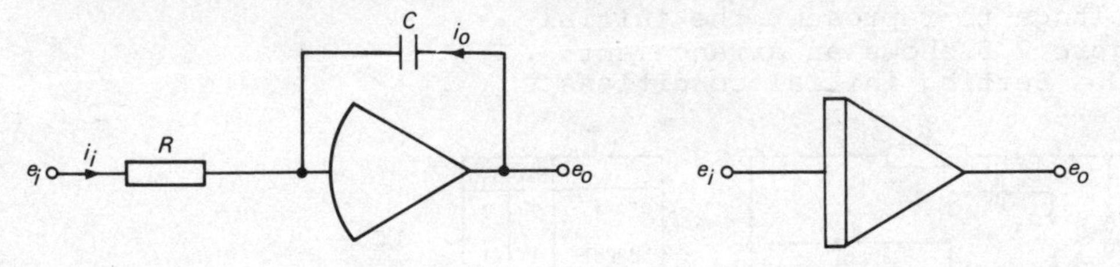

Fig.7.4 Simple integrator

an integrator is to produce the time integral of a voltage. The integrator also has an inherent sign reversal between its input and output.

The charge and the voltage across a capacitor are related by the equation

$$q = Cv \qquad (7.9)$$

Differentiating

$$\frac{dq}{dt} = C\frac{dv}{dt} \qquad (7.10)$$

Since $$i = \frac{dq}{dt} \qquad (7.11)$$

then, current through the capacitor is given by

$$i_o = C\frac{de_o}{dt} \qquad (7.12)$$

and $i_i = e_i/R$. But $i_o = -i_i$, and therefore

$$C\frac{de_o}{dt} = -\frac{e_i}{R} \quad \text{i.e.} \quad de_o = -\frac{e_i}{RC}dt$$

or $$e_o = -\frac{1}{RC}\int e_i\,dt \qquad (7.13)$$

The product RC is known as the integrator TIME CONSTANT and may be chosen to be 10, 1 or 0.1, giving gains of 0.1, 1 or 10 respectively. When it is equal to unity, the output voltage will be given by

$$e_o = -\int_0^t e_i dt - K \qquad (7.14)$$

where K is the constant of integration and represents the initial condition.

The initial condition (I.C.) is the value of the integrator output when $t = 0$. Since in many practical problems the initial value of a variable is not zero, facilities must be available to charge the feedback capacitor to the required voltage to represent the initial condition. Figure 7.5 shows an arrangement for setting I.C. Setting initial conditions

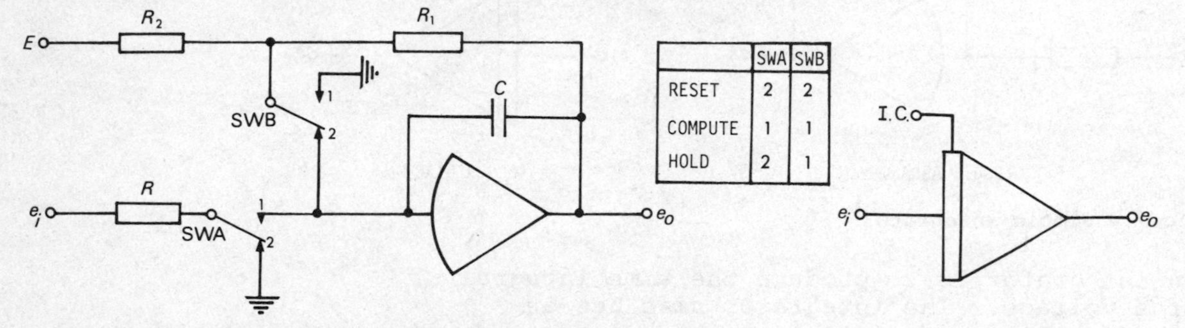

Fig.7.5 Setting initial condition

is an operation known as RESET Mode during which time the input of the amplifier is open circuited while the capacitor charges through R_2 from a d.c. supply voltage E.

7.2.4 THE SUMMING INTEGRATOR

An integrator with two or more inputs has the property of providing an output which is the time integral of the sum of its inputs. By using equations (7.7), (7.12) and (7.13), it can be shown that the output voltage of the summing integrator of Figure 7.6 is given by

$$e_o = -\frac{1}{R_1 C}\int_0^t e_1 dt - \frac{1}{R_2 C}\int_0^t e_2 dt - K \qquad (7.15)$$

R_1 and R_2 can be chosen in order to provide equal or different gains at the inputs.

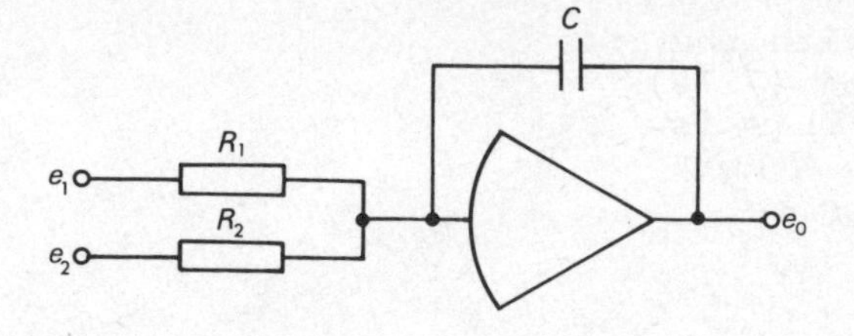

(a) CIRCUIT ARRANGEMENT

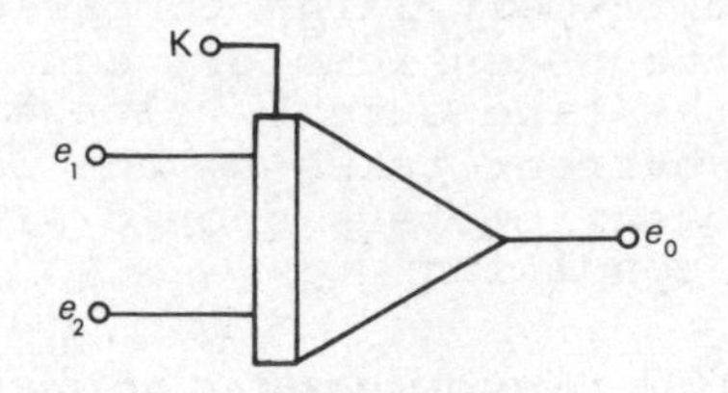

(b) BLOCK DIAGRAM

Fig.7.6 Summing integrator

7.2.5 THE DIFFERENTIATOR

Simple differentiators are avoided in analogue computing for the following reasons:

(i) They tend to oscillate.
(ii) The gain of the circuit increases with frequency; thus any random noise (high frequency) at the input will be amplified more than the low frequency input signal.
(iii) Small but rapid changes in the input often saturate the amplifier.

7.2.6 THE COEFFICIENT POTENTIOMETER

Multiplication by a constant coefficient less than unity can be achieved by using a linear potentiometer. In an analogue computer these are usually ten-turn helical potentiometers or switch attenuators with a total resistance of the order of 50 kΩ. They are usually used at the output of an element and are set manually. Sophisticated computers often use servo-set potentiometers which may be set up externally and controlled by a digital computer.

The circuit and block diagram representation of a potentiometer are shown in Figure 7.7. From the diagram:

$$e_i = iR \quad \text{and} \quad e_o = ikR$$

then,

$$\frac{e_o}{e_i} = \frac{ikR}{iR} = k$$

i.e.

$$e_o = ke_i \qquad (7.16)$$

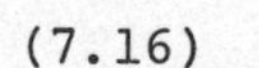

where $k < 1$ and is called the *potentiometer constant* or *coefficient*.

When the output of a potentiometer is connected to a load (the input of a summer or an integrator, for example), then unless the load

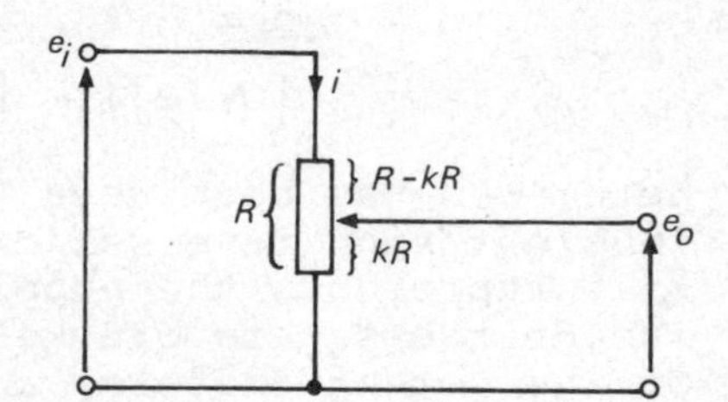

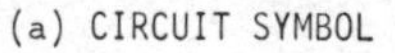

(a) CIRCUIT SYMBOL

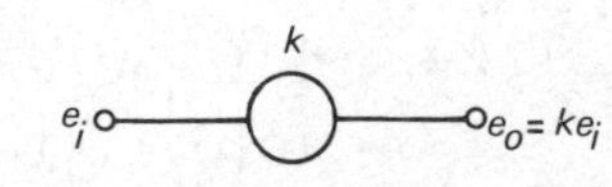

(b) BLOCK DIAGRAM

Fig.7.7 Coefficient potentiometer

resistance is extremely large compared with the resistance of the potentiometer, equation (7.16) will not give the true output voltage. This is because of the current taken by the load. For this reason, potentiometers should always be set under load conditions.

7.3 SOLUTION OF SIMULTANEOUS LINEAR EQUATIONS

Simultaneous equations are treated as two separate equations: each equation is used to solve for one variable in terms of the other variable. Two circuits are then used to generate the required outputs. The final solution is obtained by interconnecting these two circuits. As an example, consider the following simultaneous equations, given that only a ± 10 V supply is available.

$$5x - 2y = 6$$

$$-3x + 5y = 5.5$$

Solving for x in the first equation:

$$-x = -\frac{6}{5} - \frac{2}{5}y$$

$$= -1.2 - 0.4y$$

$$= -10[0.12]1 - [0.4]1y$$

Here, $-x$ has been used to account for the inherent sign reversal of the summer. The -10 represents the supply voltage, the figures inside the square brackets give the settings of the potentiometers, and the numbers immediately after the brackets indicate the gains associated with the inputs used.

Solving for y in the second equation:

$$-y = -\frac{5.5}{5} - \frac{3}{5}x$$

$$= -1.1 - 0.6x$$

$$= -10[0.11]1 - [0.6]1x$$

Separate diagrams for these equations will be:

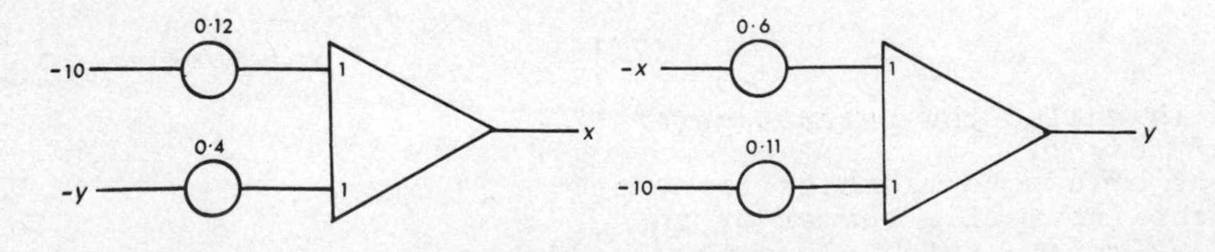

Interconnecting the two, the overall solution is shown in Figure 7.8.

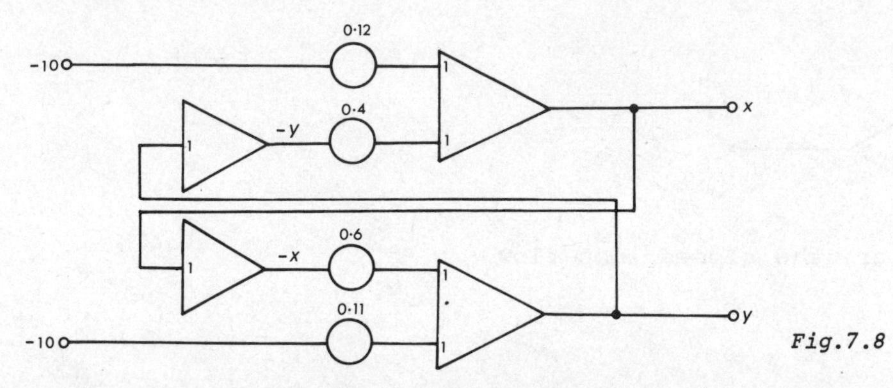

Fig.7.8

7.4 SOLUTION OF DIFFERENTIAL EQUATIONS

Integrators, inverters and potentiometers may be used to solve differential equations. The essential steps are as follows:

(1) Arrange the equation to make the highest derivative the subject on the left-hand side.
(2) Assuming the terms on the right-hand side are available, use a summer to implement this equation.
(3) Integrate until the unknown variable is obtained.
(4) Complete the circuit by feeding back signals from the appropriate points to provide the originally assumed inputs.

The above procedure is illustrated in the following two examples.

Example 7.1

The simulation of a system is expressed by the following first order differential equation:

$$4\dot{x} + 3x = 4$$

Given that $x = \frac{1}{2}$ at $t = 0$, draw an open and a closed loop flow diagram of an analogue system to solve the above equation. Compare the advantages and disadvantages of closed and open loop flow diagrams.

Solution

Rearranging the equation:

$$4\dot{x} = 4 - 3x$$

$$\dot{x} = 1 - \frac{3}{4}x$$

$$-\dot{x} = -10[0.1]1 + [0.75]1x$$

Including the I.C. for x given as ½, the open loop flow diagram will be:

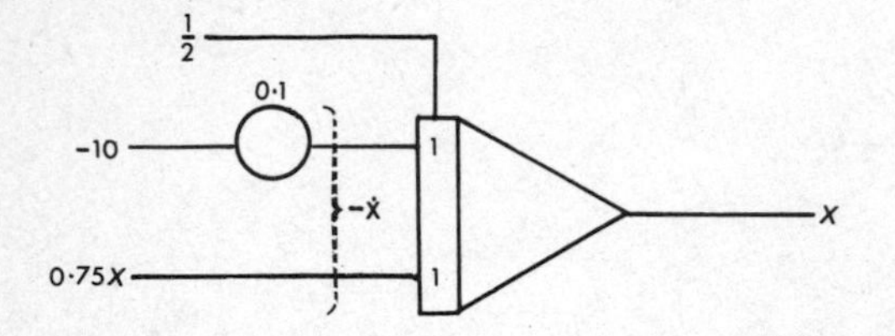

Using a potentiometer, the closed loop flow diagram will be:

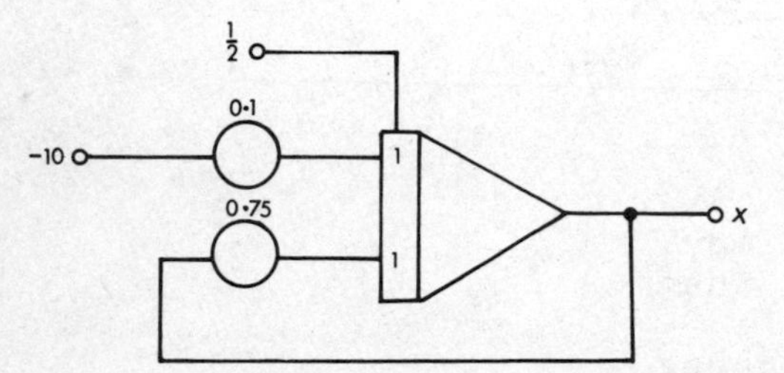

In open loop diagrams, interconnections are not shown. These diagrams are therefore easy to draw and look relatively simple. However, they are difficult to use when wiring up the computer.

Closed loop diagrams look more complicated but as they represent the exact computer wiring diagram, they are much easier to use. For checking purposes this is a great advantage, particularly with large systems.

Example 7.2

Obtain an analogue computer diagram to solve the differential equation

$$\ddot{y} + 5\dot{y} + 4y = 3$$

Solution

Rewriting the equation,

$$\ddot{y} = 3 - 5\dot{y} - 4y$$

$$= 10[0.3]1 - [0.5]10\dot{y} - [0.4]10y$$

Using a summing integrator, the terms on the right-hand side are added and integrated producing $-\dot{y}$. Further integration gives y.

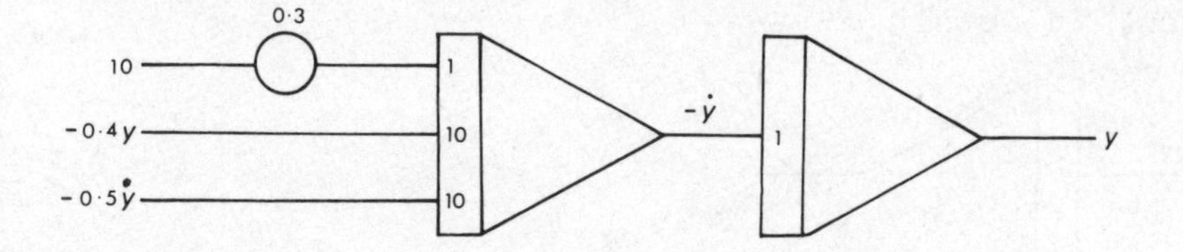

Using potentiometers, the inputs are now provided from the appropriate points on the diagram. The complete solution is shown in Figure 7.9.

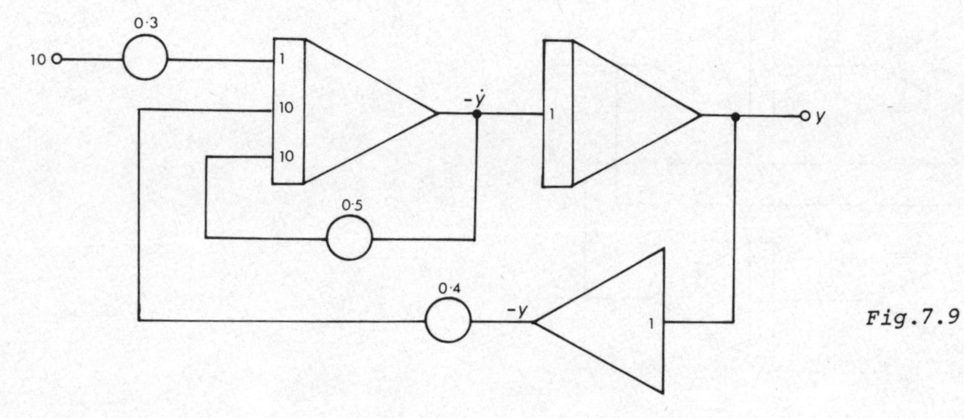

Fig.7.9

7.5 SOLUTION OF SIMULTANEOUS DIFFERENTIAL EQUATIONS

Simultaneous differential equations are treated as two separate differential equations in the manner shown in Examples 7.1 and 7.2. The two diagrams obtained are then combined to produce the final solution. Consider the following simultaneous differential equations

$$\dot{y} = x + 8$$

$$\dot{x} = y + 3$$

Rewriting:

$$\dot{y} = x + 10[0.8]1$$

$$\dot{x} = y + 10[0.3]1$$

From the first equation, to obtain y:

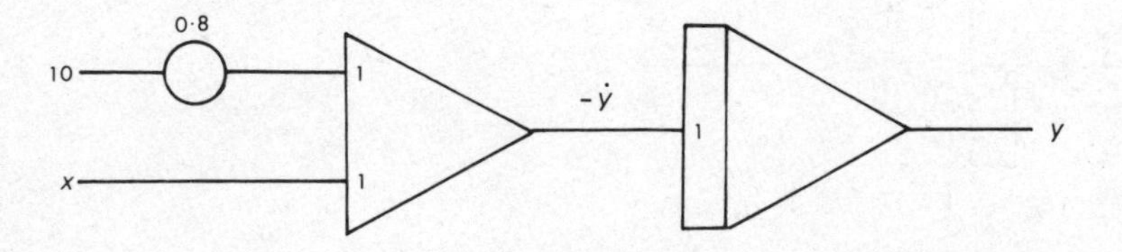

From the second equation, to obtain x:

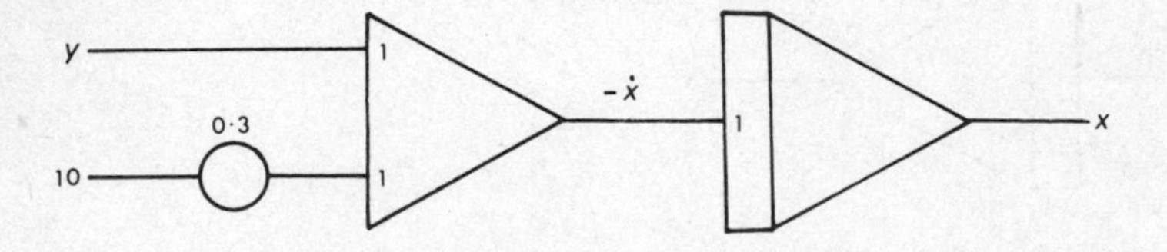

Interconnecting the two circuits, the overall solution is as shown in Figure 7.10.

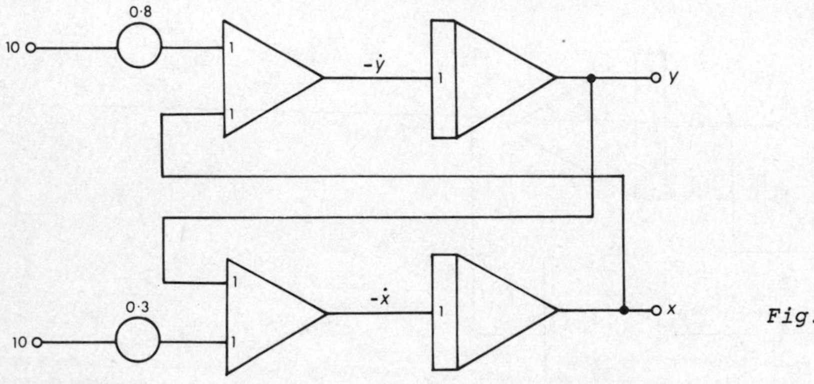

Fig.7.10

Example 7.3

Obtain an analogue computer flow diagram to solve the following second order simultaneous differential equations:

$$\ddot{y} - 4y - x = 0$$

$$\ddot{x} + 3\dot{x} + 2x = 5$$

Solution
Rearranging and introducing potentiometer coefficients:

$$\ddot{y} = [0.4]\,10y + x$$

$$\ddot{x} = 10[0.5]1 - [0.3]10\dot{x} - [0.2]10x$$

From the first equation, to obtain y:

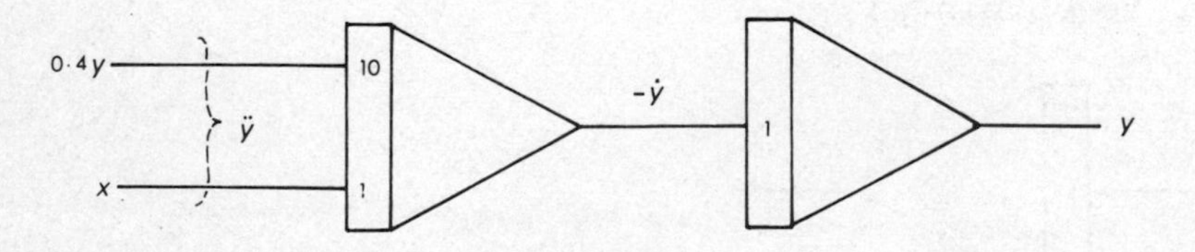

From the second equation, to obtain x:

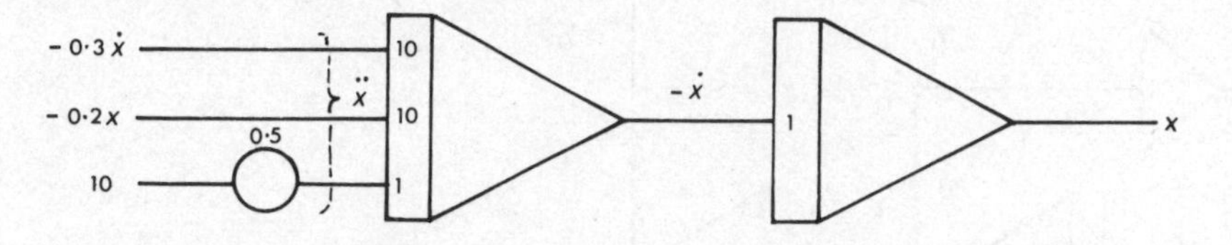

Using three further potentiometers and one inverter, the two diagrams are combined producing the overall diagram of Figure 7.11.

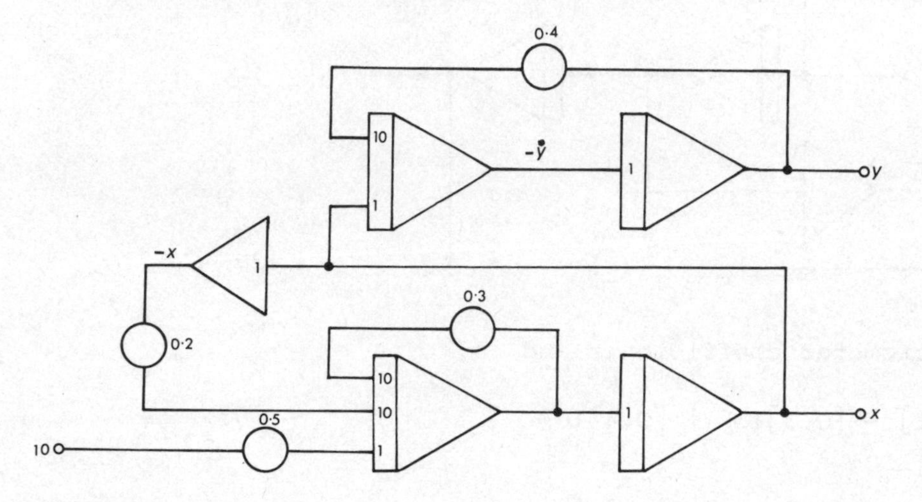

Fig.7.11

Example 7.4

Given the following differential equations:

$$\frac{d^2x}{dt^2} + 3\frac{d^2y}{dt^2} + 4\frac{dx}{dt} + 14x = 2$$

$$\frac{d^2y}{dt^2} + 5\frac{d^2x}{dt^2} + 3\frac{dy}{dt} + 7\frac{dx}{dt} + 12y = 0$$

formulate the analogue computer diagram which allows one to solve for both x and y.

Comment briefly on the factors affecting the accuracy of an analogue computer simulation.

Solution

Using the "dot" notation for derivatives, rearranging the equations:

$$-\ddot{x} = -2 + 3\ddot{y} + 4\dot{x} + 14x$$

$$-\ddot{y} = 5\ddot{x} + 3\dot{y} + 7\dot{x} + 12y$$

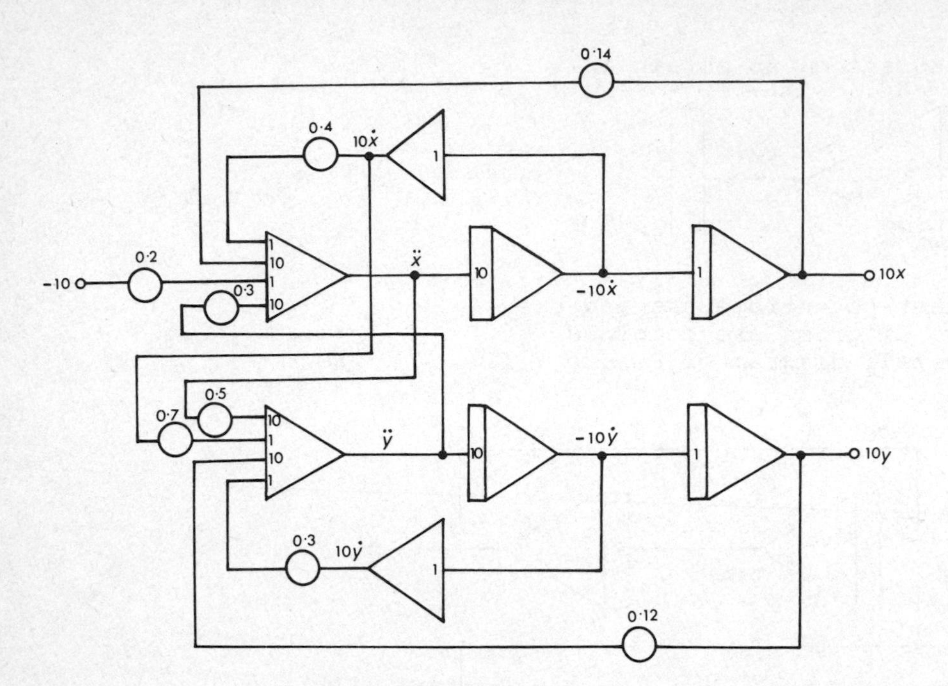

Fig.7.12

In terms of potentiometer coefficients and amplifier gains:

$$- \ddot{x} = - 10[0.2] + [0.3]10\ddot{y} + [0.4]10\dot{x} + [0.14]100x$$

$$- \ddot{y} = [0.5]10\ddot{x} + [0.3]10\dot{y} + [0.7]10\dot{x} + [0.12]100y$$

The flow diagram is shown in Figure 7.12.

The accuracy of an analogue computer simulation depends on the following:

(*a*) the quality of the computing amplifiers,
(*b*) the tolerances of components,
(*c*) the potentiometer settings,
(*d*) the accuracy of the output devices, and
(*e*) the problem itself.

7.6 AMPLITUDE SCALING

In most problems, numerical values of variables exceed the permitted operating range of amplifiers and drive them into *saturation*. Under such conditions, the amplifier is said to be *overloaded* and its output is no longer a linear function of its input.

To ensure that the maximum output voltages are kept within the permissible range, *amplitude scaling* is usually necessary. The constant which relates one unit of the variable to the computer reference voltage is called the AMPLITUDE SCALE FACTOR. The scale factor is usually expressed in terms of M.U. (Machine Unit, ± 10 V) and must be chosen so that the operating voltages are not too small (otherwise accuracy is lost) and yet are within the permitted range. The scale factor for each variable is obtained from

$$\text{Scale factor} = \frac{1 \text{ M.U.}}{\text{Maximum value of the variable}}$$

Since, in most practical problems, maximum values of variables are not available, an estimated value is used. Often it is necessary to rescale the problem after it is applied to the computer.

The process of amplitude scaling may be summarized as follows:

(1) Obtain or estimate maximum values.
(2) Choose the scale factor(s), (preferably round numbers).
(3) Substitute each variable by its scaled value and multiply each term by an appropriate factor such that the original equation remains unaltered.
(4) Write down the relationship between the variable and its derivatives.
(5) Proceed as in (3).
(6) Rewrite the scaled equations in terms of potentiometer coefficients and amplifier gains.
(7) Draw the scaled flow diagram from the scaled equations.

As an example, consider the first order differential equation $4\dot{x} + 6x = 12$. The maximum value of x occurs when $\dot{x} = 0$. Thus,

$$6x = 12 \quad \text{i.e. } x_{max} = 2$$

The scaled variable is therefore $\left(\frac{x}{2}\right)$.

(Since only x will appear at the output of an amplifier, $\dot{x}$ need not be scaled.)

Rearranging the equation:

$$-\dot{x} = \frac{6}{4}x - \frac{12}{4}$$

$$-\dot{x} = 1.5x - 3$$

but $dx/dt = \dot{x}$.

Scaling for x:

$$-\frac{d}{dt}\left(\frac{x}{2}\right) = 1.5\left(\frac{x}{2}\right) - \frac{3}{2}(1)$$

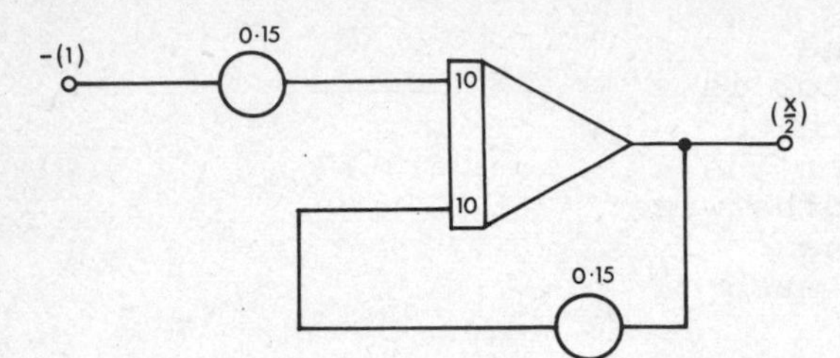

Fig.7.13

This is the scaled equation and is mathematically unaltered (effectively both sides of the equation have been divided by 2). The 1 in brackets represents one M.U.

Rewriting the equation in terms of potentiometer coefficients and amplifier gains:

$$-\frac{d}{dt}\left(\frac{x}{2}\right) = [0.15]10\left(\frac{x}{2}\right) - [0.15]10(1)$$

The scaled diagram solving for x is shown in Figure 7.13.

Example 7.5

Obtain a scaled flow diagram for the equation $4\dot{x} + x = 0$ *given that* $x_{max} = 60$ *and* $x_0 = 60$.

Solution

The scaled variable will be $\left(\frac{x}{60}\right)$. Rearranging the equation:

$$-\dot{x} = \frac{1}{4}x$$

but $dx/dt = \dot{x}$.

Scaling for x: $-\frac{d}{dt}\left(\frac{x}{60}\right) = \frac{1}{4}\left(\frac{x}{60}\right)$

Rewriting this equation in terms of potentiometer coefficient and amplifier gain:

$$-\frac{d}{dt}\left(\frac{x}{60}\right) = [0.25]1\left(\frac{x}{60}\right)$$

The I.C. x_0 is given as 60. Using the scale factor of (1/60), this is represented by 1 M.U. The scaled diagram is shown in Figure 7.14.

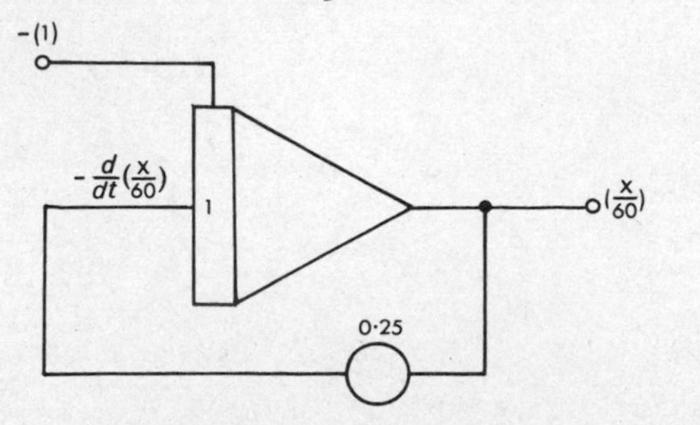

Fig.7.14

Example 7.6

Amplitude-scale the following differential equation and obtain the corresponding scaled diagram providing outputs for x, $\dot{x}$, and $\ddot{x}$:

$$2\ddot{x} + 9\dot{x} + 20x = 35$$

given that $x_{max} = 5$, $\dot{x}_{max} = 10$, $\ddot{x}_{max} = 20$ *and* $\dot{x}_0 = x_0 = 0$.

Solution

Scaled variables will be $\left(\frac{x}{5}\right)$, $\left(\frac{\dot{x}}{10}\right)$ and $\left(\frac{\ddot{x}}{20}\right)$.

Rearranging the equation:

$$-\ddot{x} = -17.5 + 4.5\dot{x} + 10x$$

Scaling for $\ddot{x}$:

$$-\left(\frac{\ddot{x}}{20}\right) = -\frac{17.5}{20}(1) + 4.5 \times \frac{1}{2}\left(\frac{\dot{x}}{10}\right) + 10 \times \frac{1}{4}\left(\frac{x}{5}\right)$$

but $d\dot{x}/dt = \ddot{x}$.

Scaling for $\dot{x}$: $\frac{d}{dt}\left(\frac{\dot{x}}{10}\right) = 2\left(\frac{\ddot{x}}{20}\right)$

Also $dx/dt = \dot{x}$.

Scaling for x: $\frac{d}{dt}\left(\frac{x}{5}\right) = 2\left(\frac{\dot{x}}{10}\right)$

Rewriting the three scaled equations in terms of potentiometer coefficients and amplifier gains:

$$-\left(\frac{\ddot{x}}{20}\right) = -[0.875]1(1) + [0.225]10\left(\frac{\dot{x}}{10}\right) + [0.25]10\left(\frac{x}{5}\right)$$

$$\frac{d}{dt}\left(\frac{\dot{x}}{10}\right) = [0.2]10\left(\frac{\ddot{x}}{20}\right)$$

$$\frac{d}{dt}\left(\frac{x}{5}\right) = [0.2]10\left(\frac{\dot{x}}{10}\right)$$

The scaled diagram is given in Figure 7.15.

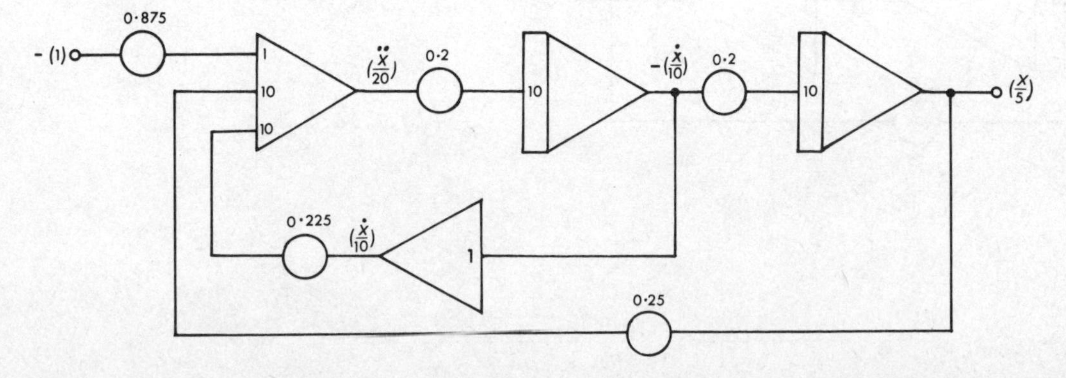

Fig.7.15

Example 7.7

A 10 μF *capacitor charged to* 500 coulombs *is discharged through a* 1 MΩ *resistor. Obtain an analogue computer diagram to give the variation of current through the resistor.*

Solution
First, the differential equation must be obtained.

For a capacitor $q = Cv$ and $i = \frac{dq}{dt} = C\frac{dv}{dt}$

Also $v = iR$ and $\frac{dv}{dt} = R\frac{di}{dt}$

Therefore $i = RC\frac{di}{dt}$

Substituting values, the differential equation will be

$$\frac{di}{dt} = 0.1i$$

Maximum value of current is equal to the initial discharge current.

$$i_{max} = i_0 = \frac{Q_0}{RC} = \frac{500}{10} = 50 \text{ A}$$

Taking a scale factor of 1/50, the scaled variable will be $\left(\frac{i}{50}\right)$. Scaled equation will be

$$\frac{d}{dt}\left(\frac{i}{50}\right) = [0.1]1\left(\frac{i}{50}\right)$$

The I.C. is 50 which is represented by one M.U. in the flow diagram of Figure 7.16.

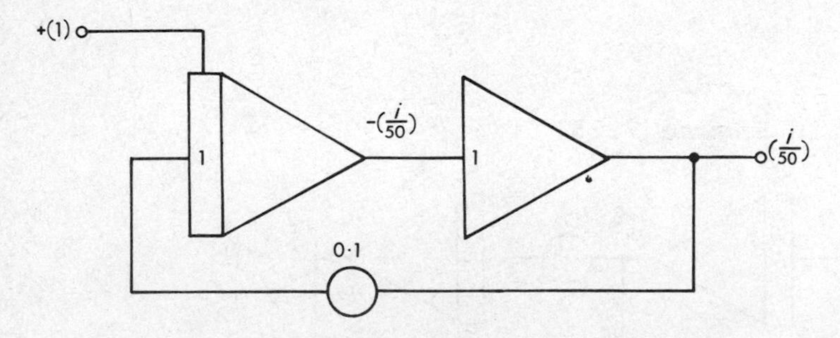

Fig.7.16

7.7 TIME SCALING

The process of computing a problem either faster or slower than "*real time*" is called *time scaling*. The factor which relates the *computer time* to the real time is called the TIME SCALE FACTOR. There are three reasons for time scaling:

(1) Physical systems may take anything from microseconds to years to undergo an operation while computer solutions normally take a few seconds.
(2) The frequency of output signals should be within the accurate range of the available recorders and output equipments.
(3) The gains associated with the inputs of integrators should always be kept below 40 (for good accuracy). Time scaling can be used to reduce the high-gain requirements of some problems.

Time scaling may be performed in two ways. The method most commonly used is one in which the time scale of the computer is changed using the relation

$$\tau = \beta t$$

where τ = computer time, t = real time, and β = time scale factor.

Then $\quad d\tau = \beta dt$

It is seen that

if $\beta > 1$, the problem is slowed down and

if $\beta < 1$, the problem is speeded up.

As an example, we will use a factor of 10 and time scale the following equation which is already amplitude scaled:

$$-\frac{d}{dt}\left(\frac{x}{5}\right) = [0.75]10\left(\frac{x}{5}\right) + [0.4]10(1)$$

The amplitude scaled flow diagram is

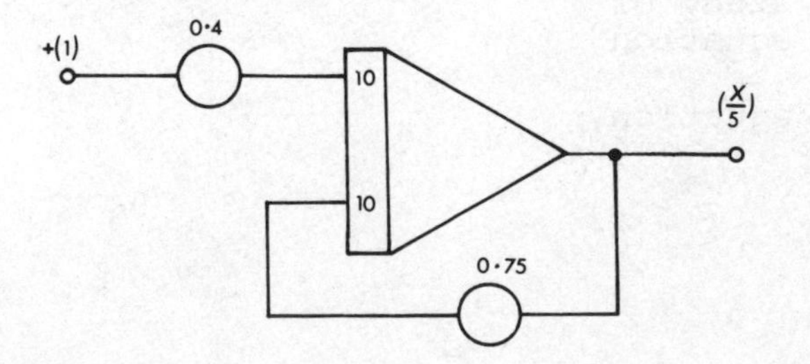

Dividing both sides by β

$$-\frac{d}{\beta dt}\left(\frac{x}{5}\right) = \frac{[0.75]10}{\beta}\left(\frac{x}{5}\right) + \frac{[0.4]10}{\beta}(1)$$

Substituting $d\tau$ for βdt and 10 for β

$$-\frac{d}{d\tau}\left(\frac{x}{5}\right) = \frac{[0.75]10}{10}\left(\frac{x}{5}\right) + \frac{[0.4]10}{10}(1)$$

Time scaled equation:

$$-\frac{d}{d\tau}\left(\frac{x}{5}\right) = [0.75]1\left(\frac{x}{5}\right) + [0.4]1(1)$$

The amplitude and time scaled flow diagram is

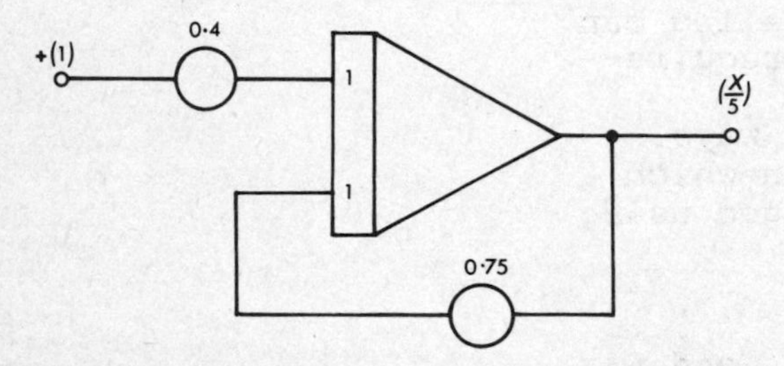

Comparison of the two diagrams shows that input gains of the integrator have been divided by a factor of 10 (since β = 10). Potentiometer settings and the amplitude scaling have not been affected. When β is not 10 (or a multiple of 10), potentiometer settings may then have to be altered.

7.7.1 ESTIMATION OF MAXIMUM VALUES

Generally speaking, there is no simple method for estimating the maximum values of the variables and their derivatives. Knowledge of the physical system, initial conditions, and type of input are often useful.

A detailed study of this topic is beyond the scope of this book. However, a brief study of the general second order differential equation will be made.

Consider a system with differential equation in the form

$$A\ddot{x} + B\dot{x} + Cx = K$$

or, rewriting:

$$\ddot{x} + \frac{B}{A}\dot{x} + \frac{C}{A}x = \frac{K}{A}$$

Comparing this with equation (5.8) we have

$$\omega_n = \sqrt{\frac{C}{A}} \quad \text{and} \quad \zeta = \frac{B}{2\sqrt{(AC)}}$$

For scaling purposes, maximum values of the variable x and its derivatives may be estimated as follows.

For a step function input and with no initial conditions ($\dot{x}_0 = x_0 = 0$):

$$x_{max} \leqslant \frac{2K}{C} \qquad \dot{x}_{max} \leqslant \frac{\omega_n K}{C} \qquad \ddot{x}_{max} \leqslant \frac{\omega_n^2 K}{C}$$

For a step function input, and with initial conditions x_0 and $\dot{x}_0$:

$$x_{max} \leqslant M + \frac{K}{C} \qquad \dot{x}_{max.} \leqslant \omega_n M \qquad \ddot{x}_{max} \leqslant \omega_n^2 M$$

where $M = \sqrt{\left\{\left(x_0 - \frac{K}{C}\right)^2 + \left(\frac{\dot{x}_0}{\omega_n}\right)^2\right\}}$

Example 7.8

The behaviour of a physical system is described by the differential equation

$$2\ddot{x} + 100\dot{x} + 14\,000x = 700 \qquad x_0 = \dot{x}_0 = 0$$

(*a*) *Estimate maximum values of x and its first derivative.*
(*b*) *Amplitude-scale the equation.*
(*c*) *Determine whether or not time scaling is necessary.*
(*d*) *Time-scale and draw the scaled flow diagram.*

Solution

(*a*) $\omega_n = \sqrt{(14\,000/2)} \simeq 83.6$ rad/s

$$x_{max} = \frac{2 \times 700}{14\,000} = 0.1$$

$$\dot{x}_{max} = \frac{\omega_n K}{C} = \frac{83.6 \times 700}{14\,000} = 4.18 \qquad \text{(take 5)}$$

(*b*) The scaled variables will be $\left(\frac{x}{0.1}\right)$ and $\left(\frac{\dot{x}}{5}\right)$.

Rewriting the equation:

$$\ddot{x} = 350 - 7000x - 50\dot{x}$$

$$\frac{d\dot{x}}{dt} = 350 - 7000x - 50\dot{x}$$

Scaling for $\dot{x}$:

$$\frac{d}{dt}\left(\frac{\dot{x}}{5}\right) = \frac{350}{5}(1) - \frac{7000}{5} \times 0.1\left(\frac{x}{0.1}\right) - 50\left(\frac{\dot{x}}{5}\right)$$

Also $dx/dt = \dot{x}$.

Scaling for x:

$$\frac{d}{dt}\left(\frac{x}{0.1}\right) = \frac{5}{0.1}\left(\frac{\dot{x}}{5}\right)$$

Rewriting the two equations in terms of potentiometer coefficients and amplifier gains:

$$\frac{d}{dt}\left(\frac{\dot{x}}{5}\right) = [0.7]100(1) - [0.14]1000\left(\frac{x}{0.1}\right) - [0.5]100\left(\frac{\dot{x}}{5}\right)$$

$$\frac{d}{dt}\left(\frac{x}{0.1}\right) = [0.5]100\left(\frac{\dot{x}}{5}\right)$$

(*c*) Inspection of these two equations shows that time scaling is necessary for two reasons:
 (i) The required input gains of 100 and 1000 are far too high.
 (ii) Recording instruments cannot operate at such a high frequency.

(*d*) The scale factor β is chosen so that the two difficulties are overcome. Dividing through by β:

$$\frac{d}{\beta dt}\left(\frac{\dot{x}}{5}\right) = \frac{[0.7]100}{\beta}(1) - \frac{[0.14]1000}{\beta}\left(\frac{x}{0.1}\right) - \frac{[0.5]100}{\beta}\left(\frac{\dot{x}}{5}\right)$$

$$\frac{d}{\beta dt}\left(\frac{x}{0.1}\right) = \frac{[0.5]100}{\beta}\left(\frac{\dot{x}}{5}\right)$$

Choosing $\beta = 50$, and using $d\tau = \beta dt$:

$$\frac{d}{d\tau}\left(\frac{\dot{x}}{5}\right) = \frac{[0.7]10}{50}(1) - \frac{[0.14]1000}{50}\left(\frac{x}{0.1}\right) - \frac{[0.5]100}{50}\left(\frac{\dot{x}}{5}\right)$$

$$\frac{d}{d\tau}\left(\frac{x}{0.1}\right) = \frac{[0.5]100}{50}\left(\frac{\dot{x}}{5}\right)$$

or $$\frac{d}{d\tau}\left(\frac{\dot{x}}{5}\right) = [0.14]1(1) - [0.28]10\left(\frac{x}{0.1}\right) - 1\left(\frac{\dot{x}}{5}\right)$$

$$\frac{d}{d\tau}\left(\frac{x}{0.1}\right) = 1\left(\frac{\dot{x}}{5}\right)$$

Figure 7.17 shows the scaled flow diagram.

Example 7.9

A position control servomechanism is represented by the second order linear differential equation:

$$J\frac{d^2\theta_a}{dt^2} + F\frac{d\theta_a}{dt} + K\theta_a = K\theta_d$$

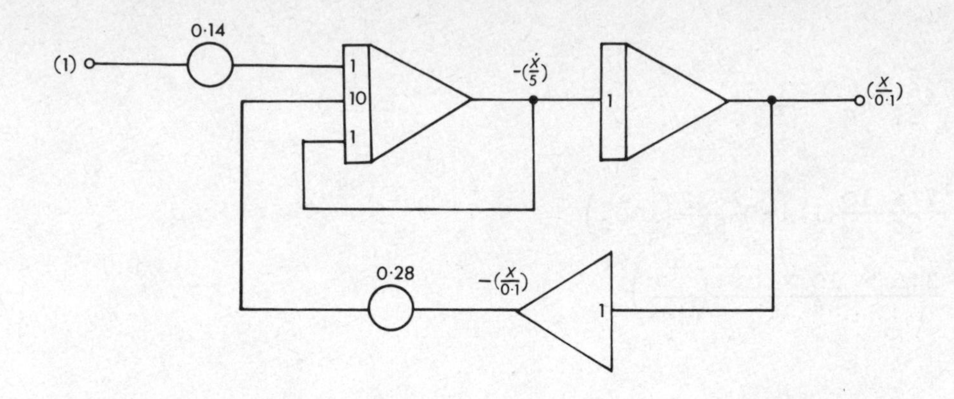

Fig.7.17

where θ_d *(a constant) is the demanded position,*

θ_a *is the actual position at some instant,*

J *is the effective moment of inertia of motor and load,*

F *is the damping torque coefficient,*

K *is the stiffness of the system, or the torque/unit error.*

The values of the constants are as follows:

J = 58 kg-m^2, F = 26.2 N-m/rad/s, and

K = 3.5 N-m/rad.

If the demanded position is given by $\theta_d = 0.2$ rad, *derive an analogue computer flow diagram to enable the actual position* θ_a *to be investigated. Take initial values* $\theta_a = \dot{\theta}_a = 0$. *Time scaling should be introduced such that the problem is speeded up by a factor of 2.*

Solution

$$\omega_n = \sqrt{\frac{K}{J}} = \sqrt{\frac{3.5}{58}} \simeq 0.246 \text{ rad/s}$$

$$|\theta_a|_{max} \simeq \frac{2K\theta_d}{K} = 2 \times 0.2 = 0.4$$

$$|\dot{\theta}_a|_{max} \simeq \omega_n |\theta_a|_{max} = 0.246 \times 0.4 = 0.098$$
$$\simeq 0.1$$

Scaled variables will be $\left(\frac{\theta_a}{0.4}\right)$ and $\left(10\dot{\theta}_a\right)$.

Substituting values:

$$58\frac{d}{dt}\dot{\theta}_a + 26.2\dot{\theta}_a + 3.5\theta_a = 3.5 \times 0.2$$

Rearranging:

$$\frac{d}{dt}\dot{\theta}_a = \frac{0.7}{58} - \frac{26.2}{58}\dot{\theta}_a - \frac{3.5}{58}\theta_a$$

Scaling for $\dot{\theta}_a$:

$$\frac{d}{dt}\left(10\dot{\theta}_a\right) = \frac{0.7 \times 10}{58}(1) - \frac{26.2}{58}\left(10\dot{\theta}_a\right)$$

$$- \frac{3.5 \times 10 \times 0.4}{58}\left(\frac{\theta_a}{0.4}\right)$$

Also $\frac{d}{dt}\theta_a = \dot{\theta}_a$

Scaling for θ_a:

$$\frac{d}{dt}\left(\frac{\theta_a}{0.4}\right) = \frac{1}{0.4 \times 10}\left(10\dot{\theta}_a\right)$$

In terms of potentiometer coefficients and amplifier gains:

$$\frac{d}{dt}\left(10\dot{\theta}_a\right) = [0.121]1(1) - [0.452]1\left(10\dot{\theta}_a\right) - [0.241]1\left(\frac{\theta_a}{0.4}\right)$$

$$\frac{d}{dt}\left(\frac{\theta_a}{0.4}\right) = [0.25]1\left(10\dot{\theta}_a\right)$$

For the problem to speed up by a factor of 2, $\beta = 1/2$.

Using $d\tau = \beta dt$ and dividing through by β:

$$\frac{d}{d\tau}\left(10\dot{\theta}_a\right) = [0.242]1(1) - [0.904]1\left(10\dot{\theta}_a\right) - [0.482]1\left(\frac{\theta_a}{0.4}\right)$$

$$\frac{d}{d\tau}\left(\frac{\theta_a}{0.4}\right) = [0.5]1\left(10\dot{\theta}_a\right)$$

The scaled flow diagram is shown in Figure 7.18.

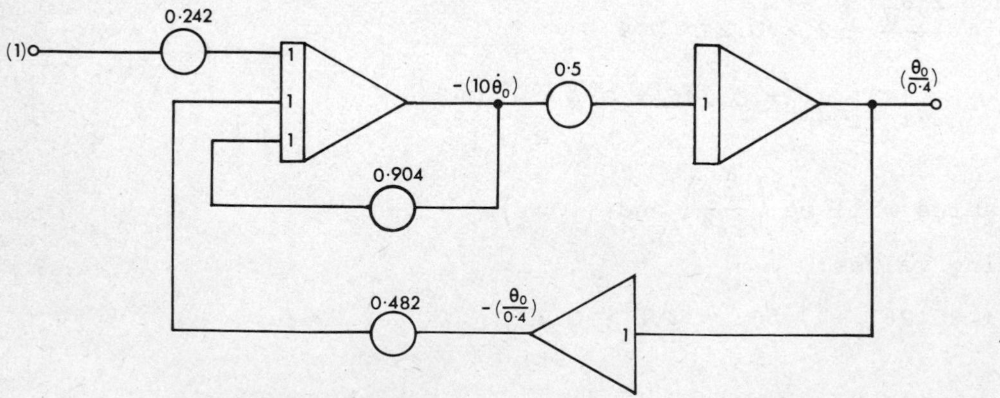

Fig.7.18

Example 7.10

The charge q on the capacitor in a particular R-L-C series circuit is given by:

$$100 = 5\frac{dq}{dt} + \frac{1}{10^3}\frac{d^2q}{dt^2} + \frac{10^6}{50}q$$

Given that $q = dq/dt = 0$ at time $t = 0$ derive a scaled flow diagram to enable q and dq/dt to be found. Employ time scaling such that "computer time" is 5000 times "real time". For amplitude scaling purposes take maximum amplitudes as follows:

$$|q|_{max} = 10^{-2} \text{ units} \quad \text{and} \quad \left|\frac{dq}{dt}\right|_{max} = 25 \text{ units}$$

Solution

Let $\frac{dq}{dt} \equiv \dot{q}$ and $\frac{d^2q}{dt^2} \equiv \frac{d}{dt}\dot{q}$

Scaled variables will be:

$$\left(\frac{q}{10^{-2}}\right) = (100q) \quad \text{and} \quad \left(\frac{\dot{q}}{25}\right)$$

Rearranging the equation:

$$\frac{d}{dt}\dot{q} = 10^5 - 5000\dot{q} - 2 \times 10^7 q$$

Scaling for $\dot{q}$:

$$\frac{d}{dt}\left(\frac{\dot{q}}{25}\right) = \frac{10^5}{25}(1) - 5000\left(\frac{\dot{q}}{25}\right) - \frac{2 \times 10^7}{25 \times 100}(100q)$$

Also $\frac{d}{dt}q = \dot{q}$

Scaling for q:

$$\frac{d}{dt}(100q) = 100 \times 25\left(\frac{\dot{q}}{25}\right)$$

Simplifying and writing in terms of potentiometer coefficients and amplifier gains:

$$\frac{d}{dt}\left(\frac{\dot{q}}{25}\right) = [0.4]10\,000(1) - [0.5]10\,000\left(\frac{\dot{q}}{25}\right) - [0.8]10\,000(100q)$$

$$\frac{d}{dt}(100q) = [0.25]10\,000\left(\frac{\dot{q}}{25}\right)$$

Time scaling:

$$\tau = 5000t \quad \text{and} \quad d\tau = 5000dt$$

$$\frac{d}{d\tau}\left(\frac{\dot{q}}{25}\right) = \frac{[0.4]10\,000}{5000}(1) - \frac{[0.5]10\,000}{5000}\left(\frac{\dot{q}}{25}\right) - \frac{[0.8]10\,000}{5000}(100q)$$

$$\frac{d}{d\tau}(100q) = \frac{[0.25]10\,000}{5000}\left(\frac{\dot{q}}{25}\right)$$

Simplifying:

$$\frac{d}{d\tau}\left(\frac{\dot{q}}{25}\right) = [0.8]1(1) - 1\left(\frac{\dot{q}}{25}\right) - [0.16]10(100q)$$

and

$$\frac{d}{d\tau}(100q) = [0.5]1\left(\frac{\dot{q}}{25}\right)$$

The scaled flow diagram is shown in Figure 7.19.

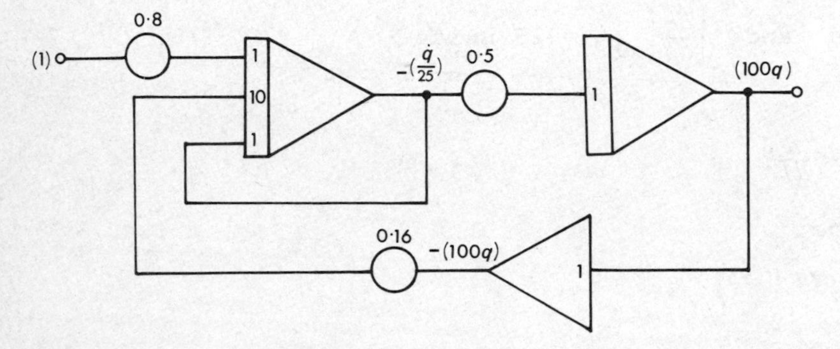

Fig.7.19

Example 7.11

Prepare a scaled computer flow diagram to obtain the solution of the following differential equation:

$$\frac{d^2x}{dt^2} + 70\frac{dx}{dt} + 6400x = 5000x_i(t)$$

where $x_i(t)$ *is a step of* 1 m.

It is given that

(i) *The problem is to be solved on a* 10 V *general purpose analogue computer.*

(ii) *The highest angular frequency in the above system is* 80 rad/s *and the solution is to be recorded on an* X-Y *recorder which gives an accurate response for angular frequencies up to* 8 rad/s.

Solution

Estimating maximum values:

$$|x|_{max} \simeq \frac{2 \times 5000}{6400} \times x_i(t) = 1.56 \quad \text{(take 2)}$$

$$|\dot{x}|_{max} \simeq \omega_n |x|_{max} = 80 \times 1.56 = 124.8$$

(take 150)

Scaled variables will be $\left(\frac{x}{2}\right)$ and $\left(\frac{\dot{x}}{150}\right)$.

Rearranging the equation and using the "dot" notation:

$$\ddot{x} = 5000 - 70\dot{x} - 6400x$$

Scaling for $\dot{x}$:

$$\frac{d}{dt}\left(\frac{\dot{x}}{150}\right) = \frac{5000}{150}(1) - 70\left(\frac{\dot{x}}{150}\right) - \frac{6400 \times 2}{150}\left(\frac{x}{2}\right)$$

Also $\quad \frac{d}{dt}x = \dot{x}$

Scaling for x:

$$\frac{d}{dt}\left(\frac{x}{2}\right) = \frac{150}{2}\left(\frac{\dot{x}}{150}\right)$$

The problem should be slowed down 10 times. Thus $d\tau = 10dt$.

Dividing through by β and substituting for dt:

$$\frac{d}{d\tau}\left(\frac{\dot{x}}{150}\right) = \frac{500}{1500}(1) - \frac{70}{10}\left(\frac{\dot{x}}{150}\right) - \frac{6400 \times 2}{1500}\left(\frac{x}{2}\right)$$

and

$$\frac{d}{d\tau}\left(\frac{x}{2}\right) = \frac{150}{20}\left(\frac{\dot{x}}{150}\right)$$

In terms of potentiometer coefficients and amplifier gains:

$$\frac{d}{d\tau}\left(\frac{\dot{x}}{150}\right) = [0.333]1(1) - [0.7]10\left(\frac{\dot{x}}{150}\right) - [0.853]10\left(\frac{x}{2}\right)$$

$$\frac{d}{d\tau}\left(\frac{x}{2}\right) = [0.75]10\left(\frac{\dot{x}}{150}\right)$$

The scaled flow diagram is shown in Figure 7.20.

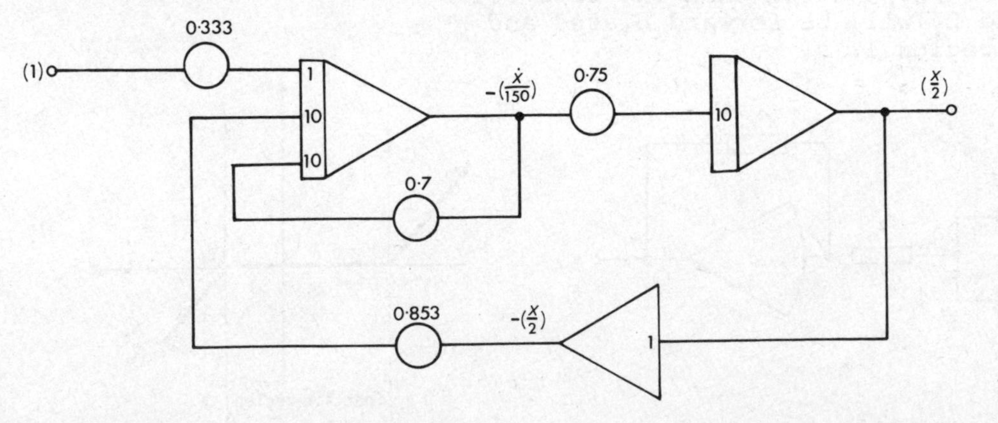

Fig.7.20

7.8 NON-LINEAR COMPUTING ELEMENTS

Non-linear elements used in analogue computing include diode function generators, multipliers, and comparators.

7.8.1 DIODE FUNCTION GENERATORS

Simulation of physical systems and the solution of many differential equations often require the use of mathematical functions such as sines, cosines, and exponentials. Other functions widely used include the *limited* function, *dead-zone*, and the *absolute value function*, shown in Figure 7.21.

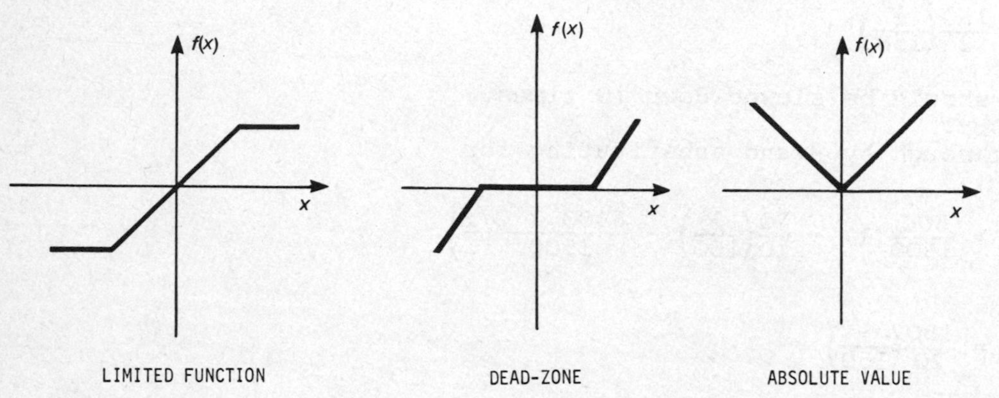

Fig.7.21

Discontinuous functions can be generated by suitable arrangements of biased diodes (acting as switches) and amplifiers. Figure 7.22 shows a few diode function generators and their input/output characteristics. The principle of operation of such circuits is rather similar and illustrated in the following examples.

Consider the DEAD-ZONE SIMULATOR of Figure 7.23 and its transfer characteristic.

When e_i is *more positive* than the bias voltage E_1, diode D_1 will be forward biased and conducting: region I.

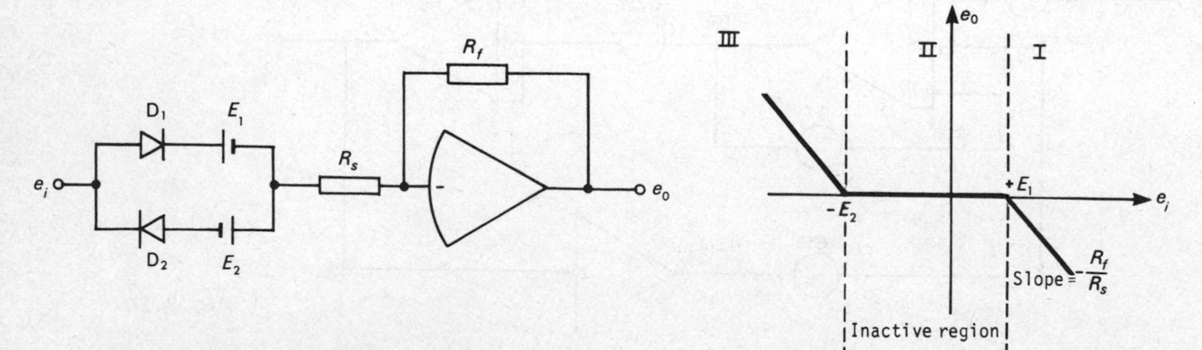

Fig.7.23 Dead-zone generator and its characteristics

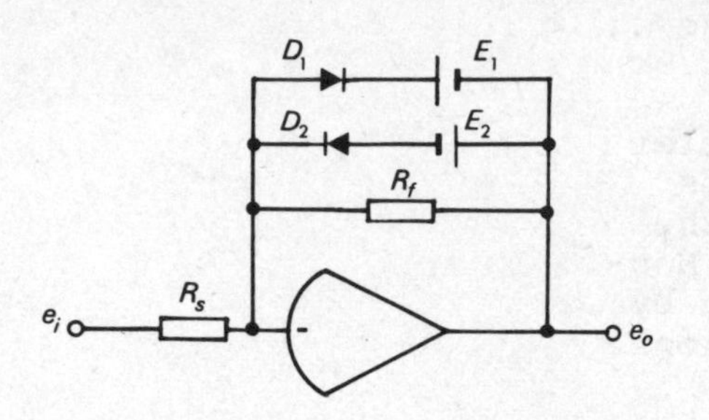

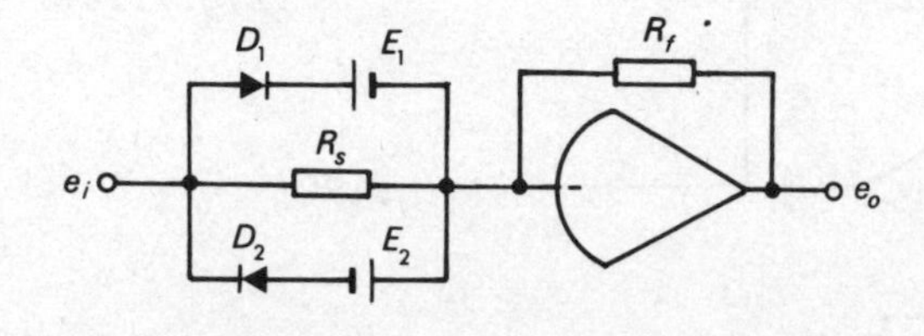

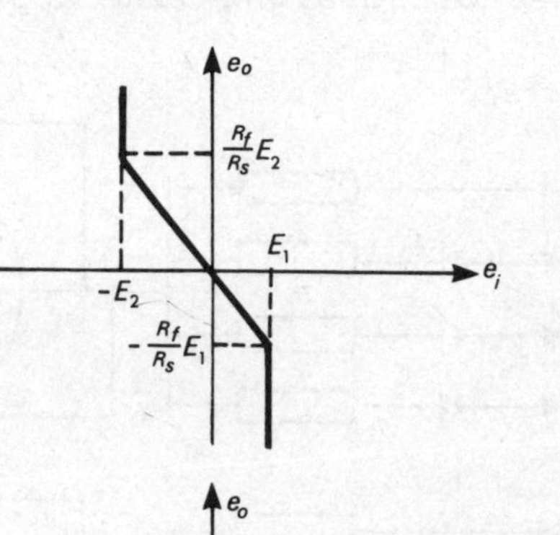

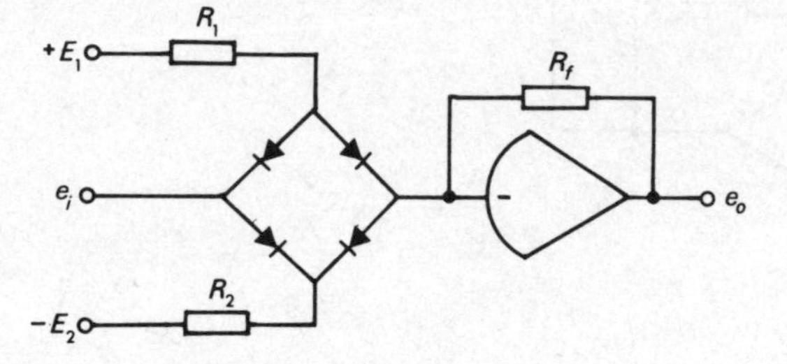

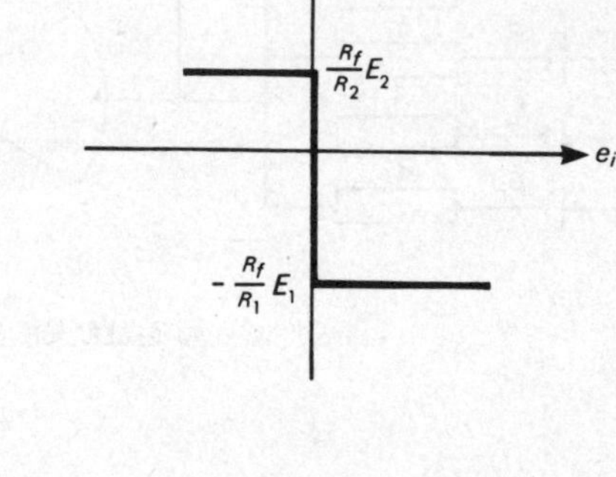

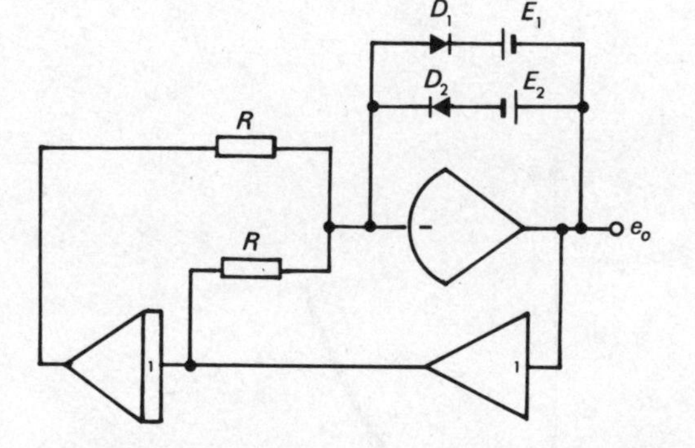

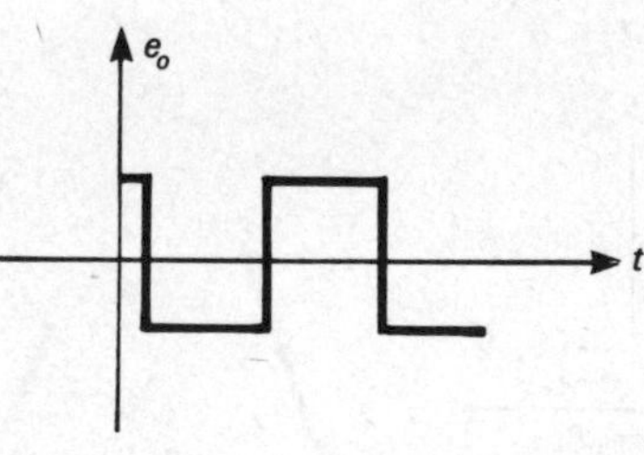

Fig.7.22

When e_i is *more negative* than E_2, diode D_2 will be forward biased and conducting: region III.

Between these limits, when $E_2 < e_i < E_1$, both diodes are reverse biased and there will be no output: region II.

The slope of the two active parts depends on the ratio of R_f/R_s and can be made different for regions I and III by inserting resistors in series with the diodes.

Figure 7.24 shows a DIODE FUNCTION GENERATOR which produces an output proportional to the square of its input. The principle of operation of the circuit is as follows.

The transfer characteristics of an amplifier depend on the feedback and series components used. If the input resistance is varied, the slope of the characteristics will change. Non-linear functions can therefore be generated by using a series of straight-line approximations.

Fig.7.24

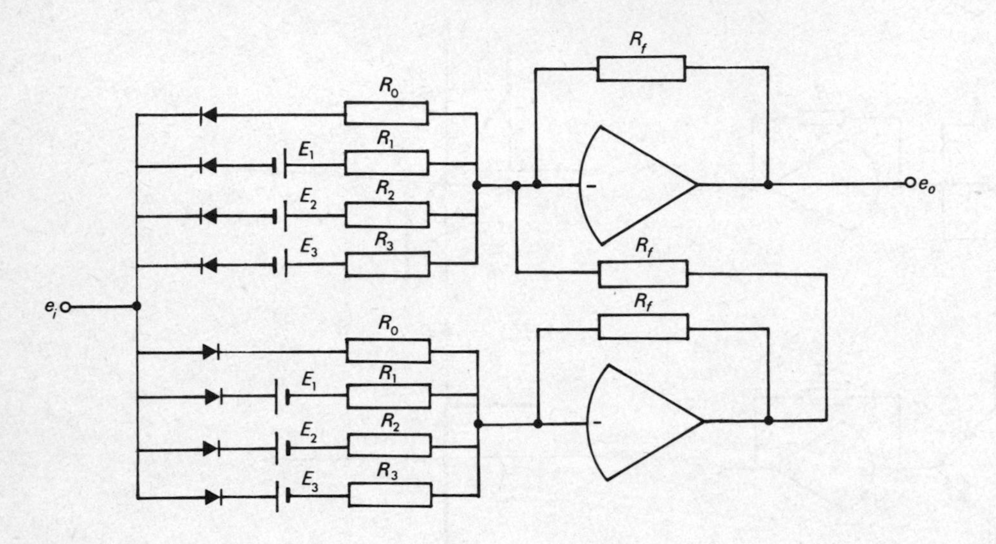

(a) SQUARE-LAW GENERATOR

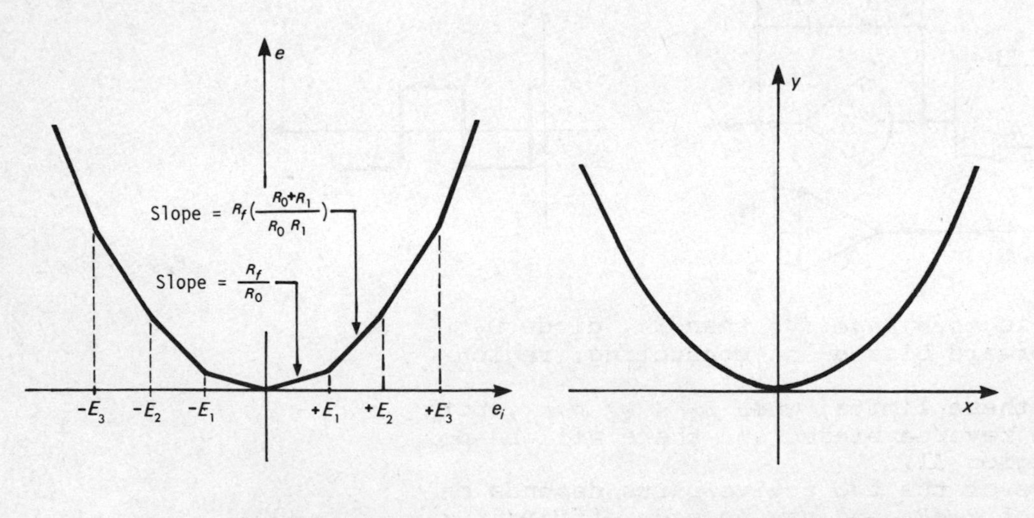

(b) GENERATED CHARACTERISTIC

(c) THEORETICAL CHARACTERISTIC

In Figure 7.24, the diodes are biased by E_1, E_2, etc. to switch on at different input levels and reduce the effective input resistance of the amplifiers. The slope of each part of the characteristics is therefore determined by the number of resistors acting in parallel.

The upper half of the circuit becomes active when the input is negative, and the lower half responds to positive values of input. It is evident that, by using a large number of biased diodes, a good approximation of the theoretical characteristics can be achieved.

7.8.2 MULTIPLIERS

Multiplication by a constant less than 1 can be obtained using a potentiometer. When two or more variables are to be multiplied, special units must be used. Multipliers play an important part in analogue computation and some of those commonly used will be considered here.

SERVOMULTIPLIER

The output of a potentiometer is given by $k \times v$ where v is the applied voltage and k is a factor depending on the position of the slider. If the applied voltage is a signal v_x and the slider is driven mechanically to a position proportional to another signal v_y, the output will then be at a voltage proportional to $v_x \times v_y$. This is the principle of servomultipliers.

The basic diagram of a servomultiplier is shown in Figure 7.25. The small servo-motor drives a set of linear potentiometers ganged together. The computer reference voltage (here ± 10 V) is applied to one of the potentiometers, the slider of which provides feedback for the servo-motor. The difference between v_y and the feedback voltage v_{fb}, error ε, is fed into the

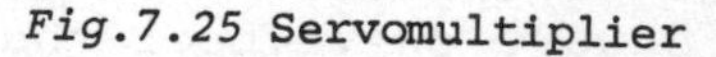

Fig.7.25 Servomultiplier

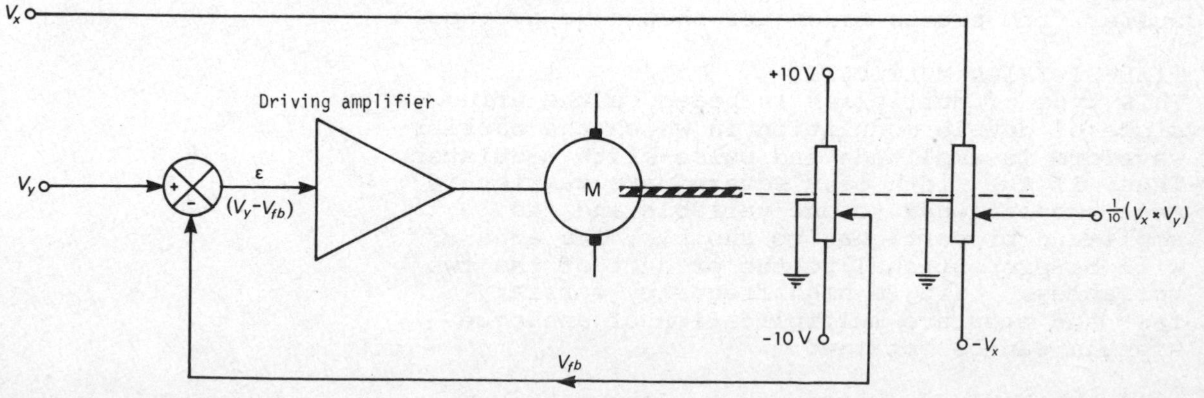

amplifier which drives the motor until the error is reduced to zero. When $v_y = v_{fb}$, the motor will stop; the position of the slider corresponds to v_y and provides an output equal to $(v_x \times v_y)/10$.

Servomultipliers are reasonably cheap and reliable, and their main feature is that several signals can be multiplied together simply by ganging further potentiometers. They are, however, slow in operation, limited in accuracy, and suitable only for very low frequencies.

QUARTER SQUARES MULTIPLIER

The operation of this type of multiplier is based on the relation:

$$\tfrac{1}{4}\left[(x + y)^2 - (x - y)^2\right] = xy$$

A block diagram implementing this relation is shown in Figure 7.26. The squaring circuits may either be square law function generators or

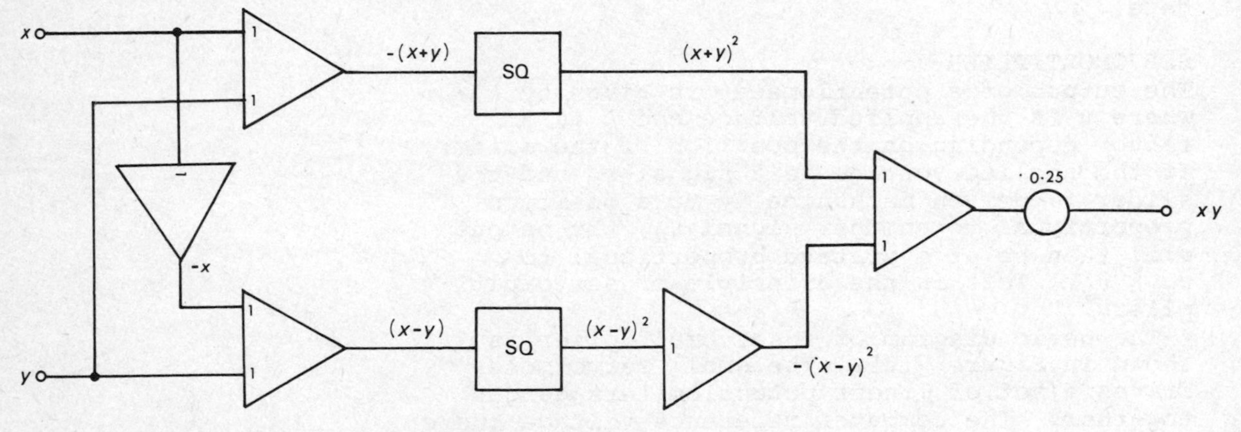

Fig.7.26 Quarter-squares multiplier

use diodes whose forward characteristic follows a square law. Due to the flatness of the square curve near the origin, small inputs suffer from excess error and should be avoided.

TIME-DIVISION MULTIPLIER

This type of multiplier is based on the principle of double modulation in which the carrier waveform is amplitude and pulse-width modulated. Thus, if the width of a square wave carrier is made proportional to one variable and its amplitude proportional to another, its area will be proportional to the product of the two variables. Using a high frequency carrier, fast and accurate multiplication of analogue signals can be obtained.

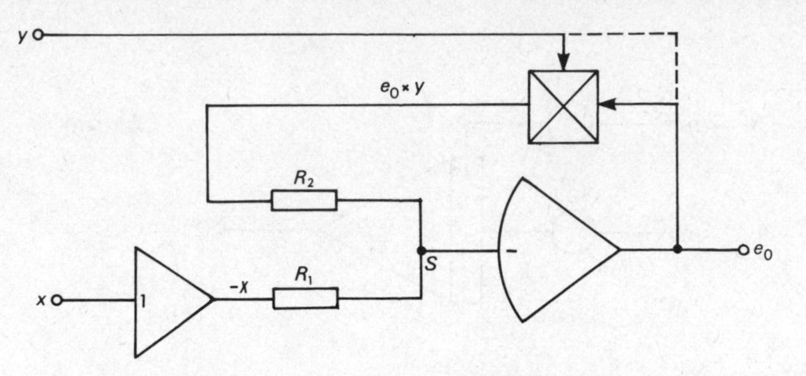

Fig.7.27 Division circuit

7.8.3 DIVIDER AND SQUARE-ROOT CIRCUIT

A division circuit may be constructed by placing the multiplier unit in the feedback loop of an amplifier as shown in Figure 7.27.

Making the usual assumptions, the sum of the currents arriving at S equals zero.

$$-\frac{x}{R_1} + \frac{y \times e_o}{R_2} = 0$$

$$y \times e_o = \frac{R_2}{R_1}x \qquad \text{i.e.} \quad e_o = \frac{R_2}{R_1} \times \frac{x}{y}$$

Thus $e_o \propto \dfrac{x}{y}$

If the y input to the multiplier is replaced by e_o, shown dotted, a square root circuit will be obtained. Proceeding as above:

$$-\frac{x}{R_1} + \frac{e_o^{\,2}}{R_2} = 0 \qquad \text{i.e.} \quad e_o^{\,2} = \frac{R_2}{R_1}x$$

Thus $e_o \propto \sqrt{x}$

7.9 GENERATION OF ANALYTIC FUNCTIONS

Analytical functions can usually be generated using integrators, inverters and potentiometers. The function is differentiated repeatedly until either:

(*a*) a constant is reached; this is fed, together with the initial condition, into a series of integrators until the required function is obtained;
or
(*b*) one of the derivatives contains the original function; this gives rise to a differential equation which can then be implemented.

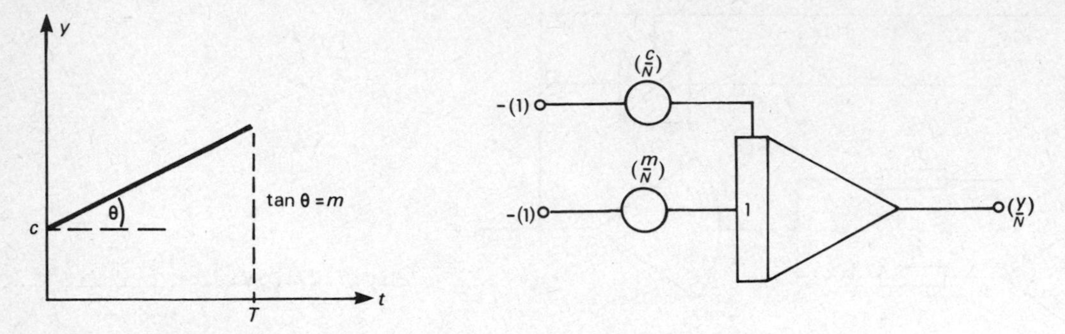

Fig.7.28 Ramp function generator

As an example, consider the ramp function shown in Figure 7.28(*a*) over a period of T seconds. The equation of this function is

$$y = mt + c$$

Differentiating with respect to t

$$\frac{dy}{dt} = m$$

The maximum value of y will be

$$y_{max} = mT + c \quad \text{(say } N\text{)}$$

The scaled variable will thus be $\left(\frac{y}{N}\right)$.

Then,

$$-\frac{d}{dt}y = -m$$

Scaled equation: $-\frac{d}{dt}\left(\frac{y}{N}\right) = -\left[\frac{m}{N}\right]1(1)$

Scaled initial condition will be $\left[\frac{c}{N}\right](1)$

The scaled diagram is shown in Figure 7.28(*b*).

Solution of many differential equations often requires the use of sinusoidal functions. As an example, consider the function

$$y = A\sin\omega t$$

Then,

$$\dot{y} = \omega A\cos\omega t$$

$$\ddot{y} = -\omega^2 A\sin\omega t$$

Therefore $\ddot{y} = -\omega^2 y$

Initial conditions: at $t = 0$, $y = \ddot{y} = 0$, $\dot{y} = \omega A$

Maximum values: $y_{max} = A$, $\dot{y}_{max} = \omega A$, $\ddot{y}_{max} = \omega^2 A$

Scaled equations will be

$$\frac{d}{dt}\left(\frac{\dot{y}}{\omega A}\right) = -\ [\omega]1\left(\frac{y}{A}\right)$$

and

$$\frac{d}{dt}\left(\frac{y}{A}\right) = [\omega]1\left(\frac{\dot{y}}{\omega A}\right)$$

Scaled initial condition for y will be:

$$\frac{\omega A}{\omega A} = (1)$$

In the flow diagram of Figure 7.29, two potentiometers each with a coefficient of ω have been used (rather than one set to ω^2) for two reasons. First, to allow for amplitude scaling, and secondly to ensure that the input condition for $\dot{y}$ is not frequency-dependent.

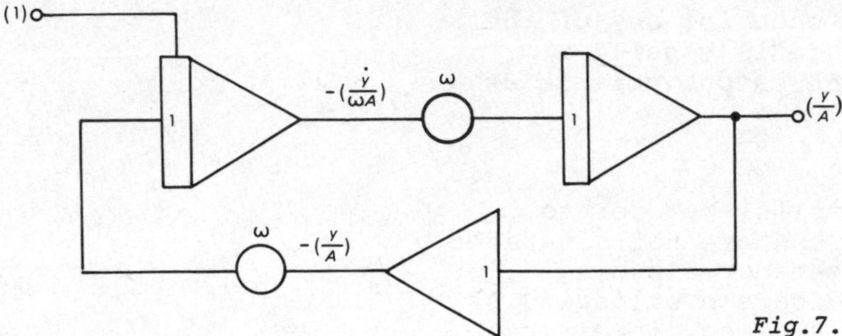

Fig.7.29 **Sinewave function generator**

7.10 CONTROL

In order to compute a dynamic problem on an analogue computer, it is necessary to control the state of the elements used by the computer. Each computer has a *control panel*, the exact detail of which varies from one make and model to another. A typical control panel comprises switches, meters, and indicators, whose operations and functions are described below.

7.10.1 MODE SWITCH

The mode switch controls a set of internal relays which put the computer into one of the following modes:

POT SET — This is the mode in which potentiometers are set to their correct value.

PROBLEM CHECK — Also known as Re-Set, is the mode in which the initial conditions are applied. At $t = 0$, the computer outputs represent the initial values of the variables.

COMPUTE — In this mode of operation the computer solves the problem applied to it.

HOLD — During this mode, the computation is "frozen" and the outputs indicate the value reached at the instant of switching.

REP-OP — In this mode the computer computes the problem for some predetermined time (say $t = 0$ to $t = 10$ s), resets to $t = 0$, and repeats the cycle as long as it is left in this mode.

7.10.2 OUTPUT SELECTION SWITCH

The output selector switch (sometimes a group of push-buttons) allows the output of any amplifier or potentiometer to be monitored on a voltmeter. This is often necessary when setting potentiometers or checking any of the outputs. The voltmeter used is usually a digital type displaying the output voltage as a fraction of one M.U.

7.10.3 OVERLOAD INDICATORS

In order to ensure that amplifiers operate within their linear range and are not saturated during a computation, buzzers or flashing lights are arranged to indicate overloading of any amplifier. Incorrect scaling, incorrect wiring, or having no feedback in an amplifier often causes overloading.

Some computers automatically go into the HOLD mode when overloading occurs. This will allow the immediate inspection of the amplifier concerned. When such a facility is not provided, after checking, the computation should be repeated.

7.11 OUTPUT DEVICES

One of the outstanding features of the analogue computer is the convenient way in which output data is presented, ready to be used without any need for decoding or processing. The results can therefore be monitored, recorded or measured as the computation proceeds. Output devices commonly used include voltmeters, oscilloscopes and X-Y plotters.

The OSCILLOSCOPE is the most useful device and can display almost any likely range of frequencies. Although in most cases sufficiently accurate results cannot be obtained, answers can be checked quickly before unnecessary recordings are made. It is ideal for test runs (to check the wiring, scaling, etc.), useful for troubleshooting, and for observing drift, noise, and high frequency oscillations.

An oscilloscope which is used in conjunction with an analogue computer should possess the following features:

(*a*) Good high frequency response.
(*b*) Long persistence screen.
(*c*) High sensitivity.
(*d*) "External X" input should be available so that the time base generated by the computer can be applied to the oscilloscope, overcoming the synchronization problem.

The main disadvantages of an oscilloscope are

(*a*) Permanent records of results are not possible.
(*b*) Very slow waveform outputs (frequencies ≈ 0.05 Hz) cannot be observed.

These difficulties together with the problem of looking for hidden details in waveforms are overcome by using PLOTTERS. X-Y plotters are electromechanical devices which give recordings of two variables against each other. They consist of two position servomechanisms controlling the motion of a pen over a chart. The two axes of motion are perpendicular to each other and correspond to the X and Y coordinates.

A simple X-Y plotter consists of a pen which is driven in the Y direction by one servo-motor mounted on the pen carriage. A second servo-motor controls the displacement of the pen carriage in the X direction. It is evident that because of the large difference in the mass of the two servo systems, the X and Y inputs will have different frequency responses.

X-Y plotters are precision devices and for good and accurate performance they should be well maintained. Their main disadvantage is that they are rather slow and unsuitable for quick checks. A special feature of an X-Y plotter is its use as a function generator. This can be achieved by replacing the pen with a pick-up head. When the head is driven over a curve drawn in conducting ink, a proportional voltage will be generated in the Y-output.

7.12 SIMULATION

One of the most important applications of the analogue computer is the simulation of feedback control systems. The usual starting point is to represent the elements of the system by blocks, with the transfer function of the various elements inserted in the blocks and thus making up a complete block diagram. Each block is then represented by a computing element (or set of elements). These elements are interconnected, in exactly the same way as the system under simulation, resulting in a computer flow diagram very similar to the block diagram of the system.

Since each element of the system is individually represented by a computer element, direct investigation of each part of the system, each parameter, and the effect of variation in control elements can be achieved often by means of one or two potentiometers.

In simulating large and complicated systems, some of the potentiometers are servo-controlled and the computer is programmed to gradually change the settings of the pots until optimum solution is reached. Sophisticated systems are often simulated and investigated by a system of *hybrid* computers which makes use of both the digital and the analogue computer.

EXERCISES

(7.1) With the aid of a diagram, explain the operation of an integrator (as used in a general purpose analogue computer) which has the facility of setting initial conditions.

(7.2) Draw and explain a diagram of an integrator which can provide all the following modes of general analogue computer control: (*a*) POT-SET, (*b*) RESET, (*c*) COMPUTE, (*d*) HOLD.

(7.3) (i) State the essential properties of an operational amplifier and show how it is used to perform:
(*a*) addition of several voltages
(*b*) integration of a voltage.
(ii) Describe, with the aid of a diagram, how an initial condition is applied to an integrator in the RESET mode.

(7.4) Obtain an analogue flow diagram to solve the simultaneous equations

$$3.5x - y = 40 \quad \text{and} \quad 1.7x + 15y = -58$$

(7.5) Obtain analogue flow diagrams to solve the following equations:

(*a*) $2\dot{x} + 3x - 5 = 0$

(*b*) $0.5\ddot{x} + 2\dot{x} + 12x = 1$

(*c*) $\dot{x} + 2y = 3$ and $3\dot{y} - 7x = 5$

(*d*) $5\ddot{x} - 2\dot{x} + y = 7$ and $3\ddot{y} + x = 0$

(*e*) $A\ddot{x} + B\dot{x} + Cy = D$ and
$E\ddot{y} + F\dot{y} + Gx = H$

(7.6) Obtain amplitude scaled flow diagrams for the following equations:

(*a*) $7\dot{x} + 2x - 17 = 0$ $x_{max} = 10$, $x_0 = 0$

(*b*) $\ddot{x} = -4x$ $x_{max} = 5$, $\dot{x}_{max} = 10$,
$x_0 = 0$, $\dot{x}_0 = 10$

(*c*) $\dot{x} = 15x + 2$ $x_0 = 0$, $\dot{x}_0 = 1$

(*d*) $\ddot{x} + 2\dot{x} + 700x = 980$ $x_0 = 0$,
$\dot{x}_0 = 15$

(*e*) $3\ddot{x} + 5\dot{x} + 35x = 450$ $x_0 = 5$,
$\dot{x}_0 = 0$

(*f*) $2\dddot{x} + 15\ddot{x} + 50\dot{x} + 55x = 100$
$x_0 = \ddot{x}_0 = 0$, $\dot{x}_0 = -5$

(7.7) A jet motor achieves a velocity v in a time t. The velocity and time are related by the following formula:

$$v = t^3 + 2t^2 + 4$$

Draw a block diagram of an analogue system with a full-scale voltage of 10 V that would calculate the acceleration in m/s^2 at selected times during the engine run. If the velocity was recorded over a period of 200 s, calculate the scale factors at each point in the block diagram.

(7.8) (*a*) Draw both the open and closed loop diagrams of the arrangement of analogue computing components that could be patched to solve the following equation

$$4\frac{dx}{dt} + 3x = 4$$

given that $x = \frac{3}{4}$ at $t = 0$

(*b*) What are the advantages and disadvantages of the two types of diagram?

(7.9) (*a*) Explain briefly why it is generally necessary to use amplitude and time scale factors when solving differential equations on an analogue computer.
(*b*) A system is described by the differential equation:

$$\frac{d^2x}{dt^2} + 60\frac{dx}{dt} + 2500x = 2000x_i(t)$$

where $x_i(t)$ = a step input of 1 m.
The estimated maximum values are:

$$x = 2\text{ m} \qquad \frac{dx}{dt} = 50\text{ ms}^{-1}$$

$$\frac{d^2x}{dt^2} = 2500\text{ ms}^{-2}$$

It is given that
(i) The problem is to be solved on a 10 V general purpose analogue computer.
(ii) The highest angular frequency in the above system is 50 rad s^{-1} and the solution is to be recorded on an X-Y plotter which gives an accurate response for angular frequencies up to 10 rad s^{-1}.

(7.10) (*a*) Why are diode function generators required in analogue computers?
(*b*) Draw a diagram of a diode function generator that will produce the function $y = x^2$ for both positive and negative values of x. With the aid of a graph illustrate the expected output of the function generator and by labelling the graph relate it to your block diagram.

(7.11) (*a*) The following analytic functions are to be generated in real time (t) by means of amplifiers and potentiometers only:

(i) Ae^{-kt} where $A = 10$ and $k = 3$

(ii) $5\cos\omega t$ where $\omega = 2\pi f$, the value of f being 5 Hz.

Derive suitable analogue computer flow diagrams, labelling all amplifier outputs and potentiometer coefficients. In both cases the function to be generated is required from time $t = 0$ onwards (t is in seconds).

(*b*) Part of a simulation problem consists of generating the function e^{kx} where x is a function of time.

Assuming that $\dot{x}$ is available as an input, establish a suitable unscaled flow diagram to generate the required function.

(7.12) Show diagrams to illustrate how you would arrange analogue computing elements to generate the following functions of time:

(i) $x = Ae^{-at}$ A and a are constants

(ii) $x = at^2 + bt + c$ a, b, c are constants (all positive)

(iii) $x = a^t$ a is a constant

(iv) Given that a voltage equivalent to (dz/dt) is available, show how you would generate the function

$$y = e^z$$

where z is not a function of time, but is an independent variable.

8 Fundamentals of Digital Systems

The use of digital computers has played a major role in the recent advances in the design and study of control systems. Many modern control systems, whether small or large, incorporate some units or elements which operate by digital signals.

This chapter introduces digital logic[†] and deals with the basic principles of operation of digital systems.

8.1 BINARY NUMBER SYSTEM

The binary number system uses the base of 2, and therefore contains two digits only, 0 and 1. A binary digit is known as a BIT, and a combination of bits is often referred to as a WORD. In *Boolean Algebra*, which is the mathematics of switching circuits, only two states can be recognized. Since the most convenient, practical, and cheapest electronic switching devices are those which can distinguish between two states, the binary number system has been adopted for digital circuits and system.

8.1.1 BINARY REPRESENTATION

A simple way of introducing binary numbers and the sequence in which they occur is to write down the decimal numbers, then cross out all those that contain a digit greater than 1.

[†] For detailed treatment of this subject reference can be made to *Solution of Problems in Computer and Control Engineering* by W.B. Bishop-Miller and B. Yousefzadeh (Pitman).

The remaining numbers form the binary numbers. Thus:

0,1,~~2~~, ~~3~~,...,~~9~~,10,11,~~12~~,...,~~99~~,100,101,~~102~~,...

~~109~~,110,111,~~112~~,~~113~~,...,~~999~~,1000,1001,~~1002~~,...

Collecting the numbers that remain, we have the binary number system:

0, 1, 10, 11, 100, 101, 110, 111, 1000,

1001, ...

Decimal numbers use the base of ten. The value of each number is governed by the relative position of its digits. The position indicates the significance or *weight* of that digit. These weights are in powers of ten. Thus, decimal number 835.6 for example, can be written as

$$8 \times 100 + 3 \times 10 + 5 \times 1 + 6 \times 0.1$$

or $8 \times 10^2 + 3 \times 10^1 + 5 \times 10^0 + 6 \times 10^{-1}$

In binary, powers of two are used. Thus, as illustrated in Table 8.1, binary number 10110.1 can be written as

$$1 \times 16 + 0 \times 8 + 1 \times 4 + 1 \times 2$$

$$+ 0 \times 1 + 1 \times \tfrac{1}{2}$$

$$= 1 \times 2^4 + 0 \times 2^3 + 1 \times 2^2 + 1 \times 2^1$$

$$+ 0 \times 2^0 + 1 \times 2^{-1}$$

$$= 16 + 0 + 4 + 2 + 0 + \tfrac{1}{2}$$

$$= 22.5 \text{ in decimal}$$

Table 8.1

...	2^4	2^3	2^2	2^1	2^0		2^{-1}	2^{-2}	...
...	16	8	4	2	1		½	¼	...
	1	0	1	1	0	•	1		

←position of binary point

Here, ½, 1, 2, 4, 8, etc. are all powers of two and represent the weights of their associated digit.

8.1.2 DECIMAL TO BINARY CONVERSION

During the conversion of a decimal number to binary, integers and fractions are dealt with separately.

INTEGERS
To convert a decimal integer into binary, the number is repeatedly divided by 2, each time recording the remainder (which can only be 0 or 1), in order from right to left.

Decimal number 53 for example will be converted to binary as follows:

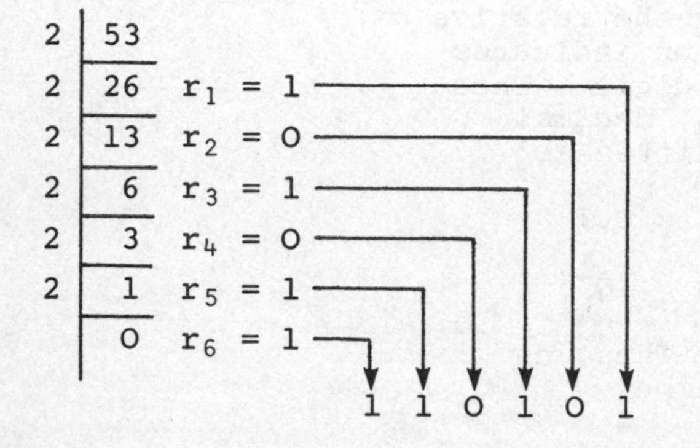

FRACTIONS
To convert a decimal fraction into binary, the number is repeatedly multiplied by 2, each time recording the integral part so obtained (which can only be 0 or 1), in order from left to right.

Decimal fraction 0.375 for example will be converted as follows:

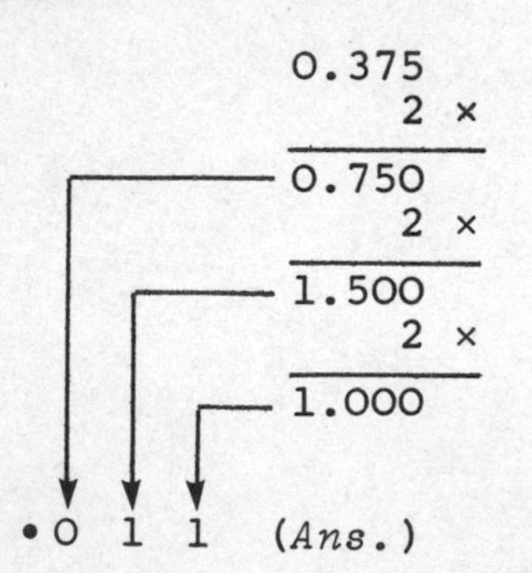

8.1.3 BINARY TO DECIMAL CONVERSION

In order to convert a binary number into its equivalent decimal number, starting from the right, each bit is multiplied by its corresponding weight. The results are then added together. Binary number 101101.01 for example is shown in Table 8.2.

Table 8.2

... 2^6	2^5	2^4	2^3	2^2	2^1	2^0	2^{-1}	2^{-2}	2^{-3} ...
... 64	32	16	8	4	2	1	0.5	0.25	0.125
	1	0	1	1	0	1 •	0	1	

From the table:

$$1 \times 32 + 0 \times 16 + 1 \times 8 + 1 \times 4 + 0 \times 2$$

$$+ 1 \times 1 + 0 \times 0.5 + 1 \times 0.25$$

$$= 32 + 8 + 4 + 1 + 0.25$$

$$= 45.25 \text{ (Ans.)}$$

8.1.4 BINARY ARITHMETIC

The arithmetic of binary numbers follows the same rules as those for decimal numbers. Table 8.3 shows the SUM and the CARRY for the addition of two bits *A* and *B*. In subtraction of two binary digits, shown in Table 8.4, where a 1 is being subtracted from a 0, the difference is 1 with a 1 borrowed (1 is effectively subtracted from 10). To compensate for this borrow, a 1 is then paid back (added) to the next bit on the left to be subtracted.

Table 8.3

A	*B*	Carry	Sum ($A + B$)
0	0	0	0
0	1	0	1
1	0	0	1
1	1	1	0

Table 8.4

A	*B*	Borrow	Difference ($A - B$)
0	0	0	0
0	1	1	1
1	0	0	1
1	1	0	0

Example 8.1

Consider the addition of two binary numbers 11011 *and* 10001.

```
Augend      11011
Addend      10001 +
            ------
Carry       10011
           -------
Sum        101100
```

Example 8.2

Consider the following binary subtraction: 10101 - 1010.

```
                10101
                 1010 -
                ------
Borrow          1010
               -------
Difference      01011
```

Table 8.5 shows the product of two bits *A* and *B*. It is seen that the product is always 0 unless both bits are 1. As illustrated below, the procedure for multiplication is exactly the same as that for decimal numbers.

```
      1101
      1001 ×
    -------
      1101
     0000
    0000
   1101
   -------
   1110101
```

Table 8.5

A	*B*	*A* × *B*
0	0	0
0	1	0
1	0	0
1	1	1

Once again, the division of two binary numbers follows the same rules as decimal division, remembering that the only two possible bits are 0 and 1.

Example 8.3

Consider the division of binary number 11001 *by* 101.

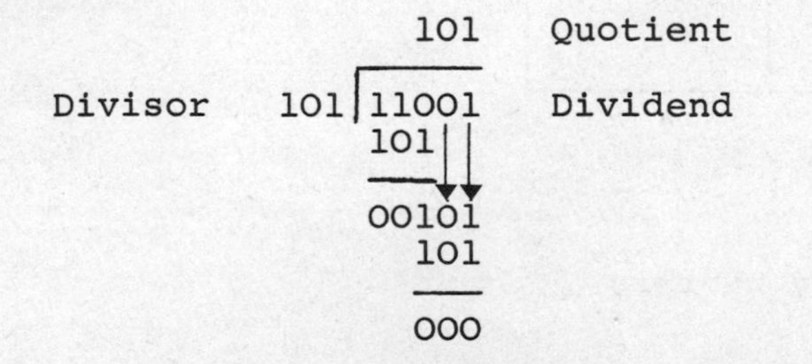

8.2 CODES

Binary numbers described in the previous section are all *pure binary* and, because their practical application is very limited, they have to be used in a coded form.

A series of simple codes are the *binary-coded-decimal* (BCD) which use four bits to represent each decimal digit. These four bits are assigned weights to obtain desired properties. The weights are chosen such that their sum is greater than 9 but does not exceed 15; one of them is a 1 and another is either a 1 or a 2.

The simplest BCD code is the 8421. This code has weights of 1, 2, 4, and 8 assigned to its bits from right to left respectively. Decimal number 95 for example, will be expressed as:

```
8 4 2 1   8 4 2 1 ←—weights
1 0 0 1   0 1 0 1 ←—binary equivalent
                    of each digit
   9         5
```

Thus 95 = 1001 0101 in 8421-BCD

= 1011111 in pure binary

Other BCD codes include the 7421, 2421, 5211, 6311, and 3321.

Then, taking decimal number 95 again:

95 = 1010 0101 in 7421-BCD

= 1111 0101 in 2421-BCD

= 1111 1000 in 5211-BCD

= 1100 0111 in 6311-BCD

= 1110 1010 in 3321-BCD

8.2.1 GRAY CODE

The *Gray* code, also known as the *reflected binary*, is an unweighted code not particularly suited to arithmetic operations, but very useful for input-output devices and analogue-to-digital convertors, and, as shown in Chapter 3, it is used in digital transducers. As shown in Table 8.6, the main characteristic of the Gray code is that each Gray code number differs from the preceding Gray code number by only one bit. The significance of this *unit-distance* code is explained in section 3.13 of Chapter 3 with the operation of shaft encoders.

DECIMAL NUMBER	PURE BINARY	GRAY CODE
0	0 0 0 0	0 0 0 0
1	0 0 0 1	0 0 0 1
2	0 0 1 0	0 0 1 1
3	0 0 1 1	0 0 1 0
4	0 1 0 0	0 1 1 0
5	0 1 0 1	0 1 1 1
6	0 1 1 0	0 1 0 1
7	0 1 1 1	0 1 0 0
8	1 0 0 0	1 1 0 0
9	1 0 0 1	1 1 0 1
10	1 0 1 0	1 1 1 1
11	1 0 1 1	1 1 1 0
12	1 1 0 0	1 0 1 0
13	1 1 0 1	1 0 1 1
14	1 1 1 0	1 0 0 1
15	1 1 1 1	1 0 0 0

Table 8.6

BINARY TO GRAY CODE CONVERSION
To convert a binary number to Gray code, the most significant bit is first recorded. Then, starting from the left, each bit is added to the bit to its right, recording the sum (disregarding any "carry"). The number obtained will be the Gray code equivalent. Taking decimal number 23 for example:

```
23 = 1  0  1  1  1      in binary
     1  1  1  0  0      in Gray code
```

GRAY CODE TO BINARY CONVERSION
To convert a Gray code number to binary, the most significant bit is first recorded. Then, starting from the left, each binary bit is generated by adding the corresponding Gray code bit and the preceding binary digit (ignoring any "carry"). Converting the Gray code number of the previous example back to pure binary:

```
1  1  1  0  0      in Gray code
1  0  1  1  1      in pure binary
```

8.3 LOGIC FUNDAMENTALS

Altogether there are five fundamental logic operations. The main three are the OR, the AND, and the NOT. The remaining two are the NOR and the NAND functions. These operations are performed by LOGIC GATES. A logic gate or element is a device or circuit which performs a pre-determined logic operation according to its logic function.

8.3.1 THE OR FUNCTION

This is one of the basic three logic operations and is designated by a plus sign +. Thus, $A + B$ will be read as A OR B. A two-input OR circuit and its corresponding logic gate symbol are shown in Figure 8.1. This circuit has an OR operation because the lamp will light if either switch A OR switch B OR if both are closed

A TRUTH TABLE which shows all the possible combinations of operations of switches A and B and the resultant state of the lamp is given in Table 8.7(a). The switches act as inputs and the lamp represents the output of the circuit. Using binary 0 and 1 for the states of the inputs and the output, the truth table for the OR function is as shown in Table 8.7(b).

Table 8.7

Switch *A*	Switch *B*	Lamp
open	open	OFF
open	closed	ON
closed	open	ON
closed	closed	ON

(*a*) Truth Table for OR circuit

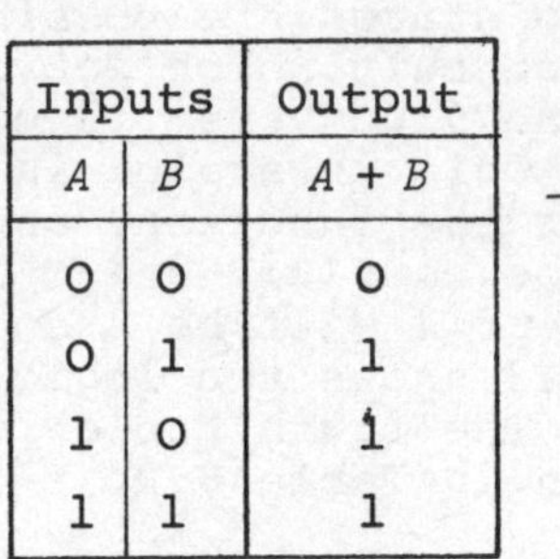

Inputs		Output
A	*B*	*A* + *B*
0	0	0
0	1	1
1	0	1
1	1	1

(*b*) Truth Table for OR function

(a) OR CIRCUIT

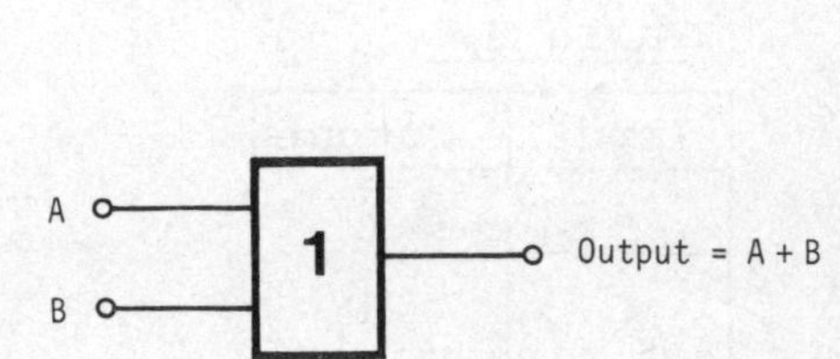

Fig.8.1 (b) OR GATE

8.3.2 THE AND FUNCTION

The AND function is designated by a dot. Thus, $A.B$ will be read as A AND B.

A two-input AND circuit and its corresponding logic gate symbol are shown in Figure 8.2. In this circuit the lamp will light only when both switches *A* AND *B* are closed. The truth tables for the operation of the AND circuit and its corresponding AND function are given in Table 8.8.

Fig.8.2

(a) AND CIRCUIT

Table 8.8

Switch *A*	Switch *B*	Lamp
open	open	OFF
open	closed	OFF
closed	open	OFF
closed	closed	ON

(*a*) Truth Table for AND circuit

Inputs		Output
A	*B*	(*A*.*B*)
0	0	0
0	1	0
1	0	0
1	1	1

(*b*) Truth Table for AND function

(b) AND GATE

8.3.3 THE NOT FUNCTION

In digital logic only two states are possible: logic-1 and logic-0. A circuit which converts a logic-1 to a logic-0 and vice versa is called an INVERTOR or *negator*. This inversion is signified by a "bar" over the function, and the function is said to be *complemented*. Thus, the invert or COMPLEMENT of A will be $\overline{A}$ and is read as NOT A. Figure 8.3 shows the logic symbol for an invertor. The truth table for the NOT function is given in Table 8.9.

Table 8.9

Input	Output
A	$\overline{A}$
0	1
1	0

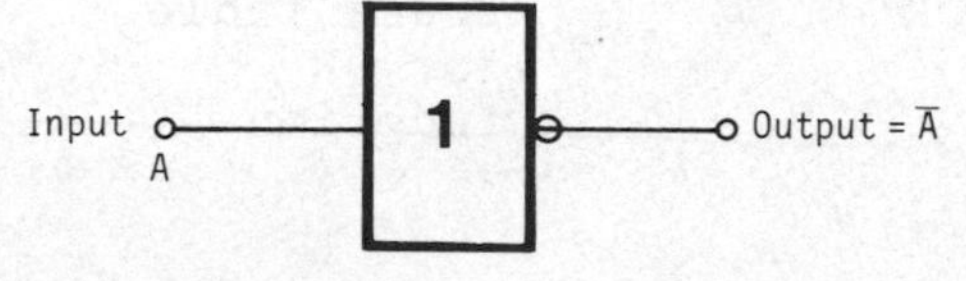

Fig.8.3 The NOT gate (inverter)

8.3.4 THE NOR FUNCTION

The NOR function may be considered as an OR function followed by a NOT function. Thus, as shown in Figure 8.4 a NOR gate performs the same operation as an OR gate followed by an invertor.

As illustrated in the truth table of Table 8.10, it is seen that the NOR operation is the logical inversion of the OR operation.

8.3.5 THE NAND FUNCTION

The NAND function is the abbreviation of NOT AND and may be considered as an AND function followed by a NOT function. Thus, as shown in Figure 8.5, the operation of a NAND gate is the same as that of an AND gate followed by an invertor.

Table 8.11 shows that the operation of a NAND gate is the logical inversion of the AND operation.

Table 8.10

Inputs		OR Output	NOR Output
A	B	$A + B$	$\overline{A + B}$
0	0	0	1
0	1	1	0
1	0	1	0
1	1	1	0

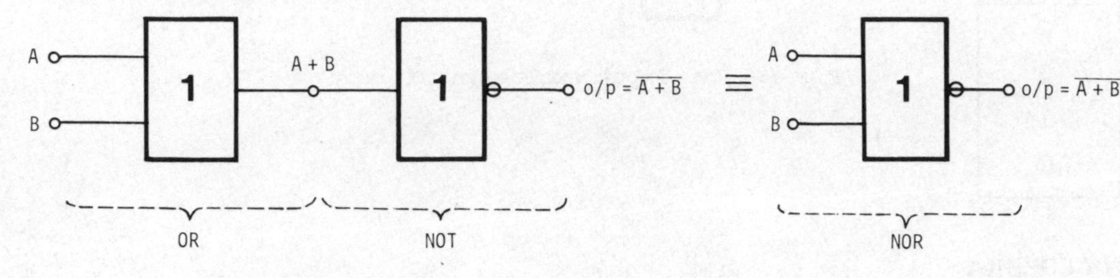

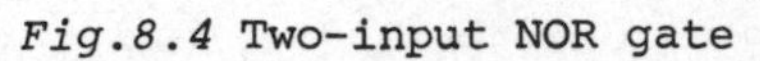
Fig.8.4 Two-input NOR gate

Table 8.11

Inputs		AND Output	NAND Output
A	B	$A.B$	$\overline{A.B}$
0	0	0	1
0	1	0	1
1	0	0	1
1	1	1	0

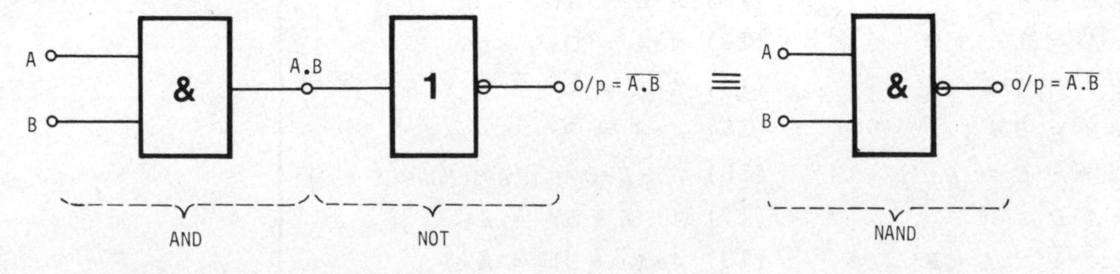

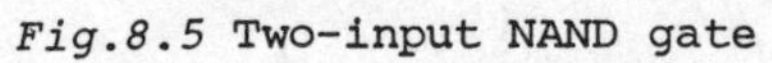
Fig.8.5 Two-input NAND gate

8.3.6 EXCLUSIVE OR

The exclusive OR function is a special case of the OR function. As illustrated in Table 8.12, it is seen that a logic-1 is generated at the output only when the two inputs are at different logic levels. The logic symbol for a two-input exclusive OR gate is shown in Figure 8.6.

Table 8.12

Inputs		Output
A	B	
0	0	0
0	1	1
1	0	1
1	1	0

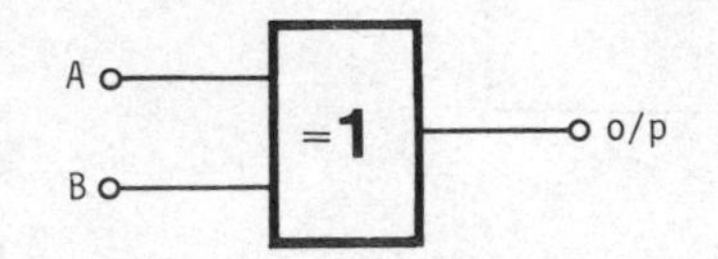

Fig.8.6 Two-input exclusive OR gate

8.4 BOOLEAN ALGEBRA

The basic theory of Boolean algebra was introduced by George Boole in 1854 and remained in the form of pure mathematics for almost a century. In 1938, Claude Shannon realised the potential of this "two-state" algebra and applied it to switching circuits. Boolean algebra has since become an essential tool in the design of digital systems.

Boolean algebra has many laws and theorems. These theorems, listed in the Table 8.13, are used to analyse and simplify logical statements and expressions.

Table 8.13

(1) $A+0 = A$	(10) $A \cdot 0 = 0$
(2) $A+1 = 1$	(11) $A \cdot 1 = A$
(3) $A+A = A$	(12) $A \cdot A = A$
(4) $A+\overline{A} = 1$	(13) $A \cdot \overline{A} = 0$
(5) $\overline{A+B} = \overline{A} \cdot \overline{B}$	(14) $\overline{A \cdot B} = \overline{A} + \overline{B}$
(6) $A+B = B+A$	(15) $A \cdot B = B \cdot A$
(7) $A \cdot B + A \cdot C = A \cdot (B+C)$	(16) $A + B \cdot C = (A+B) \cdot (A+C)$
(8) $A + A \cdot B = A$	(17) $A \cdot (A+B) = A$
(9) $A + \overline{A} \cdot B = A + B$	(18) $A \cdot (\overline{A}+B) = A \cdot B$
	(19) $\overline{\overline{A}} = A$

Theorems Number 5 and 14 are known as DE MORGANS' THEOREMS, and are of particular importance.

8.4.1 VERIFICATION OF THEOREMS BY VENN DIAGRAMS

A diagram which is drawn with overlapping circles, each circle representing one Boolean algebra variable, is called a Venn diagram. Often the circles are drawn within a rectangle which represents the Boolean algebra constant 1.

Venn diagrams may be used to prove logical identities or to simplify logic expressions. As an example, consider the identity

$$A\cdot\overline{B}+A\cdot B = A$$

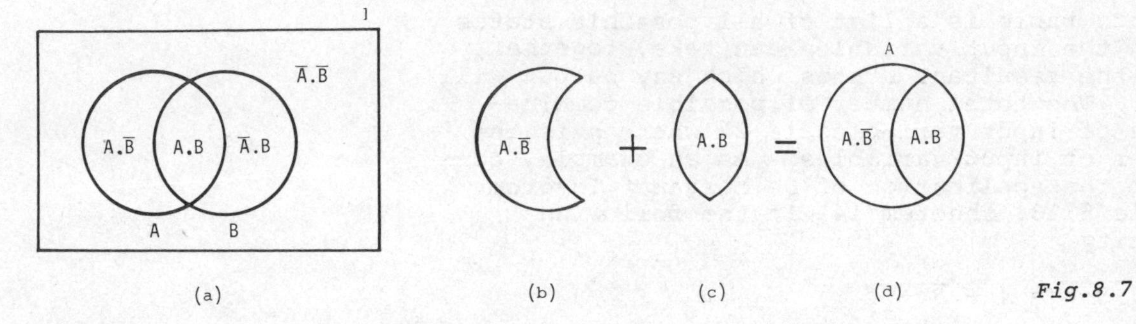

Fig.8.7

A Venn diagram is drawn with two circles, as shown in Figure 8.7(*a*). It is seen that $A\cdot\overline{B}$ is represented by the shape shown in Figure 8.7(*b*), and $A\cdot B$ is represented by the shape of Figure 8.7(*c*). The expression $A\cdot\overline{B}+A\cdot B$ is thus represented by the composite shape given in Figure 8.7(*d*), which can clearly be seen as A.

Example 8.4

Using a Venn diagram prove that $A\cdot B+A\cdot\overline{C} = A\cdot(B+\overline{C})$.

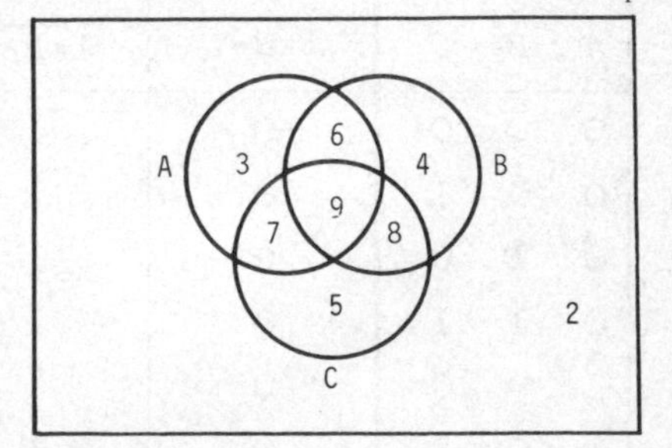

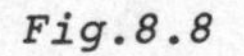

Fig.8.8

Solution

There are three variables involved, thus a three-circle diagram is drawn, and each discrete area is numbered as shown in Figure 8.8.

Variable A is composed of area 3,6,7,9.
Variable B is composed of area 4,6,8,9.
Variable $\overline{C}$ is composed of all areas outside C, thus: 2,3,4,6.

From Figure 8.8,

$$A\cdot B = (3,6,7,9)\cdot(4,6,8,9) = (6,9)$$

$$A\cdot\overline{C} = (3,6,7,9)\cdot(2,3,4,6) = (3,6)$$

Therefore

$$A \cdot B + A \cdot \overline{C} = (6,9) + (3,6) = (3,6,9) \text{ l.h.s.}$$

Further from Figure 8.8,

$$A = (3,6,7,9)$$

$$B + \overline{C} = (4,6,8,9) + (2,3,4,6) = (2,3,4,6,8,9)$$

Therefore

$$A \cdot (B + \overline{C}) = (3,6,7,9) \cdot (2,3,4,6,8,9)$$

$$= (3,6,9) \text{ r.h.s.}$$

It is proved that l.h.s. is the same as r.h.s.

8.4.2 VERIFICATION OF THEOREMS BY TRUTH TABLES

A truth table is a list of all possible states which the input variables can take, together with the resultant states which any output will take. The total number of possible combinations of input variables is 2^n where n is the number of input variables. As an example, consider the application of De Morgans' Theorem (Table 8.13, theorem 14) in the following identity

$$\overline{A \cdot B \cdot C} = \overline{A} + \overline{B} + \overline{C}$$

The truth table of Table 8.14 shows all the possible states of A, B, and C, and their corresponding complements $\overline{A}$, $\overline{B}$, and $\overline{C}$. Completing the table with the required functions, it is seen that Columns 5 and 9 are identical.

Table 8.14

A	B	C	$A \cdot B \cdot C$	$\overline{A \cdot B \cdot C}$	$\overline{A}$	$\overline{B}$	$\overline{C}$	$\overline{A} + \overline{B} + \overline{C}$
0	0	0	0	1	1	1	1	1
0	0	1	0	1	1	1	0	1
0	1	0	0	1	1	0	1	1
0	1	1	0	1	1	0	0	1
1	0	0	0	1	0	1	1	1
1	0	1	0	1	0	1	0	1
1	1	0	0	1	0	0	1	1
1	1	1	1	0	0	0	0	0

8.4.3 NAND/NOR IMPLEMENTATION

The use of De Morgans' Theorems shows that any of the five logic operations can be performed by different arrangements of either purely NAND elements or purely NOR elements. For manufacturers this is obviously desirable as bulk production of a single logic element reduces costs appreciably.

From De Morgans' Theorems:

$A \cdot B = \overline{\overline{A}} \cdot \overline{\overline{B}} = \overline{\overline{A} + \overline{B}}$ for AND operation

$A + B = \overline{\overline{A}} + \overline{\overline{B}} = \overline{\overline{A} \cdot \overline{B}}$ for OR operation

Figure 8.9 shows how different logic operations are performed using one type of element.

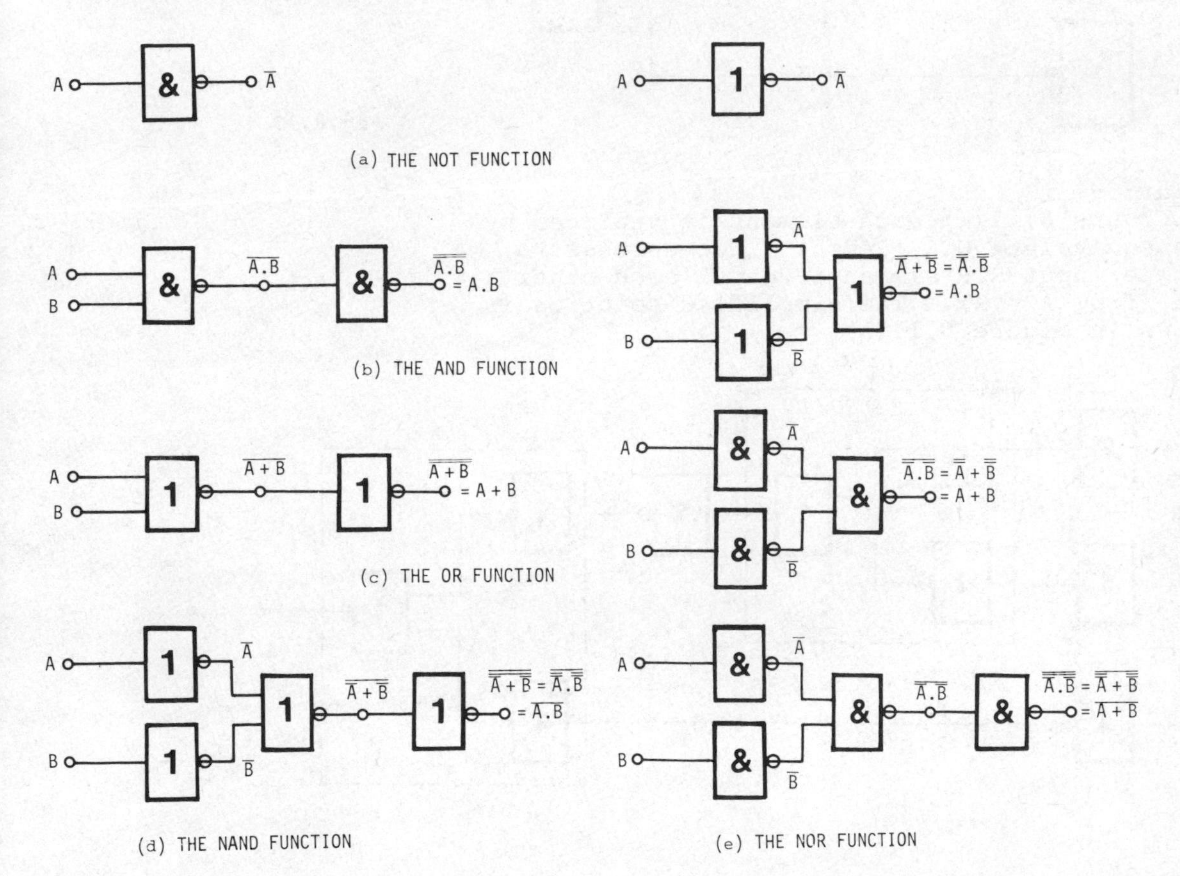

Fig.8.9

Example 8.5

Replace each element of Figure 8.10 *by its equivalent* NOR *element, and hence simplify as far as possible.*

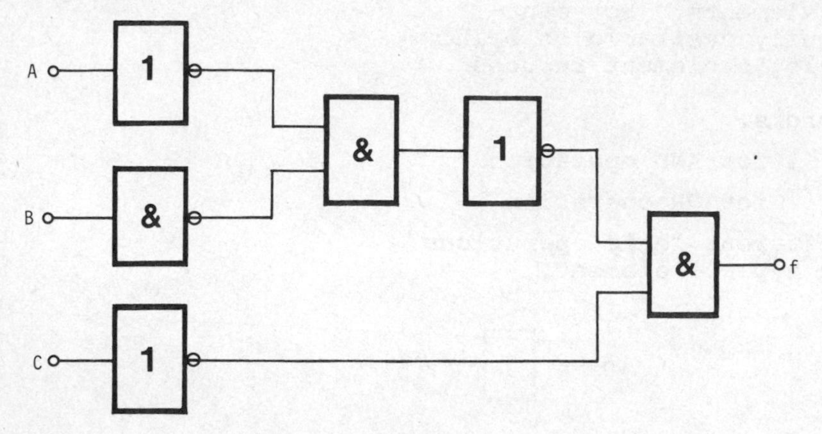

Fig.8.10

Solution

In Figure 8.11(*a*) each element is replaced by its equivalent NOR element. Two successive single input NOR elements cancel each other out (from $\overline{\overline{A}} = A$). The simplified solution is shown in Figure 8.11(*b*).

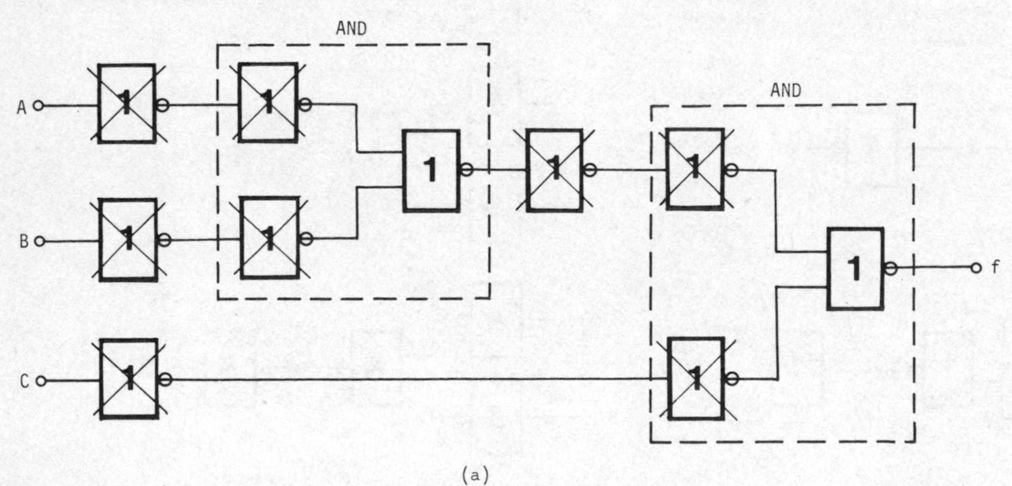

(a)

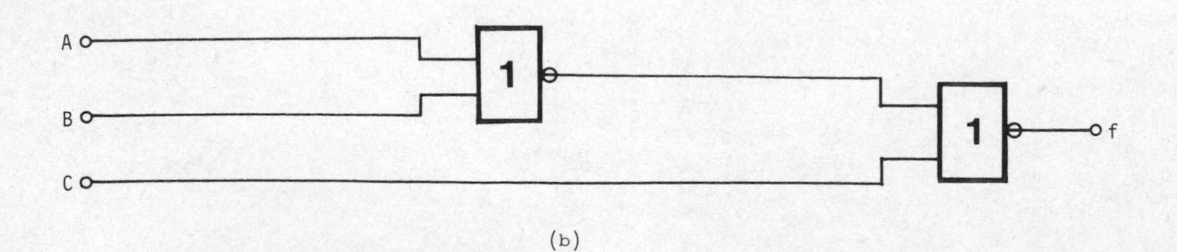

(b)

Fig.8.11

8.5 MINIMISATION OF LOGIC EXPRESSIONS

Depending on their complexity, logic expressions can be simplified by one of the following methods:

(1) Direct application of Boolean Algebra Theorems.
(2) Using mapping methods.
(3) Using Algorithmic Methods (this topic is beyond the scope of this book).

All three are based on Boolean algebra theorems.

8.5.1 LOGIC SIMPLIFICATION BY BOOLEAN ALGEBRA THEOREMS

The following examples illustrate this simple method which proves useful only when the number of variables is small and the expression is short.

Example 8.6

Simplify $A \cdot C + A \cdot \overline{C} + \overline{A} \cdot B + B \cdot \overline{B}$

Solution

$A \cdot C + A \cdot \overline{C} + \overline{A} \cdot B + 0$ by theorem 13

$A \cdot C + A \cdot \overline{C} + \overline{A} \cdot B$ by theorem 1

$A \cdot (C + \overline{C}) + \overline{A} \cdot B$ by theorem 7

$A \cdot (1) + \overline{A} \cdot B$ by theorem 4

$A + \overline{A} \cdot B$ by theorem 11

$A + B$ by theorem 9

Example 8.7

Simplify $(A + B + \overline{C}) \cdot (A + B + C)$

Solution
Removing the brackets:

$$A \cdot A + A \cdot B + A \cdot C + B \cdot A + B \cdot B + B \cdot C + \overline{C} \cdot A + \overline{C} \cdot B + \overline{C} \cdot C$$

then,

$$A + A \cdot B + A \cdot C + B \cdot A + B + B \cdot C + \overline{C} \cdot A + \overline{C} \cdot B + 0$$

$$A \cdot (1 + B + C + B + \overline{C}) + B \cdot (1 + C + \overline{C})$$

$$A \cdot (1 + B + 1) + B \cdot (1 + 1)$$

$$A \cdot (1) + B \cdot (1)$$

$$A + B$$

Example 8.8

Applying De Morgans' theorems, simplify

$$\overline{(\overline{A}+\overline{B}+\overline{C})\cdot(A+\overline{B}+\overline{C})\cdot(A+C)\cdot(B+\overline{C})}$$

Solution

Applying theorem 14, we will have

$$\overline{(\overline{A}+\overline{B}+\overline{C})} + \overline{(A+\overline{B}+\overline{C})} + \overline{(A+C)} + \overline{(B+\overline{C})}$$

Applying theorem 5, we have

$$\overline{\overline{A}}\cdot\overline{\overline{B}}\cdot\overline{\overline{C}} + \overline{A}\cdot\overline{\overline{B}}\cdot\overline{\overline{C}} + \overline{A}\cdot\overline{C} + \overline{B}\cdot\overline{\overline{C}}$$

$A\cdot B\cdot C + \overline{A}\cdot B\cdot C + \overline{A}\cdot\overline{C} + \overline{B}\cdot C$ by theorem 19

$B\cdot C\cdot(A+\overline{A}) + \overline{A}\cdot\overline{C} + \overline{B}\cdot C$ by theorem 7

$B\cdot C + \overline{A}\cdot\overline{C} + \overline{B}\cdot C$ by theorems 4 and 11

$C\cdot(B+\overline{B}) + \overline{A}\cdot\overline{C}$ by theorem 7

$C + \overline{A}\cdot\overline{C}$ by theorems 4 and 11

$C + \overline{A}$ by theorem 9

8.5.2 KARNAUGH MAPS

A very convenient and quick method of Boolean algebra simplification is the use of Karnaugh maps. This technique relies on visual judgment and skill, and is satisfactory for expressions with up to six variables.

Fig.8.12 Karnaugh map

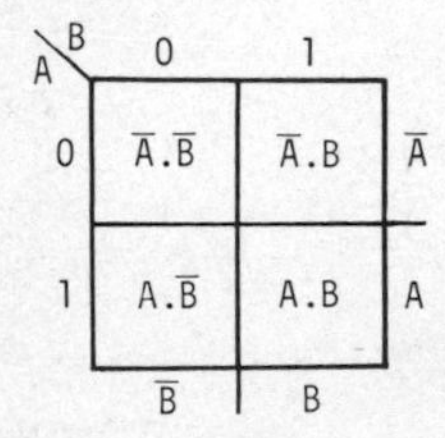

(a) TWO-VARIABLE MAP

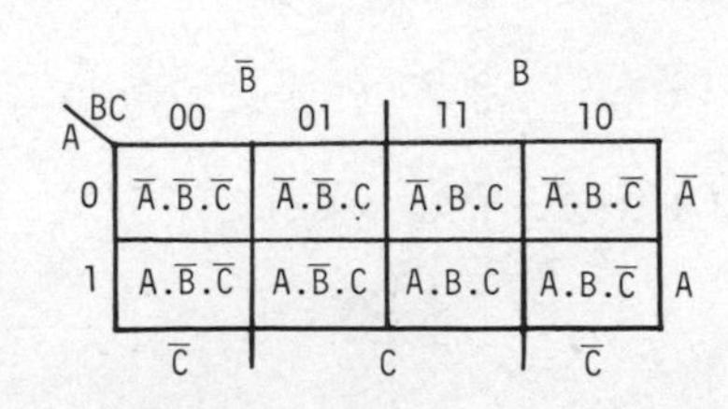

(b) THREE-VARIABLE MAP

AB \ CD	00 ($\overline{C}$)	01 ($\overline{C}$)	11 (C)	10 (C)	
00 ($\overline{A}$)	$\overline{A}.\overline{B}.\overline{C}.\overline{D}$	$\overline{A}.\overline{B}.\overline{C}.D$	$\overline{A}.\overline{B}.C.D$	$\overline{A}.\overline{B}.C.\overline{D}$	$\overline{B}$
01 ($\overline{A}$)	$\overline{A}.B.\overline{C}.\overline{D}$	$\overline{A}.B.\overline{C}.D$	$\overline{A}.B.C.D$	$\overline{A}.B.C.\overline{D}$	B
11 (A)	$A.B.\overline{C}.\overline{D}$	$A.B.\overline{C}.D$	$A.B.C.D$	$A.B.C.\overline{D}$	B
10 (A)	$A.\overline{B}.\overline{C}.\overline{D}$	$A.\overline{B}.\overline{C}.D$	$A.\overline{B}.C.D$	$A.\overline{B}.C.\overline{D}$	$\overline{B}$
	$\overline{D}$	D	D	$\overline{D}$	

(c) FOUR-VARIABLE MAP

A Karnaugh map consists of a square or rectangle divided into 2^n smaller squares, where n is the number of variables. Figure 8.12 shows a double, a treble, and a quadruple variable Karnaugh map. Each square is lettered by the expression it represents.

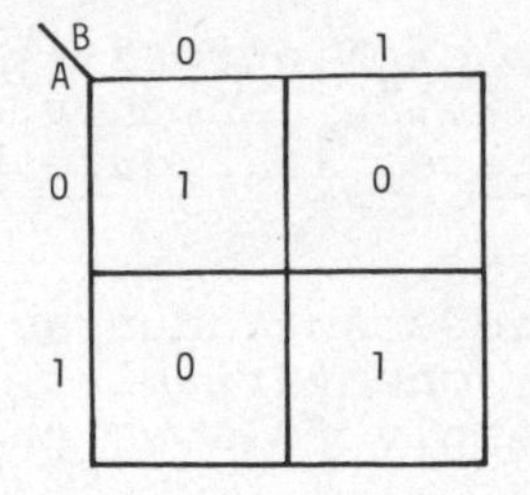

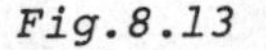
Fig.8.13

MAPPING

Before a Karnaugh map can be used to simplify a Boolean algebra expression, the expression must be mapped by inserting ones into the appropriate squares with zeros filling the remaining squares. As an example, consider the expression $f = \overline{A}\cdot\overline{B} + A\cdot B$. There are two variables and, when mapped, the Karnaugh map is as shown in Figure 8.13. Expression

$$f = A + A\cdot B + \overline{B}\cdot C\cdot D + \overline{A}\cdot B\cdot\overline{C}\cdot D$$

is mapped in the four-variable map of Figure 8.14.

AB \ CD	00	01	11	10
00	0	0	1	0
01	0	1	0	0
11	1	1	1	1
10	1	1	1	1

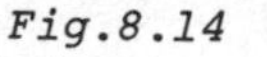
Fig.8.14

GROUPING

When the expression is mapped, the ones should be grouped together (although sometimes it is easier to group the zeros together) using the following simple rules:

(*a*) All 1s must be grouped at least *once*.
(*b*) The number of 1s falling in any group must be a power of 2, i.e. 1,2,4,8, etc.
(*c*) A grouped 1 can, if necessary, appear in more than one group.
(*d*) Only adjacent 1s can be grouped together.
(*e*) The larger the number of 1s in any one group, the better will be the minimization.
(*f*) The map should be considered as a sphere, with sides and corners being adjacent to each other.
(*g*) The number of groups formed should be kept to a bare minimum.

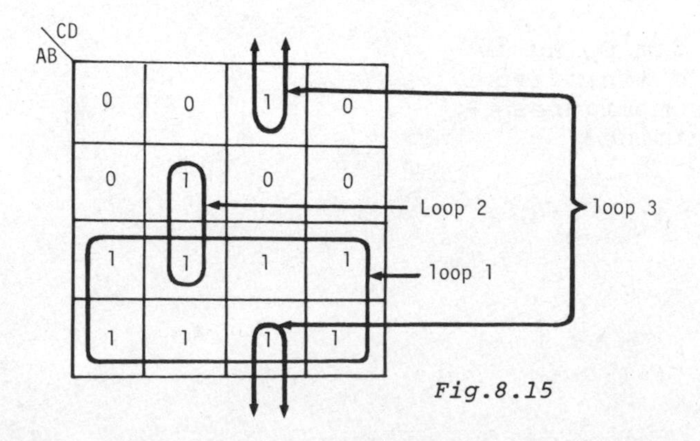

Fig.8.15

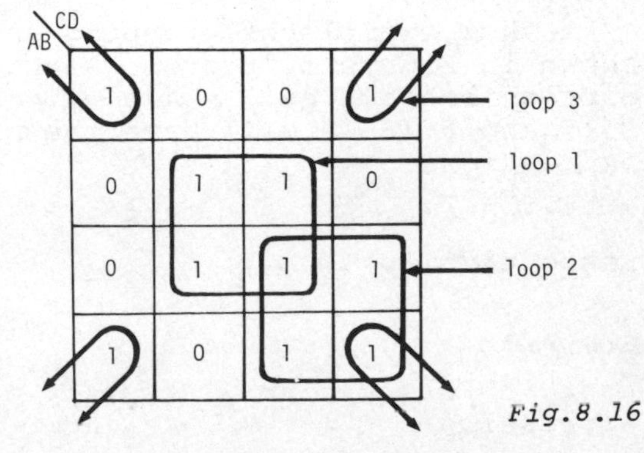

Fig.8.16

The map of Figure 8.14 is grouped as shown in Figure 8.15. Another example of grouping is illustrated in Figure 8.16.

DEMAPPING

Since adjacent squares on a Karnaugh map differ by only one variable, grouping two adjacent ones simply follows the following Boolean identity

$$A\cdot B + A\cdot\overline{B} = A\cdot(B+\overline{B})$$

$$= A\cdot 1$$

$$= A$$

This reveals that any variable which undergoes a change within the loop will be eliminated. A simple example is illustrated in Figure 8.17.

In loop 1, moving vertically, variable A changes from $\overline{A}$ to A and is thus eliminated. There is no horizontal movement and both variables B and C remain unchanged; this gives $B\cdot C$.

In loop 2, there is no vertical movement and hence A remains unchanged. Moving horizontally, we have $\overline{C}$ changing to C but $\overline{B}$ remains unchanged; this gives $A\cdot\overline{B}$.

The minimized expression from this map is therefore $f = B\cdot C + A\cdot\overline{B}$.

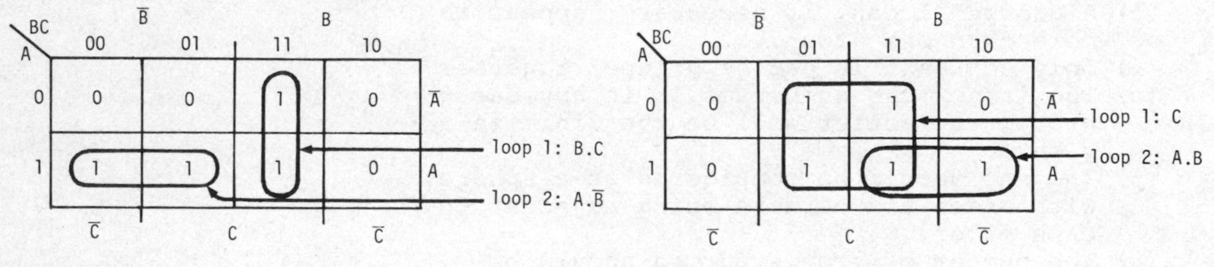

Fig.8.17

Fig.8.18

Another example of grouping and demapping is shown in Figure 8.18 from which the minimized expression will be $f = A\cdot B + C$. From Figures 8.15 and 8.16 we will have the minimized expressions

$$A + B\cdot\overline{C}\cdot D + \overline{B}\cdot C\cdot D \qquad \text{and} \qquad B\cdot D + A\cdot C + \overline{B}\cdot\overline{D}$$

respectively.

Example 8.9

Using a Karnaugh map, simplify the following expression:

$$f = \overline{A}\cdot B\cdot\overline{C}\cdot\overline{D} + \overline{A}\cdot\overline{B}\cdot C\cdot D + A\cdot\overline{B}\cdot\overline{C}\cdot D + \overline{A}\cdot B\cdot C\cdot\overline{D} + \overline{A}\cdot\overline{B}\cdot\overline{C}\cdot D + A\cdot B\cdot\overline{C}\cdot\overline{D}$$

Solution
The Karnaugh map is shown in Figure 8.19. The minimized expression is given by

$$f = B\cdot\overline{C}\cdot\overline{D} + \overline{A}\cdot\overline{B}\cdot D + \overline{A}\cdot B\cdot\overline{D} + \overline{B}\cdot\overline{C}\cdot D$$

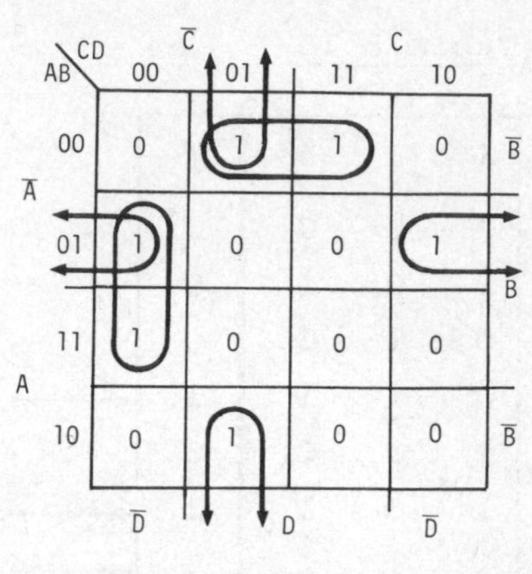

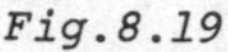
Fig.8.19

8.6 LOGIC IMPLEMENTATION

Logic expressions can be developed in several ways - from switching circuits, from truth tables, or from logical statements. These are illustrated in the following examples.

Example 8.10

Derive a logic expression from the relay switching circuit of Figure 8.20 *and minimize it as far as possible. Hence, draw the equivalent simplified logic diagram.*

Solution
From Figure 8.20

Path 1 gives $A\cdot\overline{B}$

Path 2 gives $A\cdot C$

Path 3 gives $A\cdot B$

Path 4 gives $\overline{A}\cdot\overline{C}$

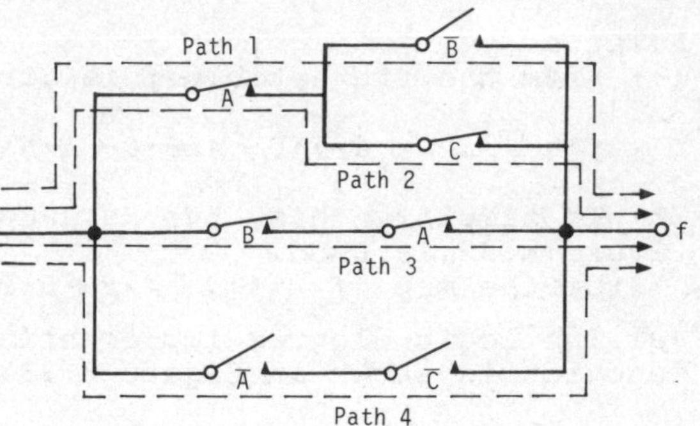

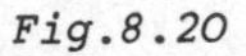
Fig.8.20

Output f is therefore given by

$$f = A\cdot\overline{B} + A\cdot C + A\cdot B + \overline{A}\cdot\overline{C}$$

$$= A\cdot(\overline{B} + C + B) + \overline{A}\cdot\overline{C}$$

$$= A\cdot(C + 1) + \overline{A}\cdot\overline{C}$$

$$= A + \overline{A}\cdot\overline{C}$$

$$= A + \overline{C} \quad \text{in minimized form.}$$

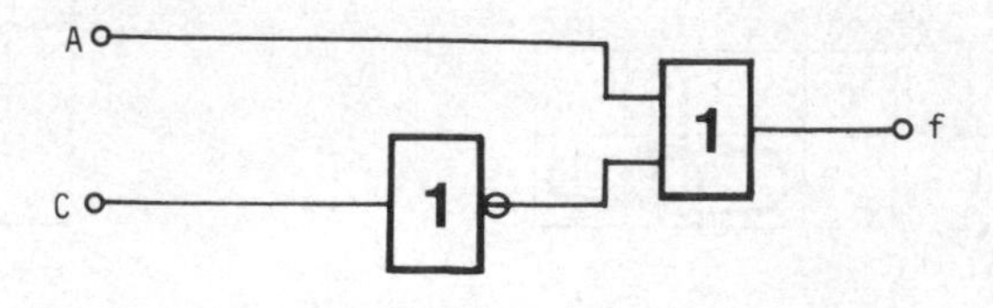

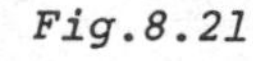
Fig.8.21

Figure 8.21 shows the logic diagram for this minimized expression.

Example 8.11

From the truth table of Table 8.15

(*a*) *Derive the logic expression for f.*
(*b*) *Simplify the expression as far as possible.*
(*c*) *Draw a logic diagram to implement this.*

Table 8.15

A	B	C	f	
0	0	0	0	
0	0	1	0	
0	1	0	0	
0	1	1	1	← $\overline{A}\cdot B\cdot C$
1	0	0	0	
1	0	1	1	← $A\cdot\overline{B}\cdot C$
1	1	0	1	← $A\cdot B\cdot\overline{C}$
1	1	1	1	← $A\cdot B\cdot C$

Solution
(*a*) From the truth table f is given by

$$f = \overline{A}\cdot B\cdot C + A\cdot\overline{B}\cdot C + A\cdot B\cdot\overline{C} + A\cdot B\cdot C$$

(*b*) To simplify this, the Karnaugh map of Figure 8.22 is used.
From the map $f = A\cdot C + A\cdot B + B\cdot C$

(*c*) The logic diagram implementing this function is shown in Figure 8.23.

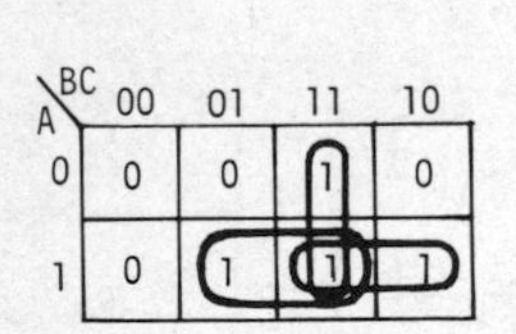

Fig.8.22

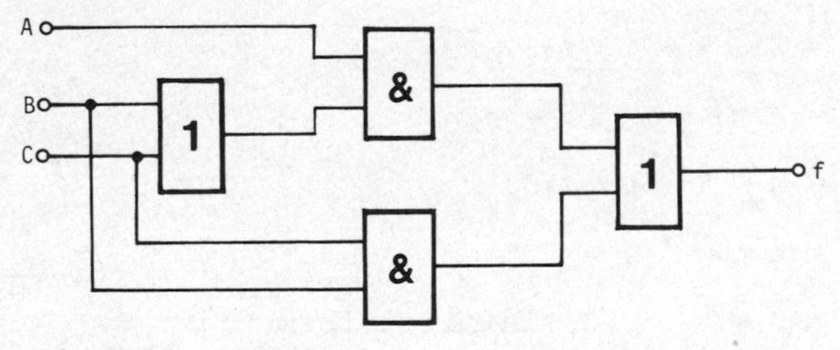

Fig.8.23

Example 8.12

A factory crane operator is to control Red and Green safety lights by four switches A, B, C and D. Design a simple logic system which will operate under the following conditions:

(*a*) *RED light on for*
(i) *switch A on, switch B off,* OR
(ii) *switch C on.*

(*b*) *GREEN light on for*
(i) *switches A and B on* AND
(ii) *switches C and D off.*

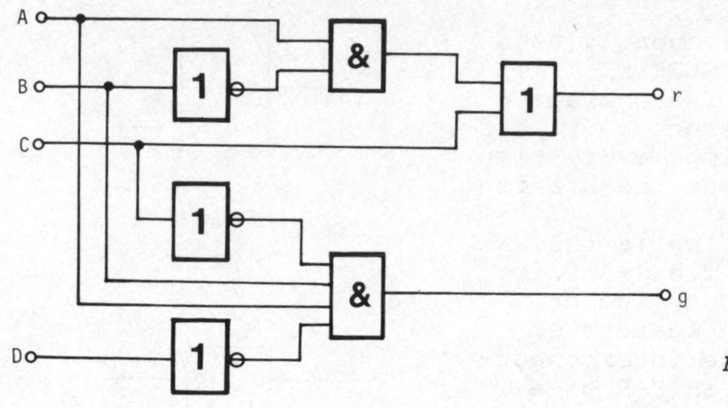

Fig.8.24

Solution
Let r be the letter representing the RED light on.
Let g be the letter representing the GREEN light on.

Then, logically speaking, we will have

$r = A \cdot \overline{B} + C$ and $g = A \cdot B \cdot \overline{C} \cdot \overline{D}$

The logic diagram implementing these is given in Figure 8.24.

8.7 DIGITAL MEMORY ELEMENTS

The simplest electronic digital memory element is the BISTABLE, commonly called the FLIP-FLOP. A flip-flop has two stable states, with the outputs at logic levels opposite to each other. The application of a control signal changes the states of the outputs over. There are several types of flip-flop, the simplest being the set-reset (S-R) flip-flop shown in Figure 8.25.

When $Q = 1$ and $\overline{Q} = 0$ the flip-flop is said to be in the SET condition

When $Q = 0$ and $\overline{Q} = 1$ it is said to be in the RESET condition.

When a logic-1 is applied to the S-input, and the R-input is at logic-0, output Q will switch to logic-1 regardless of its previous state. The flip-flop will now remain in this state (i.e. it memorizes the data given to it) even when the logic level at S is changed to logic-0. If a logic-1 is now applied to the R-input, the state of output Q will change to logic-0 ($\overline{Q}$ will go to logic-1) and will remain in this state until the flip-flop is once again set.

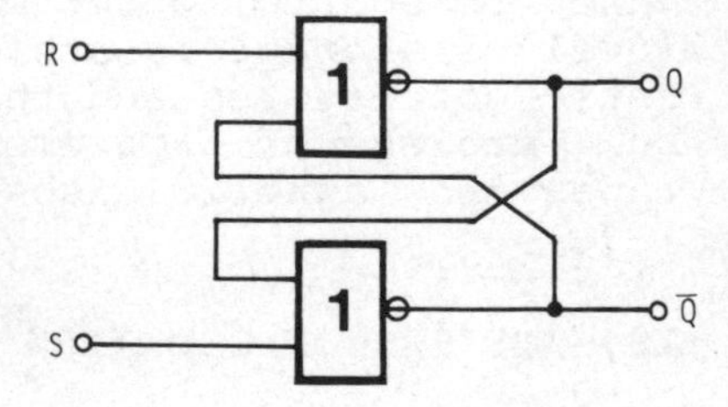

(a) S-R FLIP-FLOP USING NOR GATES

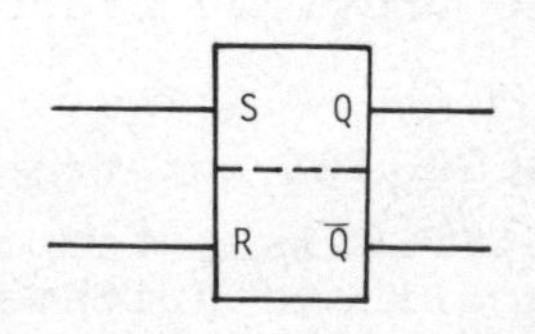

(b) CIRCUIT SYMBOL OF AN S-R FLIP FLOP

Fig.8.25

Another type of flip-flop, frequently used in digital counters, is the TRIGGER (T) flip-flop shown in Figure 8.26(*a*). The operation of this circuit is rather different as it has only one input. In a T flip-flop, every time a pulse is applied to the T-input the states of the outputs change.

The most commonly used flip-flop is the J-K type shown in Figure 8.26(*b*). The J-K flip-flop may have more than one input line at either J or K inputs. The main feature of this element is that by suitable interconnections it can be used as either an S-R or a T flip-flop.

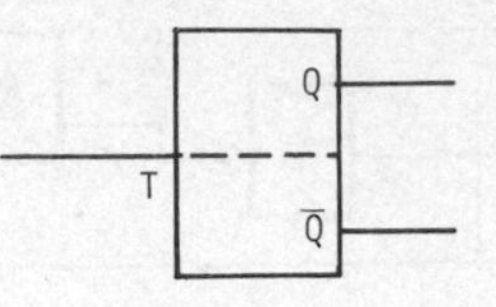

(a) T-TYPE FLIP-FLOP

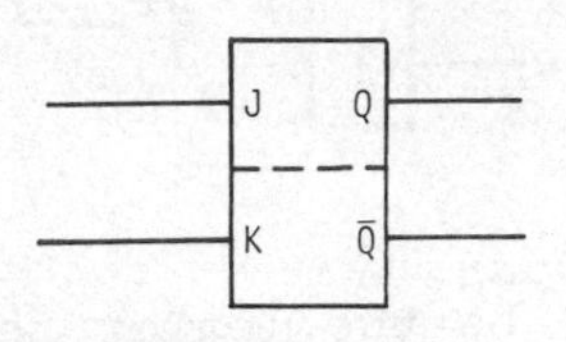

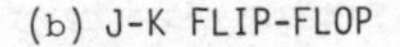
(b) J-K FLIP-FLOP

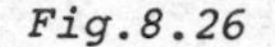
Fig.8.26

Flip-flops described above are limited in their capabilities and are normally used in a modified form. Extra internal circuitry is added to provide another input called the CLOCK INPUT. This facility, shown in Figure 8.27, allows direct control of one or a group of flip-flops operating collectively.

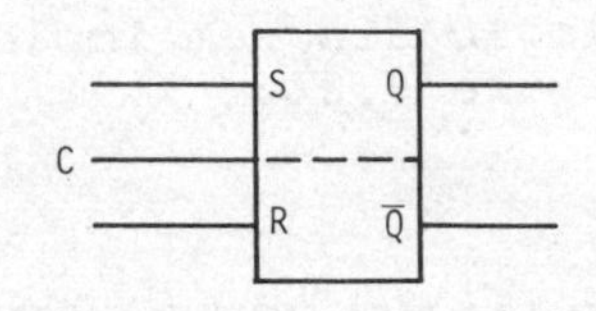

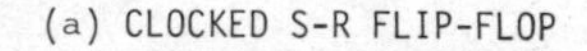
(a) CLOCKED S-R FLIP-FLOP

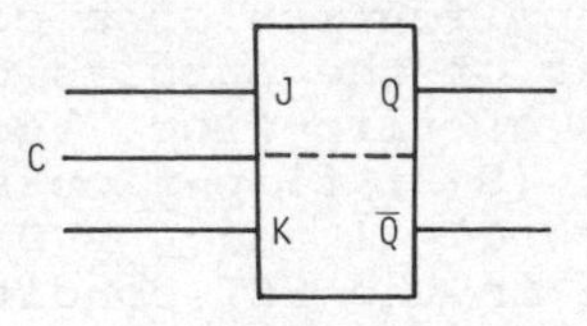

(b) CLOCKED J-K FLIP-FLOP

Fig.8.27

A simple modification of the S-R flip-flop, shown in Figure 8.28 is the DELAY (D) flip-flop. This element is widely used in counting circuits and digital read-out systems.

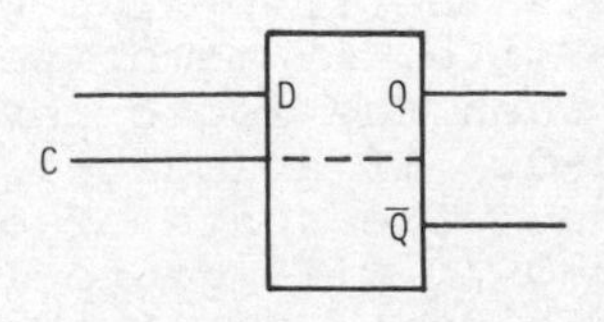

CLOCKED D-TYPE FLIP-FLOP

Fig.8.28

Flip-flops are very fast in operation but are very expensive for storing bulk information. They are generally used for storing information for very short periods and in most cases to control other circuits. The next level of storage is achieved by means of magnetic core-stores and thin film stores. For bulk storage of programs and data, magnetic tapes and magnetic discs are used.

8.8 REGISTERS AND COUNTERS

Registers and counters are a group of SEQUENTIAL logic elements commonly employed in digital systems. A sequential element is an element in which the present state of any output depends not only on present inputs but also on the previous states of inputs.

8.8.1 REGISTERS

A REGISTER is a temporary storage device used to facilitate arithmetical, logical, or transfer operations. Registers are built up from a number of flip-flops and are capable of storing related information for a desired but relatively short period. The time required to access the information contained in a register is very short and may be only a few nanoseconds.

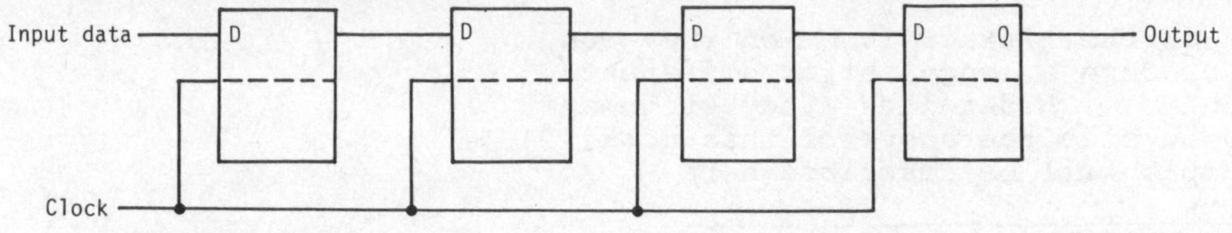

Fig.8.29 A simple shift register using D-type flip flops

The simplest but an important type of register is the SHIFT REGISTER which is a one-dimensional array of flip-flops (S-R, J-K, or D-type) arranged as shown in Figure 8.29. In such a register all flip-flops are clocked simultaneously and the application of each clock pulse causes the data to shift one place to the right (or left, depending on the connections). Thus to read the information out of a four-bit register, four clock pulses must be applied to the circuit. Shift registers are used in serial operations such as serial arithmetic where the arithmetic is carried out progressively on one pair of digits at a time.

Additional circuitry can be used to modify the mode of operation of a register. Registers commonly used in digital systems include the parallel-input type, parallel-output type and RING COUNTERS. A ring counter is a shift register whose input is derived directly from its output, thus forming a ring.

Registers are used in digital computers in

(1) the control unit to hold the program instruction while it is being executed and to store the address of the instruction reached in the program,
(2) the arithmetic unit to hold data while it has arithmetical operations performed on it.

Registers are also used to provide buffering between the central processor and the peripheral equipment, and between the internal store and the control unit.

8.8.2 COUNTERS

A particularly important group of sequential elements used in digital systems are the counter family. According to their mode of operation, counters are divided into two groups: SYNCHRONOUS or parallel and ASYNCHRONOUS or serial. Asynchronous counters are also known as *ripple-through* counters.

In the synchronous counter all states of the counter change simultaneously, while in the asynchronous counter there are small propagation delays between successive state changes.

Depending on their exact function they can be used to produce a long list of different digital counters. A detailed study of these elements is outside the scope of this book, but one example will be described here.

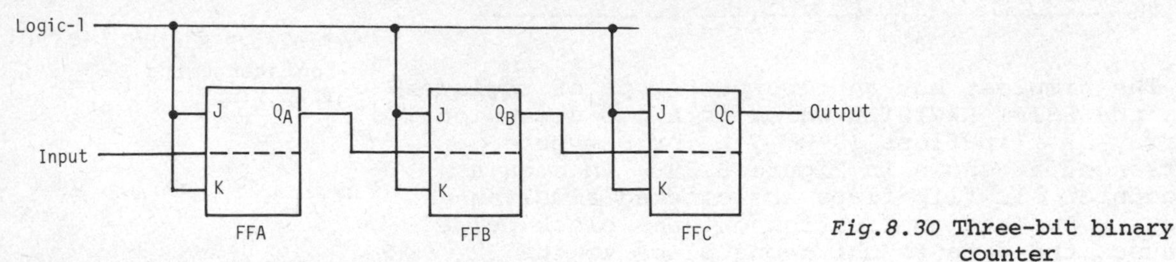

Fig.8.30 Three-bit binary counter

Consider the 3-bit counter of Figure 8.30 in which both the J and K inputs of all flip-flops are connected to a common logic-1 level. This ensures that the output of each flip-flop changes state when a pulse is applied to its input. When the counter is reset, the first input pulse to flip-flop A(FFA) causes it to set (Q_A goes to logic-1). The set output of FFA acts as the clock input to FFB. When the logic level applied to the clock input of FFB changes from 1 to 0, then FFB changes state. The set output of bistable B acts in a similar manner when applied as the clock input to FFC. Figure 8.31 shows the logic level state of the system at various time intervals, and Table 8.16 summarises the whole sequence of operation of the counter.

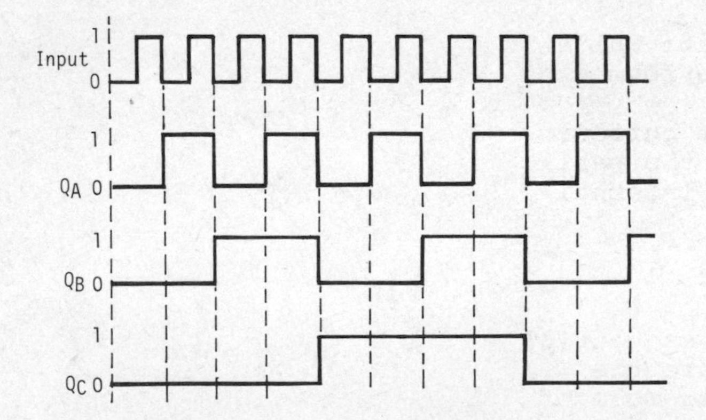

Fig.8.31

8.9 INTEGRATED CIRCUITS

With the advances made in the technology required to produce miniature discrete components, it has become possible to fabricate a complete circuit function in a single structure of a miniature form, in which the elements

Table 8.16

Input pulses	Present counter states			Next counter states		
	Q_C	Q_B	Q_A	Q_C^+	Q_B^+	Q_A^+
1	0	0	0	0	0	1
2	0	0	1	0	1	0
3	0	1	0	0	1	1
4	0	1	1	1	0	0
5	1	0	0	1	0	1
6	1	0	1	1	1	0
7	1	1	0	1	1	1
8	1	1	1	0	0	0

of the circuit cannot be physically divided without destroying the function of the circuit. Therefore integrated circuits are considered as elements which act as a system, whereas discrete components are considered as elements of a system. There are at present three forms of integrated circuits: (*a*) thin film circuits, (*b*) monolithic circuits, and (*c*) hybrid circuits.

(*a*) THIN FILM CIRCUITS

Thin film circuits have evolved from the technology used for producing high-quality discrete passive components such as resistors and capacitors. This type of integrated circuit is formed by depositing very thin films of conducting and non-conducting material upon an insulating base material called a substrate. The substrate provides structural strength for the circuit.

Passive components can be fabricated very easily in this type of integrated circuit. Resistors, for example, are formed from very narrow strips of a very thin film of material, the value of the resistor depending upon its physical dimensions. Capacitors are formed from thin layers of conducting material separated by a thin layer of insulating material. The value of the capacitor is dependent upon the physical size of its conducting plates and the thickness of the insulating thin film. Active components basically require to be welded onto the integrated circuit.

(*b*) MONOLITHIC CIRCUITS
Monolithic circuits have developed from the semiconductor technology used in producing discrete active components such as transistors. This type of integrated circuit uses diffusion processes whereby differently doped semiconducting materials are diffused into a base semiconducting substrate. As this is the normal process used in producing transistors and diodes, such components can be fabricated easily. Passive components are normally simulated by using active components formed in the integrated circuit in particular modes of operation, because otherwise the passive components would use excessive areas of integrated circuit in their fabrication. Resistors, for example, can be simulated by using active components such as field-effect transistors and using the pinch effect of the gate potential to fix the resistance between the drain and source of the field-effect device. Capacitors can be simulated by using the capacitive effect formed by the depletion region of reverse biased diodes.

(*c*) HYBRID CIRCUITS
Hybrid circuits combine the passive component fabrication advantage of thin film circuits together with the active component fabrication advantage of monolithic circuits. Production of hybrid circuits is done by adding thin film passive elements onto the surface of monolithic circuits containing the required active elements.

8.10 THE DIGITAL COMPUTER

A digital computer is an electronic machine capable of manipulating symbols, carrying out meaningful operations, and performing large amounts of arithmetical calculations at exceptionally high speeds and with high degrees of accuracy. Other special features of digital computers include long-term storage property and decision-making capability.

The digital computer system is a structured collection of software and hardware components, all contributing to a common objective. The SOFTWARE refers to all the information, operations and non-physical components necessary for effective use of the hardware. The HARDWARE consists of all the physical parts, the mechanical and electronic components built into the machine.

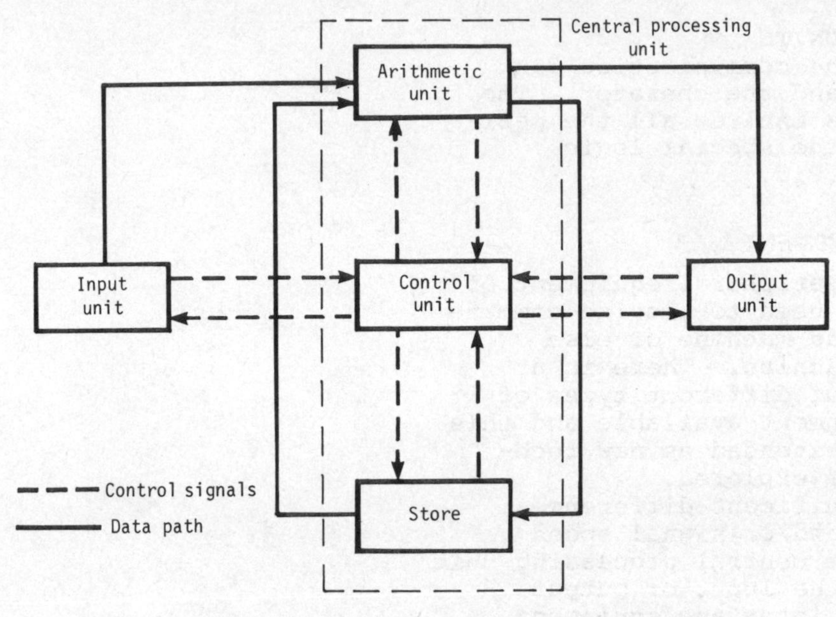

Fig.8.32 Block diagram of a simple digital computer

In its simplest form, the block diagram of a digital computer is shown in Figure 8.32. The system consists of five units, a group of which are collectively known as the *central processing unit* (C.P.U.).

THE ARITHMETIC UNIT
The arithmetic unit is used to perform arithmetical, logical or transfer operations on data.

THE STORE
The store is used to hold program instructions and the data on which operations are to be carried out. The store has two parts, the main store and the backing store. In each location of the main store, one or more program instructions or items of data are usually held. Each location or cell has allocated to it a unique number called its *address*. The backing store consists of storage units capable of holding very large quantities of information not constantly required.

THE CONTROL UNIT
The control unit controls the operation of all sections of the computer and ensures that operations are performed at the right time and that all program instructions are executed in the correct sequence.

THE INPUT AND OUTPUT UNITS
These units provide the communication link between the computer and the operator. The input and output units include all the peripheral equipment and the special logic circuits within the C.P.U.

8.10.1 THE PERIPHERAL EQUIPMENT

The input and output peripheral equipment of a digital computer is used to provide communication between man and machine or else between machine and machine. There is at present a vast range of different types of input and output equipment available and this range is continually extended as new techniques and devices are explored.

Often there are significant differences between signal level and/or signal speed encountered within the central processing unit *and* with that within the input or output equipment. Hence the interface equipment between the central processing unit and the input/output equipment often includes buffer stores and/or signal level converters. Input/output equipment may be either ON-LINE, that is directly connected via the interface to the central processing unit, or OFF-LINE, where there is no direct connection to the central processing unit.

In large-scale systems the organisation of the input/output transfers assumes such importance that a computer may be assigned solely to control these input/output transfers, thereby significantly decreasing the overall input/output transfer time.

Peripheral equipment commonly used in modern computer systems includes
punched paper-tape machines,
punched card machines,
line printers,
visual display units (V.D.U.)
character recognition devices.

PUNCHED PAPER-TAPE equipment is relatively cheap, easy to use, may provide input and output, and can be used for off-line paper-tape production or paper-tape interpretation. Because of these advantages punched paper-tape equipment is amongst the most commonly used digital computer input/output equipment.

PUNCHED CARD equipment has been used in conjunction with digital computers since the earliest days of computer technology. There is a whole range of punched card equipment such as sorters, verifiers, readers, punches, etc.; however only the readers and punches are used directly on-line.

LINE PRINTERS are one of the most important digital computer output equipment since they are used to produce rapid printed alphanumeric hard copy. There are two important types of on-line printers, the drum printer and the chain printer.

The VISUAL DISPLAY UNIT is a recent addition to the list of input/output devices available for use with a digital computer. The display unit can be used as both an input and an output device and it offers certain advantages over other input/output devices that make its use in computer aided design (C.A.D.) and interactive work highly desirable.

CHARACTER RECOGNITION devices may be used as input devices in a computer system. There are two main types available: the optical character recognition device (O.C.R.) and the magnetic ink character recognition device (M.I.C.R.). Character recognition devices offer the important advantage of directly interpreting the same documents as interpreted by the human user.

8.10.2 STORAGE EQUIPMENT

An essential property of any digital computer is the property of storage. Storage equipment used by digital computers varies considerably in capacity offered, speed of access, physical size, and cost. Systems designers use this wide variety to attempt to obtain an optimum design as regards access speed, storage capacity, and cost.

Storage equipment commonly used in conjunction with digital computers includes the

magnetic tape equipment,
magnetic disc stores,
magnetic drum stores,
ferrite core stores,
magnetic thin film stores.

MAGNETIC TAPE equipment is used to provide large capacity storage at a reasonable cost. Magnetic tape equipment is also used to provide very rapid data transfers from and to the central processing unit.

MAGNETIC DISC stores may be used as either an extension to the main memory or to provide bulk storage. Magnetic disc stores offer large-volume storage with an access time considerably shorter than that required for a magnetic tape store; however magnetic disc stores are rather more expensive than magnetic tape stores.

The MAGNETIC DRUM is also used as either an extension of the main storage or else a bulk storage device. Magnetic drum equipment is however tending to be replaced by other direct access devices such as magnetic discs.

The main store of the C.P.U. is required to be fast and should offer high capacity with random access. The FERRITE CORE store offers these facilities and is therefore used extensively to provide the main storage within the C.P.U.

8.10.3 FUTURE TRENDS

During the past 20 years, development of new ideas and techniques in computer technology has been so rapid and continuous that often production has not kept pace with new developments.

The most recent development has been in the application of superconductivity and the use of laser. Based on the work of British physicist Brian Josephson, the theory of superconductivity has been developed further, and experimental circuits that promise a new generation of smaller and faster computers are now being built in the laboratories of several major computer manufacturers. Built of Josephson junctions, these superconducting memory elements and logic devices are potential replacements for the transistorized semiconductor switches at the heart of present-day computer technology. The superconductor circuits can be refrigerated at temperatures approaching absolute zero, thus eliminating the heat problem that currently limits the speed of even the fastest computers. Since the superconductor circuits can be packed more densely than semiconductors, they will operate much faster because the distance a transmission signal has to travel is reduced. Although the theory of superconductivity has been known for years, it is only recently that the problem of building a microscopic insulation layer for such switches has been solved in the laboratory.

8.11 DIGITAL COMPUTER PROGRAMMING BASICS

A PROGRAM is a series of instructions or statements in a form acceptable to a computer. There are very many digital computer programming languages being used throughout the world today. Some of these programming languages, however, have greater prominence than others because of their widespread use. Probably the four most widely used programming languages are ALGOL, COBOL, FORTRAN and PL/1, each having particular areas of application. Although these programming languages are normally used by computer programmers, the computer is unable to operate directly upon programs written in these languages because a computer can only operate directly in its own *machine code language*. Machine code languages are referred to as *low-level languages* whereas languages such as ALGOL, COBOL, and FORTRAN are referred to as *high-level languages*.

MACHINE CODE

This is a low-level programming language that uses numeric symbols, often binary symbols only, to specify the required addresses and operation codes. All the instructions used in these programs are basic low-level instructions; it is because of this and the necessity of remembering the cumbersome numeric codes and addresses that machine code is difficult to learn and to code. The set of operation codes is normally restricted to a *single* machine type which therefore makes machine code rather inflexible. However, programs written in machine code are very efficient as the machine code operation codes are designed with the hardware of the particular computer in view. The machine code program can be run directly in the computer as the operation codes are in simple basic numeric form and do not require any translation.

MACHINE-ORIENTED LANGUAGE

This is often referred to as assembly language and is easier to learn and to code than machine code because the operation codes used are in mnemonic form. This form of low-level programming language has the advantage over machine code in that it is not restricted to use on a single computer type but normally on a *range* of computer machines. However, programs written in this form of programming language require the use of an *assembler machine* to translate the *source program* (written in machine-oriented language) into an *object program* (written in machine code) before the program can be run on the computer. The object program formed in this manner is very nearly as efficient as the equivalent program written directly in machine code as most of the translations performed by the assembler machine are one-to-one translations.

PROCEDURE-ORIENTED LANGUAGE
This is easier to learn and to code than either machine code or machine-oriented language because this type of language is more problem- or procedure-oriented than machine-oriented; that is, it is a high-level language designed with procedures and problems in mind rather than the hardware of the computer. The operation code set comprises words that are in common use in the English language and often a single operation code is equivalent to a number of machine code operation codes. Because of this, a *compiler machine* is necessary to convert the source program into an equivalent machine-oriented language by using one-to-many translations.

Before the program can be run on a computer, the object program is produced by using an assembler machine on the machine-oriented language program produced by the compiler. Procedure-oriented language programs are designed to be very flexible as these programs can be used on a great variety of different machines. The object program, however, is usually not as efficient as an object program produced by machine-oriented languages or a machine code program. However, procedure-oriented languages are very widely used as they are easy to learn and code, and it is these factors taken together which produce an overall reduction of problem solving time.

ALGOL
The name Algol is an acronym formed from ALGOrithmic Language. It is a high-level procedure-oriented language. The source program produced provides a means of defining algorithms, i.e. step-by-step procedures for general mathematical or scientific use. The instructions resemble algebraic formulae and English sentences.

FORTRAN
The name Fortran is formed from FORmula TRANslation; it is a high-level procedure-oriented language used universally. The source program is formed from combinations of algebraic formulae and English statements grouped together in a readable form. This programming language covers the same area of work as that of Algol, i.e. general mathematical and scientific problems.

COBOL
This is formed from COmmon Business Oriented Language. It is a high-level procedure-oriented language. The source program uses statements which are formed from readable English statements and it is used for general commercial problems.

PL/1
This is an acronym formed from Programming Language 1; it is a high-level procedure-oriented language. This programming language was designed to cover both the commercial and scientific areas of work by combining the problem-solving capabilities of algorithmic languages such as Algol with the data-handling capabilities of commercial languages such as Cobol.

BASIC
Basic is also a high-level scientific language. It is a "conversational" language used with an on-line computer.

8.11.1 FLOW CHARTS

A program flowchart is a pictorial representation of the solution of a problem. It is the best basis from which to form a program. Flowcharts are normally used as an in-between step, isolating the definition of the problem from the actual program itself. The program flowchart enables an overall grasp of the problem to be obtained before the actual program is written. So that program flowcharts can be widely used, standard symbols are normally employed. A set of such program flowchart symbols is shown in Figure 8.33. The use of flowcharts is illustrated in Examples 8.13 and 8.14.

FLOW LINE: a line showing the flow of control from one element to another element in a flowchart

TERMINAL ELEMENT: an element representing a terminal point in a flowchart (e.g. start, stop, etc.)

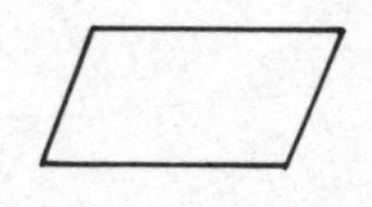

INPUT/OUTPUT ELEMENT: an element representing an input or output point in a flowchart (e.g. read, print)

PROCESS ELEMENT: an element representing a point in the flowchart where a process is carried out (e.g. multiply, shift, etc.)

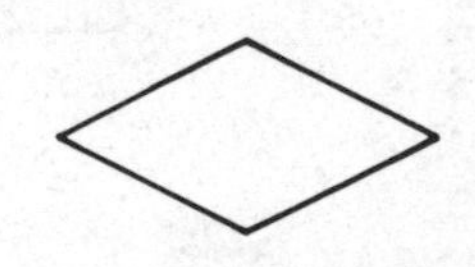

DECISION ELEMENT: an element representing a point in the flowchart where a choice between at least two alternative paths may be made

CONNECTOR: an element representing a join or common point of at least two paths

Fig.8.33
Flowchart symbols

Example 8.13

A number of people took part in a survey in connection with a proposed car park to serve a shopping centre. They were each asked three questions:

(a) Do you drive a car?
(b) Do you shop on Saturday?
(c) Do you use a car for shopping?

The resulting data have to be processed so that anyone answering yes to any two questions is put on the TOTAL *file. Anyone answering yes to (b) and yes to (c) is put on the* CAPACITY *file. Anyone answering yes to (b) and no to (c) is put on the* PEDESTRIAN *file.*

Draw a flowchart of a program to sort these data onto the correct files.

Solution

A suitable flowchart is shown in Figure 8.34.

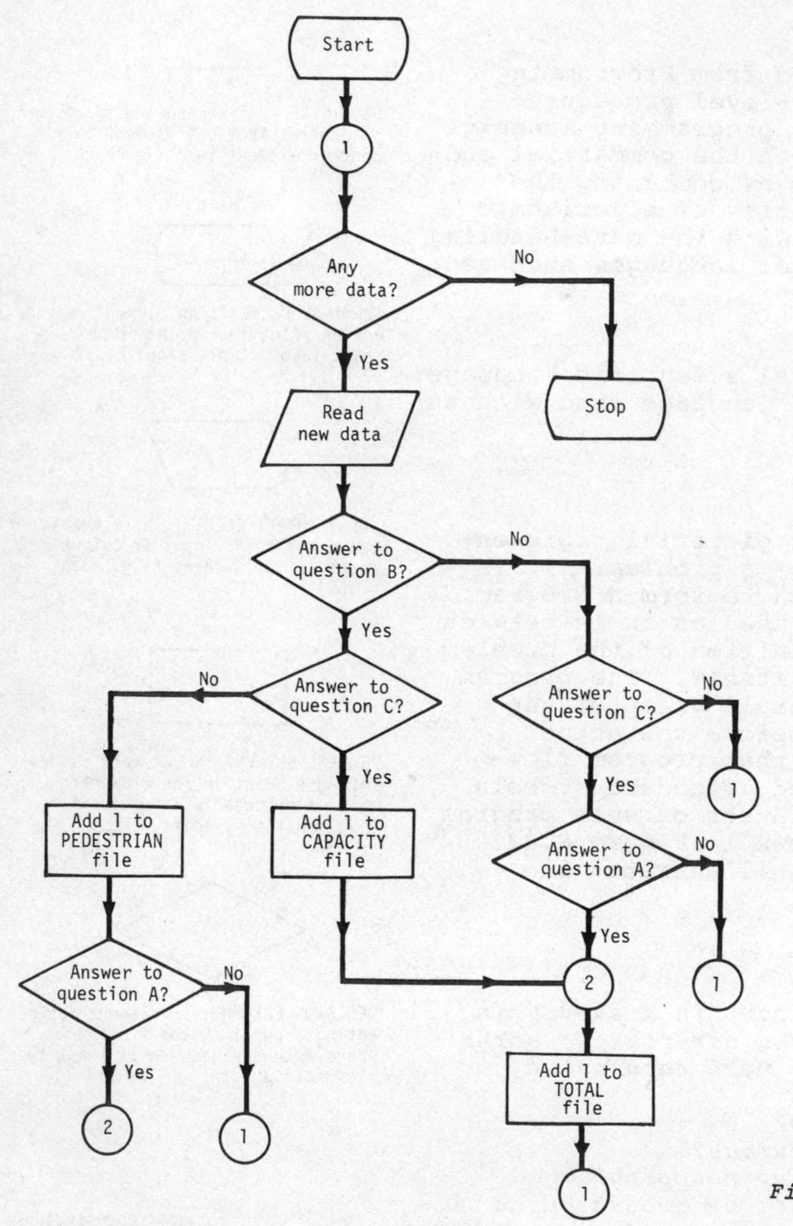

Fig.8.34

Example 8.14

Draw the flowchart of a program to sort three integers, A, B and C into ascending order. A can be found in store location 200, B in location 201, and C in location 202.

Solution

A flowchart for sorting the three integers into ascending order with the largest value in location 202 and the smallest in location 200 is given in Figure 8.35.

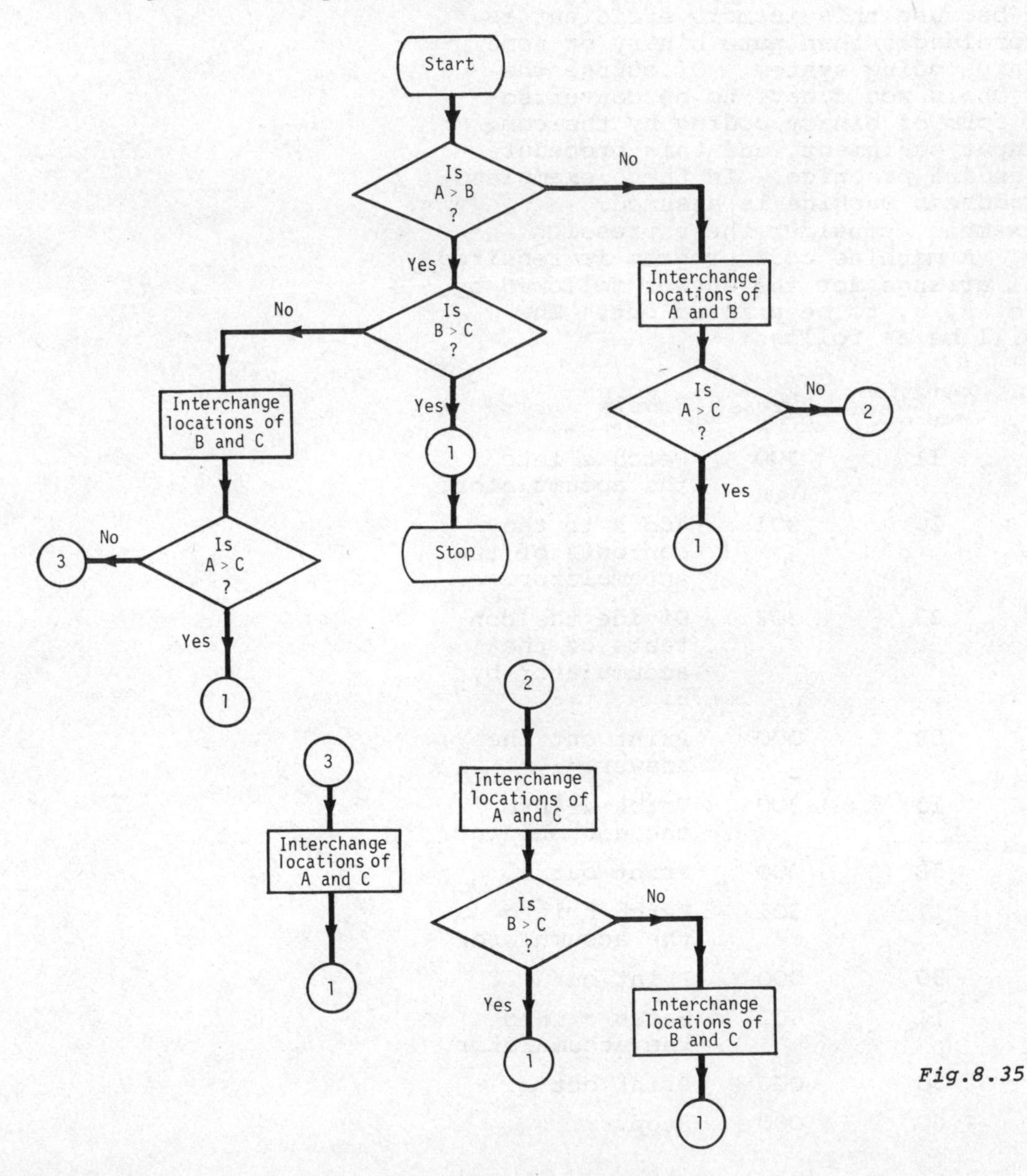

Fig.8.35

8.11.2 MACHINE CODE PROGRAMMING

As mentioned before, machine code programs are programs that use numeric symbols only to represent the required address and operation codes. Machine code is the lowest-level programming language possible and can be operated on directly by the computer without the aid of an assembler machine or a translator machine.

Three very simple but typical examples of machine code operation codes are illustrated below. The decimal number system has been used here because this is more efficient as regards wordlength than pure binary or some other binary coding system. Of course the decimal symbols would have to be converted into some form of binary coding by the computer's input equipment, and this procedure is now standard practice. In these examples a single address machine is assumed.

As an example, consider the expression $(a+b) \div c$. A machine code program is required which will arrange for the answer followed by the data a, b, c, to be printed out. The program will be as follows:

Instruction Location	*Operation Code*	*Address*	*Remarks*
00	11	300	Fetch a into the accumulator.
01	20	301	Add b to the contents of the accumulator.
02	23	302	Divide the contents of the accumulator by c.
03	50	000	Print out the answer.
04	11	300	Fetch a into the accumulator.
05	50	000	Print out a.
06	11	301	Fetch b into the accumulator.
07	50	000	Print out b.
08	11	302	Fetch c into the accumulator.
09	50	000	Print out c.
10	60	000	Stop.

In this program the first column gives the location of the instruction, the second column is taken from a predetermined set of operational codes, and the third column indicates the allocated address.

Example 8.15

Write a program using a simple machine code to perform the following calculation:

$x = 4a^2 + 2b$

where a is a binary number in location 101 *and b is a binary number in location* 102. *Place the resulting binary number x into location* 103.

Solution

Instruction Location	*Operation Code*	*Address*	*Remarks*
00	11	101	a into accumulator.
01	22	101	a^2 in accumulator.
02	22	300	$2a^2$ in accumulator.
03	20	102	$2a^2 + b$ in accumulator.
04	22	300	$2(2a^2 + b)$ in accumulator.
05	12	103	Result into location 103.
06	60	000	Stop.

In this example, data parameter 2 would have to be loaded into the computer via a tape before the program could be run.

Example 8.16

In a machine code of your own choice, write a program to perform the following operation

$a + b^2 - c^3$

where a, b and c can be found in computer store locations 100, 101 *and* 102 *respectively. Store the result in location* 200, *and if the result is negative, arrange to increase by one the number stored in location* 500. *Give a key to your code.*

Solution

Instruction Location	*Operation Code*	*Address*	*Remarks*
00	11	102	c into accumulator.
01	22	102	c^2 in accumulator.
02	22	102	c^3 in accumulator.
03	12	400	c^3 into temporary store 400.
04	11	101	b into accumulator.
05	22	101	b^2 in accumulator.
06	20	100	$a+b^2$ in accumulator.
07	21	400	$a+b^2-c^3$ in accumulator.
08	12	200	Result in location 200.
09	32	011	Jump to stop if contents of accumulator are positive.
10	01	500	Increment contents of location 500 by 1.
11	60	000	Stop.

No data tape needs to be loaded before running this program.

8.11.3 ALGOL

As previously mentioned, Algol is a procedure-oriented language and therefore is very easy to use for writing programs compared with machine-oriented languages, particularly machine code. Some of the terms and arithmetic procedures used in Algol are:
(*a*) IDENTIFIERS are names given to storage locations used in an Algol program. Identifiers are basically formed from any combination of English letters and denary digits, with the restriction that the first symbol used must be a letter. Typical examples would be: a; Sum 1; a2b.

(*b*) DECLARATIONS are statements used to define what type of format will be held in particular identifiers, i.e. identifiers must be declared before they are used. Typical examples are: <u>integer</u> a; <u>real</u> Sum 1, a2b.
(*c*) ASSIGNMENT STATEMENTS are statements which give values to variables and a typical example is x:=x+1. This means that x takes up a new value made up of the old value of x plus 1.
(*d*) STATEMENT BRACKETS are symbols or particular words used to group together and isolate particular statements from other statements. The underlined words begin and end are normally used as statement brackets, as for example:

```
begin integer a;
      read a;
      a:=a+1;
end;
```

Often statements appear within statements and the following format is then used:

```
begin
     begin
     end;
end;
```

(*e*) Numbers are split up into two forms: real and integer. A REAL NUMBER in Algol is a number containing an integral and a fractional part, for example 39.071. An INTEGER NUMBER in Algol is a number that contains an integral part only, for example 123 or -4567.

There are six basic arithmetic operations used in Algol: addition; subtraction; multiplication; division; integer division; and exponentiation.

The *addition* sign is + and can be used with real numbers and/or integer numbers.

The *subtraction* sign is - and can be associated with real numbers and/or integer numbers.

The *multiplication* sign is × although * is often used to overcome confusion with the letter x. Multiplication is defined for use with real numbers and/or integer numbers.

The *division* sign is / which is general division and can be used in association with real numbers and/or integer numbers.

The *integer division* sign is ÷ and is reserved for use with integer numbers; if the result looks like being a real number (e.g. $5 \div 2 = 2.5$), then the result becomes the next lower integer (e.g. 2).

The *exponentiation* sign is ↑ and is used for raising a real number and/or integer number to a given power, for example 2↑3 means 2 cubed.

A major difficulty arising with Algol is that no input-output statements are defined; the reason for this is that the hardware (as regards input and output) varies so much from computer to computer that it is not possible to define a set of simple input-output statements that would be universally accepted.

The input-output statements used by the Elliott company are used in this chapter as these statements are widely used in this country. The Elliott input statement explained above is the non-standard underlined word read followed by a list of variables. The Elliott output statement comprises the non-standard underlined word print followed by a list of variables. One other feature of Elliott Algol programs is that they must possess a title; this particular feature will also be employed as it can often be very convenient. The Algol programs given in Examples 8.17 and 8.18 will illustrate the use of the basic Algol features.

Example 8.17

Write an Algol program for the problem $(a+b) \div c$. The program is to arrange for the answer followed by the data a, b and c to be printed out.

Solution

```
Example Simple Calculation;
  begin real answer,a,b,c;
        read a,b,c;
        answer:=(a+b)/c;
        print answer,a,b,c;
  end;
```

Notice how simple it is to write this program. The first line comprises the program title and is terminated as are all the other statements by a semi-colon. The second line uses the underlined word begin to start the actual program, real is used to declare what type of variables are to be used (real has been used as the example does not specify what sort of numbers *a*, *b* and *c* are, therefore it is safer to treat them as reals). The non-standard underlined word read inputs 3 numbers from the computer's input equipment and places them in storage locations called a, b and c. The fourth line allows the result of the calculation to be placed in the storage location called answer. The non-standard underlined word print enables a print-out of the numeric values of answer, *a*, *b*, and then *c* to be made by the computer's output equipment. The underlined word end is used as a closing statement bracket and here is used to signify the end of the program.

Example 8.18

Write an Algol program to square each of 50 *numbers available on data tape and to print out the squares formed each time.*

Solution

```
Example Squares;
  begin real a;
        integer b;
        for b:=1 step 1 until 50 do
        begin read a;
              a:=a↑2;
              print a;
        end;
  end;
```

This program incorporates a known number of loops. The statement for b:=1 step 1 until 50 do allows for the statement between the second pair of statement brackets to be performed 50 times. The statement sets up an initial value of 1 for *b*, then proceeds to perform the required loop after which *b* is incremented by 1 (step 1), the loop is again performed and this process is repeated until *b* exceeds the value 50 (until 50).

8.12 COMPARISON OF ANALOGUE AND DIGITAL COMPUTERS

From Chapter 7 and the preceding sections it is clear that a wide range of differences exists between analogue and digital computers. In digital computers information is represented in discrete steps, whilst in analogue computers physical quantities are represented by continuous variables. Table 8.17 summarises the main differences between the two computers.

The fast operation, large-scale storage facilities and decision-making abilities of the digital computer make it suitable for use in a wide range of applications. A few examples are: scientific research, engineering problem solving, design work, systems control, routine commercial calculations, scheduling and stock control, statistical analysis, management aids, and medical diagnosis.

Table 8.17

FACTOR	ANALOGUE	DIGITAL
Representation of variables	Continuously variable voltages.	Binary numbers, in discrete steps.
Arithmetic operation	By measurement of voltages.	By addition and counting.
Accuracy	Limited by the choice of physical components and associated Input/ Output devices.	Very highly accurate.
Storage	Only short term.	Unlimited and permanent storage facilities.
Communication with computer	Program preparation, followed by connection of computing elements on the patch panel.	Program writing, followed by using either punched cards or punched tapes or directly by on-line terminal or V.D.U.
Input language	Almost universal.	Many different languages.
Output information	Graphical or by voltage indication.	Numbers, tables, graphs, diagrams, or text.
Cost	Relatively cheap.	Rather expensive.
Size	Relatively small.	Depending on type and function, can occupy a fair amount of space.
Area of application	Scientific and engineering.	Scientific, engineering and commercial.

EXERCISES

(8.1) Perform the following operations:

(*a*) Convert decimal number 598 to binary.

(*b*) Convert decimal number 0.03125 to binary.

(*c*) Convert binary number 1100110 to decimal.

(*d*) Convert binary number 111.1011 to decimal.

(*e*) Express 101.36 in 8421-BCD.

(*f*) Express 693 in 2421-BCD.

(*g*) Convert decimal number 25 to Gray Code.

(8.2) Perform the following binary operations:

(*a*) 11011 + 11101   (*b*) 1111 + 101 + 1011

(*c*) 11011 - 10111   (*d*) 10101 - 1010

(*e*) 1011 × 1101   (*f*) 1001 × 110

(*g*) 110110 ÷ 11   (*h*) 1110111 ÷ 101

(8.3) Calculate the number of different denary numbers which can be stored in a 16-bit register when coded in BCD.

(8.4) Explain what is meant by the "reflected excess-3" code. What is the advantage of this code over the "reflected binary" code?

(8.5) Using a Venn diagram, prove the following identities:

(*a*) $A\cdot B + \bar{A}\cdot B + A\cdot\bar{B} = A + B$

(*b*) $(A + A\cdot B + B\cdot\bar{C})\cdot(A\cdot C + \bar{B}\cdot C) = A\cdot C$

(8.6) Using De Morgans' Theorems and other Boolean algebra theorems show that

(*a*) $\overline{[(\overline{\bar{A}\cdot\bar{B}}) + A]} = \bar{A}\cdot\bar{B}$

(*b*) $\overline{(A\cdot B + \bar{\bar{C}})\cdot(\overline{\overline{A + B}} + C)} = C\cdot(\bar{A} + \bar{B})$

(8.7) Simplify the following expressions as far as possible:

(a) $A\cdot B + A\cdot B\cdot C + (\overline{B + C}) + A\cdot C$

(b) $B\cdot C\cdot(A + \overline{B}) + A\cdot\overline{B}\cdot(\overline{A} + C) + \overline{A}\cdot(A\cdot\overline{B} + C)$

(c) $\left[A\cdot B\cdot(\overline{B\cdot C})\right]\cdot\left[B\cdot\overline{C}\cdot(\overline{\overline{A}\cdot C})\right]$

(d) $(A + B) + (\overline{\overline{\overline{B}}\cdot\overline{\overline{C}}}) + (\overline{\overline{\overline{A}} + \overline{\overline{C}}})$

(e) $\overline{\overline{\left[\overline{X\cdot Z}\cdot(\overline{Y + \overline{\overline{Z}}})\right]}\cdot\left[(\overline{X} + Y) + X\cdot\overline{Y}\cdot Z\right]}$

(f) $(\overline{A}\cdot C + B)\cdot(B + \overline{C})\cdot(A\cdot B + C)\cdot(A + \overline{B})$

(8.8) Using Karnaugh maps simplify the following expressions:

(a) $(A + B) + (\overline{\overline{\overline{B}}\cdot\overline{\overline{C}}}) + (\overline{\overline{\overline{A}} + \overline{\overline{C}}})$

(b) $\left[A\cdot B\cdot(\overline{B\cdot C})\right]\cdot\left[B\cdot\overline{C}\cdot(\overline{\overline{A}\cdot C})\right]$

(c) $B\cdot C\cdot(A + \overline{B}) + A\cdot\overline{B}\cdot(\overline{A} + C) + \overline{A}\cdot(A\cdot\overline{B} + C)$

(d) $A\cdot B + A\cdot B\cdot C + (\overline{B + C}) + A\cdot C$

(8.9) Obtain the logic equation for the output of Figure 8.36.

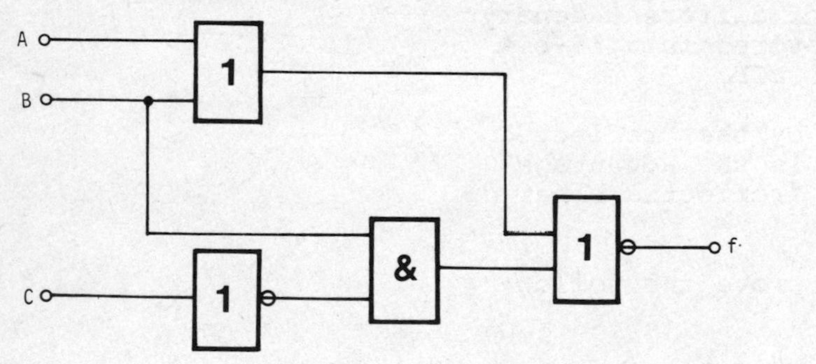

Fig.8.36

(8.10) A digital system is to be designed with *two* inputs and *three* outputs. The system should function under the following conditions:

(i) Output x indicates the absence of both inputs.
(ii) Output y indicates the presence of either input and the absence of the other.
(iii) Output z indicates the presence of both inputs.

Using logic equations and truth table:

(a) Draw the logic diagram for the system using the minimum number of gates.
(b) Modify your logic diagram to its simplest form using NAND gates only.

(8.11) Write an Algol program so that x can be calculated from

$$x = \frac{-b \pm \sqrt{(b^2 - 4ac)}}{2a}$$

(8.12) Write an Algol program to print out the larger of two numbers a and b which are available on a data tape.

9 Control Systems

A *system* may be defined as an arrangement or collection of components connected or related in such a way as to form an entire unit.

A CONTROL SYSTEM may be defined as a system whose behaviour regulates or controls either itself or another system in accordance with a command or desired input.

A control system can therefore be as large as a chemical plant, or as small and simple as an electric toaster. Each element of a control system may thus be a single device or a system containing many devices.

This chapter deals with some typical examples of control systems and various applications of control engineering.

9.1 SPEED CONTROL

Constant speed under varying load conditions is one of the requirements of many control systems. In such systems the output speed is compared with the desired speed and adjustments are made when necessary.

(1) A widely used arrangement is the VELODYNE SPEED CONTROL system of Figure 9.1 which uses a field-controlled motor with constant armature current. The tachogenerator is coupled to the output shaft and provides a voltage V_{fb} which is proportional to the speed.

The potentiometer provides a voltage V_R which sets the required speed. From the diagram

$$V_R = \varepsilon + K_s \omega_o$$

thus

$$\varepsilon = V_R - K_s \omega_o \qquad (9.1)$$

where ε is the error voltage applied to the amplifier and K_s is the tachogenerator constant.

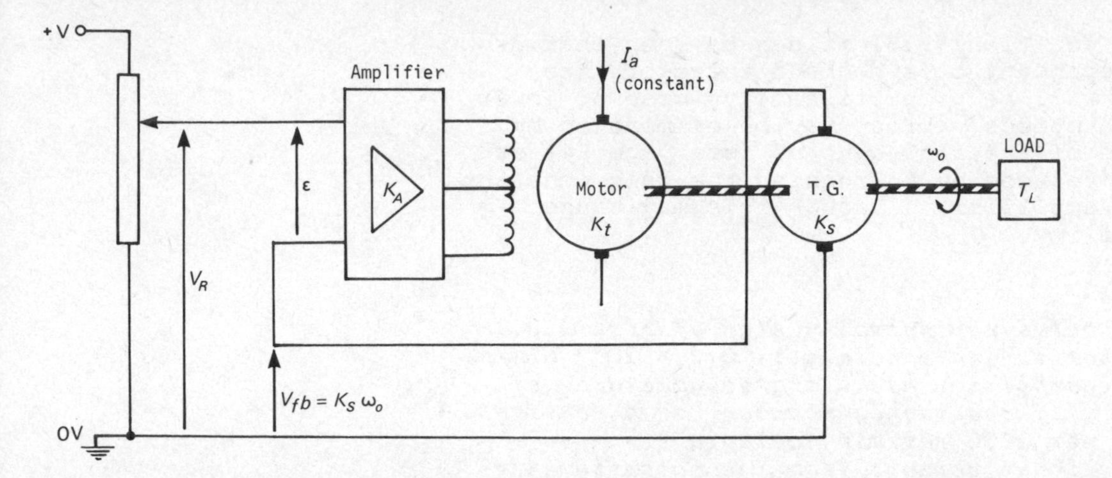

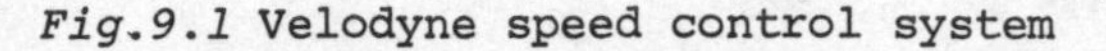
Fig.9.1 Velodyne speed control system

If, for any reason, the speed falls, the voltage fed back from the tachogenerator will fall. From equation (9.1), the error voltage applied to the amplifier will increase, and the motor will be driven faster, thus restoring the original fall.

Under steady-state condition on no-load, the torque required to maintain a constant speed is zero. Thus, from equation (9.1),

$$V_R = K_S \omega_O \tag{9.2}$$

Under load conditions, for steady output speed the motor torque must equal the load torque. Thus,

$$T_M = T_L$$

$$\varepsilon \times K_A \times K_t = T_L$$

then

$$\varepsilon = \frac{T_L}{K_A K_t} \tag{9.3}$$

where K_A is the amplifier transconductance (mS), K_t is the motor torque constant (N-m/mA), and T_L is the load torque (N-m).

A measure of the speed regulation of such a system is known as *droop* and may be defined in two ways:

(i) in terms of the error voltage applied to the amplifier, or
(ii) in terms of speed drop under rated full load conditions.

From equation (9.3) it can be seen that droop is independent of speed and therefore its effect will be proportionately worse at lower desired speeds. Droop may be eliminated by the use of *integral control* (see page 94) or may be reduced by increasing the gain constant of the amplifier, although this may cause instability.

Example 9.1

In the velodyne control system of Figure 9.1 the motor torque constant is 0.2×10^{-3} N-m/mA *and the amplifier has a transconductance of* 300 mS. *If the tachogenerator has a constant of* 2 V per 1000 rev/min, *calculate:*

(*a*) *the input voltage from the potentiometer to give a shaft speed of* 2500 rev/min *on no load,*

(*b*) *the percentage speed drop when the load torque is* 8×10^{-3} N-m.

Solution

The tachogenerator constant:

$$K_g = \frac{2}{1000} = 2 \times 10^{-3} \text{ V/rev}$$

(*a*) From equation (9.2), under no-load condition, the input voltage will be

$$V_{in} = K_g\omega = 2 \times 10^{-3} \times 2500 = 5 \text{ V}$$

(*b*) Under load condition, from equation (9.3), the error voltage will be

$$\varepsilon = \frac{T_L}{K_A \times K_M} = \frac{8 \times 10^{-3}}{300 \times 0.2 \times 10^{-3}} = \frac{2}{15} \text{ V}$$

Assuming that the on-load speed is ω', from equation (9.1),

$$\varepsilon = V_{in} - K_g\omega'$$

$$\frac{2}{15} = 5 - 2 \times 10^{-3}\omega'$$

$$\omega' = \frac{5 - 2/15}{2 \times 10^{-3}} = \frac{75 - 2}{30 \times 10^{-3}}$$

$$= 2434 \text{ rev/min}$$

The speed drop will be

$$2500 - 2434 = 66 \text{ rev/min}$$

The percentage speed drop will be

$$\frac{66}{2500} \times 100\% = 2.64\%$$

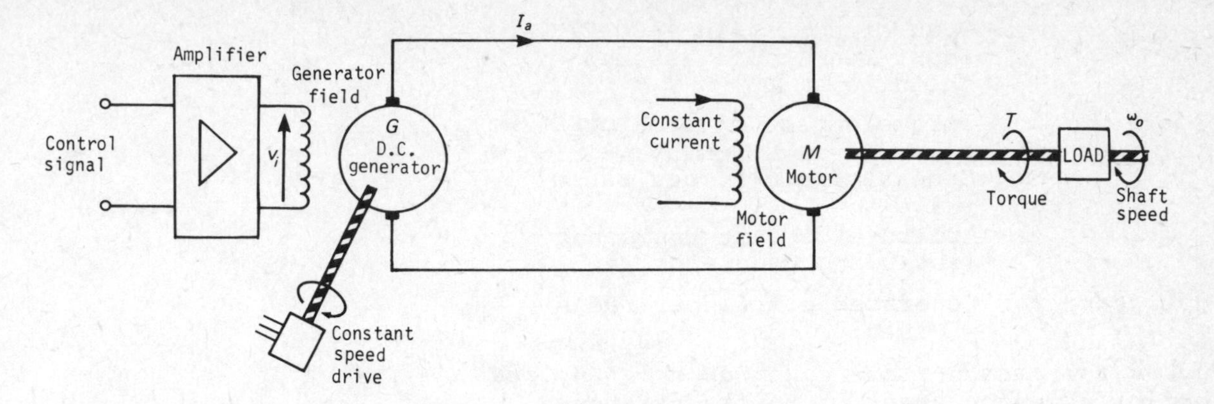

Fig.9.2 Ward-Leonard speed control system

(2) Another widely used arrangement for speed control is the WARD-LEONARD system shown in Figure 9.2.

The d.c. generator is driven by a constant-speed motor, usually an induction type, so that the generated e.m.f. will be proportional to the field current controlling it. The motor whose output speed is to be controlled has a constant field excitation and is armature-controlled by the current supplied from the d.c. generator.

Since the controlled motor has a constant field current and the generator is driven at constant speed, the motor speed is proportional to its armature current, which is in turn proportional to the generator field current and hence proportional to the controlling signal.

The generator field winding is designed to have high impedance and therefore a small current will be sufficient to produce a large magnetic field. Since output powers of the order of kilowatts are controlled by very small input powers, of the order of milliwatts, large power amplification is obtained. This arrangement is most suitable for systems in which a wide range of speed control is required with large power output.

Example 9.2

Data for the Ward-Leonard speed control system shown in Figure 9.3 is as follows:

Generator Field resistance including output resistance of amplifier = 200Ω.
Generated e.m.f. per field ampere = 2000 V.

Amplifier	Open circuit gain = 20 V/V. Input resistance assumed infinite.
Motor	Torque/ampere of armature current = 1.0 N-m. Generated e.m.f. per rad/s = 1.0 V. Armature circuit resistance (including generator) = 2.0Ω.
Tachogenerator	Generated e.m.f. per rad/s = 0.5 V.

Derive the steady-state equation relating the controlled motor speed ω_o to the reference voltage V_R and the load torque T. Find the no-load speed if V_R = 101 V. By what percentage does speed fall if a load torque T = 10 N-m is applied when V_R = 101 V?

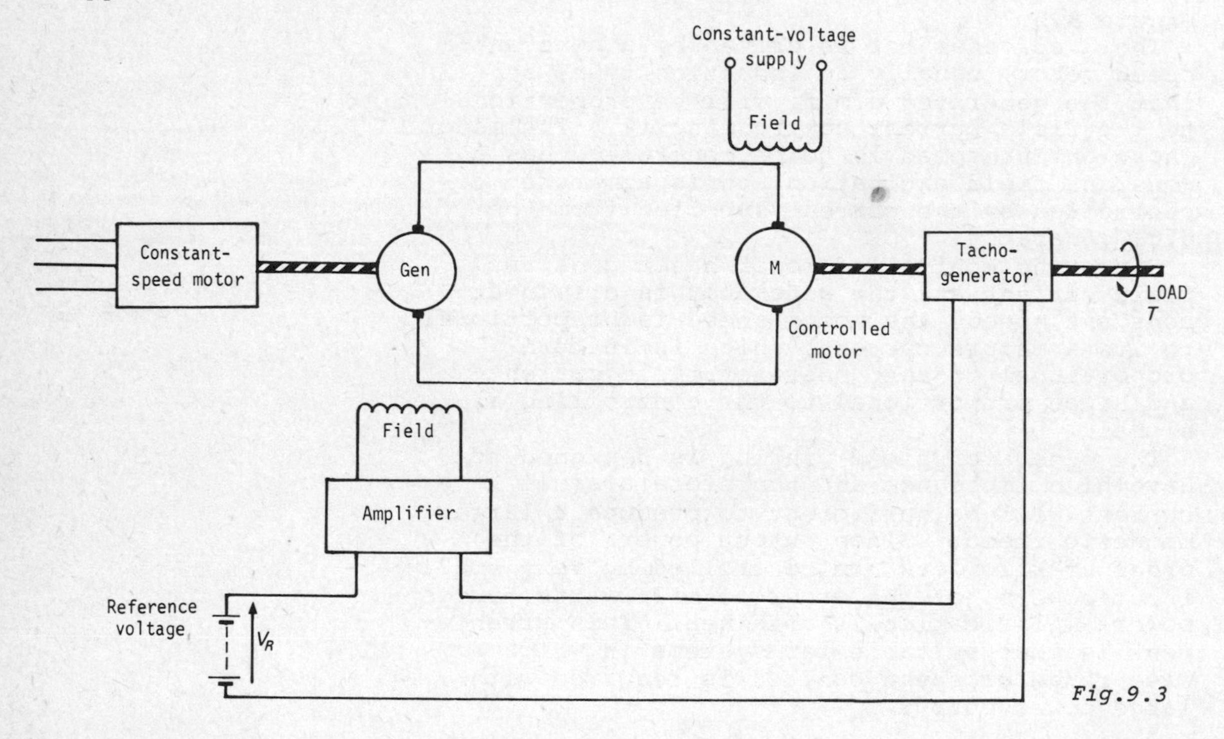

Fig.9.3

Solution

Let

ε = the error voltage applied to the amplifier

G = open circuit voltage gain of the amplifier

K_g = generator e.m.f. per ampere of field current

$R_f + R_o$ = field resistance and output resistance of the amplifier

K_t = motor torque per ampere of armature current

K_b = motor back e.m.f. per rad/s of motor speed

R_a = combined armature circuit resistance

K_s = tachogenerator voltage per rad/s of motor speed

V_R = reference voltage

T = torque

ω_o = no-load speed

Then

Error voltage $\varepsilon = V_R - K_s\omega_o$

Generated e.m.f. $E_g = K_g \times I_f$

Motor back e.m.f. $E_b = K_b \times \omega_o$

Torque produced $T = K_t \times I_a$

Field current $I_f = G \times \varepsilon/(R_f + R_o)$

For the armature circuits of the generator and motor:

$$E_g = I_a R_a + E_b$$

Substituting for E_g, I_a and E_b:

$$K_g I_f = \frac{T}{K_t} R_a + K_b \omega_o$$

Substituting for I_f:

$$\frac{G\varepsilon}{(R_f + R_o)} K_g = \frac{T}{K_t} R_a + K_b \omega_o$$

Re-arranging:

$$\omega_o = \frac{K_g G}{(R_f + R_o) K_b} \varepsilon - \frac{R_a}{K_t K_b} T$$

For simplicity, let

$$\frac{K_g G}{(R_f + R_o) K_b} \equiv M \qquad (9.4)$$

and

$$\frac{R_a}{K_t K_b} \equiv N \qquad (9.5)$$

then $\omega_o = M\varepsilon - NT$

Substituting for the error voltage:

$$\omega_o = M(V_R - K_s\omega_o) - NT$$

or

$$\omega_o(1 + K_sM) = MV_R - NT$$

Then, the steady-state equation will be

$$\omega_o = \frac{M}{1 + K_sM}V_R - \frac{N}{1 + K_sM}T \qquad (9.6)$$

From equation (9.6) it is seen that the no-load speed is given by

$$\omega_o = \frac{M}{1 + K_sM}V_R \qquad (9.7)$$

From equations (9.4) and (9.5), calculating M and N:

$$M = \frac{20 \times 2000}{200 \times 1} = 200$$

$$N = \frac{2}{1 \times 1} = 2$$

From equation (9.7), the no-load speed will be

$$\omega_o = \frac{200 \times 101}{1 + 0.5 \times 200} = 200 \text{ rad/s}$$

From equation (9.6), it is seen that under load conditions the speed drop will be

$$\frac{N}{1 + K_sM}T = \frac{2 \times 10}{1 + 0.5 \times 200} = 0.198 \text{ rad/s}$$

The percentage speed drop will be

$$\frac{0.198 \times 100}{200}\% = 0.099\%$$

9.2 POSITION CONTROL

Broadly speaking, in a position-control system an output quantity is required to *follow*, as closely as possible, the movement of an input quantity. Position control finds extensive use in military applications such as radar tracking, gun directors, missile launchers and guidance systems; in industrial and commercial applications such as machine-tool control, profile cutting, positioning of aerials, servomultipliers, automatic piloting of aircrafts; and in scientific research such as automatic handling of radioactive materials and remote control of objects such as T.V. cameras and mechanical robots.

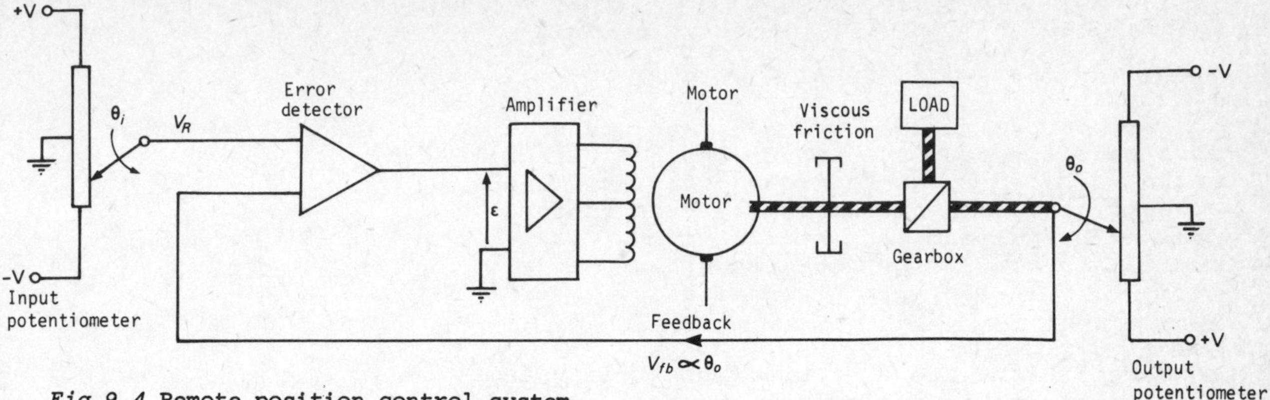

Fig.9.4 **Remote position control system**

A simple REMOTE POSITION-CONTROL SYSTEM (r.p.c.) is shown in Figure 9.4. The input potentiometer, acting as a transducer, provides a voltage proportional to the angle through which the input shaft is rotated, and the output shaft is required to rotate. The error detector is a simple analogue summer (the inputs will have signals of different polarity), and provides the error voltage ε, where

$$\varepsilon = V_R - V_{fb}$$

This error is amplified by the d.c. amplifier with a push-pull output stage which provides the current to drive the split-field type motor. A gear system is used for speed reduction; and viscous friction (a frictional force proportional to the angular velocity) is employed to damp the system (see Chapter 5). The output shaft is linked to the output potentiometer which provides a voltage proportional to the angular position of the wiper. This voltage is fed back and is compared with the input voltage.

When the input is moved through an angle θ_i, the error voltage causes the motor to rotate until the output shaft becomes in line with the input, at which time the error becomes zero and the motor will stop.

Dead zone of an r.p.c. is the range of shaft errors over which the system will not take any correcting action. This is because the motor torque produced by the error voltage is not sufficient to overcome the stiction (static friction) present. Dead zone may be reduced by increasing the gain of the system, but this has an adverse effect on the overshoot and settling time.

Example 9.3

State the essential features of a servo-mechanism.

A remote positional-control system is used to control the motion of a radar system. With the aid of a diagram, explain the function of each element used, and calculate the gear ratio needed for maximum load acceleration. The inertias of the motor and the load are 11 kg-m^2 *and* 363 kg-m^2 *respectively.*

Solution

Servomechanisms are a group of control systems which possess the following features:

(1) They are error-actuated.
(2) They have power amplification.
(3) They contain moving parts.
(4) Their operation is automatic.

A simple remote position-control system is shown in Figure 9.4. The major elements are:

(*a*) The input and output potentiometers act as transducers and convert the angles of rotation or positions of the input and output shafts into voltages.
(*b*) The input to the amplifier is the error signal. The function of the amplifier is to raise the level of this signal and actuate the motor.
(*c*) The power amplification of the system is provided by the motor which performs the heavy work, thus moving the load in accordance with the command signal.
(*d*) The gearbox matches the load to the motor and allows high speed motors to be used for good power/weight ratios.

Any deviation between the input and output potentiometers causes an error signal to be fed into the amplifier. The motor will be actuated and the resulting torque will rotate the motor in such a direction as to reduce the deviation.

$$\text{Gear ratio} \quad n = \sqrt{\frac{J_L}{J_M}} = \sqrt{\frac{363}{11}} \simeq 5.74$$

9.3 NUMERICAL CONTROL OF MACHINE TOOLS

A modern and sophisticated application of position and speed control, used in conjunction with digital computers, is the numerical control of machine tools such as milling machines, drilling machines and lathes. A detailed description of the operation of such a system is beyond the scope of this book, but a brief mention will be made here.

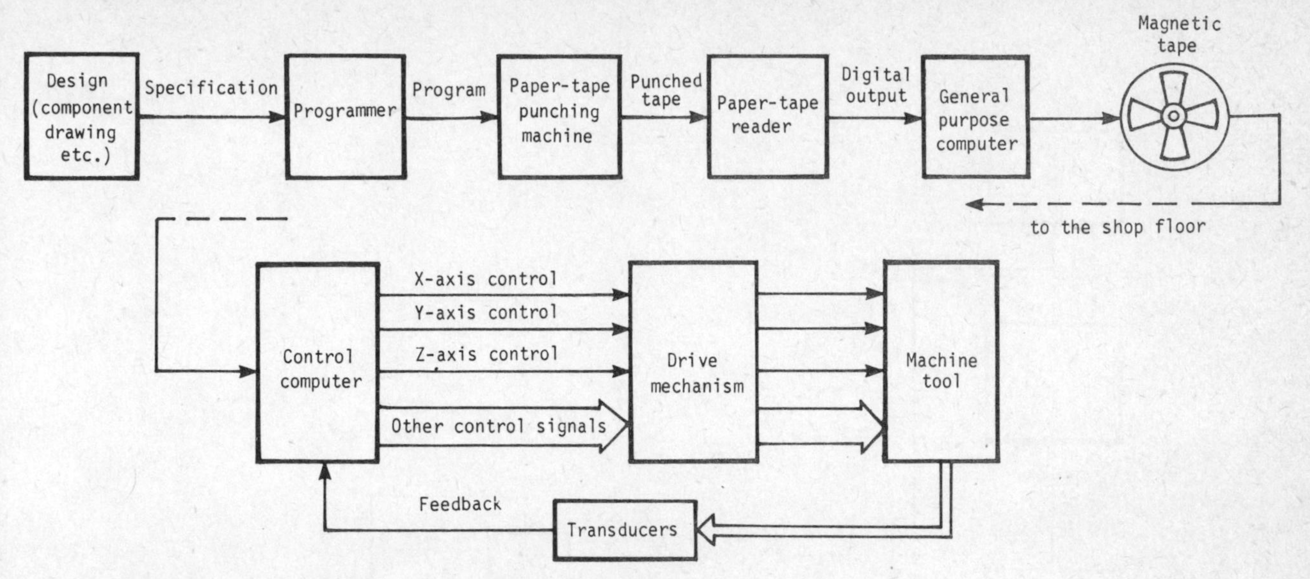

Fig.9.5 Numerical control of machine tool

A simplified block diagram showing the operational steps of a computer controlled machine tool system is shown in Figure 9.5. The drawing and design specifications are converted into a set of instructions in the form of a *program*. The program is punched on paper tapes which are then fed into the paper-tape reader. The output of the paper-tape reader is in the form of digital pulses which are fed into the general purpose digital computer. The instructions from this computer are recorded on magnetic tapes, which will be used to run the control computer. The signals from the control computer operate the drive mechanism (motors, gears, etc.) which will in turn drive the machine tool. Transducers provide feedback information for any necessary corrective action.

The complexity of the operation depends on whether the control is required in one, two or three dimensions, and also whether or not a continuous cutting path is required. For example, in a *positioning* operation a series of holes has to be drilled and it is only necessary to inform the machine of the locations of the successive holes. However, in *profiling*, when a cam has to be shaped, a continuous path has to be cut and the path of the cutting head has to be specified continuously. In addition, the *feed rate* (rate of movement) must be continuously controlled.

Numerically controlled machine tools are more expensive than conventional machines of comparable size and type. The control system

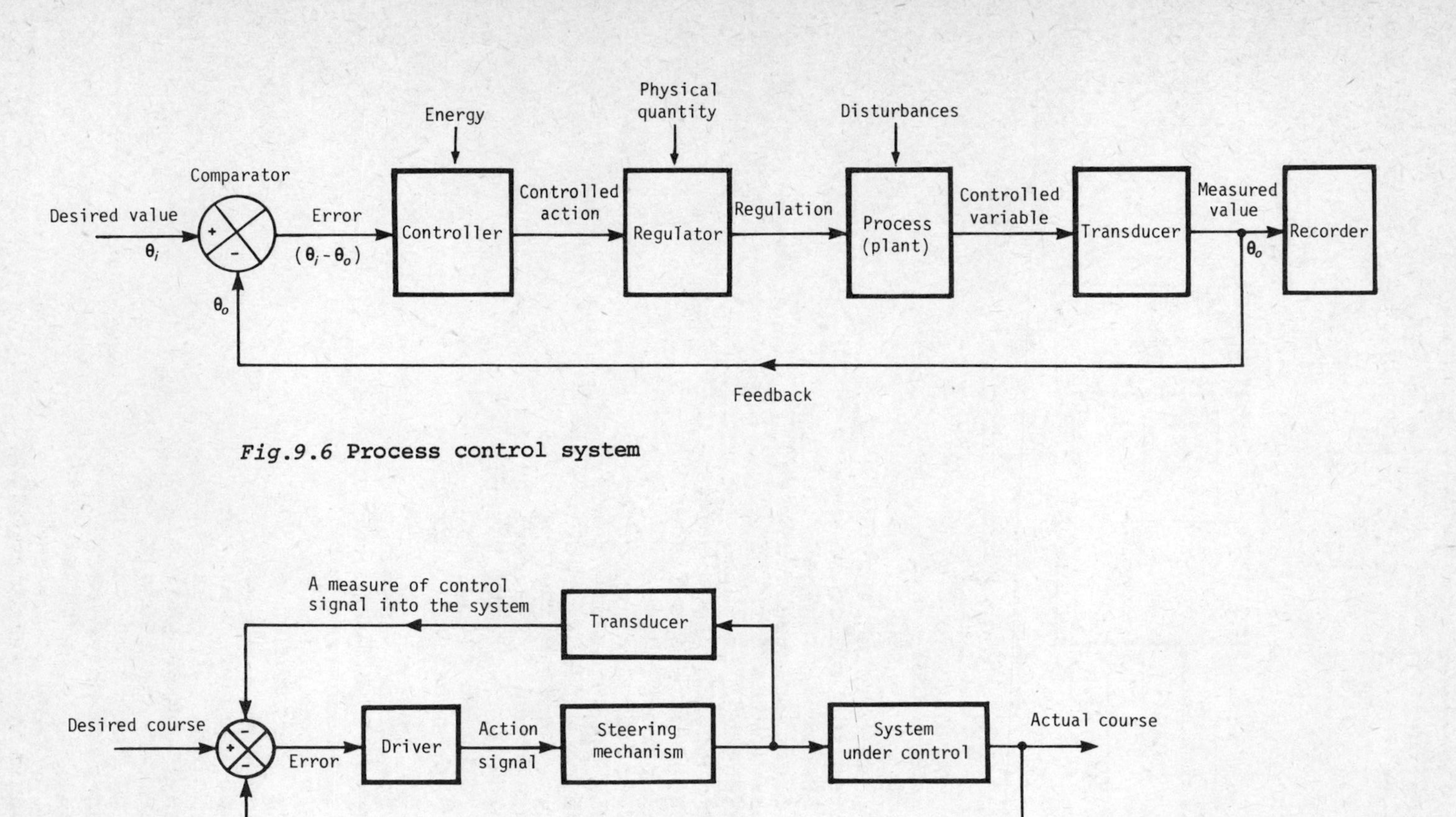

Fig.9.6 **Process control system**

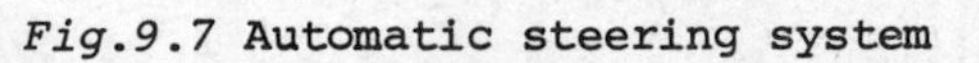

Fig.9.7 **Automatic steering system**

used in conjunction with these machines is costly and the computer program is sophisticated particularly if the control is applied to three axes. However, despite the initial disadvantage of high cost, they have many advantages, some of which are listed below:

(1) Elimination of operator errors.
(2) Lower labour costs.
(3) Reduction in the time between the receipt of a design drawing by the production engineer and the start of manufacture on the shop floor.
(4) No need for special jigs and fixtures.
(5) Flexibility in changing of component design.
(6) Accurate costing and scheduling.

9.4 PROCESS CONTROL SYSTEMS

In process control systems, quantities such as pressure, flow, liquid level, or temperature are under control. A chemical plant is an example of such a system. Figure 9.6 shows a simplified block diagram of a closed-loop process control system. The operation of the system and the function of the elements are as follows.

The *comparator unit* produces a signal which is proportional to the difference between the desired input and the actual output. The *controller*, the input of which is the error signal, may be hydraulic, pneumatic, or electric. This unit controls the action of the *regulator* in accordance with the variation in the error signal. The regulator unit is often a pneumatically or electrically controlled operated valve and regulates the *process* or *plant*. The *transducer* produces an output which is proportional to the controlled variable and is suitable to be fed into the comparator. The object of the *recorder* is to provide a continuous record of the performance of the system.

9.5 OTHER APPLICATIONS OF AUTOMATIC CONTROL

Applications of automatic control are so numerous and diversified that only a few examples can be given in a book of this nature. In all cases however, the basic principle of operation is much the same and the block diagram models remain very similar.

9.5.1 AUTOMATIC STEERING CONTROL

A simple block diagram of an automatic steering control system is shown in Figure 9.7. Here, the input to the system is the desired course and the output is the actual course. The system has two feedback signals, one derived from the steering mechanism and another taken from a measurement of the actual course. The desired course is compared with the actual course and a measure of the error is fed into the driver which will take the necessary corrective action. This type of feedback control is often used in the steering control system of big liners and the flight controls of large aircraft.

9.5.2 LATHE TRACER CONTROL

A simple lathe tracer system and its block diagram representation are shown in Figure 9.8. The template is fixed to the bed of the lathe. The cross-slide carries the tracer head and its stylus, and the cutting tool. This keeps the relative position of the cutter and the stylus fixed, enabling the tracer to act as the comparator of the system.

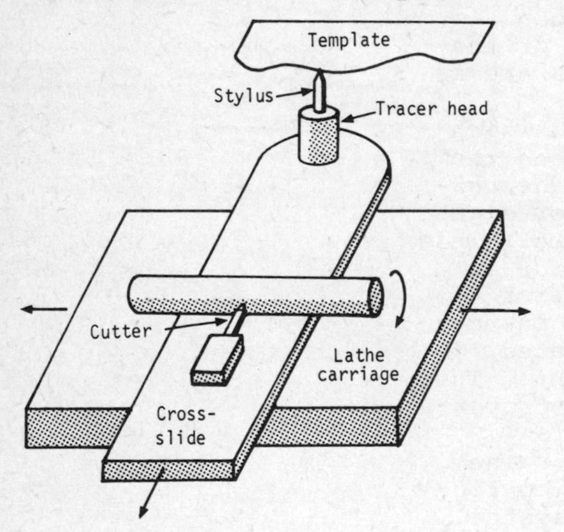

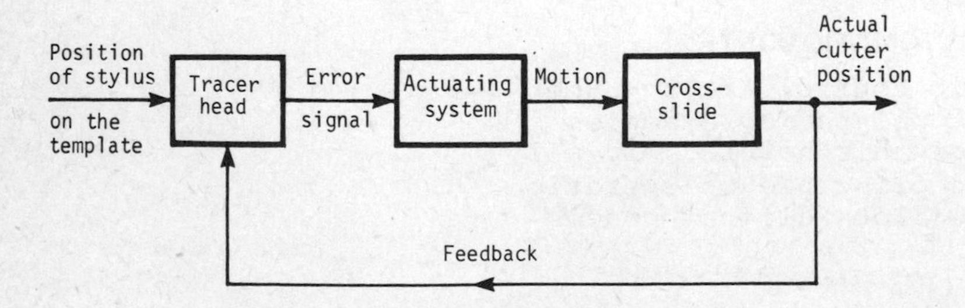

Fig.9.8 Automatic lathe tracer control

The input to the system is the desired position of the stylus on the template and the output is the shape of the object being cut. When there is a difference between the two, the tracer head initiates a signal to the actuating system which will in turn move the lathe cross-slide and hence the cutter into the correct position.

9.5.3 AUTOMATIC CONTROL OF TRAIN-BRAKE SYSTEM

The block diagram of an automatic brake system of an electrically driven train is shown in Figure 9.9. The input to the system is a stop-signal from the control station and the output is the actual stopping position of the train. As shown, the comparator receives two feedback signals, one giving speed information and the other indicating the position of the train

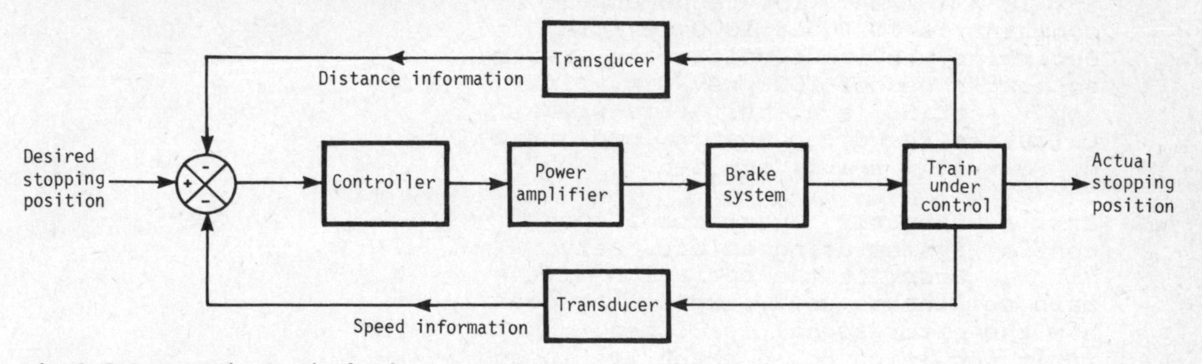

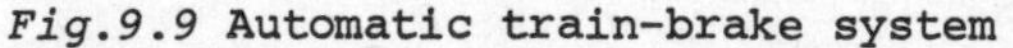
Fig.9.9 **Automatic train-brake system**

relative to the station. As the train approaches the station, the controller is activated causing the brakes to be operated. The two transducers constantly provide the necessary information which will eventually bring the train to a halt at the desired position in the station.

9.6 CONCLUSION

Other examples of automatic control include its application in the metallurgical industry, in the coal industry, in warehousing and inventory control, in biomedical experimentation such as the study of the nervous system, temperature regulation, and respiratory and cardiovascular control, in space travel and research, and many, many more.

The examples discussed in this book illustrate the diversity of the applications of automatic control systems. The potential future application of feedback control systems appears to be unlimited and there is no doubt that the theory and practice of modern control systems will play an even more important role in our technological societies.

EXERCISES

(9.1) Draw the block schematic diagram of a velodyne speed control system. Explain the operation of the system for no-load and load conditions. What is meant by the term *droop* and how is good speed regulation achieved?

(9.2) In the velodyne speed control of Figure 9.1 the amplifier transconductance is 200 mS and the motor torque constant is 5×10^{-3} N-m/mA. The tachogenerator constant is 10 V per 1000 rev/min. Determine the input voltage to give a no-load speed of 2000 rev/min. If the input setting is at half this value, calculate the droop when a load torque of 60×10^{-3} N-m is applied.

(9.3) Draw a schematic layout for a speed control system using an a.c. servo motor. Describe the operation of the main components used, explaining clearly how the error signal is formed. What fault condition could cause the speed to fall abnormally?

(9.4) Discuss briefly the factors which determine the speed of an induction motor. Explain, with the aid of appropriate diagrams, a method of controlling the speed of such a motor using a control system.

(9.5) Figure 9.2 shows a schematic arrangement of a Ward-Leonard speed control system. Assigning suitable symbols to the various system constants, deduce the steady-state equation relating the speed of the controlled motor shaft ω_o to the voltage v_i applied to the generator field for a steady load torque T.
Show, by a schematic diagram, how the system could be connected to a closed-loop speed-control system and again deduce a steady-state equation for ω_o. From this equation indicate advantages of the closed-loop system.

(9.6) A closed-loop Ward-Leonard system is to be used for the position control of a large rotatable mass. Give a circuit diagram and a description of the operation of an appropriate system. Show how switches can be incorporated to limit the angular swing of the output shaft.

(9.7) A corrosive liquid chemical is required to be maintained at a constant level in a tank for an industrial process. The tank acts as a reservoir, feeding its contents in variable quantities by gravity feed to the process. Sketch a block schematic diagram for a suitable system to control the liquid level, labelling each component, and explain its operation.

Appendix 1
Complex Number Table

For the complex number $z = 1 + jx$, the modulus M and the argument θ are such that

$1 + jx = Me^{j\theta} = M\angle\theta^\circ$, where $M = \sqrt{(1 + x^2)}$ and $\theta = \tan^{-1}x$

x	M	θ°	x	M	θ°	x	M	θ°
0.05	1.00	3	1.80	2.06	61	5.2	5.30	79
0.10	1.00	6	1.85	2.12	62	5.4	5.49	79
0.15	1.01	8	1.90	2.15	62	5.6	5.69	80
0.20	1.02	11	1.95	2.18	63	5.8	5.89	80
0.25	1.03	14	2.00	2.23	63	6.0	6.08	81
0.30	1.04	17	2.1	2.32	64	6.2	6.28	81
0.35	1.06	19	2.2	2.42	66	6.4	6.48	81
0.40	1.08	22	2.3	2.51	66	6.6	6.68	81
0.45	1.09	24	2.4	2.60	67	6.8	6.87	82
0.50	1.12	27	2.5	2.69	68	7.0	7.07	82
0.55	1.14	29	2.6	2.79	69	7.2	7.27	82
0.60	1.17	31	2.7	2.88	70	7.4	7.47	82
0.65	1.19	33	2.8	2.97	70	7.6	7.67	82
0.70	1.22	35	2.9	3.07	71	7.8	7.87	83
0.75	1.25	37	3.0	3.16	72	8.0	8.06	83
0.80	1.28	39	3.1	3.26	72	8.2	8.26	83
0.85	1.31	40	3.2	3.35	73	8.4	8.46	83
0.90	1.34	42	3.3	3.45	73	8.6	8.66	83
0.95	1.38	43	3.4	3.54	74	8.8	8.86	83
1.00	1.41	45	3.5	3.64	74	9.0	9.05	84
1.05	1.45	46	3.6	3.74	75	9.2	9.25	84
1.10	1.49	48	3.7	3.83	75	9.4	9.45	84
1.15	1.52	49	3.8	3.94	75	9.6	9.65	84
1.20	1.55	50	3.9	4.02	76	10	10.05	84
1.25	1.62	51	4.0	4.12	76	12	12.0	85
1.30	1.64	52	4.1	4.22	76	15	15.0	86
1.35	1.67	53	4.2	4.32	77	20	20.0	87
1.40	1.72	54	4.3	4.42	77	30	30	88
1.45	1.76	55	4.4	4.51	77	40	40	89
1.50	1.80	56	4.5	4.61	77	50	50	89
1.55	1.88	57	4.6	4.71	78	60	60	89
1.60	1.89	58	4.7	4.81	78	70	70	89
1.65	1.93	59	4.8	4.90	78	80	80	89
1.70	1.97	59	4.9	5.00	78	90	90	89
1.75	2.01	60	5.0	5.10	79	100	100	89
						∞	∞	90

Appendix 2
M and N Circle Data

M-Circles

M	Centre	Radius	$\sin^{-1}(1/M) = \psi$
0	Origin	0	-
0.2	+0.042	0.21	-
0.3	+0.099	0.33	-
0.4	+0.192	0.48	-
0.5	+0.333	0.67	-
0.6	+0.562	0.94	-
0.7	+0.960	1.37	-
0.8	+1.777	2.22	-
0.9	+4.26	4.74	-
1.0	$\pm\infty$	∞	90°
1.1	-5.77	5.24	65.4°
1.2	-3.27	2.73	56.5°
1.3	-2.45	1.88	50.3°
1.4	-2.04	1.46	45.6°
1.5	-1.80	1.20	41.8°
1.6	-1.64	1.03	38.7°
1.7	-1.53	0.900	36.0°
1.8	-1.47	0.842	33.7°
1.9	-1.39	0.729	31.8°
2.0	-1.33	0.666	30.0°
2.25	-1.24	0.550	26.4°
2.50	-1.19	0.476	23.6°
3.00	-1.12	0.375	19.5°
3.50	-1.10	0.340	16.6°
4.00	-1.07	0.266	14.5°
5.00	-1.04	0.208	11.5°

N-Circles

ϕ	N	Centre	Radius
-5°	-0.087	-5.75	5.95
-10°	-0.176	-2.84	2.88
-20°	-0.364	-1.37	1.46
-30°	-0.577	-0.866	1.00
-40°	-0.839	-0.596	0.778
-50°	-1.19	-0.420	0.656
-60°	-1.73	-0.289	0.577
-70°	-2.75	-0.182	0.532
-80°	-5.67	-0.087	0.506

Answers to Exercises

Chapter 1

(1.1) $R_2/(R_1 + R_2)$

(1.4) $(j\omega)^2/[(j\omega)^2 + (3/RC)j\omega + 1/R^2C^2]$

(1.5) 4%

(1.7) $[G_1G_2(G_3 + G_4)] \Big/ \{1 + G_1G_2[(G_3 + G_4)H_2 - H_1]\}$

(1.8) $(G_1G_2R + G_2N_1 + G_1G_2H_1N_2)/(1 - G_1G_2H_1H_2)$

Chapter 4

(4.1) 9000

Chapter 5

(5.2) 50%

(5.3)(b) $\theta_0(t) = 1 + 1.155e^{-t}\sin(1.732t - 60^\circ)$

(5.4) $F = 3.22$ N-m/rad/s, $K = 1.24$ N-m/rad

(5.5) $\zeta = 1/3$; the transient response is thus oscillatory.

(5.7) 3000 N-m/rad/s, 3 rad/s

(5.10) $n^2J\ddot{\theta}_o + [F + (n^2K_AK_tK_s)]\dot{\theta}_o + nK_AK_tK_r\theta_o = nK_AK_tK_r\theta_i$,
$\omega_n = 10$ rad/s, 176 N-m/rad/s, 0.04 rad

(5.12) 0.15°

(5.14) $J\ddot{\theta}_o + F\dot{\theta}_o + K_AK_tK_r\theta_o = K_AK_tK_r\theta_i$, $K_A = 1.5$

(5.15) $0.588 < \zeta < 0.707$, $M_{pf} = 1.05$, $\omega_{rf} = 15.72$ rad/s

Chapter 6

(6.1) (*a*) $25s^2 + 40s + 2 = 0$ (*b*) $s^3 + 8s^2 + 25s + 20 = 0$
(*c*) $s^3 + 2s^2 + 5s + 10 = 0$ (*d*) $s^2 + 2\zeta\omega_n s + \omega_n^2 = 0$

(6.2) (*a*) Stable, (*b*) Unstable, (*c*) Unstable, (*d*) Stable

(6.3) $0 < K < 96$

(6.4) Stable

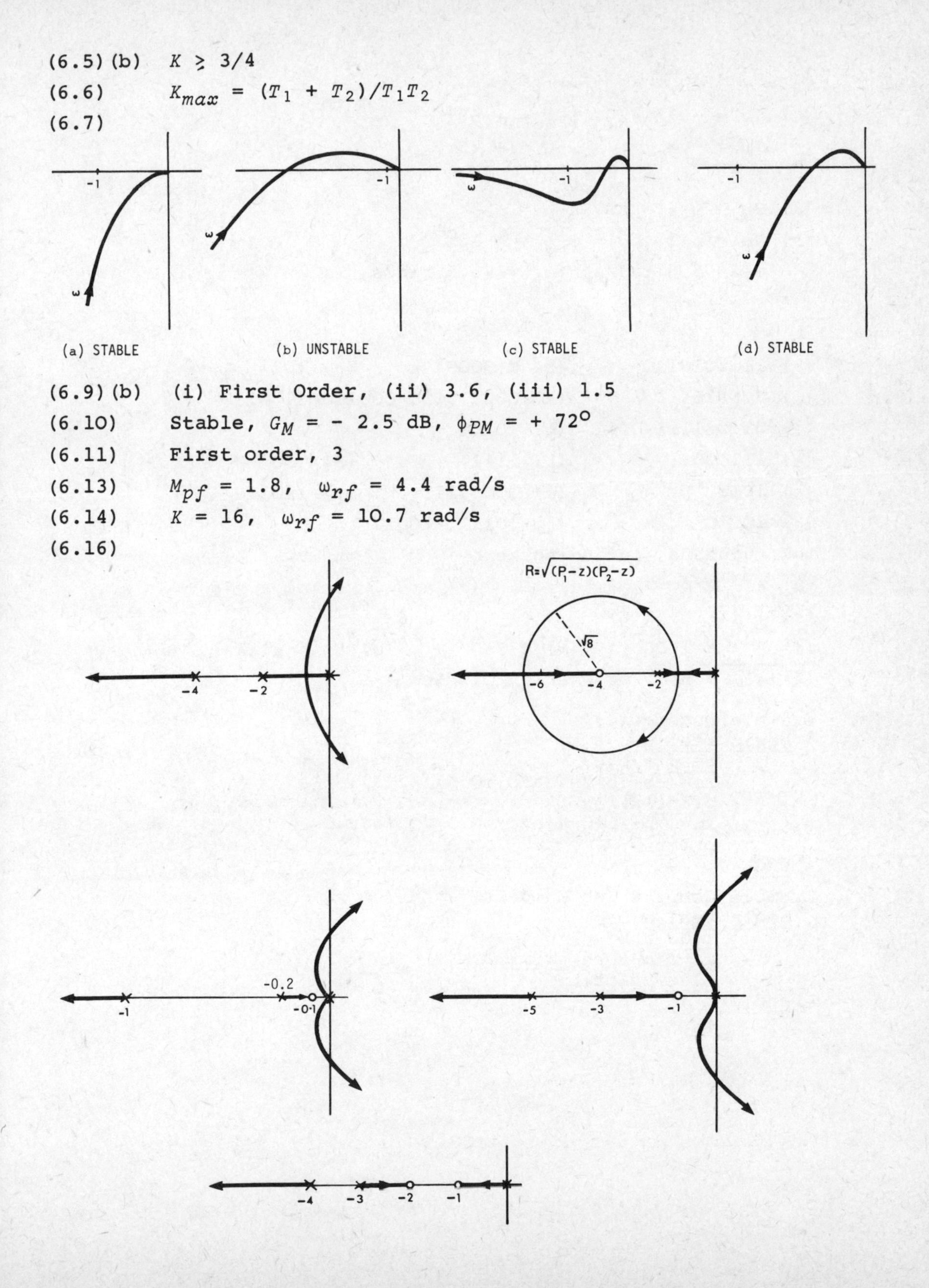

(6.5)(b) $K \geq 3/4$

(6.6) $K_{max} = (T_1 + T_2)/T_1T_2$

(6.7)

(a) STABLE (b) UNSTABLE (c) STABLE (d) STABLE

(6.9)(b) (i) First Order, (ii) 3.6, (iii) 1.5

(6.10) Stable, $G_M = - 2.5$ dB, $\phi_{PM} = + 72^{\circ}$

(6.11) First order, 3

(6.13) $M_{pf} = 1.8$, $\omega_{rf} = 4.4$ rad/s

(6.14) $K = 16$, $\omega_{rf} = 10.7$ rad/s

(6.16)

(6.17) $K = 500, \quad \zeta = 0.4$

(6.19) $G_M = -3$ dB, $\phi_{PM} = +8^{\circ}$

(6.20) (*a*) $G_M = -38$ dB, $\phi_{PM} = +58^{\circ}$

(*b*) $G_M = -\infty$, $\phi_{PM} = +45^{\circ}$

(6.21) $K = 1.25$

(6.25) (*a*) $G_M = -2.5$ dB, $\phi_{PM} = +5^{\circ}$

(*b*) $G_M = -27$ dB, $\phi_{PM} = +45^{\circ}$

$\omega_1 = 5.6$ rad/s, $\omega_{\pi} = 25.5$ rad/s

Chapter 8

(1) (*a*) 1001010110 (*b*) 0.00001 (*c*) 102

(*d*) 7.6875 (*e*) 0001 0000 0001.0011 0110

(*f*) 0110 1111 0011 (*g*) 10101

(2) (*a*) 111000 (*b*) 11111 (*c*) 100

(*d*) 1011 (*e*) 10001111 (*f*) 110110

(*g*) 10010 (*h*) 10111.110011

(3) Ten thousand, including zero

(7) (*a*) $A + \overline{B}\cdot\overline{C}$ (*b*) C (*c*) $A\cdot B\cdot\overline{C}$ (*d*) $A + B + C$

(*e*) $\overline{Y}\cdot(\overline{X}\cdot Z + X\cdot\overline{Z})$ (*f*) $A\cdot B$

(8) (*a*) $A + B + C$ (*b*) $A\cdot B\cdot\overline{C}$ (*c*) C (*d*) $A + \overline{B}\cdot\overline{C}$

(9) $\overline{(A + B) + B\cdot\overline{C}}$ or when simplified $\overline{A}\cdot\overline{B}$

(11)

```
Example Quadratic;
  begin real a,b,c,x;
        read a,b,c;
        x:=(-b+(b↑2-4*a*c)↑0.5)/(2*a);
        print x;
        x:=(-b-(b↑2-4*a*c)↑0.5)/(2*a);
        print x;
  end;
```

(12)

```
Example Compare Two Numbers;
  begin real a,b;
        read a,b;
        if a>b then print a
        else print b;
  end;
```

Index

Preface

One of the physician's most important skills is the ability to apply basic science in a clinical setting. To help develop this skill, *Medical Biochemistry* combines the chemical, physiologic and pathologic perspectives on human biochemistry.

Anyone involved in medical education has grumbled at some point that, despite the enormous amount of knowledge taught during the preclinical years, students are often unable to recall relevant information or apply it clinically. This is now being addressed in the new medical curricula, which focus on understanding principles rather than ingesting fact after fact.

In the early stages of planning this book, we decided *Medical Biochemistry* would provide students with the biochemistry they need to know when they practice medicine. We have been particular about maintaining scientific rigor and, at the same time, have made conscious decisions to reduce the number of facts irrelevant to clinical practice including the molecular weights of enzyme molecules, thermodynamic parameters of reactions, and the numerous intermediates in some metabolic pathways.

In each chapter core knowledge is enriched by Advanced Concepts which point the reader to new developments or more specialist topics. The information is presented in a clinical context: the core is illustrated by clinical examples to simulate a ward-round environment. Special effort has been made to select cases which illustrate important concepts in metabolism, but also to provide down-to-earth examples of clinical situations, which the future physician will confront on a daily basis. For such cases, a problem-solving approach has been taken and clinical laboratory data are given for interpretation. This combination of the metabolic aspects of biochemistry with its application to the diagnosis and monitoring of disease is critical in providing a context for learning biochemistry.

Another issue in medical education is the need to integrate knowledge from separate basic science disciplines. This integration involving molecular, anatomical and physiologic information is essential for developing a perspective on the function of the human body in health and disease. We have avoided presenting biochemistry as a collection of metabolic pathways, which operate in isolated systems. Instead, the reader will find chapters dedicated to the function of the lungs, kidneys and liver, to the regulation of acid-base and electrolyte balance, and to biochemical endocrinology.

All in all, we hope that *Medical Biochemistry* will help students to develop an understanding of biochemistry as an everyday science, a science that is as useful in the laboratory as in the hospital ward or the physician's office.

Marek Dominiczak
John Baynes

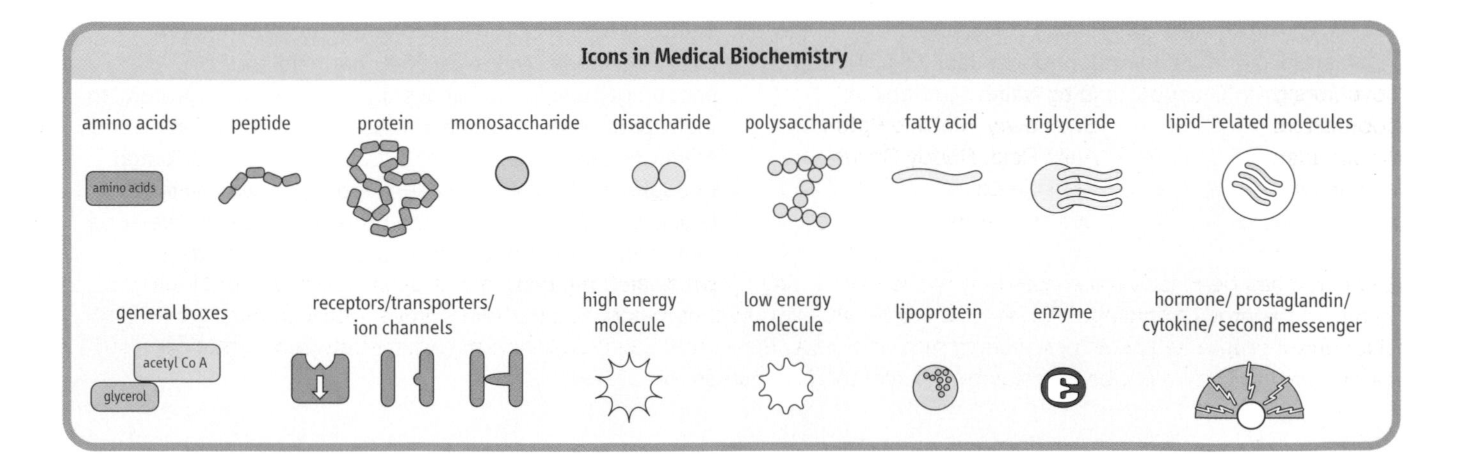

Dedication

To those who have made it possible: my parents, John and Marie Baynes; my mentors Ed Heath and Finn Wold; my wife, Suzanne Thorpe; and to the more than 1 000 medical students whom I have taught, and who have taught me, at the University of South Carolina School of Medicine in Columbia.

John Baynes

To Anna, my parents, and to Peter Jacob, who at the age of thirteen graciously accepted that "the book" was important.

To my teachers, H Gemmell Morgan and Stefan Angielski, who showed me how exciting laboratory medicine can be.

Marek Dominiczak

Acknowledgements

We wish to thank all the contributors for their efforts, for many helpful suggestions and for their patience in following the editorial philosophy and for complying with sometimes impossible deadlines. Special thanks go to the administrative staff at our laboratories: Jacky Gardiner and Anne Cooney in Glasgow, and to Wilma Murdock in Columbia.

It was a privilege to work with the publishing team at Mosby international: Louise Crowe, Michele Campbell and Cathryn Waters. We are grateful for their constant support, enthusiasm and skill in directing the production of this book. We thank Dianne Zack for her help and encouragement in the initial stages. We are also grateful to the superb team of artists at Mosby for their expertise and creativity and to many other individuals who contributed throughout. The team was such that there was never a feeling that the creation of the book was a chore. We hope that at least some of the enthusiasm that we had permeated the text.

Prof. Neill

Medical Biochemistry

John Baynes PhD
Carolina Distinguished Professor
Department of Chemistry and Biochemistry
and School of Medicine
University of South Carolina
Columbia, USA

Marek H Dominiczak
MD, FRCPath, FRCP (Glasg.)
Head of Department of Biochemistry
Western Infirmary and Gartnavel General
Hospital
Honorary Clinical Senior Lecturer
University of Glasgow
Glasgow, UK

London Edinburgh New York Philadelphia Sydney Toronto

Published in 1999 by Mosby, an imprint of Harcourt Brace and Company Limited.

ISBN: 0 7234 3012 8

Reproduction by Prospect Litho, Basildon, England
Printed and bound by Grafos SA, Arte sobre papel, Barcelona, Spain

Cataloging-in-Publication Data:
Catalogue records for this book are available from the US Library of Congress and the British Library

Editor	Louise Crowe
Development Editor	Michele Campbell
Project Manager	Cathryn Waters
Designer	Greg Smith
Layout	Rob Curran
Illustration	Mike Saiz
	Danny Pyne
Illustrators	Paul Burnson
	Rob Dean
	Milvia Romici
	Mick Ruddy
	Marion Tasker
Cover design	Greg Smith
Copyeditors	Sue Lowry, Melanie Paton
Proofreaders	Anita Reid, Roddy Crews
Production	Siobhan Egan
Index	Liza Weinkove

Great care has been taken to ensure that normal levels, drug dosages and test procedures described in this book are correct at the time of printing. However, the reader should note that, before administering drugs, dosages must be verified with current drug data sheets provided by manufacturers. Similarly, the test protocols may differ between institutions, and their implementation must be authorized by senior medical personnel.

Contributors

Gary Bannon PhD
Department of Biochemistry and
Molecular Biology
School of Medicine
University of Arkansas for Medical
Sciences
Little Rock, AR
USA

Graham Beastall PhD FRCPath
Department of Clinical Biochemistry
Glasgow Royal Infirmary
Glasgow, UK

Iain Broom MBChB MIBiol FRCPath
Department of Clinical Biochemistry
University Medical School
Aberdeen, UK

John C Chamberlain BM PhD MRCPath
Department of Clinical Chemistry
Royal Liverpool University Hospital
Liverpool, UK

Alan Elbein PhD
Department of Biochemistry and
Molecular Biology
University of Arkansas for Medical
Sciences, Little Rock, AR
USA

Alex Farrell MBChB MRCGP DCH DRCOG
Department of Immunology
Western Infirmary
Glasgow, UK

Bill Fraser MD FRCPath
Department of Clinical Chemistry
Royal Liverpool University Hospital
Liverpool, UK

Junichi Fujii PhD
Department of Biochemistry
Osaka University Medical School
Osaka
Japan

Maggie Harnett PhD
Department of Immunology
Western Infirmary
Glasgow, UK

George Helmkamp PhD
Department of Biochemistry and
Molecular Biology
University of Kansas Medical Center
Kansas City, KS
USA

Andrew Jamieson MBChB (Hons) PhD MRCP (UK)
Department of Endocrinology
Southern General Hospital
Glasgow, UK

Alan F Jones MBBChir DPhil MRCP MRCPath
Department of Clinical Chemistry
Birmingham Heartlands Hospital
Birmingham, UK

Gur P Kaushal PhD
Department of Medicine and
Biochemistry and Molecular Biology
University of Arkansas for Medical
Sciences
Little Rock, AR
USA

Gordon Lowe MBChB MD FRCP
Department of Medicine
Glasgow Royal Infirmary
Glasgow, UK

Masatomo Maeda PhD
Department of Pharmaceutical
Science
Osaka University
Osaka
Japan

Allen Rawitch PhD
Department of Biochemistry and
Molecular Biology
University of Kansas Medical Center
Kansas City, KS
USA

David Shapiro PhD MRCPath
Department of Biochemistry
Gartnavel General Hospital
Glasgow, UK

William Stillway PhD
Department of Biochemistry
Medical University of South Carolina
Charleston, SC
USA

Mirosława Szczepańska-Konkel PhD
Department of Clinical Biochemistry
Medical University
Gdańsk
Poland

Naoyuki Taniguchi MD PhD
Department of Biochemistry
Osaka University Medical School
Osaka
Japan

Edward Thompson PhD MD DSc FRCPath MRCP
Department of Neuroimmunology
Institute of Neurology
London, UK

Robert Thornburg PhD
Department of Biochemistry and
Biophysics
Iowa State University
Ames, IA
USA

Contents

Contents

Contents

Contents

Contents

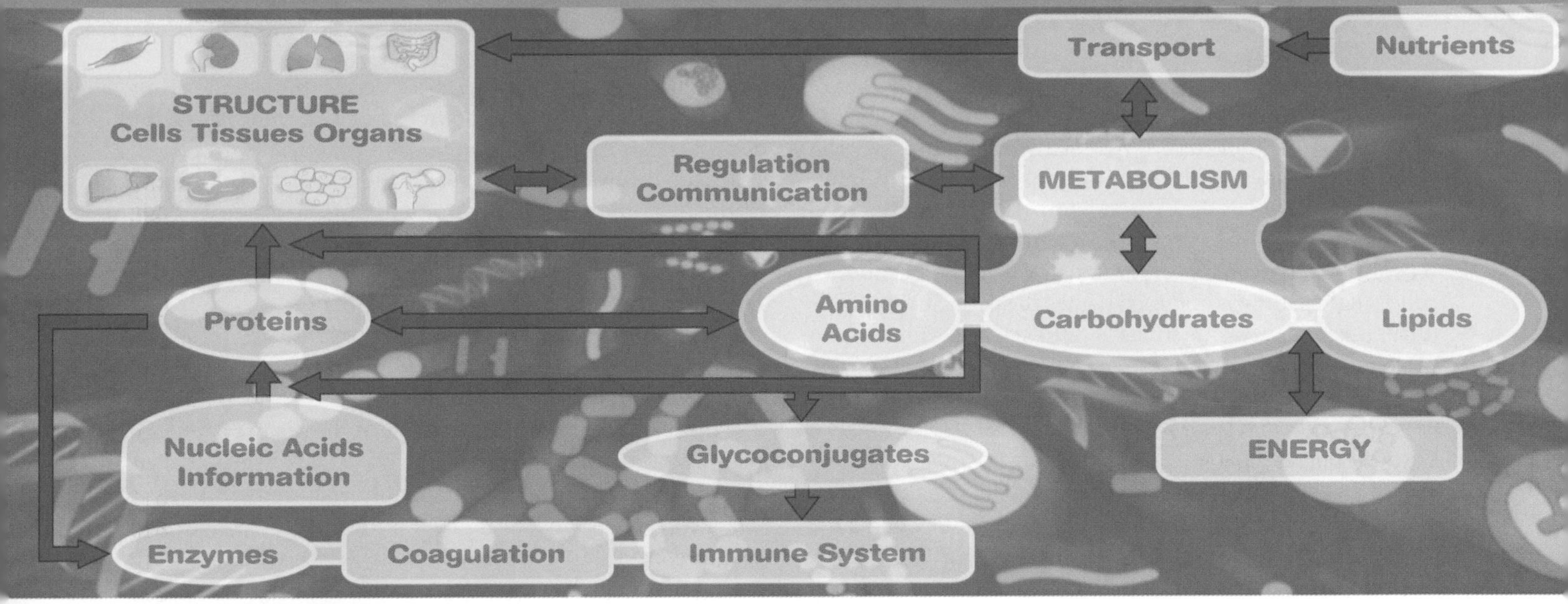

1 Introduction

The story of biochemistry – an overview

The study of human biochemistry opens our eyes to how the body works, and provides a basis for understanding what can, and often does, go wrong. From a physician's point of view, biochemistry provides not only a description of how the system works, but also a foundation for understanding how to improve its operation (by appropriate nutrition, exercise, preventive medicine), how to diagnose problems and, where possible, how to remedy them. Current therapies include recombinant proteins, such as human insulin or erythropoietin synthesized by bacteria, and future therapies will include genetic engineering, involving gene rather than organ transplants. To understand how the human body works, and the basis of the therapies for its maintenance and healing, it is essential to understand not only the 'nuts and bolts', but also the functional interactions between metabolic pathways, organs, and tissues. This, in a broad sense, is the realm of physiologic biochemistry.

The living organism communicates with its environment

It is useful to consider the human organism from two points of view: as a tightly controled, internally integrated metabolic system, and as a flexible, open system that communicates with its environment. Interactions between these systems are essential for the maintenance of our internal, homeostatic environment. We regularly consume fuel (food) and water, and we constantly take up oxygen in inspired air and transport it to tissues for oxidative metabolism. We use the energy from metabolism of foods to perform work and to maintain body temperature – respiration is a controled combustion reaction. We exhale or excrete the primary metabolic products, carbon dioxide, and water. Water represents approximately 50% of our total body mass; we control its loss, its electrolyte and metabolite concentrations, and use it as the common medium for biochemical reactions. Carbon dioxide, before elimination, is used for buffering the pH of body fluids.

Most metabolism occurs within the complex ecosystem of the cell, the components of which include the nucleus, cell membrane, endoplasmic reticulum (rough and smooth), Golgi apparatus, mitochondrion, lysosomes, peroxisomes, and the elements of the cytoskeleton. The compartmentalization of metabolic processes is important for several reasons: to protect the organism from autodigestion, to concentrate pathways and metabolites in space, and to enable different pathways, such as synthesis and turnover of proteins, to operate at the same time.

Medical Biochemistry combines metabolic and clinical aspects

The main pathways of carbohydrate and lipid metabolism are used as routes of access to other processes in biochemistry. At the beginning of every chapter, there is a simplified flow chart giving an overview of the many aspects of biochemistry (Fig. 1.1); the specific area to which that particular chapter relates is highlighted, to help the reader orientate him/herself in the overall picture.

We start with an introduction to the structure and function of proteins, lipids, and biological membranes. Proteins are the building-blocks and catalysts of biochemical systems: as structural units, they hold us together and form the architectural framework of our tissues; as enzymes, together with helper molecules known as coenzymes and cofactors, they catalyze and control biochemical reactions. We describe the elements of the homeostatic environment, such as pH, oxygen tension, inorganic ion and buffer concentrations, in which human metabolism takes place. Therapeutic efforts to maintain the stability of this environment are a significant part of clinical medicine. The chapter on blood plasma is particularly important, because plasma is an accessible 'window' on metabolism and serves as a source of clinical information for the diagnosis and management of disease. The enzymatic reactions involved in blood coagulation, and the complexity of the immune system, illustrate the sophistication of our defenses against disturbances in this environment.

Biological membranes contain and compartmentalize biochemical processes

Biological membranes perform important roles in the compartmentalization of metabolites, metabolic pathways and metabolite transport. Bioenergetics, the science of energy recovery and utilization in biological systems, is introduced through the function of mitochondrial membranes in oxidative phosphorylation. This is the process, involving oxygen consumption or respiration, by which we harness the energy of fuels, trapping it as adenosine triphosphate (ATP). ATP is the 'common currency of metabolism' for exchange of metabolic energy, transducing the energy from fuel metabolism for use in work, transport, and biosynthesis.

Fuel metabolism – the pathway from foods to ATP

The process of fuel metabolism is introduced with the digestive activity of the gut, followed by discussion of our nutritional requirements and the role of vitamins and minerals in our diet. The metabolism of fuels is introduced through glycolysis, an ancient, anaerobic metabolic pathway for glucose metabolism and energy production. Glycolysis proceeds through identical steps both in our brain cells and in the anaerobic bacteria in our intestines; it transforms glucose to pyruvate, setting the stage for oxidative metabolism in the mitochondrion. This pathway provides an opportunity

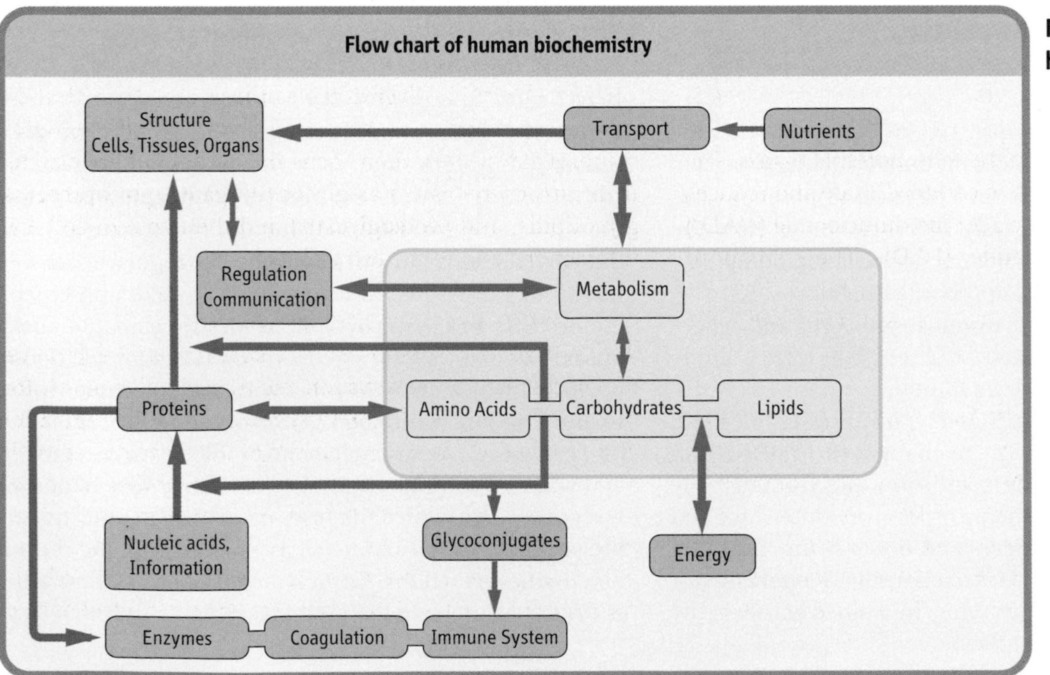

Fig. 1.1 The flow chart of human biochemistry.

to introduce the mechanisms of regulation of metabolic pathways by small-molecule allosteric effectors, by reversible chemical modification of key enzymes, and by control of gene expression.

Glucose is not only our major carbohydrate fuel, but also the circulating form of carbohydrate in blood, and several chapters explain how its concentration is regulated. This is because maintenance of a normal concentration of blood glucose, only one-fifth of a teaspoon of sugar in a liter of blood (100 mg/dL; 1 g/L; 5 mmol/L) is essential for our survival. When blood glucose decreases to less than 45 mg/dL (2.5 mmol/L), we may fall into a hypoglycemic coma; when it remains consistently greater than 125 mg/dL (7 mmol/L), we become diabetic and are at risk for a range of chronic diabetic complications – vascular disease, renal failure, blindness. We also discuss the metabolism of glycogen, the storage form of glucose in liver and muscle. In this chapter, we introduce the function of hormones in the regulation of metabolic pathways, and describe how organs use hormones to communicate with one another and how hormones activate, inactivate, and coordinate intracellular metabolic activities.

Both oxidative metabolism and antioxidant protection are essential for our survival

Oxygen is not required for conversion of glucose to pyruvate during glycolysis, but it is needed for oxidative metabolism of pyruvate to carbon dioxide and water – which is essential for maximal extraction of the energy available from glucose. Conversely, oxygen can also be toxic, causing oxidative stress and damage. We discuss both the advantageous and disadvantageous features of oxygen, emphasizing how we harness it usefully and, at the same time, protect ourselves from its more damaging effects through antioxidant defenses.

In aerobic cells, pyruvate is transformed into a key metabolite, acetyl coenzyme A (acetyl CoA), which is the common intermediate in the energy metabolism of carbohydrates, lipids, and amino acids. It enters the central metabolic engine of the cell, the tricarboxylic acid cycle (TCA cycle, citric acid cycle, Krebs cycle) in the mitochondrial matrix. The TCA cycle oxidizes acetyl CoA to carbon dioxide and reduces the coenzymes, nicotinamide adenine dinucleotide (NAD^+) and flavin adenine nucleotide (FAD). These reduced nucleotides are the substrates for oxidative phosphorylation in the mitochondrion; their oxidation provides the energy for synthesis of ATP.

The fast–feed cycle: catabolism versus anabolism

As we work our way through energy metabolism, a complex web of interactions between biochemical pathways becomes evident. The process of glycolysis, in addition to its role in setting the stage for energy metabolism, where we began, produces metabolites that serve as the starting point for synthesis of amino acids, proteins, lipids, and nucleic acids.

An important principle that evolves is the partial reversibility of the main pathways of carbohydrate and lipid metabolism – catabolism versus anabolism. The direction of metabolism is constantly shifting with the fast–feed cycle. In the fed state, the active pathways are glycolysis, glycogen synthesis, lipogenesis, and protein synthesis, rejuvenating tissues and storing the excess of metabolic fuel. In the fasting state, which begins only a few hours after our last meal, the direction of metabolism is reversed: glycogen and lipid stores are degraded, protein is converted into glucose by the pathway of gluconeogenesis, and other biosynthetic processes are slowed down. We place particular emphasis on the mechanisms for adaptation to the extreme changes in energy status induced by feeding, changing diets, fasting, and starvation. We also consider the integration of fuel metabolism within and among tissues, storage of fuel in tissues and its transport in plasma between tissues, differences in fuel preferences and the metabolic capacity of individual tissues, and the derangements in fuel metabolism that occur in diabetes and atherosclerosis.

Biochemistry is often perceived by students as excessively complicated. It becomes much less so, once one develops an overall view and a mental picture of how the different parts of metabolism interact. We have developed a general (and necessarily simplified) scheme which looks not unlike the drawing of the London Underground! It is included as an Appendix at the end of the book. Its purpose is that the reader, after going through different chapters on specific subjects, can refer to it to place a particular aspect/pathway/class of substances in the context of the whole system.

Tissue diversity and specialization

To illustrate the diversity of biochemistry in specialized tissues, we describe the mechanism of muscle contraction, the roles of the lung and kidneys in acid–base and electrolyte balance, that of the liver in biosynthesis and detoxification, and the processes of bone metabolism and biological mineralization. We then focus on the role of specialized microstructures, such as glycoconjugates (glycoproteins, glycolipids, and proteoglycans) and their role in cell–cell interactions and in the extracellular matrix.

Underlying it all: the human genome

We then turn our attention to the structure of the genome, the mechanism of conservation and transfer of genetic information, the control of protein synthesis, and the regulation of gene expression. These pathways are complex – protein synthesis is controled by information encoded in deoxyribonucleic acid (DNA) and transcribed into ribonucleic acid (RNA), then translated into peptides which fold into functional protein molecules. Of importance is not only the spectrum of expressed proteins, but also the control of their temporal expression during development, adaptation, and aging. Information presented in these chapters offers many opportunities for understanding treatments and strategies for inhibiting bacterial and viral infections, including acquired immunodeficiency syndrome (AIDS). This is followed by a discussion of applications of recombinant DNA and polymerase chain reaction (PCR) technology in the clinical laboratory and in molecular medicine. The concluding chapters of the book deal with integrated topics, such as the function of the immune system, biochemical endocrinology, and the specialized biochemistry of nerve tissue. In the last chapter, we deal with the issues of cell growth and the failure of biochemical systems in cancer.

Due to the fast-changing knowledge and general re-thinking of health care delivery around the world, major changes are taking place in medical education today. The modern curricula emphasize the integration of basic and clinical disciplines and introduce clinical contact at an early stage. The amount of facts students need to learn is being decreased, as young doctors are equipped with skills to continue learning throughout their whole professional career.

Our most important aim is to help the reader apply the understanding of biochemistry to clinical practice. This is why the chapters contain clinical examples which usually illustrate a single practical point. We hope that after reading the book, you will conclude that biochemistry becomes even more enjoyable when linked to problems you confront at the bedside.

Further reading

Dominiczak MH. Teaching and training laboratory professionals for the 21st century. *Clin Chem Lab Med* 1998;**36**:133–36.

Jolly B, Rees L, eds. *Medical education in the millenium*. Oxford: Oxford University Press; 1998;1–268.

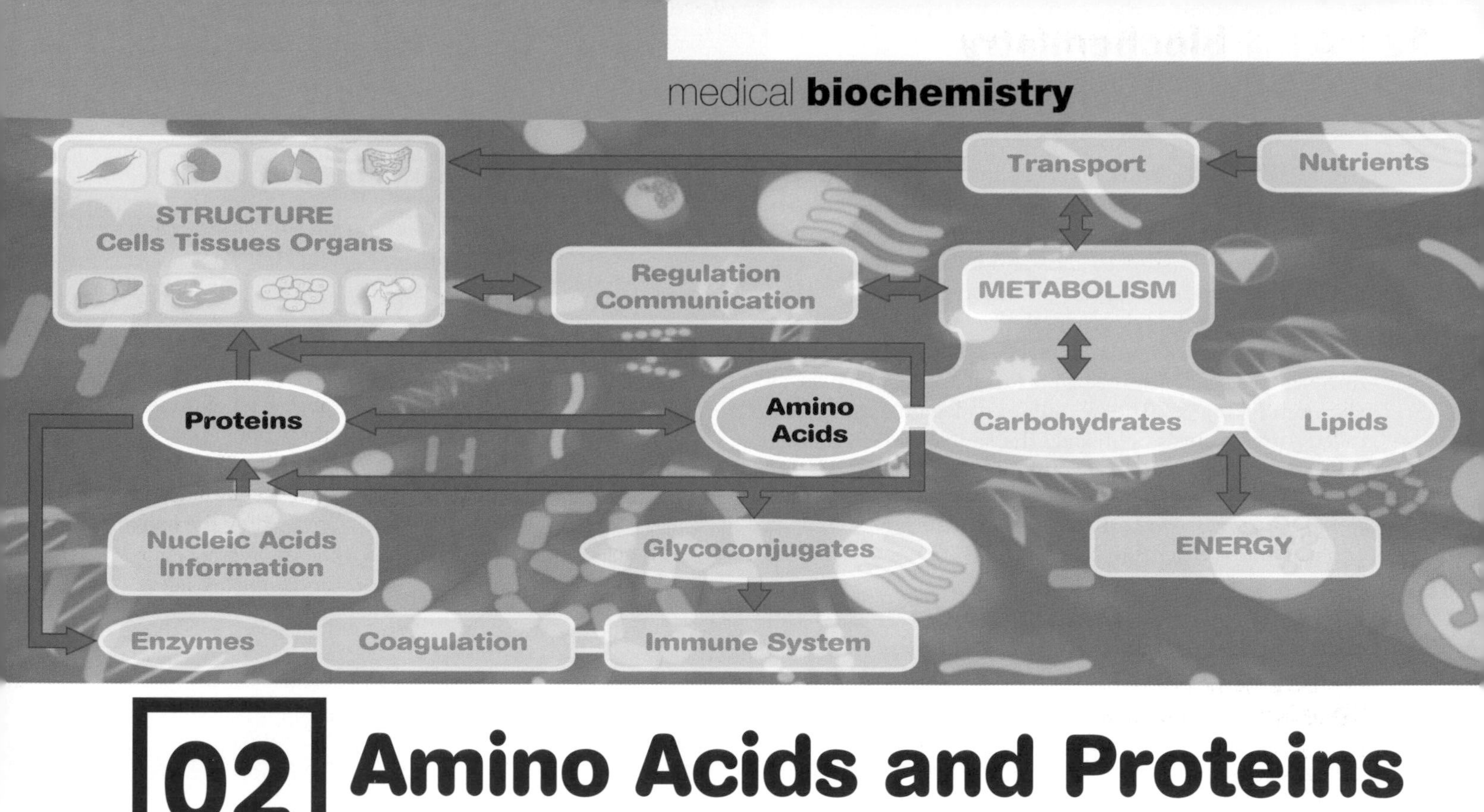

02 Amino Acids and Proteins

Introduction

Proteins are the primary structural and functional polymers in living systems. They have a broad range of activities, including catalysis of metabolic reactions and transport of vitamins, minerals, oxygen, and fuels. Some proteins make up the structure of tissues, while others function in nerve transmission, muscle contraction, and cell motility, others in blood clotting and immunologic defences, and others as hormones and regulatory molecules. Proteins are synthesized as a sequence of amino acids linked together in a linear polyamide (polypeptide) structure, but they assume complex three-dimensional shapes in performing their function. There are approximately 300 amino acids present in various animal, plant, and microbial systems, but only 20 amino acids are coded by DNA to appear in proteins. Many proteins also contain modified amino acids and accessory components termed prosthetic groups.

Amino acids

Stereochemistry: configuration at the α-carbon, D- and L-isomers

Each amino acid has a central carbon, called the α-carbon, to which four different groups are attached (Fig. 2.1):

- a basic amino group ($-NH_2$)
- an acidic carboxyl group ($-COOH$)
- a hydrogen atom ($-H$)
- a distinctive side chain ($-R$).

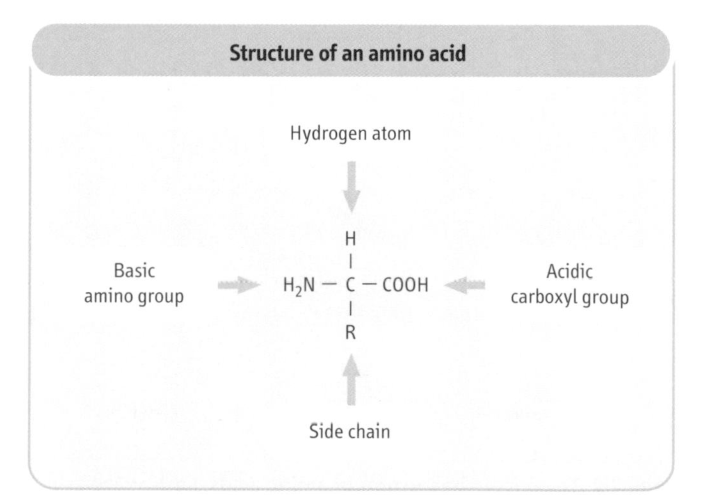

Fig. 2.1 Four different groups are attached to the α-carbon of an amino acid.

One of the 20 amino acids, proline, is not an α-amino acid but an α-imino acid (see below). Except for glycine, all amino acids contain at least one asymmetric carbon atom (the α-carbon atom), giving two isomers that are optically active, i.e. can rotate plane-polarized light. These isomers, referred to as enantiomers, are said to be chiral, a word derived from the Greek word for hand. Such isomers are nonsuperimposable mirror images and are analogous to left and right hands, as shown in Figure 2.2. The two amino acid configurations are called D (for dextro or right) and L (for laevo or left). All amino acids in proteins are of the L-configuration, because proteins are biosynthesized by enzymes that insert only L-amino acids into the chains.

Classification of amino acids based on chemical structure

The properties of each amino acid are dependent on its side chain (–R); the side chains are the functional groups that are the major determinants of the conformation and function of proteins, as well as the electrical charge on the molecule. Knowledge of the properties of these side chains is important for understanding methods of analysis, purification, and identification of proteins. Amino acids with charged, polar, or hydrophilic side chains are usually exposed on the surfaces of proteins. The nonpolar hydrophobic residues are usually buried in the hydrophobic interior of a protein and are out of contact with water. The 20 amino acids in proteins encoded by DNA are listed in Figure 2.3 and are classified according to their side-chain functional groups.

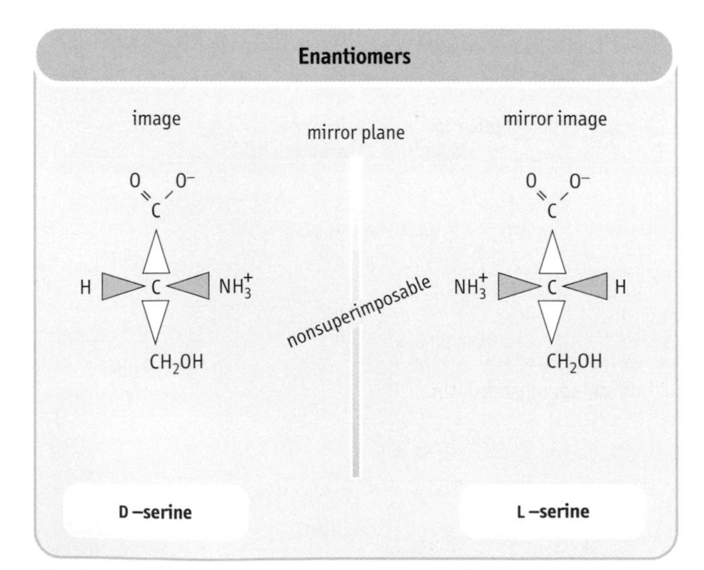

Fig. 2.2 The mirror-image pair of amino acids. Each amino acid represents nonsuperimposable mirror images. The mirror-image stereoisomers are called enantiomers.

The 20 α–amino acids specified by the genetic code

Amino acids	Structure of R moiety
Aliphatic amino acids	
glycine (Gly, G)	$-H$
alanine (Ala, A)	$-CH_3$
valine (Val, V)	$-CH(CH_3)_2$
leucine (Leu, L)	$-CH_2-CH(CH_3)_2$
isoleucine (Ile, I)	$-CH(CH_3)-CH_2-CH_2$
Sulfur-containing amino acids	
cysteine (Cys, C)	$-CH_2-SH$
methionine (Met, M)	$-CH_2-CH_2-S-CH_3$
Aromatic amino acids	
phenylalanine (Phe, F)	$-CH_2-$ (benzene ring)
tyrosine (Tyr, Y)	$-CH_2-$ (benzene ring) $-OH$
tryptophan (Trp, W)	$-CH_2-$ (indole ring, N H)
Imino acid	
proline (Pro, P)	CH_2 / CH_2 CH_2 / $-HN-CH-COOH$ (Whole structure)
Neutral amino acids	
serine (Ser, S)	$-CH_2-OH$
threonine (Thr, T)	$-CH(CH_3)-OH$
asparagine (Asn, N)	$-CH_2-CONH_2$
glutamine (Gln, Q)	$-CH_2-CH_2-CONH_2$
Acidic amino acids	
aspartic acid (Asp, D)	$-CH_2-COOH$
glutamic acid (Glu, E)	$-CH_2-CH_2-COOH$
Basic amino acids	
histidine (His, H)	$-CH_2-$ (imidazole ring, N NH)
lysine (Lys, K)	$-CH_2-CH_2-CH_2-CH_2-NH_2$
arginine (Arg, R)	$-CH_2-CH_2-CH_2-NH-C(=NH)-NH_2$

Fig. 2.3 The 20 amino acids found in proteins. The three-letter and single-letter abbreviations in common use are given in parentheses.

Aliphatic amino acids

Glycine, alanine, valine, leucine, and isoleucine, referred to as aliphatic amino acids, have saturated hydrocarbons as side chains. Alanine, which has a relatively simple structure and is the most abundant amino acid in most proteins, is a representative example. All of these amino acids are hydrophobic in nature.

Aromatic amino acids

Phenylalanine, tyrosine, and tryptophan have aromatic side chains. The nonpolar aliphatic and aromatic amino acids contribute to the internal hydrophobic interactions of the proteins. Tyrosine and tryptophan, which have some polarity, may be located near the surface. The hydroxyl group (–OH) of tyrosine is weakly acidic. Reversible phosphorylation of the hydroxyl group of tyrosine in some enzymes is important in the regulation of metabolic pathways. The aromatic amino acids are responsible for the ultraviolet absorption of most proteins, which have absorption maxima between 275 and 285 nm. Tryptophan has a greater absorption in this region than the other two aromatic amino acids. The molar absorption coefficient of a protein is useful in determining the concentration of a protein in solution, based on spectrophotomctry. Typical absorption spectra of aromatic amino acids and a protein are shown in Figure 2.4.

Neutral polar amino acids

Neutral polar amino acids contain hydroxyl groups or amide side-chain groups. Serine and threonine contain hydroxyl groups. These amino acids sometimes exist at the catalytic sites of certain kinds of enzyme called hydrolases (see Chapter 5). Serine and threonine are also involved in O-glycosidic linkage of sugar chains to form glycoproteins (see Chapter 24). These amino acids are also important residues because serine/threonine kinases phosphorylate these amino acid residues in enzymes that regulate energy metabolism and fuel storage in the body (see Chapter 12). Asparagine and glutamine have amide-bearing side chains. These are polar, but uncharged under physiological conditions.

Acidic amino acids

Aspartic and glutamic acids contain carboxylic acids on their side chains and are ionized at pH 7.0, and, as a result, carry negative charges on their β- and γ-carboxyl groups respectively. When in the ionized state, these amino acids are referred to as aspartate and glutamate, respectively.

Basic amino acids

The side chains of lysine and arginine are fully protonated at neutral pH and, therefore, positively charged. Lysine contains a primary amino group (NH_2) attached to the terminal ε-carbon of the side chain. The ε-amino group of

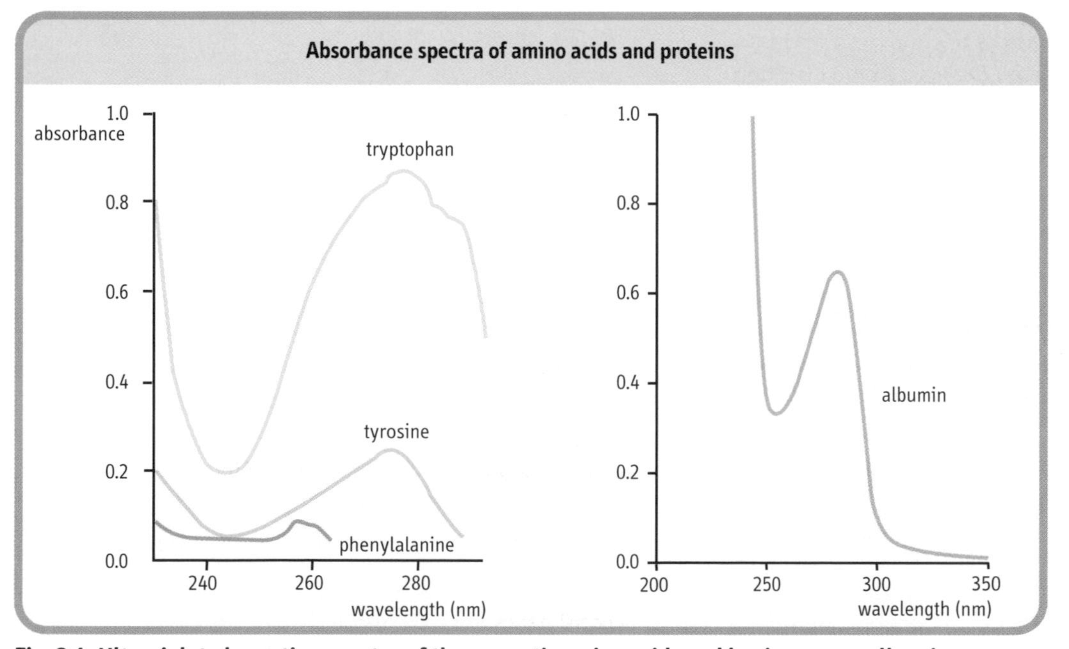

Fig. 2.4 Ultraviolet absorption spectra of the aromatic amino acids and bovine serum albumin.
Aromatic amino acids such as tryptophan, tyrosine, and phenylalanine have absorbance maxima at ~280 nm. Each purified protein has a distinct molecular absorption coefficient at around 280 nm. According to Beer's law, $A_\lambda = \varepsilon_\lambda \times c \times l$ where A_λ = absorbance at wavelength λ, ε_λ = absorption coefficient at wavelength λ, c = concentration, and l = light path length of the cuvette measured in cm.

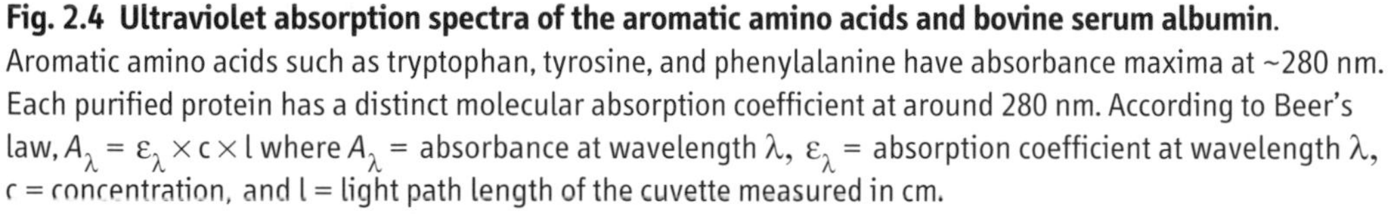

A bovine serum albumin solution (1 mg dissolved in 1 ml of water) has an absorbance of 0.67 at 280 nm using a 1 cm cuvette. The absorption coefficient of proteins is often expressed as $E^{1\%}$ (10 mg/ml solution). For albumin, $E^{1\%}_{280\,mm} = 6.7$ when l = 1 cm. Although proteins vary in their Trp, Tyr, and Phe content, measurements of absorbance at 280 nm are useful for estimating the concentration of protein in solutions.

lysine has a p$K_a \approx 11$. Arginine is the most basic amino acid (p$K_a \approx 13$), and its guanidine group exists as a protonated guanidinium ion at pH 7.0.

Histidine (p$K_a \approx 6$) has an imidazole ring as the side chain and functions as a general acid–base catalyst in many enzymes. The protonated form of imidazole is called an imidazolium ion.

Sulfur-containing amino acids

Cysteine and its oxidized form, cystine, are sulfur-containing amino acids and of low polarity. Cysteine plays an important role in stabilization of protein structure, since it can participate in formation of a disulfide bond with other cysteine residues to form cystine residues, crosslinking protein chains. Two regions of a single polypeptide chain, remote from each other in the sequence, may be covalently linked through a disulfide bond (intrachain disulfide bond). Disulfide bonds are also formed between two polypeptide chains (interchain disulfide bond), forming covalent protein dimers. These bonds can be reduced by enzymes, or by reducing agents such as 2-mercaptoethanol or dithiothreitol to form cysteine residues. Methionine is the third sulfur-containing amino acid and contains a nonpolar methyl thioether group in its side chain.

Imino acids

Proline is different from other amino acids in that its side-chain pyrrolidine ring includes both the α-amino and α-carboxyl groups. This structure forces a 'bend' in a polypeptide chain, sometimes causing abrupt changes in the direction of the chain. Chemically speaking, proline is not an α-amino ($-NH_2$) acid but, rather, an α-imino (–NH–) acid.

Classification of amino acids based on the polarity of the amino acid side chains

Figure 2.5 depicts the functional groups of amino acids and their polarity (hydrophilicity). Polar amino acid side chains can be involved in hydrogen bonding to water and other polar groups. The hydrophobic nature of side chains contributes to protein folding by hydrophobic interactions.

Ionization state of an amino acid

Amino acids are amphoteric molecules, that is, they have both basic and acidic groups. Monoamino and monocarboxylic acids are ionized in different ways in solution, depending on the solution pH. At pH 7, the 'zwitterion' $^{+}H_3N{-}CH_2{-}COO^{-}$ is the dominant species of glycine in solution, and the overall molecule is therefore electrically neutral. At acid pH, the α-amino group is protonated and positively charged, yielding $^{+}H_3N{-}CH_2{-}COOH$, while at alkaline pH glycine exists primarily as the anionic $H_2N{-}CH_2{-}COO^{-}$ species.

At a pH intermediate between the pK_a (see below) of the amino and carboxyl groups, known as the isoelectric point (pI), the zwitterionic form of the amino acid has no net charge.

Henderson–Hasselbach equation and pK_a

As described above, all amino acids contain ionizable groups that act as weak acids or bases, giving off or taking on protons when the pH is altered. The general dissociation of a weak acid is given by the equation:

Summary of the functional groups of amino acids and their polarity

Amino acids	Functional group	Hydrophilic (polar) or hydrophobic (apolar)	Examples
acidic	carboxyl, –COOH	polar	Asp, Glu
basic	amine, $-NH_2$ imidazole, guanidino,	polar polar polar	Lys His Arg
neutral	amides, $-CONH_2$ hydroxyl, –OH	polar polar	Asn, Gln Ser, Thr,
aliphatic	hydrocarbon	apolar	Ala, Val, Leu, Ile, Met
aromatic	C-rings	apolar	Phe, Trp, Tyr

Fig. 2.5 Polarity of functional groups of amino acids.

Nonprotein amino acids

Some amino acids that occur in free or combined states, but not in proteins, play important roles in metabolism. Measurement of abnormal amino acids in urine (aminoaciduria) is useful for clinical diagnosis (see Chapter 18). In plasma, free amino acids are usually found in the order of 10 to 100 μmol/L, including many that are not found in protein. Citrulline, for example, is an important metabolite of L-arginine and a product of nitric oxide synthase, an enzyme that produces nitric oxide, an important signaling molecule. Urinary amino acid concentration is usually expressed as μmol/g creatinine. Creatinine is an amino acid derived from muscle, and is excreted in relatively constant amounts per unit body mass per day. Thus, the creatinine concentration in urine, normally about 1 mg/mL, can be used to correct for urine dilution. The most abundant amino acid in urine is glycine, which is present as 400–2000 μg/g creatinine.

(1) $HA \rightleftharpoons H^+ + A^-$

where HA is the protonated form (conjugate acid or associated form) and A^- is the unprotonated form (conjugate base, or dissociated form).

The dissociation constant (K_a) of a weak acid is defined as:

(2) $$K_a = \frac{[H^+][A^-]}{[HA]}$$

The hydrogen ion concentration [H^+] of a solution of a weak acid is calculated as follows. Equation (2) can be rearranged to give:

(3) $$[H^+] = K_a \times \left(\frac{[HA]}{[A^-]}\right)$$

Equation (3) can be expressed in terms of a negative logarithm:

(4) $$-\log[H^+] = -\log K_a - \log\left(\frac{[HA]}{[A^-]}\right)$$

since pH equals the negative logarithm of [H^+], i.e. $-\log[H^+]$ and pK_a equals the negative logarithm of the dissociation constant for a weak acid, i.e. $-\log K_a$.

The Henderson–Hasselbalch equation, which is derived from equation (4) and is shown below, is useful in dealing with acid–base equilibrium systems and is often used for making buffer solutions, as described below.

(5) $$pH = pK_a + \log\left(\frac{[A^-]}{[HA]}\right)$$

When the conjugate base and acid are present at equal concentrations, their ratio is 1, and the log is 0, so pH = pK_a. Buffers are solutions that minimize a change in [H^+] upon addition of acid or base. A buffer has maximal buffering capacity at its pK_a when the acidic and basic forms are present at equal concentrations. The acidic, protonated form reacts with added base, and the basic unprotonated form neutralizes added acid, as shown below for an amino compound:

$$RNH_3^+ + OH^- \rightleftharpoons RNH_2 + H_2O$$

$$RNH_2 + H^+ \rightleftharpoons RNH_3^+$$

An alanine solution has maximal buffering capacity at pK_a values 2.4 and 9.8, and minimal buffering capacity at its pI, as shown in Figure 2.6. When dissolved in water, alanine exists as a dipolar ion, or zwitterion, in which the carboxyl group is unprotonated ($-COO^-$) and the amino group is protonated ($-NH_3^+$). The pH of the solution is 6.0, half-way between the pK_a of the amino and carboxyl groups. The titration curve of alanine by NaOH is shown in Figure 2.6.

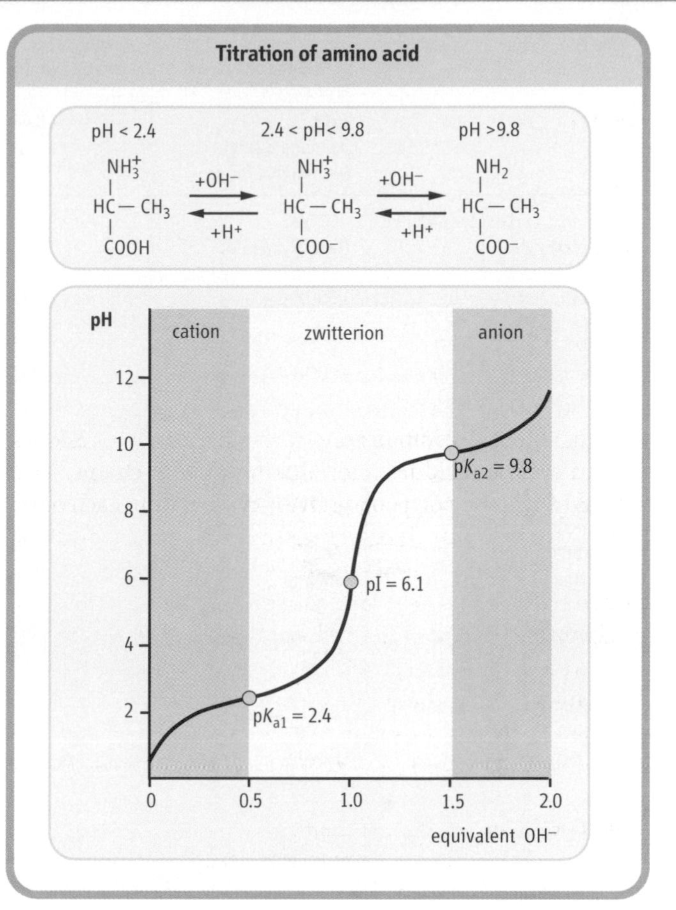

Fig. 2.6 Titration curve of alanine. The curve shows the number of equivalents of NaOH consumed by alanine in titrating the solution from pH 0 to pH 12. Alanine contains two ionizable groups: an α-carboxyl group and an α-amino group. As NaOH is added, these two groups are titrated. The pK_a of the α-COOH is 2.4, whereas that of the α-NH_3^+ group is 9.8. At very low pH, the predominant ion species of alanine is:

$$^+H_3N\text{–}CH(CH_3)\text{–}COOH$$

the fully protonated form. At the mid point in the first stage of the titration (pH 2.4), equimolar concentrations of proton-donor and proton-acceptor species are present, providing good buffering power.

$$^+H_3N\text{–}CH(CH_3)\text{–}COOH \qquad H_2N - CH(CH_3) - COO^-$$

The second stage of the titration corresponds to the removal of a proton from the -NH_3^+ group of alanine. The pH at the mid point of this stage is 9.8, equal to the pK_a for the –NH_3^+ group. The titration is complete at a pH of about 12, at which point the predominant form of alanine is:

$$H_2N - CH(CH_3) - COO^-$$

The pH at which a molecule has no net charge is known as its isoelectric point, pI, and is calculated as:

$$pI = [(pK_{a1} + pK_{a2})/2] = [(2.4 + 9.8)/2] = 12.2/2 = 6.1.$$

p*K* values and ionized groups in proteins

Group	Acid (conjugate acid) (protonated form)	Base + H^+ (conjugate base) (unprotonated form)	pKa
terminal carboxyl residue (α-carboxyl)	–COOH (carboxylic acid)	$-COO^- + H^+$ (carboxylate)	3.0–5.5
aspartic acid (β-carboxyl)	–COOH	$-COO^- + H^+$	3.9
glutamic acid (γ-carboxyl)	–COOH	$-COO^- + H^+$	4.3
histidine (imidazole)	 HN⊕NH (imidazolium)	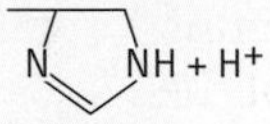 N NH + H^+ (imidazole)	6.0
terminal amino (α-amino)	$-NH_3^+$ (amino)	$-NH_2 + H^+$ (amine)	8.0
cysteine (sulfhydryl)	–SH (thiol)	$-S^- + H^+$ (thiolate)	8.3
tyrosine (phenolic hydroxyl)	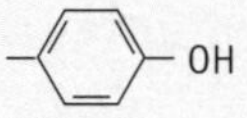 –C₆H₄–OH (phenol)	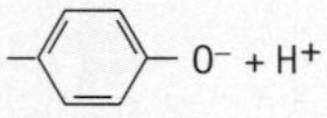 –C₆H₄–$O^- + H^+$ (phenolate)	10.1
lysine (ε-amino)	$-NH_3^+$ (ε-amino)	$-NH_2 + H^+$ (ε-amine)	10.5
arginine (guanidino)	$-NH{=}C(NH_2)_2^{\oplus}$ (guanidinium)	$-NH-C(=NH)NH_2 + H^+$ (guanidino)	12.5

Fig. 2.7 Ionized groups and p*K* values in proteins. pK_a indicates the approximate value, because it depends on temperature, buffer, etc.

The side chains of seven amino acids in proteins readily ionize. p*K* values for those ionized groups are shown in Figure 2.7.

Peptides and proteins

Primary structure of proteins

Proteins are macromolecules with a backbone formed by polymerization of amino acids in a polyamide structure. The amide bonds in protein, known as peptide bonds, are formed by linking α-carboxyl groups of one amino acid to the α-amino groups of the next amino acid by amide bonds. During the formation of a peptide bond, a molecule of water is eliminated, as shown in Figure 2.8. The amino acid units on a peptide chain are referred to as amino acid residues. A peptide chain consisting of three

Glutathione

Glutathione (GSH) is a tripeptide with the sequence L-γ-glutamyl-L-cysteinylglycine, as shown in Fig. 2.9. If the thiol group of the cysteine is oxidized, the disulfide GSSG is formed. GSH is the major peptide present in the cell. In the liver, the concentration of GSH is in the millimolar range. GSH plays a major role in the maintenance of proteins in their reduced forms and in antioxidant defenses (Chapter 11). The enzyme γ-glutamyl transpeptidase is involved in the metabolism of glutathione and is a marker for some liver diseases, including hepatocellular carcinoma and alcoholic liver diseases.

amino acid residues is called a tripeptide. For example, glutathione (Fig. 2.9) which has two peptide linkages is a tripeptide.

The terms dipeptide, tripeptide, oligopeptide (fewer than 50 amino acids), and polypeptide are often used by protein chemists. By convention, the amino terminus is taken as the first residue, and the sequence of amino acids is written from left to right. When writing the peptide sequence, one uses either the three-letter or the one-letter abbreviations of amino acids, such as Asp-Arg-Val-Tyr-Ile-His-Pro-Phe-His-Leu or D–R–V–Y–I–H–P–F–H–L (see Fig. 2.3 for abbreviations). This peptide is angiotensin, a peptide hormone that affects blood pressure. The amino acid residue having a free amino group at the end of the peptide, Asp, is called the N-terminal amino acid (amino terminus), whereas the residue having a free carboxyl group, Leu, is called the C-terminal amino acid (carboxyl terminus). The primary structure of a protein is defined by the linear sequence of amino acid residues. Proteins contain between 50 and 2000 amino acid residues. The mean molecular mass of an amino acid residue is about 110 dalton units (Da). Therefore the molecular mass of most proteins is between 5500 and 220 000 Da. Human carbonic anhydrase B, an enzyme that plays a major role in acid–base balance (see Chapter 22), is a protein with a molecular mass of 29 000 Da or 29 kilodaltons (kDa).

The charge and polarity characteristics of a peptide chain

The amino acid composition of a peptide chain has a profound effect on its physical and chemical properties. Proteins rich in aliphatic or aromatic amino groups are relatively insoluble in water and more soluble in cell membranes. Proteins rich in polar amino acids are more water soluble. Although the α-amino and α-carboxyl amino acids are involved in amide bonds and are neutral, nonionic species, amino acids with side-chain acidic (Glu, Asp) or basic (Lys, His, Arg) groups will confer charge and buffering capacity to a protein. The balance between acidic and basic side chains in a protein determines its isoelectric point (pI) and net charge in solution. Proteins rich in lysine and arginine are basic in solution and cationic at neutral pH, while acidic proteins, rich in aspartate and glutamate residues, are acidic and anionic. Because of their side-chain functional groups, all proteins become more positively charged at acidic pH and more negatively charged at basic pH. Proteins are an important part of the buffering capacity of blood and biological fluids.

Secondary structure

The secondary structure of a protein refers to the local structure of the polypeptide chain. This structure is determined by hydrogen bond interactions between the carbonyl oxygen group of one peptide bond and the amide hydrogen of another nearby peptide bond. There are two types of secondary structure, the α-helix and the β-pleated sheet.

The α-helix

The α-helix is a rod-like structure with the peptide chain tightly coiled and the side chains of amino acid residues extending outward from the axis of the spiral. Each amide carbonyl group is hydrogen-bonded to the amide hydrogen of a peptide bond that is four residues away along the same chain. There are on average 3.6 amino acid residues per turn of the helix, and the helix winds in a right-handed manner in almost all natural proteins, i.e. turns in a clockwise fashion around the axis (Fig. 2.10).

The β-pleated sheet

If the H-bonds are formed between peptide bonds in different chains, the chains become arrayed parallel or antiparallel to one another in what is commonly called a β-pleated sheet. The β-pleated sheet is an extended structure as opposed to the coiled α-helix. It is pleated because the carbon-carbon (C-C) bonds are tetrahedral and cannot exist in a planar configuration. If the polypeptide chain runs in the same direction, it forms a parallel β-sheet (Fig. 2.11), but in the opposite direction, it forms an antiparallel structure. The β-turn or β-bend refers to the segment in which the polypeptide abruptly reverses direction. Glycine (Gly) and proline (Pro) residues often occur in β-turns on the surface of globular proteins.

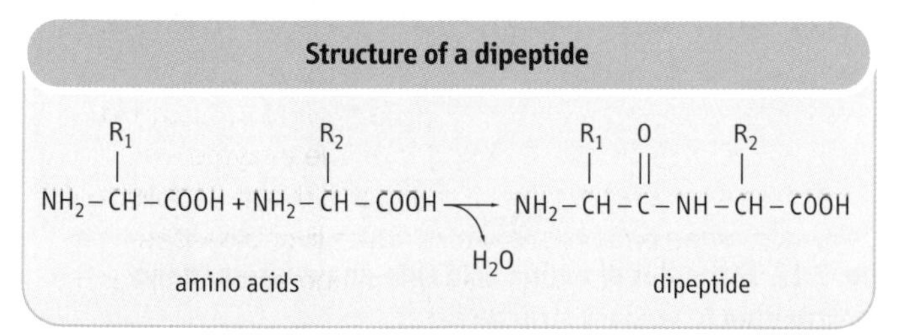

Fig. 2.8 Structure of a peptide bond.

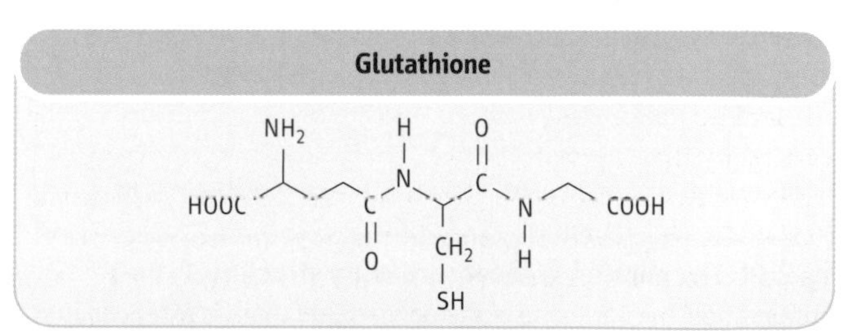

Fig. 2.9 Structure of glutathione (GSH).

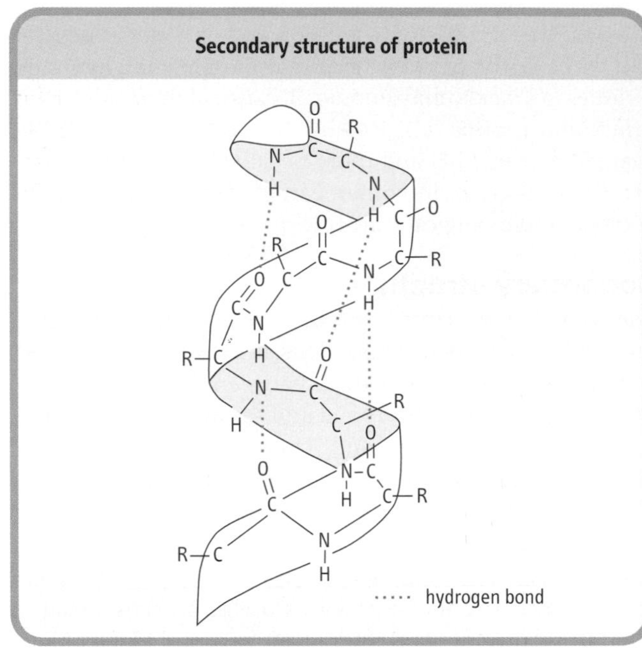

Fig. 2.10 **An α-helical secondary structure.** Hydrogen bonds between 'backbone' amide –NH– and –CO– groups stabilize the helix.

Tertiary structure

The three-dimensional, folded and biologically active conformation of a protein is referred to as its tertiary structure. This structure reflects the overall shape of the molecule. The tertiary structures of over 1000 proteins have been determined by X-ray crystallography and nuclear magnetic resonance spectroscopy. The folded conformation of proteins that contain more than 200 residues consists of several smaller folded units designated as domains.

The three-dimensional tertiary structure of a protein is stabilized by interactions between side-chain functional groups, covalent disulfide bonds, hydrogen bonds, salt bridges, and hydrophobic interactions (Fig. 2.12). The side chains of tryptophan and arginine serve as hydrogen donors, whereas asparagine, glutamine, serine, and threonine can serve as both hydrogen bond donors and acceptors. Lysine, aspartic acid, glutamic acid, tyrosine, and histidine also can serve as both donors and acceptors in the formation of ion pairs (salt bridges). Two opposite-charged amino acids, such as glutamate with a γ-carboxyl group and lysine with an ε-amino group, may form a salt bridge, primarily on the surface of proteins.

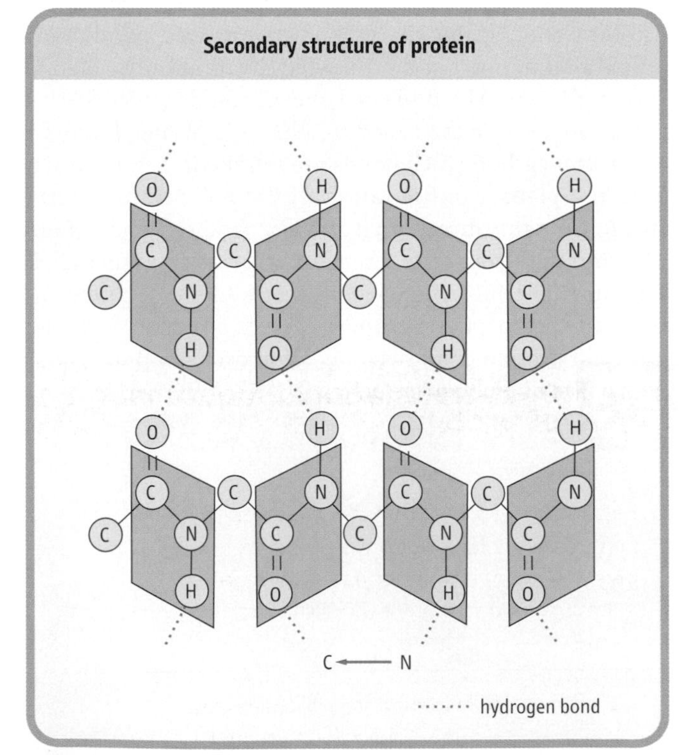

Fig. 2.11 **The parallel β-sheet secondary structure.** In the β-configuration, the backbone of the polypeptide chain is extended into a zigzag structure. When the zigzag polypeptide chains are arranged side by side, they form a structure resembling a series of pleats.

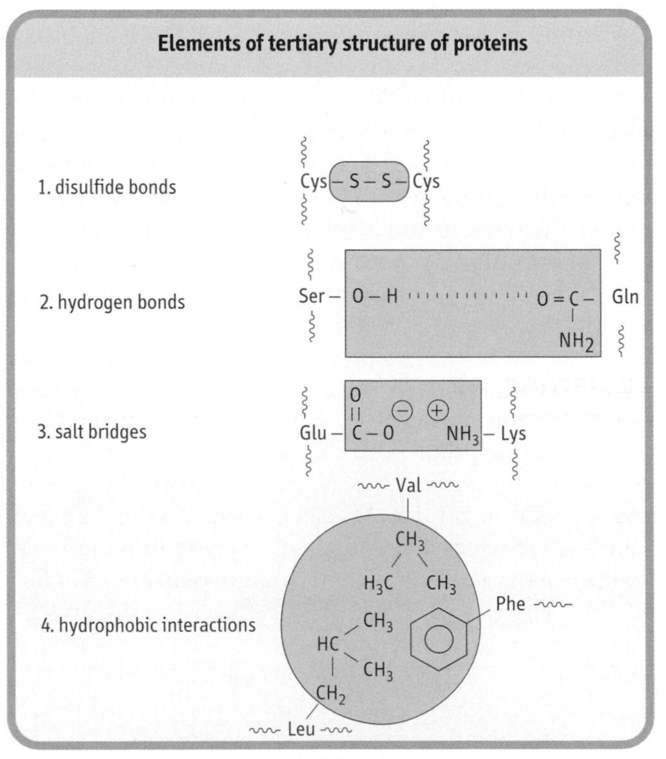

Fig. 2.12 **Examples of amino acid side-chain interactions contributing to tertiary structure.**

Compounds such as urea and guanidine hydrochloride frequently cause denaturation or loss of tertiary structure when present at high concentrations such as, for example, 8 mol/L urea. These reagents are called chaotropic agents. They are thought to form hydrogen bonds with the amino acid residues of the peptide chain, thereby disrupting the internal hydrogen bonds that stabilize the native structure. They also alter the solvent properties of water so that hydrophobic interactions in the protein are weakened. In some cases, the denaturation may be partially or almost completely reversed by dialysis or dilution, which lowers the concentration of the chaotropic agent. The refolding of proteins to their natural conformation emphasizes that much of the secondary and tertiary structures of a protein is encoded in the primary sequence.

Collagen

Human genetic defects involving collagen illustrate the close relationship between amino acid sequence and three-dimensional structure. Collagen is the most abundant protein family in the mammalian body, representing about a third of body proteins. It is a major component of connective tissue such as cartilage, tendons, the organic matrix of bones, and the cornea of the eye.

Comment. Collagen contains 35% Gly, 11% Ala, and 21% Pro plus Hyp (hydroxyproline). The amino acid sequence in collagen is generally a repeating tripeptide unit, Gly-Xaa-Pro or Gly-Xaa-Hyp, where Xaa can be any amino acid. This repeating sequence adopts a left-handed helical structure with three residues per turn. Three of these helices wrap around one another with a right-handed twist. The resulting three-stranded molecule is referred to as tropocollagen. Tropocollagen molecules self-assemble into collagen fibrils and are packed together to form collagen fibers. There are metabolic and genetic disorders related to collagen. Scurvy, osteogenesis imperfecta and Ehlers–Danlos syndrome (Chapter 26) result from defects in collagen synthesis.

Quaternary structure

Quaternary structure refers to a complex or an assembly of two or more separate peptide chains that are held together by non-covalent or, in some cases, covalent interactions. In general, most proteins larger than 50 kDa consist of more than one chain and are referred to as dimeric, trimeric, or multimeric proteins. Many multisubunit proteins are composed of different kinds of functional subunits, such as the regulatory and catalytic subunits. Hemoglobin is a tetrameric protein (see Chapter 4), and beef heart mitochondrial ATPase has 10 protomers (see Chapter 8). The smallest unit is referred to as a monomer or subunit. Figure 2.13 indicates the structure of the dimeric protein Cu,Zn-superoxide dismutase. Figure 2.14 is an overview of the primary, secondary, tertiary, and quaternary structures of a tetrameric protein.

Post-translational modification of proteins

Modifications of proteins formed during syntheses include such reactions as glycosylation, phosphorylation, amidation, sulfation, acetylation, and mixed disulfide formation. Among these, protein glycosylation, which is catalyzed by enzymes called glycosyltransferases, is one of the most abundant modifications of proteins (see Chapter 24).

Creutzfeldt–Jacob disease

A 56-year-old male cattle rancher presented with epileptic cramp and dementia and was diagnosed as having Creutzfeldt–Jacob disease, a well-known human prion disease. The prion diseases, also known as transmissible spongiform encephalopathies, are neurodegenerative diseases that affect both humans and animals. This disease in sheep and goats is designated as scrapie, and in cows as spongiform encephalopathy (mad cow disease). The diseases are characterized by the accumulation of an abnormal isoform of a host-encoded protein, prion protein-cellular form (PrPC), in affected brains.

Comment. Prions appear to be composed only of PrPSc (Scrapie form) molecules, which are abnormal conformers of the normal, host-encoded protein. PrPC has a high α-helical content and is devoid of β-pleated sheets, whereas PrPSc has a high β-pleated sheet content. The conversion of PrPC into PrPSc involves a profound conformational change. The progression of infectious prion diseases appears to involve an interaction between PrPC and PrPSc, which induces a conformational change of the α-helix-rich PrPC to the β-pleated sheet-rich conformer of PrPSc. PrPSc-derived prion disease may be genetic or infectious. The amino acid sequences of different mammalian PrPC are similar, and the conformation of the protein is virtually the same in all mammalian species. The central protein of the transmissible agent, or prion, was discovered by Dr Stanley B Prusiner, the Nobel Prize winner in Physiology in 1997.

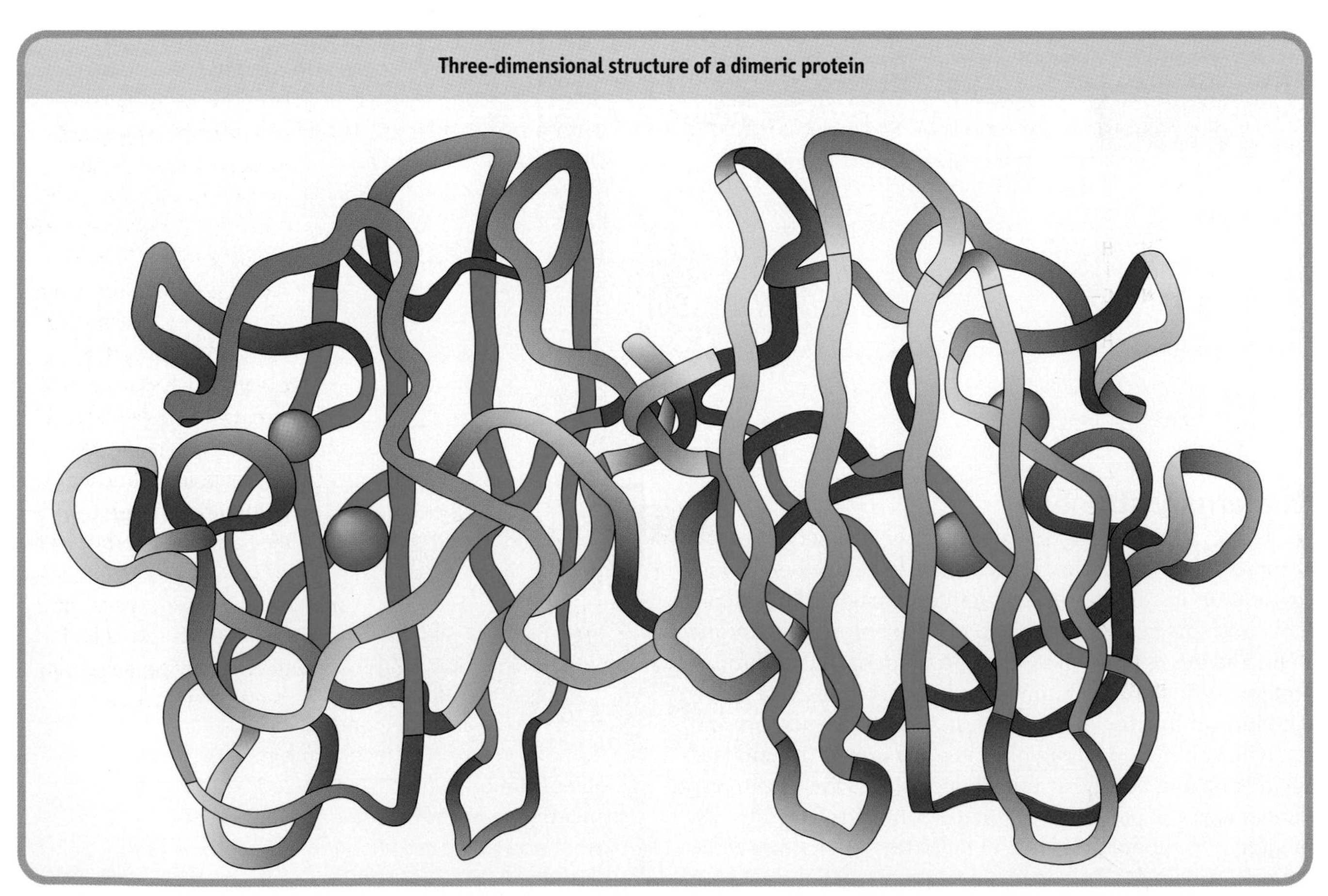

Fig. 2.13 Quaternary structure of Cu,Zn-superoxide dismutase from spinach. Cu,Zn-superoxide dismutase has a dimeric structure, with a monomer molecular mass of 16 000 kDa. Each subunit consists of eight antiparallel β-sheets called a β-barrel structure, in analogy with geometric motifs found on native American and Greek weaving and pottery. Courtesy of Dr Y Kitagawa.

Purification and characterization of proteins

There are thousands of different kinds of proteins in cells, and it is essential to purify proteins in order to understand their structural and functional properties. Protein-purification procedures take advantage of separations based on charge, size, binding properties, and solubility. The complete characterization of the protein requires an understanding of its amino acid composition, and its complete primary, secondary, and tertiary structure, and for multimeric proteins, its quaternary structure.

In order to characterize a protein, it is first necessary to purify the protein by separating it from other components in complex biological mixtures. The source of the proteins is commonly blood or tissues, or microbial cells such as bacteria and yeast. First the cells or tissues are disrupted by grinding or homogenization in buffered isotonic solutions, commonly at physiologic pH and at 4°C to minimize protein denaturation during purification. The 'crude extract' containing organelles such as nuclei, mitochondria, lysosomes, microsomes, and cytosolic fractions can then be fractionated by high-speed centrifugation or ultracentrifugation. Proteins that are tightly bound to the other biomolecules or membranes may be solubilized using organic solvent or detergent.

Salting-out (ammonium sulfate fractionation)

The solubility of a protein is dependent on the concentration of dissolved salts, and the solubility may be increased by the addition of salt at a low concentration (salting-in) or decreased by high salt concentration (salting-out). When ammonium sulfate, one of the most soluble salts, is added to a solution of a protein, some proteins precipitate at a given salt concentration while others do not. Human serum immunoglobulins are precipitable by 33–40% saturated $(NH_4)_2SO_4$, while albumin remains soluble. Saturated ammonium sulfate is about 4.1 mol/L. Most proteins will precipitate from an 80% saturated $(NH_4)_2SO_4$ solution.

Separation on the basis of size

Dialysis and ultrafiltration

Small molecules, such as salts, can be removed from protein solutions by dialysis or ultrafiltration. Dialysis is performed by adding the protein–salt solution to a semipermeable membrane tube (commonly a nitrocellulose or collodion membrane). When the tube is immersed in a dilute buffer solution, small molecules will pass through and large protein molecules will be retained in the tube, depending on the pore size of the dialysis membrane. This procedure is

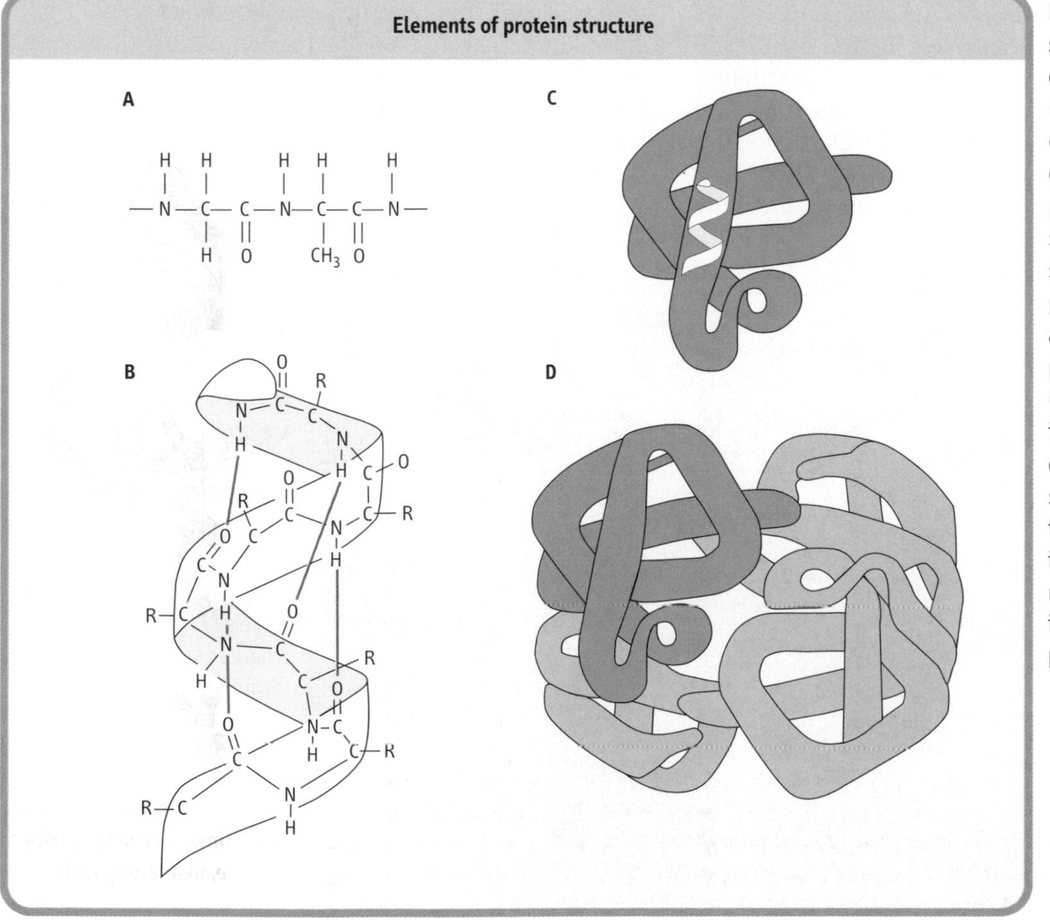

Fig. 2.14 Primary, secondary, tertiary, and quaternary structures. (A) The primary structure is composed of a linear sequence of amino acid residues of proteins. (B) The secondary structure indicates the local spatial arrangement of polypeptide backbone yielding an extended α-helical or β-pleated sheet structure. (C) The tertiary structure illustrates the three-dimensional conformation of a subunit of the protein; while the quaternary structure (D) indicates the assembly of multiple polypeptide chains into an intact, tetrameric protein.

particularly useful for removal of $(NH_4)_2SO_4$ during protein purification, since the salts will interfere with the interaction of proteins with ion-exchange columns. Figure 2.15 illustrates the dialysis of proteins.

Ultrafiltration is a rapid and convenient technique for the simultaneous concentration of proteins and the removal of small molecule contaminants. For example, if a membrane with an appropriate pore size is used, the protein solution can be concentrated by applying pressure. Water passes through the membrane with small molecules (the ultrafiltrate), and the protein is concentrated in the retentate. Dilute buffer may be added batch wise or continually to the retentate to achieve both concentration and dialysis of the protein solution. A similar membrane can be used in a centrifuge tube to concentrate proteins by low-speed centrifugation. Using membranes with defined pore sizes, proteins of different sizes can also be separated by ultrafiltration.

Gel-filtration (molecular sieving)

Gel-filtration, or gel-permeation, chromatography uses a column of insoluble but highly hydrated polymers such as dextrans, agarose, or polyacrylamide. Gel-filtration chromatography depends on the differential migration of dissolved solutes through gels that have pores of defined sizes. If one introduces a small volume of human serum albumin dissolved in 0.145 mol/L NaCl to the top of the small-pore gel-filtration column and then allows it to elute slowly through the column, the albumin molecules will elute rapidly from the column, because they are restricted largely to the space surrounding the gel particles. In contrast, the salt will diffuse into and out of the particles, accessing a larger volume of the column, and eluting after the albumin. The salt is said to be retained or retarded by the column. This technique is frequently used for protein purification and for desalting protein solutions. Figure 2.16 indicates the principle of gel filtration. There are commercially available gels made from carbohydrate polymer beads designated as dextran (Sephadex series), polyacrylamide (Bio-Gel P series), and agarose (Sepharose series), respectively. The gels vary in pore size, and one can chose the gel-filtration materials according to the fractionation ranges desired.

Separation on the basis of charge

Ion-exchange chromatography

Ion-exchange chromatography takes advantage of the differential affinity of charged ions or molecules in solution for immobilized charged substances. When the charged ion or molecule with one or more positive charges exchanges with another positively charged component that was bound to a negatively charged immobilized phase, the process is called cation exchange. The inverse process is called anion exchange. The cation exchanger, carboxymethyl-cellulose ($–O–CH_2–COO^-$), and anion exchanger, diethylaminoethyl (DEAE) cellulose ($–O–C_2H_4–NH^+[C_2H_5]_2$, are frequently used for the purification of proteins. Consider purifying a protein mixture containing albumin and immunoglobulin, at pH 7.5. Albumin is negatively charged, having a pI of 4.8. Immunoglobulin is positively charged, since it has a pI above pH 7.5. If the mixture is applied to a DEAE column, the albumin sticks to the DEAE column whereas the immunoglobulin passes through the column. The albumin that had been retained on the column can then be eluted by a salt concentration gradient in which the salt anions compete with protein binding sites on the column, or by decreasing the pH so that the albumin becomes positively charged. Figure 2.17 illustrates the principle of ion-exchange chromatography. As with gel permeation chromatography, proteins may become separated from one another, in this case, based on small differences in their pI.

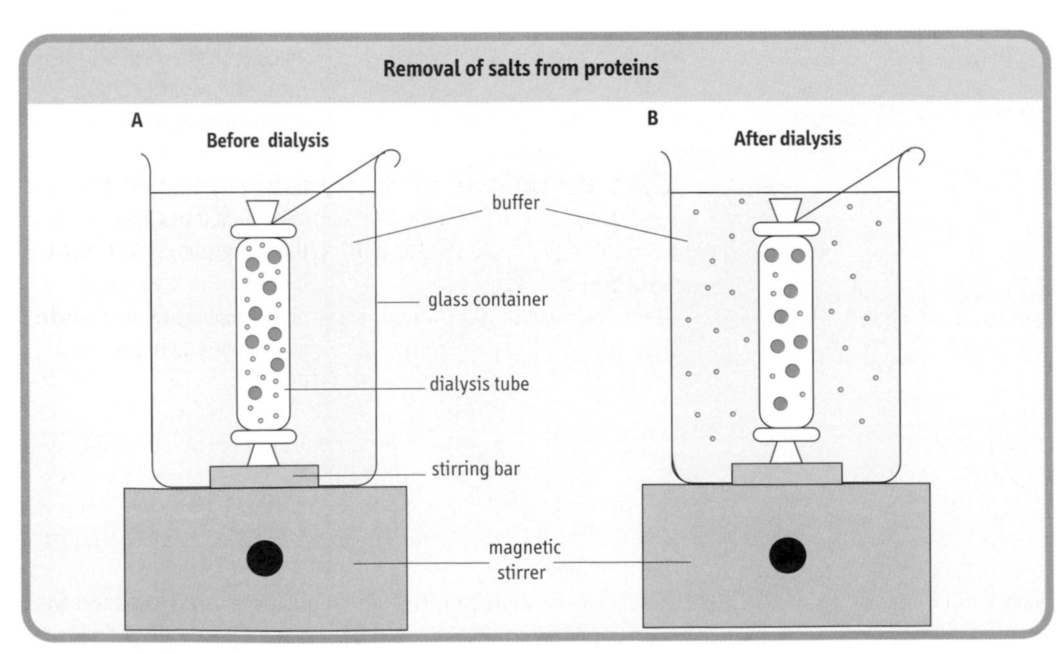

Fig. 2.15 Dialysis of proteins. Protein and low-molecular-mass compounds are separated by dialysis on the basis of size. (A) A protein solution with salts is placed in a dialysis tube in a beaker and dialyzed with stirring against an appropriate buffer. (B) The protein is retained in the tube whereas salts will pass through the membrane.

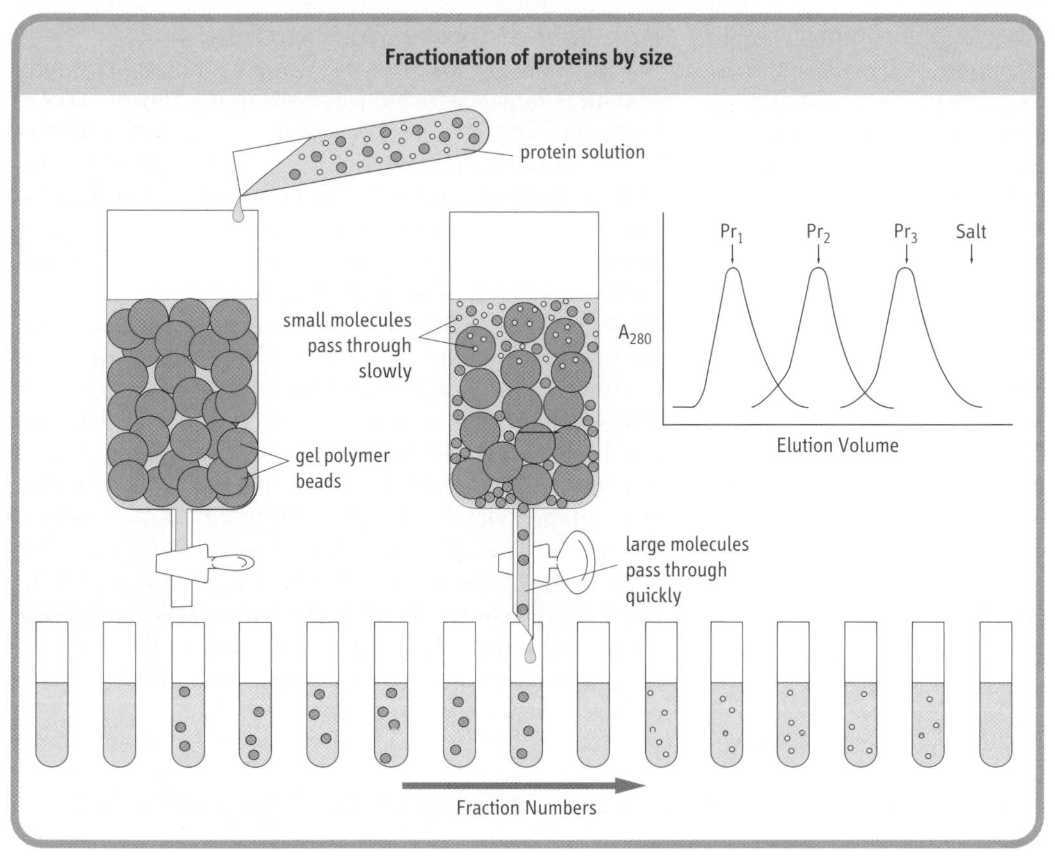

Fig. 2.16 Gel filtration of proteins. Proteins with different molecular sizes are separated by gel filtration based on their relative size. The smaller the protein, the more readily it exchanges into polymer beads, whereas larger proteins may be completely excluded. Larger molecules flow more rapidly through this column, leading to fractionation on the basis of molecular size. The chromatogram on the right shows a theoretical fractionation of three proteins, Pr_1–Pr_3 of decreasing molecular weight.

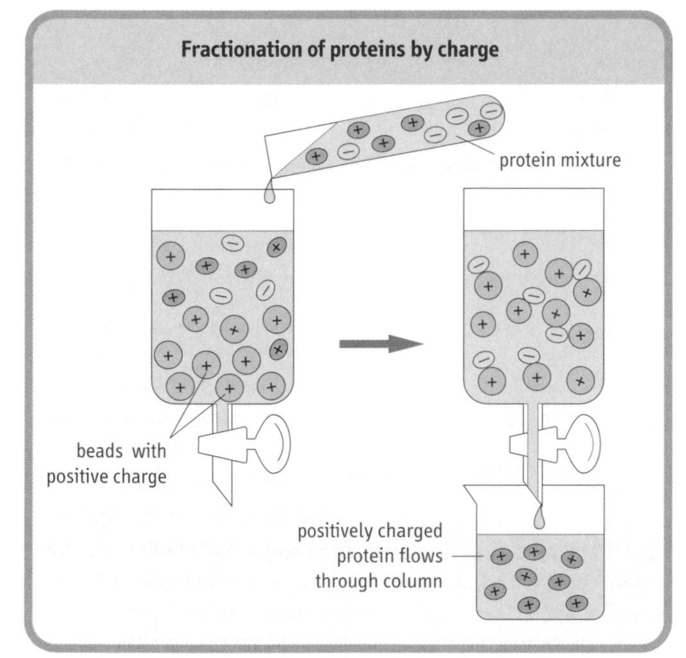

Fig. 2. 17 Ion-exchange chromatography. Mixtures of proteins can be separated by ion-exchange chromatography according to their net charges. Beads that have positive-charge groups are called anion exchangers, whereas those having negative-charge groups are cation exchangers. This figure depicts an anion-exchange column. Negatively charged protein binds to positively charged beads, and positively charged protein flows through the column.

Affinity chromatography

Affinity chromatography is a convenient and specific method for purification of proteins. For example, one can chemically bind a ligand that interacts with or binds to a specific protein in a complex mixture. The protein of interest will be selectively and specifically bound to the ligand while the others wash through the column. The bound protein can then be eluted by a high salt concentration, mild denaturation, or by a soluble form of the ligand or ligand analogs. This will be discussed further in Chapter 5.

Determinations of purity and molecular weight of proteins by sodium dodecyl sulfate–polyacrylamide gel electrophoresis (SDS–PAGE)

Electrophoresis is applicable for the separation of a wide variety of charged molecules, including amino acids, polypeptides, proteins, and DNA. When a current is applied to molecules in dilute buffers, those with a net negative charge at the selected pH migrate toward the anode and those with a net positive charge toward the cathode. A porous support, such as paper, cellulose acetate, or polymeric gel, is commonly used to minimize diffusion and convection.

Like chromatography, electrophoresis may be used for preparative fractionation of proteins at physiologic pH. A denaturing detergent, SDS, is also used in a PGE system to separate and resolve protein subunits according to size. The protein preparation is treated with an excess of thiol, such as β-mercaptoethanol, and SDS. Under these conditions, the thiol reduces all disulfide bonds in proteins, and the SDS binds to the protein, resulting in total disruption of the secondary, tertiary, and quaternary structure of the protein. The anionic denatured protein chains are then resolved by electrophoresis in a polyacrylamide gel. Because the binding of SDS is proportional to the length of a peptide chain, the relative mobility of each anion, representing individual, denatured peptide chains, is proportional to the molecular mass of the polypeptide chain. On SDS–PAGE, a purified protein preparation can readily be analyzed for homogeneity after staining with sensitive and specific dyes such as Coomassie Blue, or with a silver staining technique, as shown in Figure 2.18.

The typical steps in the purification procedure of a protein are summarized in Figure 2.19.

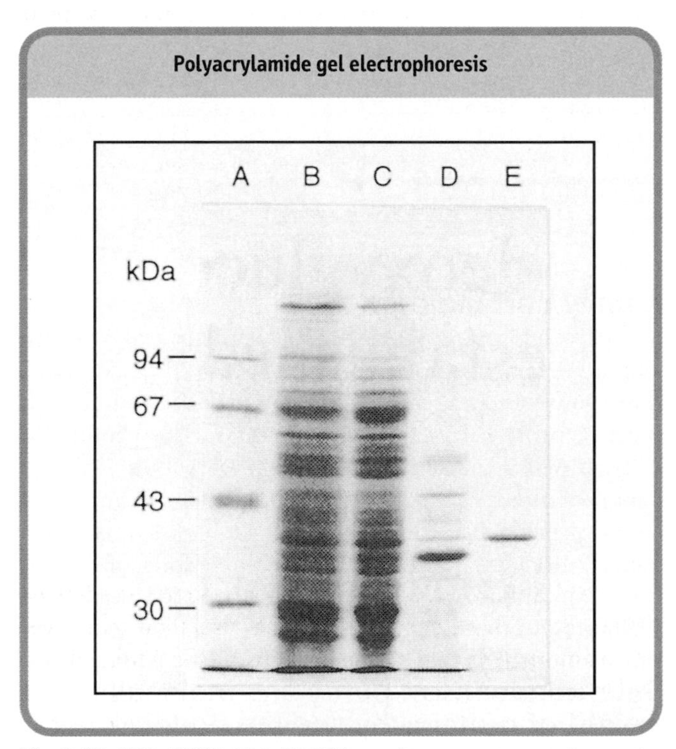

Fig. 2.18 SDS–PAGE. SDS–PAGE is used to separate proteins on the basis of their molecular weights. Larger molecules are retarded in the gel matrix, whereas the smaller ones move more rapidly. Lane A contains standard proteins with known molecular masses (indicated in kDa on the left side). Lanes B, C, D, and E show results of SDS–PAGE analysis of the protein at various stages in purification: B = total protein isolate; C = ammonium sulfate precipitate; D = fraction from gel-permeation chromatography; E = purified protein from ion exchange chromatography.

Analysis of the protein structure

For the determination of its amino acid composition, a protein is subjected to hydrolysis by using 6 mol/L HCl at 110°C in a sealed and evacuated tube for 24–48 h. Under these conditions, tryptophan, cysteine, and most of the cystine are destroyed. Glutamine and asparagine are quantitatively deaminated to give glutamate and aspartate respectively. Recovery of serine and threonine is incomplete and decreases with increasing time of hydrolysis. Alternative hydrolysis procedures may be used for measurement of tryptophan, while cysteine and cystine may be converted to an acid-stable cysteic acid prior to hydrolysis. Following hydrolysis, the free amino acids are separated on an automated amino acid analyzer using an ion-exchange column, or by reversed-phase high-performance liquid chromatography (HPLC). The amino acids are reacted with chromogenic or fluorogenic reagents, such as ninhydrin or dansyl chloride, Edman's reagent (see below), or *o*-phthalaldehyde. These techniques allow the measurement of as little as 1 pmol of each amino acid. A typical elution pattern of amino acids in a purified protein is shown in Figure 2.20.

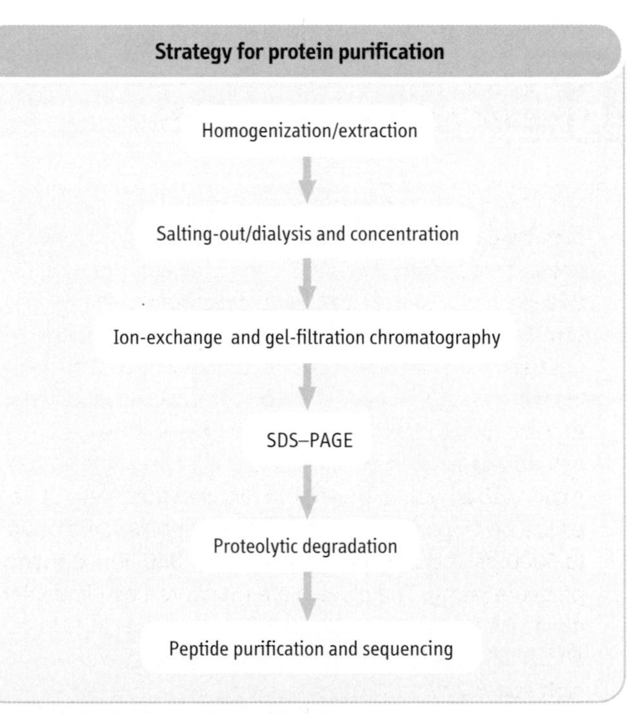

Fig. 2.19 Typical protein purification sequence. Purification of a protein involves a sequence of steps in which contaminating proteins are removed, based on difference in size, charge and hydrophobicity. Purification is monitored by SDS–PAGE. The primary sequence of the protein is determined by automated Edman degradation of peptides. The three-dimensional structure of the protein may be determined by X-ray crystallography.

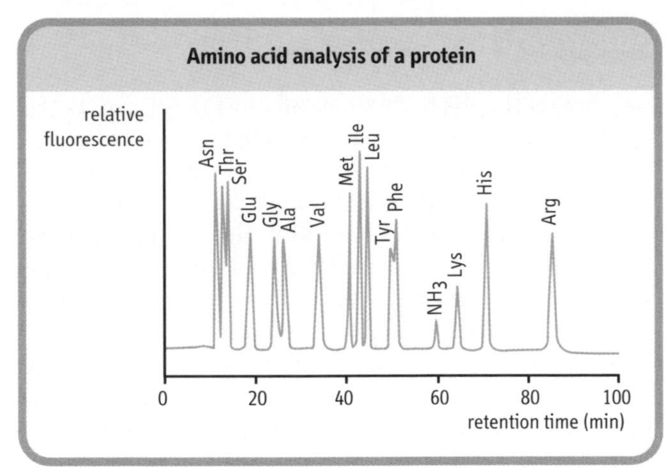

Fig. 2.20 Typical chromatogram from an amino acid analysis. Amino acids in proteins were separated by cation-exchange chromatography. The hydrolysate was applied to the cation exchange column in a dilute buffer at acidic pH (~3.0), at which all amino acids are positively charged. The amino acids are then eluted by a gradient of increasing pH and salt concentrations. The most anionic (acidic) amino acids elute first, followed by the neutral and basic amino acids. Amino acids are derived by postcolumn reaction with a fluorogenic compound, such as *o*-phthalaldehyde.

High performance liquid chromatography (HPLC)

HPLC is a powerful chromatographic technique for high-resolution separation of proteins, peptides, and amino acids. The principle of the separation may be based on the charge, size, or hydrophobicity of proteins. The narrow columns are packed with a noncompressible matrix of fine beads coated with a thin layer of a stationary phase. A protein mixture is applied to the column, and then the components are eluted by either isocratic or gradient chromatography. The small particle size, typically 5–10 μm requires that the mobile phase be forced through the column at pressures of up to 5000 psi (pounds per square inch – 350 atmospheric pressure [atm]). The eluates are monitored by ultraviolet absorption, refractive index, or fluorescence detectors. This technique gives high-resolution separation with high specificity and high sensitivity and is the most common technique for purification of proteins and peptides. Most commonly, the peptides are bound to a reversed-phase column, such as an octadecyl (C-18) silica column, then eluted with a gradient of increasing concentrations of organic solvents.

Determination of the primary structure of proteins

Information on protein primary structure is essential for understanding the functional properties, the identification of the family to which the protein belongs, as well as characterization of mutant proteins that cause disease.

The determination of the amino acid sequence of a purified protein is relatively straightforward. Since the data on sequences of 'overlapping peptides' rather than a single long peptide give useful information, various cleavage techniques are employed. A protein may be cleaved first by proteolytic digestion by endoproteases, such as trypsin, V8 protease, or lysyl endopeptidase, to obtain peptide fragments. Trypsin cleaves peptide bonds on the C-terminal side of arginine and lysine residues, provided the next residue is not proline. Lysyl endopeptidase is also frequently used to cleave at the C-terminal side of lysine. Cleavage by chemical reagents such as cyanogen bromide is also useful. Cyanogen bromide cleaves on the C-terminal side of methionine residues. Before cleavage, proteins with cysteine and cystine residues are reduced by 2-mercaptoethanol (MeSH) and then treated with iodoacetate to form carboxymethylcysteine residues. This avoids spontaneous formation of inter- or intradisulfide formations during analyses.

$$R_1 - S - S - R_2 + 2MeSH$$

$$\downarrow$$

$$R_1SH + R_2SH + MeSSMe$$

$$\downarrow$$

$$RSH + ICH_2COOH$$

$$\downarrow$$

$$RS - CH_2COO^-$$

The cleaved peptides are then subjected to reversed-phase HPLC to purify the peptide fragments, and then sequenced on an automated protein sequencer, using the Edman degradation technique (Fig. 2.21).

The sequence of overlapping peptides is then used to obtain the primary structure of the protein. Mass spectrometry may also be used to obtain both the molecular mass and sequence of polypeptides simultaneously. Both techniques can be applied directly to proteins or peptides recovered from SDS–PAGE. Once the partial amino acid sequence is obtained, one can determine the nucleotide sequence of the DNA that encodes this polypeptide segment. After chemically synthesizing this DNA, it can be used to identify and isolate the gene containing its nucleotide sequence (see Chapter 33).

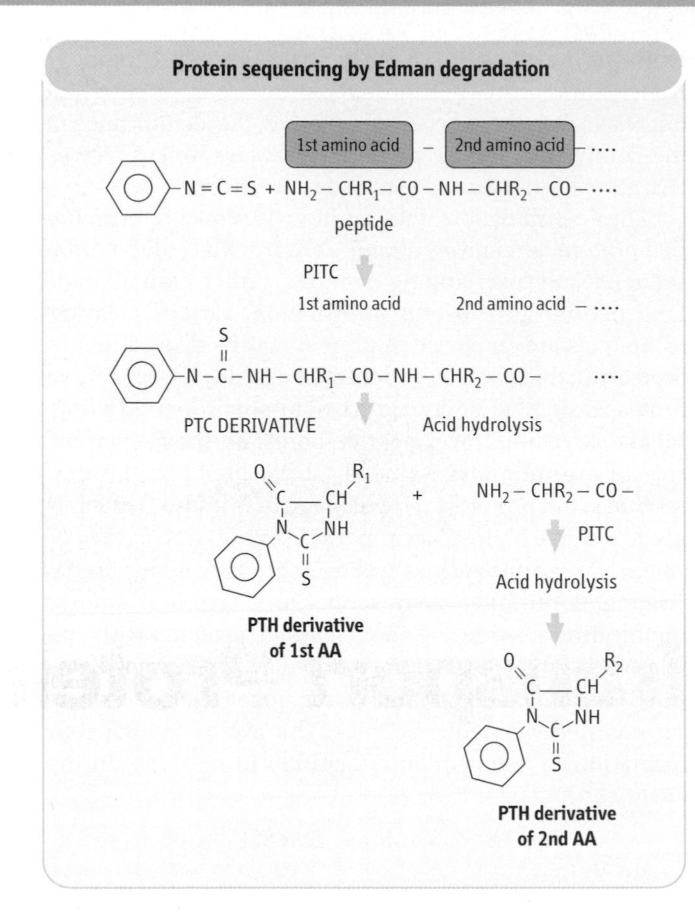

Fig. 2.21 Steps in Edman degradation. The Edman degradation method sequentially removes one residue at a time from the amino end of a peptide. Phenyl isothiocyanate (PITC) reacts with the N-terminal amino group of the immobilized peptide to form a phenylthiocarbamyl derivative (PTC amino acid) in alkaline solution. Acid treatment removes the first amino acid as the phenylthiohydantoin (PTH) derivative, which is identified by HPLC. The peptide chain minus the N-terminal amino acid is then reacted again with fresh reagent in an automated procedure.

Summary

There are thousands of different proteins in cells, and each protein has a different structure and function. Purification and characterization of proteins is essential for elucidating their structure and function. By taking advantage of differences in their size, solubility, charge, and binding capacity, proteins can be purified to homogeneity using various chromatographic and electrophoretic techniques. The molecular mass and purity of a protein, and its subunit composition, can be determined by SDS–PAGE. The primary structure can be determined by hydrolysis of a protein and automated Edman degradation. Deciphering the primary and three-dimensional structures of a protein leads to an understanding of structure–function relationships in proteins.

Further reading

Eigen M. Prionics or the kinetic basis of prion diseases. *Biophysical Chemistry* 1996;**63**:A1–A18.

Eisenhaber F, Persson B, Argos P. Protein structure prediction: recognition of primary, secondary and teriary structural features from amino acid sequence. *Crit Rev Biochem Mol Biol* 1995;**30**:1–94.

Jaenicke R. Stability and folding of ultrastable proteins: eye lens crystallins and enzymes from thermophiles. *FASEB Journal* 1996;**10**:84–92.

Murzin AG. Structural classification of proteins: new superfamilies. *Curr Opin Struct Biol* 1996;**6**:386–394.

Scopes RK. Protein purification in the nineties. *Biotech Appl Biochem* 1996;**23**:197–204.

Vogel G. Prusiner recognized for once-heretical prion theory (news). *Science* 1997;**278**:214.

Winston RL, Fitzgerald MC. Mass spectrometry as a readout of protein structure and function. *Mass Spectrom Rev* 1997;**16**:165–179.

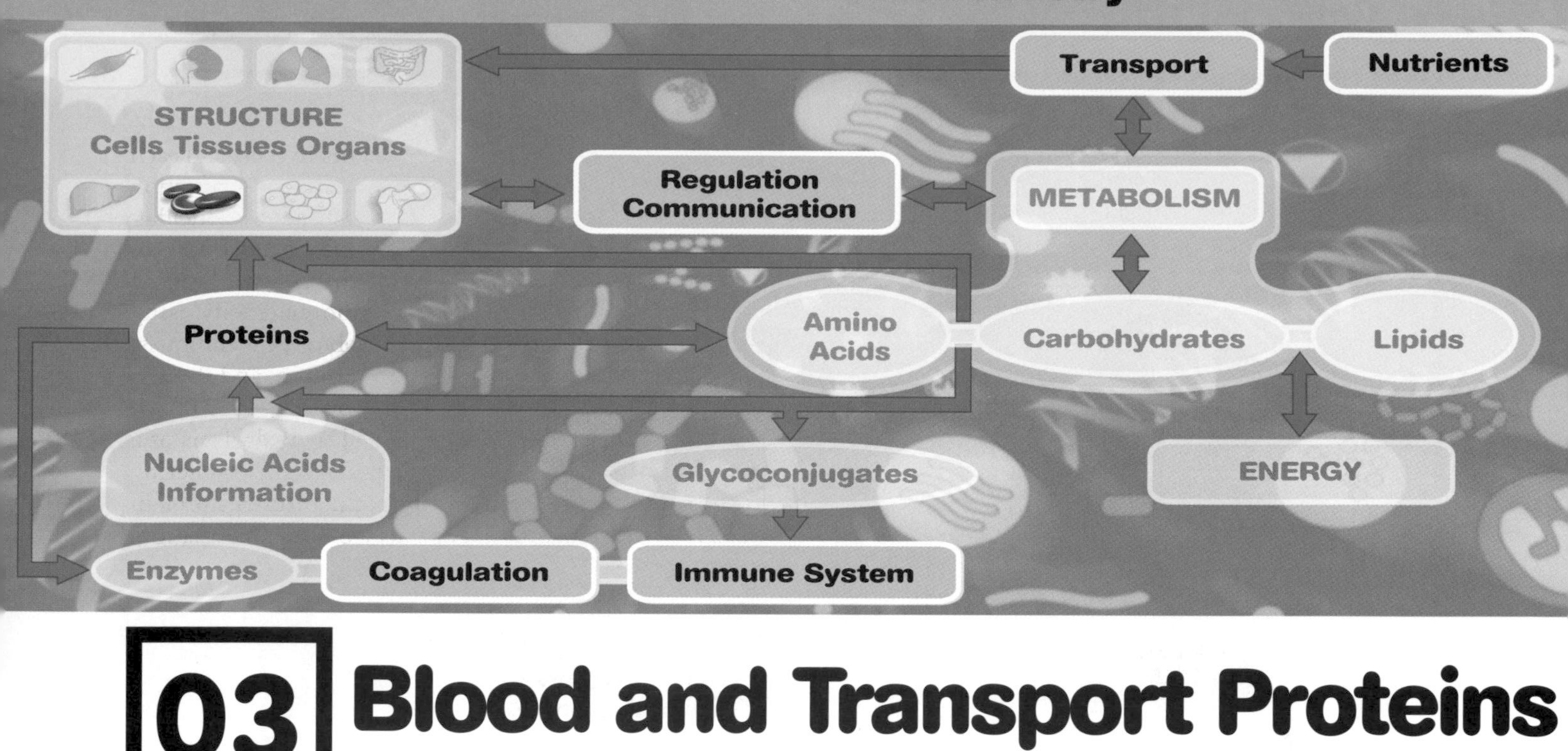

03 Blood and Transport Proteins

Blood functions as a transport and distribution system for the body, delivering essential nutrients to tissues and at the same time removing waste products. It is composed of an aqueous solution containing molecules of varying sizes and a number of cellular elements. Some of the components of blood perform important roles in the defence against external insult and repair of damaged tissues.

Plasma and serum

The formed elements of blood are suspended in an aqueous solution that is termed plasma. Plasma is obtained by centrifuging blood that has been treated with an anticoagulant to prevent clotting. In laboratory practice, the most common anticoagulants are lithium, heparin and ethylene diamine tetra-acetic acid (EDTA). When blood is collected for transfusion, citrate is used as an anticoagulant, as this preserves procoagulants and its effects are readily reversible by calcium. If blood is allowed to clot and is then centrifuged, the aqueous solution produced is termed serum. During clotting, fibrinogen is converted to fibrin as a result of proteolytic cleavage by thrombin, and so a major difference between plasma and serum is the absence of fibrinogen in serum.

Formed elements

There are three major cellular components circulating in the bloodstream:

Erythrocytes

are not classified as true cells, as they do not possess nuclei and intracellular organelles. They are cellular remnants, containing specific proteins and ions, which can be present in high concentrations. Erythrocytes are the end-product of erythropoiesis in the bone marrow, which is under the control of erythropoietin produced by the kidney (Fig. 3.1). Hemoglobin is synthesized in the erythrocyte precursor cells – erythroblasts and reticulocytes – under a tight control dictated by the concentration of heme, the synthesis of which involves the chelation of reduced ferrous iron (Fe^{2+}) by four nitrogen atoms in the centre of a porphyrin ring (see Chapter 27). The main functions of erythrocytes are the transport of oxygen and the removal of carbon dioxide and hydrogen; as they lack cellular organelles, they are not capable of protein synthesis and repair. As a result, erythrocytes have a finite life span of 60–120 days before being trapped and broken down in the spleen.

Leukocytes

are cells of which the main function is to protect the body from infection. Most leukocytes are produced in the bone marrow, some are produced in the thymus, and others mature within several tissues (Fig. 3.2)(see Chapter 34). Leukocytes can control their own synthesis by secreting into the blood signal peptides that can subsequently act on the bone marrow stem cells. In order to function correctly, leukocytes have the ability to migrate out of the bloodstream into surrounding tissues.

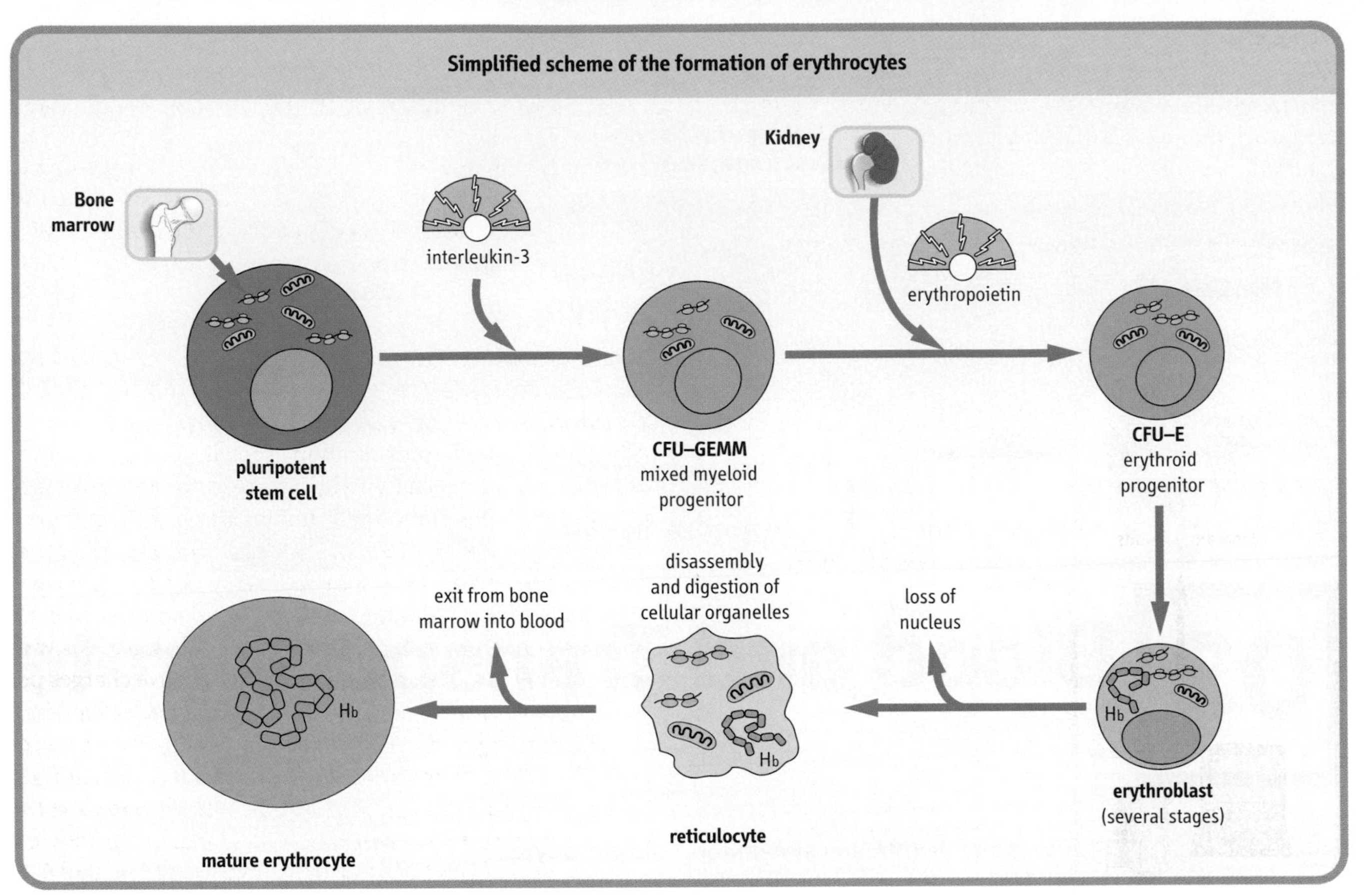

Fig. 3.1 Formation of erythrocytes. In an average day, 10^{11} erythrocytes are formed. Hemoglobin is synthesized in the erythrocyte and reticulocyte before the loss of ribosomes and mitochondria. CFU, colony-forming unit. GEMM: granulocyte, erythroid, monocyte, megakaryocyte. CFU-E, colony-forming unit erythroid.

Platelets

are not true cells, but are membrane-bound fragments derived from megakaryocytes residing in the bone marrow. They have a key role in the process of blood clotting (see Chapter 6).

Leukocytes (white blood cells)

Leukocyte group	Subgroup	Function
granulocytes	neutrophils	destroy small organisms
	basophils	secrete histamine, mediating inflammatory response, and platelet activating factor
	eosinophils	destroy parasites allergic reaction
lymphocytes	B lymphocytes	synthesize antibody
	T lymphocytes	participate in the specific immune response
monocytes	macrophages	destroy invading organisms

Fig. 3.2 Classification and functions of leukocytes.

Plasma and serum

The role of the clinical laboratory

The clinical laboratory performs a large number of biochemical analyses on body fluids, which can give answers to specific clinical questions about an individual patient. Such analyses are usually requested to aid in the diagnosis or treatment of specific conditions. Blood and urine samples form the majority of specimens. Whereas measurements are sometimes performed on whole blood, serum or plasma are preferred for most analyses of molecules and ions. In general, the time devoted to the analysis of each sample is relatively short, but the entire process from a request for analysis to receipt of a result can take several hours, and involves many steps. Throughout the process, constant checking, attention to detail, and quality assurance are performed to ensure that the results produced are of the highest quality and are clinically valid. An outline of the workflow through the laboratory is shown in Figure 3.3.

Plasma proteins

Plasma proteins can be broadly classified into two groups: those, including albumin, that are synthesized by the liver, and the immunoglobulins, which are produced by plasma cells of the bone marrow, usually as part of the immune response.

A number of plasma proteins have the ability to bind certain ligands with a high affinity and specificity. These proteins can then act as a reservoir for the ligand and help control its distribution and availability by transporting it to tissues throughout the body. Binding to a protein can also render a toxic substance less harmful to the tissues. Major binding proteins and their ligands are shown in Figure 3.4.

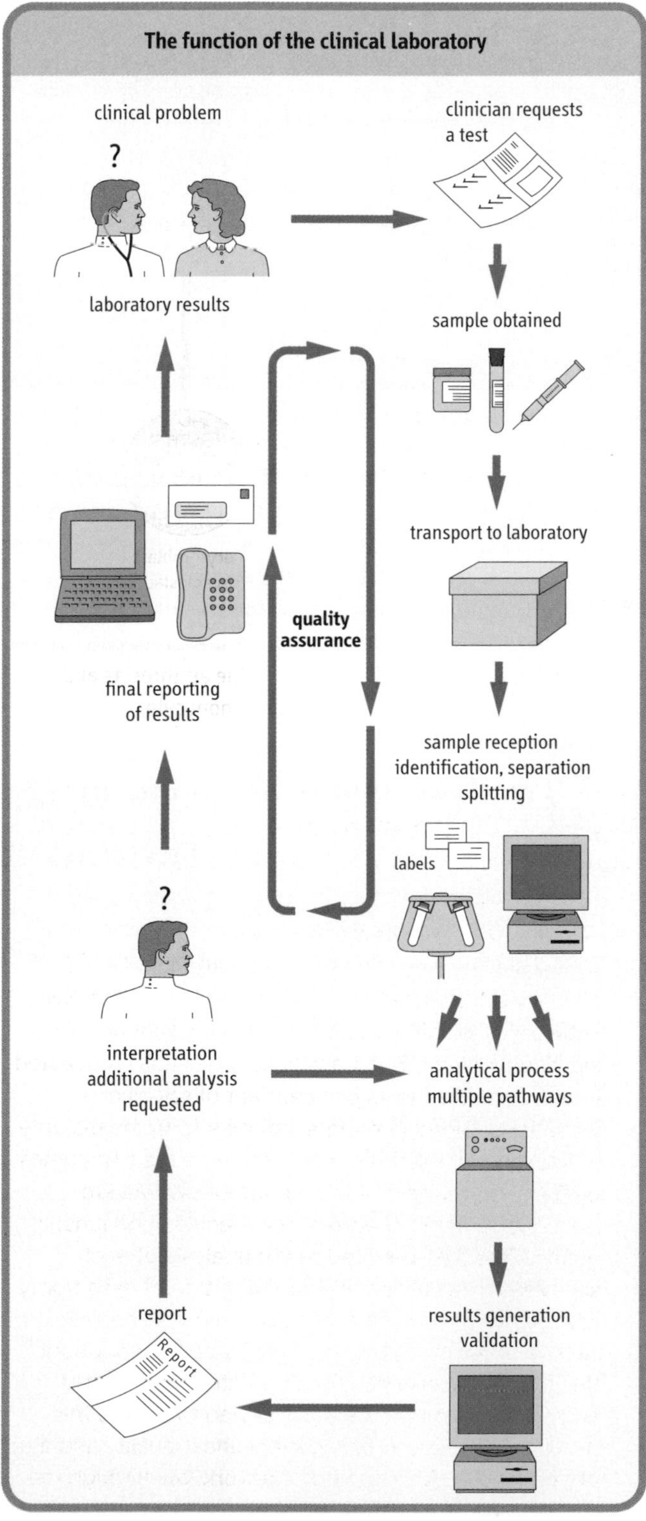

Fig. 3.3 Flow diagram indicating the steps involved in the generation of results from the clinical laboratory.

Albumin

In addition to its functions as a protein reserve in nutritional depletion and as an osmotic regulator, albumin is a major transport protein

Albumin, the predominant plasma protein having no known enzymatic or hormonal activity, accounts for approximately 50% of the protein found in human plasma, and is present normally at a concentration of 35–45 g/L. It is easy to isolate, and has been extensively studied. With a molecular weight of about 66 kDa, albumin is one of the smallest plasma proteins and, given its highly polar nature, dissolves easily in water. At pH 7.4, it is an anion with 200 negative charges per molecule; this gives it a vast capacity for nonselective binding of many ligands. The presence of large amounts of albumin in the body (4–5 g/kg body weight), with at least 38% being present intravascularly, also helps to explain the critical role that it has in exerting colloid osmotic pressure.

The rate of synthesis of albumin (14–15 g daily) is critically dependent on nutritional status, especially the extent of amino acid deficiencies. The half-life of albumin is about 20 days, and degradation appears to occur by pinocytosis in all tissues.

Transport proteins and their ligands

Proteins	Ligands
Cation binding	
Albumin	divalent and trivalent cations, e.g. Cu^{2+}, Fe^{3+}
Ceruloplasmin	Cu^{2+}
Transferrin	Fe^{3+}
Hormone binding	
Thyroid-binding globulin (TBG)	Thyroxine (T_4), Tri-iodothyronine (T_3)
Cortisol-binding globulin (CBG)	Cortisol
Sex hormone-binding globulin (SHBG)	Androgens (testosterone), estrogens (estradiol)
Hemoglobin/protoporphyrin binding	
Albumin	Heme, bilirubin, biliverdin
Haptoglobin	Hemoglobin dimers
Free fatty acid binding	
Albumin	Free fatty acids, steroids

Fig. 3.4 Transport proteins and their ligands. Almost all plasma proteins bind ligands, and this is a major function of many proteins. Albumin can bind many molecules weakly and nonspecifically, but other proteins bind tightly to specific molecules – for example, transferrin is specific for ferric iron (Fe^{3+}).

Nephrotic syndrome

A 44-year-old lady was admitted to hospital because of weakness, anorexia, recurrent infections, bilateral leg edema, and deteriorating respiratory function. Her plasma albumin concentration was 19 g/L (normal range 35–45 g/L) and her urinary protein excretion 20 g/24 h (normal value <0.15 g/24 h). Renal biopsy confirmed the diagnosis of acute glomerulonephritis.

Comment. This lady had the classic triad of the nephrotic syndrome: hypoalbuminemia, proteinuria, and edema. The nephritis has resulted in damage to the glomerular basement membrane, with resultant leak of albumin. Continued loss of albumin exceeds the synthetic capacity of the liver, and results in hypoalbuminemia; consequently, the capillary osmotic pressure is significantly reduced. This leads to both peripheral (leg) edema and pulmonary edema (breathlessness). With increasing glomerular damage, proteins of larger molecular mass, such as immunoglobulins and complement, are lost.

Albumin is the primary plasma protein responsible for the transport of hydrophobic fatty acids, bilirubin, and drugs

Albumin demonstrates a unique ability to solubilize, in aqueous phase, a heterogeneous range of substances that include the long-chain fatty acids, sterols, and several synthetic compounds. The transport of long-chain fatty acids underpins much of the body's distribution of energy-rich substrates. Through binding, consequently solubilizing, and ultimately transporting fatty acids such as stearic acid, oleic acid, and palmitic acid, albumin enables the transport of these hydrophobic molecules in the predominantly hydrophilic milieu of the plasma. Associative studies have demonstrated the presence of numerous fatty acid binding sites on the albumin molecule, with variable affinities. The highest-affinity sites are believed to lie within the globular segments of the albumin molecule (Fig. 3.5).

In addition to binding fatty acids, albumin has an important role in binding bilirubin, thereby rendering it, not only water soluble and transportable from the reticulo-endothelial system to the liver, but also temporarily nontoxic.

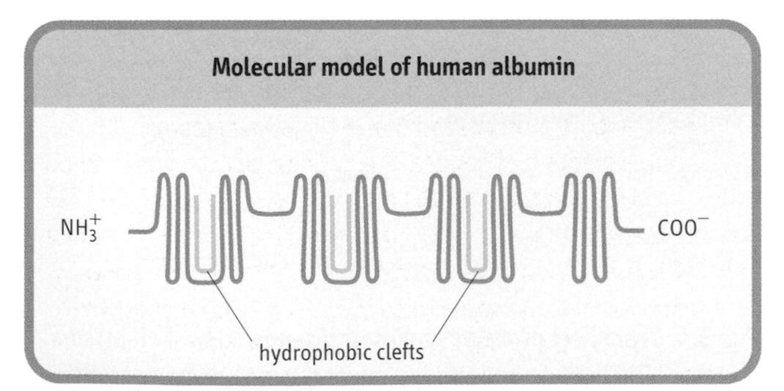

Fig. 3.5 Molecular model of human albumin. The hydrophobic clefts are globular segments of albumin that are able to bind fatty acids with high affinity.

The presence of sites within the albumin molecule that are capable of binding a variety of drugs, including salicylate, barbiturates, sulfonamides, penicillin, and warfarin, is of great pharmacologic relevance. Chiefly, such interactions are weak and the ligands become easily displaced by competitors for the binding site. Given such binding, not only does albumin play a part in drug solubilization, but it may also determine the proportion of free, and thus pharmacologically active, drug available in the plasma.

Anticoagulants and albumin

A 45-year-old man with a long-standing history of coronary artery disease underwent coronary artery bypass grafting and received anticoagulant medication after the procedure. Oral anticoagulation was monitored by measurement of the prothrombin time (time required for clot formation effected by extrinsic clotting factors) (see Chapter 6), standardized on a common scale as an International Normalized Ratio (INR). For the majority of indications, INR is maintained in the range 2–3. The patient's condition was initially stabilized by means of a dose of 3 mg of warfarin daily, with an INR of 2.8. Later, he was prescribed aspirin. His INR subsequently increased rapidly to 4.0, and returned to its previous value only on reduction of the dose of warfarin to 2 mg daily.

Comment. Both aspirin and warfarin compete for binding to albumin. The aspirin displaced additional warfarin from albumin binding and resulted in a greater free warfarin concentration. It was possible to achieve adequate anticoagulation with a lower dose of warfarin.

Transport of metal ions

The ability of proteins to bind and transport metal ions is of major importance

Iron is an essential element for many metabolic processes, and is an important component of the heme proteins, myoglobin, hemoglobin, and cytochromes. Within the plasma, iron is transported bound to transferrin as ferric ions (Fe^{3+}) and is released from the protein after it has bound to specific cell receptors and the resulting complex has been internalized. The iron is then deposited in storage sites as ferritin or hemosiderin, or is used in synthesis of heme proteins. The binding of ferric ions to transferrin protects against the toxic effects of these ions. In inflammatory reactions, the iron–transferrin complex is degraded by the reticuloendothelial system without a corresponding increase in their synthesis; this results in low plasma concentrations of transferrin and iron.

- **Ferritin** is the major iron storage compound found in almost all cells of the body. It acts as the reserve of iron in the liver and bone marrow. The concentration of ferritin in plasma is proportional to the amount of stored iron, and so measurement of plasma ferritin is one of the best indicators of iron deficiency;
- **Hemosiderin** is formed from deproteinized ferritin and is found in the liver, spleen, and bone marrow. It is insoluble in aqueous solutions, and forms aggregates that slowly release iron when deficiency exists;
- **Ceruloplasmin** is the major transport protein for copper, an essential trace element. Ceruloplasmin helps export copper from the liver to peripheral tissues, and is essential

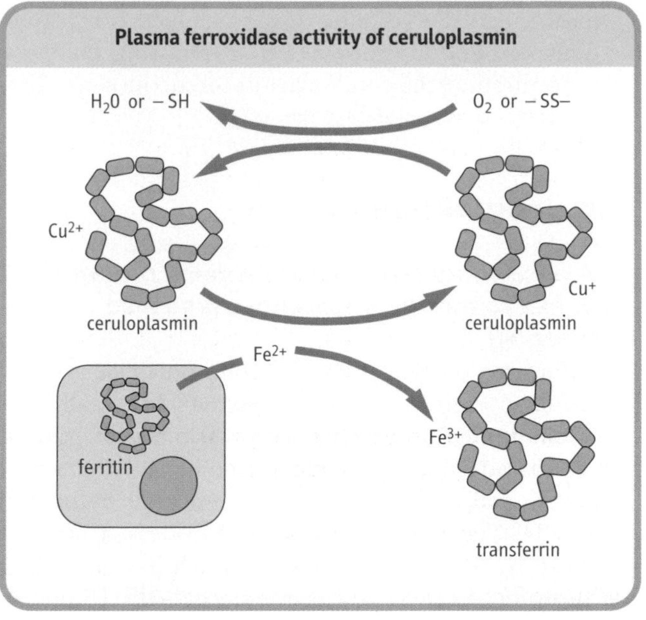

Fig. 3.6 Oxidation of Fe^{2+} by ceruloplasmin permits the binding and transport of iron by plasma transferrin. Ceruloplasmin Cu^{2+} is regenerated by reaction with oxygen or oxidized thiol groups.

Hemolysis and free haemoglobin

Handling of free hemoglobin

When erythrocytes are prematurely hemolyzed, they release hemoglobin into the plasma, where it dissociates into dimers that are bound to haptoglobin. The hemoglobin–haptoglobin complex is metabolized more rapidly than haptoglobin alone in the cells of the liver and reticular system, producing an iron–globulin complex and bilirubin. This prevents the loss of iron in the urine. When excessive hemolysis occurs, the plasma haptoglobin concentration can become very low. If hemoglobin has been broken down into heme and globin, the free heme can be bound by hemopexin; unlike haptoglobin, which is an acute phase protein, hemopexin is not affected by acute phase reactions. The heme–hemopexin complex is taken up by liver cells, where iron is bound to ferritin. A third complex, called methemalbumin, can form between oxidized heme and albumin. These mechanisms have evolved to allow the body not only to scavenge iron and prevent major losses, but also to complex free heme, which is toxic to many tissues.

Wilson's disease

A 14-year-old girl was admitted as an emergency. She was jaundiced (yellow pigmentation to sclera and skin), had abdominal pain, was disorientated, had an enlarged, tender liver, and was in acute liver failure. Previous history revealed behavior disturbance, difficulty with movement in the recent past, and truancy from school. Her ceruloplasmin concentration was 50 μmol/L (normal range 200–450 μmol/L [20–45 mg/dL]), serum copper was 8 μmol/L (normal range 12–25 μmol/L [80–160 μg/dL]), urinary excretion of copper was 2.2 μmol/24 h (normal range 0.2–1.6 μmol/24 h [13–25 μg/dL]), and a liver biopsy established the diagnosis of Wilson's disease.

Comment. This case highlights the importance of measurement of ceruloplasmin. In Wilson's disease, a deficiency of ceruloplasmin results in low plasma concentrations of copper. The metabolic defect is in the excretion of copper in bile and its reabsorption in the kidney; copper is deposited in liver, brain, and kidney. Liver symptoms are present in patients of younger age, and cirrhosis and neuropsychiatric problems are manifest in those who are older. Detection of low plasma concentrations of ceruloplasmin and copper, increased urinary excretion of copper, and markedly increased concentrations of copper in the liver confirms the diagnosis.

for the regulation of the oxidation–reduction reactions, transport, and utilization of iron (Fig. 3.6). Increased concentrations of ceruloplasmin occur in active liver disease and in tissue damage.

Immunoglobulins

Antigens/immunoglobulins: proteins produced in response to foreign substances (see also Chapter 34)

The immune system may be conceptualized as two independent entities, served by separate lymphoid cells: thymically derived T lymphocytes oversee immunoregulation and cell-based immune function, whereas B lymphocytes synthesize and secrete antibodies (immunoglobulins)(see Chapter 34). These immunoglobulin antibodies are proteins, produced by the immune system, which have a defined specificity for a foreign particle (immunogen) that stimulated their synthesis. Not all foreign substances entering the body can elicit this response, however; those that do are called immunogens, whereas any agent that can be bound by an antibody is termed an antigen.

The immunoglobulins are a uniquely diverse group of molecules, recognizing and reacting with a wide range of specific antigens (epitopes) and giving rise to a series of effects that result in the eventual elimination of the presenting antigen. Some immunoglobulins have additional effector functions: for example, IgG is involved in complement activation.

Structure

Immunoglobulins share a common Y-shaped structure of two heavy and two light chains

The basic immunoglobulin is a Y-shaped molecule containing two identical units termed heavy (H) chains, and two identical, but smaller, units termed light (L) chains. Several H chains exist, and the nature of the H chain determines the class of immunoglobulin: IgG, IgA, IgM, IgD, and IgE are characterized by γ, α, μ, δ, and ε H chains, respectively. L chains are of only two types κ and λ, and both types may be found in any one class of immunoglobulin, although obviously not within the same molecule. Each polypeptide chain within the immunoglobulin is characterized by a series of globular regions, which have considerable sequence homology and, in evolutionary terms, are probably derived from protogene duplication.

The N-terminal domains of both H and L chains contain a region of variable amino acid sequence (the V region); together, these determine antigenic specificity. Both H and L chains are required for full antibody activity, as the physically apposed H and L V regions form a functional pocket into which the epitope fits; this is termed the antibody recognition (Fab_2) region. The domain immediately adjacent to the V region is much less variable, in both H and L chains.

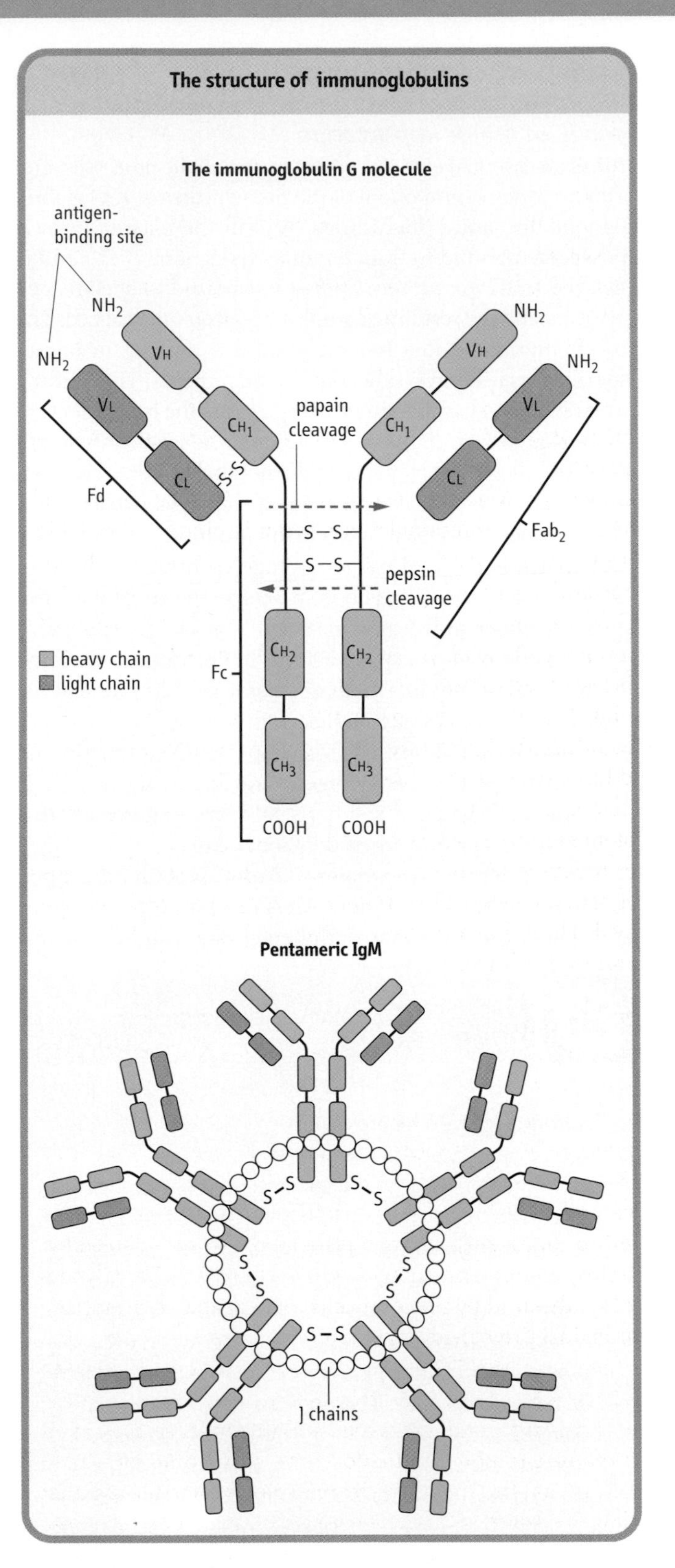

Fig. 3.7 Diagrammatic representation of the basic structure of immunoglobulin, and that of pentameric IgM. V, variable region; C, constant region; H, heavy chain; L, light chain; J chain, joining chain; Fab_2, fragment generated by pepsin cleavage of the molecule; Fc, Fd, fragments generated by papain proteolysis.

The remainder of the H chain consists of a further constant region (Fc region) consisting of a hinge region and two additional domains. This constant region is responsible for immunoglobulin functions other than epitope recognition, such as complement activation. This basic structure of immunoglobulins is depicted in Figure 3.7. When antigen binds to the immunoglobulin, conformational changes are transmitted through the hinge region of the antibody, to the Fc region, which is then said to have become activated.

Major immunoglobulins

IgG is the most common immunoglobulin, protecting tissue spaces and freely crossing the placenta

IgG (with an overall molecular mass of 160 kDa) consists of the basic 2H2L immunoglobulin subunit joined by a variable number of disulfide bonds. The γ H chains have several antigenic and structural differences, allowing classification of IgG into a number of subclasses according to the type of H chain present; however, functional differences between the subclasses are minor.

IgG circulates in high concentrations in the plasma, accounting for 75% of immunoglobulin present in adults, and has a half-life of 22 days. It is present in all extracelluar fluids, and appears to eliminate small, soluble antigenic proteins through aggregation and enhanced phagocytosis by the reticuloendothelial system. From weeks 18–20 of pregnancy, IgG is actively transported across the placenta and provides humoral immunity for the fetus and neonate before maturation of the immune system.

IgA is found widely in secretions and presents an antiseptic barrier, protecting mucosal surfaces

Having an H chain similar to the γ chain of IgG, α chains possess an extra 18 amino acids at the C terminus. The extra peptide sequence enables the binding of a joining or J chain. This short (129-residue) acidic glycopeptide, synthesized by plasma cells, allows dimerization of secretory IgA. IgA is often found in noncovalent association with the so-called secretory component, a highly glycosylated 71-kDa polypeptide, synthesized by mucosal cells and capable of protecting IgA against proteolytic digestion.

IgA represents 7–15% of plasma immunoglobulins and has a half-life of 6 days. It is found, in particular in the dimerized form, in parotid, bronchial, and intestinal secretions. It is a major component of colostrum. IgA would appear to function as the primary immunologic barrier against pathogenic invasion of mucous membranes. It can promote phagocytosis, cause eosinophilic degranulation, and activate complement via the alternate pathway.

IgM is confined to the intravascular space and helps eliminate circulating antigens and microorganisms

IgM, the final major class of immunoglobulin, is polyvalent, with a high molecular mass. It has a basic form similar to that of IgA, having the extra H-chain domain that allows for J-chain binding, and is thus capable of polymerization. IgM normally circulates as a pentamer (with a molecular mass of 971 kDa) linked by disulfide bonds and the J chain (see Fig. 3.7).

IgM accounts for 5–10% of plasma immunoglobulins and has a half-life of 5 days. With its polymeric nature and high molecular mass, most IgM is found confined to the intravascular space, although lesser amounts may be found in secretions, usually in association with secretory component. It is the first antibody to be synthesized after an antigenic challenge.

Minor immunoglobulins

IgD is the surface receptor for antigen in B lymphocytes

IgD differs from the standard immunoglobulin structure chiefly by its high carbohydrate content of numerous oligosaccharide units, resulting in an increased molecular mass of 190 kDa. δ chains are characterized by having only a single interconnecting disulfide bridge, and an elongated hinge region that is particularly susceptible to proteolysis.

Accounting for less than 0.5% of circulating plasma immunoglobulin mass, IgD has a role that remains elusive, although, as a surface component of the mature B cell, it probably has some role in antigenic reception. Rare cases of isolated IgD deficiency seem to be associated with no obvious pathology.

IgE is present only in trace amounts, and acts to bind antigen and promote a release of vasoactive amines from mast cells

Similar to IgM in its unit structure, IgE has ε H chains that consist of five, rather than four, domains, but J-chain binding and polymers do not occur. The extended H chain helps to explain the high molecular mass of IgE – about 200 kDa.

IgE has a high affinity for binding sites on mast cells and basophils. Antigenic binding at the Fab_2 region induces crosslinking of the high-affinity receptor, granulation of the cell, and release of vasoactive amines. By this mechanism, IgE plays a major part in allergy/atopy and mediates antiparasitic immunity.

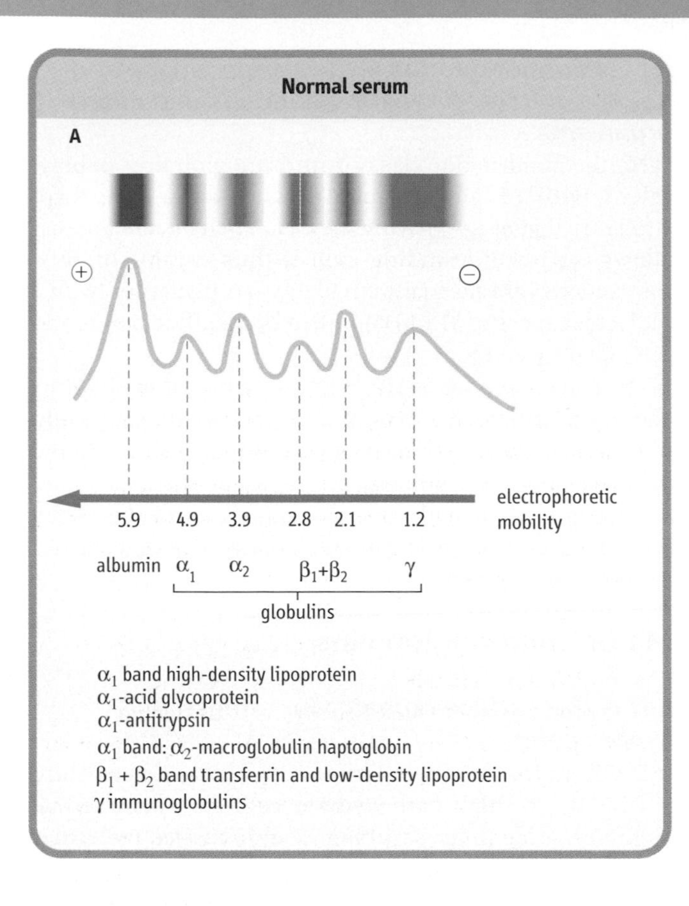

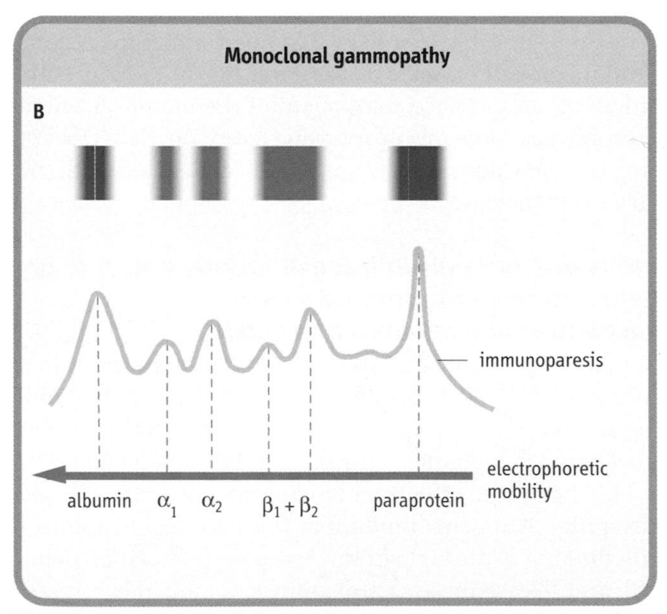

Fig. 3.8 Comparison of electrophoretic appearance of normal serum and that of monoclonal immunoglobulins. (A) Normal serum. (B) Monoclonal gammopathy: a strongly stained band is present in the γ globulin region on electrophoresis, and there is an associated reduction of staining in the remainder of the γ-region (immunoparesis).

Monoclonal gammopathy

A 65-year-old man presented with a sudden onset of low back pain. A radiograph revealed a crush fracture of the second lumbar vertebra, and discrete areas in which the bone was thin (lytic lesions); in the skull, these had a characteristic 'punched out' appearance. Serum electrophoresis demonstrated the presence of a monoclonal immunoglobulin. This proved to be an IgG immunoglobulin and, on electrophoresis, excess free κ chains (Bence–Jones protein) were found in the patient's urine.

Comment. Patients with monoclonal gammopathies present mostly after the age of 50 years; multiple myeloma affects men and women with equal incidence. The clinical features are due to both the malignant proliferation of monoclonal plasma cells and the synthesis and secretion of antibody by these cells. Bone lesions affect the skull, vertebrae, ribs, and pelvis. There may be generalized osteoporosis and pathologic fractures. In up to 20% of cases, no plasma protein is detected, although usually Bence–Jones proteins are present. Such cases are commonly associated with suppression of production of other immunoglobulins (immunoparesis). The presence of excess light chains may cause renal failure as a result of the deposition of Bence–Jones proteins in the renal tubules. Other common findings in myelomatosis include anemia and hypercalcemia.

Monoclonal immunoglobulins

Monoclonal immunoglobulins are the product of a single B cell, and arise from benign or malignant transformations of B cells

Monoclonal immunoglobulins result from the proliferation of a single B cell clone, which thus produces identical antibodies. Usually, these are structurally normal molecules, but sometimes they may be in some way fragmented or truncated. The absolute physical identity of the monoclonal immunoglobulins leads to a single band in gel electrophoresis, revealed by protein staining as a single, dense band in the gamma region (the paraprotein band) (Fig. 3.8).

Monoclonal immunoglobulins are associated with diverse malignant pathologies such as myeloma and Waldenstrom's macroglobulinemia, and also from more benign transformations that are usually termed monoclonal gammopathies of uncertain significance (MGUS).

The acute phase response and C-reactive protein

The acute phase response is a nonspecific response to the stimulus of tissue injury or infection; it affects several organs and tissues

During the acute phase response, there is a characteristic pattern of change in certain proteins – in particular, a marked increase in specific protein synthesis (predominantly in the liver), along with a decrease in the plasma concentration of some other proteins (Fig. 3.9). An increase in the synthesis of proteins such as proteinase inhibitors (α_1-antitrypsin), coagulation proteins (fibrinogen, prothrombin), complement proteins, and C-reactive protein (CRP) is of obvious clinical benefit. Albumin, prealbumin, and transferrin are examples of proteins that have a decreased synthesis during the acute phase response, and are thus termed 'negative acute phase reactants'.

CRP is a major component of the acute phase response. It is synthesized in the liver and is constructed of five polypeptide subunits, giving a molecular weight of around 130 kDa. It is present in only minute quantities (<1 mg/L in normal serum) and is believed to mediate binding of foreign polysaccharides, phospholipids, and complex polyanions, and also activating complement via the classical pathway (see Chapter 34).

Acute phase response

A 45-year-old woman suffered severe lower limb injuries in a road traffic accident. After her admission to hospital, biochemical profiling revealed slightly decreased concentrations of total serum protein (58 g/L) (normal 63–86 g/L) and serum albumin (35 g/L) (normal 37–56 g/L). Serum electrophoresis revealed an increase in the α_1 and α_2 protein fractions. Four days after her operation, the patient's condition deteriorated, and she developed an increased temperature, sweating, and confusion. An acute infection was diagnosed and treatment with appropriate antibiotics was commenced. CRP concentrations peaked 5 days after the operation (Fig. 3.10).

Comment. Increased concentrations of α_1 and α_2 proteins (which include $\alpha_1$antitrypsin, $\alpha_1$acid glycoprotein, and haptoglobin), together with a decrease in serum albumin concentration, suggest an 'acute phase response'. This response is also associated with an increase in CRP, the erythrocyte sedimentation rate (ESR), and increased plasma viscosity. Treatment of the infection and a therapeutic response can be assessed by the resulting decrease in plasma CRP concentration.

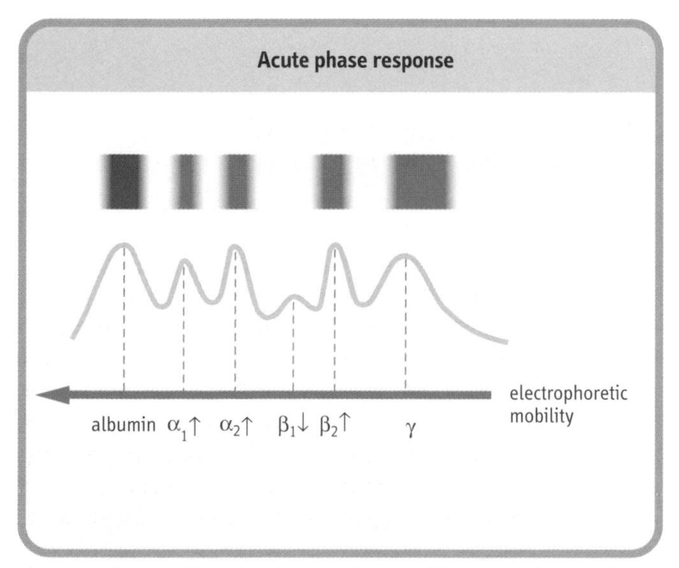

Fig. 3.9 Electrophoretic pattern observed in serum during the acute phase response. Albumin is decreased, $\alpha_1 + \alpha_2$ increased, β_1 decreased, β_2 increased and a mild increase in γ.

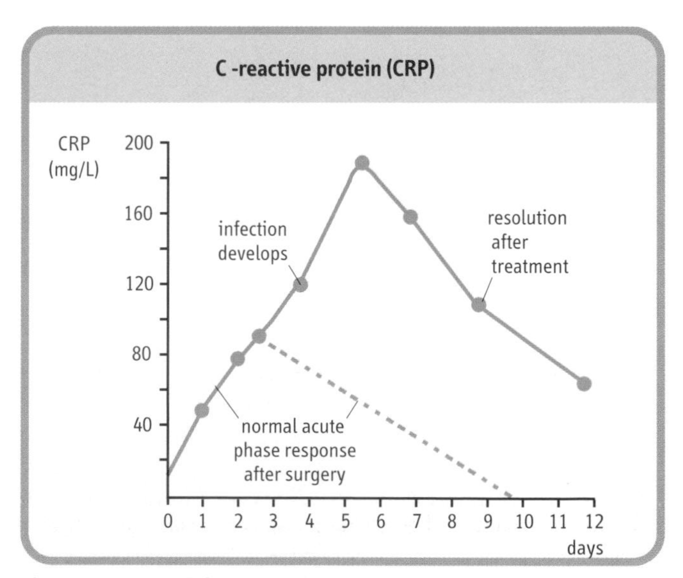

Fig. 3.10 CRP and the acute phase reaction to surgery. The concentration of CRP increases as part of the acute phase response to surgery, and a further increase may be observed if recovery is complicated by infection. (The dotted line represents normal response to surgery).

Summary

This chapter has described the major components and functions of blood. The formed elements – erythrocytes, leucocytes and platelets are suspended in an aqueous solution (plasma) and have several specialised functions such as transport of oxygen, destruction of external agents and clotting of blood. Plasma contains many proteins broadly classified into those synthesized in liver (e.g. albumin) and immunoglobulins. Albumin can function as a major transport protein for several ligands – trace metals, hormones, bilirubin and free fatty acids. Other proteins are more specific in their transport function binding particular ligands e.g. ceruloplasmin and Cu^{2+}, TBG and thyroid hormones. Immunoglobulins are unique molecules, with a common structure, that participate in the defence against antigens that may enter or attempt to enter the body. Five classes of immunoglobulin exist with different specific protective functions.

Changes in the concentration of plasma proteins can give important clinical information. A characteristic pattern with decreased albumin, prealbumin and transferrin and increased α1-antitrypsin, fibrinogen and C-reactive protein is indicative of the acute phase response. Serum and urine protein electrophoresis is an important way of identifying the presence of monoclonal immunoglobulins.

Further reading

Hoffbrand AV, Pettit JE, eds. *Clinical haematology*. London: Mosby-Wolfe; 1994;**1–34**:280–289.

Bataille R, Harousseau JL. Multiple myeloma. *N Engl J Med* 1997;**336**:1657–1664.

Natelson S, Natelson EA, eds. *Principles of applied clinical chemistry, vol 3: Plasma proteins in nutrition and transport*. New York: Plenum Press; 1980.

Russell-Jones DL, Umpleby AM, eds. *Baillière's clinical endocrinology and metabolism: protein metabolism, vol 10, number 4*. London: Baillière Tindall; 1996.

Thompson D, Milford-Ward A, Whicher JT. The value of acute phase protein measurements in clinical practice. *Ann Clin Biochem* 1992;**29**:123–131.

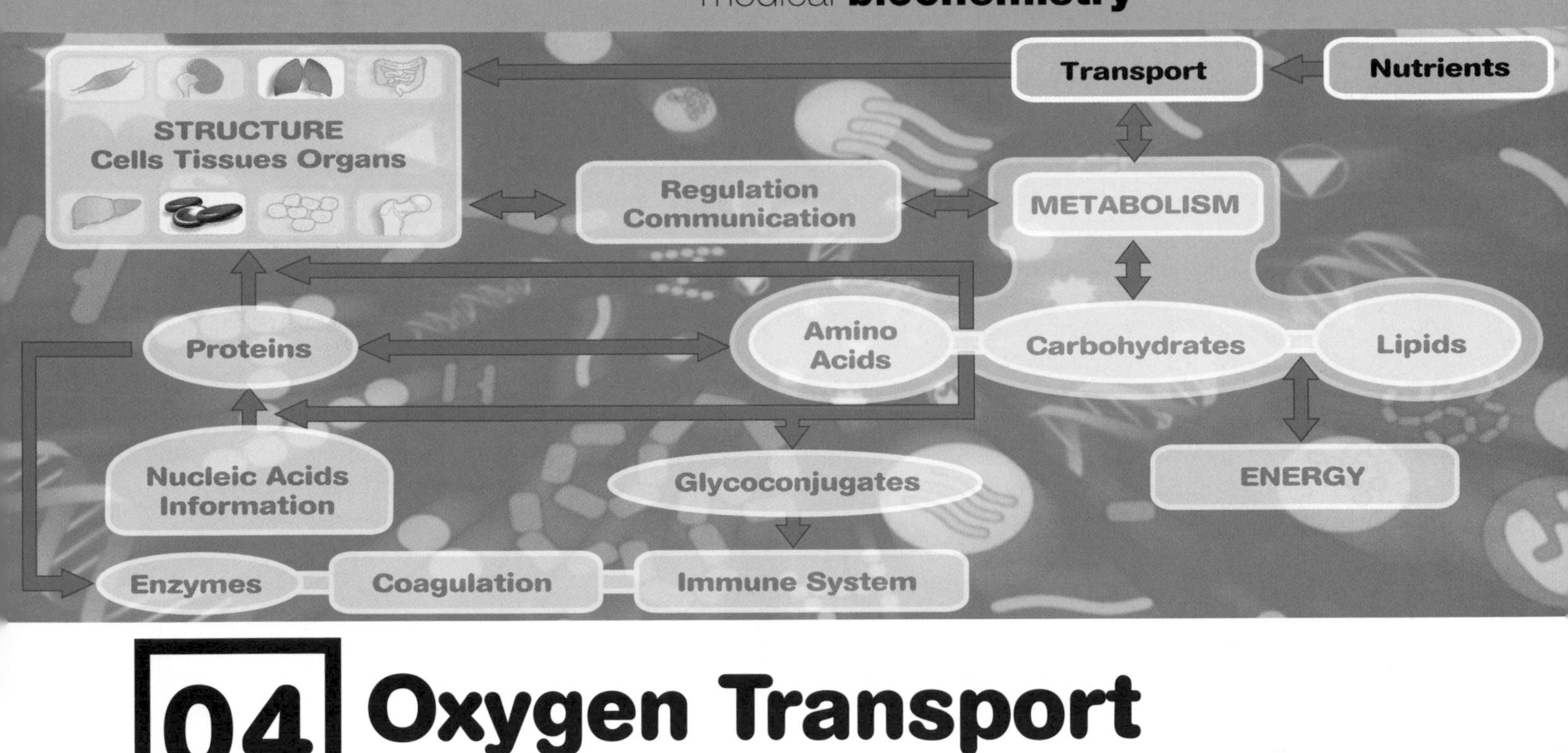

04 Oxygen Transport

Introduction

Humans are aerobic organisms. Our lungs extract oxygen (O_2) from air and deliver carbon dioxide (CO_2) in exhaled gases. The inspired O_2 leads to a more efficient utilization of metabolic fuels, such as glucose and fatty acids, whereas the expired CO_2 is a major product of cellular metabolism. Living systems contain proteins that interact with O_2 and, consequently, increase its solubility in water and sequester it from further reaction. In mammals, these proteins are myoglobin (Mb) and hemoglobin (Hb). Mb, found primarily in skeletal and striated muscle, serves to store O_2 in the cytoplasm and deliver it on demand to the mitochondrion; Hb, restricted to the erythrocytes, is responsible for the movement of O_2 between the lungs and other tissues. This chapter presents the molecular features of heme, the biochemical and physiologic relationships between the structures of Mb and Hb and their interaction with O_2, and the pathologic aspects of selected Hb mutations.

Properties of O_2

In mixtures of gases each component makes a specific contribution, known as its partial pressure, that is directly proportional to its concentration. It is customary to use the partial pressure of a gas as a measure of its concentration in physiologic fluids. For atmospheric O_2 at a barometric pressure of 760 mmHg (1 Atmosphere), the partial pressure, pO_2, is 150–160 mmHg [1 mmHg = 133 Pascal (Pa)]. The amount of O_2 in solution is, in turn, directly proportional to its partial pressure. Thus, in arterial blood the maximum pO_2 is 100 mmHg (13.3 kPa), equivalent to a concentration of dissolved O_2 of 0.13 mmol/L (4.2 mg/L).

Due to its limited solubility in aqueous solvents, O_2 is transported in blood and stored in muscle in complexes with the proteins Hb and Mb respectively. Hb is a tetrameric protein with four O_2-binding sites. In arterial blood with an Hb concentration of 150 g/L (2.3 mmol/L), the contribution of Hb-bound O_2 is about 8.6 mmol/L (275 mg/L). The overall effect is a dramatic 60-fold increase in the O_2 content of this physiologic fluid, yielding almost 200 mL dissolved O_2/L of blood.

Structure of the heme prosthetic group

Heme is the O_2-binding molecule common to Mb and Hb; it is a porphyrin molecule to which an iron atom (Fe^{2+}) is coordinated (Fig. 4.1). The Fe–porphyrin prosthetic group is, with the exception of two propionate groups, hydrophobic and planar. Heme becomes an integral component of the globin proteins during polypeptide synthesis; it is the heme molecule that gives globin proteins their characteristic red-brown color. Globins increase the aqueous solubility of the otherwise poorly soluble, hydrophobic heme molecule. Once sequestered inside a hydrophobic pocket created by the folded globin polypeptide, heme encounters a protective environment that minimizes the highly favored oxidation (rusting) of Fe^{2+} to Fe^{3+} in the presence of O_2. Such an environment is essential for globins to bind and release O_2. Should the iron atom become oxidized to Fe^{3+}, generating the proteins metmyoglobin or methemoglobin, heme can no longer interact with O_2, and O_2 transport is compromised.

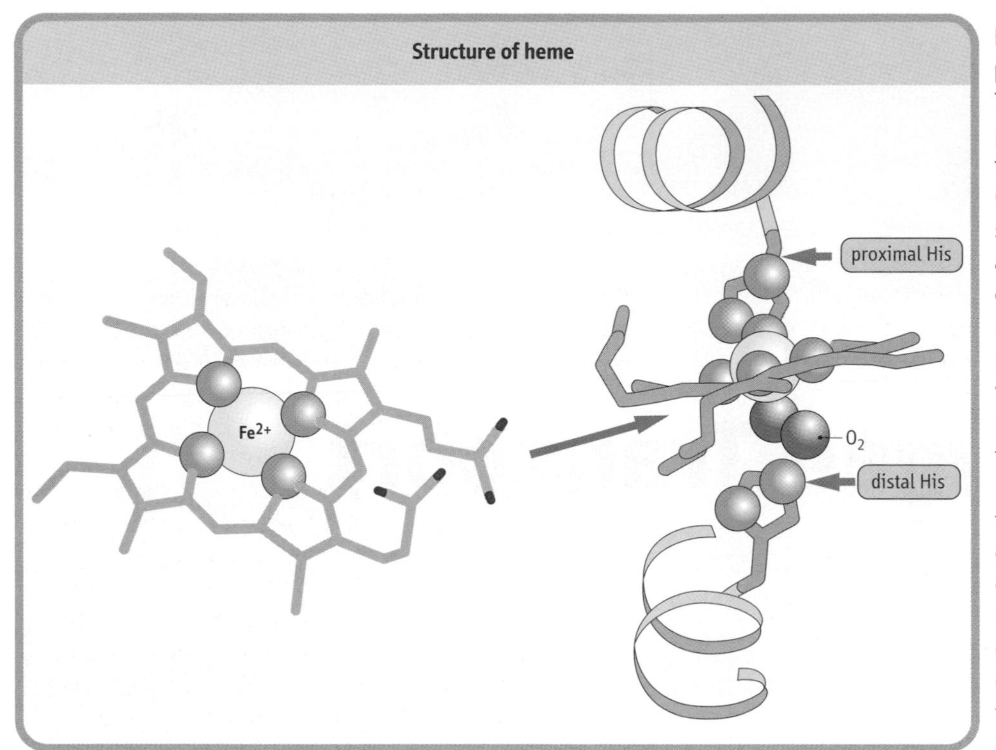

Fig. 4.1 Heme is a complex of porphyrin and iron. Left (top view): The carbon framework of protoporphyrin IX, a conjugated tetrapyrrole ring, is depicted in gray; O_2 molecules are red. Iron (yellow sphere) is normally chelated by six atoms exhibiting octahedral coordination geometry; pyrrole nitrogen atoms (blue spheres) provide four of these. Right (side view): In the globin structure, the planar heme is positioned between the proximal and distal histidines (His); only the former has an imidazole nitrogen (blue sphere) close enough to bond with iron. The α-helices that contain these histidines are shown in pink. For globins in the oxygenated state, O_2 occupies the sixth and final position; in deoxygenated globins, the sixth position remains vacant.

General characteristics of mammalian globin proteins

Globins constitute an ancient family of soluble metalloproteins whose structure and function have been preserved for several million years among leguminous plants, certain invertebrates, and vertebrates. Mb and Hb are examples of globins found in all mammals. Mb consists of a single globin polypeptide and a heme prosthetic group (Fig. 4.2). Hb is a tetrameric assembly of closely related globin subunits. The globin polypeptide is a single chain of approximately 150 amino acids. Each globin contains one noncovalently bound heme prosthetic group. The most significant aspect of the secondary structure of globins is the high proportion of α-helix: over 75% of the amino acids are associated with eight helical segments containing as few as six and as many as 28 residues. These α-helices are organized into a tightly packed, nearly spherical, globular tertiary structure (Fig. 4.2).

Polar amino acids are located almost exclusively on the exterior surface of globin polypeptides, and contribute to the remarkably high solubility of these proteins (for example, 5.2 mmol/L [335 g/L] Hb in the erythrocyte, or >30% protein). Amino acids that are both polar and hydrophobic, such as threonine, tyrosine, and tryptophan, are oriented with their polar functions toward the protein's exterior. Hydrophobic residues

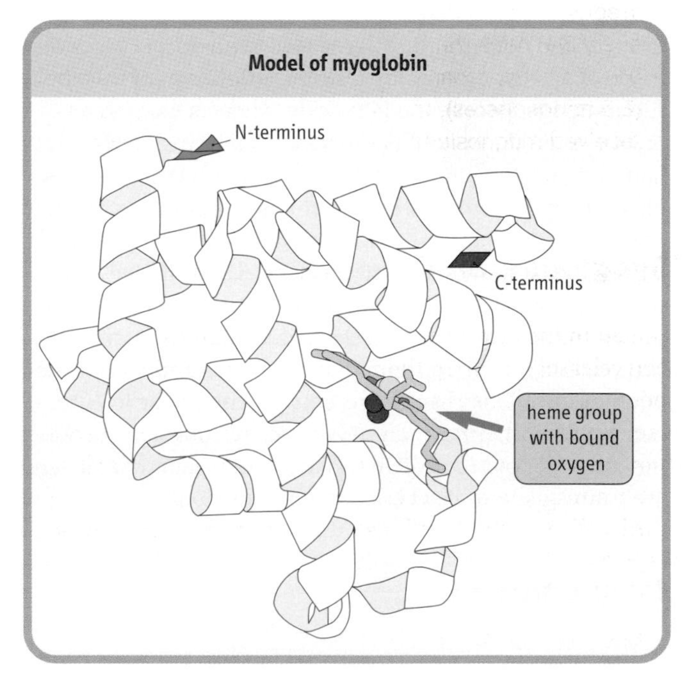

Fig. 4.2 Mb is a compact globular protein. In this depiction of mammalian Mb only the globin polypeptide backbone is shown, to emphasize the high proportion of secondary structure (exclusively α-helix). The N-terminus is blue; the C-terminus is red. The heme group, with bound O_2, is illustrated as a 'ball-and-stick' structure.

are buried within the interior, where they stabilize the folding of the polypeptide and binding of the iron–porphyrin prosthetic group. The only exceptions to this general distribution of amino acid residues in globins are the two histidines that play indispensable roles in the heme-binding pocket. The side chains of these histidines are oriented perpendicular to and on either side of the planar heme prosthetic group. One histidine has an imidazole nitrogen that is close enough to bond directly to the Fe^{2+} atom: this is the *proximal histidine*. On the opposite side of the heme plane is the other histidine: this is the *distal histidine*. The distal histidine is too far from the heme iron for direct bonding; rather, it confers important geometrical constraints on the sixth coordination site. The alignment of heme and distal histidine permits O_2 to bind favorably to the Fe^{2+} atom and, under normal conditions, restricts interaction with carbon monoxide (CO).

Hyperbaric O_2 therapy in treatment of acute CO poisoning

A 22-year-old pregnant female, carrying a fetus of 31 weeks of gestational age, was transported to the maternity clinic of a university hospital for suspected CO poisoning. The patient was experiencing headache, nausea, and visual abnormalities. She stated that her workplace had been undergoing repairs to the heating and ventilation systems during the past 2 weeks, and on the day of her hospital visit the fire department had evacuated the building after detecting a high level of CO (200 ppm).

Vital signs were blood pressure of 116/68 mmHg, pulse rate of 100, and respiratory rate of 24. Noteworthy in the patient's evaluation was a carboxyhemoglobin component of 1.5% of total Hb (normal = 0.5–1.0). External fetal monitoring indicated a fetal heart rate of 135, with occasional, moderate irregularities. Uterine contractions were occurring every 3–5 min. The patient underwent treatment in the hospital's hyperbaric O_2 chamber: 30 min at 250 kPa (2.5 atmospheres), then 60 min at 200 kPa. She also received magnesium sulfate intravenously to resolve the premature contractions. The patient was discharged 2 days later. She delivered a healthy female infant at 38 weeks of gestational age who, on examination at birth and at 6 weeks of age, exhibited no apparent sequelae to her *in utero* exposure to CO.

Comment. Like O_2, CO binds to heme prosthetic groups. Because the affinity of globin-bound heme for CO is more than 10^4 times that of O_2, prolonged exposure of hemoglobin (Hb) to CO would be virtually irreversible ($t_{1/2}$ = 4–5 h) and lead to highly toxic levels of carboxyhemoglobin. Hyperbaric O_2 is the treatment of choice for severe or complicated CO poisoning. The administration of 100% O_2 at 200–300 kPa creates arterial and tissue pO_2 values of 2000 mmHg and 4000 mmHg respectively. The immediate result is a reduction in the $t_{1/2}$ of carboxyhemoglobin to less than 20 min. Hyperbaric O_2 is also used in the treatment of decompression sickness, arterial gas embolism, radiation-induced or ischemic tissue injury, and severe hemorrhage.

Myoglobin: an O_2-storage protein

Located in the cytosol of muscle cells, Mb binds O_2 that has been released by Hb in the tissue capillaries and has subsequently diffused across cellular membranes. This stored O_2 is readily available to organelles, particularly the mitochondrion, that carry out oxidative metabolism. With its single ligand-binding site, the reaction of Mb with O_2, the binding affinity K_a and the fractional O_2 saturation Y are defined respectively as:

$$Mb + O_2 \rightleftharpoons Mb \bullet O_2$$

$$K_a = \frac{[Mb \bullet O_2]}{[Mb][O_2]}$$

$$Y = \frac{[Mb \bullet O_2]}{[Mb \bullet O_2] + [Mb]}$$

Combining these two equations, expressing the concentration of O_2 in terms of its partial pressure pO_2, and substituting the term P_{50} for K yields the equation for the O_2 saturation curve of Mb:

$$Y = \frac{pO_2}{(pO_2 + P_{50})}$$

By definition, the constant P_{50} is the value of pO_2 at which $Y = 0.5$ or half the ligand sites are occupied (saturated by O_2). A low value of P_{50} corresponds to a high affinity for O_2. In a plot of Y versus pO_2, the equation for ligand binding by Mb describes a hyperbola (Fig. 4.3) with $P_{50} \approx 5$ mmHg.

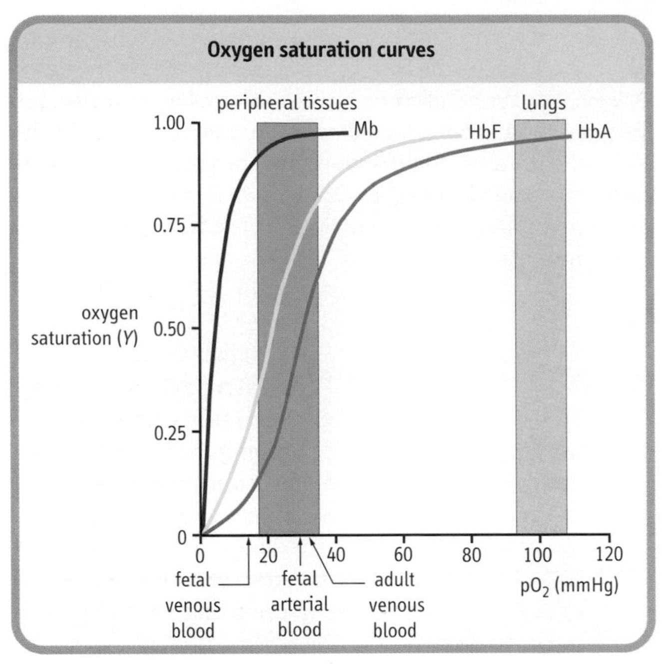

Fig. 4.3 Mb and Hb have different O_2 saturation curves. The fractional saturation (*Y*) of O_2-binding sites is plotted against the concentration of O_2 [pO_2 (mmHg)]. Curves are shown for Mb, fetal Hb (HbF), and adult Hb (HbA). Also indicated, by arrows and shading, are the normal levels of O_2 measured in various adult and fetal blood samples.

In the capillary beds of muscle tissues, pO_2 values are in the range of 20–40 mmHg (2.7–5.3 kPa). Predictably, working muscles exhibit lower pO_2 values than muscles at rest. Mb has a P_{50} of about 2 mmHg (0.3 kPa). With its high affinity for O_2, Mb readily becomes saturated. As O_2 is consumed during aerobic metabolism in muscle tissues, O_2 dissociates from Mb and diffuses into mitochondria, the powerplants of the muscle cell.

Hemoglobin: an O_2-transport protein

Hb, the principal O_2-transport protein in blood, is localized in the erythrocytes (red blood cells). As a delivery vehicle, Hb must be able to bind O_2 efficiently as it enters the lung alveoli during respiration and to release O_2 with similar efficiency to cells as erythrocytes circulate through tissue capillaries. This remarkable duality of function is achieved by cooperative interactions among the globin subunits of Hb.

Quaternary structure of human hemoglobin

Human Hb is a tetramer of two α-globin and two β-globin sub-units: $\alpha_2\beta_2$. These subunits are organized in a tetrahedral array, a geometry that predicts several types of

Blood component replacement following hemorrhage or anemia

Acute or chronic loss of blood and its components accompanies major traumatic or medical hemorrhage and many forms of anemia. A diminished blood volume (hypovolemia) may seriously affect cardiac and vascular function, while a significant decrease in circulating erythrocytes (anemia) will compromise delivery of O_2 to tissues. Increasingly, these conditions are being corrected by the administration of blood components, rather than whole blood transfusions. Crystallite and colloid infusions and/or packed red cells provide rapid and effective resuscitation, while decreasing the risk of excessive anticoagulants, donor alloantigens, and other transfusion complications. There remains the option of providing other blood components, such as platelets and specific clotting factors.

During the restoration of hematologic imbalances, several parameters are usually monitored. Two of these are hematocrit (Hct), the percentage of blood volume occupied by erythrocytes, and the concentration of hemoglobin (Hb). Normal values differ with gender and age: 37–54% for Hct and 120–180 g/L (1.9–2.8 mmol/L) for Hb, with newborns and males at the higher end. An acute bleeding episode may reduce Hb to 39 g/L (0.6 mmol/L). Within hours of replacement therapy, however, the patient should demonstrate an increase in Hb of 10–12 g/L (0.16–0.19 mmol/L) per unit of packed red cells and a corresponding improvement in Hct. Anemia may also respond to erythrocyte concentrates. In determining the cause of anemia, measurements of the size and hemoglobin content of individual erythrocytes are particularly useful. These parameters are mean corpuscular volume (MCV) and mean corpuscular hemoglobin (MCH). Their normal values are 80–100 fL (f, femto = 10^{-15}) for MCV and 27–33 pg Hb (p, pico = 10^{-12}) for MCH. Another diagnostically informative term is mean corpuscular hemoglobin concentration (MCHC), a measure of the average hemoglobin concentration in erythrocytes. MCHC, expressed as g/L, may be calculated two ways: MCH/MCV or (Hb•100)/Hct.

subunit–subunit interactions (Fig. 4.4). Experimental analysis of the quaternary structure indicates multiple noncovalent interactions between each pair of dissimilar subunits, that is, at the α–β interfaces. In contrast, there are few interactions between identical subunits, at the α–α or β–β interfaces. Thus, Hb is more appropriately considered a dimer of the heterodimer: $(\alpha\beta)_2$. The actual number and nature of contacts differ in the presence or absence of O_2 and allosteric effectors. Strong associations within the αβ heterodimer and at the interface between the two heterodimers are now recognized as major factors determining O_2 binding and release.

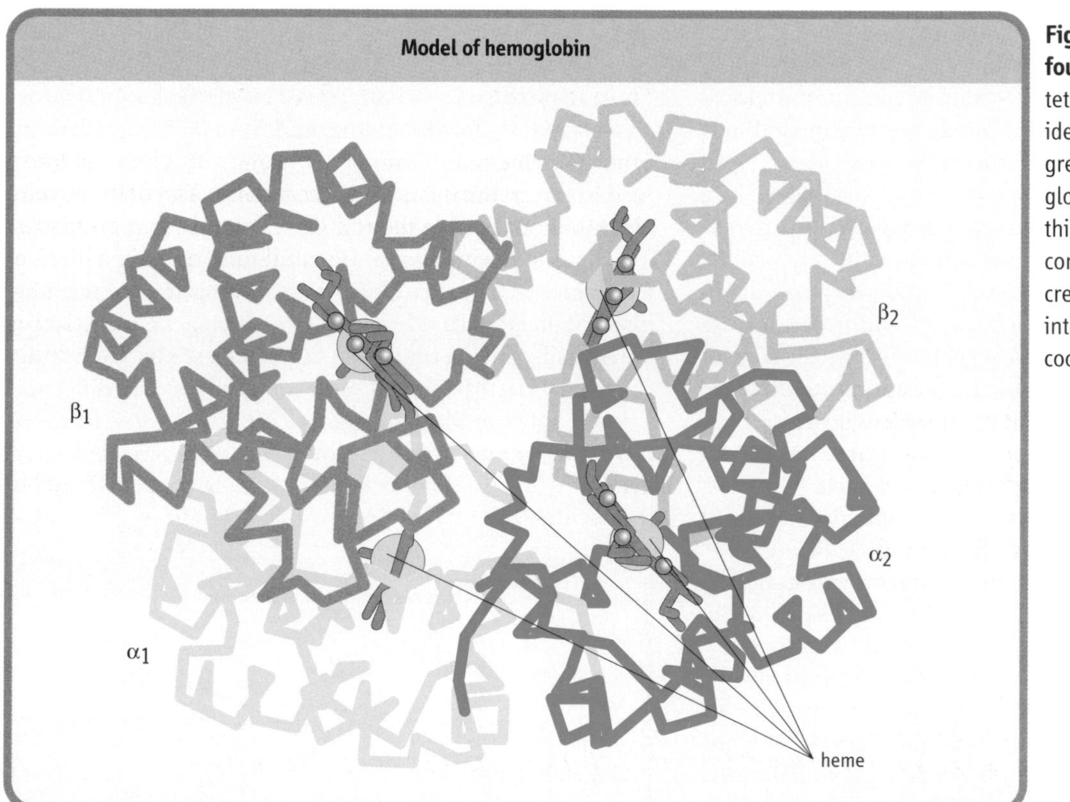

Fig. 4.4 Hb is a tetramer of four globin subunits. Hb is a tetrahedral complex of two identical α-globins (α_1 and α_2, greens) and two identical β-globins (β_1 and β_2, reds). With this geometry each globin contacts the other three, creating subunit interfaces and interactions that define cooperativity.

Interactions with O_2

Hb can bind up to four molecules of O_2. With its multiple ligand-binding sites, the binding affinity and the fractional saturation of Hb are more complex functions than those of Mb. Consequently, the equation for the fractional O_2 saturation curve is modified to:

$$Y = \frac{pO_2^n}{pO_2^n + P_{50}^n}$$

where n is the Hill coefficient. In a plot of *Y* versus pO_2 when $n > 1$, the equation for ligand binding describes a sigmoid (S-shaped) curve (Fig. 4.3).

The Hill coefficient, determined experimentally, is a measure of cooperativity among ligand-binding sites, i.e. the extent to which the interaction of O_2 with one subunit influences the interactions of O_2 with other subunits. For fully cooperative binding, n is equal to the number of sites, an indication that binding at one site maximally enhances binding at other sites in the same molecule. In the absence of cooperativity, even with multiple sites, the Hill coefficient would be 1; this is what is observed for Mb and those Hb mutants that have lost functional subunit–subunit contacts. In practice, a Hill coefficient between 1 and n, the theoretical maximum, is commonly observed.

The normal value for adult Hb ($n = 2.7$) is an indication of strongly cooperative ligand binding. Hb has an O_2 affinity of 27±2 mm Hg (~4kPa), significantly less than that of Mb. The steepest slope of the saturation curve for Hb lies in a range of pO_2 that is found in most tissues (Fig. 4.3). Thus, relatively small changes in pO_2 will result in considerably larger changes in Hb:O_2 interaction. Slight shifts of the curve in either direction will also dramatically influence O_2 affinity.

Transition between deoxygenated and oxygenated hemoglobin

As deoxygenated Hb becomes oxygenated, significant structural changes take place throughout the protein molecule.

For example, in the heme-binding pocket, as a consequence of O_2 coordination to iron and a new orientation of atoms in the heme structure, the proximal histidine and the helix to which it belongs shift their positions (see Fig. 4.1). This subtle conformational change triggers major structural realignments elsewhere within the subunit. In turn, these tertiary structural changes are transmitted, even amplified, in the overall quaternary structure. One αβ heterodimer rotates and slides relative to the other. Because of the inherent asymmetry of the Hb tetramer, these combined motions result in quite dramatic changes in the location of specific amino acid residues. Existing noncovalent bonds are broken and new ones are created at the heterodimer interfaces.

Contact between the two heterodimers is stabilized by a mixture of hydrogen and electrostatic bonds. Approximately 30 amino acids participate in the noncovalent interactions that characterize the deoxygenated and oxygenated Hb conformations. These two quaternary conformations are known as the T- and R-states respectively. In the T-state (tense), the noncovalent interactions between the heterodimers are stronger; in the R-state (relaxed), these noncovalent interactions are, in summation, weaker. O_2 affinity is lower for the T-state and higher for the R-state.

Although the molecular structures of both deoxygenated and fully oxygenated Hb have been studied extensively, progress toward the isolation and characterization of partially oxygenated Hb has been slow. Nevertheless, in defining the theoretical transition between the T- and R-states, a number of models have been developed. At one extreme is the model in which each subunit *sequentially* responds to O_2 binding with a conformational change, thereby permitting hybrids of the T- and R-states. At the opposite extreme is the model in which all four subunits switch *concertedly*. Hybrid states are forbidden and O_2 binding shifts the equilibrium between T- and R- states. The majority of evidence now points toward a sequential model. A sequential model also better explains some of the complex allosteric phenomena. Nevertheless, two questions remain unanswered:

- Do the α- and β-subunits differ in ligand affinity?
- Which subunit binds the first (or releases the last) molecule of O_2?

Interactions with allosteric effectors

Allosteric proteins and effectors

Hb is one of the best-studied examples of an allosteric protein, a protein that exhibits changes in ligand (or substrate) affinity under the influence of small molecules. These small molecules, called allosteric effectors, bind to proteins at sites that are spatially distinct from the ligand-binding sites. Allosteric proteins are typically multisubunit proteins. An allosteric effector may exert either a positive influence on ligand interaction (increased affinity) or a negative influence (decreased affinity). Hb is modified negatively by a number of chemically different allosteric effectors, including H^+, CO_2, and 2,3-bisphosphoglycerate (2,3-BPG) (Fig. 4.5). In keeping with the literal translation of allostery ('other shape'), interactions between Hb and its allosteric effectors are, in general, accompanied by changes in tertiary and/or quaternary structure.

Bohr effect

The O_2 affinity of Hb is exquisitely sensitive to pH, a phenomenon called the Bohr effect. It is most readily described as a right shift in the O_2 saturation curve with decreasing pH. Thus, an increased concentration of H^+ (decreased pH) favors an increased P_{50} for O_2 binding, i.e. favors O_2 release; the converse also holds.

To understand the Bohr effect at the level of protein structure and to appreciate the role of H^+ as an allosteric effector, it is important to recall that Hb is a highly charged molecule. Experimental evidence suggests that the principal residues involved in the Bohr effect are the N-terminal amino group of the α-chain, and the imidazole side chains of His^{122} α and the C-terminal His^{146} β. The pK values of these residues differ sufficiently between the deoxygenated and oxygenated forms of Hb to cause more extensive protonation of the T-state (deoxygenated Hb). These interactions, in turn, are linked directly to the decrease in O_2 affinity of the T-state of Hb.

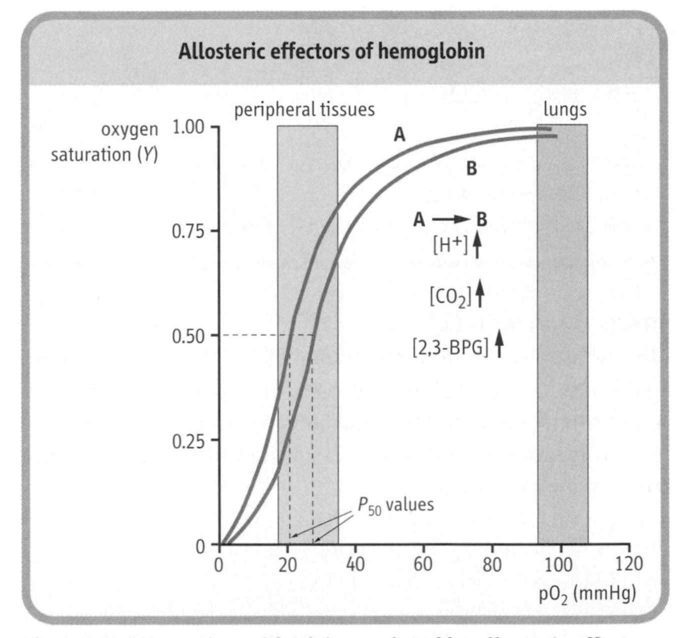

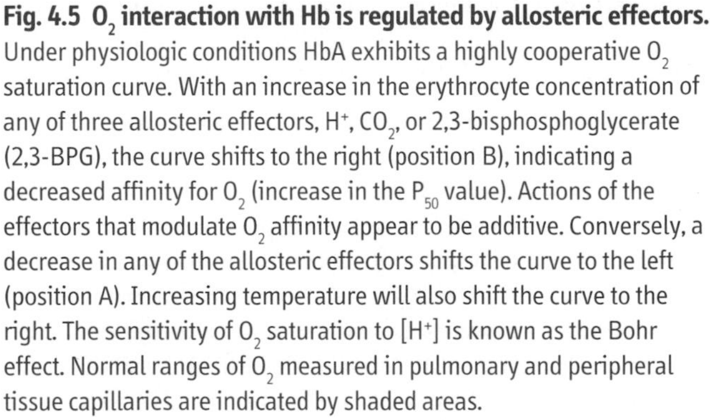

Fig. 4.5 O_2 interaction with Hb is regulated by allosteric effectors. Under physiologic conditions HbA exhibits a highly cooperative O_2 saturation curve. With an increase in the erythrocyte concentration of any of three allosteric effectors, H^+, CO_2, or 2,3-bisphosphoglycerate (2,3-BPG), the curve shifts to the right (position B), indicating a decreased affinity for O_2 (increase in the P_{50} value). Actions of the effectors that modulate O_2 affinity appear to be additive. Conversely, a decrease in any of the allosteric effectors shifts the curve to the left (position A). Increasing temperature will also shift the curve to the right. The sensitivity of O_2 saturation to $[H^+]$ is known as the Bohr effect. Normal ranges of O_2 measured in pulmonary and peripheral tissue capillaries are indicated by shaded areas.

When Hb binds O_2, protons dissociate from the weak acid functions. Conversely, in acidic media, protonation of the conjugate bases inhibits O_2 binding. In their circulation between pulmonary alveoli and peripheral tissue capillaries, erythrocytes encounter markedly different conditions of pO_2 and pH. The high pO_2 in the lungs promotes ligand saturation, yet it also forces protons from the Hb molecule to stabilize the R-state. In the capillary bed, particularly in metabolically active tissues, the pH is slightly lower, due to the production of acidic metabolites. O_2-saturated Hb, upon entering this environment, will acquire some 'excess' protons, shift toward the T-state, and release O_2 for uptake by the tissue's cells. This interdependent relationship between O_2 saturation and pH is largely responsible for meeting the needs of active, aerobic metabolism, a concept discussed in Chapter 8.

Effect of CO_2

Closely related to the Bohr effect is the ability of CO_2 to alter the O_2 affinity of Hb. Like the negative allosteric effect of H^+, increasing levels of CO_2 decrease the affinity for O_2. Accordingly, a right shift in the ligand saturation curve occurs as pCO_2 increases. It should be emphasized that the allosteric effector is, in fact, CO_2, not HCO_3^-. CO_2 reacts reversibly with the N-terminal amino groups of the globin polypeptides to form carbamino–Hb:

$$Hb-NH_3^+ + CO_2 \rightleftharpoons Hb-NH-COO^- + 2H^+$$

This transient chemical modification of Hb is not only a specialized example of allosteric control; it also represents one means of delivering CO_2 to the lungs for clearance from the body.

There is a strong physiologic correlation between pCO_2 and O_2 affinity. CO_2 is a major product of mitochondrial oxidation and, like H^+, will be particularly abundant in metabolically active tissues. Upon diffusing into the blood, a small portion of CO_2 reacts with oxygenated Hb, shifts the equilibrium toward the T-state, and thereby promotes the dissociation of bound O_2. The vast majority of peripheral-tissue CO_2, however, is hydrated in the presence of erythrocyte carbonic anhydrase to carbonic acid (H_2CO_3), a weak acid that dissociates readily to H^+ and HCO_3^-:

$$CO_2 + H_2O \rightleftharpoons H_2CO_3 \quad \text{enzyme-catalyzed reaction}$$

$$H_2CO_3 \rightleftharpoons H^+ + HCO_3^- \quad \text{acid–conjugate base reaction}$$

Interestingly, from both the carbamination reaction and hydration/dissociation reactions involving CO_2, an additional pool of protons is generated, protons that become available to participate in the Bohr effect and facilitate O_2–CO_2 exchange. On its return to the lungs, blood transports two forms of CO_2: carbamino–Hb and the H_2CO_3/HCO_3^- acid–conjugate base pair. Blood and Hb are now exposed to a low pCO_2, and through mass action the carbamination reaction is reversed and binding of O_2 is again favored. Similarly, in the pulmonary capillaries, erythrocyte carbonic anhydrase converts H_2CO_3 into CO_2 and H_2O, products whose gaseous forms are expelled into the atmosphere, as discussed further in Chapter 22. Clearly the concentrations of H^+ and CO_2 are often tightly associated.

Hyperventilation, numbness, and dizziness

A college student with severe muscle spasms in her arms, numbness in her extremities, some dizziness, and respiratory difficulty was brought to the student health center. The patient had been vigorously working out in an attempt to relieve the stress of upcoming examinations when she suddenly began to experience forced, rapid breathing. Suspecting hyperventilation, a health care worker began to reassure the student and helped her regain normal respiration by first cupping her hands over her nose and mouth and then nasal breathing at a slowed rate. After 20 minutes the spasms ceased, feeling returned to her fingers, and the lightheadedness resolved.

Comment. Alveolar hyperventilation is an abnormally rapid, deep, and prolonged breathing pattern that leads to respiratory alkalosis, i.e. a profound decrease in pCO_2 and an increase in blood pH that can be attributed to the increased loss of CO_2 and H_2CO_3 from the body. With decreased [CO_2] and [H^+], two allosteric effectors of O_2 binding and release, the affinity of Hb for O_2 increases sufficiently to reduce the delivery of O_2 to peripheral tissues, including the central nervous system. Another characteristic of alkalosis is a decreased level of ionized calcium in plasma, a situation that contributes to muscle spasms and cramps. In general, hyperventilation may be triggered by hypoxemia, pulmonary and cardiac diseases, metabolic disorders, pharmacologic agents, and anxiety.

Effect of 2,3-disphosphoglycerate

An organic phosphate compound, 2,3-disphosphoglycerate; is a third important modulator of O_2 affinity. An alternative intermediate in the glycolytic pathway (Chapter 11), this molecule is synthesized in human erythrocytes. Like H^+ and CO_2, it is an indispensable negative allosteric effector that, when bound to Hb, causes a marked increase in P_{50}. Indeed, if it were not for the high erythrocyte concentration of 2,3-BPG (4.1 ± 0.5 mmol/L [1.1 ±0.1 g/L], nearly equal to that of Hb), the O_2 saturation curve of Hb would approach that of Mb!

A central cavity is formed among the four subunits within the tetrameric structure of Hb. At one end of this cavity, where the two β subunits are juxtaposed, there is an allosteric binding site for one molecule of 2,3-BPG, generated from the alignment of multiple positive charges. One critical consequence of the conformational differences between the T- and R-states is that only deoxygenated Hb interacts with the negatively charged 2,3-BPG. The effector binding site, created in the region where the β-globins contact each other, is a cluster of the N-terminal amino group (Val^1 β) and the side chains of His^2 β, Lys^{82} β, and His^{143} β. Electrostatic interactions stabilize the complex between the effector and Hb.

The importance of 2,3-BPG as an allosteric effector is underscored by observations that its erythrocyte concentration is responsive to various physiologic and pathologic conditions. When there is a chronic tissue deprivation of O_2 (decreased pO_2), the level of 2,3-BPG increases. Such compensatory increases have been described for certain anemias, during cardiac failure, and on adaptation to high altitudes. The net result is a greater stabilization of the deoxygenated, low-affinity T-state and further shift of the saturation curve to the right, thereby facilitating release of more O_2 to the hypoxic tissues. Under most circumstances, the rightward shift has an insignificant effect on the O_2 saturation of Hb in the lungs.

Normal Hb variants

Over 95% of the Hb found in adult humans is HbA, with the $\alpha_2\beta_2$ globin chain composition. A minor normal Hb is HbA_2; this tetramer accounts for up to 4% of the total and has an $\alpha_2\delta_2$ polypeptide composition. Functionally, these two adult Hbs are indistinguishable. Not surprisingly, mutations of the gene encoding δ-globin are without clinical consequence.

Another minor Hb is fetal Hb, HbF; its subunits are α-globin and γ-globin. While it accounts for no more than 1% of adult Hb, HbF predominates in the fetus during the second and third trimesters of gestation. The most striking difference between HbF and HbA is the sensitivity to 2,3-BPG. Comparison of the primary structures of the β- and γ- polypeptides reveals a replacement of His^{143} β by serine in γ-globin. Consequently, two of the cationic groups that participate in the binding of the anionic allosteric effector are no longer present. Predictably, the interaction of 2,3-BPG with HbF is considerably weaker, resulting in an increased affinity for O_2 (P_{50} = 19 mmHg for HbF vs 27 mmHg for HbA) and a greater stabilization of the oxygenated R-state. The direct benefit of this structural and functional change in the HbF isoform is a more efficient transfer of O_2 from maternal HbA to fetal HbF (Fig. 4.3). The HbF variant, barely detectable in most adults, often increases up to 15–20% in individuals with mutant adult Hbs, such as sickle cell disease. This is an example of the body's compensatory response to a pathologic abnormality. Evaluation of many Hb variants is performed by electrophoretic analysis.

Sickle cell disease

Sickle cell disease is caused by an inherited structural abnormality in the β-globin polypeptide. Clinically, an individual with sickle cell disease presents with intermittent episodes of hemolytic and painful vaso-occlusive crises, the latter leading to severe pain in bones, chest, and abdomen. There is also likely to be impaired growth, increased susceptibility to infections, and multiple organ damage. Sickle cell disease has a prevalence of 40% in some regions of equatorial Africa; among African–Americans heterozygotic carriers number 8%, and homozygote newborns are 0.17%. The Hb molecule in sickle cell disease (HbS) has been

Electrophoretic detection of hemoglobin variants and mutants

Hemoglobin is a protein composed of globin polypeptides and heme prosthetic groups. The summation of acid and conjugate base functions in this molecule impart a net charge, one that varies with the pH of the aqueous solvent. The pH at which the net charge is 0 (positive charges = negative charges) is defined as the isoelectric point, pI, of the protein; the pI of HbA is 6.95. Separation of proteins in a pH gradient generated by an electric field and based on differences in pI is called isoelectric focusing. Commercial kits are available that permit rapid identification of all significant Hb variants and many of the more common mutants, including HbS and HbC, as illustrated in Figure 4.6. To be detected by this method, a mutant must exhibit a net gain or loss of charged residues when compared with HbA. The volume of hemolysate required (<100 μL) makes this technique suitable for neonate and adult blood samples. Quantification is performed by scanning densitometry. The sickle cell trait sample illustrated here (Fig. 4.6) contains 55% HbA, 45% HbS, and 10% HbF.

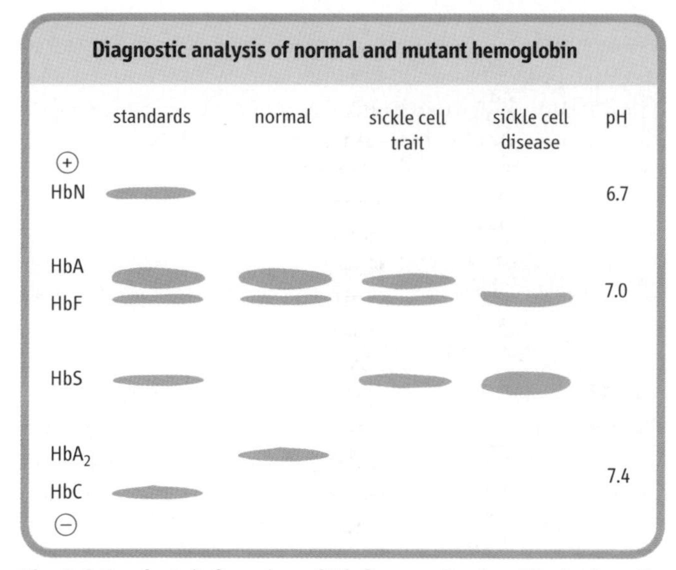

Fig. 4.6 Isoelectric focusing of Hb from patients with sickle cell disease and trait compared with normal.

studied biochemically and biophysically for nearly 50 years and sickle cell anemia has become the paradigm of a molecular disease.

HbA remains a true solute at rather high concentrations, largely as a result of an exterior surface that is compatible with and nonreactive to nearby Hb molecules. HbS, the variant most commonly associated with sickle cell disease, cannot tolerate high protein concentrations when deoxygenated. The poorly soluble deoxygenated HbS becomes arrayed into long, filamentous polymers that readily precipitate, distorting erythrocyte morphology to the characteristic sickle shape. The mutation is Glu^6 β→Val: a surface-localized charged amino acid is replaced by a hydrophobic residue. Valine on the mutant β-globin (β_S) subunit fits into a complementary pocket (sometimes called a 'sticky patch') formed on the β-globin subunit of another Hb molecule, a pocket that becomes exposed only upon the release of bound O_2 in tissue capillaries. In the homozygous individual with sickle cell disease (HbS/HbS), these complementary interactions increase exponentially, and a complex process of nucleation and polymerization occurs. In contrast, in the heterozygous individual (HbA/HbS), the extent of macromolecular growth is much more restricted. In solution, HbS has interactions with O_2 (P_{50} value, Hill coefficient) and allosteric effectors that are equivalent to those for HbA.

Sickled erythrocytes exhibit less deformability; they no longer move freely through the microvasculature and often block blood flow. Moreover, these cells lose water, become fragile, and have a considerably shorter life span, leading to anemia. Except during extreme physical exertion, the heterozygous individual (sickle cell trait) appears normal. For reasons that remain to be elucidated, heterozygosity is associated with an increased resistance to malaria, specifically growth of the infectious agent *Plasmodium falciparum* in the erythrocyte. This observation represents an example of a selective advantage that the HbA/HbS heterozygote exhibits over either the HbA/HbA normal or the HbS/HbS homozygote.

Other hemoglobinopathies

Nearly 1000 mutations in the genes encoding the α- and β-globin polypeptides have been documented. As with most mutational events, the majority of these lead to few, if any, clinical problems. There are, however, over several hundred mutations that give rise to abnormal Hb and pathohematologic phenotypes. Hb mutants, usually named after the location (hospital, city, or geographical region) in which the

Analgesic management of sickle cell vaso-occlusive crises

Vaso-occlusive pain is the most common problem reported by individuals with sickle cell disease. It is also the most frequent reason for emergency-room treatment and hospital admission of individuals with sickle cell disease. Episodes of vaso-occlusive pain are unpredictable and are often excruciating and incapacitating. The origin of this progressive pain involves altered rheologic and hematologic properties of erythrocytes attributable to HbS polymerization and aggregation, coupled with an inflammatory leukocytosis and elevation of plasma acute-phase proteins (Chapter 3). The pain lasts an average of 4–10 days. To provide relief to the individual experiencing vaso-occlusive pain, nonnarcotic, narcotic, and adjuvant analgesics are used alone or in combination. Obviously, the severity and duration of the pain dictate the most appropriate analgesic regimen. Several recent studies suggest additional options for the patient and physician:

- continuous intravenous infusion of a nonsteroidal anti-inflammatory drug (ketoralac) reduced the requirement for opioid (meperidine)
- continuous epidural administration of local anesthetic (lidocaine) and opioid (fentanyl) effectively decreased pain that was unresponsive to conventional measures.

abnormal protein was first identified, may be classified according to the type of structural change and altered function and the resulting clinical characteristics (Fig. 4.7). While many of these mutants have predictable phenotypes, there are others that are surprisingly pleiotropic in their impact on multiple properties of the Hb molecule. With few exceptions, Hb variants are inherited as autosomal recessive traits. Occasionally, double heterozygotes are identified.

Examples and classification of hemoglobinopathies

Hemoglobin	Mutation	Biochemical change	Classification	Clinical consequences
Hb Helsinki (rare)	Lys ^{82} β⇒Met	tetramer exhibits reduced binding of 2,3-bisphosphoglycerate	increased O_2 affinity	mild polycythemia (increased number of erythrocytes)
Hb Titusville (rare)	Asp ^{94} α ⇒Asn	heterodimer interface is altered to stabilize the T-state	decreased O_2 affinity	mild cyanosis (dark blue-purple skin coloration from deoxygenated blood)
HbM Boston (rare)	His ^{58} α ⇒Tyr	substitution of distal His causes structural perturbation in the heme pocket and decreased Bohr effect	ferric heme (methemoglobin)	cyanosis of skin and mucous membranes
HbS (common)	Glu ^{6} β ⇒Val	polymerization of deoxygenated protein occurs in microvasculature	altered solubility	hemolytic and vaso-occlusive crises; severe pain
Hb Gun Hill (rare)	Δ $β^{91-95}$	deletion of a critical Leu, required for heme contact, leads to misfolded protein	unstable protein	formation of Heinz bodies (inclusions of denatured hemoglobin); jaundice (yellow coloration of integument and sclera from heme metabolites); pigmented urine
Hb Agnana (rare)	$β^{94}$ [+TG] ⇒ 156 amino acids	frameshift causes elongation to an unstable, nonsense protein	abnormal synthesis	severe anemia and other complications, consistent with a thalassemia (impaired synthesis of globin polypeptides)

Fig. 4.7 Classification of hemoglobinopathies.

A patient with mutant hemoglobin

During a complete physical examination related to diabetes mellitus and cataracts, a 66-year-old male presented with cyanosis of the lips. The patient was found to have mild hypertension, left ventricular hypertrophy, and other electrocardiographic abnormalities. Since childhood, he has followed a restricted exercise program because of suspected heart disease. Laboratory findings indicated increases in hematocrit (Hct), erythrocyte count, and Hb concentration. Platelet and leucocyte counts were normal, as was pulmonary function. The P_{50} of the patient's whole blood was 58 mmHg (normal = 27 ± 2); the Hill coefficient was 1.6 (normal = 2.7 ± 0.2). The possibility of an abnormal Hb was considered. Biochemical analysis of his Hb revealed a mutant β-globin polypeptide in which Asn^{102} β was replaced by Thr. Electrophoretic analysis indicated that the patient was heterozygous for Hb Kansas.

Comment. Not only does this mutant Hb exhibit a dramatically reduced affinity for O_2 (increased P_{50}), but subunit cooperativity is also markedly disrupted (decreased n). Examination of the tertiary structures of deoxygenated and oxygenated Hb provides an explanation: $Asn^{102}β$ hydrogen bonds to $Asp^{94}α$ in the heterodimer interface, a bond that only exists in the R-state. These residues are too far apart in the T-state to create a noncovalent interaction. In Hb Kansas the absence of this critical oxygenated Hb-stabilizing bond shifts the equilibrium between the T- and R-states toward deoxygenation. Moreover, the altered heterodimer contact results in less communication among the four ligand-binding sites and a loss of subunit cooperativity. Like the overwhelming majority of hemoglobinopathies, Hb Kansas is very rarely seen in a clinical setting. Yet, the phenotypes of these mutations continue to be invaluable in understanding protein structure–function relationships.

Summary

This chapter describes two important proteins that reversibly interact with O_2. Mb, a tissue oxygen storage molecule, and Hb, a blood oxygen transport molecule, both use an ancient heme-containing polypeptide domain motif to sequester O_2 and increase its solubility. These proteins must function efficiently in rather different biochemical environments to sustain aerobic metabolism. As a tetramer of globins, Hb is one of the best-characterized examples of cooperativity in ligand interactions. With its wide variety of effector molecules, Hb is also a prototype of an allosteric protein. Conformational changes in both the tertiary and quaternary structures characterize the transition between deoxygenated and oxygenated states. Mutations to globin genes lead to a spectrum of structural and functional variants, among which are fetal Hb and sickle cell disease.

Further reading

Bunn HF. Pathogenesis and treatment of sickle cell disease. *New Engl J Med* 1997;**337**:762–769.

Frantzen F. Chromatographic and electrophoretic methods for modified hemoglobins. *J. Chromatol Biomed Sci. & Appl.* 1997;**699**:269–286.

Hsia CCW. Respiratory function of hemoglobin. *N Engl J Med* 1998;**338**:239–247.

Kroeger KS. Structures of hemoglobin-based blood substitutes: insights into the function of allosteric proteins. *Structure* 1997; **5**:227–237.

Morita K, Fukuzawa J, Onodera S, Kawamura Y, Sasaki N, Fujisawa K, Ohba Y, Miyaji T, Hayashi Y, Yamazaki N. Hemoglobin Kansas found in a patient with polycythemia. *Ann Hematol* 1992;**65**:229–231.

Silverman RK, Montano J. Hyperbaric oxygen treatment during pregnancy in acute carbon monoxide poisoning. *J Reprod Med* 1997;**42**:309–311.

05 Catalytic Proteins–Enzymes

Principles and classification of enzymes

Almost all biological functions are supported by chemical reactions catalyzed by enzymes, biological catalysts. Efficient metabolism is controlled by orderly, sequential, and branching metabolic pathways. Enzymes accelerate chemical reactions under physiologic conditions, 37°C and neutral pH. However, an enzyme cannot alter the equilibrium of a reaction, but can only accelerate the reaction rate, by decreasing the activation energy of the reaction (Fig. 5.1). Enzymes are involved in most forms of metabolism, and regulation of their activities allows metabolism to adapt to rapidly changing conditions.

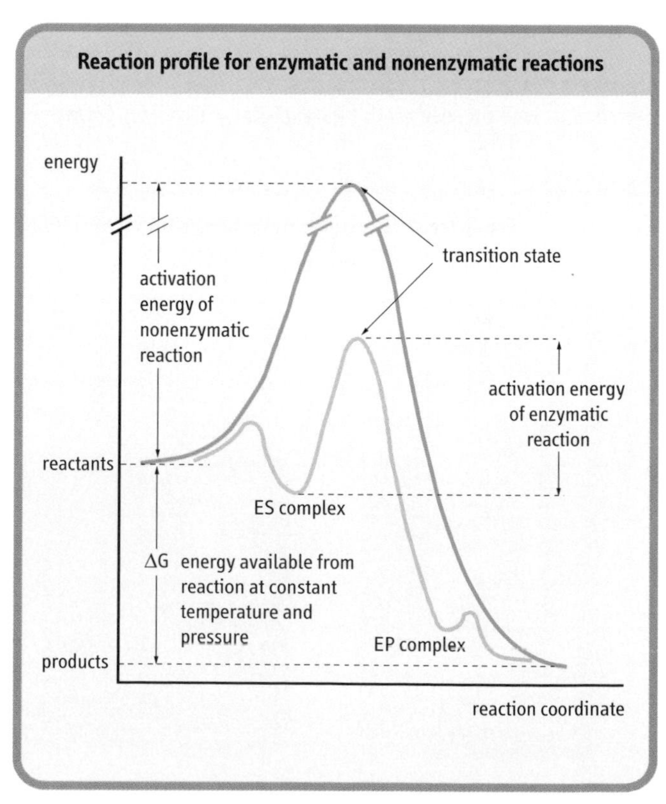

Fig. 5.1 Reaction profile for enzymatic and nonenzymatic reactions. The basic principles of an enzyme-catalyzed reaction are the same as any chemical reaction. When a chemical reaction proceeds, the substrate must gain activation energy to reach a point called the transition state of the reaction, at which the energy level is maximum. Since the transition state of the enzyme-catalyzed reaction has a lower energy than that of the uncatalyzed reaction, the reaction can proceed faster. ES complex, enzyme–substrate complex; EP complex, enzyme–product complex.

Factors affecting enzymatic reactions

Effect of temperature

In the case of an inorganic catalyst, the reaction rate increases with the temperature of the system. Enzymes function as catalysts at body temperature, but also display a temperature optimum *in vitro*, because the three-dimensional structure of a protein involves weak bonding interactions, such as hydrogen bonds, it can be disrupted by high temperature.

Effect of pH

Similarly, every enzyme has a pH optimum, because ionizable amino acids, such as histidine, glutamate, and cysteine, participate in the catalytic reaction. Since the pH in our body is in the range of 6.8–7.4, many enzymes function optimally in this range as well. There are some exceptional enzymes, such as pepsin, with a pH optimum of 1.5–2.0, which is secreted by gastric cells and functions in gastric juice. Changes in pH affect the ionic charge of amino acid side chains of enzymes, and can have a dramatic effect on their catalytic activity. In addition, various molecules, including substrates, products, intermediates, and regulatory molecules, also affect the rate of enzymatic reactions.

Definition of enzyme activity

The activity of an enzyme is obtained by determining the rate of an enzyme-catalyzed reaction under defined conditions. Reaction rate or velocity (v) is generally expressed as the rate of conversion of substrate to product per min, i.e. mol/min. Since the catalytic activity of an enzyme is normally independent of reaction volume, i.e. it is unaffected by dilution, substrate turnover per unit of time under defined conditions (pH, buffer, temperature) is commonly used for defining enzyme-catalyzed reactions. The amount of enzyme activity that catalyzes conversion of 1 mole of substrate into 1 mole of product per second is expressed as the katal (1 kat = 1 mol/s). However, the katal is inconvenient for expressing the actual enzyme activity, because it is generally a very small number. The much larger international unit (1 IU = 1 μmol/min) is more convenient and is commonly used as the standard unit of activity. The specific activity of an enzyme, expressed as μmol/min/mg of protein, or IU/mg of protein, indicates the amount of enzyme in a protein sample, and is useful for estimating the purity of an enzyme. The higher the specific activity of an enzyme, i.e. the more units/mg of protein, the higher its purity or homogeneity.

Reaction specificity and substrate specificity are determined by the active-site structure

Most enzymes are highly specific for both the type of reaction catalyzed and the nature of the substrate(s). Reaction specificity, i.e. the reaction that the enzyme catalyzes, is determined chemically by the amino acid residues in the catalytic center of the enzyme. In general, the active site of the enzyme is composed of the catalytic site and the substrate-binding site. Substrate specificity is determined by the structure, charges, polarity, and hydrophobicity of the substrate-binding site. This is because the substrate must bind in the active site as the first step in the reaction, setting the stage for catalysis. Highly specific enzymes such as catalase and urease, which degrade H_2O_2 and urea respectively, catalyze only one type of reaction, but some enzymes have broader substrate specificity. The serine proteases are a typical example of such a group of enzymes. These are a family of closely related enzymes, such as the pancreatic enzymes, chymotrypsin, trypsin, and elastase, that contain a reactive serine residue in the catalytic site. They catalyze the hydrolysis of peptide bonds on the carboxyl side of a limited range of amino acids in protein. Although they have similar structures and catalytic mechanisms, their substrate specificities are quite different (Fig. 5.2).

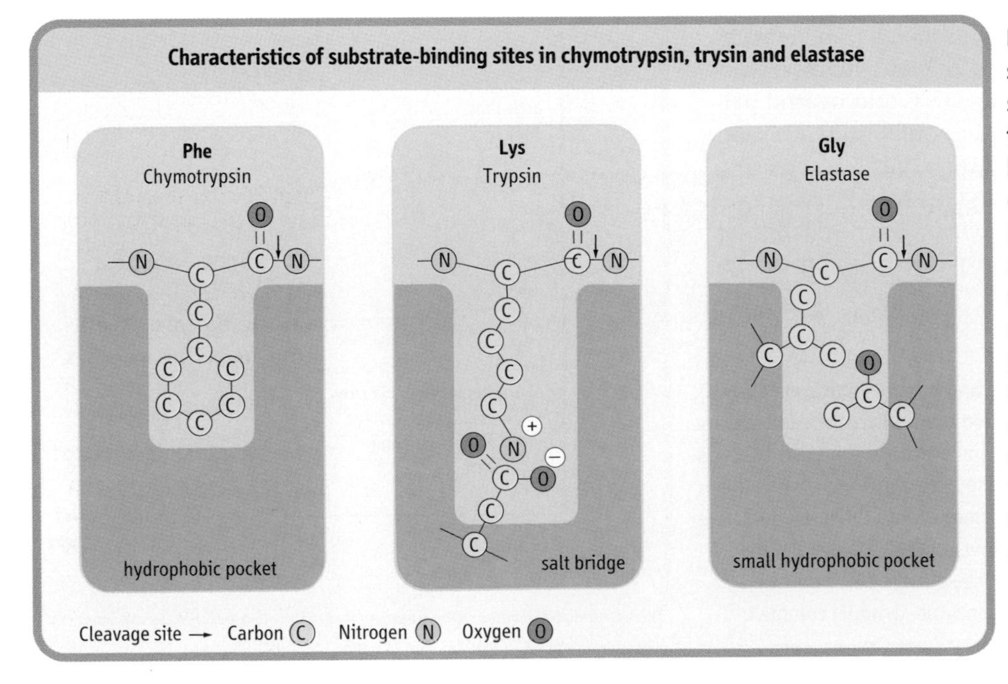

Fig. 5.2 Characteristics of the substrate-binding sites in the serine proteases chymotrypsin, trypsin, and elastase, which hydrolyse peptide bonds. In chymotrypsin a hydrophobic pocket binds aromatic amino acid residues such as phenylalanine (**Phe**). In trypsin, the negative charge of the aspartate residue in the active site promotes cleavage to the carboxyl side of positively charged lysine (**Lys**) and arginine (Arg) residues. In elastase, valine and threonine block the binding site and permit binding of amino acids with small side chains, such as glycine (**Gly**).

Isozymes are enzymes that catalyze the same reaction, but differ in their primary structure and/or subunit composition. Levels of some tissue-specific enzymes and isoenzymes in serum are measured for diagnostic purposes (Figs 5.3 and 5.4).

Nomenclature

A systematic classification is required to organize the different enzymes that catalyze the many thousands of reactions which take place in our body. All enzymes are assigned a four-digit enzyme (EC) number. The first digit indicates membership of one of the six major classes of enzymes shown in Figure 5.5. The next two digits indicate substrate subclasses and sub-subclasses. For example, the transfer of reducing equivalents from one redox system to another is catalyzed by the oxidoreductases (Class 1). The transfer of other functional groups from one substrate to another is catalyzed by the transferases (Class 2). The hydrolases (Class 3) catalyze

Tissue specificity of LDH isoenzymes

A 56-year-old female was admitted to an intensive care unit. The patient had suffered from a slight fever for 1 week, and had some chest pain, and difficulty breathing for the past 24 h. No abnormality was found on chest X-ray or by electrocardiography. However, a blood test showed white blood cells 12,100/mm^3 (4000–9000/mm^3 = normal), red blood cells 240 × 10^4/mm^3 (380–500 × 10^4/mm^3 = normal), hemoglobin 8.6 g/dL (11.8–16.0 g/dL = normal), lactate dehydrogenase (LDH) 1400 IU/L (200–400 IU/L = normal). Levels of other enzymes were normal. Based on the blood tests, the LDH isozyme profile and other data, the patient was eventually diagnosed with malignant lymphoma.

Comment. LDH is a tetrameric enzyme composed of two different 35 kDa subunits. The heart contains mainly the H type, and skeletal muscle and the liver the M type, which are encoded by different genes. Five types of tetramers, called isozymes, can be formed from these subunits: H_4 (LDH_1), H_3M_1 (LDH_2), H_2M_2 (LDH_3), H_1M_3 (LDH_4), and M_4 (LDH_5). Since isozyme distributions differ among tissues, it is possible to diagnose tissue damage by assaying total LDH activity and then by isozyme profiling (Fig 5.3).

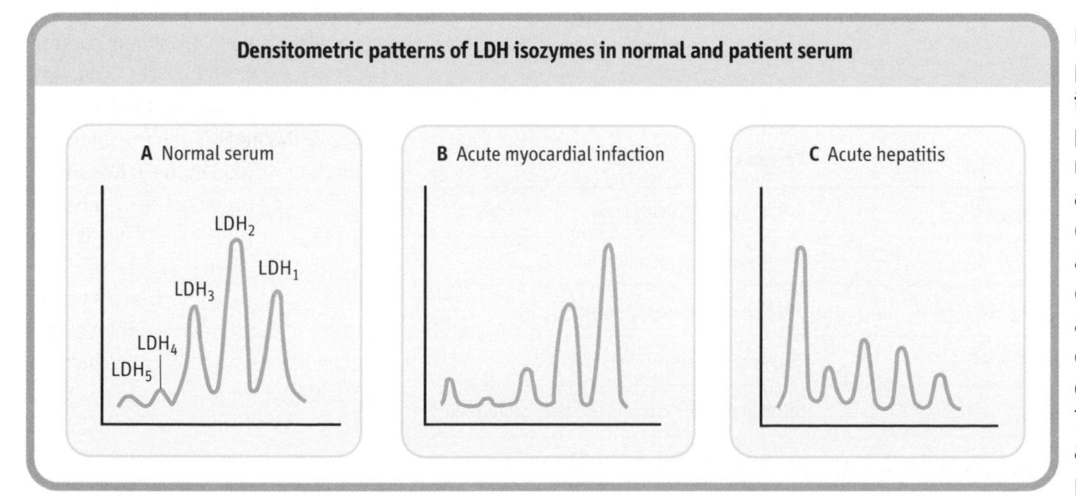

Fig. 5.3 Densitometric patterns of the LDH isozymes in serum of patients diagnosed with myocardial infarction or acute hepatitis. Isozymes, differing slightly in charge, are separated by electrophoresis on cellulose acetate, visualized using a chromogenic substrate, and quantified by densitometry. Total serum LDH activity is also increased in these patients.

group transfer, but the acceptor molecule is exclusively a water molecule. Reactions involving the addition or removal of H_2O, NH_3 or CO_2 are catalyzed by lyases (Class 4), also called synthases. Isomerases (Class 5) catalyze isomerization reactions by rearranging atoms within a molecule, and thus do not affect the composition of the substrate. Ligases, also called synthetases (Class 6), catalyze energy-dependent synthetic reactions.

Some enzymes used for clinical diagnosis of disease

Enzyme	Tissue source (s)	Diagnostic use
AST	heart, skeletal muscle, liver, brain	myocardial infarction
ALT	liver	hepatitis hepatoma (ALT>AST)
amylase	pancreas, salivary gland	acute pancreatitis, bile duct obstruction
CPK	skeletal muscle, heart, brain	muscular dystrophy, myocardial infarction
GGT	liver	hepatitis, malignant tumor
LDH	heart (LDH_1), liver (LDH_5)	myocardial infarction, hepatitis
lipase	pancreas	acute pancreatitis, bile duct obstruction
alkaline phosphatase	osteoblast	some bone diseases, bone tumor
acid phosphatase	prostate	prostate cancer

Fig. 5.4 Enzymes used for clinical diagnosis. AST = aspartate amino transferase; ALT = alanine amino transferase; CPK = creatine phosphokinase; GGT = gamma glutamyl transferase; LDH = lactate dehydrogenase.

Roles of coenzymes

Helper molecules referred to as coenzymes play an essential part in enzyme-catalyzed reactions. Enzymes with covalently or noncovalently bound coenzymes are referred to as holoenzymes. A holoenzyme without a coenzyme is termed an apoenzyme. Coenzymes are divided into two categories. Soluble coenzymes bind to the protein moiety of the enzyme, undergo a chemical change, and are ultimately released. Prosthetic groups are tightly bound to and remain associated with the enzyme during the entire catalytic cycle. Most coenzymes are vitamin derivatives. Derivatives of the B vitamins, niacin and riboflavin, act as coenzymes and are involved in oxidoreductase reactions. The structure and function of coenzymes will be described in later chapters. Some enzymes require inorganic (metal) ions, frequently termed cofactors, for their activity, e.g. blood-clotting enzymes that require Ca^{2+} and oxidoreductases, which use iron, copper, and manganese.

Enzyme kinetics

The Michaelis–Menten equation: a simple model for an enzymatic reaction

Enzyme reactions are multistep in nature and comprise several partial reactions. In 1913, long before the structure of proteins was known, Michaelis and Menten developed a simple model for examining the kinetics of enzyme-catalyzed reactions. The Michaelis–Menten model assumes that the substrate S binds to the enzyme E, forming an essential intermediate, the enzyme–substrate complex (ES), which decomposes to E + product (P). The model also assumes that E, S, and ES are all in rapid equilibrium with one another, so that a steady concentration of ES is rapidly achieved.

$$E + S \rightleftharpoons ES \rightarrow E + P$$

The catalytic constant, k_{cat}, also known as the turnover number, is defined as the number of substrate molecules that can be converted per enzyme molecule per unit time. The

Enzyme classification

Class	Reaction	Enzymes
1. Oxidoreductases	$A_{red} + B_{ox} \rightarrow A_{ox} + B_{red}$	dehydrogenase, peroxidase
2. Transferases	$A{-}B + C \rightarrow A + B{-}C$	hexokinase, transaminase
3. Hydrolases	$A{-}B + H_2O \rightarrow A{-}H + B{-}OH$	alkaline phosphatase, trypsin
4. Lyases	$A(XH){-}B \rightarrow A{-}X + B{-}H$	carbonic anhydrase, dehydratases
5. Isomerases	$A \rightleftharpoons Iso{-}A$	triose phosphate isomerase, phosphoglucomutase
6. Ligases	$A + B + ATP \rightarrow A{-}B + ADP + Pi$	pyruvate carboxylase, DNA ligase

Fig. 5.5 Classification of enzymes.

proportion of ES, compared with the total enzyme molecules $[E]_t$, i.e. the ratio $[ES]/[E]_t$, limits the velocity of an enzyme, v, so that:

$$v = k_{cat}[ES]$$

Since E, S, and ES are all in chemical equilibrium, the enzyme achieves maximal velocity, V_{max}, at very high (saturating) substrate concentrations [S] when $[ES] \approx [E]_t$.

For the dissociation of the ES complex, the law of mass action yields:

$$K_d = \frac{[E][S]}{[ES]}$$

Given that,

$$[E]_t = [E] + [ES]$$

it can be shown that,

$$\frac{[ES]}{[E]_t} = \frac{[S]}{(K_m + [S])}, \text{ where } K_m \approx K_d$$

Consequently, v is given by:

$$v = \frac{(k_{cat}[E]_t[S])}{(k_m + [S])}$$

Since $k_{cat}[E]_t$ corresponds to the maximum velocity, V_{max}, that is attained at high (saturating) substrate concentrations, we obtain the Michaelis–Menten equation:

$$v = \frac{(V_{max}[S])}{(K_m + [S])}$$

Analysis of the above equations indicates that the Michaelis constant, K_m, has units of concentration and corresponds to the substrate concentration at which v is 50% of the maximum velocity, i.e. $[ES] = 1/2\ [E]_t$ and $v = V_{max}/2$.

The Michaelis–Menten model is based on the assumptions that:

- E, S, and ES are in rapid equilibrium
- there are no forms of the enzyme present other than E and ES
- the conversion of ES into E + P is irreversible.

Similar types of kinetic models have been developed for describing the kinetics of multisubstrate, multiproduct enzymes.

Isozymes: hexokinase and glucokinase

Hexokinase catalyzes the first step in glucose metabolism in all cells, namely the phosphorylation reaction of glucose by adenosine triphosphate (ATP) to form glucose 6-phosphate (G-6-P):

glucose + ATP → glucose 6-phosphate + ADP

This enzyme has a low K_m for glucose (0.1 μmol/L) and is inhibited allosterically by its product, G-6-P. Since normal glucose levels in blood are about 5 mmol/L and intracellular levels are approximately 0.2 μmol/L, hexokinase efficiently catalyzes the reaction under normal conditions. This is beneficial for tissues such as the brain whose energy supply is mainly dependent on blood glucose. Hepatocytes, which store glucose as glycogen, and pancreatic β-cells, which regulate glucose consumption in tissues and its storage in liver by secreting insulin, contain an isozyme called glucokinase. Glucokinase catalyzes the same reaction as hexokinase, but has a higher K_m for glucose (10 mmol/L) and is not inhibited by the product, G-6-P. Since glucokinase has a much higher K_m than hexokinase, glucokinase phosphorylates glucose only when blood glucose levels become higher, for example following a meal (see Fig. 5.6). One of the physiologic roles of glucokinase in the liver is to provide G-6-P for the synthesis of glycogen, a storage form of glucose.

Use of the Lineweaver–Burk and Eadie–Hofstee plots for estimating K_m and V_{max}

In a plot of reaction rate versus substrate concentration, the rate approaches the maximum velocity (V_{max}) asymptotically (Fig. 5.7A). It is difficult to obtain reliable values for V_{max}, and, as a result, K_m, by simple extrapolation. To solve this problem, several linear transformations of the Michaelis–Menten equation have been developed.

Lineweaver–Burk plot

The Lineweaver–Burk, or double reciprocal, plot is obtained by taking the reciprocal of the steady-state Michaelis–Menten equation. By rearranging the equation, we obtain:

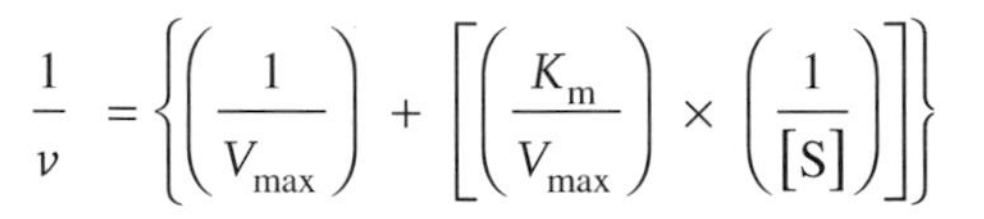

$$\frac{1}{v} = \left\{\left(\frac{1}{V_{max}}\right) + \left[\left(\frac{K_m}{V_{max}}\right) \times \left(\frac{1}{[S]}\right)\right]\right\}$$

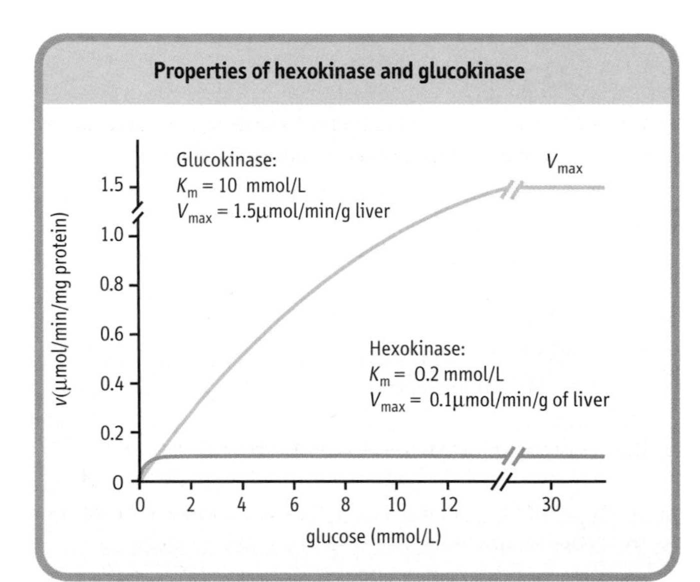

Fig. 5.6 Properties of hexokinase and glucokinase in liver. Hexokinase and glucokinase catalyze the same reaction, phosphorylation of glucose to glucose 6-phosphate (G-6-P). They exhibit different kinetic properties and have different tissue distribution and physiologic function.

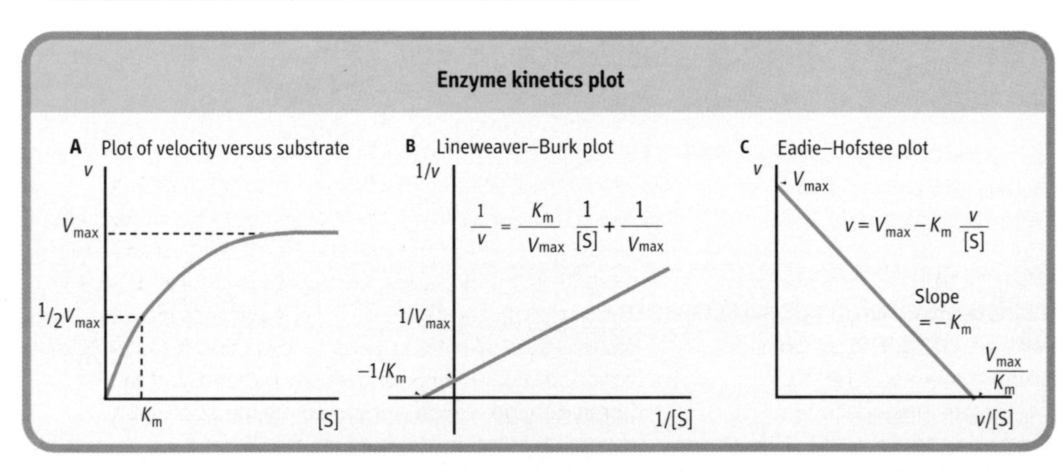

Fig. 5.7 (A) Michaelis–Menten plot of velocity (*v*) versus substrate concentration ([S]); (B) Lineweaver–Burk plot; and (C) Eadie–Hofstee plot.

Measurement of enzyme activity in clinical samples

In clinical laboratories, enzyme activity is measured in the presence of saturating substrate(s) and coenzyme concentrations. Initial kinetics are normally recorded to minimize reverse reactions, i.e. conversion of product to substrate in reversible reactions. Under these conditions, $v \approx V_{max}$ and enzyme activity is directly proportional to enzyme concentration. Enzyme activity is expressed in I U/mL of plasma, serum, cerebrospinal fluid, etc. For interlaboratory comparisons, the conditions for the enzyme assay must be standardized, e.g. the substrate and coenzyme concentrations used, the buffer, buffer concentration, ionic species and ionic strength, pH and temperature.

This equation yields a straight line:

$$y = mx + b.$$
$$(y = 1/v,\ x = 1/[S],\ m = \text{slope},\ b = y \text{ intercept})$$

A graph of $1/v$ versus $1/[S]$ (Fig. 5.7B) has a slope of K_m/V_{max}, a $1/v$ intercept of $1/V_{max}$, and a $1/[S]$ intercept of $-1/K_m$. Although the Lineweaver–Burk plot is widely used for kinetic analysis of enzyme reactions, reciprocals of the data are calculated, and a small experimental error can result in a large error in the graphically determined values of K_m and V_{max}. An additional disadvantage is that important data obtained at high substrate concentrations are concentrated into a narrow region near the $1/v$ axis.

Eadie–Hofstee plot

A second, widely used linear form of the Michaelis–Menten equation is the Eadie–Hofstee plot (Fig. 5.7C). This is described by the equation:

$$v = V_{max} - \left\{ K_m \times \left(\frac{v}{[S]} \right) \right\}$$

A plot of v versus $v/[S]$ has a v intercept of V_{max}, a $v/[S]$ intercept of V_{max}/K_m, and a slope of $-K_m$. The Eadie–Hofstee plot involves only one reciprocal and does not compress the data at high substrate concentrations.

Enzymes are inhibited in different ways

Among numerous substances affecting metabolic processes, enzyme inhibitors are particularly important. Many drugs, either naturally occurring or synthetic, act as enzyme inhibitors. Metabolites of these compounds may also inhibit enzyme activity. Most enzyme inhibitors act reversibly, but there are also irreversible inhibitors that permanently modify the target enzyme. Using Lineweaver–Burk plots, it is possible to distinguish three forms of reversible inhibition: competitive, uncompetitive, and noncompetitive inhibition.

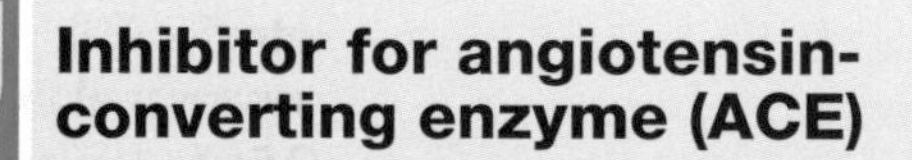

Inhibitor for angiotensin-converting enzyme (ACE)

A 50-year-old man was admitted to hospital suffering from general fatigue, a stiff shoulder, and headache. The patient was 1.8 m tall and weighed 84 kg. The patient's blood pressure was 196/98 mmHg (normal = 140/90 mmHg) and his pulse was 74. He was diagnosed as hypertensive. The patient was given captopril, an angiotension converting enzyme (ACE) inhibitor. After 5 days' treatment, his blood pressure returned to normal levels.

Comment. Renin in the kidney converts angiotensinogen into angiotensin I, which is then proteolytically cleaved to angiotensin II by ACE. Angiotensin II increases renal fluid and electrolyte retention, contributing to hypertension. Inhibition of ACE activity is therefore an important target for hypertension treatment. Captopril inhibits ACE competitively, decreasing blood pressure.(See also Chapter 21.)

Competitive inhibitors cause an apparent increase in K_m, without changing V_{max}

An enzyme can be inhibited competitively by substances that are similar in chemical structure to the substrate (Fig. 5.8).

These compounds compete with substrate for the active site of the enzyme causing an apparent increase in K_m, but no change in V_{max}. The kinetic scheme for competitive inhibition, is:

$$E \begin{array}{l} + S \rightleftharpoons ES \rightarrow E + P \\ + I \rightleftharpoons EI \end{array}$$

Methanol poisoning can be cured by ethanol administration

A 46-year-old male presented to the emergency room 7 h after consuming a large quantity of bootleg alcohol. He could not see clearly and complained of abdominal and back pain. Laboratory results indicated severe metabolic acidosis, a serum osmolality of 465 mmol/L (reference range 285-295 mmol/L), and serum methanol level of 4.93 g/L. By aggressive treatment, including an ethanol drip, bicarbonate, and hemodialysis, he survived and regained his eyesight.

Comment. Methanol poisoning is uncommon but extremely hazardous. Ethylene glycol poisoning exhibits similar clinical characteristics. The most important initial symptom of methanol poisoning is visual disturbance. Laboratory evidence of methanol poisoning includes severe metabolic acidosis and increased plasma solute concentration. Methanol is slowly metabolized to formaldehyde, which is then rapidly metabolized to formate by alcohol dehydrogenase. Formate accumulates during methanol intoxication and is responsible for the metabolic acidosis in the early stage of intoxication. In later stages lactate may also accumulate, due to formate inhibition of respiration. Ethanol is metabolized by alcohol dehydrogenase, which binds ethanol with much higher affinity than either methanol or ethylene glycol. Ethanol is thus a useful agent to inhibit competitively the metabolism of methanol and ethylene glycol to toxic metabolites. Early treatment with ethanol, together with alkali to combat acidosis and hemodialysis to remove methanol and its toxic metabolites, yields a good prognosis.

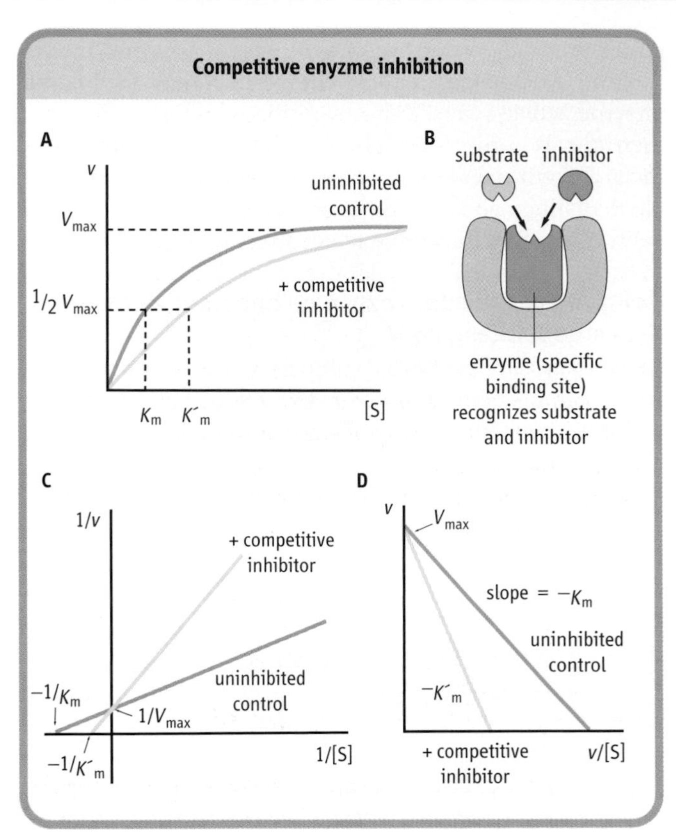

Fig. 5.8 Competitive enzyme inhibition. (A) Plot of velocity versus substrate concentration; (B) mechanism of competitive inhibition; (C) Lineweaver–Burk plot in the presence of a competitive inhibitor; and (D) Eadie–Hofstee plot in the presence of a competitive inhibitor. K'_m is apparent K_m in presence of inhibitor.

The inhibition constant (K_i) is the dissociation constant of the enzyme–inhibitor complex (EI). The lower the K_i, the more efficient the inhibition of enzyme activity.

K_m and V_{max} decrease in uncompetitive inhibition

An uncompetitive inhibitor binds only to the enzyme–substrate complex and not to the free enzyme. The equation shows the kinetic scheme for uncompetitive inhibition. In this case, the K_i is the dissociation constant for the enzyme–substrate–inhibitor complex (ESI).

$$E + S \rightleftharpoons ES \begin{cases} + I \rightleftharpoons ESI \\ \searrow E + P \end{cases}$$

The inhibitor causes a decrease in V_{max} because a fraction of the enzyme–substrate complex is diverted by the inhibitor to the inactive ESI complex. Bound inhibitor also affects the dissociation of substrate, causing an apparent decrease in K_m, i.e. the affinity for substrate is increased.

Insecticide poisoning

A 55-year-old man was spraying an insecticide containing organic fluorophosphates in a rice field. The patient suddenly developed a frontal headache, eye pain, and tightness in his chest, typical signs of over-exposure to toxic organic fluorophosphates. The patient was treated with an intravenous injection of 2 mg of atropine sulfate, and gradually recovered.

Comment. Organic fluorophosphates form covalent phosphoryl–enzyme complexes with acetylcholinesterase and irreversibly inhibit the enzyme. Acetylcholinesterase terminates the action of acetylcholine during neuromuscular activity by hydrolyzing the acetylcholine to acetate and choline. Inhibition of this enzyme prolongs the action of acetylcholine, leading to constant neuromuscular stimulation. Atropine competitively blocks acetylcholine binding at the neuromuscular junction.

Enzyme inhibition: transition-state inhibition and suicide substrate

Enzymes catalyze reactions by inducing the transition state of the reaction. It should therefore be possible to construct molecules that bind very tightly to the enzyme by mimicking the transition state of the substrate. Transition states themselves cannot be isolated, because they are not a stable arrangement of atoms, and some bonds are only partially formed or broken. But for some enzymes, analogues can be synthesized that are stable, but still have some of the structural features of the transition state.

Penicillin (Fig. 5.9) is a good example of a transition state analog. It inhibits the transpeptidase that crosslinks bacterial cell-wall peptidoglycan strands, the last step in cell-wall synthesis in bacteria. It has a strained 4-membered lactam ring that mimics the transition state of the normal substrate. When penicillin binds to the active site of the enzyme, its lactam ring opens, forming a covalent bond with a serine residue at the active site. Penicillin is a potent irreversible inhibitor of bacterial cell-wall synthesis, making the bacteria osmotically fragile and unable to survive in the body.

V_{max} decreases in noncompetitive inhibition

A noncompetitive inhibitor can bind either to the free enzyme or to the enzyme–substrate complex. Thus, noncompetitive inhibition is more complex than other types of inhibition. The next equation shows the kinetic pattern observed for noncompetitive inhibition.

$$\begin{array}{ccc} E + S & \rightleftharpoons & ES \rightarrow E + P \\ + & & + \\ I & & I \\ \upharpoonleft\downharpoonright & & \upharpoonleft\downharpoonright \\ EI & & ESI \end{array}$$

Many drugs and poisons inhibit enzymes irreversibly

Disulfiram (Antabuse®) is a drug used in the treatment of alcoholism. Alcohol is metabolized in two steps to acetic acid. The first enzyme, alcohol dehydrogenase, yields acetaldehyde, which is then converted into acetic acid by aldehyde dehydrogenase. The latter enzyme has an active site cysteine that is irreversibly modified by disulfiram, resulting in accumulation of acetaldehyde in the blood. People who take disulfiram become sick because of accumulation of acetaldehyde in tissues, leading to alcohol avoidance.

Many nerve gases irreversibly inhibit acetylcholinesterase, an enzyme that plays an important role in the transmission of nerve impulses. Diisopropylfluorophosphate (DFP), one of these agents, reacts with a critical serine residue at the active site of acetylcholinesterase and serine proteases.

Alkylating reagents, such as iodoacetamide, irreversibly inhibit the catalytic activity of some enzymes by modifying essential cysteine residue.

In many cases, irreversible inhibitors are used to identify active-site residues involved in enzyme catalysis and to gain insight into the mechanism of enzyme action. By sequencing the protein, it is possible to identify the specific amino acid residue modified by the inhibitor and involved in catalysis.

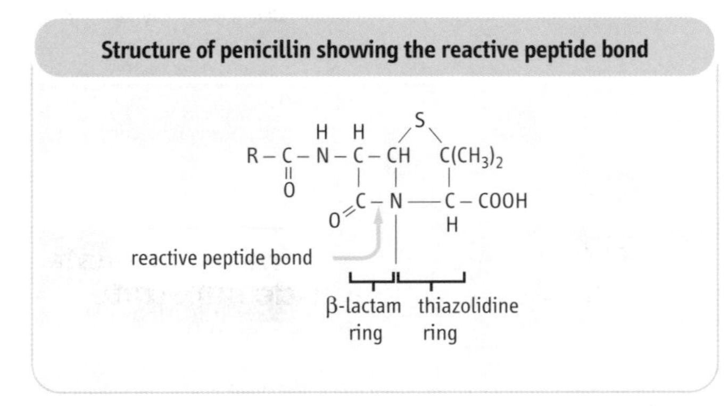

Fig. 5.9 Structure of penicillin showing the reactive peptide bond in the β-lactam ring.

Regulation of enzyme activity

In multistep metabolic pathways, the slowest step limits the overall rate of reaction. It is therefore most efficient to regulate the metabolic pathway by controlling key enzymes that are involved in this 'rate-limiting' step. Generally, five independent mechanisms are involved in these processes:

- The expression of the enzyme protein from the corresponding gene changes in response to the cell's changing environment or metabolic demands.
- Enzymes may be irreversibly activated or inactivated by proteolytic enzymes.

Cyclic AMP (cAMP)-dependent regulation of protein kinase A (PKA)

Many hormonal stimuli trigger the formation of cAMP by cyclization of ATP by adenylate cyclase. cAMP serves as an intracellular messenger, mediating hormonal signals. Most of its effects in eukaryotic cells are mediated by the activation of a single protein kinase, called PKA. This kinase alters the activities of target proteins by phosphorylating specific serine or threonine residues. The enzyme consists of two types of subunits: a 49 kDa regulatory (R) subunit, which has high affinity for cAMP, and a 38 kDa catalytic (C) subunit. In the absence of cAMP, the R and C subunits form an R_2C_2-complex that is enzymatically inactive. The binding of two molecules of cAMP to each of the regulatory subunits leads to the dissociation of R_2C_2 into an R_2-subunit and two C subunits. These free catalytic subunits are then enzymatically active.

Each R chain contains the sequence arginine-arginine-glycine-alanine-isoleucine, which matches the consensus sequence for phosphorylation by this kinase except for the presence of alanine in place of serine. In the R_2C_2-complex, this pseudosubstrate sequence of R occupies the catalytic site of the C subunit, thereby preventing the entry of protein substrates. The binding of cAMP to the R-chains allosterically moves the pseudosubstrate sequences out of the catalytic sites. The released C-chains are then free to bind and phosphorylate substrate proteins. This is an unusual type of allosteric control since ligand binding promotes dissociation of protein subunits.

- Enzymes may be reversibly activated and inactivated by covalent modification, such as phosphorylation.
- Allosteric regulation modulates the activity of key enzymes through reversible binding of small molecules at sites distinct from the active site in a process that is relatively rapid and, hence, the first response of cells to changing conditions.
- The degradation of enzymes by intracellular proteases in the lysosome or by proteosomes in the cytosol also determines the lifetimes of the enzymes and consequently enzyme activity over a much longer period of time.

Catalytic mechanism of serine protease

In enzyme reactions, ionizable amino acids, such as histidine and cysteine, participate in the enzyme-catalyzed reaction. In the serine protease family, a specific serine residue forms the catalytic center, with aid from other amino acids. Ser^{195}, His^{57}, and Asp^{102} form a 'catalytic triad' in chymotrypsin (Fig. 5.10). When the proton of the hydroxyl group of Ser^{195} is hydrogen bonded to $histidine^{57}$, the more nucleophilic oxygen atom of $serine^{195}$ is able to attack the carboxyl carbon atom of the peptide bond in the substrate. The role of the carboxylate group of Asp^{102} is to stabilize the positively charged form of His^{57} in the transition state.

Trypsin and elastase, two other digestive enzymes, are similar to chymotrypsin, in many respects. About 40% of the amino acid sequences of these three enzymes are identical, and their three-dimensional structures are very similar. All three enzymes contain a serine–histidine–aspartate–catalytic triad, and are inactivated by the binding of fluorophosphates such as diisopropylfluorophosphate to the serine residue in this triad.

Proteolytic activation of digestive enzymes

Some enzymes are stored in a specific area or compartment, such as exocytotic vesicles in cells, in inactive precursor forms termed proenzymes or zymogens. This type of enzyme includes the digestive enzymes, which are stored as inactive zymogens in the pancreas. The zymogens are secreted in pancreatic juice following a meal and are activated in the gastrointestinal tract by conversion of trypsinogen into trypsin by the action of intestinal enteropeptidase. This enzyme, located on the inner surface of the duodenum, hydrolyzes an N-terminal peptide from the inactive trypsinogen. Rearrangement of the tertiary structure

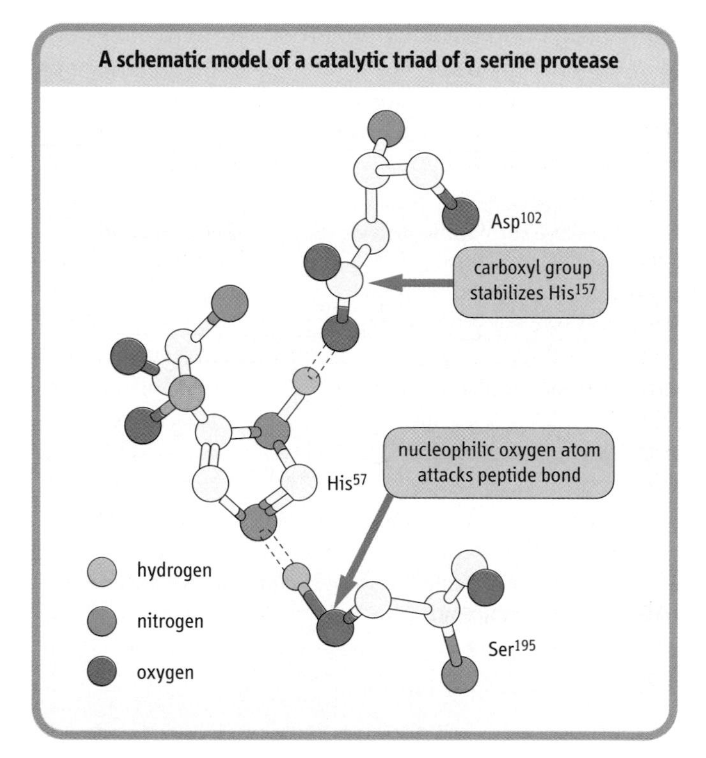

Fig. 5.10 The catalytic triad of serine protease.

Hemophilia is caused by a defect in a zymogen activation

A child was admitted to hospital with muscle bleeding affecting the femoral nerve. Laboratory findings indicated a blood-clotting disorder, hemophilia A, resulting from deficiency of Factor VIII. Factor VIII was administered to the patient to restore blood-clotting activity.

Comment. Formation of a blood clot results from a cascade of zymogen-activation reactions. Over a dozen different proteins, known as blood-clotting factors, are involved. In the final step, the blood clot is formed by conversion of a soluble protein, fibrinogen (Factor I), into an insoluble, fibrous product, fibrin, which forms the matrix of the clot. This last step is catalyzed by the serine protease, thrombin (Factor IIa). Hemophilia is a disorder of blood clotting caused by a defect in one of the sequence of clotting factors. Hemophilia A, the major (85%) form of hemophilia, is caused by a defect of clotting Factor VIII. (See Chapter 6).

yields the proteolytically active form of trypsin. The active trypsin then proteolytically digests other zymogens, such as procarboxypeptidase, proelastase, and chymotrypsinogen, as well as other trypsinogen molecules. Similar proteolytic cascades are observed during blood clotting and fibrinolysis (dissolution of clots).

Since the pancreas is an important organ for controlling blood glucose, the unregulated activation of these enzymes would cause inflammation of the pancreas (pancreatitis) and could lead to diabetes.

Allosteric regulation of rate-limiting enzymes in metabolic pathways

The substrate saturation curve for an 'isosteric' (single shape) enzyme is hyperbolic (see Fig. 5.7A). Allosteric enzymes often show sigmoidal plots of reaction velocity versus substrate concentration [S] (Fig. 5.11). An allosteric effector molecule binds to the enzyme at a site that is distinct and physically separate from the substrate-binding site, but which affects the overall substrate binding and/or reaction velocity. When the substrate is the effector molecule, this is referred to as a homotropic effect, whereas if it is different, it is referred to as a heterotropic effect. Homotropic effects are observed when the reaction of one substrate molecule with an enzyme affects the reaction of the second substrate molecule with a different active site on a multimeric protein. The interaction between subunits makes the binding of substrate cooperative and results in a sigmoidal curve in the plot of v versus [S]. The effect is essentially identical with that described for the binding of O_2 to hemoglobin, except that in the case of enzymes, substrate binding leads to an enzyme-catalyzed reaction.

Positive and negative cooperativity

Positive cooperativity indicates that the reaction of a substrate with one active site makes it easier for another substrate to react at another active site. Negative cooperativity means that the reaction of a substrate with one active site makes it more difficult for a substrate to react at the other active site. Since the affinity of the enzyme changes with substrate concentration, it cannot be described by simple Michaelis–Menten kinetics. Instead, it is characterized by substrate concentration giving a half-maximal rate, $[S]_{0.5}$, and the Hill coefficient (H). The H values are larger than 1 for enzymes with positive cooperativity and less than 1 for those with negative cooperativity. For most allosteric enzymes, intracellular substrates are near the $[S]_{0.5}$ so that the enzyme's activity responds to slight changes in substrate concentration.

The model most often invoked to rationalize allosteric behavior was established by Monod, Wyman, and Changeaux, the so-called concerted model (Fig. 5.12). As with O_2 binding to Hb, in the absence of substrate, the enzyme has a low affinity for substrate, and is in the T-state (tense state). The other conformation of the enzyme is the R-state (relaxed state). Binding of allosteric effector molecules shifts the fraction of enzyme from one state to the other. While enzymes are shifted to the R-state by the binding of positive allosteric effector molecules, they are stabilized in the T-form by negative allosteric effector molecules. In this model, all the active sites in the R-state are the same and all have higher substrate affinity than in the T-state. Because the transition between the T- and R-states occurs at the same time for all subunits, this is called the concerted model. An alternative model, the so-called sequential model, has also been proposed by Koshland, Nèmethy, and Filmer. It postulates that each subunit changes independently to a different conformation and that different subunits may have different affinities for substrate. It is now recognized that both models are applicable depending on the enzyme.

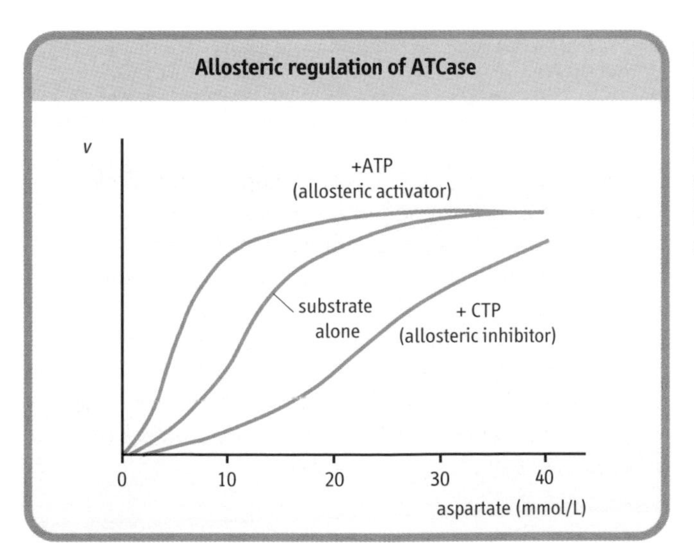

Fig. 5.11 Plot of velocity (v) versus substrate concentration in the presence of an allosteric activator or allosteric inhibitor. Aspartate transcarbamoylase (ATCase) is an example of an allosteric enzyme. Aspartate (substrate) homotropically regulates ATCase activity, providing sigmoidal kinetics. CTP, an end product, heterotropically inhibits, but ATP, a precursor, heterotropically activates ATCase. This enzyme is described in more detail in Chapter 28.

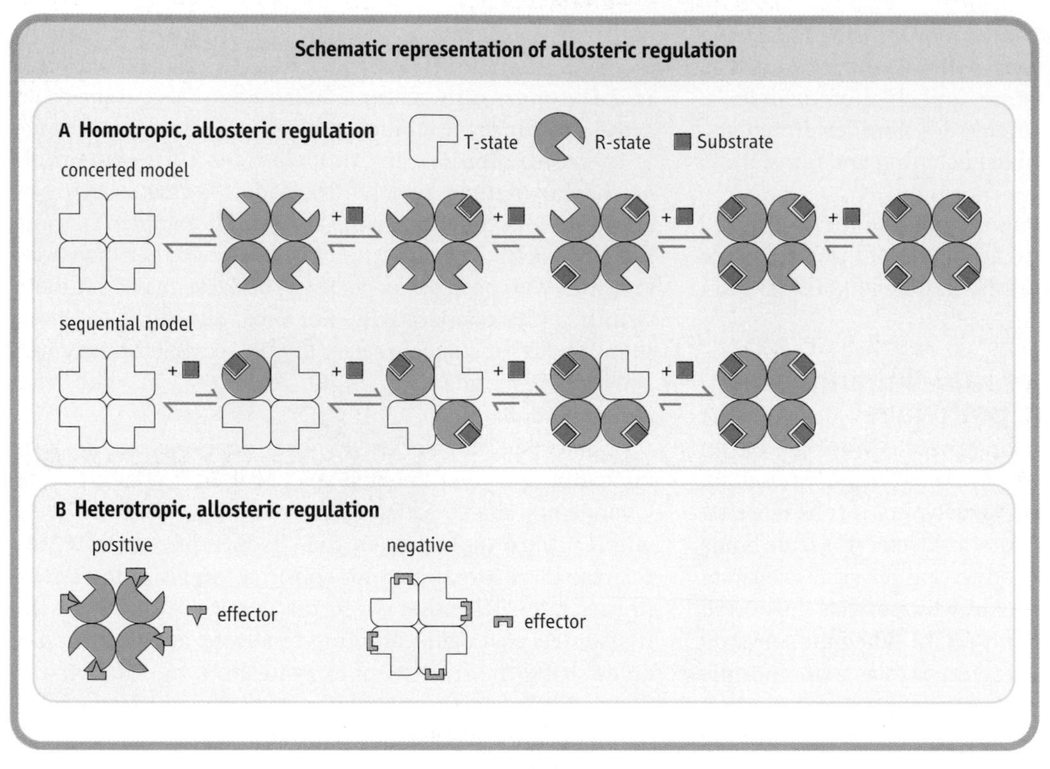

Fig. 5.12 Allosteric regulation. (A) In homotropic regulation, the substrate acts as an allosteric effector. In the concerted model, all of the subunits convert from the T- into the R- state at the same time; in the sequential model, they change one by one, with each substrate-binding reaction. (B) In heterotropic regulation, the effector is distinct from the substrate, and binds at a structurally different site on the enzyme.

Summary

Most metabolism is catalyzed by biological catalysts – enzymes. Their catalytic activities are apparent at body temperature, and they are strictly regulated by several mechanisms. Both covalent and noncovalent modifications are involved in this regulation and allow for efficient metabolic control. Enzymes can be inhibited by synthetic as well as endogenous compounds. Allosteric effectors convert enzymes into more or less active forms. Kinetic analyses of enzymatic reaction are beneficial for evaluating the biological role of enzymes and for elucidating their reaction mechanisms. In addition, assays of enzymes in blood are useful for diagnosis of some diseases. Uncontrolled enzymatic activity, however, sometimes causes serious diseases. In such cases, appropriate inhibitors are quite useful for therapeutic purposes.

Further reading

Flexner C. HIV Protease inhibitors. *New Engl J Med* 1998;**338:** 1281–1292.

Holwerda BC. Herpes virus proteases: targets for novel antiviral drugs. *Antiviral Research* 1997;**35:** 1-21.

Huijgen HB, Sanders GT, Koster RW, Vreeken J, Bossuyt PM. The clinical value of lactate dehydrogenase in serum: a quantitative review. *Eur J Clin Chem Clin Biochem* 1997;**35**:569–579.

Kruse JA. Methanol poisoning. *Intensive Care Med* 1992;**18**:391–397.

Ma AO, Gibbons GH. New insights on renovascular hypertension. *Curr Opin Cardiol* 1994;**9**:598–605.

Martinowitz UP, Schulman S. Continuous infusion of factor concentrates: review of use in hemophilia A and demonstration of safety and efficacy in hemophilia B. *Acta Haematol* 1995;**94**:35–42.

Pan P, Woehl E, Dunn F. Protein architecture, dynamics and allostery in tryptophan synthase channeling. *Trends Biochem Sci* 1997; **22**:22–27.

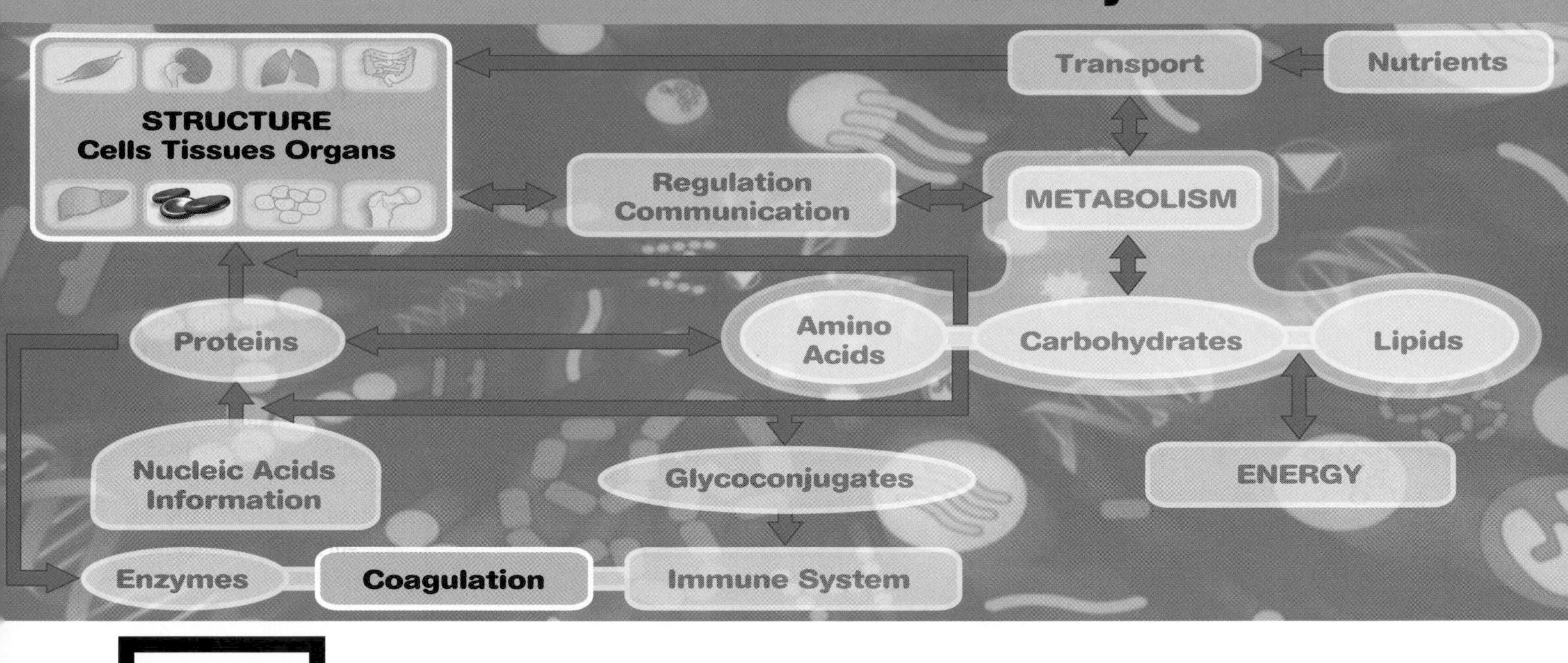

06 Hemostasis and Thrombosis

Circulation of the blood within the cardiovascular system is essential for transportation of gases, nutrients, minerals, metabolic products, and hormones between its organs. It is also essential that blood should not leak excessively from blood vessels when they are injured by the traumas of daily life. Animal evolution has therefore resulted in the development of an efficient, but complex, series of hemodynamic, cellular, and biochemical mechanisms that limit such blood loss by forming platelet–fibrin plugs at sites of vessel injury (hemostasis). Genetic disorders that result in loss of individual protein functions, and therefore in excessive bleeding (e.g. hemophilias), have played an important part in the identification of many of the biochemical mechanisms in hemostasis.

It is essential also that these hemostatic mechanisms are appropriately controled by inhibitory mechanisms, otherwise an exaggerated platelet–fibrin plug may produce local occlusion of a major blood vessel (artery or vein) at its site of origin (thrombosis), or may break off and block a blood vessel downstream (embolism). Arterial thrombosis is the major cause of heart attacks, stroke, and the need for limb amputations in developed countries, but venous thrombosis and embolism are also major causes of death and disability. Clinical use of antithrombotic drugs (antiplatelet, anticoagulant, and thrombolytic agents) is now widespread in developed countries, and requires an understanding of how they interfere with hemostatic mechanisms to exert their antithrombotic effects.

Hemostasis

Hemostasis means 'the arrest of bleeding'

After tissue injury that ruptures smaller vessels (including everyday trauma, injections, surgical incisions, and tooth extractions), a series of interactions between the vessel wall and the circulating blood normally occur, resulting in cessation of blood loss from injured vessels within a few minutes (hemostasis). Hemostasis results from effective sealing of the ruptured vessels by a hemostatic plug composed of blood platelets and fibrin. Fibrin is derived from circulating fibrinogen, whereas platelets are small cell fragments that circulate in the blood and have an important role in the initiation of hemostasis.

Hemostasis requires the effective, coordinated function of blood vessels, platelets, coagulation factors, and the fibrinolytic system

Figure 6.1 provides an overview of hemostatic mechanisms, and illustrates some of the interactions between blood vessels, platelets, and the coagulation system in hemostasis; each of these components of hemostasis also interacts with the fibrinolytic system. The initial response of small blood vessels to injury is arteriolar vasoconstriction, which temporarily reduces local blood flow. Flow reduction transiently reduces blood loss, and may also promote formation of the platelet–fibrin plug. Activation of blood platelets is followed by their adhesion to the vessel wall at the site of injury, and their subsequent aggregation to each other, building up an occlusive platelet mass that forms the initial (primary) hemostatic plug. This platelet plug is friable and, unless subsequently stabilized by fibrin, will be washed away by local blood pressure when vasoconstriction reverses.

Vascular injury also activates coagulation factors, which interact sequentially to form thrombin, which converts circulating soluble plasma fibrinogen to insoluble, crosslinked fibrin. This forms the subsequent (secondary) hemostatic plug, which is relatively resistant to dispersal by blood flow or fibrinolysis. There are two pathways of the activation of

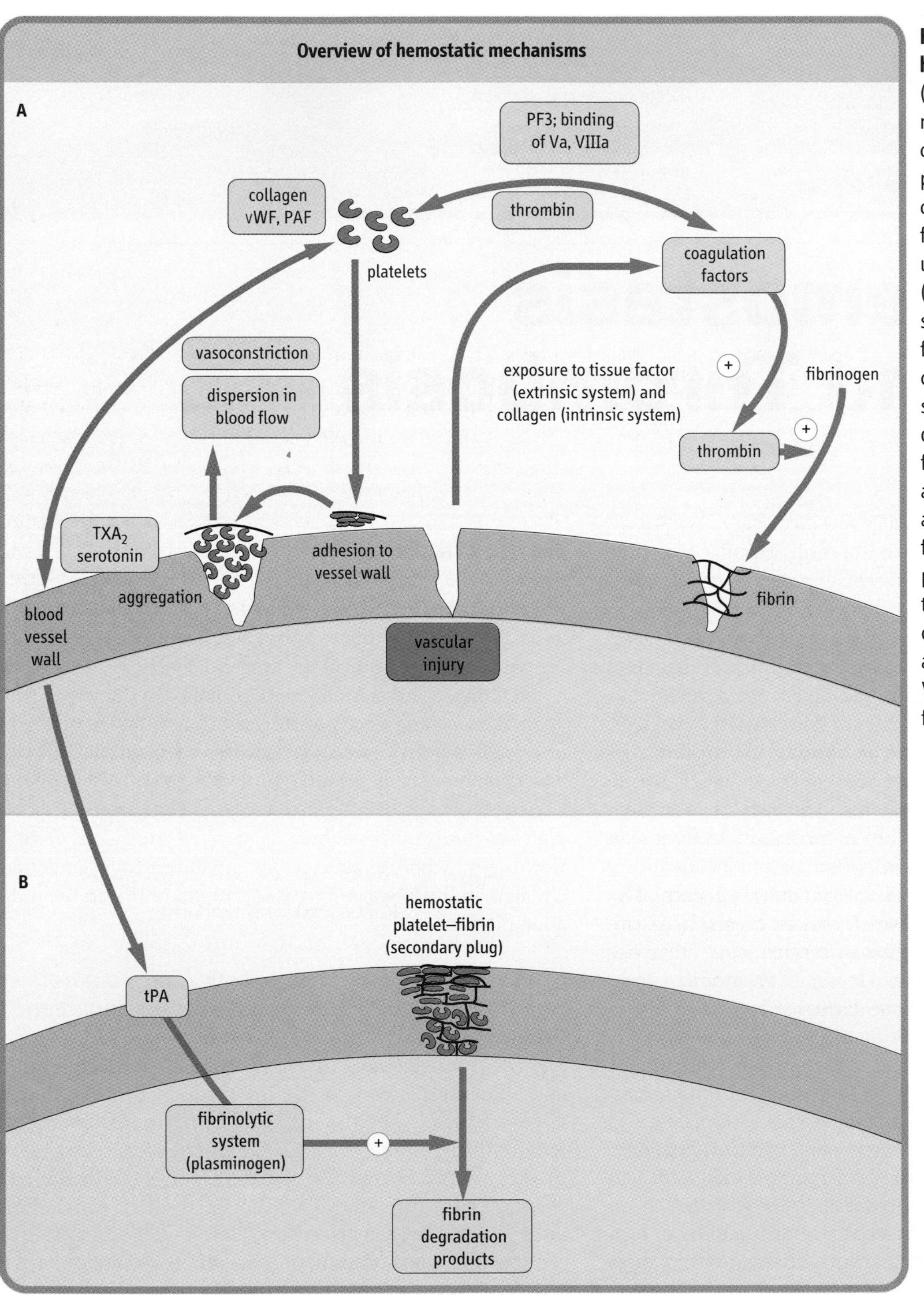

Fig. 6.1 Overview of hemostatic mechanisms. (A) Vascular injury sets in motion a series of events that culminate in formation of a primary plug of platelets. This can be dispersed by blood flowing through the vessel unless the plug is stabilized. (B) The primary plug is stabilized by a network of fibrin (formed from crosslinked fibrinogen). The secondary plug is stable and is degraded only when the fibrinolytic system has been activated. PAF, platelet activating factor; PF3, platelet factor 3; tPA, tissue-type plasminogen activator; TXA_2, thromboxane A_2; Va, activated coagulation factor V; VIIIa, activated coagulation factor VIII; vWF, von Willebrand factor.

Causes of excessive bleeding

	Congenital	Acquired
vessel wall	disorders of collagen synthesis (Ehlers–Danlos syndrome)	vitamin C deficiency (scurvy) corticosteroid excess
platelets	vWF deficiency (von Willebrand disease) platelet GPIb-IX deficiency (Bernard–Soulier syndrome) platelet GPIIb-IIIa deficiency (Glanzmann's thrombasthenia)	antiplatelet drugs (e.g. aspirin) defective formation of platelets excessive destruction of platelets
coagulation	coagulation factor deficiencies (hemophilias)-factor VIII factor IX factor XI fibrinogen etc.	vitamin K deficiency (factors II, VII, IX, X) oral anticoagulants (vitamin K antagonists, e.g. warfarin) liver disease disseminated intravascular coagulation (DIC)
fibrinolysis	antiplasmin deficiency PAI-1 deficiency	fibrinolytic drugs (e.g. tPA, urokinase, streptokinase)

Fig. 6.2 Congenital and acquired causes of excessive bleeding. GPIb-IX, GPIIb-IIIa, glycoprotein receptors Ib-IX and IIb-IIIa; PAI-1, plasminogen activator inhibitor type 1.

coagulation factors: the extrinsic pathway which is initiated by the exposure of the flowing blood to tissue factor, released from subendothelial tissue; and the intrinsic pathway which has an important amplification role in generating thrombin and fibrin.

The lysis of fibrin is equally important to health as is its formation

Hemostasis is a continuous process throughout life, and would result in excessive fibrin formation and vascular occlusion if unchecked. Evolution has therefore produced a fibrinolytic system; this is activated by local fibrin formation, resulting in local generation of plasmin, an enzyme which digests fibrin plugs (in parallel with tissue repair processes), thus maintaining vascular patency. Digestion of fibrin results in generation of circulating fibrin degradation products (FDP). These are detectable in plasma of healthy individuals at low concentration, which illustrates that fibrin formation and lysis are continuing processes in health.

Excessive bleeding may result from defects in each of the components of hemostasis, which may be caused by disease (congenital or acquired) or by antithrombotic drugs (Fig. 6.2).

The vascular, platelet, coagulation and fibrinolytic components of hemostasis will now be discussed in turn.

Vessel wall

Vascular injury has a key role in initiating local formation of the platelet–fibrin plug and in its subsequent removal by the fibrinolytic system

All blood vessels are lined by a flat sheet of endothelial cells, which have important roles in the interchange of chemicals, cells, and microbes between the blood and the body tissues. Endothelial cells in the smallest blood vessels (capillaries) are supported by a thin layer of connective tissue, rich in collagen fibers, called the intima. In veins, a thin layer (the media) of contractile smooth muscle cells allows some venoconstriction: for example, superficial veins under the skin constrict in response to surface cooling. In arteries and arterioles, a well-developed muscle layer allows powerful vasoconstriction, including the vasoconstriction after local injury that forms part of the hemostatic response. Larger vessels also have a supportive connective tissue outer layer (the adventitia). Figure 6.3 illustrates the structure of the vessel wall.

Intact normal endothelium does not initiate or support platelet adhesion or blood coagulation. This thromboresistance is partly due to endothelial production of two potent vasodilators and inhibitors of platelet function: prostacyclin (prostaglandin I_2, PGI_2) and nitric oxide, otherwise known as endothelium-derived relaxing factor (EDRF) (see Chapter 17).

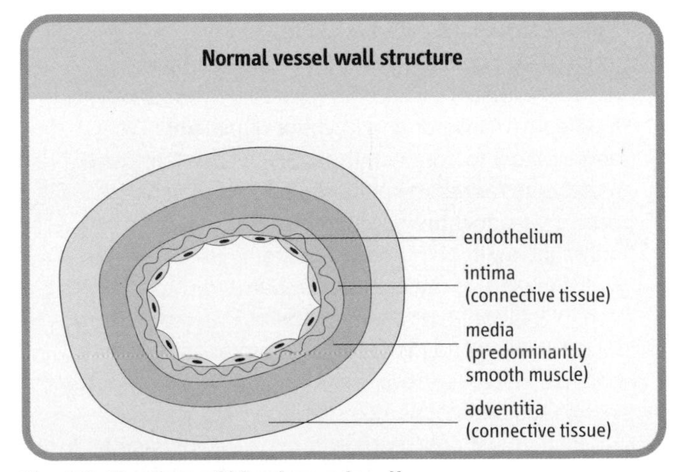

Fig. 6.3 Structure of blood vessel wall.

Prostacylin and nitric oxide are generated in the vessel wall

Biochemical mediators of vasoconstriction and vasodilatation

The diameters of arteries and arterioles throughout the body continuously alter to regulate blood flow according to local and general metabolic and cardiovascular requirements. Control mechanisms include neurogenic (sympathetic/adrenergic; see Chapter 38) and myogenic pathways, and local biochemical mediators, including prostacyclin (PGI_2) and nitric oxide.

Prostacyclin is the major arachidonic acid metabolite formed by vascular cells. It is a potent vasodilator, and also a potent inhibitor of platelet aggregation. It has a short half-life in plasma (3 minutes).

Nitric oxide is also a potent vasodilator formed by vascular endothelial cells, with a short half-life. It was initially termed endothelium-derived relaxing factor (EDRF). In common with that of prostacyclin, its generation by endothelial cells is enhanced by many biochemicals, and also by blood flow and shear stress. In the normal circulation, nitric oxide appears to have a key role in flow-mediated vasodilatation. It is synthesized by two distinct forms of endothelial nitric oxide synthase (eNOS): constitutive and inducible. Constitutive eNOS rapidly provides relatively small amounts of nitric oxide for short periods, related to vascular flow regulation. The beneficial effects of nitrate drugs in hypertension and angina may partly reflect their effects on this pathway (see Chapter 21). Inducible eNOS is stimulated by cytokines in inflammatory reactions, and releases large amounts of nitric oxide for long periods. Its suppression by glucocorticoids may partly account for their anti-inflammatory effects.

Both prostacyclin and nitric oxide appear to exert their vasodilator actions by diffusing locally from endothelial cells to vascular smooth muscle cells (see Fig. 6.3), where they stimulate guanylate cyclase, resulting in increased formation of cyclic guanosine 3′,5′-monophosphate (cGMP) and relaxation of vascular smooth muscle, probably via alteration of the intracellular calcium concentration (see Chapter 36).

The vasoconstriction that occurs after vascular injury is partly mediated by two platelet activation products: serotonin (5-hydroxytryptamine), and thromboxane A_2 (TXA_2; see Fig. 6.1), a product of platelet prostaglandin metabolism. In addition, after a vascular injury that disrupts the endothelial cell lining, flowing blood is exposed to subendothelial collagen, which activates the intrinsic pathway of blood coagulation. The endothelial cell damage also exposes flowing blood to subendothelial tissue factor, which activates the extrinsic pathway of blood coagulation (see Fig. 6.1).

Exposure of flowing blood to collagen as a result of endothelial damage also stimulates platelet activation. Platelets bind to collagen via von Willebrand factor (vWF), which is released from the endothelial cells. vWF in turn binds both to collagen fibers and to platelets (via a platelet membrane glycoprotein receptor, GPIb-IX). Platelet

Platelets produce thromboxane A_2

Thromboxane A_2 and aspirin

It has already been noted that PGI_2, the major arachidonic acid metabolite formed by vascular cells, is a potent vasodilator and inhibitor of platelet aggregation. In contrast, the major arachidonic acid metabolite formed by platelets is TXA_2, which is a potent vasoconstrictor and stimulates platelet aggregation. In common with prostacyclin, TXA_2 has a short half-life. In the late 1970s, Salvador Moncada and John Vane contrasted the effects of PGI_2 and TXA_2 on blood vessels and platelets, and hypothesized that a balance between these two compounds was important in the regulation of hemostasis and thrombosis.

Congenital deficiencies of cyclo-oxygenase or thromboxane synthase (the enzymes involved in TXA_2 synthesis) result in a mild bleeding tendency. Ingestion of even low doses of acetylsalicylic acid (aspirin) irreversibly acetylates cyclo-oxygenase and suppresses TXA_2 synthesis and platelet aggregation for several days, resulting in a mild bleeding tendency. Bleeding is especially likely from the stomach, as a result of the formation of stomach ulcers secondary to the inhibition of cytoprotective gastric mucosal prostaglandins by aspirin. Although, in persons at high risk of thrombosis, this bleeding tendency is outweighed by a reduction in risk of thrombosis, aspirin is contraindicated in individuals with a history of bleeding disorders or existing stomach or duodenal ulcers.

Platelet activation exposes glycoprotein receptors

Platelet membrane receptors and their ligands, vWF and fibrinogen

Platelets have a key role in hemostasis and thrombosis, through adhesion to the vessel wall and subsequent aggregation to form a platelet-rich hemostatic plug or thrombus. These processes involve exposure of specific membrane glycoprotein receptors after platelet activation by several compounds (see Fig. 6.4).

Platelet receptor GPIb-IX plays a key part in the adhesion of platelets to subendothelium. It binds vWF, which also interacts with specific subendothelial receptors, including those on subendothelial collagen. Congenital deficiencies of GPIb-IX (Bernard–Soulier syndrome) or, more commonly, of vWF, result in a bleeding tendency. In contrast, high plasma concentrations of vWF are associated with increased risk of thrombosis. For patients at high risk of thrombosis, therapeutic strategies directed against vWF (for example anti-vWF antibodies) are currently being developed, to reduce the thrombotic risk.

Another receptor, GPIIb-IIIa has a key role in platelet aggregation. After platelet activation, hundreds of thousands of GPIIb-IIIa receptors can be exposed in a single platelet. These receptors interact with fibrinogen or vWF, which bind platelets together, forming a hemostatic or thrombotic plug. Congenital deficiency of GPIIb-IIIa (the rare Glanzmann's thrombasthenia) causes a severe bleeding disorder; in contrast, deficiencies of either fibrinogen or vWF cause a milder bleeding disorder, because these two ligands can substitute for each other. High plasma concentrations of fibrinogen are associated with increased risk of thrombosis, partly because of its platelet-binding activity. For patients at high risk of thrombosis, inhibitors of the GPIIb-IIIa receptor (such as antireceptor antibodies) are being developed and are proving to be clinically effective.

activating factor (PAF) from the vessel wall may also activate platelets in hemostasis (see Fig. 6.1).

Because collagen has a key role in the structure and the hemostatic function of small blood vessels, vascular causes of excessive bleeding include congenital or acquired deficiencies of collagen synthesis (see Fig. 6.2). Congenital disorders include the rare Ehlers–Danlos syndrome (see Chapter 2). Acquired disorders include the relatively common vitamin C deficiency, scurvy (see Chapter 10), and excessive exogenous or endogenous corticosteroids.

Platelets

Blood platelets form the initial hemostatic plug in small vessels, and the initial thrombus in arteries and veins

Platelets are circulating, anuclear microcells of mean diameter 2–3 μm. They are fragments of bone marrow megakaryocytes, and circulate for about 10 days in the blood. The concentration of platelets in normal blood is 150–400×10^9/L.

Platelets can be activated by several chemical agents, including adenosine diphosphate (ADP, released by platelets, erythrocytes, and endothelial cells), epinephrine, collagen, thrombin, and PAF; by immune complexes (generated during infections); and by high physical shear stresses (shear stress is the tangential force applied to the cells by the flow of blood). Most of the chemical agents appear to act by binding to specific receptors on the platelet surface membrane (see Fig. 6.4). After receptor stimulation, several pathways of platelet activation can be initiated, resulting in several biological activities:

- **change in platelet shape**– from a disc to a sphere with extended pseudopodia – facilitates aggregation and coagulant activity;
- **release of several compounds involved in hemostasis** from intracellular granules – for example ADP, serotonin, TXA_2, and vWF;
- **aggregation**, via exposure of GPIb-IX membrane receptor and linking by vWF (under high shear conditions), and via exposure of another membrane glycoprotein receptor, GPIIa-IIIb and linking by fibrinogen (under low shear conditions);
- **adhesion to the vessel wall** via exposure of the GPIb-IX membrane receptor, through which vWF binds platelets to subendothelial collagen.

Finally, stimulation of the platelet membrane receptor triggers the activation of platelet membrane phospholipases, which hydrolyze membrane phospholipids, releasing arachidonic acid. Arachidonic acid is metabolized by cyclooxygenase and thromboxane synthase to TXA_2, a potent but labile (half-life 30 seconds) mediator of platelet activation and vasoconstriction (see Fig 6.4).

Platelet-related bleeding disorders

Congenital defects in platelet adhesion/aggregation can cause lifelong excessive bleeding

A simple screening test – measurement of the skin bleeding time (normal range, 2–10 minutes) – is sufficient to detect congenital defects of platelet adhesion/aggregation, in which the time is characteristically prolonged. The most common such defect is von Willebrand disease (see Fig. 6.2), a group

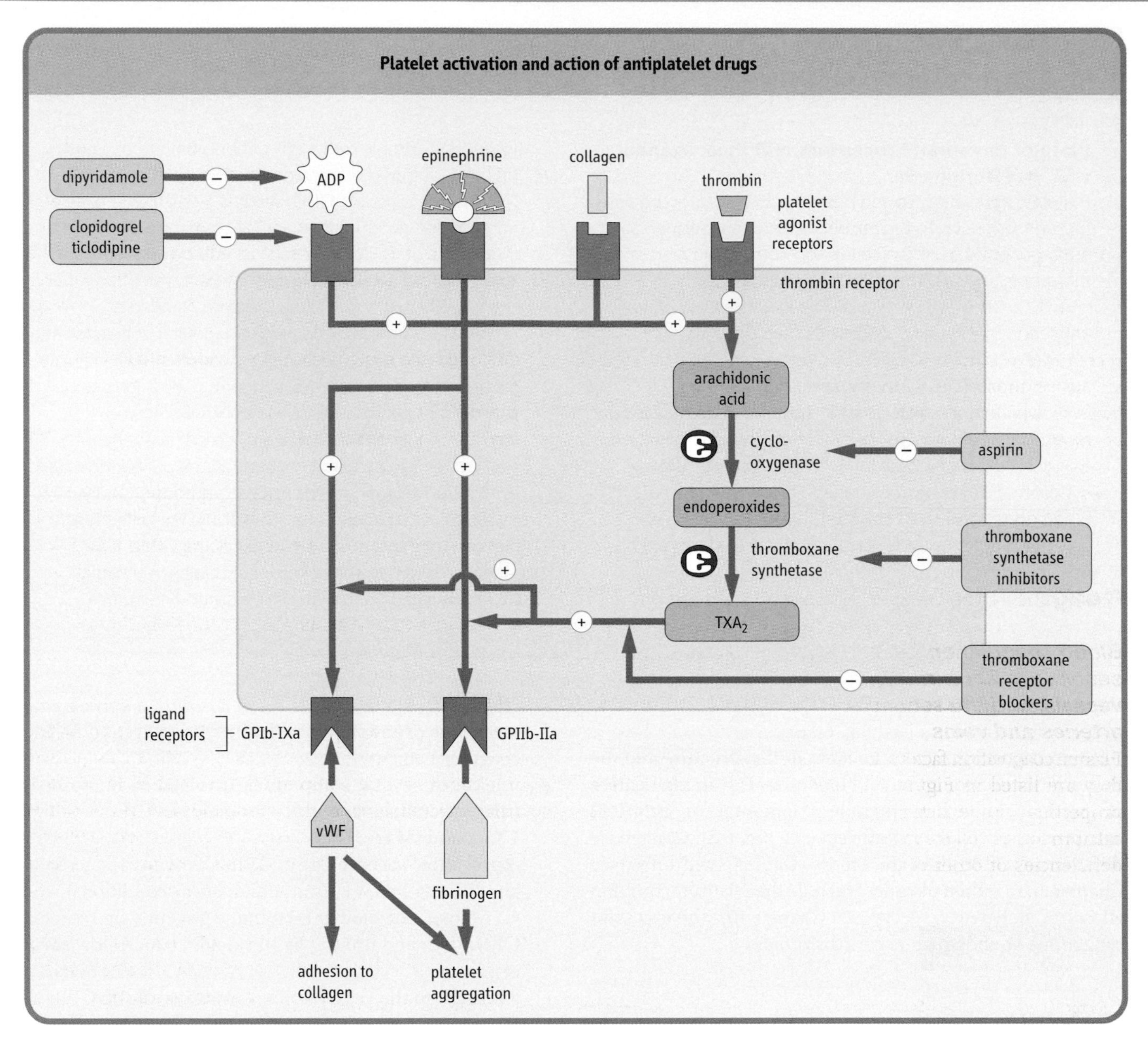

Fig. 6.4 Pathways of platelet activation and mechanisms of action of antiplatelet drugs. Stimulation of platelet agonist receptors results in exposure of platelet ligand receptors, partly through the platelet prostaglandin (cyclo-oxygenase) pathway. Ligand receptors bind vWF and fibrinogen in platelet adhesion/aggregation.

of autosomal dominant disorders that result in low plasma concentrations of vWF multimers. These multimers are composed of subunits (molecular weight 250 000) that are released from endothelial cells (and platelet granules) and circulate in plasma at a concentration of 1 mg/dL (Fig. 6.5). Not only does vWF have an important role in platelet hemostatic function, it also carries coagulation factor VIII (antihemophilic factor) in the circulation and delivers it to sites of vascular injury. Hence, plasma concentrations of factor VIII may also be low in von Willebrand disease. Treatment of this disease is to increase the low plasma levels of vWF activity, usually by means of desmopressin (a synthetic analogue of vasopressin – see Chapter 21 – which releases vWF from endothelial cells into plasma), but sometimes with human plasma concentrates.

Less common congenital bleeding disorders include GPIb-IX deficiency (Bernard–Soulier syndrome), GPIIb-IIIa deficiency (Glanzmann's thrombasthenia), and fibrinogen deficiency (because fibrinogen bridges GPIIb-IIIa receptors of adjacent platelets).

Acquired disorders of platelets include a low platelet count (thrombocytopenia), which may be the result either of defective formation of platelets by bone marrow megakaryocytes (as in marrow neoplasia, or aplasia), or of excessive destruction of platelets (e.g. by antiplatelet antibodies, or in splenomegaly or disseminated intravascular coagulation, DIC).

Antiplatelet drugs

Antiplatelet drugs are used in the prevention or treatment of arterial thrombosis; their sites of action are illustrated in Figure 6.4. As described above, aspirin inhibits cyclo-oxygenase and hence reduces the formation of TXA_2. Because it also has the effect of reducing the formation of PGI_2, which itself has antiplatelet activity, agents acting more specifically as thromboxane synthase inhibitors or thromboxane receptor antagonists have also been investigated as potential antiplatelet agents, but do not appear to be more effective than aspirin. Dipyridamole acts by reducing the availability of ADP, and ticlopidine and clopidogrel inhibit the ADP receptor (Fig. 6.4). These drugs have antithrombotic effects similar to those of aspirin, but cause less gastric bleeding because they do not interfere with synthesis of prostaglandins in the stomach; dipyridamole has an effect additional to that of aspirin. Recently, GPIIb-IIIa antagonists (e.g. antibodies) have been used; however, as in their congenital equivalent (Glanzmann's thrombasthenia), severe bleeding may occur.

Coagulation

Blood coagulation factors interact to form the secondary, fibrin-rich, hemostatic plug in small vessels, and the secondary fibrin thrombus in arteries and veins

Plasma coagulation factors are identified by roman numerals: they are listed in Figure 6.5, together with some of their properties. Tissue factor was formerly known as factor III, calcium ion as factor IV; factor VI does not exist. Congenital deficiencies of other coagulation factors (I–XIII) result in excessive bleeding, which illustrates their physiologic importance in hemostasis. The exception is factor XII deficiency, which does not increase the bleeding tendency, despite prolonging blood clotting times *in vitro*; the same is true for its cofactors, prekallikrein or high-molecular-weight kininogen (HMWK). A possible explanation for this is given below.

Figure 6.6 illustrates the currently accepted scheme of blood coagulation. Since the early 1960s, this has been accepted as a 'waterfall' or 'cascade' sequence of interactive pro-enzyme to enzyme conversions, each enzyme activating the next pro-enzyme in the sequence(s). Activated factor enzymes are designated by the letter 'a' – for example, factor XIa. Traditionally, the scheme has been divided into three parts:

- **the intrinsic pathway,**
- **the extrinsic pathway,**
- **the final common pathway.**

These are described below. They are distinguished on the basis of the nature of the initiating factor and its corresponding test in the clinical hemostasis laboratory; hence, three tests of coagulation are performed on citrated, platelet-poor plasma in clinical laboratories:

- **activated partial thromboplastin time (APTT),**
- **prothrombin time,**
- **thrombin time.**

Platelet-poor plasma is used in these tests because the platelet count influences clotting time results. To obtain the platelet-poor plasma, citrate anticoagulant is added to blood to sequester calcium ions reversibly, and the blood is centrifuged at 2000 g for 15 minutes. The coagulation time tests are initiated by adding calcium and appropriate initiating agents.

Coagulation factors and their properties

Factor	Synonyms	Molecular weight	Plasma concentration (mg/dL)
I	Fibrinogen	340 000	200–400
II	Prothrombin	70 000	10
III	Tissue factor (thromboplastin)	44 000	0
IV	Calcium ion	40	9–10
V	Proaccelerin, labile factor	330 000	1
VII	Serum prothrombin conversion accelerator (SPCA), stable factor	48 000	0.05
VIII	Antihemophilic factor (AHF)	330 000	0.01
(vWF)		(250 000)n	1
IX	Christmas factor	55 000	0.3
X	Stuart–Prower factor	59 000	1
XI	Plasma thromboplastin antecedent (PTA)	160 000	0.5
XII	Hageman factor	80 000	3
XIII	Fibrin-stabilizing factor (FSF)	320 000	1–2
Prekallikrein	Fletcher factor	85 000	5
High-molecular-weight kininogen (HMWK)	Fitzgerald, Flaujeac Williams factor, contact activation cofactor	120 000	6

Fig. 6.5 Coagulation factors and some of their properties. n indicates number of subunits.

The intrinsic pathway

The term 'intrinsic' implies that no extrinsic factor such as tissue factor or thrombin is added to the blood, other than a contact with nonendothelial 'surface'. The clinical test of this pathway is the activated partial thromboplastin time (APTT), also known as the kaolin–cephalin clotting time (KCCT) because kaolin (microparticulated clay) is added as a standard 'surface' and cephalin (brain phospholipid extract) as a substitute for platelet phospholipid. The normal range of the APTT is about 30–50 seconds; prolongations are observed in deficiencies of factors XII (or its cofactors, prekallikrein or HMWK), XI, IX (or its cofactor, factor VIII), X (or its cofactor, factor V), prothrombin (factor II), or fibrinogen (factor I) (see Fig. 6.2). The test is used to exclude the common congenital hemophilias (deficiencies of factors VIII, IX, or XI; see Fig. 6.5), and to monitor heparin treatment (see box). Hemophilias caused by factor VIII or IX deficiency occur in about 1 in 10 000 males; inheritance is X-linked recessive, transmitted by carrier females. Treatment is usually with factor VIII or IX concentrates.

The extrinsic pathway

The term 'extrinsic' refers to the effect of tissue factor, which (after combining with coagulation factor VII) greatly accelerates coagulation, by activating both factor IX and factor X (Fig. 6.6). Tissue factor is a polypeptide that is expressed in all cells other than endothelial cells. The clinical test of this pathway is the prothrombin time (PT), in which tissue factor is added to plasma. The normal range is about 10–15 seconds; prolongations are observed in deficiencies of factors VII, X, V, II, or I. In clinical practice, the test is used to diagnose both the rare congenital defects of these factors and, much more commonly, to diagnose acquired bleeding disorders, resulting from:

- **vitamin K deficiency** (e.g. malabsorption, obstructive jaundice; see Chapters 10 and 27), which reduces hepatic synthesis of factors II, VII, IX, and X. Treatment is by injections of vitamin K;
- **oral anticoagulants** (e.g. warfarin) which are vitamin K antagonists, reducing hepatic synthesis of these factors. Excessive bleeding in patients taking warfarin can be treated by stopping the drug, giving vitamin K, or replacing factors II, VII, IX, and X with fresh frozen plasma or concentrates;
- **liver disease**, which reduces hepatic synthesis of these factors. For example, the prothrombin time is a good marker of liver failure after acetaminophen (paracetamol) overdose (see Chapter 27). Treatment is by replacing factors II, VII, IX, and X with fresh frozen plasma or concentrates.

The final common pathway

The third part of coagulation is tested clinically by the thrombin time, in which exogenous thrombin is added to plasma. The normal range of values is about 10–15 seconds, but prolongations are observed in fibrinogen deficiency. This may be congenital, or due to acquired consumption of fibrinogen in DIC, or may occur after administration of fibrinolytic drugs (see below). Treatment is with fresh frozen plasma or fibrinogen concentrates.

Thrombin

It is currently believed that activation of blood coagulation is usually initiated by vascular injury, causing exposure of flowing blood to tissue factor, which results in activation of factors VII and IX. Subsequently, activation of factors X and II (prothrombin) occurs preferentially at sites of vascular injury, and platelet activation, which provides procoagulant activity (platelet factor 3, PF3) as a result of exposure of negatively-charged platelet surface membrane lipoproteins, such as phosphatidylserine. This is accompanied by the exposure, on activated platelets, of high-affinity binding sites for several activated coagulation factors (especially factors Va and VIIIa), and provision of platelet phospholipid, which further catalyzes coagulation activation. As a result of these biochemical interactions (see Figs 6.1 and

Classical hemophilia: congenital factor VIII deficiency

A 3-year-old boy was admitted from the Emergency Room of his local hospital because of extensive bruising after a fall down a few stairs (see Fig. 6.7). A routine coagulation screen test showed a greatly prolonged APTT of more than 150 seconds (normal range, 30–50 seconds). Assay of coagulation factor VIII showed a very low level; the vWF level was normal. Enquiry of his (asymptomatic) mother revealed a family history of excessive bleeding, in her brother and in their father.

Comment. Because of this typical history of an X-linked recessive bleeding disorder, a low coagulation factor VIII level, and a normal vWF level, a diagnosis of classical hemophilia (congenital factor VIII deficiency) was made. The family were referred to the local Hemophilia Center and counseled about the risks of further affected sons and carrier daughters. The child was treated with intravenous factor VIII concentrate for the presenting bleed, and for future bleeds, injuries, or surgery.

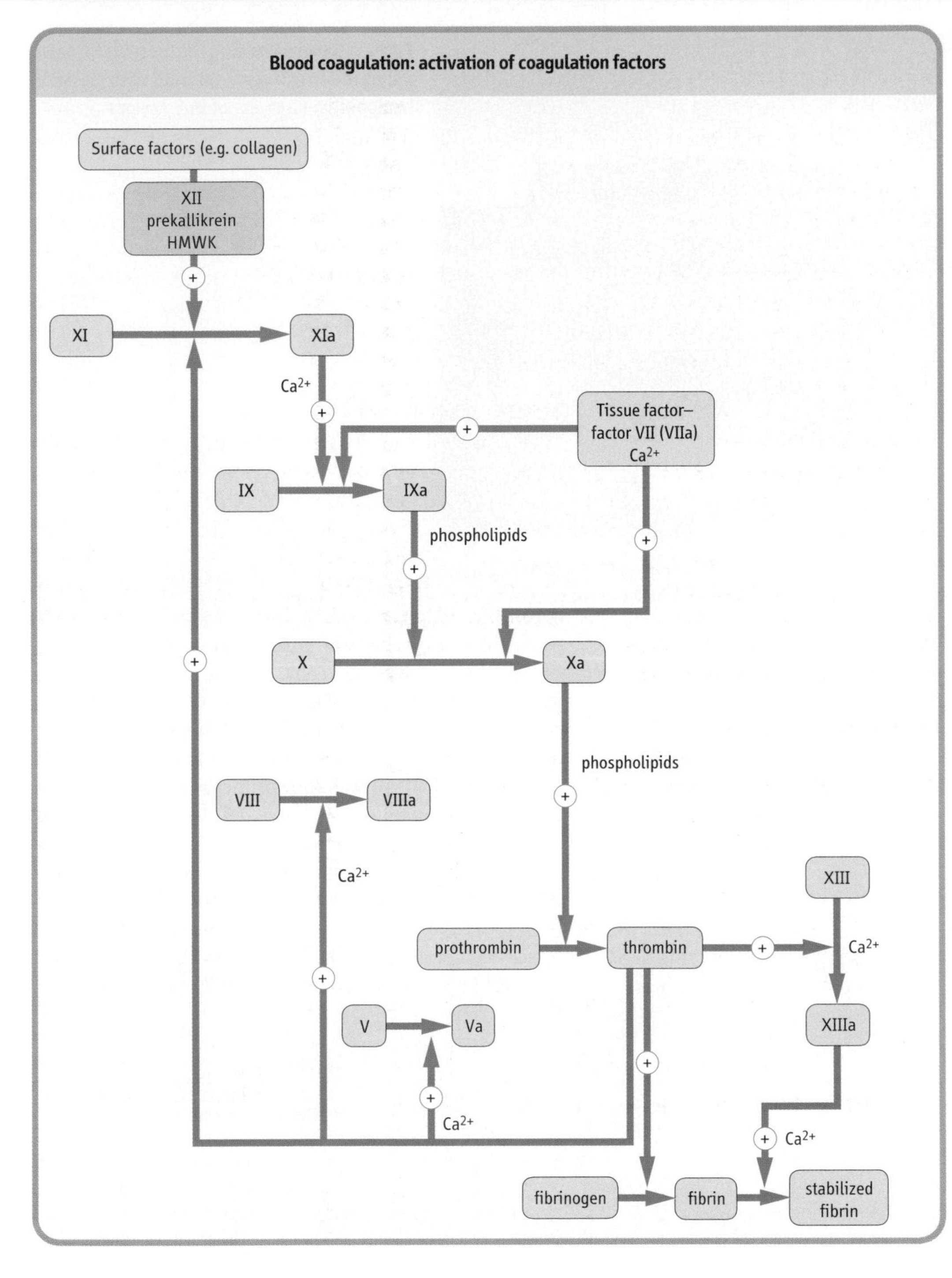

Fig. 6.6 Blood coagulation. After the initiation of blood coagulation, the coagulation factor pro-enzymes are activated sequentially: activated factor enzymes are designated by the letter 'a'. The gray box indicates contact factors that have no apparent function in *in vivo* hemostasis. Phospholipids are supplied *in vivo* by platelets. HMWK, high-molecular-weight kininogen.

6.6), thrombin and fibrin formation are efficiently localized at sites of vascular injury.

Thrombin has a central role in hemostasis

Not only does thrombin convert circulating fibrinogen to fibrin at sites of vascular injury (producing the secondary, fibrin-rich, hemostatic plug); it also activates factor XIII (transglutaminase), which crosslinks such fibrin, rendering it resistant to dispersion by local blood pressure or by fibrinolysis (see Figs 6.1 and 6.6). Furthermore, thrombin stimulates its own generation in a positive feedback cycle in two ways:

- **it catalyzes activation of factor XI**; this may explain why congenital deficiencies of factor XII, prekallikrein, or HMWK are not associated with excessive bleeding (Fig. 6.6);
- **it catalyzes activation of factors VIII and V.**

Thrombin also activates platelets (see Fig. 6.4).

Now that the central role of thrombin in hemostasis and thrombosis has been recognized, there is current interest in the development of direct antithrombins as antithrombotic drugs; these include examples such as hirudin (originally obtained from the medicinal leech, *Hirudo medicinalis*) and its synthetic derivatives.

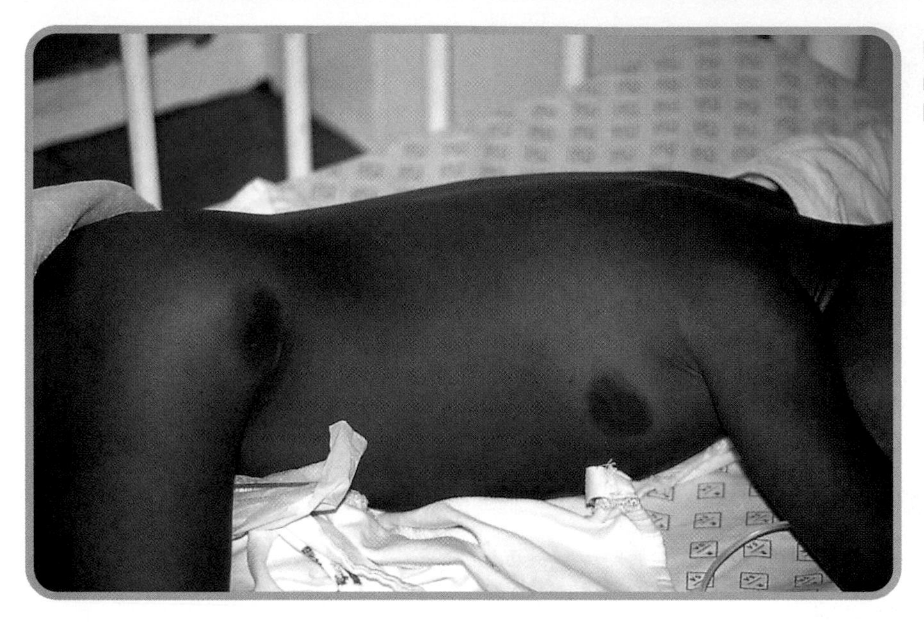

Fig. 6.7 Severe bruising that resulted from a minor fall in a 3-year-old child with classic hemophilia. (Courtesy of Dr S Taylor.)

Coagulation inhibitors are essential to prevent excessive thrombin formation and thrombosis

Three systems of coagulation inhibitors have been identified (see Figs 6.8 and 6.9):

- **antithrombin:** this is a protein synthesized in the liver. Its activity is catalyzed by the antithrombotic drug, heparin, and by heparin-like endogenous glycosaminoglycans (GAGs) that are present on the surface of vascular endothelial cells. It inactivates not only thrombin, but also factors IXa and Xa (see Fig. 6.8). Congenital antithrombin deficiency results in increased risk of venous thromboembolism.
- **protein C and its cofactor, protein S:** these are vitamin-K-dependent proteins, synthesized in the liver. When thrombin is generated, it binds to thrombomodulin (molecular weight 74,000), which is present on the surface of vascular endothelial cells. The thrombin–thrombomodulin complex activates protein C, which forms a complex with its cofactor, protein S. This complex selectively degrades factors Va and VIIIa by limited proteolysis (Fig. 6.8). Hence, this pathway forms a negative feedback upon thrombin generation. Congenital deficiencies of protein C or protein S result in increased risk of venous thromboembolism; a further cause of increased risk of venous thromboembolism is a mutation in coagulation factor V (factor V Leiden), which confers resistance to its inactivation by activated protein C. This mutation is common, occurring in about 3% of the population in Western countries.

Heparin treatment is ineffective in antithrombin deficiency

A 40-year-old man was admitted from the Emergency Room of his local hospital because of acute pain and swelling of his left leg, after recent major surgery. Ultrasound imaging of the leg confirmed occlusion of the left femoral vein by a recent thrombus. He was prescribed anticoagulant therapy with heparin at standard doses. Baseline APTT was normal, at 32 seconds (normal range, 30–50 seconds). After the patient received treatment with heparin, the APTT was prolonged only slightly (38 seconds). At this time, the patient volunteered a strong family history of 'clots in the legs' at a young age. A thrombophilia screening test was performed, and showed a low plasma antithrombin level.

Comment. The patient was treated with intravenous antithrombin concentrate, which increased his plasma antithrombin levels to within the normal range. At the same doses of heparin, his APTT was prolonged to therapeutic values (80 seconds – about twice the mean reference level). The clinical response was satisfactory, and the patient's medication was subsequently changed from intravenous heparin anticoagulation to oral warfarin.

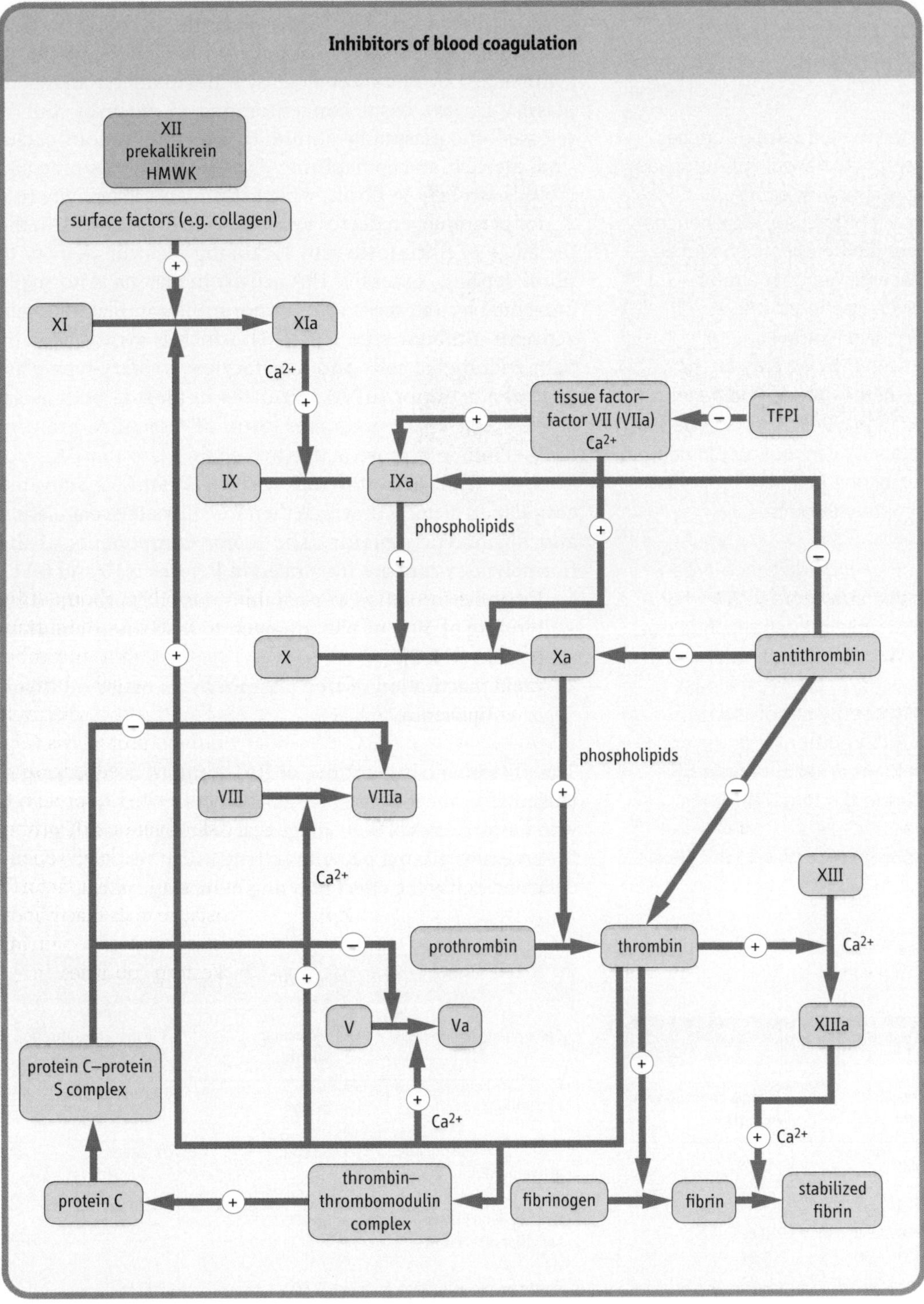

Fig. 6.8 Sites of action of blood coagulation inhibitors. TPFI, tissue factor pathway inhibitor.

- **tissue factor pathway inhibitor (TFPI):** this protein is synthesized in endothelium and the liver; it circulates bound to lipoproteins. It inhibits the tissue factor–VIIa complex (Fig. 6.8), which may explain the severe bleeding in hemophilia caused by deficiency in factor VIII or IX (failure to sustain thrombin and fibrin formation). Conversely, deficiency of TFPI does not appear to increase the risk of thrombosis.

Fibrinolysis

The fibrinolytic system also limits excessive fibrin formation

The coagulation system acts to form fibrin; the fibrinolytic system acts to limit excessive formation of fibrin (both intra- and extravascular) through plasmin-mediated fibrinolysis. Circulating plasminogen binds to fibrin via lysine-binding

Thrombolytic treatment in myocardial infarction

Occlusion of a coronary artery by a thrombus causes death of that part of the heart muscle which is supplied by the artery (myocardial infarction). In acute myocardial infarction, the patient typically experiences sudden, severe, heavy chest pain. Patients should be sent urgently to the hospital emergency room for consideration for thrombolytic treatment with a plasminogen activator drug, given intravenously. Prompt thrombolysis dissolves the coronary artery thrombus, reduces the size of the infarct, and hence reduces the risk of complications, which include death and heart failure. Aspirin is also given routinely in acute myocardial infarction, to inhibit the platelet component of the developing coronary artery thrombus.

Comment. Thrombolytic drugs include tissue-type and urinary-type plasminogen activators (tPA and uPA) (Fig. 6.11), produced by recombinant gene technology, or their synthetic variants. Worldwide, the most commonly used thrombolytic drug is streptokinase (a plasminogen activator produced by streptococci), because of its lower cost. All thrombolytic drugs can cause bleeding (see Fig. 6.2), as a result of lysis of hemostatic plugs in addition to the target thrombi.

sites; it is converted to active plasmin by plasminogen activators. Tissue-type plasminogen activator (tPA) is synthesized by endothelial cells; it normally circulates in plasma in low basal concentrations (5 ng/mL), but is released into plasma by stimuli that include venous occlusion, exercise, and epinephrine. Together with plasminogen, it binds strongly to fibrin, which stimulates its activity (the K_m for plasminogen decreases from 65 to 0.15 μmol/L in the presence of fibrin), thereby localizing plasmin activity to fibrin deposits. Excessive tPA activity in plasma is normally prevented by an excess of its major inhibitor, plasminogen activator inhibitor type 1 (PAI-1), which is synthesized by both endothelial cells and hepatocytes. Urinary-type plasminogen activator (uPA) circulates in plasma both as an active single-chain precursor form, uPA (scuPA, pro-urokinase) and as a more active two-chain form (tcuPA, urokinase). One activator of scuPA is surface-activated coagulation factor XII, which therefore links the coagulation and fibrinolytic systems. The major components of the fibrinolytic system are illustrated in Figures 6.10 and 6.11.

Excessive formation of plasmin is normally prevented by:

- **binding of 50% of plasminogen to histidine-rich glycoprotein (HRG);**
- **rapid inactivation of free plasmin by its major inhibitor, α_2 antiplasmin.**

The physiologic importance of PAI-1 and α_2 antiplasmin is illustrated by the increased bleeding tendency that is associated with the rare cases of their congenital deficiencies (see Fig. 6.2); the excessive plasma plasmin activity which results from the deficiencies has the effect of lysing hemostatic plugs.

Properties of coagulation inhibitors

Inhibitor (synonym)	Molecular weight	Plasma concentration (mg/dL)
Anthithrombin (antithrombin III)	65 000	18–30
Protein C	56 000	0.4
Protein S	69 000	2.5
Tissue factor pathway inhibitor, TFPI (lipoprotein-associated coagulation inhibitor, LACI)	32 000	0.1

Fig. 6.9 Coagulation inhibitors.

Properties of fibrinolytic system components

Component (synonym)	Molecular weight	Plasma concentration (mg/dL)
Plasminogen	92 000	0.2
Tissue-type plasminogen activator, tPA	65 000	5 (basal)
Urinary-type plasminogen activator type 1, uAI-1	54 000	20
Plasminogen activator inhibitor type 1	48 000	200
Antiplasmin (α_2 antiplasmin)	70 000	700

Fig. 6.10 Components of the fibrinolytic system, and some of their properties.

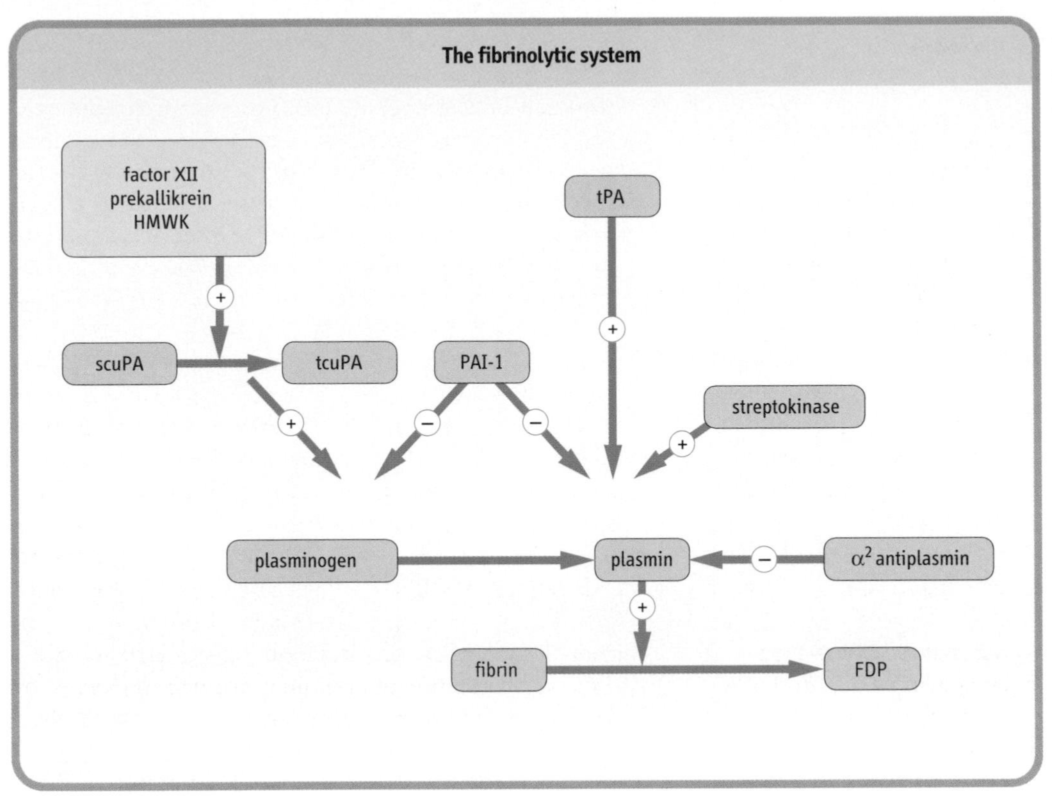

Fig. 6.11 The fibrinolytic system. Plasminogen can be activated to plasmin by uPA (urokinase), tPA, or streptokinase. uPA and tPA are inhibited by plasminogen activator inhibitor type 1 (PAI-1). Plasmin is inhibited by antiplasmin. Plasmin degrades fibrin to Fibrinogen degradation products (FDP). HMWK, high-molecular-weight kininogen.

Summary

Hemostasis constitutes a number of processes which guard the body against blood loss. Injury to the blood vessel wall sets in motion complex phenomena which involve blood platelets and a cascade of coagulation factors classified into two systems, intrinsic and extrinsic. The integrity of both systems may be tested by simple laboratory tests. Deficiencies of factors participating in the coagulation cascade, and/or disordered platelet function result in bleeding disorders. Eventually, blood clots are degraded by the fibrinolytic system. The process of fibrinolysis prevents thrombotic phenomena and there normally is a balance between haemostatic and fibrinolytic tendency. Therapeutic clot dissolution using infusion of enzymes such as streptokinase is now an established treatment for acute myocardial infarction.

Further reading

Poller L, Ludlam CA, eds. *Recent advances in blood coagulation*. Edinburgh: Churchill Livingstone; 1997.

Ratnoff OD, Forbes CD, eds. *Disorders of hemostasis*, 3rd edn. Philadelphia: Saunders; 1996.

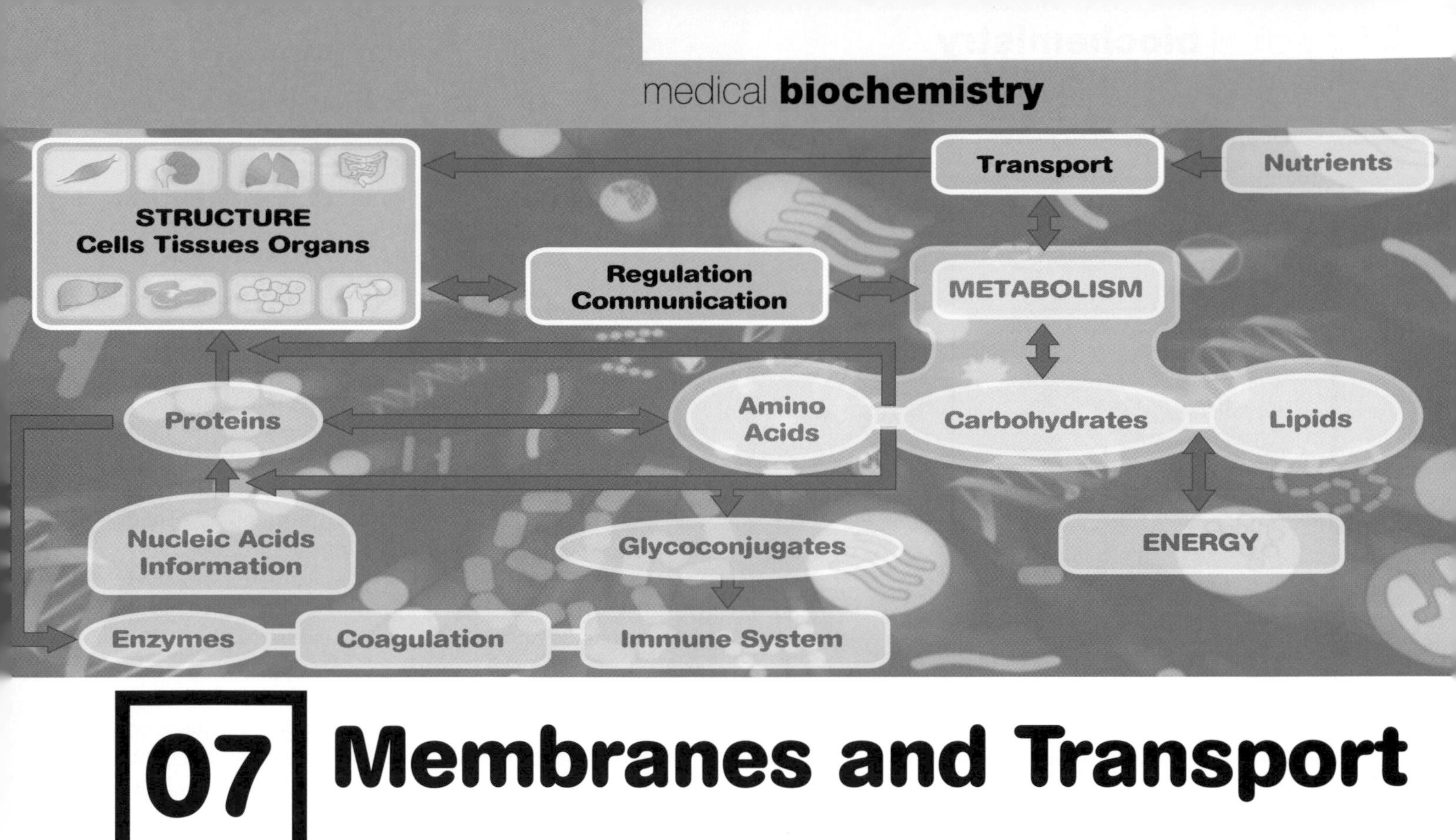

07 Membranes and Transport

Cells are surrounded by membranes composed of proteins and lipids; intracellular organelles are also compartmentalized by membranes

Biomembranes are thin films consisting of proteins and lipids. They are not rigid or impermeable but, rather, are highly mobile and dynamic structures. Membranes are the gatekeepers of the cell. They control not only access of inorganic ions, nutrients, and biological compounds, but also the entry of drugs and the exit of waste products. Integral transmembrane proteins have important roles in transporting these molecules, and often maintain concentration gradients across the membranes. K^+, Na^+, and Ca^{2+} concentrations in the cytoplasm are maintained at 140, about 10, and 10^{-4} mmol/L (546, 23, and 0.0007 mg/dL), respectively, by the transporter proteins, whereas those outside (in the blood) are 5, 145, and 1–2 mmol/L (20, 333, and 7–14 mg/dL), respectively. The driving force for transport is directly or indirectly provided by energy released from adenosine triphosphate (ATP). We will begin this chapter with a description of the structure and properties of lipids, and then explain how lipids and proteins interact to form biomembranes. The transport properties of membranes will be illustrated by several important examples.

Membrane organization

Lipid composition of biomembranes

Lipids are small, nonpolar molecules that can be extracted into organic solvents. Most of our fat (triglyceride) is stored in the form of cytoplasmic droplet in adipose tissue (see Chapter 25). In cell membranes, the major and common class of lipids are phospholipids. These are composed of an L-glycerol backbone with long-chain aliphatic 'fatty' acids attached at the C-1 and C-2 positions in ester linkage. Phosphoric acid is linked as an ester to position C-3, and an amine-containing polar head group is further linked to the phosphate moiety (Fig. 7.1).

Variations in the size and degree of unsaturation of the fatty acid components in phospholipids affect the fluidity of biomembranes

Fatty acids in biological systems normally contain an even number of carbon atoms – a property that stems from their synthesis in two-carbon units. C-16 and C-18 fatty acids are the most common components of phospholipids, and nearly 50% of the fatty acids in biomembranes are unsaturated, containing one or more carbon–carbon double bonds. The

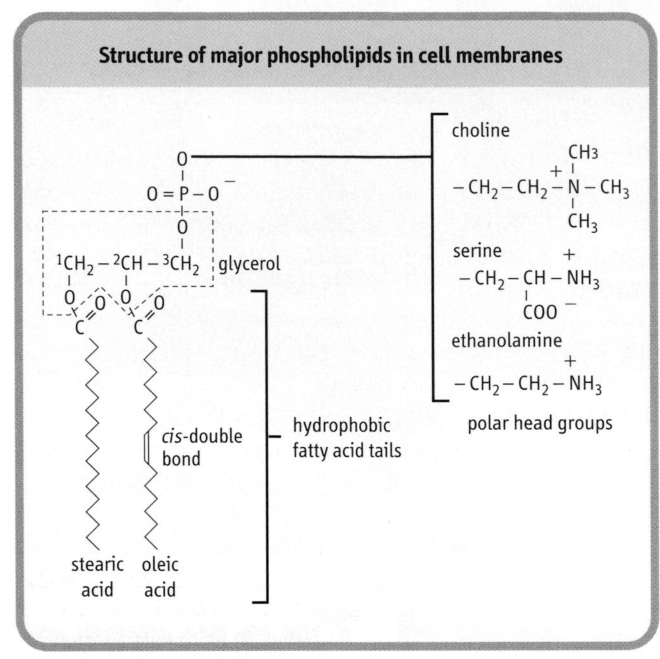

Fig. 7.1 Structure of major phospholipids in cell membranes. Two of the –OH groups in glycerol (at the C-1 and C-2 positions) are linked to fatty acids, while the third –OH group is phosphorylated. The phosphate is further linked to one of the variety of small polar head groups (such as choline, serine, and ethanolamine).

double bonds in these unsaturated fatty acids are all in the cis configuration. This places a 'kink' in their structure and interferes with their molecular packing, so that they have lower melting points. In contrast, saturated fatty acids pack readily into semicrystalline arrays and thus have a relatively high melting point, being solid and waxy at room temperature (Fig. 7.2).

Both oils and fats are triglycerides: triolein (glycerol trioleate: olive oil) is a liquid, whereas tristearin (glycerol tristearate: animal fat) is a solid. In general, saturated fatty acids are attached at the C-1 position, and unsaturated fatty acids at the C-2 position of the glycerol in phospholipids.

Phospholipids form the basic structure of membranes

Phospholipids are amphipathic molecules, because they are composed of both hydrophobic fatty acids and hydrophilic or polar head groups. The characteristic head groups of membrane phospholipids are choline, serine, and ethanolamine (Fig. 7.1). When they are hydrated, phospholipids form micelles and lamellar structures, and, under suitable conditions, they spontaneously organize into single bilayer structures (Fig. 7.3) – not only lamellar structures, but also closed vesicular structures termed liposomes. Liposomes having defined lipid compositions are being evaluated clinically as drug carrier and delivery systems.

In addition to the major phospholipids described above (Fig. 7.1), there are a number of other important lipids in biomembranes. These include phosphatidylinositol, cardiolipin, sphingolipids (sphingomyelin and glycolipids), and cholesterol, which are described in detail in later chapters. These lipids have special functions in processes such as cellular recognition and signal transduction.

Current membrane structural model

The membrane can be pictured as a mosaic of globular proteins embedded in a fluid-like phospholipid bilayer

The protein to lipid ratio differs among various biomembranes, ranging from about 80% (dry weight) lipid in the myelin sheath that insulates nerve cells, to 30% lipid in the mitochondrial inner membrane. Lipid composition also differs among membranes. Glycolipids, for example, are found primarily on the outer surface of the plasma membrane. However, the model of biomembrane structure that is

Naturally occurring fatty acids

Carbon atoms	Chemical formula	Systematic name	Common name	Melting point (°C)
Saturated fatty acids				
12	$CH_3(CH_2)_{10}COOH$	*n*-dodecanoic	lauric	44.2
14	$CH_3(CH_2)_{12}COOH$	*n*-tetradecanoic	myristic	53.9
16	$CH_3(CH_2)_{14}COOH$	*n*-hexadecanoic	palmitic	63.1
18	$CH_3(CH_2)_{16}COOH$	*n*-octadecanoic	stearic	69.6
20	$CH_3(CH_2)_{18}COOH$	*n*-eicosanoic	arachidic	76.5
Unsaturated fatty acids				
16	$CH_3(CH_2)_5CH{=}CH(CH_2)_7COOH$		palmitoleic	–0.5
18	$CH_3(CH_2)_7CH{=}CH(CH_2)_7COOH$		oleic	13.4
18	$CH_3(CH_2)_4CH{=}CHCH_2CH{=}CH(CH_2)_7COOH$		linoleic	–5
18	$CH_3CH_2CH{=}CHCH_2CH{=}CHCH_2CH{=}CH(CH_2)_7COOH$		linolenic	–11
20	$CH_3(CH_2)_4CH{=}CHCH_2CH{=}CHCH_2CH{=}CHCH_2CH{=}CH(CH_2)_7COO$		arachidonic	–49.5

Fig. 7.2 Structure and melting point of naturally occurring fatty acids. Melting point increases with the chain length of the fatty acid and decreases with the number of its double bonds.

generally accepted is the fluid mosaic model proposed by Singer & Nicolson in the early 1970s. This model represents the membrane as a fluid-like phospholipid bilayer into which globular proteins are embedded (Fig. 7.3). The polar head groups of the phospholipids are exposed on the external surface of the membrane, with the fatty acyl chains oriented to the inside of the membrane. Whereas membrane lipids and proteins easily move on the membrane surface (lateral diffusion), 'flip-flop' movement of lipids between the outer and inner bilayer leaflets rarely occurs without the aid of an integral membrane enzyme, flippase. Although this model is basically correct, there is also growing evidence that many membrane proteins have limited mobility and are anchored in place by attachment to cytoskeletal proteins.

Membrane proteins are classified as integral (intrinsic) membrane proteins and peripheral (extrinsic) membrane proteins. The former are embedded deeply in the lipid bilayer and some of them traverse the membrane several times (transmembrane protein), whereas peripheral membrane proteins are bound to membrane lipids and proteins by electrostatic interactions (Fig. 7.3). Ion transfer systems are typical examples of integral transmembrane proteins (Fig. 7.3). Most of their transmembrane portions are α-helical and composed of amino acid residues with nonpolar side chains; about 20 amino acid residues forming six to seven α-helical turns are enough to traverse a membrane of 5 nm (50 Å) thickness. The protein membrane domains interact with the hydrophobic tail of the lipid molecules, the transmembrane portions, or both, of the same or other integral membrane proteins.

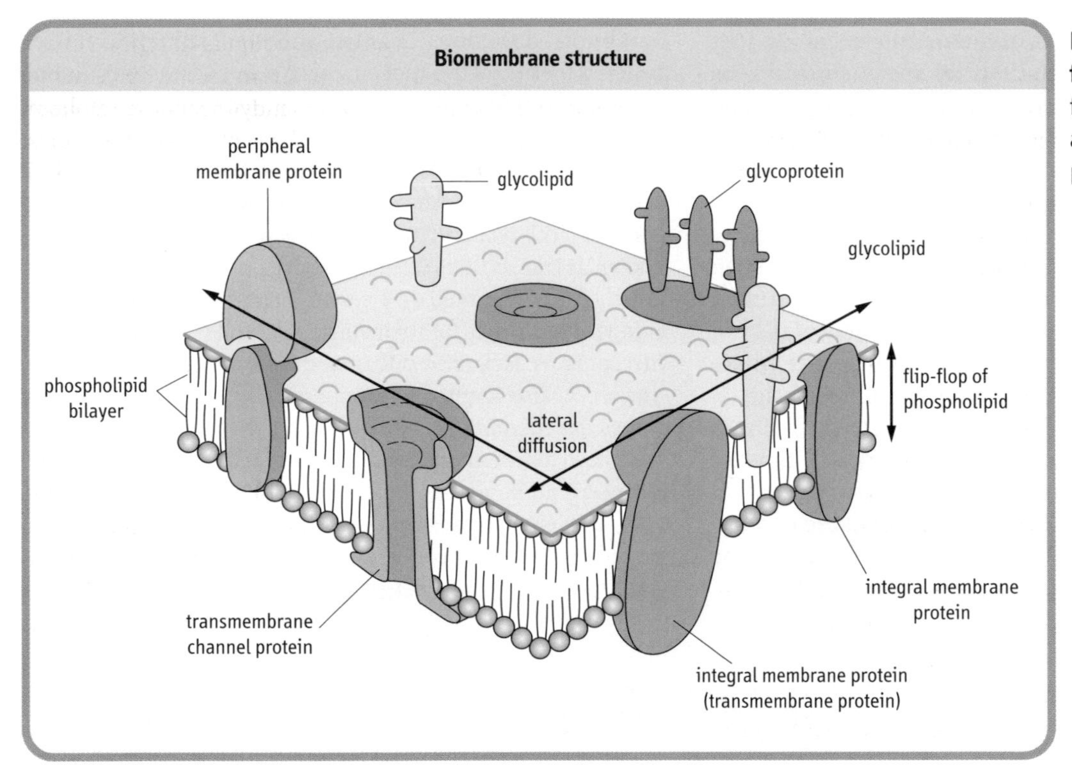

Fig. 7.3 Fluid mosaic model for biomembranes. In the fluid mosaic model, proteins are embedded in a fluid phospholipid bilayer.

Amphipathic compounds

Membrane perturbation by amphipathic compounds

Amphipathic compounds have distinct polar and nonpolar moieties. They include many anesthetics and tranquilizers. The pharmacologic activities of such compounds are dependent on their ability to interact with membranes and perturb membrane structure. A number of antibiotics and natural products, such as bile salts and fatty acids, are also amphipathic. While effective at therapeutic concentrations, some of these drugs exhibit detergent-like action in moderate to high concentrations and disrupt the bilayer structure, resulting in membrane leakage.

Structural and metabolic role of membranes

Membranes are more than barriers

A major role of membranes is to maintain the structural integrity and barrier function of cells and organelles. However, membranes are not rigid or impermeable: they are fluid, and their components move around, are metabolized, and are subject to metabolic turnover. The turnover of membrane components is especially important for the cellular response to information from inside and outside the cell: recognition, transfer, amplification, and signal transduction processes occur in or on the membranes. Furthermore, both small and large molecules must pass through biomembranes. With few exceptions, these transport processes are mediated by specific membrane proteins.

Phospholipids not only provide a fluid environment, but also regulate the activities of membrane enzymes. Particular phospholipids are required for specific membrane structures, such as curved regions and junctions with adjacent membranes. The inside surface of the membrane is more suited to phosphatidylethanolamine and phosphatidylserine, in which the polar heads are small and the hydrocarbons are more spread out, because of their larger contents of polyunsaturated fatty acids. As a result of such differing requirements, phospholipids are distributed asymmetrically between outer and inner leaflets of the cell membrane: phosphatidylcholine and sphingomyelin are abundant in the outer leaflet, whereas phosphatidylethanolamine and phosphatidylserine are found in the inner leaflet. Such differences are actively maintained by flippase (see Fig. 7.3). Cell damage often leads to loss of this membrane lipid asymmetry. It is believed that exposure of phosphatidylserine in the outer leaflet of the plasma membrane is a signal for red blood cell turnover.

Types of transport processes

Simple diffusion through the phospholipid bilayer

When substrates inside or outside the cell are transported according to their concentration gradient, movement of the substrates across the membrane is called diffusion or passive transport. This process includes two systems, simple diffusion and facilitated diffusion (or facilitated transport) (Fig. 7.4). Small, nonpolar molecules (such as O_2, CO_2, N_2, and benzene) and uncharged polar molecules (such as urea, ethanol, and small organic acids) move through membranes by simple diffusion without the aid of membrane proteins (Fig. 7.5A). In simple diffusion, hydrophobicity of the molecules is an important factor for their transport across the membrane, as the interior of the phospholipid bilayer is hydrophobic. Although water molecules can be transported by simple diffusion, channel proteins (aquaporins) are believed to control the movement of water across most membranes, especially in the kidney for concentration of the urine.

Transport mediated by membrane proteins

Transport of larger, polar molecules, such as amino acids or sugars, into a cell requires the involvement of membrane proteins known as transporters (also called porters, permeases, translocases, or carrier proteins). These transport proteins are as specific as are enzymes for their substrates, and work by one of two mechanisms: facilitated diffusion or active transport. The term 'carrier' is also applied to ionophores, which move passively across the membrane together with the bound ion (Fig. 7.6). Facilitated diffusion catalyzes the movement of a substrate down a concentration gradient and does not require energy. In contrast, active transport is a process in which substrates are transported against their concentration gradient. Active transport must be coupled to an energy-producing reaction (see Fig. 7.5A).

Transport systems of biomembranes

Type			Transport protein	Energy coupling	Specificity	Saturability	Rate (molecules/ transport protein/s)
passive transport or diffusion	simple diffusion		–	–	–	–	
	facilitated diffusion*	transporter	+	–	+	+	~10^2
		channel	+	–	+	+	10^7–10^8
active transport	primary		+	+	+	+	10^2–10^4
	secondary	symporter	+	+	+	+	
		antiporter	+	+	+	+	10^0–10^2**
		uniporter	+	+	+	+	

Fig. 7.4 Classification of transport systems of biomembranes. Transport systems are classified according to the role of transport proteins and energy coupling. *Facilitated diffusion down a concentration gradient without energy coupling is known as uniport. **The Cl^-/HCO_3^- antiporter seems to be an exception, as its transport rate is significantly high, at 10^5 mol/s.

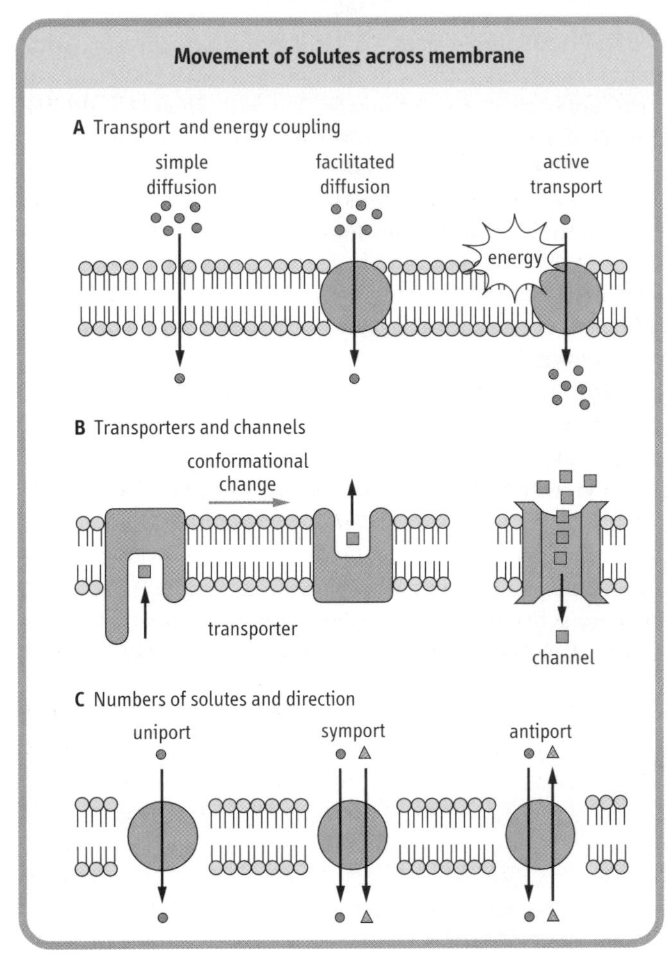

Fig. 7.5 Various examples of solute movement across membranes.

The transport of glucose into cells is mediated by substrate-specific transport proteins. The transport of blood glucose is generally by facilitated diffusion, as the intracellular concentration of glucose is typically less than 5 mmol/L (90 mg/dL). In contrast, the transport of glucose from the intestine into blood involves both facilitated diffusion and active transport processes. Active transport is especially important for maximal recovery of sugars from the intestine when the intestinal concentration of glucose is less than that in the blood.

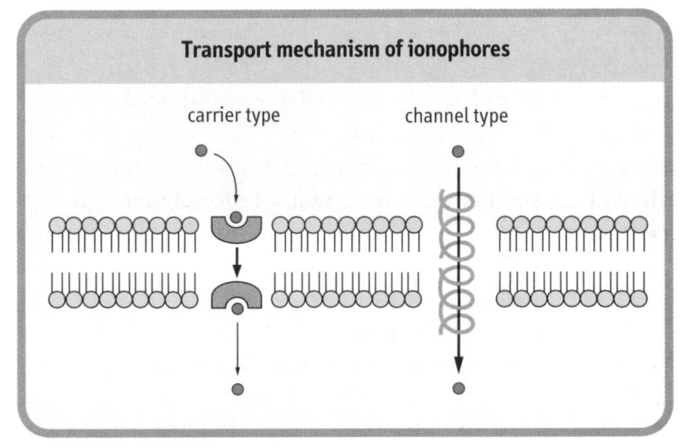

Fig. 7.6 Mobile ion carriers and channel-forming ionophores. Ionophores permit net movement of ions only down their electrochemical gradients.

Membrane permeability

Antibiotics inducing ion permeability

Peptide antibiotics act as ionophores and increase the permeability of membranes to specific ions; bactericidal effects of ionophores are attributed to disturbance of the ion transport systems of bacterial membranes. Ionophores permit net movement of ions only down their electrochemical gradients. There are two classes of ionophores: mobile ion carriers (or caged carriers) and channel formers (Fig. 7.6). Valinomycin is a typical example of a mobile ion carrier. It is a cyclic peptide with a lipophilic exterior and ionic interior. It dissolves in the membrane and diffuses between the inner and outer surfaces. K^+ binds in the central core of valinomycin, and the complex diffuses across the membrane, releasing the K^+ and gradually dissipating the K^+ gradient. A β-helical gramicidin A molecule, a linear peptide with 15 amino acid residues, forms a pore. The head-to-head dimer of gramicidin A makes a transmembrane channel that allows movement of monovalent cations (H^+, Na^+, and K^+). These ionophores are toxic, but are used as tools for laboratory experiments.

Polyene antibiotics such as amphotericin B and nystatin exert their cytotoxic action by rendering the membrane of the target cell permeable to ions and small molecules. Formation of a sterol–polyene complex is essential for the cytotoxic function of these antibiotics, as they display a selective action against organisms in which the membranes contain sterols. Thus they are active against yeasts, a wide variety of fungi, and other eucaryotic cells, but have no effect on bacteria. Because their affinity toward campesterol, a fungal membrane component, is higher than that for cholesterol, these antibiotics have been used for the treatment of topical infections of fungal origin.

Saturability and specificity are important characteristics of the transport system in which proteins participate

The rate of facilitated diffusion is generally much greater than that of simple diffusion: transport protein catalyzes the transport process. In contrast to simple diffusion, in which the rate of transport is directly proportional to the substrate concentration, facilitated diffusion is a saturable process, having a maximum transport rate, T_{max} (Fig. 7.7). When the concentration of extracellular molecules (transport substrates) becomes very high, the T_{max} is achieved by saturation of the transporter proteins with substrate. The kinetics of facilitated diffusion for substrates can be described by the same equation that is used for enzyme catalysis (the Michaelis–Menten equation):

$$S_{out} + \text{transporter} \overset{K_t}{\rightleftharpoons} (S \bullet \text{transporter complex}) \rightarrow S_{in}$$

where K_t is the dissociation constant of substrate (S) and transporter, and S_{out} is the concentration of transport substrate. Then the transport rate, t, is calculated approximately to be:

$$t = \frac{T_{max}}{\left(1 + \frac{K_t}{S_{out}}\right)}$$

where the K_t is the concentration that gives the half-maximal rate.

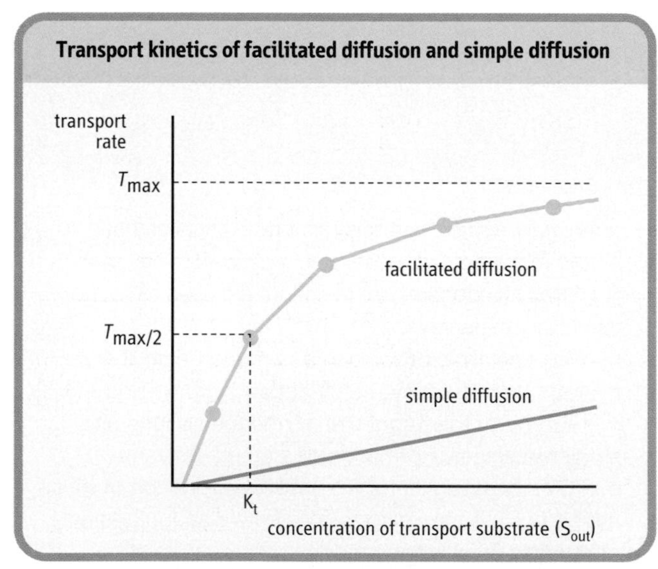

Fig. 7.7 Comparison of the transport kinetics of facilitated diffusion and simple diffusion. The rate of transport of substrate is plotted against the concentration of substrate in the extracellular medium. In common with enzyme catalysis, transporter-catalyzed uptake has a maximum transport rate, T_{max} (saturable). K_t is the concentration at which the rate of substrate uptake is half maximal. For simple diffusion, the transport rate is slower and directly proportional to substrate concentration.

The transport process is usually highly specific: each transporter transports only a single species of molecules or structurally related compounds. The red blood cell glucose transporter has high affinity for D-glucose, but low affinity for the related sugars, D-mannose and D-galactose (10–20 times lower). L-Glucose is not transported, because its affinity is more than 1000 times lower than that of the D-form.

Active transport processes

ATP is the universal energy currency for transport processes

ATP is a high-energy product of metabolism and is often described as the 'energy currency' of the cell. The phosphoanhydride bond of ATP releases free energy when it is hydrolyzed to produce adenosine diphosphate (ADP) and inorganic phosphate. Such energy is used for syntheses of large and small cellular molecules, cellular movement, and uphill transport of molecules against concentration gradients. Primary active transport systems use ATP directly to drive transport; secondary active transport uses an electrochemical

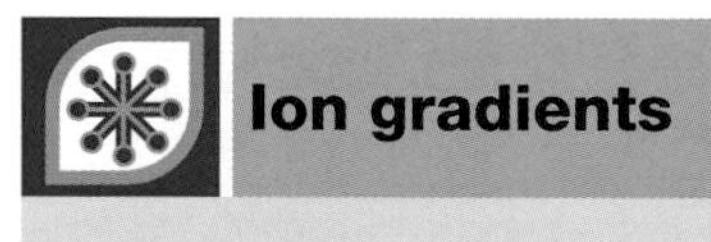

Ion gradients

Concentration gradient and electrochemical gradient of ions

The permeability of most nonelectrolytes through membranes can be analyzed by assuming that the rate-limiting step is the diffusion within the lipid bilayer. The relative rate of simple diffusion of a molecule across the membrane is therefore proportional to the concentration gradient across the bilayer and to the hydrophobicity of the molecule. For charged molecules and ions, transport across the membrane must be facilitated by a transporter or channel, and is driven by the electrochemical gradient.

This gradient is the combination of the voltage gradient (membrane potential) and the concentration gradient of the ion across the membrane. These forces may act in the same direction or in opposite directions. In the case of Na^+ ions, the concentration difference between outside and inside the cell is about a factor of 10, being maintained by the Na^+/K^+-ATPase. The Na^+/K^+-ATPase is electrogenic, pumping out three Na^+ and pumping in two K^+ ions, generating an inside-negative membrane potential. K^+ leaks out through K^+ channels, down its concentration gradient, further increasing the voltage gradient. The concentration gradient of Na^+ ions and the voltage gradient power the uptake of other molecules against their concentration gradient by symport with Na^+.

gradient of Na^+ and H^+, or membrane potential produced by primary active transport processes. Sugars and amino acids are generally transported into cells by secondary active transport systems.

The most important primary active transport systems are ion pumps (ion transporting ATPases or pump ATPases)

The pump ATPases are classified into four groups (Fig. 7.8).

Coupling factor ATPases (F-ATPases) in mitochondrial, chloroplast, and bacterial membranes (see Chapter 8) hydrolyze ATP and transport hydrogen ions (H^+). This ATPase is also called H^+-ATPase (H^+-transporting ATPase). In mitochondria, the 'powerhouses' of the cell, the F_1-ATPase works in the opposite direction, synthesizing ATP from ADP and phosphate as protons move down a concentration gradient generated across the inner membrane by oxidation reactions during metabolism. The proton gradient energetically couples the oxidation and phosphorylation reactions in a process known as oxidative phosphorylation (see Chapter 8). The product, ATP, is released into the mitochondrial matrix, then transported to the outside (the cytoplasmic side) through an ATP–ADP translocase in the mitochondrial inner membrane (an example of an antiport system shown in Figure 7.5C). The translocase allows one molecule of ADP to enter only if one molecule of ATP exits simultaneously.

Cytoplasmic vesicles, such as lysosomes, endosomes, and secretory granules, are acidified by the V-type (vacuolar) H^+-ATPase in their membranes. Acidification by this V-ATPase is important for the activity of lysosomal enzymes that have acidic pH optima, and for the accumulation of drugs and neurotransmitters in secretory granules. The V-ATPase also acidifies the extracellular environments of osteoclasts and kidney epithelial cells. F- and V-type ATPases are structurally similar, and seem to be derived from a common ancestor. The ATP-binding catalytic subunit and the subunit forming the H^+ pathway are conserved between these ATPases.

P-ATPases form phosphorylated intermediates that drive ion translocation: the 'P' refers to the phosphorylation. These transporters have an active-site aspartate residue which is reversibly phosphorylated by ATP during the transport process. The P-type Na^+/K^+-ATPase in various tissues and the Ca^{2+}-ATPase in the sarcoplasmic reticulum have important roles in maintaining cellular ion gradients. Na^+/K^+-ATPases create an electrochemical gradient of Na^+ that produces the driving force for uptake of nutrients. The discharge of this electrochemical gradient is also fundamental to the process of nerve transmission (see Chapter 38).

Various primary active transporters in eucaryotic cells

Group	Member	Location	Substrate(s)	Functions
F-ATPase (coupling factor)	H^+-ATPase	mitochondrial inner membrane	H^+	ATP synthesis generation of electrochemical gradient of H^+
V-ATPase (vacuolar)	H^+-ATPase	cytoplasmic vesicles (lysosome, secretory granules) plasma membranes (ruffled border of osteoclast, kidney epithelial cell)	H^+	activation of lysosomal enzymes accumulation of neurotransmitters turnover of bone acidification of urine
P-ATPase (phosphorylation)	Na^+/K^+-ATPase	plasma membranes (ubiquitous, but abundant in kidney and heart)	Na^+ and K^+	generation of electrochemical gradient of Na^+ and K^+
	H^+/K^+-ATPase	stomach (parietal cell in gastric gland)	H^+ and K^+	acidification of stomach lumen
	Ca^{2+}-ATPase	sarcoplasmic reticulum and endoplasmic reticulum	Ca^{2+}	Ca^{2+} sequestration into sarcoplasmic (endoplasmic) reticulum
	Ca^{2+}-ATPase	plasma membrane	Ca^{2+}	Ca^{2+} excretion to outside of the cell
	Cu^{2+}-ATPase	plasma membrane and cytoplasmic vesicles	Cu^{2+}	Cu^{2+} absorption from intestine and excretion from liver
ABC transporter (ATP binding cassette)	P-glycoprotein	plasma membrane	various drugs	excretion of harmful substances multidrug resistance for anticancer drugs
	MRP	plasma membrane	glutathione conjugate	detoxification multidrug resistance for anticancer drugs
	CFTR*	plasma membrane	Cl^-	outward rectifying chloride channel
	TAP	endoplasmic reticulum	peptide	presentation of peptides for immune response

Fig. 7.8 Primary active transporters in eucaryotic cells. Various examples of primary active transporters (ATP-powered pump ATPases) are listed, together with their location. The F-ATPase is reversible, whereas others catalyse unidirectional ATP hydrolysis reactions. *Some of ABC transporters function as channels or channel regulators. MRP, multidrug resistance-associated protein; CFTR, cystic fibrosis transmembrane conductance regulator; TAP, transporter associated with antigen presentation.

The ATP-binding cassette (ABC) transporters comprise the fourth active transporter family. 'ABC' is the abbreviation for '**A**TP-**b**inding **c**assette', referring to an ATP-binding region in the transporter. Members of this transporter family have four domains: two highly hydrophobic domains, each of which traverses the membrane six times, forming the translocation pathway for substrate, and two cytoplasmic catalytic domains containing the ATP-binding region. In some cases, the transporter is formed by a combination of two halves, each polypeptide carrying a transmembrane region and an ATP-binding region (transporter associated with antigen presentation, TAP, in Figure 7.8). In bacteria, the four domains are often formed by separate polypeptides.

ABC transporters are physiologic and clinically important (see Fig. 7.8). P-glycoprotein ('P' = permeability) and MRP (multidrug resistance-associated protein) pump out many anticancer drugs from cancer cells, causing a problem in cancer treatment. They are thought to have a physiological role in excretion of toxic metabolites and xenobiotics. TAP transporters, a class of ABC transporters associated with antigen presentation, are required for initiating the immune response against foreign proteins; they are located in endoplasmic reticulum and are involved in antigen peptide transport from the cytosol into endoplasmic reticulum. Some ABC transporters are present in peroxisomal membrane. They appear to be involved in the transport of peroxisomal enzymes necessary for oxidation of very-long-chain fatty acids.

Uniport, symport, and antiport are examples of secondary active transport

Secondary active transport is classified into three types: uniport (monoport), symport (cotransport) and antiport (countertransport) (see Fig. 7.5C). Transport substrates move in the same direction during symport, and in opposite directions during antiport. The movement of one substrate against its concentration gradient can be driven by movement of another substrate (usually cations such as Na^+ and H^+) down a gradient. Uniport of charged substrates may also be electrophoretically driven by the membrane potential of the cell. The proteins participating in these active transport systems are termed uniporters, symporters, and antiporters, respectively. Selected examples of these transport systems are presented below.

Transport by channels and pores

Channels are often pictured as tunnels across the membrane, in which binding sites for substrates (ions) are accessible from either side of the membrane at the same time (Fig. 7.5B). Conformational changes are not required for the translocation of substrates entering from one side of the membrane to exit on the other side. Both voltage changes and ligand binding induce conformational changes in channels that have the effect of opening or closing the channels – processes known as voltage or ligand 'gating'. Movement of molecules through channels is fast (10^7–10^8/s) in comparison with the rates achieved by transporters (Fig. 7.4).

Menkes and Wilson diseases

X-linked Menkes disease is a lethal disorder that occurs in 1 in 100,000 newborn infants and is characterized by abnormal and hypopigmented hair, a characteristic facies, cerebral degeneration, connective tissue and vascular defects, and death by the age of 3 years. A copper-transporting P-ATPase that is expressed in all tissues except liver is defective in this disease (see Fig. 7.8). In patients with Menkes disease, copper enters the intestinal cells, but is not transported further, resulting in severe copper deficiency. Subcutaneous administration of copper histidine may be an effective treatment if started early.

The gene for Wilson disease also encodes a copper-transporting P-ATPase and is 60% identical with that of the Menkes gene. It is expressed in liver, kidney, and placenta. Wilson disease occurs in 1 in 35 000–100 000 newborns and is characterized by failure to incorporate copper into ceruloplasmin in the liver and failure to excrete copper from the liver into bile, resulting in toxic accumulation of copper in the liver and also in the kidney, brain, and cornea. Liver cirrhosis, progressive neurologic damage, or both, occur during childhood to early adulthood. Chelating agents such as penicillamine are used for treatment of patients with this disease. Oral zinc treatment may be useful for decreasing the absorption of dietary copper.

Comment. Copper is an essential trace metal and an integral component of many enzymes. However, it is toxic in excess, because it binds to proteins and nucleic acids, enhances the generation of free radicals, and catalyzes oxidation of lipids and proteins in membranes.

The terms 'channel' and 'pore' are generally used interchangeably. However, 'pore' is used most frequently to describe somewhat nonselective structures that discriminate between substrates primarily on the basis of size. Thus only sufficiently small molecules will pass freely through the pores. The term 'channel' is usually applied to describe ion channels.

There are three examples of pores that are important for cellular physiology

The gap junction between endothelial, muscle, and neuronal cells is a cluster of small pores, in which two cylinders of six connexin subunits in plasma membranes join each other to form a pore about 1.2–2.0 nm (12–20 Å) in diameter. Molecules smaller than about 1 kDa can pass between cells through gap junctions. Such cell–cell communication is important for physiologic coupling, for example in the concerted contraction of uterine muscle during labor and delivery. These pores are usually maintained in an open state, but will close when cell membranes are damaged or when the metabolic rate is depressed.

Nuclear pores have a functional radius of about 9.0 nm (90 Å) through which proteins and nucleic acids enter and leave the nucleus.

A third class of pores is important for protein sorting. Mitochondrial proteins encoded by nuclear genes are transported to this organelle through pores in the outer mitochondrial membrane. Nascent polypeptide chains of secretory proteins and plasma membrane proteins also pass through pores in the endoplasmic reticulum membrane.

Examples of transport systems and their coupling

Characteristics of glucose transporters (uniporters)

Glucose transporters are essential for facilitated diffusion of glucose into cells. The glucose transporter family comprises five members, named GLUT-1 to GLUT-5. They are transmembrane proteins similar in size, all having about 500 amino acid residues and 12 transmembrane helices. GLUT-1, in red blood cells, has a K_m of 15–20 mmol/L); most of the GLUT-1 molecules are active under fasting conditions (glucose concentration of 5 mmol/L; 90 mg/dL). In contrast, pancreatic islet β-cells express GLUT-2, with a K_m of more than 10 mmol/L (180 mg/dL). In response to the intake of food and resulting increase in blood glucose concentration, GLUT-2 molecules mediate an increase in the cellular uptake of glucose, leading to insulin secretion (see Chapter 20). Cells in insulin-sensitive tissues such as muscle and adipose have GLUT-4. Insulin stimulates translocation of GLUT-4 from intracellular vesicles to the plasma membrane, facilitating glucose uptake during meals. While GLUT-1 distributes ubiquitously in various tissues and GLUT-2 in liver, intestine and kidney, GLUT-3 and GLUT-5 are found in brain and testis, and intestinal epithelial cells, respectively.

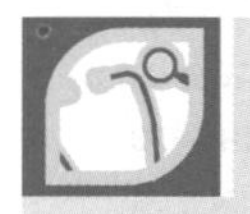

Cystic fibrosis

Cystic fibrosis is the most common potentially lethal autosomal recessive disease of white populations, affecting 1 in 2500 newborns. Cystic fibrosis is usually manifested as exocrine pancreatic insufficiency, an increase in the concentration of chloride ions (Cl^-) in sweat, male infertility, and airway disease. The last of these leads to progressive lung dysfunction, which is the major cause of morbidity and mortality in cystic fibrosis.

Comment. Cystic fibrosis is caused by mutations in the gene encoding a Cl^- channel named CFTR (cystic fibrosis transmembrane conductance regulator). CFTR is classified as an ABC transporter, as it has characteristic ATP-binding regions (ATP-binding cassette) in its amino acid sequence (see Fig. 7.8). ATP binding to the CFTR is required for channel opening. The lack of this channel activity in epithelia of CF patients is believed to be central to the pathogenesis of the disease. Liposome-mediated CFTR gene transfer has resulted in a significant reduction in the numbers of bacteria bound to ciliated airway epithelial cells.

Sulfonylurea therapy

Sulfonylureas such as tolubutamide and glibenclamide stimulate insulin secretion, which decreases blood glucose concentration. These drugs bind to a component of an ATP-sensitive K^+ channel (K_{ATP}-β, called the sulfonylurea receptor, SUR_1) in pancreatic islet β-cells and close the K_{ATP} channel, stimulating insulin secretion in diabetes. Defective K_{ATP} channels, which are unable to transport K^+, induce low blood glucose concentration – a condition called familial persistent hyperinsulinemic hypoglycemia of infancy (PHHI) that occurs in 1 per 50 000 persons – as a result of loss of K^+-channel function and continuous insulin secretion.

Comment. The K_{ATP} is composed of two subunits, which participate in regulation of insulin secretion in pancreatic islet β-cells. When the blood concentration of glucose increases, glucose is transported into the β-cell through a glucose transporter (GLUT-2) and metabolized, resulting in an increase in cytoplasmic ATP concentration. The ATP binds to the regulatory subunit of the K^+ channel, K_{ATP}-β causing structural change of a K_{ATP}-α subunit, which closes the K_{ATP} channel. This induces depolarization of the plasma membrane (decreased voltage gradient across the membrane) and activates voltage-dependent Calcium (Ca^{2+}) channels (VDCCs). The entry of Ca^{2+} stimulates exocytosis of vesicles that contain insulin. The binding of sulfonylureas to K_{ATP} (or an associated protein) on the outside of the plasma membrane is thought to mimic the regulatory effect of intracellular ATP.

Ca^{2+} transport and mobilization in muscle

Striated muscle (skeletal and cardiac) is composed of bundles of multinucleated muscle cells (see Chapter 19). Each cell is packed with bundles of actin and myosin filaments (myofibrils) that produce contraction. During muscle contraction, nerves at the neuromuscular junction stimulate local depolarization of the membrane by opening voltage-dependent Na^+ channels. The depolarization spreads rapidly into invaginations of the plasma membrane called the transverse (T) tubules, which extend around the myofibrils.

Voltage-dependent Ca^{2+} channels (VDCC) located in the T tubules of skeletal muscle change their conformation as a result of membrane depolarization, and directly activate a Ca^{2+}-release channel in the sarcoplasmic reticulum membrane, a network of flattened tubules that surrounds each myofibril in the muscle-cell cytoplasm. The escape of Ca^{2+} from the lumen (interior compartment) of the sarcoplasmic reticulum increases the cytoplasmic concentration of Ca^{2+} (depolarization-induced Ca^{2+} release) from 10^{-4} mmol/L (0.0007 mg/L) to about 10^{-2} mmol/L (0.07 mg/dL), triggering ATP hydrolysis by myosin and muscle contraction. A Ca^{2+}-ATPase in the sarcoplasmic reticulum then uses ATP to transport Ca^{2+} back out of the cell into the lumen, decreasing the cytoplasmic concentration of Ca^{2+}, and the muscle relaxes (Fig. 7.9).

In heart muscle, VDCCs permit the entry of a small amount of Ca^{2+}, which induces release of Ca^{2+} from the lumen of the sarcoplasmic reticulum (Ca^{2+}-induced Ca^{2+} release). Not only the sarcoplasmic reticulum Ca^{2+}-ATPase, but also an Na^+/Ca^{2+}-antiporter and a plasma membrane Ca^{2+}-ATPase (Fig. 7.8) are responsible for pumping out cytoplasmic Ca^{2+} from heart muscle (Fig. 7.9). The rapid restoration of ion gradients allows for rhythmic contraction of the heart.

Role of Na^+/K^+-ATPase in glucose uptake

Glucose is almost completely reabsorbed from ultrafiltrate in the proximal tubules of the kidney

The kidneys constitute an ultrafiltration system that removes small molecules from blood. However, glucose, amino acids,

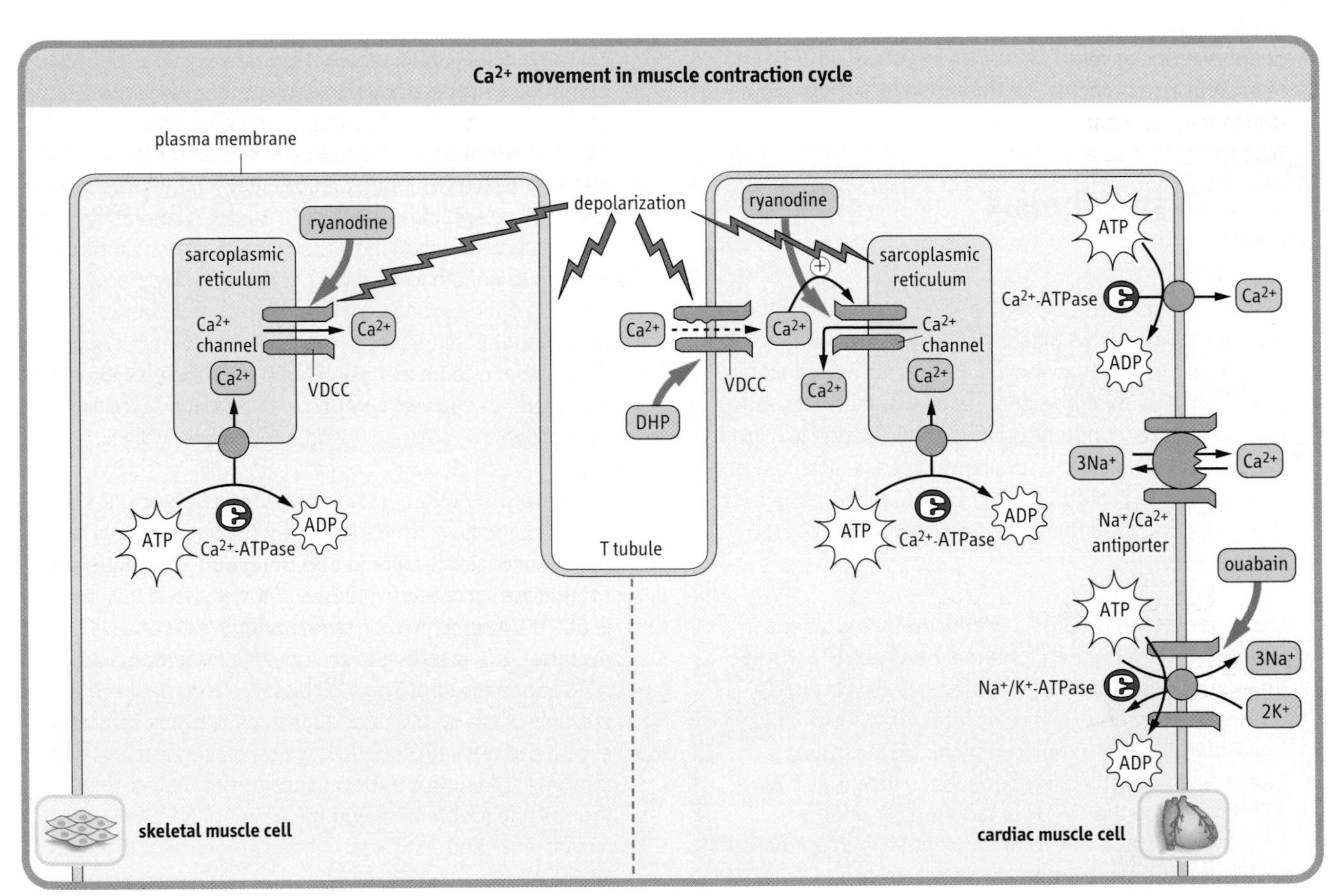

Fig. 7.9 Roles of transporters in muscle contraction. Ca^{2+} movements in skeletal (left) and cardiac (right) muscle cells are shown. Thick arrows indicate the binding sites for inhibitors. In skeletal muscle, VDCCs directly activate release of Ca^{2+} from the sarcoplasmic reticulum. The increased cytoplasmic Ca^{2+} concentration triggers muscle contraction. A Ca^{2+}-ATPase in the sarcoplasmic reticulum transports Ca^{2+} back into the lumen, decreasing the cytoplasmic Ca^{2+} concentration, and the muscle relaxes. In heart muscle, VDCCs allow entry of a small amount of Ca^{2+}, which induces release of Ca^{2+} from the lumen of sarcoplasmic reticulum. Two types of Ca^{2+}-ATPases and a Na^+/Ca^{2+}-antiporter are responsible for pumping cytoplasmic Ca^{2+} out of the muscle cell. The Na^+/Ca^{2+}-antiporter uses the sodium (Na^+) gradient produced by Na^+/K^+-ATPase to antiport Ca^{2+}.

many ions, and other nutrients in the ultrafiltrate are almost completely reabsorbed in the proximal tubules, by symport processes. Glucose is reabsorbed primarily by the Na^+-coupled glucose symporter, sodium glucose transporter 2 ($SGLT_2$) (one-to-one stoichiometry), into epithelial cells located in the renal proximal tubule. Much smaller amounts of glucose are recovered by $SGLT_1$ in a later segment of the tubule, which couples transport of one molecule of glucose to two sodium ions. The concentration of Na^+ in the filtrate is 140 mmol/L (322 mg/dL), while that inside the epithelial cells is 30 mmol/L (69 mg/dL), so that Na^+ flows 'downhill' along its gradient, dragging glucose 'uphill' against its concentration gradient. The low intracellular concentration of Na^+ is maintained by a Na^+/K^+-ATPase on the opposite side of the tubular epithelial cell, which antiports three cytoplasmic sodium ions and two extracellular potassium ions, coupled with hydrolysis of a molecule of ATP. Glucose accumulated in the epithelial cell is further transported downhill across the membrane to the blood by GLUT-2 (facilitated diffusion).

Dietary glucose is also transported into intestinal epithelial cells against a concentration gradient. $SGLT_1$, Na^+/K^+-ATPase, and GLUT-2 also function in this process (Fig. 7.10).

Various drugs inhibit transporters in muscle

Phenylalkylamine (verapamil), benzothiazepine (diltiazem), and dihydropyridine (DHP; nifedipine) are Ca^{2+}-channel blockers that inhibit VDCCs (Fig. 7.9). Ryanodine inhibits the Ca^{2+}-release channel in the sarcoplasmic reticulum. These drugs are used to inhibit both the increase in cytoplasmic Ca^{2+} concentration and the force of muscle contraction. In contrast, cardiac glycosides such as ouabain and digoxin increase heart muscle contraction and are used for treatment of congestive heart failure. They act by inhibiting the Na^+/K^+-ATPase that generates the Na^+ concentration gradient used to drive export of Ca^{2+} by the Na^+/Ca^{2+} antiporter. Snake venoms such as α-bungarotoxin, and tetrodotoxin from the puffer fish inhibit voltage-dependent Na^+ channels. Lidocaine, a Na^+-channel blocker, is used as a local anesthetic and antiarrhythmic drug. Inhibition of Na^+ channels represses transmission of the depolarization signal.

Proton pump in the stomach

The lumen of the stomach is highly acidic (pH ≈1) because of the presence of a proton pump (H^+/K^+-ATPase; P-ATPase in Fig. 7.8) that is specifically expressed in gastric parietal cells. The gastric proton pump is localized intracellularly in vesicles in the resting state. Stimuli such as histamine and gastrin induce fusion of the vesicles with the plasma membrane (Fig. 7.11A). The pump antiports two cytoplasmic protons and two extracellular potassium ions, coupled with hydrolysis of a molecule of ATP, thus it is called an H^+/K^+-ATPase. Cl^- is secreted through a Cl^- channel, producing hydrochloric acid (HCl) (gastric acid) in the lumen (Fig. 7. 11B).

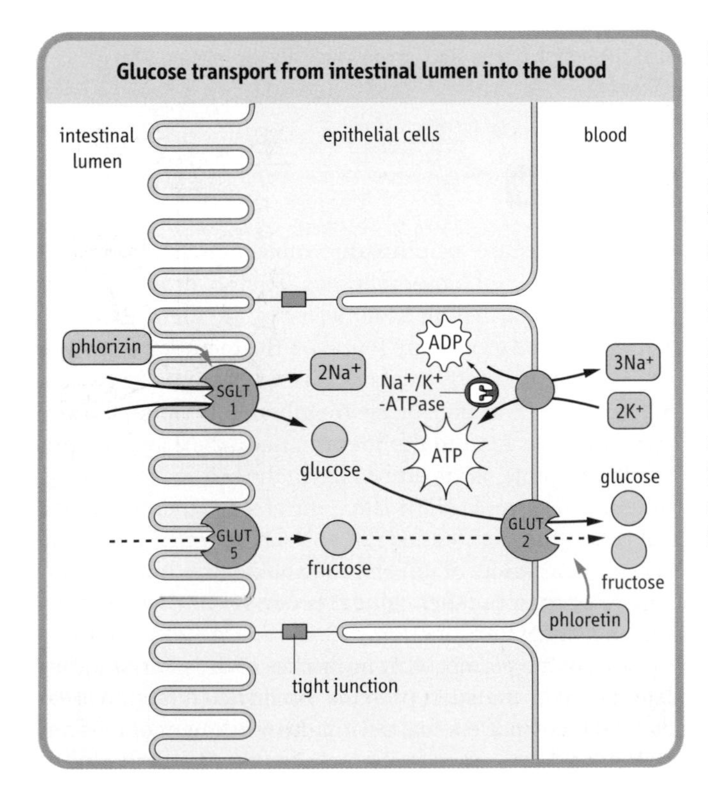

Fig. 7.10 Transport of glucose across an intestinal epithelial layer. Glucose is pumped into the cell through the Na^+-coupled glucose symporter ($SGLT_1$), and passes out of the cell by facilitated diffusion mediated by the GLUT-2 uniporter. The Na^+ gradient for glucose symport is maintained by the Na^+/K^+-ATPase, which keeps the intracellular concentration of Na^+ low. $SGLT_1$ is inhibited by phlorizin and GLUT-2 by phloretin. Phloretin-insensitive GLUT-5 catalyzes the uptake of fructose by facilitated diffusion. The fructose is then exported through GLUT-2. A defect of $SGLT_1$ causes glucose/galactose malabsorption. Adjacent cells are connected by impermeable tight junctions, which prevent solutes from crossing the epithelium. However, leakage of salts (Na^+ and Cl^-) through tight junctions induces diarrhea as a result of inhibition of water absorption. Diarrhea is also induced by laxatives such as phenolphthalein, which is an irritant cathartic for the colon. Thick arrows indicate the binding sites for inhibitors.

Inhibiting the gastric proton pump and eradication of *Helicobacter pylori*

Chronic strong acid secretion by the gastric proton pump injures the stomach and the duodenum, leading to gastric and duodenal ulcers. Proton pump inhibitors such as omeprazole are delivered to parietal cells from the circulation after oral administration. Omeprazole is a prodrug: it accumulates in the acidic compartment, as it is a weak base, and is converted to the active compound under the acidic conditions in the gastric lumen. The active form covalently modifies cysteine residues located in the extracytoplasmic domain of the proton pump. H_2 blockers (receptor antagonists) such as cimetidine and ranitidine indirectly inhibit acid secretion by competing with histamine for its receptor (Fig. 7.11).

Comment. Infection of the stomach by *Helicobacter pylori* also causes ulcers and is associated with an increased risk of gastric adenocarcinoma. Recently, antibiotic treatment has been introduced to eradicate *Helicobacter pylori*. Interestingly, antibiotic treatment together with omeprazole is much more effective, possibly because of an increased stability of the antibiotic under the weakly acidic condition produced by proton pump inhibition.

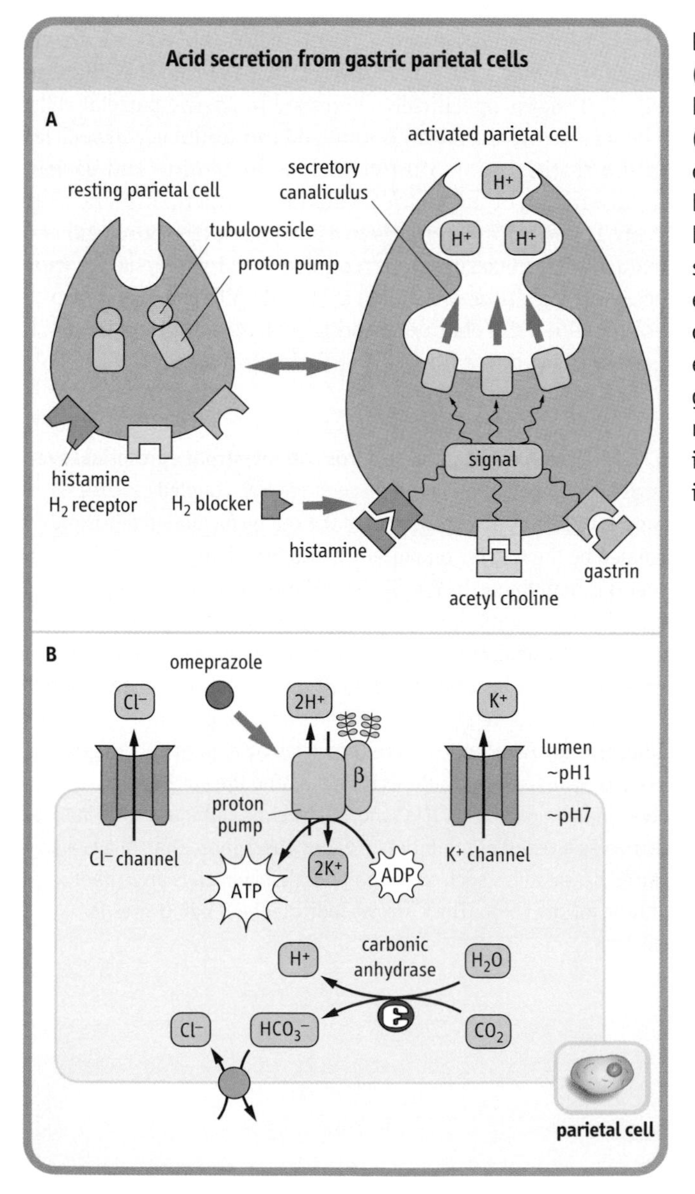

Fig. 7.11 of acid secretion from gastric parietal cells.
(A) Acid secretion is stimulated by extracellular signals and accompanied by morphologic changes in parietal cells [from resting (left) to activated (right)]. The proton pump (H^+/K^+-ATPase) moves to the secretory canaliculus (plasma membrane) from cytoplasmic tubulovesicles. H_2 blockers compete with histamine at the histamine H_2 receptor. (B) Ion balance in the parietal cell. The H^+ transported by the proton pump are supplied by carbonic anhydrase. Bicarbonate, the other product of this enzyme, is antiported with Cl^- and the Cl^- are secreted through a Cl^- channel. The potassium (K^+) imported by the proton pump are again excreted by a K^+ channel. The proton pump has catalytic α- and glycosylated β- subunits. Omeprazole covalently modifies cysteine residues located in the extracytoplasmic domain of α- subunit and inhibits the proton pump. Thick arrows indicate the binding sites for inhibitors.

Summary

Phospholipids are amphipathic molecules that form the bilayer structure of biomembranes. Their hydrophobic fatty acid tails are located on the interior of the membrane and their polar head groups are found on the membrane surface. The unsaturated fatty acids bound to phospholipids contribute to the fluid state of the membrane. Peripheral membrane proteins bind to the membrane surface and integral membrane proteins are embedded in the bilayer. Transporter proteins, an example of the latter, traverse the membrane several times. Biomembranes not only act as permeability barriers and mediators of ion and metabolite flux, but also have important roles in other cellular processes such as recognition and signal transduction.

Most of the permeability properties of the membrane are determined by transport proteins. Facilitated diffusion is catalyzed by transporters that permit the movement of ions and molecules down concentration gradients, whereas uphill or active transport requires energy. Primary active transport is catalyzed by pump ATPases that use energy produced by ATP hydrolysis. Secondary active transport uses electrochemical gradients of Na^+ and H^+, or membrane potential produced by primary active transport processes. Uniport, symport, and antiport are examples of secondary active transport. Protein-mediated transport is a saturable process with high substrate specificity.

Numerous substrates such as ions, nutrients, small organic molecules including drugs and peptides, and proteins are transported by various transporters. All of these transporters are indispensable for homeostasis. The expression of unique sets of transporters is important for specific cell functions such as muscle contraction, nutrient and ion absorption by intestinal epithelial cells and reabsorption by kidney cells, and secretion of acid from gastric parietal cells.

Further reading

Crystal RG. The gene as the drug. *Nature Medicine* 1995;**1**:15–17.
Dunne MJ, Kane C, Shepherd RM, *et al.* Familial persistent hyperinsulinemic hypoglycemia of infancy and mutations in the sulfonylurea receptor. *N Engl J Med* 1997;**336**:703–706.
Jan LY, Jan YN. Membrane permeability. *Curr Opin Cell Biol* 1994;**6**:569–570, 571–615.
Singer SJ, Nicholson GL. The fluid mosaic model of the structure of cell membranes. *Science* 1972;**175**:720–731.
Thomson AB, Wild G. Adaptation of intestinal nutrient transport in health and disease. Part I. *Dig Dis Sci* 1997;**42**:453–469.
Tümer Z, Horn N. Menkes disease: recent advances and new aspects. *J Med Genet* 1997;**34**:265–274.
Epstein FH, Ackerman MJ, Clapham DE. Ion channels – basic science and clinical disease. *N Engl J Med* 1997;**336**:1575–1586.

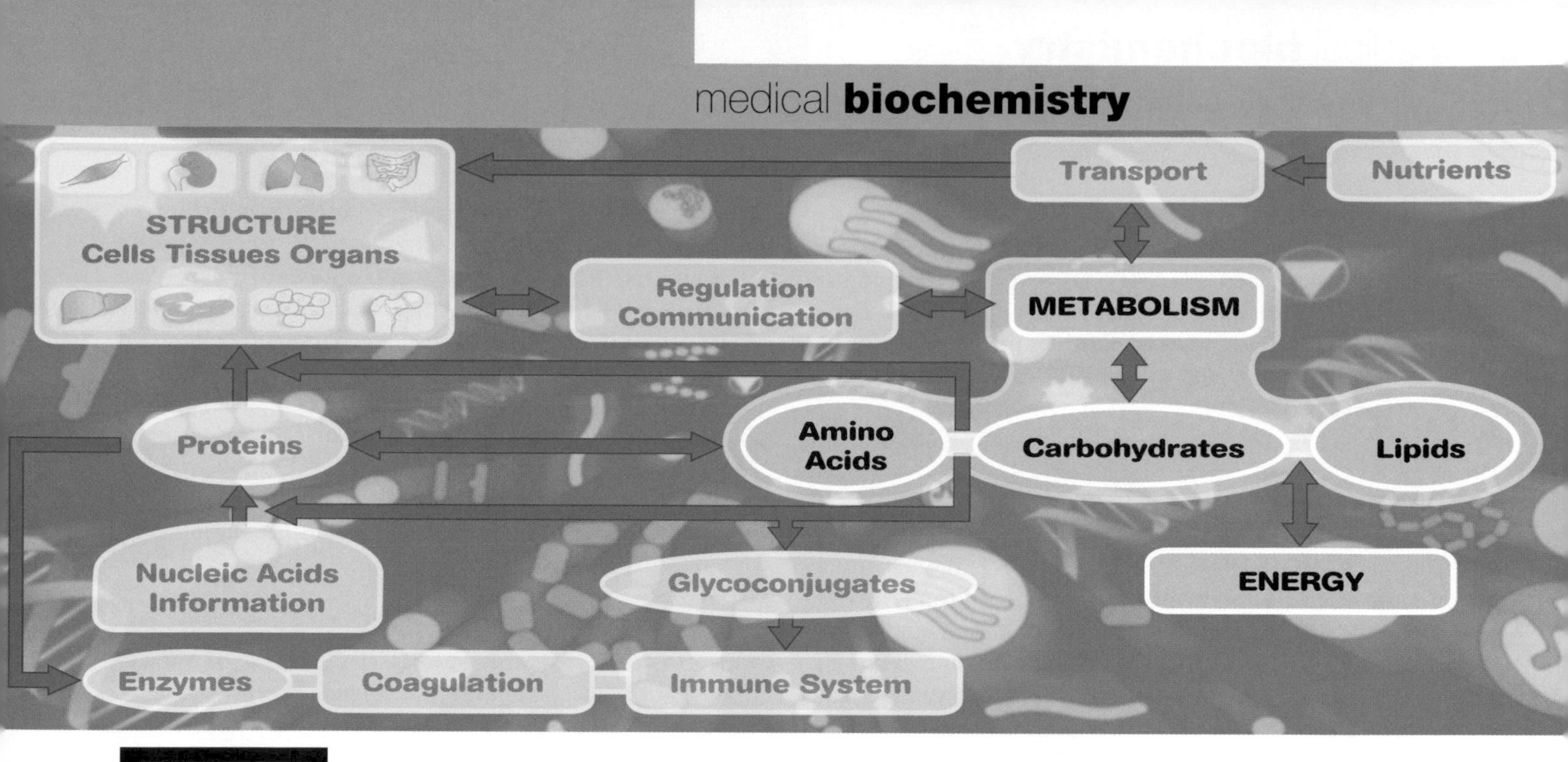

8 Bioenergetics and Oxidative Metabolism

Oxidation as a source of energy

Oxidation of metabolic fuels is essential to life. In higher organisms, dietary fuels, primarily carbohydrates and lipids, are metabolized to carbon dioxide, the fully oxidized form of carbon, and water, the fully reduced form of oxygen. During the process of **oxidative phosphorylation**, electrons are removed from fuels, transferred to oxidation–reduction (redox) coenzymes, and then transferred – by a sequence of redox reactions – to oxygen. The energy provided by the oxidation reactions is conserved as the high-energy phosphorylated compound, adenosine triphosphate (ATP), which is then used to perform work, such as transport of metabolites, physical activity, and biosynthesis. This chapter will provide an introduction to the concept of free energy – that is, the energy available from chemical reactions in biological systems, and the transduction of this energy from fuels into useful work. The pathways and specific molecules through which electrons are transported to oxygen and the mechanism of generation of ATP will be described and related to the structure of the mitochondrion, the powerhouse of the cell and the major source of cellular ATP.

Energy content of foods

The science of thermodynamics may be applied directly to nutrition and metabolism. Normal nutrition, and disorders such as obesity, diabetes, and cancer, all require an understanding of thermodynamics, and for any controled diet, it is important that the energy content of foods be known. The commonly accepted energy values for the four major food categories are shown in Figure 8.1; alcohol is included, because it is a significant dietary component for some people.

The calorific value of foods

Metabolic fuel	Energy content (kcal/g)
fats	9
carbohydrates	4
proteins	4
alcohol	7

Fig. 8.1 Energy content of the major classes of food. Note that the thermodynamic term, kcal (energy required to increase the temperature of 1 g of water by 1°C) is equivalent to the common nutritional Calorie (with a capital C), i.e. 1 Cal = 1 kcal, 1 kcal = 4.187 kJ.

Stages of fuel oxidation

The oxidation of fuels can be divided into two general stages. In the first stage, reduced coenzymes are produced in numerous reactions during the oxidation of fuels, and in the second stage, ATP is synthesized in a defined common pathway, using the free energy provided by oxidation of the reduced coenzymes (Fig. 8.2).

Free energy

The Gibbs' free energy (ΔG) of a reaction is the maximum amount of useful energy that can be obtained from the reaction at constant temperature and pressure. The units of free energy are J/mol (kcal/mol). It is not possible to measure the absolute free energy content of a substance, but when reactant A reacts to form product B, the free energy change in this reaction ΔG, can be determined. For the reaction A → B:

$$\Delta G = G_B - G_A$$

where G_A and G_B are the free energy of A & B, respectively. All reactions in biologic systems are considered to be reversible reactions, so that the free energy of the reverse reaction, B → A, is numerically equivalent, but opposite in sign to that of the forward reaction.

It is clear that, if there is a greater concentration of B than of A at equilibrium, the reaction A → B is favorable – that is, it tends to move forward from a standard state in which A and B are present at equal concentrations. In this case, the reaction is said to be a spontaneous or exergonic reaction, and the free energy of this reaction is defined as negative: that is, $\Delta G < 0$, indicating that energy is liberated by the reaction. Conversely, if the concentration of A is greater than that of B at equilibrium, the forward reaction is termed unfavorable, nonspontaneous or endergonic, and the reaction has a positive free energy: that is, B tends to form A, rather than A to form B. In this case, energy input would be required to push the reaction A → B forward from its equilibrium position to the standard state in which A and B are present at equal concentration. The total free energy available from a reaction depends on both its tendency to proceed forward from the standard state (ΔG) and the amount (moles) of reactant converted to product.

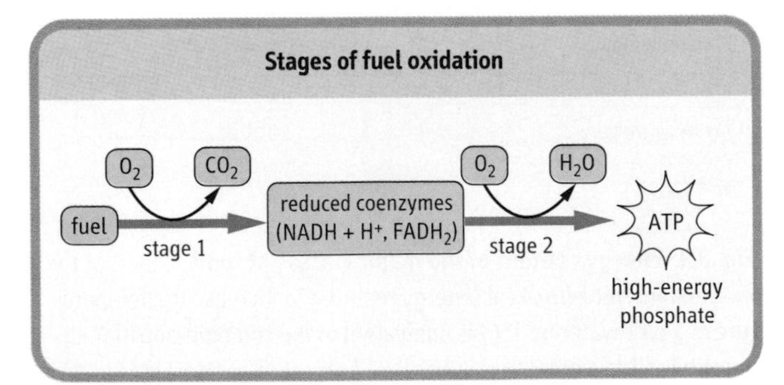

Fig. 8.2 Fuel oxidation. NADH, reduced nicotinamide adenine dinucleotide; $FADH_2$, reduced flavin adenine dinucleotide.

The free energy of metabolic reactions is related to their equilibrium constants

Thermodynamic measurements are based on standard-state conditions in which reactant and product are present at 1 molar concentrations or 1 atmosphere of pressure. Standard free energies, represented by the symbol ΔG° are determined or calculated for reactions that take place at 25°C (298° K). ΔG°′, the biologic free energy change in reaction is measured at pH 7.0. The free energy available from a reaction, in which 1 mole of substrate is converted to product, may be calculated from its equilibrium constant by the Gibbs equation:

$$\Delta G^{\circ\prime} = -RT \ \ln Ke_q$$

where T is absolute temperature (°kelvin), and K_{eq} is the natural logarithm of the equilibium constant for the reaction; and R is the gas constant:

$$R = -8.4 \ JK^{-1} mol^{-1} (-2 \ cal \ K^{-1} \ mol^{-1})$$

Several common metabolic intermediates that you will encounter in your studies are listed in Figure 8.3, along with the equilibrium constants and free energies for their hydrolysis reactions. Those intermediates with free energy changes equal to or greater than that of ATP, the central energy transducer of the cell, are considered to be high-energy compounds, and generally have anhydride bonds. The lower-energy compounds listed are all phosphate esters and, in comparison, do not yield as much free energy on hydrolysis. The hydrolysis reaction of glucose-6-phosphate (G-6-P) is written as:

$$G\text{-}6\text{-}P + H_2O \rightarrow \text{glucose} + \text{phosphate}$$
$$\Delta G^{\circ\prime} = -13.8 \ kJ/mol \ (-3.3 \ kcal/mol)$$

Thermodynamics of hydrolysis reactions

Metabolite	K_{eq}	ΔG°′ (kcal mol⁻¹)	kJ/mol⁻¹
phosphoenolpyruvate	1.2×10^{11}	– 14.8	– 61.8
phosphocreatine	9.6×10^8	– 12.0	– 50.2
1,3-bisphosphoglycerate	6.8×10^8	– 11.8	– 49.3
pyrophosphate	9.7×10^5	– 8.0	– 33.4
acetyl coenzyme A	4.1×10^5	– 7.5	– 31.3
ATP	2.9×10^5	– 7.3	– 30.5
G-1-P	5.5×10^3	– 5.0	– 20.9
F-6-P	7.0×10^2	– 3.8	– 15.9
G-6-P	3.0×10^2	– 3.3	– 13.8

Fig. 8.3 Equilibrium constants and standard free energy of hydrolysis of various metabolic intermediates. F-6-P, fructose-6-phosphate; G-1-P, glucose-1-phosphate; G-6-P, glucose-6-phosphate.

This reaction has a negative free energy and occurs spontaneously. The reverse reaction, synthesis of G-6-P from glucose and phosphate, would require input of energy.

Coupling with ATP

Living systems must transfer energy from one molecule to another without losing all of it as heat. Some of the energy must be conserved in chemical form and used to drive non-spontaneous biosynthetic reactions. Most of the energy obtained from the oxidation of metabolic fuels is channeled into the synthesis of ATP, a universal energy transducer in living systems that is often referred to as the common currency of metabolic energy, because it is used to drive so many energy-requiring reactions. ATP consists of the purine base, adenine, the five-carbon sugar, ribose, and α, β, and γ phosphate groups (Fig. 8.4). The two anhydride linkages are said to be high-energy bonds, because their hydrolysis yields a large negative change in free energy.

ATP is commonly used to drive biosynthetic reactions

The free energy of a high-energy bond, such as the phosphate anhydride bonds in ATP, can be used to drive or push forward reactions that would otherwise be unfavorable. For example, the first step in the metabolism of glucose is the synthesis of G-6-P. As shown in Figure 8.3, this is not a favorable reaction: the hydrolysis ($\Delta G^{\circ\prime} = -3.3$ kcal/mol), rather than synthesis ($\Delta G^{\circ\prime} = +3.3$ kcal/mol) of G-6-P is the favored reaction. However, as shown below, the synthesis of G-6-P (reaction I) can be energetically coupled to the hydrolysis of ATP (reaction II), yielding a 'net reaction' III that is favorable for synthesis of G-6-P:

Reaction

I glucose + phosphate → G-6-P
$\Delta G^{\circ\prime} = +13.8$ kJ/mol (+3.3 kcal/mol)

II ATP + H_2O → ADP + phosphate
$\Delta G^{\circ\prime} = -30.5$ kJ/mol (−7.3 kcal/mol)

III Net reaction:
glucose + ATP→ G-6-P + ADP
$\Delta G^{\circ\prime} = -12.6$ kJ/mol (−3 kcal/mol)

This is possible because of the high free energy or 'group transfer potential' of ATP. The physical transfer of the phosphate from ATP to glucose occurs in the active site of a kinase enzyme, such as glucokinase. This motif, in which ATP is used to drive biosynthetic reactions, transport processes, or muscle activity, occurs commonly in metabolic pathways.

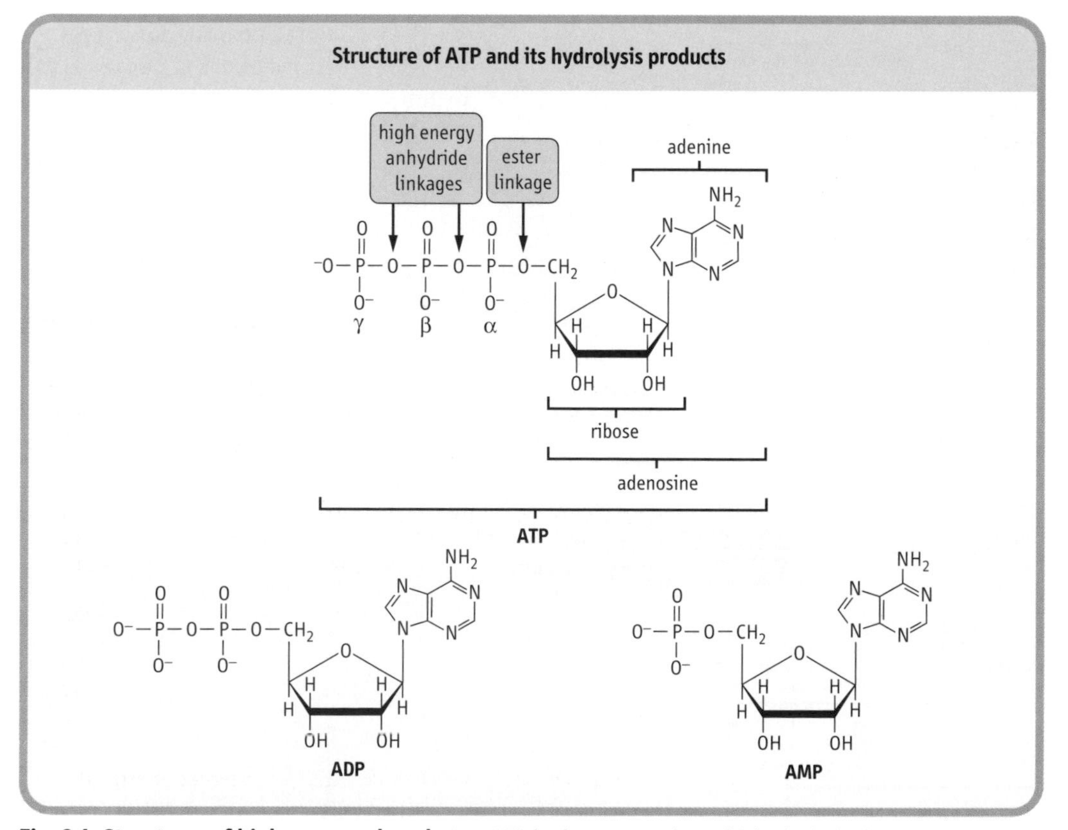

Fig. 8.4 Structures of high-energy phosphates. ATP is shown, together with its hydrolysis products, adenosine diphosphate (**ADP**) and adenosine monophosphate (**AMP**).

NAD^+, FAD and FMN are the major redox coenzymes

The major redox coenzymes involved in transduction of energy from fuels to ATP are nicotinamide adenine dinucleotide (NAD^+), flavin adenine dinucleotide (FAD) and flavin mononucleotide (FMN). During energy metabolism, electrons are transferred from carbohydrates and fats to these coenzymes, reducing them to NADH, $FADH_2$ and $FMNH_2$. In each case, two electrons are transferred, but the number of protons transferred differs. NAD^+ accepts a hydride ion (H^-) that consists of one proton and two electrons; the remaining proton is released into solution (Fig. 8.5). FAD and FMN accept two electrons and two protons.

Mitochondrial synthesis of ATP from reduced coenzymes

Metabolism of carbohydrates begins in the cytoplasm, through a pathway known as glycolysis (see Chapter 11), whereas energy production from fats occurs exclusively in the mitochondrion. Mitochondria are subcellular organelles, about the size of bacteria. They are essential for aerobic metabolism in eukaryotes. Their main function is to oxidize metabolic fuels and conserve free energy by synthesis of ATP. This is accomplished in several steps:

- **fuels are oxidized** to produce reduced nucleotides; this occurs in the mitochondrial matrix;
- **electron transport:** the electrons from reduced nucleotides are transferred by a sequence of reactions to oxygen, forming water; the free energy available from the oxidation of reduced coenzymes is used to pump protons out of the mitochondrion;
- **the proton gradient is discharged** through an enzyme that uses the free energy of the proton gradient to synthesize ATP from ADP and phosphate,

Mitochondria are surrounded by a dual membrane system (Fig. 8.6). The inner membrane is pleated or invaginated, forming structures known as cristae, and is impermeable to most small molecules and ions, including metabolites, ATP, coenzymes, phosphate, and protons. Transporter proteins facilitate the movement of these molecules across the inner membrane. The inner membrane also contains all the components involved in oxidative phosphorylation – the process by which reduced nucleotides are oxidized and ATP is synthesized.

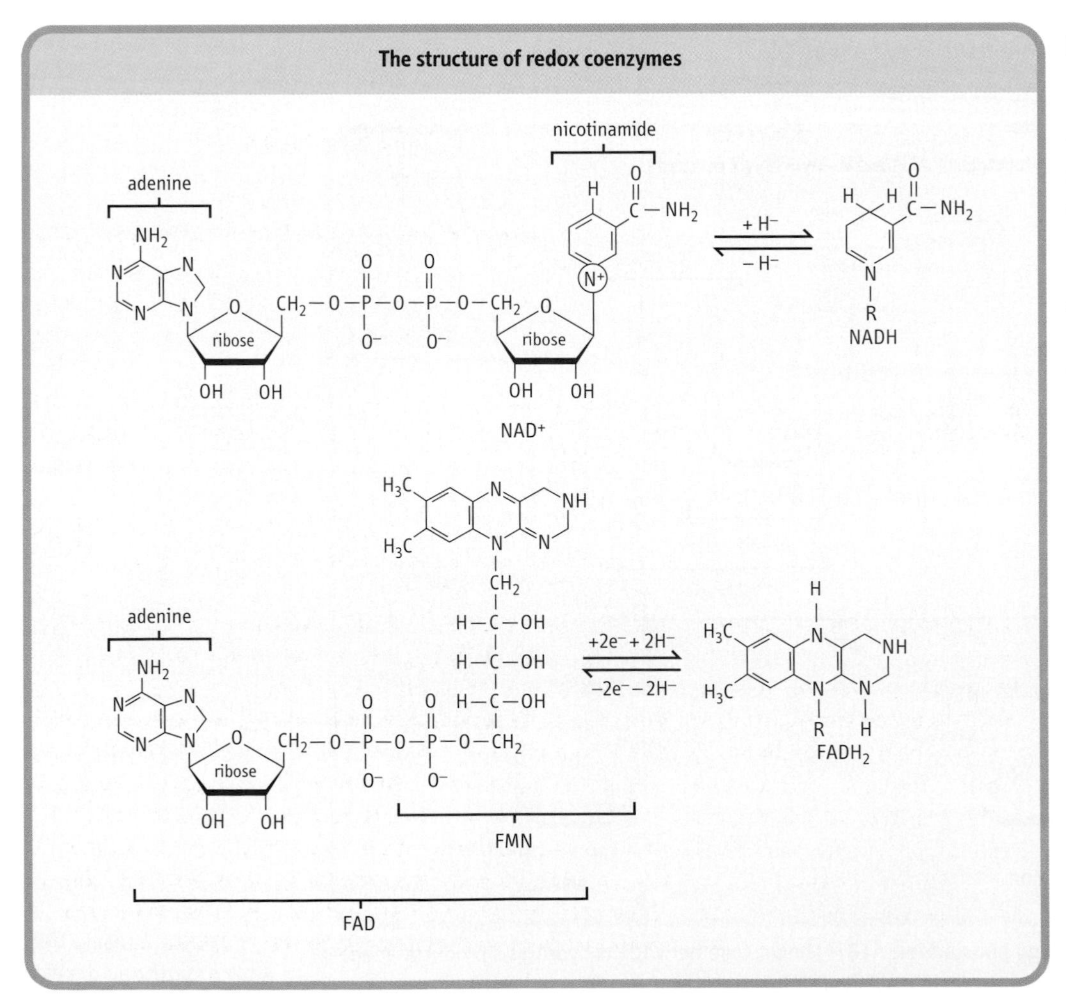

Fig. 8.5 Redox coenzymes. NAD^+ and its reduced form, NADH, consists of adenine, two ribose units, two phosphates, and nicotinamide; FAD and $FADH_2$, and flavin mononucleotide (FMN) and $FMNH_2$ contain riboflavin and a ribitol group. The nicotinamide and riboflavin components of these coenzymes are reversibly oxidized and reduced during electron transfer (redox) reactions.

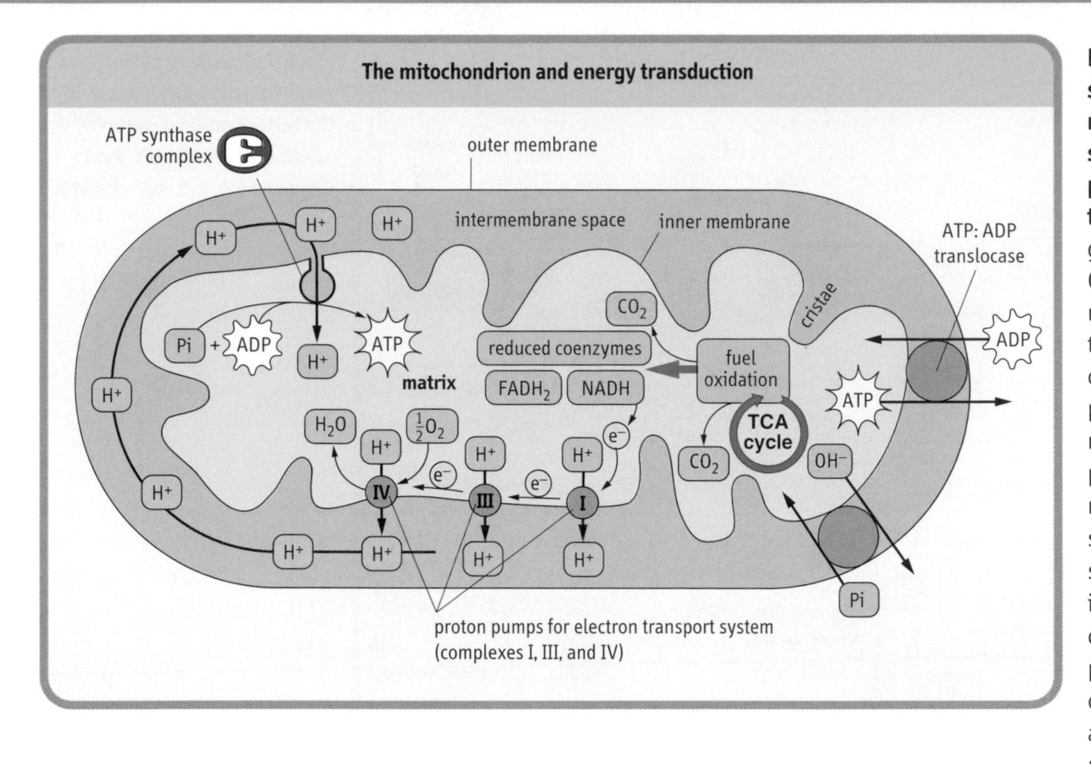

Fig. 8.6 Generalized structure of the mitochondrion and summary of major pathways of energy transduction. Fuel oxidation generates NADH and $FADH_2$. Oxidation of reduced nucleotides via the electron transport system reduces oxygen to water and pumps protons out of the mitochondrion. Influx of protons back into the mitochondrion powers the synthesis of ATP by ATP synthase. Mitochondrial ATP is then exchanged for cytoplasmic ADP and phosphate. e^-, negatively charged electron; $^1/_2O_2$, one atom of oxygen; Pi, inorganic phosphate.

Transduction of energy from reduced coenzymes to high-energy phosphate

The oxidation of reduced nucleotides by the electron transport system produces a large amount of free energy. When the oxidation of one mole of NADH is coupled to the reduction of $\frac{1}{2}$ mole of oxygen to form water, the energy produced is sufficient, theoretically, to synthesize eight moles of ATP:

$$NADH + H^+ + \tfrac{1}{2}O_2 \rightarrow NAD^+ + H_2O$$
$$\Delta G^{\circ\prime} = -220\ kJ/mol\quad(-52.4\ kcal/mol)$$

and thence, as $\Delta G^{\circ\prime}$ for ATP is −30.5kJ/mol (see Fig. 8.3),

$$-220/-30.5 = 7.2$$

The free energy of oxidation of NADH produces, in effect, an electrical current that is used to pump protons to the outside of the inner mitochondrial membrane by three electron transport complexes (see below). These protons can re-enter the mitochondrion only through an ATP synthase complex. In the same way that water pressure behind a dam is used for production of electrical energy, the proton gradient across the inner mitochondrial membrane is a source of free energy for ATP synthesis: the ATP synthase complex can be said to constitute a molecular motor – a turbine powered by a current of protons.

The mitochondrial electron transport system

The entire electron transport system, also known as the electron transport chain or respiratory chain, is located in the inner mitochondrial membrane. It consists of several large protein complexes and two small, independent components, ubiquinone and cytochrome *c*. Electrons are conducted in a defined sequence from reduced coenzymes through this system to oxygen, and the free energy changes drive the transport of protons from the matrix to the intermembrane space (Fig. 8.7).

There are four sites of entry of electrons to the electron transport system: one for NADH (Complex I) and three for $FADH_2$ (Complex II). These pathways meet at the small, lipophilic molecule, ubiquinone (coenzyme Q_{10}), at the beginning of the common electron transport pathway, which consists of Complex III, cytochrome *c*, and Complex IV. Protons are pumped from the matrix into the intermembrane space by only three of the Complexes (I, III, and IV). The final electron acceptor, at the end of the chain, is molecular oxygen, which is reduced to water. For each two moles of electrons transported through Complexes I, III, and IV, a sufficient number of protons is pumped for the synthesis of approximately one mole of ATP. If electron transport begins with an electron pair from NADH, approximately three moles of ATP

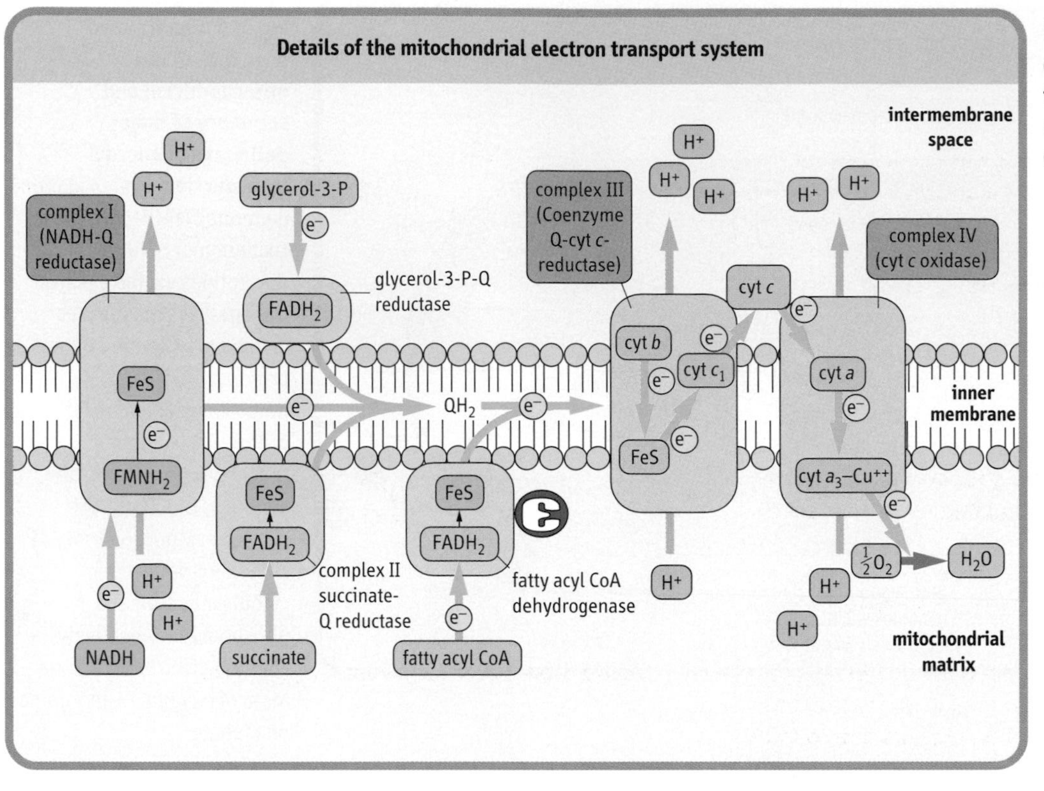

Fig. 8.7 A section of the electron transport system in the mitochondrial inner membrane. Acyl CoA, Acyl Coenzyme A; cyt, cytochrome.

Iron–sulfur complexes

Iron–sulfur complexes participate in redox reactions

Iron is an important constituent of heme proteins, such as hemoglobin, myoglobin, cytochromes, and catalase, but it is also associated with iron–sulfur (FeS) complexes or nonheme iron proteins that function as electron transporters in the mitochondrial electron transport system. The Fe_2S_2 and Fe_4S_2 types are shown in Figure 8.8. In each case, the iron–sulfur center is bound to a peptide through cysteine residues. The FeS complexes undergo reversible distortion and relaxation during redox reactions. The redox energy is said to be stored in the 'conformational energy' of the protein.

are synthesized, whereas an electron pair from $FADH_2$ yields about two moles of ATP, because the proton-pumping capability of Complex I is bypassed by Complex II.

Flavoproteins are components of complexes I and II

Complex I, also called NADH-Q reductase or NADH dehydrogenase, is a flavoprotein containing FMN. It oxidizes mitochondrial NADH, and transfers electrons through FMN and iron–sulfur (FeS) complexes to ubiquinone, providing enough energy via the proton gradient to synthesize about one mole of ATP. Three flavoproteins in Complex II transfer electrons from oxidizable substrates via $FADH_2$ to ubiquinone (Q_{10}) (Fig. 8.7):

- **succinate – Q reductase** oxidizes succinate to fumarate and reduces FAD to F ADH_2
- **glycerol-3-phosphate (glycerol-3-P)** – Q_{10} reductase, a part of the glycerol-3-P shuttle (see below), oxidizes cytoplasmic glycerol-3-P to dihydroxyacetone phosphate (DHAP) and reduces FAD to $FADH_2$
- **fatty acyl Coenzyme A (acyl CoA)** catalyzes the first step in the mitochondrial oxidation of fatty acids and also produces $FADH_2$

Ubiquinone (coenzyme Q_{10}) transfers electrons to complex III

Ubiquinone is virtually ubiquitous in living systems. It is a small, lipid-soluble compound found in the inner membrane of animal and plant mitochondria and in the plasma membrane of bacteria. The primary form of mammalian

Iron deficiency leads to anaemia

A 45-year-old woman complains of tiredness and appears pale. She is a vegetarian and is experiencing a monthly menstrual flow that is heavy and prolonged. Her hematocrit is 0.32 (reference range 0.41–0.46) and her hemoglobin concentration 90 g/L (normal range 120–160 g/L).

Comment. Iron deficiency anemia is a common nutritional problem and is especially common in menstruating and pregnant women because of their increased dietary requirement for iron. Men require about 1 mg iron/day, menstruating women about 2 mg/day, and pregnant women about 3 mg/day. Iron is required to maintain normal amounts of hemoglobin, the cytochromes, and iron–sulfur complexes that are central to oxygen transport and energy metabolism. All these processes are impaired in iron deficiency. Heme iron, which is found in meats, is absorbed much more readily than inorganic iron such as that found in egg yolks, vegetables, and nuts, and with the trend toward eating less meat, it follows that the incidence of iron deficiency is increasing.

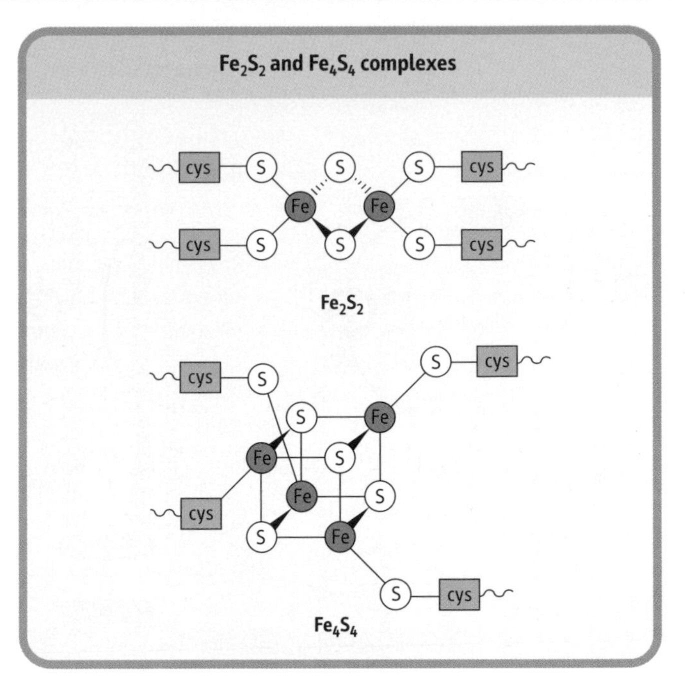

Fig. 8.8 Iron–sulfur complexes. Cys, cysteine.

Transfer of electrons from NADH into mitochondria

Electron shuttles

NADH is produced in the cytosol during carbohydrate metabolism. Because it cannot cross the inner mitochondrial membrane, it cannot donate electrons to the electron directly to the electron transport system. Transferring electrons from NADH into mitochondria, rather than physical transfer of the NADH itself, solves this problem. The glycerol-3-P shuttle is the simpler of the two known redox shuttles (Fig. 8.9). It transfers the electrons, but not the NADH itself, from cytoplasm to mitochondrion by reducing FAD to $FADH_2$. A characteristic feature of these shuttles is that they are powered by cytoplasmic and mitochondrial isoforms of the same enzyme. Cytoplasmic glycerol-3-P dehydrogenase catalyzes reduction of DHAP to glycerol-3-P. The cytoplasm-derived glycerol-3-P is oxidized back to DHAP by another isozyme of glycerol-3-P dehydrogenase in the inner mitochondrial membrane, which reduces FAD to $FADH_2$. The electrons then enter the electron transport system via Complex II, reducing ubiquinone. The yield of ATP from cytoplasmic NADH by this pathway is two moles, rather than the maximum of three moles available from mitochondrial NADH via the NADH-Q reductase complex (Complex I).

Many cells utilize the glycerol 3-P shuttle, but heart and liver utilize the malate–aspartate shuttle, which yields three moles of ATP per mole of NADH. This shuttle is more complicated because the substrate, malate, can cross the inner mitochondrial membrane, but the membrane is impermeable to the product, oxaloacetate–there is no oxaloacetate transporter. The exchange is therefore accomplished by interconversion between keto - and amino acids, involving cytoplasmic and mitochondrial glutamate and α-ketoglutarate, and isozymes of glutamate–oxaloacetate transaminase (aspartate aminotransferase).

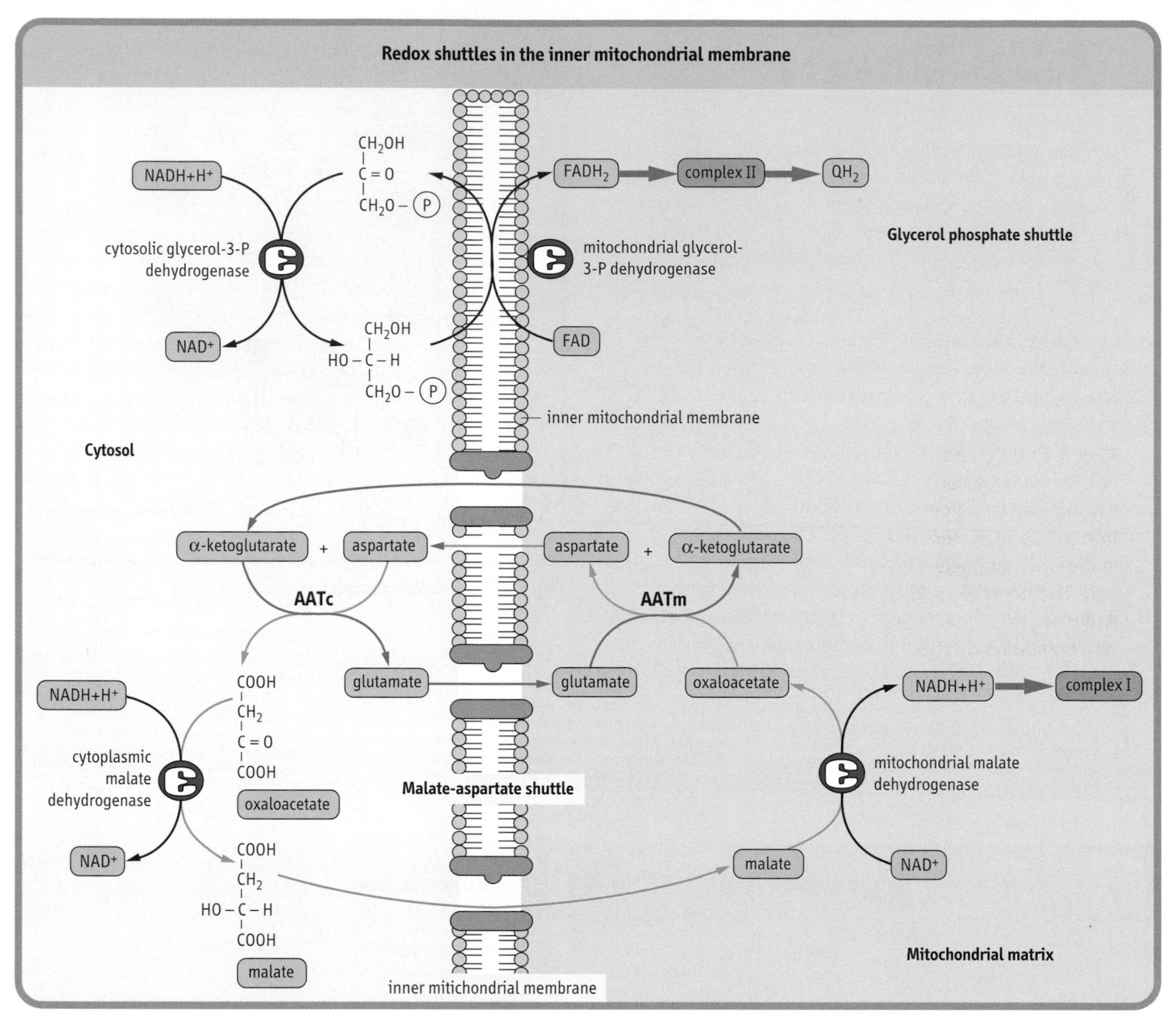

Fig. 8.9 The glycerol phosphate and malate–aspartate shuttles. AAt, asparate amino transferase; m, mitochondrion; c, cytoplasmic.

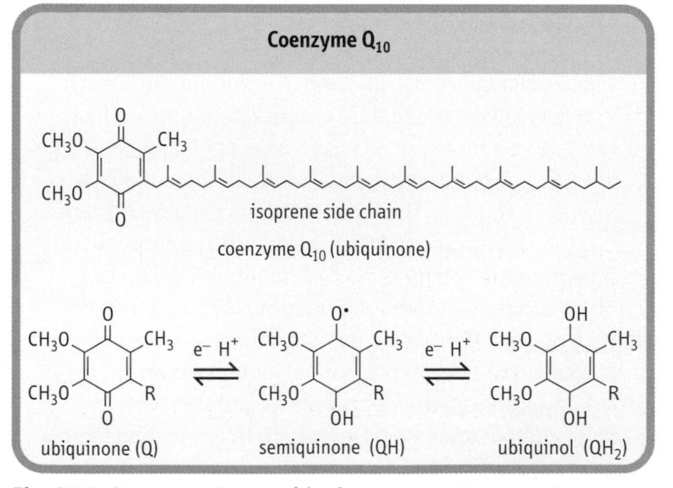

Fig. 8.10 Coenzyme Q_{10}, or ubiquinone, accepts one or two electrons, transferring them from flavoproteins to Complex III.

ubiquinone contains a side chain of 10 isoprene units and is designated CoQ_{10}. It diffuses in the inner membrane, accepts electrons from mitochondrial flavoproteins, and transfers them to Complex III (QH_2–cytochrome *c* reductase) (Fig. 8.10).

Complex III (ubiguinone QH_2) – Cytochrome *c* reductase

Ubiquinone funnels electrons from flavoproteins in Complexes I and II to Complex III, an oligomer of about eight peptides containing cytochrome *b*, an FeS center, and cytochrome c_1. This enzyme complex, also known as ubiquinone–cytochrome *c* reductase, oxidizes ubiquinone and reduces cytochrome *c*. Transport of two electrons to cytochrome *c* yields sufficient free energy change and protons pumped to synthesize one mole of ATP.

A rare coenzyme Q_{10} deficiency

A 4-year-old boy presents with seizures, progressive muscle weakness, and encephalopathy. Accumulation of lactate, a product of anaerobic metabolism of glucose, in the cerebrospinal fluid (CSF) suggested a defect in mitochondrial oxidative metabolism. Muscle mitochondria were isolated for study. The activities of the individual Complexes I, II, III, and IV were normal, but the combined activities of I+III and II+III were significantly decreased. Treatment with coenzyme Q_{10} improved the muscle weakness, but not the encephalopathy.

Comment. Severe muscle weakness, encephalopathy, or both, may be caused in so called mitochondrial myopathies by mitochondrial defects involving the electron transport system. The finding of increased lactate in the CSF suggests a defect in oxidative phosphorylation. The decreased activities of Complexes I+III and II+III suggested a deficiency in coenzyme Q_{10}, which was confirmed by direct measurements.

Cytochromes

Cytochromes, found in the mitochondrion and endoplasmic reticulum, are proteins that contain heme groups, but which are not involved in oxygen transport (Fig. 8.11). The core structure of the heme is a tetrapyrrole ring similar to that of hemoglobin, differing only in the composition of the side chains. The heme group of cytochromes *b* and c_1 is known as iron protoporphyrin IX and is the same heme that is found in hemoglobin, myoglobin, and catalase. Cytochrome *c* contains heme C, covalently bound to the protein through cysteine residues. Cytochromes *a* and a_3 contain heme A, which, in common with ubiquinone, contains an isoprene side chain. In hemoglobin and myoglobin, heme must remain in the ferrous (Fe^{2+}) state; in cytochromes, the heme iron is reversibly reduced and oxidized between the Fe^{2+} and Fe^{3+} states as electrons are shuttled from one protein to another.

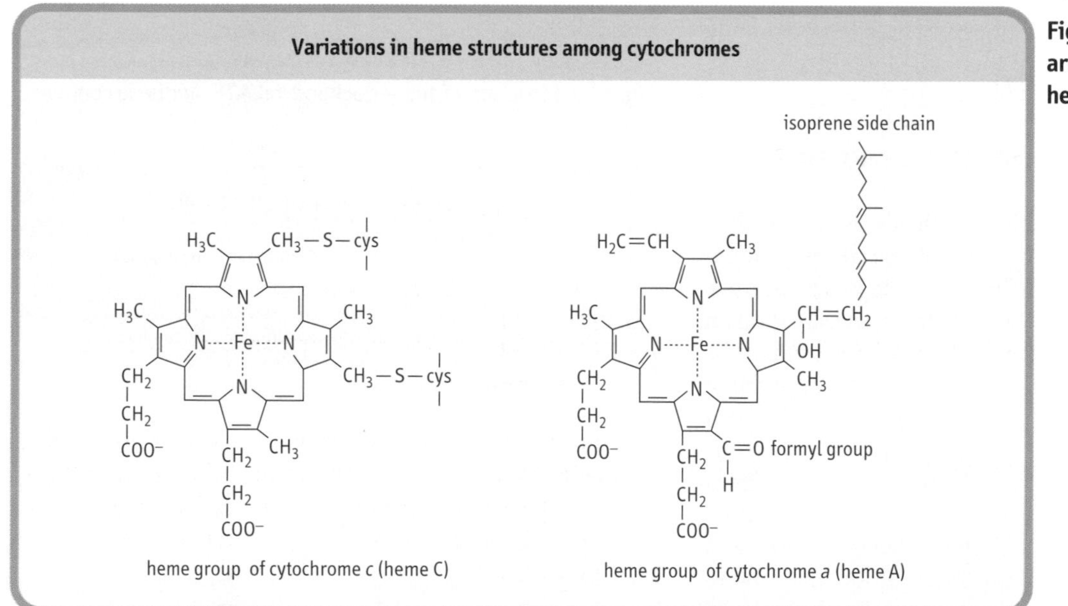

Fig. 8.11 The cytochromes are proteins that contain heme groups.

Cytochrome *c*

Cytochrome *c*, a small heme protein that is loosely bound to the outer surface of the inner membrane, shuttles electrons from Complex III to Complex IV. The binding of cytochrome *c* to Complexes III and IV is largely electrostatic, involving a number of lysine residues on the protein surface. Reduction of ferricytochrome *c* (Fe^{3+}) to ferrocytochrome *c* (Fe^{2+}) by cytochrome c_1 leads to a change in the three-dimensional structure of the protein, promoting transfer of electrons to cytochrome *a* in Complex IV (see Fig. 8.7).

Complex IV

Complex IV known as cytochrome *c* oxidase, or cytochrome oxidase, oxidizes the mobile cytochrome *c*, and conducts electrons through cytochromes *a* and a_3, finally reducing oxygen to water. Copper is a common component of this and other oxidase enzymes. Poisonous

Copper deficiency in neonates

Copper is required in trace amounts for optimal human nutrition. Although copper deficiency is rare in adults, premature infants have low stores of copper and may suffer from its deficiency. This may lead to anemia and cardiomyopathy, because of failure to synthesize adequate amounts of cytochrome c oxidase and other enzymes, including several cuproenzymes involved in the synthesis of heme.

Comment. Copper deficiency can impair ATP production by inhibiting the terminal reaction of the electron transport chain, leading to pathology in the heart, where energy demand is high. Dietary formulae for premature infants must contain adequate copper; cow's milk alone is unsuitable, because it is low in copper.

compounds, such as azide and cyanide (discussed below) react with the copper and inhibit this and other oxidase enzymes. In common with Complexes I and III, the cytochrome oxidase complex pumps protons out of mitochondria, providing for the synthesis of an additional mole of ATP.

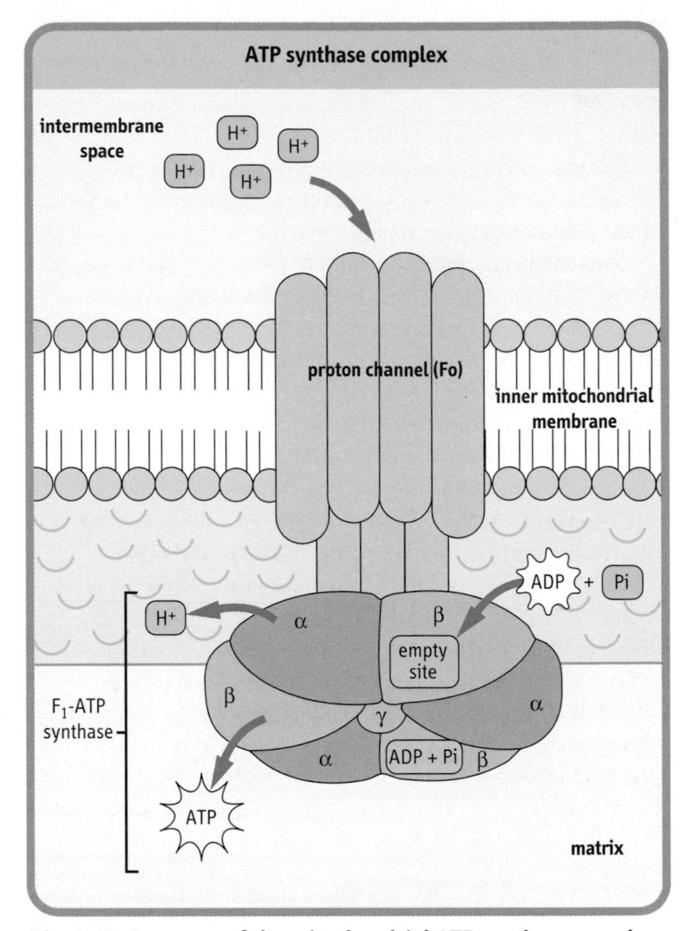

Fig. 8.12 Structure of the mitochondrial ATP synthase complex.

The ATP-synthetic proton gradient

According to the **chemiosmotic hypothesis**, mitochondrial electron transport is coupled to ATP synthesis through a proton gradient. Protons are pumped out of the matrix into the intermembrane space during electron transport, creating an electrochemical potential across the inner membrane. The outside of the mitochondrion becomes more acidic and more positively charged than the matrix. The synthesis of ATP, a nonspontaneous reaction, is driven by the flux of protons back into the matrix along the electrochemical gradient through ATP synthase. Thus the mitochondrion is an ATP battery in which the energy for ATP synthesis is stored as a proton gradient. To operate, it requires a closed inner membrane system, impermeable to protons, except through the ATP synthase complex.

The ATP synthase complex is an example of rotary catalysis

The ATP synthase complex, also called F_0F_1-ATP synthase, and F_0F_1–ATPase consists of two major complexes; (F = coupling factor; see Inhibitors of ATP synthase). The inner membrane component, termed F_0 because of its sensitivity to oligomycin (see below), contains the proton channel and a stalk piece through which protons can flow back into the

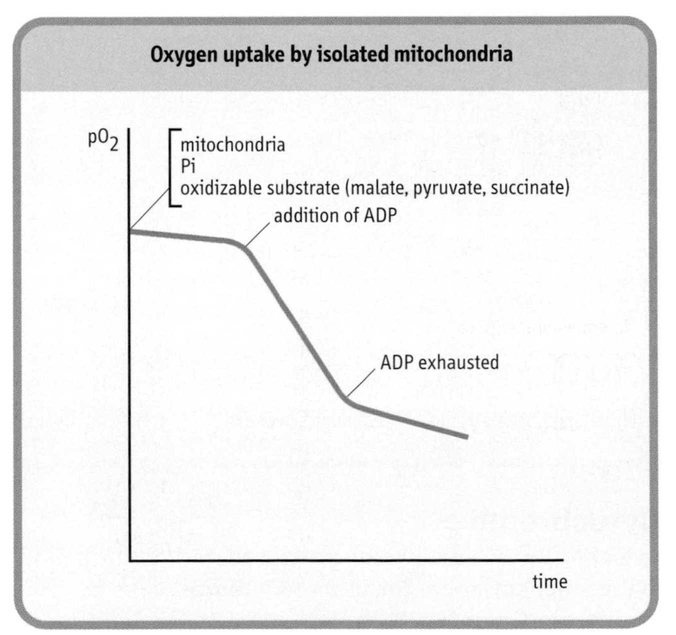

Fig. 8.13 Effect of ADP on the uptake of oxygen by isolated mitochondria. This may be studied in an isolated (sealed) system with an oxygen electrode and a recording device. The graph shows a typical recording of oxygen consumption (pO_2, partial pressure of oxygen) by normal mitochondria on introduction of ADP.

mitochondrion. The second, F_1-ATP synthase, complex is bound to F_0 through the stalk, projecting into the matrix (Fig. 8.12). It consists of a central γ-subunit surrounded by alternating α- and β-subunits with a stoichiometry of α_3/β_3. There are three nucleotide-binding sites, located mostly on the β-subunits. The central γ-subunit physically rotates in response to the proton flux.

Rotation of the γ-subunit induces conformational changes in the α_3/β_3-subunits such that the nucleotide-binding sites alternate between three states: one state is empty, the second binds ADP and inorganic phosphate (Pi), and the third releases ATP. The subunits are asymmetrical, because each is in a different conformation at any given moment. By an unknown mechanism, the protons induce rotation of the γ-subunit, which induces release of ATP, setting the stage for repetition of the cycle. This complex is actually a proton-driven motor, and it is an example of rotary catalysis.

P:O ratios and respiratory control

The P:O ratio is a measure of the number of high-energy phosphates (i.e. amount of ATP) synthesized per atom of oxygen ($\frac{1}{2}O_2$) consumed, or per mole of water produced. As shown earlier, the theoretical yield of ATP from one mole of NADH is about 8 moles; however, by actual measurement with isolated mitochondria, the P:O ratio for oxidation of metabolites that yield NADH is about 3 and the ratio for those that yield $FADH_2$ is about 2. The P:O ratio can be calculated from the amount of ADP used to synthesize ATP and the amount of oxygen taken up by mitochondria. For example, if 2 mmol of ADP is added, and 0.5 mmol of oxygen (1.0 atom of oxygen) is taken up, the P:O ratio is 2.0.

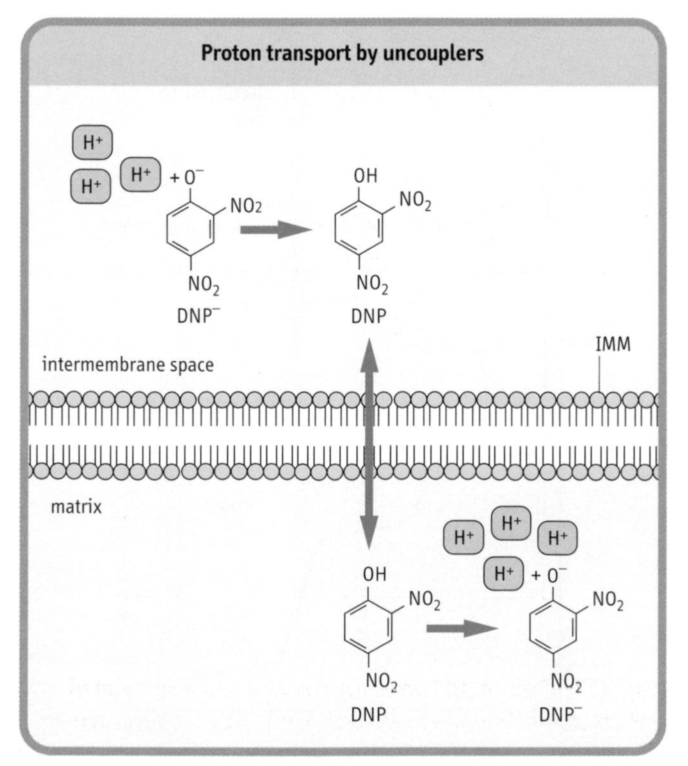

Fig. 8.14 Uncouplers transport protons into the mitochondrion, dissipating the proton gradient. DNP, 2,4-dinitrophenol.

'Respiratory control' is the mechanism by which oxidation and phosphorylation are coupled and controled by mitochondrial concentrations of ADP

Normally, the processes of oxidation and phosphorylation are tightly coupled: substrates are oxidized, electrons are transported, and oxygen is consumed only when synthesis of ATP is required. Thus resting mitochondria consume oxygen at a slow rate, which can be greatly stimulated by addition of ADP (Fig. 8.13). Oxygen uptake declines again when the concentration of ADP is depleted and ATP synthesis terminates. Mitochondria can become partially uncoupled if the inner membrane loses its structural integrity; they are said to be 'leaky', because protons can diffuse through the inner membrane by pathways not involving the ATP synthase. This occurs if isolated mitochondria are treated with mild detergents that introduce holes in the inner membrane, or even if they have been stored for a period of time. Such mitochondria become uncoupled and lose respiratory control; their P:O ratio declines.

The mechanism of respiratory control probably depends on the requirement for ADP and Pi binding to the ATP synthase complex: in the absence of ADP and Pi, protons cannot enter the mitochondrion through this complex. The electrochemical gradient then reaches a high level, which feeds back and shuts down the electron transport chain and oxygen consumption. This happens because the free energy of the electron transport reactions is sufficient to generate a pH gradient of only 2 units across the membrane. When the pH gradient exceeds this amount, the pumps cannot provide sufficient energy to pump additional protons, and they shut down.

Uncouplers

Uncouplers of oxidative phosphorylation dissipate the proton gradient by transporting protons back into mitochondria, bypassing the ATP synthase. This stimulates respiration, because the system makes a futile attempt to restore the proton gradient by oxidizing more fuel and pumping more protons out of mitochondria. A typical uncoupler, such as 2,4-dinitrophenol (DNP) (Fig. 8.14), is a weak acid or base, and is hydrophobic. Its ionization constant, near pH 7, is ideal to enable it to accept a proton on the outer, more acidic side of the inner mitochondrial membrane. Because of its hydrophobicity, it may then freely diffuse through the inner mitochondrial membrane. On the matrix side, when it encounters a higher, less acidic pH, the proton is released,

Thermogenin: a natural uncoupler

Brown adipose tissue is abundant in newborn and hibernating animals. The brown color is due to the high concentration of mitochondria, rich in cytochromes. Mitochondria of brown adipose tissue are uncoupled by a specialized inner membrane protein called uncoupling protein (UCP) or thermogenin, which transports protons back into the matrix, bypassing ATP synthase. Because the oxidative energy is released as heat, rather than as ATP, thermogenin is believed to be involved in regulation of body temperature during hibernation and protection of vital internal organs during variations in body temperature in the newborn. Clearly, uncouplers could also be used to reduce body weight without exercise: fat stores would be rapidly oxidized, and the free energy would be dissipated as heat. In fact, DNP has been used as a drug in patients seeking to lose weight. Unfortunately, hyperthermia and toxic side effects were excessive, and its use has been banned by the Food and Drug Administration in the USA.

effectively discharging the pH gradient. Other uncouplers include preservatives and antimicrobial agents, such as pentachlorophenol and *p*-cresol.

Inhibitors of oxidative metabolism

Electron transport system inhibitors

Inhibitors of electron transport selectively inhibit Complexes I, III, or IV, interrupting the flow of electrons through the respiratory chain. This results in a block in proton pumping, ATP synthesis, and oxygen uptake. Several inhibitors are poisons that could be encountered in the practice of medicine.

Rotenone inhibits Complex I (NADH-Q reductase)

Rotenone is an insecticide that inhibits Complex I of the electron transport system. Because malate and lactate are oxidized by NAD, their oxidation will be inhibited by rotenone. However, substrates yielding $FADH_2$ can still be oxidized, because Complex I is bypassed and electrons can be donated to ubiquinone through Complex II. Addition of ADP to a suspension of mitochondria supplemented with malate and phosphate (Fig. 8.15) markedly stimulates oxygen uptake as ATP synthesis occurs. Oxygen uptake is markedly inhibited by rotenone, but, when succinate is added, ATP synthesis and oxygen consumption resume until the supply of ADP is exhausted. Rotenone inhibition of Complex I causes reduction of all components prior to the point of inhibition, because they cannot be oxidized, whereas those after the point of inhibition become fully oxidized. Analysis of this crossover point indicates that Complex I precedes Complex III and Complex IV in the electron transport chain. Similar analyses were important for determining the complete sequence of the electron transport complexes.

Antimycin A inhibits Complex III (QH_2-cytochrome c reductase)

The inhibition of Complex III by antimycin A prevents transfer of electrons from either Complex I or Complex II to cytochrome *c*. In this case, components preceding Complex III in the process become reduced, and those after it become

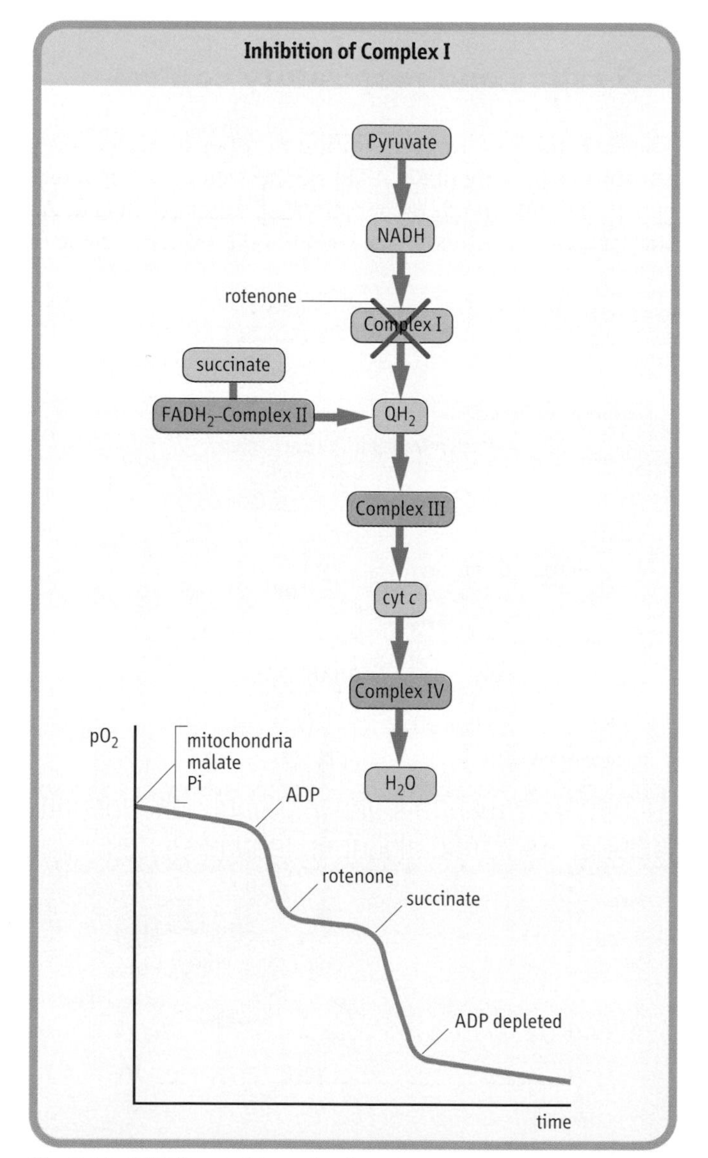

Fig. 8.15 Inhibitors of Complex I, such as rotenone, retard oxygen uptake by mitochondria when NADH-producing substrates are being oxidized.

oxidized. The oxygen uptake curve (Fig. 8.16) shows that the stimulation of respiration by ADP is inhibited by antimycin A, but that the addition of succinate does not relieve the inhibition. Ascorbic acid can reduce cytochrome *c*, and addition of ascorbic acid restores respiration, illustrating that Complex IV is unaffected by antimycin A.

Cyanide and carbon monoxide inhibit Complex IV

Both cyanide and carbon monoxide bind to and inhibit Complex IV (cytochrome *c* oxidase) (Fig. 8.17). Because Complex IV is the terminal electron transfer complex, its inhibition cannot be bypassed. All components preceding Complex IV become reduced. As it is not possible to pump protons, uncouplers such as DNP have no effect.

Inhibitors of ATP synthase

The proton channel of ATP synthase is F_0 protein, so named because it is sensitive to oligomycin that blocks the transport of protons back into mitochondria. Oligomycin inhibits respiration, but, in contrast to electron transport inhibitors, it is not a direct inhibitor of the electron transport system. It causes a build-up of protons outside the mitochondrion, because the proton pumping system is still intact, but the proton channel is blocked. The addition of DNP after oxygen uptake has been inhibited by oligomycin illustrates this point: DNP dissipates the proton gradient and stimulates oxygen uptake as the electron transport system attempts to re-establish the proton gradient (Fig. 8.18).

Inhibitors of the ATP:ADP translocase

Most ATP is synthesized in the mitochondrion, but used in the cytosol for biosynthetic reactions. Newly synthesized mitochondrial ATP and spent cytosolic ADP are exchanged by a mitochondrial ATP:ADP translocase, representing about 10% of the protein in the inner mitochondrial membrane. This translocase can be inhibited by unusual plant and mold

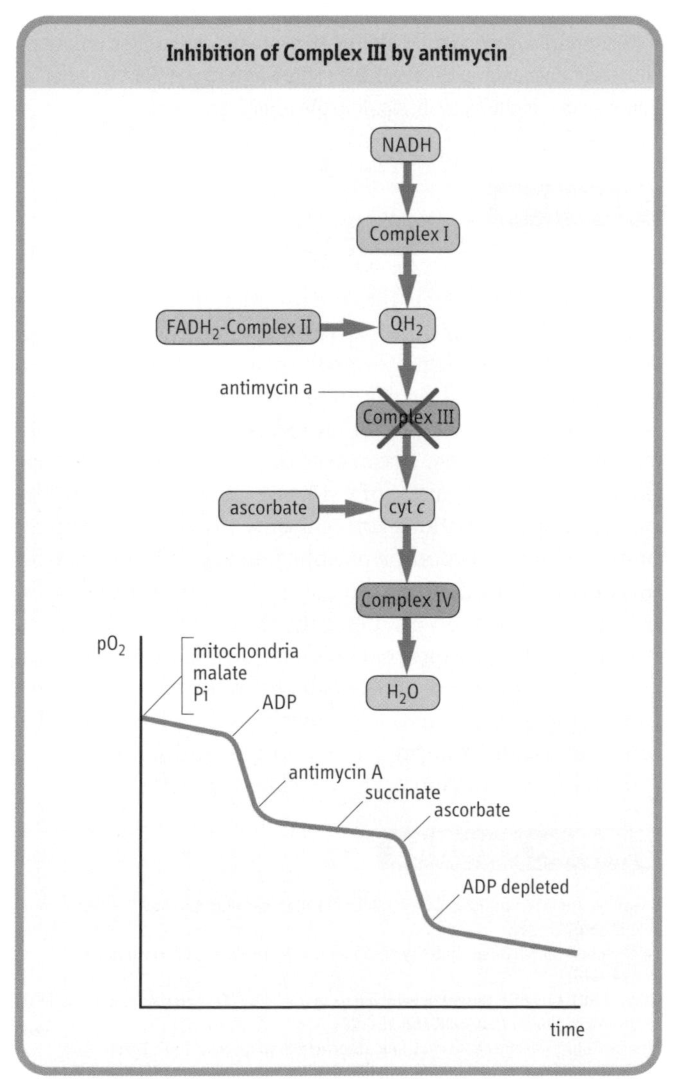

Fig. 8.16 Antimycin A inhibits Complex III, blocking transfer of electrons from both Complex I and Complex II.

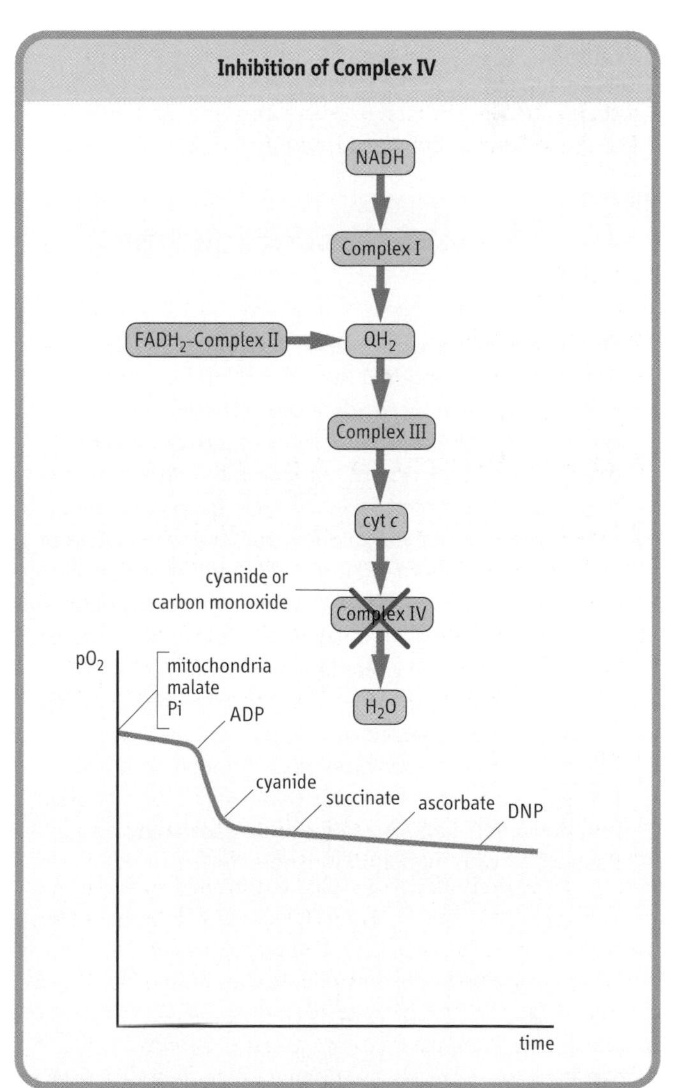

Fig. 8.17 The inhibition of Complex IV interrupts the transfer of electrons in the final step of electron transport. Electrons cannot be transferred to oxygen, and the synthesis of ATP is halted.

Cyanide and carbon monoxide are mitochondrial poisons

Both cyanide and carbon monoxide bind to hemoglobin and inhibit oxygen transport. They also inhibit electron transport and production of ATP.

Comment. Cells respond to cyanide or carbon monoxide poisoning by switching to anaerobic metabolism, resulting in lacticacidosis and ultimate death, unless immediate measures are taken. Carbon monoxide poisoning may be treated with the administration of oxygen. In both cyanide and carbon monoxide poisoning, methylene blue can be administered: it alleviates the inhibition of Complex IV by accepting electrons from Complex III (cytochrome *c* reductase), allowing both Complex I and Complex III to pump protons, so that ATP can continue to be synthesized. Cyanide can also be converted to the relatively harmless thiocyanate ion by the administration of thiosulfate.

The mitochondrial genome

Disorders of mitochondrial function
Mitochondria contain proteins specified both by nuclear and by mitochondrial DNA. The mitochondrial genome (mtDNA) is maternally inherited. Defects of Complexes I, II, III, and IV, cytochrome *c*, ATP synthase, and biosynthesis of ubiquinone have been described. Most mitochondrial mutations result in accumulation of lactic acid, a product of anaerobic metabolism of glucose, and impaired ATP production, which may result in cell death, especially in skeletal and cardiac muscles and nerve tissue, which are heavily dependent on oxidative metabolism. Thus mitochondrial diseases are frequently characterized by myopathies, cardiomyopathies, and encephalopathies. Although controversial, some evidence indicates that Huntington's chorea and Parkinson's and Alzheimer's diseases may also be associated with mitochondrial defects.

toxins, such as bongkrekic acid and atractyloside. Their effects are similar to those of oligomycin *in vitro* – a proton gradient builds up and electron transport stops, but respiration can be reactivated by uncouplers.

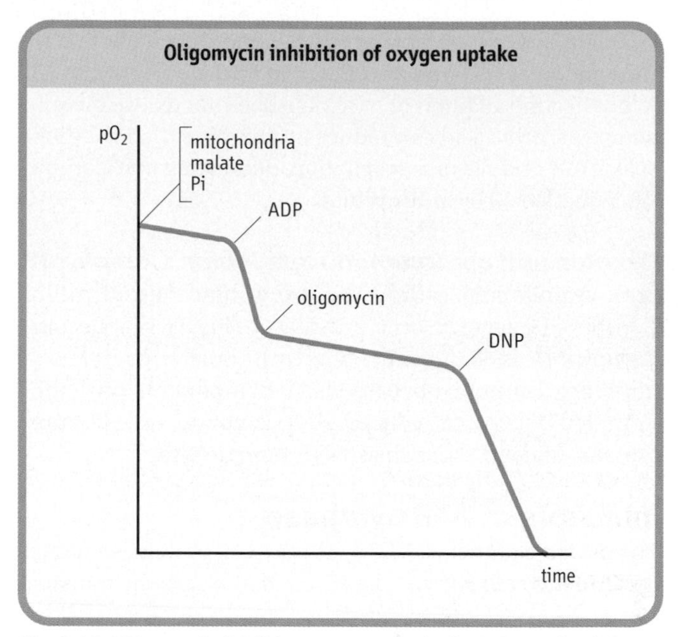

Fig. 8.18 Oligomycin inhibits oxygen uptake in ATP-synthesizing mitochondria. Oligomycin inhibits ATP synthase and oxygen uptake in coupled mitochondria. However DNP stimulates oxygen uptake after oligomycin inhibition, by dissipating the proton gradient.

Summary

The mitochondrion is a complex factory for transducing redox energy from fuels into the production of ATP. During a normal day, we use and resynthesize our body weight in ATP! Most of the enzymes of energy metabolism are located in the mitochondrial matrix and provide reduced coenzymes for oxidative phosphorylation by enzymes of the inner mitochondrial membrane. Because of the role of the proton gradient in the mechanism of oxidative phosphorylation, the mitochondrion functions correctly only when its inner membrane is intact. Numerous poisons and toxins interfere with the normal function of mitochondria, acting on electron transport, the proton gradient, the ATP synthase complex, and ATP:ADP translocase. Understanding the sites of action and effects of these inhibitors has been important in determining the mechanism of oxidative phosphorylation and also offers an excellent approach for understanding the overall operation and function of the mitochondrion.

Further reading

Boyer PD. The ATP synthase – a splendid molecular machine. *Ann Rev Biochem* 1997;**66:**717–749.
Fairweather-Tait SJ. Bioavailability of copper. *Eur J Clin Nutr* 1997;**51**(suppl 1): S24–S26.
Graeber MB, Muller U. Recent developments in the molecular genetics of mitochondrial disorders. *J Neurol Sci* 1998;**153:**251–263.
Khan S. Rotary chemiosmotic machines. *Biochem Biophys Acta* 1997;**322:**86–105.
Marin-Garcia J, Ananthakrishnan R, Goldenthal MJ. Hypertrophic cardiomyopathy with mitochondrial DNA depletion and respiratory enzyme defects. *Pediatr Cardiol* 1998;**19:**266–268.
Noji H, Yasuda R. Direct observation of the rotation of F1-ATPase. *Nature* 1997;**386:**299–302.

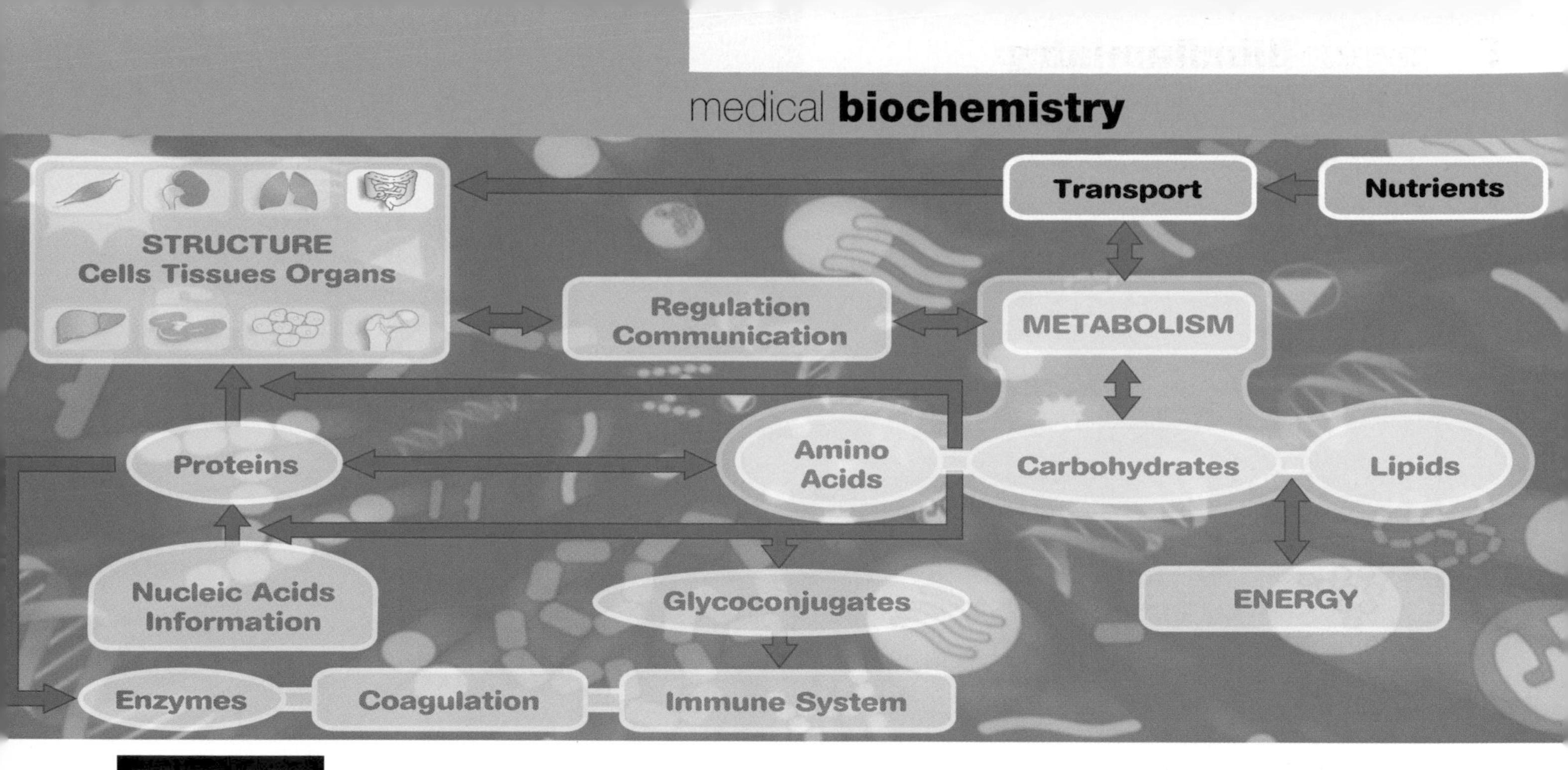

9 Function of the GI Tract in Digestion and Absorption of Fuels

Introduction

The survival of an organism depends on its ability to transduce energy derived from food and stored as chemical energy, to electrical and mechanical energy

Our main food is taken into the gastrointestinal (GI) tract and this organ, or combination of organs, is responsible for the breaking down of complex macromolecules to simple monomeric or dimeric units – the process of digestion. These smaller, simpler molecules are then transferred across the GI tract for further delivery to the remainder of the body's tissues – the process of absorption. The macromolecules encountered as primary fuel sources are of three types:

- carbohydrates,
- lipids,
- proteins.

Carbohydrates and lipids have primarily a fuel function as part of food intake but also have a nonfuel function in the body. Protein, on the other hand, is primarily for nonfuel purposes but can, under certain circumstances, serve as a fuel source. The composition of foods varies widely in the proportions of protein, carbohydrate and fats. In addition, some materials ingested, particularly of plant origin, are indigestible and constitute what is termed 'fiber'.

In addition to digestion, the GI tract also acts as a transport organ and has a metabolic function where food is altered before being transported

The GI tract is effectively a large coiled tube whose interior is extracorporeal (i.e. outside the body). Its function therefore is to transfer fuel and other materials from outside to inside the body (Fig. 9.1). The gut can be organized into various anatomical areas, each with a specific function related to digestion and absorption: the stomach and duodenum dealing with the initial process of mixing and digestion; the jejunum continuing with the digestive processes, and beginning the process of absorption; the ileum involved in the absorption of digested foodstuffs with the large bowel being involved in the absorption of fluid and electrolytes (Fig. 9.2). Along the lengths of the gut various fluids, electrolytes and proteins are added to aid in the mixing, hydration and digestion of the food. There is thus the capacity to sequester fluid, electrolytes and proteins in the gut with effective loss from the body.

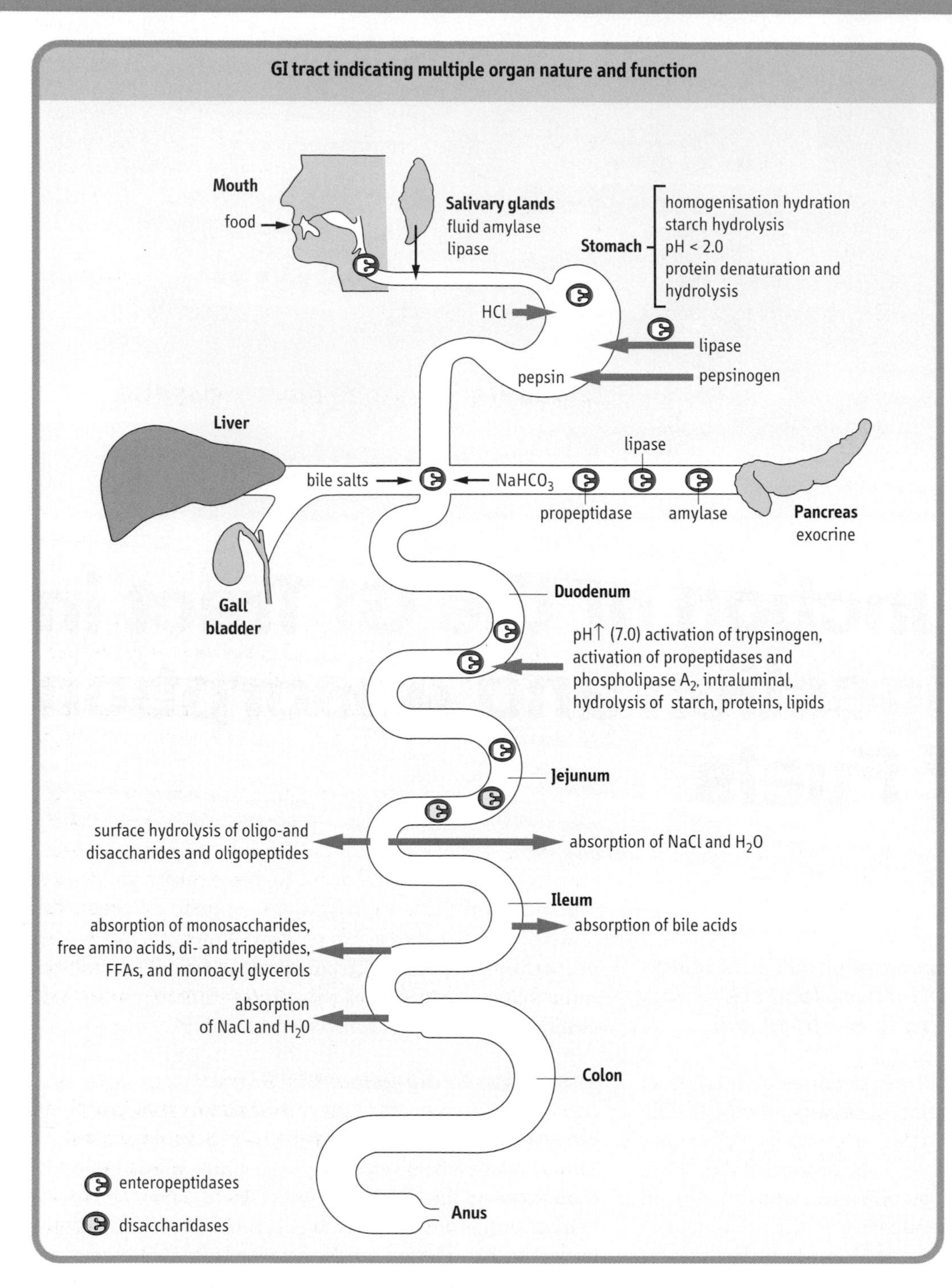

Fig. 9.1 The GI tract includes several organs and functions. FFAs, free fatty acids.

The gut also serves a major metabolic function and can alter the food intake rather than simply passing all digested food to the other organs. It will, for example, treat the simple monosaccharide, glucose, differently when it receives it from the lumen or via the mesenteric blood supply. Glucose derived from the lumen is transferred directly to the liver unaltered, whereas glucose received via the blood supply is metabolized to lactate prior to passage to the liver. In addition, the amino acid glutamine, which is derived from the protein constituent of food, is used by the enterocytes as a major fuel source and does not enter the portal blood supply (see below).

General principles of digestion

Digestion is a sequential, ordered process with physiologic links between each stage

The process of digestion is characterized by several specific stages in characteristic sequence, allowing the interaction of fluid and emulsifying agents, acid–base changes, enzymes and specific membrane substrate contacts. This, in turn, requires physiologic links and signals to the liver, pancreas, gall bladder and salivary glands. The processes that are involved are outlined in Figure 9.1 and can be summarized as:

- lubrication and homogenization of food with fluids secreted by glands of the GI tract, starting in the mouth;
- secretion of enzymes whose prime function is to break down macromolecules to a mixture of oligomers, dimers and monomers by the process of hydrolysis;
- within different regions of the GI tract, secretion of electrolytes, acids or bases occurs to optimize the environmental conditions for enzymic hydrolysis specific to that region of the GI tract;
- secretion of bile acids to aid in the solubilization of dietary lipid (as an emulsion), thus allowing appropriate enzymic hydrolysis and absorption;
- within the jejunum, further hydrolysis of oligomers and dimers is continued by membrane-bound surface enzymes;
- specific transport of digested material into the enterocyte and thence to blood or lymph.

Signs and symptoms of GI maldigestion/malabsorption require considerable pathology in terms of structural/functional relationships

In general, there is considerable reserve function in all aspects of digestion and absorption. Minor degrees of functional loss may go unnoticed by the individual, allowing pathology to progress for some time before being noted by the individual and diagnosed by the physician. Numerous sections of the gut are involved in the process and each area contains specialized glands and unique surface epithelial properties, as outlined in Figure 9.2.

Gastrointestinal (GI) tract organization by functional requirements

Gastrointestinal organ	Primary function in absorption of foodstuffs
salivary glands	production of fluid and digestive enzymes for homogenization, lubrication and digestion of carbohydrate (amylase) and lipid (lingual lipases)
stomach	secretion of HCl and proteases to initiate hydrolysis of proteins.
pancreas	secretion of HCO_3^-, proteases, lipases and starches to continue digestion of protein/lipids
liver/gall bladder	secretion and storage of bile acids for release to small bowel
small bowel	final intraluminal digestion of foodstuffs, membrane digestion of carbohydrate dimers and specific absorptive pathways for digested material
large bowel	absorption of fluid and electrolytes and products of bacterial action in the colon

Fig. 9.2 Organization of the GI tract in relation to functional requirements.

Each of the organs involved in the process of digestion/absorption has the capacity to increase its activity several fold in response to specific stimulation; this adds considerably to the gut's reserve capacity. For pancreatic disease to become manifest, 90% of the function has to be destroyed. In addition, areas of the gut can accommodate loss of function of one particular organ. For example, both the pancreas and small intestine can take over total loss of gastric digestion, and lingual lipases can accommodate, in part, some loss of pancreatic lipase production.

Digestive enzymes and zymogens

Most digestive enzymes in the gut are secreted as inactive precursors

With the exception of amylase (in saliva) and lingual (tongue-associated) lipases, digestive enzymes secreted into the gut lumen are present as inactive precursors termed zymogens. The process of secretion of all gut enzymes is similar in all organs involved (i.e. salivary glands, gastric mucosa and pancreas). These organs contain specialized cells for the synthesis, packaging and transport of enzymes to the cell surface, and thence to the intestinal lumen. Since the lumen is extracorporeal, these secretions are therefore termed exocrine that is, 'secreting', to the outside.

Enzymes involved in protein digestion (proteases) and the lipase, phospholipase A_2, are synthesized as inactive zymogens and are only activated on release to the gut lumen. In general, these enzymes once released and in their active form, can activate their own precursors. Activation of their precursors can occur by either change in pH (e.g. pepsinogen in the stomach is converted to pepsin, the active enzyme, at pH <4.0) or by the action of specific enteropeptidases bound to the duodenum mucosal membrane (see Fig. 9.1).

All digestive enzymes are hydrolases

The action of all digestive enzymes is a process of hydrolysis. The products of such hydrolytic procedures are oligomers, dimers and monomers of the parent macromolecule. Carbohydrates thus form mixtures of disaccharides and monosaccharides. Proteins are broken down to a mixture of amino acids, di- and tripeptides. Lipids, on the other hand, are treated somewhat differently – being broken down to a mixture of fatty acids (FA), glycerol, and mono- and diacyl glycerols (Fig. 9.3).

Carbohydrate digestion and absorption

Dietary carbohydrates are present as mono-, di- and polysaccharides and provide the major source of our daily energy requirements

Dietary carbohydrate consists of mainly plant and animal starches (polysaccharides), the disaccharides sucrose and

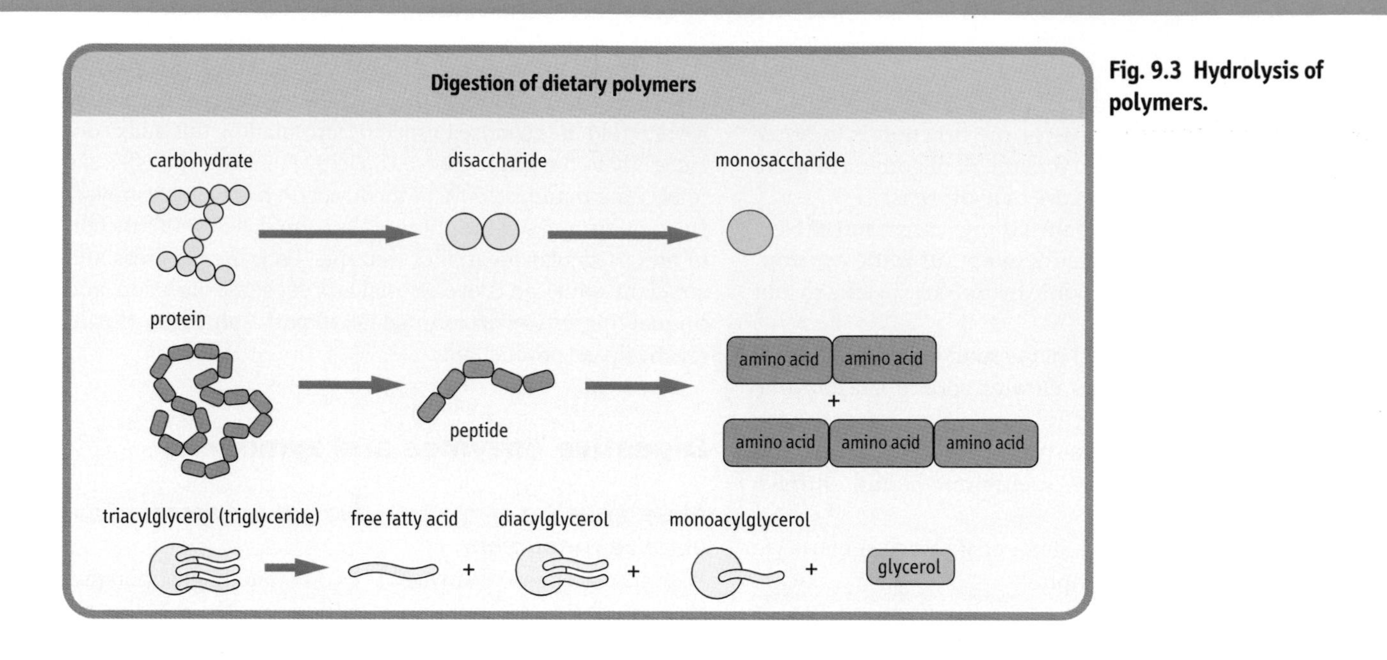

Fig. 9.3 Hydrolysis of polymers.

lactose, and a series of monosaccharides, in particular glucose and fructose (Fig. 9.4).

The latest recommendations in terms of healthy eating suggest carbohydrate should provide more than 50% of our daily energy requirements. The nature of the carbohydrates providing this should also be in the form of complex carbohydrates and not as simple sugars (i.e. glucose, fructose or disaccharides such as sucrose). Monosaccharides arising as constituents of the diet in themselves or by a process of digestion of di- and polysaccharides contained in the diet are glucose, fructose and galactose – the latter derived mainly from dairy products. These simple sugar monomers require no further digestion to be absorbed across the GI tract.

Both disaccharides and polysaccharides (starch and glycogen) require hydrolytic cleavage prior to absorption

Disaccharides are acted upon by membrane-bound disaccharidases on the intestinal mucosal surface. Starch and glycogen require the additional hydrolytic capacity of the enzyme amylase found in the secretions of the salivary glands and pancreas (Fig. 9.5).

Starch is a plant polysaccharide, whereas glycogen is the animal equivalent. Both contain a mixture of linear chains of glucose molecules linked by $\alpha1\rightarrow4$ glycosidic bonds (amylose) and by branched glucose chains with $\alpha1\rightarrow6$ linkages (amylopectin) but glycogen contains more branches than starch. The digestion of these polysaccharides is promoted by endosaccharidases and amylase produced by the salivary glands and pancreas. This process is carried out within the gut lumen where amylase is found unbound to the enterocyte mucosal membrane.

Polysaccharide digestion

During the eating process and the homogenization that occurs with mastication in the mouth and the action of gastric folds, dietary polysaccharides become hydrated. Hydration of polysaccharides is essential for the appropriate action of amylase. This enzyme is specific for internal $\alpha1\rightarrow4$-glycosidic linkages and is totally inert towards $\alpha1\rightarrow6$ linkages. In addition amylase does not act on $\alpha1\rightarrow4$ linkages of glycosyl residues serving as branching units (see Fig. 9.5).

The cleaved units thus formed are the trisaccharide maltotriose, the disaccharide maltose and an oligosaccharide with one or more $\alpha1\rightarrow6$ branches and containing on average eight glycosyl units termed the 'α-limit dextrin'. These compounds are then further cleaved to glucose units by oligosaccharidase and α-glucosidase, the latter removing single glucose residues from $\alpha1\rightarrow4$-linked oligosaccharides (including maltose) from the nonreducing end of the oligomer. A sucrase–isomaltase complex, secreted as a single polypeptide precursor molecule and activated to two separate active polypeptide enzymes one of which (isomaltase) is responsible for the hydrolytic cleavage of $\alpha1\rightarrow6$ glycosidic linkages. The final product of digestion of starches is thus glucose but through a complex series of enzyme reactions. The initial digestion involves amylase, which occurs free in the lumen, whereas the final processes involve α-glucosidases and isomaltase, which are attached to the enterocyte mucosal membrane.

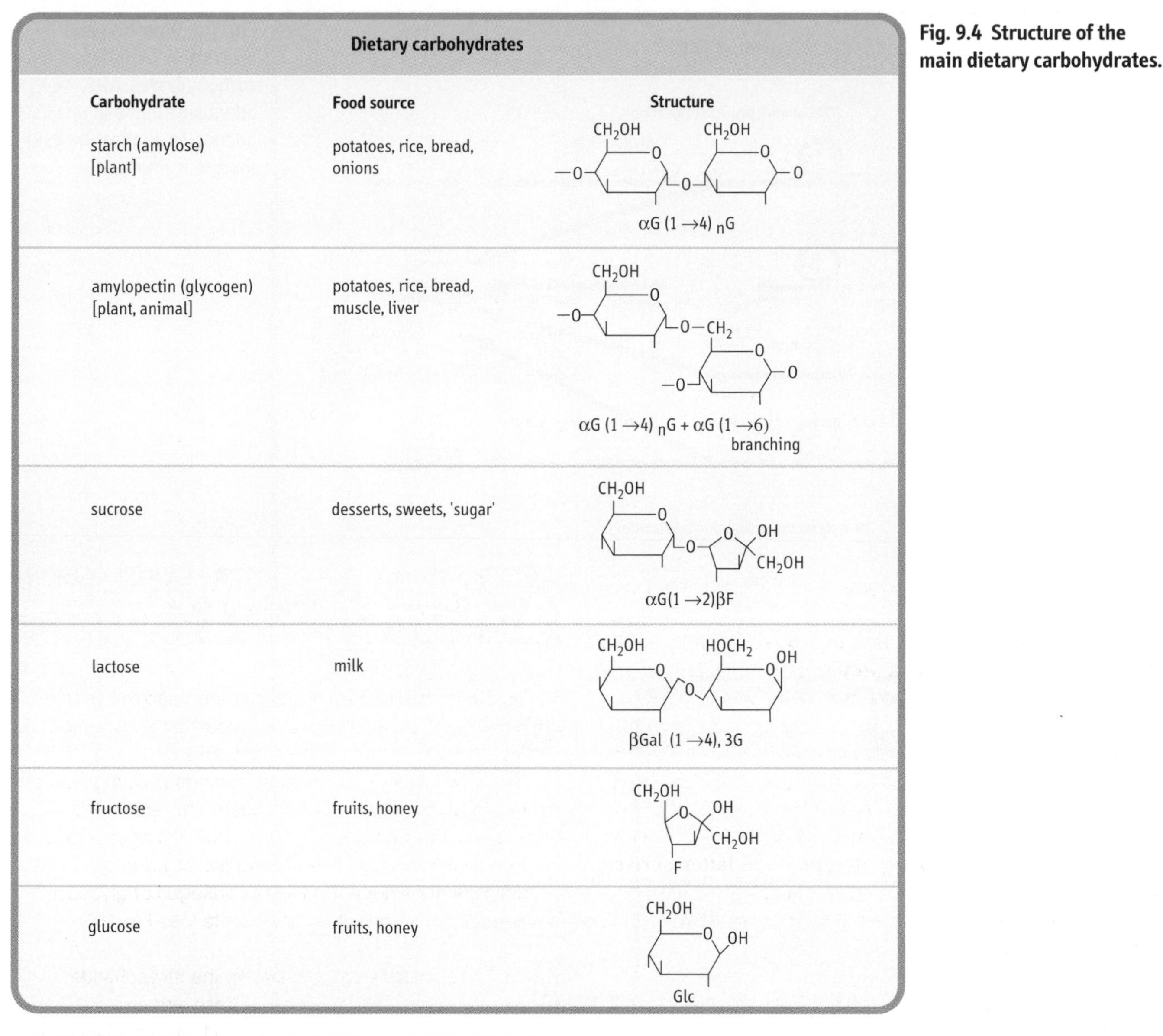

Dietary carbohydrates

Carbohydrate	Food source	Structure
starch (amylose) [plant]	potatoes, rice, bread, onions	αG (1 →4) $_n$G
amylopectin (glycogen) [plant, animal]	potatoes, rice, bread, muscle, liver	αG (1 →4) $_n$G + αG (1 →6) branching
sucrose	desserts, sweets, 'sugar'	αG(1 →2)βF
lactose	milk	βGal (1 →4), 3G
fructose	fruits, honey	F
glucose	fruits, honey	Glc

Fig. 9.4 Structure of the main dietary carbohydrates.

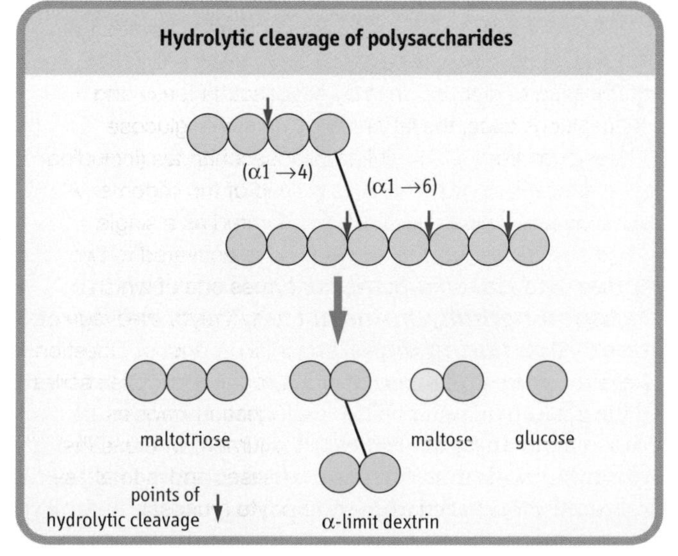

Fig. 9.5 Action of amylase.

The products of this enzyme-mediated process are the disaccharide maltose, the trisaccharide maltotriose and a branched unit, termed the α-limit dextrin. All of these products are then further hydrolyzed by enzymes bound to the enterocyte mucosal membrane, the L-glucosidases, with the final formation of the monosaccharide glucose (Fig. 9.6A).

Dietary disaccharides such as lactose, sucrose and trehalose are hydrolyzed to their constituent monomeric sugars by a series of specific disaccharidases, which are attached to the small intestinal brush-border membrane. The catalytic domains of these proteins are free in the lumen to react with their specific substrates, while their noncatalytic, structural domain(s) are attached to the enterocyte membrane.

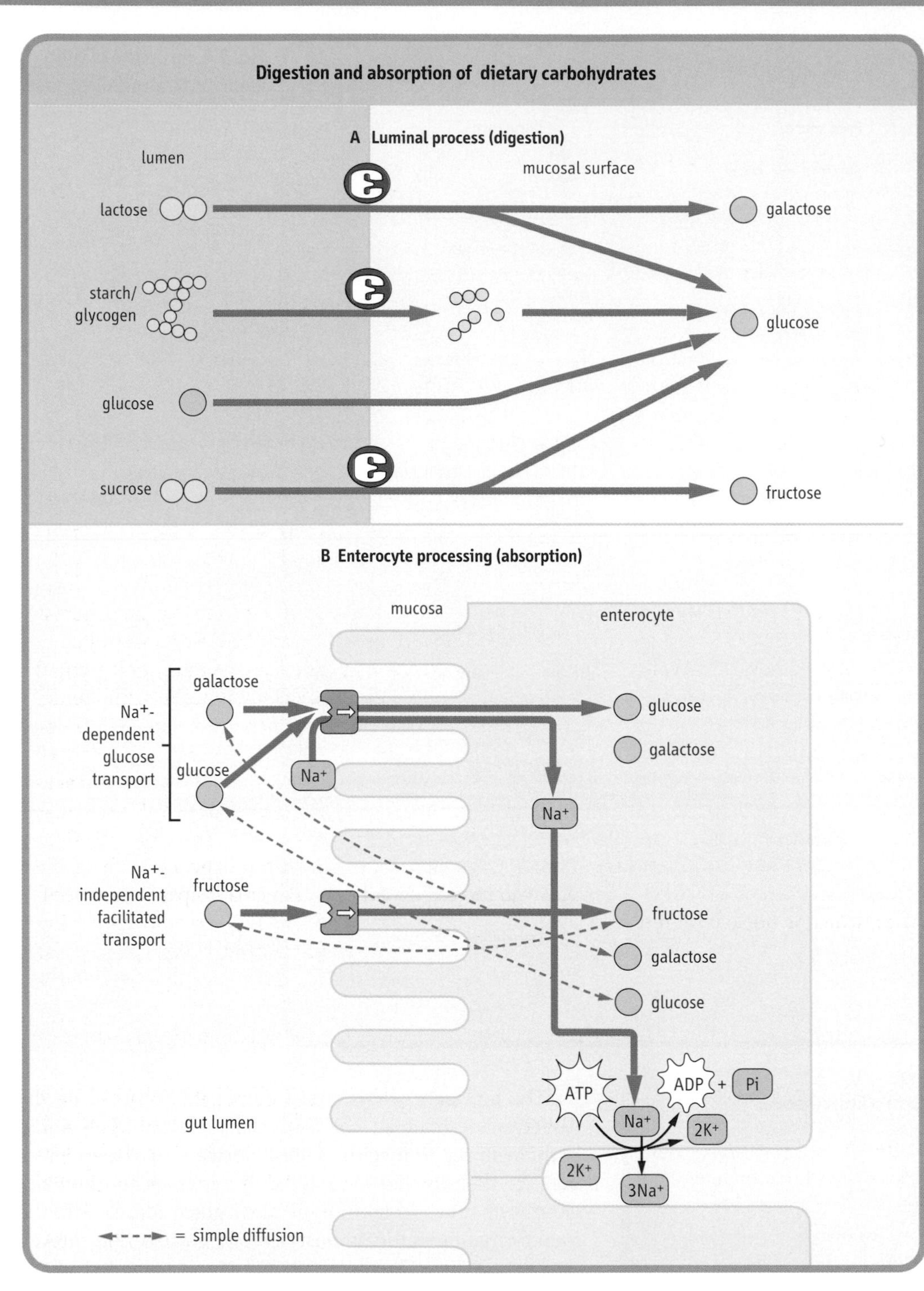

Fig. 9.6 Digestion and absorption of dietary carbohydrates. ADP, adenosine diphosphate; ATP, adenosine triphosphate; Pi inorganic phosphate.

With the exception of lactase, all disaccharidases are inducible

The greater the amount of a disaccharide found in the diet or produced by digestion, the greater is the amount of that specific disaccharidase produced by the enterocyte. The rate-limiting step in absorption of dietary disaccharides is thus the transport of resultant monomeric sugars. Lactase is a noninducible brush-border disaccharidase and therefore the rate-limiting factor in its absorption is the hydrolysis of lactose itself and not the transport of glucose and galactose.

Active and passive transport systems exist to transport carbohydrates across the brush-border membrane during digestion

Since the process of digestion adds to the osmotic load within the gut lumen, water will be pulled from the vascular compartment to the gut. Increased activity of brush-border hydrolysis will thus increase the osmotic load, while increased monosaccharide transport across the enterocyte brush border will decrease the osmotic load. As discussed above for most oligo- and disaccharidases transport of the resulting monomers is rate-limiting and thus compensatory mechanisms exist to avoid pooling of fluid in the gut. As monomeric sugar concentrations rise in the gut lumen, increasing osmolality, there is a compensatory decrease in the activity of brush-border disaccharidases. There is thus both control of osmotic load and prevention of fluid shifts.

Glucose, fructose and galactose are the primary monosaccharides resulting from the digestion of dietary carbohydrate. The absorption of these sugars and other minor monosaccharides is via specific carrier-mediated mechanisms (Fig. 9.6B), all of which demonstrate substrate specificity and stereospecificity, are capable of demonstrating saturation kinetics and can be inhibited specifically. In addition, all monosaccharides can cross the brush-border membrane by a simple diffusion process, although this is extremely slow. At least two carrier-mediated transport mechanisms for monosaccharides exist – a Na^+-dependent co-transporter and a Na^+-independent transporter. At the brush-border membrane both glucose and galactose are transported by the Na^+ dependent glucose transporter. This membrane-linked protein binds with glucose (galactose) and Na^+ at separate sites and transports both into cytosol. The Na^+ is thus transported down its concentration gradient carrying glucose along against its concentration gradient. This transporter mechanism is linked to Na^+-dependent ATPase, which removes Na^+ from the cell in exchange for K^+, with the concomitant hydrolysis of ATP. The transport of glucose (galactose) is thus an indirect active process.

Fructose is transported across the brush-border membrane by a Na^+-independent facilitated diffusion process involving another specific membrane-associated protein, possibly glucose transporter (GLUT-5), which is present on the serosal side of the enterocyte and is thought to be responsible for moving glucose (galactose) from the enterocyte to the capillaries.

Digestion and absorption of lipids

Approximately 90% of fat intake in the diet is as triacylglycerols (TAG) also termed triglycerides with the remainder consisting of cholesterol, cholesterol ester, phospholipid and free fatty acids (FFA). The hydrophobic nature of fats excludes water-soluble digestive enzymes. Fat globules also present limited surface area for enzyme action.

Lipids cannot be broken down by water-soluble digestive enzymes, instead, an emulsification process takes place

The change in the physical structure of lipids begins in the stomach where the heat helps to liquefy lipids, and peristaltic movements aid in the formation of a lipid emulsion. This emulsification process is also aided by the salivary and gastric lipases. The initial rate of hydrolysis is slow due to the separate aqueous and lipid phases and relatively small lipid–water interface. Once hydrolysis begins, however, the water–immiscible TAGs are degraded to fatty acids, which act as surfactants, breaking down lipid globules to smaller particles increasing their surface and facilitating more rapid hydrolysis. The lipid phase therefore becomes dispersed throughout the aqueous phase as an emulsion.

Additional dietary factors also act as surfactants. These include phospholipids, FFAs and monoacyl glycerols. These dietary components aid in the emulsification process and promote the binding of the acid-stable lipases to the interface. This, in turn, facilitates the hydrolysis of TAG and lipid emulsification.

In the duodenum, pancreatic enzymes and bile salts act on the lipid emulsion

The lipid emulsion is ejected from the stomach into the duodenum where dietary lipid undergoes its major digestive process using enzymes secreted by the pancreas, with solubilization being further aided by the release of bile salts from the gall bladder. The major enzyme secreted by the pancreas is pancreatic lipase. This enzyme is, however, inactivated in the presence of bile salts normally secreted into the small intestine during lipid digestion. This inhibitory phenomenon is overcome by the concomitant secretion of co-lipase by the pancreas. Co-lipase binds to both the water-lipid interface and to pancreatic lipase simultaneously anchoring and activating the enzyme. As indicated in Figure 9.7, very little dietary TAG is completely hydrolyzed to glycerol and FFAs. The second and third fatty acids are hydrolyzed from the TAGs with increasing difficulty. Pancreatic lipase aims specifically for the sn-1 and sn-3 bonds of TAGs, thus producing mainly 2-monoacylglycerol (2-MAG) for absorption into the enterocyte.

Bile salts are essential for solubilizing lipids during the digestive process

Without bile salts acting as detergents, the digested lipid would not be in a suitable form for absorption from the gut. The structure of bile acids is demonstrated by cholic acid (Fig. 9.8). This molecule is planar with both a hydrophobic and hydrophilic surface. The hydrophobic region of bile acids is formed by the upper surface of the fused-ring system, while the carboxyl group and all hydroxyl groups are on the opposite surface, conveying hydrophilicity. Bile acids, or actually bile salts at the alkaline pH of the intestines, reversibly form aggregates at concentrations above a critical level, the critical micellar concentration. Such aggregates are termed 'micelles' and their constituent bile acids are in equilibrium with free bile acids. Micelles are thus equilibrium structures with well-defined sizes, considerably smaller than lipid emulsion droplets. The size of these micelles is dependent on bile acid concentration and the ratio of bile acid to lipid content.

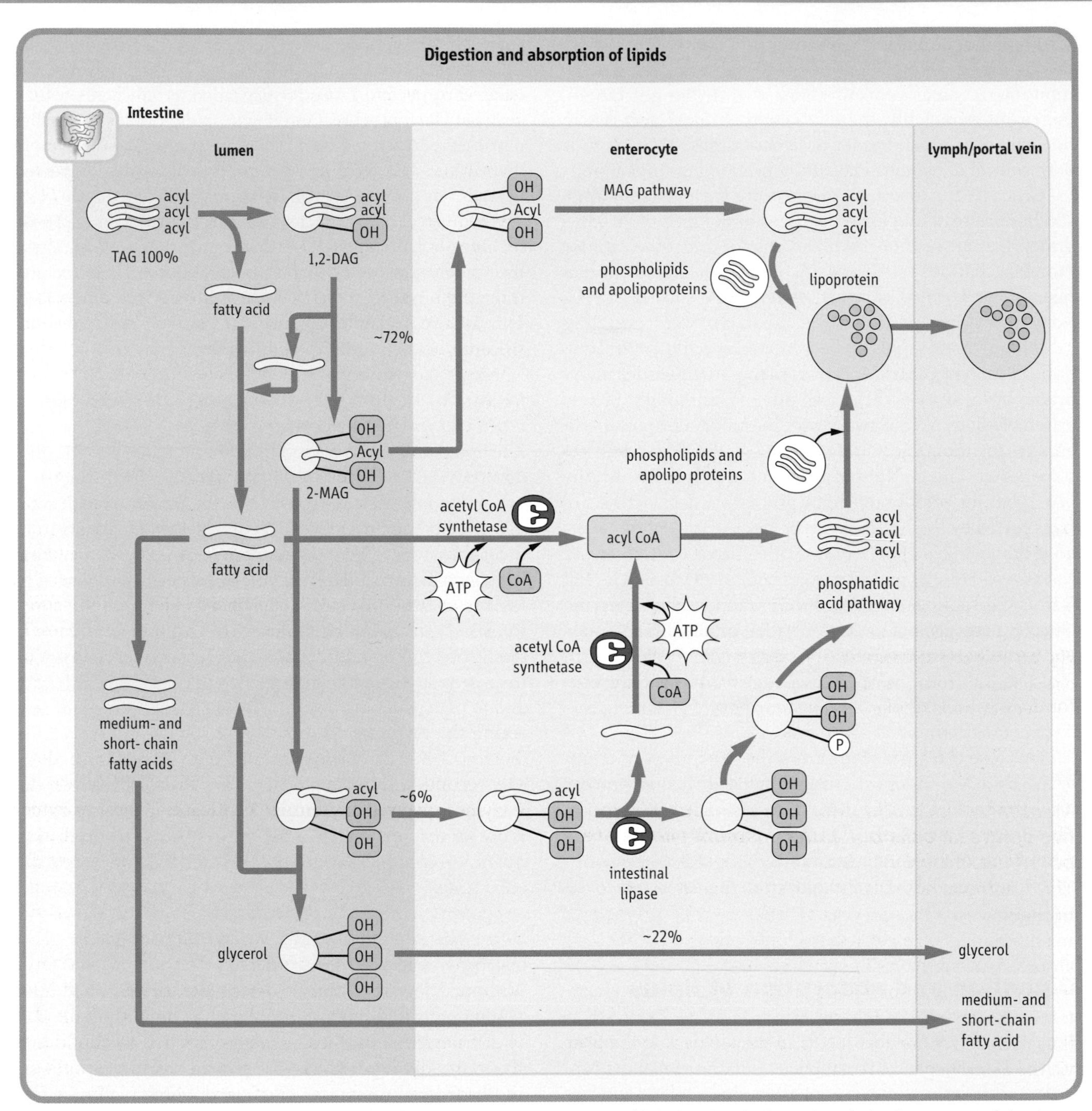

Fig. 9.7 Digestion and absorption of dietary lipids. The percentages can vary widely, but they indicate the relative importance of the three routes indicated. TAG, triacylglycerol; DAG, diacylglycerol; MAG, monoacylglycerol; CoA, coenzyme A. This diagram does not take into account the solubilization factors involved or micelle formation (see below).

The fat emulsion droplets turn into micellar structures

This facilitates the transport of fat through the aqueous environment of the gut. Bile salt micelles can solubilize other lipids and these mixed micelles have disc-like shapes. During digestion of TAGs in the luminal emulsion, the lipid digest is transferred from the fat emulsion droplets to micellar structures. The micelles mediate the transport of lipid digest through the aqueous environment of the gut lumen to the brush border of the enterocyte, where the digest is transferred to the intestinal epithelial cell.

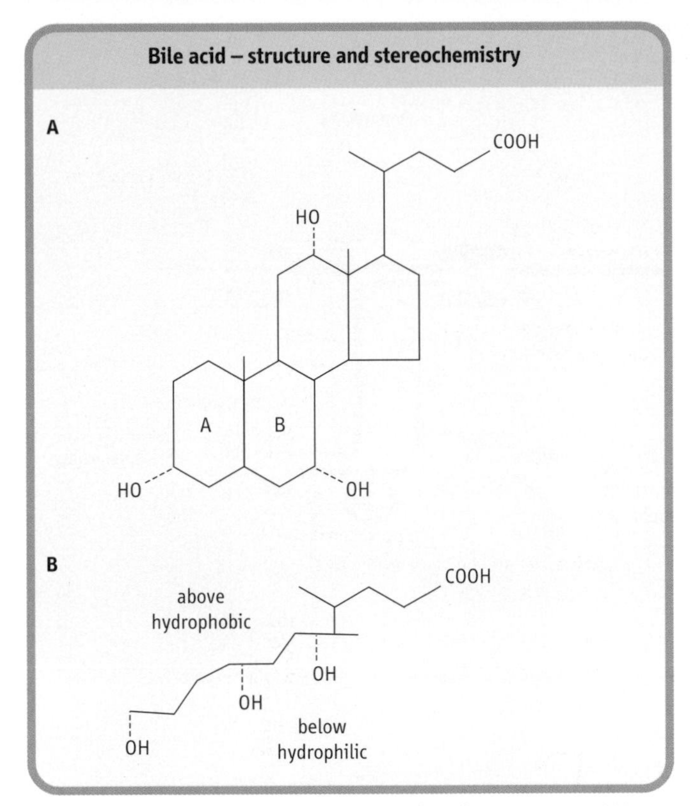

Fig. 9.8 Bile acid structure and stereochemistry. Cholic acid is one of the main components of bile acid. (A) Structure of cholic acid. (B) Stereochemical structure of cholic acid.

Most fatty acids and 2-MAG are absorbed into the epithelial cells but water-insoluble lipids are poorly absorbed into the small intestine

The absorption of lipids into the epithelial cell of the small intestine occurs by a process of diffusion through the plasma membrane. Almost 100% of fatty acids and 2-MAG are absorbed, these being slightly water-soluble. Water-insoluble lipids are poorly absorbed – only 30–40% of dietary cholesterol is absorbed. The bile salts pass on to the ileum where they themselves are absorbed and passed back to the liver, via the enterohepatic circulation.

The fate of fatty acids entering the enterocyte is dependent on their chain length

Medium and short-chain fatty acids (less than 10 carbon atoms) pass directly through the cell to the hepatic portal blood supply. In contrast, fatty acids of more than 12 carbon atoms are bound to a fatty acid binding protein and transferred to the rough endoplasmic reticulum for resynthesis into TAGs. The glycerol for this process is provided by absorbed 2-MAG (the MAG pathway; see Fig. 9.7), hydrolysis of 1-MAG producing free glycerol, or via glycerol-3-phosphate produced during glycolysis (the phosphatidic acid pathway; see Fig. 9.7). Glycerol produced in the intestinal lumen is not reutilized in the enterocyte for TAG synthesis but passes directly to the portal system.

Activation of fatty acids is required for TAG synthesis

Fatty acid activation is accomplished by producing acyl CoA derivatives using acyl CoA synthase. All long-chain fatty acids absorbed by the intestinal epithelial cell are thus reutilized to form TAG before being transferred to the lymphatic system as chylomicrons. Chylomicrons are large, lipid-rich (99% lipid, 1% protein) particles synthesized within the enterocyte on the rough endoplasmic reticulum. They are released into the intercellular space by exocytosis and finally leave the intestine via the lymphatics. The protein component, apolipoprotein B48 is essential for the final release of chylomicrons from the enterocyte (see Chapter 17). This mechanism for lipid transport has evolved to avoid lipid overload of the liver after a meal.

Digestion and absorption of proteins

The protein load received by the gut is derived from two primary sources; 70–100 g dietary protein and 35–200 g endogenous protein, the latter either as secreted enzymes and proteins in the gut or from intestinal epithelial cell turnover. Only 1–2 g nitrogen, equivalent to 6–12 g protein, are lost in the feces on a daily basis. Thus, the digestion and absorption of protein is extremely efficient.

Peptidases hydrolyse proteins

Proteins, like other dietary macromolecules, are broken down by hydrolysis of specific peptide bonds and hence the enzymes involved are termed 'peptidases'. These enzymes can either cleave internal peptide bonds (i.e. endopeptidases) or cleave off one amino acid at a time from either the –COOH or $-NH_2$ terminal of the polypeptide (i.e. they are exopeptidases subclassified into carboxypeptidases and aminopeptidases, respectively). The endopeptidases cleave the large polypeptides to smaller oligopeptides, which can be acted upon by the exopeptidases to produce the final products of protein digestion, amino acids, di- and tripeptides, which are then absorbed by the enterocytes. Depending on the source of the peptidases, the protein digestive process can be divided into gastric, pancreatic and intestinal phases (Figs 9.9 and 9.10).

The process of protein digestion begins in the stomach

In the stomach, secreted HCl reduces the pH to 1–2 with, consequent denaturation of dietary proteins. This denaturation process makes proteins more accessible to protease activity by the unfolding of the polypeptide chain. In addition, pepsins are secreted by the chief cells of the gastric mucosa. These acid proteases are released as the inactive precursors, pepsinogen A and B, and are activated by either an intramolecular reaction (autoactivation) at a pH less than 5 or by active pepsin (autocatalysis). At a pH greater than 2, the liberated peptide remains bound to pepsin and acts as an inhibitor of

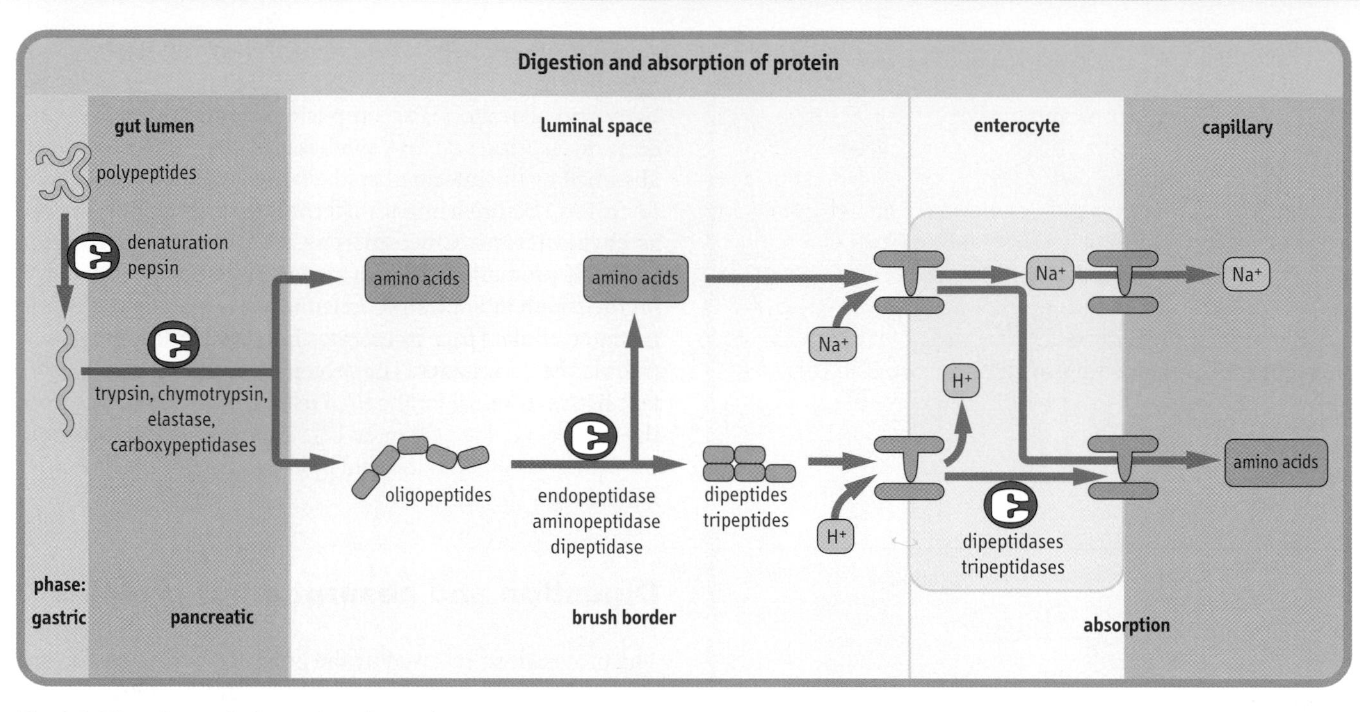

Fig. 9.9 Digestion and absorption of proteins.

Protein digestive enzymes

Source	Zymogen	Activation	Substrate	End product
stomach				
fundus	pepsinogen A	HCl	protein	peptides
pylorus	pepsinogen B	pH 1–2 autoactivation	protein	peptides
pancreas	trypsinogen	enteropeptidase trypsin	protein peptides	polypeptides dipeptides
	chymotrypsinogen	trypsin	protein, peptides	as for trypsin
	proelastase	trypsin	protein peptides	polypeptides dipeptides
	procarboxy-peptidases	trypsin	polypeptides at –COOH end	small peptides amino acids
	Enzyme			
small intestine (no inactive precursor)	aminopeptidase	not applicable	polypeptide at $-NH_2$ end dipeptides	small peptides amino acids
	dipeptidases			amino acids
	endopeptidases		polypeptides	small peptides dipeptides

Fig. 9.10 Enzymes responsible for protein digestion.

pepsin activity. This inhibition is removed by either a drop in pH to 2 or less or by further pepsin action. The major products of pepsin digestion of protein are large peptide fragments and some free amino acids. These gastric protein digests stimulate cholecystokinin release in the duodenum, initiating the release of the main digestive enzymes by the pancreas.

The proteolytic enzymes released from the pancreas are also released as inactive zymogens and depend on the action of duodenal enteropeptidase to convert trypsinogen to the active trypsin. This enzyme is then capable of activating both itself from a precursor state and all other pancreatic zymogens, thus activating chymotrypsin, elastase and carboxypeptides A and B. Since the prime role of trypsin is activating other pancreatic enzymes, its activity is controlled within the pancreas and pancreatic ducts by a low molecular weight inhibitory peptide.

Pancreatic proteases have different substrate specificity with respect to peptide bond cleavage

Trypsin cleaves at lysine and arginine, chymotrypsin at aromatic amino acids and elastase at smaller hydrophobic amino acids. The combined effect of these pancreatic enzymes is to produce an abundance of free amino acids and small molecular weight peptides of two to eight residues in length.

In association with protease secretion, the pancreas also produces copious amounts of sodium bicarbonate ($NaHCO_3$). This results in neutralization of the acid contents of the stomach, promoting pancreatic alkaline protease activity.

Endopeptidases, dipeptidases and aminopeptidases complete the digestion of proteins

The final digestion of di- and oligopeptides is dependent on membrane-bound small intestinal endopeptidases, dipeptidases and aminopeptidases. The end-products of this surface enzyme activity are free amino acids, di- and tripeptides which can then be absorbed across the enterocyte membrane by specific carrier-mediated transport. Di- and tripeptides are further hydrolyzed to their constituent amino acids inside the enterocyte. The final transfer is therefore of free amino acids across the contraluminal plasma membrane into the portal blood system.

Active transport of amino acids into intestinal epithelial cells

These mechanisms are similar to those described for glucose uptake. At the brush-border membrane Na^+-dependent symporters for amino acid uptake are functional with consequent ATP-linked pumping out of Na^+ at the contraluminal membrane. This is an indirect, active process. A similar H^+-dependent symporter is present on the brush-border surface for di- and tri-peptide active transport into the cell. Na^+-independent transporters are present on the contraluminal surface, thus allowing amino acid-facilitated transport to the hepatic portal system.

From both genetic and transporter studies, at least six specific symporter systems have been identified for the uptake of L-amino acids from the intestinal lumen:

- neutral amino acid symporter for amino acids with short or polar side-chains (Ser, Thr, Ala);
- neutral amino acid symporter for aromatic or hydrophobic side-chains (Phe, Tyr, Met, Val, Leu, Ileu);
- imino acid symporter (Pro, OH-Pro);
- basic amino acid symporter (Lys, Arg, Cys);
- acidic amino acid symporter (Asp, Glu);
- β-amino acid symporter (β-Ala, Tau).

These transporter systems are also present in the renal tubules and defects in their constituent protein structure can lead to disease (e.g. Hartnup disease). Pathologies can thus be produced in both the kidney and intestine.

Lactose intolerance

A 15-year-old African–American came across to the UK on an exchange visit for 2 months. After 2 weeks in the UK, he complained of abdominal discomfort, a feeling of being bloated, increased passage of urine and, more recently, the development of diarrhoea. His only change in diet noted at the time was the introduction of yoghurt into his diet. The young man had developed a considerable liking for this foodstuff and was consuming 1–2 large cartons per day. A lactose tolerance test was performed, whereby the young man was given 50 g lactose in an aqueous vehicle to drink. Plasma glucose levels did not rise by more than 1 mmol/L (18 mg/dL) over the next 2 hours, with sampling at 30-minute intervals. A diagnosis of lactose intolerance was made.

Comment. Lactose intolerance is a physiologic change resulting from acquired lactase deficiency. Lactase activity decreases with increasing age in children but this decline in activity is genetically predetermined and demonstrates ethnic variation. Lactase deficiency in the adult black population varies from 45–95%. If symptoms of malabsorption occur after the introduction of milk to adult diets, the diagnosis of acquired lactose deficiency should be considered. A diagnosis is made by challenging the small bowel with lactose and monitoring the rise in plasma glucose. An increase of more than 1.7 mmol/L (30 mg/dL) is considered normal. A rise of less than 0.1 mmol/L (20 mg/dL) is diagnostic of lactose deficiency. A rise of 1.1–1.7 mmol/L (20–30 mg/dL) is inconclusive.

Coeliac disease

A 22-year-old man presented with a history suggestive of severe malabsorption syndrome, with a history of weight loss, diarrhea, abdominal bloating and anemia. Laboratory features included hemoglobin of 90 g/L (9 g/dL) (reference range 135–180 g/L; 13.5–18 g/dL), low serum iron, serum ferritin, folate in the red cells and serum B_{12}. Biopsy of his small bowel demonstrated flattening of the mucosal surface, villous atrophy and disappearance of microvilli. A diagnosis of gluten-induced enteropathy (coeliac disease) was made. All wheat products were removed from the patient's diet, with concomitant improvement in his symptomatology and recovery of gut function.

Comment. Coeliac disease is characterized by severe malabsorption and specific diagnostic features exhibited by the intestinal mucosa. Since the absorptive surface is markedly reduced, the resulting indigestion/malabsorption are severe. The histologic changes are due to the interaction of gluten, the principal protein of wheat, with the epithelium. There is evidence to suggest that the deficit is located within the mucosal cells of the intestine and permits polypeptides, resulting from peptic and tryptic digestion of gluten, not only to exert local harmful effects within the intestine but also to be absorbed and to induce an antibody response. Circulating antibodies to wheat gluten and its fractions are frequently present in cases of coeliac disease.

Summary

The absorption and digestion of foods are essential processes which make the metabolic fuels available to the organism. Defects in these mechanisms result in a variety of malabsorption and food intolerance syndromes. From a clinical point of view it is essential to understand the nature of the carbohydrate, protein and fat intake in the diet. Certain carbohydrates and proteins cannot be properly digested by some individuals and lead to the development of disease and the expression of GI stress (see Lactose intolerance and coeliac disease). Elimination of these foods from the diet is, in essence, the treatment of that particular pathology. The commonest of these food-associated pathologies is lactose intolerance and gluten (wheat protein) sensitivity, the latter also known as coeliac disease.

Further reading

Bronner F. Calcium absorption – a paradigm for mineral absorption. *Journal of Nutrition* 1998;**128:**917–20.

Bruno MJ, Haverkort EB, Tytgat GN, van Leeuwen DJ. Maldigestion associated with exocrine pancreatic insufficiency: implications of gastrointestinal physiology and properties of enzyme preparations for a cause-related and patient-tailored treatment. *American Journal of Gastroenterology* 1995;**90:**1383–1393.

Duerksen DR, Nehra V, Bistrian BR, Blackburn GL. Appropriate nutritional support in acute and complicated Crohn's disease. *Nutition* 1998;**14:**462–465.

Kerner JA Jr. Formula allergy and intolerance. *Gastroenterology Clinics of North America* 1995;**24:**1–25.

Lentze MJ. Molecular and cellular aspects of hydrolysis and absorption. *American Journal of Clinical Nutrition* 1995;**61(Suppl 4):**946S–951S.

Rose RC. Intestinal absorption of water soluble vitamins. *Proceedings of the Society for Experimental Biology and Medicine* 1996;**212:**191–198.

Ushijima K, Riby JE, Kretchmer N. Carbohydrate malabsorption. *Pediatrics Clinics of North America* 1995;**42:**899–915.

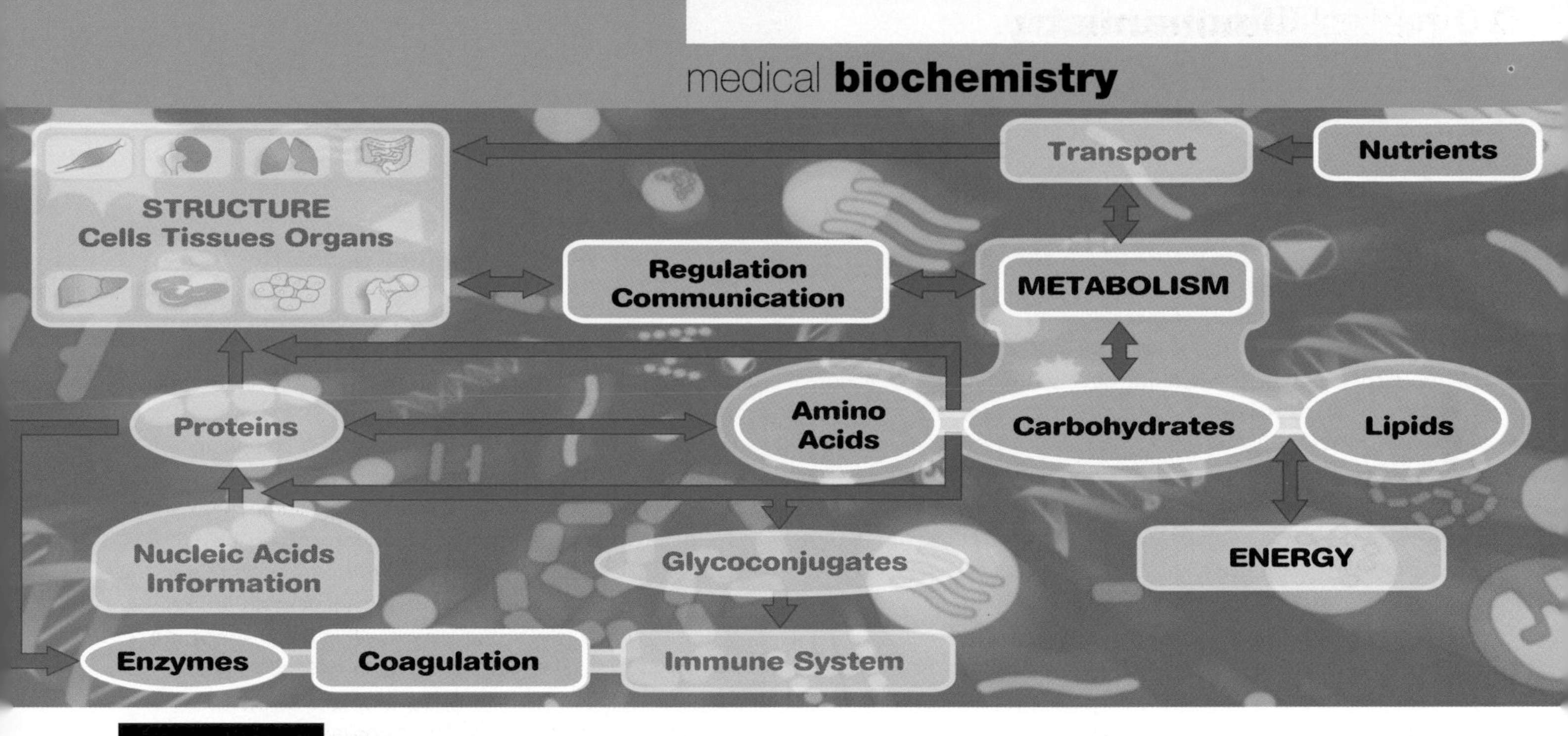

10 Vitamins, Minerals and Nutrition

Introduction

Vitamins can be subdivided into fat- and water-soluble

Vitamins, especially the water-soluble group, are an inherent part of functional protein molecules, often enzymic in nature. These vitamins act as coenzymes in the specific reactions catalyzed, e.g. riboflavin in oxidoreductase reactions or biotin in decarboxylation reactions.

Fat-soluble vitamins include vitamins A, D, E, and K; while water-soluble vitamins include vitamins B_1, B_2, B_3, B_5, B_6, B_{12}, folate, biotin and ascorbic acid (vitamin C).

Certain trace metals are essential nutrients

More recently, the importance of certain trace metals as essential nutrients has also been recognized. Again, many of these function as part of protein molecules or metalloenzymes. Proteins without their trace metal prosthetic groups, e.g. Zn, Mn, and Mg are devoid of their biological function.

Other trace elements are cytotoxic

In addition to the essential trace elements, certain other trace metals find their way into the food chain and can be toxic to the cell (e.g. Cd, Hg, and Al). This list of toxic metals also includes essential trace metals elements when taken in large amounts, e.g. Cu, Zn, and Mn, etc. The essential requirements of these trace metals in the prevention of disease and their association with certain pathologies only came to light with the development of suitable methods of biological analyses, for example atomic absorption spectrometry and mass spectrometry.

Recommended dietary allowances

Absolute or relative deficiencies of vitamins and trace metals lead to characteristic diseases or syndromes

To prevent the development of pathologies caused by vitamin or trace metal deficiencies, certain levels of intake have been recommended for continual health of the population. These minimum intake values are termed recommended daily allowances (RDAs). RDAs are not absolute, but vary from population to population, e.g. as a function of age, sex, pregnancy, etc. They also depend, to some extent, on the level of macronutrient intake.

Although single deficiency states may occur, poor diets are often characterized by multiple nutrient deficiencies. Specific vitamin-associated pathologies are also well recognized.

Assessment of micronutrient status

Micronutrients are not normally deficient as a single item

Assessment of nutrient status is fraught with difficulties since malnutrition, or undernutrition, is associated with multiple nutrient deficiencies, each one having functional implications and all being interrelated. Micronutrient assessment is even more difficult in situations where subclinical levels of nutrient deficit exist. Measurement of circulating vitamin levels are inappropriate, as, in the case of water-soluble vitamins, these are associated with the most recent intake and do not reflect current vitamin status. This is especially so, when one considers that most of these vitamins are intimately linked to enzyme structure.

Measurement of enzyme function associated with particular water-soluble vitamins has been suggested as the most appropriate way to assess adequate nutritional status. This is usually carried out as stimulation tests, i.e., enzyme activity is measured in the absence and presence of the vitamin as reagent. Deficit is recognized by the stimulation of enzyme activity in the presence of added vitamins.

Likewise the determination of simple circulating concentrations of fat-soluble vitamins is also inappropriate. These vitamins are associated with body fat and are often stored in specific tissues with circulating concentrations kept relatively constant. For example, vitamin A is stored in the liver and transported by specific binding proteins in the serum. A drop in level of a specific nutrient in the diet is thus not necessarily indicative of nutrient deficit. Furthermore, a drop in level of a nutrient within a particular tissue, e.g. blood or plasma, need not indicate a deficiency or an increased requirement: it could simply reflect a metabolic adjustment to stress or change in physiologic status, such as pregnancy. Similar situations relate to trace metals, where the circulating levels of these nutrients bear little relation to nutrient status. For evaluation of trace element toxicity, tissues other than blood may need to be assayed before a definite diagnosis of metal poisoning can be made.

Fat-soluble vitamins

Ample reserves of fat-soluble vitamins are stored in the tissues, as they are not readily absorbed from food

Fat-soluble vitamins are not so readily absorbed or extracted from the diet as water-soluble vitamins but ample reserves are stored in tissues. With the exception of vitamin K they are, in fact, not coenzymes. Indeed vitamins A and D act more like hormones. Their presence in storage sites is related to their toxicity in unstored forms: both vitamins A and D can be markedly toxic if present in the circulation in large amounts. This is not true of either vitamin E or K or the vitamin A precursor, carotenes. The latter, like vitamin E, are stored in adipose tissue.

Vitamin A

Vitamin A is found in animals as retinol, retinal, and retinoic acid; its provitamin, β-carotene, is found in plants

Vitamin A is a generic term for a collection of three vitamers, retinol, retinal and retinoic acid, all of which are found in animals. The term 'retinoids' has been used to define these three compounds and other associated synthetic compounds with vitamin A-like activity. Vitamin A also exists as a provitamin, β-carotene, in plants and is converted to all-*trans* retinol by the action of β-carotene dioxygenase in the small bowel. Further metabolism in the enterocyte produces retinol and retinoic acid (Fig 10.1). These compounds are then transported to the liver where vitamin A is stored as retinol palmitate.

The stores of vitamin A in the liver comprise approximately 1 year's supply

Liver, egg yolk, butter and milk are good sources of preformed vitamin A. Dark-green and yellow vegetables are good sources of β-carotene. The conversion of carotenoids to vitamin A is rarely 100% and the potency of foods is described in retinol equivalents (RE) such that:

1 RE = 1 μg of retinol or 6 μg β-carotene or 12 μg of other carotenes

Vitamin A may confer a protective action against cancer and cardiovascular disease

More recently β-carotene has received attention in its role as an antioxidant. In this role it is thought to prevent the development of diseases in which the action of oxygen free radicals is implicated. Large clinical trials and epidemiologic studies are in progress in both the USA and Europe to evaluate the protective role of vitamin A against cancer and cardiovascular disease.

Since normal epithelial cell growth and differentiation depends on retinoids, and many human tumors arise from epithelial cells (carcinomas), vitamin A has been proposed to play a protective role. Some epidemiologic studies have demonstrated an inverse relationship between the vitamin A content of the diet and the risk of cancer. Some investigators have also shown a protective role of added vitamin A against known carcinogens. Considerably more work in this area is required as the direct effect of retinoids on gene expression and regulation is far from clear.

Vitamin A is stored in the liver but transport is required to its sites of action

Owing to the fat-soluble nature of the vitamin, specific transport mechanisms are involved, both in the blood and tissue sites of action. Transport is effected by specific proteins – serum retinol binding protein (SRBP) and cytosolic binding proteins (CRBP). In addition, retinoic acid is thought to be transported to cells either bound to albumin or a specific retinoic acid binding protein (RABP). Complex

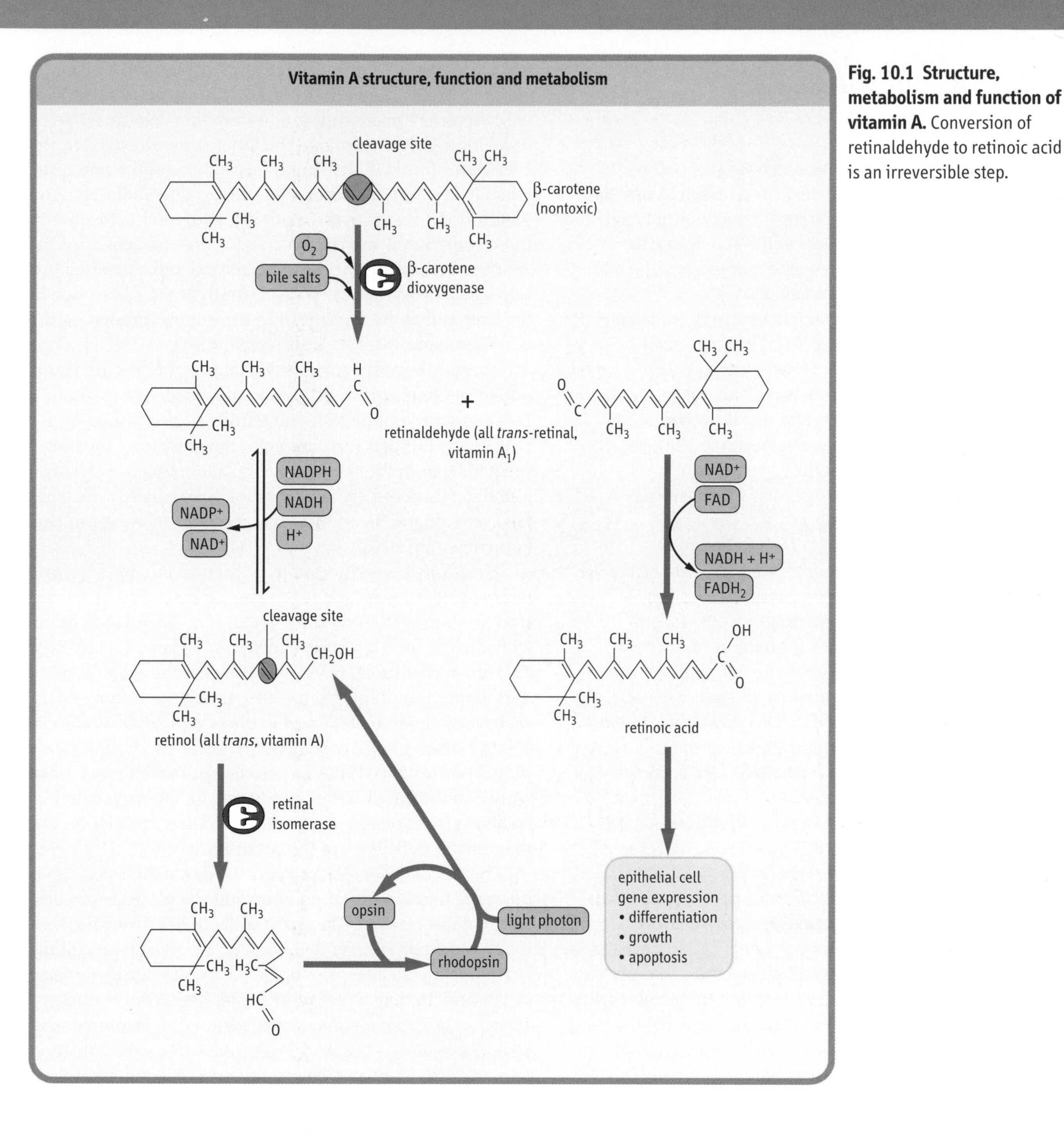

Fig. 10.1 Structure, metabolism and function of vitamin A. Conversion of retinaldehyde to retinoic acid is an irreversible step.

mechanisms and other tissue proteins are involved in the molecular trafficking of retinol to the nucleus of the cell.

Vitamin A is necessary for vision mediated by the rod cells, so deficiency often presents as 'night blindness'

The visual pigment, rhodopsin, is found in the rod cells of the retina and is formed by the binding of 11-*cis*-retinal to the apoprotein opsin. When rhodopsin is exposed to light, it is bleached, retinal dissociates and is isomerized and reduced to all-*trans*-retinol (see Fig. 10.1). This reaction is accompanied by a conformational change and elicits a nerve impulse perceived by the brain as light. Rod cells are responsible for vision in poor light. Vitamin A deficiency often presents as defective night vision or 'night blindness'.

Corneal softening and opacity also result from vitamin A deficiency

Vitamin A also affects growth and differentiation of epithelial cells; its deficiency thus produces defective epithelialization and keratomalacia – a condition affecting the cornea producing softening and opacity. Severe vitamin A deficiency leads to progressive keratinization of the cornea and possibly permanent blindness. Vitamin A deficiency is the commonest cause of blindness in the world.

Vitamin A deficiency

A 47-year-old woman with a long history of Crohn's disease (a chronic inflammatory bowel disease) has had to be fed for some months using intravenous nutrition. Initially, this treatment included intravenous fat and fat-soluble vitamins. As a result of complications in the administration of the intravenous feed, the fat component was removed and more energy supplied using a carbohydrate source. Prior to her starting intravenous feeding she had been receiving supplements of oral vitamins: these had been discontinued. Three months after the alteration of her intravenous feeding regimen she began to complain of being unable to see appropriately in dim light. Measurement of her serum vitamin A indicated a level well below the reference range.

Comment. Intravenous feeding solutions are highly purified and micronutrients must be added. The removal of fat from the prescription precluded the administration of fat-soluble vitamins intravenously. This was not noted at the time and the patient proceeded to develop symptoms of vitamin A deficiency, i.e. night blindness. This could have been avoided by providing a separate infusion of fat emulsion one or two days per week to act as a carrier for the fat-soluble vitamins.

Although severe vitamin A deficiency is rare in the USA and Europe, subclinical deficiency may lead to increased susceptibility to infection

Severe vitamin A deficiency is seen only in the developing world. In the USA and Europe deficiencies are rare, but subclinical states of deficiency may contribute to increased susceptibility to cancer and infection. Although rare in the general population, vitamin A deficiency is fairly common in patients with severe liver disease or fat malabsorption.

Vitamin A excess is toxic

Excess vitamin A administration results in toxicity, with symptoms including bone pain, dermatitis, hepatosplenomegaly, nausea and diarrhea. It is virtually impossible to develop vitamin A toxicity by ingesting normal foods; however, toxicity may result from use of pure vitamin A supplements. Increased intake of vitamin A is also associated with teratogenicity and should be avoided during pregnancy.

Vitamin D

Vitamin D is the only vitamin that is not usually required in the diet of man

Vitamin D should be reclassified as a hormone since it is only under conditions of inadequate exposure to sunlight that dietary intake is required. Vitamin D is, in fact, a group of closely related sterols produced by the action of ultraviolet light on certain provitamins, ergosterol in plants and 7-dehydrocholesterol in animals. The latter is synthesized in the liver and is found in the skin. The products of the photolytic reaction are ergocalciferol (vitamin D_2) and cholecalciferol (vitamin D_3) respectively. Both are equipotent. Neither of these compounds are responsible for the end-activity of the vitamin, both being further metabolized and converted to a series of hydroxylated derivatives, firstly at the 25-position in the liver and at the 1-position in the kidney, producing the active compound 1,25-dihydroxyvitamin D_3 (1,25$(OH)_2D_3$).

Vitamin D_3 and its hydroxylated metabolites are transported in the plasma bound to a specific globulin, Vitamin D-binding protein (DBP). The affinity of provitamin D_3 for DBP is low but that for D_3 is high, thus ensuring the movement of D_3 from the skin to the circulation. Vitamin D_3 is also found in the diet where its absorption is associated with other fats (see Chapter 9), and it is transported to the liver in chylomicrons (Fig 10.2).

The 25-hydroxylation step is carried out by a hepatic microsomal enzyme and is the rate-limiting step in conversion of vitamin D_3 to its active metabolite. Regulation of the 25-hydroxylation step is dependent on the hepatic content of 25-hydroxyvitamin D_3 (25$(OH)D_3$). This is the major form of the vitamin found in both the liver and in the circulation, in each case bound to DBP, and levels of 25$(OH)D_3$ in the circulation reflect hepatic stores of the vitamin. A significant proportion of 25$(OH)D_3$ is excreted in the bile and reabsorbed in the small bowel, producing an enterohepatic circulation. Disturbance in the enterohepatic circulation can thus lead to deficiency of this vitamin.

The main site for further hydroxylation at the 1-position is the renal tubules, although bone and the placenta can also carry out this reaction. The enzyme, 25$(OH)D_3$ 1α hydroxylase is a mitochondrial enzyme producing the most potent of the vitamin D metabolites and the only naturally occurring form of vitamin D that is active at physiologic concentrations. 1,25$(OH)_2D_3$ can maintain normal serum Ca^{2+} in animals that do not have kidneys or parathyroid glands. Its production, however, is closely regulated by parathyroid hormone, phosphate and itself.

The renal tubules, cartilage, intestine and placenta also contain a 24-hydroxylase, producing the inactive 24,25-dihyroxyvitamin D_3, (24,25$[OH]_2D_3$). The level of the 24,25$(OH)_2D_3$ is reciprocally related to the level of the 1,25$(OH)_2D_3$ in the circulation.

Most of the vitamin intake is via milk and other fortified foodstuffs

Fish oils and egg yolks are naturally rich sources of vitamin D; other natural foodstuffs, including liver, are poor sources of the vitamin. The vitamin is released from chylomicrons in the liver by DBP and hydroxylated at the 25-position for storage.

1,25$(OH)_2D_3$ is produced by action of the liver and kidney and is transported in plasma, bound to DBP. Since it

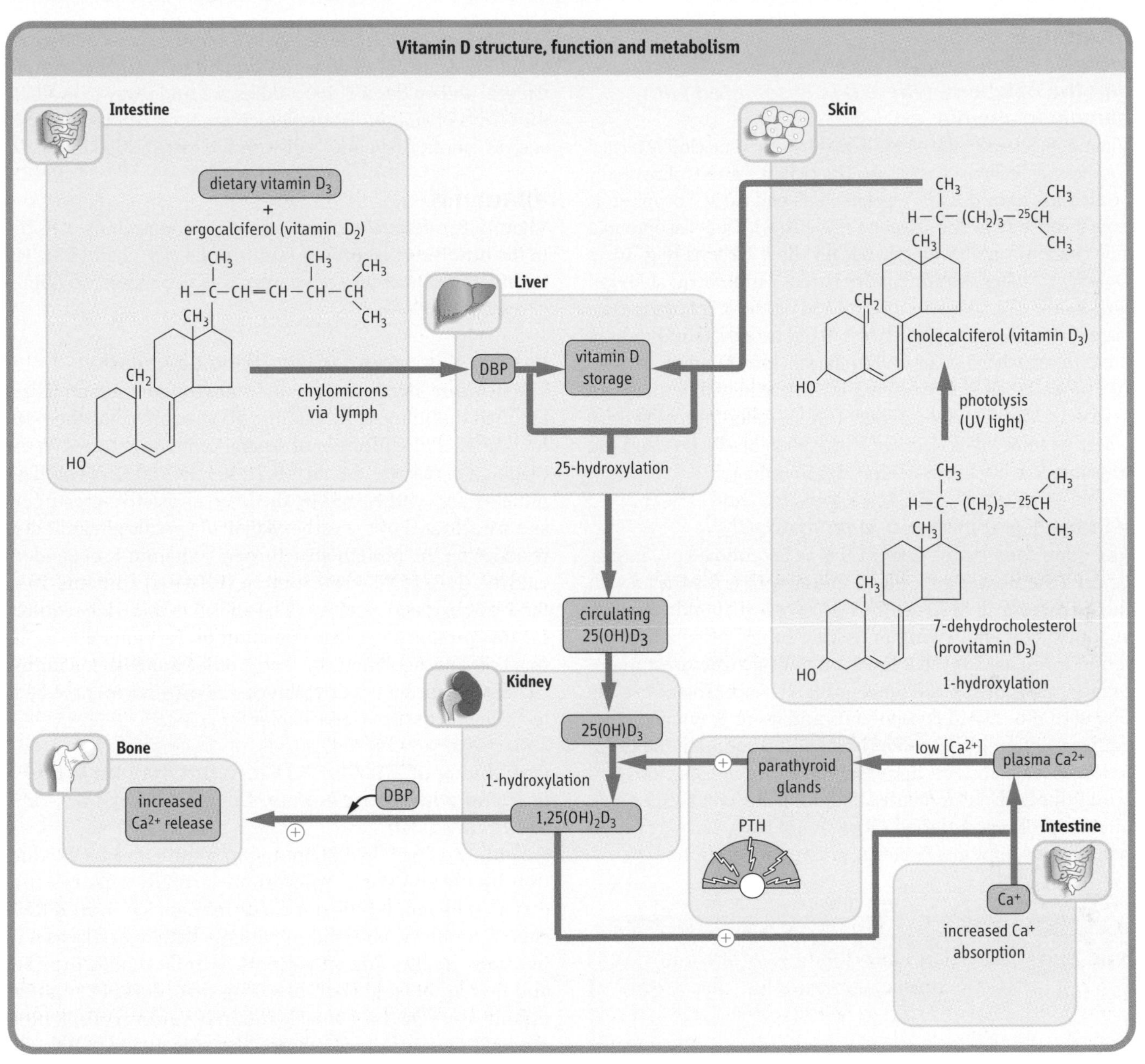

Fig. 10.2 Structure, function and metabolism of vitamin D. Note that excess $25(OH)D_3$ can mimic $1,25(OH)_2D_3$ but at greater physiologic concentrations. PTH, parathyroid hormone.

affects Ca^{2+} transport and metabolism at a distance, vitamin D may be described as a hormone and should be considered as such. In the intestinal epithelial cells, it binds to a cytoplasmic receptor like other steroid hormones (see Chapter 16) and this ligand–protein complex is transported to the nucleus where the active hormone $1,25(OH)_2D_3$ induces gene expression affecting calcium metabolism (see Chapter 23).

Deficiency of vitamin D produces rickets in children and osteomalacia in adults

Rickets is characterized by the production of soft pliable bones due to defective mineralization secondary to calcium deficiency, and is associated with continued osteon and cartilage formation. The characteristic bowing of the leg bones and the formation of the rickety rosary around costochondral junctions results. Demineralization of pre-existing bones takes place in the adult, increasing susceptibility to fractures. Vitamin D deficiency is also characterized by low circulating concentrations of calcium in association with increased serum alkaline phosphatase activity.

Vitamin D toxicity states are seen at levels of intake greater than 10 times the RDA

Vitamin D excess leads to enhanced calcium absorption and bone reabsorption, leading to hypercalcemia and metastatic calcium deposition. There is also a tendency to develop kidney stones from the hypercalciuria secondary to the hypercalcemia.

Vitamin E

Vitamin E deficiency is rare except in pregnancy and the newborn, where it is associated with hemolytic anemia

Vitamin E occurs in the diet as a mixture of several closely related compounds, called tocopherols. The richest sources of naturally occurring vitamin E are vegetable oils and nuts. Tocopherols have a substituted chromanone nucleus, with a polyisoprenoid side chain of variable length; usually three carbons (Fig. 10.3). Deficiency states in humans are virtually unknown, although low vitamin E intake in pregnancy and newborn infants is associated with hemolytic anemia. This is usually found only in preterm infants fed on formula milk with low vitamin E content. In European folklore, vitamin E is associated with fertility and sexual activity. This is certainly true in other animal species where vitamin E is associated with sperm production and egg implantation, but this is not the case in humans.

Vitamin E is a membrane antioxidant

The main function of vitamin E is as an antioxidant, in particular a membrane antioxidant, and as such is associated with the membrane lipid structure. It is the most abundant natural antioxidant and, owing to its lipid solubility, vitamin E is found associated with all lipid-containing structures: membranes, lipoproteins and fat deposits. It is absorbed from the diet with other lipid components and there is no facilitated pathway or specific transport protein. It is found in the circulation associated with lipoproteins. Fat malabsorption will reduce the body fat content of vitamin E and in this situation, after a prolonged period, some clinicians have reported neurologic symptoms related to vitamin E deprivation.

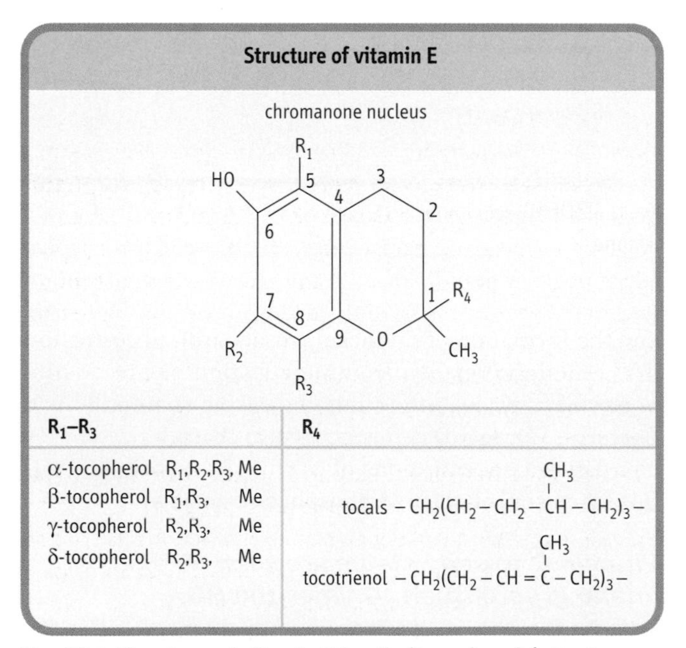

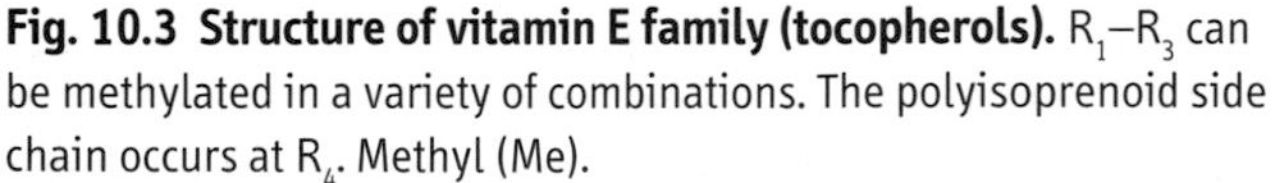

Fig. 10.3 Structure of vitamin E family (tocopherols). R_1–R_3 can be methylated in a variety of combinations. The polyisoprenoid side chain occurs at R_4. Methyl (Me).

Like the lack of deficiency symptoms, there is a corresponding dearth of evidence in support of vitamin E toxicity. Clinical and epidemiologic studies are underway to evaluate the effects of vitamin E supplementation on chronic diseases such as cancer, Alzheimer's disease and cardiovascular disease.

Vitamin K

Vitamin K refers to a group of related compounds, varying in the number of isoprenoid units in its side chain. Like vitamin E, the absorption of vitamin K is dependent on appropriate fat absorption.

Vitamin K is necessary for blood coagulation

The structure, nomenclature and source of the vitamin Ks are outlined in Figure 10.4. Vitamin K is required for the post-translational modification of several proteins required in the coagulation cascade, i.e. factors II, VII, IX and X. All of these proteins are synthesized by the liver as inactive precursors and are activated by the carboxylation of specific glutamic acid residues on the protein structure by a vitamin K-dependent enzyme (Fig. 10.5). Prothrombin (factor II) contains 10 of these carboxylated residues (Gla) and all of these are required for this protein's specific chelation of Ca^{2+} ions during its function in coagulation. Recently, other proteins containing vitamin K-dependent Gla residues, e.g. osteocalcin, have been identified in tissues (see Chapter 6).

Deficiency of vitamin K is rare but may be found in those with liver disease, fat malabsorption, or in the newborn

Vitamin K is widely distributed in nature and its production by the intestinal microflora virtually ensures that dietary deficiency does not occur in man. Vitamin K deficiency, however, is found in patients with liver disease, in newborn infants and in patients with fat malabsorption, and it is associated with bleeding disorders. Premature infants are especially at risk and may suffer from hemorrhagic disease of the newborn. The placenta is inefficient at passing maternal vitamin K to the fetus and immediately after birth the circulating concentration normally drops, but recovers on absorption of food; this is possibly delayed in preterm infants. In addition, the gut of the newborn is sterile, so that the intestinal microflora does not provide a source of vitamin K for several days after birth.

Specific inhibitors of vitamin K-dependent carboxylase reactions are used in the treatment of thrombosis-related disease, e.g. patients with deep vein thrombosis and/or pulmonary thromboembolism, or patients with atrial fibrillation. These are drugs of the dicoumarin group, e.g. warfarin, which inhibit the action of vitamin K – probably via the mechanisms involved in the regeneration of the active hydroquinone. This drug is also used as rat poison and vitamin K is thus the antidote for human poisoning by this agent.

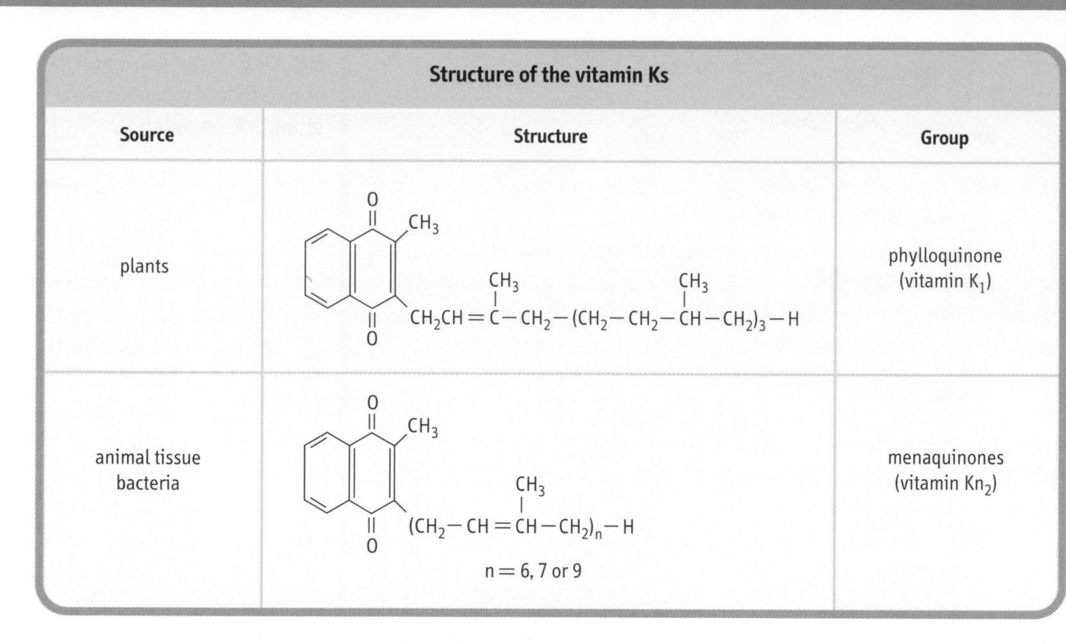

Structure of the vitamin Ks		
Source	Structure	Group
plants	$CH_2CH=C(CH_3)-CH_2-(CH_2-CH_2-CH(CH_3)-CH_2)_3-H$ (2-methyl-1,4-naphthoquinone ring)	phylloquinone (vitamin K_1)
animal tissue bacteria	$(CH_2-CH=C(CH_3)-CH_2)_n-H$ (2-methyl-1,4-naphthoquinone ring); n = 6, 7 or 9	menaquinones (vitamin Kn_2)

Fig. 10.4 The structure, group, nomenclature and source of the vitamin Ks.

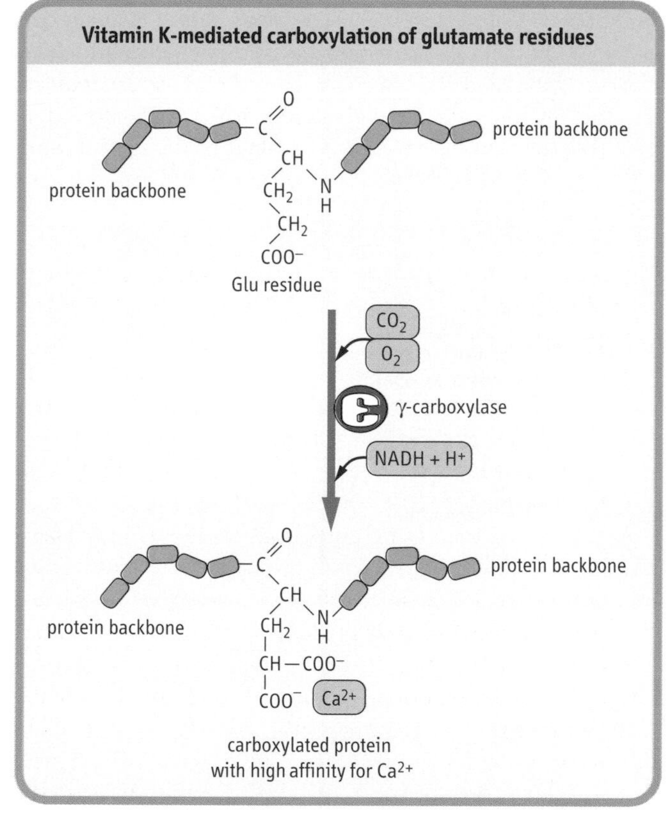

Fig. 10.5 Vitamin K-mediated carboxylation of glutamate residues (Glu). This reaction produces carboxylated residues, which are required for Ca^{2+} chelation.

Water-soluble vitamins

With the exception of vitamin B_{12} there is no storage capacity for water-soluble vitamins

As a consequence of the absence of storage, all water-soluble vitamins must be supplied as regular constituents of the diet. Any excess intake of these vitamins results in their excretion in the urine. In contrast to the fat-soluble vitamins there is no common toxicity associated with overconsumption of these vitamins, either enterally or parentally.

B-complex vitamins

B-complex vitamins act as coenzymes in all metabolic pathways

B-complex vitamins are essential to the normal metabolism of all cells and are involved as coenzymes in all metabolic pathways. The B-complex group of vitamins is outlined in Figure 10.6 and their associated deficiency states are also presented. Deficiency of a single B vitamin is rare, patients often presenting with multiple deficiency states.

Thiamine (vitamin B_1)

The active form of thiamine is thiamine pyrophosphate

Thiamine, in its active form as thiamine pyrophosphate, is essential for decarboxylation and transfer reactions and for normal carbohydrate energy metabolism (see Chapter 8). Thiamine is required for the transketolase reaction in the hexose monophosphate pathway (see Chapter 20) and erythrocyte activity of this enzyme is the most frequently used test to assess thiamine status.

Thiamine deficiency initially causes loss of appetite, constipation and nausea but may progress to impaired nerve cell function

The pathways in which thiamine is required are well characterized, but the failure of these in deficiency states and the signs and symptoms of deficiency are not clearly related. Loss of appetite, constipation and nausea are early symptoms of thiamine deficiency, progressing to depression, peripheral neuropathy and instability, the latter related to impaired nerve cell function. These are the signs and symptoms seen in the elderly or in low-income groups with poor diet. Further deterioration in thiamine status results in mental confusion

Vitamin B-complex			
Vitamin	**Structure**	**Deficiency disease**	**Food source**
thiamine (vit B_1)	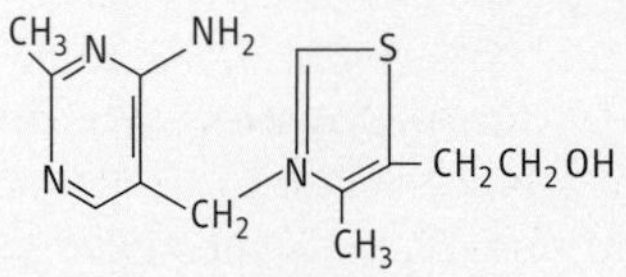	beri-beri	seeds, nuts, wheatgerms, legumes, lean meat
riboflavin (vit B_2)	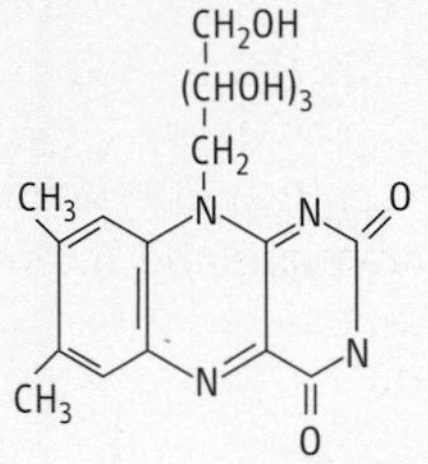	(pellagra)	meats, nuts, legumes
niacin (vit B_3)	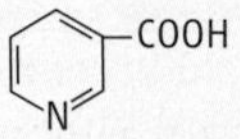	pellagra	meats, nuts, legumes
panthothenic acid (vit B_5)	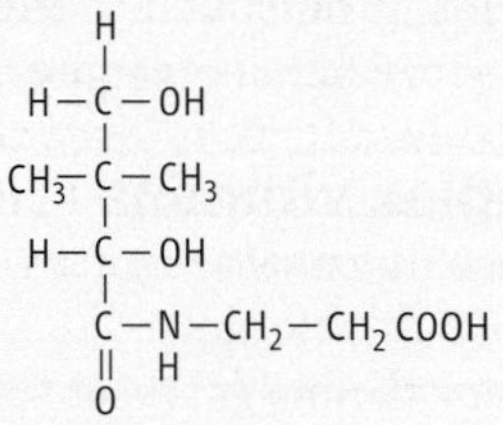		yeast, grains, egg yolk, liver
pyridoxine (vit B_6)	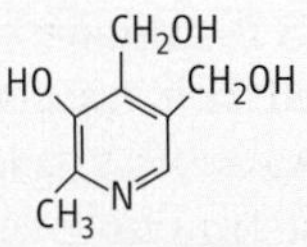	neurologic disease	yeast, liver, wheatgerm, nuts beans, bananas
biotin	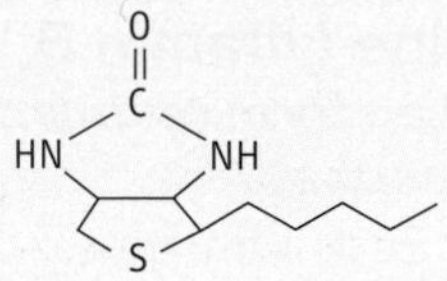	widespread injury	egg-white
folate	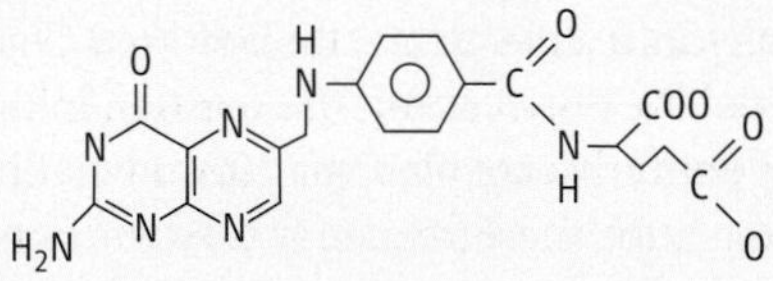	anemia	yeast, liver, leafy vegetables
cobalamin (vit B_{12})	complex	pernicious anemia	liver, kidney, egg, cheese

Fig. 10.6 Structure, sources and deficiency diseases of vitamin B complex.

(loss of short-term memory), ataxia and loss of eye coordination. This combination, often seen in alcoholics, is termed Wernicke–Korsakoff psychosis. Severe thiamine deficiency results in beri-beri, either 'dry' (without fluid retention), or 'wet' – the latter being associated with myocardial dysfunction and high output cardiac failure with edema. Beri-beri is characterized primarily by advanced neuromuscular symptoms, and is usually associated with populations relying exclusively on polished rice for food. Wet beri-beri is particularly associated with alcoholism.

The greater the caloric intake, the larger the requirement for B vitamins

Diseases that are associated with large caloric requirements will require greater intakes of thiamine and other complex vitamins. Increased energy requirements, in particular from carbohydrates, require increased amounts of vitamin as a result of increased enzyme activity. Such situations are found in certain states of infection, post-trauma and in burns.

Riboflavin (vitamin B_2)

As with other water-soluble vitamins, riboflavin is attached to the sugar alcohol, ribitol. The molecule is colored, fluorescent, decomposes in visible light but is heat-stable. It is found in the oxidoreductases as flavin mononucleotide (FMN) and flavin adenine dinucleotide (FAD), and is required for the energy metabolism of both sugars and lipids (see Chapter 11). The activation of riboflavin is via an ATP-dependent enzyme system resulting in the production of FMN and FAD.

Lack of riboflavin in the diet causes a generally nonfatal deficiency syndrome of inflammation of the corners of the mouth (angular stomatitis), the tongue (inflammation of glossitis) and scaly dermatitis. A degree of photophobia may also exist. Owing to its light sensitivity, riboflavin deficiency may occur in newborn infants, with jaundice, who are treated by phototherapy. Erythrocyte enzyme activity measurements (glutathione reductase) are used to determine nutritional status of the patient with regard to riboflavin. Hypothyroidism is known to affect the conversion of riboflavin to flavin mononucleotide (FMN) and flavin adenine dinucleotide (FAD).

Niacin (vitamin B_3)

Niacin is a generic name for nicotinic acid or nicotinamide, either of which is an essential nutrient

Niacin is active as part of the coenzyme nicotinamide adenine dinucleotide (NAD^+) or nicotinamide adenine dinucleotide phosphate ($NADP^+$) in oxidoreductases. The active form of the vitamin required for synthesis of NAD^+ or $NADP^+$ is nicotinate, and therefore nicotinamide must be deaminated before becoming available for synthesis of these coenzymes.

It is possible to synthesize niacin from tryptophan and hence, in the truest sense, niacin is not a vitamin. The conversion is, however, extremely inefficient and, for all intents and purposes, cannot supply sufficient amounts of the vitamin. In addition, the conversion requires thiamine, pyridoxine and riboflavin, and therefore on marginal diets such a synthesis would be problematical.

Severe niacin deficiency can produce dermatitis, diarrhea and dementia

Niacin deficiency produces initially a superficial glossitis but more severe deficiency states progress to pellagra, which is characterized by dermatitis, diarrhea and dementia. Certain drugs, e.g. isoniazid, also predispose to niacin deficiency. In the modern world pellagra is a medical curiosity and does not figure as a current deficiency disease.

Pyridoxine (vitamin B_6)

Vitamin B_6 exists in three active forms known as vitamers: pyridoxine, pyridoxal, and pyridoxamine, and their corresponding phosphates

Pyridoxine is the major form of vitamin B_6 in the diet, and pyridoxal phosphate is the active form of the vitamin. All forms of the vitamin are absorbed from the gut, during which some hydrolysis of the phosphates occurs. Most tissues, however, contain pyridoxal kinase, thus resynthesizing the active phosphorylated forms required for the synthesis, catabolism and interconversion of amino acids (see Chapter 2).

As with other B-complex vitamins, assessment of pyridoxine status is based on the measurement of erythrocyte enzymes, in this case, aspartate amino transferase. Owing to the central role of vitamin B_6 in amino acid metabolism, requirements for this vitamin increase with protein intake. Since energy transduction mechanisms are involved in amino acid metabolism some of the signs and symptoms of pyridoxine deficiency are similar to the other B vitamins involved in energy transduction pathways.

Pyridoxine is also required for synthesis of the neurotransmitters, serotonin and noradrenaline (see Chapter 38), and also for the synthesis of sphingosine, a component of sphingomyelin and sphingolipids (see Chapter 25). Vitamin B_6 deficiency is related to irritability, nervousness and depression in its mild form, progressing to peripheral neuropathy, convulsions and coma in severe deficiency. The vitamin is also required in heme biosynthesis (see Chapter 27) and severe deficiency is thus also associated with a sideroblastic anemia.

The antituberculosis drug, isoniazid, by binding to pyridoxine, and the oral contraceptive pill, by increasing the synthesis of enzymes requiring the vitamin, interfere strongly with pyridoxine and deficiencies may occur. Peripheral neuropathy in association with isoniazid is also well recognized. The debate concerning the contraceptive pill continues but it is generally accepted that somewhat higher doses of pyridoxine than the RDA are required to maintain appropriate enzyme function.

Biotin

Biotin is normally synthesized by intestinal flora; its deficiency can lead to depression, hallucinations, muscle pain and dermatitis

Biotin is a coenzyme in multienzyme complexes involved in carboxylation reactions in a wide range of metabolic pathways. The majority of the requirements for biotin are met from synthesis in the bowel by intestinal bacteria. Biotin deficiency therefore results from defects in its utilization.

The consumption of raw eggs can cause biotin deficiency because the egg-white protein, avidin, combines strongly with biotin, preventing its absorption and thus inducing biotin deficiency. Certain inherited enzyme deficiencies can also lead to apparent biotin deficiency syndrome, owing to single or multiple carboxylase deficiency.

Symptoms of biotin deficiency include depression, hallucinations, muscle pain and dermatitis. Children with multiple decarboxylase deficiency also demonstrate immunodeficiency disease.

Folic acid

DNA synthesis indirectly requires folic acid

Rapidly dividing cells have high requirements for this vitamin since its role is in the synthesis of purines and thiamine required for DNA synthesis (see Chapter 39). On the basis of selective toxicity in rapidly growing cells, e.g. bacteria and cancer cells, this function of folate has also formed the basis for drug developments including antibiotics and anticancer agents, e.g. trimethoprim and methotrexate respectively.

Megablastic anemia can result from folate deficiency

Folate functions in 1-carbon addition reactions in numerous pathways including the synthesis of choline, serine, methionine and nucleic acids. It is, however, the latter two that achieve medical or clinical significance, and failure to synthesize them in deficiency states accounts for the signs and symptoms of megaloblastic anemia, i.e. the presence of enlarged blast cells in the bone marrow. Deficiency of folate is one of the commonest vitamin deficiencies and the hematologic abnormalities associated with this cannot be distinguished from those of vitamin B_{12} deficiency (see below). The neurologic changes are also similar. Where there are suboptimal intakes a generalized disorder of cellular metabolism occurs. The block in synthesis slows down the production of erythrocytes, causing the appearance of macrocytic erythrocytes with fragile membranes and a tendency to hemolyze. A macrocytic anemia thus ensues in association with a megaloblastic bone marrow.

There are many causes of folate deficiency, including inadequate intake, impaired absorption, impaired metabolism and increased demand. The most common example of increased demand is during pregnancy and lactation. Folic acid requirements increase dramatically as the blood volume and number of cells increase. By the third trimester, folic acid requirements have doubled. However, megaloblastic anemias in pregnancy, other than multiple pregnancy, are rare. The general practice is to provide folate supplements during pregnancy. Folate deficiencies are seen in the elderly as a result of poor diet and poor absorption.

Vitamin B_{12}

Vitamin B_{12} (cobalamin) has a complex ring structure similar to the porphyrin system of heme (see Chapter 4) but is more hydrogenated. The iron of the heme system is replaced by a cobalt ion (Co^{3+}) at the center. In addition, and essential for the chelation of the cobalt ion, a dimethylbenzimidazole ring is also part of the active molecule (Fig 10.7).

Vitamin B_{12} is the only vitamin not synthesized by any member of the plant or animal kingdoms

Vitamin B_{12} is synthesized solely by bacteria. Vitamin B_{12} is thus absent from all plants but is concentrated in the livers of animals in three forms: methylcobalamin; adenosylcobalamin; and hydroxycobalamin. Liver is therefore a useful source of this vitamin and has been used in the treatment of deficiency states in the past.

It is impossible to consider the function of vitamin B_{12} in isolation from folate

That the roles of vitamin B_{12} and folate are interrelated is exemplified by the fact that deficiency of either produces the same signs and symptoms of disease. The reaction involving both these vitamins is the conversion of homocysteine to methionine – a methylation reaction (Fig. 10.8).

Vitamin B_{12} is required in only one further reaction, that is the isomerization of methylmalonyl Coenzyme A (CoA) to succinyl CoA. The coenzyme form of the vitamin

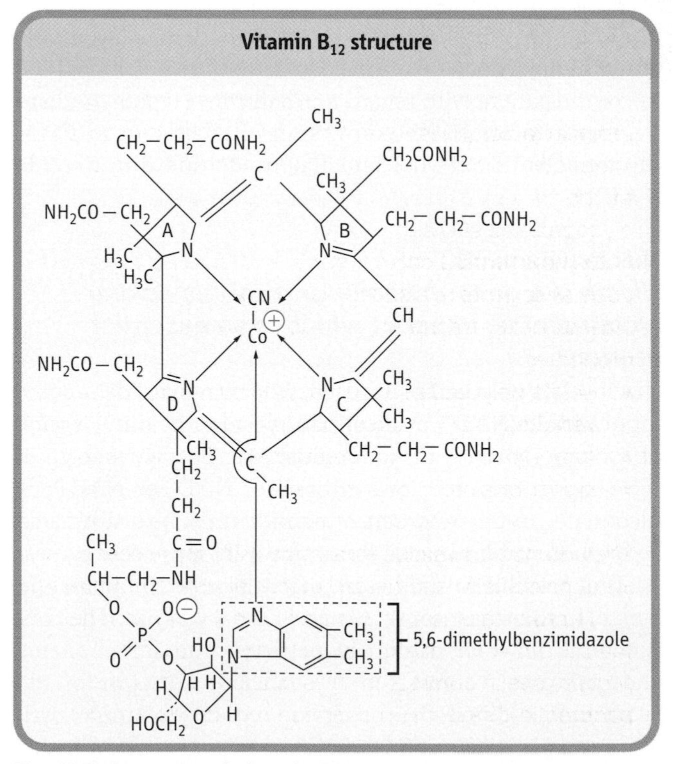

Fig. 10.7 Structure of vitamin B_{12}. The structure shown indicates the presence of a cyano-group (CN) attached to the cobalt: this is an artifact of extraction but is also the most stable form of the vitamin and indeed is the commercially available product for treatment. The cyano-group does require removal for conversion to the active form of the vitamin. This is also the only known function of cobalt in mammalian systems.

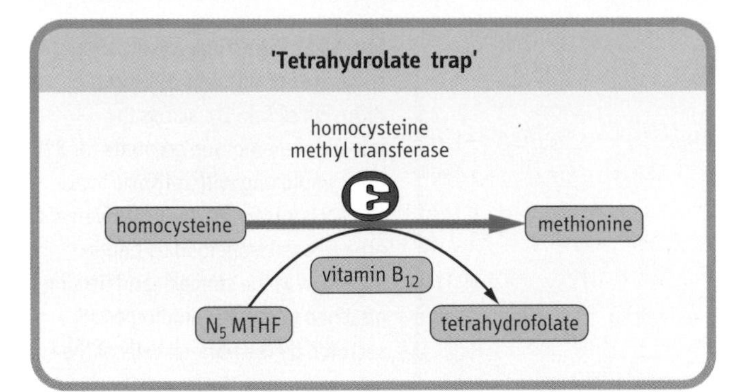

Fig. 10.8 Vitamin B_{12} and folate are involved together in the conversion of homocysteine to methionine. An absence of cobalamin (vitamin B_{12}) leads to cessation of the reaction and build up of N^5 methyltetrahydrofolate (N_5MTHF), known as the 'tetrahydrafolate trap'.

in this case is 5′-deoxyadenosyl cobalamin. Specific mechanisms exist for the absorption and transport of cobalamin (Fig. 10.9).

The megaloblastic anemia characteristic of vitamin B_{12} deficiency is probably due to a secondary deficiency of reduced folate, in other words, as a consequence of the accumulation of N_5-methyltetrahydrofolate; hence the folate/B_{12}-associated syndrome. A neurologic presentation also exists and can present in the absence of anemia. This is known as subacute combined degeneration of the cord. This neurologic disorder is probably secondary to a relative deficiency of methionine in the cord.

Since vitamin B_{12} is required in only two reactions, deficiency of this vitamin will result in an accumulation of methylmalonic acid and homocysteine. This will therefore result in methylmalonic aciduria and homocystinuria.

Vitamin B_{12} deficiency can occur through several mechanisms. The one most commonly seen is that termed 'pernicious anemia', which is due to lack of intrinsic factor (IF) in the stomach and hence absorption is prevented, for other reasons, this will obviously cause failure of production of IF, which should be remembered in cases of gastric surgery. A similar situation will also arise upon surgical removal of the ileum, e.g. as in Crohn's disease (see Chapter 9). Vegans are at risk of developing a true dietary deficiency since the vitamin is found only in foods of animal origin, unless the vegetable diet is contaminated with microorganisms, such as yeasts. Vitamin B_{12} is also secreted in the bile and there is a marked enterohepatic circulation of vitamin B_{12}.

Vitamin B_{12} transport proteins

Intrinsic factor (IF) is a highly specific glycoprotein. Other cobalamin-binding proteins, R-proteins, secreted by the salivary glands and stomach, are also glycoproteins and along with *trans*-cobalamin (TC)I and III are now termed cobalaphilins. The third type of cobalamic protein, also a glycoprotein, is TCII. All three classes of B_{12}-transport proteins have similar properties:

- single polypeptide chain (340–375 amino acid residues),
- single binding site for cobalamin,
- glycoproteins.

They do not, however, cross-react with each other immunologically, and are coded for by different genes.

R-proteins bind cobalamin more tightly at acid pH than IF but are normally degraded by pancreatic proteinases in contrast to IF, which is not. Thus, in pancreatic disease, R-proteins are not degraded and cobalamin is not available to bind to IF, with loss of absorptive capacity for this vitamin.

In the final absorption process, a specific site on the IF molecule binds with the ileal receptor in the presence of Ca^{2+} and at neutral pH. As the IF-B_{12} complex crosses the ileal mucosa, IF is released and the B_{12} is transferred to a plasma transport protein TCII. Other cobalamin-binding proteins, e.g. TCI and possibly TCIII, exist in the plasma and liver. In the latter, these provide excellent storage forms of the vitamin, a situation that is unique in terms of water-soluble vitamins, is more akin to that of vitamin A and E.

Once cobalamin is bound to TCII in portal blood, it disappears from the plasma in a few hours. The major circulating vitamer is methylcobalamin with only a trace of hydroxycobalamin. In the liver, 5′-deoxyadenosyl cobalamin accounts for 70% of the total and methylcobalamin for only 3%.

The TCII-cobalamin complex delivers exogenous cobalamin to the tissues, where it binds to specific cell-surface receptors and enters the cell by a process of endocytosis, ultimately releasing the cobalamin as hydroxycobalamin. Conversion of hydroxycobalamin to methylcobalamin occurs in the cytosol for participation in the homocysteine–methionine conversion. 5-deoxyadenosyl cobalamin is derived by a mitochondrial process from hydroxycobalamin to allow the mitochondrial reaction of the conversion of methylmalonyl CoA to succinyl CoA. TCII is also thought to be necessary for the delivery of vitamin B_{12} to the tissues of the central nervous system (CNS).

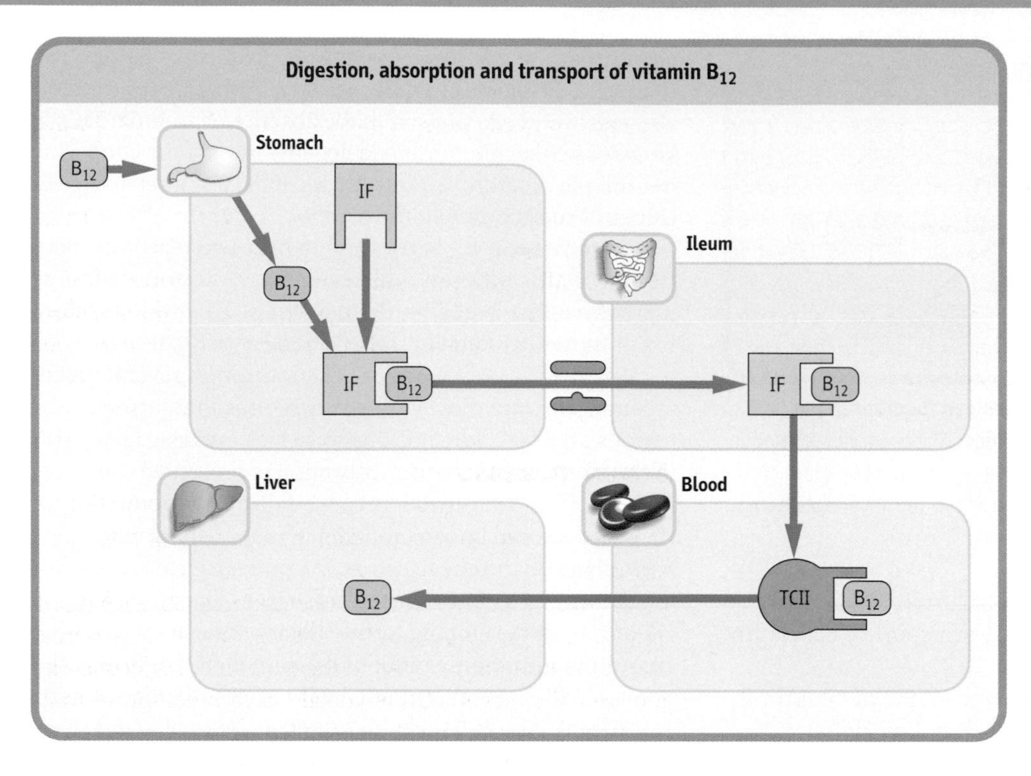

Fig. 10.9 Digestion, absorption and transport of vitamin B_{12}. Simple diffusion of free B_{12} across the intestinal membrane accounts for 3% but complexing with intrinsic factor (IF) accounts for 97%. B_{12} derivatives are released from food by peptic digestion in the stomach and become attached to specific binding on IF, secreted by the parietal cells of the gastric mucosa. IF-B_{12} complex is required for absorption by specific receptor sites on the ileal mucosa. The rate-limiting factor in this process is the number of ileal receptor sites. Other transport proteins are involved in the delivery or storage of the cobalamins, the so-called *trans*-cobalamin proteins I, II and III (TCI, II and III) and R-proteins, the latter secreted by the salivary glands and gastric mucosa.

Disturbances of this circulation can therefore have major effects on vitamin B_{12} status (Fig. 10.10).

To give folate alone in a case of vitamin B_{12} deficiency will aggravate the neuropathy. Therefore, if supplements are required during investigation of the cause of megaloblastic anemia, folate needs to be given together with vitamin B_{12}, after blood and bone marrow are taken

Vitamin C

While most of the animal kingdom synthesizes vitamin C, man has to ingest it

Vitamin C, also known as ascorbic acid, is an essential nutrient in man, the higher primates, the guinea pig and fruit-eating bats. In all other animals, a specific pathway exists for its synthesis. Studies on the dietary role of this compound in the etiology of disease should therefore be confined to the above species. The synthetic pathway and structure of vitamin C is shown in Figure 10.11.

Mechanisms of B_{12} deficiency

Mechanism	Time to develop clinical deficiency (years)
vegan diet	10–12
intrinsic factor failure	1–4
ileal dysfunction	rapid

Fig. 10.10 Mechanism of development of B_{12} deficiencies and time to develop signs and symptoms of deficiency.

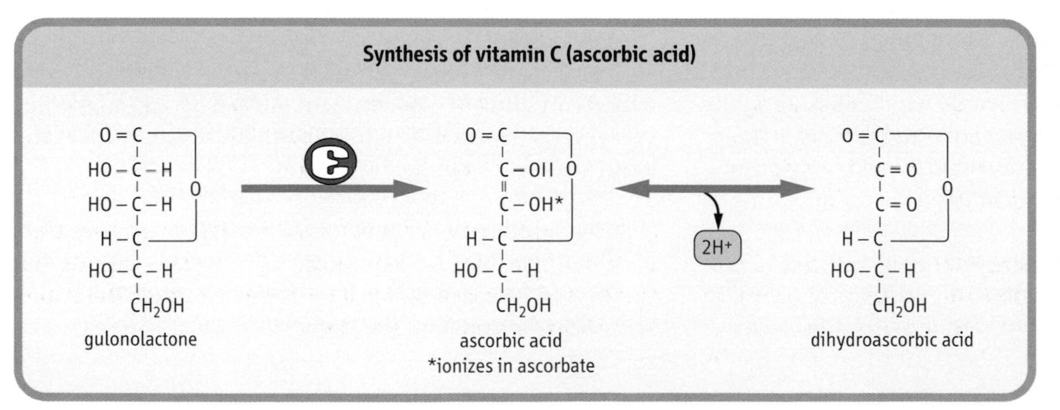

Fig. 10.11 Structure and synthesis of vitamin C (ascorbic acid). Note that the enzyme that converts gulonolactone to ascorbic acid is absent in man, higher primates, the guinea pig and the fruit-eating bat.

Vitamin C is a reducing agent

Vitamin C serves as a reducing agent and the active vitamin is ascorbic acid, which is oxidized during the transfer of reducing equivalents to dihydroascorbic acid, which in itself can act as a source of the vitamin. The mechanism of action of ascorbic acid relative to its many activities is far from clear. However, in a large proportion of synthetic degradative pathways in which it is involved, the prime function of this compound is to main metal cofactors in their lower valence state, e.g. Fe^{2+} and Cu^{+}. This is certainly the case in its role in the synthesis of collagen where it is required specifically for the hydroxylation of proline (see Chapter 26). Vitamin C participates in:

- collagen biosynthesis,
- degradation of tyrosine,
- adrenaline synthesis,
- bile acid formation,
- steroidogenesis,
- absorption of iron,
- bone mineral metabolism.

Vitamin C is an antioxidant

Its role as a general antioxidant has, in recent years, achieved greater attention. In this capacity it is thought to be involved in the prevention of atherosclerosis and coronary heart disease. Despite numerous studies the relationship between vitamin C and atherosclerosis remains unclear. Another area associated with its antioxidant properties is in the prevention and treatment of cancer. Again confusion still remains as to the importance of increased intakes of this vitamin in the area of cancer prevention and therapy.

Except in the elderly population, vitamin C deficiency resulting in the full clinical picture of scurvy is rare: in mild cases of vitamin C deficiency immune function may be decreased

Scurvy is related to the defective collagen synthesis associated with vitamin C deficiency. It is characterized by subcutaneous and other hemorrhages, muscle weakness, soft, swollen, bleeding gums, osteoporosis and poor wound healing and anemia. The osteoporosis results from the inability to maintain bone matrix in association with demineralization. This latter aspect results in the appearance of Looser's zones on radiography, especially in the hands. Milder forms of vitamin C deficiency are more common and the manifestation of such include easy bruising and the formation of petechiae (small, pinpoint hemorrhages under the skin) both due to increased capillary fragility. Immune function is also compromised in cases of mild vitamin C deficiency.

This reduction in immuncompetence is the basis of providing megadoses of the vitamin in preventing the common cold and also its role in cancer prevention. No clear evidence exists, however, to substantiate these claims first made by Linus Pauling in 1970. Certainly vitamin C is required for normal leukocyte function and leukocyte vitamin C levels drop precipitously after stress that is related to either trauma or infection.

There is no clear evidence that vitamin C taken in excess is toxic

Theoretically, however, since vitamin C is metabolized to oxalate, there is a risk of the development of renal oxalate stones in susceptible individuals, although this has not been substantiated in practice.

Trace metals

The active components of some proteins are metal ions

Metal ions are required as active components of several proteins. The most obvious of these is iron and its function as part of the proteins involved in the transfer of molecular oxygen (see Chapter 4). Other metals have been found to be essential for normal biological function. These include metals previously thought to be toxic; indeed, environmental excesses of these result in toxicity. Such metals include chromium, selenium, manganese, copper and zinc, and are now termed essential trace metals.

Zinc

Zinc is involved in carbohydrate and energy metabolism

Zinc is an integral part of numerous enzymes associated with carbohydrate and energy metabolism, both protein degradation and synthesis, nucleic acid synthesis, intercellular transport functions and protection from oxidative damage. Its primary effects, however, are most obviously seen in the maintenance of skin integrity and its involvement in wound healing. It is also important for maintaining both appropriate exocrine and endocrine pancreatic function. Spermatogenesis is a zinc-dependent process based on the metal's role in testosterone metabolism.

Absorption of zinc from the diet is an active process and shares gut transport mechanisms with copper and iron

On absorption, zinc is found bound to the protein metallothioneine, a cysteine-rich protein, which is also associated with the binding of other divalent metal ions, e.g. copper. Its synthesis is dependent on the amount of zinc in the diet and may play a role in reducing toxicity.

Zinc excess may interfere with copper absorption

Zinc is probably the least toxic of the trace metals but increased oral doses of zinc interfere with copper absorption leading to deficiency of the latter.

Zinc deficiency affects growth and sexual development in children; skin lesions may also be present

Zinc deficiency is not uncommon: in children it is characterized by growth retardation, skin lesions and impairment of sexual development. A specific inherited metabolic defect in the absorption of zinc from the gut was identified in the 1970s, termed 'acrodermatitis enteropathica' with the clinical appearance of severe skin lesions, diarrhea, and loss of hair (alopecia). Zinc deficiency also leads to impairment in taste and smell and to delayed or failed wound healing.

Increased losses of zinc occur in major burns and in patients with renal damage. Loss of zinc in the latter is due to its association with plasma albumin, lost as part of the nephropathy. Substantial amounts of zinc may be lost during dialysis. Failure to replace zinc during intravenous feeding in situations where there is frequently an increased demand also produces signs and symptoms of deficiency. The metabolic response to trauma by increasing metallothioneine synthesis also results in a reduction of serum zinc levels.

Measurement of serum zinc concentrations is the usual method of assessing zinc status. Nevertheless, many conditions and environmental factors can affect this, including inflammation, stress, cancer, smoking, steroid administration and hemolysis. More recently, erythrocyte metallothioneine levels have proved more appropriate in assessing zinc status since amounts of this protein do not alter with physiologic state.

Zinc deficiency

A 34-year-old man who required total intravenous feeding had been receiving the same prescription for some 4 months, with no assessment of his trace metal status. During this time, he continued to have major gastrointestinal losses and intermittent pyrexia. Initially, he developed a rash across his face, head and neck, with accompanying hair loss and, by the end of the 4-month period, was clearly zinc-deficient. He had a widespread acne-type rash and was virtually devoid of hair. His serum zinc concentration at that time was less than 1 mmol/L (range: 13–21 mmol/L).

Comment. Patients with major catabolic illness and increased gastrointestinal losses have markedly increased zinc requirements. The zinc-depleted state the patient developed would aggravate his illness: (a) by preventing healing of his gastrointestinal lesions; and (b) by making him more susceptible to infection due to defects in his immune competence. Patients receiving intravenous feeding need to have their micronutrient status checked regularly and prescriptions altered if required.

Copper

Copper is involved in scavenging superoxide and other oxygen free radicals

Copper is associated with several oxygenases. These include cytochrome oxidase and superoxide dismutase, the latter also requiring zinc for activity. One of the main roles of copper, especially in superoxide dismutase, but also in association with the plasma copper-carrying protein ceruloplasmin, is the scavenging of superoxide and other oxygen free radicals. Copper is also required for appropriate crosslinking of collagen, being an essential component of lysyl oxidase.

Absorption of copper from the gut is also associated with metallothioneine and excess of one interferes with the absorption of the other. Copper availability in the diet is less affected by dietary constituents than zinc, although high fiber intake reduces availability by complexing with copper.

Copper deficiency produces an anemia; skin and hair may also be affected

Copper deficiency is most likely to occur from reduced intake or excess loss, e.g. in renal dialysis. Copper deficiency manifests itself as a microcytic (small erythrocytes) microchromic (pale erythrocytes) anemia that is resistant to iron therapy. There is also a reduction in the number of leukocytes in the blood (neutropenia) and degeneration of vascular tissue with bleeding (due to defects in elastin and collagen production). Skin depigmentation and alteration in hair structure also occur in severe deficiency.

Copper excess can interfere with iron and zinc levels

Copper when taken orally is generally nontoxic but, in large doses, it accumulates in tissues and can interfere with other metal ions, specifically iron and zinc. Chronic excessive intake, however, results in cirrhosis. Acute toxicity is manifested by marked hemolysis and damage to both liver and brain cells. This latter situation is seen in the inherited metabolic defect of Wilson's disease, where the liver's capacity to synthesize cerulomin is compromised. This results in a reduced excretion of copper, with chronic accumulation of copper in the tissues.

Summary

Vitamins and minerals play an important role in nutrition. Nutritional deficiencies may develop both due to insufficient intake, and also in the course of clinical conditions which impair their absorption from the gut. Vitamin and trace metal supplements are particularly important in patients with gastrointestinal disorders who remain on artificial diets and on parenteral nutrition.

Numerous other trace metals are required for normal biologic function, for example, manganese, molybdenum, vanadium, nickel, and even cadmium. The latter is probably better known for its renal toxic effects and is seen especially in shipyard workers exposed to this metal over long periods of time.

With the passage of time, other metals and other functions of known essential minerals will be realized as techniques for separation and analysis develop. This in turn may lead to a better understanding of the epidemiology of certain diseases which may have, at least in part, an environmental aetiology.

Further reading

Banhegyi G. Braun L. Csala M. Puskas F. Mandl J. Ascorbate metabolism and its regulation in animals. *Free Radical Biology & Medicine* 1997;**23:**793–803.

Buchman AL. Vitamin supplementation in the elderly: a critical evaluation. *Gastroenterologist* 1996;**4:**262-275.

Kapadia CR. Vitamin B12 in health and disease: part I--inherited disorders of function, absorption, and transport. *Gastroenterologist*; 1995 **3:**329–44.

Linder MC. *Nutritional Biochemistry and metabolism,* 2nd ed. Elsevier; 1992.

Murray RK, Granner DK, Mayes PA, Rodwell VW. *Harpers Biochemistry,* 2nd ed. Appleton and Lange; 1996.

Roeckel IE. Dickson LG. Understanding iron absorption and metabolism, aided by studies of hemochromatosis. *Annals of Clinical & Laboratory Science* 1998;**28:**30–33.

Wapnir RA. Copper absorption and bioavailability. *American Journal of Clinical Nutrition* 1998; **67 (Suppl 5):**1054S-1060S.

Viteri FE. Iron supplementation for the control of iron deficiency in populations at risk. *Nutrition Reviews* 1997;**55:**195–209.

Wasserman RH. Fullmer CS. Vitamin D and intestinal calcium transport: facts, speculations and hypotheses. *Journal of Nutrition* 1995;**125(Suppl 7):**1971S–1979S.

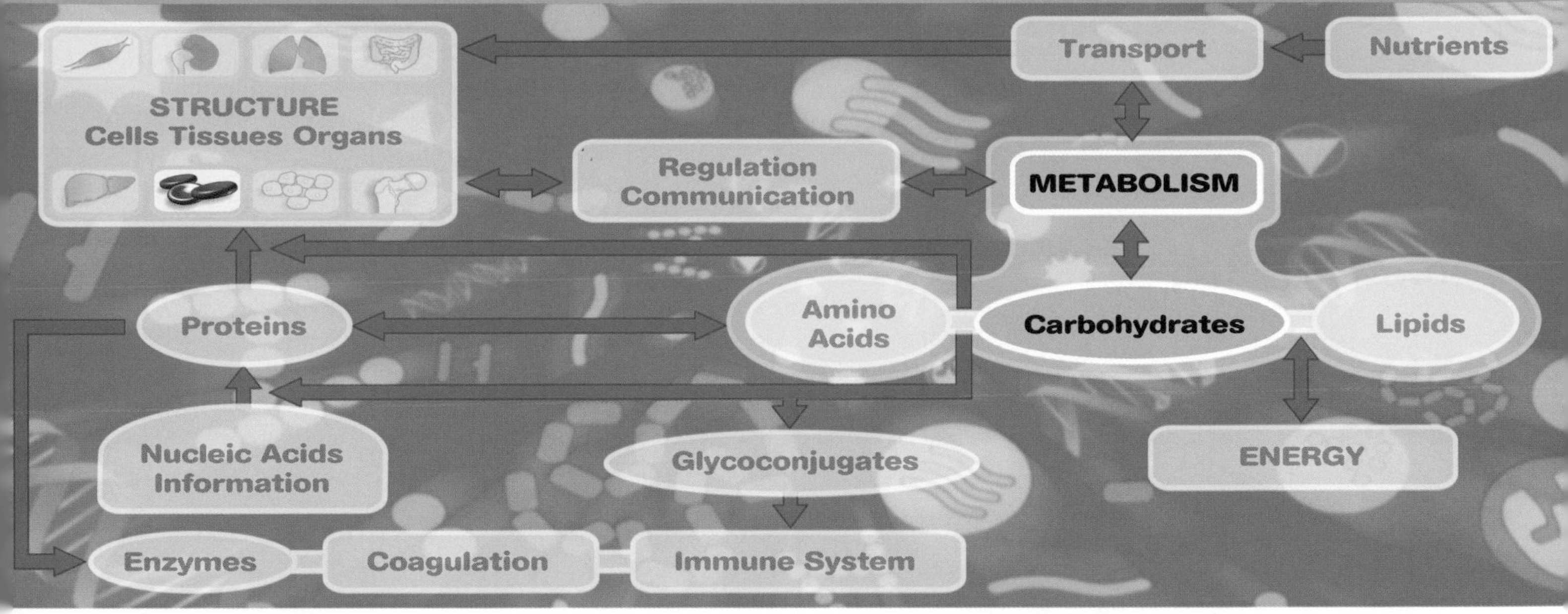

11 Anaerobic Metabolism of Glucose in the Red Cell

Introduction

Erythrocytes

The red blood cell (RBC), or erythrocyte, represents 40–45% of blood volume and over 90% of the formed elements (erythrocytes, leucocytes, and platelets) in blood. The RBC is, both structurally and metabolically, the simplest cell in the body – the end product of the maturation of bone-marrow reticulocytes. During its maturation, the RBC loses all its subcellular organelles. Without nuclei, it lacks the ability to synthesize DNA or RNA. Without ribosomes or an endoplasmic reticulum, it cannot synthesize or secrete protein. Because it cannot oxidize fats, a process requiring mitochondrial activity, the RBC relies exclusively on glucose as a fuel – it derives this glucose from blood, yielding adenosine 5′-triphosphate (ATP) by a pathway termed glycolysis, i.e. carbohydrate (*glyco*) splitting (*lysis*). The metabolism of glucose in the RBC is entirely anaerobic, consistent with the primary role of the RBC in oxygen transport and delivery, rather than its utilization.

Glycolysis

Glucose is the major carbohydrate on Earth, the backbone and monomer component of cellulose and starch. Glycolysis is the ubiquitous, central metabolic pathway for glucose metabolism in the biosphere. It is found not only in all mammalian cells, but even in yeast and bacteria. It is a relatively inefficient bioenergetic pathway, recovering only about 5% of the energy available by oxidative metabolism of glucose in cells containing mitochondria. However, even in aerobic cells with mitochondria, such as brain, liver, and muscle, oxidative metabolism of glucose proceeds through glycolysis. Pyruvate, a three-carbon acid, is the end product of glycolysis; 2 moles of pyruvate are formed per mole of glucose. In cells with mitochondria and oxidative metabolism, pyruvate is converted completely into CO_2 and H_2O – glycolysis in this oxidative setting is termed *aerobic* glycolysis. In RBCs, which lack mitochondria and oxidative metabolism, pyruvate is reduced to lactic acid, a three-carbon hydroxyacid, the product of *anaerobic* glycolysis. Each mole of glucose yields 2 moles of lactate, which are then excreted into blood. Two lactic acid molecules contain exactly the same number of carbons, hydrogen, and oxygen as 1 molecule of glucose; however, there is sufficient free energy available from the cleavage and rearrangement of glucose molecules to produce 2 moles of ATP per mole of glucose converted into lactate. Most of this ATP is used by the RBC to maintain electrochemical and ion gradients across its plasma membrane. About 10% of glucose is diverted to the pentose phosphate pathway to protect hemoglobin (Hb) and other cell constituents against oxidative damage, and another 10–20% to the synthesis of 2,3-bisphosphoglycerate (2,3-BPG), an allosteric regulator of the O_2 affinity of Hb. The commitment to synthesis of 2,3-BPG is a special function of RBC metabolism, but glycolysis and the pentose phosphate pathway are found together in all mammalian cells.

Glycolysis

Overview (Fig 11.1)

Glucose enters the RBC by facilitated diffusion, *via* the insulin-independent glucose transporter, GLUT-1. The glucose concentration in the RBC is not significantly different from that in plasma. Glycolysis proceeds through a series of phosphorylated intermediates, starting with the synthesis of glucose-6-phosphate (G-6-P). During this process, which involves 10 distinct enzymatically catalyzed steps, two molecules of ATP are expended (*investment* stage) to build up a nearly symmetric intermediate, fructose-1,6-bisphosphate (F-1,6-BP), which is then cleaved (*splitting* stage) to two three-carbon triose phosphates. These are eventually converted into lactate during the *yield* stage of glycolysis. The *yield* stage includes both redox and phosphorylation reactions, leading to formation of four molecules of ATP during the conversion of the triose phosphates into lactate. Two moles of ATP are formed from each mole of triose phosphate, yielding a net 2 moles of ATP per mole of glucose converted into lactate.

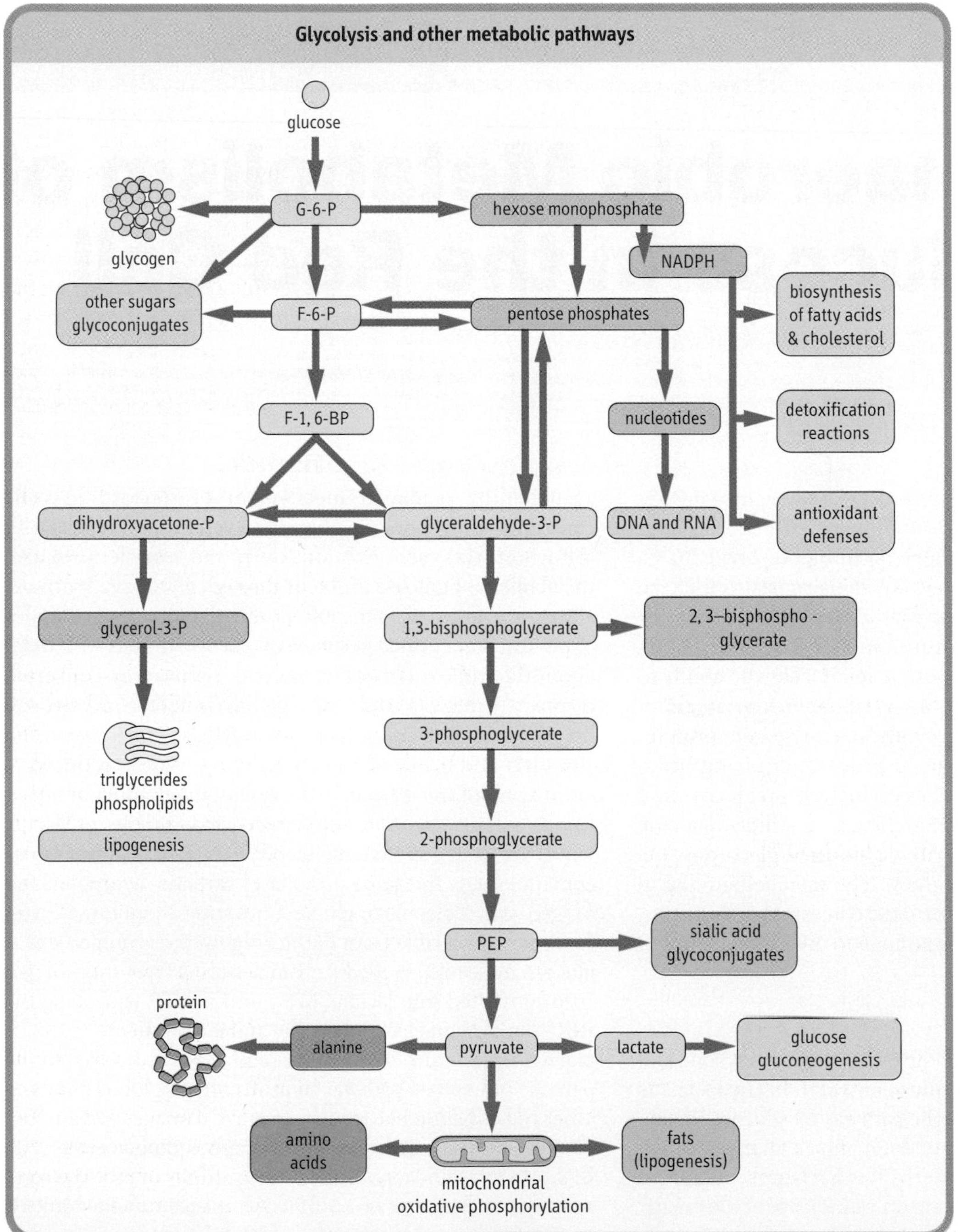

Fig. 11.1 Interactions between glycolysis and other metabolic pathways. The green colored boxes indicate intermediates involved in the pathway of glycolysis. Other boxes illustrate some of the metabolic interactions between glycolysis and other metabolic pathways in the cell.

Glycolysis is a relatively inefficient pathway for extracting energy from glucose: the yield of 2 moles of ATP per mole of glucose is only about 5% of the 36–38 ATPs that are formed by oxidation of glucose completely to CO_2 and H_2O in the brain. One might ask why this 10-step pathway is so complex – couldn't it have been done in fewer steps? The answer is, from a metabolic point of view, that glycolysis is a central pathway, and most of the intermediates provide a branch point to other metabolic pathways. Thus, through various glycolytic intermediates, the metabolism of glucose intersects with the metabolism of fats and proteins, as well as with alternative pathways of metabolism of glucose and other sugars. Some of these metabolic interactions are shown in Fig. 11.1.

Glucose utilization in the red cell

In a 70 kg person, there are about 5 L of blood and a little over 2 kg (2 L) of RBCs. These cells constitute about 3% of total body mass and consume ≈ 20 g (0.1 mole) of glucose per day, representing about 10% of total body glucose metabolism. The RBC has the highest specific rate of glucose utilization of any cell in the body, approximately 10 g of glucose/kg of tissue/day, compared with ≈ 2.5 g of glucose/kg of tissue/day for the whole body. In the RBC, about 90% of this glucose (≈ 8 g or 0.1 mole) is metabolized via glycolysis, yielding ≈ 0.2 mole of lactate (≈ 18 g/day). Despite its high rate of glucose consumption, the RBC has one of the lowest rates of ATP synthesis of any cell in the body, ≈ 0.2 mole of ATP (100 g) per day, reflecting the fact that most of its glucose metabolism is carried out by anaerobic glycolysis, which produces lactate and traps only a fraction of the energy available from complete combustion of glucose to CO_2 and H_2O.

The investment stage of glycolysis

Glucose-6-phosphate

The first step in the commitment of glucose to glycolysis is the phosphorylation of glucose to G-6-P, catalyzed by the enzyme hexokinase (Fig. 11.2, top). The formation of G-6-P from free glucose and inorganic phosphate is energetically unfavorable, so that a molecule of ATP must be expended or *invested* in the phosphorylation reaction. The product, G-6-P, is now trapped in the RBC, along with other phosphorylated intermediates in glycolysis, because there are no transport systems for sugar phosphates in the plasma membranes of mammalian cells. Hexokinase has a low K_m (0.1 mmol/L) for its substrates, and is normally saturated with both ATP and glucose, present in the RBC at 1–2 and 4–5 mmol/L concentration, respectively.

Fructose-6-phosphate

The second step in glycolysis is the conversion of G-6-P into fructose 6-phosphate (F-6-P) by phosphoglucose isomerase. Isomerases catalyze freely reversible equilibrium reactions, in this case an aldose–ketose interconversion. The product, F-6-P, can now be phosphorylated at C-1 by phosphofructokinase-1 (PFK-1) to yield the pseudosymmetric intermediate, fructose 1,6-bisphosphate (F-1,6-BP). PFK-1 requires ATP as a substrate and, like hexokinase, catalyzes an essentially irreversible reaction (K_{eq} ≈ 500). Both hexokinase and PFK-1 are important regulatory enzymes in glycolysis, but PFK-1 is the critical step. This reaction commits glucose to glycolysis, the only pathway for metabolism of F-1,6-BP in the red cell.

The splitting stage of glycolysis

In the splitting stage of glycolysis, F-1,6-BP is cleaved in the middle by a reverse-aldol reaction (Fig. 11.2, bottom). The aldolase reaction is a freely reversible equilibrium reaction, yielding dihydroxyacetone phosphate and glyceraldehyde 3-phosphate, from the top and bottom half of the molecule, respectively. Only the glyceraldehyde 3-phosphate continues through the *yield* stage of glycolysis, but triosephosphate isomerase catalyzes the interconversion of dihydroxyacetone phosphate into glyceraldehyde 3-phosphate, so that both halves of the glucose molecule are metabolized to lactate. Retro-aldol reactions require a carbonyl group adjacent to the cleavage site, providing a chemical rationale for shifting the carbonyl group at C-1 of G-6-P to the C-2 position of F-6-P.

The yield stage of glycolysis

The yield stage of glycolysis produces 4 moles of ATP, yielding a net of 2 moles of ATP per mole of glucose converted into lactate (Fig. 11.3). The synthesis of ATP is accomplished by kinases that catalyze *substrate-level phosphorylation*, a process in which a high-energy phosphate compound is formed, and then transfers its phosphate to ATP. Since only two phosphates are available at this point, one per mole of triose phosphate, two additional phosphates must be recruited from the soluble, cytoplasmic pool. These phosphates are trapped in the form of an acyl phosphate in the first step of the *yield* stage. The reaction is catalyzed by glyceraldehyde 3-phosphate dehydrogenase (G3PDH), yielding 1,3-biphosphoglycerate (1,3-BPG), simultaneously with reduction of NAD^+ to NADH. The energy for the reduction of NAD^+ and the formation of the acyl phosphate group is provided by the oxidation of the aldehyde group of glyceraldehyde 3-phosphate to the carboxylic acid group in 1,3-BPG.

Glyceraldehyde-3-phosphate dehydrogenase

The G3PDH reaction provides an interesting illustration of the role of enzyme-bound intermediates in the formation of high-energy phosphates. How does the oxidation of an aldehyde and the reduction of NAD^+ lead to the formation of an acylphosphate bond in 1,3-BPG? How does the phosphate enter the picture, and become activated to a high-energy state? The inhibition of G3PDH by iodoacetamide,

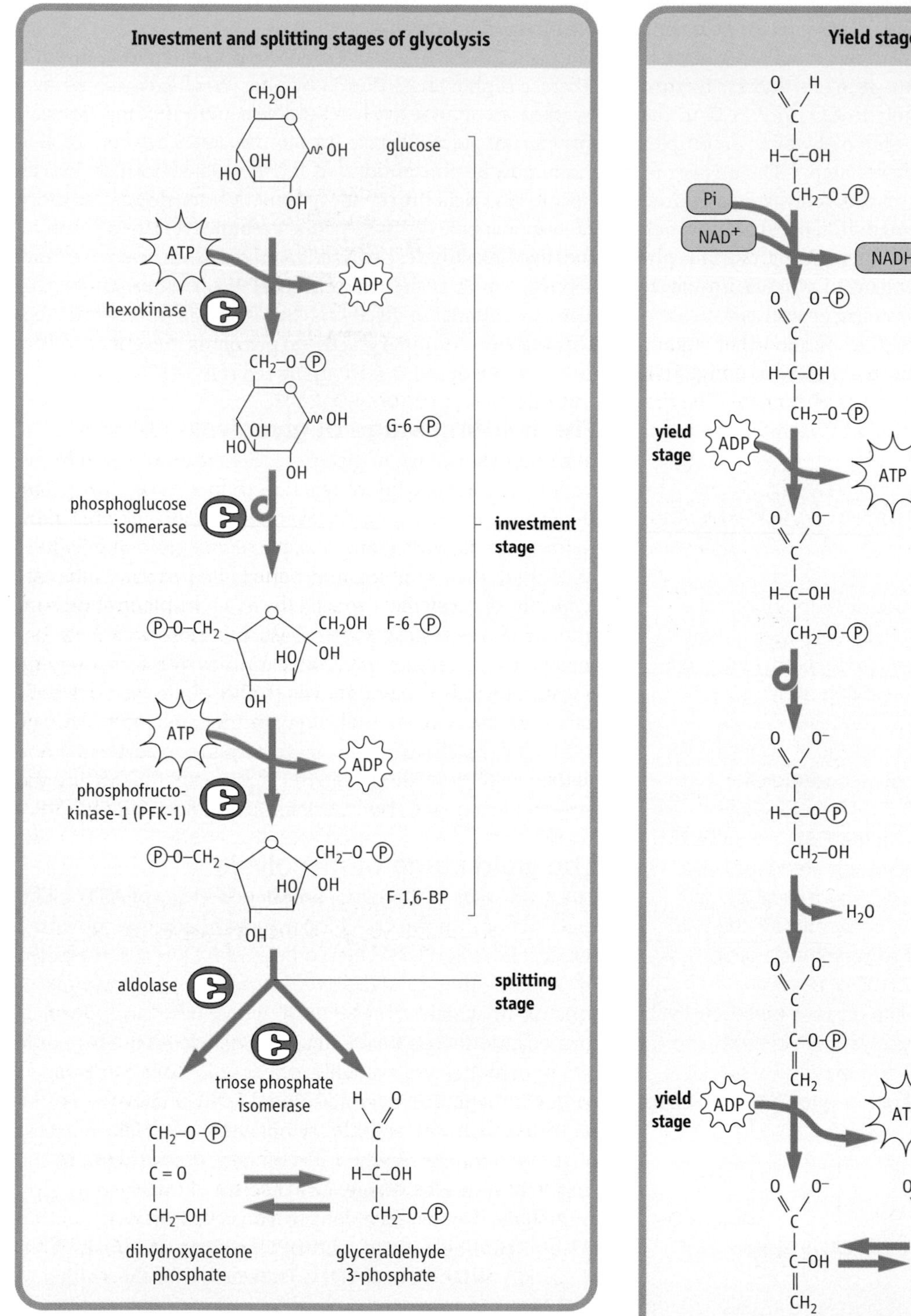

Fig. 11.2 The *investment* and *splitting* stages of glycolysis. Note the consumption of ATP at the hexokinase and phosphofructokinase-1 reactions.

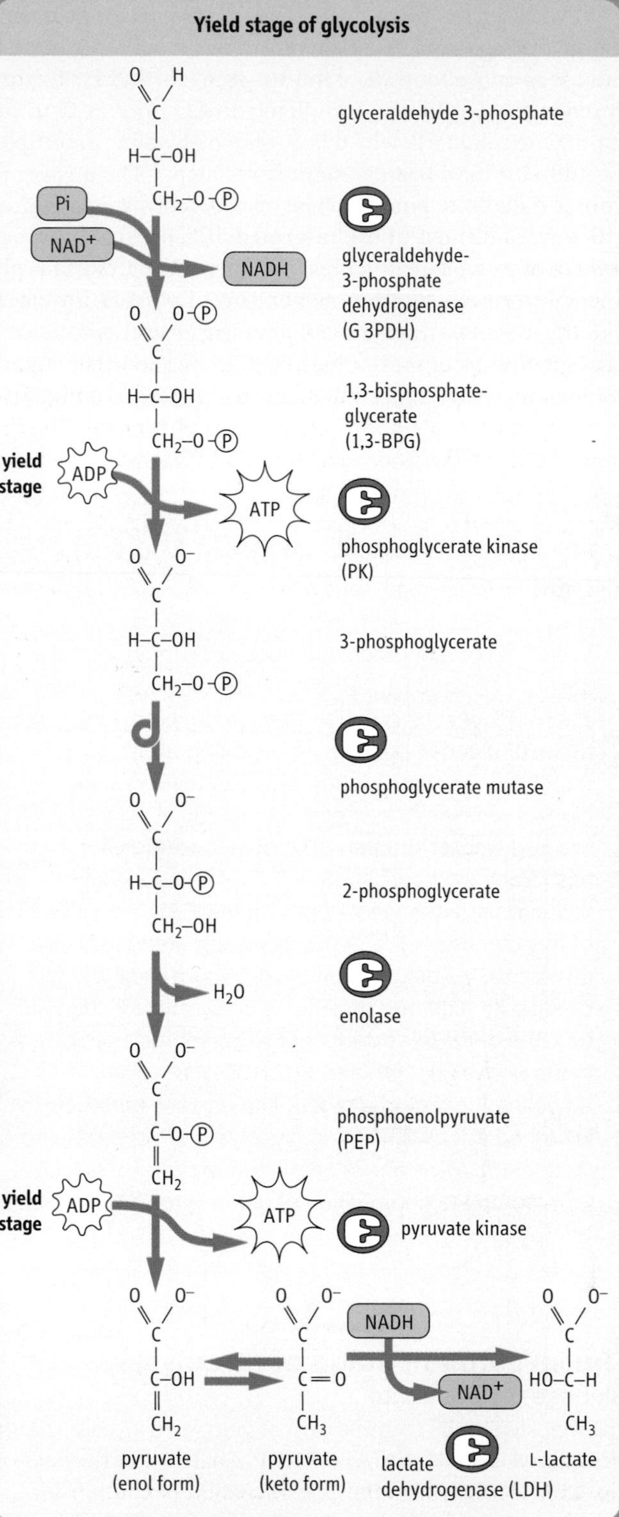

Fig. 11.3 The *yield* stage of glycolysis. Substrate-level phosphorylation reactions catalyzed by phosphoglycerate kinase and pyruvate kinase produce ATP, using the high-energy compounds 1,3-bisphosphoglycerate and phosphoenolpyruvate, respectively. Note that the NADH produced during the glyceraldehyde 3-phosphate dehydrogenase reaction is converted back into NAD^+ during the lactate dehydrogenase reaction, permitting continued glycolysis in the presence of only catalytic amounts of NAD^+.

p-chloromecuribenzoate and *N*-ethylmaleimide pointed to involvement of an active-site sulfhydryl residue, leading to the proposed mechanism of action of this enzyme, described in Fig. 11.4.

Substrate-level phosphorylation

Phosphoglycerate kinase catalyzes transfer of the phosphate group from the acyl phosphate of 1,3-BPG to ADP. This *substrate-level phosphorylation* reaction yields the first ATP produced in glycolysis. The remaining phosphate group in 3-phosphoglycerate is an ester phosphate and does not have enough energy to phosphorylate ADP, so a series of isomerization and dehydration reactions are enrolled to convert the ester phosphate into a high-energy enol phosphate. The first step is to move the phosphate to C-2 of glycerate, converting 3-phosphoglycerate into 2-phosphoglycerate, catalyzed by the enzyme phosphoglycerate mutase (see Fig. 11.3). This enzyme has an active-site histidine residue, and a phosphohistidine intermediate is formed on the enzyme during catalysis. 2-Phosphoglycerate then undergoes a dehydration reaction, catalyzed by enolase, to yield phosphoenolpyruvate (PEP), a high-energy phosphate compound. PEP is then used by pyruvate kinase to phosphorylate ADP, yielding pyruvate and the second ATP, again by *substrate-level phosphorylation*. It seems strange that the high-energy phosphate bond in PEP can be formed by a simple sequence of isomerization and dehydration reactions. However, the thermodynamic driving force for these reactions is probably derived from charge–charge repulsion between the phosphate and carboxylate groups of 2-phosphoglycerate and the isomerization of enolpyruvate to pyruvate following the phosphorylation reaction.

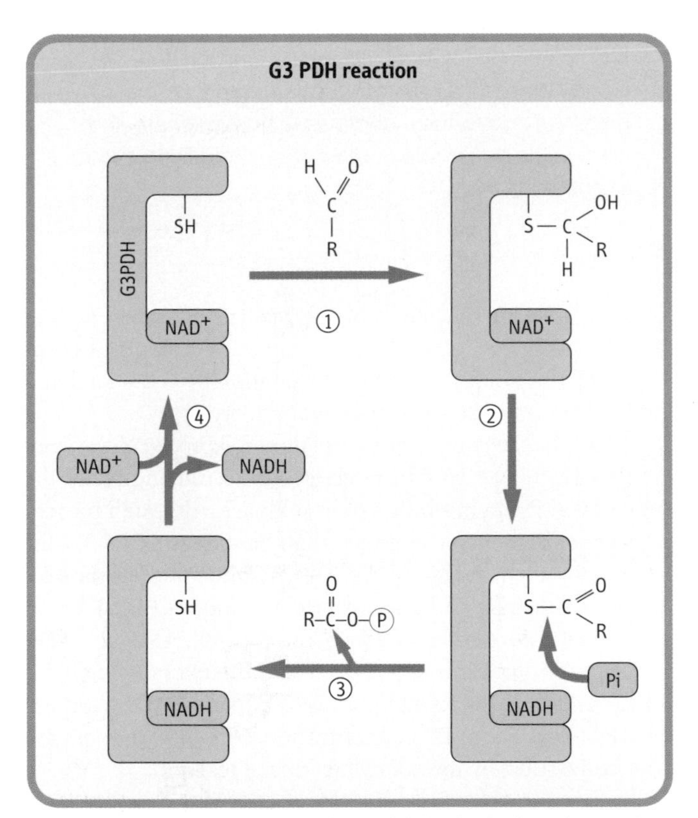

Fig. 11.4 Mechanism of the glyceraldehyde-3-phosphate dehydrogenase (G3PDH) reaction. In Step 1, an active-site sulfhydryl group of G3PDH forms a thiohemiacetal adduct with glyceraldehyde-3-phosphate. In Step 2, the thiohemiacetal is oxidized to a thioester by NAD^+, also bound in the active site of the enzyme. In Step 3, phosphate enters the active site and, in a phosphorylase reaction, displaces the thiol group, yielding 1,3-bisphosphoglycerate and regenerating the sulfhydryl group. In Step 4, the enzyme exchanges NADH for NAD^+, completing the catalytic cycle.

Lactate dehydrogenase (LDH)

Phosphoglycerate kinase and pyruvate kinase catalyze the ATP-generating reactions of glycolysis, yielding 2 moles of ATP per mole of triose phosphate, for a total of four moles of ATP per mole of F-1,6-BP. After adjustment for the ATP invested in the hexokinase and PFK-1 reactions, the net energy yield is 2 moles of ATP per mole of glucose converted into pyruvate. Two molecules of pyruvate have exactly the same number of carbons and oxygens as 1 molecule of glucose; however, there is a deficit of four hydrogens – each pyruvate has four hydrogens, a total of eight hydrogens for two pyruvates, compared with 12 in a molecule of glucose. The 'missing' four hydrogens remain in the form of 2NADH and $2H^+$ formed on reduction of NAD^+ in the G3PDH reaction. Since NAD^+ is present in only catalytic amounts in the cell and is an essential cofactor for glycolysis, there must be a mechanism for reoxidation of NADH and

Inhibition of substrate-level phosphorylation by arsenate

Arsenic is just below phosphorus in the Periodic Chart of the Elements, and it might be expected to share some of the properties and reactivity of phosphate. In fact, arsenate has pK_a values similar to those of phosphate and can actually be used by G3PDH, producing 1-arsenato-3-phosphoglycerate. However, the acyl–arsenate bond is unstable and hydrolyzes rapidly in water. Because the high-energy acyl-phosphate bond is discharged nonenzymatically, ATP is not generated by substrate-level phosphorylation. While arsenate does not inhibit any of the enzymes of glycolysis, it dissipates the redox energy available from the G3PDH reaction and prevents the formation of ATP by substrate-level phosphorylation at the phosphoglycerate kinase reaction. In effect, arsenate *uncouples* the energy available from oxidation of G3PDH for the phosphorylation of adenosine diphosphate (ADP). In the presence of arsenate, the net yield of ATP from anaerobic glycolysis drops to zero moles of ATP per mole of glucose converted to lactate.

regeneration of NAD^+. This is accomplished under anaerobic conditions by the lactate dehydrogenase (LDH) reaction, in which pyruvate is reduced to lactate by NADH, regenerating NAD^+. In mammals, all cells have LDH, and lactate is the end product of glycolysis under anaerobic conditions. Under aerobic conditions, mammalian cells use molecular O_2 and mitochondrial reactions to oxidize NADH to NAD^+ and pyruvate to CO_2 and H_2O, so that lactate is not formed. Despite the capacity of oxidative metabolism, however, some cells may at times 'go glycolytic', forming lactate, e.g. in muscle during O_2 debt and in phagocytes in poorly perfused, infected tissues.

Some microorganisms have an alternative mechanism for oxidizing NADH formed during glycolysis. During fermentation in yeast, for example, the pathway of glycolysis is identical with that in the RBC, except that pyruvate is converted into ethanol. The pyruvate is first decarboxylated by pyruvate carboxylase to acetaldehyde, releasing CO_2. The NADH produced in the G3PDH reaction is then re-oxidized by an alcohol dehydrogenase, regenerating NAD^+ and producing ethanol (Fig. 11.5). Ethanol is a toxic compound, and yeast die when the ethanol concentration in their medium reaches about 12%, which is the approximate concentration of alcohol in natural wines.

Regulation of glycolysis

To the best of our knowledge, RBCs consume glucose at a fairly steady rate. They are not physically active like muscle and do not require energy for transport of O_2 or CO_2. Glycolysis in red cells appears to be regulated simply by the energy needs of the cell, i.e. the requirement for ATP, which is relatively constant. The balance between ATP consumption and production is controlled allosterically at three sites, the hexokinase, PFK-1, and pyruvate kinase reactions. Based on measurements of the V_{max} of the various enzymes in RBC lysates *in vitro*, hexokinase is

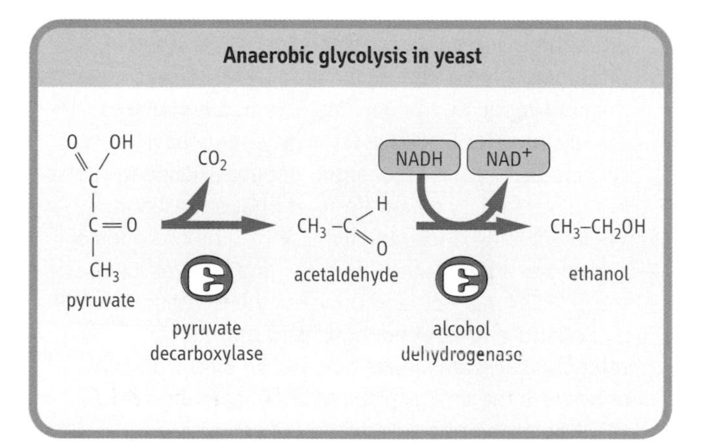

Fig. 11.5 Formation of ethanol by anaerobic glycolysis during fermentation. Pyruvate is decarboxylated by pyruvate decarboxylase, yielding acetaldehyde and CO_2. Alcohol dehydrogenase uses NADH to reduce acetaldehyde to ethanol, regenerating NAD^+ for glycolysis.

Inhibition of enolase by fluoride

Measurement of blood glucose concentration is used for the diagnosis of diabetes. Frequently these measurements are made in the clinical laboratory more than 1 h after the collection of the blood sample. Because RBCs can metabolize glucose to lactate, even in a sealed, anaerobic vial, the glucose in blood will be consumed, with concomitant production of lactate, which will lead to acidification of the blood sample. These reactions proceed in RBCs at room temperature, so that both blood glucose concentration and pH will decrease during storage. How can this be prevented? This is readily achieved by adding an inhibitor of glycolysis. Sulfhydryl reagents would work – they are inhibitors of G3PDH; however, most blood samples are collected with a small amount of a much cheaper reagent, sodium fluoride, in the sample-collection vial. Fluoride is a strong competitive inhibitor of enolase, blocking glycolysis and lactate production in the RBC. It is an unusual competitive inhibitor, since fluoride bears little resemblance to 2-phosphoglycerate. In this case, fluoride forms a complex with phosphate and Mg^{2+} in the active site of the enzyme, blocking access of substrate.

present at the lowest activity of all glycolytic enzymes. Its V_{max} is in fact only five times the rate of glucose consumption by the RBC. It is subject to product inhibition by G-6-P so that it probably never operates at maximal activity.

PFK-1 has the dominant regulatory role in glycolysis, controlling the flux of F-6-P through glycolysis and, indirectly, the level of G-6-P and inhibition of hexokinase. Although present at 20 times higher concentration than hexokinase, PFK-1 activity is uniquely sensitive to the energy status of the cell. ATP is both a substrate and an allosteric inhibitor of PFK-1, exerting fine control over the activity of the enzyme. AMP and ADP relieve the inhibition by ATP, so that the overall activity of PFK-1 and thus the rate of glycolysis depends on the ratio of (AMP + ADP) to ATP concentrations. These products are interconvertible by the adenylate kinase reaction:

$$2\,ADP \rightleftharpoons ATP + AMP$$

When ATP is consumed and ADP increases, AMP is formed. The increasing ADP and AMP concentrations relieve the inhibition of PFK-1 by ATP, activating glycolysis. The phosphorylation of ADP during glycolysis and then of AMP by the adenylate kinase reaction gradually restores the ATP concentration or 'energy charge' of the cell, and, as the AMP concentration declines, the rate of glycolysis decreases to a steady-state level.

As shown in Figure 11.6, the concentration of ATP in the RBC (1.5 mmol/L), is poised at the steep point in the concentration–response curve for inhibition of PFK-1. AMP relieves this inhibition, but is present at 50-fold lower concentrations (0.03 mmol/L). Thus, small changes in ATP concentration in the RBC yield much larger relative changes in AMP concentration. The activity of PFK-1 is exquisitely sensitive to changes in cellular AMP concentration and the overall energy status of the cell.

In addition to regulation at hexokinase and PFK-1, pyruvate kinase in liver is activated by F-1,6-BP, the product of the PFK-1 reaction. This process, known as feed-forward regulation, may be important in the RBC to limit the accumulation of reactive triose phosphate intermediates in the cytosol (see the discussion of the glyoxylase pathway below).

Each of the three enzymes involved in regulation of glycolysis – hexokinase, PFK-1, and pyruvate kinase – has characteristic features of regulatory enzymes: they are dimeric or tetrameric enzymes, are present at low V_{max} in comparison with other enzymes in the pathway, and catalyze irreversible reactions. The regulation of glycolysis in liver, muscle, and other tissues is more complicated, because of variations in their rates of fuel consumption and the interplay between carbohydrate and lipid metabolism during aerobic metabolism. In these tissues, the amount and activity of the regulatory enzymes are also regulated by covalent modification and other allosteric effectors.

Synthesis of 2,3-bisphosphoglycerate

2,3-Bisphosphoglycerate (2,3-BPG) (Fig. 11.7) is an important byproduct of glycolysis in the RBC, sometimes reaching 5 mmol/L concentration, comparable with the molar concentration of Hb in the RBC. It is in fact the major phosphorylated intermediate in the erythrocyte, present at even higher concentrations than ATP (1–2 mmol/L) or inorganic phosphate (1 mmol/L). 2,3-BPG is a negative allosteric effector of the O_2 affinity of Hb. It decreases the O_2 affinity of deoxyhemoglobin, promoting the release of O_2 in peripheral tissue, without significantly affecting O_2 saturation at the higher O_2 tension in the lungs. Its presence in the RBC explains the observation that the O_2 affinity of purified HbA is greater than that of whole RBCs. 2,3-BPG concentration increases in the RBC during adaptation to higher altitudes and in anemia, promoting the release of O_2 to tissues when the O_2-transport capacity of blood is decreased. Fetal Hb is less sensitive than adult Hb to the effects of 2,3-BPG, so that 2,3-BPG in maternal erythrocytes is one factor promoting efficient transfer of O_2 across the placenta from HbA to HbF.

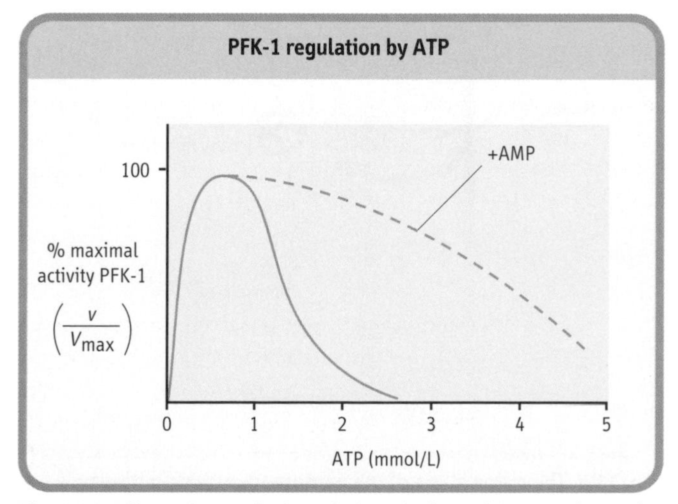

Fig. 11.6 Allosteric regulation of phosphofructokinase-1 (PFK-1) by ATP. AMP is a potent activator of PFK-1 in the presence of ATP.

The pentose phosphate pathway

Overview

The pentose phosphate pathway is a cytosolic pathway present in all cells, so named because it is the primary pathway for formation of pentose phosphates for synthesis of nucleotides, DNA, and RNA. This pathway branches from glycolysis at the level of G-6-P, thus its alternative designation, the hexose monophosphate pathway. The pentose phosphate pathway is also described as a shunt, rather than a pathway, because, when pentoses are not needed for biosynthetic reactions, the pentose phosphate intermediates are cycled back into the mainstream of glycolysis by conversion into F-6-P and glyceraldehyde-3-phosphate. This rerouting is especially important in the RBC and in nondividing cells, where there is limited need for synthesis of DNA and RNA. In addition to pentose phosphates, the major product of the pentose phosphate pathway is NADPH. In nucleated cells with active lipid biosynthesis, such as lactating mammary glands, the adrenal cortex, and the liver, the NADPH is used in redox reactions

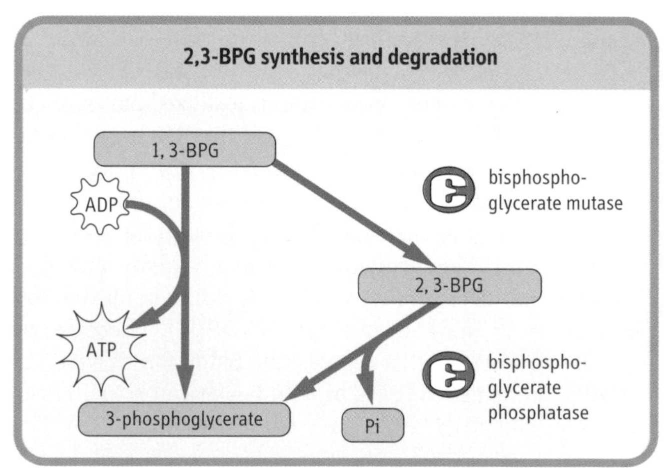

Fig. 11.7 Pathway for biosynthesis and degradation of 2,3-bisphosphoglycerate (2,3-BPG). BPG mutase catalyzes the conversion of 1,3-BPG into 2,3-BPG. The same enzyme has bisphosphoglycerate phosphatase activity, which hydrolyzes the 2-phosphate group, yielding 3-phosphoglycerate. Note that this pathway bypasses the phosphoglycerate kinase reaction, so that the overall yield of ATP per mol of glucose is decreased.

required for biosynthesis of fatty acids, cholesterol, steroid hormones, and bile salts. The liver also uses NADPH for hydroxylation reactions involved in the detoxification and excretion of drugs. The RBC has little biosynthetic activity, but still shunts about 10% of glucose through the pentose phosphate pathway, in this case almost exclusively for the production of NADPH. The NADPH is used primarily for the reduction of a cysteine-containing tripeptide, glutathione, an important intermediate in antioxidant defenses.

The pentose phosphate pathway is divided into an irreversible *redox* stage, which yields both NADPH and pentose phosphates, and a reversible *interconversion* stage, in which excess pentose phosphates are converted into glycolytic intermediates. Both stages are important in the RBC, since it needs NADPH for reduction of glutathione, but has limited need for *de novo* synthesis of pentoses.

The redox stage of the pentose phosphate pathway: synthesis of NADPH

NADPH is synthesized by two dehydrogenases in the first and third reactions of the pentose phosphate pathway (Fig. 11.8). The first mole of NADPH is produced in the G-6-P dehydrogenase (G6PDH) reaction, in which the aldehyde group of G-6-P is oxidized to an acid, yielding 6-phosphogluconic acid lactone, a cyclic sugar acid. The lactone is hydrolyzed to 6-phosphogluconic acid, catalyzed by the enzyme lactonase. Oxidative decarboxylation of 6-phosphogluconate, catalyzed by 6-phosphogluconate dehydrogenase, then yields the ketose sugar, ribulose 5-phosphate, plus 1 mole of CO_2 and the second and final mole of NADPH per mole of G-6-P.

G-6-P and 6-phosphogluconate dehydrogenase maintain a cytoplasmic ratio of NADPH:NADP$^+$ ≈100. Interestingly, because NAD$^+$ is required for glycolysis, the ratio of NADH:NAD$^+$ in the cytoplasm is nearly the inverse, less than 0.01. Although the total concentrations (oxidized plus reduced forms) of NAD(H) and NADP(H) in the RBC are similar (25 μmol/L), the cell maintains these two redox systems with similar redox potentials at such different set-points in the same cell by isolating their metabolism through the specificity of cytoplasmic dehydrogenases. The glycolytic enzymes (G3PDH and LDH) use only NAD(H), while pentose phosphate pathway enzymes use only NADP(H). There are no enzymes in the RBC that catalyze the reduction of NAD$^+$ by NADPH, so that high levels of both NAD$^+$ and NADPH can exist simultaneously in the same compartment.

The interconversion stage of the pentose phosphate pathway (Fig. 11.9)

In cells with active nucleic acid synthesis, ribulose 5-phosphate may be isomerized to ribose 5-phosphate for synthesis of ATP and other ribonucleotides (Fig. 11.9, top right). However, in nondividing cells, the majority of pentose phosphates are routed back to glycolysis. This is accomplished by a series of equilibrium reactions in which 3 moles of ribulose 5-phosphate are converted into 2 moles of F-6-P and 1 mole of glyceraldehyde 3-phosphate. Certain restrictions are imposed on the interconversion reactions, i.e. they may be carried out only by transfer of two or three carbon units between sugar phosphates. Each reaction must also involve a ketose donor and an aldose receptor. Isomerases and epimerases provide the five-carbon aldose and ketose phosphate substrates for the *interconversion* stage. Transketolase, a thiamine-dependent enzyme, catalyzes the two-carbon

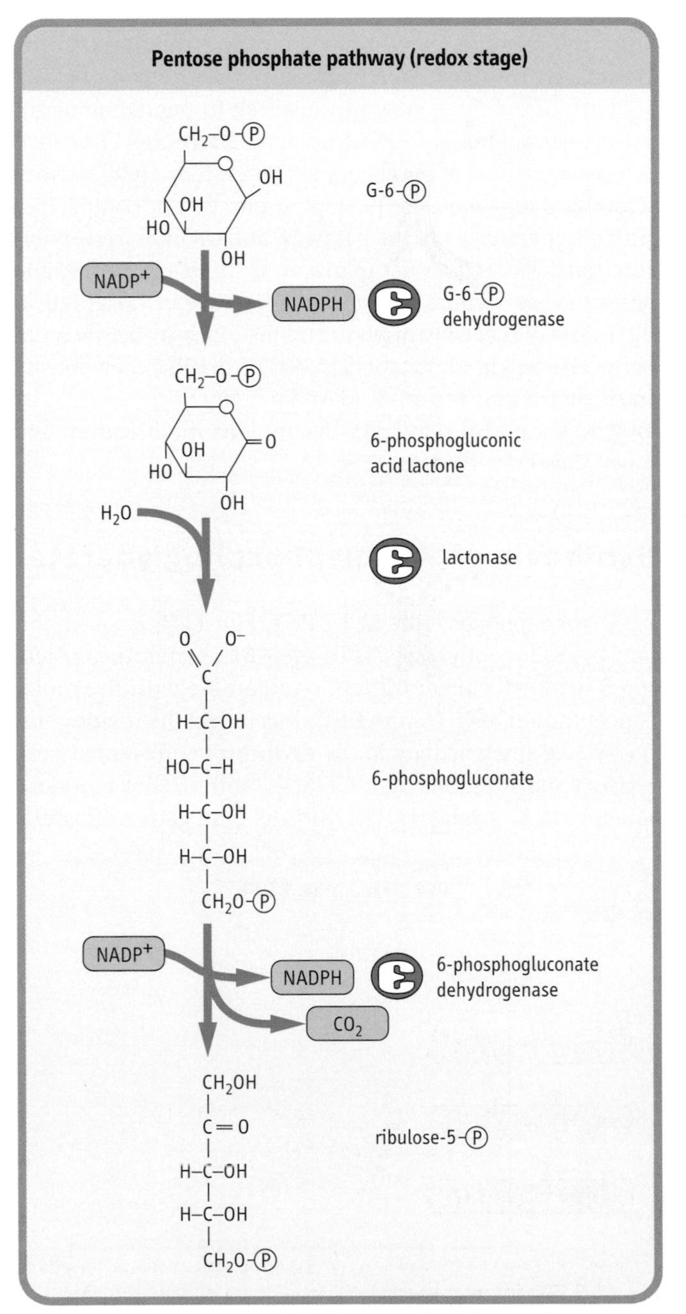

Fig. 11.8 The redox stage of the pentose phosphate pathway. A sequence of three enzymes forms 2 moles of NADPH per mole of G-6-P, which is converted into ribulose 5-phosphate, with evolution of CO_2.

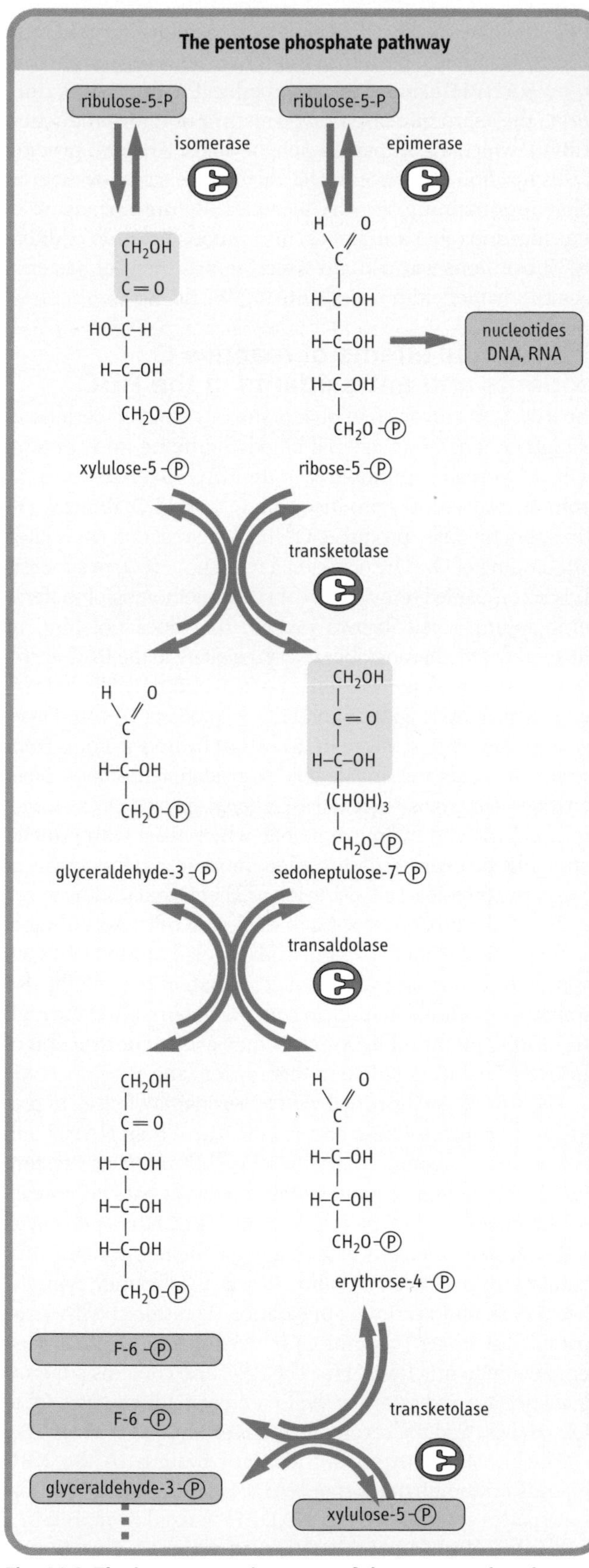

Fig. 11.9 The interconversion stage of the pentose phosphate pathway. The carbon skeletons of three molecules of ribulose-5-phosphate are shuffled around to form two molecules of F-6-P and one molecule of glyceraldehyde 3-phosphate.

transfer reactions. Transaldolase acts similarly to the aldolase in glycolysis, except a three-carbon unit is transferred to another sugar, rather than released as a free triose phosphate in the aldolase reaction.

As shown in Fig. 11.9, two molecules of ribulose 5-phosphate, the first pentose product of the *redox* stage, are converted into separate products: one molecule is isomerized to the aldose sugar, ribose-5-phosphate, and the other is epimerized to xylulose 5-phosphate. Transketolase then catalyzes transfer of two carbons from xylulose-5-phosphate to ribose-5-phosphate, yielding a seven-carbon ketose sugar, sedoheptulose-7-phosphate, and the three-carbon glyceraldehyde-3-phosphate. Transaldolase then catalyzes a three-carbon transfer from sedoheptulose-7-phosphate to glyceraldehyde-3-phosphate, yielding the first glycolytic intermediate, F-6-P, and a residual erythrose-4-phosphate. A third mole of pentose phosphate now enters the pathway as ribulose-5-phosphate and donates two carbons to erythrose-4-phosphate in a second transketolase reaction, yielding a second mole of F-6-P and 1 mole of glyceraldehyde-3-phosphate, both of which enter glycolysis. The net effect of these reactions is shown below.

Ribulose-5-P	⇌	Ribose-5-P
2 Ribulose-5-P	⇌	2 Xylulose-5-P
Xylu-5-P + Rib-5-P	⇌	Glyc-3-P + Sed-7-P
Sed-7-P + Glyc-3-P	⇌	Ery-4-P + Fru-6-P
Xylu-5-P + Ery-5-P	⇌	Glyc-3-P +Fru-6-P
3 Ribulose-5-P	⇌	**Glyceraldehyde-3-P + 2 F-6-P**

Thus, the three five-carbon sugar phosphates formed in the redox stage of the pentose phosphate pathway are converted into two six-carbon and one three-carbon glycolytic intermediates. In the RBC, these glycolytic intermediates normally continue through glycolysis to lactate, illustrating that glucose is only temporarily shunted away from the mainstream of glycolysis.

Under conditions of oxidative stress, such as drug-induced hemolytic anemia (see Glucose-6-phosphate dehydrogenase deficiency), there is greatly increased demand for NADPH. In these cases F-6-P produced in the *interconversion* stage of the pentose phosphate cycle may be converted back to G-6-P by the phosphoglucose isomerase reaction and recycled through the pentose phosphate pathway to enhance the production of NADPH.

Antioxidant defenses and reactive oxygen species (ROS)

O_2 is a strong oxidizing agent, and oxygenated RBCs are rich not only in O_2, but also in easily oxidizable substrates, such as heme iron and membrane lipids. Fortunately, molecular O_2 is relatively inert. It can be mixed with hydrogen gas, a strong reducing agent, but no reaction will occur until a flame or other source of heat is applied to raise the energy of the system.

Because of its inertness, transition metal ions, such as iron and copper, are required for activation of O_2 in biological systems – otherwise metabolic oxidation reactions using molecular O_2 as oxidant would be as slow as the combustion reaction between hydrogen and O_2 at body temperature and pH, i.e. negligible. In addition to metalloproteins involved in O_2 transport, such as Hb in blood and myoglobin (Mb) in muscle, all oxidases and oxygenases that use molecular O_2 for metabolism are metalloenzymes. In some cases these enzymes produce partially reduced forms of O_2 that are much more reactive than O_2 itself, such as the superoxide radical anion ($O_2^{\bullet-}$), its protonated form, the hydroperoxy radical ($HOO^\bullet$, $pK_a \approx 4.5$), and hydrogen peroxide (H_2O_2). These reactive oxygen species (ROS) are precursors to strongly oxidizing species such as the hydroxyl radical ($OH^\bullet$) and metal–oxo complexes (Fig. 11.10).

Measurement of blood glucose

In the clinical laboratory plasma and urinary glucose are commonly measured by enzymatic methods using automated analyzers and enzymes of glycolysis or the pentose phosphate pathway. However, in some circumstances glucose concentration in body fluids needs to be measured outside of the laboratory. This is done using glucose reagent strips. These strips are impregnated with enzymes and reagents, such as glucose oxidase, which generates H_2O_2, and a peroxidase which uses H_2O_2 to oxidize a substrate and produce a colored product. The color produced is directly related to the glucose concentration in the sample. The color reaction is assessed either visually or using a reflectance meter. Such methods are not as accurate or precise as laboratory methods, but are commonly used where rapid or frequent measurements of blood glucose are required. They are especially useful to diabetic patients who must check their glucose levels regularly in a non-laboratory setting.

The beneficial effects of reactive O_2

As outlined in Figure 11.11, a sequence of reactions producing ROS is initiated during phagocytosis and killing of bacteria by macrophages. First, NADPH oxidase in the macrophage plasma membrane is activated to produce $O_2^{\bullet-}$, which is then converted into H_2O_2 by superoxide dismutase:

$$2O_2^{\bullet-} + 2H^+ \rightarrow H_2O_2 + O_2$$

The H_2O_2 is used by another macrophage enzyme, myeloperoxidase, to oxidize chloride ion, ubiquitous in body fluids, to hypochlorous acid (HOCl), which, together with H_2O_2, mediates bactericidal activity by oxidative degradation of microbial lipids, proteins, and DNA. The consumption of O_2 by NADPH oxidase is responsible for the 'respiratory burst', the sharp increase in O_2 consumption for production of ROS, which accompanies phagocytosis. The end product of this reaction sequence, HOCl, is also the active oxidizer in chlorine-containing laundry bleaches. Before the advent of penicillin and other antibiotics, intravenous infusion of dilute HOCl solutions was actually used for treatment of bacterial sepsis in battlefield hospitals during World War I.

The harmful effects of reactive O_2: oxidants and antioxidants in the RBC

The RBC is not involved in phagocytosis. However, because of its high O_2 tension in arterial blood and heme iron content, ROS are formed continuously in the RBC. Oxidases, such as xanthine oxidase, may produce both $O_2^{\bullet-}$ and H_2O_2 directly. Hb also spontaneously produces $O_2^{\bullet-}$ in a side reaction associated with binding of O_2. The occasional reduction of O_2 to superoxide is accompanied by oxidation of Hb to methemoglobin (ferrihemoglobin), a rust-brown protein that does not bind or transport O_2. Methemoglobin may precipitate in the RBC, forming inclusions known as Heinz bodies, and may also release heme, which reacts with $O_2^{\bullet-}$ and H_2O_2 to produce $OH^\bullet$ and reactive iron–oxo species. These ROS abstract hydrogen atoms from membrane lipids and initiate lipid peroxidation reactions. Lipid peroxides decompose to reactive carbonyl species, such as malondialdehyde and hydroxynonenal, which react with protein, damaging the integrity of the cell membrane and the activity of transporter proteins, collapsing ion gradients and leading to cell death. Similar processes occur in nucleated cells. According to the free radical theory of aging, oxidative damage to DNA contributes to cellular aging. Ultraviolet radiation in sunlight also produces O_2 radicals, leading to suntans or burns, and the high flux of ROS produced by X-rays is the basis for destruction of tumor tissue during radiation therapy for cancer.

The RBC is well-fortified with antioxidant defenses to protect itself against oxidative stress. First, there is an NADH and cytochrome b_5-dependent methemoglobin reductase system (Fig. 11.12), which reduces methemoglobin back to normal ferrohemoglobin. Normally, less than 1% of Hb is present as methemoglobin. Persons with congenital methemoglobinemia, resulting from methemoglobin reductase deficiency, typically have a dark and cyanotic appearance. Treatment with large doses of ascorbate (vitamin C) is used to reduce their methemoglobin to functional Hb. The RBC also contains superoxide dismutase and catalase; the first enzyme dismutates $O_2^{\bullet-}$ to H_2O_2 and O_2, and the second dismutates H_2O_2 to H_2O and O_2.

The most important antioxidant system in the RBC depends on the formation of NADPH by the pentose phosphate pathway. Most of this NADPH is used to maintain a high intracellular concentration of reduced glutathione (GSH). GSH is a tripeptide (γ-glutamyl-cysteinyl-glycine [Fig. 11.13]). It is present at ≈ 2 mmol/L in the RBC, 99% in the reduced form, GSH. The reducing activity of its sulfhydryl group is essential for its biological activity.

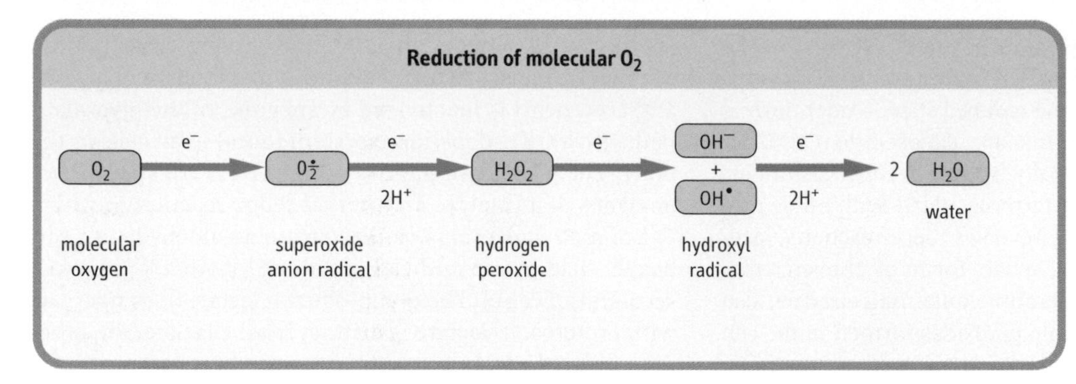

Fig. 11.10 Molecular oxygen and partially reduced, reactive forms of oxygen. Reduction of molecular O_2 in a series of one-electron steps yields superoxide, hydrogen peroxide, hydroxyl radical, and water. The intermediate, activated forms of oxygen are known as reactive oxygen species (ROS).

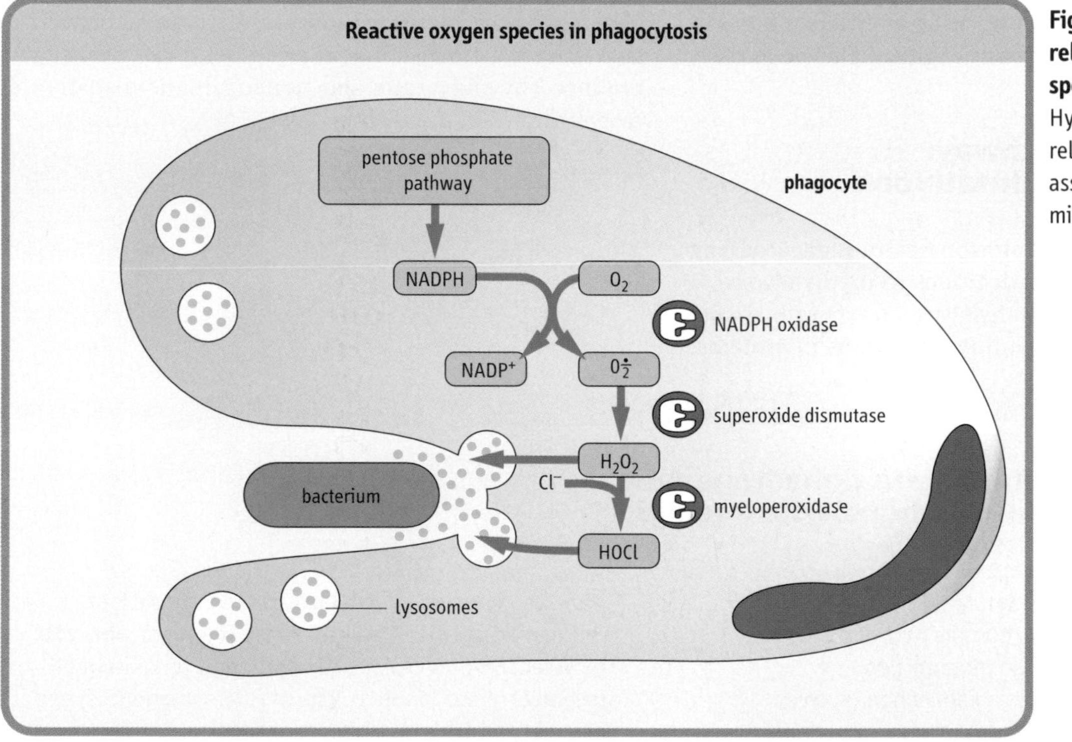

Fig. 11.11 Generation and release of reactive oxygen species during phagocytosis. Hydrolytic enzymes are also released from lysosomes to assist in degradation of microbial debris.

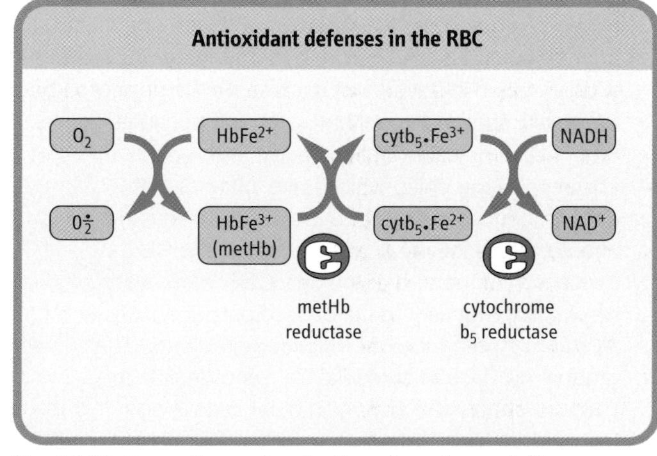

Fig. 11.12 Formation and reduction of methemoglobin.

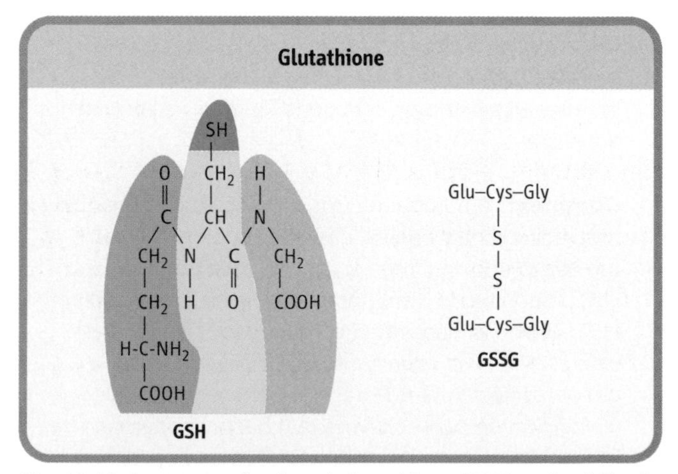

Fig. 11.13 Structure of reduced glutathione (GSH) and oxidized glutathione (GSSG).

GSH is involved in several antioxidant systems in the RBC. It acts as a sulfhydryl buffer, maintaining -SH groups in proteins and enzymes in the reduced state. Under normal circumstances, when proteins are exposed to O_2, their sulfhydryl groups may gradually oxidize to form disulfides, either intramolecularly or intermolecularly with other proteins. GSH nonenzymatically reverses these reactions, leading to regeneration of the active form of the enzyme. Glutathione peroxidase, a selenium-containing enzyme, also reduces both H_2O_2 and organic peroxides formed in the cell (Fig. 11.14). During the course of its protective functions, GSH is oxidized to the disulfide form, GSSG, which is then regenerated by the action of the NADPH-dependent enzyme, glutathione reductase. Selenium is required in trace amounts in the diet and is often described as an antioxidant nutrient because of its role in the active site of glutathione peroxidase.

The glyoxalase pathway: a special role for glutathione

A small fraction of triose phosphates produced during metabolism spontaneously degrades to methylglyoxal, a reactive dicarbonyl sugar. Methylglyoxal reacts with amino, guanidino, imidazole, and sulfhydryl groups in proteins, leading to enzyme inactivation and protein crosslinking. Methylglyoxal is also formed during metabolism of glycine and acetone. It is inactivated by enzymes of the glyoxalase pathway, a GSH-dependent system found in all cells in the body. The glyoxalase pathway (Fig. 11.15) consists of two enzymes that catalyze an internal redox reaction in which C-1 of methylglyoxal is oxidized from an aldehyde to a carboxylic acid group and C-2 is reduced from a ketone to a secondary alcohol. The end product D-lactate does not react with proteins. (D-lactate is distinct from L-lactate, the product of glycolysis, but may be converted into L-lactate, for further metabolism.) Levels of methylglyoxal and D-lactate are increased in blood of diabetic patients, because levels of glucose and glycolytic intermediates, including triose phosphates, are increased intracellularly in diabetes. The glyoxalase system also inactivates other dicarbonyl sugars produced by enzymatic and nonenzymatic oxidation of carbohydrates in the body.

Glucose-6-phosphate dehydrogenase deficiency causes hemolytic anemia

Just prior to a planned departure to the tropics, a patient visits his physician, complaining of weakness, and noting that his urine had recently become unexplainably dark. Physical examination revealed slightly jaundiced (yellow, icteric) sclera. Laboratory tests indicated a low hematocrit, a high reticulocyte count, and a significantly increased blood level of bilirubin. The patient had been quite healthy during a previous visit a month ago when he received immunizations and prescriptions for drugs related to his travel plans.

Comment. A number of drugs, particularly primaquine and related antimalarials, undergo redox reactions in the cell, producing large quantities of superoxide and H_2O_2. Superoxide dismutase converts superoxide into H_2O_2, which is inactivated by glutathione peroxidase, using NADPH as coenzyme. Some persons have a genetic defect in G-6-P dehydrogenase, typically yielding an unstable enzyme that has a shorter half-life in the RBC or an enzyme that is unusually sensitive to inhibition by NADPH. In either case, because of the decreased activity of this enzyme and insufficient production of NADPH under stress, the cell's ability to recycle GSSG to GSH is impaired, and drug-induced oxidative stress leads to lysis of RBCs (hemolysis) and hemolytic anemia. Bilirubin, a product of heme metabolism also accumulates in blood. If the hemolysis is severe enough, Hb spills over into the urine, resulting in hematuria and dark-colored urine. There are over 200 known mutations in G-6-P dehydrogenase, yielding a wide variation in severity of disease. RBCs appear to be especially sensitive to oxidative stress, because, unlike other cells, they cannot synthesize and replace their enzymes. Older cells, which have lower G-6-P dehydrogenase activity, are therefore particularly affected. The activity of all enzymes in the RBC declines with the age of the cell. Cell death eventually results from inability of the cells to produce sufficient ATP for maintenance of cellular ion gradients. The gradual increase in cytosolic Ca^{2+} and decline in pentose phosphate activity in older cells is one mechanism leading to crosslinking of membrane proteins and turnover of the RBC in the spleen.

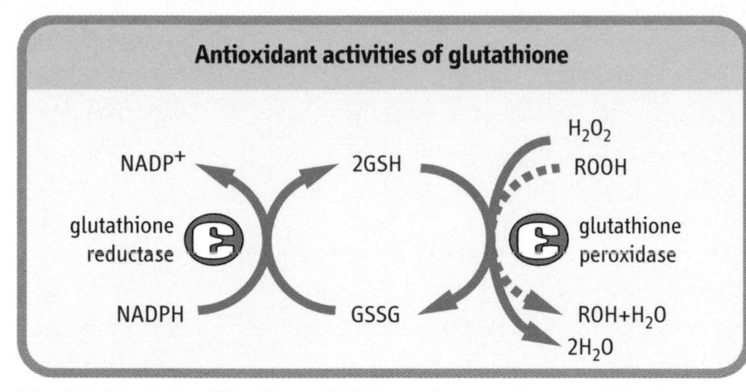

Fig. 11.14 Detoxification of H_2O_2 and organic peroxides by glutathione peroxidase.

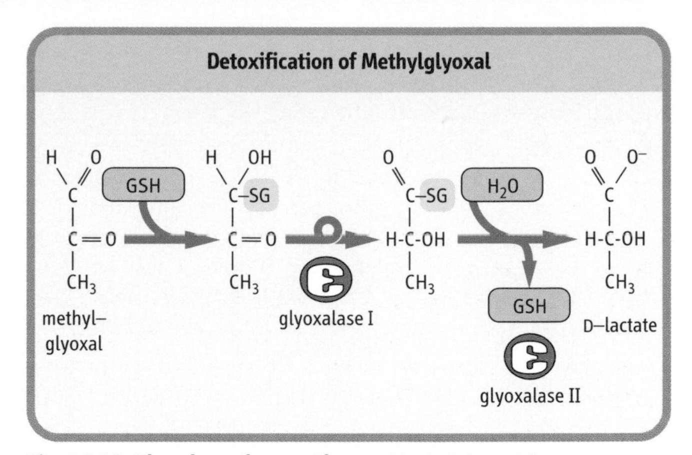

Fig. 11.15 The glyoxalase pathway. Methylglyoxal forms a hemithioacetal adduct with GSH. The hemithioacetal rearranges to a thioester derivative in the first step of the pathway, followed by hydrolysis to release D-lactate. GSH is regenerated in this pathway.

Summary

This chapter deals with two ancient, central metabolic pathways common to all cells in the body, glycolysis and the pentose phosphate pathway. The RBC, which lacks mitochondria and the capability for oxidative metabolism and obtains all of its ATP energy by glycolysis, is used as a model for introduction to these pathways. Anaerobic glycolysis in the RBC provides a limited amount of ATP by anaerobic metabolism of the six-carbon sugar, glucose, to two molecules of the three-carbon hydroxyacid, lactate. Through a series of sugar phosphate intermediates, glycolysis provides metabolites for branch points to numerous other metabolic pathways, including the pentose phosphate pathway. This pathway provides pentoses for synthesis of DNA and RNA in nucleated cells, and NADPH for biosynthetic reactions. In all cells, but especially in the RBC, this pathway also has an essential role in providing NADPH for reduction of glutathione, an important cofactor for antioxidant defense systems that protect the cell against oxidative stress.

Further reading

Arya R, Layton DM, Bellingham AJ. Hereditary red cell enzymopathies. *Blood Reviews* 1995;**9**:165–175.

Brooks C. Neonatal hypoglycemia. *Neonatal Network* 1997;**16**:15–21.

Brown SP, Keith WB. The effect of acute exercise on levels of erythrocyte 2,3-bisphosphoglycerate: a brief review. *J Sports Med* 1993;**11**:479–484.

Fothergill-Gilmore LA, Michels PA. Evolution of glycolysis. *Prog Biophys Mol Biol* 1993;**59**:105–235.

Kletzien RF, Harris PK, Foellmi LA. Glucose-6-phosphate dehydrogenase: a 'housekeeping' enzyme subject to tissue-specific regulation by hormones, nutrients and oxidant stress. *FASEB J* 1994;**8**:174–181.

Knull H, Minton AP. Structure within eukaryotic cytoplasm and its relationship to glycolytic metabolism. *Cell Biochemistry and Function* 1996;**14**:237–248.

Thornalley PJ. The glyoxalase system: new developments towards functional characterization of a metabolic pathway fundamental to biological life. *Biochem J* 1990;**269**:1–11.

Thornalley PJ. The glyoxalase system in health and disease. *Mol Aspects Med* 1993;**14**:287–371.

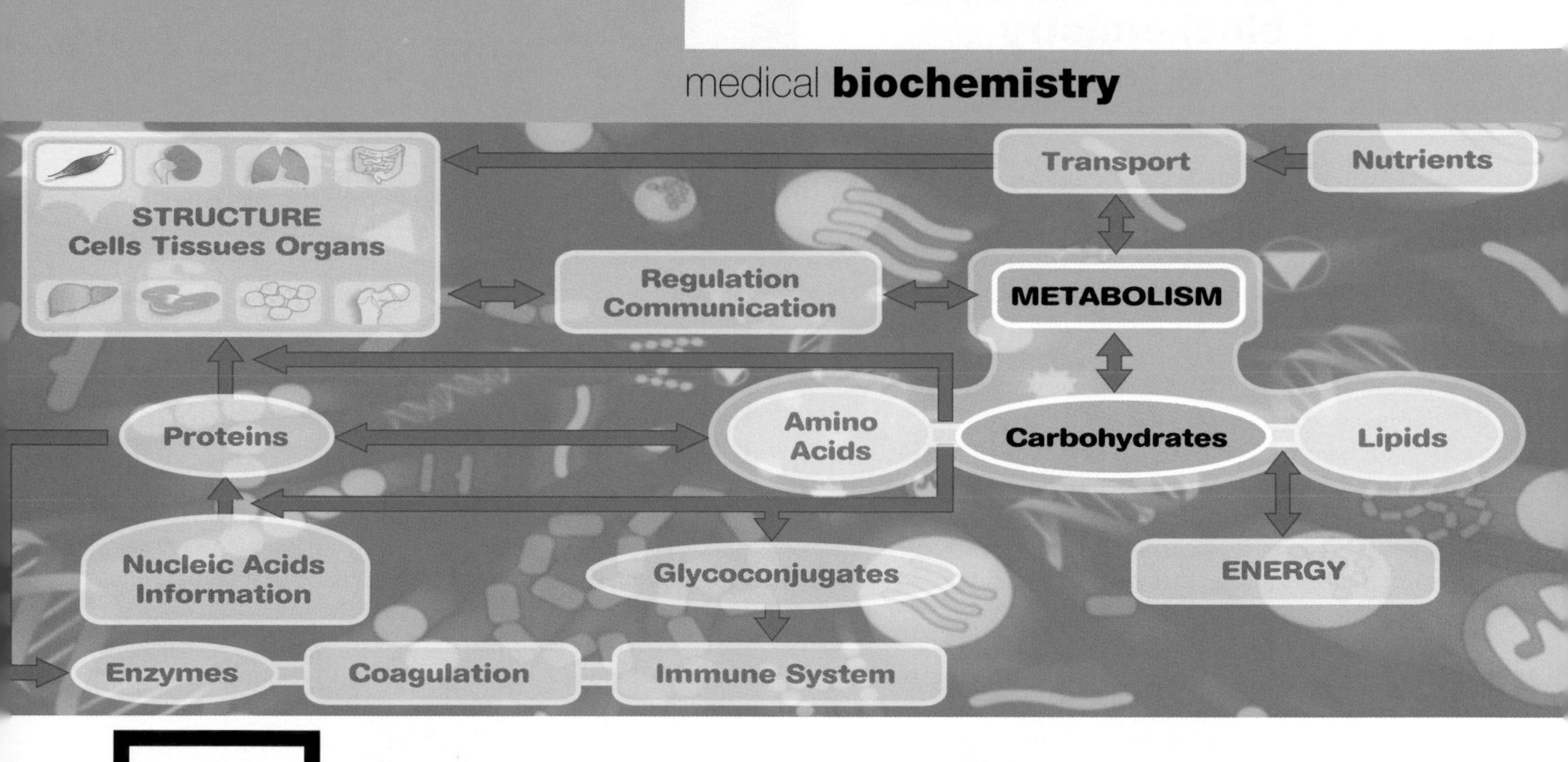

12 Carbohydrate Storage and Synthesis in Liver and Muscle

Introduction

The red cell and the brain have an absolute requirement for blood glucose for energy metabolism. These cells consume about 80% of the 200 g of glucose consumed in the body per day. There is a maximum of about 10 g of glucose in the total plasma and extracellular fluid volume, so that blood glucose must be replenished constantly. Otherwise, hypoglycemia develops rapidly and compromises brain function, leading to confusion and disorientation and possibly life-threatening coma at blood glucose concentrations below 2.5 mmol/L (45 mg/dL). We absorb glucose from our intestines for only 2–3 h following a carbohydrate-containing meal, so there must be a mechanism for maintenance of blood glucose between meals.

Glycogen, a polysaccharide storage form of glucose in liver, is our first line of defense against declining blood glucose concentration. During and immediately following a meal, glucose is converted in liver into the storage polysaccharide glycogen (a process known as glycogenesis). Glycogen is gradually degraded between meals, by the pathway of glycogenolysis, releasing glucose to maintain blood glucose concentration. However, total hepatic glycogen stores are barely sufficient for maintenance of blood glucose concentration during a 12-h fast.

During sleep, when we are not eating, there is a gradual shift from glycogenolysis to *de novo* synthesis of

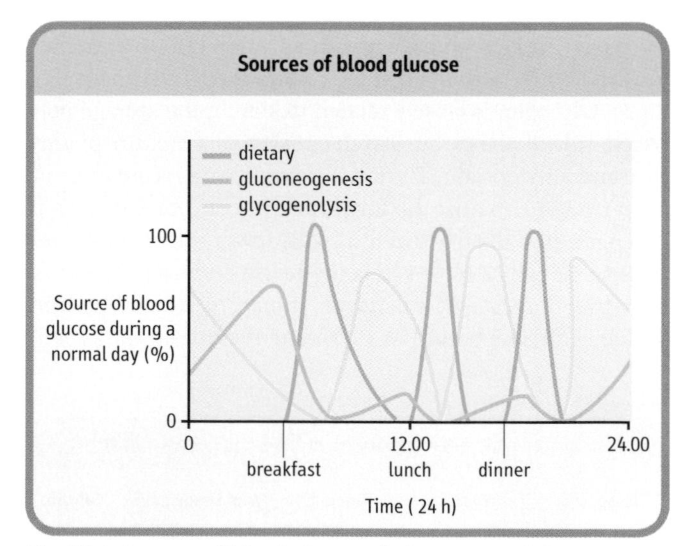

Fig. 12.1 Sources of blood glucose during a normal day. Between meals, blood glucose is derived primarily from hepatic glycogen. Depending on the frequency of snacking, glycogenolysis and gluconeogenesis may be more or less active during the day. Late in the night, following depletion of a major fraction of hepatic glycogen, gluconeogenesis becomes the primary source of blood glucose.

glucose, also an hepatic pathway, known as gluconeogenesis (Fig. 12.1). Gluconeogenesis is essential for survival during fasting or starvation, when glycogen stores are negligible. The liver uses amino acids from muscle protein as the primary precursor of glucose, but also makes use of lactate from glycolysis and glycerol from fat catabolism. Lipids, mobilized from adipose tissue, provide the energy for gluconeogenesis.

Glycogen is also stored in muscle, but this glycogen is not available for maintenance of blood glucose. Glucose, derived in part from glycogen, especially during bursts of physical activity, is essential for muscle energy metabolism, even though muscle relies primarily on fats as a source for energy.

The tissue concentration of glycogen is higher in liver than in muscle, but because of the relative masses of muscle and liver, the majority of glycogen in the body is stored in muscle (Fig. 12.2). Although hepatic glycogen stores, typically 50–100 g, are less than the approximately 200 g of glucose required for daily survival, this is not an insignificant amount of glucose, representing about 200–300 g (1/2 lb) of glycogen, including water of hydration.

This chapter describes the pathways of biosynthesis (glycogenesis) and mobilization (glycogenolysis) of glycogen in liver and muscle and the pathway of gluconeogenesis in liver.

Structure of glycogen

Glycogen is a branched polysaccharide of glucose, a homoglucan. It contains only two types of glycosidic linkages, chains of α1→4-linked glucose residues with α1→6 branches spaced about every 4–6 residues along the α1→4 chain (Fig. 12.3). Glycogen is closely related to starch, the storage polysaccharide of plants, but starch consists of a mixture of amylose and amylopectin. The amylose component contains only linear α1→4 chains; the amylopectin component is more glycogen-like in structure but with fewer α1→6 branches, about one per 12 α1→4-linked glucose residues. The gross structure of glycogen is dendritic in nature, expanding from a core sequence bound to a tyrosine residue in the protein glycogenin and developing into a final structure resembling a head of cauliflower. The many glucose molecules on the surface of the glycogen molecule provide ready access for enzymes involved in release of glucose from the glycogen polymer.

Glucose and glycogen stores in the body (70 kg adult)

Tissue	Type	Amount	% of tissue mass	Calories
liver	glycogen	75 g	3–5 %	300
muscle	glycogen	250 g	0.5–1.0%	1000
blood and extracellular fluid	glucose	10 g	—	40

Fig. 12.2 Tissue distribution of carbohydrate energy reserves (70 kg adult).

Pathway of glycogenesis from blood glucose in liver

The liver is rich in the high-capacity, low-affinity (K_m >10 mmol/L) glucose transporter GLUT-2, making it freely permeable to glucose delivered at high concentration in portal blood during and following a meal. The liver is also rich in glucokinase, an enzyme that is specific for glucose and converts it into glucose 6-phosphate (G-6-P). Glucokinase (GK) is inducible by continued consumption of a high-carbohydrate diet. It has a high K_m, about 5–7 mmol/L, so that it is poised to increase in activity as portal glucose increases above the normal 5 mmol/L (100 mg/dL) blood glucose concentration. Unlike hexokinase, GK is not inhibited by G-6-P, so that the concentration of G-6-P increases rapidly in liver following a carbohydrate-rich meal, forcing glucose into all of the major pathways of glucose metabolism: glycolysis, the pentose phosphate pathway, and glycogenesis. Glucose is channeled into glycogen, providing a carbohydrate reserve for maintenance of blood glucose during the postabsorptive state. Excess G-6-P in liver, beyond that needed to replenish glycogen reserves, is then funneled into glycolysis, both for energy production and for conversion into triglycerides, which are exported for storage in adipose tissue. Glucose

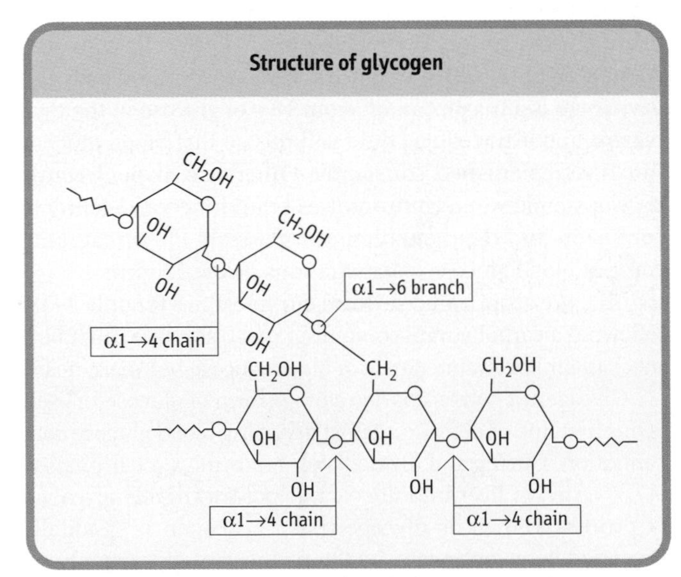

Fig. 12.3 Close-up of the structure of glycogen. The figure shows α1→4 chains and an α1→6 branch point. Glycogen is stored as granules in liver and muscle cytoplasm. Most of the glycogenic and glycogenolytic enzymes are bound to these granules, assuring rapid changes in glycogen metabolism in response to allosteric and hormonal stimuli.

that passes through the liver leads to an increase in peripheral blood glucose concentration following carbohydrate-rich meals. This glucose is used in muscle for synthesis and storage of glycogen and in adipose tissue as a source of glycerol for triglyceride biosynthesis.

The pathway of glycogenesis from glucose (Fig. 12.4, left) involves four steps:

- conversion of G-6-P into glucose 1-phosphate (G-1-P) by phosphoglucomutase;
- activation of G-1-P to the sugar nucleotide uridine diphosphate (UDP)-glucose by the enzyme UDP-glucose pyrophosphorylase;
- transfer of glucose to glycogen in α1→4 linkage by glycogen synthase, a member of the class of enzymes known as glycosyl transferases;
- when the α1→4 chain exceeds eight residues in length, glycogen branching enzyme, a transglycosylase, transfers some of the α1→4-linked sugars to an α1→6 branch, setting the stage for continued elongation of both α1→4 chains until they, in turn, become long enough for transfer by branching enzyme.

Glycogen synthase is the regulatory enzyme for glycogenesis, rather than UDP-glucose pyrophosphorylase, because UDP-glucose is also used for synthesis of glycoproteins, glycolipids, and other sugars. Pyrophosphate (PPi), the other product of the pyrophosphorylase reaction, is rapidly hydrolyzed to inorganic phosphate by pyrophosphatase.

Pathway of glycogenolysis in liver

As with most metabolic pathways, separate enzymes, sometimes in separate subcellular compartments, are required for the forward and reverse pathways. The pathway of glycogenolysis (Fig. 12.4, right) begins with removal of the abundant, external α1→4-linked glucose residues in glycogen. This is accomplished not by a hydrolase, but by glycogen phosphorylase, an enzyme that uses cytosolic phosphate and releases glucose from glycogen in the form of G-1-P, which is converted into G-6-P by phosphoglucomutase. In liver the glucose is

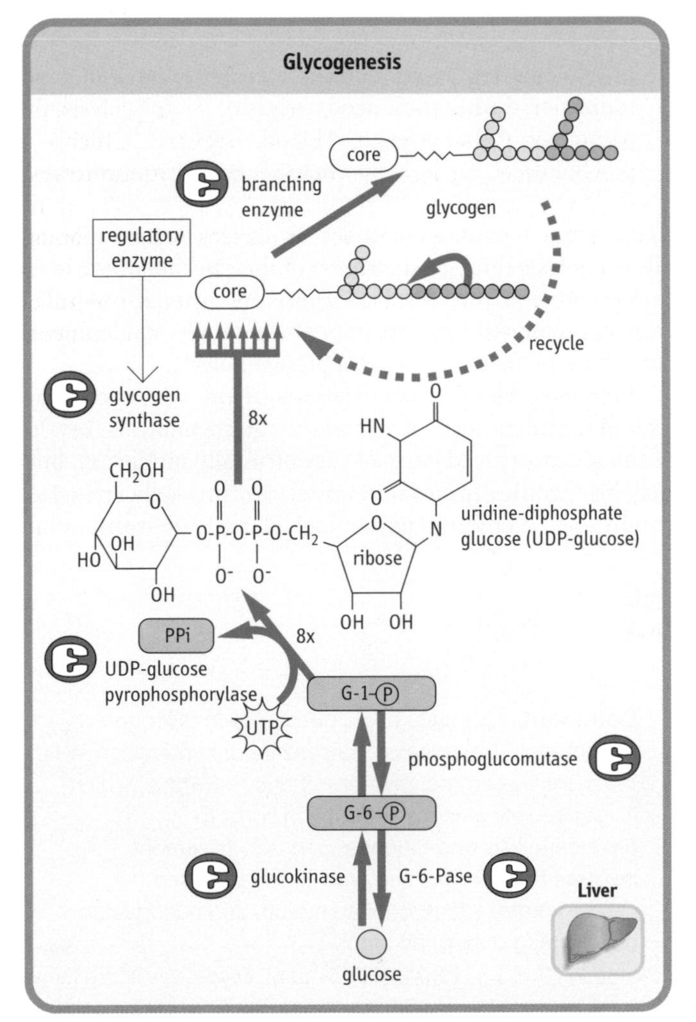

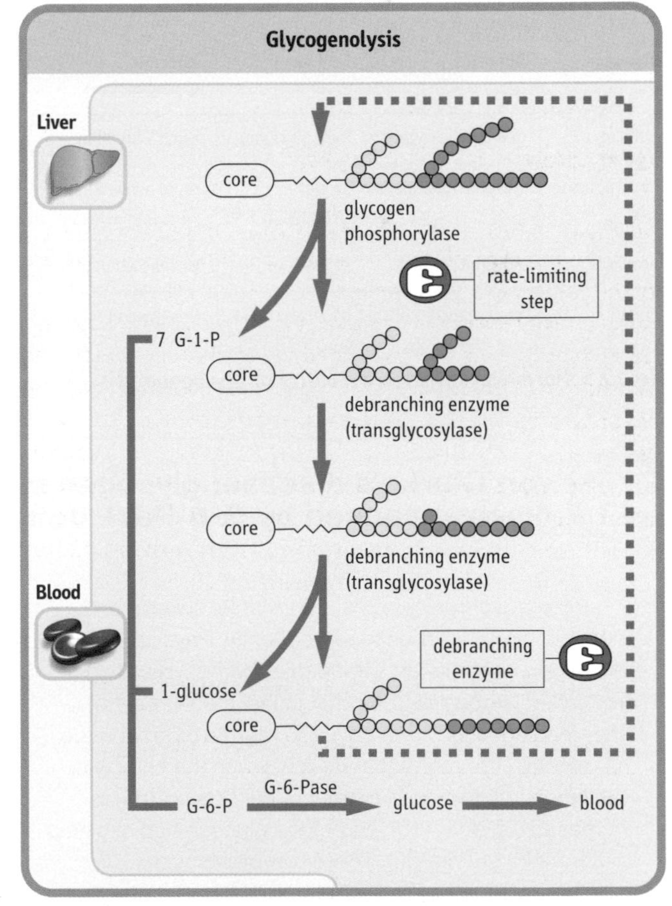

Fig. 12.4 Pathways of glycogenesis (left) and glycogenolysis (right). The branching arrays at the top of the figure are meant to illustrate the three-dimensional array of glycogen branching. This structure places a substantial fraction of the total glucose molecules on the periphery of the molecule, immediately available for glycogen phosphorylase activity.

released from G-6-P by glucose-6-phosphatase (G-6-Pase), and the glucose exits via the GLUT-2 transporter into blood. The rate-limiting, regulatory step in glycogenolysis is catalyzed by phosphorylase, the first enzyme in the pathway.

Phosphorylase is specific for α1→4 glycosidic linkages; it cannot cleave α1→6 linkages. Further, this enzyme cannot approach the branching glucose residues efficiently. Thus, as shown in Figure 12.4, phosphorylase cleaves the external glucose residues until the branches are three or four residues long, then debranching enzyme, which has both transglycosylase and glucosidase activity, moves a short segment of glucose residues bound to the α1→6 branch to the end of an adjacent α1→4 chain, leaving a single glucose residue at the branch point. This glucose is then removed by the exo-1,6-glucosidase activity of branching enzyme, allowing glycogen phosphorylase to proceed with degradation of the α1→4 chain until another branch point is approached, setting the stage for a repeat of the transglycosylase and glucosidase reactions. About 90% of the glucose is released from glycogen as G-1-P, and the remainder, the α-1→6 branching residues, as free glucose.

Hormonal control of glycogenolysis

Hormone	Source	Initiator	Effect on glycogenolysis
glucagon	pancreatic α-cells	hypoglycemia	rapid activation
epinephrine	adrenal medulla	stress, hypoglycemia	rapid activation
cortisol	adrenal cortex	stress	chronic activation
insulin	pancreatic β-cells	hyperglycemia	inactivation

Fig. 12.5 Hormones involved in control of glycogenolysis.

Hormonal regulation of hepatic glycogenolysis

The study of glycogen metabolism is best approached by first addressing the regulation of glycogenolysis. Glycogenolysis in liver is activated in response to a demand for blood glucose, either because of its utilization during the post-absorptive state or in preparation for increased glucose utilization in response to stress. There are three major hormonal activators of glycogenolysis: glucagon, epinephrine (adrenaline), and cortisol (Fig. 12.5). Glucagon is a peptide hormone (3500 Da), secreted from the α-cells of the endocrine pancreas. Its primary function is to activate hepatic glycogenolysis for maintenance of normoglycemia. It has a short half-life in plasma, about 5 min, as a result of receptor binding, renal filtration, and proteolytic inactivation in liver. Glucagon concentration in plasma therefore changes rapidly in response to the need for blood glucose. Blood glucagon increases between meals, decreases during a meal, and is chronically increased during fasting or on a low-carbohydrate diet.

Glycogenolysis is also activated in response to both acute and chronic stress. The stress may be:

- physiologic, e.g. in response to increased blood glucose utilization during prolonged exercise;
- pathologic, e.g. as a result of blood loss;
- psychological, e.g. in response to acute or chronic threats.

Acute stress, regardless of its source, causes an activation of glycogenolysis through the action of the catecholamine hormone, epinephrine, released from the adrenal medulla. During prolonged exercise, both glucagon and epinephrine contribute to the stimulation of glycogenolysis.

Increased blood concentrations of the adrenocortical steroid hormone cortisol also induce glycogenolysis. Levels of the glucocorticoid cortisol vary diurnally in plasma, but may be chronically elevated under continuously stressful conditions, including psychological and environmental

von Gierke's disease: glycogen storage disease caused by G-6-Pase deficiency

A baby girl is chronically cranky, irritable, sweaty, and lethargic, and demands food frequently. Physical evaluation indicates an extended abdomen, resulting from an enlarged liver. Blood glucose, measured 1 h after feeding, was 3.5 mmol/L (70 mg/dL) (normal value ~5 mmol/L [100 mg/dL]). After 4 h, when the child was exhibiting irritability and sweating, her heart rate was increased (pulse = 110), and blood glucose had declined to 2 mmol/L (40 mg/dL). These symptoms were corrected by feeding. A liver biopsy showed massive deposition of glycogen particles in the liver cytosol.

Comment. This child has a deficiency in glycogen mobilization. Because of the severity of hypoglycemia, the most likely mutation is in hepatic G-6-Pase, which is required for glucose production by both glycogenolysis and gluconeogenesis. Treatment involves frequent feeding with slowly digested carbohydrate, e.g. uncooked starch, and nasogastric drip feeding during the night.

(cold) stress. Glucagon serves as a general model for the mechanism of action of hormones that act by way of cell-surface receptors. Cortisol, which acts at the level of gene expression, will be discussed later in Chapter 32.

Mechanism of action of glucagon

Glucagon binds to a hepatic plasma-membrane receptor and initiates a cascade of reactions that lead to mobilization of hepatic glycogen. On the inside of the plasma membrane there is a class of signal-transduction proteins, known as G-proteins, that bind guanosine triphosphate (GTP) and guanosine diphosphate (GDP), nucleotide analogs of ATP and ADP. GDP is bound in the resting state. Binding of glucagon to the plasma-membrane receptor stimulates exchange of GDP for GTP on the G-protein, and the G-protein then undergoes a conformational change that leads to dissociation of one of its subunits, which then binds to and activates the plasma membrane enzyme, adenylate cyclase. This enzyme converts cytoplasmic ATP into cyclic-3′,5′-AMP (cAMP), a soluble mediator that is described as the 'second messenger' for action of glucagon (and other hormones). Cyclic AMP binds to the cytoplasmic enzyme protein kinase A (PKA), causing dissociation of inhibitory (regulatory) subunits from the catalytic subunits of the heterodimeric enzyme, relieving inhibition of PKA (see Chapter 5), which then initiates a series of protein-phosphorylation reactions.

The pathway for activation of glycogen phosphorylase (Fig. 12.6) involves phosphorylation of many molecules of phosphorylase kinase by PKA, which then phosphorylates and activates many molecules of glycogen phosphorylase. The net effect of the sequential steps, beginning with activation of many molecules of adenylate cyclase by G-protein, is a 'cascade amplification' system, not unlike that of a series of amplifiers in a radio or stereo set, resulting in a massive increase in signal strength within seconds after recognition of glucagon binding to the hepatocyte plasma membrane. Phosphorylation of phosphorylase initiates glycogenolysis, leading to production of G-6-P in liver, which is then hydrolyzed to glucose and exported into blood. One other target of PKA is Inhibitor-1, a protein phosphatase inhibitor protein, which is activated by phosphorylation. Phosphorylated Inhibitor-1 inhibits cytoplasmic phosphoprotein phosphatases, which would otherwise reverse the phosphorylation of enzymes and quench the response to glucagon (see Fig. 12.6).

Glycogenolysis and glycogenesis are opposing pathways. Theoretically, G-1-P produced by phosphorylase could be rapidly activated to UDP-glucose and reincorporated into glycogen. To prevent this wasteful or futile cycle, PKA also acts directly on glycogen synthase, in this case inactivating the enzyme. Thus, the activation of glycogenolysis is coordinated with inactivation of glycogenesis. Other hepatic pathways, including protein, cholesterol, fatty acid, and triglyceride biosynthesis, and glucose synthesis (gluconeogenesis) and utilization (glycolysis) are also inhibited by phosphorylation of key regulatory enzymes, focusing liver metabolism in response to glucagon on the provision of glucose to blood for maintenance of vital body functions (see Chapter 20).

Perhaps in order to balance the cascade of events amplifying the response to glucagon, there are multiple, redundant mechanisms to insure rapid termination of the hormonal response. In addition to the slow GTPase activity of the G_α-subunit there is also a phosphodiesterase activity in the cell that

G-proteins

G-proteins are trimeric, plasma-membrane, guanosine-nucleotide-binding proteins that are involved in signal transduction for a wide variety of hormones (Fig. 12.6). In some cases they stimulate (G_s) and, in other cases, they inhibit (G_i) protein kinases and protein phosphorylation. G-proteins are closely associated with hormone receptors in plasma membranes and consist of three subunits, α, β, and γ; the G_α-subunit binds GDP in the resting state. Following hormone binding (ligation), the receptor recruits G-proteins, stimulating exchange of GDP for GTP on the G_α-subunit. GTP binding leads to release of the β- and γ-subunits, and the α-subunit is then free to bind to and activate adenylate cyclase. The hormonal response is amplified following receptor binding, because a single receptor can activate many α-subunits. Hormonal responses are also turned off at the level of receptors and G-proteins by two mechanisms:

- the G_α-subunit has a sluggish guanosine triphosphate phosphatase (GTPase) activity that hydrolyzes GTP, with a half-time measured in minutes, so that it dissociates from, and thereby ceases to activate, adenylate cyclase;
- phosphorylation of the hormone receptor by protein kinases decreases its affinity for the hormone, a process described as desensitization.

These effects dampen the cellular response and require higher levels of extracellular hormone for a continued high level response to the hormone. Chronic high levels of circulating hormone may lead to hormone resistance.

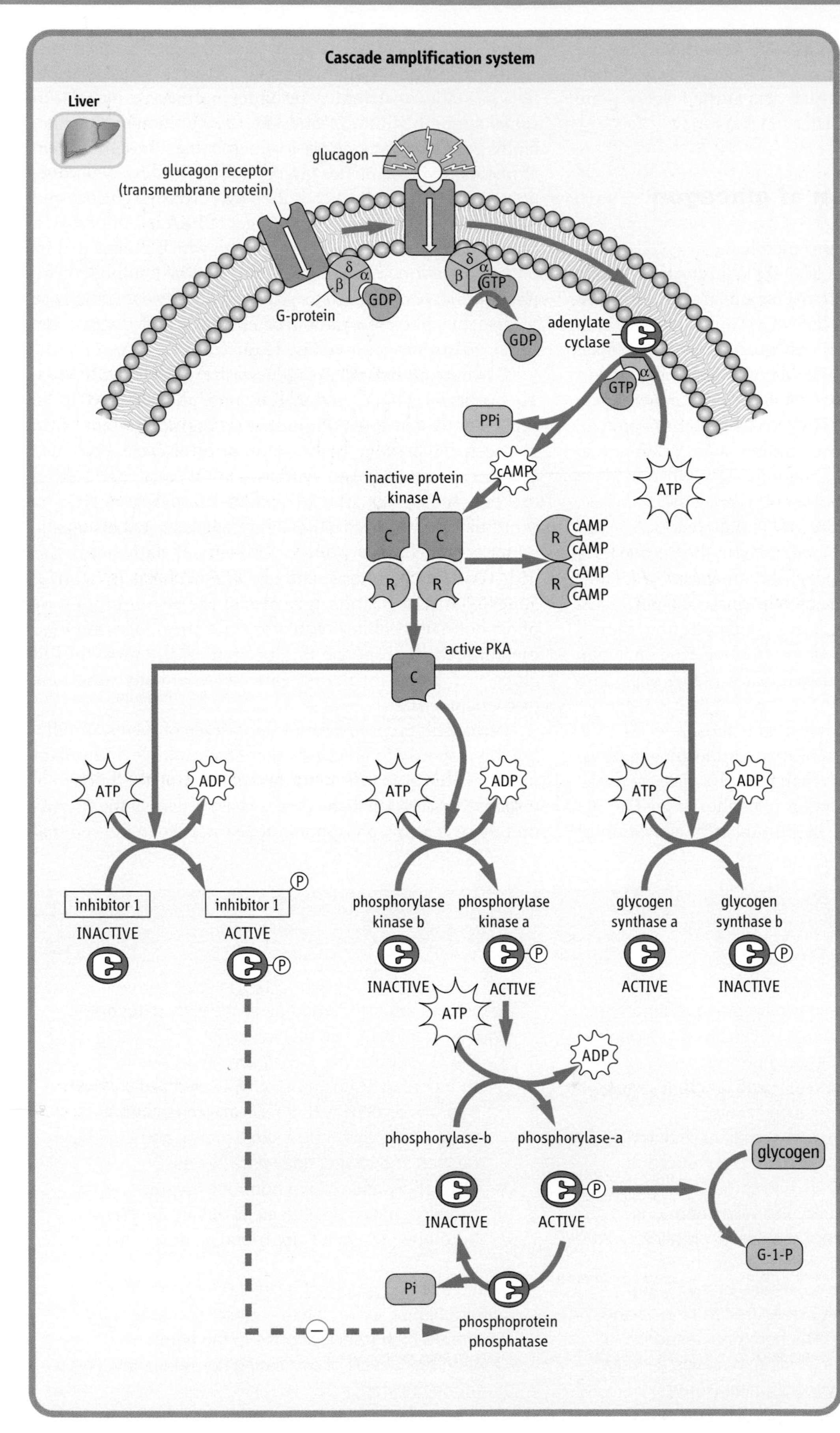

Fig. 12.6 Mobilization of hepatic glycogen by glucagon. A cascade of reactions amplifies the hepatic response to glucagon binding to its plasma-membrane receptor. cAMP is known as the second messenger of glucagon action. PKA indirectly activates phosphorylase via phosphorylase kinase and directly inactivates glycogen synthase. C, catalytic subunits; R, regulatory (inhibitory) subunits.

hydrolyzes cAMP to AMP, permitting reassociation of the inhibitory and catalytic subunits of PKA, decreasing its protein kinase activity. There are also phosphoprotein phosphatases that remove the phosphate groups from the active, phosphorylated forms of phosphorylase kinase and phosphorylase. The decrease in cAMP concentration and PKA activity also leads to decreased phosphorylation of Inhibitor-1, permitting increased activity of phosphoprotein phosphatases. Thus, an array of mechanisms act in concert to insure that hepatic glycogenolysis declines rapidly in response to declining blood glucagon concentration (Fig. 12.7). As shown in Figure 12.8, there are numerous types of glycogen-storage diseases affecting both liver and muscle glycogen metabolism.

Protein kinase A is very sensitive to small changes in cAMP concentration

As illustrated in Figure 12.6, cAMP-dependent PKA is a tetrameric enzyme with two different types of subunits (R_2C_2); a catalytic C-subunit that has protein kinase activity and a regulatory R-subunit that inhibits the protein kinase activity. The R-subunit has a sequence of amino acids that would normally be recognized and phosphorylated by the C-subunit, except that this sequence in R contains an alanine, rather than a serine or threonine, residue. Binding of 2 molecules of cAMP to each R-subunit results in conformational changes that lead to dissociation of a $(cAMP_2–R)_2$ dimer from the C-subunits. The monomeric, active C-subunits then proceed to phosphorylate serine and threonine residues in target enzymes. This is not a typical allosteric regulatory mechanism, but the complete activation of PKA involves cooperative binding of 4 molecules of cAMP to two R subunits. PKA is fully activated at submicromolar concentrations of cAMP, so that it is exquisitely sensitive to small changes in adenylate cyclase activity.

Mechanisms of termination of hormonal response to glucagon
hydrolysis of GTP on G_α-subunit
hydrolysis of cAMP by phosphodiesterase
protein phosphatase activity

Fig. 12.7 Several mechanisms are involved in terminating the hormonal response to glucagon.

Glycogen-storage diseases

Type	Name	Enzyme deficiency	Structural or clinical consequences
I	von Gierke's	G-6-Pase	severe postabsorptive hypoglycemia, lactic acidemia, hyperlipidemia
II	Pompe's	lysosomal α-glucosidase	glycogen granules in lysosomes
III	Cori's	debranching enzyme	altered glycogen structure, hypoglycemia
IV	Andersen's	branching enzyme	altered glycogen structure
V	McArdle's	muscle phosphorylase	excess muscle glycogen deposition, exercise-induced cramps and fatigue
VI	Hers'	liver phosphorylase	hypoglycemia, not as severe as Type 1

Fig. 12.8 Major classes of glycogen-storage diseases.

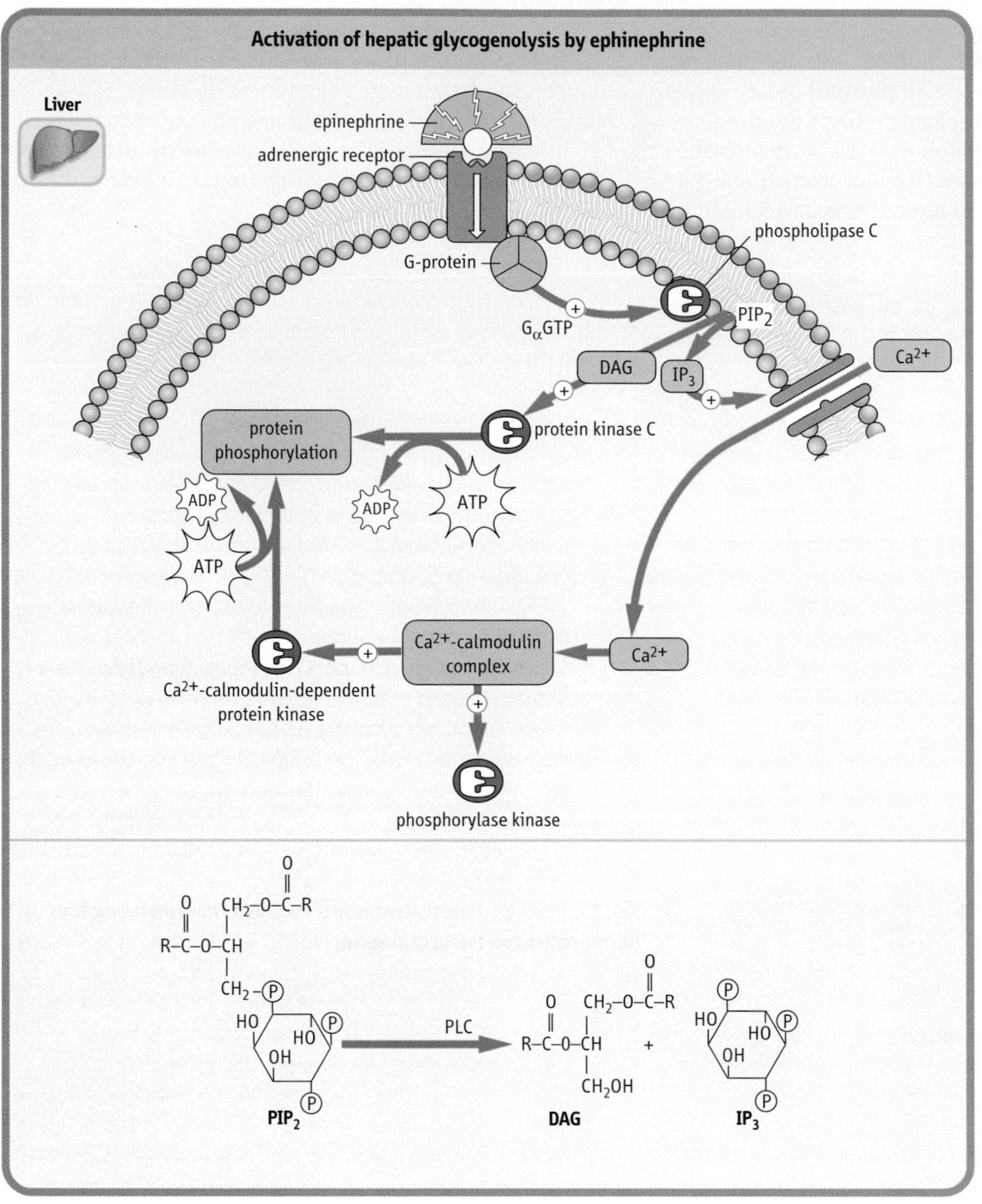

Fig. 12.9 Mechanism of activation of glycogenolysis in liver via the α-adrenergic receptor. Diacylglycerol (DAG) and inositol trisphosphate (IP_3) are second messengers mediating the adrenergic response. PIP_2 (phosphatidylinositol biphosphate).

Mobilization of hepatic glycogen by epinephrine

Epinephrine works through several distinct receptors on different cells. The best studied of these receptors are the α- and β-adrenergic receptors; they recognize different conformations or aspects of the structure of epinephrine, bind epinephrine with different affinities, and work by different mechanisms. During severe hypoglycemia, glucagon and epinephrine work together to magnify the glycogenolytic response in liver. However, even when blood glucose is normal, epinephrine is released in response to real or perceived threats, causing an increase in blood glucose to support a 'fight or flight' response. Caffeine in coffee and theophylline in tea are inhibitors of phosphodiesterase, and also cause an increase in cAMP and blood glucose. Like epinephrine, caffeine, administered in the form of a few strong cups of coffee, can also make us alert and responsive.

Epinephrine action on hepatic glycogenolysis proceeds by two pathways. One of these, through the epinephrine β-receptor, is essentially identical with that for glucagon, involving a plasma-membrane epinephrine-specific receptor, G-proteins, and cAMP. The epinephrine response augments the effects of glucagon during severe hypoglycemia, and also explains, in part, the rapid heartbeat, sweating, tremors, and anxiety associated with hypoglycemia. Epinephrine also

Maximal inhibition of glycogen synthase is achieved only through sequential action of several kinases

When both glucagon and epinephrine are acting on liver, the activation of glycogenolysis and inhibition of glycogenesis is mediated by at least three kinases: PKA, PKC, and Ca^{2+}-calmodulin-activated protein kinase. All three of these protein kinases phosphorylate key serine and threonine residues in regulatory enzymes. These and other protein kinases work in concert with one another in a process known as sequential or hierarchical phosphorylation, leading to phosphorylation of up to nine amino acid residues in glycogen synthase. Maximal inhibition of glycogen synthase is achieved only through the sequential activity of several kinases. In some cases, certain serine or threonine residues must be phosphorylated in a specific sequence by cooperative action of different kinases, i.e. phosphorylation of one site by one enzyme requires prior phosphorylation of another site by a separate enzyme.

McArdle's disease: glycogen storage disease which reduces capacity to exercise

A 30-year-old man consults his physician because of chronic arm and leg muscle pains and cramps during exercise. He indicates that he has always had some muscle weakness and, for this reason, was never active in scholastic sports, but the problem didn't become severe until he recently enrolled in an exercise program to improve his health. He also noted that the pain generally disappeared after about 15–30 min, and then he could continue his exercise without discomfort. His blood glucose concentration was normal during exercise, but serum creatine kinase (MM isoform) was elevated, suggesting muscle damage. Blood glucose declined slightly during 15 min of exercise, but unexpectedly blood lactate also declined, rather than increased, even when he was experiencing muscle cramps. A biopsy indicated an unusually high level of glycogen in muscle, suggesting a glycogen-storage disease.

Comment. This patient suffers from McArdle's disease, a rare deficiency of muscle phosphorylase activity. The actual enzyme deficiency must be confirmed by enzyme assay, since a number of other mutations could also affect muscle glycogen metabolism. During the early periods of intense exercise, the muscle obtains most of its energy by metabolism of glucose, derived from glycogen. During cramps, which normally occur during oxygen debt, much of the pyruvate produced by glycolysis is excreted into blood as lactate. In this case, however, the patient did not excrete lactate, suggesting a failure to mobilize muscle glycogen. His recovery after about 0.5 h results from epinephrine-mediated physiologic responses that provide fuels, both glucose and fatty acids, from blood, overcoming the deficit in muscle glycogenolysis. Treatment of McArdle's disease usually involves exercise avoidance or, if necessary, carbohydrate consumption prior to exercise. Otherwise, the course of the disease is uneventful.

works simultaneously through an α-receptor, but by a different mechanism. Binding to α-receptors also involves G-proteins, common elements in hormone signal transduction, but in this case the G-protein is specific for activation of a membrane isozyme of phospholipase C (PLC), which is specific for cleavage of a membrane phospholipid, phosphatidylinositol bisphosphate (PIP_2) (Fig. 12.9). Both products of PLC action, diacylglycerol (DAG) and inositol trisphosphate (IP_3), act as second messengers of epinephrine action. DAG activates protein kinase C (PKC), which, like PKA, initiates a series of protein-phosphorylation reactions. IP_3 promotes the transport of Ca^{2+} into the cytosol. Ca^{2+} then binds to the cytoplasmic protein calmodulin, which binds to and activates phosphorylase kinase, leading to phosphorylation and activation of phosphorylase, providing glucose for blood. A Ca^{2+}-calmodulin-dependent protein kinase and other enzymes are also activated, either by phosphorylation or by association with the Ca^{2+}-calmodulin complex. A range of metabolic pathways are activated in response to stress, especially those involved in the mobilization of energy reserves.

Glycogenolysis in muscle

The tissue localization of hormone receptors provides tissue specificity to hormone action. Only those tissues with glucagon receptors respond to glucagon. Muscle may be rich in glycogen, even during hypoglycemia, but it lacks both the glucagon receptor and G-6-Pase. Therefore muscle glycogen cannot be mobilized to replenish blood glucose. Muscle glycogenolysis is activated in response to epinephrine through the β-adrenergic receptor (cAMP-mediated), providing a supply of carbohydrate for the energy needs of muscle. This occurs not only during 'fight or flight' situations, but also during prolonged exercise. There are also two important hormone-independent mechanisms for activation of glycogenolysis in muscle (Fig. 12.10). First, the influx of Ca^{2+} into the muscle cytoplasm in response to nerve stimulation activates the basal, unphosphorylated form of phosphorylase kinase by action of the Ca^{2+}-calmodulin complex. This hormone-independent activation of phosphorylase provides for rapid activation of glycogenolysis during short bursts of exercise, even in the absence of epinephrine action. A second mechanism for activation of muscle glycogenolysis involves direct allosteric activation of phosphorylase by AMP. Increased usage of ATP during a rapid burst of muscle activity leads to

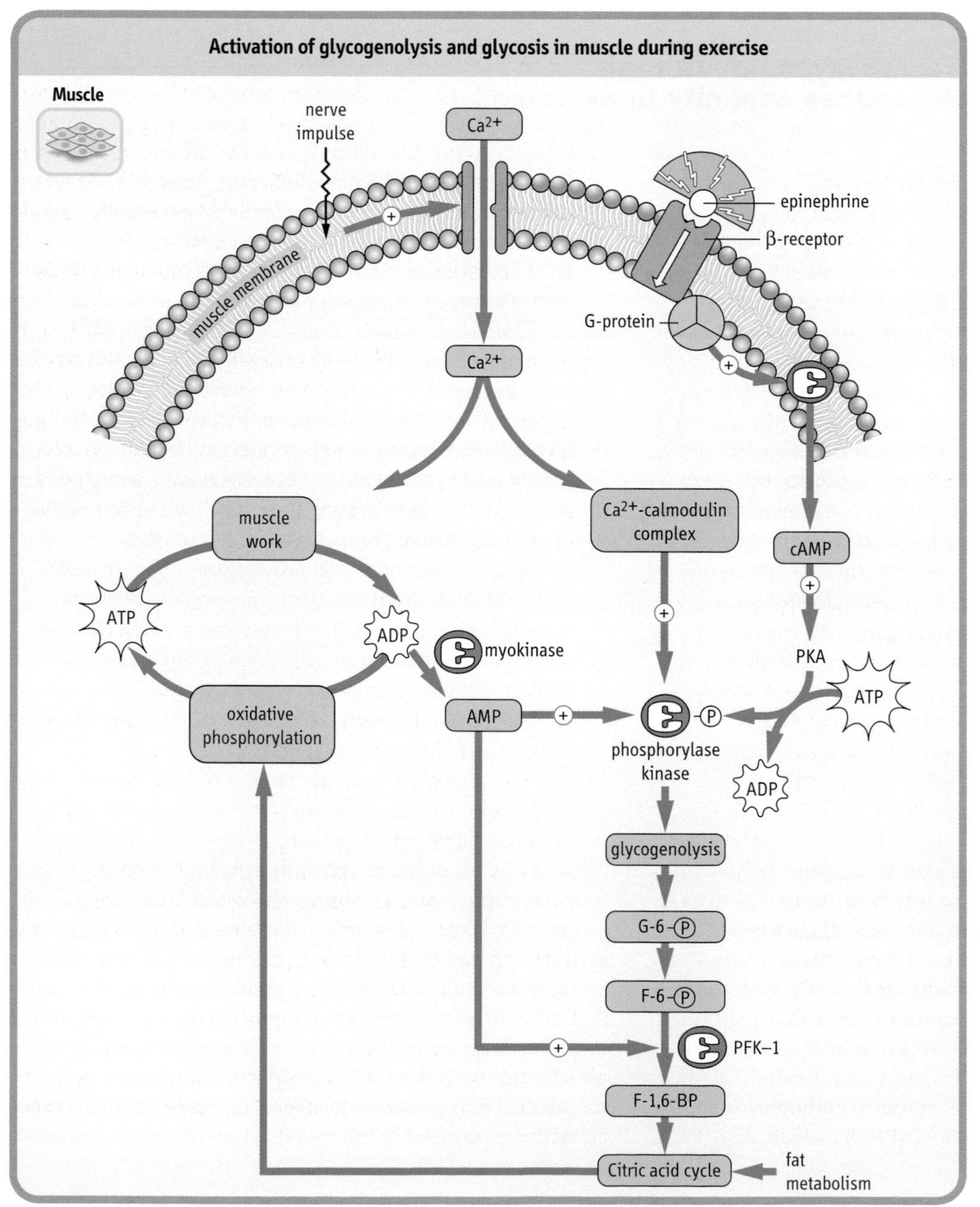

Fig. 12.10 Regulation of protein kinase A in muscle. PFK-1, phosphofructokinase-1.

rapid accumulation of ADP, which is converted in part into AMP by action of the enzyme myokinase (adenylate kinase), which catalyzes the reaction

$$2\ ADP \rightleftharpoons ATP + AMP$$

AMP activates both the basal and phosphorylated forms of phosphorylase, enhancing glycogenolysis either in the absence or presence of hormonal stimulation. AMP also relieves inhibition of phosphofructokinase-1 (PFK-1) by ATP (see Chapter 11), stimulating the utilization of glucose through glycolysis for energy production. The stimulatory effects of Ca^{2+} and AMP assure that the muscle can respond to its energy needs, even in the absence of hormonal input.

Regulation of glycogenesis

Glycogenesis, and energy storage in general, occur during and immediately following meals. Glucose and other carbohydrates, rushing into the liver from the intestines via the portal circulation and are efficiently trapped, to make glycogen. Excess glucose proceeds to the peripheral circulation, where it is taken up into muscle and adipose tissue for energy reserves or storage. We normally eat sitting down, rather than during exercise, so that utilization and storage of energy are temporally compartmentalized functions in our lives. Energy storage is under the control of the polypeptide hormone insulin, which is stored in β-cells in the pancreatic islets of Langerhans. Insulin is secreted into blood following a meal, tracking blood glucose concentration. It has two primary functions in carbohydrate metabolism: first, insulin reverses the actions of glucagon in phosphorylation of proteins, turning off glycogen phosphorylase and activating glycogen synthase, promoting glucose storage; second, it stimulates the uptake of glucose into muscle and adipose tissue, facilitating synthesis and storage of glycogen and triglycerides. Insulin also acts at the level of gene expression stimulating the synthesis of enzymes involved in carbohydrate metabolism and storage and conversion of glucose into triglycerides.

Protein tyrosine phosphorylation, rather than serine and threonine phosphorylation, is a characteristic feature of insulin and growth factor activity. Insulin binding to its transmembrane receptor (Fig. 12.11) stimulates aggregation of receptors and promotes tyrosine kinase activity in the intracellular domain of the receptor. The insulin receptor kinase activity autophosphorylates its tyrosine residues, enhancing its protein tyrosine kinase activity and phosphorylating tyrosine residues in other intracellular effector proteins, which activate secondary pathways. Among these are kinases that phosphorylate serine and threonine residues on proteins, but at sites and on proteins distinct from those phosphorylated by PKA and PKC. Insulin-dependent activation of GTPase, phosphodiesterase and phosphoprotein phosphatases also checks the action of glucagon, which is typically present in high concentration in the blood at mealtimes, i.e. several hours since the last meal.

The liver also appears to be directly responsive to ambient blood glucose concentration, increasing glycogen synthesis following a meal, even in the absence of hormonal input. Thus, the increase in hepatic glycogenesis begins more rapidly than the increase in insulin concentration in blood, and perfusion of liver with glucose solutions *in vitro*, in the absence of insulin, also leads to inhibition of glycogenolysis and activation of glycogenesis. This appears to occur by direct allosteric inhibition of phosphorylase by glucose and secondary stimulation of protein phosphatase activity.

Child born of malnourished mother may have hypoglycemia

A baby girl is born at 39 weeks of gestation to a young, malnourished mother. The child is also thin and weak at birth and, within 1 h after birth, is showing signs of distress, including rapid heartbeat and respiration. Her blood glucose was 3.5 mmol/L (63 mg/dL) at birth, and declined rapidly to 1.5 mmol/L (27 mg/dL) by 1 h, when she was becoming unresponsive and comatose. Her condition was markedly improved by infusion of a glucose solution, followed by a carbohydrate-rich diet. She improved gradually over the next 2 weeks before discharge from the hospital.

Comment. During development *in utero*, the fetus obtains glucose exogenously, from the placental circulation. But, following birth, the child relies at first on mobilization of hepatic glycogen, and then on gluconeogenesis for maintenance of blood glucose. Because of the malnourished state of the mother, the child was born with negligible hepatic glycogen reserves. Thus, she was unable to maintain blood glucose homeostasis postpartum and rapidly declined into hypoglycemia, initiating a stress response. After surviving the transient hypoglycemia, she probably still lacks adequate muscle mass to provide a sufficient supply of amino acids for gluconeogenesis. Infusion of glucose, followed by a carbohydrate-rich diet, will address these deficits, but may not correct more serious damage from prolonged malnutrition during fetal development.

Most, if not all, cells in the body are responsive to insulin in some way, but the major sites of insulin action, on a mass basis, are muscle and adipose tissue. These tissues normally have low levels of cell-surface glucose transporters, restricting the entry of glucose – they rely mostly on lipids for energy metabolism. In muscle and adipose tissue, insulin-receptor tyrosine kinase activity induces movement of glucose transporter-4 (GLUT-4) from intracellular vacuoles to the cell surface, increasing glucose transport into the cell. The glucose is then used in muscle for synthesis of glycogen, and in adipose tissue to produce glyceraldehyde 3-phosphate which is converted to glycerol 3-phosphate for synthesis of triglycerides (Chapter 15).

Large child born of a diabetic mother

A baby boy, born of a poorly controlled, chronically hyperglycemic, diabetic mother, was large and chubby (macrosomic) at birth (5 kg), but appeared otherwise normal. He declined rapidly, however, and within 1 h showed all of the symptoms of hypoglycemia, similar to the previously described case of the baby girl born of a malnourished mother. The difference, in this case, was that the boy was obviously on the heavy side, rather than thin and malnourished.

Comment. This child has experienced a chronically hyperglycemic environment during uterine development. He adapted by increasing endogenous insulin production, which has a growth hormone-like activity, resulting in macrosomia. At birth, when placental delivery of glucose ceases, he has a normal blood glucose concentration and a substantial supply of hepatic glycogen. However, his high blood-insulin concentration promotes glucose uptake into muscle and adipose tissue. The resultant insulin-induced hypoglycemia leads to a stress response, which was corrected by glucose infusion. His ample body mass will provide a good reservoir for synthesis of blood glucose from muscle protein when gluconeogenesis is activated at 1–2 days postpartum.

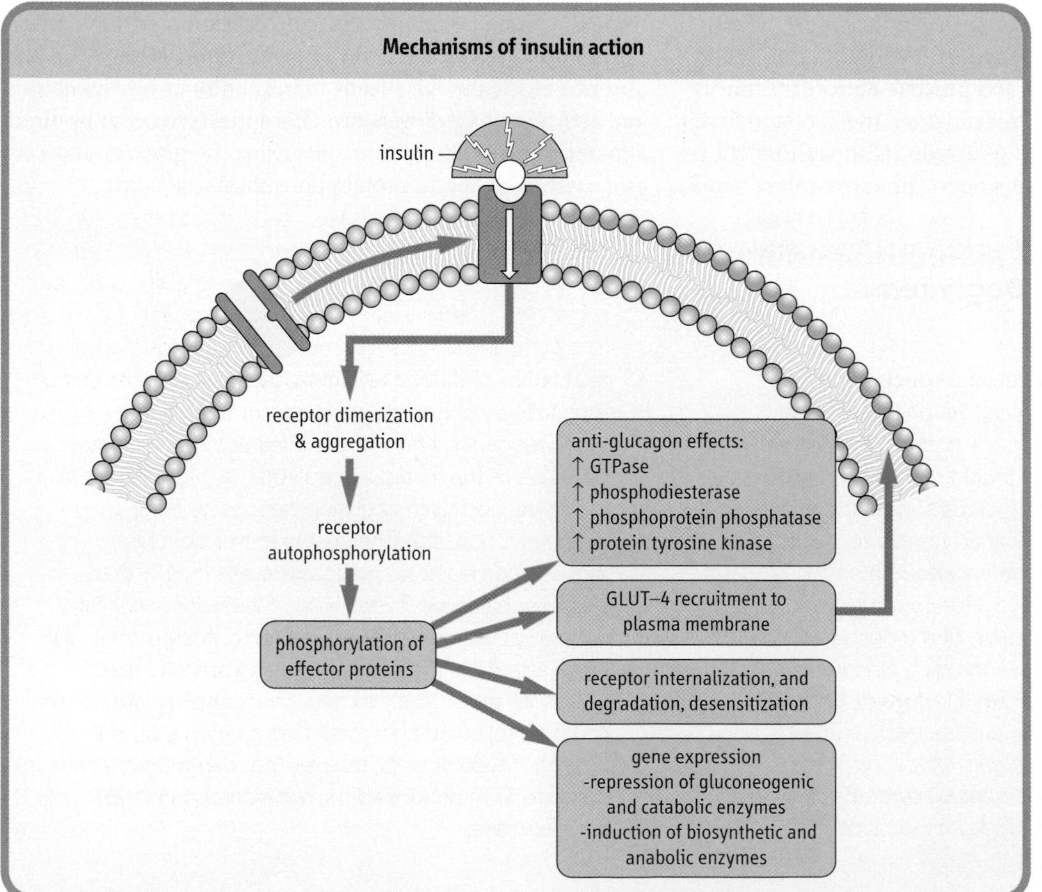

Fig. 12.11 Regulatory effects of insulin on hepatic and muscle carbohydrate metabolism.

Gluconeogenesis

During fasting and starvation, when hepatic glycogen is depleted, gluconeogenesis is essential for maintenance of blood glucose homeostasis. Gluconeogenesis requires both a source of energy and a source of carbons for formation of the backbone of the glucose molecule. The energy is provided by metabolism of fats released from adipose tissue. The carbon skeletons are provided from three primary sources:

- lactate produced in tissues such as the red cell by anaerobic glycolysis;
- amino acids derived from muscle protein;
- glycerol released from triglycerides during lipolysis in adipose tissue.

Among these, muscle protein is the major precursor of blood glucose – the rate of gluconeogenesis is often limited by the availability of substrate, including the rate of proteolysis in muscle. During prolonged fasting, malnutrition or starvation, we lose both adipose and muscle mass. The fat is used both for the general energy needs of the body and to support gluconeogenesis, while most of the amino acids in protein are converted into glucose.

Gluconeogenesis from lactate

Gluconeogenesis is conceptually the opposite of anaerobic glycolysis, but proceeds by a slightly different pathway, involving both mitochondrial and cytosolic enzymes (Fig. 12.12). Lactate is the end product of anaerobic glycolysis – blood lactate is derived primarily from anaerobic glycolysis in red cells and exercising muscle. Gluconeogenesis is the only metabolic pathway in the body that uses lactate, and it converts the lactate back into glucose, using, in part, the same glycolytic enzymes involved in conversion of glucose into lactate. The lactate cycle involving the liver, red cells, and muscle, known as the Cori cycle, is discussed in Chapter 20.

Alcohol excess can lead to hypoglycemia

A middle-aged, emaciated, chronic alcoholic man collapsed in a bar at about 11 a.m., and was transported to the emergency room by ambulance. Another patron noted that the man had had only a few shots of vodka, and did not appear to be unusually drunk, although he was a little confused, at the time that he fainted. The bartender suggested that the man might have had a heart attack. Physical examination revealed a somewhat clammy skin, unusual for a winter morning, rapid breathing, and a rapid heartbeat. Laboratory tests indicated a blood glucose of 2.5 mmol/L (50 mg/dL), in the hypoglycemic range, and a blood alcohol level of 0.2%, above the legal limit for intoxication. Subsequent tests indicated a normal level of creatine phosphokinase, an enzyme measured for early diagnosis of myocardial infarction, high serum asparate aminotransferase activity, indicative of ongoing liver damage (cirrhosis), a slightly acidic blood pH (7.29 versus normal 7.35), low pCO_2, and high blood lactate. The man responded to an infusion of a glucose solution, regained consciousness, had brunch, and a few hours later, after a miraculous recovery, was referred to a counselor for treatment. What happened?

Comment. This patient probably had not eaten breakfast before starting his morning binge. His glycogen stores were negligible, so he was dependent on gluconeogenesis for maintenance of blood glucose concentration, but gluconeogenesis may be compromised both by liver damage and by the limited muscle mass available to mobilize amino acids for gluconeogenesis. The consumption of alcohol places additional stress on gluconeogenesis, since alcohol is metabolized primarily in the liver. The two-step metabolism of alcohol is relatively unregulated, leading to a rapid increase in hepatic NADH:

$$CH_3CH_2OH \xrightarrow[\text{alcohol dehydrogenase}]{NAD^+ \to NADH} CH_3CHO \xrightarrow[\text{aldehyde dehydrogenase}]{NAD^+ \to NADH} CH_3COOH$$

The increase in hepatic NADH shifts the equilibrium of the LDH reaction toward lactate, limiting gluconeogenesis from pyruvate derived from lactate (or alanine), leading to accumulation of lactic acid in blood (lactic acidemia). It also shifts cytosolic oxaloacetate toward malate, reducing gluconeogenesis from citric acid cycle intermediates, and shifts dihydroxyacetone phosphate toward glycerol-3-phosphate, reducing gluconeogenesis from glycerol. Thus, the redox imbalance induced by alcohol consumption leads to a large increase in NADH in the cytoplasm, at the expense of NAD^+, inhibiting the flux of all major substrates into gluconeogenesis. The low blood glucose leads to a stress response (rapid heart beat, clammy skin), an effort to enhance stimulation of gluconeogenesis by combined action of glucagon and epinephrine. The rapid breathing is a physiologic response to metabolic acidosis, resulting from the excess of lactic acid.

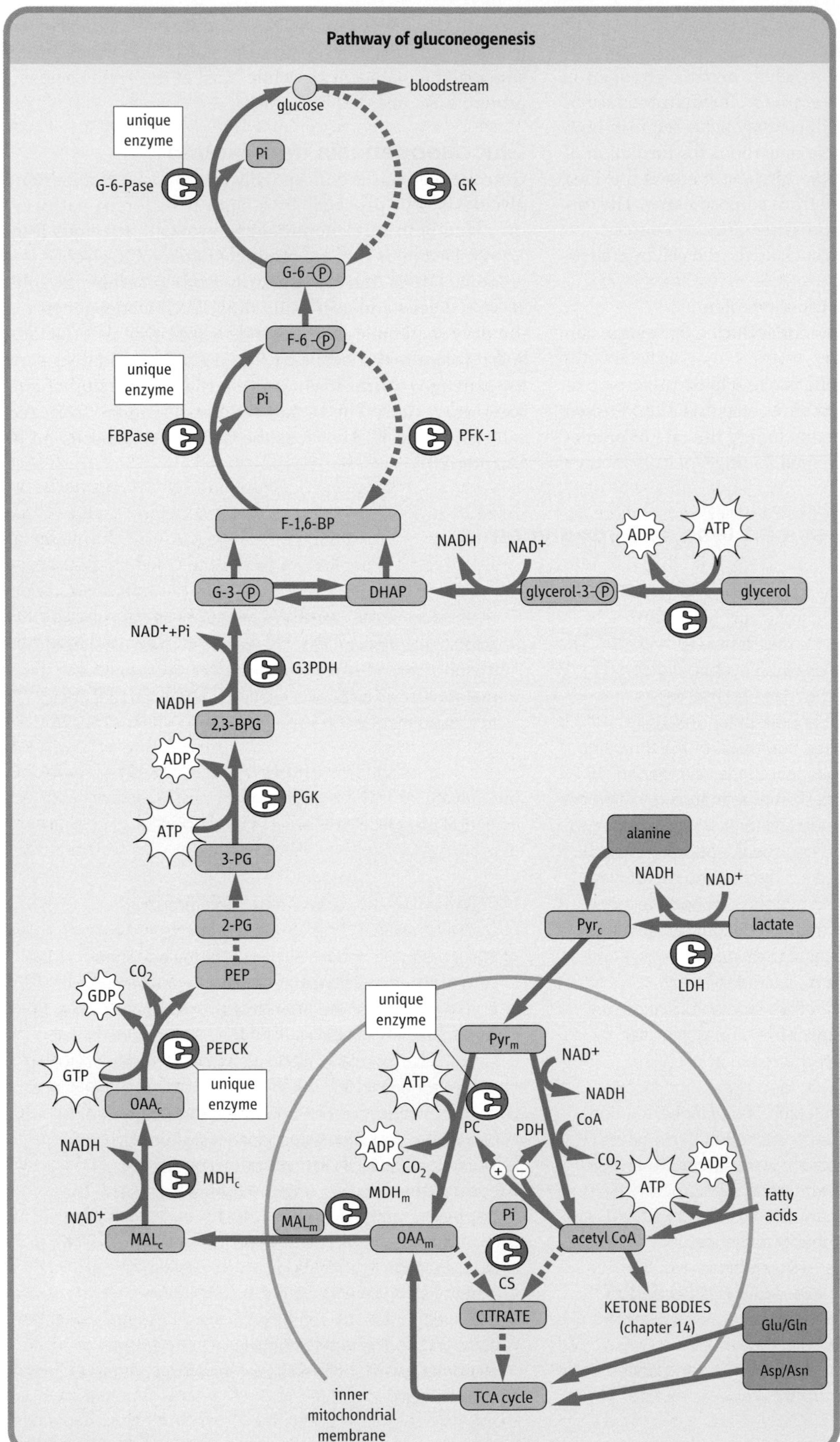

Fig. 12.12 Pathways of Gluconeogenesis. Gluconeogenesis is the reverse of glycolysis. Unique enzymes overcome the irreversible kinase reactions of glycolysis.
Compartments:
C, cytoplasmic; IMM, inner mitochondrial membrane; M, mitochondrial.
Enzymes:
CS, citrate synthase; FBPase, fructose-1,6-bisphosphatase; G6Pase, glucose-6-phosphatase. GK, glucokinase MDH, malate dehydrogenase; PC, pyruvate carboxylase; PDH, pyruvate; dehydrogenase; PEP, phosphoenolpyruvate; PEPCK, PEP carboxykinase; PGK, phosphoglycerate kinase;
Substrates:
2,3-BPG, bisphosphoglycerate; DHAP, dihydroxyacetone phosphate; F-1,6-BP, fructose-1,6-bisphosphate; Glyc-3-P, glyceraldehyde 3-phosphate; MAL, malate; OAA, oxaloacetate; Pyr, pyruvate; 3-PG, 3-phosphoglycerate.
Solid lines: active during gluconeogenesis.
Dotted lines: inactive during gluconeogenesis.

A critical problem in the reversal of glycolysis is overcoming the irreversibility of three kinase reactions: glucokinase (GK), phosphofructokinase-1 (PFK-1), and pyruvate kinase (PK). The fourth kinase in glycolysis, phosphoglycerate kinase (PGK), catalyzes a freely reversible, equilibrium reaction, transferring a high-energy acyl phosphate in 1,3-biphosphoglycerate to an energetically similar pyrophosphate bond in ATP. To circumvent the three irreversible reactions, the liver uses four unique enzymes: pyruvate carboxylase (PC) in the mitochondrion and phosphoenolpyruvate carboxykinase (PEPCK) in the cytoplasm to bypass PK, fructose-1,6-bisphosphatase (F-1,6-BPase) to bypass PFK-1, and G-6-Pase to bypass GK (see Fig. 12.12). Because these enzymes are located in different compartments, substrates, including pyruvate and malate, must be shuttled between the mitochondrion and the cytosol.

Gluconeogenesis from lactate involves, first, its conversion into PEP, a process requiring investment of two ATP equivalents because of the high energy of the enol-phosphate bond in PEP. Lactate is first converted into pyruvate by lactate dehydrogenase (LDH), then enters the mitochondrion, where it is converted into oxaloacetate by PC, using biotin and ATP. Oxaloacetate is reduced to malate for export from the mitochondrion, then reoxidized to oxaloacetate by cytosolic malate dehydrogenase. The cytosolic oxaloacetate is then decarboxylated by PEPCK, using GTP as a cosubstrate, yielding PEP. The energy for synthesis of PEP from oxaloacetate is derived from both the GTP and the decarboxylation of oxaloacetate.

Glycolysis may now proceed backwards from PEP until it reaches the next irreversible reaction, PFK-1. This enzyme is bypassed by a simple hydrolysis reaction, catalyzed by FBPase without production of ATP, as would be required by reversal of the PFK-1 reaction. Similarly, the bypass of GK is accomplished by hydrolysis of G-6-P by G-6-Pase, without production of ATP. The free glucose is then released into blood.

Gluconeogenesis is fairly efficient – the liver can make a kilogram of glucose per day by gluconeogenesis, and actually does so in poorly controlled, hyperglycemic diabetic patients. Gluconeogenesis from pyruvate is also moderately expensive, requiring a net expenditure of 4 moles of ATP per mole of pyruvate converted into glucose. This ATP is provided by oxidation of fatty acids (Chapter 14).

Gluconeogenesis from amino acids and glycerol

Most amino acids are glucogenic, i.e. following deamination their carbon skeletons can be converted into glucose. Alanine and glutamine are the major amino acids exported from muscle for gluconeogenesis. Their relative concentrations in venous blood from muscle exceed their relative concentration in muscle protein, indicating considerable reshuffling of muscle amino acids to provide gluconeogenic substrates. As discussed in more detail in Chapter 18, alanine is converted directly into pyruvate by the enzyme alanine aminotransferase, and then gluconeogenesis proceeds as described for lactate. Other amino acids are converted into tricarboxylic acid cycle (TCA cycle) intermediates, then to malate for gluconeogenesis. Aspartate, for example, is converted into oxaloacetate by aspartate amino transferase, and glutamate into α-ketoglutarate by glutamate dehydrogenase. Other glucogenic amino acids are converted by less direct routes into alanine or intermediates in the tricarboxylic acid cycle for gluconeogenesis. Excess amino groups generated during the transamination reactions are converted into urea, via the urea cycle and the urea is excreted in urine (Chapter 18).

Glycerol enters gluconeogenesis at the level of triose phosphates (see Fig. 12.12). Following release of glycerol and fatty acids from adipose tissue, the glycerol is taken up into liver and phosphorylated by glycerol kinase, then enters the gluconeogenic pathway as dihydroxyacetone phosphate. Only the glycerol component of fats can be converted into glucose. As discussed in Chapter 14, metabolism of fatty acids involves their conversion in two carbon oxidation steps to form acetyl CoA, which is then metabolized in the tricarboxylic acid cycle by condensation with oxaloacetate to form citrate. While the carbons of acetate are theoretically available for gluconeogenesis, two molecules of CO_2 are eliminated during conversion of citrate into malate. Thus, although energy is produced during the tricarboxylic acid cycle, the two carbons invested for gluconeogenesis from acetyl CoA are lost as CO_2. For this reason, acetyl CoA, and therefore, even-chain fatty acids, cannot serve as a substrate for gluconeogenesis. However, odd-chain and branched-chain fatty acids yield propionyl CoA, which can serve as a minor precursor for gluconeogenesis. Propionyl CoA is first carboxylated to methylmalonyl CoA, which undergoes racemase and mutase reactions to form succinyl CoA, a tricarboxylic acid cycle intermediate (see Chapter 15). Succinyl CoA is converted into malate, exits the mitochondrion and is oxidized to oxaloacetate. Following decarboxylation by PEPCK, the three carbons of propionate appear intact in PEP for gluconeogenesis.

Regulation of gluconeogenesis

Like glycogen metabolism in liver, gluconeogenesis is regulated primarily by hormonal mechanisms. In this case, the regulatory process involves counter-regulation of glycolysis and gluconeogenesis, largely by phosphorylation/dephosphorylation of enzymes, under control of glucagon and insulin. The primary control point is at the regulatory enzymes PFK-1 and F-1,6-BPase, which, in liver, are exquisitely sensitive to the allosteric effector fructose 2,6-bisphosphate (F-2,6-BP). F-2,6-BP is an activator of PFK-1 and an inhibitior of F-1,6-BPase. As shown in Figure 12.13, F-2,6-BP is synthesized by an unusual, bifunctional enzyme, phosphofructokinase-2/fructose-2,6-bisphosphatase (PFK-2/F-2,6-BPase) which has both kinase and phosphatase activities. In the phosphorylated state, under the influence of glucagon, this enzyme displays F-2,6-BPase activity, reducing the level of F-2,6-BP, which simultancously decreases the stimulation of glycolysis (PFK-1) and relieves inhibition of gluconeogenesis (F-1,6-BPase). The coordinate, allosterically-mediated decrease in PFK-1 and increase in F-1,6-BPase activity ensures that glucose made by gluconeogenesis is not consumed by glycolysis in a futile cycle, but released into blood by G-6-Pase. Similarly, any flux of glucose from glycogen, also induced by glucagon, is diverted to blood, rather than to glycolysis, by inhibition of PFK-1. Pyruvate kinase (PK) is also inhibited by phosphorylation by protein kinase A(PKA), providing an additional site for inhibition of glycolysis.

When glucose enters the liver following a meal, insulin mediates the dephosphorylation of PFK-2/F-2,6-BPase, turning on its PFK-2 activity. The resultant increase in F-2,6-BP activates PFK-1 and inhibits F-1,6-BPase activity. Gluconeogenesis is inhibited, and glucose entering the liver is then incorporated into glycogen or routed into glycolysis and lipogenesis. Thus, liver metabolism following a meal is focused on synthesis and storage of both carbohydrate and lipid energy reserves, which it will later use, in the postabsorptive state, for maintenance of blood glucose and fatty acid homeostasis.

Gluconeogenesis is also regulated in the mitochondrion by acetyl CoA. The influx of fatty acids from adipose tissue, stimulated by glucagon to support gluconeogenesis, leads to an increase in hepatic acetyl CoA, which is both an inhibitor of pyruvate dehyrogenase (PDH) and an essential allosteric activator of pyruvate carboxylase (PC) (see Figs 12.12 and 12.14). In this way, fat metabolism inhibits the oxidation of pyruvate and favors gluconeogenesis in liver. In muscle, the utilization of glucose is limited both by the low level of GLUT-4 in the plasma membranes and by inhibition of PDH by acetyl CoA. Active fat metabolism and high levels of acetyl CoA in muscle promote the excretion of a significant fraction of pyruvate as lactate, even in the resting state.

Conversion of fructose and galactose to glucose

As discussed in detail in Chapter 24, fructose is metabolized almost exclusively in the liver. It enters glycolysis at the level of triose phosphates, bypassing the regulatory enzyme, PFK-1, so that large amounts of pyruvate may be forced on the mitochondrion for use in energy metabolism or fat biosynthesis. During a gluconeogenic state, this fructose may also proceed toward G-6-P, providing a convenient source of blood glucose. Gluconeogenesis from galactose is equally efficient, since G-1-P, derived from galactose 1-phosphate (Chapter 24), is readily isomerized to G-6-P by phosphoglucomutase. These sugars are good sources of glucose, independent of glycogenolysis and gluconeogenesis.

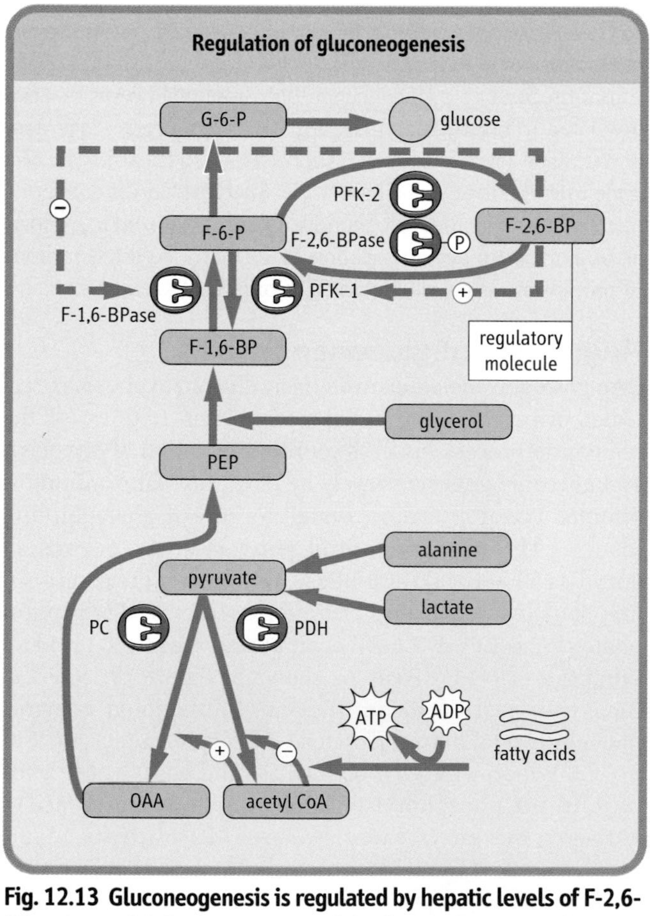

Fig. 12.13 Gluconeogenesis is regulated by hepatic levels of F-2,6-BP and acetyl CoA. The upper part of the diagram focuses on the reciprocal regulation of F-1,6-BPase and PFK-1 by F-2,6-BP and the lower part on the reciprocal regulation of pyruvate dehydrogenase (PDH) and pyruvate carboxylase (PC) by acetyl CoA.

General features of hormone action

- tissue specificity, determined by receptor distribution
- multistep, cascade amplification
- intracellular second messengers
- coordinate counter-regulation of opposing pathways
- augmentation and/or opposition by other hormones
- multiple mechanisms of termination of response

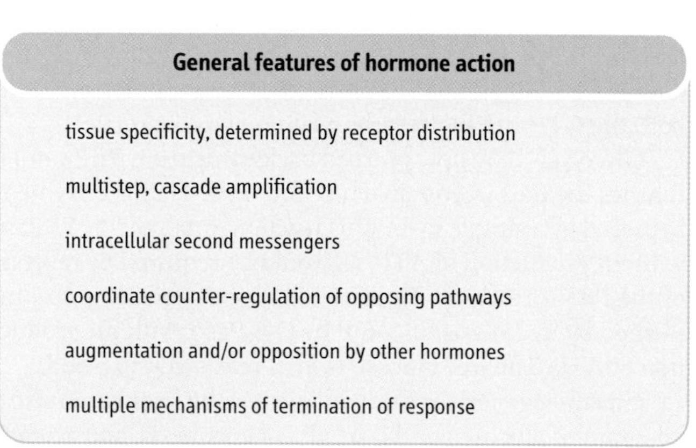

Fig. 12.14 Features of hormone action. The multihormonal regulation of gluconeogenesis illustrates fundamental principles of hormone action.

Summary

Glycogen is stored in two tissues in the body for different reasons: in liver for short-term maintenance of blood glucose homeostasis, and in muscle as a source of energy. Glycogen metabolism in these tissues responds rapidly to both allosteric and hormonal control. In liver, the balance between glycogenolysis and glycogenesis is regulated by the balance between concentrations of glucagon and insulin in the circulation, which controls the state of phosphorylation of enzymes. Phosphorylation of enzymes under the influence of glucagon directs glycogen mobilization and is the most common condition in the liver, e.g. during sleep. Increases in blood insulin during and after meals promote dephosphorylation of the same enzymes, leading to glycogenesis.

Insulin also promotes glucose uptake into muscle and adipose tissue for glycogen and triglyceride synthesis following a meal. Epinephrine also controls phosphorylation of liver enzymes, enabling a burst in hepatic glycogenolysis and an increase in blood glucose for stress responses. Muscle is responsive to epinephrine, but not to glucagon; in this case the glucose produced by glycogenolysis is used for energy metabolism. Muscle glycogenolysis is also responsive to intracellular Ca^{2+} and AMP concentrations, providing a mechanism for coupling glycogenolysis to energy consumption during exercise. The actions of insulin, glucagon, and epinephrine illustrate many of the fundamental principles of hormone action (Fig. 12.14).

Gluconeogenesis takes place primarily in liver, and is designed for maintenance of blood glucose during the fasting state. It is essential after 12 h of fasting, when the majority of hepatic glycogen has been consumed. The major substrates for gluconeogenesis are lactate, amino acids, and glycerol, while fatty acid metabolism provides the necessary energy. The major control point is at the level of phosphofructokinase-1 (PFK-1), which is activated by the allosteric effector F-2,6-BP. The synthesis of F-2,6-BP is under control of the bifunctional enzyme, PFK-2/F-2,6-BPase, whose kinase and phosphatase activities are regulated by phosphorylation/dephosphorylation, under hormonal control by insulin and glucagon. During fasting and active gluconeogenesis, glucagon mediates phosphorylation and activation of the phosphatase activity of this enzyme, leading to a decrease in the level of F-2,6-BP, and a corresponding decrease in glycolysis. Oxidation of pyruvate is also inhibited in the mitochondrion by inhibition of PDH by acetyl CoA, derived from fat metabolism .

Further reading

Brooks C. Neonatal hypoglycemia. *Neonatal network* 1997;**16**:15–21.

Donovan CM, Sumida KD. Training enhanced hepatic gluconeogenesis: the importance for glucose homeostasis during exercise. *Med Sci Sports Exercise* 1997;**29**:628–634.

Fischer EH. Cellular regulation by protein phosphorylation: a historical review. *Biofactors* 1997;**6**:367–374.

Hargreaves M. Interactions between muscle glycogen and blood glucose during exercise. *Exercise Sports Sci Rev* 1997;**25**:21–39.

Hawley JA, Schabort EJ, Noakes TD, Dennis SC. Carbohydrate loading and exercise performance. An update. *Sports Med* 1997;**24**:73–81.

Kurland IJ, Pilkis SJ. Covalent control of 6-phosphofructo-2-kinase/fructose-2,6-bisphosphatase. *Protein Sci* 1995;**4**:1023–1037.

Melendez R, Melendez–Hevia E, Cascante M. How did glycogen structure evolve to satisfy the requirement for rapid mobilization of glucose? A problem of physical constraints in structure building. *J Mol Evol* 1997;**45**:446–455.

Mizock BA. Alterations in carbohydrate metabolism during stress: a review. *Am J Med* 1995;**98**:75–84.

Triomphe TJ. Gycogen storage disease: a basic understanding and guide to nursing care. *J Pediatric Nursing* 1997;**12**: 230–249

Tarui S. Glycolytic defects in muscle: aspects of collaboration between basic science and clinical medicine. *Muscle and Nerve* 1995;**3**: S2–9.

Van den Berghe G. Disorders of gluconeogenesis. *J Inh Metab Dis* 1996;**19**:470–477.

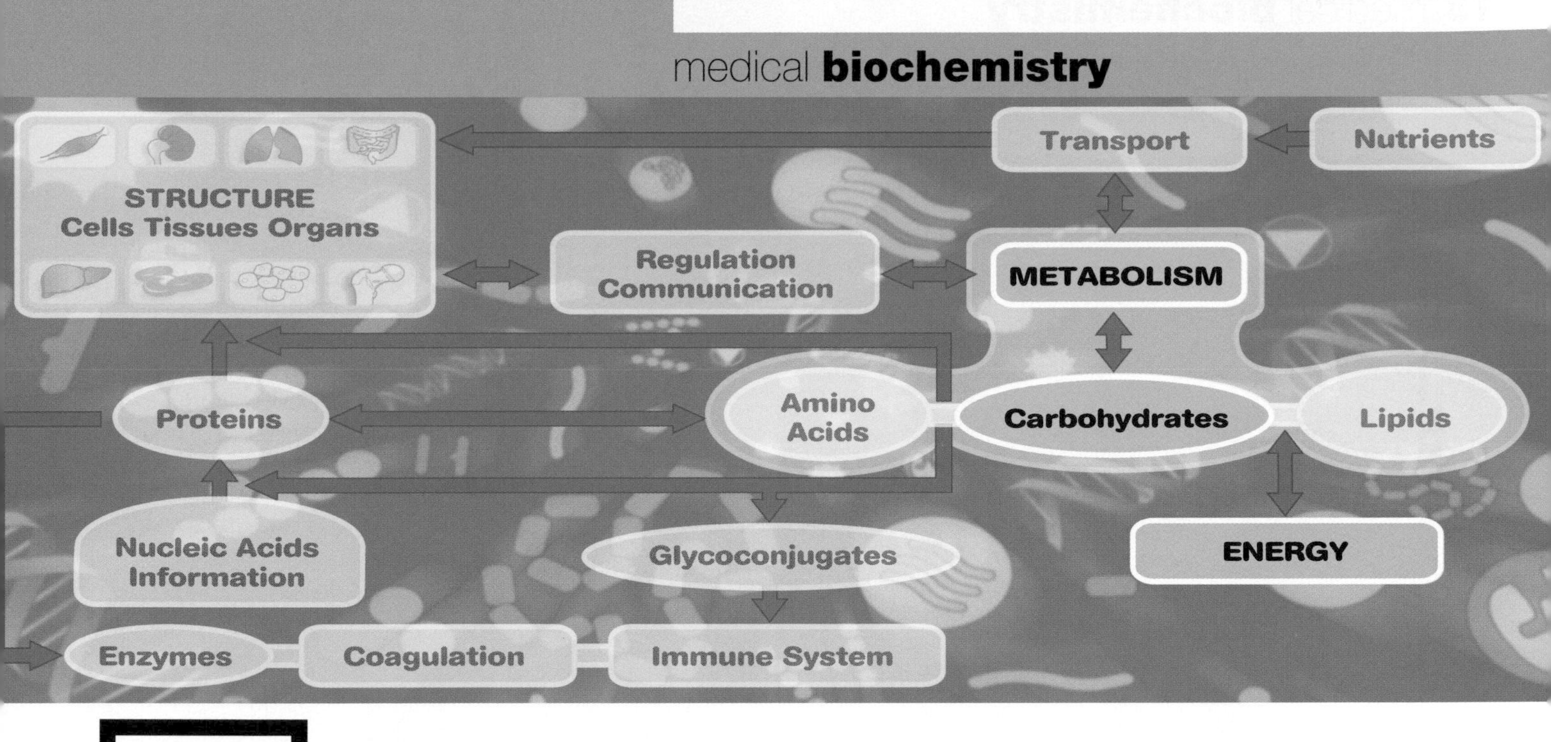

13 The Tricarboxylic Acid Cycle

Introduction

The TCA cycle is a common pathway for metabolism of all fuels

Located in the mitochondrion, the tricarboxylic acid cycle (TCA cycle), also known as the Krebs or citric acid cycle, is a common pathway for metabolism of all fuels. It oxidatively strips electrons from fat, carbohydrate and protein fuels, producing the majority of the reduced coenzymes that are used for the generation of adenosine triphosphate (ATP) in the electron transport chain. Although the TCA cycle does not use oxygen in any of its reactions, it requires oxidative metabolism in the mitochondrion for re-oxidation of reduced coenzymes. The TCA cycle has two major functions, energy production and biosynthesis (Fig. 13.1).

Energy production

The four oxidative steps in the TCA cycle provide free energy for ATP synthesis

Acetyl Coenzyme A (acetyl CoA) is the starting metabolite for the TCA cycle. It is a common end-product of metabolism of carbohydrates, fatty acids and amino acids and is oxidized to produce reduced coenzymes of the four oxidation reactions in the TCA cycle. Three produce reduced nicotinamide adenine dinucleotide (NADH) and one produces reduced flavin adenine dinucleotide ($FADH_2$). These provide free energy for synthesis of ATP by the electron transport system and oxidative phosphorylation (see Chapter 8). One high-energy phosphate, guanosine triphosphate (GTP) is produced in the cycle by substrate-level phosphorylation. Most of the body's carbon dioxide is produced by decarboxylation reactions in the TCA cycle.

Biosynthesis

The TCA cycle provides a common ground for interconversion of fuels and metabolites

The TCA cycle (Fig. 13.1) participates in the synthesis of glucose from amino acids and lactate during fasting and starvation (see Chapter 12). It is also involved in the conversion of carbohydrates to fat for storage following a carbohydrate-rich meal (see Chapter 15). It is the source of most of the nonessential amino acids in the body, such as aspartate and glutamate, which are made directly from TCA cycle intermediates. One TCA cycle intermediate, succinyl Coenzyme A (succinyl CoA), serves as a precursor to porphyrins for heme biosynthesis. Many biosynthetic reactions proceeding from the TCA cycle require the input of carbons from intermediates other than acetyl CoA. Reactions that supply carbons to the cycle are known as anaplerotic reactions.

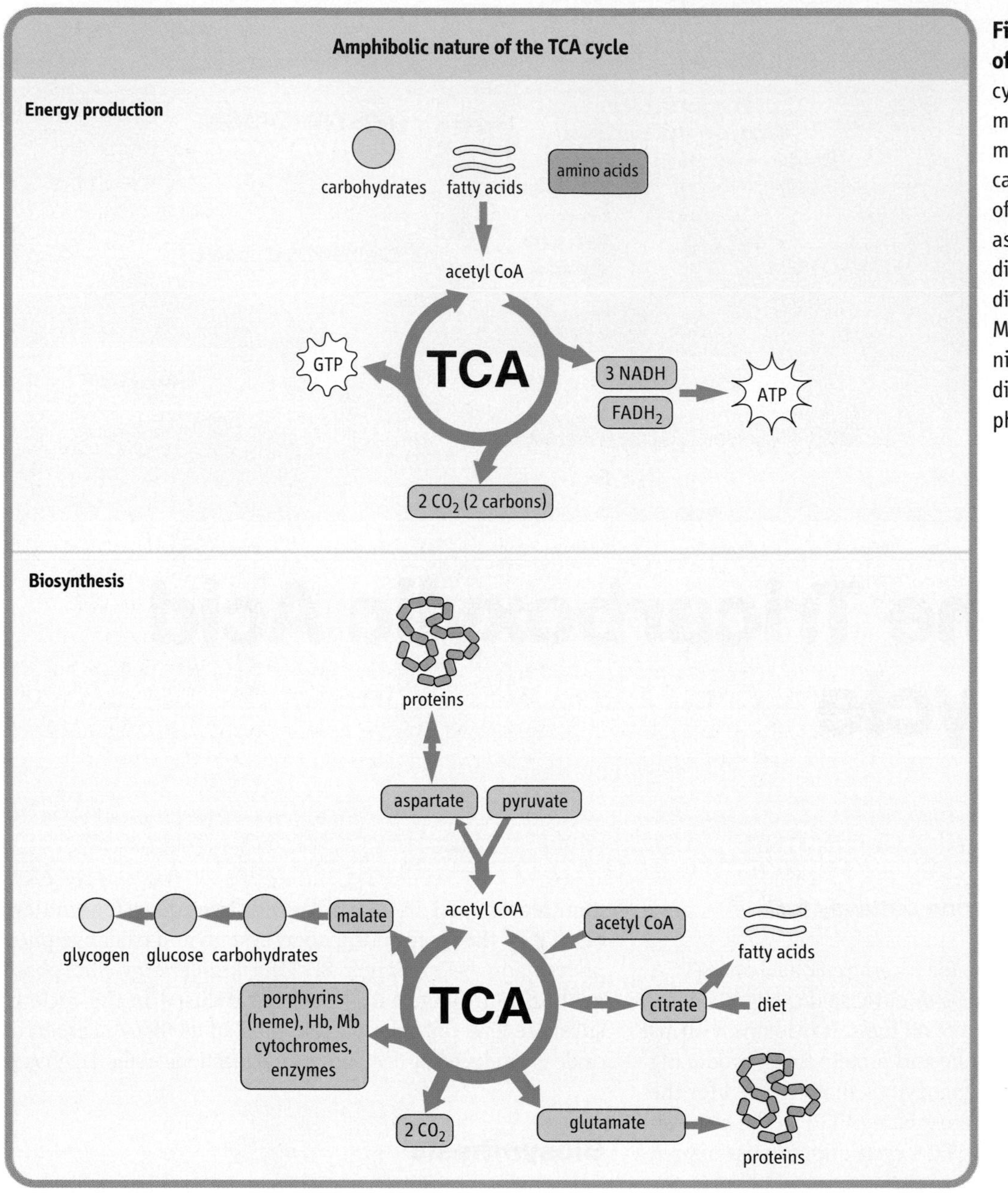

Fig. 13.1 Amphibolic nature of the TCA cycle. The TCA cycle provides energy and metabolites for cellular metabolism. Because of the catabolic and anabolic nature of the TCA cycle, it is described as amphibolic. FAD, flavin dinucleotide; GDP, guanosine diphosphate; Hb, hemoglobin; Mb, myoglobin; NAD, nicotinamide adenine dinucleotide; Pi, inorganic phosphate.

The TCA cycle is located in the mitochondrion

All TCA cycle enzymes are located in the mitochondrion. Compartmentalization of this cycle within the mitochondrion is important metabolically, because it allows identical intermediates to be used for entirely different purposes inside and outside mitochondria. Acetyl CoA, for example, cannot cross the inner mitochondrial membrane. The main fate of mitochondrial acetyl CoA is oxidation in the TCA cycle, but in the cytoplasm, it is used for biosynthesis of fatty acids and cholesterol.

Metabolic defects in the TCA cycle are rare

Metabolic defects involving enzymes of the TCA cycle are rare, because normal functioning of the TCA cycle is absolutely essential to sustain life. Products of energy-producing pathways, such as pyruvate from glycolysis and acetyl CoA from fatty acid oxidation, must be metabolized in the TCA cycle for efficient production of ATP. Any defect in the TCA cycle will severely inhibit energy metabolism and ATP production. Cells deprived of ATP die rapidly.

Acetyl CoA is a common product of many catabolic pathways

The TCA cycle begins with acetyl CoA, which has three major metabolic precursors (Fig. 13.2). Carbohydrates undergo glycolysis to yield pyruvate, which can be taken up by mitochondria and oxidatively decarboxylated to acetyl CoA by the enzyme complex, pyruvate dehydrogenase. Fats are converted to free fatty acids, which are taken up by cells and transported into mitochondria, where they undergo oxidation to acetyl CoA. Lastly, proteolysis of tissue proteins releases constituent amino acids, most of which are metabolized to acetyl CoA and TCA-cycle intermediates.

The first version of the TCA cycle, proposed by Krebs in 1937, began with pyruvic acid, not acetyl CoA. Pyruvic acid was decarboxylated and condensed with oxaloacetic acid through an unknown mechanism to form citric acid. The key intermediate, acetyl CoA, was not identified until years later. It is tempting to begin the TCA cycle with pyruvic acid, unless it is recognized that lipids and amino acids are also major metabolic sources of acetyl CoA. It is for this reason that the TCA cycle is said to begin with acetyl CoA, not pyruvic acid.

Pyruvate provides both acetyl CoA and oxaloacetate (OAA) for the synthesis of citrate –the first step of the TCA cycle (Fig. 13.3). Acetyl CoA is produced from pyruvate by pyruvate dehydrogenase, one of the gatekeepers controlling the flux of carbons into the TCA cycle. OAA is formed from pyruvate through another mitochondrial enzyme, pyruvate carboxylase. Acetyl CoA and OAA condense to form citrate but, because the OAA is regenerated in the cycle, small amounts of OAA will catalyze the oxidation of larger amounts of acetyl CoA or pyruvate. In this way, the level of OAA in the mitochondrion helps govern TCA cycle activity.

Deficiencies in pyruvate metabolism and TCA cycle

Lacticacidemia in infancy, resulting from excessive anaerobic metabolism of carbohydrates, is the most common metabolic feature of disorders in pyruvate metabolism or TCA cycle enzymes. Tissues dependent on aerobic metabolism, such as brain and muscle are most severely affected, so that the clinical picture includes impaired motor function, neurologic disorders and mental retardation.

Comment. These diseases are rare, but deficiencies in pyruvate carboxylase and all of the components of the pyruvate dehydrogenase complex (PDH) have been described, including the associated kinase and phosphatase enzymes. One such disorder involves E3 (dihydrolipoyl dehydrogenase), the subunit that is common to α-ketoacid dehydrogenases, including PDH and α-ketoglutarate dehydrogenase. Blood levels of lactate, pyruvate, α-ketoglutarate, alanine and branched-chain amino acids are elevated in these patients.

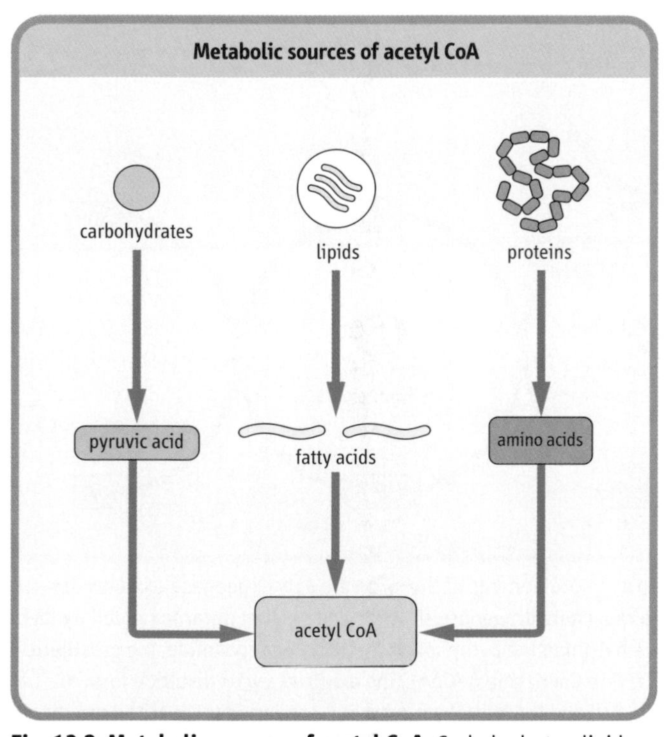

Fig. 13.2 Metabolic sources of acetyl CoA. Carbohydrates, lipids and amino acids are precursors of mitochondrial acetyl CoA for operation of the TCA cycle.

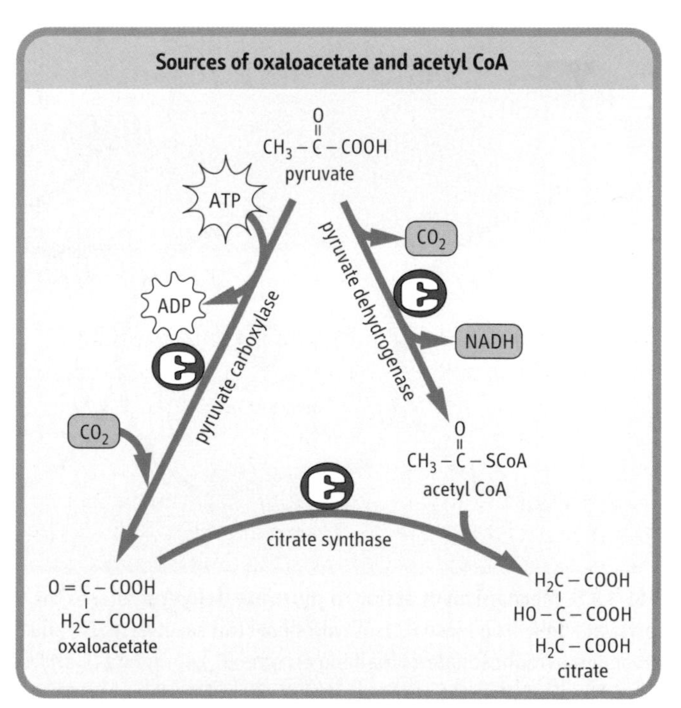

Fig. 13.3 Sources of oxaloacetate and acetyl CoA. Acetyl CoA and OAA are derived from pyruvate through the catalytic action of pyruvate dehydrogenase and pyruvate carboxylase, respectively. ADP, adenosine diphosphate.

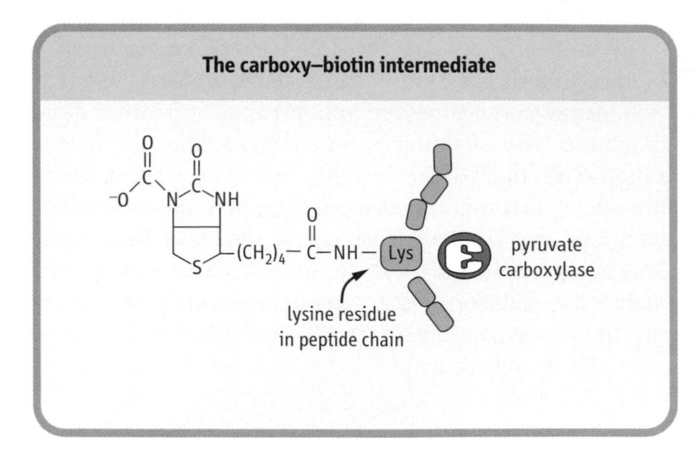

Fig. 13.4 Role of biotin in synthesis of oxaloacetate from pyruvate. Pyruvate carboxylase catalyzes carboxylation of pyruvate to oxaloacetate. The coenzyme, biotin, is covalently bound to pyruvate carboxylase, and transfers the carbon originating from CO_2 to pyruvate.

Pyruvate carboxylase

Pyruvate may be converted to four different metabolites in one step

Pyruvate is at a crossroads in metabolism. It may be converted in one step to lactate (lactate dehydrogenase), alanine (alanine aminotransferase), acetyl CoA (pyruvate dehydrogenase) or oxaloacetate (pyruvate carboxylase). Depending on metabolic circumstances, pyruvate may be routed toward gluconeogenesis (see Chapter 12), fatty acid biosynthesis (see Chapter 15) or the TCA cycle itself, so that more high-energy phosphate can be generated. Pyruvate carboxylase, like most other carboxylases, uses CO_2 and the coenzyme biotin (Fig. 13.4), one of the water-soluble B vitamins. ATP is used to drive the carboxylation reaction. The enzyme is a tetramer of identical subunits. Each subunit contains an allosteric site that binds acetyl CoA, an essential positive heterotropic modifier of enzyme activity. Pyruvate carboxylase has an absolute requirement for acetyl CoA – the enzyme does not work in its absence. An abundance of mitochondrial acetyl CoA acts a signal for the generation of additional oxaloacetate.

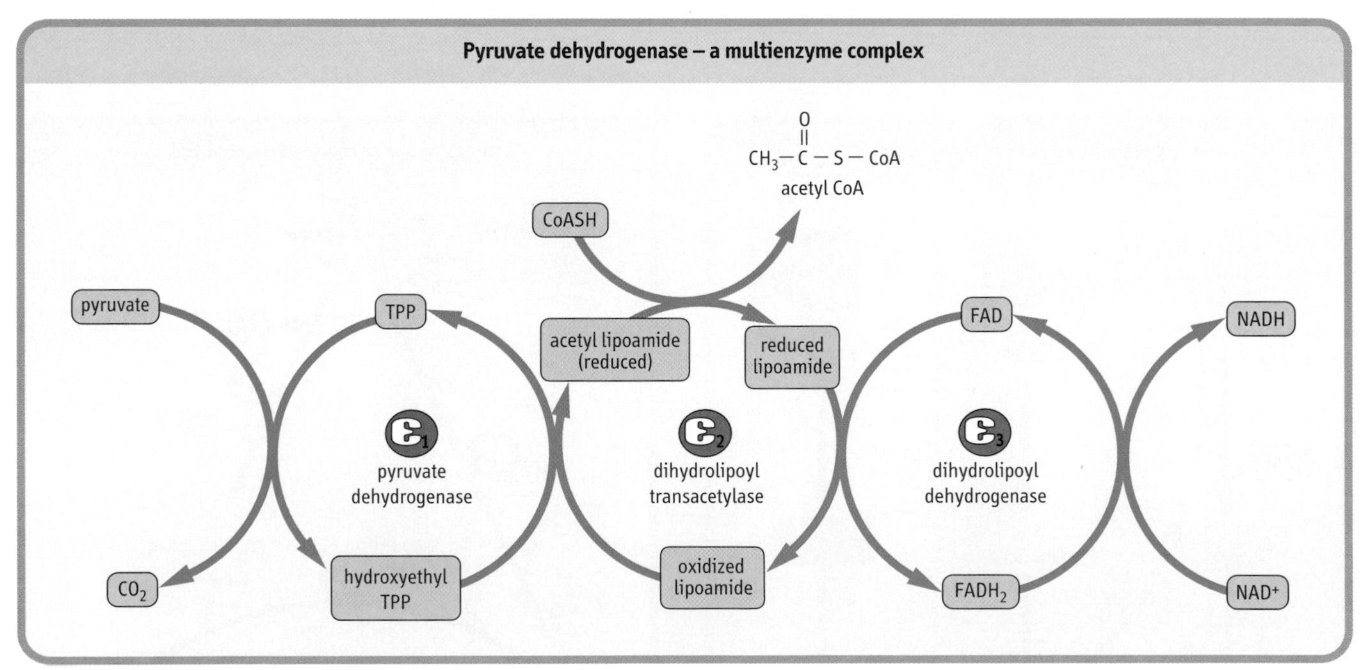

Fig. 13.5 Mechanism of action of pyruvate dehydrogenase. The three enzyme components of the pyruvate dehydrogenase complex are pyruvate dehydrogenase (E_1), dihydrolipoyl transacetylase (E_2) and dihydrolipoyl dehydrogenase (E_3). Pyruvate is first decarboxylated by the thiamine pyrophosphate-containing enzyme (E_1), forming CO_2 and hydroxyethyl-thiamine pyrophosphate (HETPP). Lipoamide, the prosthetic group on E_2, serves as a carrier in the transfer of the 2-carbon unit from HETPP to Coenzyme A (CoA). The oxidized, cyclic disulfide form of lipoamide accepts the hydroxyethyl group from HETPP. The lipoamide is reduced and the hydroxyethyl group converted to an acetyl group during this transfer reaction, forming acetyldihydrolipoamide. Following transfer of the acetyl group to CoA, E_3 reoxidizes the lipoamide, using FAD, and the $FADH_2$ is, in turn, oxidized by NAD^+, yielding NADH.

The role of lipoamide in the pyruvate dehydrogenase reaction

Lipoic acid is linked to the ε-amino group of a lysine residue of the protein by an amide bond, providing a long, flexible arm that can transfer the lipoyl derivatives among different active sites (Fig. 13.6). After formation of acetyldihydrolipoamide by E_1, the arm of E_2 moves the acetyldihydrolipoamide group to the CoA binding site, where acetyl CoA is formed. The reduced dihydrolipoamide arm then moves to E_3 where the lipoamide is oxidized back to lipoamide. The $FADH_2$ is then reoxidized to FAD by NAD^+, releasing NADH and regenerating the active enzyme.

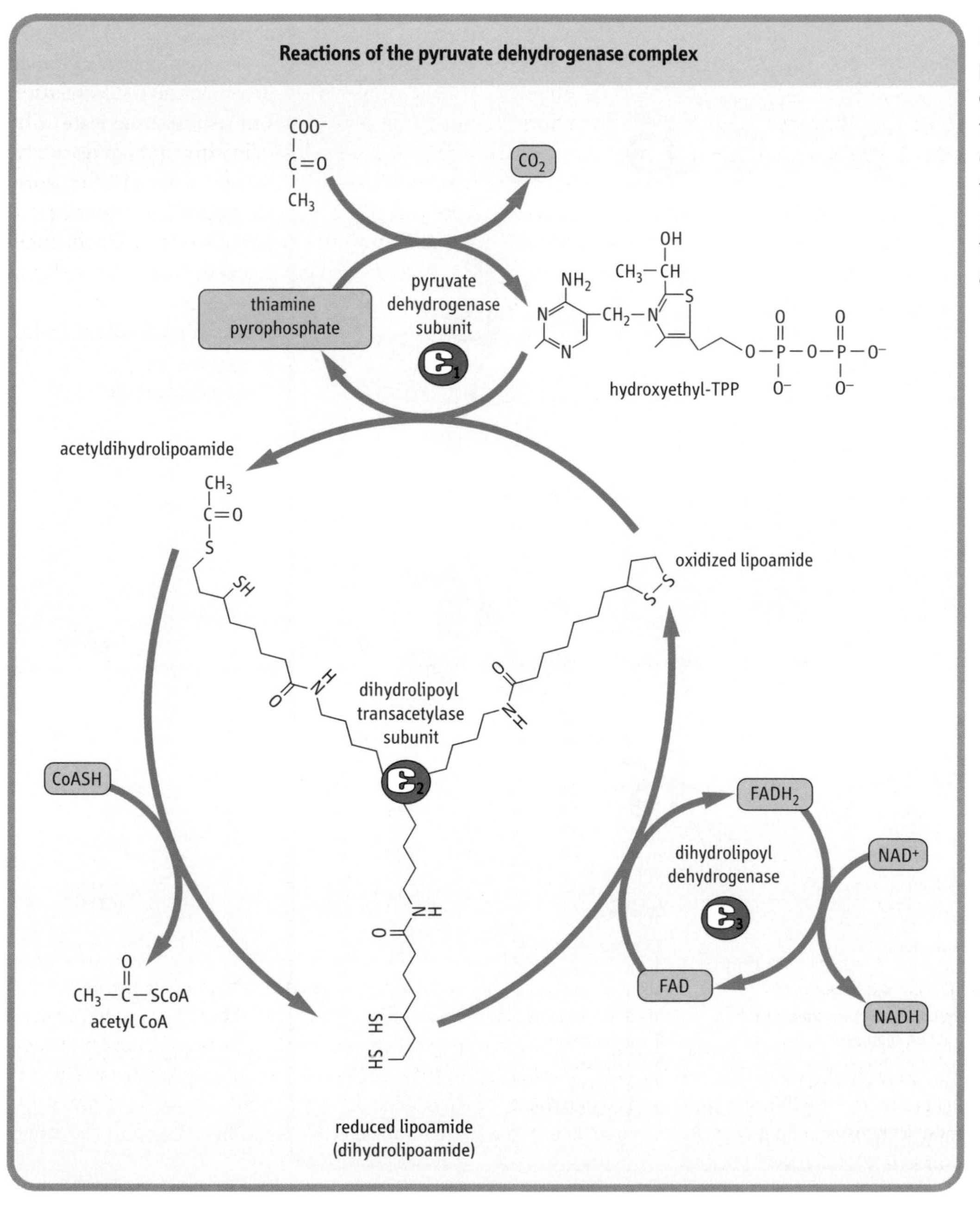

Fig. 13.6 Lipoic acid in the pyruvate dehydrogenase complex. Lipoamide moves from one active site to another on the transacetylase subunit in a 'swinging arm' mechanism. The structures of thiamine pyrophosphate (TPP) and lipoamide are shown.

The pyruvate dehydrogenase complex

Pyruvate dehydrogenase is one of several α-ketoacid dehydrogenases having analogous reaction mechanisms. These include α-ketoglutarate dehydrogenase in the TCA cycle and α-ketoacid dehydrogenases associated with the catabolism of leucine, isoleucine and valine. Pyruvate dehydrogenase catalyzes the conversion of pyruvate to acetyl CoA (see Fig. 13.3). It is a multienzyme complex, consisting of three enzymes, pyruvate dehydrogenase (E_1), dihydrolipoyl transacetylase (E_2) and dihydrolipoamide dehydrogenase (E_3). Five coenzymes are also required for its activity: thiamine pyrophosphate; lipoamide (lipoic acid bound in amide linkage to protein); CoA, FAD, and NAD^+. The three enzymes of the complex are bound to one another, and the intermediates, tethered to the protein, traverse minimal distances between each catalytic step. One PDH complex from the bacterium *Escherichia coli* consists of 60 polypeptide chains with the transacetylase component (E_2) at the core and the two dehydrogenase components (E_1 and E_3) on the outside. This

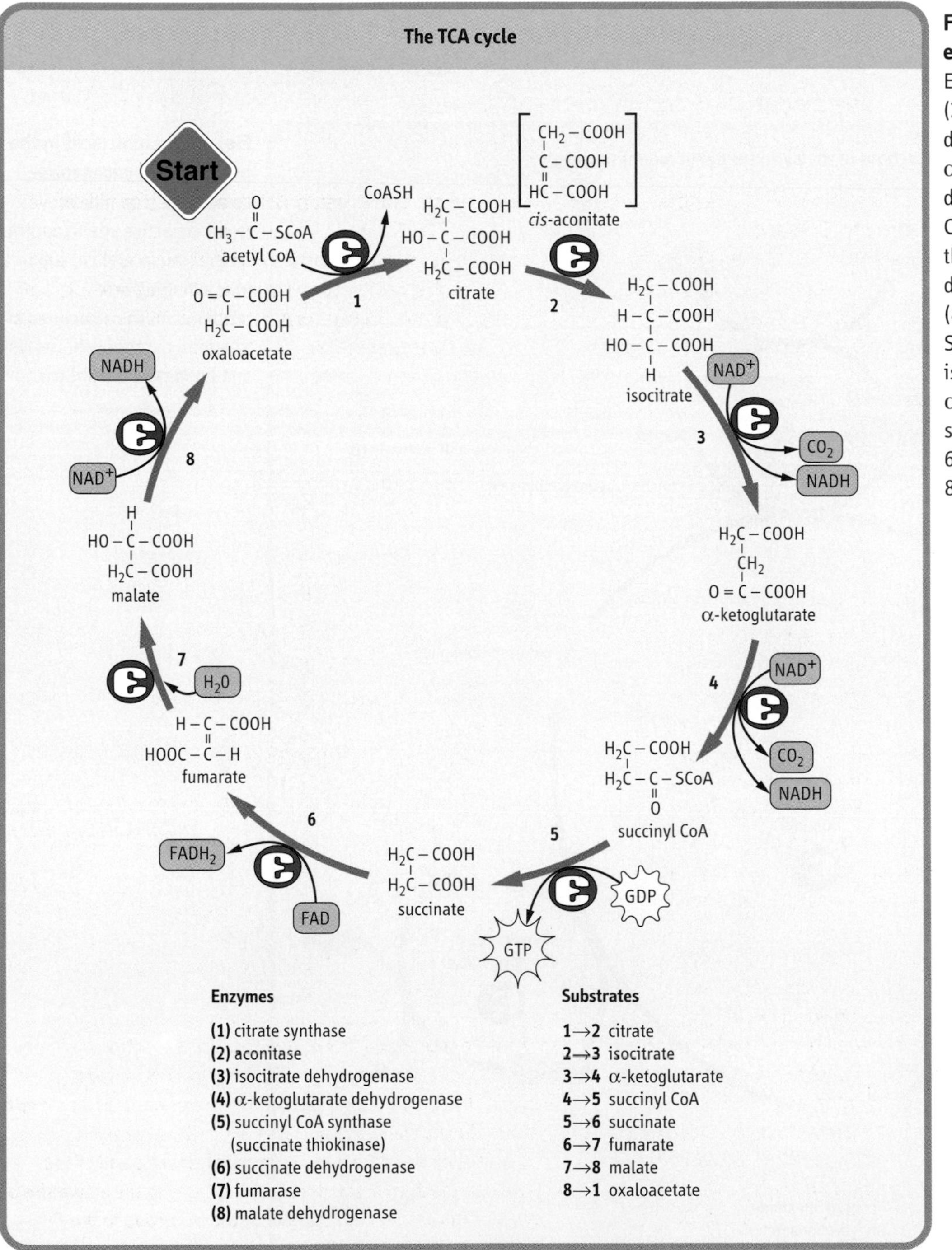

Fig. 13.7 Intermediates and enzymes of the TCA cycle. Enzymes: (**1**) citrate synthase; (**2**) aconitase; (**3**) isocitrate dehydrogenase; (**4**) α-ketoglutarate dehydrogenase; (**5**) succinyl CoA synthase (succinate thiokinase); (**6**) succinate dehydrogenase; (**7**) fumarase; (**8**) malate dehydrogenase. Substrates: 1→2 citrate; 2→3 isocitrate; 3→4 α-ketoglutarate; 4→5 succinyl CoA; 5→6 succinate; 6→7 fumarate; 7→8 malate; 8→1 oxaloacetate.

enzyme is at a central point in carbohydrate metabolism and, as might be expected, is subject to allosteric and covalent regulation. The steps in the pyruvate dehydrogenase reaction are summarized in Figures 13.5 and 13.6.

Several coenzymes required by the pyruvate dehydrogenase complex are synthesized from vitamin precursors. Thiamine pyrophosphate is made from vitamin B_1, thiamine. Riboflavin and nicotinamide (niacin) are required for the synthesis of FAD and NAD^+. Pantothenic acid is a vitamin component of CoA. Lipoic acid, although it is not a vitamin, is often provided in dietary supplements. Deficiencies in any of these coenzymes would have obvious effects on energy metabolism. Increases in cellular pyruvate and α-ketoglutarate are found in beri-beri (see Chapter 10). This results from thiamine deficiency and is characterized by cardiac and skeletal muscle weakness and neurologic disease.

Enzymes and reactions of the TCA cycle

The TCA cycle is a sequence of eight enzymatic reactions (Fig. 13.7). It begins with condensation of acetyl CoA with OAA to form citrate. The OAA is regenerated on completion

Toxicity of fluoroacetate – a suicide substrate

Fluoroacetate, originally isolated from plants, is a potent toxin. It is activated as fluoroacetyl CoA and then condenses with oxaloacetate to form fluorocitrate. Death results from inhibition of the TCA cycle by 2-fluorocitrate, a strong inhibitor of aconitase. Fluoroacetate is an example of a 'suicide substrate,' a compound that is not toxic *per se*, but is metabolically activated to a toxic product. Thus, the cell is said to commit suicide by converting an apparently harmless substrate to a lethal toxin. Similar processes are involved in the activation of many environmental procarcinogens to carcinogens that induce mutations in DNA.

Stereospecificity of enzymes

Aconitase catalyzes isomerization at the oxaloacetate end of the citrate molecule. However, citrate has no asymmetric centers; it is achiral. How does aconitase know 'which end is up?' The answer lies in the nature of citrate binding to the active site of aconitase, a process known as three-point attachment. As shown in Figure 13.8, because of the geometry of the active site of aconitase, there is only one way for citrate to bind. This 'three-point binding' places the oxaloacetate carbons in the proper orientation for the isomerization reaction, while the carbons derived from acetyl CoA are excluded from the active site. Although citrate is a symmetric or achiral molecule, it is termed 'prochiral' because it is converted to an asymmetric, chiral molecule, isocitrate. Similar types of three-point binding processes are involved in transaminase reactions that produce exclusively L-amino acids from ketoacids. The reduction of the nicotinamide ring by NAD(H)-dependent dehydrogenases is also stereospecific. Some dehydrogenases place the added hydrogen exclusively on the front face of the nicotinamide ring (viewed with the amide group to the right), while others add hydrogen only to the back face (Fig. 13.9).

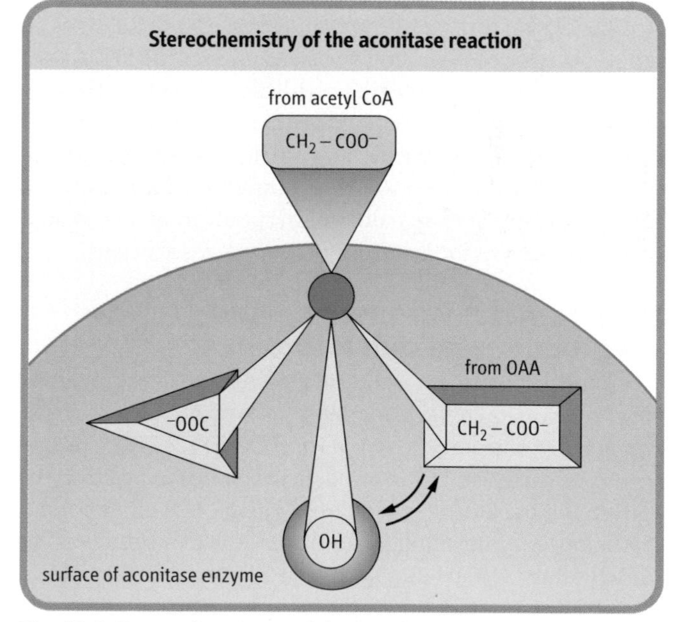

Fig. 13.8 Stereochemistry of the aconitase reaction. Aconitase converts achiral citrate to a specific chiral form of isocitrate. Binding of the C-3 hydroxyl (OH) and carboxylate (COO^-) groups of citrate on the enzyme surface places the carboxymethyl ($-CH_2-COO^-$) group, derived from the oxaloacetate end of the molecule, in touch with the third binding locus in the active site of aconitase. This assures the transfer of the OH group to the CH_2 group derived from the OAA group, indicated by arrows, rather than that derived from acetyl CoA.

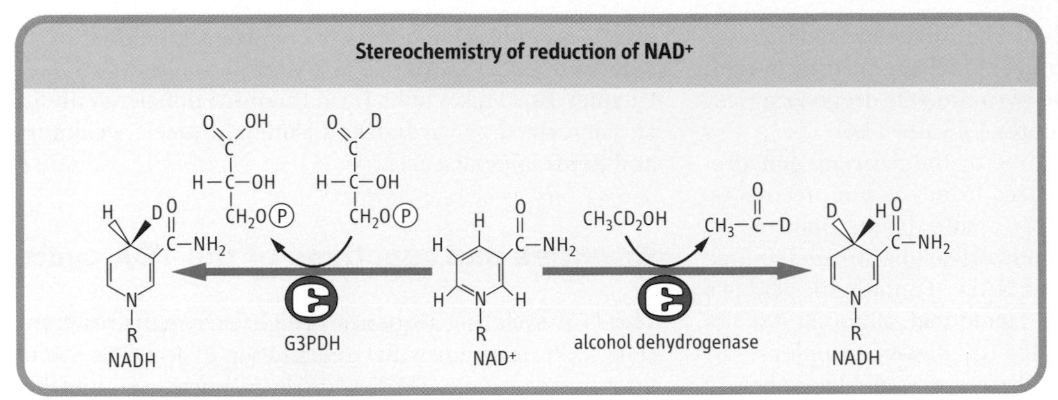

Fig. 13.9 Stereochemistry of the reduction of NAD⁺ by dehydrogenases. Alcohol dehydrogenase places the hydrogen ion on the front face of the nicotinamide ring, while glyceraldehyde-3-phosphate dehydrogenase (G3PDH) places the hydrogen on the back face of the ring. The two positions can be discriminated using deuterated (D) substrates.

of the cycle. Of the four oxidations in the cycle, two involve decarboxylation. Three of these produce NADH and one produces $FADH_2$. GTP, a high-energy phosphate, is also produced in one step by substrate-level phosphorylation.

Citrate synthase

Citrate synthase begins the TCA cycle by catalyzing the condensation of acetyl CoA and OAA to form citric acid. The reaction is driven by cleavage of the high-energy thioester bond of citroyl CoA, an intermediate in the reaction. A later TCA cycle enzyme, succinyl CoA synthetase, utilizes the high-energy thioester bond in succinyl CoA to produce GTP, a high-energy phosphate.

Aconitase

Aconitase is an iron-sulfur protein (Chapter 8). It catalyzes an intramolecular redox reaction that isomerizes citrate to isocitrate through the enzyme-bound intermediate *cis*-aconitate. The two-step reaction is reversible and involves dehydration followed by hydration. Although citrate is a symmetric molecule, aconitase works specifically on the OAA end of citrate, not the end derived from acetyl CoA. Such stereochemical specificity occurs because of the geometry of the active site of aconitase (see below).

Isocitrate dehydrogenase and α-ketoglutarate dehydrogenase

Isocitrate dehydrogenase and α-ketoglutarate dehydrogenase catalyze two sequential oxidative decarboxylation reactions in which NAD^+ is reduced to NADH and CO_2 is released. The first of these enzymes, isocitrate dehydrogenase, catalyzes the conversion of isocitrate to α-ketoglutarate. It is an important regulatory enzyme, inhibited under energy-rich conditions by high levels of NADH and ATP, and is activated when NAD^+ and ADP are produced by metabolism. Inhibition of this enzyme following a carbohydrate meal causes intramitochondrial accumulation of citrate, which is then exported to the cytosol for lipogenesis (see Chapter 15). Citrate is also an important allosteric modifier of enzymes, inhibiting phosphofructokinase-1 (see Chapter 11) and activating acetyl CoA carboxylase (see Chapter 15).

The second dehydrogenase enzyme, α-ketoglutarate dehydrogenase, catalyzes the oxidative decarboxylation of α-ketoglutarate, yielding NADH, CO_2 and succinyl CoA, a high-energy thioester compound. Like pyruvate dehydrogenase, this enzyme contains three subunits having the same designations as pyruvate dehydrogenase (E_1, E_2 and E_3). E_3 is identical in the two complexes and is encoded by the same gene. The reaction mechanisms and the cofactors thiamine pyrophosphate, lipoate, CoA, FAD and NAD^+, are the same. Both enzymes begin with an α-keto acid, pyruvate or α-ketoglutarate, and both form CoA esters, acetyl CoA or succinyl CoA, respectively.

At this point, the net carbon yield of the TCA cycle is zero, i.e. two carbons were introduced as acetyl CoA and two have been liberated in the form of CO_2. Note, however, that because of the asymmetry of the aconitase reaction, neither of the CO_2 molecules produced in this first round trip through the TCA cycle originates from the carbons of the acetyl CoA – they are derived from the OAA end of the citrate molecule. Both of the carbons that originated from acetyl CoA remain in the TCA cycle intermediates, and may appear in compounds produced in biosynthetic reactions branching from the TCA cycle, including glucose, aspartic acid and heme. However, because of the loss of two CO_2 molecules at this point, there is no net synthesis of these metabolites from acetyl CoA. Humans cannot perform net synthesis of glucose from acetyl CoA derived from fat metabolism. Malate is an important intermediate in gluconeogenesis (see Chapter 12), and conversion of a mole of acetyl CoA to malate involves the loss of two moles of CO_2. Thus, there is no net transfer of carbons from acetyl CoA to glucose.

Succinyl CoA synthetase

Succinyl CoA synthetase (succinate thiokinase) catalyzes the conversion of energy-rich succinyl CoA to succinate and free CoA. The free energy of hydrolysis of the thioester bond of acetyl CoA is conserved by formation of GTP from GDP and inorganic phosphate (Pi). Since the respiratory chain is not involved, this is a substrate-level phosphorylation reaction, like the reactions catalyzed by phosphoglycerate kinase and pyruvate kinase in glycolysis (see Chapter 11). GTP is used by enzymes such as phosphoenolpyruvate carboxykinase (PEPCK) in gluconeogenesis (see Chapter 12), but is also readily equilibrated with ATP by the enzyme nucleoside diphosphate kinase as shown:

$$GTP + ADP \rightleftharpoons GDP + ATP$$

The next three reactions in the TCA cycle illustrate a common theme in metabolism for introducing a carbonyl group into a molecule:

- introduction of an ethyleniec double bond,
- addition of water across the double bond to form an alcohol derivative, then
- oxidation of the alcohol to form a ketone.

This same sequence occurs, in the form of enzyme-bound intermediates, during conversion of citrate to α-ketoglutarate, above, and in the oxidation of fatty acids (see Chapter 14).

Succinate dehydrogenase

Succinate dehydrogenase is a flavoprotein containing the prosthetic group FAD. As described in Chapter 8, this enzyme is embedded in the inner mitochondrial membrane where it is a part of Complex II (succinate-Q reductase). The reaction involves oxidation of succinate to the *trans*-dicarboxylic acid, fumarate, with reduction of FAD to $FADH_2$.

Fumarase

Fumarase stereospecifically adds water across the *trans* double bond of fumarate to form the α-hydroxy acid L-malate.

Malate dehydrogenase

Malate dehydrogenase catalyzes the oxidation of L-malate to oxaloacetate, producing NADH and completing one round trip through the TCA cycle. The oxaloacetate may then react with acetyl CoA continuing the cycle of reactions.

Energy yield from the TCA cycle

During the course of the TCA cycle, each molecule of acetyl CoA generates sufficient reduced nucleotide coenzymes for synthesis of ~11 moles ATP by oxidative phosphorylation.

$$3\ NADH \rightarrow 9\ ATP,\ 1\ FADH_2 \rightarrow 2\ ATP$$

Together with the GTP synthesized by substrate-level phosphorylation in the succinyl CoA synthetase (succinate thiokinase) reaction, a total of ~12 ATP equivalents are available per mole of acetyl CoA. Complete metabolism of a mole of glucose through glycolysis and the TCA cycle yields ~36–38 moles ATP (Fig. 13.10). In contrast, only 2 moles of ATP (net) are recovered by anaerobic glycolysis in which glucose is converted to lactate (see Chapter 11).

Regulation of the TCA cycle

There are several levels of control of the TCA cycle. In general, the overall activity of the cycle depends on the availability of NAD^+ for the dehydrogenase reactions. This, in turn, is linked to the rate of NADH consumption for oxidative phosphorylation, which depends on the rate of ATP utilization and production of ADP. Thus, as ATP is used, ADP is produced, and NADH is consumed for oxidative phosphorylation, and NAD^+ is produced. The TCA cycle is activated, and fuels are consumed, and more NADH is produced for oxidative phosphorylation. The mitochondrial level of NAD^+ provides the link between work (ATP utilization) and fuel consumption.

The malonate block

The malate dehydrogenase reaction played an important role in the elucidation of the cyclic nature of the TCA cycle. Addition of tricarboxylic acids (citrate, aconitate and α-ketoglutarate) was known to catalyze pyruvate metabolism – we now know that this is the result of formation of catalytic amounts of OAA from these intermediates. In 1937, Krebs found that malonate, the 3-carbon dicarboxylic acid homologue of succinate and competitive inhibitor of succinate dehydrogenase, blocked metabolism of pyruvate by using minced muscle preparations. He also showed that malonate inhibition of pyruvate metabolism led to accumulation not only of succinate, but also of citrate and α-ketoglutarate, suggesting that succinate was a product of pyruvate metabolism and that the tricarboxylic acids might be intermediates in this process. Interestingly, fumarate and oxaloacetate also stimulated pyruvate oxidation and led to accumulation of citrate and succinate during malonate block, suggesting that the 3- and 4-carbon acids might combine to form the tricarboxylic acids. The experiments with fumarate indicated that there were two paths between fumarate and succinate, one involving reversal of the succinate dehydrogenase reaction, which was inhibited during malonate block, and the other involving conversion of a series of organic acids to succinate. These observations, combined with Krebs' experience a few years earlier in characterization of the urea cycle (see Chapter 18), led to the description of the TCA cycle.

Energetics of glucose oxidation

Reaction	Mechanism	moles ATP / mol Glc
hexokinase	phosphorylation	–1
phosphofructokinase	phosphorylation	–1
G3PDH	NADH, oxidative phosphorylation	+6 (+4)*
phosphoglycerate kinase	substrate-level phosphorylation	+2
pyruvate kinase	substrate-level phosphorylation	+2
pyruvate dehydrogenase	NADH, oxidative phosphorylation	+6
isocitrate dehydrogenase	NADH, oxidative phosphorylation	+6
α-ketoglutarate dehydrogenase	NADH, oxidative phosphorylation	+6
succinyl CoA synthetase	substrate-level phosphorylation (GTP)	+2
succinate dehydrogenase	$FADH_2$, oxidative phosphorylation	+4
malate dehydrogenase	NADH, oxidative phosphorylation	+6
TOTAL		**38 (36)***

Fig. 13.10 ATP yield from glucose during oxidative metabolism. The yields of ATP shown are approximate. Recent work suggests that the actual yields of ATP from NADH and $FADH_2$ are closer to 2.5 and 1.5, respectively, yielding approximately 30 moles of ATP per mole of glucose. The oxidation of glucose in a bomb calorimeter yields 686 kcal/mole, while the synthesis of ATP requires 7.3 kcal/mole. Aerobic metabolism of glucose is therefore about 35% efficient (686 kcal/mole glucose ÷ 7.3 kcal/mole ATP = 94 theoretical moles of ATP/mol glucose; 36/94 = 38%). *The ATP yield depends on the route of transport of redox equivalents to the mitochondrion – six ATP by the malate aspartate shuttle, and four ATP by the glycerol phosphate shuttle (see Chapter 8).

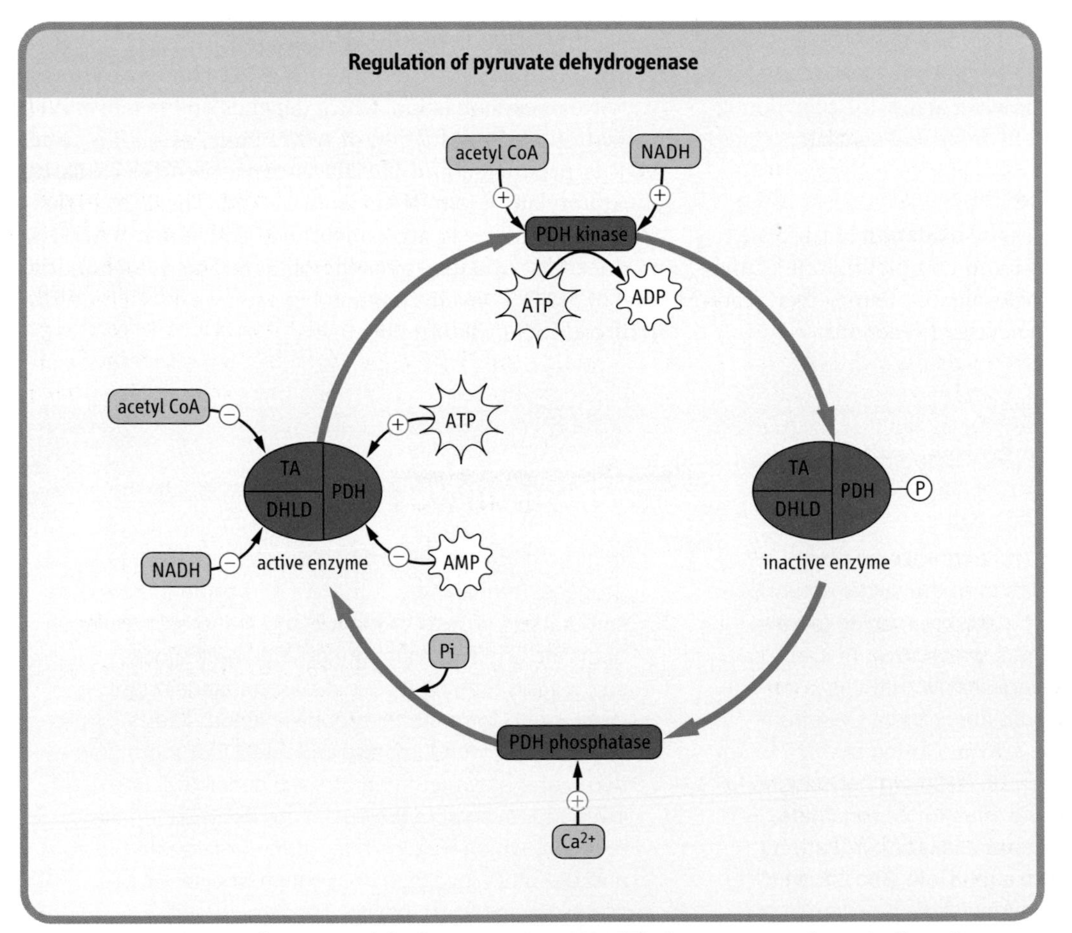

Fig. 13.11 Regulation of pyruvate dehydrogenase. Pyruvate dehydrogenase regulates the flux of pyruvate into the TCA cycle. NAD(H), ATP and acetyl CoA exert both allosteric and covalent control of enzyme activity. PDH, pyruvate dehydrogenase subunit; TA, dihydrolipoyl transacetylase; DHLD, dihydrolipoamide dehydrogenase subunit. Ca^{2+} also affects PDH phosphatase activity, in response to changes in intracellular Ca^{2+} during muscle contraction (see Chapter 19).

In addition to its responsiveness to energy consumption, there are several regulatory enzymes that affect the activity of the TCA cycle. The activity of pyruvate dehydrogenase, and therefore the supply of acetyl CoA from glucose, lactate and alanine, is regulated by allosteric and covalent modifications (Fig. 13.11). The products of the pyruvate dehydrogenase reaction, NADH and acetyl CoA, act as negative allosteric effectors of the enzyme. In addition, the pyruvate dehydrogenase complex has associated kinase and phosphatase enzymes that modulate the degree of phosphorylation of regulatory serine residues in the enzyme. NADH and acetyl CoA activate the kinase, which phosphorylates and inactivates the enzyme. In contrast, the substrates of the reaction, pyruvate, CoA and NAD^+ promote dephosphorylation and activate the enzyme. This is an important regulatory process during fasting and starvation, when gluconeogenesis is essential to maintain blood glucose concentration (see Chapter 12). Active fat metabolism during fasting leads to increased NADH and acetyl CoA in the mitochondrion, which leads to inhibition of pyruvate dehydrogenase and blocks the utilization of pyruvate for energy metabolism. Pyruvate, lactate and alanine are directed toward the synthesis of glucose (see Chapter 12). Conversely, insulin stimulates pyruvate dehydrogenase by activating the phosphatase where dietary carbohydrates are in excess.

OAA has a catalytic role in stimulating the entry of acetyl CoA into the TCA cycle and, at times, the availability of OAA appears to regulate the activity of the cycle. This occurs especially during fasting when levels of ATP and NADH, derived from fat metabolism, are increased in the mitochondrion. The increase in NADH shifts the malate:oxaloacetate equilibrium toward malate, directing TCA cycle intermediates toward malate, which is exported to the cytosol for gluconeogenesis (see Chapter 12). Meanwhile, acetyl CoA derived from fat metabolism is directed toward ketone bodies (see Chapter 14) because of the lack of OAA, which is required for efficient activity of the TCA cycle.

Isocitrate dehydrogenase is also a major regulatory enzyme within the TCA cycle. It is subject to allosteric inhibition by ATP and NADH. During consumption of a high carbohydrate diet under resting conditions, the level of carbohydrate-derived substrates increases, while the demand for ATP is limited. Under these circumstances, the accumulation of ATP and NADH inhibits isocitrate dehydrogenase, so that citrate accumulates in the mitochondrion. The citrate is then exported to the cytosol for synthesis of fatty acids, which are exported from the liver and stored in adipose tissue as triglycerides (see Chapter 14).

Anaplerotic reactions

As noted above (see Fig. 13.1), TCA cycle intermediates may be siphoned off for various biosynthetic processes. Removal of succinyl CoA for heme biosynthesis could gradually deplete mitochondrial OAA. Theoretically, the TCA cycle would cease to function if the intermediates were not replenished. Anaplerotic, meaning 'filling up', reactions provide the TCA cycle with intermediates, maintaining the activity of the cycle. Pyruvate carboxylase is the classical example of an anaplerotic reaction. It converts pyruvate to OAA, which is required for initiation of the cycle. A cytoplasmic enzyme, malic enzyme, also converts pyruvate to malate (see Chapter 15), which can then enter the mitochondrion as a substrate for the TCA cycle. Pyruvate may also react with aspartate or glutamate in transaminase reactions, producing alanine and the TCA cycle intermediates, OAA and α-ketoglutarate, respectively. Several other 'glucogenic' amino acids (see Chapter 18) may also serve as a source of pyruvate or TCA cycle intermediates, guaranteeing that the cycle is never stalled because of a lack of intermediates.

Summary

The TCA cycle is a common, central pathway for fuel catabolism in the body. Its major products are CO_2 and reduced coenzymes. It is located in the mitochondrion, in close association with the enzymes of oxidative phosphorylation, which use these coenzymes for production of ATP. The activity of the cycle is regulated by allosteric effectors and covalent modifications co-ordinating fuel consumption with the energy needs of the body. In addition to its role in energy metabolism, the TCA cycle also produces several intermediates through which carbons are exchanged between amino acids, carbohydrates and lipids.

Further reading

Medina JM, Tabernero A, Tovar JA, Martin-Barrientos J. Metabolic fuel utilization and pyruvate oxidation during the postnatal period. *J Inherited Metabolic Dis* 1996; **19**:432–442.

Patel MS, Naik S, Wexler ID, Kerr DS. Gene regulation and genetic defects in the pyruvate dehydrogenase complex. *J Nutrition* 1995; **125**:1753S–1757S.

Robinson BH, MacKay N, Chun K, Ling M. Disorders of pyruvate carboxylase and the pyruvate dehydrogenase complex. *J Inherited Metabolic Dis* 1996; **19**:452–462.

Rustin P, Bourgeron T, Parfait B, Chretien D, Munnich A, Rotig A. Inborn errors of the Krebs cycle: a group of unusual mitochondrial diseases in humans. *Biochem Biophys Acta* 1997; **1361**:185–197.

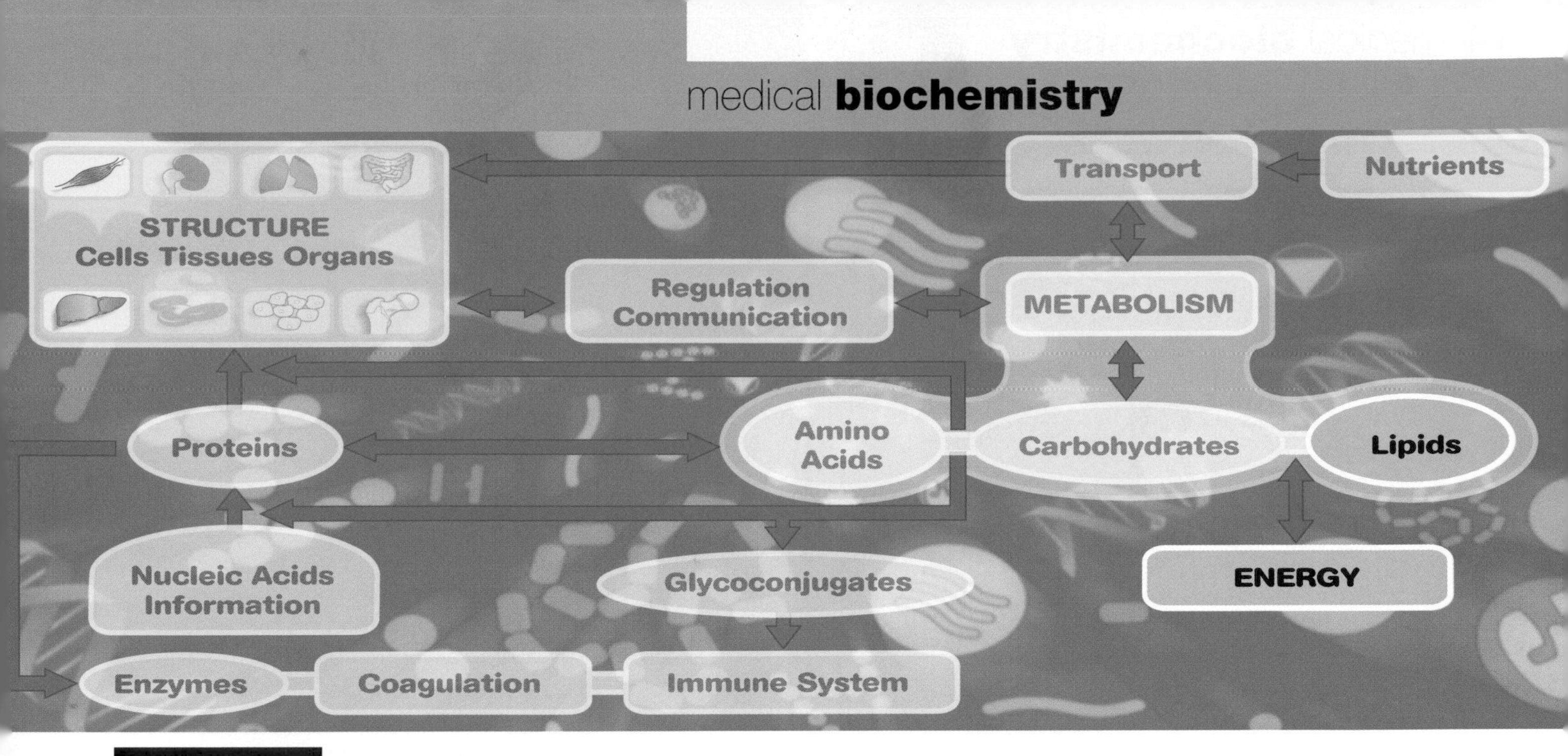

14 Oxidative Metabolism of Lipids in Liver and Muscle

Fats are normally the major source of energy in liver and in muscle, and in human tissues in general, except for red cells of blood and brain. Triglycerides are the storage and transport form of fats; fatty acids are the immediate source of energy. They are released from adipose tissue, transported in association with plasma albumin, and equilibrate into cells for metabolism. The catabolism of fatty acids is entirely oxidative; after they have been transported through the cytoplasm, their oxidation proceeds in both the peroxisome and the mitochondrion, primarily by a cycle of reactions known as β-oxidation. Carbons are released, two at a time, from the carboxyl end of the fatty acid; the major end-products are acetyl coenzyme A (CoA) and the reduced forms of the nucleotides, flavin adenine dinucleotide and nicotinamide adenine dinucleotide ($FADH_2$ and NADH). In muscle, the acetyl CoA is metabolized via the tricarboxylic acid (TCA) cycle and oxidative phosphorylation to produce ATP; in liver, it is shunted largely to the synthesis of ketone bodies (ketogenesis), which are water-soluble lipid derivatives that, like glucose, are exported for use in other tissues. Fat metabolism is controled primarily by the rate of triglyceride hydrolysis (lipolysis) in adipose tissue; this is regulated by hormonal mechanisms involving insulin and glucagon, epinephrine, and cortisol. These hormones coordinate the metabolism of carbohydrate, lipid and protein throughout the body (see Chapter 20).

Activation of fatty acids and their transport to the mitochondrion

Fatty acids do not exist in free form in the body – they are soaps, and would dissolve membranes

In blood, fatty acids are bound to albumin, which is present at a concentration of approximately 0.5 mmol/L (35 mg/mL) in plasma. Each molecule of albumin can bind six to eight fatty acid molecules. In the cytosol, fatty acids are bound to a series of fatty-acid-binding proteins and enzymes. As the priming step for their catabolism, the fatty acids are activated to their CoA derivative, using adenosine triphosphate (ATP) as the energy source (Fig. 14.1). The carboxyl group is first activated to an enzyme-bound, high-energy acyl adenylate intermediate, formed by reaction of the carboxyl group of the fatty acid with ATP. The acyl group is then transferred to CoA by the same enzyme, fatty acyl CoA synthetase. This enzyme is commonly known as fatty acid thiokinase, because ATP is consumed in the formation of the thioester bond in acyl CoA.

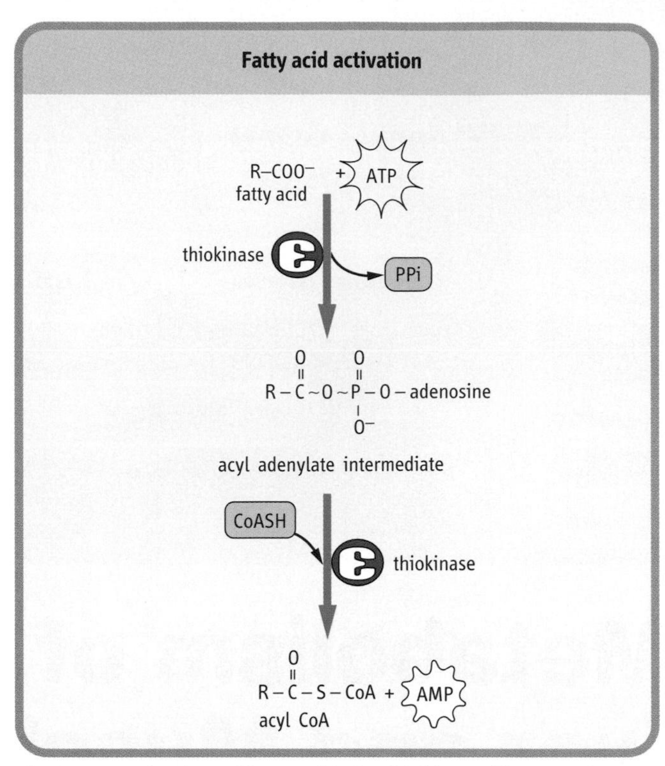

Fig. 14.1 Activation of fatty acids by thiokinase (fatty acyl CoA synthetase). ATP forms an enzyme-bound acyl adenylate intermediate, which is discharged to form acyl CoA. AMP, adenosine monophosphate; CoASH, coenzyme A; PPi, inorganic pyrophosphate.

The length of the fatty acid dictates where it is activated to CoA

Short- and medium-chain fatty acids (Fig. 14.2) can cross the mitochondrial membrane by passive diffusion, and are activated to their CoA derivative within the mitochondrion. Very-long-chain fatty acids are shortened to long-chain fatty acids in peroxisomes. Long-chain fatty acids (16 ± 4 carbons) are the major components of storage triglycerides and dietary fats. They are activated to their CoA derivatives in the cytoplasm and are transported into the mitochondrion via the carnitine shuttle.

Metabolism of fatty acids

Size class	Number of carbons	Site of catabolism	Membrane transport
short chain	2–4	mitochondrion	free diffusion
medium chain	4–12	mitochondrion	diffusion
long chain	12–20	mitochondrion	carnitine cycle
very long chain	>20	peroxisome	unknown

Fig. 14.2 Metabolism of the four classes of fatty acids.

Peroxisomes

Role of peroxisomes in fatty acid oxidation
Peroxisomes are subcellular organelles found in all nucleated cells. They are the principal sites of metabolism of hydrogen peroxide (H_2O_2) in the cell, and account for nearly 20% of oxygen consumption in hepatocytes. They are able to conduct β-oxidation of medium- to very-long-chain fatty acids by a pathway similar to mitochondrial oxidation, but with significant differences: the action of acyl CoA dehydrogenase, for example, produces H_2O_2, rather than $FADH_2$. Peroxisomes are also relatively inefficient at catabolism of short-chain fatty acids, so products such as hexanoyl- and octanoylcarnitine are exported, to be catabolized in the mitochondrion. Peroxisomes are also much more efficient at oxidizing medium-chain α,ω-dicarboxylic acids produced by microsomal or peroxisomal α-oxidation of fatty acids.

Peroxisomes are believed to have a role in production of acetyl CoA for anabolic reactions during energy-rich conditions – that is, when mitochondrial β-oxidation would be inhibited. They also have a special role in the metabolism of very-long-chain fatty acids and phytanic acids (see below). Zellweger syndrome, resulting from the absence of peroxisomes, is characterized by accumulation of long chain fatty acids and pristanic acids in plasma and tissues.

The carnitine shuttle

CoA is a large, polar, nucleotide derivative, and cannot penetrate the mitochondrial inner membrane. Thus, for the transport of long-chain fatty acids, the fatty acid is first transferred to the small molecule, carnitine, by carnitine palmitoyl transferase-I (CPT-I), located in the outer mitochondrial membrane. An acyl carnitine transporter or translocase in the inner mitochondrial membrane then facilitates transfer of the fatty acid into the mitochondrion, where CPT-II regenerates the acyl CoA, releasing free carnitine. The carnitine shuttle (Fig. 14.3) operates by an antiport mechanism in which free carnitine and the acyl carnitine derivative move in opposite directions across the inner mitochondrial membrane. The shuttle is an important site in the regulation of fatty acid oxidation. As discussed in the next chapter, the carnitine shuttle is inhibited after the ingestion of carbohydrate-rich meals, preventing the catabolism of newly synthesized fatty acids and favoring their export from the liver for storage in adipose tissue.

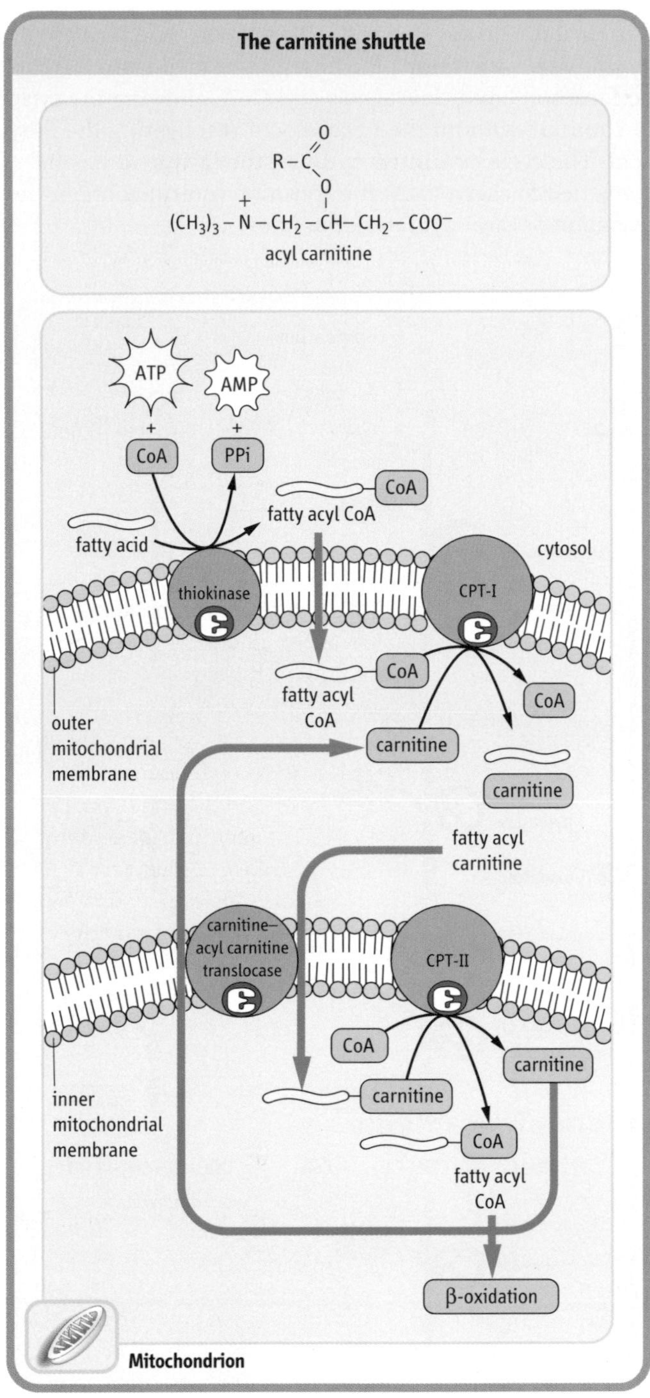

Fig. 14.3 Transport of long-chain fatty acids into the mitochondrion. The three components of the carnitine pathway include extra- and intramitochondrial CPTs and the carnitine–acyl carnitine translocase.

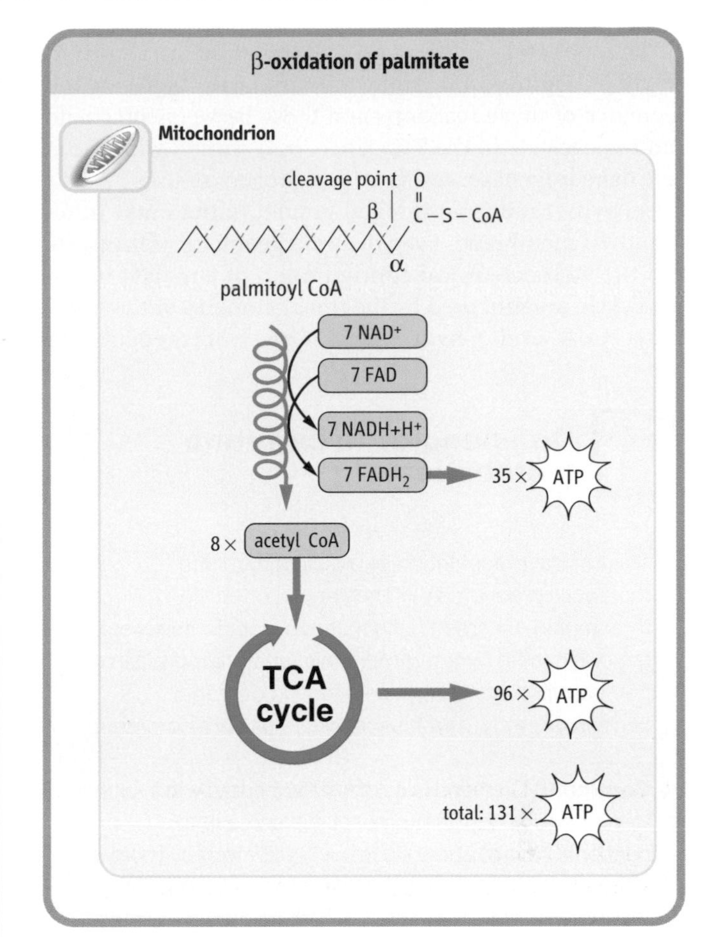

Fig. 14.4 Overview of β-oxidation of palmitate. In a cycle of reactions, the carbons of the fatty acyl CoA are released in two-carbon acetyl CoA units; the yield of ATP from this β-oxidation is nearly equivalent to that from complete oxidation of glucose. In liver, the acetyl CoA units are then used for synthesis of ketone bodies, and in other tissues they are metabolized in the TCA cycle to form ATP. The complete oxidation of palmitate yields a net 129 moles of ATP, after correction for the two-mole equivalents of ATP invested at the thiokinase reaction. The overall production of ATP per gram of palmitate is about twice that per gram of glucose, because glucose is already partially oxidized in comparison with palmitate. For this reason, as illustrated in Figure 14.5, the caloric value of fats is about twice that of sugars.

Oxidation of fatty acids

Mitochondrial β-oxidation

Fatty acyl CoAs are oxidized in a cycle of reactions involving oxidation of the β-carbon to a keto (C = O) group; hence the terminology, β-oxidation (Fig. 14.4). The oxidation is followed by cleavage between the α- and β-carbons by a thiolase reaction. One mole each of acetyl CoA, $FADH_2$, and $NADH+H^+$ is formed during each cycle, along with a fatty acyl CoA with two fewer carbon atoms. For a 16-carbon fatty acid, such as palmitate, the cycle is repeated seven times, yielding eight moles of acetyl CoA (Figs 14.4, 14.5), plus seven moles of $FADH_2$ and seven moles of $NADH+H^+$. This process occurs in the mitochondrion, and the reduced nucleotides are used directly for synthesis of ATP by oxidative phosphorylation.

The four steps in the cycle of β-oxidation are shown in detail in Figure 14.6. Note the similarity between the sequence of these reactions and those between succinate and oxaloacetate in the TCA cycle. In common with succinate dehydrogenase, acyl CoA dehydrogenase uses FAD as a coenzyme, and is an integral protein in the inner mitochondrial membrane. Even the *trans* geometry of fumarate and the stereochemical configuration of L-malate in the TCA cycle are mirrored by the *trans* geometry of the *trans*-enoyl CoA and L-hydroxyacyl CoA intermediates in β-oxidation. The last step of the β-oxidation cycle is catalyzed by thiolase, which traps the energy obtained from the carbon–carbon bond cleavage as acyl CoA, allowing the cycle to continue without the necessity of reactivating the fatty acid. The cycle continues until all the fatty acid has been converted to acetyl CoA, the common intermediate in the oxidation of carbohydrates and lipids.

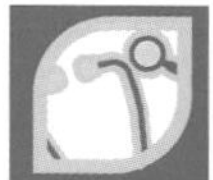

Deficiencies in carnitine metabolism

The clinical presentation of deficient carnitine metabolism occurs in infancy and is often life threatening. Characteristic features include hypoketotic hypoglycemia, hyperammonemia, and altered plasma free carnitine concentration. Hepatic damage, cardiomyopathy, and muscle weakness are common.

Comment. Carnitine is synthesized primarily in liver and kidney, and is normally present in plasma in a concentration of about 50 μmol/L (0.8 mg/dL). There are high-affinity uptake systems for carnitine in most tissues, including the kidney, which resorbs carnitine from the glomerular filtrate, limiting its excretion in urine. Homozygous deficiencies in carnitine transport, CPTs-I and -II, and the translocase result in defects in long-chain fatty acid oxidation. Plasma and tissue carnitine concentrations decrease to <1 μmol/L in carnitine transport deficiency, because of both defective uptake into tissues and excessive loss in urine. On the other hand, plasma free carnitine may exceed 100 μmol/L (2 mg/dL) in CPT-I deficiency. In both translocase and CPT-II deficiency, total plasma carnitine may be normal, but is mostly in the form of acyl carnitine esters of long-chain fatty acids – in the former case, because they cannot be transported into the mitochondrion, and in the latter because of backflow out from mitochondria. These diseases are treated by carnitine supplementation, by frequent high-carbohydrate feeding, and by avoidance of fasting.

Energy yield from glucose and palmitate

Substrate	Molecular weight	Net ATP yield (mol/mol)	mol ATP / g	Caloric value (Cal/g)
glucose	180	36–38	0.2	4
palmitate	256	129	0.5	9

Fig. 14.5 Comparative energy yield from glucose and palmitate.

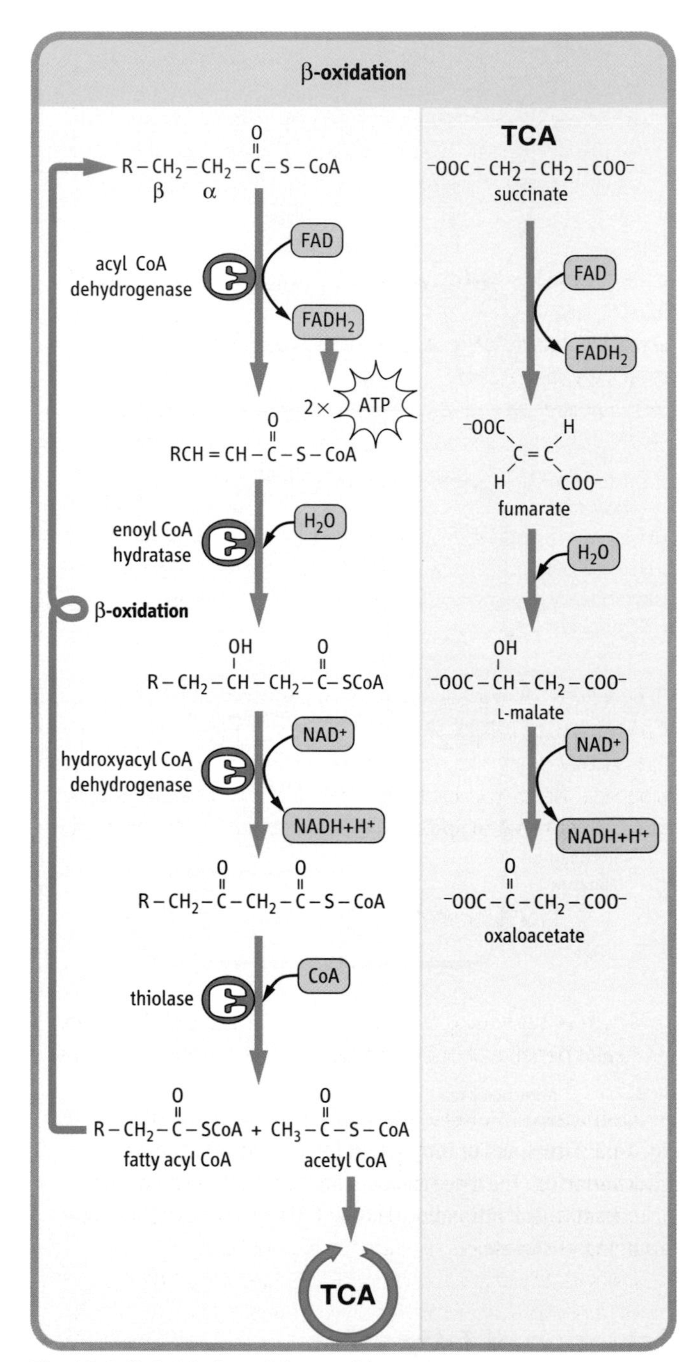

Fig. 14.6 β-Oxidation of fatty acids. Oxidation occurs in a series of steps at the carbon that is β to the keto (C=O) group. Thiolase cleaves the resultant β-ketoacyl CoA derivative to give acetyl CoA and a fatty acid with two fewer carbon atoms, which then re-enters the β-oxidation cascade. Note the similarity between these reactions and those of the TCA cycle, shown on the right.

Impaired oxidation of medium-chain fatty acids

Fatty acyl CoA dehydrogenase
Fatty acyl CoA dehydrogenase is not a single enzyme, but a family of enzymes with chain-length specificity for oxidation of short-, medium- and long-chain fatty acids. Medium-chain fatty acyl CoA dehydrogenase (MCAD) deficiency is an autosomal recessive disease characterized by hypoketotic hypoglycemia. It presents in infancy, and is characterized by high concentrations of medium-chain carboxylic acids, acyl carnitines, and acyl glycines in plasma and urine. Hyperammonemia may also be present, as a result of liver damage. Concentrations of hepatic mitochondrial medium-chain acyl CoA derivatives are also increased, limiting β-oxidation and formation of acetyl CoA for ketogenesis. The inability to metabolize fats during fasting is life threatening. MCAD deficiency is treated by frequent feeding, avoidance of fasting, and carnitine supplementation. Deficiencies in short- and long-chain fatty acid dehydrogenases have also been described, and have similar clinical features.

Alternative pathways of oxidation of fatty acids

Unsaturated fatty acids yield less $FADH_2$ when they are oxidized

Unsaturated fatty acids are already partially oxidized, so less $FADH_2$, and correspondingly less ATP, is produced by their oxidation. The double bonds in polyunsaturated fatty acids have *trans* geometry and occur at three-carbon intervals, whereas the intermediates in β-oxidation have *cis* geometry and the reactions proceed in two-carbon steps. The metabolism of unsaturated fatty acids therefore requires several additional enzymes, both to shift the position and to change the geometry of the double bonds.

Odd-chain fatty acids gain access to the TCA cycle via propionyl CoA

The oxidation of fatty acids with an odd number of carbons proceeds from the carboxyl end, like that of normal fatty acids, except that propionyl CoA is formed by the last thiolase cleavage reaction. The propionyl CoA is converted to succinyl CoA by a multistep process involving three enzymes and the vitamins biotin and cobalamin (Fig. 14.7). The succinyl CoA enters directly into the TCA cycle.

Branched-chain fatty acids must be catabolized, to acetyl CoA and propionyl CoA via α-oxidation

Phytanic acids are branched-chain polyisoprenoid lipids found in plant chlorophylls. Because the β-carbon of phytanic acids is at a branch point, it is not possible to oxidize this carbon to a ketone. The first and essential step in catabolism of phytanic acids is microsomal α-oxidation to pristanic acid, releasing the (α-)carbon-1 as carbon dioxide. Thereafter, as shown in Figure 14.8, acetyl CoA and propionyl CoA are released alternately and in equal amounts. Refsum's disease is a rare neurologic disorder, characterized by accumulation of phytanic acid deposits in nerve tissues as a result of a genetic defect in α-oxidation.

Butter or margarine?

There is continuing debate among nutritionists about the health benefits of butter versus those of margarine in foods.

Comments. Butter is rich in saturated fatty acids and cholesterol, which are risk factors for atherosclerosis. Margarine contains no cholesterol, and is richer in unsaturated fatty acids. However, the unsaturated fatty acids in margarine are mostly the unnatural *trans*-fatty acids formed during the partial hydrogenation of vegetable oils. *Trans*-fatty acids affect plasma lipids in the same fashion as long-chain saturated fats, suggesting that there are comparable risks associated with the consumption of butter or of margarine. The resolution of this issue is complicated by the fact that various forms of margarine, for example soft-spread and hard-block types, vary significantly in their content of *trans*-fatty acids.

Ketogenesis in liver

Gluconeogenesis in fasting and starvation

The liver uses fatty acids as the source of energy for gluconeogenesis during fasting and starvation

Fats are a rich source of energy and, under conditions of fasting or starvation, liver mitochondrial concentrations of fat-derived ATP and NADH are high, inhibiting the isocitrate dehydrogenase reaction and shifting the oxaloacetate–malate equilibrium toward malate. TCA cycle intermediates that are formed from amino acids released from muscle as part of the response to fasting and starvation (see Chapter 19) are converted to malate in the TCA cycle, but the malate then leaves the mitochondrion, to take part in gluconeogenesis. The resulting low level of oxaloacetate in the mitochondrion limits the activity of the TCA cycle, resulting, in particular, in an inability to metabolize acetyl CoA in the TCA cycle. The liver, meanwhile, obtains sufficient energy to support gluconeogenesis simply via the enzymes of β-oxidation, which generate both $FADH_2$ and NADH.

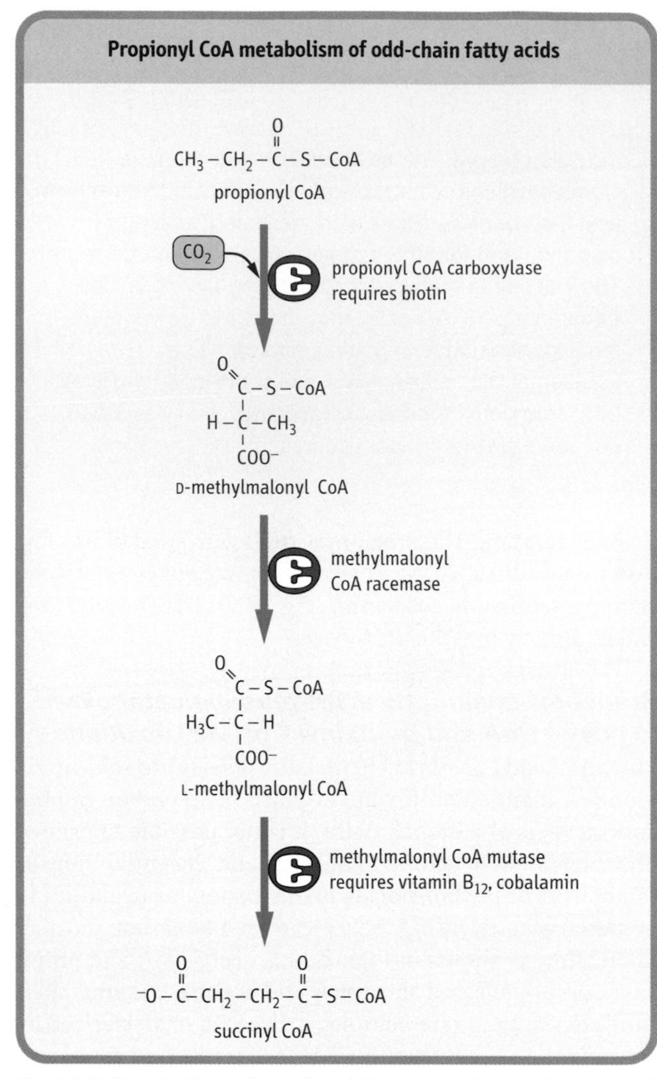

Fig. 14.7 Metabolism of propionyl CoA to succinyl CoA. Propionyl CoA from odd-chain fatty acids is a minor source of carbons for gluconeogenesis. The intermediate, methylmalonyl CoA, is also produced during catabolism of branched-chain amino acids. Defects in methylmalonyl CoA mutase or deficiencies in vitamin B_{12} lead to methylmalonic-aciduria.

What does the liver do with the excess acetyl CoA that accumulates in fasting or starvation?

The problem of dealing with excess acetyl CoA is a critical one, because CoA is present only in catalytic amounts in tissues, and free CoA is required to initiate and continue the cycle of β-oxidation. To recycle the acetyl CoA, the liver uses a pathway known as ketogenesis, in which the CoA component is regenerated and the acetate group appears

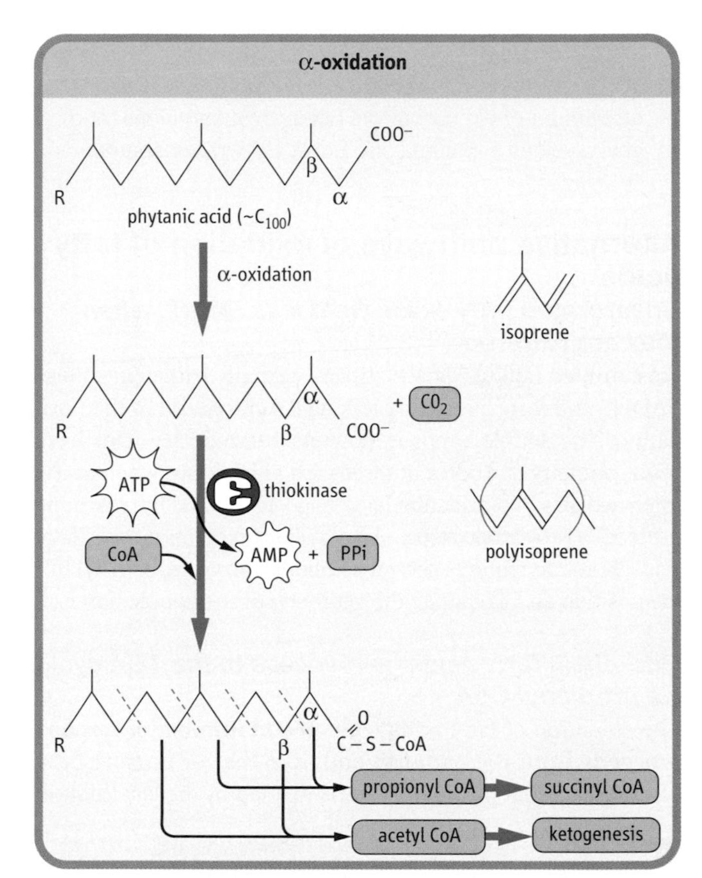

Fig. 14.8 α-Oxidation of branched-chain phytanic acids. The first carbon of phytanic acids is removed as carbon dioxide. In subsequent cycles of β-oxidation, acetyl CoA and propionyl CoA are released alternately.

Alternative pathways of fatty acid oxidation and associated disorders

Dicarboxylic aciduria and ω-oxidation of fatty acids

Several disorders of lipid catabolism, including alterations in the carnitine shuttle, acyl CoA dehydrogenase deficiencies, and Zellweger syndrome, (a defect in peroxisome biogenesis are associated with the appearance of medium-chain dicarboxylic acids in urine; both odd- and even-chain dicarboxylic acids may be involved. When fatty acid β-oxidation is impaired, fatty acids are oxidized, one carbon at a time, from the ω-carbon by microsomal cytochrome P450-dependent hydroxylases and dehydrogenases. These dicarboxylic acids are substrates for peroxisomal β-oxidation, which continues to the level of 6–10-carbon dicarboxylic acids, which are then excreted in urine.

in blood in the form of three water-soluble lipid-derived products: acetoacetate, β-hydroxybutyrate, and acetone. The pathway of formation of these 'ketone bodies' (Fig. 14.9) involves the synthesis and decomposition of hydroxymethylglutaryl (HMG) CoA in the mitochondrion. The liver is unique in its content of HMG CoA synthase and lyase, but is deficient in enzymes required for metabolism of ketone bodies, which explains their export from the liver.

Ketone bodies are taken up in extrahepatic tissues, including skeletal and heart muscle, where they are converted to CoA derivatives for metabolism (Fig. 14.10). Ketone bodies are an efficient source of energy (Fig. 14.11) during fasting and starvation, and appear to be used in muscle in proportion to their plasma concentration. During starvation, the brain also converts to the use of ketone bodies for more than 50% of its energy metabolism, sparing

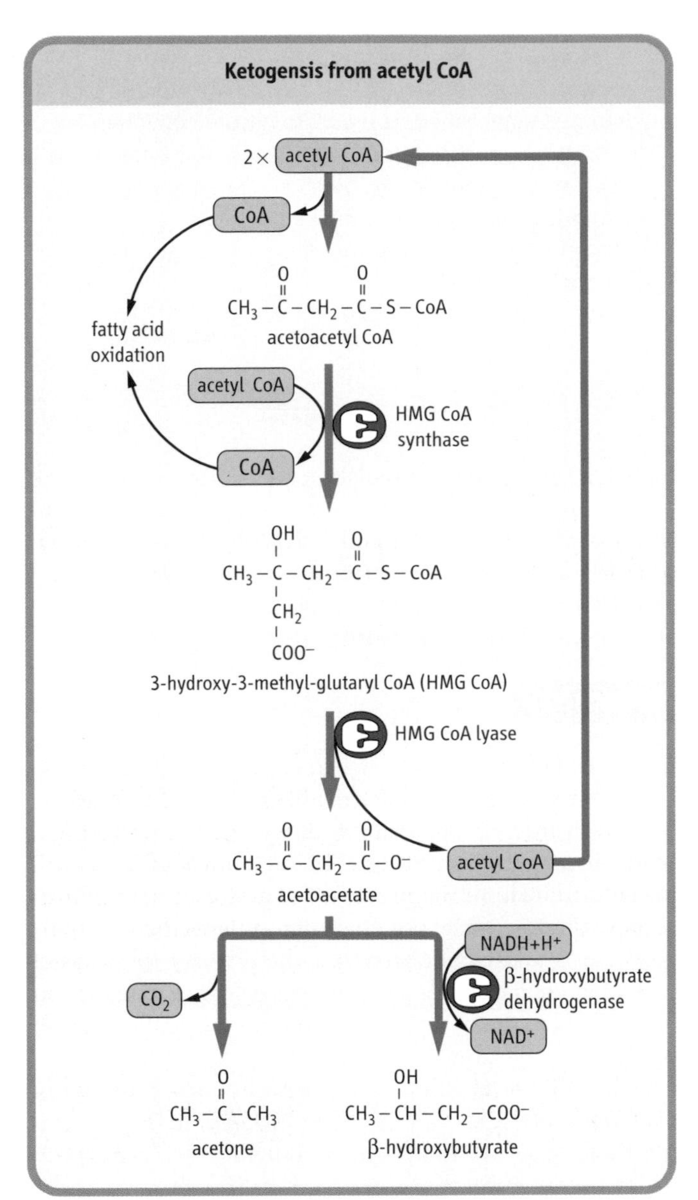

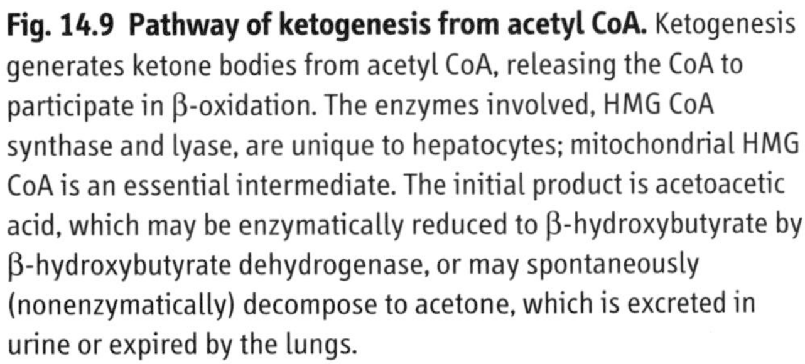
Fig. 14.9 Pathway of ketogenesis from acetyl CoA. Ketogenesis generates ketone bodies from acetyl CoA, releasing the CoA to participate in β-oxidation. The enzymes involved, HMG CoA synthase and lyase, are unique to hepatocytes; mitochondrial HMG CoA is an essential intermediate. The initial product is acetoacetic acid, which may be enzymatically reduced to β-hydroxybutyrate by β-hydroxybutyrate dehydrogenase, or may spontaneously (nonenzymatically) decompose to acetone, which is excreted in urine or expired by the lungs.

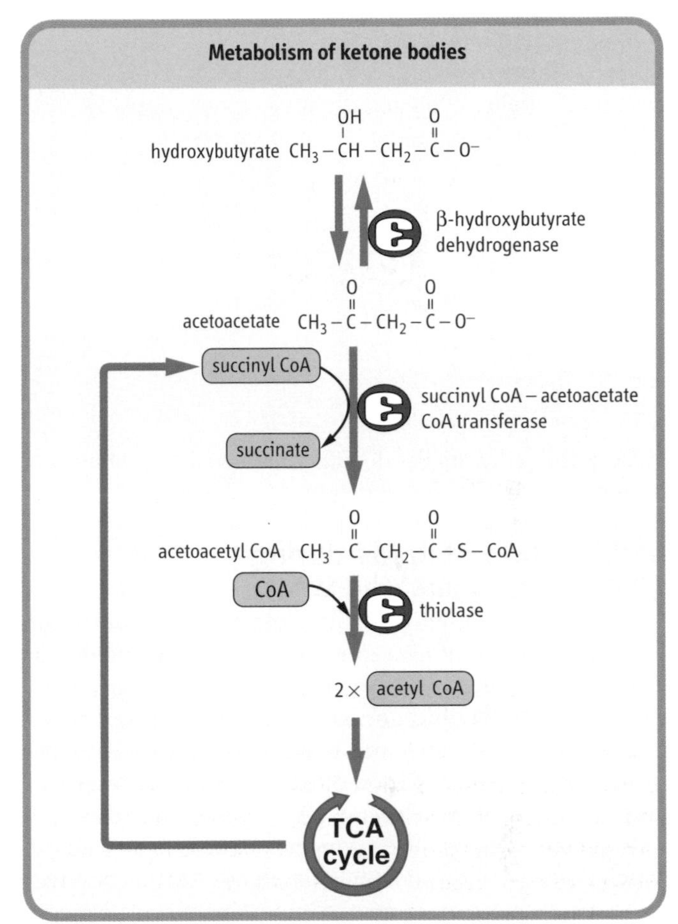

Fig. 14.10 Metabolism of ketone bodies in peripheral tissues. Succinyl CoA acetoacetate CoA transferase activates the transfer of acetoacetate to acetoacetyl CoA. A thiokinase-type enzyme may also directly activate acetoacetate in some tissues.

Plasma concentrations of fatty acids and ketone bodies

Substrate	Plasma concentration (mmol/L) Normal	Fasting	Starvation
fatty acids	0.6	1.0	1.5
acetoacetate	<0.1	0.2	1–2
β-hydroxybutyrate	<0.1	1	5–10

Fig. 14.11 Plasma concentrations of fatty acids and ketone bodies in different nutritional states.

Keto-Stix and weight-loss programs

The appearance of ketone bodies in the urine is an indication of active fat metabolism and gluconeogenesis. Ketonuria may also occur normally in association with a high-fat, low-carbohydrate diet. Some weight-loss programs encourage gradual reduction in carbohydrate and total caloric intake until ketone bodies appear in urine. Dieters are urged to maintain this level of caloric intake, checking urinary ketones regularly to confirm the consumption of body fat.

Comment. Keto-Stix and similar 'dry chemistry' tests are convenient test strips for urinary ketone bodies. They contain a chemical reagent, such as nitroprusside, which reacts with acetoacetate in urine to form a lavender color, graded on a scale with a maximum of '4+' (see Chapter 21 and Fig. 21.12). A reaction of '1+' (representing 5–10 mg ketone bodies/100 mL) or '2+' (10–20 mg/100 mL) on the test strip was established as a goal to assure continued fat metabolism, and therefore weight loss. This type of diet is discouraged today, because the appearance of ketone bodies in the urine indicates greater concentrations in the plasma, and may cause metabolic acidosis.

glucose and reducing the demand on degradation of muscle protein for gluconeogenesis (see also Chapter 20).

Mobilization of lipids during gluconeogenesis and work

Insulin, glucagon, epinephrine, and cortisol control the direction and rate of glycogen and glucose metabolism in liver. During fasting and starvation, hepatic gluconeogenesis is activated by glucagon and requires the coordinated degradation of proteins and release of amino acids from muscle, and the degradation of triglycerides and release of fatty acids from adipose tissue. This process, known as lipolysis, is controlled by the enzyme hormone-sensitive lipase, which is activated by phosphorylation by cAMP-dependent protein kinase A in response to increasing plasma concentrations of glucagon (see Chapter 20). The activation of hormone-sensitive lipase has predictable effects – increasing the concentration of free fatty acids, glycerol, and ketone bodies in plasma during fasting and starvation (Fig. 14.12); similar effects are observed in response to epinephrine during the stress response. Epinephrine activates both glycogenolysis in the liver and lipolysis in adipose tissue, so that both fuels, glucose and fatty acids, increase in blood during stress. Cortisol exerts a more chronic effect on lipolysis and also causes insulin resistance. Cushing's syndrome (see Chapter 35), in which there are high blood concentrations of cortisol, is characterized by hyperglycemia, muscle wastage, and redistribution of fat from glucagon-sensitive adipose depots to atypical sites, such as the cheeks, upper back, and trunk.

Summary

Unlike carbohydrate fuels, which enter the body primarily as glucose or sugars that are readily converted to glucose, lipid fuels are heterogeneous with respect to chain length, branching, and unsaturation. The catabolism of fats is primarily a mitochondrial process, but also occurs in peroxisomes. Using a variety of chain-length-specific transport processes and catabolic enzymes, the primary pathways of catabolism of fatty acids involve their oxidative degradation in two-carbon units – a process known as β-oxidation, which produces acetyl CoA. In muscle, the acetyl CoA units are used for ATP production in the mitochondria, whereas in liver the acetyl CoA is catabolized to ketone bodies, primarily acetoacetate and β-hydroxybutyrate, that are exported for energy metabolism in peripheral tissue.

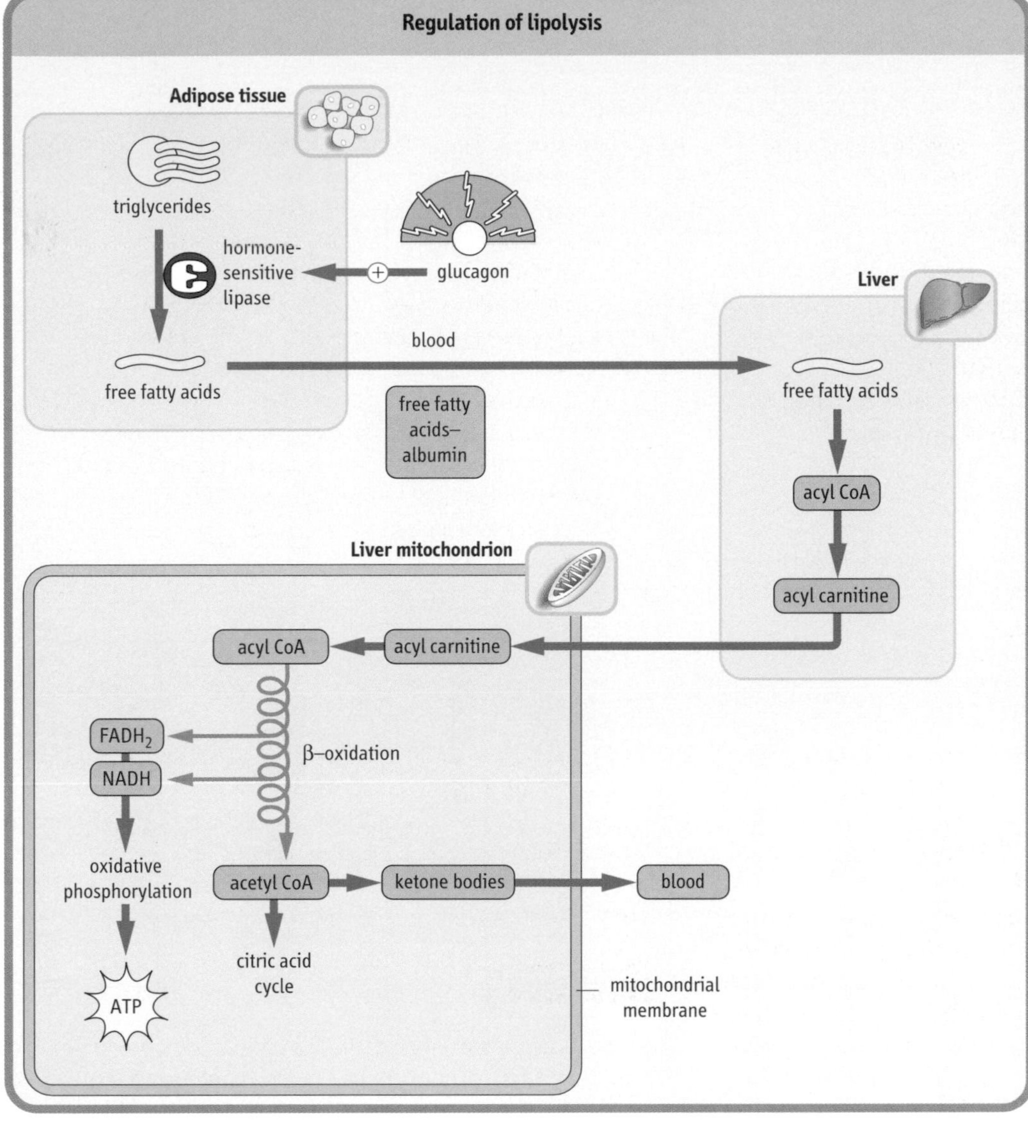

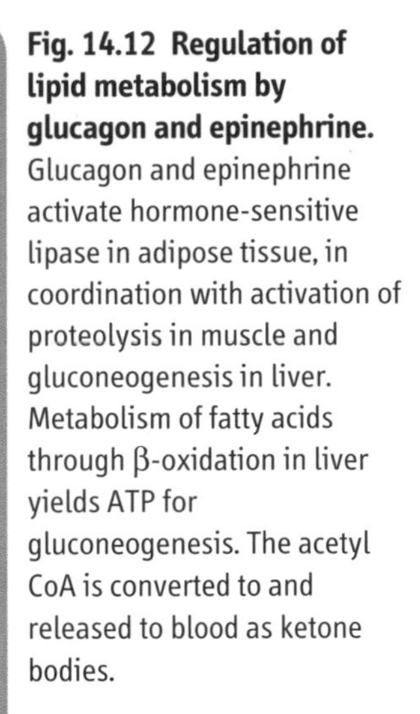

Fig. 14.12 Regulation of lipid metabolism by glucagon and epinephrine. Glucagon and epinephrine activate hormone-sensitive lipase in adipose tissue, in coordination with activation of proteolysis in muscle and gluconeogenesis in liver. Metabolism of fatty acids through β-oxidation in liver yields ATP for gluconeogenesis. The acetyl CoA is converted to and released to blood as ketone bodies.

Further reading

Atar D, Spiess M, Mandinova A, Cierpka H, Noll G, Luscher TF. Carnitine – from cellular mechanisms to potential clinical applications in heart disease. *Eur J Clin Invest* 1997;**27**:973–976.

Eaton S, Bartlett K, Pourfarzam M. Mammalian mitochondrial β-oxidation. *Biochem J* 1996;**320**:345–357.

McGarry JD, Brown NF. The mitochondrial carnitine palmitoyltransferase system. From concept to molecular analysis. *Eur J Biochem* 1997;**244**:1–14.

Mitchell GA, Kassovska-Bratinova S, Boukaftane Y, *et al.* Medical aspects of ketone body metabolism. *Clin Invest Med* 1995;**18**:193–216.

Pollitt RJ. Disorders of mitochondrial long-chain fatty acid oxidation. *J Inherit Metab Dis* 1995;**18**:473–490.

Singh I. Biochemistry of peroxisomes in health and disease. *Mol Cell Biochem* 1997;**167**:1–29.

Swink TD, Vining EP, Freeman JM. The ketogenic diet: 1997. *Adv Pediatr* 1997;**44**:297–329.

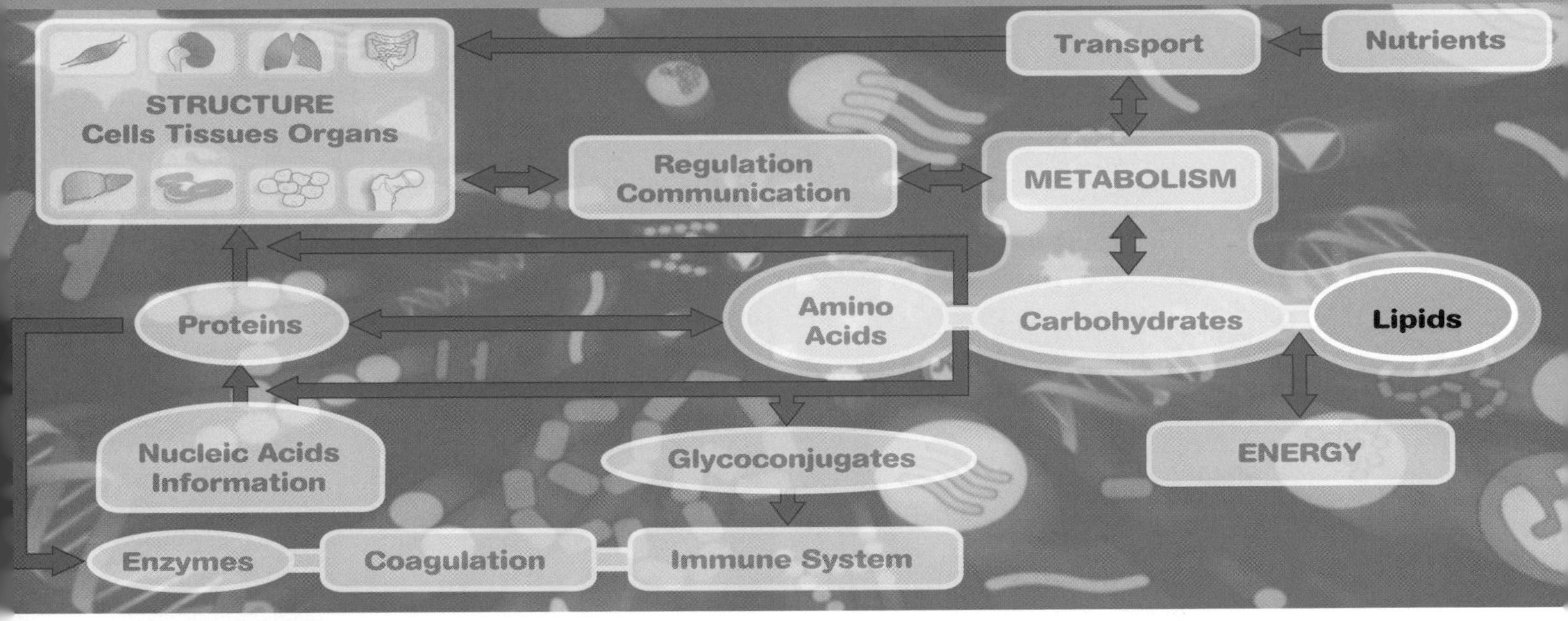

15 Biosynthesis and Storage of Fatty Acids in Liver and Adipose Tissue

The majority of fatty acids required by man are supplied in the diet; however, pathways for their *de novo* synthesis from 2-carbon compounds are present in numerous tissues, e.g. liver, brain, kidney, mammary gland and adipose tissue. In general, the pathway of *de novo* synthesis is primarily used in conditions of excess energy intake – specifically, excess intake of carbohydrate, in which the carbohydrate is converted to fatty acids in the liver and stored as triglyceride (TG, also known as triacylglycerol [TAG]) in adipose tissue. In man, adipose tissue is not an important site of synthesis of fatty acids (lipogenesis); the main organ of lipogenic activity is the liver. Lipogenesis does not appear to be a critical requirement in humans, and therefore no life-threatening illnesses associated with its malfunction have been identified. It does, however, have an important bearing on the development of obesity, and is inhibited in type 1 diabetes mellitus.

The pathway for lipogenesis is not simply the reverse of oxidation of fatty acids (see Chapter 14). Not only does lipogenesis require a completely different set of enzymes, it is also located in a different cellular compartment, the cytosol, and uses nicotinamide dinucleotide phosphate ($NADP^+$), as opposed to the nicotinamide dinucleotide (NAD^+) that is required for β-oxidation.

Fatty acid synthesis

The synthesis of fatty acids in mammalian systems can be considered as a two-stage process, both stages requiring acetyl Coenzyme A (acetyl CoA) and both stages using multifunctional proteins in multienzyme complexes:

- **stage 1**: involves acetyl CoA carboxylase,
- **stage 2:** involves fatty acid synthetase.

Acetyl CoA carboxylase

In the first stage of fatty acid biosynthesis, acetyl CoA, generally derived from carbohydrate metabolism, is converted to malonyl CoA under the action of the enzyme, acetyl CoA carboxylase (Fig. 15.1). This is a biotin-dependent enzyme with distinct enzyme functions and a carrier protein function: a biotin carboxylase, a transcarboxylase, and a biotin–carboxyl-carrier protein. The enzyme is synthesized in an inactive protomer form, each protomer containing the above subunits, a molecule of biotin, and a regulatory allosteric site for the binding of citrate (a Krebs cycle metabolite) or of palmitoyl CoA, the end-product of the fatty acid biosynthetic pathway. The reaction itself takes place in two stages: the carboxylation of biotin, involving adenosine triphosphate (ATP), followed by the transfer of this carboxyl

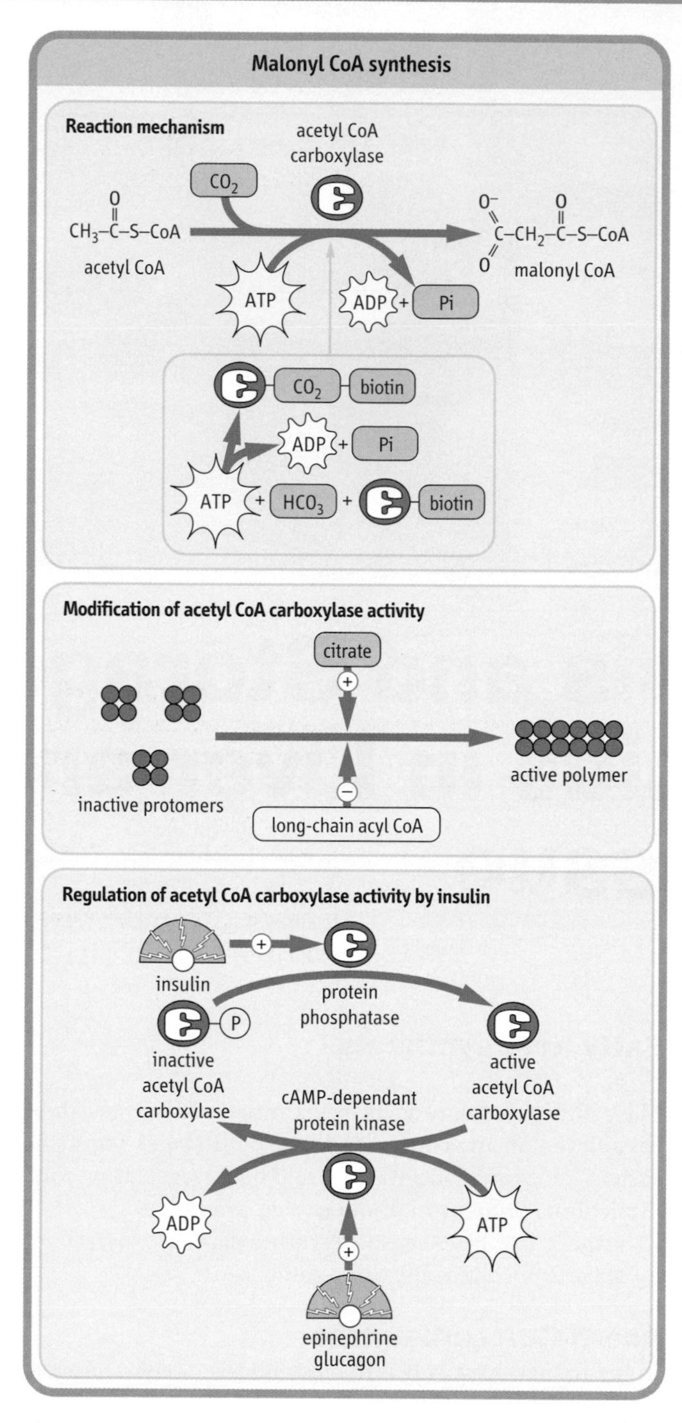

Fig. 15.1 Conversion of acetyl CoA to malonyl CoA. Acetyl CoA carboxylase is covalently attached to biotin, which is carboxylated, with conversion of ATP to adenosine diphosphate (ADP) and inorganic phosphate (Pi). The enzyme requires the presence of citrate for polymerization to its active form. Enzyme activity is also regulated by insulin, independently of the citrate-stimulated polymerization. cAMP, cyclic adenosine monophosphate.

group to acetyl CoA to produce the end-product, malonyl CoA, releasing the free enzyme–biotin complex.

This crucial lipogenic enzyme is subject to both short- and long-term control mechanisms

Acetyl CoA carboxylase requires the presence of citrate or isocitrate, which allows the polymerization of the protomer that produces the active form of the enzyme. This polymerization process is also inhibited by palmitoyl CoA at the same allosteric site. The stimulatory and inhibitory effects of these two molecules are entirely logical: under conditions of high citrate concentration, energy storage is desirable, but decreased synthesis of fatty acids is appropriate when the product (palmitoyl CoA) accumulates. There is an additional control mechanism involving hormone-dependent protein phosphatase/kinase (see Fig. 15.1): phosphorylation, which inhibits activity of the enzyme, and dephosphorylation, which activates it, is promoted by glucagon. These effects are independent of the effects of citrate or palmitoyl CoA.

The carboxylation of acetyl CoA to malonyl CoA is the committed step of fatty acid synthesis. It is therefore quite clear why this enzyme is under such strict regulation by two independent mechanisms, certainly in the short term. Longer-term control mechanisms also exist and these relate to the induction or repression of enzyme synthesis effected by diet: synthesis of acetyl CoA carboxylase is upregulated under conditions of high-carbohydrate/low-fat intake, whereas starvation or high-fat/low-carbohydrate intake leads to downregulation of synthesis of the enzyme.

Fatty acid synthetase

The second stage of fatty acid biosynthesis is also accomplished by a multienzyme complex, fatty acid synthetase. This enzyme system is more complex than acetyl CoA carboxylase. The protein contains seven distinct enzyme activities and an acyl-carrier protein, the latter replacing CoA. The structure of this molecule is shown in Figure 15.2: it consists of a dimer of identical polypeptides arranged head to tail. Each chain contains all seven enzyme activities and an acyl-carrier protein; however, the structure–function relationships are shared between the two polypeptide chains.

Six cycles of synthetase activity precede a final thioesterification to palmitate

The reaction proceeds by the addition of malonyl CoA units after an initial priming of the cysteine (–Cys–SH) group with acetyl CoA under the action of acetyl transacylase (Fig. 15.3). Malonyl CoA then combines with the –SH of the pantetheine group that is attached to the acyl-carrier protein of the other subunit, under the action of malonyl transacylase. Next, 3-ketoacyl synthase catalyzes the reaction between the previously attached acetyl group and the malonyl residue, liberating carbon dioxide and forming the 3-ketoacyl-enzyme complex. This frees the cysteine residue that had been occupied by acetyl CoA. The 3-ketoacyl group subsequently undergoes sequential reduction, dehydration, and reduction to form a

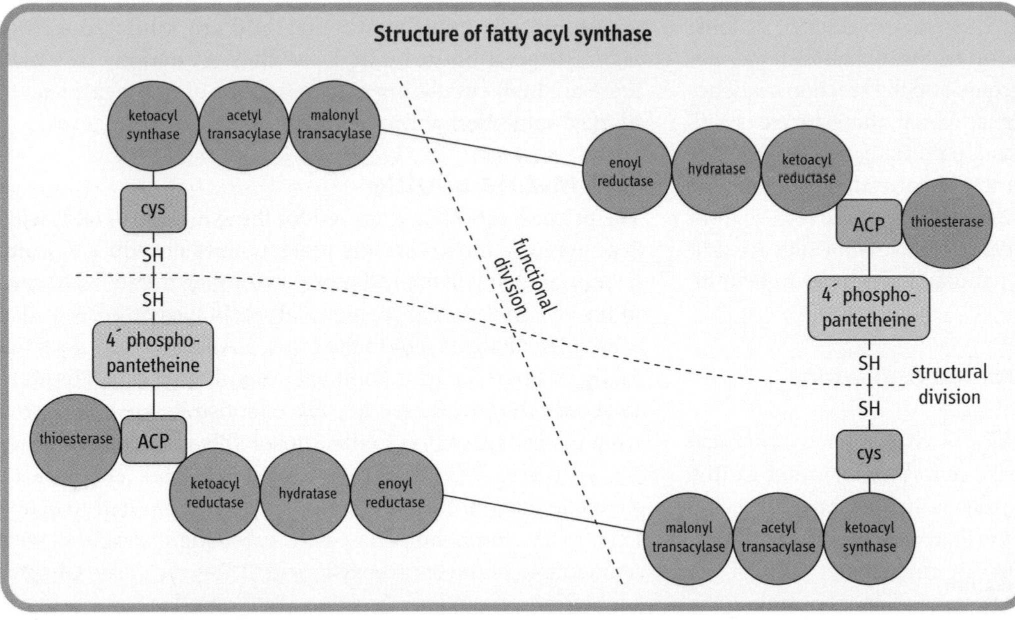

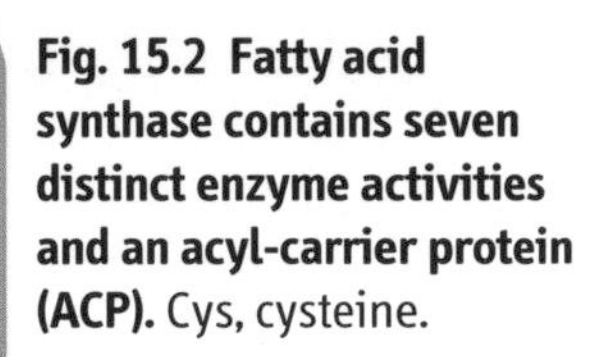

Fig. 15.2 Fatty acid synthase contains seven distinct enzyme activities and an acyl-carrier protein (ACP). Cys, cysteine.

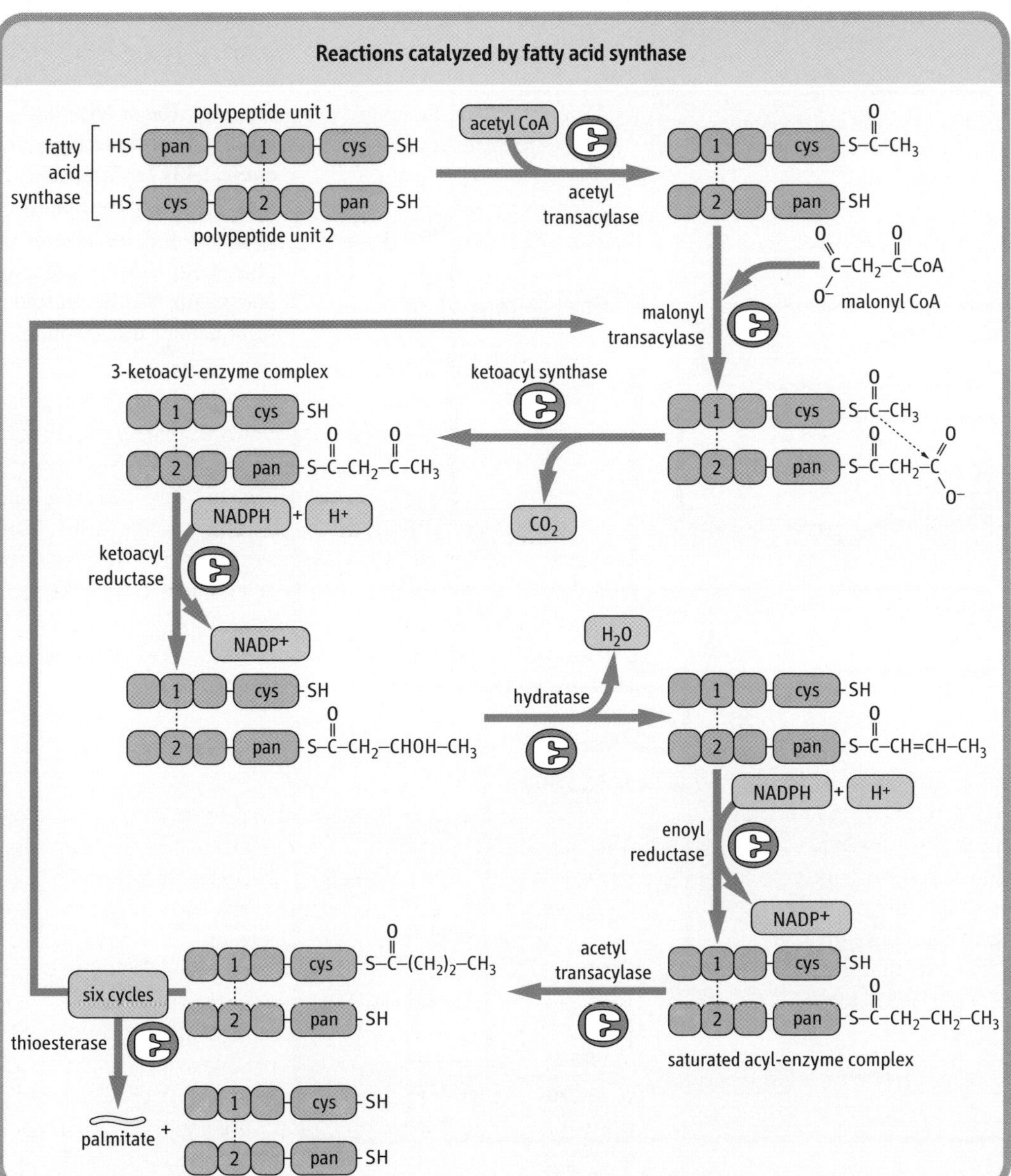

Fig. 15.3 Second stage of fatty acid biosynthesis. The cyclical part of the reaction is completed six times to release one 16-carbon palmitate molecule. NADPH, reduced nicotinamide dinucleotide phosphate; pan, pantetheine.

saturated acyl enzyme complex. The next molecule of malonyl CoA then displaces the acyl group from the pantotheine–SH group to the now free cysteine group, and the reaction sequence is repeated through six cycles, after which, thioesterase is activated and the 16-carbon fatty acid, palmitate, is released. The presence of a 16-carbon chain in the saturated acyl-enzyme complex activates the thioesterase resulting in the release from the enzyme complex, free palmitate. The two –SH sites are now free, allowing another cycle of palmitate synthesis to be initiated.

This stage, also, has associated regulatory mechanisms

In common with the acetyl CoA carboxylase system, fatty acid synthase activity is also regulated, both by substrate flux (the presence of phosphorylated sugars) via an allosteric effect, and by induction and repression of the enzyme. Alterations in total enzyme protein are effected by the nutritional state of the individual; consequently, this is the main factor controling the rate of lipogenesis. Rates are thus greatest in the presence of a high-carbohydrate/low-fat diet, and are inhibited during fasting/starvation or by high-fat diets. Situations in which there are high circulating concentrations of fatty acids lead to marked inhibition of lipogenesis.

The malate shuttle

The primary molecule required for the synthesis of fatty acids is acetyl CoA. However, this metabolite is a product of mitochondrial metabolism and hence is compartmentalized within the cell; in a similar fashion, fatty acid biosynthesis is also compartmentalized, but to the cytosol. As acetyl CoA does not easily traverse the mitochondrial membrane, a mechanism exists whereby the 2-carbon units of substrate are transferred from the mitochondria to the cytosol: this is the malate shuttle, involving the malate–citrate antiporter (Fig. 15.4). Pyruvate derived from glycolysis is decarboxylated to acetyl CoA in the mitochondria, with subsequent reaction with oxaloacetate in the tricarboxylic acid (TCA) cycle (see Chapter 13) to form citrate. Translocation of a molecule of citrate to the cytosol via the TCA antiporter is accompanied by transfer of

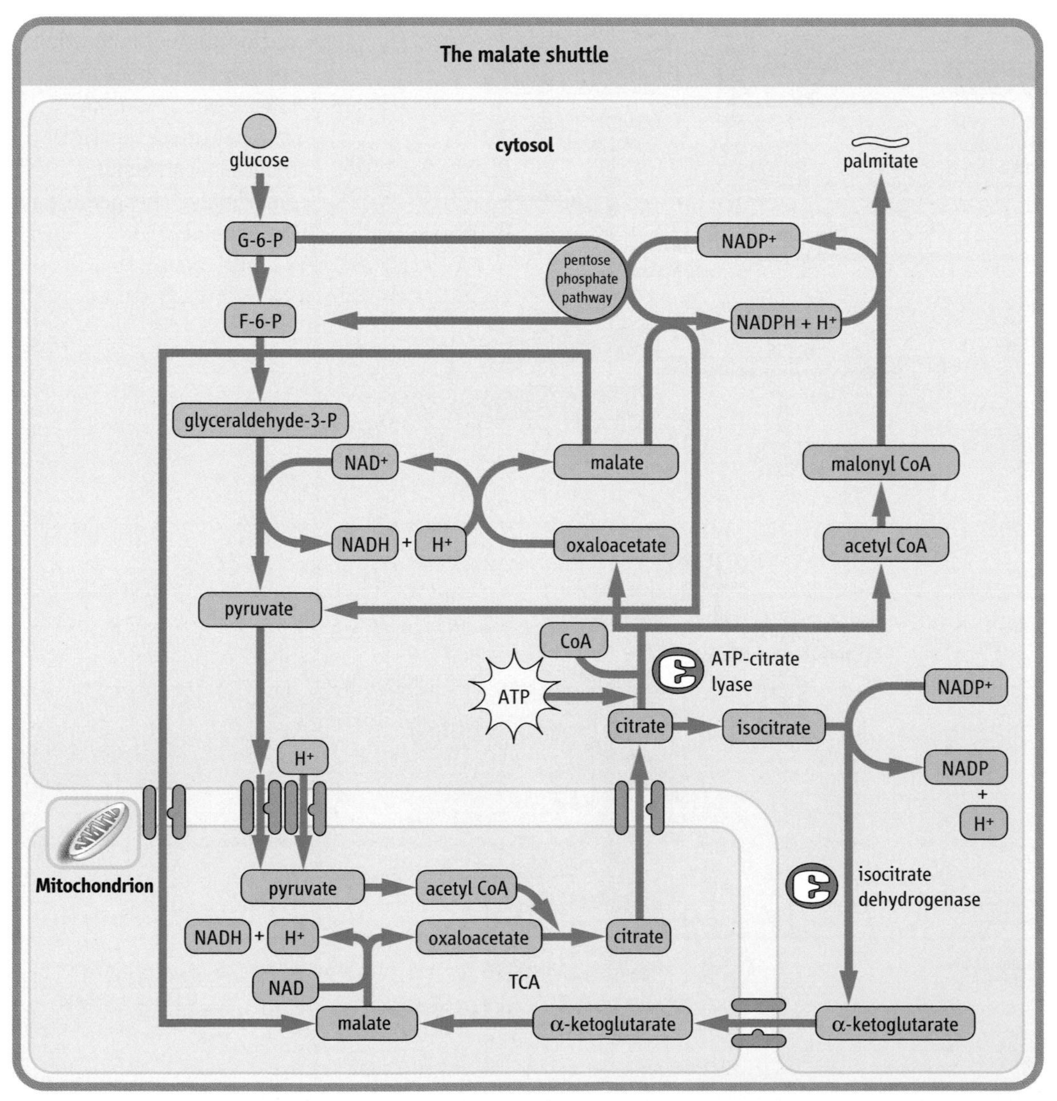

Fig. 15.4 The provision of acetyl CoA and reducing equivalents for fatty acid biosynthesis in the malate shuttle. F-6-P, fructose-6-phosphate; G-6-P, glucose-6-phosphate; NADH, reduced nicotinamide dinucleotide.

Lipid abnormalities in alcoholism

A 36-year-old woman attending a routine well-woman clinic was found to have serum concentrations of TG 73.0 mmol/L (6388 mg/dL) and cholesterol 13 mmol/L (503 mg/dL). On questioning about her alcohol intake, after some initial prevarication, she admitted to drinking three bottles of vodka and six bottles of wine per week. When she discontinued her alcohol intake, her TG concentrations decreased to 2 mmol/L (175 mg/dL) and her cholesterol concentration decreased to 5.0 mmol/L (193 mg/dL). Three years later, the woman presented with an enlarged liver and return of her lipid abnormality. Liver biopsy indicated alcoholic liver disease, with hepatic steatosis (infiltration of the liver cells with fat).

Comment. In alcoholic individuals, the metabolism of alcohol produces increased hepatic nicotinamide adenine dinucleotide (NADH) and hydrogen ions (H^+). Increased NADH + H^+/NAD^+ ratios inhibit the oxidation of fatty acids. Fatty acids reaching the liver either from dietary sources or by mobilization from adipose tissue are therefore re-esterified with glycerol to form TG. In the initial stages of alcoholism, these are packaged with apolipoproteins and exported as very-low-density lipoproteins (VLDL). Increased concentrations of VLDL, and hence of serum TG, are often present in the early stages of alcoholic liver disease. As the liver disease progresses, there is a failure to produce the apolipoproteins and export the fat as VLDL; accumulation of TG in the liver cells thus ensues.

a molecule of malate to the mitochondrion. Citrate in the cytosol, in the presence of ATP and CoA, undergoes cleavage to acetyl CoA and oxaloacetate, effected by citrate lyase. This makes acetyl CoA available for carboxylation to malonyl CoA and the synthesis of fatty acids; the oxaloacetate is converted to malate via malate dehydrogenase, and then to pyruvate by the $NADP^+$-dependent malic enzyme. The reduced nicotinamide adenine phosphate (NADPH) so formed becomes available for lipogenesis.

The synthesis of fatty acids is further linked to glucose metabolism through the pentose phosphate pathway. This pathway is the main provider of NADPH required for lipogenesis and the activities of the two pathways are closely linked.

Fatty acid elongation

Palmitate released from fatty acid synthetase provides the basis of supply of other long-chain fatty acids required by the body,

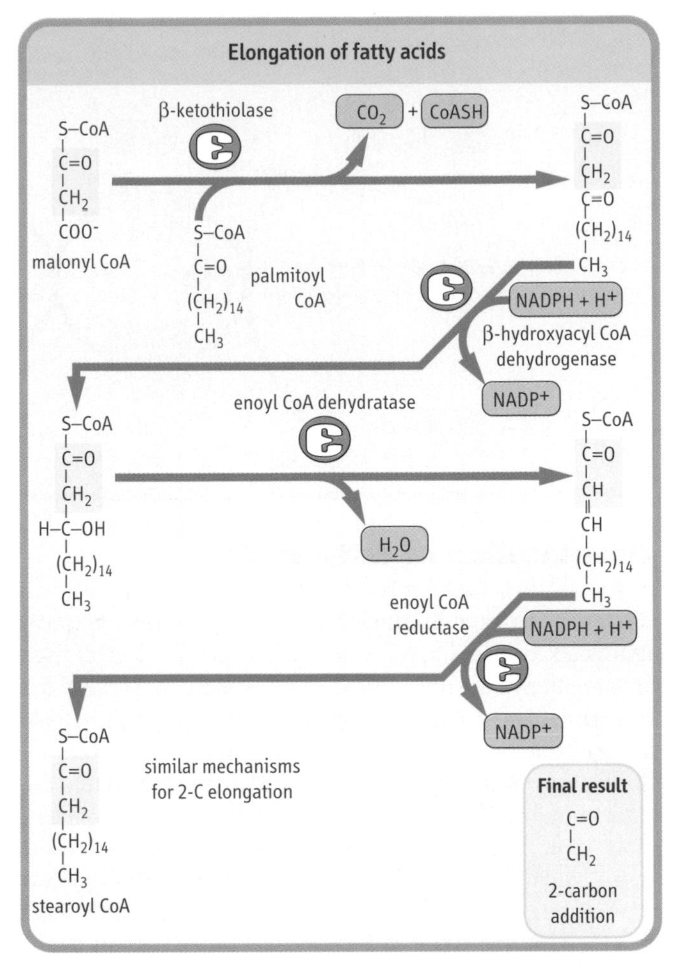

Fig. 15.5 Fatty acid elongation occurs on the endoplasmic reticulum under the influence of another enzyme complex, fatty acid elongase.

with the exception of certain essential fatty acids (see below). Chain elongation of palmitate occurs by the addition of 2-carbon fragments derived from malonyl CoA (Fig. 15.5). This process occurs on the endoplasmic reticulum by the action of yet another multienzyme complex: fatty acid elongase. The system of reactions occurring during chain elongation is similar to those involved in fatty acid synthesis, except that the fatty acid is attached to CoA, rather than to the pantetheine residue of acyl-carrier protein.

The substrates for the cytosolic fatty acid elongase include the saturated fatty acids from 10-carbon upwards, in addition to unsaturated fatty acids. Very-long-chain (22–24-carbon) fatty acids are produced in the brain, and elongation of stearoyl CoA in the brain increases rapidly during myelination, producing fatty acids required for the synthesis of sphingolipids.

Mitochondria are also able to elongate fatty acids using a different system, which is NADH-dependent and uses acetyl CoA as the source of 2-carbon fragments. It is simply the reverse of β-oxidation (see Chapter 13) and the substrates for chain elongation are short- and medium-chain fatty acids containing fewer than 16 carbon atoms.

During fasting and starvation, elongation of fatty acids is greatly reduced.

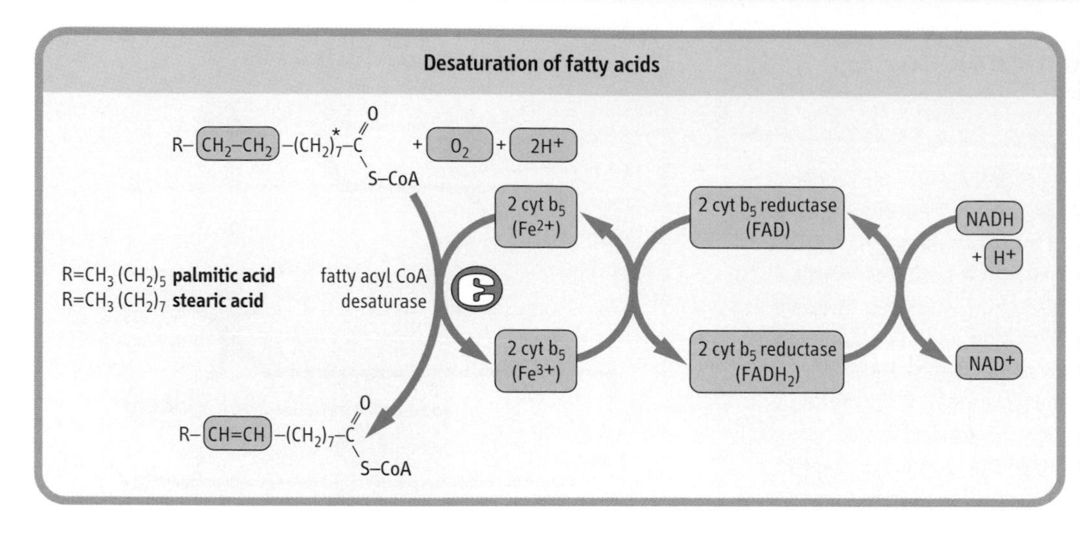

Fig. 15.6 Desaturation of fatty acids. *Human desaturases cannot introduce a double bond between carbon 10 and the methyl (ω) end of the fatty acid. This $(CH_2)_7$ represents the limit at which a double bond can be introduced. cyt b_5, cytochrome b_5; FAD, flavin adenine dinucleotide; $FADH_2$, reduced flavin adenine dinucleotide; Fe^{3+}, ferric iron.

Desaturation of fatty acids

The body has a requirement for mono- and polyunsaturated fatty acids, in addition to saturated fatty acids. Some of these are supplied in the diet; indeed, it is essential for survival that two particular unsaturated fatty acids, linoleic and linolenic, are available in the diet. However, a system for desaturation does exist, and requires molecular oxygen, NADH, and cytochrome b_5. The process of desaturation, like that of chain elongation, occurs on the endoplasmic reticulum, and results in the oxidation of both the fatty acid and NADH (Fig. 15.6).

In man, the desaturase system is unable to introduce double bonds between carbon atoms beyond carbon 10 and the ω (terminal methyl) carbon atom. The most common desaturation reactions to occur are between carbon atoms 9 and 10 (annotated as Δ^9 desaturations), e.g. those with palmitic acid producing palmitoleic acid (C-16:1 Δ^9), and those with stearic acid producing oleic acid (C-18:1, Δ^9).

The annotation C-18:1Δ,^{9}ω C-16:1,Δ^9 indicates the presence of a double bond in an 18-carbon chain ω-16-carbon chain fatty acid at position between C-9 and C-10 in each fatty acid. C-18:2,$\Delta^{9,12}$ would thus indicate two double bonds in an 18-carbon fatty acid between carbons C-9 and C-10, and between carbons C-12 and C-13.

Essential fatty acids

Because of the inability of the human desaturase system to introduce double bonds beyond C-10, and the fact that two series of fatty acids – having double bonds three carbons from the methyl end (ω-3 fatty acids) and six carbons from the methyl end (ω-6 fatty acids) – are required for the synthesis of eicosanoids, these ω-3 and ω-6 fatty acids must be present in the diet or synthesized from other fatty acids in the diet. Mainly, they are obtained from dietary plant oils: these oils contain the ω-6 fatty acid, linoleic acid (C-18:2, $\Delta^{9,12}$) and the ω-3 fatty acid, linolenic acid (C-18:3, $\Delta^{9,12,15}$). Linoleic acid is converted by a series of elongation and desaturation reactions to arachidonic acid (C-20:4, $\Delta 5,^{8,11,14}$), the precursor for the synthesis of prostaglandin and other eicosanoids in man. Elongation and desaturation of linolenic acid produces

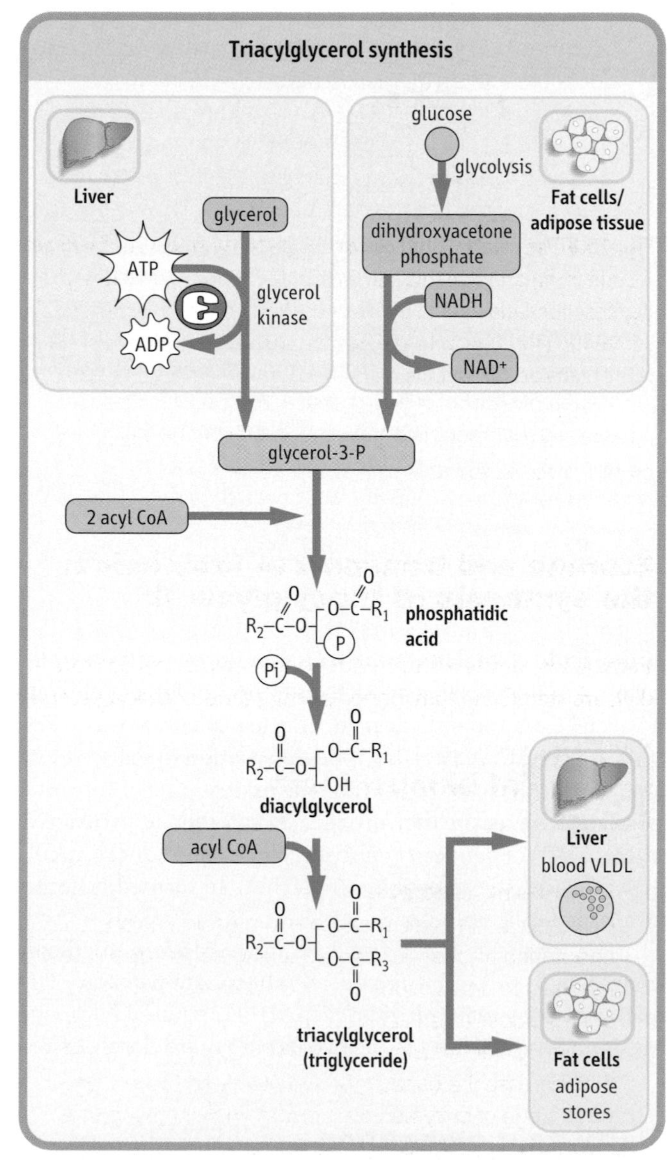

Fig. 15.7 TAGs are synthesized in adipose tissue and the liver. The source of glycerol-3-P is different in the two tissues, because there is no glycerol kinase in adipose tissue.

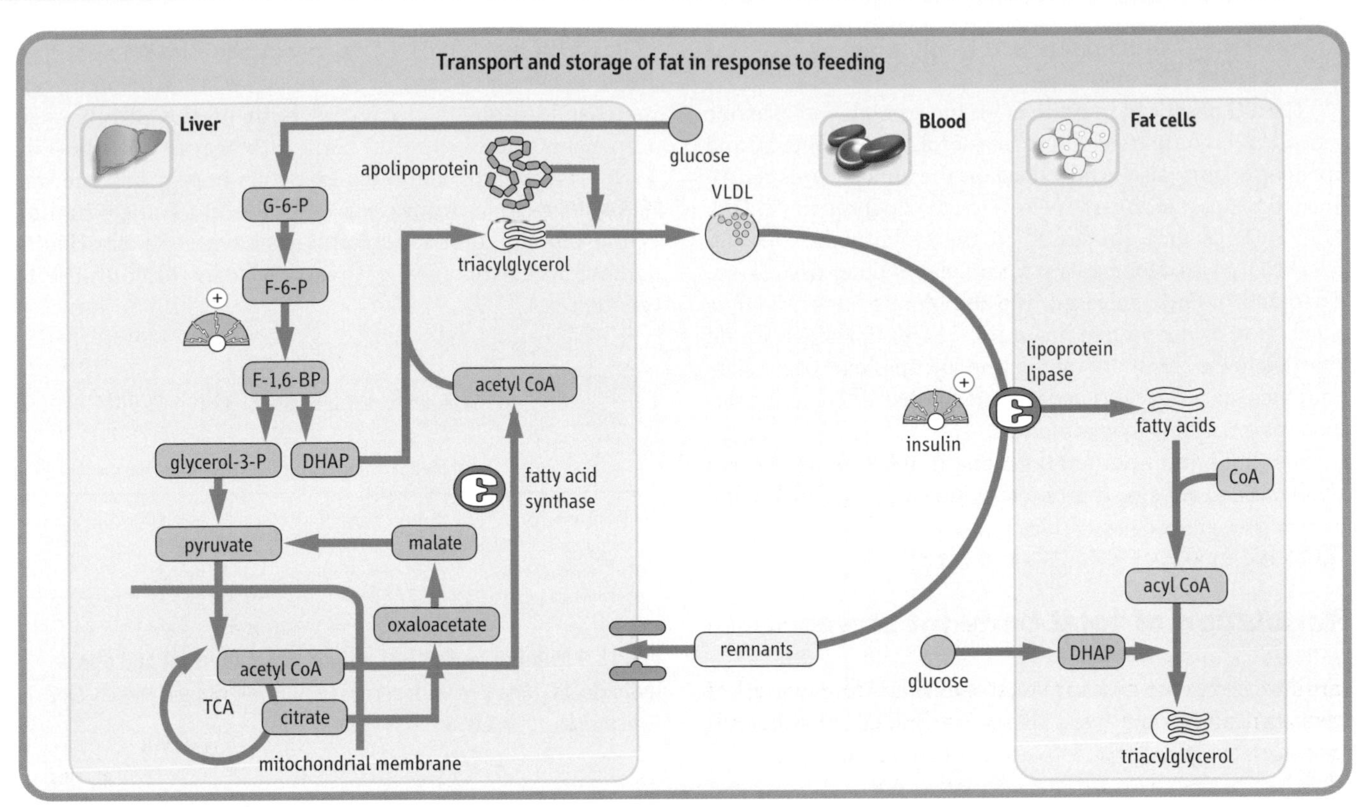

Fig. 15.8 The relationship between the biosynthesis of fatty acid in the liver, its export as VLDL, and increased storage of fat as TAG. Insulin is important in this pathway. In the liver, it stimulates glycolysis, thereby increasing pyruvate production. By stimulating the dephosphorylation of the pyruvate dehydrogenase complex and activating this enzyme, it promotes the production of acetyl CoA, thus stimulating the TCA cycle and increasing the concentration of citrate; in turn, through stimulation of acetyl CoA carboxylase, this increases the rate of fatty acid biosynthesis. F-1,6-BP, fructose-1,6 biphosphate; IDL, intermediate-density lipoprotein.

eicosapentaenoic acid (EPA; C-20:5, $\Delta 5,^{8,11,14,17}$), used as a precursor for a different series of eicosanoids.

Storage and transport of fatty acids: the synthesis of triacylglycerols

Fatty acids, derived from endogenous synthesis or from the diet, are stored and transported as triglycerides (triacylglycerols). In both liver and fat, TGs are produced by a pathway involving glycerol-3-phosphate (glycerol-3-P) and then phosphatidic acid as intermediates (Fig. 15.7). However, the glycerol-3-P is of different origin in the two tissues: in the liver, glycerol itself provides the source, via the action of glycerol kinase, but in adipose tissue, which lacks this enzyme, glucose is the source, via glycolysis and the immediate precursor of glycerol-3-P is dihydroxyacetone phosphate (DHAP) (Fig. 15.7). The storage of fatty acids in

Fat biosynthesis

Long-term adaptation

Changes in the level of enzyme expression as an adaptation to changes in food intake are key regulatory mechanisms by which the appropriate ability to store energy substrates is achieved.

Being well fed allows the individual to induce enzymes that increase the capacity to synthesize fat in the liver. A wide range of enzymes are induced, including enzymes involved in glycolysis, e.g. glucokinase and pyruvate kinase, as well as enzymes linked to increased production of the reduced forms of NADP (G-6-P dehydrogenase, 6-phosphogluconate dehydrogenase, and malic enzyme). For more direct and rapid rates of fatty acid biosynthesis, there is increased expression of citrate lyase, acetyl CoA carboxylase, fatty acid synthetase, and Δ^9 desaturase.

In the same way that those enzymes are induced in the well-fed state, so also there is a concomitant decrease or repression of synthesis of the key enzymes involved in gluconeogenesis. Phosphoenolpyruvate carboxykinase, G-6-Pase, and some aminotransferases are reduced in amount, either by reduction in synthesis, or by increased degradation (see Chapter 20).

adipose tissue can therefore occur only when glycolysis is activated in the fed state.

The TG produced in the liver on the smooth endoplasmic reticulum is complexed with cholesterol, phospholipids, and apolipoproteins (also synthesized on the endoplasmic reticulum), for export to form very-low-density lipoprotein (VLDL). The VLDL is then processed in the Golgi apparatus and released into the bloodstream for uptake by other tissues (see Chapter 17). Once released into the bloodstream, VLDL is acted upon by lipoprotein lipase (LPL). This enzyme is found attached to the basement membrane glycoproteins of capillary endothelical cells and is active both against VLDL and, after food intake, against chylomicrons.

Insulin is an important hormone in relation to fatty acid synthesis and storage. It promotes glucose uptake both in the liver and in adipose tissue (Fig. 15.8).

Regulation of total body fat stores

Until recently, the mechanisms involved in the control of body fat (adiposity) were little recognized. It was well understood that increased energy intake without appropriate increase in energy expenditure was associated with increased adiposity, in terms both of the numbers of adipocytes and of their fat content – that is, obesity (Fig. 15.9) – but the mechanisms by which energy balance was regulated were far from clear. With recent advances in molecular biology, those mechanisms have been clarified to some extent, and new pathways of control continue to emerge.

Number and size of adipocytes

	Cell size (mg lipid/cell)	Total cell number ($\times 10^9$)
normal weight	0.66 → 0.06	26 → 6.8
juvenile-onset obesity	0.90 → 0.05	85 → 6.9
adult-onset obesity	0.98 → 0.14	62 → 4.2
obese – treated	0.45 → 0.05	62 → 5.3

Fig. 15.9 Number and size of adipocytes in normal and obese individuals. (Adapted with permission from Grimber, Hirsch. *Ciba Symposium* 1972;**8**:350.)

Obesity

A 48-year-old ex-Army infantryman (height 1.91 m) presented with the problem of increasing weight over the previous 8 years since he left the Army. At the time of his retirement from active service, he weighed 95 kg (209 lb); at presentation he weighed 193 kg (424.6 lb). His current occupation was that of truck driver. For health reasons, he had also stopped smoking some 5 years previously. He denied any change in food intake since leaving the Army, but admitted to taking little or no exercise. Detailed enquiry indicated that his daily dietary intake provided between 12 600 and 16 800 kJ (3000 and 4000 kcal), with a fat intake approaching 40%. The patient was initially placed on a healthy eating plan, with fat intake reduced to 35% of total calories. He was advised to exercise and proceeded to swim three or four times per week. His weight immediately began to decrease, rapidly at first, and then at 3–4 kg (6.6–8.8 lb) each month until it stabilized at 145–150 kg (319–330 lb). He was then placed on a high-protein/low-carbohydrate/low-fat diet, which induced a return of weight loss that continued for a further year, resulting in a final weight of 93 kg (204.6 lb).

Comment. The development of obesity is a major problem worldwide, and is associated with failure of energy balance mechanisms and regulation of food intake. Clinical obesity is now clearly defined in terms of height and weight through the body mass index (BMI), which is calculated as the weight in kilograms divided by the (height in meters)^2:

$$\text{BMI (kg/m}^2) = \frac{\text{weight (kg)}}{(\text{height [m]})^2}$$

BMI 25–30 kg/m^2 is classified as overweight or grade I obesity, BMI >30 kg/m^2 is clinical or grade II obesity, and BMI >40 kg/m^2 is classified as morbid or grade III obesity. Any imbalance between energy input and energy output such that the former exceeds the latter will result in increasing weight.

Medical complications of obesity are widespread and are associated with both chronic and acute disease. The most important of the associated diseases is type 2 diabetes mellitus: 80% of this type of diabetes is associated with the obese state. Often, obesity-associated illnesses include coronary heart disease, hypertension, stroke, arthritis, and gall bladder disease. Treatment of obesity is therefore paramount in the prevention of many of these conditions.

The control of energy balance, and hence adiposity, is achieved through the integration of several factors, some of which are genetically linked, whereas others are related to the environment and to behavior:

- **genetically linked factors:** it is clear that appetite control involves a number of recently discovered protein messengers and receptors;
- **environmental factors** leading to the development of obesity include the relative abundance of food and the type of food, e.g. the energy-dense foods currently in vogue.

The leptin system is the best developed of the control mechanisms

Leptin is a small-molecular-weight protein, produced by white adipose tissue, that has numerous metabolic effects. One of these is at the hypothalamus, where the interaction of leptin with its receptor molecule leads to suppression of food intake through the release of corticotropin-releasing hormone and suppression of neuropeptide Y.

Many other newly identified proteins are now known to be involved in the control of whole-body energy and fat metabolism, but the integration of these mechanisms remains to be elucidated. Such protein molecules and their appropriate receptors are the focus of considerable research activity, in an effort to counteract the growing problem – indeed an endemic – of obesity.

Summary

Fat synthesis and storage are essential components of fuel metabolism in the body, but excess accummulation of fat leads to obesity which is becoming a growing problem, indeed a modern pandemic. Lipogenesis takes place in the cytosol and utilizes multienzyme complexes of fatty acyl carboxylase and fatty acyl synthetase. These are subject to complex regulation. The malate shuttle facilitates the transfer of two-carbon units for the cytosolic assembly of acetyl-CoA for use in lipogenesis. The essential unsaturated fatty acids, linoleic and linolenic, must be supplied: linolenic acid is converted to the arachidonic acid, which in turn is the precursor of prostaglandins. Transport of fatty acids between tissues involves binding of free fatty acids to albumin, and a sophisticated system of lipoproteins, which transport water-insoluble tiracylglycerols (triglycerides). They are described in Chapter 17.

Further reading

Auwerx J. Staels B. Leptin. *Lancet* 1998;**351:**737–742.

Campfield LA. Smith FJ. Burn P. Strategies and potential molecular targets for obesity treatment. *Science* 1998;**280:**1383–1387.

Friedman JM. Leptin, leptin receptors, and the control of body weight. *Nutrition Reviews* 1998;**56:**s38-46; discussion s54–75.

Rosenbaum M. Leibel RL. Hirsch J. Obesity. *New England Journal of Medicine* 1997;**337:** 396–407.

Semenkovich CF. Regulation of fatty acid synthase (FAS). *Progress in Lipid Research* 1997:**36:**43–53.

Towle HC. Kaytor EN. Shih HM. Regulation of the expression of lipogenic enzyme genes by carbohydrate. *Annual Review of Nutrition* 1997;**17:**405–433.

Weigle DS. Kuijper JL. Obesity genes and the regulation of body fat content. Bioessays 1996;**18:**867–874.

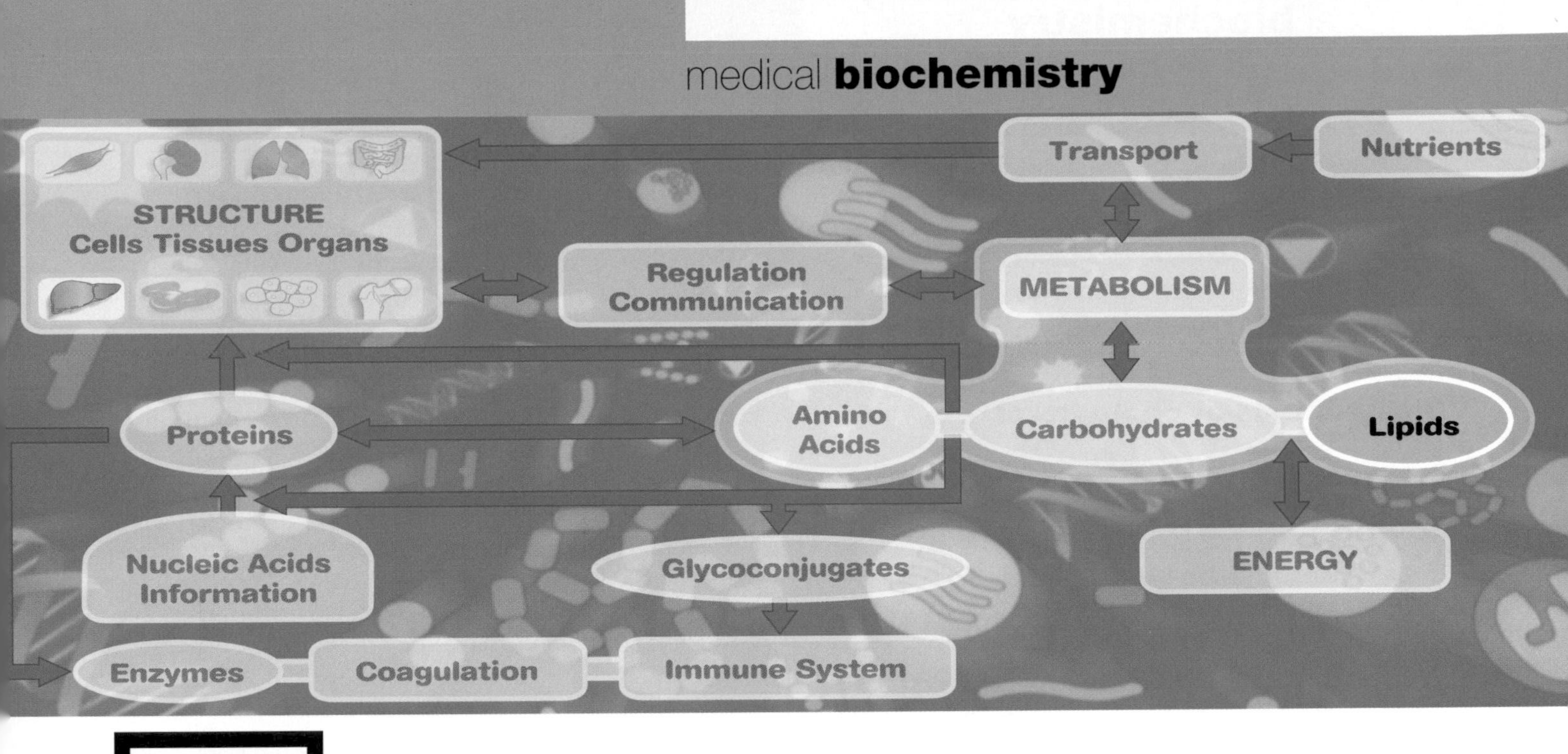

16 Biosynthesis of Cholesterol in Liver

Cholesterol is one molecule with many functions

Cholesterol is a lipid that is an essential component of mammalian cell membranes. In addition, it is the precursor of three important classes of biologically active compounds, the bile acids, the steroid hormones and vitamin D. Disorders of cholesterol metabolism play a role in the etiology of cardiovascular disease and cholesterol is a major component of gall stones.

The typical daily Western diet contains approximately 500 mg (1.2 mmol) of cholesterol, mainly in meat, eggs and dairy products. Under normal circumstances, some 70% of this is absorbed from the gut. A similar amount of cholesterol is synthesized *de novo* each day, mainly in the liver. For this reason both dietary control and manipulation of the processes of cholesterol biosynthesis and metabolism are important in the prevention of cardiovascular disorders.

Structure of cholesterol

The structure of cholesterol is shown in Figure 16.1. It has a molecular weight of 386 Da and contains 27 carbon atoms, of which 17 are incorporated into four fused rings (the perhydrocyclopentano-phenanthrene nucleus), two are in angular methyl groups attached at the junctions of rings AB and CD and eight are in the peripheral side chain. Cholesterol is almost entirely composed of carbon and hydrogen atoms; there is a solitary hydroxyl group attached to carbon atom 3. It is also almost completely saturated having just one double bond between carbon atoms 5 and 6. In three-dimensional terms the ring structure of cholesterol is approximately planar. The two angular methyl groups and the side chain project above this plane and so are in the β-configuration.

This structure gives cholesterol a low solubility in water (approximately 5 μmol/L). Only about 30% of circulating cholesterol occurs in the free form, the majority is esterified through the hydroxyl group to a wide range of long-chain fatty acids including oleic and linoleic acids. Cholesterol esters are even less soluble in water than free cholesterol and so it is perhaps surprising to discover cholesterol circulating in plasma in concentrations of about 5 mmol/L (200 mg/dL). The

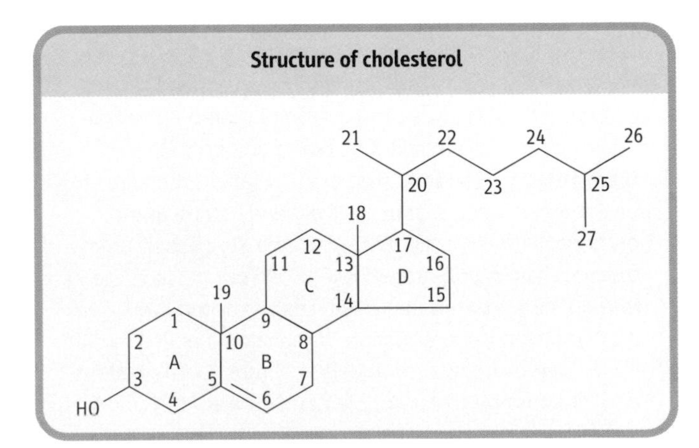

Fig. 16.1 The structure of cholesterol. A–D is the conventional notation used to describe the four rings. Numbers 1–27 are used to describe the carbon atoms.

apparent paradox is explained by the presence of a range of lipoproteins which bind and thereby solubilize the cholesterol molecule (see Chapter 17). Within these lipoproteins, the hydrophobic cholesterol esters are located in the core of the molecule, with free cholesterol in the outside layer.

Biosynthesis of cholesterol

Acetyl coenzyme A is the starting point in the biosynthesis of cholesterol

Virtually all human cells have the capacity to make cholesterol. In quantitative terms, however, the liver is the major site of cholesterol biosynthesis with the intestine, adrenal cortex and gonads also making a significant contribution. An examination of the structure of cholesterol makes it clear that generation of the many carbon–carbon and carbon–hydrogen bonds requires a source of carbon atoms, a source of reducing power and the expenditure of significant amounts of energy. Acetyl coenzyme A (acetyl CoA) provides a high-energy starting point. Acetyl CoA may be derived from several sources including the β-oxidation of long-chain fatty acids, the dehydrogenation of pyruvate and the oxidation of ketogenic amino acids such as leucine and isoleucine. The reducing power is provided by reduced nicotinamide dinucleotide phosphate (NADPH), which is generated by the enzymes of the pentose phosphate pathway (see Chapter 11).

Additional energy is provided by the breakdown of adenosine triphosphate (ATP). Overall the production of 1 mole of cholesterol requires 18 moles of acetyl CoA, 36 moles of ATP and 16 moles of NADPH. All the biosynthetic reactions occur within the cytoplasm although some of the enzymes required are bound to membranes of the endoplasmic reticulum.

HMG CoA reductase is an example of a rate-limiting enzyme

The control of metabolic regulation of any pathway is usually achieved by modulation of the activity of one key enzyme – known as the rate-limiting enzyme. This enzyme often catalyzes the committed step – the first one that may be identified as being unique to that pathway. It is of interest that the regulation of cholesterol biosynthesis occurs at a relatively early stage in the process, with the enzyme that uses a six-carbon atom molecule as its substrate. HMG CoA reductase is the rate-limiting enzyme that catalyzes the committed step that results in the production of mevalonic acid. Hepatic HMG CoA reductase synthesis is stimulated by fasting and inhibited by dietary cholesterol intake. HMG CoA reductase activity is controled by covalent modification induced by cholesterol feedback and by several metabolic hormones.

Mevalonic acid is the first unique compound

Three molecules of acetyl CoA are converted into the six-carbon atom mevalonic acid (Fig. 16.2). The first two steps are condensation reactions leading to the formation of 3-hydroxy-3-methylglutaryl CoA (HMG CoA). These reactions, catalyzed by acetoacetyl CoA thiolase and HMG CoA synthase, are common to the formation of ketone bodies, although the latter process occurs within mitochondria rather than the cytosol. These reactions are also favored energetically since they involve cleavage of a thioester bond and liberation of free coenzyme A. However, the key reaction in the early stages of cholesterol biosynthesis is that catalyzed by the microsomal enzyme HMG CoA reductase, which leads to the irreversible formation of mevalonic acid.

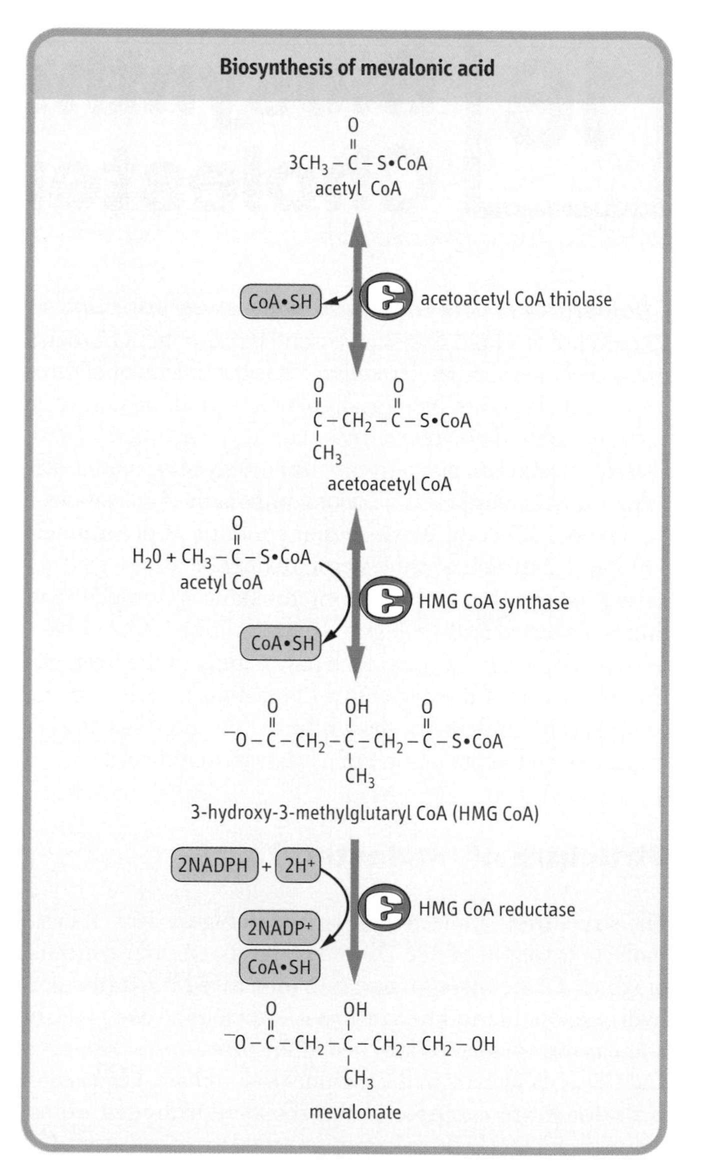

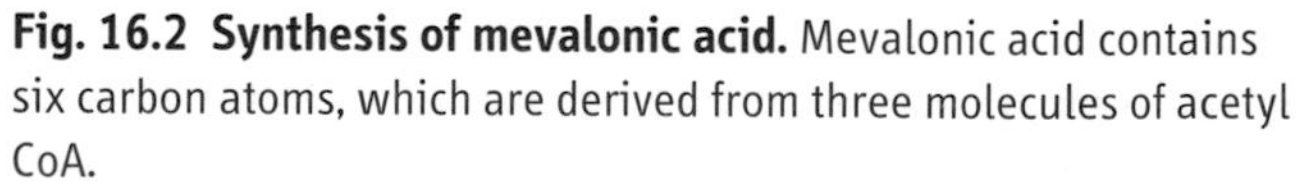
Fig. 16.2 Synthesis of mevalonic acid. Mevalonic acid contains six carbon atoms, which are derived from three molecules of acetyl CoA.

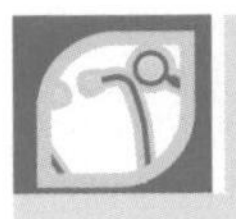

HMG CoA reductase inhibitors

Despite strict dietary control a 50-year-old man, from a family with a history of cardiovascular disease, had a serum cholesterol result of 8.0 mmol/L (desirable levels are <5.0 mmol/L). He started to take pravastatin (an inhibitor of HMG CoA reductase) and 3 months later his cholesterol was 5.5 mmol/L.

Comment. Partial inhibition of the rate-limiting enzyme of cholesterol biosynthesis may be expected to bring about a lowering of plasma cholesterol in a subject who has a controlled dietary intake. This has proved to be the case. A family of competitive inhibitors of HMG CoA reductase, known as 'statins', have been developed following the original discovery that mevastatin, isolated from *Penicillium citrinum*, had enzyme-inhibiting properties. These drugs bring about a 20–40% reduction in low density lipoprotein (LDL) cholesterol. It is interesting to note that in addition to inhibiting HMG CoA reductase activity, the 'statins' also appear to increase the number or functional activity of the LDL receptors, thereby increasing the clearance of several lipoproteins.

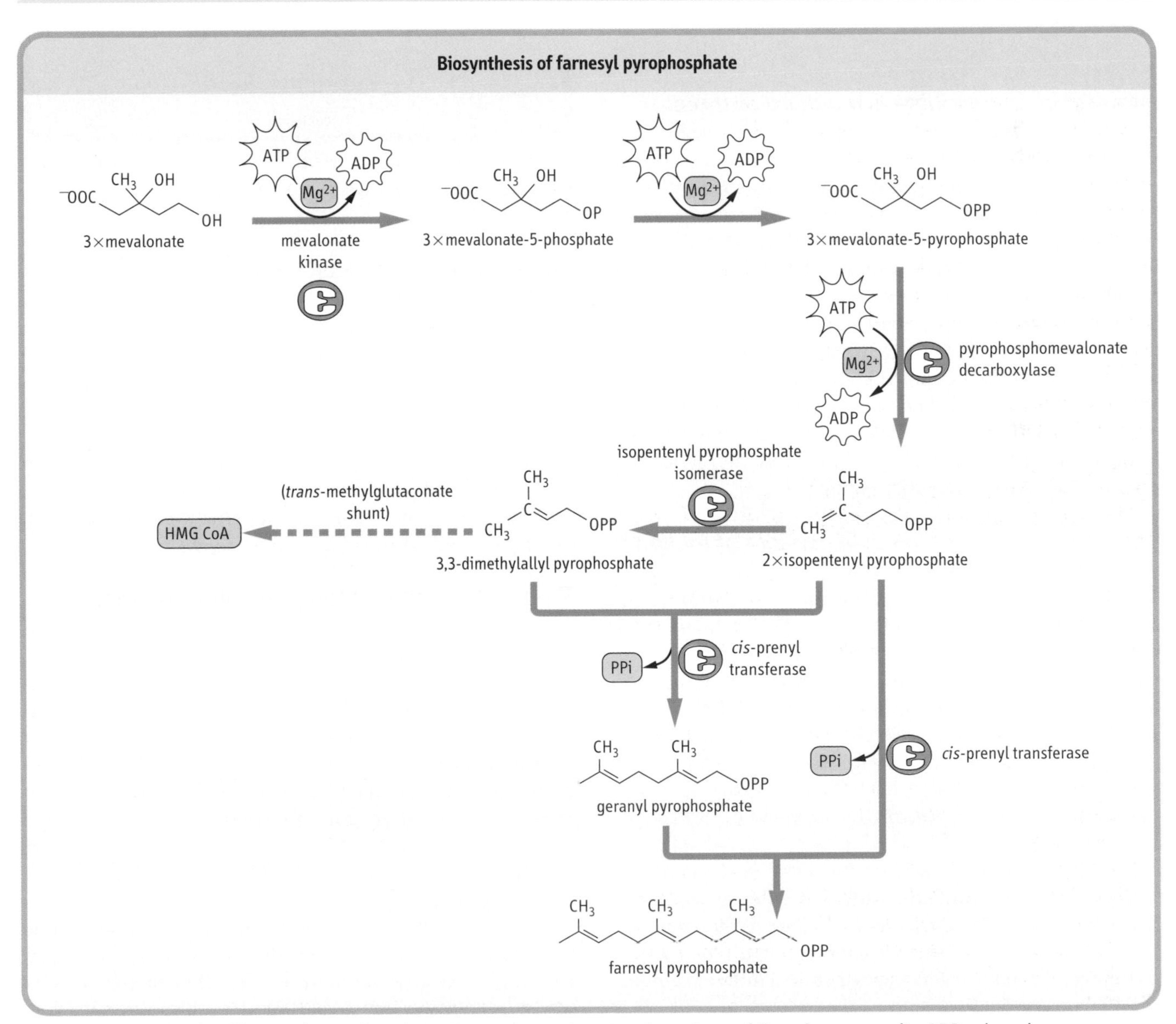

Fig. 16.3 Biosynthesis of farnesyl pyrophosphate. Farnesyl pyrophosphate is made up of three isoprene units. ADP, adenosine diphosphate; Mg^{2+}, magnesium; PPi, pyrophosphate.

The transmethylglutaconate shunt – a secondary point of control

It was once believed that the production of mevalonic acid led to the inevitable formation of farnesyl pyrophosphate. However, it is now known that dimethylallyl pyrophosphate, one of the isoprene units formed from mevalonate (see Fig 16.3), can be dephosphorylated and broken down back into acetoacetate and acetyl CoA, which may then be diverted into other metabolic pathways such as fatty acid biosynthesis. Thus, high-energy compounds once destined to be converted into cholesterol may be redeployed for a higher priority need. This mechanism is known as the transmethylglutaconate shunt.

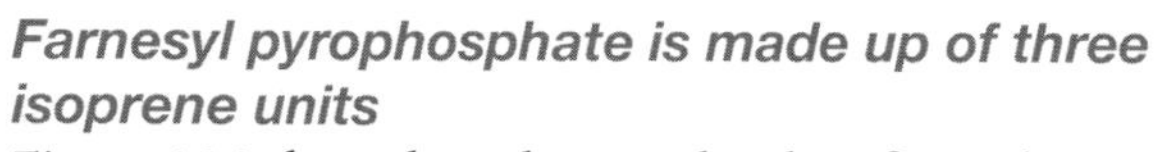

Farnesyl pyrophosphate is made up of three isoprene units

Figure 16.3 shows how three molecules of mevalonic acid are each decarboxylated into five-carbon atom isoprene units, which are condensed sequentially to produce the 15-carbon atom molecule farnesyl pyrophosphate. The first two reactions require kinase enzymes and ATP to generate the pyrophosphate moiety. Decarboxylation results in the isomeric isoprene units, isopentenyl pyrophosphate and dimethylallyl pyrophosphate, which condense together to form geranyl pyrophosphate. A further condensation with isopentenyl pyrophosphate produces farnesyl pyrophosphate. As well as being an intermediate in cholesterol biosynthesis, farnesyl pyrophosphate is the branching point for the synthesis of dolichol and ubiquinone.

Squalene is a linear molecule capable of folding into a ring formation

Squalene synthase is a complex enzyme present in the endoplasmic reticulum that facilitates the condensation at the pyrophosphate end of two molecules of farnesyl pyrophosphate. Several intermediates are involved but the resulting product is squalene, a 30-carbon atom hydrocarbon containing six double bonds which enable it to fold into a ring similar to the steroid nucleus (Fig. 16.4).

Lanosterol is the product of squalene cyclization in mammals

Before ring closure, squalene is converted to squalene 2,3-oxide by a mixed function oxidase in the endoplasmic reticulum. Thereafter, cyclization occurs under the action of the enzyme oxidosqualene: lanosterol cyclase (Fig. 16.5). It is interesting to note that in plants there is a different product of squalene cyclization, known as cycloartenol, which is further metabolized to a range of phytosterols, including β-sitosterol, rather than to cholesterol.

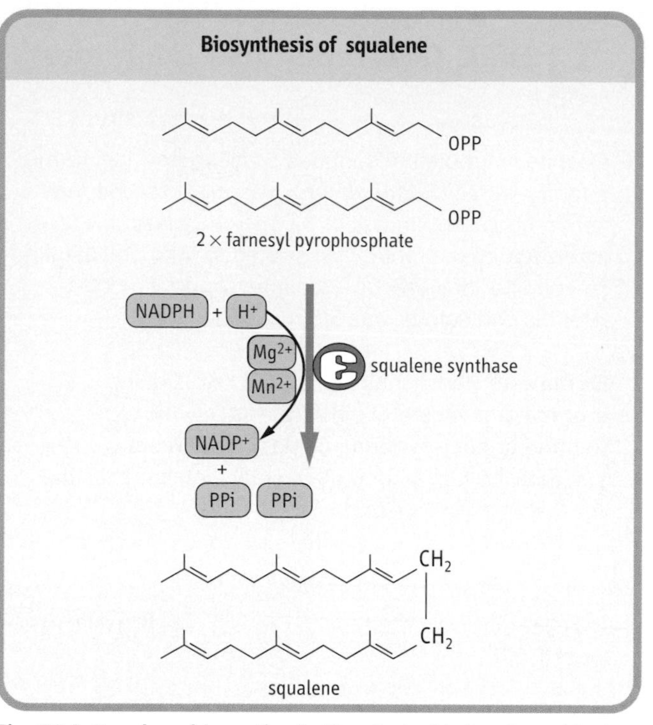

Fig. 16.4 Squalene biosynthesis. The six double bonds enable the structure to fold into a ring similar to the steroid nucleus. Mn^{2+}, manganese ion.

The final stages of cholesterol biosynthesis occur on a carrier protein

Squalene, lanosterol and all the intermediates to cholesterol are hydrophobic molecules. In order for the final steps of the pathway to occur in an aqueous medium, the intermediates react while bound to a squalene and sterol-binding protein. The exact sequence of the conversion from the 30-carbon lanosterol into the 27-carbon cholesterol is not well understood, although it involves three decarboxylation reactions, an isomerization and a reduction (see Fig. 16.5). NADPH is consumed in four of these reactions.

Regulation of cholesterol biosynthesis

There are many factors that contribute to the sophisticated process of regulation of cholesterol biosynthesis (Fig. 16.6). Under normal circumstances there is an inverse relationship between dietary cholesterol intake and cholesterol biosynthesis. This relationship ensures a relatively constant daily supply of cholesterol but it explains why dietary restriction is only likely to achieve a 15% reduction in circulating cholesterol concentrations.

Intracellular free cholesterol is derived from several pools

Immediately after absorption from the gut, cholesterol is transported to the liver in the form of chylomicron remnants. At the same time the cholesterol pool is transported in plasma as part of several lipoproteins (see Chapter 17). For the purposes of considering the regulation of cholesterol biosynthesis, assumptions will be made that

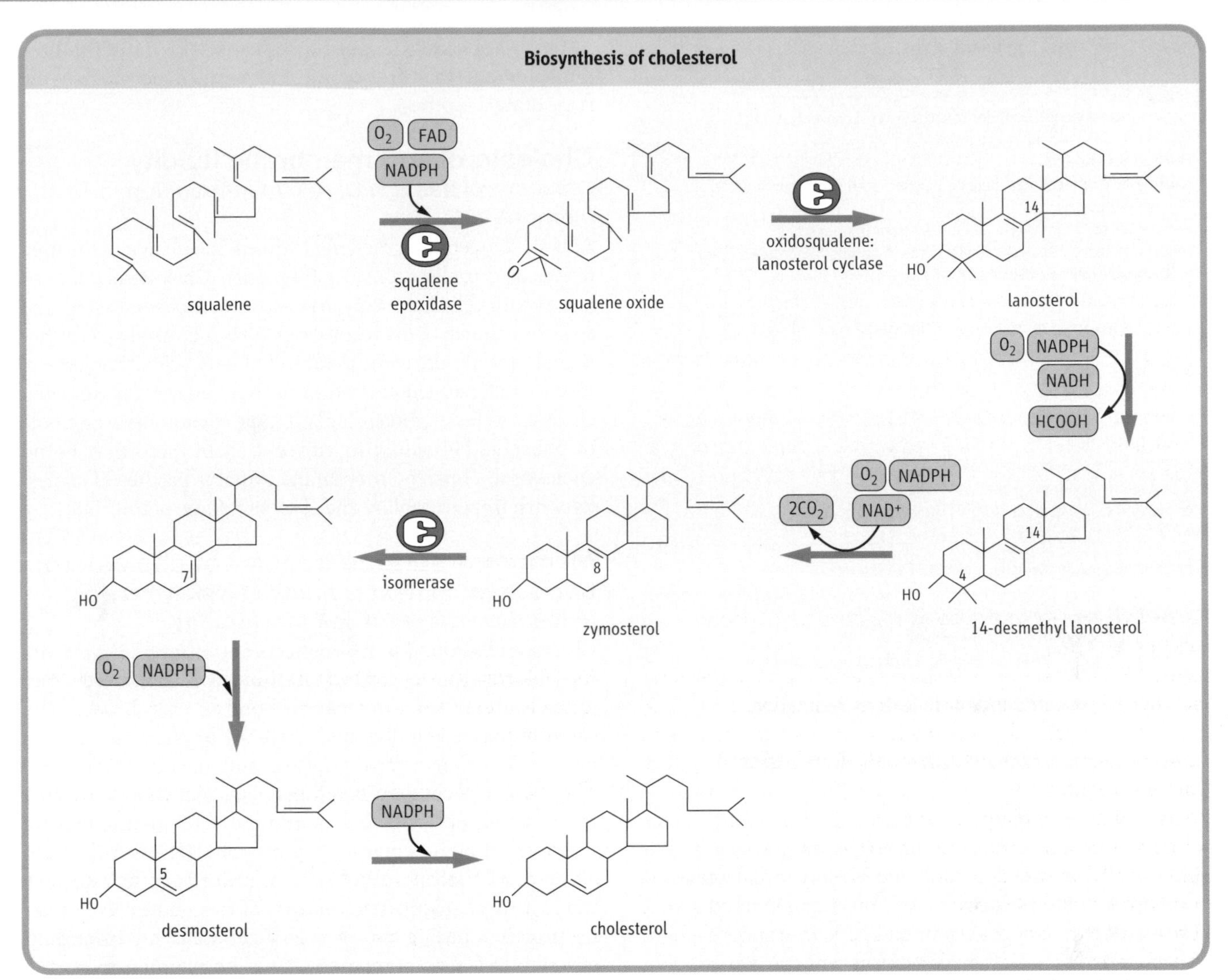

Fig. 16.5 Cholesterol biosynthesis. These reactions occur while bound to a squalene- and sterol-binding protein. FAD, flavine adenine dinucleotide; NADH, reduced nicotinamide adenine dinucleotide.

this cholesterol is present, mainly as cholesterol ester, linked to apolipoprotein B100 within low-density lipoprotein (LDL) and that LDL cholesterol is a stable marker both of the absorption and further metabolism of dietary cholesterol and also of *de novo* synthesis.

The free cholesterol within the hepatocyte may be derived from several pools:

- it is the product of biosynthesis within the cell;
- it may be liberated from stored intracellular cholesterol ester by hydrolysis;
- it may be derived from the chylomicron remnants that bind to the LDL receptor-related protein (LRP) and so permit entry of cholesterol and cholesterol ester;
- it may enter via the LDL receptor-mediated route. LDL binds to its specific receptor on the hepatocyte surface and is internalized in the form of clathrin-coated vesicles. These vesicles are modified prior to being acted upon by lysosomal enzymes, which separate the LDL from its receptor, remove cholesterol from its apolipoprotein and hydrolyze cholesterol esters into free cholesterol;
- some free cholesterol may enter the cell directly from extracellular LDL.

HMG CoA reductase is the key regulatory enzyme

A significant rise in the intracellular free cholesterol concentration brings about regulation in four ways:

- it causes a reduction both in the activity and synthesis of HMG CoA reductase, thus limiting further cholesterol synthesis. The exact mechanism by which this is achieved remains unclear but may involve intracellular oxygenation of cholesterol to more potent enzyme inhibitors;
- it results in downregulation of LDL receptors to limit the LDL receptor-mediated pathway;
- the intracellular esterification of free cholesterol into cholesterol ester is increased;
- cholesterol crosses the plasma membrane and binds to HDL, promoted by lecithin:cholesterol acyl transferase (see Fig. 16.6).

Factors influencing cholesterol regulation
Factors increasing intracellular free cholesterol concentration
de novo biosynthesis hydrolysis of intracellular cholesterol esters by the enzyme cholesterol ester hydrolase dietary intake of cholesterol and uptake from chylomicrons receptor-mediated uptake of cholesterol-containing lipoproteins (LDL) direct uptake of free cholesterol from lipoproteins
Factors decreasing intracellular free cholesterol concentration
inhibition of cholesterol biosynthesis downregulation of the LDL receptor intracellular esterification of cholesterol by acyl Coenzyme A:cholesterol acyl transferase release of cholesterol to lipoproteins (HDL) promoted by lecithin: cholesterol acyl transferase conversion to bile acids or steroid hormones
Factors influencing the activity of HMG CoA reductase
intracellular concentration of HMG CoA intracellular concentration of cholesterol hormones: insulins, tri-iodothyronine (+), glucagon, cortisol (–)

Fig. 16.6 Factors influencing cholesterol regulation.

There is a circadian rhythm of cholesterol biosynthesis

Hepatic synthesis of cholesterol is at a peak at about 6 hours after dark and at a minimum some 6 hours after exposure to light. This rhythm is the result of corresponding changes in HMG CoA reductase activity. The mechanism of control of HMG CoA reductase activity in these circumstances is poorly understood, although dietary patterns will play a part.

Hormones also regulate HMG CoA reductase activity

Several hormones affect the activity of HMG CoA reductase and so influence cholesterol biosynthesis. Insulin and tri-iodothyronine increase HMG CoA reductase activity, while glucagon and cortisol have the opposite effect.

Role of cholesterol

Cholesterol is an essential component of cell membranes

Mammalian cell membranes are composed of lipid and protein with small amounts of carbohydrate. The basic structural characteristics of membranes are described in Chapter 7 but, in outline, the protein components of membranes are embedded in and may span a hydrophobic phospholipid bilayer. It is now recognized that membranes are very fluid structures in which both the lipid and protein molecules may move or undergo conformational change so as to allow the specific passage of certain molecules while maintaining a generally impermeable barrier between the intra- and extracellular aqueous phases. The more fluid the phospholipid bilayer becomes, the more permeable the membrane becomes.

Cholesterol and membrane fluidity

Cholesterol has a vital role in influencing membrane fluidity

At body temperature, the long hydrocarbon chains of the lipid bilayer are capable of considerable motion. Cholesterol is located between these hydrocarbon chains to form a loose crosslink and so reduce fluidity. This relative rigidity is increased still further if cholesterol is adjacent to saturated fatty acids. Cholesterol forms clustered regions within the lipid bilayer. In areas of a cholesterol cluster, there may be 1 mole of cholesterol per mole of phospholipid, while in adjacent areas there may be no cholesterol. Thus, the membrane contains patches of cholesterol-rich impermeability and cholesterol-free permeability.

Membrane surface transfer of cholesterol

Cholesterol content is found in widely varying amounts in different cell membranes

Cholesterol is found in the highest concentrations in plasma membranes (up to 25% of the lipid content), while it is virtually absent from inner mitochondrial membranes. The cholesterol is held in the lipid bilayer by physical interactions between the planar steroid nucleus and the fatty acid chains. The absence of covalent bonding means that cholesterol may transfer in and out of the membrane. Outside the plasma membrane high-density lipoprotein (HDL) is the main acceptor of cholesterol released from cells by this surface transfer mechanism. The removal of free cholesterol either by protein binding or by esterification is an important stimulus to further transmembrane cholesterol transport.

Bile acids

Structure of the bile acids

Quantitatively the most important metabolic products of cholesterol are the bile acids

In man there are four main bile acids (Fig. 16.7). These bile acids all have 24 carbon atoms with the terminal three carbon atoms of the cholesterol side chain removed during synthesis. They all have a saturated steroid nucleus and differ only in the number and position of the additional hydroxyl groups. It is worthy of note that all these hydroxyl groups have the α-configuration (below the plane of the nucleus) and this means that isomerization of the 3β-hydroxyl group of cholesterol must occur.

Biosynthesis and secretion of the bile acids

Biosynthesis of bile acids occurs within the liver parenchymal cells

Biosynthesis occurs within the liver parenchymal cells to produce cholic and chenodeoxycholic acids. The rate-limiting step in the biosynthesis is the microsomal 7-hydroxylase enzyme,

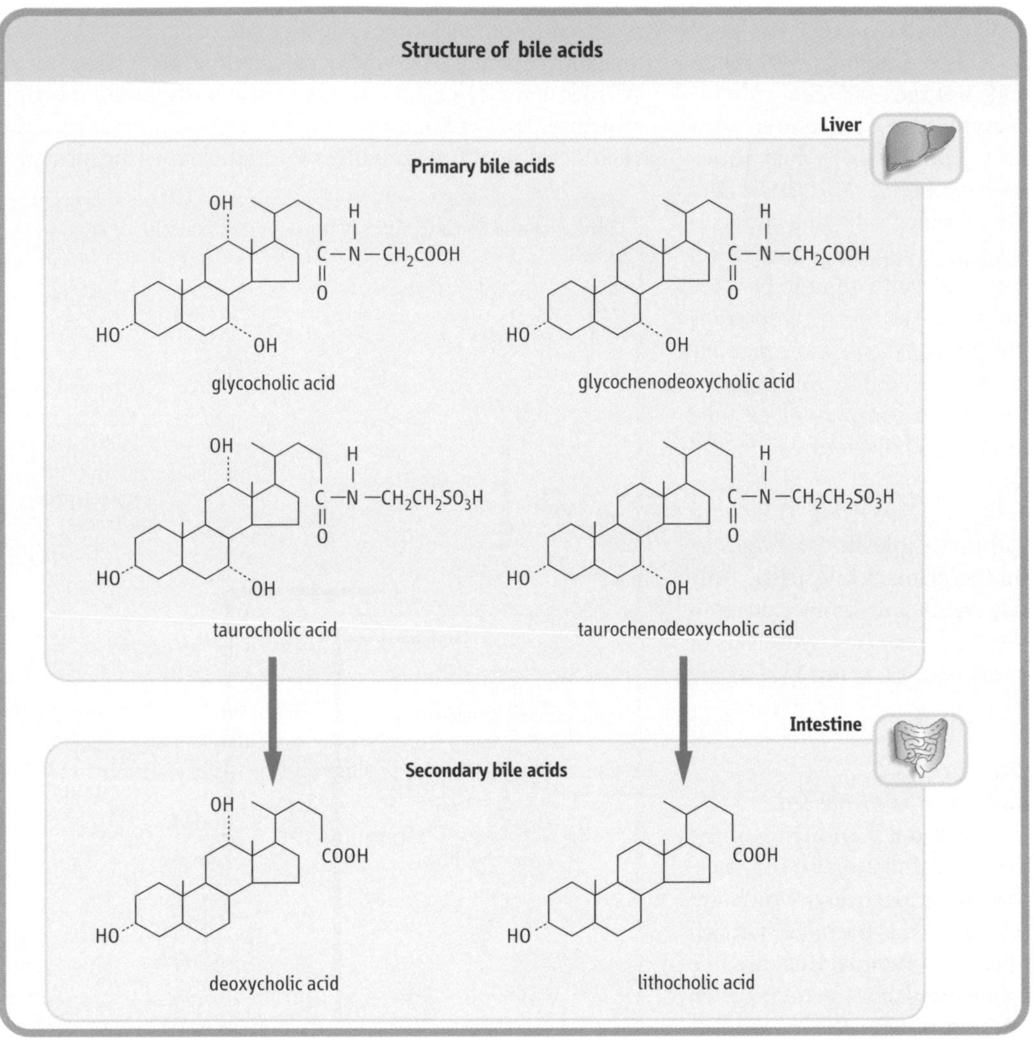

Fig. 16.7 Structure of the bile acids. Primary bile acids are synthesized in the liver and secondary bile acids in the intestine.

Gall stones

A 45-year-old woman complains of acute abdominal pain and occasional vomiting following fatty food. The only biochemical abnormality is a modestly raised alkaline phosphatase at 400 U/L (<280 U/L). A cholecystogram demonstrates the presence of gall stones.

Comment: Gall stones occur in up to 20% of the population of Western countries. The condition is characterized by severe abdominal pain resulting from the formation of cholesterol-rich stones within the gall bladder. Cholesterol is present in high concentrations in bile, being 'solubilized' in micelles that also contain phospholipids and bile acids. When the liver secretes bile with a cholesterol to phospholipid ratio greater than 1:1 it is difficult to solubilize all the cholesterol in micelles, thus there is a tendency for the excess to crystallize around any insoluble nuclei. This is compounded by further concentration of the bile in the gall bladder by reabsorption of water and electrolytes. The condition may be managed conservatively by reducing the dietary cholesterol and by increasing the availability of bile acids that will assist with cholesterol solubilization in the bile and excretion via the gut. Alternative treatment includes disintegration of stones by shock waves (lithotripsy) and surgery. In this case the elevated alkaline phosphatase is a marker of partial blockage of the bile duct (cholestasis).

which requires oxygen, NADPH and cytochrome P450. Prior to secretion these primary bile acids are conjugated through the carboxyl group forming amide linkages with either glycine or taurine (see Fig. 16.7). In man there is a 3:1 ratio in favor of glycine conjugates. The secreted products are thus principally glycocholic, glycochenodeoxycholic, taurocholic and taurochenodeoxycholic acids. At physiologic pH, the bile acids are mainly ionized and so they occur as sodium or potassium salts. The terms 'bile acids' and 'bile salts' tend to be used interchangeably. As their name suggests, these compounds are secreted from the liver via the bile canaliculi and larger bile ducts, either directly into the duodenum or for storage in the gall bladder. They are an important component of bile, together with water, phospholipids, cholesterol, salts and excretory products such as bilirubin.

Deoxycholic and lithocholic acids (see Fig. 16.7) are secondary bile acids formed within the intestine through the action of bacterial enzymes on the primary bile acids. Only a proportion of primary bile acids are converted into secondary bile acids, a process which requires hydrolysis of the amide link to glycine or taurine prior to removal of the 7α-hydroxyl group.

Functions of bile acids

Bile acids assist the digestion of dietary fat

The secretion of bile from the liver and the emptying of the bile duct are controlled by the gastrointestinal hormones, hepatocrinin and cholecystokinin, respectively, which are released when partially digested food passes from the stomach to the duodenum. Once secreted into the intestine, the bile acids act as detergents assisting the emulsification of ingested lipids into very small globules; this aids the enzymatic digestion and absorption of dietary fat (see Chapter 9).

Bile acids are recirculated via the enterohepatic circulation

Up to 30 g of bile acids pass from the bile duct into the intestine each day but only 2% of this (approximately 0.5 g) is lost through the feces. Most is deconjugated and reabsorbed. Passive reabsorption of bile acids occurs in the jejunum and colon but the majority is reabsorbed in the ileum by a process of active transport. The reabsorbed bile acids are transported in blood via the portal vein, noncovalently bound to albumin, and resecreted into the bile. This is the enterohepatic circulation and it explains why bile contains both primary and secondary bile acids.

Steroid hormones

Structure of the steroid hormones

Systematic nomenclature is used to describe the structure of steroid hormones

Cholesterol is the precursor of all of the steroid hormones. Mammals produce many steroid hormones, some of which differ only by a double bond or by the orientation of a hydroxyl group. Consequently, it has been necessary to employ systematic nomenclature to detail exact structures. There are three broad groups of steroid hormones (Fig. 16.8). The corticosteroids have 21 carbon atoms in the basic pregnane ring structure. Loss of the remaining two carbon atoms from the cholesterol side chain produces the androstane ring structure

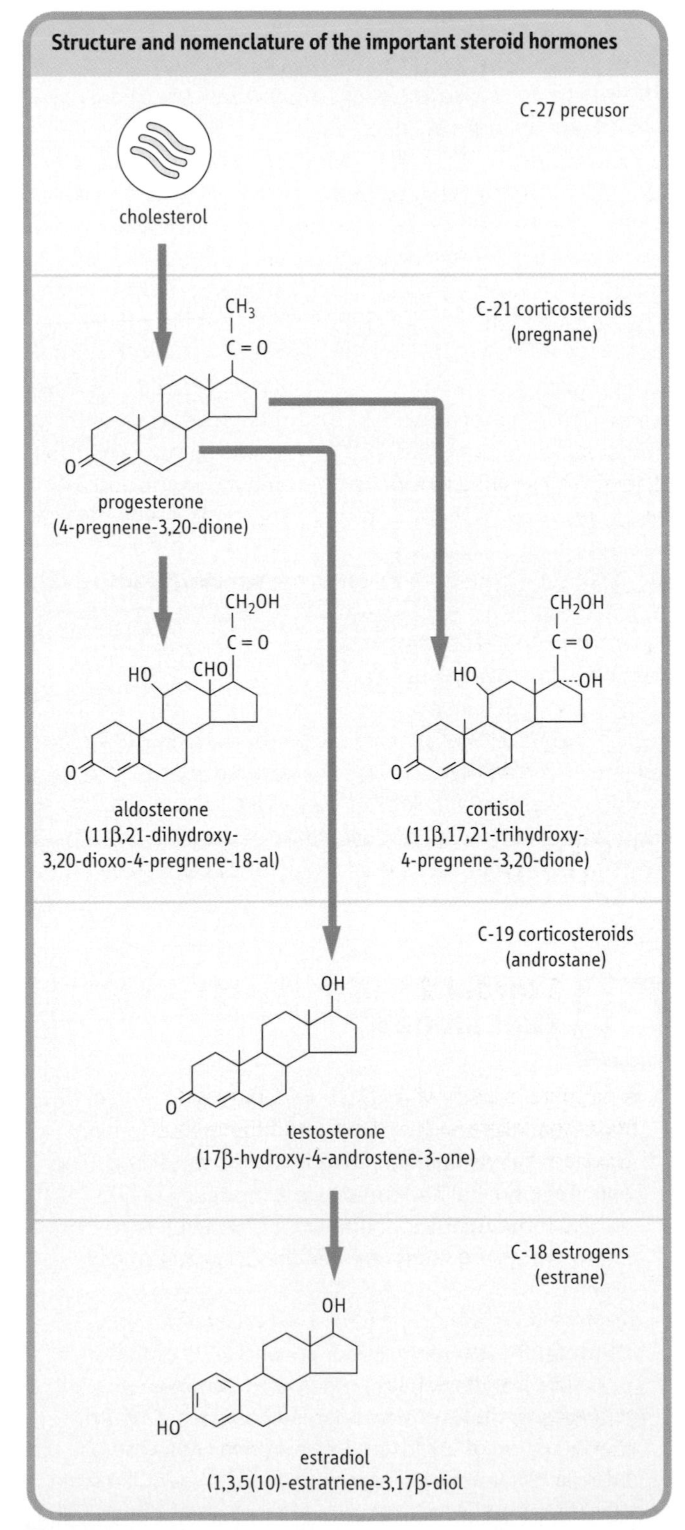

Fig. 16.8 The structures of the most important human steroid hormones. Their trivial and systematic (in parentheses) names are shown.

and the group of hormones known as the androgens. Finally, loss of the angular methyl group at carbon atom 19 as part of the aromatization of the A ring results in the estrane structure found in the estrogens. The presence and position of double bonds and the position and orientation of hydroxyl or other functional groups on the basic nucleus may then be described.

Biosynthesis of the steroid hormones

Conversion of cholesterol into steroid hormones occurs in only three organs

The biosynthesis of the steroid hormones occurs within the adrenal cortex, the testis in man and the ovary in woman. It is normal practice to consider the corticosteroids as the products of the adrenal cortex, the androgens as the products of the testis and the estrogens as the products of the ovary. A simplified pathway to the steroids depicted in Figure 16.8 is shown in Chapter 35. It is the relative abundance and activity of the steroidogenic enzymes in each of the three organs that determines the major secreted product. However, this is not absolute and all three organs are capable of secreting small amounts of steroids from a group other than their major product. In pathologic situations, such as a defect in steroidogenesis or a steroid-secreting tumor, a very abnormal pattern of steroid secretion may be observed.

Cytochrome P450 mono-oxygenase enzymes control steroidogenesis

Most of the enzymes involved in converting cholesterol into steroid hormones are cytochrome P450 proteins that require oxygen and NADPH. In its simplest form, this enzyme complex catalyzes the replacement of a carbon–hydrogen bond with a carbon–hydroxyl bond; hence, the collective term mono-oxygenase. Hydroxylation of adjacent carbon atoms is the forerunner to cleavage of the carbon–carbon bond. Comparison of the structure of cholesterol (see Fig. 16.1) with those of the steroid hormones (see Fig. 16.8) demonstrates that the biosynthetic pathway is largely made up of cleavage of carbon–carbon bonds and hydroxylation reactions. These enzymes have their own nomenclature in which the symbol P450 is followed by a specific suffix. Thus, P450scc refers to the side-chain cleavage enzyme and P450c18 to the enzyme that hydroxylates carbon atom 18. A list of relevant enzymes can be found in Chapter 35.

Biosynthesis of the corticosteroids

The cellular substructure of the adrenal cortex is arranged in three different layers. The inner two layers (zona fasciculata and zona reticularis) are responsible for the synthesis of cortisol, the main glucocorticoid, and the adrenal androgens. The outer layer (zona glomerulosa) is responsible for the synthesis of aldosterone, the main mineralocorticoid. Although many of the steps are similar they are controled by very different mechanisms and this has led to the suggestion that the adrenal cortex may be considered as two separate endocrine organs. It should be emphasized that steroidogenesis is a complex process regulated by several hormonal and nonhormonal factors.

The biosynthesis of cortisol depends on stimulation by pituitary adrenocorticotropic hormone (ACTH) which binds to its plasma membrane receptor and triggers a range of intracellular events which cause hydrolysis of cholesterol esters stored in lipid droplets and activation of the P450scc cholesterol 20,22-desmolase enzyme. This is the rate-limiting step of steroidogenesis which converts C-27 cholesterol into pregnenolone, the first of the C-21 pregnane family of corticosteroids. Thereafter, conversion to cortisol requires a dehydrogenation–isomerization and three sequential

Steroid 21-hydroxylase deficiency

A neonate is born with ambiguous genitalia. Within 48 hours the infant is dehydrated and distressed. Biochemical investigation reveals:

Na^+ 115 mmol/L (135–145 mmol/L)
K^+ 7.0 mmol/L (3.5–5.0 mmol/L)
17–hydroxyprogesterone 550 nmol/L (<50 nmol/L)

Comment. This baby has the most severe form of steroid 21-hydroxylase deficiency, the commonest of a range of conditions, known as congenital adrenal hyperplasia, that are characterized by defects in the activity of one of the enzymes in the steroidogenic pathway. As the name implies, all these conditions have a genetic basis, which leads to a failure to produce cortisol (+/– aldosterone). As a result there is reduced negative feedback on the pituitary production of ACTH, which continues to stimulate the adrenal gland to produce steroids upstream of the enzyme block. These include 17-hydroxyprogesterone, which is further metabolized to testosterone (see Fig. 16.8) resulting in androgenization of a female neonate. The electrolyte abnormality (salt loss) is the result of mineralocorticoid deficiency and requires urgent treatment with fluids and steroids. Long-term maintenance therapy with hydrocortisone and a mineralocorticoid will suppress ACTH and androgen production.

A less severe form of this condition (partial enzyme deficiency) occurs in young women who present with menstrual irregularity and hirsutism, both consequences of adrenal androgen excess.

hydroxylation reactions at C-17, -21 and -11 under the control of cytochrome P450 enzymes. Control of the rate of cortisol biosynthesis is achieved by negative feedback by cortisol on the secretion of ACTH.

The main stimulus to the synthesis of aldosterone is not ACTH but angiotensin II produced as a result of the enzymes of the renin–angiotensin system (see Chapter 21). Potassium is an important secondary stimulus. Angiotensin II, by binding to its receptor, and potassium work cooperatively to activate the same first step in the pathway by which cholesterol is converted into pregnenolone. The zona glomerulosa lacks the enzyme P450c17 17α-hydroxlase but has abundant amounts of P450c18 18-hydroxylase which is the first of a two-stage reaction, which results in the 18-aldehyde group found in aldosterone.

Biosynthesis of the androgens

The conversion of corticosteroids into androgens requires the P450c17 17–20 lyase/desmolase enzyme and a substrate that contains a 17α-hydroxyl group. This breaks the C-17–C-20 bond to yield the androstane ring structure. This enzyme is abundant in the Leydig cells of the testis and in the granulosa cells of the ovary. In these cases, however, the stimuli to the rate-limiting cholesterol side-chain cleavage step are luteinizing hormone (LH) and follicle stimulating hormone (FSH), respectively. Thus the same biosynthetic step, controled by almost identical enzymes, is controled by two different hormones in two different tissues. This is an excellent example of specificity arising from hormone–cell receptor interaction.

Biosynthesis of the estrogens

The conversion of androgens into estrogens involves removal of the angular methyl group at C-19 under the action of P450c19 19-aromatase. The A ring undergoes two dehydrogenations as part of the reaction, and the characteristic 1,3,5(10)-estratriene nucleus results. This aromatase enzyme is found most abundantly in the granulosa cells of the ovary although it is worthy of note that an enzyme in adipose tissue can also convert some testosterone into estradiol.

Function and elimination of the steroid hormones

Steroid hormones act via nuclear receptors

All the steroid hormones act by binding to nuclear receptors and stimulating transcription which results in specific protein synthesis (see Chapter 36). All the receptors belong to a superfamily of hormone receptors with specificity being achieved in a small hormone-binding domain. The biological actions of the steroid hormones are diverse and are best considered as part of the trophic hormone system to which they belong. This system is described in Chapter 35.

Steroid hormones are excreted in the urine

Most steroid hormones are excreted via the kidney. In general there are two main steps in this process. Firstly, the biological potency of the steroid must be removed and this is achieved by a series of reduction reactions. Secondly, the lipid steroid molecule must be rendered water-soluble by conjugation to a glucuronide or sulfate moiety, usually through the hydroxyl group at C-3. These steps occur predominantly within the liver. As a result there are many different steroid hormone conjugates in urine, some of them present in high concentrations. Urinary steroid profiling by gas chromatography–mass spectrometry typically identifies more than 30 such steroids and the relative concentrations of these may be used to pinpoint specific defects in the steroidogenic pathway.

Vitamin D_3

Intake/manufacture of vitamin D_3

Vitamin D_3 is also known as cholecalciferol and, as this name implies, it is derived from cholesterol. Small amounts of the fat-soluble vitamins D occur in food, (e.g. fish liver oil, egg yolk), but the majority of cholecalciferol is manufactured in the Malpighian layer of the epidermis of the skin. Cholesterol is converted to 7-dehydrocholesterol, which acts as the substrate for a unique nonenzymatic photolysis reaction in which ultraviolet rays from sunlight mediate the opening of the B-ring of cholesterol (Fig. 16.9) so destroying the steroid nucleus. This reaction is inversely related to the amount of pigment in the skin and directly related to the amount of sunlight exposure.

Transport and conversion of vitamin D_3

Vitamin D is converted from vitamin to hormone within a single family of compounds

Cholecalciferol is transported in plasma bound to a specific vitamin D-binding protein. In the liver it undergoes hydroxylation at C-25 to produce 25-hydroxycholecalciferol. This step, which is not regulated, is catalyzed by a cytochrome P450 mono-oxygenase in the endoplasmic reticulum. 25-hydroxycholecalciferol is then transported in plasma to the kidney, bound to the same specific protein. Production of the potent hormone, 1,25-dihydroxycholecalciferol (also known as calcitriol), requires the contribution of a 1α-hydroxylase enzyme, which is located within the mitochondria of the renal proximal convoluted tubule. This is a complex cytochrome P450 enzyme system which controls the rate-limiting step in hormone production, yet another mono-oxygenase reaction (Fig. 16.9).

1,25-Dihydroxycholecalciferol

1,25-Dihydroxycholecalciferol is an important calciotropic hormone

1,25-dihydroxycholecalciferol acts in the same way as the steroid hormones. It diffuses into its target cell and binds to a nuclear receptor that is a member of the steroid superfamily of hormone receptors (see Chapter 35). The main target cell is in the intestinal mucosa where hormone binding results in transcription of a range of proteins, including calcium-binding protein (CBP), which is rich in amino acids containing carboxyl side chains, allowing it to bind calcium

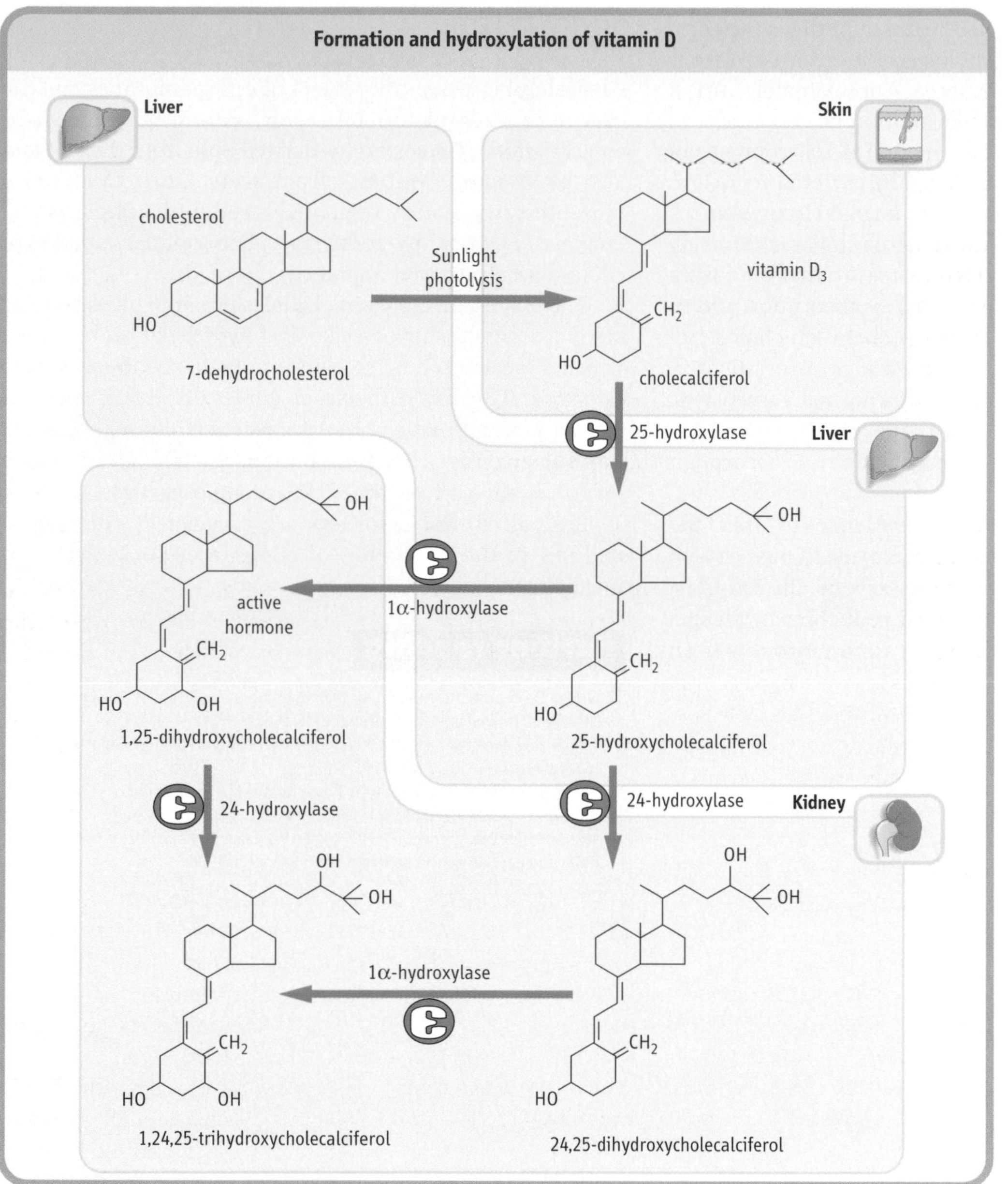

Fig. 16.9 Formation of vitamin D_3 (cholecalciferol).

ions electrostatically. As a result of a complex series of actions, of which CBP is only a part, there is an increased transfer of calcium and phosphate ions from the intestinal lumen, across the mucosal cell and into the circulation.

Homeostasis is achieved by regulation of the renal 1α-hydroxylase enzyme

Like the steroid hormones 1,25-dihydroxycholecalciferol is subject to tight feedback regulation. Low-calcium diets and hypocalcaemia result in marked increases in 1α-hydroxylase activity. This effect requires parathyroid hormone (PTH), which is also released in response to hypocalcemia. Low phosphorus diets and hypophosphatemia also induce 1α-hydroxylase activity but these provide a weaker stimulus than hypocalcemia. 1,25-dihydroxycholecalciferol is an important regulator of its own production. High concentrations inhibit renal 1α-hydroxylase and stimulate the formation of a biologically inactive 24,25-dihydroxycholecalciferol.

Cholesterol metabolism and excretion

Cholesterol cannot be metabolized by mammalian cells

Somewhat surprisingly, cholesterol cannot be digested in the gut or broken down by mammalian cells into carbon dioxide and water. Removal from the body is thus dependent on transfer into the gut prior to excretion via the feces. As the section on bile acids indicated there is a considerable flux of cholesterol, either directly or as bile acid metabolites, from the liver into bile and then into the duodenum via the common bile duct. About 1g of cholesterol is eliminated from the body each day through the feces. Approximately 50% is excreted after conversion to bile acids. The remainder is excreted as the isomeric saturated neutral sterols coprostanol (5β-) and cholestanol (5α-) produced by bacterial reduction of the cholesterol molecule. Under normal circumstances the sophisticated pattern of regulation of vcholesterol metabolism (see Fig. 16.6) is controled to ensure that the losses through fecal excretion are balanced by absorption and *de novo* synthesis. Given the relationship between cholesterol excess (particularly in plasma) and coronary heart disease, one therapeutic option is to disturb this normal homeostatic relationship in order to ensure that more cholesterol is excreted than is produced by the combination of absorption and synthesis.

There is now accumulating evidence to link the prevalence of atherosclerosis and coronary heart disease with hypercholesterolemia. Furthermore, several clinical trials have demonstrated that a sustained reduction in plasma cholesterol leads to reduced mortality and morbidity from coronary heart disease (see Chapter 17).

Summary

Cholesterol is a vital constituent of cell membranes and the precursor molecule for bile acids, steroid hormones and cholecalciferol. Cholesterol is derived both from the diet and also by *de novo*. synthesis from acetyl CoA. Cholesterol biosynthesis is strictly regulated through the rate limiting enzyme, HMG CoA reductase. The cellular uptake of cholesterol is also well regulated.

The further metabolism of cholesterol into bile acids and steroid hormones involves several hydroxylation reactions which are catalyzed by cytochrome P450 mono-oxygenase enzymes. The biosynthesis of cholecalciferol involves a unique nonenzymatic photolysis reaction although the rate limiting enzyme is the formation of the biologically active form is again a cytochrome P450 mono-oxygenase.

Several clinical disorders are associated with abnormalities in the regulation of cholesterol homeostasis or metabolism.

Further reading

Betteridge DJ, Khan M. Review of new guidelines for management of dyslipidaemia. *Baillière's Clin Endocrinol and Metab* 1995; **9**: 867–890.

Griffin BA. Low-density lipoprotein heterogeneity. *Baillière's Clinical Endocrinology and Metabolism* 1995; **9**: 687–704.

Kirchmair R, Ebenbichler CF, Patsch JR. Post-prandial lipaemia. *Baillière's Clinical Endocrinology and Metabolism* 1995; **9**: 705–720.

Wilson JD, Foster DW, Kronenburg HM, Larsen PR (Eds). *Williams Textbook of Clinical Endocrinology*, 9th edn. Philadelphia: WB Saunders, 1998.

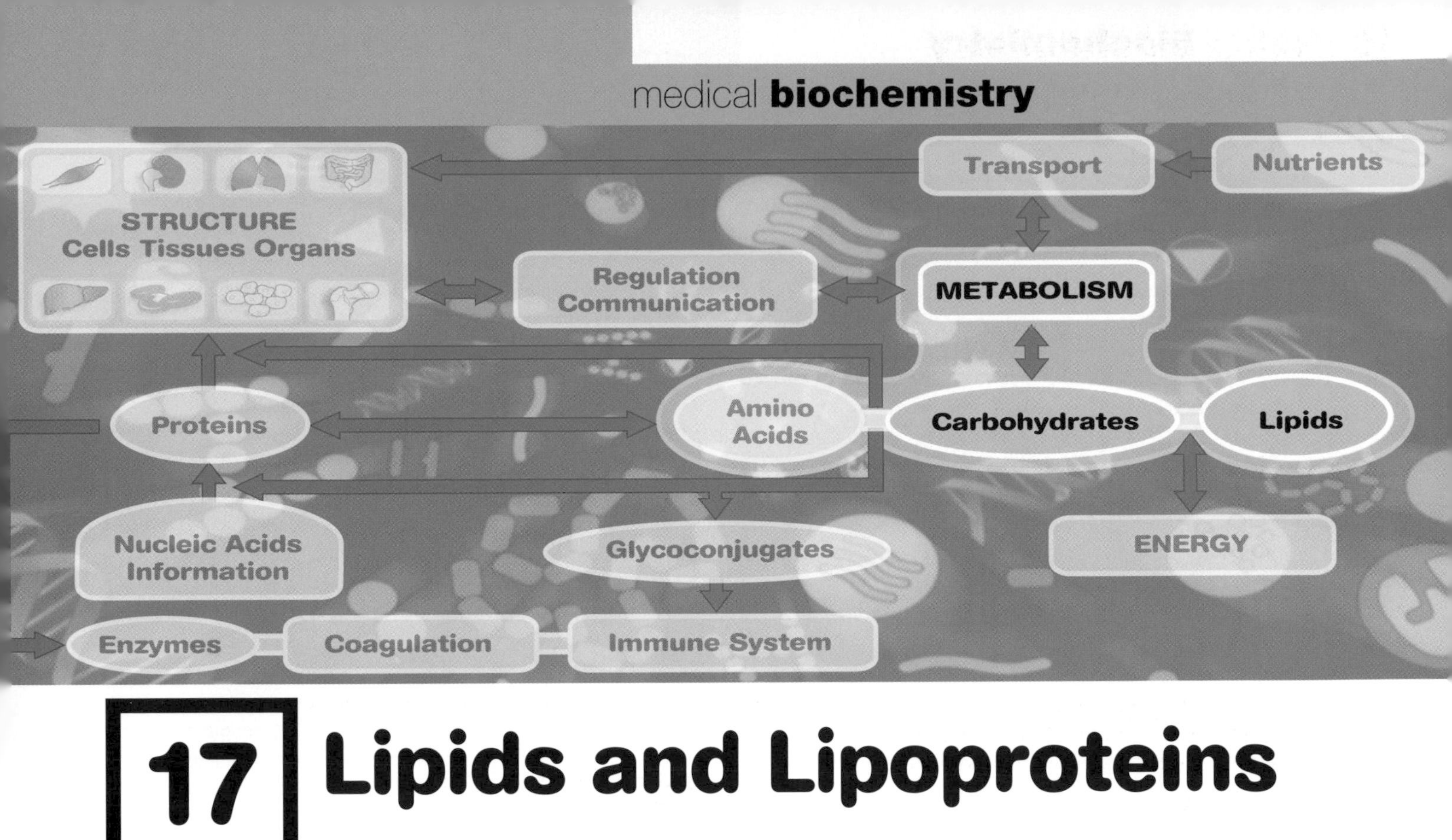

17 Lipids and Lipoproteins

Lipoproteins

This chapter discusses the metabolism of lipids and lipoproteins. Lipoproteins are important because they provide means for fat transport between different organs and tissues. Their clinical importance is the role they play in the development of atherosclerosis: a phenomenon that underlies a range of diseases of the cardiovascular system, such as coronary heart disease and stroke. These diseases are presently the most important cause of mortality in the Western populations.

The transport of fatty acids and triglycerides between tissues

Lipid metabolism is a part of the body energy metabolism. The lipids directly used for energy production are fatty acids. They are synthesized primarily in the liver and intestine but are stored mainly in adipose tissue as glycerol esters, triglycerides. In the fasting state, fatty acids are released from triglycerides for use in the liver and muscle (see Chapter 20). Also, dietary fat needs to be transported from the intestine to the peripheral tissues and liver. Free fatty acids travel in plasma bound to albumin, but triglycerides are too hydrophobic for such transport. They are therefore packaged, together with cholesterol and proteins (apoproteins), by hepatocytes and enterocytes into particles known as lipoproteins. These lipoproteins are secreted into plasma, where they undergo stepwise enzymatic hydrolysis, releasing fatty acids to tissues. While in the plasma, lipoproteins exchange components; this changes their size and shape. The conformation of constituent apolipoproteins also changes, and once they 'fit' into cellular receptors, the whole lipoprotein particles are taken up by cells (internalized).

Cholesterol: a structural component and a steroid hormone precursor

Cholesterol is an essential constituent of cell membranes. It is also a precursor of bile acids and steroids, including vitamin D. A cell can either synthesize cholesterol or acquire it from its environment. Cholesterol is synthesized from acetyl CoA (see Chapter 16). The rate-limiting step in its synthesis is the reduction of 3-hydroxy-3-methylglutaryl CoA (HMG CoA) to mevalonate by HMG CoA reductase. The activity of HMG CoA reductase is inversely related to the amount of cholesterol within the cell.

Cells can also acquire cholesterol from outside the cell through plasma-membrane low-density-lipoprotein (LDL) receptors, which mediate lipoprotein internalization into the cell. The cell satisfies its need for cholesterol by a balance between synthesis and import (Fig. 17.1).

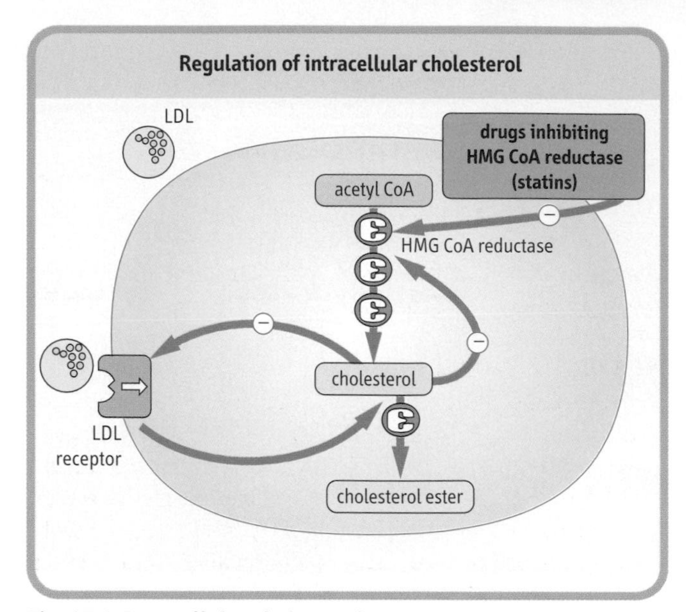

Fig. 17.1 Intracellular cholesterol concentration is precisely regulated. The intracellular cholesterol regulates both the activity of a key enzyme in its synthesis, HMG CoA reductase, and also the number of LDL receptors on the cell membrane. Statins are the drugs that inhibit HMG CoA reductase.

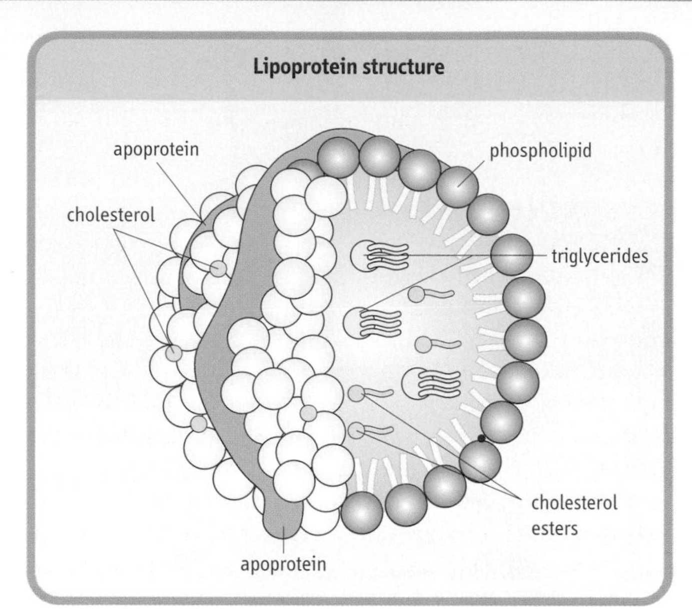

Fig. 17.2 The lipoprotein particle. The external monolayer of a lipoprotein particle contains free cholesterol, phospholipids, and apoproteins. Cholesterol esters and triglycerides are located in the particle core.

Statins: drugs inhibiting cholesterol synthesis

Statins such as simvastatin or pravastatin are competitive inhibitors of HMG CoA reductase, a rate-limiting enzyme in the pathway of cholesterol synthesis. The inhibition of this enzyme results in a decrease in intracellular cholesterol levels. This triggers the regulatory system: an increase in the expression of LDL-receptors on the cell surface. An increase in the number of cellular receptors leads to increased cellular uptake of LDL and, consequently, a lower plasma cholesterol concentration. Statins are now widely used in the treatment of lipid disorders.

Plasma lipoproteins

A lipoprotein particle contains a hydrophilic core of cholesterol esters and triglycerides (Fig. 17.2). Amphipathic phospholipids and free cholesterol, together with proteins (apoproteins) form an outer layer. Some proteins, such as apoprotein B (apoB), are deeply embedded into the particles. Others, such as apoC, are in looser contact on the surface and can be easily exchanged between different particles.

Lipoprotein particles form a continuum of size and density in plasma (Fig. 17.3). They can be classified according to their hydrated density into chylomicrons, very-low-density lipoproteins (VLDL), remnant particles (these include intermediate-density lipoproteins [IDL]), LDL, and high-density lipoproteins (HDL). VLDL and remnant particles are triglyceride-rich, whereas LDL is cholesterol-rich.

With a decreasing triglyceride content, the density of lipoprotein particles increases and the size decreases, from chylomicrons through VLDL, remnant particles, IDL, LDL, to HDL, which are the heaviest.

Lipoproteins can also be characterized by their electrophoretic mobility. On electrophoresis α-lipoproteins (HDL) migrate furthest, followed by pre-β-lipoproteins (VLDL) and β-lipoproteins (LDL). Chylomicrons remain at the origin of the electrophoretic strip. The electrophoretic pattern of lipoprotein separation has been the basis for the, now classic, phenotypic classification of dyslipidemias into five types (Fig. 17.4). One should remember, however, that a lipid disorder can present as different phenotypic patterns (see Fig. 17.11). Recently, lipid disorders have been classified either as defined genetically determined disorders or simply according to plasma levels of cholesterol, triglycerides, and HDL-cholesterol (see below).

Apoproteins (apolipoproteins)

Protein components of lipoprotein particles are called apoproteins. Each type of lipoprotein contains a characteristic set of apoproteins. Apoproteins interact with cellular receptors and therefore determine the metabolic fate of lipoprotein particles. They also serve as activators and inhibitors of enzymes involved in lipoprotein metabolism. The main apoproteins are listed in Figure 17.5. The five most important ones are apoA, apoB, apoC, apoE, and apo(a). Apoproteins A (AI and AII) are present in HDL. Apoprotein B plays a key role in the metabolism of non-HDL lipoproteins: its variant called apoB48 controls the metabolism of chylomicrons, and another, apoB100, the metabolism of VLDL, IDL, and LDL. Apoproteins C act as enzyme activators and inhibitors rather than structural components of lipoproteins. They are loosely associated with the surface of lipoproteins, and exchange between chylomicrons, VLDL, IDL and HDL. ApoE plays an essential role in the metabolism of remnant particles and is important in the nervous system. Lipoprotein (a) [Lp(a)] has a structural similarity to plasminogen, a key protein in the coagulation cascade.

The apoprotein content of lipoprotein classes

Chylomicrons

Chylomicrons transport dietary triglycerides from the intestine to the peripheral tissues. Their main apoprotein is apoB48, which is synthesized in the intestine and is a truncated (N-terminal 48%) form of apoB. Chylomicrons also contain apoproteins A, C, and E.

VLDL, IDL, and LDL

VLDL transport triglycerides from the liver to the periphery. The main apoprotein of VLDL is apoB100, but they also contain apoproteins C and E. As VLDL loses triglycerides, it transforms into IDL, which is either taken up by the liver, or transforms further into LDL. Both transformations involve successive loss of triglycerides and of all the apoproteins except apoB100. The removal of triglycerides decreases the particle size and increases its density. The changes in size and shape of these particles affect the conformation of apoproteins on their surface. This in turn affects their metabolic channeling. For example, in a relatively large VLDL particle,

The lipoprotein classes

Particle	Density (kg/L)	Main component	Apoproteins	Diameter (μm)
chylomicrons	<0.95	TG	B48 (A, C, E)	75–1200
VLDL	0.95–1.006	TG	B100 (A, C, E)	30–80
IDL	1.006–1.019	TG & cholesterol	B100, E	25–35
LDL	1.019–1.063	Cholesterol	B100	18–25
HDL	1.063–1.210	Protein	AI, AII (C, E)	5–12

Fig. 17.3 The characteristics of the main lipoprotein classes. TG, triglyceride; VLDL, very-low-density lipoproteins; IDL, intermediate-density lipoproteins; HDL, high-density lipoproteins. When separated by electrophoresis VLDL are called pre-beta lipoproteins, LDL, beta lipoproteins and HDL alpha lipoproteins.

Phenotypic classification of dyslipidemia

Dyslipidemia type (Fredrickson)	Increased electrophoretic fraction (lipoproteins)	Increased cholesterol	Increased triglyceride
I	chylomicrons	yes	yes
IIa	beta (LDL)	yes	no
IIb	pre-beta & beta (VLDL & LDL)	yes	yes
III	'broad beta' band (IDL)	yes	yes
IV	pre-beta (VLDL)	no	yes
V	pre-beta (VLDL) plus chylomicrons	yes	yes

Fig. 17.4 Classification of dyslipidemia. This is a phenotypic classification of dyslipidemia developed by Fredrickson on the basis of electrophoretic separation of serum lipoproteins. Note that this classification is based only on the lipid patterns observed in plasma. For genetic classification refer to Figure 17.11.

Apoproteins			
Apoprotein	**Structural function**	**Receptor**	**Effect on enzyme activity**
AI	HDL	scavenger receptor B1 (SRB1) putative HDL receptor	LCATactivator
AII	HDL	HDL receptor?	LCAT cofactor
(a)	lp(a)	plasminogen receptor?	probably interferes with fibrinolysis
B48	chylomicrons	LRP	HTGL?
B100	VLDL, IDL, LDL	LDL receptor	–
CI, CII	–	–	LPL activation
CIII	–	–	LPL inhibition
E	remnant particles	LDL receptor	–

Fig. 17.5 The function of apoproteins. Both apoE and apoB bind to LDL receptor, which is sometimes referred to as apoB/E receptor. LCAT, lecithin:cholesterol acyltransferase; LRP, LDL receptor-related protein; HTGL, hepatic triglyceride lipase; LPL, lipoprotein lipase.

apoB100 and apoE are extended on the surface, and remain in a nonreceptor-binding conformation. As VLDL shrinks to transform into a remnant particle (VLDL remnants and IDL), apoE assumes a conformation that allows binding to the LDL receptor. This is how the remnants are taken up by the liver. Subsequently, when IDL convert into LDL, it is apoB100 that assumes the conformation allowing its binding to the LDL receptor.

HDL

The main apoproteins of HDL are apoAI and apoAII. They are synthesized in the intestine and liver. HDL also contains apoC and apoE, which it actively exchanges with other lipoproteins, particularly with triglyceride-rich ones.

Lipoprotein-metabolism enzymes and transfer proteins

Lipoprotein lipase (LPL) and hepatic triglyceride lipase (HTGL)

Two hydrolases, lipoprotein lipase (LPL) and hepatic triglyceride lipase (HTGL), are important in lipoprotein metabolism. LPL binds to heparan sulfate proteoglycans on the surface of the vascular endothelial cells, and HTGL is associated with liver plasma membranes. LPL digests triglycerides in chylomicrons and VLDL and releases fatty acids and glycerol for cell metabolism or storage. LPL may hydrolyze as much as 10 g triglycerides per hour. LPL can also anchor lipoproteins to the cell surface and to the extracellular matrix. HTGL also digests lipoproteins, but its substrates are particles already partially digested by LPL: it facilitates conversion of IDL into LDL.

Lecithin:cholesterol acyltransferase (LCAT)

Another important enzyme in lipoprotein metabolism is lecithin:cholesterol acyltransferase (LCAT), a glycoprotein with a molecular mass of 67 kDa associated with HDL particles. LCAT esterifies cholesterol remaining on the surface of HDL. The esterification of cholesterol changes its polarity and moves it to the inside of the particle, which assumes a spherical shape. The LCAT gene is expressed primarily in the liver. ApoAI is an activator of LCAT.

Cholesterol ester transfer protein (CETP)

The exchange of components between different lipoproteins is facilitated by the cholesterol ester transfer protein (CETP), a 74 kDa protein synthesized in the liver. CETP transfers cholesterol esters from HDL to VLDL or LDL, in exchange for triglycerides.

Lipoprotein (a)

Lipoprotein (a) is a particle consisting of an LDL particle containing apoB100 linked through a disulfide bond to another apoprotein, apo(a) (Fig. 17.6). Apo(a) is a glycoprotein with a considerable size polymorphism. The molecular mass of its isoforms ranges between 300 and 800 kDa. The isoforms are designated S, F, S1, S2, S3, and S4. Apo(a) possesses a protease domain and a number of repeating sequences of approximately 80–90 amino acids in length, stabilized by disulfide bonds into a triple-loop structure called kringles (this is because their conformation resembles the shape of Danish pastry, kringle). One of the kringles, kringle IV, is repeated 35 times within the apo(a) sequence. The number of kringle IV repeats determines the differences in size of isoforms.

Lipoprotein (a) is assembled in the liver and has a pre-beta mobility on electrophoresis. Its density spans the LDL and HDL range (1.04–1.125 g/mL). Its concentration in plasma ranges widely between 0.2 and 120 mg/dL. Both the apo(a) size and its plasma concentration are determined by the apo(a) gene locus.

Apolipoprotein (a) exhibits a considerable sequence homology with plasminogen. Because of this, although it does not possess plasminogen's protease activity, it may interfere with the action of plasminogen, impairing the process of clot resolution (fibrinolysis).

Measurement of lipoproteins

There is no method so far for the measurement of the concentration of entire lipoprotein particles in plasma. Instead, we measure a component of a lipoprotein, such as cholesterol or an apoprotein, and use this as a marker of particle concentration. For clinical purposes we measure total cholesterol concentration in plasma. Since most of the cholesterol present in plasma is in LDL, the total cholesterol level is an approximation of LDL concentration. Other important variables are the triglycerides and the concentration of HDL-cholesterol. All lipoprotein subfractions can be measured in specialist laboratories (see Fig. 17.7).

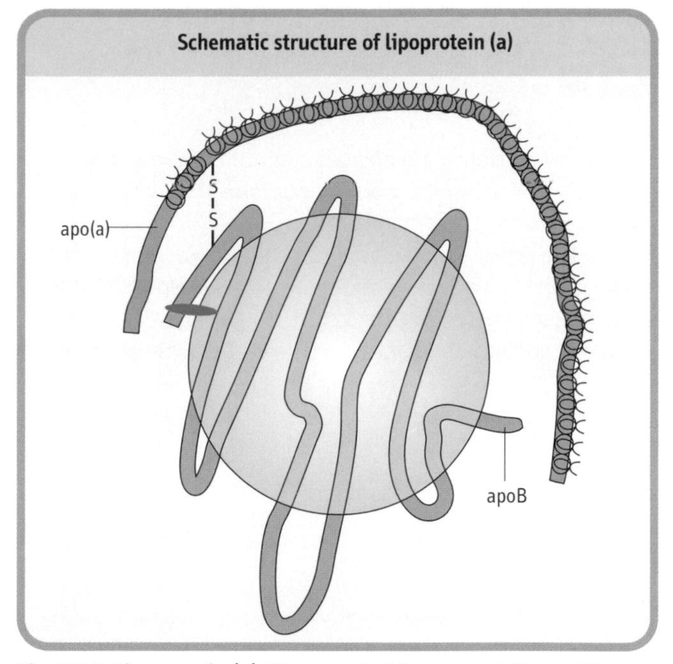

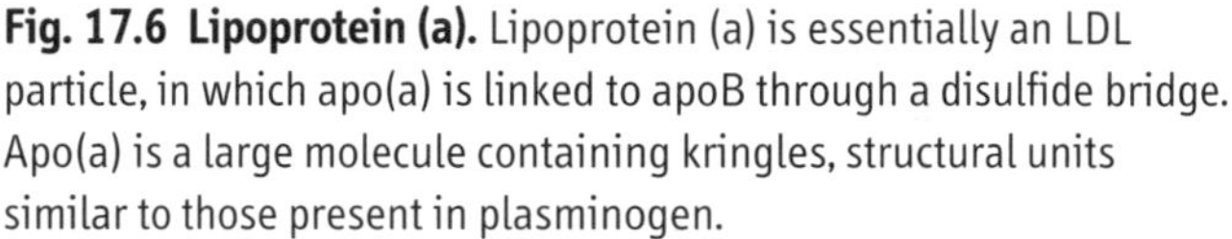

Fig. 17.6 Lipoprotein (a). Lipoprotein (a) is essentially an LDL particle, in which apo(a) is linked to apoB through a disulfide bridge. Apo(a) is a large molecule containing kringles, structural units similar to those present in plasminogen.

Ultracentrifugation

In all clinical laboratories simple centrifuges are used to separate serum or plasma from red blood cells. In these machines, a moderate centrifugal force, 2000–3000 *g*, is used to sediment the blood cells. When much larger centrifugal forces (40 000–100 000 *g*) are applied to solutions, the centrifugation, now termed ultracentrifugation, becomes an important separation method for particles and molecules.
Ultracentrifugation is extensively used in lipid research. When the centrifugal force is applied to a solution, particles that are heavier than the surrounding solvent will sediment, and those lighter than the solvent will float to the surface at a rate proportional to the applied centrifugal force and to the particle size. The formula below summarizes factors that affect the particle movement:

$$v = \frac{\left[d^2(Pp - Ps)g\right]}{18\mu}$$

where v = sedimentation rate, d = diameter, Pp = particle density, Ps = solvent density, μ = viscosity of the solvent, and g = gravitational force.

In a technique known as flotation ultracentrifugation, plasma containing lipoproteins is overlayered with a solution of defined density, e.g. 1.063 kg/L, the density of VLDL. After several hours of centrifugation (depending on the type of rotor, the rotor speeds would be in the range of 40 000 rev/min), the VLDL float to the surface, where they can be harvested. Other density solutions can be used to separate other lipoproteins. Variations of the ultracentrifugation technique, such as density-gradient centrifugation, can be applied to separate a plasma sample into several 'bands' containing different lipoprotein fractions. Ultracentrifugation is also extensively used in protein and nucleic acid biochemistry.

The heterogeneity of LDL

The main lipoprotein classes are not homogenous. VLDL, LDL and HDL contain particles that differ in size and composition.

The major subfractions of LDL can be separated by gradient polyacrylamide gel electrophoresis (PAGE) and are designated LDL 1, LDL 2, and LDL 3. LDL 1 and 2 are large and light particles. LDL 3 are smaller and denser. Small dense LDL are more susceptible to oxidation, have lower affinity toward the LDL receptor, and penetrate the arterial intima more easily than larger LDL. Thus, a person may have a normal cholesterol concentration in plasma but may still develop atherosclerosis because his/her lipoproteins are more atherogenic than usual.

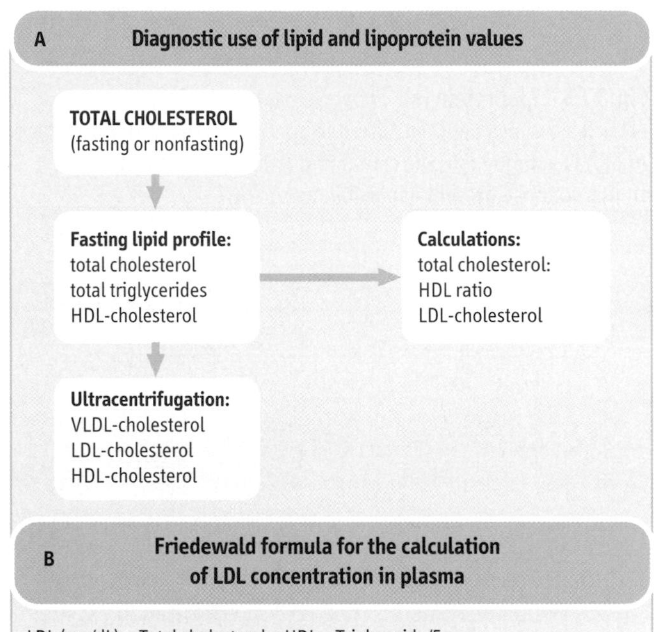

Fig. 17. 7 (A) Three steps in the diagnostic use of lipid and lipoprotein measurements. The measurement of total cholesterol is a first-line screening test. The next step, fasting lipid profile, provides information about the main lipid markers of cardiovascular risk: levels of total cholesterol, LDL-cholesterol, and HDL-cholesterol. Ultra-centrifugation analysis of lipoprotein subfractions and measurement of apoproteins are specialist tests. The total-cholesterol-to-HDL ratio is a useful parameter that indicates a balance between cholesterol transport to and from peripheral tissues. A ratio of total cholesterol: HDL-cholesterol above 5 indicates an increased cardiovascular risk. **(B) The calculation of LDL-cholesterol level in plasma.** The LDL-cholesterol level can be calculated from the value of total cholesterol, triglycerides, and HDL-cholesterol using the Friedewald formula shown above.

Laboratory testing for lipid disorders

There are three levels of testing in the diagnosis of lipoprotein disorders (Fig. 17.7[A]). The first level is the screening for total cholesterol. This can be done on either a fasting or nonfasting sample. The second level consists of measurements of total cholesterol, triglycerides, and HDL in serum. Such measurements should be performed after a 12-h fast. The LDL concentration is usually calculated from the total cholesterol and HDL concentration (Fig. 17.7[B]). The final step, performed only in specialist lipid laboratories, is lipoprotein analysis by ultracentrifugation.

Lipoprotein metabolism

Lipoprotein metabolism proceeds through the following stages:

- The assembly of lipoprotein particles. Chylomicrons are assembled in the gut, and VLDL in the liver.
- Off-loading of triglycerides from lipoproteins. The enzymes LPL and HTGL participate in this process. The resulting particles that emerge from chylomicrons and VLDL are called remnant particles (remnants).
- The binding, internalization, and degradation of remnant particles by cellular receptors.
- Transformation of some remnants into LDL, their binding to the receptor, and cellular internalization.
- Exchange of components between different types of particles. This is facilitated by the action of LCAT enzyme and transport proteins such as CETP.

The transport of dietary fat

Chylomicrons transport dietary lipids and are assembled in the gastrointestinal tract. The dietary triglycerides are first acted upon by intestinal (pancreatic) lipases and are absorbed as monoacylglycerol, free fatty acids, and some glycerol (see Chapter 9). In the enterocytes, triglycerides are resynthesized and, together with phospholipids and cholesterol, are reassembled with apoB48 into chylomicrons. Chylomicrons are secreted into the lymph and reach plasma through the thoracic duct. They normally appear in plasma only after fat-containing meals. In the peripheral tissues, such as muscle and adipose tissue, they are acted upon by LPL, which offloads their triglycerides. After this, the particle becomes smaller and is now called a chylomicron remnant. The remnants are transported to the liver, bind to the LDL receptor-related protein (LRP), and are internalized. The half life of chylomicrons is short, less than 1 h (Fig. 17.8).

Transport of endogenously synthesized lipids

Endogenously synthesized triglycerides are transported by VLDL assembled in the liver. The VLDL assembly occurs on a backbone of apoB100. During this process, the binding of triglycerides to apoB is facilitated by microsomal triglyceride transfer protein (MTP). Triglyceride-rich VLDL exchanges triglycerides for cholesterol esters with HDL. In the peripheral tissues, similarly to chylomicrons, VLDL is acted upon by LPL, which hydrolyses its triglycerides and converts the particle into a VLDL remnant. The remnant is smaller, contains relatively more cholesterol than VLDL and contains apoE. Now, apoE becomes responsible for the metabolism of a remnant. At this stage, a proportion of remnants are internalized in the liver. Some, however, are further hydrolyzed by HTGL to yield IDL. IDL, by losing still more triglycerides, transform into LDL. By this stage, all apoproteins, except apoB100, have been lost from the particles, and apoB100 acquires a conformation allowing it to bind to the LDL receptor.

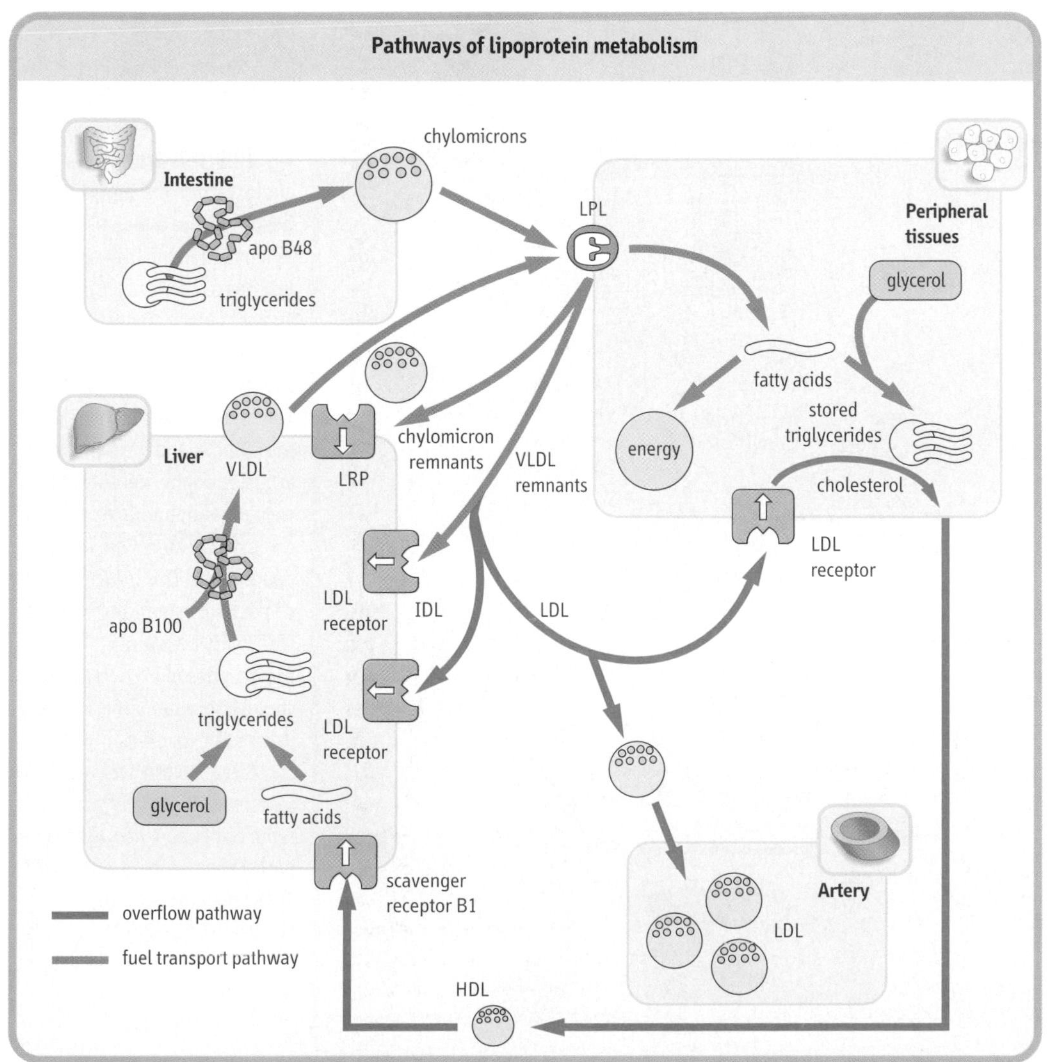

Fig. 17.8 Lipoprotein metabolism: the fuel transport pathway and the overflow pathway. **The fuel transport pathway:** chylomicrons transport triglyceride to the periphery, and their remnants are metabolized in the liver. VLDL transport fuel from the liver to peripheral tissues, and the remnants also return to the liver. Part of VLDL remnants and IDL are further converted into LDL, which enters the overflow pathway. **The overflow pathway:** LDL travels in blood from the liver through the peripheral tissues and back to the liver. On its way it can enter the arterial wall. Cholesterol is transported from the cells back to the liver by HDL particles. The lipid hypothesis of atherosclerosis states that the amount of lipids deposited in the arteries is proportional to their plasma concentration.

LDL metabolism

The cholesterol-rich LDL can be taken up either by the liver (approximately 80% of particles follow this path) or by the peripheral tissues. Both routes involve uptake by the LDL receptor (see Fig. 17.8). After internalization, the LDL–receptor complex is digested by lysosomal enzymes. The cholesterol is released and esterified within the cell by acyl-cholesterol acyltransferase (ACAT) and the receptor recycles back to the membrane. The amount of cholesterol released within the cell regulates the rate of endogenous cholesterol synthesis by inhibiting the HMG CoA reductase. Also, a high intracellular cholesterol concentration decreases the LDL-receptor expression. Conversely, low intracellular cholesterol increases the expression of the LDL receptor (see Fig. 17.1).

Apoprotein E (apoE)

Apoprotein E plays a key role in the metabolism of remnant lipoproteins. Its synthesis is controled by three major alleles, e2, e3, and e4. It is a relatively small protein, with a molecular mass of 34 kDa. There are three major isoforms: E2, E3, and E4. ApoE is recognized both by the LDL receptor and by the LRP. The E2 isoform has much lower affinity (1%) to the receptor than others. This, in the homozygous state, slows down the uptake of remnant particles and results in a dyslipidemia known as familial dyslipidemia (type 3 hyperlipidemia). Also, apoE seems to be particularly important for lipid metabolism in the nervous system where it is synthesized by brain astrocytes. The presence of an e4 allele is associated with Alzheimer's disease, a disorder responsible for at least 50% of all observed cases of dementia.

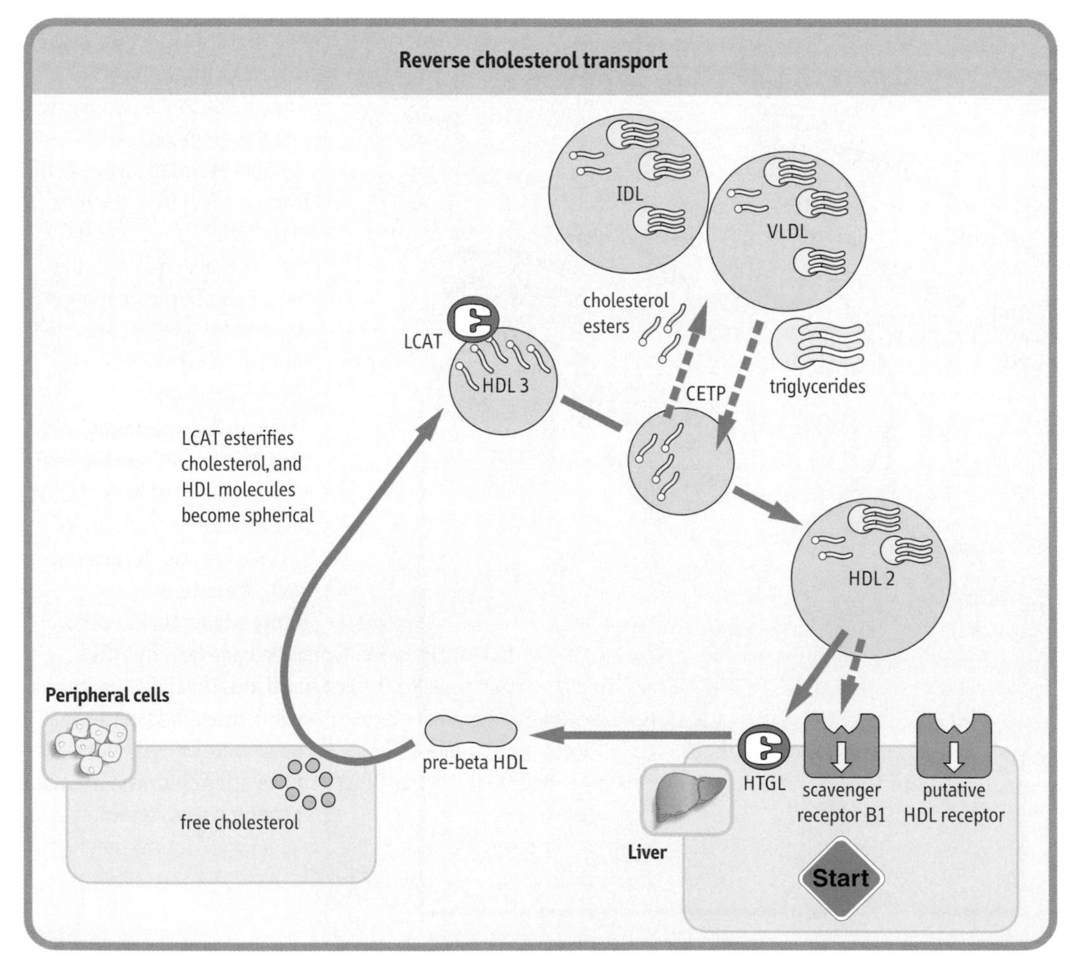

Fig. 17.9 HDL transports cholesterol from peripheral tissues to the liver. HDL are assembled in the liver and intestine as discoid particles. They scavenge cholesterol from cell membranes. LCAT associated with HDL esterifies cholesterol. The esters move to the inside of the particle, and the particle acquires a spherical shape. HDL exchanges apoproteins and cholesterol esters with other lipoproteins. This is facilitated by cholesterol ester transfer protein (CETP). HDL acquires triglycerides in exchange for cholesterol esters. This increases its size. Off-loading of cholesterol in the liver results in a decrease in particle size: redundant apoA-rich, lipid-poor particles emerge (pre-beta HDL). These now take part in cholesterol transport.

Cholesterol removal from cells (reverse cholesterol transport)

The VLDL–IDL–LDL cascade results in the delivery of endogenous lipids from the liver to peripheral cells. The removal of cholesterol from peripheral tissues is also possible, and is mediated by HDL. HDL transports cholesterol from the peripheral tissues to the liver. HDL is assembled in the liver and intestine as a small disc-like particle. It acquires cholesterol in peripheral tissues. Then, the cholesterol is esterified by LCAT associated with the particle. Cholesterol esters move to the hydrophobic center of the particle, and the particle becomes spherical. It is now designated HDL 3. In addition, in a process facilitated by CETP, cholesterol esters are exchanged for triglycerides with particles such as VLDL, chylomicrons, and remnants. The acquired triglyceride increases the particle size. HDL 3 now changes into HDL 2. This triglyceride-rich HDL becomes a substrate for HTGL, which removes its triglycerides in the liver. HDL size decreases, and redundant surface material is released as so-called pre-beta HDL. Pre-beta HDL continues to scavenge excess cholesterol from cell membranes (Fig. 17.9).

The concept of two main pathways of lipoprotein metabolism

Putting it all together, chylomicrons and VLDL metabolism constitute the triglyceride distribution network in the body. We can call it the fuel transport pathway of lipoprotein metabolism (Fig. 17.8). After the fuel distribution is complete, the LDL is generated from the products of the fuel transport pathway. The transformation of remnants into LDL, its transport either to liver or to peripheral tissues, its internalization, and the removal of cholesterol from cells by HDL can be regarded as the 'overflow pathway' of lipoprotein metabolism (Fig.17.8). This concept is helpful in understanding the role of lipids in the development of atherosclerosis.

The most important consequence of an overload of either the fuel transport pathway or the overflow pathway is damage to the vascular wall. Normally the apoE-containing remnants are quickly metabolized in the fuel transport pathway and no excess of LDL appears in the circulation. However, if dietary intake of cholesterol is high, or if there is a decrease in the number of LDL (apoB/E) receptors, more particles enter the overflow pathway and more LDL appear in plasma.

LDL stays in the circulation for several days. During that time it may enter the vascular wall. While in plasma, LDL is relatively protected against oxidation by plasma antioxidants such as vitamins C and E, beta-carotene, and also by antioxidants contained in lipoprotein particles themselves, such as alpha-tocopherol (vitamin E) and ubiquinol. However, the antioxidant levels in the extracellular space are much lower: therefore, once LDL exits plasma, its phospholipids and fatty acids become susceptible to oxidation. Oxidized LDL is strongly atherogenic (see below).

Lipoprotein receptors

The LDL receptor allows cells to acquire cholesterol from outside the cells. The receptor gene is located on chromosome 19; the mature receptor protein is 839 amino acids long and spans the cell membrane (Fig. 17.10). The receptor was discovered by Goldstein and Brown, who jointly received the Nobel Prize for their work. The LDL receptor is also known as the apoB/E receptor, because it has an affinity to two apoproteins: apoE and apoB100. ApoE binds to LDL receptor with a higher affinity than apoB.

ApoB48 does not bind to the LDL receptor. Chylomicrons use a different hepatic receptor, which belongs to LDL-receptor gene family, the LRP. A third lipoprotein receptor, VLDL receptor, has a high degree of homology with the LDL receptor and preferentially binds apoE-containing VLDL. Its function in humans is still uncertain.

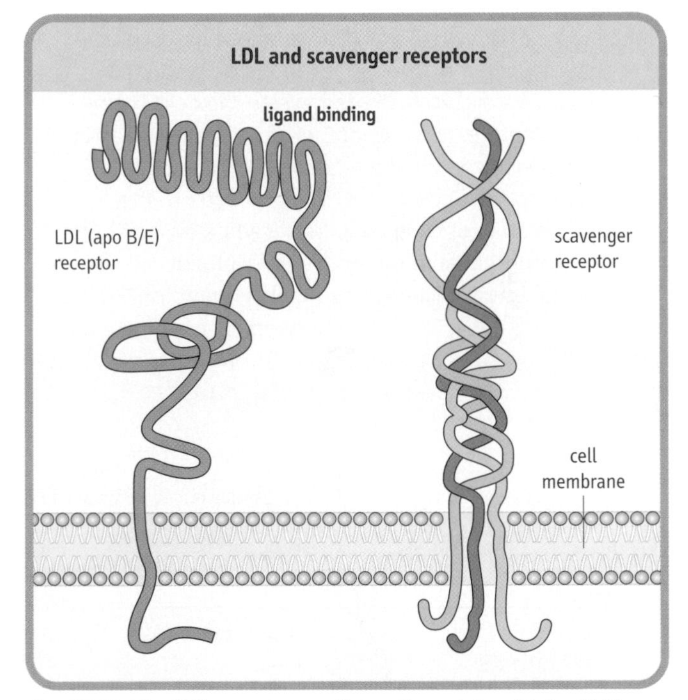

Fig. 17.10 The LDL receptor mediates the uptake of intact LDL, and the scavenger receptor internalizes modified LDL. Both receptors span the cell membranes. The LDL receptor is regulated by the intracellular cholesterol content, while the scavenger receptor remains unregulated. The so-called scavenger receptor type A, which is illustrated here, is present on macrophages and possesses a collagen-like structure. Scavenger receptor type B1 participates in HDL-metabolism.

The scavenger receptor

Scavenger receptors are membrane receptors with a relatively broad specificity. They are present on the phagocytic cells such as macrophages in the vascular wall, in pulmonary alveoli and peritoneum, and also in the Kupffer cells in the liver (see Fig. 17.10). These receptors differ from the LDL receptor in that they are not subject to feedback regulation. There are two classes of scavenger receptor: A and B. Of these, class A receptors are known as collagenous receptors, because they contain a collagen-like triple-helical structure. While intact LDL does not interact with the scavenger receptor, chemically modified (acetylated or oxidized LDL) is readily bound. Class B scavenger receptor interacts with HDL particles in the liver.

Dyslipidemias

The defects in various components of lipoprotein metabolism lead to changes in plasma lipoprotein levels and to disorders known as hyperlipidemias or, better, dyslipidemias. Figure 17.11 lists the clinically most important genetic dyslipidemias.

Vascular endothelium secreting nitric oxide is important for the function of blood vessel walls

The ability of blood vessels to dilate (vasodilatation) and to constrict (vasoconstriction) is important for the regulation of tissue and organ blood flow. Blood-vessel walls are lined with a monolayer of cells known as the endothelial cells (endothelium). Endothelial surface prevents platelet adhesion and blood-clot formation. It also forms a mechanical barrier between the blood and the rest of the vessel wall. Substances can penetrate endothelium either through junctions between the endothelial cells, or by transgressing the cells themselves. Damage to the endothelium, either mechanical or functional, impairs the function of a blood vessel. Damage to the endothelium is a key factor in the earliest stages of atherosclerosis (see below).

The endothelium controls vasodilatation by secreting endothelium-derived relaxing factor (EDRF), nitric oxide (NO). NO is synthesized from L-arginine by endothelial NO synthase (eNOS).The activity of eNOS is controled by the intracellular calcium concentration. There are two isoenzymes of eNOS: eNOS that is constitutively (constantly) expressed in the endothelium, and inducible (iNOS) found in vascular smooth-muscle cells and in macrophages. NO causes vasodilatation via stimulation of guanylate cyclase and cyclic GMP production (see Chapter 36).

Current data suggest that low levels of NO may contribute to development of high blood pressure (hypertension). NO may also be protective against atherosclerosis, because it reduces monocyte adhesion to the endothelium and proliferation of the vascular smooth-muscle cells, phenomena that contribute to the formation of atherosclerotic plaque. Drugs such as glyceryl trinitrate, used in angina pectoris, act by releasing NO to vascular smooth-muscle cells.

The most important genetic dyslipidemias.

Dyslipidemia	Frequency/ inheritance	Defect	Plasma lipid pattern	Increased cardiovascular risk
familial hypercholesterolemia	1:500 autosomal dominant	LDL receptor	hypercholesterolemia or mixed hyperlipidemia (IIa or IIb)	yes
familial combined hyperlipidemia	1:50 autosomal dominant	overproduction of apoB100	hypercholesterolemia or mixed hyperlipidemia (IIa or IIb)	yes
familial dysbetalipoproteinemia (type III hyperlipidemia)	1:5000 autosomal recessive	presence of E2/E2 isoform defective remnant binding to LDL receptor	mixed hyperlipidemia (III)	yes

Mixed hyperlipidemia = increased plasma cholesterol and triglycerides

Fig 17.11 The three clinically most important dyslipidemias are familial hypercholesterolemia, familial combined hyperlipidemia, and familial dysbetalipoproteinemia.

Atherogenesis

The key event in early atherosclerosis is damage to the endothelium (Fig. 17.12), which may be caused by oxidized lipoproteins, hypertension, and cigarette smoking. First the damage is functional only. The endothelium becomes more permeable to lipoproteins and allows migration of cells to the underlying layer, the intima. LDL penetrate the vascular wall and deposit in the intima, where they may undergo oxidation. Oxidized LDL stimulate endothelial expression of adhesion molecules, such as vascular cell-adhesion molecule 1 (VCAM-1) and monocyte chemotactic protein-1 (MCP-1), which attract monocytes, white blood cells that are the precursors of phagocytic macrophages, to the arterial wall. Monocytes enter the wall and transform into macrophages under the influence of macrophage monocyte-colony-stimulating factor (M-CSF). Oxidized LDL facilitates this process by inhibiting the mobility of macrophages, thus immobilizing them in the subendothelial space. Oxidized apoB100 can no longer bind to the LDL receptor. Instead, oxidized LDL

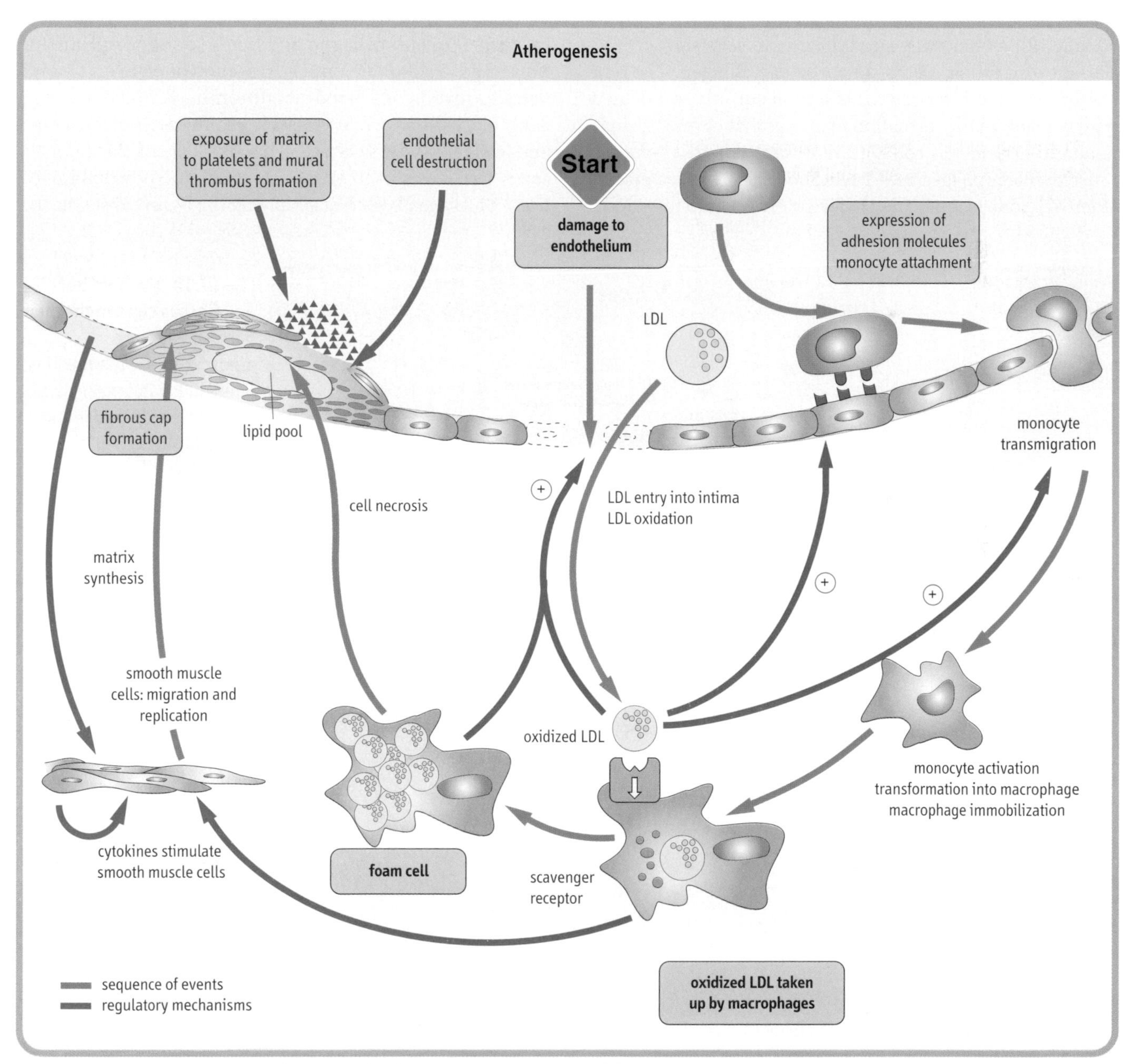

Fig. 17.12 Atherogenesis results from complex interactions among several factors. The main stages of atherogenesis are indicated by boxes. The sequence of events is described in the text.

is ingested by macrophages through the scavenger receptor. Since the scavenger receptor is not regulated, the cells become overloaded with lipid and become 'foam cells'. Conglomerates of foam cells form fatty streaks, yellow patches visible in the arterial wall. Dying foam cells release lipid that forms pools within the arterial wall.

Further damage occurs when surrounding smooth muscle and endothelial cells start to secrete a range of small peptides regulating cell growth known as cytokines, and growth factors such as platelet-derived growth factor (PDGF), interleukin-1 (IL-1), and tumor necrosis factor (TNF), which stimulate smooth-muscle cells to proliferate and to migrate toward the lumen side of the arterial wall (see also Chapters 36 and 39). At that time smooth-muscle cells start synthesizing extracellular matrix, in particular collagen. The relocation of smooth-muscle cells and accumulation of new matrix result in the formation of a 'cap' that covers the lipid pool and consists of collagen-rich fibrous tissue, smooth muscle, macrophages, and T lymphocytes. This is a mature atherosclerotic plaque. The plaque protrudes into the arterial lumen, grows slowly over years, and may finally obstruct the artery. This decreases blood flow in the affected vessel. In the coronary arteries, which supply the blood to the heart muscle, plaque formation leads to a chest pain that the patients feel on effort (angina pectoris). If the narrowing occurs in the arteries supplying the legs, it causes characteristic pain on walking (intermittent claudication).

Macrophages and T cells reside preferably at the edges of the plaque (Fig. 17.13). Here, macrophages secrete metalloproteinases, enzymes that degrade extracellular matrix. In addition, T cells produce γ-interferon, which inhibits collagen synthesis in the smooth-muscle cells. These processes weaken the plaque cap and may result in its breakage. The breakage exposes collagen and lipids to the blood stream. This leads to adherence and aggregation of platelets and initiates formation of a blood clot (thrombus), which may suddenly block the artery. Such blockage interrupts blood supply and leads to tissue necrosis. This blockage of the coronary artery manifests itself as a heart attack (myocardial infarction). Blockage in arteries supplying the brain causes stroke.

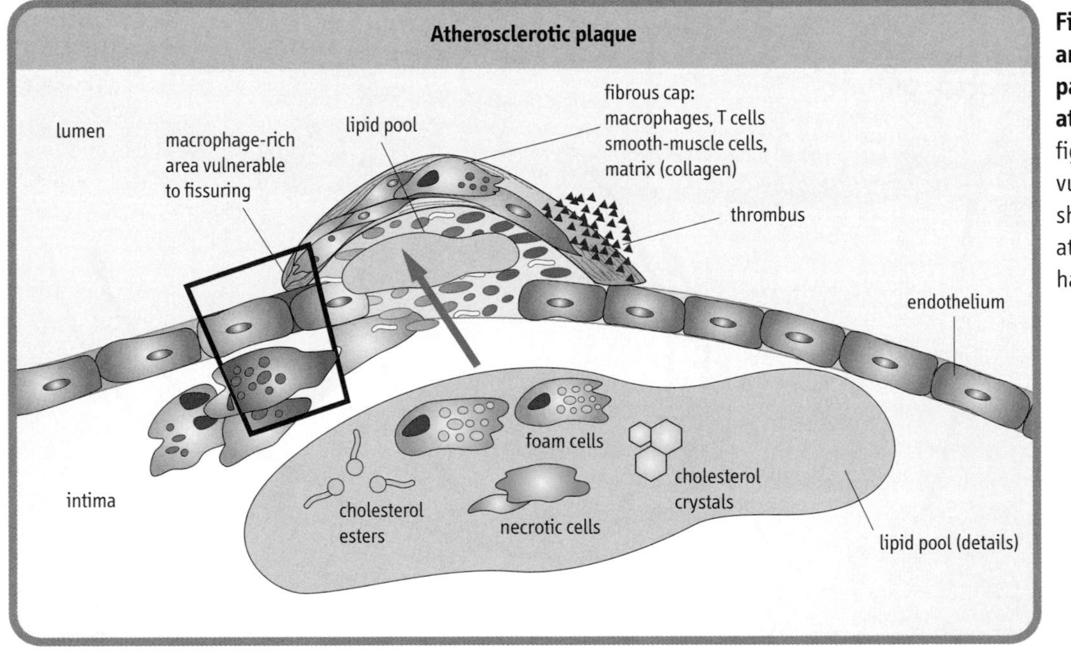

Fig. 17.13 The lipid pool and fibrous cap are the main parts of a mature atherosclerotic plaque. This figure illustrates plaque areas vulnerable to breakage and shows the thrombus forming at the place where the plaque has broken.

Antioxidants and atherosclerosis

In animals, antioxidants such as the drug probucol, vitamin E, or butylated hydroxytoluene inhibit the formation of atherosclerotic-like changes. Antioxidants such as beta-carotene and α-tocopherol (vitamin E) protect LDL against oxidation. In humans, treatment with antioxidants increases the resistance of LDL to oxidation.

Epidemiologic studies indicate that taking antioxidants such as vitamins E and C or beta-carotene may decrease the risk of cardiovascular disease. However, the results of clinical trials are not yet conclusive. More research is needed to establish whether antioxidants should be taken to prevent atherosclerosis-related diseases.

The assessment of cardiovascular risk

Epidemiologic studies, such as the Multiple Risk Factor Intervention Trial (MRFIT) in the USA, clearly show that there is a relationship between the levels of lipids and lipoproteins in plasma and the incidence of coronary heart disease (Fig. 17.14). The risk of cardiovascular disease is related to plasma levels of total cholesterol and LDL-cholesterol. The risk increases when the total cholesterol level is above 5.2 mmol/L (200 mg/dL). The desirable level of LDL appears to be below 4.1 mmol/L (160 mg/dL), with even lower levels suggested for individuals who have coronary disease or multiple cardiovascular risk factors. The magnitude of risk associated with high LDL can be further modified by other factors, such as increased concentration of Lp(a).

There is also an inverse relationship between the cardiovascular risk and HDL concentration. The majority of clinical studies also suggest that an increased triglyceride level is associated with an increased risk for coronary heart disease. The interpretation of the fasting lipid profile is summarized in Fig. 17.15.

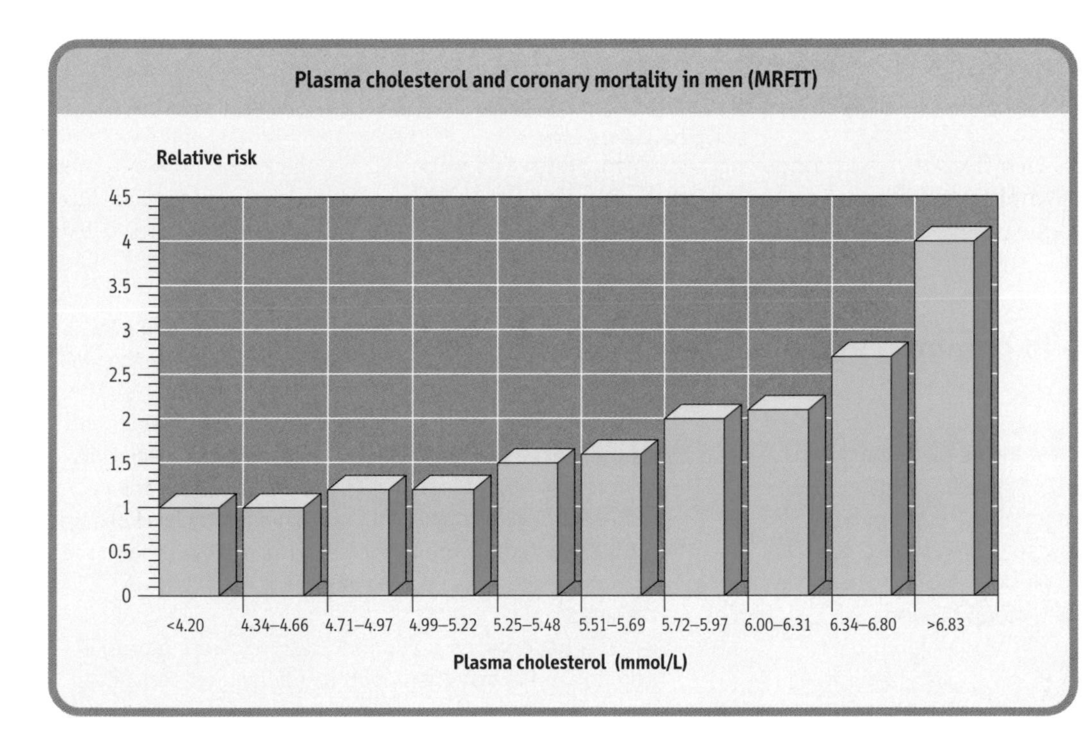

Fig. 17.14 Total cholesterol levels in plasma in relation to the number of deaths from heart disease. Note – converting plasma lipid values into SI units: cholesterol: to convert mg/dL into mmol/L, mutliply by 0.02586; triglycerides: to convert mg/dL into mmol/L, multiply by 0.01129. The data are from the Multiple Risk Factor Intervention Trial (MRFIT). Stammler *et al.* JAMA 1986;**256:**2823.

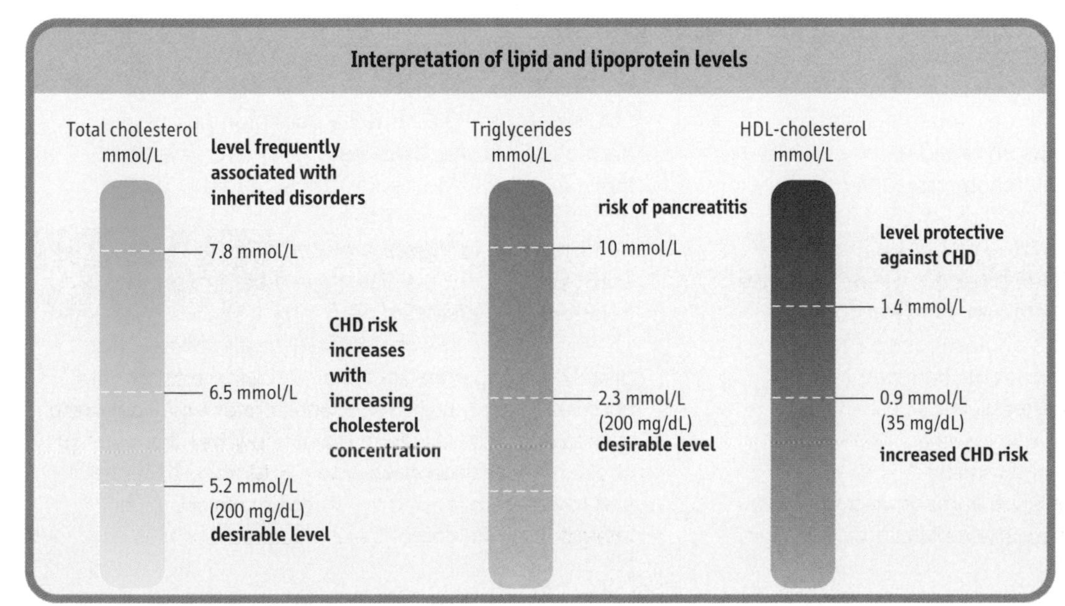

Fig. 17.15 The interpretation of cholesterol, triglyceride, and HDL levels in serum. Note – converting plasma lipid values into US units: total cholesterol, LDL-cholesterol, and HDL-cholesterol: to convert mmol/L into mg/dL, multiply by 38.67; triglycerides: to convert mmol/L into mg/dL, multiply by 87.5. CHD, coronary heart disease.

Cardiovascular risk factors and their management

Risk factor	Comment	Remedy
male sex	the difference in cardiovascular risk between sexes equalizes in post menopausal women	
age		
smoking		smoking cessation
high plasma cholesterol (high LDL-cholesterol)	2–3% decrease in risk for 1% decrease of total plasma cholesterol.	diet low in saturated fats and, where appropriate, cholesterol-lowering drugs
low plasma HDL		smoking cessation, regular exercise
hypertension	major risk factor for stroke and a risk factor for CHD	control blood pressure: diet and drugs
obesity		attain ideal weight
sedentary lifestyle		regular moderate exercise
diabetes	cardiovascular disease is the main cause of death in diabetes	diet and drugs (insulin in type 1 diabetes) Lipid-lowering if dyslipidemia present

Fig 17.16 Cardiovascular risk factors and their management. The plasma lipids are not the only factors that determine the risk of cardiovascular disease.

Dyslipidemia is common in diabetes mellitus

Mr B is 67 years old and has type 2 diabetes. His cholesterol is 6.9 mmol/L (265 mg/dL), triglycerides 1.9 mmol/L (173 mg/dL), and HDL 0.9 mmol/L (35 mg/dL). Fasting blood glucose is 8.5 mmol/L (153 mg/dL) and glycated hemoglobin$_{1c}$ (HbA$_{1c}$) 6%. He is on a diet and takes an oral hypoglycemic agent (sulfonylurea derivative)

Comment. Diabetes carries a 2–3 times increased risk of coronary heart disease. This patient's diabetes is well-controled. In spite of this his cholesterol level remains high, thus the need for lipid-lowering drug treatment. Low HDL is relatively common in type 2 diabetes.

The patient was prescribed a lipid lowering drug in addition to his treatment with a sulfonylurea derivative.

Familial hypercholesterolemia: an inherited lipid disorder associated with heart attacks at young age

A 32-year-old heavy smoker developed a sudden crushing chest pain. He was admitted to the casualty department. Myocardial infarction was confirmed by ECG changes and by high creatine kinase MB isoenzyme concentration. On examination the patient had tendon xanthoma on hands and thickened Achilles tendons. He also had a prominent family history of coronary heart disease. His cholesterol was 10.0 mmol/L (390 mg/dL), triglycerides 2 mmol/L (182 mg/dL) and HDL 1.0 mmol/L (38 mg/dL).

Comment. This patient has familial hypercholesterolemia (FH), an autosomal dominant disorder characterized by a decreased number of LDL receptors. FH carries a very high risk of premature coronary disease, and heterozygotic individuals develop heart attacks as early as the 3rd or 4th decade of life. The frequency of FH homozygotes in Western populations is approximately 1:500. FH is the most important inherited lipid disorder.

This patient was immediately treated with a clot-dissolving drug (thrombolysis). Subsequently he underwent coronary artery bypass graft and was then treated with lipid-lowering drugs, on which his cholesterol level decreased to 4.8 mmol/L (185 mg/dL) and triglyceride level to 1.7 mmol/L with HDL increasing to 1.1 mmol/L (42 mg/dL).

Plasma lipids are not the only factors that determine the development of atherosclerotic disease. Other contributing factors are listed in Figure 17.16. Specially designed risk tables allow an estimate of the overall risk of atherosclerotic event, such as heart attack in persons with several risk factors (Fig. 17.17). Because the presence of other risk factors increases the risk associated with dyslipidemia, the desirable cholesterol levels are lower in individuals with other risk factors, such as diabetes or hypertension.

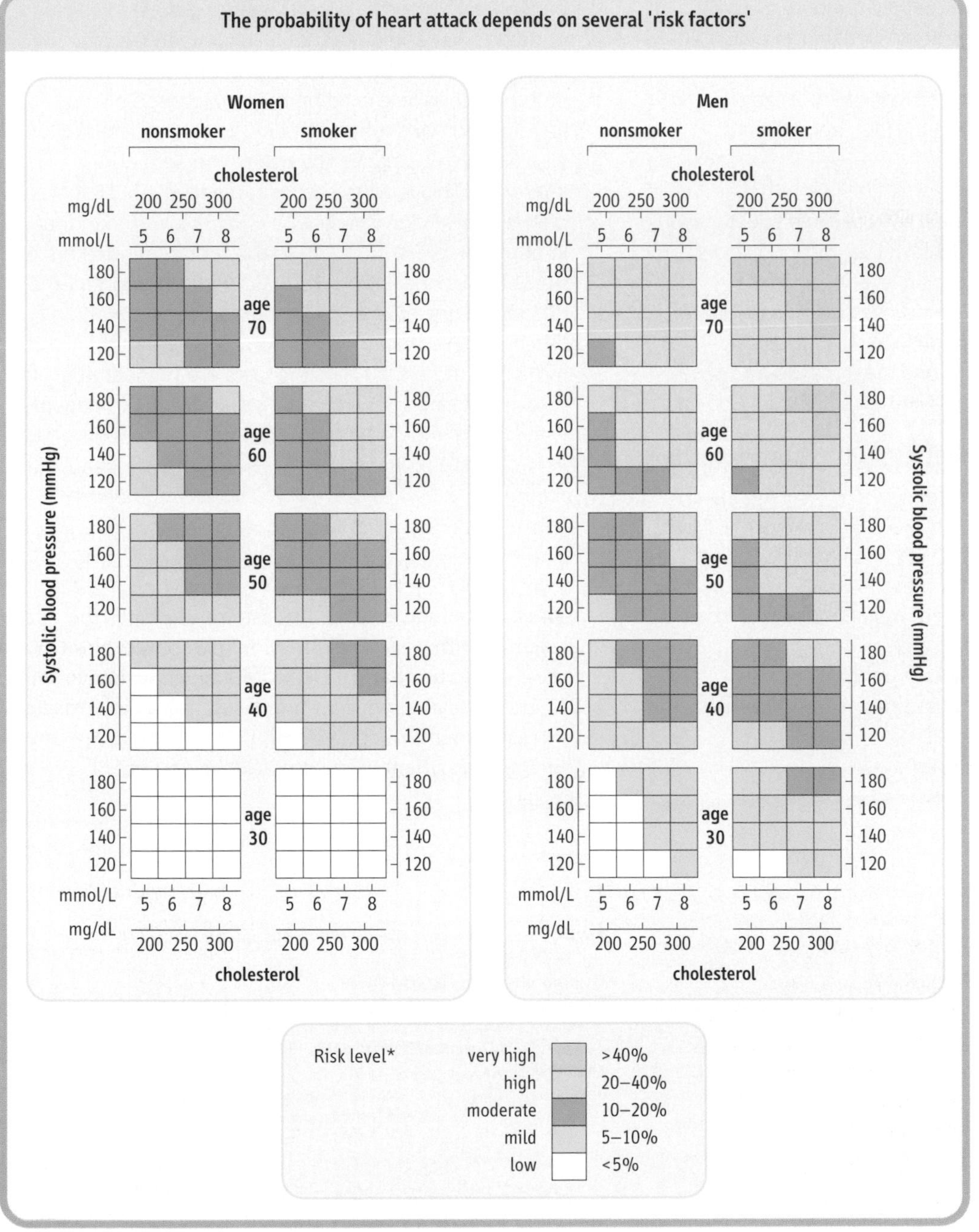

Fig. 17.17 Risk factors have an additive effect on the risk of myocardial infarction. This is the risk table developed by the European Atherosclerosis Society, European Cardiac Society, and the European Society of Hypertension. It illustrates that the presence of risk factors such as hypertension and smoking increases the risk associated with cardiovascular elevated cholesterol levels. The presence of several risk factors at the lower level can be associated with the risk equivalent to that of a single risk factor at the higher level. The table also shows that the risk of myocardial infarction increases with age, and that, at the equivalent age, the risk is higher in men than in women. *The table presents an estimated risk of having a heart attack within 10 years. New recommendations were published in 1998, (see Further reading).

The effect of lifestyle change

A 57-year-old man was referred to the lipid clinic because of hypertriglyceridemia. His triglycerides were 6 mmol/L (545 mg/dL), cholesterol was 5 mmol/L (192 mg/dL), and HDL was 1 mmol/L (39 mg/dL). He was obese, drank 30 units of alcohol per week, and had a sedentary lifestyle.

After two unsuccessful attempts at dieting, he finally managed to lose 7 kg of weight over the subsequent 6 months, cut down drinking to below 20 U/week, and took up regular walking. By 12 months later, his triglycerides were 2.5 mmol/L (227 mg/dL), cholesterol 4.8 mmol/L (186 mg/dL), and HDL 1.2 mmol/L (46 mg/dL).

Comment. Lifestyle change may result in a spectacular change of lipid profile. To achieve this, the individuals must be committed to changing lifestyle and to maintaining the change over a prolonged period of time. In practice this is often very difficult.

Note: 1 unit of alcohol is one measure (60 mL) of liquor, one glass (170 mL) of wine, or a half-pint (300 mL) of beer.

Prevention of coronary heart disease

The risk of heart attack can be decreased by eliminating risk factors through smoking cessation, control of high blood pressure, and control of plasma cholesterol (Fig. 17.16). Cholesterol and LDL levels can be lowered with a low-cholesterol diet. In addition, there are drugs that lower plasma cholesterol concentration either by inhibiting intracellular cholesterol synthesis (HMG CoA reductase inhibitors or statins), by stimulation of LPL (derivatives of fibric acid, fibrates), or by preventing cholesterol absorption in the gut (bile acid binding resins). In people who have already had a heart attack, cholesterol reduction significantly decreases the risk of the second infarction. This is why the desirable level of LDL-cholesterol in individuals with CHD is lower than that in other people. According to the recommendations of the U.S. National Cholesterol Education Program, the desirable level of LDL-cholesterol in people with no CHD and less than 2 risk factors is <4.1 mmol/L (<160 mg/dL). If more than 2 risk factors are present, the level should be <3.4 mmol/L (<130 mg/dL). In patients with CHD it should be below 2.6 mmol/L (<100mg/dL).

Summary

The different lipoproteins present in plasma have distinctive functions:

- chylomicrons transport dietary lipids;
- VLDL carry endogenously synthesized lipids from the liver to the peripheral tissues;
- LDL carry mostly cholesterol;
- HDL remove cholesterol from cells.

The intracellular level of cholesterol regulates the rate of cholesterol synthesis and expression of LDL receptors. High levels of cholesterol and LDL in plasma, and low levels of HDL, are associated with an increased risk of atherosclerosis. Measurements of cholesterol, LDL, and HDL are important in the assessment of risk in the development of coronary heart disease.

Further reading

Dominiczak MH, ed. Seminars in clinical biochemistry. Glasgow:University of Glasgow 1997, 415pp.
Dominiczak MH. Lipids: the story so far. *Br J Cardiol* 1997;**4:**425–429.
Howlett R. Nobel award stirs up debate on nitric oxide breakthrough. *Nature* 1998;**395:**625–26.
Libby P. Molecular bases of the acute coronary syndromes. *Circulation* 1995;**91**:2844–2850.
Prevention of Coronary heart disease in clinical practice. Summary of Recommendations of the Second Joint Task Force of European and other Societies on Coronary Prevention, 1998.
Ross R. The pathogenesis of atherosclerosis: a perspective for the 1990s. *Nature* 1993;**362**:801–809.
Stammler *et al. JAMA* 1986;**256**:2823.
Summary of the second report of the National Cholesterol Education Program (NCEP). Expert Panel on Detection, Evaluation and Treatment of High Blood Cholesterol in Adults (Adult Treatment Panel II). *JAMA* 1993;**269**:3015–23.
Internet sites:
American Heart Association: www.americanheart.org
National Heart, Lung and Blood Institute: www.nhlbi.nih.gov/nhlbi/nhlbi.htm/

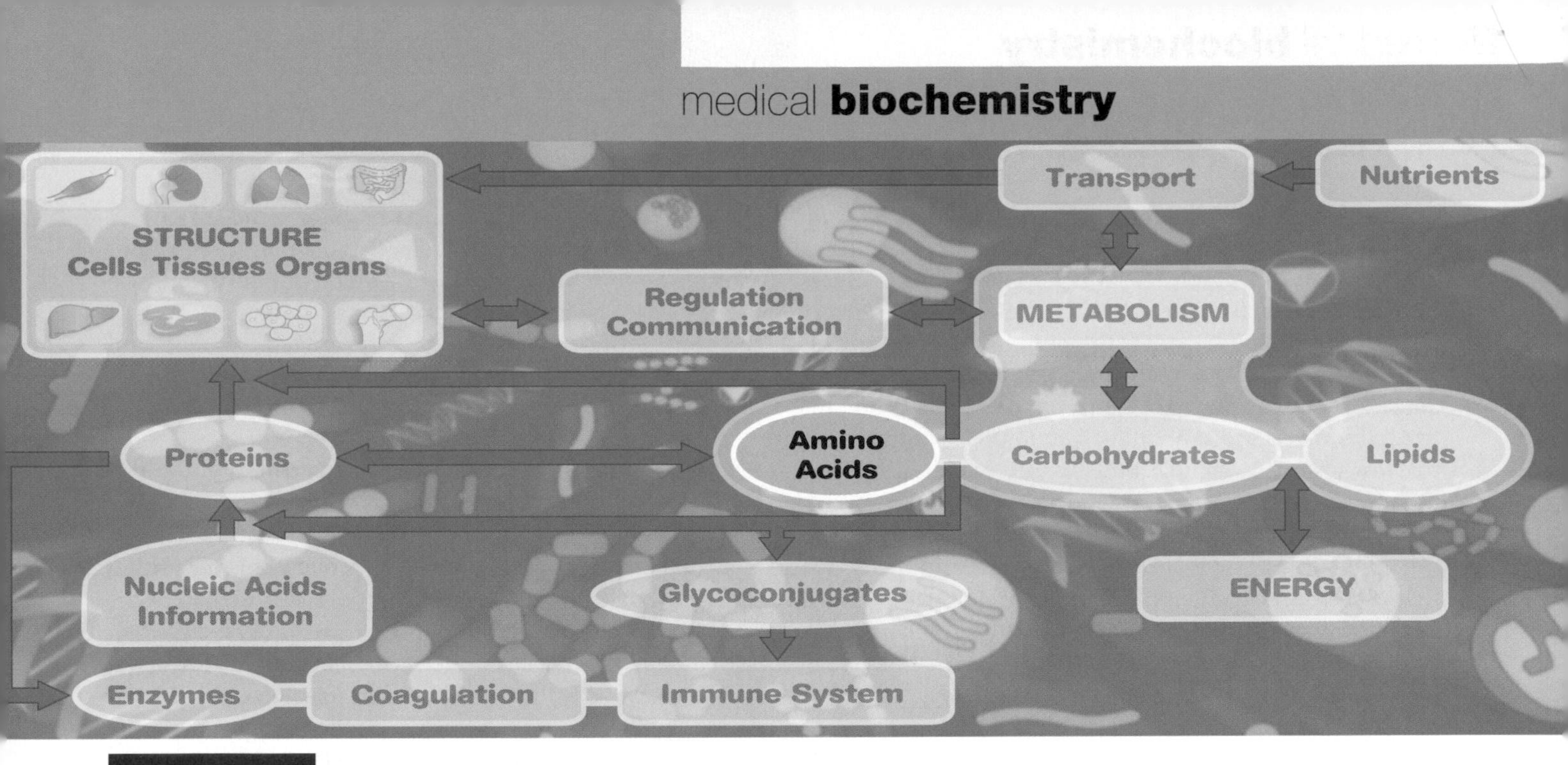

18 Biosynthesis and Degradation of Amino Acids

Amino acid metabolism – overview

Amino acids have several functions in addition to their roles as building blocks of peptides and proteins, and as precursors of neurotransmitters and hormones. Their carbon skeletons may be used to produce energy in oxidative metabolism by the end stages of glycolysis and the tricarboxylic acid (TCA) cycle. The carbon skeletons of some amino acids can be used to produce glucose through gluconeogenesis in the liver, thereby providing a metabolic fuel for tissues that require or prefer glucose; such amino acids are designated as glucogenic or glycogenic amino acids. In addition, the carbon skeletons of certain amino acids can produce the equivalent of acetyl CoA or acetoacetate and are termed ketogenic, indicating that they can be metabolized to give immediate precursors of lipids or ketone bodies. Conversely, the carbon skeletons of many amino acids may themselves be derived from metabolites in central pathways and thus allow the biosynthesis of some of, but not all, the amino acids in humans. Amino acids that can be synthesized in this way are therefore not required in the diet (non essential amino acids), whereas amino acids having carbon skeletons that cannot be derived from normal human metabolism must be supplied in the diet (essential amino acids).

Nitrogen metabolism

The nitrogen component of amino acids, the α-amino groups, must be removed before the carbons can be used in other metabolic pathways. There are several ways in which this can be achieved:

- **transamination** (the transfer of the amino group to a suitable keto acid acceptor); this reaction requires pyridoxal phosphate and involves a pyridoxamine intermediate;
- **oxidative deamination** (the oxidative removal of the amino group, also resulting in keto acids); the amino acid oxidases are flavoproteins, and produce ammonia;
- **removal of a molecule of water by a dehydratase**, e.g. serine or threonine dehydratase; this reaction produces an unstable, imine intermediate that is hydrolyzed to yield an α-keto acid and ammonia.

In contrast, in order to complete the synthesis of nonessential amino acids, amino groups must be added to the appropriate carbon skeletons. This generally occurs through the transamination of an α-keto acid corresponding to that specific amino acid.

Dietary protein is necessary as a source for essential amino acids; however, in an individual consuming adequate amounts of protein, a significant quantity of amino acid is utilized for energy metabolism. The carbons derived from amino acids may also be converted to fat (triacylglycerol) or carbohydrate (glycogen) for storage. The resulting excess nitrogen must be metabolized and excreted. As the primary form in which the nitrogen is removed from amino acids is ammonia, and because free ammonia is quite toxic, humans and most higher animals rapidly convert the ammonia derived from amino acid catabolism to urea, which is neutral, less toxic, very soluble, and excreted in the urine. Thus the primary nitrogenous excretion product in humans is urea. Animals that excrete urea are termed ureotelic. Urea is by far the primary form of excreted nitrogen in humans, but, in healthy individuals, small amounts of nitrogen (~15% of total nitrogen excretion) are excreted in the form of uric acid, creatinine, and ammonium ion. In an average individual, more than 80% of the excreted nitrogen is in the form of urea (25–30 g/24 hours).

Amino acids from body proteins and the diet

Relationship to central metabolism

Although body proteins represent a significant proportion of potential energy reserves (Fig. 18.1), under normal circumstances they are not available to contribute to energy production. In an extended fast, however, muscle protein is degraded to amino acids for the synthesis of essential proteins, and to keto acids for gluconeogenesis to maintain blood glucose concentration. Unlike carbohydrates and lipids, amino acids do not have a dedicated storage form equivalent to glycogen or fat.

Protein and weight-loss diets

Protein and weight-loss diets

High-protein diet plans that claim that the dieter can consume as many calories as they wish in the form of protein are not only poor choices because they do not work, they are also dangerous, because of the high nitrogen load they create. Thus the best diet for weight loss is one that contains reduced calories, but an adequate amount of high-quality protein, to ensure that the need for essential or required amino acids is met. In spite of the many claims of fashionable diets and over-the-counter products, real weight loss can be achieved only through the consumption of fewer calories than are required for basal and exercise-related metabolism. Unfortunately, most of the effective medications that have been developed for appetite suppression have significant side effects.

In addition to its role as an important source of carbon skeletons for oxidative metabolism and energy production, dietary protein must provide adequate amounts of those amino acids which we cannot make, to support normal protein synthesis. The relationships of body protein and dietary protein to central amino acid pools and to central metabolism are shown in Figure 18.2.

Digestion and absorption of dietary protein

In order for dietary protein to contribute to either energy metabolism or pools of essential amino acids, the protein must be digested (broken down to the level of free amino acids or small peptides) and absorbed across the gut. After any remaining di- and tripeptides are broken down in enterocytes, the free amino acids are released into the portal vein and carried to the liver for energy metabolism or biosynthesis. Some amino acids are also redistributed to other tissues, to meet similar needs.

Amino acid degradation

Metabolism of the carbon skeleton and the amino group are coordinated

Before the carbon skeletons of most amino acids are metabolized, the α-amino group is removed. The principal mechanism for removal of amino groups from the common amino acids is via transamination, or the transfer of amino groups from the amino acid to a suitable α-keto acid acceptor. Several enzymes, called amino transferases (or transaminases), are capable of removing the amino group from most amino acids and producing the corresponding α-keto acid. Amino transferase enzymes utilize pyridoxal phosphate, a cofactor derived from the B-vitamin pyridoxine, as a key component in their catalytic mechanism. The ability of pyridoxal phosphate to form Schiff base adducts with amino groups is central to the action of all of these enzymes. These structures and the net reaction catalyzed by amino transferases are shown in Figure 18.3.

Storage forms of energy in the body

Stored fuel	Tissue	Amount (g)†	Energy (kJ)	(kcal)
glycogen	liver	70	1176	280
glycogen	muscle	120	2016	480
free glucose	body fluids	20	336	80
triacylglycerol	adipose	15 000	567 000	135 000
protein	muscle	6000	100 800	24 000

†in a 70-kg individual

Fig. 18.1 Proteins represent a substantial energy reserve in the body. (Adapted with permission from Cahill GF Jr, *Clin Endocrinol Metab* 1976;**5**:398.)

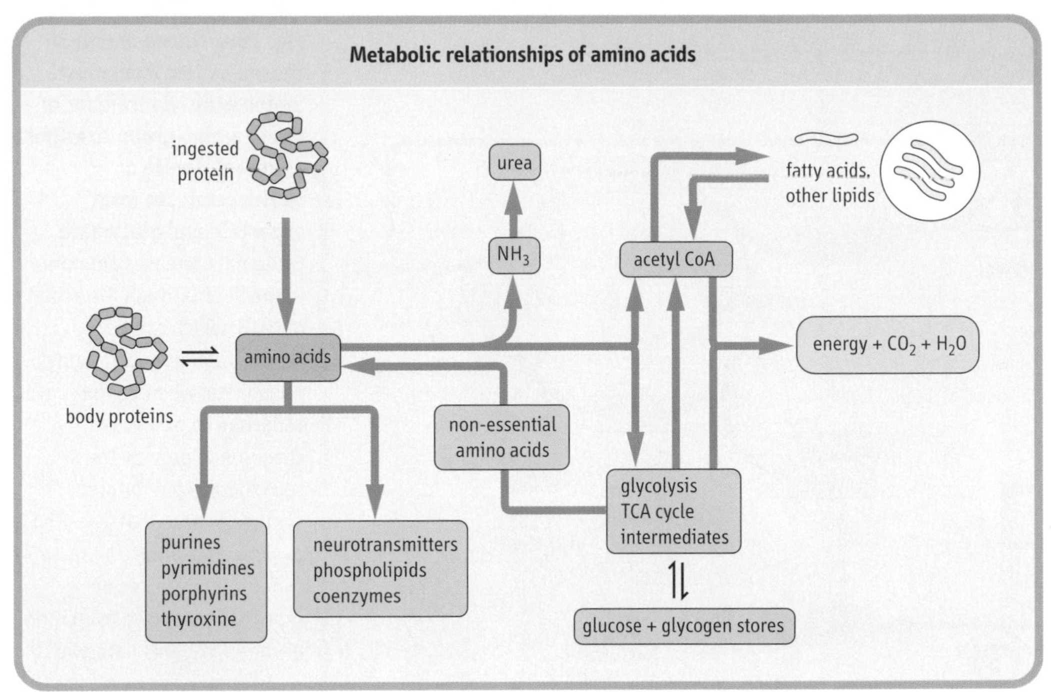

Fig. 18.2 The pool of free amino acids is derived from the degradation and turnover of body proteins, and from the diet. The amino acids are precursors of important biomolecules, including hormones, neurotransmitters, and newly synthesized proteins, and also serve as a carbon source for central metabolism and energy production.

Nitrogen atoms are incorporated into urea from two sources

The transfer of an amino group from one keto acid carbon skeleton to another may seem to be unproductive and not useful in itself; however, when one considers the nature of the primary keto acid acceptors that participate in these reactions (α-ketoglutarate and oxaloacetate) and their products (glutamate and aspartate), the logic of this metabolism becomes clear. Nitrogen atoms are incorporated into urea from two sources (Fig. 18.4). Ammonia produced

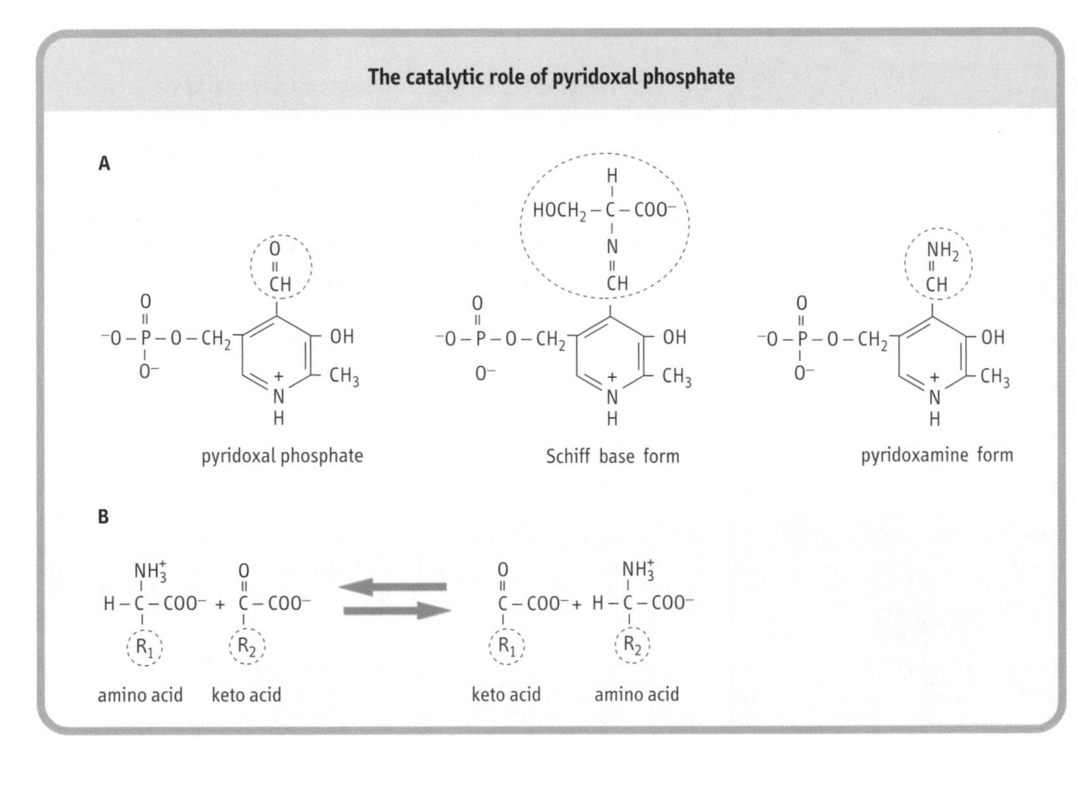

Fig. 18.3 Amino transferases or transaminases use pyridoxal phosphate as a cofactor and involve a pyridoxamine intermediate capable of donating an amino group to a suitable α-keto acid. (A) Structures of the components involved. The cofactor, pyridoxal phosphate, is used in a variety of enzyme-catalyzed reactions involving both amino and keto compounds, including transamination and decarboxylation (e.g. histidine decarboxylase) reactions. (B) Transamination involves both a donor amino acid, designated R_1, and an acceptor α-keto acid, designated R_2. The products are an α-keto acid derived from the carbon skeleton of R_1 and an α-amino acid derived from the carbon skeleton of R_2.

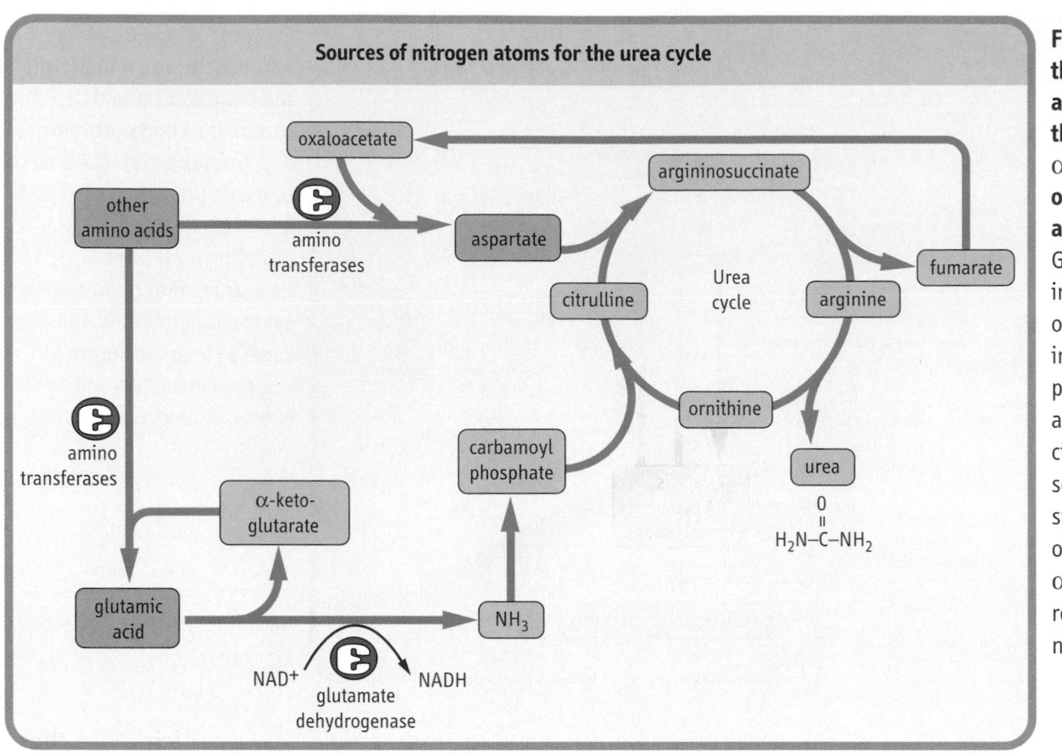

Fig. 18.4 Nitrogen enters the urea cycle from most amino acids via transfer of the α-amino group to either α-ketoglutarate or oxaloacetate, to form aspartate and glutamate. Glutamate releases ammonia in the liver through the action of GDH, which can be incorporated into carbamoyl phosphate, whereas the aspartate combines with citrulline to provide the second nitrogen for urea synthesis. Note that oxaloacetate and α-ketoglutarate can be repeatedly recycled to channel nitrogen into this pathway.

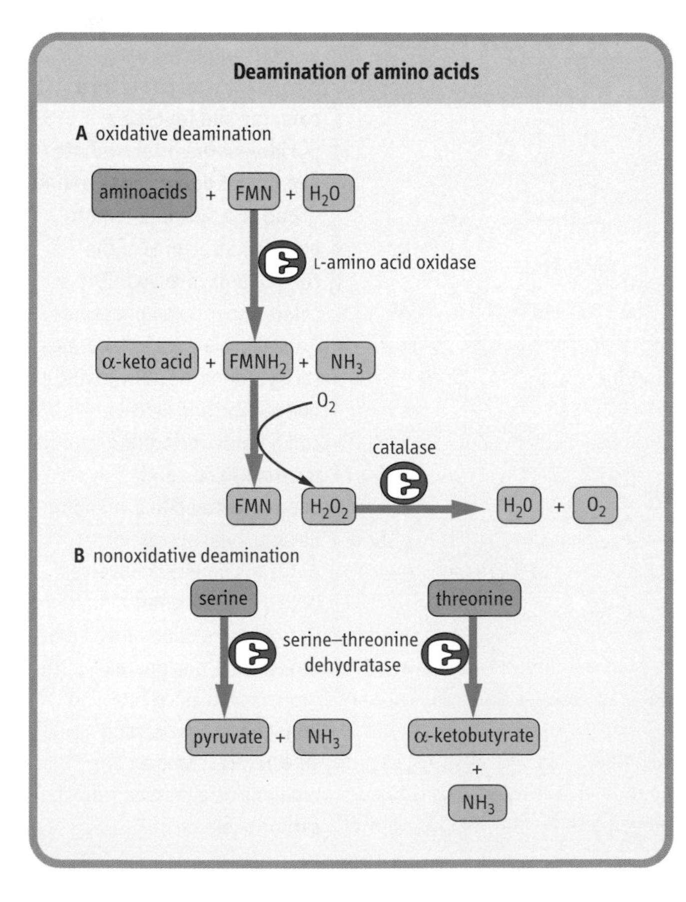

Fig. 18.5 The primary route of amino group removal is via transamination, but there are additional enzymes capable of removing the α-amino group. (A) Amino acid oxidases such as L-amino acid oxidase produce ammonia and an α-keto acid directly, using flavin mononucleotide (FMN) as a cofactor. The reduced form of the flavin must be regenerated using molecular oxygen, and this reaction is one of several that produce H_2O_2. The peroxide is decomposed by catalase.

(B) A second means of deamination is possible only for hydroxyamino acids (serine and threonine), through a dehydratase mechanism that involves a dehydration followed by the readdition of water and loss of the amino group as ammonia.

primarily from glutamate (via the glutamate dehydrogenase (GDH) reaction, see Fig. 18.6) enters the urea cycle as carbamoyl phosphate. The second nitrogen is contributed to urea by aspartic acid. The fumarate formed in this process may be recycled via the TCA cycle to oxaloacetate, which can accept another amino group to reform aspartate or participate in either the TCA cycle or gluconeogenesis (see Fig. 18.8 and Chapter 8). Thus the funneling of amino groups from other amino acids into glutamate and aspartate provides the nitrogen for urea synthesis in a form appropriate for the urea cycle (see below). Although there are other pathways that may lead to the release of amino groups from some amino acids through the action of amino acid oxidases or a dehydratase mechanism (Fig. 18.5), they represent minor contributions to the flow of amino groups from amino acids to urea.

Glutamine and alanine are key transporters of amino groups between various tissues and the liver

In addition to the role of glutamate as a carrier of amino groups to GDH, glutamate serves as a precursor of glutamine, a process that consumes a molecule of ammonia. This is an important point because glutamine, along with alanine (see Chapter 2) is a key transporter of amino groups between various tissues and the liver, and is present in greater concentrations than most other amino acids in blood. The three forms of the same carbon skeleton, α-ketoglutarate, glutamate, and glutamine, are interconverted via amino transferases, glutamine synthetase, glutaminase, and GDH (Fig. 18.6). Thus glutamine can serve as a buffer for ammonia utilization, as a source of ammonia, and as a carrier of amino groups. Because ammonia is quite toxic, a balance must be maintained

Alanine transport role

Alanine and interorgan carbon and nitrogen flow

Much of the carbon flow that occurs between skeletal muscle – and several other tissues – and the liver is facilitated by the release of alanine into the blood by peripheral tissues, and its uptake by the liver. The alanine taken into the liver is converted to pyruvate and the nitrogen component is incorporated into urea. The pyruvate can be used in gluconeogenesis to produce glucose, which may be released into the blood for transport back to peripheral tissues. In this 'glucose–alanine cycle', three carbons are transported to the liver for gluconeogenesis, along with a nitrogen atom. The functioning of this cycle allows the net conversion of amino acid carbons to glucose, the elimination of amino acid nitrogen as urea, and the return of carbons to the peripheral tissues in the form of glucose. This cycle works in a fashion similar to the Cori cycle (Chapter 12), in which lactate is released into the blood by skeletal muscle and used for gluconeogenesis in the liver, the key difference being that alanine also carries a nitrogen atom to the liver. It is of significance that alanine and glutamine are released in approximately equal quantities from skeletal muscle and represent almost 50% of the amino acids released by skeletal muscle into the blood – an amount that far exceeds the proportion of these amino acids in muscle proteins.

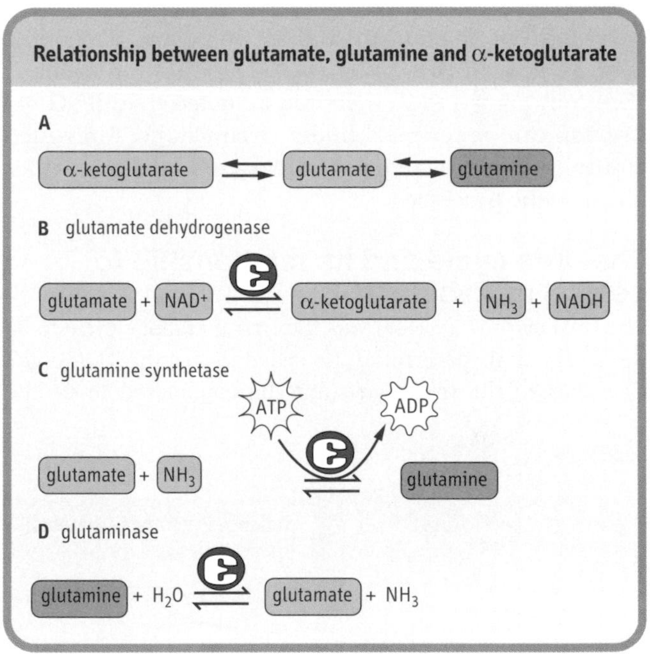

Fig. 18.6 The several forms of the carbon skeleton of glutamic acid have key roles in the metabolism of amino groups. (A) Three forms of the same carbon skeleton. (B) The GDH reaction is a reversible reaction that can produce glutamate from α-ketoglutarate or convert glutamate to α-ketoglutarate and ammonia, to enable urea synthesis. The latter reaction is extremely important in the synthesis of urea, as amino groups are fed to α-ketoglutarate via transamination from other amino acids. (C) Glutamine synthetase catalyzes an energy-requiring reaction that is very important in overall amine metabolism, as glutamine has a key role in the transport of amino groups from one tissue to another, and as a buffer against high concentrations of free ammonia. (D) The second half of the glutamine transport system for nitrogen is the enzyme, glutaminase, which hydrolyzes glutamine to glutamate and ammonia. This hydrolysis is important in the kidney in the management of proton transport and pH control.

Monosodium glutamate reaction

A healthy 30-year-old woman experienced the sudden onset of headache, sweating, and nausea after eating at an ethnic restaurant. She felt weak, and experienced some tingling and a sensation of warmth in her face and upper torso. The symptoms passed after about 30 minutes and she experienced no further problems. Upon visiting her doctor the next day, she learned that some individuals react to foods containing high levels of the food additive monosodium glutamate, the sodium salt of glutamic acid.

Comment. The flu-like symptoms that develop, previously described as 'Chinese Restaurant Syndrome', have been attributed to central nervous system (CNS) effects of glutamate or its derivative, the inhibitory neurotransmitter, γ-amino butyric acid (GABA). Interestingly, studies have shown that this phenomenon causes no CNS damage and that, although bronchospasm may be triggered in individuals with severe asthma, the symptoms are generally brief and completely reversible.

between its production and utilization. A summary of the sources and pathways that use or produce ammonia is shown in Figure 18.7. It should be noted that the GDH reaction can be reversed under circumstances in which amino groups are required for amino acid and other biosynthetic processes.

The urea cycle and its relationship to central metabolism

The urea cycle (Fig. 18.8) was the first metabolic cycle to be well-defined; its description preceded that of the TCA cycle. The start of the urea cycle may be considered to be the synthesis of carbamoyl phosphate from an ammonium ion and bicarbonate in liver mitochondria. This reaction, shown below, requires two molecules of adenosine triphosphate (ATP), and is catalyzed by the enzyme, carbamoyl phosphate synthetase I (CPS I), which is found in high concentrations in the mitochondrial matrix.

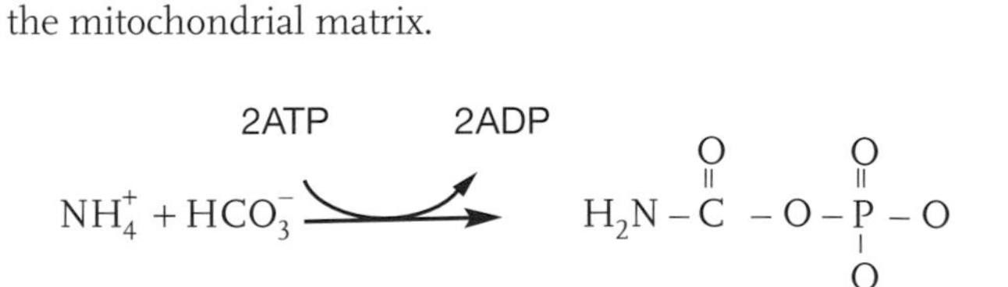

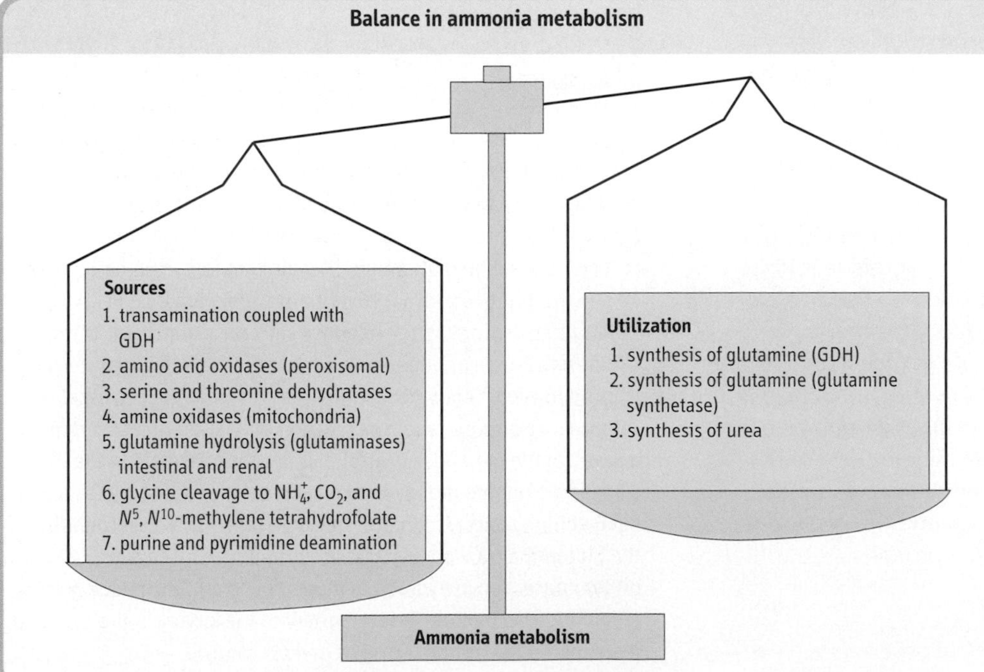

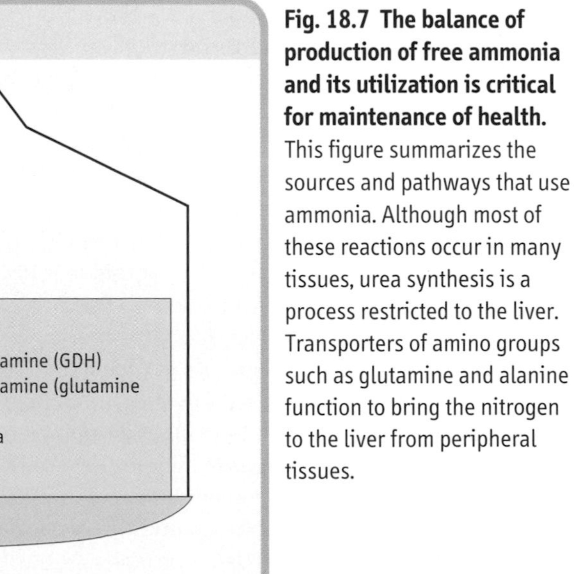

Fig. 18.7 The balance of production of free ammonia and its utilization is critical for maintenance of health. This figure summarizes the sources and pathways that use ammonia. Although most of these reactions occur in many tissues, urea synthesis is a process restricted to the liver. Transporters of amino groups such as glutamine and alanine function to bring the nitrogen to the liver from peripheral tissues.

Parkinson's disease

An otherwise healthy, 60-year-old man noticed that he occasionally observed a tremor in his left arm when relaxing and watching television. He also noticed occasional muscle cramping in his left leg, and his spouse noticed that he would occasionally develop a mask-like stare. A visit to his doctor for a complete physical examination and a subsequent visit to a neurologist confirmed a diagnosis of Parkinson's disease, and he was prescribed a medication that contained L-dihydroxyphenylalanine and a monoamine oxidase inhibitor. L-Dihydroxyphenylalanine is a precursor of the neurotransmitter dopamine, while monoamine oxidase is the enzyme responsible for the oxidative deamination and degradation of dopamine. His symptoms improved immediately, although, with the passage of time, he experienced significant side effects from the medication, including an inability to initiate movements of major muscle groups.

Comment. This condition, which occurs commonly in the elderly, can also occur in younger individuals. It is a progressive disease caused by the death of dopamine-producing cells in the substantia nigra and the locus ceruleus. Although medications can markedly reduce the symptoms, the disease is progressive, and may result in severe disability from loss of motor function. The very high doses of medication required to enable the drugs to cross the blood/brain barrier, and the use of an inhibitor of monoamine oxidase, which, in addition to dopamine, has a role in the metabolism of other neurotransmitters, are likely to be responsible for the many side effects that result from treatment. Because of this, some efforts have been made to develop transplantation of dopamine-producing tissues into the brain of affected individuals, but with limited success to date.

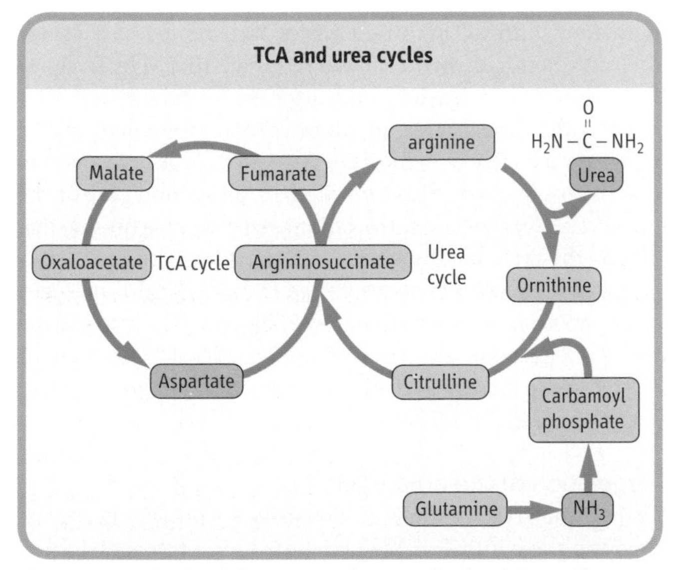

Fig. 18.8 Analysis of the urea cycle reveals that it is really two cycles, with the carbon flow split between the primary urea synthetic process and the recycling of fumarate to aspartate, which involves parts of the citric acid cycle.

Hereditary hyperammonemia

An apparently healthy 5-month-old female infant was brought to a pediatrician's office by her mother, with a complaint of periodic bouts of vomiting and a failure to gain weight. The mother also reported that the child would oscillate between periods of irritability and lethargy. Subsequent examination and laboratory results revealed an abnormal electroencephalogram, a markedly increased concentration of plasma ammonia (323 μmol/L [550 μg/dL]; the normal range is 15–88 μmol/L [25–150 μg/dL]), and greater than normal concentrations of glutamine, but low concentrations of citrulline. Orotate (a pyrimidine nucleotide precursor) was noted to be excreted in the urine.

Comment. The infant was admitted to hospital and treated with intravenous phenylacetate and benzoate along with arginine. The infant improved rapidly, and was discharged from hospital on a low-protein diet with arginine supplementation. Subsequent biopsy of the patient's liver indicated a level of ornithine transcarbamoylase activity in the tissue that was about 10% of normal.

Carbamoyl phosphate synthesis

The enzyme carbamoyl phosphate synthetase I (CPS I), which participates in urea synthesis, is found in the mitochondrion and primarily in the liver; a second enzyme, CPS II, is found in the cytosol and in virtually all tissues. Although the product of both these enzymes is the same, namely carbamoyl phosphate, the enzymes are derived from different genes and function in urea synthesis (CPS I) or pyrimidine biosynthesis (CPS II). Additional differences between the two enzymes include their source of nitrogen (NH_3 in the case of the former and glutamine in the latter) and their requirement for *N*-acetylglutamate (required by the former but not by the latter). Under normal circumstances, CPS I and II function independently and in different cellular compartments; however, when the urea cycle is blocked as a result of a deficiency in ornithine transcarbamoylase, the accumulated mitochondrial carbamoyl phosphate spills over into the cytosolic compartment and may stimulate excess pyrimidine synthesis, which is reflected in a build-up of orotic acid in the blood and urine.

Ammonia toxicity

Ammonia encephalopathy

All the mechanisms involved in ammonia toxicity – the encephalopathy in particular – are not well-defined. It is clear, however, that when its concentration builds up in the blood and other biological fluids, ammonia diffuses into cells and across the blood/brain barrier. This increase in ammonia causes an increased synthesis of glutamate from α-ketoglutarate and increased synthesis of glutamine. Although this is a normal detoxifying reaction in cells, when concentrations of ammonia are significantly increased, supplies of α-ketoglutarate in cells of the CNS may be depleted, resulting in inhibition of the TCA cycle and the production of ATP. There may be additional mechanisms accounting for the bizarre behavior observed in individuals with high blood concentrations of ammonia. One could speculate that either glutamate, as a major inhibitory neurotransmitter, or its derivative, γ-amino butyric acid (GABA), may also have some role in causing the CNS effects.

The mitochondrial enzyme, CPS I, is unusual in that it requires *N*-acetylglutamate as a cofactor. It is one of two carbamoyl phosphate synthetase enzymes that have key roles in metabolism. The second, CPS II, is found in the cytosol, does not require *N*-acetylglutamate, and is involved in pyrimidine biosynthesis.

Ornithine transcarbamoylase catalyzes the condensation of carbamoyl phosphate with the amino acid, ornithine, to form citrulline. In turn, aspartate is condensed with citrulline to form argininosuccinate. This step is catalyzed by argininosuccinate synthetase and requires ATP; the reaction cleaves the ATP to adenosine monophosphate (AMP) and inorganic pyrophosphate (PPi) (2 ATP equivalents). The formation of argininosuccinate brings to the complex the second nitrogen atom destined for urea. Argininosuccinate is in turn cleaved by argininosuccinase, to arginine and fumarate. The arginine produced in this series of reactions is then broken down by arginase, to a molecule of urea and one of ornithine. The ornithine can then be used to reinitiate this cyclic pathway, while the urea diffuses into the blood, is transported to the kidney, and excreted in urine. The net process of urea synthesis is summarized below:

$$CO_2 + NH_4^+ + 3ATP + aspartate + 2H_2O \Rightarrow\Rightarrow\Rightarrow$$
$$Urea + 2ADP + 2Pi + AMP + PPi + fumarate$$

The urea cycle is split between the mitochondrial matrix and the cytosol

The first two steps in the urea cycle occur in the mitochondrion, and citrulline then diffuses into the cytosol, where the cycle is completed with the release of urea and the regeneration of ornithine. Ornithine must be transported back across the mitochondrial membrane to reinitiate the cycle. Carbons from fumarate, released in the argininosuccinase step, may re-enter the mitochondrion after hydration to malate and be recycled via enzymes in the TCA cycle, to oxaloacetate and ultimately to aspartate, thus completing the second part of the urea cycle. Because the same carbon skeleton of ornithine that begins the cycle is returned after the arginase reaction, this pathway is sometimes also referred to as the ornithine cycle. Urea synthesis occurs virtually exclusively in the liver and the role of the enzyme, arginase, in other tissues is believed to be related more closely to ornithine requirements in those tissues than to the production of urea.

Regulation of the urea cycle

The primary regulation of the urea cycle appears to be through the control of the concentration of *N*-acetylglutamate, the required allosteric activator of CPS. High concentrations of arginine stimulate *N*-acetylation of glutamate. In addition, concentrations of the enzymes of this pathway increase or decrease in response to a high- or low-protein diet.

Defects in any of the enzymes of the urea cycle have serious consequences. Infants born with defects in any of the first four enzymes in this pathway may appear normal at birth, but rapidly become lethargic, lose body temperature, and may have difficulty breathing. Blood concentrations of ammonia increase quickly, followed by cerebral edema. A deficiency of arginase produces less severe symptoms, but is nevertheless characterized by increased concentrations of blood arginine and at least a moderate increase in blood ammonia. In individuals with high blood concentrations of ammonia, hemodialysis must be used, often followed by intravenous administration of sodium benzoate and phenyllactate. These compounds can condense with glycine and glutamine, respectively, to form water-soluble adducts, trapping the ammonia in a nontoxic form that can be excreted in the urine.

The concept of nitrogen balance

Because nitrogen metabolism is quite dynamic and there is no significant storage form of nitrogen or amino compounds in humans, a careful balance is maintained through the control of nitrogen excretion in relation to that ingested in the diet. In a normal, healthy diet, the protein content exceeds the amount required to supply essential and nonessential amino acids for protein synthesis, and the amount of nitrogen excreted is approximately equal to that taken in. Such a healthy adult would be said to be 'in nitrogen balance'. When there is a need to increase protein synthesis, such as in recovering from trauma or in a rapidly growing child, the amount of nitrogen excreted is less than that consumed in the diet, and the individual would be in 'positive nitrogen balance'. The converse is true in protein malnutrition: because of the need to synthesize essential body proteins, other proteins, such as muscle protein or hemoglobin, are degraded and more nitrogen is lost than is consumed in the diet. Such an individual would be said to be in 'negative nitrogen balance'. Fasting and starvation are also characterized by negative nitrogen balance, as body protein is degraded to amino acids and their carbon skeletons are used for gluconeogenesis. The concept of nitrogen balance reminds us of the continuous turnover in the normal human body of amino acids, proteins, and some nucleic acids.

Metabolism of the carbon skeletons of amino acids

Metabolic intermediates

Metabolism of the carbon skeletons of amino acids interfaces with carbohydrate and lipid metabolism

When one examines the metabolism of the carbon skeletons of the 20 common amino acids, there is an obvious interface with carbohydrate and lipid metabolism. Virtually all the carbons can be converted into intermediates in the glycolytic pathway, the TCA cycle, or lipid metabolism. The first step in this process is the transfer of the α-amino group by transamination to α-ketoglutarate or oxaloacetate, providing glutamate and aspartate, the sources for the nitrogen atoms of the urea cycle (Fig. 18.9). The single exception to this is lysine, which does not undergo transamination. Although the details of pathways for the various amino acids vary, the general rule is that there is loss of the amino group, followed by either direct metabolism in a central pathway (glycolysis, the TCA cycle, or ketone body metabolism), or one or more additional conversions to yield an intermediate in one of the central pathways. Examples of amino acids that follow the former scheme include alanine, glutamate, and aspartate, which yield pyruvate, α-ketoglutarate and oxaloacetate, respectively, upon removal of their amino group. The branched-chain amino acids, leucine, valine, and isoleucine, and the aromatic amino acids, tyrosine, tryptophan, and phenylalanine are examples of the latter, more complex scheme.

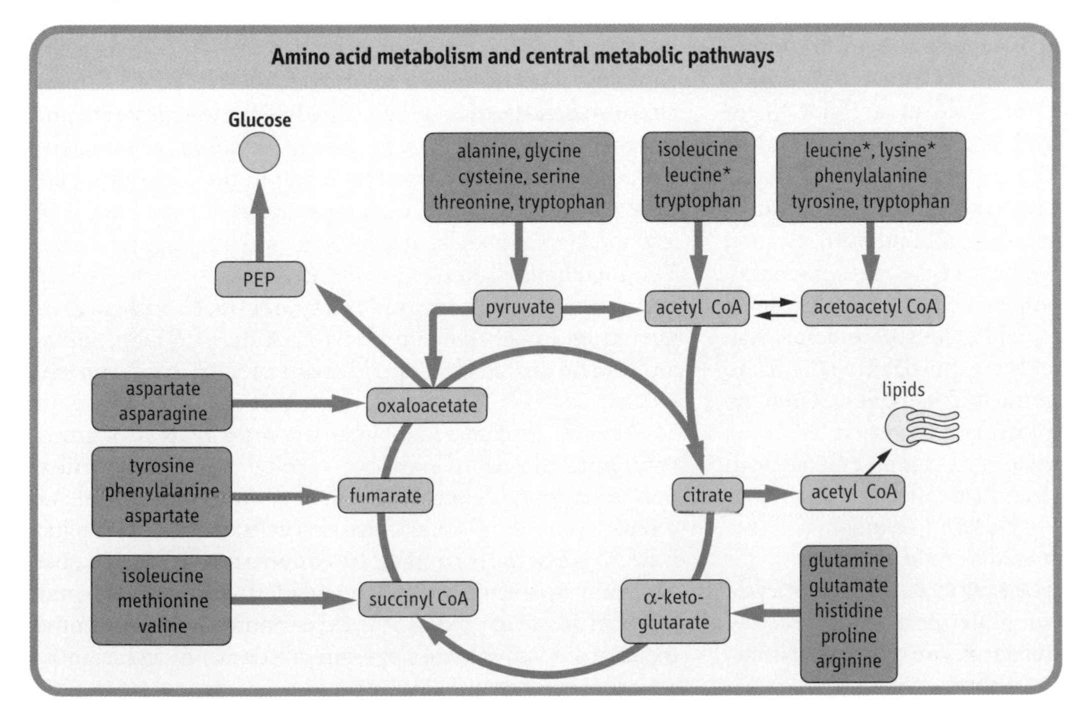

Fig. 18.9 This figure summarizes the interactions between amino acid metabolism and central metabolic pathways. *These amino acids are ketogenic only. PEP, phosphoenolpyruvate.

Enzymes of the urea cycle		
Enzyme	**Reaction catalyzed**	**Remarks**
carbamoyl phosphate synthetase	formation of carbamoyl phosphate from ammonia and CO_2	requires *N*-acetylglutamate as a cofactor, uses 2 ATP, located in the **mitochondrion**, deficiency leads to high blood concentrations of ammonia and related toxicity
ornithine transcarbamoylase	formation of citrulline from ornithine and carbamoyl phosphate	releases P_i, an example of a transferase, located in the **mitochondrion**, deficiency leads to high blood concentrations of ammonia and orotic acid, as carbamoyl phosphate is shunted to pyrimidine biosynthesis
argininosuccinate synthetase	formation of argininosuccinate from citrulline and aspartate	requires ATP, which is cleaved to AMP + PP_i – an example of a ligase, located in the **cytosol**, deficiency leads to high blood concentrations of ammonia and citrulline
argininosuccinase	cleavage of argininosuccinate to arginine and fumarate	an example of a lyase, located in **cytosol**, deficiency leads to high blood concentrations of ammonia and citrulline
arginase	cleavage of arginine to ornithine and urea	an example of a hydrolase, located in the **cytosol** and primarily in the liver, deficiency leads to moderately increased blood ammonia and high blood concentrations of arginine

Fig. 18.10 Five enzymes catalyze the urea cycle in lines. The first enzyme, CPS-1, which fixes NH_4^+ as cobalamin phosphate, is the regulatory enzyme and is sensitive to the allosteric effector, *N*-acetylglutamate.

Amino acids may be either glucogenic or ketogenic

Depending on the point at which the carbons from an amino acid enter central metabolism, that amino acid may be considered to be either glucogenic or ketogenic (i.e. possessing the ability to increase the concentrations of either glucose or ketone bodies, respectively, when fed to an animal). Those amino acids that feed carbons into the TCA cycle at the level of α-ketoglutarate, succinyl CoA, fumarate, or oxaloacetate, and those that produce pyruvate can all give rise to the net synthesis of glucose via gluconeogenesis and are hence designated glucogenic. Those amino acids that feed carbons into central metabolism at the level of acetyl CoA or acetoacetyl CoA are considered ketogenic. It should be borne in mind that, because of the nature of the TCA cycle, no net flow of carbons can occur between acetate or its equivalent to glucose via gluconeogenesis (see Chapter 20). Several amino acids, primarily those with more complex or aromatic structures, can yield fragments that may be both ketogenic and glucogenic. An examination of Figure 18.9 indicates that, while the majority of amino acids are glucogenic, several are considered to be both glucogenic and ketogenic. Only the amino acids leucine and lysine are regarded as being exclusively ketogenic and, because of its complex metabolism and lack of ability to undergo transamination, some authors do not consider lysine to be exclusively ketogenic. These classifications may be summarized as follows:

- **glucogenic amino acids** (yield pyruvate, or a TCA cycle intermediate): aspartic acid, glutamic acid, asparagine, glutamine, histidine, proline, arginine, glycine, alanine, serine, cysteine, methionine, valine;
- **ketogenic amino acids** (yield acetoacetate or acetyl CoA): leucine, lysine;
- **both glucogenic and ketogenic amino acids** (yield pyruvate, or a TCA cycle intermediate, in addition to acetoacetate or acetyl CoA): phenylalanine, tyrosine, tryptophan, isoleucine, threonine.

Metabolism of the carbon skeletons of selected amino acids

Leucine is an example of a ketogenic amino acid. Its catabolism begins with transamination to produce 2-ketoisocaproate. The metabolism of 2-ketoisocaproate requires oxidative decarboxylation by a dehydrogenase complex to produce isovaleryl CoA. The further metabolism of isovaleryl CoA involves several steps that result in the formation of 3-hydroxy-3-methylglutaryl CoA, a precursor of both acetyl CoA and the ketone bodies, acetoacetate and 3-hydroxybutyrate. The metabolism of leucine and the other branched-chain amino acids is summarized in Figures 18.11 and 18.12. It should be noted that propionyl CoA derived from either amino acid degradation or odd-chain fatty acids is converted to succinyl CoA.

Alanine, aspartate and glutamate are examples of glucogenic amino acids. In each case, through either transamination or oxidative deamination, the resulting α-keto acid is a direct precursor of oxaloacetate via central metabolic pathways. Oxaloacetate can then be converted to PEP, and subsequently to glucose via gluconeogenesis. Other glucogenic amino acids reach the TCA cycle or related metabolic intermediates through several steps, after the removal of the amino group (see Fig. 18.9).

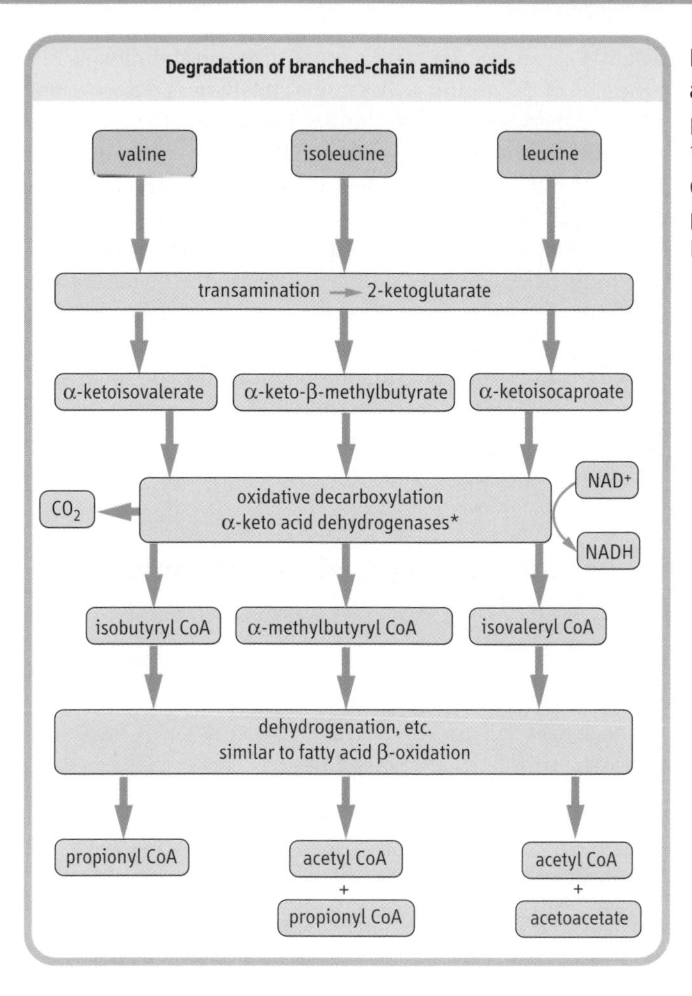

Fig. 18.11 Metabolism of the branched-chain amino acids produces acetyl CoA and acetoacetate. In the case of valine and isoleucine, propionyl CoA is produced and metabolized, in two steps, to succinyl CoA. *In common with pyruvate dehydrogenase and alpha ketoglutarate dehydrogenase (αKGDH), these enzymes involve thiamine pyrophosphate (thiamine-PP), lipoic acid, flavin adenine dinucleotide, NAD^+ and CoA.

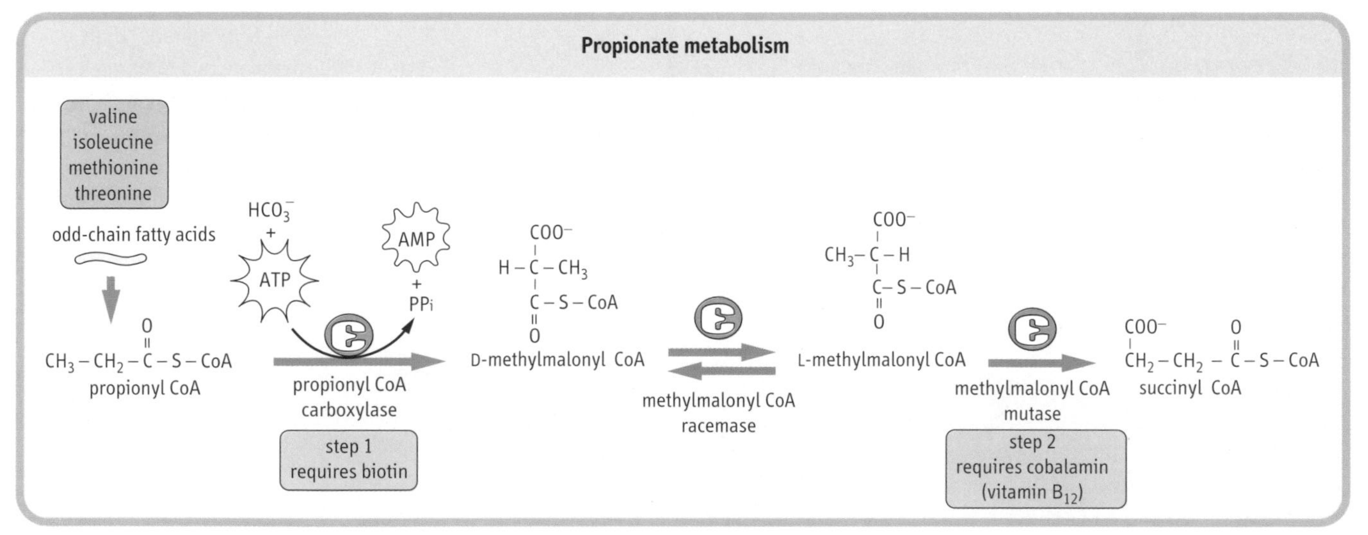

Fig. 18.12 The conversion of propionyl CoA to succinyl CoA is a two-step process. The initial step involves the addition of a carboxyl group by a biotin-requiring carboxylase. Methylmalonyl CoA mutase is a cobalamin (vitamin B_{12})-containing enzyme and a deficiency of vitamin B_{12} leads to a number of symptoms, including methylmalonic aciduria.

Tryptophan is a good example of an amino acid that yields both glucogenic and ketogenic precursors. After cleavage of its heterocyclic ring and a complex set of reactions, the core of the amino acid structure is released as alanine (a glucogenic precursor), while the balance of the carbons are ultimately converted to glutaryl CoA (a ketogenic precursor). A summary of the key points in the catabolism of the aromatic amino acids is given in Figure 18.13.

Amino acid biosynthesis

Evolution has left our species without the ability to synthesize almost half the amino acids that are essential building blocks and precursors of a variety of critical molecules

Humans use 20 amino acids to build peptides and proteins that are essential to the many functions of their cells. Biosynthesis of the amino acids involves biosynthesis of the carbon skeletons for their corresponding α-keto acids, followed by addition of the amino group via transamination. However, humans are capable of carrying out the biosynthesis of the carbon skeletons of only about half of those α-keto acids. Amino acids that we cannot synthesize are termed essential amino acids, and are required in the diet. While almost all of the amino acids can be classified as clearly essential or nonessential, a few require further qualification. For example, although cysteine is not generally considered an essential amino acid because it can be derived from the nonessential amino acid, serine, its sulfur must come from the required or essential amino acid, methionine. Similarly, the amino acid, tyrosine, is not required in the diet, but must be derived from the essential amino acid, phenylalanine. This relationship between phenylalanine and tyrosine will be discussed further in considering the inherited disease, phenylketonuria (PKU). Figures 18.14 and 18.15 list the nonessential and essential amino acids, and the source of the carbon skeleton in the case of those not required in the diet.

Inherited diseases of amino acid metabolism

In addition to the deficiencies in the urea cycle, specific defects in the metabolism of the carbon skeletons of various amino acids were among the first disease states to be associated with simple inheritance patterns. These observations gave rise to the concept of the genetic basis of inherited

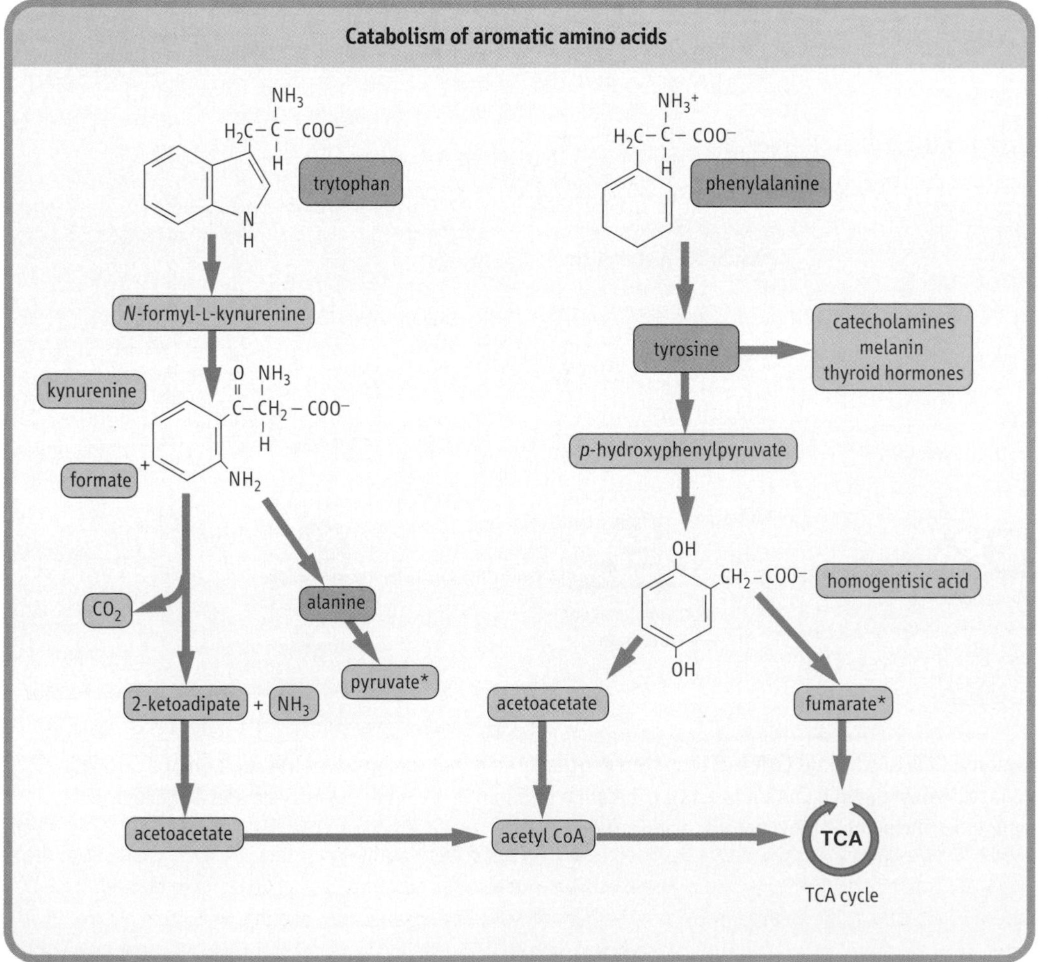

Fig. 18.13 This figure summarizes the catabolism of the aromatic amino acids, illustrating the pathways that lead to ketogenic and glucogenic precursors derived from both tyrosine and tryptophan. *Both pyruvate and fumarate can lead to net glucose synthesis. They constitute the gluconeogenic portions of the metabolism of these amino acids.

Origin of nonessential amino acids

Amino acid	Source in metabolism, etc.
alanine	from pyruvate via transamination
aspartic acid, asparagine, arginine, glutamic acid, glutamine, proline	from intermediates in the citric acid cycle
serine	from 3-phosphoglycerate (glycolysis)
glycine	from serine
cysteine*	from serine; requires sulfur derived from methionine
tyrosine*	derived from phenylalanine via hydroxylation

Fig. 18.14 Origins of non-essential amino acids (i.e. those not required in a normal diet). *These are examples of nonessential amino acids that depend on adequate amounts of an essential amino acid (i.e. one that is required in the diet).

Essential amino acids

Mnemonic	Amino acid*	Notes or comments
P	phenylalanine	required in the diet also as a precursor of tyrosine
V	valine	one of three branched-chain amino acids
T	threonine	metabolized like a branched-chain amino acid
T	tryptophan	its complex heterocyclic side chain can not be synthesized in humans
I	isoleucine	one of three branched-chain amino acids
M	methionine	provides the sulfur for cysteine and participates as a methyl donor in metabolism; the homocysteine is recycled
H	histidine	its heterocyclic side chain cannot be synthesized in humans
A	arginine	whereas arginine can be derived from ornithine in the urea cycle in amounts sufficient to support the needs of adults, growing animals require it in the diet
L	leucine	a pure ketogenic amino acid
L	lysine	neither of the nitrogens of lysine can undergo transamination

Fig. 18.15 Essential amino acids (i.e. those required in the diet). *Some texts suggest that the essential amino acids can be memorized by using a mnemonic for the first letters of the names of the required amino acids: PVT TIM HALL.

metabolic disease states, also known as 'inborn errors of metabolism'. Garrod considered a number of disease states that appeared to be inherited in a Mendelian pattern, and proposed a correlation between these abnormalities and specific genes, in which the disease state could be either dominant or recessive. Dozens of inborn errors of metabolism have now been described, and the molecular defect has been described for many of them. Three classical inborn errors of metabolism will be discussed in some detail here.

Phenylketonuria

PKU is an inborn error of metabolism resulting from a deficiency of the enzyme, phenylalanine hydroxylase. The hydroxylation of phenylalanine is a required step in both the normal degradation of the carbon skeleton of this amino acid and the

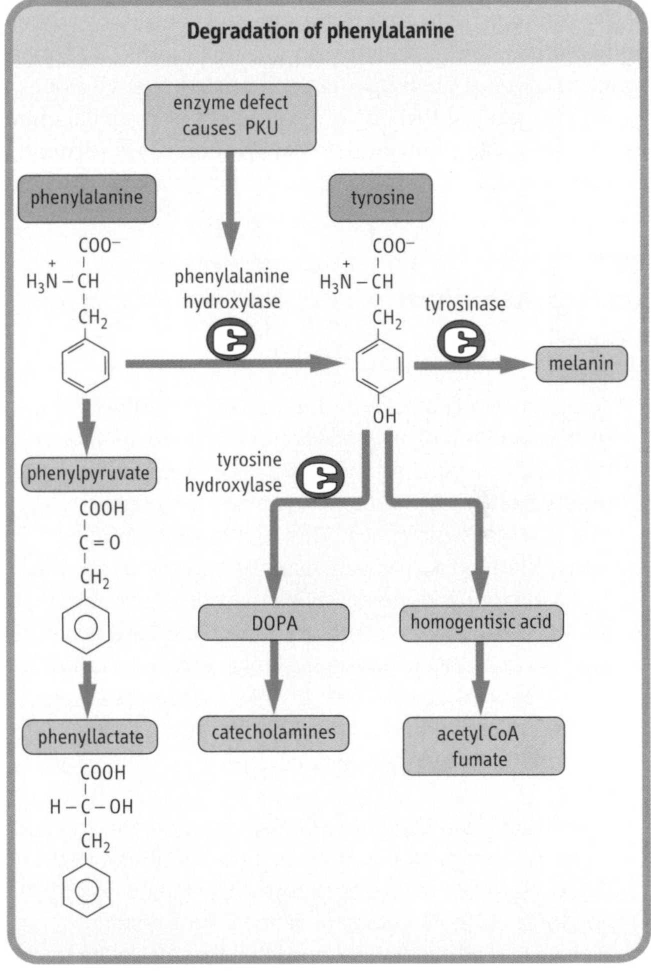

Fig. 18.16 In order to enter normal metabolism, phenylalanine must be hydroxylated by the enzyme, phenylalanine hydroxylase. A defect in this enzyme leads to PKU, an accumulation of high concentrations of both phenylalanine and abnormal metabolites such as phenylpyruvate and phenyllactate. Tyrosine can participate in three basic pathways, which lead, respectively, to the formation of central metabolites (acetyl CoA and fumarate), catecholamines, and a reactive quinone that leads to melanin. DOPA, dihydroxyphenylalanine.

synthesis of tyrosine (Fig. 18.16). When untreated, this metabolic defect leads to excessive urinary excretion of phenylpyruvate and phenyllactate, and severe mental retardation. In addition, individuals with PKU tend to have very light pigmentation, unusual gait, stance, and sitting posture, and a high frequency of epilepsy. In the USA, this autosomal recessive defect occurs in about 1 in 16 000 live births in white populations and in about 1 in 50 000 in the black population. Because of this, and the ability to prevent the most serious consequences of the defect through diet control, newborns in most developed countries are routinely tested for blood concentrations of phenylalanine.Fortunately, with early detection and the use of a diet restricted in phenylalanine but supplemented with tyrosine, most of the mental retardation can be avoided. Mothers who are homozygous for this defect have a very high probability of bearing children with congenital defects and mental retardation unless their phenylalanine concentrations can be decreased through diet control. The developing fetus is very sensitive to the toxic effects of high concentrations of phenylalanine and related phenylketones.

While classical PKU involves a defect in phenylalanine hydroxylase, not all hyperphenylalaninemias are related to this enzyme. There are other cases in which a defect occurs in the biosynthesis or reduction of a required tetrahydrobiopterin cofactor. Although such defects do lead to increased concentrations of phenylalanine, they cannot be treated by diet alone, probably because of a requirement for the cofactor in other pathways that are related to the biosynthesis of neurotransmitters.

Alkaptonuria (black urine disease)

A second inherited defect in the phenylalanine–tyrosine pathway involves a deficiency in the enzyme that catalyzes the oxidation of homogentisic acid (an intermediate in the metabolic breakdown of tyrosine and phenylalanine). In this condition, which occurs in 1 in 1 000 000 live births, homogentisic acid accumulates and is excreted in urine. This compound autooxidizes on standing or on treatment with alkali, and gives the urine a dark color. Unfortunately, individuals with alkaptonuria ultimately suffer from deposition of dark (ochre-colored) pigment in cartilage tissue, with subsequent tissue damage, including severe arthritis; the onset of these symptoms is generally in the 3rd or 4th decade of life. This autosomal recessive disease was the first of several that Garrod considered in proposing his initial hypothesis for inborn errors of metabolism. Although alkaptonuria is relatively benign compared with PKU, little is available in the way of treatment other than relief of symptoms. High doses of

Albinism

A full-term infant, born to a normal and healthy mother and father, was observed to have a marked lack of pigmentation. The infant, who appeared to be otherwise normal, had blue eyes and very light blond, almost white, hair. This lack of pigmentation was confirmed as classical albinism on the basis of a family history and the establishment of a lack of the enzyme, tyrosinase, which is responsible for a two-step process involving the hydroxylation of tyrosine to dihydroxyphenylalanine (DOPA) and a subsequent further oxidation to a quinone, which leads to the formation of melanin in melanocytes.

Comment. As a separate DOPA-producing enzyme, tyrosine hydroxylase, is involved in biosynthesis of the catecholamine neurotransmitters and Albinos do not appear to have neurological deficits. As a result of their lack of pigmentation, however, they are quite sensitive to damage from sunlight and must take added precautions against ultraviolet radiation from the sun. Albinos have normal eyesight, in spite of the lack of pigmentation, but are generally very sensitive to bright light. (See Figure 18.15 for an outline of this pigment-forming pathway.)

Mixed-function oxidases

Phenylalanine hydroxylase

The hydroxylation of phenylalanine is a critical step in the metabolism of that amino acid, and in the biosynthesis of tyrosine and its several important derivatives. Phenylalanine hydroxylase is an example of a 'mixed-function oxidase', an enzyme that utilizes a reduced form of a cofactor and molecular oxygen to carry out a hydroxylation reaction. In this case, the reduced cofactor is tetrahydrobiopterin. In this reaction, the oxygen donates one of its atoms to the hydroxylation and the other to water, while the reduced cofactor is oxidized (in this case, to dihydrobiopterin). The net result is the hydroxylation of the substrate, the production of a molecule of water, and the oxidation of the cofactor. In order for this reaction to continue, the dihydrobiopterin must be reduced to its tetrahydro form again, and this requires a second enzyme, dihydrobiopterin reductase, which uses NADH to drive the reduction. The further hydroxylation of tyrosine, in the pathway that leads to catecholamines, requires a similar mixed-function oxidase, tyrosine hydroxylase.

ascorbic acid have been used in some patients, to help reduce the deposition of pigment on collagen, but the progress of the disease has not been significantly affected by this strategy.

Maple syrup urine disease (MSUD)

The normal metabolism of the branched-chain amino acids, leucine, isoleucine, and valine, involves loss of the a-amino group, followed by an oxidative decarboxylation of the resulting α-keto acid. This decarboxylation step is catalyzed by branched-chain keto acid decarboxylase, a multienzyme complex that is associated with the inner membrane of the mitochondrion. In approximately 1 in 300 000 live births in the general US population, a defect in this enzyme leads to accumulation of the keto acids corresponding to these branched-chain amino acids in the blood, and then to branched-chain ketoaciduria. When untreated or unmanaged, this condition may lead to both physical and mental retardation of the newborn and a distinct maple syrup odor of the urine. This defect can be partially managed with a low-protein or modified diet, but not in all cases. In some instances, supplementation with high doses of thiamine pyrophosphate, a cofactor for this enzyme complex, has been helpful.

Signaling molecules

Derivatives of amino acids have important roles as signaling molecules

In addition to their role as building blocks for peptides and proteins, several of the common amino acids serve as precursors of amino acid derivatives that function as neurotransmitters or hormones. Some of the amino acids may be used as neurotransmitters directly, for example glycine, aspartate, and glutamate, whereas others may be converted to neurotransmitters or hormones through modification. Tyrosine is notable in that it serves as a precursor of several neurotransmitters and of the thyroid hormones. The pathways of biosynthesis and mechanisms of action of neurotransmitters are discussed in Chapter 38.

Summary

In this chapter, we have seen that the metabolism of amino acids is integrally related to central metabolism. The degradative or catabolic metabolism of amino acids generally begins with the removal of the α-amino group, which is transferred to α-ketoglutarate and oxaloacetate, and ultimately excreted in the form of urea. The resulting carbon skeletons are converted to an intermediate or intermediates that enter central metabolism at various points. Because carbon skeletons corresponding to the various amino acids can be derived from or feed into the glycolytic pathway, the TCA cycle, fatty acid biosynthesis, and gluconeogenesis, amino

Selenocysteine

In addition to the 20 common amino acids found in proteins, a 21st amino acid has recently been discovered and shown to be an essential residue in several enzyme systems. This amino acid, selenocysteine, is required at the active sites of the antioxidant enzyme, glutathione peroxidase (see Chapter 10), 5′-deiodinases, glycine reductase, and several other enzymes. This rare and unusual amino acid is incorporated into these proteins by a transfer ribonucleic acid (tRNA) that bears a UCA anticodon and which is initially aminoacylated with serine. The tRNA is then modified to a selenocysteine-bearing species through the action of selenophosphatase. Selenocysteine has unique properties, and there is at least one report that the substitution of selenocysteine with cysteine resulted in a marked decrease in enzyme activity. It is because of the need for selenocysteine that trace amounts of selenium are required in the diet.

Cystinuria

A 21-year-old man came to the emergency room with severe pain in his right side and back. Subsequent examination and evaluation indicated a kidney stone, and increased concentrations of cystine, arginine, and lysine in the urine. This patient exhibited the charateristic symptoms of cystinuria.

Comment. Cystinuria is caused by a defect in one of the amino acid transport systems that is responsible for moving cysteine and several other amino acids across epithelial cell membranes. In this case, cysteine, which is normally reabsorbed by means of this system as fluids are filtered through the kidney glomerulus, cannot be reabsorbed and remains in the urine. The cysteine spontaneously oxidizes to its disulfide form, cystine. Because cystine has very limited solubility, it tends to precipitate in the urinary tract, forming kidney stones. The condition is generally treated by restricting the dietary intake of methionine (a biosynthetic precursor of cysteine), encouraging high fluid intake to keep the urine dilute, and, more recently, with various drugs that may convert urinary cysteine to a more soluble compound that will not precipitate.

acid metabolism should not be considered as an isolated system. Although amino acids are not stored like glucose (glycogen) or fatty acids (triacylglycerols), they have an important and dynamic role, not only in providing the building blocks for the synthesis and turnover of protein, but also in normal energy metabolism, providing a carbon source for gluconeogenesis when needed and an energy source of last resort in starvation. In addition, amino acids provide precursors for the biosynthesis of a variety of small signaling molecules, including neurotransmitters and hormones. The severe consequences of abnormal metabolism evident in inherited diseases such as phenylketonuria(PKU) and maple syrup urine disease (MSUD) afford an indication of the importance of normal amino acid metabolism.

Further reading

Brusilow SW, Maestri NE. Urea cycle disorders: diagnosis, pathophysiology and therapy. *Adv Pediatr* 1996;**43**:127–170.

Cark BJ. After a positive Guthrie test – what next? Dietary management for the child with phenylketonuria. *Eur J Clin Nutr* 1992;**46**:S33–S39.

Chuang DT, Shih VE. Disorders of branched chain amino acid and keto acid metabolism. In: Scriver CR, Beaudet AL, Sly WS, Valle D, eds. *The metabolic and molecular basis of inherited disease, 7th ed, Vol 1*. New York: McGraw-Hill; 1995:1239–1277.

Garrod AE, Inborn errors in metabolism, Oxford University Press; 1909.

Le Boucher J, Cynober L. Protein metabolism in burn injury. *Ann Nutr Metab* 1997;**41**:69–82.

Levy PA, Miller JB, Shapira E. The advantage of phenylalanine to tyrosine ratio for the early detection of phenylketonuria. *Clin Chim Acta* 1998;**270**:177–181.

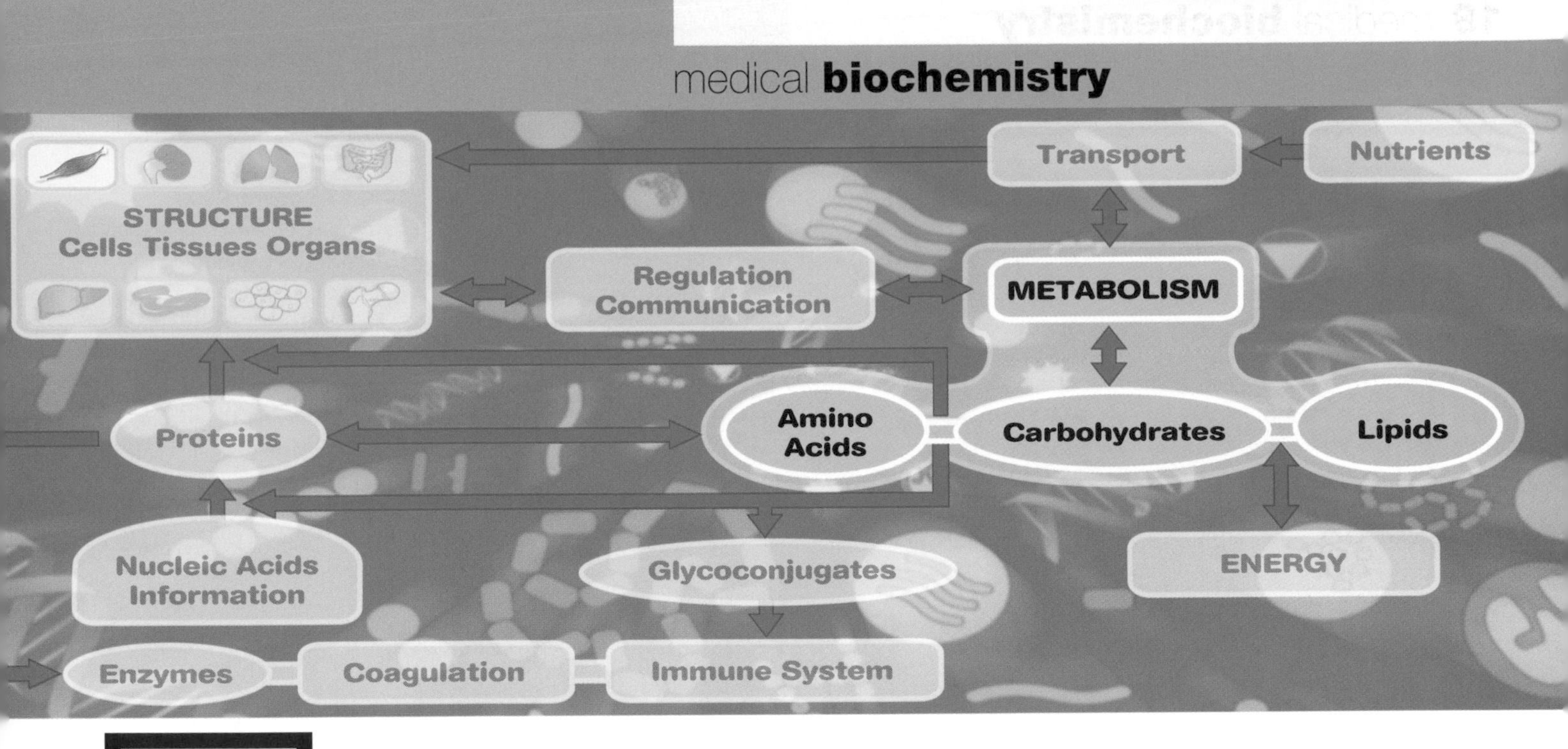

19 Muscle: Energy Metabolism and Contraction

Introduction

Muscle is the major consumer of fuel and ATP in the body

Muscle accounts for about 40% of our body mass and is the major consumer of body fuels and adenosine triphosphate (ATP) during a normal day. It exemplifies the relationship between protein structure and function and illustrates the conservation of the chemical energy of ATP in the form of the conformational energy of muscle protein. Fatty acids and ketone bodies, provided by diet, adipose tissue and liver, are its major sources of energy, but glucose mobilization from glycogen is essential for short bursts of activity and continued fat utilization during prolonged exercise.

ATP is used for muscle contraction

Muscle uses ATP for both maintenance of ion gradients and for muscle contraction. It is rich in creatine phosphate, a storage form of high-energy phosphate used for rapid regeneration of ATP. Muscle consists largely of two filamentous proteins, actin and myosin. These proteins consume ATP in the presence of Ca^{2+}, leading to muscle contraction. The Ca^{2+} content of the muscle cytoplasm (sarcoplasm) is normally very low, 10^{-7} mol/L or less, but increases rapidly by ~100-fold in response to neural stimulation, leading to contraction. Mechanochemical coupling during contraction involves interdigitation of actin and myosin chains, a process described by the sliding-filament model of muscle contraction. This chapter will deal with muscle from three points of view: its structure, the mechanism of mechanochemical coupling and its energy metabolism.

Structure of muscle

In skeletal and cardiac muscle, the structure of muscle is hierarchical

Skeletal and cardiac muscle have a hierarchical structure (Fig. 19.1). At the macroscopic level, these muscles consist of bundles (fasciculi) of elongated, multinucleated fiber cells (myofibers), contributing to the striated (lined, furrowed) appearance of skeletal muscle. The myofiber cells contain bundles of myofibrils, which are, in turn, composed of myofilament proteins (Fig. 19.2).

There is a repeating pattern of light- and dark-staining regions in the myofibril

Electron microscopic analysis indicates that, like fibrous collagen in connective tissue, the myofibril has a periodic structure (Fig.19.3), with alternating light- and dark-staining regions known as the I (isotropic)- and A (anisotropic)-bands, respectively. At the center of the I-band is a discrete,

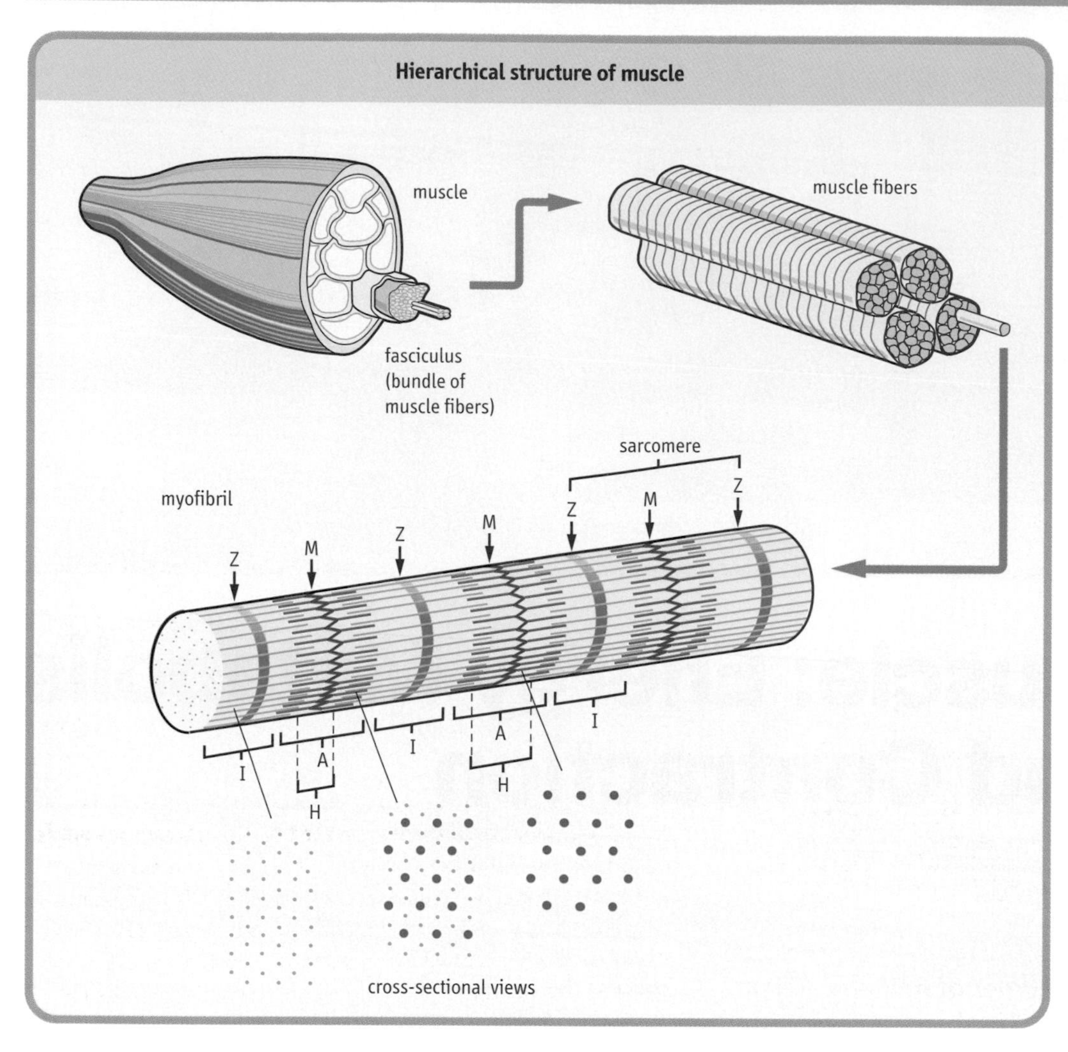

Fig. 19.1 Hierarchical structure of skeletal muscle, showing an exploding view of fasciculi, myofibers, myofibrils and myofilament proteins. The cross-sectional views show the I-band (thin, actin filaments), the H-zone (thick, myosin filaments), and the dark-staining regions of the A-band corresponding to the region of overlap of actin and myosin filaments.

Elements of muscle structure

Microscopic unit	Fasciculus: bundle of muscle cells
cellular unit	myofiber cell: long, multinucleated cell
subcellular unit	myofibril: composed of myofilament proteins
functional unit	sarcomere: contractile unit, repeating unit of the myofibril
myofilament components	proteins: primarily actin and myosin

Fig. 19.2 The structural elements of muscle arranged in descending order of size.

darker staining Z-line, while the center of the A-band has a lighter staining H-zone with a central M-line. The functional (contracting) unit of the myofibril, known as the sarcomere (Fig. 19.3; see also Fig. 19.1), extends from one Z-line to the next. The sarcomere may decrease by as much as 70% in length during muscle contraction.

During contraction, actin and myosin filaments slide past each other; this leads to reduction in the length of the sarcomere

Myofilaments are composed primarily of two filamentous proteins, actin and myosin. Cross-sectional analysis of the myofibril (Fig. 19.2C) indicates that the actin and myosin filaments extend in opposite directions from both sides of the Z- and M-lines, respectively, and overlap and slide past one another during the contractile process. The Z- and M-lines are, in effect, base plates for anchoring the actin and myosin filaments. Increased overlap of these filaments during contraction causes a decrease in length of the H-zone (myosin only) and I-bands (actin only) of the sarcomere.

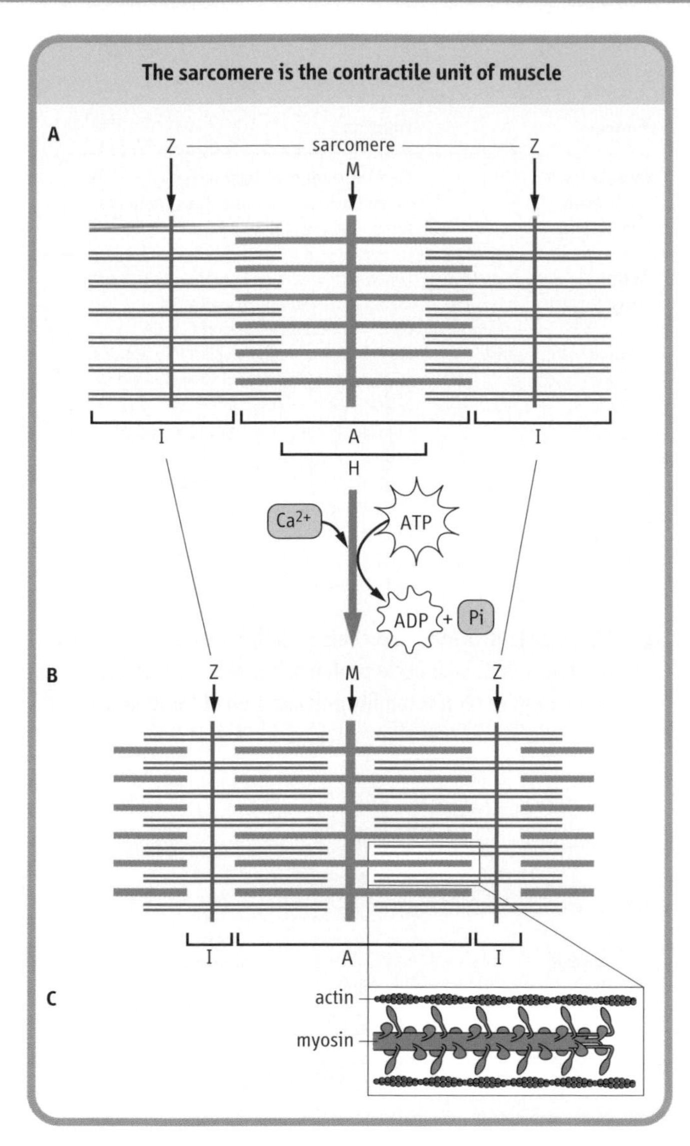

Fig. 19.3 Schematic structure of the sarcomere, indicating the distribution of actin and myosin in the A- and I-bands. (A) relaxed sarcomere; (B) contracted sarcomere; (C) magnification of contracted sarcomere, illustrating the polarity of the arrays of myosin molecules. Increased overlap of actin and myosin filaments during contraction, accompanied by a decrease in the length of the I-band, illustrates the sliding-filament model of muscle contraction.

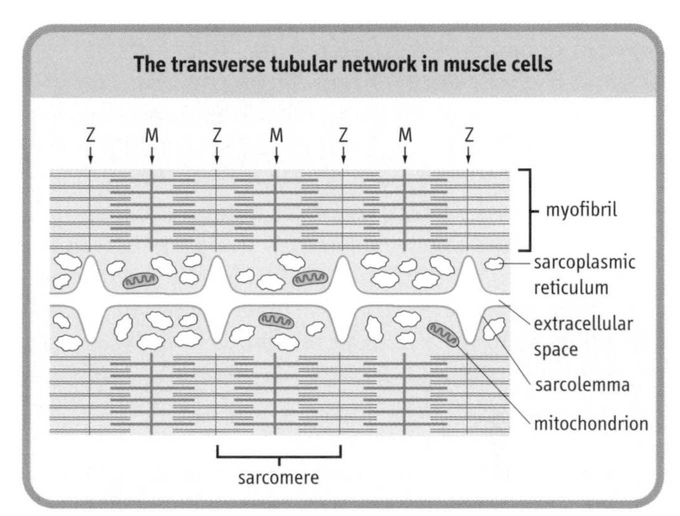

Fig. 19.4 Side-view of the transverse tubular network in muscle cells. Transverse tubules are invaginations of the sarcolemma, which are in intimate contact with the sarcoplasmic reticulum (SR). The SR is a continuous, tubular compartment in close association with the myofibrils. The transverse tubules are extensions of the sarcolemma around the Z-line. They transmit the depolarizing nerve impulse to terminal regions of the SR, coordinating Ca^{2+} release and contraction of the myofibril.

The sarcoplasmic reticulum surrounds, and is in intimate contact with the myofiber bundles

Myofiber bundles in muscle cells are enveloped in a specialized membrane structure, known as the sarcoplasmic reticulum (SR), derived from the endoplasmic reticulum of the muscle cell. The SR is rich in a Ca^{2+}-ATPase, which pumps Ca^{2+} into the SR, maintaining cytosolic Ca^{2+} in the muscle cell at submicromolar (~10^{-7} mol/L) concentrations. At the same time, the concentration of Ca^{2+} in the SR is in the mmol/L range, much of the Ca^{2+} being bound to the SR protein, calsequestrin. The plasma membrane of the myofibril cell, known as the sarcolemma, invaginates in and around the myofibrils at the Z-lines, forming a series of transverse tubules that interact with the SR (Fig. 19.4). During muscle contraction, the myofibril cell responds to a neural input, initiating a wave of depolarization of the Na^+/K^+-gradient across the muscle plasma membrane, which is transmitted to the SR, causing a voltage-gated opening of Ca^{2+}-channels. Ca^{2+} is released from the SR, and the concentration of Ca^{2+} in the muscle cytoplasm (sarcoplasm) increases by as much as 100-fold, initiating ATP hydrolysis and muscle contraction. Through intimate contact with the SR, the meshwork of transverse tubules derived from the plasma membrane ensures that Ca^{2+} is released throughout the sarcoplasm, coordinating the contractile response. The contractile response is terminated by ATP-dependent uptake of Ca^{2+} into the SR, a process requiring typically less than 100 ms.

Muscle proteins

Actin and myosin translate the chemical energy of ATP into the mechanical action of muscle

Muscle contraction is mediated by changes in the conformations and interactions of actin and myosin. The actomyosin complex transforms the chemical energy of ATP into the mechanical action of muscle. Myosin is one of the largest proteins in the body, with a molecular mass of approximately 500 kDa, and accounts for more than one-half of muscle protein; it has ATPase activity in the presence of Ca^{2+}. Actin is composed of smaller subunits (42 kDa), known as G-actin (globular), but polymerizes into a filamentous array (F-actin). These two myofilament proteins represent about 90% of total muscle protein. Other actomyosin-associated proteins (Fig. 19.5) are required for assembly of muscle proteins and coordination of the contractile response.

Muscle proteins and their functions

Protein	Function
Myosin	Ca^{2+}-dependent ATPase activity
C-protein	assembly of myosin into thick filaments
M-protein	binding of myosin filaments to M-line
Actin	G-actin polymerizes to filamentous F-actin
tropomyosin	stabilization and propagation of conformational changes of F-actin
troponins-C, I and T	modulation of actin–myosin interactions
α-and β-actinins	stabilization of F-actin and anchoring to Z-line
nebulin	possible role in determining length of F-actin filaments
titin	control of resting tension and length of the sarcomere
desmin	organization of myofibrils in muscle cells
dystrophin	reinforcement of cytoskeleton and muscle cell plasma membrane

Fig. 19.5 Muscle proteins and their functions. Actin and myosin account for over 90% of muscle proteins, but several associated proteins are required for assembly and function of the actomyosin complex.

Myosin

Myosin comprises two heavy and four light chains; it also contains two hinge regions

Under the electron microscope, myosin appears as an elongated protein with two globular heads. Structurally, it consists of two heavy and four light chains. The heavy chains form an extended α-helical coiled-coil structure, and the light chains are bound to one end of each heavy chain, forming globular domains. Structural analysis by limited proteolysis indicates that there are two flexible hinge regions in the molecule (Fig. 19.6). One is about two-thirds of the way along the helical chain and divides the molecule into light meromyosin (LMM: helical region) and heavy meromyosin (HMM: short helical tail plus globular domains). The other hinge is between the short helical and globular domains of HMM. Thick filaments are formed by self-association of LMM helices, up to 400 myosin molecules per thick filament. The filaments extend outward from the M-line toward the Z-line of each myofibril (see Figs 19.3 and 19.6).

Myosin light chains have Ca^{2+}-dependent ATPase activity and are involved in reversible interactions with actin

The myosin light chains in the globular domain are homologous to calmodulin and have Ca^{2+}-dependent ATPase activity. These chains are also involved in reversible interactions with actin. ATP binding to the myosin head groups reduces their affinity for actin. Hydrolysis of the bound ATP to ADP and inorganic phosphate (Pi) results in more than a 1000-fold increase in the binding affinity of the myosin head groups for actin; however, this interaction requires Ca^{2+}-dependent structural changes in the actin filaments. Rigor mortis occurs after death as a result of the inability of muscle to regenerate ATP, which is required to maintain the low Ca^{2+} concentration in the sarcoplasm. The hydrolysis of ATP on myosin and the increase in sarcoplasmic Ca^{2+} leads to tight interactions between myosin and actin, forming rigid muscle tissue.

Actin

F-actin is a polymer of G-actin subunits; in muscle there are about twice as many actin as myosin chains

F-actin is formed by polymerization of G-actin subunits in a head-to-tail manner, with two polymer chains coiling around one another to form the F-actin myofilament (see Fig. 19.6). These molecules extend in opposite directions from the Z-line, overlapping with the myosin chains

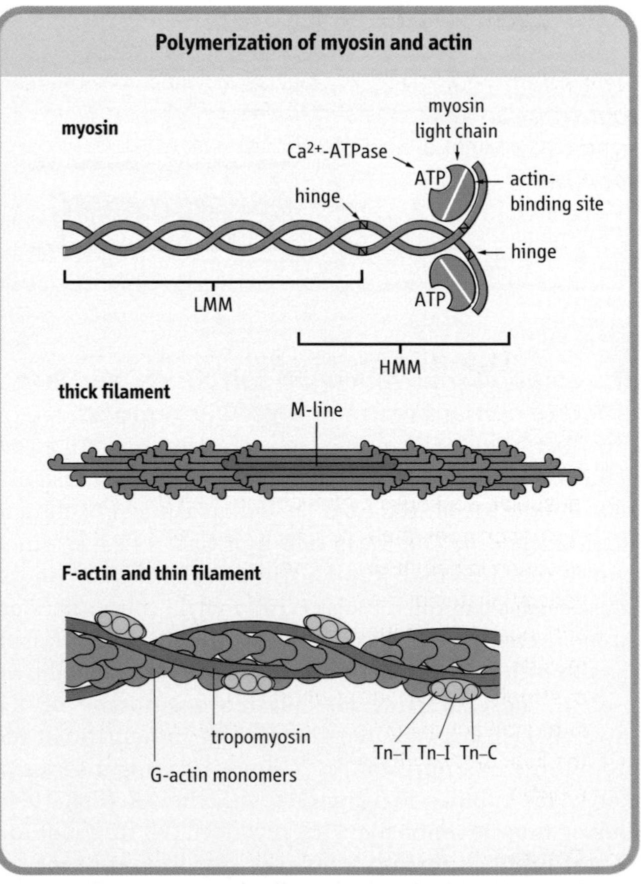

Fig. 19.6 Polymerization of myosin and actin into thick and thin filaments. Tn-C, calcium-binding troponin; Tn-I, troponin inhibitory subunit; Tn-T, tropomyosin binding troponin.

extending from the M-line. There are approximately twice as many actin as myosin chains in muscle, yielding an array in which each myosin molecule is associated with six actin molecules and each actin with three myosin molecules (see Fig. 19.1 for a cross-sectional view).

Tropomyosin and troponins

Tropomyosin stabilizes F-actin and coordinates conformational changes among actin subunits during contraction

Tropomyosin is a fibrous protein that extends along the grooves of F-actin, each tropomyosin molecule contacting about seven G-actin units. It has a role in stabilizing F-actin and coordinating conformational changes among actin subunits during contraction. In the absence of Ca^{2+}, tropomyosin blocks the myosin binding site on actin. A complex of troponin proteins is bound to tropomyosin: Tn-T (tropomyosin-binding), Tn-C (calcium-binding) and Tn-I (inhibitory subunit). Troponins modulate the interaction between actin and myosin. Ca^{2+} binding to Tn-C, a calmodulin-like protein, induces changes in Tn-I, which are then transduced to tropomyosin, moving it out of the myosin-binding site and permitting actin–myosin interactions.

The contractile response

The sliding-filament model of muscle contraction describes a series of chemical and structural changes in the actomyosin complex

The general features of the sliding-filament model of muscle contraction are described by a series of chemical and structural changes in the actomyosin complex. The contractile response is powered by reversible 'cross-bridge' interactions between myosin head groups and actin. A cycle of binding of myosin to actin, conformational changes in the hinge regions of myosin, and release of myosin occurs, with the conformational change providing the 'power stroke' for muscle contraction. This sequence of reactions is summarized in Figure 19.7.

The sliding-filament model explains muscle contraction as the result of:

- the reversible interaction between actin and myosin.
- conformational change in the hinge regions of myosin induced by hydrolysis of ATP and relaxed by dissociation of ADP and Pi.

The latter process, the dissociation of ADP and Pi, rather than the hydrolysis of ATP, is the rate-limiting step in myosin ATPase activity and muscle contraction. Myosin is a sluggish ATPase, and actin was so-named because it accelerates the rate of myosin ATPase activity by over 100-fold. Otherwise, the rate of muscle contraction would be prohibitively slow. The stability of the contracted state is maintained by multiple and continuous actin–myosin interactions, so that slippage is minimized until Ca^{2+} is removed from the sarcoplasm, allowing the muscle to relax.

Cardiac and smooth muscle

Cardiac muscle is striated and contracts rhythmically under involuntary control

Cardiac muscle is striated, but contracts rhythmically and continuously. Unlike skeletal muscle, it is not a voluntary

Duchenne muscular dystrophy

A young boy was brought to the clinic because of muscular weakness, noticeably in his legs. Physical evaluation confirmed muscle atrophy, and a 20-fold elevation in serum creatine (phospho) kinase (CK) activity. Histology revealed muscle loss, some necrosis and increased connective tissue and fat volume in muscle. A tentative diagnosis of Duchenne muscular dystrophy (DMD) was confirmed by immunoelectrophoretic (Western blot) analysis showing the lack of the cytoskeletal protein dystrophin in muscle.

Comment. Dystrophin is a high-molecular-weight cytoskeletal protein that reinforces the plasma membrane of the muscle cell and mediates interactions with the extracellular matrix. In its absence, the plasma membrane of muscle cells is damaged during the contractile process, leading to muscle cell death. The dystrophin gene is located on the X-chromosome and is unusually long, nearly 2.5×10^6 base pairs. Mutations are relatively common, the frequency of DMD being approximately 1 in 3500 male births. DMD is a progressive myodegenerative disease, commonly leading to confinement to a wheelchair by puberty, with death by age 20 years from respiratory or cardiac failure. Dystrophin is completely absent in DMD patients. A milder form of the disease, known as Becker muscular dystrophy, has milder symptoms and is characterized by expression of an altered dystrophin protein and survival into the fourth decade.

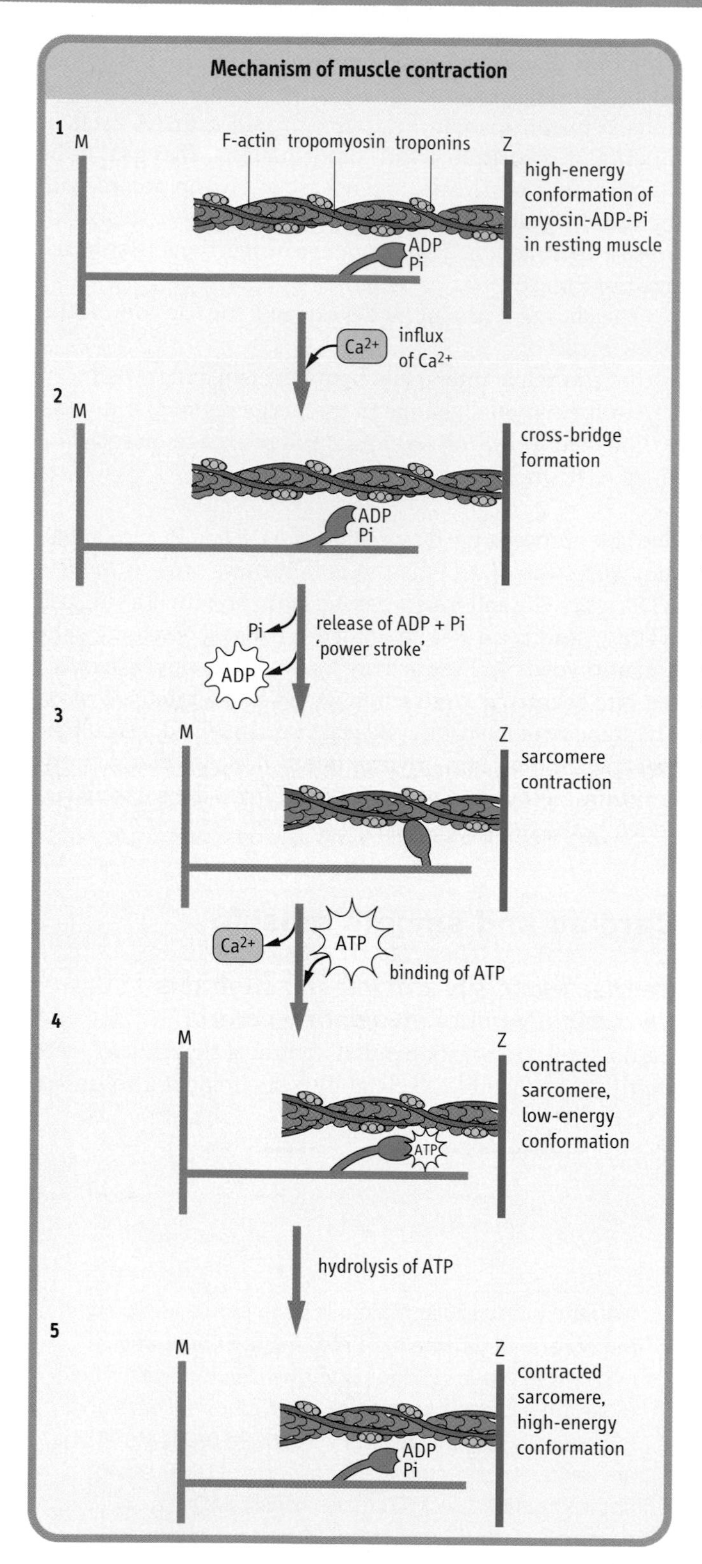

Fig. 19.7 Proposed stages in muscle contraction, according to the sliding-filament model.
(**1**). In resting, relaxed muscle, Ca^{2+} concentration is ~10^{-7} mol/L. The head group of myosin chains contains bound ADP and Pi, and is extended forward along the axis of the myosin helix in a **high-energy** conformation. Although the myosin-ADP-Pi complex has a high affinity for actin, binding of myosin to actin is inhibited by tropomyosin, which blocks the myosin-binding site on actin at low Ca^{2+} concentration.
(**2**). When muscle is stimulated, Ca^{2+} enters the sarcoplasm through voltage-gated Ca^{2+} channels (see Chapter 7). Ca^{2+} binding to Tn-C causes a conformational change in Tn-I, which is transmitted through Tn-T to tropomyosin. Movement of tropomyosin exposes the myosin-binding site on actin. Myosin-ADP-Pi binds to actin, forming a cross-bridge.
(**3**). Release of Pi, then ADP, from myosin, catalyzed by actin, causes a major conformational change in myosin, producing the 'power stroke', which moves the actin chain about 10 nm (100 Å) in the direction opposite the myosin chain, increasing their overlap and causing muscle contraction.
(**4**). Binding of ATP to myosin and uptake of Ca^{2+} from the sarcoplasm leads to dissociation of the actomyosin cross-bridge.
(**5**). The ATP is hydrolyzed, and the free energy of hydrolysis of ATP is conserved as the high-energy conformation of myosin, setting the stage for continued muscle contraction.

muscle. The general mechanism of contraction of heart muscle is similar to that in skeletal muscle with some exceptions:

- the SR is less well developed, and the transverse tubule network, an extension of the plasma membrane, is more developed in the heart. Thus, the heart is more dependent on, and actually requires, extracellular Ca^{2+} for its contractile response;
- cardiac muscle is more responsive to hormonal regulation. For example, cyclic adenosine monophosphate (cAMP)-dependent protein kinases phosphorylate transport proteins and Tn-I, mediating changes in the force of contraction in response to epinephrine.

Smooth muscle contraction is also modulated by Ca^{2+} but it lacks the Ca^{2+}-sensitive troponin complex. Instead, the interaction between myosin and actin is modulated by the state of phosphorylation of smooth muscle isoforms of myosin light chain. The phosphorylation of myosin and the affinity of myosin for actin is under the control of a Ca^{2+}-calmodulin-sensitive myosin light chain kinase and protein phosphatase. The kinase is inactivated by phosphorylation by cAMP-dependent protein kinase A, contributing to smooth muscle relaxation in response to β-adrenergic stimulation.

Isoforms of actin and myosin are not solely confined to muscle cells

Isoforms of actin and myosin are also found in the cytoskeleton of nonmuscle cells, where they have roles in diverse processes, e.g. cell migration, vesicle transport during endocytosis and exocytosis, maintenance or changing of cell shape, and anchorage of intracellular proteins to the plasma membrane.

Muscle energy metabolism

Muscle consists of two types of striated muscle cells; fast-twitch and slow-twitch fibers

Two general types of striated, skeletal muscle are readily distinguished, even by someone who is not a biochemist;

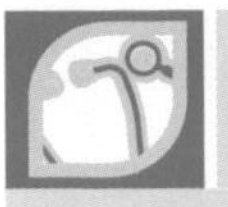

Malignant hyperthermia

About 1 in 150 000 patients treated with halothane (gaseous halocarbon) anesthesia or muscle relaxants, responds with excessive skeletal muscle rigidity and severe hyperthermia with a rapid onset, up to 2°C (4°F) within 1 hour. Unless treated rapidly, cardiac abnormalities may be life-threatening; mortality from this condition exceeds 10%. This genetic disease results from excessive or prolonged release of Ca^{2+} into the SR, most commonly the result of mutations in the Ca^{2+}-release channels within the SR. Excessive release of Ca^{2+} leads to a prolonged increase in sarcoplasmic Ca^{2+} concentration. Muscle rigidity results from Ca^{2+}-dependent consumption of ATP, and hyperthermia results from increased metabolism to replenish the ATP. As muscle metabolism becomes anaerobic, lacticacidemia and acidosis may develop. The cardiac abnormalities result from hyperkalemia, caused by release of potassium ions from muscle; as supplies of ATP are exhausted, muscle is unable to maintain ion gradients across its plasma membrane. Treatment of malignant hyperthermia includes use of muscle relaxants, e.g. dantrolene, an inhibitor of the ryanodine-sensitive Ca^{2+}-channel (see Chapter 7), to inhibit Ca^{2+}-release from the SR. Supportive therapy involves cooling, administration of oxygen, correction of blood pH and electrolyte imbalances and also treatment of cardiac abnormalities.

these are the white meat and red meat of fowl. The white meat, characteristic of the breast muscle of chicken, is white because of its relative lack of mitochondria, lower blood flow and myoglobin content. It also has a lower fat content and is richer in glycogen, so that it has a somewhat sweeter taste. The breast muscle of a chicken is rich in anaerobic 'fast-twitch' fibers, which normally perform rhythmic, voluntary, low-stress work and rely on glycogen and anaerobic glycolysis for short bursts of additional energy during stressful situations. Chicken breast muscle is not capable of sustaining prolonged flight; chickens tend to run and squawk a lot in response to fright, but can fly for only short distances at a time.

In contrast, the breast muscle of a migrating goose is a red, greasy meat. It is rich in aerobic, 'slow-twitch' fibers, and is red-brown in color because it is well perfused with blood and rich in mitochondria (cytochromes) and myoglobin. This voluntary muscle uses fats for prolonged strenuous activity during annual migrations, and fat metabolism requires mitochondria. Fat is the preferred fuel store in wings for flight energy, not only because fat provides 9 Cal/g, compared with 4 Cal/g for carbohydrates but also because fat is stored in droplets of pure fat, whereas carbohydrate, stored as glycogen, is stored with water of hydration, reducing its energy content to 1–2 Cal/g of glycogen. If the migrating goose stored energy in its wings as glycogen, rather than fat, the additional weight would make it nearly impossible for the goose to get off the ground! Like the wing muscle of the duck, skeletal and cardiac muscle in humans is well perfused with blood, rich in mitochondria, and relies largely on fats for metabolism. In addition to fat depots in adipose tissues, animals store significant quantities of fat in their muscle tissue, accounting for the high fat content of roast beef and steak.

For short bursts of energy, skeletal muscle relies on its ATP stores and an additional reserve of the high-energy storage compound, creatine phosphate (creatine-P), to regenerate ATP rapidly during the first few minutes prior to full activation of glycogenolysis. Creatine is synthesized from arginine and glycine (Fig. 19.8), and is phosphorylated reversibly to creatine-P by the enzyme creatine (phospho)kinase (CK or CPK). CK is a dimeric protein and exists as three isozymes: the MM (muscle), BB (brain) and MB isoforms. The MB isoform is enriched in cardiac tissue.

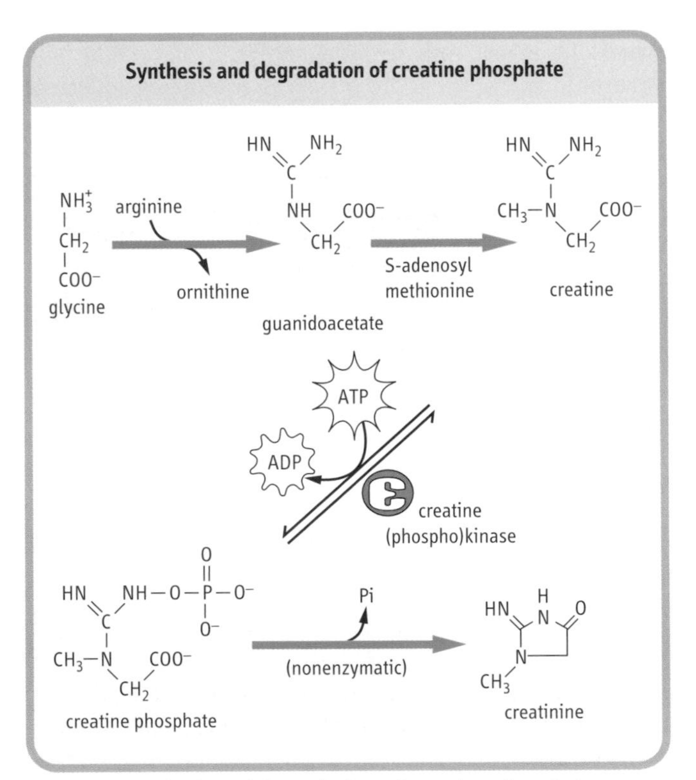

Fig. 19.8 Synthesis and degradation of creatine phosphate (creatine-P). Creatine is synthesized from glycine and arginine precursors. Creatine-P is unstable and undergoes slow, spontaneous degradation to Pi and creatinine, the cyclic anhydride form of creatine, which is excreted from the muscle cell into plasma and then into urine.

Assay of creatinine to assess renal function and urine dilution

Since creatine-P concentration is relatively constant per unit muscle mass, the production of creatinine is also relatively constant during the day. It is also eliminated in urine at a relatively constant amount per hour, primarily by glomerular filtration, and to a lesser extent by tubular secretion. Since its concentration in urine varies with the dilution of the urine, levels of metabolites in random urine samples are often normalized to the urinary concentration of creatinine. Otherwise, a 24-h collection would be required to assess daily excretion of a metabolite. Normal creatinine concentration in plasma is about 1 mg/dL (60–120 μmol/L). Increases in plasma creatinine concentration are commonly used as an indicator of renal failure.

Resting muscle obtains most of its energy from fat metabolism

The level of creatine-P in resting muscle is several-fold higher than that of ATP (Fig. 19.9). Thus, ATP concentration remains relatively constant during the initial stages of exercise. It is replenished not only by the action of CK but also by adenylate kinase (myokinase) as follows:

Creatine phosphokinase: creatine-P + ADP $\rightarrow$ creatine + ATP
Adenylate kinase: 2 ADP $\rightleftharpoons$ ATP + AMP

Muscle at rest obtains most of its acetyl CoA, and therefore its ATP, from fat metabolism. However, during the initial stages of exercise, muscle glycogenolysis, followed by both anaerobic and aerobic glycolysis, is the major source of

Changes in energy resources in working muscle

Metabolite	Metabolite concentration (mmol/kg dry weight)		
	resting	3 minutes	8 minutes
ATP	27	26	19
creatine-P	78	27	7
creatine	37	88	115
lactate	5	8	13
glycogen	408	350	282

Fig. 19.9 Concentrations of energy metabolites in human leg muscle during bicycle exercise. These experiments were conducted during ischemic exercise, which exacerbates the decline in ATP concentration. They illustrate the rapid decline in creatine-P and the increase in lactate from anaerobic glycolysis of muscle glycogen. Data are adapted from Timmons JA *et al. J Clin Invest* 1998; **101**:79–85.

Assay of myocardial enzymes and troponins to diagnose myocardial infarction

Myocardial infarction (MI) is the result of blockage of blood flow to the heart. Tissue damage results in leakage of intracellular enzymes into blood (Fig. 19.10). Among these are glycolytic enzymes (LDH, see Chapter 5); however, measurements of myoglobin, total plasma CK and CK-MB isozymes are most frequently used for the diagnosis and management of MI. Myoglobin is a small protein (17 000 kDa) and rises most rapidly in plasma, within 2 hours following MI. Although it is sensitive, it lacks specificity for heart tissue. It is cleared rapidly by renal filtration and returns to normal within 1 day. Since plasma myoglobin also increases following skeletal muscle trauma, it would not be useful for diagnosis of MI, e.g. following an automobile accident. Total plasma CK and the CK-MB isozyme begin to rise within 3–10 hours following an MI, and reach a peak value of up to 25 times normal after 12–30 hours; they may remain elevated for 3–5 days. Total CK may also increase as a result of skeletal muscle damage but the measurement of CK-MB provides specificity for cardiac damage.

Comment. Enzyme-linked immunosorbent assays (ELISA) for the myocardial troponins are now being used widely for the diagnosis and management of MI. These assays depend on the presence of unique isoforms of troponin subunits in the adult heart. Tn-T concentration in plasma increases within a few hours after a heart attack, peaks at up to 300 times normal plasma concentration, and may remain elevated for 1–2 weeks. An assay for a specific isoform in an adult heart, Tn-T_2, is essentially 100% sensitive for diagnosis of MI and yields fewer than 5% false-positive results. Significant increases in plasma Tn-T are detectable even in patients with unstable angina and transient episodes of ischemia in the heart.

energy. Ca^{2+} entry into muscle activates phosphorylase kinase, catalyzing the conversion of phosphorylase b to phosphorylase a. AMP also allosterically activates muscle phosphorylase and phosphofructokinase-1, accelerating glycolysis from muscle glycogen. During the first 15–30 minutes of exercise, there is a gradual shift to aerobic metabolism of fatty acids. The glycogen reserves in muscle are sufficient to support the energy needs of muscle during exercise for only about 1 hour, however they are preserved by the changeover to fat metabolism. As exercise continues, epinephrine activates hepatic gluconeogenesis, providing an additional source of glucose for muscle, and also contributes to cAMP-mediated activation of glycogenolysis in muscle.

Long-term muscle performance (stamina) depends on levels of muscle glycogen

Marathon runners typically 'hit the wall' at about the time that their muscle glycogen is fully depleted. This occurs because there is a continuing requirement for a basal level of glycogen and carbohydrate metabolism in muscle, even when fats are the primary source of muscle energy. Pyruvate, derived from glucose, is an important source of oxaloacetate, produced by the pyruvate carboxylase reaction. This oxaloacetate is required for efficient activity of the tricarboxylic acid cycle and metabolism of acetyl CoA derived from fats.

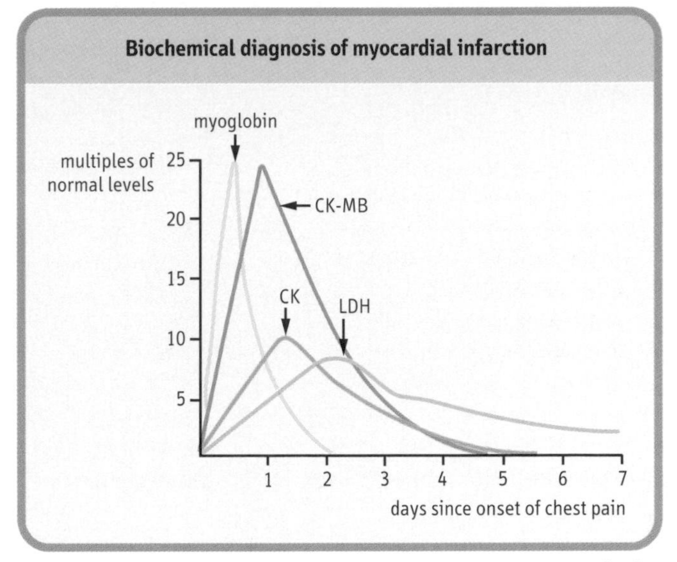

Fig. 19.10 Biochemical diagnosis of myocardial infarction (MI). Various marker enzymes increase in plasma following MI. LDH, lactate dehydrogenase. Adapted from Pettigrew AR, Pacanis A. Diagnosis of myocardial infarction. In: Dominiczak MH, (ed.), *Seminars in Clinical Biochemistry.* University of Glasgow Computer Publishing Unit, Glasgow, 1997.

Summary

Muscle is the major consumer of fuels and ATP in the body. Both glycolysis and lipid metabolism are essential for muscle activity. Reliance on these pathways varies with the muscle and its exercise history. White muscle relies largely on glycogen and anaerobic glycolysis for short bursts of energy. Red muscle is aerobic tissue; at rest, it uses fats as its primary source of energy. During the initial phases of exercise, it relies on glycogenolysis and glycolysis, then gradually converts to fat metabolism for long-term energy production. The ATP produced in muscle drives both the maintenance of ion gradients, which is an electrochemical process, and muscle contraction, which is a mechanochemical process. Contraction is described by a 'sliding-filament' model in which hydrolysis of ATP is coupled to changes in the conformation of myosin. Relaxation of the high-energy conformation of myosin during interaction with actin produces a 'power stroke', resulting in increased overlap of the actin–myosin filaments and shortening of the sarcomere, the contractile unit in muscle.

Further reading

Abraham RB, Adnet P, Glauber V, Perel A. Malignant hyperthermia. *Postgrad Med J* 1998; **74**: 11–17.

Horowitz A, Menice CB, Laporte R, Morgan KG. Mechanisms of smooth muscle contraction. *Physiolog Rev* 1996; **76**: 967–1003.

Huxley HE. Getting to grips with contraction: the interplay of structure and biochemistry. *Trends Biochem Sci* 1998; **23**: 84–87.

Keffer JH. Myocardial markers of injury. Evolution and insights. *Am J Clin Pathol* 1997; **105**: 305–320.

Loke J, MacLennan DH. Malignant hyperthermia and central core disease: disorders of Ca^{2+} release channels. *Am J Med* 1998; **104**: 470–486.

Maughan RJ. Creatine supplementation and exercise performance. *Intl J Sports Nutr* 1995; **5**: 94–101.

Mercer DW. Role of cardiac markers in evaluation of suspected myocardial infarction. Selecting the most clinically useful indicator. *Postgraduate Medicine* 1997; **102**: 113–117, 121–122.

Oplatka A. Are rotors at the heart of all biological motors? *Biochem Biophys Res Commun* 1998; **246**: 301–306.

Petrof BJ. The molecular basis of activity-induced muscle injury in Duchenne muscular dystrophy. *Molec Cell Biochem* 1998; **179**: 111–123.

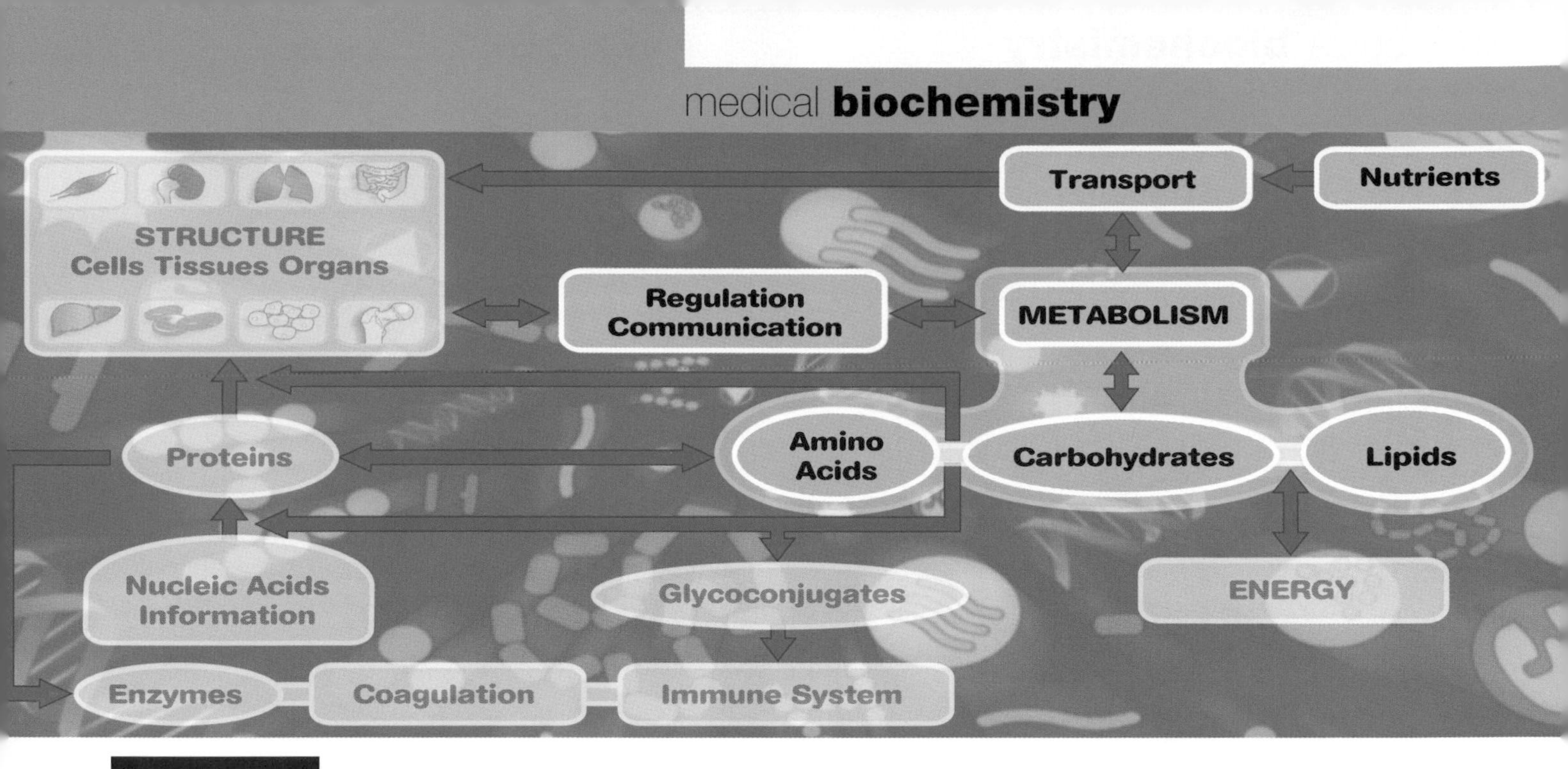

20 Glucose Homeostasis and Fuel Metabolism

The metabolic fuels

The existence of an organism depends on the continuous provision of energy for metabolic processes. This chapter describes how the body handles energy substrates (metabolic fuels) under different circumstances.

The main metabolic fuels are glucose and fatty acids

The most important metabolic fuels are glucose and fatty acids. In normal circumstances, glucose is the only fuel the brain can use. Glucose is also preferentially utilized by muscle during the initial stages of exercise. The amount of glucose present in the extracellular fluid is minute – only about 20 g (<1 oz) (equivalent of 80 kcal) (335 kJ). To ensure the continuous provision of glucose to the brain and other tissues, metabolic fuels are stored for use in times of need. Carbohydrates are stored as glycogen. The amount of available glycogen stored is not large; approximately 75 g (approx 2.5 oz) in the liver and 400 g (< 1 lb) in the muscles (about 1900 kcal (7955 kJ) altogether). Liver glycogen can remain the main supplier of glucose for no longer than 16 h. To safeguard the continuous supply of glucose over longer periods, the body transforms noncarbohydrate compounds into glucose during gluconeogenesis.

Long-chain fatty acids are the ideal storage fuel

The caloric value of fats 9 kcal/g (37.6 kJ/g) is higher than that of either carbohydrates 4 kcal/g (16.7 kJ/g)or proteins 4 kcal/g, and therefore long-chain fatty acids are an ideal storage fuel. The body has a virtually unlimited capacity for the accumulation of fats. A 70 kg (154 lb) man will have approximately 15 kg (33 lb) of fat stored as adipose tissue triglycerides, equivalent to over 130 000 kcal (544 300 kJ). Fatty acids can support the body's energy needs over prolonged periods of time. In extreme circumstances, humans can fast for as long as 60–90 days and obese persons may survive for over a year without food.

Amino acids can be used as a fuel during fasting, illness, or injury

Amino acids normally serve as substrates for the synthesis of the body's own proteins, rather than as a source of energy. However, during a prolonged fast, or after illness or injury, proteins are degraded and the constituent amino acids are converted into glucose. Excess amino acids provided with food are normally converted to carbohydrates either for storage or for energy metabolism. The main metabolic pathways and key metabolites are listed in Fig. 20.1.

Utilization and storage of metabolic fuels

Pathways	Main substrates	End products
Anabolic		
gluconeogenesis	lactate, alanine, glycerol	glucose
glycogen synthesis	G-1-P	glycogen
protein synthesis	amino acids	proteins
lipogenesis	acetyl CoA, glycerol	fatty acids, triglycerides
Catabolic		
glycolysis	glucose	pyruvate, ATP
tricarboxylic acid cycle	pyruvate	NADPH + H^+, $FADH_2$, CO_2, H_2O, ATP
glycogenolysis	glycogen	G-1-P, glucose
pentose phosphate pathway	G-6-P	NADPH + H^+, pentoses, CO_2
lipolysis	triglycerides→fatty acids	glycerol, acetyl CoA
proteolysis	proteins	amino acids→glucose, amino acids→ketones

Fig. 20.1 Principal anabolic and catabolic pathways, and their main substrates and end products. Note that key metabolites, such as pyruvate and acetyl CoA, serve as links between different pathways.

The organ–fuel interactions

At rest, the brain uses approximately 20% of all oxygen (O_2) consumed by the body. As mentioned above, glucose is normally the brain's only fuel, but during starvation the brain can use ketones as an alternate energy source.

Gluconeogenesis occurs primarily in the liver

When the glucose content of the extracellular fluid decreases, glycogen is mobilized within seconds, providing a short-term supply of endogenous glucose. Subsequently, this supply is complemented by gluconeogenesis, the other source of endogenous glucose. Gluconeogenesis takes place primarily in the liver, with the kidneys contributing during a prolonged fast. The substrates for gluconeogenesis originate from anaerobic glycolysis (lactate) and the breakdown of either muscle protein (alanine) or adipose tissue triglycerides (glycerol).

Muscle handles carbohydrate quite differently to the liver. In contrast to the liver, muscle does not have glucose-6-phosphatase (G-6-Pase) and cannot release glucose into the circulation. Instead, it uses glycogen for its own energy needs. Muscle does, however, contribute to endogenous glucose production by releasing lactate, a product of anaerobic glycolysis, which is transported to the liver, where it enters gluconeogenesis. Muscle can use both glucose and fatty acids as energy sources. During intensive exercise, glucose is the preferred fuel. Fatty acids are the main energy source at rest and during prolonged exercise (see Chapter 19).

Glucose homeostasis

In the fasting state, glucose turnover in a 70 kg (154 lb) individual is approximately 2 mg/kg/min (200 g/24 h). The plasma glucose concentration reflects the balance between intake (glucose absorption from the gut), tissue utilization (glycolysis, pentose phosphate pathway, tricarboxylic acid (TCA) cycle, glycogen synthesis) and endogenous production (glycogenolysis and gluconeogenesis). Glucose homeostasis is controlled primarily by the anabolic hormone insulin and also by several insulin-like growth factors. Several catabolic hormones (glucagon, catecholamines, cortisol, and growth hormone) oppose the action of insulin; they are known as anti-insulin or counter-regulatory hormones (Fig. 20.2).

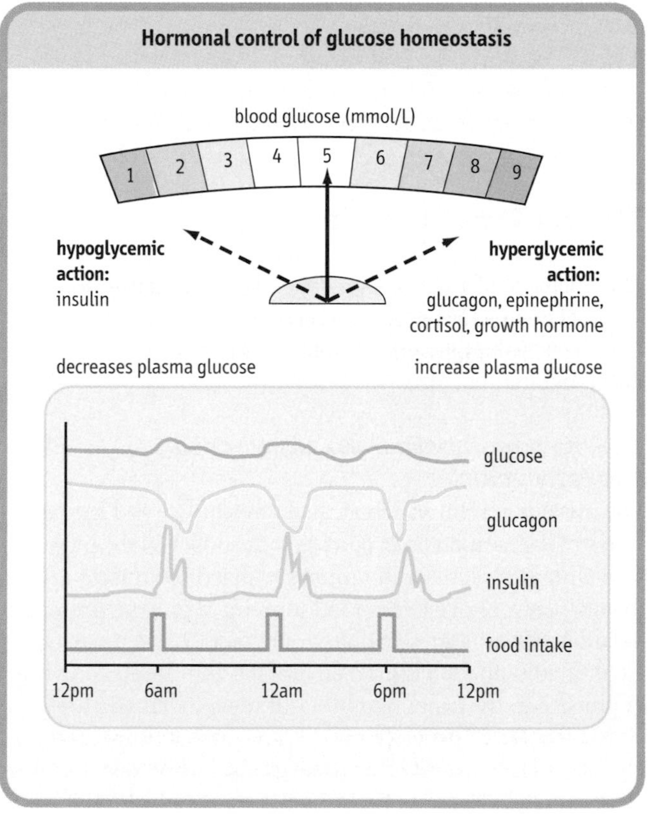

Fig. 20.2 Plasma glucose concentration is a result of the balance between the hypoglycemic action of insulin and the hyperglycemic action of anti-insulin hormones. The lower part of the figure illustrates the daily patterns of insulin, glucagon, and plasma glucose concentrations. Glucose concentration throughout the day remains in a relatively narrow range. To obtain glucose concentrations in mg/dL, multiply by 18.

Insulin and glucagon are the main hormones responsible for controlling plasma glucose levels

Insulin is secreted from the pancreas in response to the increase in plasma glucose following a meal. Insulin decreases the plasma glucose concentration by promoting the uptake of glucose into tissues, intracellular glucose metabolism, and glycogen synthesis. Anti-insulin hormones stimulate both the release of glucose from glycogen stores and its *de novo* synthesis, thus causing an increase in glucose concentration in plasma (hyperglycemia). The fine balance between insulin and glucagon action is a key factor in the control of fuel metabolism. Insulin and glucagon are both secreted from the same anatomic location – the pancreatic islets of Langerhans. Insulin is secreted by β-cells, which constitute approximately 70% of all islet cells, and glucagon is secreted by the α-cells. The glucose level acts as a signal that initiates the islet hormonal response. Glucose stimulates the secretion of insulin and suppresses the secretion of glucagon.

The metabolic effects of insulin

In a highly coordinated fashion, insulin promotes an anabolic state – storage of carbohydrate and lipids, and synthesis of protein. Insulin acts on three main target tissues – the liver, muscle, and adipose tissue (Fig. 20.3). As the first step in its action, insulin binds to a four-subunit protein membrane receptor. The β-subunit of the receptor contains a transmembrane protein with an adenosine triphosphate (ATP)-binding site and has latent tyrosine kinase activity. The binding of insulin to the receptor activates the tyrosine kinase, which autophosphorylates the receptor. Receptor autophosphorylation in turn initiates further intracellular phosphorylations. Tyrosine kinase can phosphorylate other proteins such as the insulin receptor substrate-1 (IRS-1). In turn, IRS-1 generates signals which affect glucose transport and glycogen synthesis. Other pathways of intracellular signaling such as the G-protein pathway are also involved in insulin action (see also Chapter 36).

In the liver, insulin stimulates both glycolysis and glycogen synthesis. It also suppresses lipolysis and promotes the synthesis of long-chain fatty acids (lipogenesis). The lipids are then packaged into very-low-density lipoproteins (VLDL), which are secreted into the blood. In the peripheral tissues, insulin induces lipoprotein lipase, an enzyme that offloads triglycerides from either hepatic VLDL or dietary chylomicrons by hydrolyzing them into glycerol and fatty acids (see Chapter 17). Insulin also stimulates triglyceride synthesis from glycerol and fatty acids in adipose tissue. In muscle, insulin increases glucose transport, glucose metabolism, and glycogen synthesis. Insulin also increases cellular uptake of amino acids and stimulates protein synthesis.

Plasma glucose after myocardial infarction

A 66-year-old woman was admitted to the cardiology ward after suffering a heart attack (myocardial infarction). Her random glucose level was 10.5 mmol/L (189 mg/dL). An oral glucose tolerance test (OGTT) carried out the next day showed a fasting blood glucose of 6.5 mmol/L (117 mg/dL), a 1-h level of 10.8 mmol/L (195 mg/dL), and a 2-h postglucose level of 9.0 mmol/L (162 mg/dL).

Comment. This response to a glucose load would normally be consistent with glucose intolerance. However, this patient underwent a major stress, myocardial infarction, which is associated with a counter-regulatory hormone response. The OGTT should not be performed immediately after major stress. The patient should be tested again in 3 months' time.

Insulin and C-peptide

Insulin synthesis and secretion

Insulin consists of two peptide chains linked by two disulfide bonds (Fig. 20.4). The α-chain contains 21 amino acids and the β-chain 30 amino acids. The molecular weight of insulin monomer is 5500 Da. The precursor of insulin within the β-cells of the islet of Langerhans is the single chain preproinsulin. During insulin synthesis a 24-amino-acid signal sequence is first cleaved from preproinsulin by a peptidase, yielding proinsulin.

Proinsulin and C-peptide

Proinsulin consists of the insulin sequence interspersed by a connecting peptide (C-peptide). At the final stage of insulin synthesis, proinsulin is split into insulin and C-peptide (Fig. 20.4), both of which are then released from the cell. C-peptide is released in an amount equimolar to insulin. This is exploited in the clinical laboratories to assess β-cell function in patients treated with exogenous (therapeutically injected) insulin. In these patients, endogenous insulin cannot be measured directly, because the exogenous insulin would interfere in the assay. In such circumstances, C-peptide measurement provides an assessment of β-cell function.

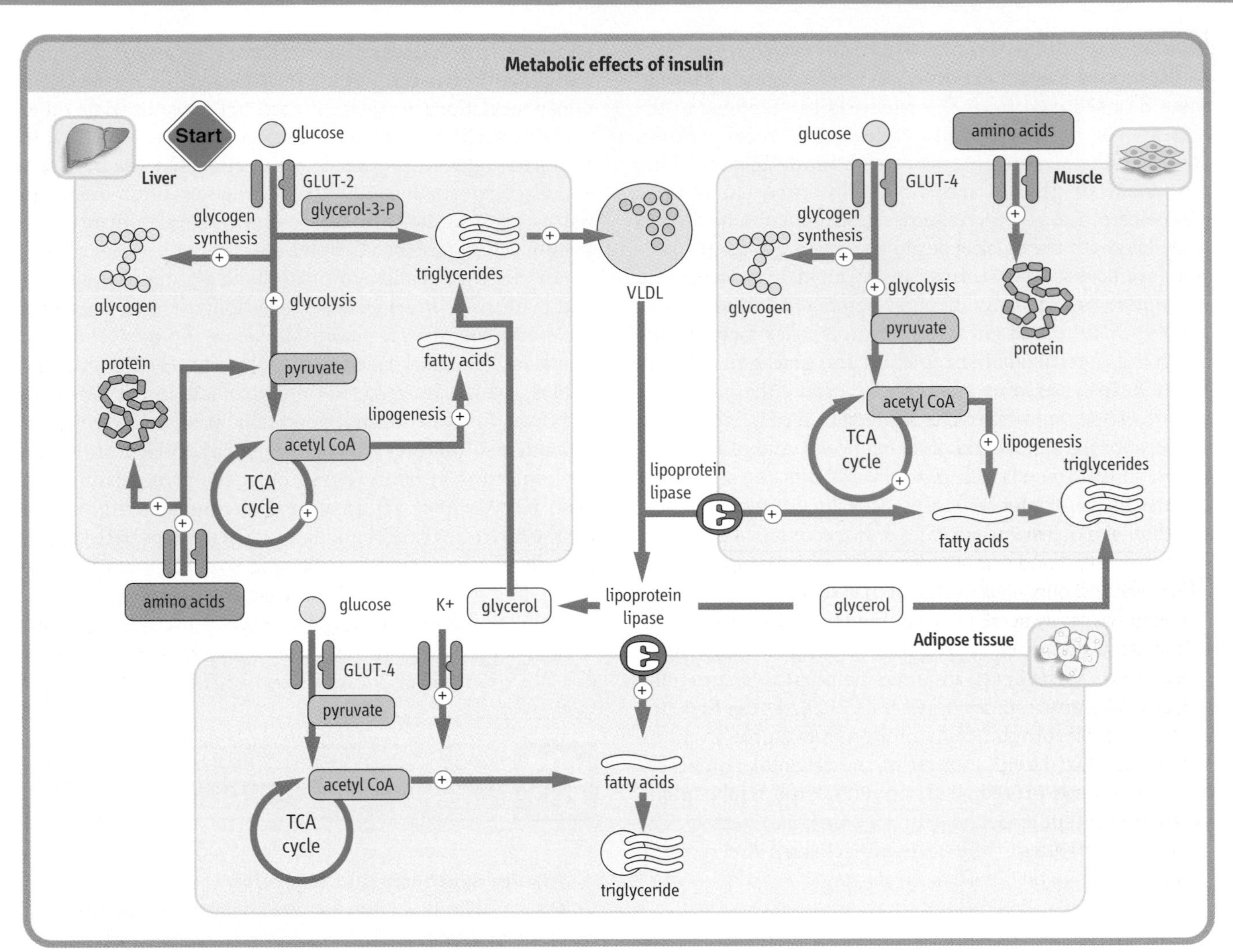

Fig. 20.3 The main insulin target tissues are liver, muscle, and adipose tissue. Insulin affects carbohydrate, lipid, and protein metabolism, and also the cell potassium uptake. Glucose transport mediated by GLUT-4 transporter in muscle and adipose tissue is insulin-dependent. The glucose transporter in liver (GLUT-2) is insulin-independent. Please note the transport of triglycerides between tissues use very low density lipoproteins (VLDL) particles as a vehicle. (See also Chapter 17.)

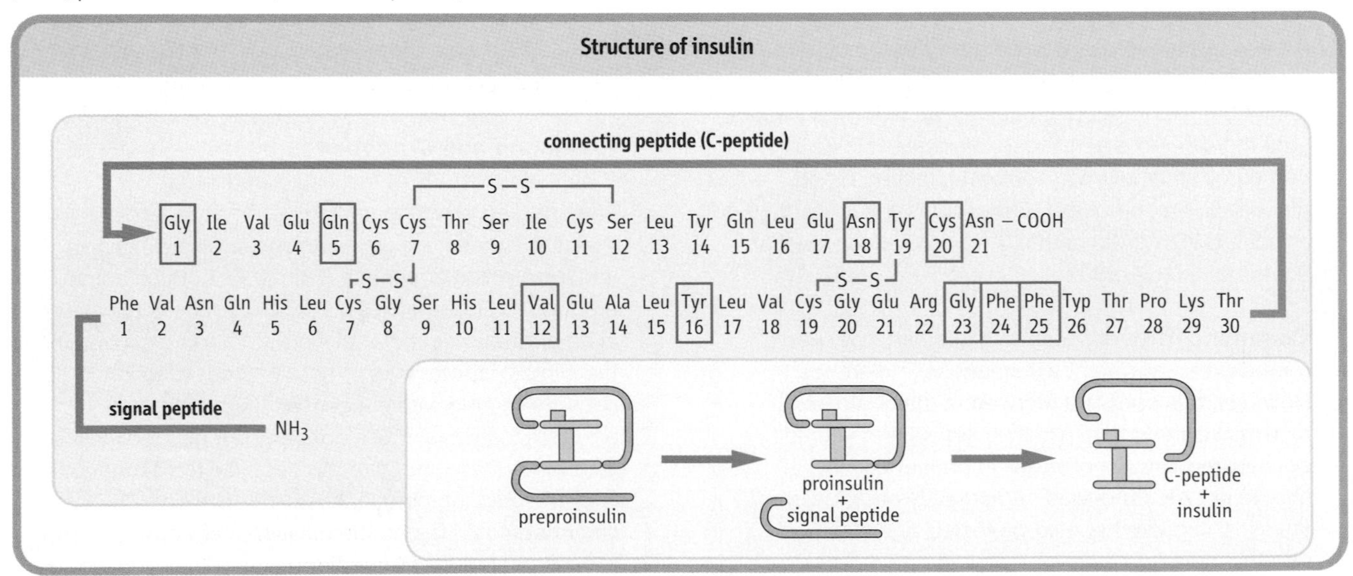

Fig. 20.4 The insulin molecule. C-peptide connects the α- and β-chains of insulin. Boxes indicate amino acid residues active in binding to the insulin receptor.

Insulin secretion

Stimulation of insulin secretion by glucose

The glucose concentration in the vicinity of the β-cell is sensed by the β-cell glucose transporter GLUT-2. Glucose is carried into the cell by GLUT-2, where it is phosphorylated into glucose 6-phosphate (G-6-P) by glucokinase which also is a part of the glucose-sensing mechanism. Increased availability of G-6-P increases the rate of glucose utilization and ATP production in the β-cell. This changes the flux of of ions across the cell membrane, depolarizes the cell and increases the concentration of cytoplasmic free calcium (see Chapter 36). The final result is insulin exocytosis. Insulin secretion from the β-cell after glucose stimulation is biphasic. The first phase of insulin secretion occurs within 10–15 min of stimulation and is the release of preformed insulin. The second phase, which lasts up to 2 hours, is the release of newly synthesized insulin (Fig. 20.5). Insulin secretion is also stimulated by gastrointestinal hormones and some amino acids, such as leucine, arginine, and lysine. Gastrointestinal hormones, such as glucose-dependent insulinotropic peptide (GIP), cholecystokinin, glucagon-like peptide-1 (GLP-1) and vasoactive intestinal peptide (VIP), are secreted following ingestion of foods and potentiate insulin secretion. Thus, the insulin response to orally administered glucose is greater than to an intravenous infusion.

The metabolic effects of glucagon

Glucagon is a small, single chain, 29-amino-acid peptide, with a molecular weight of 3485 Da. Overall, glucagon focuses energy metabolism on the endogenous production of glucose. Its main effect is the mobilization of the fuel reserves for maintenance of the blood glucose level between meals. Glucagon inhibits glucose-utilizing pathways and the storage of metabolic fuels. It acts rapidly on the liver to stimulate glycogenolysis, and to inhibit glycogen synthesis, glycolysis, and lipogenesis (Fig. 20.6). Gluconeogenesis and ketogenesis are then activated.

As with insulin, the first stage of glucagon action is its binding to a specific membrane receptor. The signaling pathway of the glucagon receptor is much better understood than that of insulin. The glucagon–receptor complex causes the binding of guanosine 5′-triphosphate (GTP) to a G-protein complex (see also Chapter 36). This binding stimulates the dissociation of G-protein subunits, one of which (G_α) stimulates adenylate cyclase. Adenylate cyclase converts ATP into a second messenger – cyclic AMP (cAMP). cAMP in turn activates cAMP-dependent protein kinase which, through phosphorylation of regulatory enzymes, controls the activity of key enzymes in carbohydrate and lipid metabolism (Figs 20.7 and 20.8).

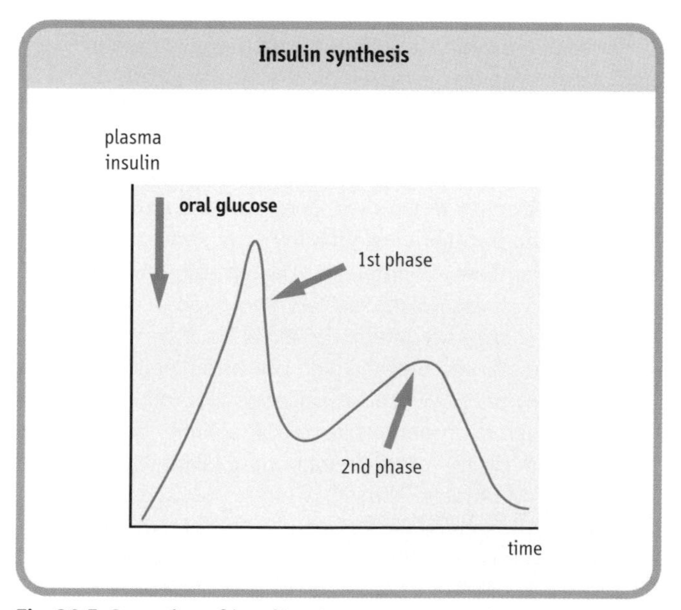

Fig. 20.5 Secretion of insulin. The pattern of insulin secretion after glucose load. Note the biphasic pattern of insulin secretion.

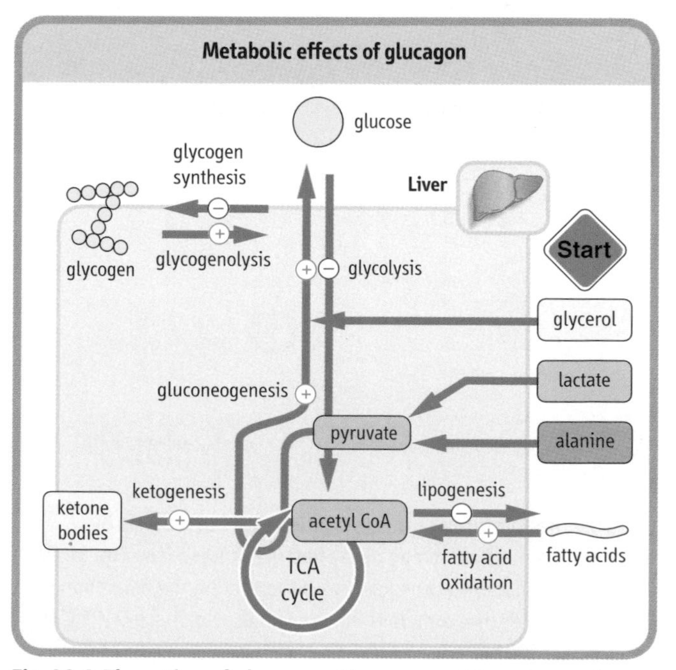

Fig. 20.6 The action of glucagon. Glucagon mobilizes glucose from every available source. Glucagon also increases lipolysis, and ketogenesis from acetyl CoA.

Epinephrine (adrenaline) has effects similar to glucagon in the liver but works through a different receptor (the β-adrenergic receptor – see Chapter 38). Epinephrine promotes an increase in blood glucose in response to stress, even when blood glucose is normal and glucagon is low. This increases the availability of glucose to red blood cells and the brain during stress (see below).

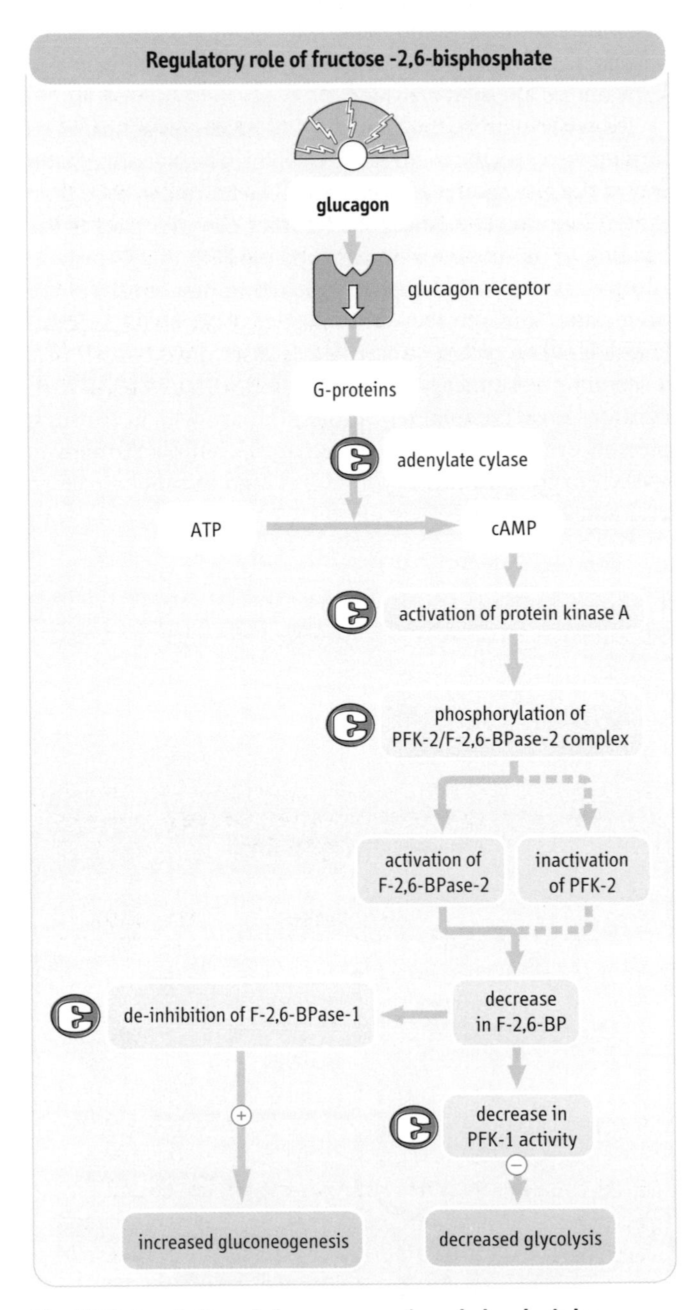

Fig. 20.7 Regulation of gluconeogenesis and glycolysis by glucagon through action on phosphofructokinase (see Fig. 20.8). Glucagon regulates gluconeogenesis by controling the bifunctional enzyme complex that contains both phosphofructokinase-2 (PFK-2) and fructose 2,6-biphosphatase-2 (F-2,6-BPase-2) activity. This changes the concentration of fructose-2,6-biphosphate (F-2,6-BP), which in turn regulates the activity of 'mainline' enzymes: PFK-1 and F-2,6-BPase-1. See also Chapter 11.

Muscle does not possess glucagon receptors. Glycogenolysis in muscle is stimulated primarily by epinephrine in response to stress.

Fine-tuning of fuel metabolism

Hormones determine the general direction of metabolism by the induction or suppression of key enzymes (Fig. 20.9). Such regulation comes into effect over days or weeks. This mechanism responds to diet, and also to stress and disease. For example, activities of hepatic enzymes differ for persons remaining on a chronically high-fat diet compared to a high-carbohydrate diet.

In addition, there is a short term regulation of pathways, which is also affected by hormones: Figure 20.8 illustrates a concerted regulation of glycogen breakdown, gluconeogenesis and lipolysis provided by hormone-driven phosphorylation of key enzymes. Other mechanisms operating in the short term include substrate interactions, allosteric effectors, the cell energy level and redox potential (see also Chapters 11, 12 and 15).

Disorders of glycogen metabolism

Glycogen storage diseases

Glycogen is degraded in response to a falling blood glucose concentration. A defect in the glycogenolytic pathway can lead to insufficient glucose supply and may cause hypoglycemia. This happens in patients with inherited deficiencies of enzymes controling glycogen metabolism. Seven types of glycogen storage disease are known. They are very rare, but provide an excellent insight into human energy metabolism (see Chapter 12). The symptoms of glycogen storage diseases vary and depend on the site of the enzyme defect. For instance, type 1 glycogen storage disease (von Gierke's disease) is a deficiency of G-6-Pase, which leads to a fasting hypoglycemia unresponsive to epinephrine and glucagon. On the other hand, patients with type V disease (McArdle's disease), which is caused by muscle phosphorylase deficiency, do not experience hypoglycemia, but have a limited ability to perform strenuous exercise.

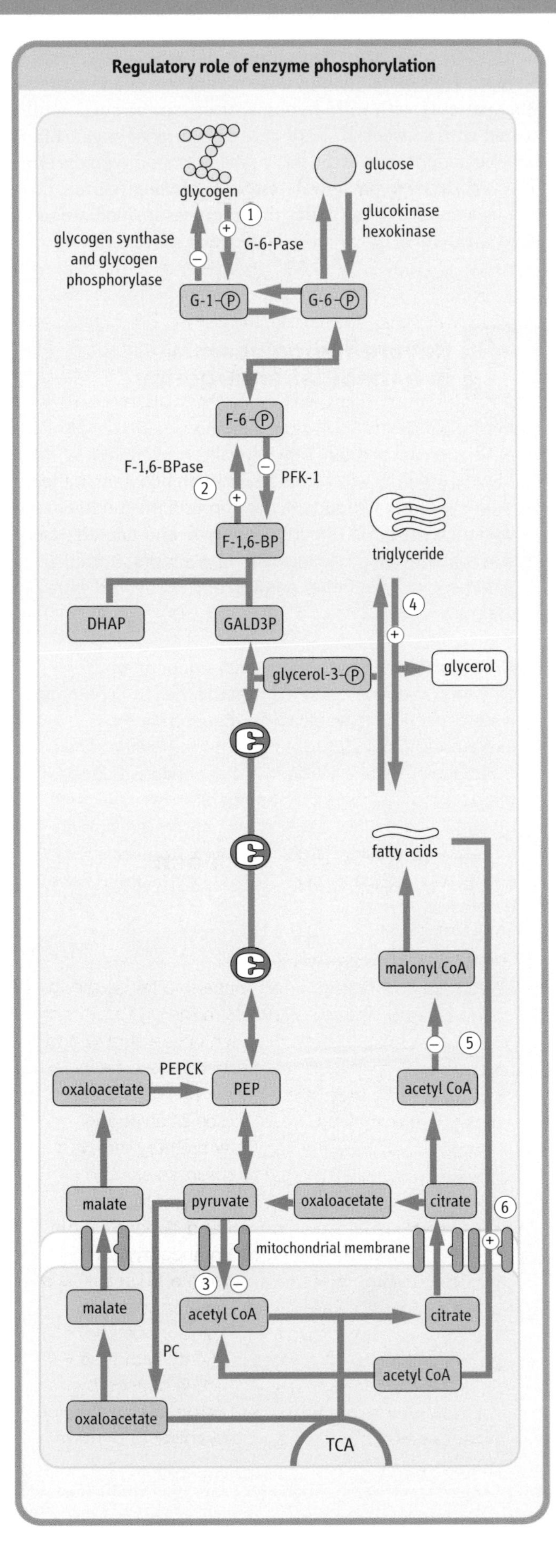

Fig. 20.8 Covalent modification of key enzymes by phosphorylation is an important mechanism coordinating pathways of carbohydrate and lipid metabolism. Phosphorylation of enzymes is triggered by anti-insulin hormones, such as glucagon and epinephrine.

Effect on glycogen breakdown. Phosphorylation activates hepatic glycogen phosphorylase and inactivates glycogen synthase, thus favoring glycogen degradation (1).

Regulation of gluconeogenesis. Phosphorylation of the PFK-2/F-2,6-BPase-2 complex decreases F-2,6-BP formation (see Fig. 20.7). This slows down glycolysis and accelerates gluconeogenesis (2). In addition, since F-1,6-BP allosterically activates pyruvate kinase lower down the glycolytic pathway, its decrease also slows down glycolysis at this stage (3).

Effect on lipolysis. Glucagon stimulates lipolysis by phosphorylation of hormone-sensitive lipase (4) and by inhibiting acetyl CoA carboxylase, an enzyme converting acetyl CoA to malonyl CoA (5), which normally inhibits carnitine-palmitoyl transferase-1. A decrease in the intracellular level of malonyl CoA de-inhibits the carnitine shuttle (6) and facilitates the entry of fatty acids into mitochondria (see also Chapter 14). DHAP, dihydroxyacetone phosphate; GALD3P, glyceraldehyde-3-phosphate; PEP, phosphoenolpyruvate.

Reciprocal effects of insulin and glucagon on key enzymes of carbohydrate metabolism

Enzyme	Effect of glucagon	Effect of insulin
G-6-Pase	+	–
F-1,6-BPase	+	–
PEPCK	+	–

Fig. 20.9 Enzyme induction and repression by insulin and glucagon. Insulin affects the synthesis of key enzymes of glycolysis and gluconeogenesis. On a high-carbohydrate diet, insulin induces gene transcription for the glycolytic enzymes glucokinase, PFK, pyruvate kinase, and glycogen synthase. At the same time, insulin represses the key enzymes of gluconeogenesis, pyruvate carboxylase (PC), phosphoenolpyruvate carboxykinase (PEPCK), F-1,6-BPase, and G-6-Pase.

Glucagon effects on enzyme synthesis oppose those of insulin. On a high-fat diet, glucagon represses the synthesis of glucokinase, PFK-1, and pyruvate kinase, and induces the transcription of PEPCK, F-6-Pase, and G-6-Pase.

Hypoglycemia

Hypoglycemia (a low concentration of blood glucose) is defined as a blood glucose concentration below 2.5 mmol/L (45 mg/dL) (see Fig. 20.12). A decrease in plasma glucose concentration stimulates the body defense mechanisms, primarily in the sympathetic nervous system. Epinephrine and glucagon are released, resulting in a stress response, the manifestations of which may include sweating, trembling, increased heart rate, and a feeling of hunger. If blood glucose continues to fall, brain function is compromised owing to lack of glucose (neuroglycopenia). The patient becomes confused and may lose consciousness. Profound hypoglycemia can be fatal.

Hypoglycemia in healthy individuals is usually mild and may occur during exercise after a period of fasting, or by inhibition of gluconeogenesis as a result of alcohol ingestion. Alcohol increases the intracellular NADH + H^+/NAD^+ ratio, which favors conversion of pyruvate to lactate and reduces the pool of pyruvate available for gluconeogenesis. Hypoglycemia may be a feature of endocrine syndromes characterized by a low cortisol level caused by an insufficient amount of adrenocorticotrophic hormone (ACTH; see Chapter 35). Hypoglycemia may be caused by a rare insulin-secreting tumor of β-cells – insulinoma. Other causes of hypoglycemia are listed in Figure 20.10.

Hypoglycemia is the most common complication of diabetes

Remember that hypoglycemia is the most common complication of diabetes. The diabetic patient develops hypoglycemia if the balance between insulin dose, glucose supply with meals, and physical activity, is disrupted. Thus, hypoglycemia may occur as a result of taking too much insulin or injecting the usual amount of of insulin but missing a meal. Exercise increases tissue glucose uptake independently of insulin. To prevent hypoglycemia, diabetic patients must decrease their usual insulin dose before strenuous exercise. Most patients with mild hypoglycemia can be successfully treated with a sweet drink or several lumps of sugar. Many diabetic patients sense the early symptoms of hypoglycemia and carry candy to prevent them. Severe hypoglycemia, however, is a medical emergency that requires immediate treatment with either intravenous glucose or glucagon.

Severe hypoglycemia is a medical emergency

A 12-year-old diabetic boy was playing with his friends. He received his normal insulin injection in the morning but continued playing through the lunch time without a meal. He became confused and fainted. He was instantly given an injection of glucagon from the emergency kit his father carried, and recovered within minutes.

Comment. An immediate improvement after glucagon injection confirms that this boy's symptoms were caused by hypoglycemia, caused by the exogenous insulin and insufficient food intake. Spectacular recovery from hypoglycemia was due to the action of glucagon. In the hospital, hypoglycemic patients who cannot eat or drink are treated with an intravenous infusion of glucose. An intramuscular glucagon injection is an emergency measure that can be applied at home.

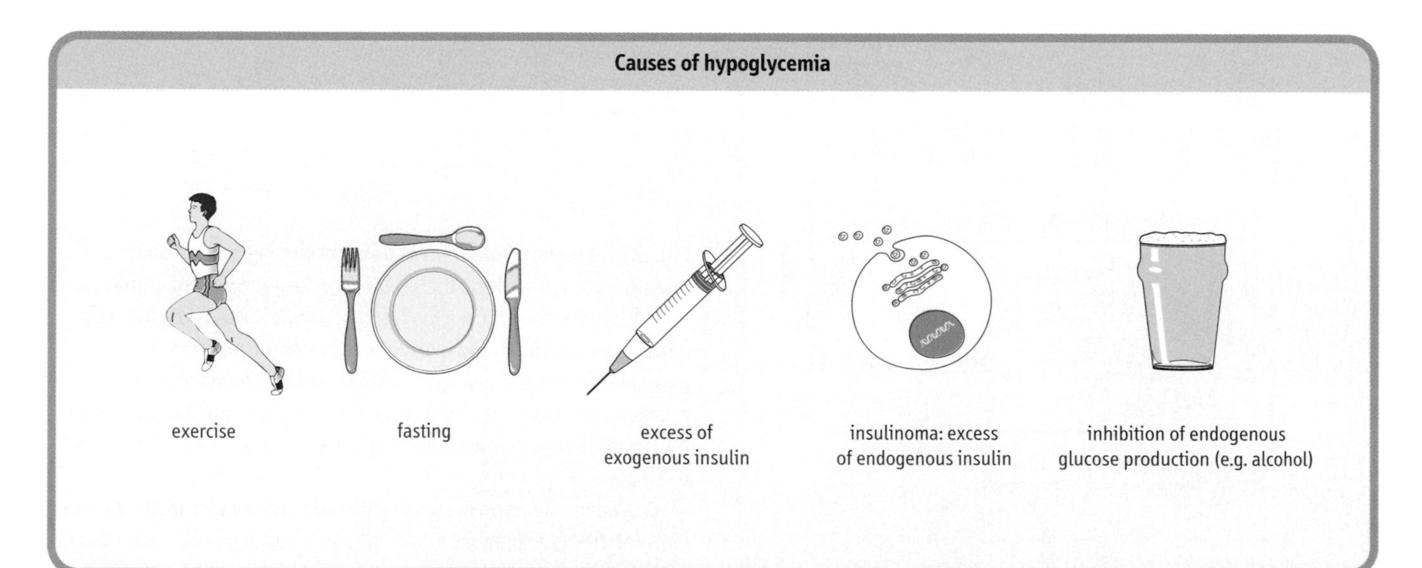

Fig. 20.10 Causes of hypoglycemia.

The assessment of fuel metabolism status

The measurement of plasma glucose is an important clinical test

How does one assess the status of fuel resources in a patient? The most important laboratory test is the measurement of plasma glucose. The physician differentiates between the normal glucose concentration (normoglycemia), too high a level (hyperglycemia – usually associated with diabetes), and too low a level (hypoglycemia). The blood glucose concentration increases after ingestion of food, therefore it is important to relate the time of blood sampling to the last meal. The interpretation of both fasting and postmeal (postprandial) glucose concentrations is important.

The fasting plasma glucose is relatively stable. Normal fasting (no caloric intake for approx 10 h) plasma glucose remains below 6.1 mmol/L (<110 mg/dL). A concentration above 6.1 mmol/L but below 7.0 mmol/L (126 mg/dL) is defined as impaired fasting glucose (IFG). The level of 7.0 mmol/L (126 mg/dL) or above, if confirmed, indicates diabetes. A plasma glucose measured irrespective of the meal

Insulin resistance

Sites of insulin resistance

Many obese individuals have mildly increased plasma glucose levels, together with normal or even increased plasma insulin concentrations. Patients with type 2 diabetes (see below) sometimes present with hyperglycemia, together with an 'inappropriately' high plasma insulin concentration. This apparent inconsistency is due to tissue insensitivity to insulin, known as insulin resistance. In insulin resistance, despite the presence of insulin, glucose is not efficiently utilized by tissues. Within a target cell, insulin resistance may be caused by defects at several levels (Fig. 20.11). Insulin-receptor binding could be compromised due to a downregulation of receptors, as in obesity. Rarely, a mutation in the insulin-receptor gene may cause extreme resistance. Resistance is also caused by the presence of antireceptor antibodies. Defects in receptor signaling (i.e. in tyrosine kinase activity or protein phosphorylation) may also cause insulin resistance. Insulin resistance may be caused by a defective enzyme, such as glycogen synthase or pyruvate dehydrogenase.

Insulin resistance syndrome

In type 2 diabetes, insulin resistance is a major factor contributing to hyperglycemia. Insulin resistance is also important in obesity. Interestingly, insulin resistance may have a broader clinical significance than just contribution to the diabetic syndrome: it is observed in many patients with arterial hypertension. Insulin-resistant individuals, who have hypertension and abnormalities of lipid metabolism, are also at a high risk of early coronary heart disease. Such a syndrome is known as 'syndrome X' or 'insulin resistance syndrome'.

Sites of insulin resistance

Site of resistance	Possible defect	Role in diabetes
prereceptor	insulin receptor antibodies, abnormal molecule	rare
receptor	decreased number or affinity of insulin receptors	not important in diabetes
postreceptor	defects in signal transduction: defective tyrosine phosphylation (?), reduced IRS-1 level, decreased phosphatidylinositol-3′ kinase decreased activity of key enzymes such as pyruvate dehydrogenase or glycogen synthase	most probable site of insulin resistance in diabetes
glucose transport	deficient or defective glucose transporters	to be established

Fig. 20.11 Insulin resistance. Cellular insulin resistance is caused by defects that may arise at receptor or postreceptor level. Insulin resistance is important in the development of type 2 diabetes mellitus, and probably also in the development of cardiovascular disease.

Insulin resistance increases the risk of heart attack

A 45-year-old man was referred to the cardiology outpatient clinic for an investigation of chest discomfort which he felt when climbing steep hills and, in his own words, 'when stressed or excited'. The patient was 170 cm tall and weighed 102 kg (224 lb). His blood pressure was 160/98 mmHg (upper limit of normal = 140/90 mmHg), triglyceride concentration was 4 mmol/L (364 mg/dL) (desirable level< 2.3 mmol/L [200 mg/dL]), and fasting plasma glucose was 6.5 mmol/L (117 mg/dL). His resting ECG was normal but an ischemic pattern was observed during exercise. His plasma insulin response to the oral glucose load was higher than normal.

Comment. This obese man presented with arterial hypertension, hypertriglyceridemia, glucose and impaired fasting. The impaired fasting of glucose in this case was due to peripheral insulin resistance. Such a cluster of abnormalities carries a high risk of development of coronary heart disease. Indeed, this man did have evidence of cardiac ischemia at a relatively early age.

time (random plasma glucose) is useful for diagnosis of hypoglycemia or severe hyperglycemia, but is of little use when the abnormality is mild. In a nondiabetic person random plasma glucose below 2.5 mmol/L (45 mg/dL) is regarded as dangerously low, and is termed hypoglycemia.

The criteria for the diagnosis of diabetes have been developed from epidemiologic data. The definition of diabetes is based on the evidence that people with glucose levels above the diagnostic level are more likely to suffer from the diabetic complications. The criteria for the diagnosis of diabetes described here are those proposed by the American Diabetic Association in 1997. The previous criteria, accepted by WHO set the cut-off point of fasting blood glucose diagnostic for diabetes as equal or above 7.8 mmol/L (140 mg/dL).

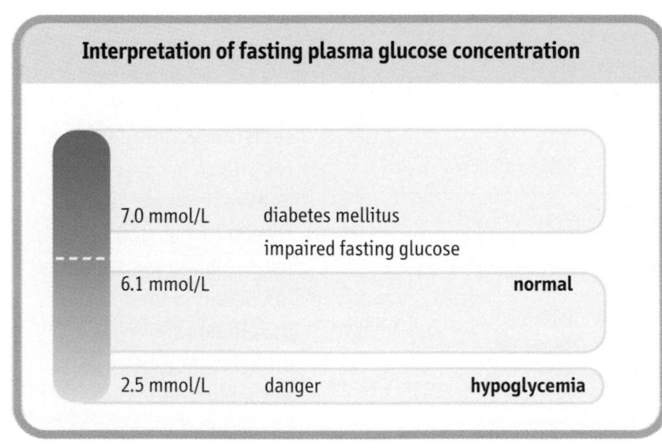

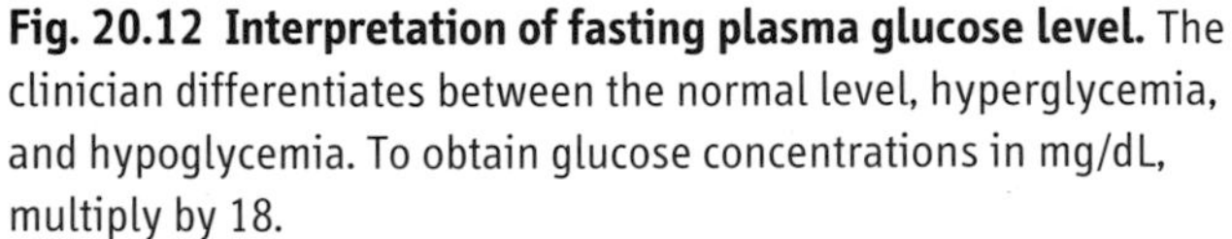

Fig. 20.12 Interpretation of fasting plasma glucose level. The clinician differentiates between the normal level, hyperglycemia, and hypoglycemia. To obtain glucose concentrations in mg/dL, multiply by 18.

Measurements of glucose and ketone levels in the urine are also indicators of metabolic control in diabetes

Other useful tests are measurements of glucose and ketones in urine. At normal plasma concentration, all the glucose filtered through the renal glomeruli is reabsorbed in the proximal tubule, and none appears in the urine. The urinary threshold for glucose is approximately 10.0 mmol/L (180 mg/dL). At higher plasma glucose concentrations the capacity of the renal tubular transport system is exceeded, and glucose filters into the urine (glucosuria). Rarely, the renal glucose threshold is lower than normal, and glucosuria can be detected at lower blood glucose levels. Thus, the measurement of urine glucose is a relatively insensitive test for diabetes.

Testing for the presence of ketones in urine (ketonuria) is clinically important. A high level of ketones signifies a high rate of lipolysis. This occurs in healthy individuals during prolonged fasting or on a high-fat diet. In a diabetic patient, ketonuria is a warning sign of metabolic decompensation.

Plasma lactate measurements may be important in rare circumstances. A high plasma lactate level indicates increased anaerobic metabolism, and usually is a marker of inadequate tissue oxygenation (hypoxia; see also Chapter 4).

Currently, metabolites, such as free fatty acids or glycerol, are rarely measured in clinical practice. This may change in the future as we continue to learn more about the importance of fatty acids in diabetes.

Metabolism after a meal and in the fasting state (terminology)

Normally an individual oscillates between a high-insulin/low-glucagon state and a low-insulin/high-glucagon state. The high-insulin/low-glucagon state occurs during the ingestion of food and for several hours after a meal (absorptive, postprandial or fed state). The low-insulin/high-glucagon state occurs on fasting. The early period of fasting, between 6 and 12 h, is called the postabsorptive state. Beyond that, it is either 'prolonged fasting' or 'starvation'.

Postprandial (absorptive) state

Following a meal, insulin release is stimulated and glucagon release is inhibited

Constituents of a meal stimulate insulin release and suppress the secretion of glucagon. This changes the metabolism of the liver, adipose tissue, and muscle (Fig. 20.13). After a meal, glucose utilization by the brain remains unchanged, but there is a large increase in glucose uptake in the insulin-dependent tissues, mainly skeletal muscle. Glucose oxidation and glycogen synthesis are stimulated in the liver, adipose tissue, and muscle; lipolysis is inhibited. Glucose taken up by the liver is

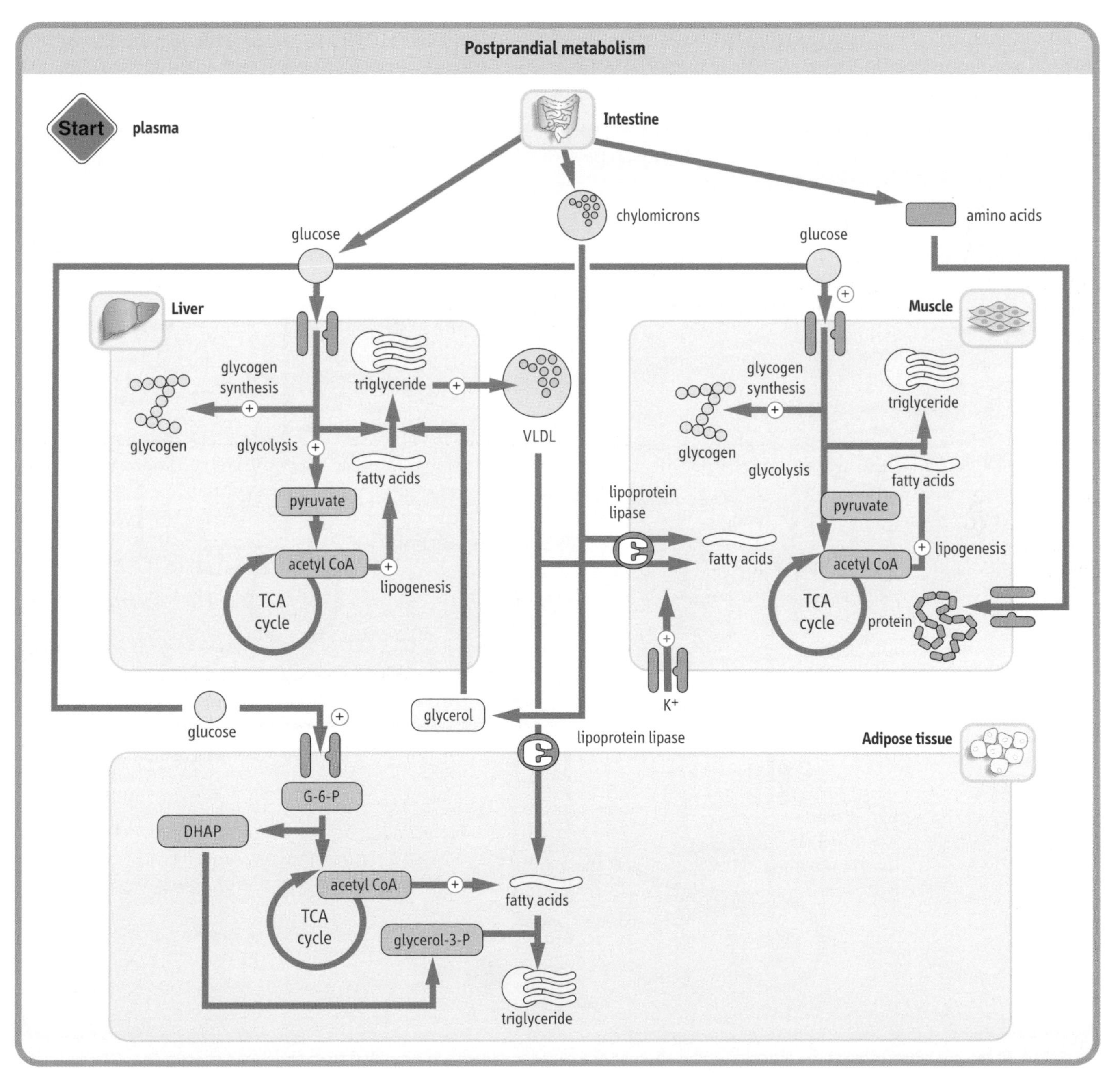

Fig. 20.13 Postprandial metabolism. In the postprandial state insulin directs metabolism towards storage and synthesis (anabolism). DHAP, dihydroxyacetone phosphate.

immediately phosphorylated by hexokinase into G-6-P at low concentrations, and by an inducible enzyme, glucokinase, at higher glucose concentrations. Excess glucose is directed into the pentose phosphate pathway to yield NADPH + H^+, which is essential for various reductive biosynthetic processes, such as lipogenesis and cholesterol synthesis.

Intestinal absorption of fat results in the assembly of large chylomicron particles by the intestinal cells. Chylomicrons loaded with triglycerides are released into the lymphatic system and reach the circulation through the thoracic duct. Chylomicron triglycerides are hydrolyzed by lipoprotein lipase, an insulin-inducible enzyme present on the peripheral endothelium, which hydrolyzes triglycerides to glycerol and free fatty acids (see Chapter 17). The fatty acids are taken up by adipose tissue and are stored as triglycerides. Fatty acids are used as a fuel in muscle. Triose phosphate produced from glycolysis is reduced to glycerol-3-phosphate (glycerol-3-P), which is used for the synthesis of triglycerides. In the liver and adipose tissue, insulin suppresses lipolysis and increases intracellular fatty acid synthesis. VLDL are assembled for the transport of lipids synthesized *de novo* in the liver. Insulin stimulates amino acid uptake and protein synthesis in the liver, muscle, and adipose tissue. Protein degradation in these tissues decreases.

Postabsorptive state

In clinical jargon the postabsorptive state is often referred to as 'fasting'. During the postabsorptive state, glucose metabolism approaches a steady state (i.e. hepatic glucose production, largely from glycogenolysis, equals its tissue uptake).

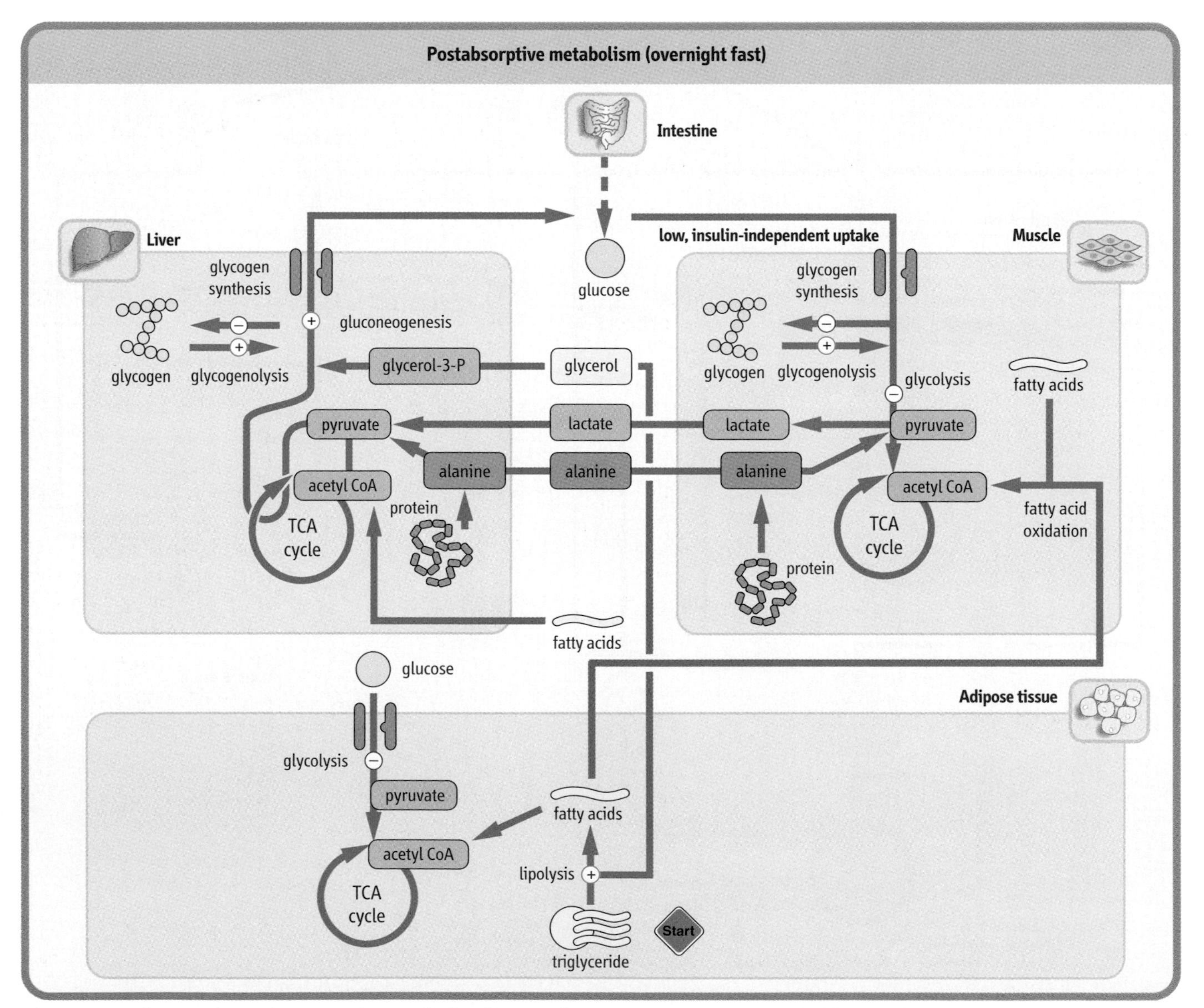

Fig. 20.14 In the postabsorptive state, glucose, in the absence of exogenous supply, is provided from endogenous sources (glycogenolysis and gluconeogenesis). Alanine and lactate cycles are operative (see Fig. 20.15 for details) Note there are three main gluconeogenic substrates: alanine, lactate, and glycerol.

After an overnight fast, insulin secretion decreases and glucagon secretion increases. This leads to a decrease in glycogen synthesis and to an increase in glycogenolysis (Fig. 20.14). The liver gradually becomes a glucose-producing organ.

In the postabsorptive state, approximately 80% of all glucose is taken up by insulin-independent tissues. Of this, 50% is taken up by the brain and 20% by red blood cells. In the postabsorptive state insulin-dependent tissues use little glucose – muscle and adipose tissue together are only responsible for approximately 20% of total glucose utilization.

After a 12-h fast, 65–75% of endogenous glucose is derived from glycogen, and the rest originates from gluconeogenesis. The contribution of gluconeogenesis increases with time. Muscle contributes to gluconeogenesis by releasing lactate which, after being oxidized to pyruvate, enters gluconeogenesis. The formed glucose is recycled to the skeletal muscle. This sequence of reactions is known as the Cori cycle. Also, low insulin level also stimulates proteolysis. The two main amino acids released from muscle are alanine and glutamate. A cycle analogous to that described by Cori, but involving alanine (the glucose–alanine cycle), also operates between muscle and the liver. Alanine released from muscle, enters gluconeogenesis in the liver, after being converted to pyruvate by transamination (Fig. 20.15).

Glucagon activates lipolysis in adipose tissue

In adipose tissue, glucagon action on hormone-sensitive lipase activates lipolysis. This releases the third main gluconeogenic substrate, glycerol, and also provides free fatty acids for energy metabolism in muscle. Activation of lipolysis secondarily stimulates ketogenesis from acetyl CoA in the liver, yielding acetoacetate, hydroxybutyrate, and the product of spontaneous decarboxylation of acetoacetate, acetone. All three metabolites are known as 'ketone bodies', and are oxidized in heart and skeletal muscle.

Blood samples for laboratory tests

The fast–fed cycle should be remembered when ordering laboratory tests. Blood sampling during the postabsorptive state provides the best assessment of basal metabolism. This is why patients are often asked not to eat anything 8–12 h before coming to the outpatient clinics to have blood taken for tests.

Plasma glucose response to the ingestion of glucose

The oral glucose tolerance test

Blood glucose response to a standard carbohydrate load is also a basis for the oral glucose tolerance test (OGTT). For a meaningful result it is essential that the test is performed under standard conditions.

The patient should attend in the morning, after fasting for approximately 10 h. To ensure as 'clean' a basal state as possible, the patient should sit throughout the test, since exercise affects the body's handling of glucose. The patient should not be in any way stressed, and the test should not be performed during or immediately after an acute illness (compare the effects of stress below).

Fasting plasma glucose is measured first. Next, the patient is given a standard quantity of glucose to drink (75 g in 300 mL of water) and glucose is measured again after 30, 60, 90, and 120 min (Fig. 20.16).

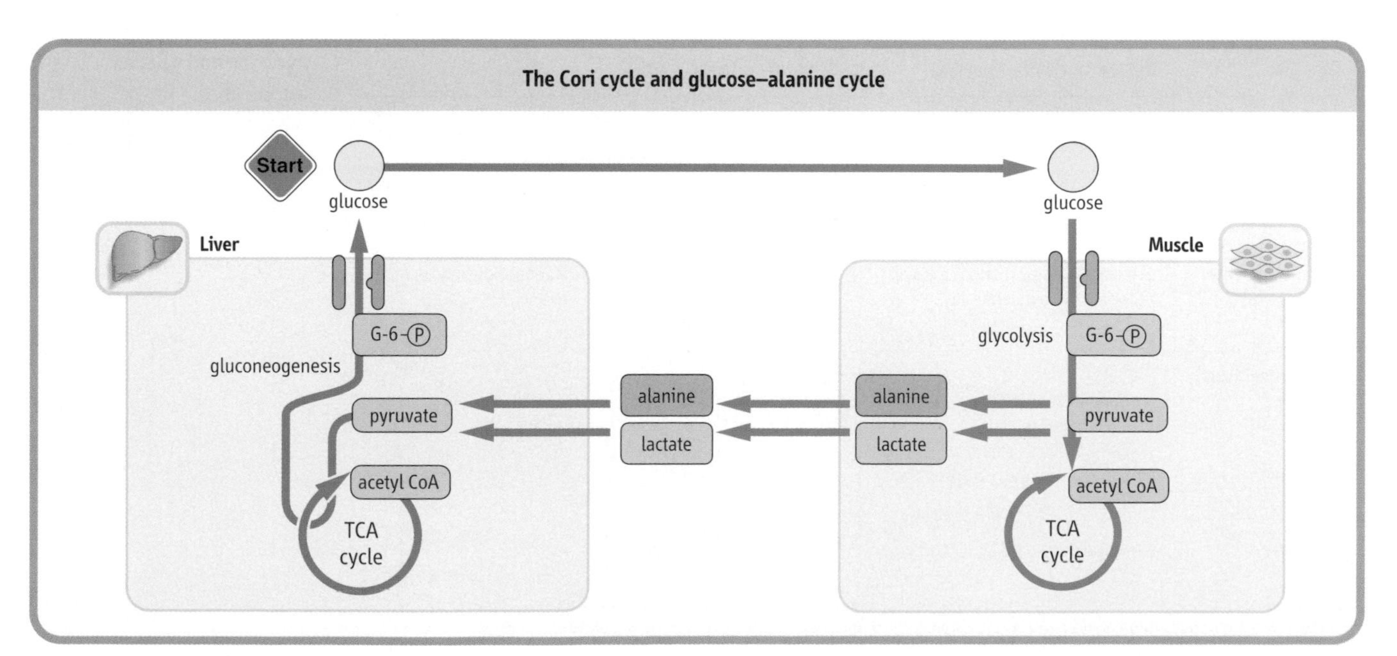

Fig. 20.15 The Cori (glucose–lactate) cycle allows recycling of lactate back to glucose, but does not contribute to the *de novo* synthesis of glucose.

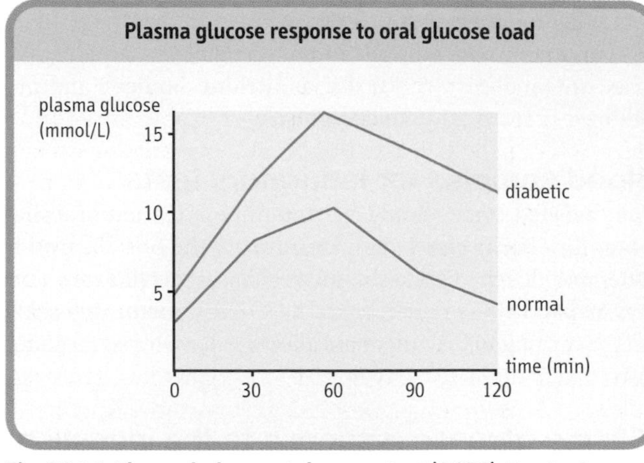

Fig. 20.16 The oral glucose tolerance test (OGTT). To obtain glucose concentrations in mg/dL, multiply by 18.

Normally during the OGTT, plasma glucose rises to a peak concentration after approximately 60 min and returns to a near fasting state within 120 min. If it remains above 11.1 mmol/L (200 mg/dL/min) in the 120-min sample, it indicates the presence of diabetes, even if the fasting blood glucose is normal (see below). Some individuals with normal fasting blood glucose, have a postload glucose level somewhere between the normal and diabetic concentration. Subjects with levels between 6.1 and 7.8 mmol/L (≥100 mg/dL and ≤140 mg/dL) 2 h after glucose load are classified as having impaired glucose tolerance. Although they are not diabetic, they are at an increased risk of developing diabetes in the future. The criteria for the interpretation of the OGTT are given in Figure 20.17. OGTT used to be regarded as a reference test for the diagnosis of diabetes. Current evidence suggests that the simpler measurement of fasting plasma glucose is equally accurate.

Prolonged fast and starvation

The brain can adapt to using ketone bodies in times of starvation

Prolonged fasting is a chronic low-insulin, high-glucagon state. There is also a decrease in thyroid hormone levels and concomitant lethargy. Early in the fasting period, free fatty acid levels increase in plasma and they become a major energy source. At the same time, because oxaloacetate is being directed towards gluconeogenesis, low levels of oxaloacetate in mitochondria tend to limit TCA cycle activity (see Chapter 13). Consequently, large amounts of acetyl CoA produced by β-oxidation of fatty acids in the liver are channeled into ketogenesis instead of the TCA cycle. There is an increase in circulating ketone bodies, which are oxidized in the muscle. The muscle releases amino acids, primarily alanine and glutamine, which serve as substrates for gluconeogenesis. In prolonged fasting, gluconeogenesis in the kidney becomes a significant source of endogenous glucose. As the excessive loss of protein is dangerous and can lead to respiratory muscle weakness and pneumonia, the body minimizes the use of proteins as a gluconeogenic substrate by becoming almost totally dependent on fat as an energy source (Fig. 20.18). The amount of glucose transporter-4 (GLUT-4) in the adipose tissue and muscle decreases. As the plasma concen-

Diagnosis of diabetes mellitus and glucose intolerance

Condition	Diagnostic criteria (mmol/L)	Diagnostic criteria (mg/dL)	Comments
normal fasting plasma glucose	below 6.1	below 110	
impaired fasting glucose (IFG)	equal or above 6.1 but below 7.0	equal or above 110 but below 126	
impaired glucose tolerance (IGT)	plasma glucose 2 h after 75 g load 7.8 or above, but below 11.1	plasma glucose 2 h after 75 g load 140 or above, but below 200	diagnosed during OGTT
diabetes mellitus*			
criterion 1	random plasma glucose 11.1 or above**	random plasma glucose 200 or above**	
criterion 2	fasting plasma glucose 7.0 or above	fasting plasma glucose 126 or above	
criterion 3	2 h value during OGTT 11.1 or above	2 h value during OGTT 200 or above	

**If one of the criteria is fulfilled, diagnosis is provisional. The diagnosis needs to be confirmed next day using a different criterion.*
***if accompanied by symptoms (polyuria, polydypsia, unexplained weight loss). These are the criteria proposed by the American Diabetes Association in 1997 (see further reading).*

Fig. 20.17 Diagnostic criteria for diabetes mellitus and glucose intolerance.

tration of ketone bodies continues to rise, the brain adapts to use them as a fuel. Some brain tissue also converts to the glycolytic metabolism of glucose, rather than complete oxidation to CO_2. This decreases the requirement for endogenous glucose synthesis and spares muscle protein as the produced lactate enters the Cori cycle which also operates between the brain and the liver.

Stress and metabolic response to injury

Stress is not only 'fight and flight' response but also trauma, injury, surgery, burns, or infection. All of these are associated with a metabolic response, characterized by increased oxygen consumption and hypermetabolism, in which the sympathetic nervous system plays a major role (Fig. 20.19). This time the main anti-insulin hormones are catecholamines (primarily epinephrine) and glucagon; cortisol is also important.

In times of stress, the brain has priority for fuel

In the first phase of the stress response there is vasoconstriction, which limits blood loss after injury. Fuels are mobilized from all available sources, but the provision of glucose for the brain has priority; high concentrations of epinephrine and glucagon stimulate glycogenolysis and gluconeogenesis. This leads to moderate hyperglycemia. Decreased peripheral uptake of glucose enhances the hyperglycemic effect. Later, the metabolic rate increases and energy is provided primarily from the oxidation of fatty acids and from protein. Gluconeogenesis from muscle-derived amino acids increases. A negative nitrogen balance develops and peaks approximately 2–3 days post injury.

Stress induces insulin resistance in muscle, adipose tissue, and liver, probably at a postreceptor level (see Fig. 20.11). The insulin-dependent transport of glucose in adipose tissue and skeletal muscle decreases, possibly through the inhibition of insulin's effect on GLUT-4. Interestingly, there is an increase in insulin-independent glucose uptake, particularly

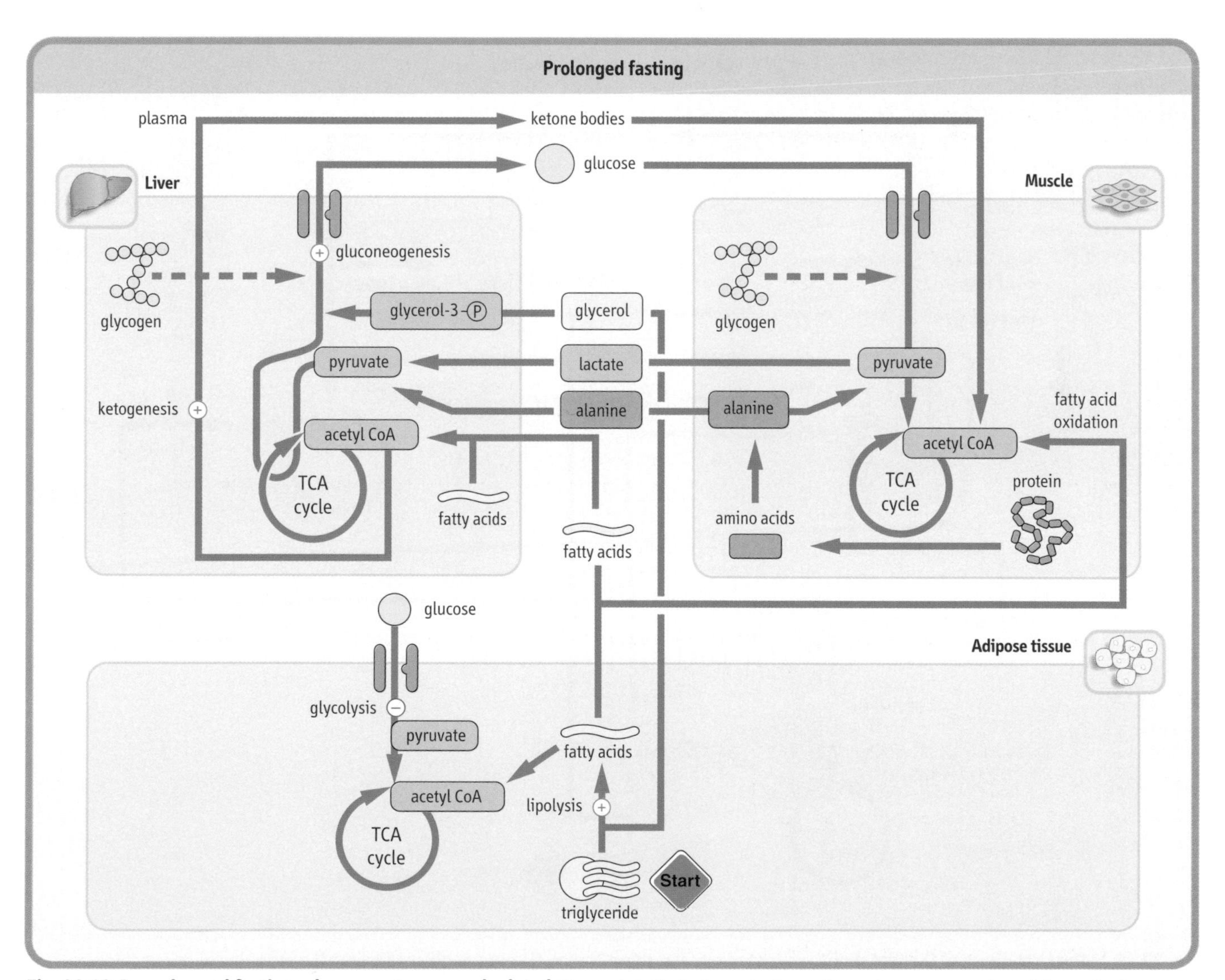

Fig. 20.18 In prolonged fasting, glycogen stores are depleted. The metabolic fuel supply depends on gluconeogenesis and, in particular, on lipolysis. Ketone bodies become an important fuel, which spares the muscle protein.

in muscle. This is caused by the action of tumor necrosis factor (TNF) and other cytokines, such as interleukin-1 (IL-1) (see Chapter 39). TNF also stimulates muscle glycogen breakdown. Glucocorticoids contribute to the stress response by inhibiting glucose transport into peripheral cells, and play a permissive role in the stimulation of gluconeogenesis by glucagon and catecholamines. Glucocorticoids facilitate gluconeogenesis by induction of G-6-Pase and PEPCK genes (see Fig. 20.9). Cytokines, such as IL-6, also contribute through their effect on PEPCK. They stimulate lipolysis in adipose tissue and contribute to muscle proteolysis.

Overall, the blood glucose concentration increases and there is an increase in lactate concentration. In the liver, lactate is converted to pyruvate via lactate dehydrogenase.

Thus, the metabolic response to stress is characterized by suppression of anabolic pathways (glycogen synthesis, lipogenesis), increased catabolism (glycogenolysis, lipolysis, and proteolysis), increased insulin-independent peripheral glucose uptake, and insulin resistance. Clinically there is fever, tachycardia (increased heart rate), tachypnea (increased respiratory rate), and leukocytosis (increased number of white blood cells). Prolonged stress may lead to organ failure.

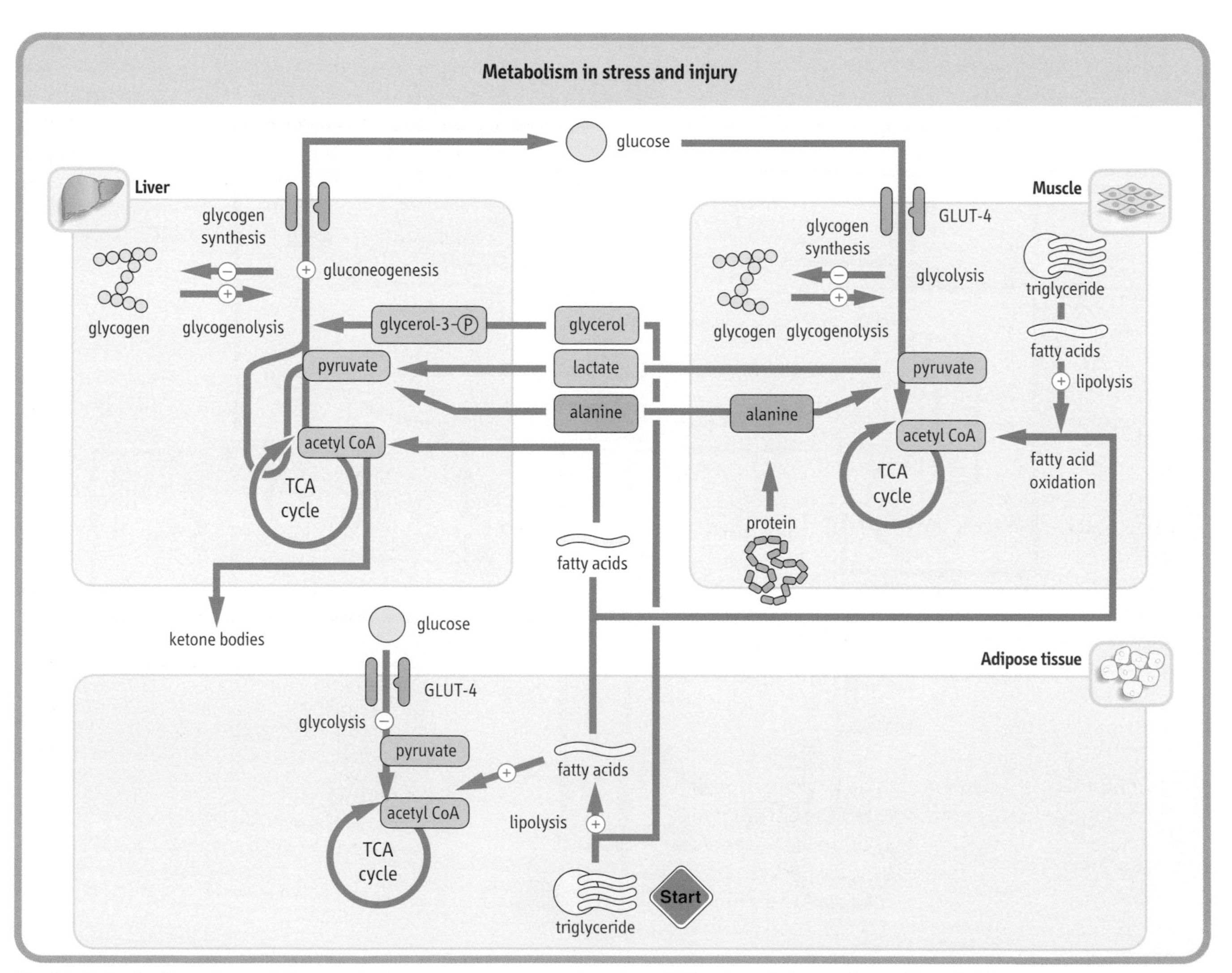

Fig. 20.19 In the first phase of the metabolic response to stress, there is mobilization of glucose from all available sources. Later, metabolic fuels are provided from fatty acids and from protein catabolism. The definition of metabolic stress includes response to injury, trauma, surgery, burns, and infection.

Catabolic state after road traffic accident

A 60-year-old man was admitted to an Intensive Care Unit following multiple trauma in a road traffic accident. He developed respiratory failure and was intubated. On the third day he was still unable to eat. His fasting blood glucose was 6.7 mmol/L (121 mg/dL), he had mild ketonuria (+), and his 24-h urine collection revealed the excretion of 600 mg of α-amino nitrogen/24 h (normal 50–300 mg/24 h).

Comment. This patient is hypercatabolic as a result of the stress response to injury. He is not able to take food. His high urinary nitrogen loss indicates excessive catabolism of muscle protein. He requires nutritional support. The patient was prescribed an intravenous nutrition regimen containing 2000 kcal (8374 kJ) as 50% glucose solution, 1000 kcal (4187 kJ) as lipid emulsion, and amino acid solution containing the equivalent of 18 g of nitrogen/day.

Major stress also affects water and electrolyte metabolism

A 65-year-old woman underwent a partial surgical excision of the stomach (gastrectomy). After surgery she was given a standard intravenous fluid replacement. The volume of fluids to be replaced was calculated on the basis of fluid lost in urine, through gastric drainage, and included an allowance for an insensible loss (water loss with breath and sweat). In spite of carefully calculated fluid volume and a normal renal function, the patient developed tissue swelling (became edematous) and her laboratory results were consistent with water overload.

Comment. The response to stress, such as major surgery, includes the stimulation of the secretion of antidiuretic hormone from the posterior pituitary (see Chapter 35). This causes increased water reabsorption by the kidney and a consequent water retention. This needs to be taken into account when prescribing fluid therapy in the postoperative period
See also Chapters 21 and 22.

Stress response and laboratory tests

The understanding of the stress response is important for the physician as it affects the results of laboratory tests. Stressed patients present with mild hyperglycemia, which should not be confused with diabetes mellitus. Injury, infection, or trauma are also associated with the so-called acute phase response, in which a variety of proteins, such as α_1-antitrypsin, C-reactive protein (CRP), haptoglobin, α_1-acid glycoprotein, complement proteins, and others, are produced. In fact, the measurements of CRP are important in the monitoring of the progress of therapy in patients with severe infections and inflammatory disorders.

Diabetes mellitus: a disorder of fuel metabolism

Diabetes mellitus is a group of metabolic diseases characterized by hyperglycemia leading to long-term complications (Fig. 20.20). It is a common disease, which affects 1–2% of Western populations. There are two main forms of diabetes. Of all diabetic patients, 10% have type 1 and 90% have type 2 (Fig. 20.21). Type 1 patients are unable to produce insulin and must receive exogenous insulin to survive. On the other hand, type 2 patients have at least partially preserved insulin secretion, but are often insulin-resistant. Some patients may have no clinical symptoms at all, with the diagnosis made exclusively on the basis of laboratory results. The diagnostic criteria are summarized in Figure 20.17.

Classification of diabetes

Syndrome	Comments
type 1	autoimmune destruction of β-cells
type 2	β-cell failure and insulin resistance
other types	genetic defects of β-cells (e.g. mutations of glucokinase gene). Rare insulin resistance syndromes. Diseases of exocrine pancreas. Endocrine diseases (acromegaly, Cushing's syndrome). Drugs and chemical-induced diabetes. Infections (e.g. mumps). Rare syndromes with the presence of antireceptor antibodies. Diabetes accompanying other genetic diseases (e.g. Down syndrome)
gestational diabetes	any degree of glucose intolerance diagnosed in pregnancy

Fig. 20.20 The most important forms of diabetes are type 1 (insulin-dependent) diabetes and type 2 (noninsulin-dependent) diabetes. Approximately 80% of all diabetic patients have type 2 diabetes.

Comparison of type 1 and type 2 diabetes mellitus

	Type 1	Type 2
Onset	usually below 20 years of age	usually over 40 years of age
Insulin synthesis	absent: immune destruction of β-cells	preserved: combination of impaired β-cell function and insulin
Plasma insulin concentration	low or absent	low, normal, or high
Genetic susceptibility	yes, inheritance associated with HLA antigens	not associated with HLA, important polygenic inheritance
Islet cell antibodies at diagnosis	yes	no
Obesity	uncommon	common
Ketoacidosis	yes	possible after major stress

Fig. 20.21 Type 1 and type 2 diabetes represent two syndromes of different pathogenesis. The table compares the characteristics of type 1 (formerly known as insulin-dependent diabetes, IDDM, or juvenile diabetes) with type 2 diabetes (formerly known as noninsulin-dependent diabetes, NIDDM, or maturity-onset diabetes).

Type 1 diabetes

Type 1 diabetes develops in young people, with the peak incidence at approximately 12 years of age. It is caused by autoimmune destruction of pancreatic β-cells. The precipitating cause is still unclear. It could be that a viral infection initiates the chain of autoimmune reactions. Alternatively, a cytokine response to viral infection, or to another insult, could attract monocytes and macrophages that infiltrate and destroy the pancreatic islets.

In addition to the inflammatory infiltration of the islets, a proportion of patients have antibodies against β-cell proteins. These are often present before the diagnosis of diabetes. Autoantibodies to insulin are also seen in some individuals.

Type 2 diabetes

Type 2 diabetes usually develops in patients who are over 40 years old and are typically obese. The pathogenesis of type 2 diabetes involves the impairment of insulin secretion and insulin resistance. It is unclear which of these mechanisms is the primary factor, but the type 2 diabetic patient usually has both impaired β-cell function and a degree of insulin resistance. The response of the diabetic β-cell to the glucose stimulus is suboptimal and, after glucose stimulation, there is no first phase of insulin secretion (see Fig. 20.5).

Genetics of diabetes

Susceptibility to type 1 diabetes appears to be inherited; there are 20 different regions of the human genome that are associated with sensitivity to type 2 diabetes. However, no 'diabetes gene' has been discovered to date. The best investigated susceptibility gene is located on chromosome 6 in the major histocompatibility complex (MHC) that codes for immune system recognition molecules known as histocompatibility antigens (HLA) (see also Chapter 34). Susceptibility to type 2 diabetes is associated with HLA types DR4 and DQw8 , whereas DR2 and DQw1.2 appear to suppress the tendency to develop diabetes. The sibling of a type 1 diabetic patient has a 10% chance of developing diabetes by the age of 50 years. This is much higher than the risk of diabetes in the general population.

The situation in type 2 diabetes is less clear. No doubt there is a strong hereditary component to this disease; monozygotic twins are concordant for type 2 diabetes in 90–95% of cases, and first-degree relatives of type 2 diabetes individuals have a 40% chance of developing type 2 diabetes. By contrast, in individuals with no diabetic relatives such risk amounts to only 10%. No type 2 diabetes susceptibility genes have been identified to date. Such genes could be coding for any of the components of fuel metabolism – hormones, receptors, ion channels, transporters, or indeed enzymes – associated with carbohydrate and lipid pathways.

Metabolism in diabetes

Uncontrolled diabetes leads to a life-threatening diabetic ketoacidosis

Persons with type 1 diabetes do not have any, or have only trace amounts of, insulin in plasma. They also have an increased plasma glucagon concentration. Lack of insulin results in the inability of glucose to enter insulin-dependent tissues, such as adipose tissue and muscle. It also contributes to the relative excess of glucagon. As a consequence, glycolysis and lipogenesis are inhibited, and glycogenolysis, lipolysis, ketogenesis, and gluconeogenesis are stimulated (Fig. 20.22). The key event in diabetes is that the liver becomes a producer of glucose. Increased endogenous glucose production, together with impaired glucose transport, lead to fasting hyperglycemia. Simultaneously, unopposed lipolysis produces an excess of acetyl CoA. Ketogenesis is stimulated. In a grossly decompensated patient, ketonemia and ketonuria develop (Fig. 20.23). Overproduction of acetoacetic and β-hydroxybutyric acids decrease the pH of blood, which is normally between 7.37 and 7.44 (see Chapter 22), and causes metabolic acidosis. In a type 1 diabetic patient, ketoacidosis can develop very quickly, even after missing a single insulin dose. In type 2 diabetes, ketoacidosis is relatively rare but may be precipitated by a major stress, such as myocardial infarction.

As glucose is osmotically active, renal excretion of a large amount of glucose leads to loss of water (osmotic diuresis). Poorly controlled diabetic patients complain of having to drink large quantities of fluid (polydypsia) and of passing large volumes of urine (polyuria). The resulting fluid loss eventually leads to dehydration. Diabetic ketoacidosis is a life-threatening condition.

Long-term complications of diabetes

Diabetic ketoacidosis, a sudden (acute) metabolic disturbance, is only one part of the diabetic syndrome. The other is the slow development of changes in small (microangiopathy) and large

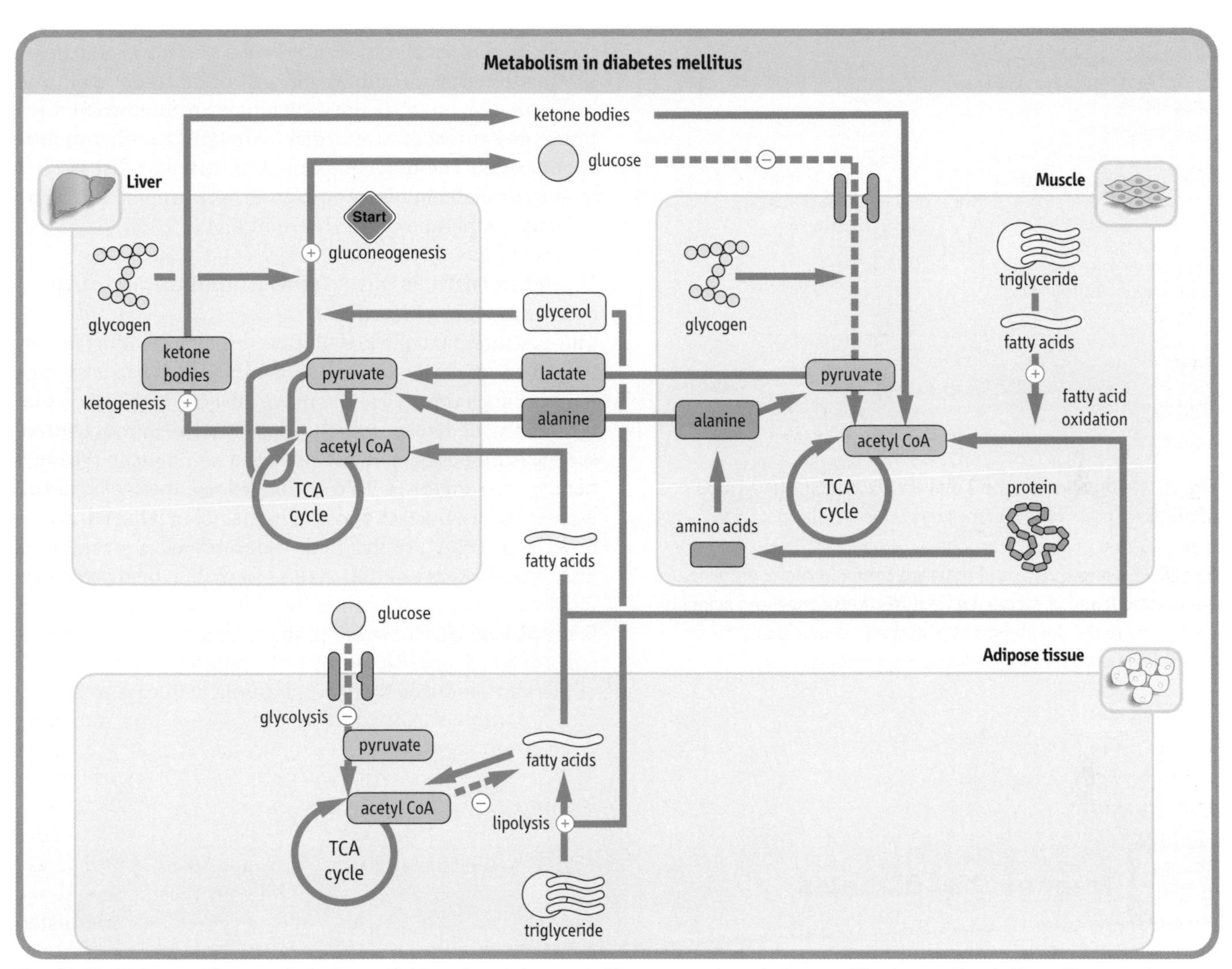

Fig. 20.22 Diabetes affects carbohydrate, lipid, and protein metabolism. Hyperglycemia is caused by the combined effect of increased endogenous glucose production by the liver and impaired peripheral glucose uptake.

Diabetic ketoacidosis affects body potassium balance

Insulin increases potassium uptake by cells. Lack of insulin leads to release of potassium, particularly from skeletal muscle. Since uncontroled diabetes is accompanied by an osmotic diuresis, the released potassium is excreted through the kidney. Most diabetic patients admitted to hospital with ketoacidosis are potassium-depleted. Exogenous insulin given to such patients stimulates the entry of potassium into cells. This further depletes the plasma potassium pool and can lead to very low plasma potassium levels (hypokalemia). Hypokalemia is dangerous, owing to its effects on cardiac muscle. Thus, except for patients with very high potassium levels, potassium supplementation needs to be considered in the treatment of diabetic ketoacidosis. See Chapters 21 and 22.

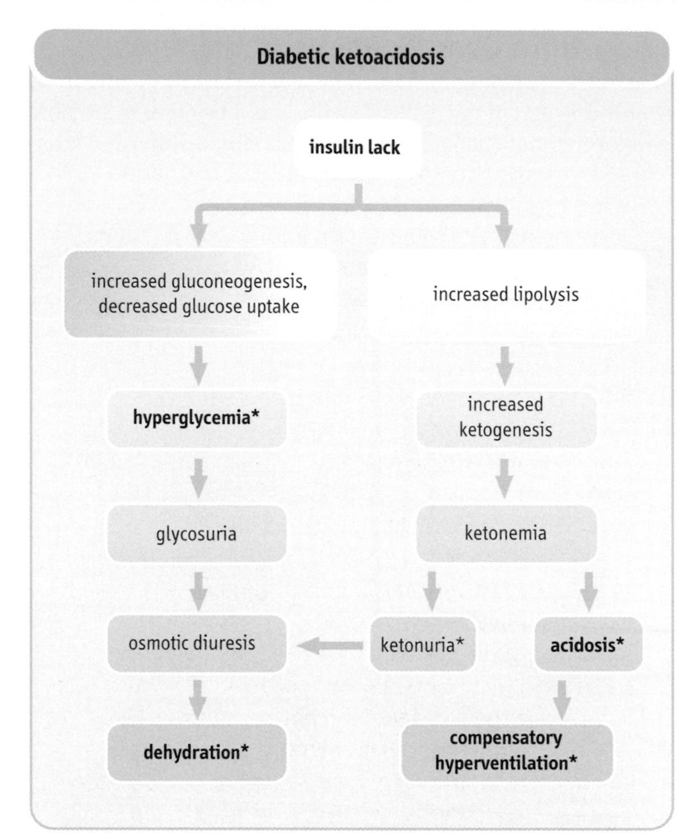

Fig. 20.23 Consequences of hyperglycemia (osmotic diuresis, dehydration) and of increased lipolysis (ketonemia and acidosis) contribute to the development of diabetic ketoacidosis. *Indicates the most important clinical and laboratory findings.

(macroangiopathy) arteries. Although they develop slowly, in the long term these changes are as dangerous as ketoacidosis. Diabetic complications lead to diabetic kidney failure (diabetic nephropathy), blindness (caused by diabetic retinopathy), and to the impairment of nerve function (diabetic neuropathy). Diabetic patients also develop lens opacities (cataracts) more frequently than non-diabetic persons. Currently, diabetes is the main cause of blindness in the Western world and one of the main causes of kidney failure. In addition, owing to macroangiopathy, diabetic individuals are at a two to three times greater risk of myocardial infarction than their nondiabetic peers. The diabetic peripheral vascular disease is a major cause of foot ulcers and lower limb amputations. Late diabetic complications are shown in Figure 20.24.

Glycated HbA_{1c} is a measure of diabetic control over a period of time

The measurement of blood glucose remains the most important laboratory test in diabetes. Its inherent drawback is that it may change very quickly. A major advance in the laboratory monitoring of diabetic patients has been the introduction of the measurement of glucose-modified hemoglobin (glycated hemoglobin or HbA_{1c}). As erythrocytes age, there is a gradual conversion of a fraction of native hemoglobin (HbA) to its glycated form, HbA_{1c}, so that in an older red cell, a greater fraction of HbA exists as HbA_{1c}. HbA_{1c} formation increases more rapidly during hyperglycemia (Fig. 20.27) and is proportional to the plasma glucose concentration. As the formation of glucose adducts at physiologic pH is virtually irreversible, the rate of glycation induced by hyperglycemia will leave a 'record'

Treatment of diabetes

Patients who have type 1 diabetes are treated with daily subcutaneous insulin injections throughout life. Diabetic patients in whom blood glucose is difficult to control are treated with several injections per day, or sometimes, with a constant insulin infusion, delivered by a programmable, portable pump. The rate of infusion is increased at meal times to help with disposal of exogenous glucose. Diet and exercise are also important in the management of diabetes.

Emergency treatment of diabetic ketoacidosis addresses four issues: insulin lack, dehydration, potassium depletion, and acidosis. The ketoacidotic patient requires insulin infusion to reverse the metabolic effect of the excess of anti-insulin hormones, and the infusion of fluids to treat dehydration. Intravenous fluids normally contain potassium supplements to prevent a decrease in plasma potassium hypokalemia. This treatment is usually sufficient to control the metabolic acidosis; however, when the acidosis is severe, infusion of an alkaline solution (sodium bicarbonate) may be required.

Type 2 diabetic patients do not usually require insulin treatment because insulin synthesis is at least partly preserved. Instead, the treatment relies on diet and oral hypoglycemic agents. Drugs, such as sulfonylurea derivatives stimulate insulin secretion. Another class of compounds, biguanides (e.g. metformin) reduce hyperglycemia by increasing peripheral glucose uptake.

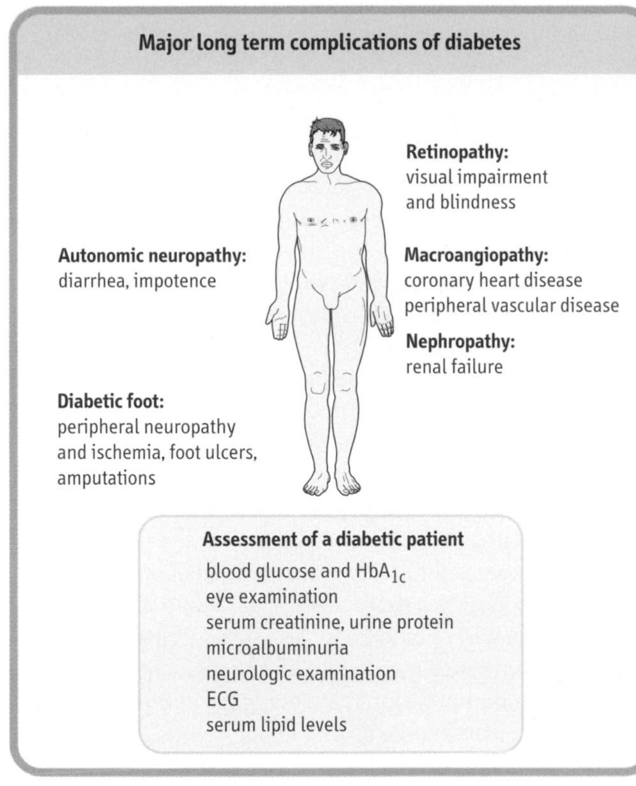

Fig. 20.24 Late complications of diabetes mellitus. These include abnormalities of small arteries (microangiopathy: diabetic retinopathy and nephropathy), large arteries (diabetic macroangiopathy: coronary heart disease and peripheral vascular disease), and also in diabetic neuropathy, which results from a combination of vascular and structural tissue changes.

Typical presentation of diabetic ketoacidosis

A 15-year-old girl is admitted to the Accident and Emergency department. She is confused and her breath smells of acetone. She has dry skin and tongue, which are the signs of dehydration. She also takes quick, deep breaths (hyperventilates). Her random blood glucose is 18.0 mmol/L (324 mg/dL) and ketones are present in the urine. Her serum potassium concentration is 3.5 mmol/L (normal = 3.5–5.0 mmol/L) and her arterial blood pH is 7.20 (normal = 7.37–7.44).

Comment. This is a typical presentation of diabetic ketoacidosis. Hyperventilation is a compensatory response to acidosis (see Chapter 22). The patient needs to be treated as a medical emergency. She will receive an intravenous infusion containing physiologic saline with potassium supplements to replace lost fluid, and an insulin infusion.

Long-term complications of diabetes

The pathogenesis of diabetic complications
Many aspects of the development of late complications remain unclear. The factors presently regarded as important are the nonenzymatic modification of proteins by glucose and associated with this oxidate process (glycoxidation), and abnormalities in the polyol pathway.

The polyol pathway
Glucose can be reduced to sorbitol by the action of aldose reductase (Fig. 20.25). Sorbitol is further oxidized by sorbitol dehydrogenase to fructose. Since aldose reductase has a high k_m for glucose, the pathway is not very active at normal glucose levels. In hyperglycemia, however, glucose levels in insulin-independent tissues, such as the red blood cells, nerve, and lens, increase and consequently there is an increase in the activity of the polyol pathway. Like glucose, sorbitol exerts an osmotic effect. This is thought to play a role in the development of diabetic cataracts. In addition, the high level of sorbitol decreases cellular uptake of another alcohol, myoinositol, which in turn causes a decrease in the activity of plasma membrane Na^+/K^+ ATPase. This in turn affects nerve function and, along with hypoxia and reduced nerve blood flow, contributes to the development of diabetic neuropathy. Drugs inhibiting aldose reductase improve the peripheral nerve function in diabetes.

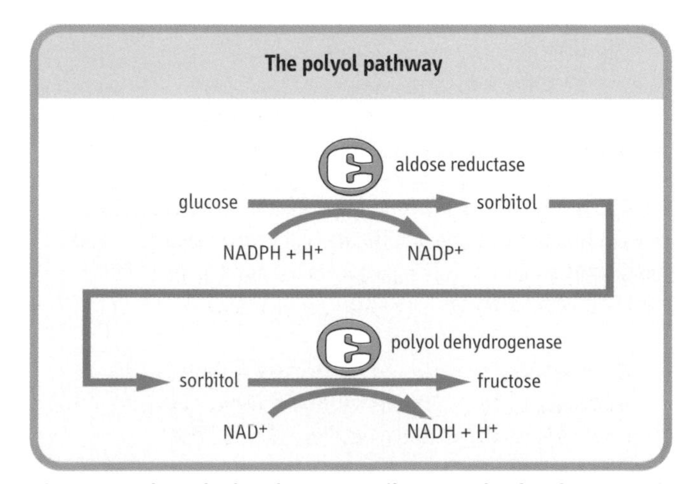

Fig. 20.25 The polyol pathway contributes to the development of diabetic neuropathy. This pathway may be inhibited by inhibitors of its rate-limiting enzyme, aldose reductase.

Long-term complications of diabetes: the role of protein glycation

Protein modification by glucose

The entry of glucose into brain and peripheral nerve tissue, kidney, intestine, lens, and red blood cells does not depend on insulin action. Consequently, during hyperglycemia, the intracellular level of glucose in these cells is high. This promotes the nonenzymatic attachment of glucose to protein molecules (protein glycation).

Glucose–protein adducts transform further in a sequence of nonenzymatic reactions (Fig. 20.26), collectively known as the Maillard reaction (so-called after the French chemist, Louis Camille Maillard). Amino acid residues involved include the α-amino terminal amino acid and ε-amino groups of lysine residues. First, a labile Schiff base is formed. This spontaneously transforms to ketoamine through the Amadori rearrangement. Glycated hemoglobin, hemoglobin A_{1c} (HbA_{1c}), is the most widely studied Amadori product. Other proteins, such as albumin and collagen, can also form Amadori products. Glycation changes the electrical charge of proteins and can affect their functions, such as binding to membrane receptors. For instance, glycation of apolipoprotein B slows down the rate of receptor-dependent metabolism of low-density lipoproteins (LDL). (See Chapter 17.)

Amadori products transform further to form protein crosslinks known as advanced glycation end products (AGE). AGE crosslink long-lived body proteins such as tissue collagen or a nerve protein, myelin. AGE formation 'stiffens' the extracellular matrix and decreases the elasticity of, for instance, the arterial wall. Their formation also affects the function of endothelial cells, phagocytes (macrophages), and smooth muscle cells in the wall of blood vessels. Through these mechanisms AGE may contribute to the development of the late complications of diabetes and probably other vascular diseases.

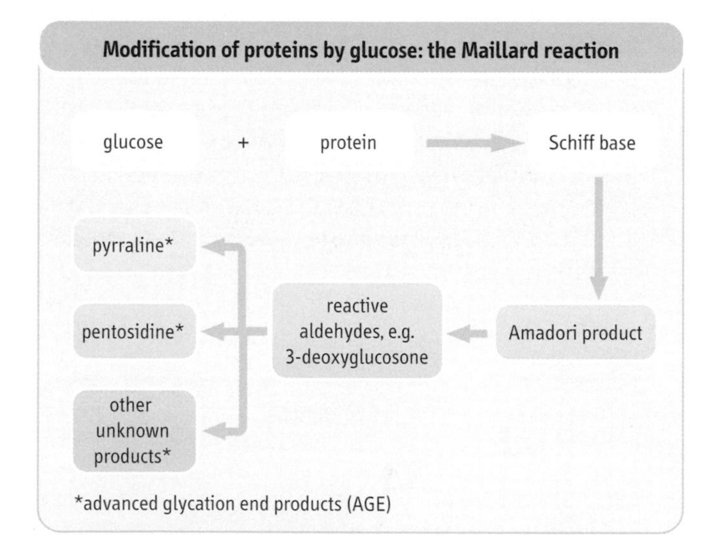

Fig. 20.26 The Maillard (browning) reaction is an ubiquitous process that leads to the modification of protein structure and function by sugars. This is important in the development of the late complications of diabetes.

throughout the rest of the erythrocyte's life, even if hyperglycemia is successfully treated. Thus, HbA_{1c} concentration in blood reflects the time-averaged level of glycemia over the 3–6 weeks preceding the measurement. The normal concentration of HbA_{1c} is 4–6% of total HbA. Levels below 7% indicate acceptable control of diabetes. Higher levels suggest poor control (Fig. 20.28).

Only a few laboratory are useful in the assessment of long-term complications of diabetes

These are primarily tests of renal function, such as the measurement of the concentration of urea and creatinine in plasma (see Chapter 21). Urinary protein excretion (proteinuria) above 300 mg/day is the most important diagnostic test of the progress of diabetic nephropathy. Minimal amounts of albumin present in urine (microalbuminuria, above 200 mg/day) predict the development of nephropathy. Microalbuminuria is only detectable by an assay which is more sensitive than the conventional method for the measurement of albumin in serum.

The importance of good glycemic control

Maintaining near-normal blood glucose levels prevents the development of late complications of diabetes. A recently completed landmark clinical study, the Diabetes Control and Complications Trial (DCCT), has shown that the development of late complications of diabetes in type 1 diabetes is related to long term glycemia. This study has also shown that in patients who have complications, good control of glycemia delays further development of retinopathy, nephropathy, and neuropathy. Similar results were obtained for type 2 diabetic patients during the U.K. Prospective Diabetes Study (UKPDS) completed in 1998. Thus, the aim of treatment of diabetes should be the achievement of blood glucose levels as close to normal as possible, without precipitating hypoglycemia (see Further reading).

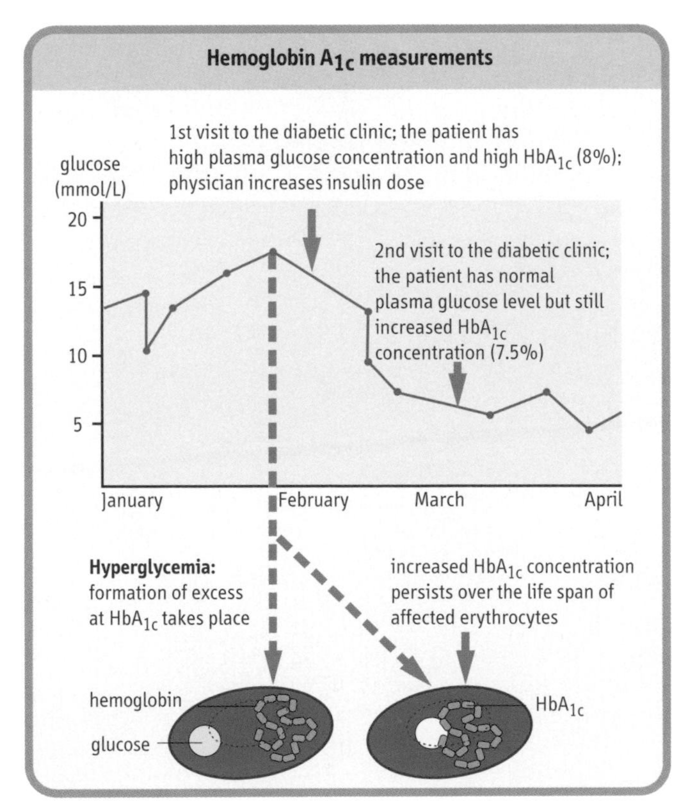

Fig. 20.27 The concentration of HbA_{1c} in blood reflects the time-averaged glycemia over the 3–6 weeks preceding its measurement. To obtain glucose concentrations in mg/dL, multiply by 18.

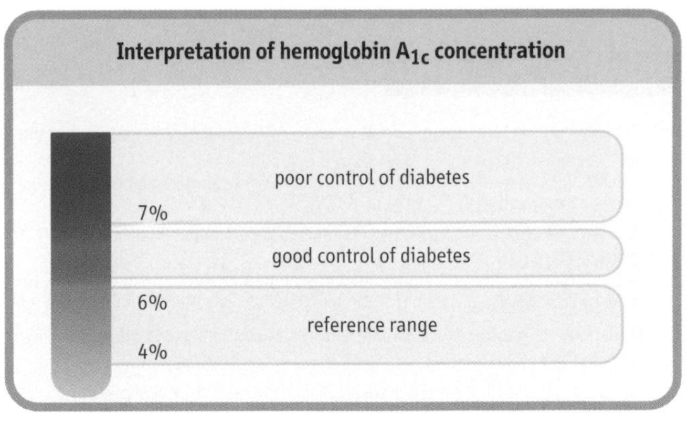

Fig. 20.28 The HbA_{1c} concentration allows the physician to assess the quality of glycemic control in a diabetic patient.

HbA_{1c} identifies patients who do not comply with treatment

Diabetes

A 15-year-old insulin-dependent boy visits a Diabetic Clinic for a check-up. He tells the doctor that he complies with all the dietary advice and never misses insulin. Indeed, his random blood glucose is 6 mmol/L (108 mg/dL), but his HbA_{1c} concentration is 11% (normal 4–6%). He has no glycosuria or ketones in his urine.

Comment. Blood and urine glucose results indicate good control of this boy's diabetes at the time of measurement, whereas the HbA_{1c} level suggests poor control over the last 3–6 weeks. The probability is that he only complied with treatment days before he was due to come to the clinic. This is not uncommon in adolescents, who find it hard to accept the necessity to adjust their lifestyle to the requirements of diabetes treatment.

Summary

This chapter has described the metabolic inter-relationships between key tissues and organs – liver, adipose tissue, and skeletal muscle. These interactions preserve glucose homeostasis and, in healthy persons, maintain blood glucose concentration within a narrow range. Disruption of this homeostatic system results in potentially life-threatening conditions – hypoglycemia and diabetes mellitus. The metabolite concentrations in blood change during the fast–feed cycle, and are influenced by stress and disease. The measurement of plasma glucose concentration is part of a routine assessment of every patient admitted to hospital. In diabetic patients, a broader range of tests are performed. These include the measurements of glucose, ketones, and HbA_{1c}, and tests of renal function, including microalbuminuria.

Further reading

Ashcroft FM, Ashcroft SJH, eds. Insulin. Molecular biology to pathology. (Oxford: Oxford University Press;1992): 1–418.

Atkinson MA, Maclaren NK. The pathogenesis of insulin-dependent diabetes mellitus. *New Engl J Med* 1995;**331**:1428–1436.

Kahn CR, Weir GC, eds. *Joslin's Diabetes Mellitus, 13th edn*. Philadelphia: Lea and Febiger, 1994.

Mizock BA. Alterations in carbohydrate metabolism during stress: a review of literature. *Am J Med* 1995;**98**:75–84.

Moller DE, Flier JS. Insulin resistance-mechanisms, syndromes and implications. *New Engl J Med* 1991;325:938–948.

Turner RC, Hattersley AT, Shaw JTE, Levy JC. Type 2 diabetes: clinical aspects of molecular studies. *Diabetes* 1995;**44**:1–10.

Service FJ. Hypoglycemic disorders. *New Engl J Med* 1995;**332**:1144–1152.

The Diabetes Control and Complications Trial Research Group. The effect of intensive treatment of diabetes on the development and progression of long-term complications of insulin-dependent diabetes mellitus. *N Engl J Med* 1993;**329**:977–986.

The Expert Committee on the diagnosis and classification of diabetes mellitus. Report of the Expert Committee on the diagnosis and classification of diabetes mellitus. *Diabetes Care* 1997;**20**:1183–1197.

U.K. Prospective Diabetes Study (UKPDS) Group. Intensive Blood–glucose control with sulphonylureas or insulin compared with conventional treatment and risk of complications in patients with type 2 diabetes (UKPDS 33). *Lancet* 1998;**352**:837–53.

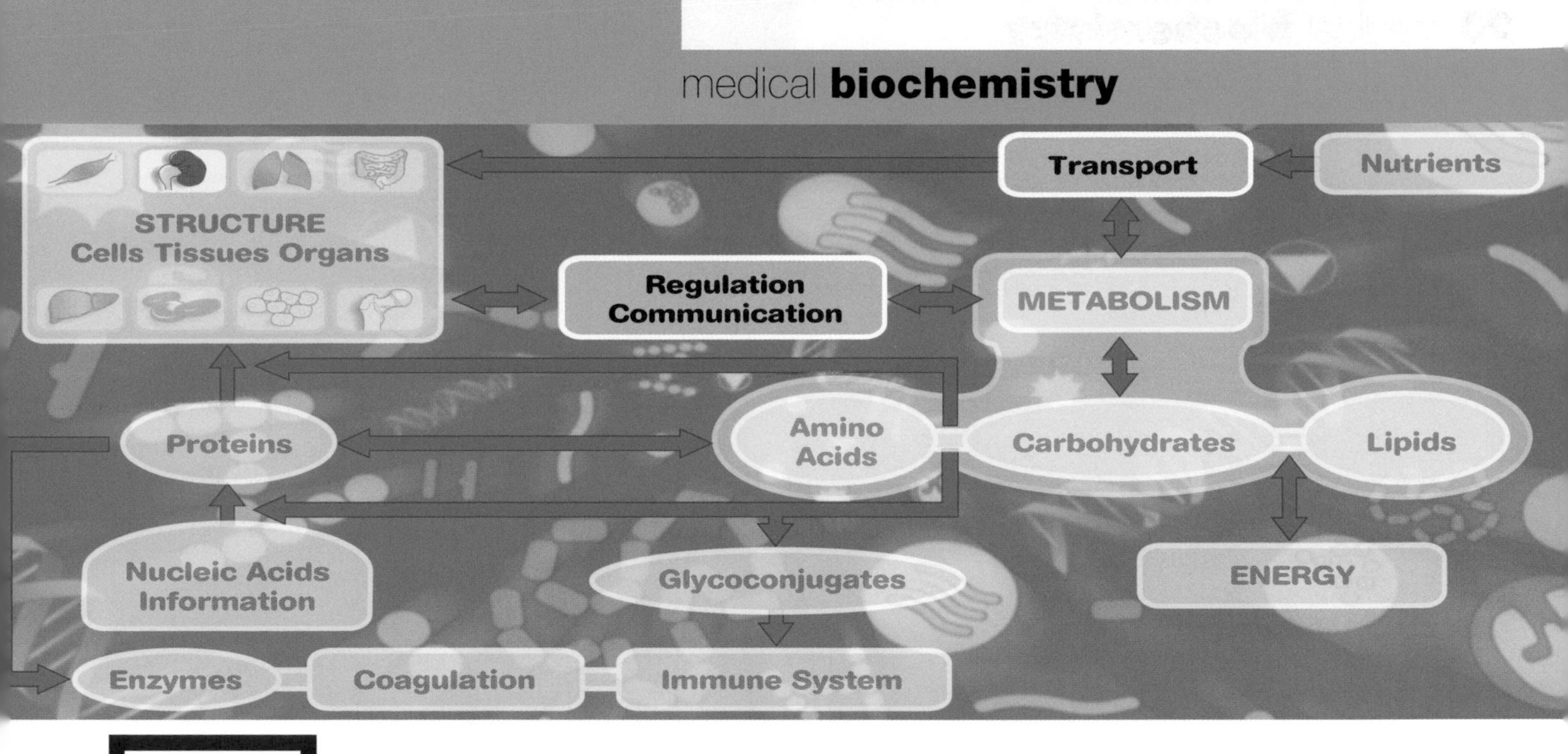

21 Water and Electrolyte Balance: Kidney Function

One can view a human body as a conglomerate of cells suspended in the extracellular fluid. The solutions in the body are compartmentalized, the most important compartments being intracellular and extracellular. The different organs are connected by a transport system comprising the blood vessels, which constantly pump nutrients through tissues and remove metabolic waste. Disorders affecting water and electrolyte balance are important in many areas of medicine.

Body water

Water constitutes about 60% of the body weight

The water content of the body changes with age: it is about 75% in the newborn and decreases to less than 50% in older individuals. Water content is greatest in brain tissue (about 90%) and least in adipose tissue (10%).

Various hydrophilic ions and neutral molecules are soluble in water, proteins form colloids, and lipoproteins that are dispersed in water as pseudomicelles. Water itself participates as substrate and product in many chemical reactions in glycolysis, the tricarboxylic acid cycle, and the respiratory chain.

Both a relative deficiency and an excess of water impair the function of tissues and organs. The stability of subcellular structures and activities of numerous enzymes are dependent on adequate cell hydration.

Body water compartments

Approximately two-thirds of total body water is in the intracellular fluid (ICF), and one-third is distributed outside cells, in the extracellular fluid (ECF). ECF consists of interstitial fluid and lymph (15% body weight), plasma (3% body weight), and transcellular fluids, which include gastrointestinal fluid, urine, CSF, and others (Fig. 21.1).

The capillary wall, which separates plasma from the interstitial fluid, is freely permeable to water and electrolytes, but restricts the flow of proteins

These properties of the capillary wall mean that, whereas ECF and plasma have a similar distribution of low-molecular-weight ions and molecules, the concentration of protein is four to five times greater in plasma than in the interstitial fluid.

The total plasma concentration of cations is about 150 mmol/L, of which sodium is responsible for approximately

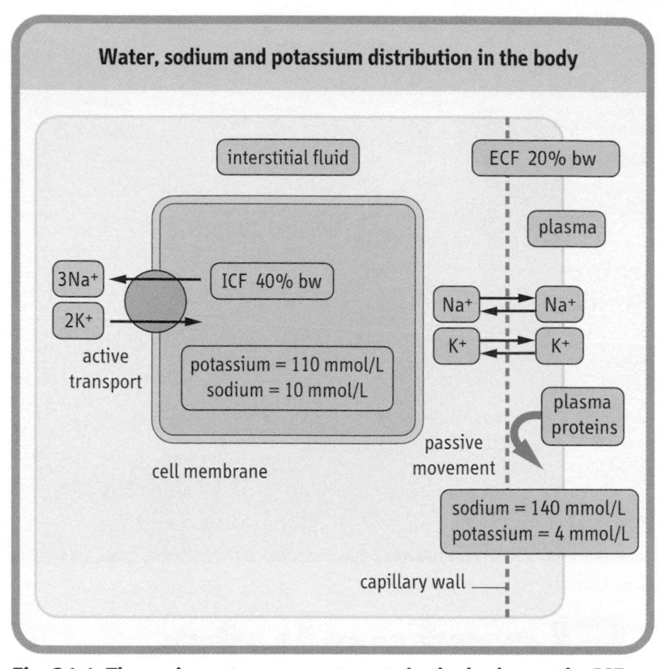

Fig. 21.1 The main water compartments in the body are the ICF and the ECF. ECF includes interstitial fluid and plasma. A gradient of sodium and potassium concentration is maintained across cell membranes by Na^+/K^+-ATPase, which pumps sodium from ICF to the ECF, and potassium in the opposite direction. Sodium is a major contributor to the osmolality of the ECF and a main determinant of the distribution of water between ECF and ICF. In contrast, distribution of water between plasma and interstitial fluid is determined by the oncotic pressure exerted by plasma proteins. bw, body weight.

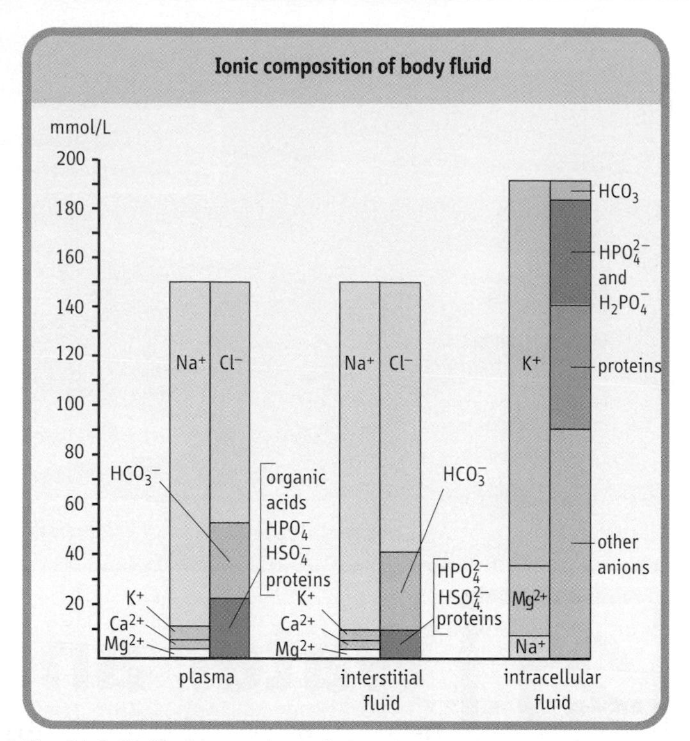

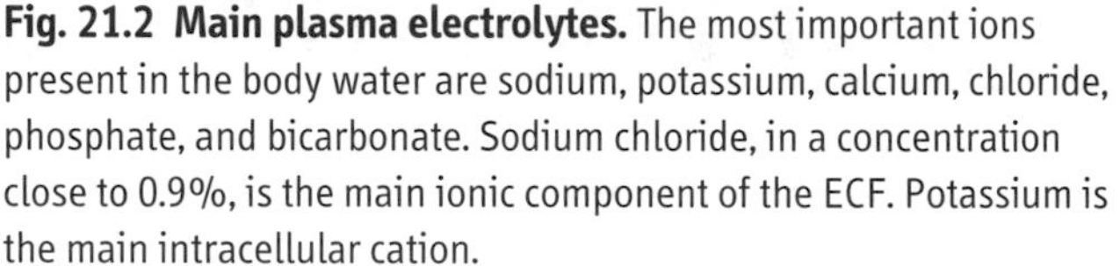

Fig. 21.2 Main plasma electrolytes. The most important ions present in the body water are sodium, potassium, calcium, chloride, phosphate, and bicarbonate. Sodium chloride, in a concentration close to 0.9%, is the main ionic component of the ECF. Potassium is the main intracellular cation.

In addition to the electrolytes shown here, plasma glucose and urea contribute to plasma osmolality. Normally their contribution is small, because they are present in plasma at relatively low molar concentrations (about 5 mmol/L each). The contribution of glucose to osmolality becomes important in diabetes, when its concentration increases: because glucose is confined to the ECF, glucose excess induces movement of water from the cells to the ECF. The concentration of urea increases in renal failure; however, because it can cross cell membranes freely, it does not contribute to water movement between ECF and ICF.

140 mmol/L. The most abundant plasma anions are chloride, with an average concentration of 100 mmol/L, and bicarbonate, with an average concentration of 25 mmol/L (Fig. 21.2). The remaining anions constitute the so-called anion gap, which includes phosphate, sulphate, protein, and organic acids such as lactate, citrate, pyruvate, acetoacetate, and 3-hydroxybutyrate. In practice, the anion gap is estimated according to the formula:

$$\text{anion gap} = \{[Na^+] + [K^+]\} - \{[Cl^-] + [HCO_3^-]\}$$

Calculated this way, the anion gap is about 12 mmol/L. It may increase severalfold in disorders where inorganic and organic anions, accumulate, e.g. renal failure or diabetic ketoacidosis.

Potassium is the main cation in the ICF. The concentration of potassium in the ICF is about 110 mmol/L, which is about 30-fold greater than that in the ECF. The concentrations of sodium and chloride in the ICF are only 10 mmol/L and 4 mmol/L, respectively. Anions present in the ICF include proteins, phosphate and other substances that cannot diffuse through the cell membranes.

Water diffuses freely across most cell membranes, whereas the movement of ions and neutral molecules is restricted

Transport of small molecules across membranes is possible because of the existence of specific transport proteins, including ion pumps. Of the several ion pumps, the sodium/potassium ATPase pump (Na^+/K^+ATPase), also referred to as sodium–potassium pump, shows the greatest transport activity. Its action is based on the hydrolysis of one ATP molecule, which drives the transfer of three sodium ions from the cell to the outside, and two potassium ions from the outside into the cell. This pump maintains both chemical and electrical potential gradients between the inside and the outside of the cell (Fig. 21.3); for most cells, the membrane potential ranges from 50 to 90 mV, being negative inside the cell. The electrochemical gradient is a source of energy for transport of many substances, in particular for the cotransport of sodium ions with glucose, amino acids, and phosphate.

Osmolality and volume of body fluids

The relative volume of ECF and ICF depends on the amount of osmotically active substances in these compartments

All molecules dissolved in the body water contribute to the osmotic pressure, which is proportional to the molal concentration of the solution. One millimole of a substance dissolved in 1 kg H_2O at 37°C exerts an osmotic pressure of approximately 19 mmHg. Under physiologic conditions, the average concentration of all osmotically active substances in the ECF is 290 mmol/kg H_2O, and this remains in equilibrium with the ICF.

A change in the concentration of osmotically active ions in either of the water compartments creates a gradient of

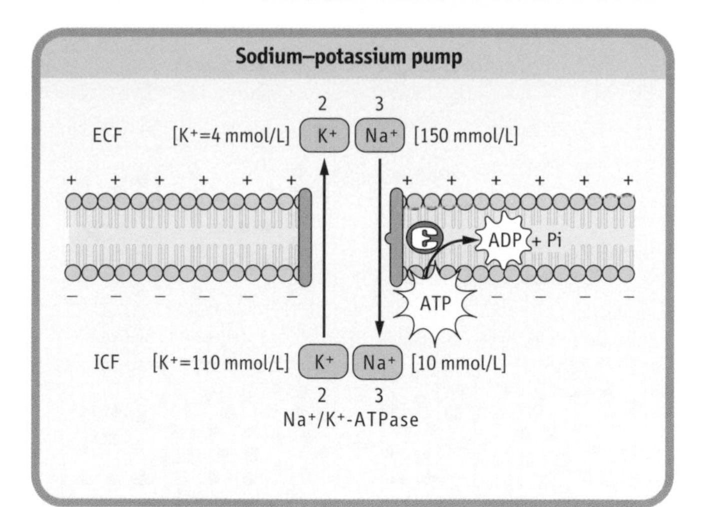

Fig. 21.3 The enzyme, Na^+/K^+-ATPase, is responsible for maintaining sodium and potassium concentration gradients across cell membranes. It also has a key role in the reabsorption of sodium in the kidney tubules.

The composition of body fluids determines abnormalities resulting from their loss from the body

Clinical problems caused by the loss of body fluids

It is possible to predict problems that may arise from the loss of body fluids. For instance, sweat contains less sodium than extracellular fluid, therefore excessive sweating leads to a predominant loss of water, 'concentrates' extracellular fluid sodium, and leads to an increased plasma concentration of sodium. In contrast, the sodium content of the intestinal fluid is similar to that of plasma, with considerable amounts of potassium. The loss of intestinal fluid (as occurs, for instance, in severe diarrhea) results in dehydration and hypokalemia, but may not lead to large changes in plasma sodium concentration (Fig. 21.4).

Electrolyte content of body fluids

	Sodium (mmol/L)	Potassium (mmol/L)	Bicarbonate (mmol/L)	Chloride (mmol/L)
plasma	140	4	25	100
gastric juice	50	15	0–15	140
small intestinal fluid	140	10	variable	70
feces in diarrhea	50–140	30–70	20–80	variable
bile, pleural, and peritoneal fluids	140	5	40	100
sweat	12	10	—	12

Fig. 21. 4 The electrolyte composition of body fluids. Loss of fluid that has an electrolyte content similar to that of plasma leads to dehydration with normal plasma electrolyte concentrations. In contrast, dehydration may be accompanied by hyperatremia when the sodium content of the lost fluid is less than that of plasma (e.g. sweat). Overhydration is usually accompanied by hyponatremia. (Adapted with permission from Dominiczak MH, ed. *Seminars in clinical biochemistry,* 2nd ed. Glasgow: Glasgow University, 1997.)

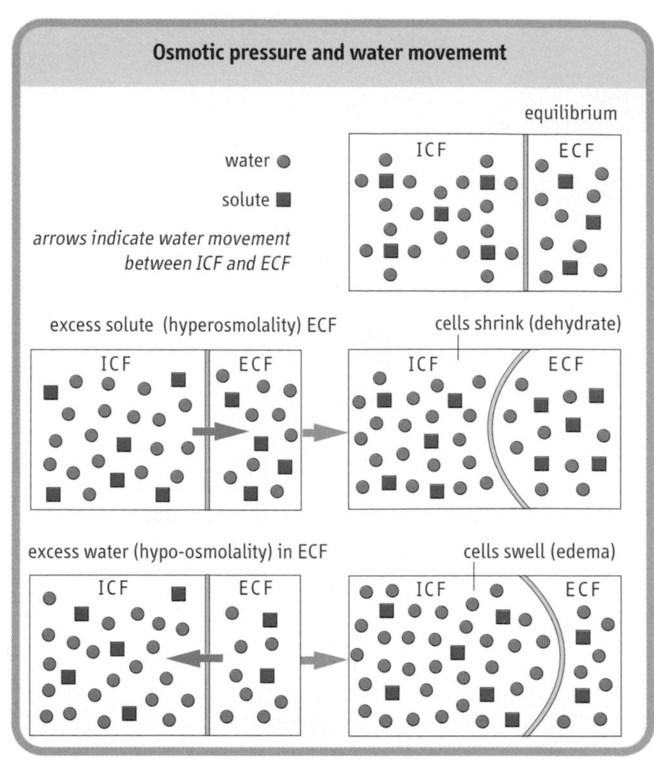

Fig. 21.5 Osmotic pressure controls the movement of water between different compartments. An increase in ECF osmolality leads to movement of water from the cells into the ECF, and to cellular dehydration. In contrast, when ECF osmolality decreases, water moves from the ECF into the cells. This leads to expansion of the cell volume (edema). Blue arrows indicate the direction of water movement. Creatinine is mg/dL = μmol/L × 0.0113. Urea is mg/dL = mmol/L × 6.02. In several countries the measurement of urea nitrogen is used instead of blood urea. The normal range of an adult is 2.9–8.9 mmol/L (8–25mg/dL).

osmotic pressure and, consequently, movement of water between compartments. Water diffuses from a compartment of low osmolality to one of high osmolality until the osmotic pressures are identical in both of them (iso-osmolality). Thus the relative volume of ECF and ICF depends on the amount of osmotically active substances in these compartments; the sodium ion is the most important one in the ECF. Glucose, present in plasma in a concentration about 5 mmol/L (90 mg/dL), does not contribute much to osmolality, but in diabetes it may reach very high concentrations (in extreme circumstances above 50 mmol/L [900 mg/dL]) that profoundly affect the osmolality (see Chapter 20) (Fig. 21.5).

Within the ECF, the distribution of water between intravascular and extravascular compartments depends on the plasma protein concentration

Proteins, particularly albumin, induce their own osmotic pressure (about 3.32 kPa [25 mmHg]), known as the oncotic pressure. This is balanced by the hydrostatic pressure, which forces the fluid out of the capillaries. In the arterial

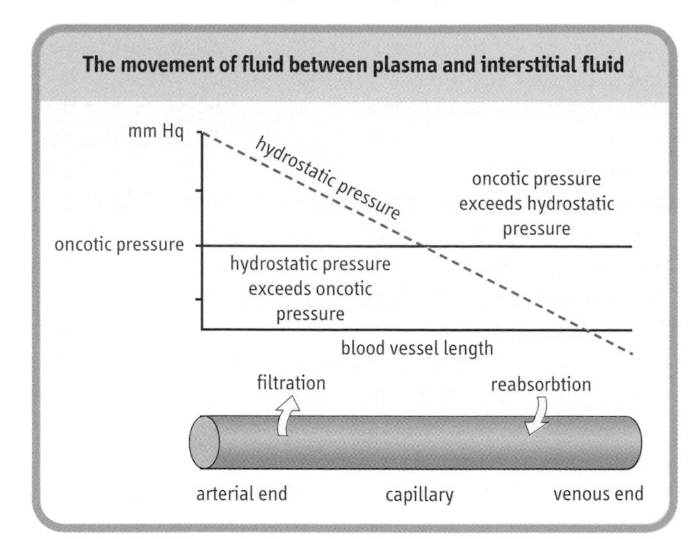

Fig. 21.6 The balance between oncotic and hydrostatic pressures determines the movement of fluid between plasma and interstitial fluid.

part of capillaries, the hydrostatic pressure prevails over the oncotic pressure. As a result, plasma water and low-molecular-weight substances filter into the extravascular space. In contrast, in the low-pressure venous part of capillaries, oncotic pressure prevails over hydrostatic pressure, and draws the fluid from the extravascular space into the vascular lumen (Fig. 21.6). A reduction in plasma oncotic pressure, for example as occurs when the plasma concentration of albumin decreases, increases movement of plasma to the extravascular space and results in edema.

Regulation of cell osmolality and volume

When there is no osmotic gradient across the cell membrane, the mechanism responsible for maintaining the cell fluid volume is the sodium–potassium pump. Inhibition of the pump, for example by ouabain, is followed by an increase in the concentration of sodium and decrease in that of potassium

Edema results from a loss of protein

A 17-year-old girl is referred to a nephrologist after she noticed an increased puffiness of her face over a month or so. Testing her urine shows protein excretion of 12 g/day. Reference value for urinary protein excretion is less than 0.15g/day.

Comment. In this patient, renal biopsy showed inflammation of the kidney cortex (glomerulonephritis). The damage to the renal filtration barrier led to proteinuria, with the urinary protein loss causing a decrease in the plasma oncotic pressure and a retention of water in the ECF, causing edema.

inside cells, and thus a decrease in membrane potential (depolarization). To maintain electrical neutrality, chloride ions enter the cells as counterions to sodium. The amount of ions entering the cell is greater than that leaving it. This generates an osmotic gradient and results in the movement of water into cells.

The activity of the sodium–potassium pump is regulated by the intracellular sodium concentration

If the intracellular concentration of sodium increases, the activity of the sodium–potassium pump also increases, and sodium ions are extruded from the cell itself. In this way, cells protect themselves from changes in the fluid volume. However, the cell volume may also be controled by generation of osmotically active substances within the cell. For instance, the brain cells adapt to increased ECF osmolality by increasing their amino acid concentration. Also, cells in the renal medulla, which are exposed to a hyperosmotic environment (see below), produce an osmotically active alcohol, sorbitol. They also concentrate the amino acid, taurine, in response to osmotic stress.

The water balance

The body constantly exchanges water with the environment. In a steady state, the intake of water equals its loss. The main source of water intake is oral and the main source of loss is with urine. We also lose water through sweat and the lung; this is called the 'insensible' water loss and amounts to approximately 500 mL daily, but can increase to several liters after sweating during intensive exercise, and during fever, which causes hyperventilation (Fig. 21.7).

Diarrhea may lead to severe fluid loss

A 4-year-old child is admitted to the Pediatric Unit after 2 days of severe diarrhea. Clinically, the child is dehydrated. Her sodium concentration is 145 mmol/L, her creatinine concentration 50 μmol/L (0.57 mg/dL), and her urea concentration 5.2 mmol/L (31 mg/dL). Treatment with an intravenous infusion of sodium chloride and antibiotics is started, and the girl recovers within 2 days. Children may have different reference values than adults:

- **sodium:** 133-145 mmol/L
- **creatinine:** 20-80 μmol/L (0.23-0.90mg/dL)
- **urea:** 2.5-6.5 mmol/L (16.2-39 mg/dL)

Comment. This child was severely dehydrated, but there was little evidence of abnormality in her biochemistry tests. This is because she lost, through the intestine, a large amount of fluid having an electrolyte concentration similar to that of plasma. Therefore, in spite of volume loss, there was no change in electrolyte concentrations.

The kidney

The main function of the kidneys is to maintain the composition, osmolality, and volume of the ECF. The control of the acid–base balance is closely associated with this function (see Chapter 22). Kidneys remove products of metabolism such as urea, uric acid, and creatinine, and retain valuable metabolites such as glucose, amino acids, and proteins. They also metabolize and remove numerous drugs and toxins from the body. Kidney function is regulated by endocrine systems involving peptide and steroid hormones; also, the kidneys themselves are an endocrine organ: they produce renin, 1,25 $(OH)_2$ vitamin D (see chapter 23), and erythropoietin, which affect blood pressure, calcium homeostasis, and the production of erythrocytes, respectively.

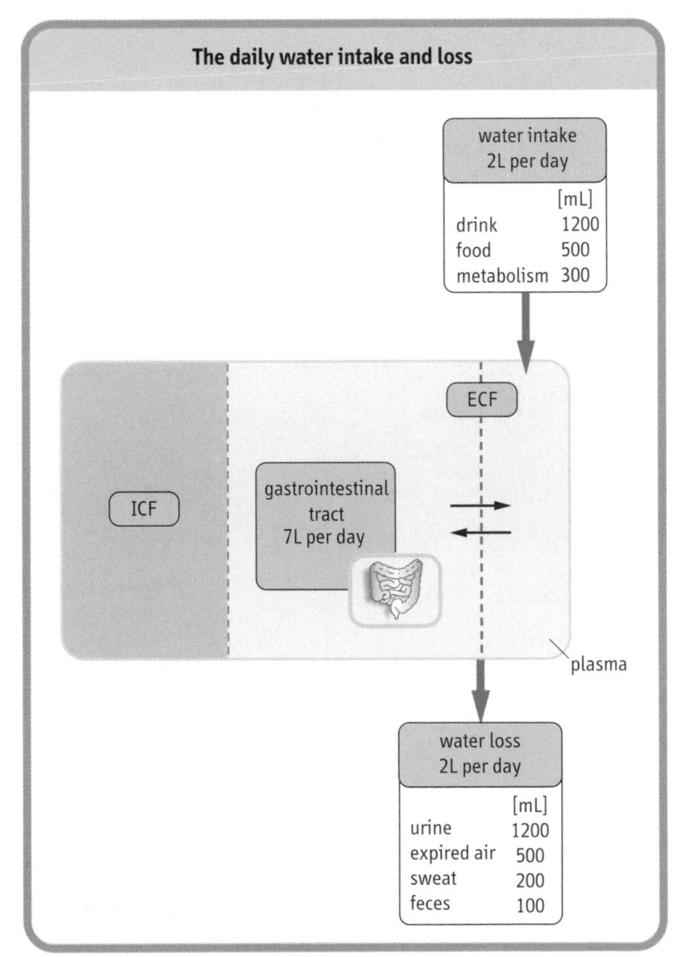

Fig. 21.7 The daily water balance in an adult person. In a steady state, intake of water equals output: water is obtained from the diet and from oxidative metabolism, and is lost through the kidneys, skin, lungs, and intestine. Note how much water enters and leaves the gastrointestinal tract daily; this explains why severe diarrhea leads to dehydration (see above).

The nephron is the functional unit of the kidney

Each kidney consists of approximately 1 million tiny structures called nephrons (Fig. 21.8). Glomeruli located in the kidney cortex connect the cardiovascular system to the excretory tubules. An incoming small artery (afferent arteriole) branches out inside the glomerulus capsule (Bowman's capsule) into bundles of capillary vessels that combine again into the outgoing (efferent) arteriole. The glomerular capsule itself transforms into a thin, long tube, the renal tubule; the first part of this is the proximal tubule, which becomes a thin loop of Henle, and then thickens again to form the distal tubule and collecting duct. The nephron is twisted around, so that the distal tubule lies close to the glomerulus. This results in an important functional junction: a group of cells in the distal tubule, the macula densa, together with the so-called juxtaglomerular cells, belonging to the afferent arteriole, form the juxtaglomerular apparatus, which responds to changes in arterial blood pressure by secreting a proteolytic enzyme, renin (see below). The macula densa also senses the tubular chloride concerentration.

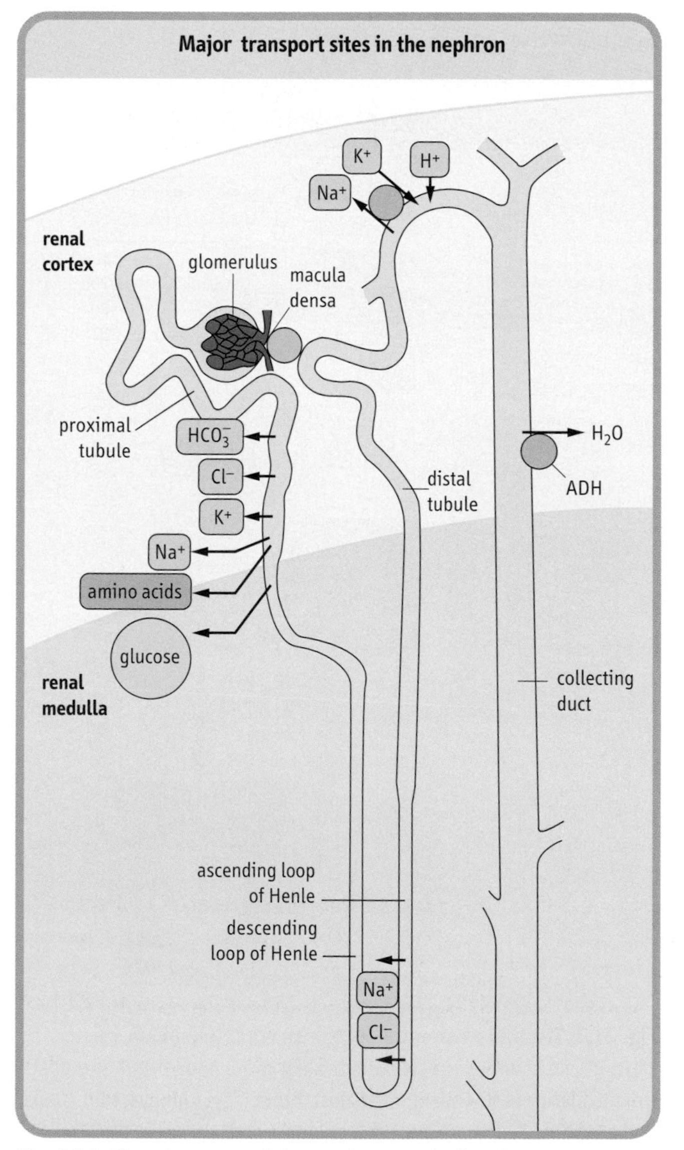

Fig. 21.8 The structure of the nephron and of major transport sites within it.

Energy metabolism and transport function of the kidney

Most of the metabolic processes in the kidneys are aerobic, and oxygen consumption in the kidneys is consequently high: it is approximately equal to that of cardiac muscle, and is threefold greater than that of the brain. This high metabolic activity is required to maintain the tubular reabsorption: about 70% of the consumed oxygen is used to support active sodium transport, which in turn determines the reabsorption of glucose and amino acids. The partial pressure of oxygen (pO_2) is greater in the kidney cortex than in the medulla; because of this low (pO_2) in the medulla and the high activity of Na^+/K^+-ATPase there, cells of the distal segments of the nephron are particularly vulnerable to a decreased oxygen supply.

In the renal cortex, the main energy substrates used are fatty acids, lactate, glutamate, citrate, and ketone bodies, whereas in the renal medulla, oxidative metabolism is the main source of energy. Renal tubular cells with a high Na^+/K^+-ATPase activity possess multiple mitochondria close to the plasma membrane, so that the ATP released to the cytosol is easily accessible.

The formation of urine

The excretory function of the kidneys includes filtration of the plasma in the glomeruli, transport of water and solutes from the tubular lumen back to the blood (tubular reabsorption), and transport of substances from tubular cells to the lumen (tubular secretion).

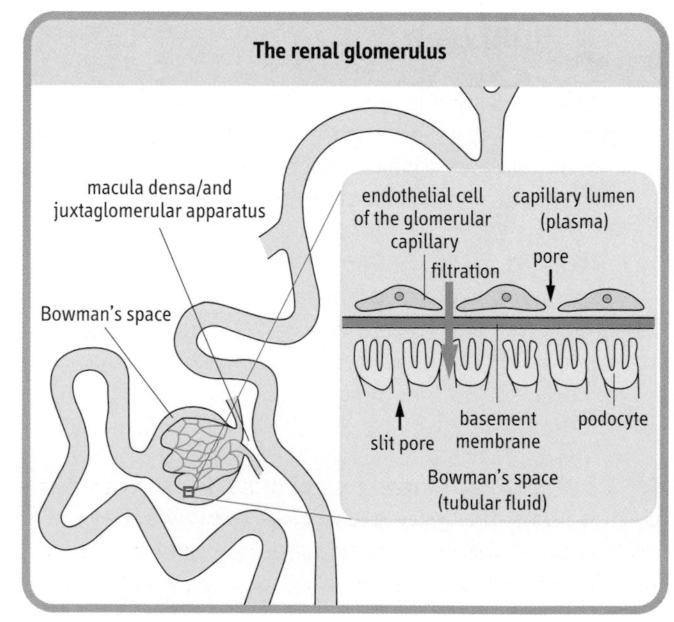

Fig. 21.9 The glomerular filtration barrier. This consists of the endothelial cells, the basement membrane, and the podocytes. Macula densa cells sense the chloride concentration in the distal tubule and adjust the diameters of different arterioles accordingly, thus regulating the glomerular blood flow.

Plasma flowing through glomerular capillaries is filtered into the Bowman's space. This glomerular filtration depends on the available filtration surface and on the permeability of the filtration barrier, which includes endothelial cells with characteristic pores (fenestrations) that line the glomerular blood vessels, the basement membrane, and epithelial cells (podocytes) that have characteristic foot processes (Fig. 21.9). The fenestrations in the endothelial layer and the spaces between foot processes of the podocytes cause the capillary wall to act as a sieve that filters water and small molecules. Filtration of larger molecules is limited by their size, shape, and electric charge: at pH 7.4, most plasma proteins are negatively charged, and so is the filtration barrier; this hinders filtration of even small proteins such as myoglobin (molecular mass 17 kDa), and almost totally prevents filtration of albumin (69 kDa).

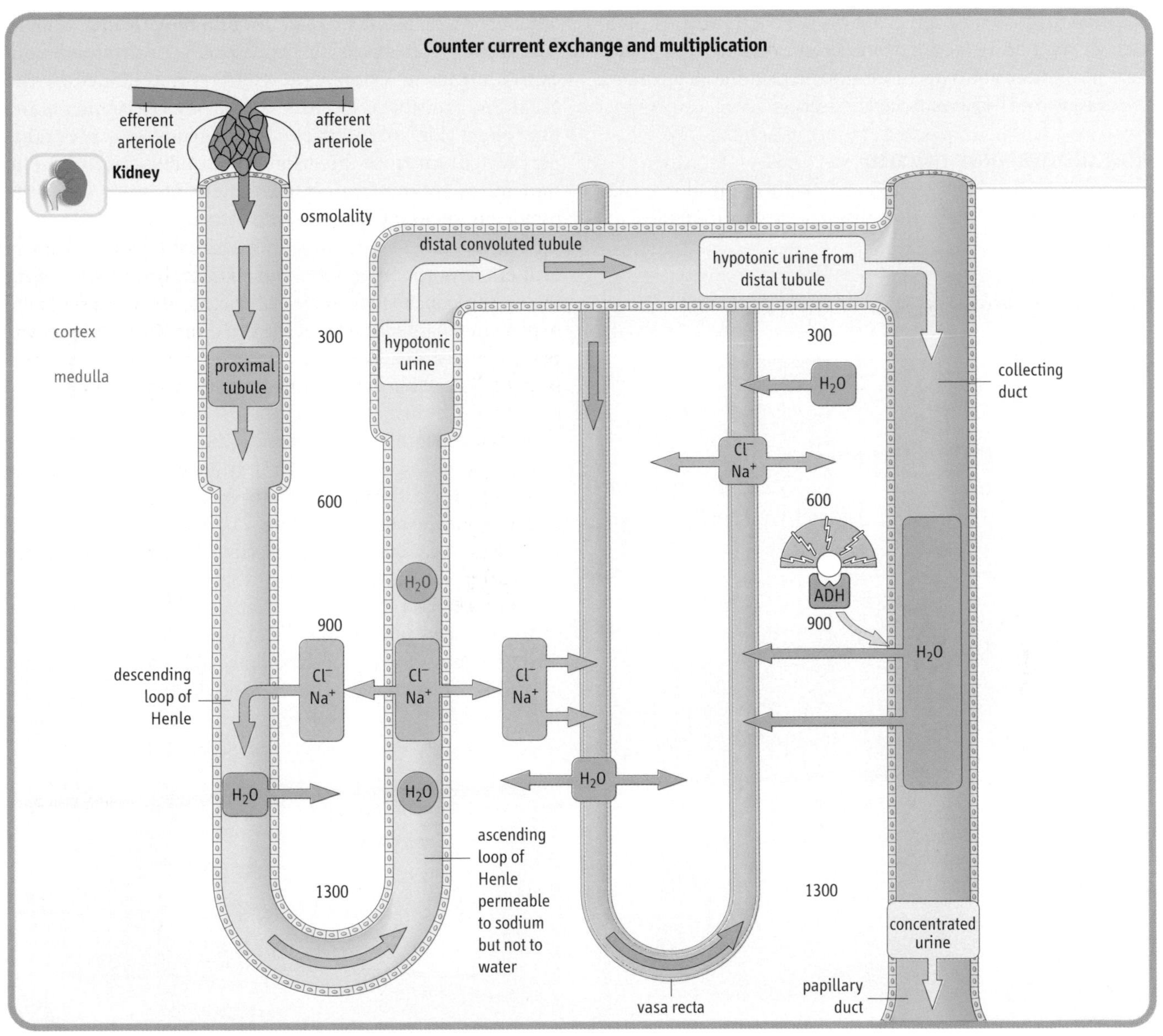

Fig. 21.10 The countercurrent mechanism is essential for the formation of urine, and for the efficient reabsorption of water in the distal tubule: it enables the kidney to form a high osmolality environment in the medulla. The descending and ascending arms of the loop of Henle have different permeability characteristics. In the ascending arm, sodium and chloride ions are pumped out into the interstitial fluid. They then diffuse freely into the lumen of the descending limb, creating a functional 'loop', which perpetuates the increase in osmolality of the filtrate reaching the ascending limb. This is called 'countercurrent multiplication'.

As a result of this, the osmolality of the renal cortex is similar to that of plasma (300 mmol/L), whereas in the medulla it is as great as 1300 mmol/L. The high osmolality of the medulla facilitates the reabsorption of water in the collecting ducts. This is known as 'countercurrent exchange'. Reabsorbed water diffuses into the medullary blood vessels (vasa recta). The amount of water absorbed is controlled by antidiuretic hormone (ADH).

Countercurrent multiplication and countercurrent exchange act in concert to enable the efficient reabsorption of water in the collecting duct.

Glomerular filtration is driven by the hydrostatic pressure in the glomerular capillaries, which amounts to approximately 50 mmHg. The forces counteracting this are the oncotic pressure of the plasma and the hydrostatic back-pressure (approximately 10 mmHg) of the filtrate in the glomerular capsule.
Countercurrent multiplication leads to the build-up of osmotic gradient in the medulla through the different permeability characteristics of descending and ascending loop of Henle. Countercurrent exchange is the process of 'return' of reabsorbed water and electrolytes to blood vessels (*vasa vecta*). Note movements of water, sodium and chloride as the blood vessel traverses hyperosmolar medulla.

The glomerular filtrate

The volume, composition, and osmolality of the glomerular filtrate change during its flow through the renal tubules.

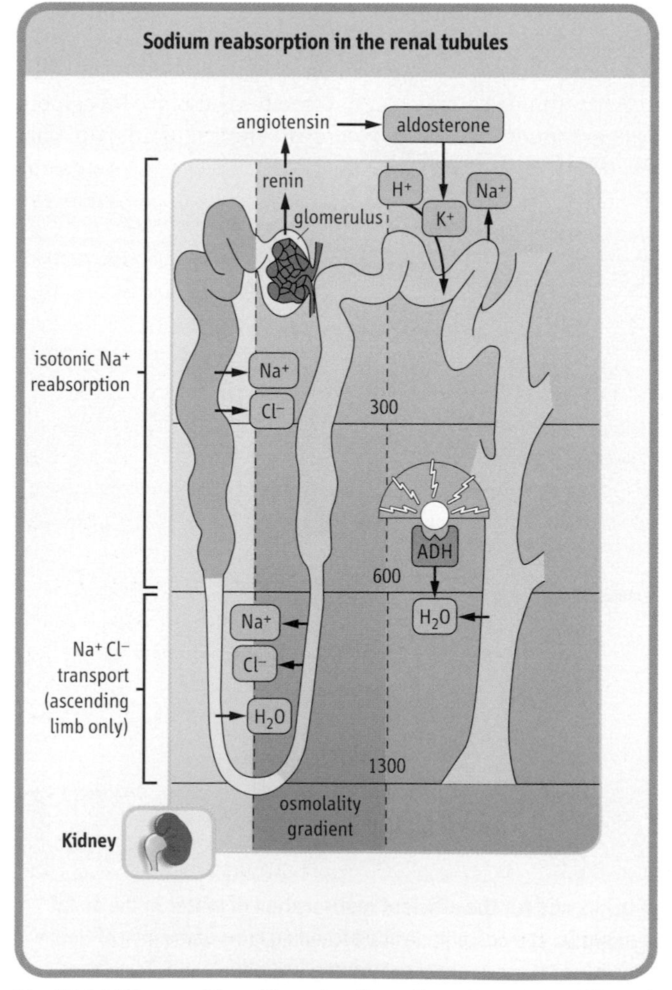

Fig. 21.11 The renal handling of sodium. More than 80% of filtered sodium is actively reabsorbed in the proximal tubule. In addition, sodium and chloride ions are reabsorbed in the ascending limb of the loop of Henle (see Fig. 21.10). On the other hand, in the distal tubule, sodium reabsorption is stimulated by aldosterone and is coupled with the secretion of hydrogen and potassium ions. Aldosterone causes sodium retention and an increased secretion of potassium.

The amount of blood that flows through the kidneys amounts to about 20% of the cardiac output. Quantitatively, the main components of the glomerular filtrate are sodium and accompanying anions, mainly chloride, phosphate, and bicarbonate. Approximately 70% of the filtrate is reabsorbed in the proximal tubule.

Sodium is reabsorbed by several mechanisms: through specific ion channels, in exchange with the hydrogen ion, and in cotransport with glucose, amino acids, phosphate, and other anions. The entry of sodium into the proximal tubular cells is passive. It is possible because of its relatively low concentration in the cytoplasm, which is maintained by the Na^+/K^+-ATPase that is localized on the basolateral membrane (the 'blood side') of tubular cells and catalyzes active transport of sodium from the cells to the interstitial fluid. The transfer of sodium causes movement of water from the tubule lumen to the surrounding extracellular space.

The isotonic fluid leaving the proximal tubule undergoes further reabsorption of water and sodium. In the descending limb of the loop of Henle, water diffuses from the tubules to the hyperosmotic interstitium of the renal medulla, and sodium and chloride diffuse back into the glomerular lumen. In contrast, the ascending limb is impermeable to water, but actively reabsorbs sodium and chloride; this generates hyperosmolality in the renal medulla. These permeability characteristics of the descending and ascending limbs of the loop of Henle enable an efficient increase in the osmolality of the medulla; this is known as the countercurrent system (Fig. 21.10).

In the distal tubule and the collecting duct, still more sodium is reabsorbed by active transport that is controled by aldosterone. In addition, the tubular secretion of hydrogen and potassium ions takes place in the distal tubule. Tubular fluid that leaves the loop of Henle is hypotonic, and is about to become urine (Fig. 21.10 and 21.11).

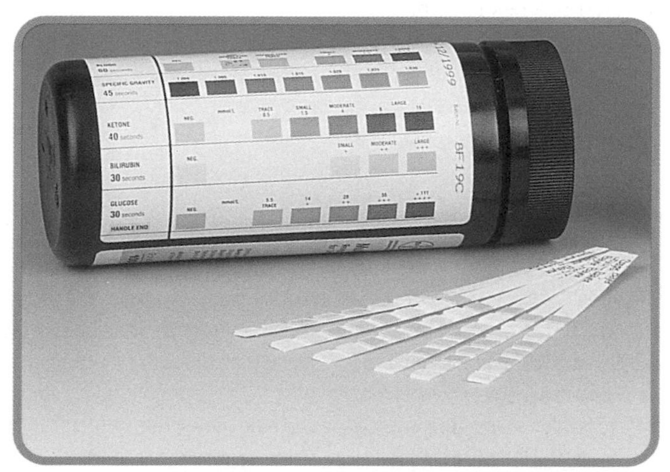

Fig. 21.12 The testing of urine (urinalysis) is performed using ready-made dry tests; these are strips coated with reagents immobilized on a plastic support. Dipping the strip in the urine initiates a reaction yielding a colored product. The readout is against a standarized color scale.

Diabetes often leads to the impairment of renal function

A 37-year-old woman with a 22-year history of type 1 diabetes comes for a routine visit to the Diabetic Clinic. She has poor diabetic control, with a glycated hemoglobin (HbA_{1c}) value of 8%. A quantitative measurement of albumin in urine reveals a concentration of (40 mg/L), indicating microalbuminuria. Reference values are:

- **HbA_{1c}:** desirable value below 6%.
- **microalbuminuria:** less than 30 mg/L.

Comment. This patient has an impaired renal function as a result of glomerular damage in diabetes. The presence of microalbuminuria predicts future overt diabetic renal disease.

The urine

Kidneys are able to excrete from 0.5 L to more than 10 L of urine daily; the average daily volume is 1–2 L. The minimum volume necessary to remove the products of metabolism is approximately 0.5 L/24 h.

Urine osmolality varies from about 80 to 1200 mmol/L. The osmolality of the glomerular filtrate is about 300 mmol/L, therefore the maximal urine concentration is approximately fourfold.

In spite of the fact that, for many substances, the capacity of the tubular transport systems is limited, only small amounts of amino acids (0.7 g/24 h), and almost no glucose, are normally present in the urine.

Each substance reabsorbed in the renal tubules has its specific renal transport maximum (T_{max}). When the function of renal tubules is impaired, the T_{max} for substances such as amino acids decreases, and aminoaciduria occurs. Aminoaciduria may also result from the accumulation of amino acids such as phenylalanine, leucine, isoleucine, and valine in the plasma, as a result of specific metabolic defects. In such cases, aminoaciduria occurs because of an excessive load of the amino acid that exceeds the reabsorptive capacity of the renal tubules.

The analysis of urine helps to diagnose hereditary defects in amino acid transport (aminoacidurias), and is also important in conditions such as renal disease and diabetes

The substances measured in urine by the clinical laboratory include proteins, amino acids and glucose, and the number of red blood cells and leukocytes (Fig. 21.12).

Protein is normally detectable in the urine in trace amounts only, but increases significantly with loss of glomerular selectivity; the presence of significant amounts of protein in urine is an important sign of renal disease. Minimal amounts of albumin present in the urine (microalbuminuria) predict the development of diabetic nephropathy (see Chapter 20). Larger proteins such as immunoglobulins appear in the urine when the glomeruli are damaged: the presence of immunoglobulin light chains (Bence–Jones protein) occurs in multiple myeloma. In patients with hemolytic anemia, the urine may contain free hemoglobin. Myoglobin appears when there is widespread muscle damage (rhabdomyolysis). The measurement of ketones in the urine is important in monitoring the glycemic control particularly in type 1 diabetes (Chapter 20).

The renin–angiotensin system

An outline of this system is presented in Figure 21.13.

Renin is released from the juxtaglomerular cells in response to decreased renal perfusion pressure. It is a protease that uses angiotensinogen as its substrate. Angiotensinogen, a glycoprotein of more than 400 amino acids, is synthesized in the liver and has a variable structure and molecular weight. The renin liberates angiotensin I from angiotensinogen.

Angiotensin I is a short, 10-amino-acid peptide. It is the substrate for angiotensin-converting enzyme (ACE). ACE is responsible for the removal of two amino acids from angiotensin I, to produce the most potent known vasocon-

Endocrine causes of hypertension

Disorders of aldosterone secretion

Hyperaldosteronism is a common finding in hypertension. However, primary hyperaldosteronism is rare, and occurs as a result of abnormal adrenal activity. It may be the result of a single adrenal tumor, adenoma (Conn's syndrome). In most cases, however, the hyperaldosteronism is secondary to increased secretion of renin. Pheochromocytomas are catecholamine-secreting tumors that are a cause of hypertension of about 0.1% of hypertensive patients. It is important to make the diagnosis of pheochromocytoma, because it can be surgically removed, and it is a potentially dangerous condition.

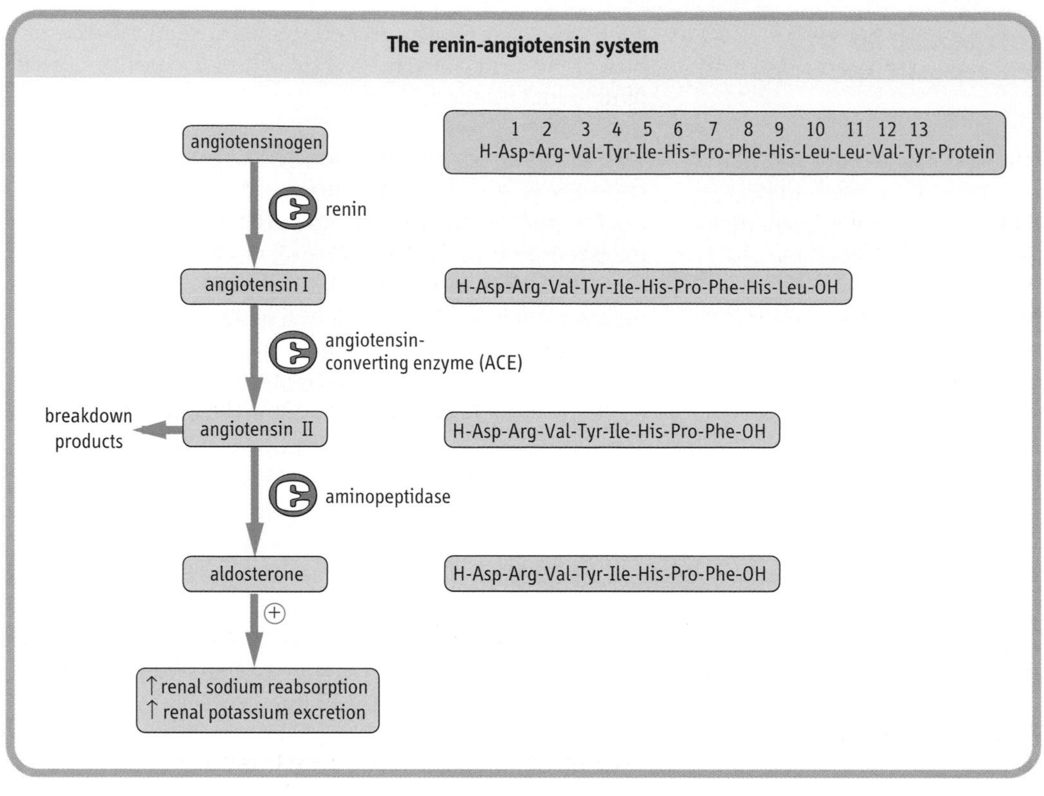

Fig. 21.13 The renin–angiotensin–aldosterone system.

Arterial hypertension is a common disease

Hypertension is inappropriately increased arterial blood pressure above the desirable level of systolic blood pressure 140 mmHg and diastolic pressure 90mmHg (140/90 mmHg). According to the World Health Organization, up to 20% of the population of the developed world may suffer from the condition. Traditionally, hypertension has been classified as 'essential' (primary) or 'secondary'. A cause of essential hypertension has not yet been identified, although it is known to involve multiple genetic and environmental factors including neural, endocrine, and metabolic components. Hypertension is associated with an increased risk of stroke and myocardial infarction. The diuretics such as bendrofluazide and drugs which affect the renin–angiotensin system have a key role in the modern treatment of hypertension.

Renin–angiotensin–aldosterone system and cardiac failure

A 65-year-old man with a history of myocardial infarction presents with increasing fatigue, shortness of breath and ankle edema. Imaging of the heart using an ultrasound technique (echocardiogram) shows impaired function of the left ventricle. The patient's serum measurements reveal: sodium 140 mmol/L, potassium 3.5 mmol/L, protein 30 g/dL, creatinine 80 μmol/L (0.90 μg/dL), and urea 7.5 mmol/L (45 mg/dL). Reference value:

- **serum protein concentration:** 35–45 g/L

Comment. This man presents with symptoms and signs of cardiac failure. The impaired function of the heart leads to a decreased blood flow through the kidney and to activation of the renin–angiotensin system and the stimulation of aldosterone secretion. Aldosterone, in turn, causes an increased reabsorption of sodium in the kidney. This leads to water retention and 'dilution' of the ECF. The resulting decrease in albumin concentration causes edema.

Diuretics are drugs used for treatment of edema, cardiac failure and hypertension

Diuretics are drugs that stimulate water and sodium excretion. Diuretics act by inhibiting the reabsorption of sodium and chloride in the renal tubules. Thiazide diuretics, e.g. bendrofluazide, decrease sodium reabsorption in the distal tubules by blocking sodium and chloride cotransport. So-called loop diuretics, such as frusemide, inhibit sodium reabsorption in the ascending loop of Henle. Spironolactone, a competitive inhibitor of aldosterone, inhibits sodium–potassium exchange in the distal tubules, and also decreases potassium excretion; it is known as a potassium-sparing diuretic. Finally, an osmotic diuresis may be induced by the administration of the sugar alcohol, mannitol. The net effect of diuretics is to increase urine volume and induce the loss of water and sodium. They are important in the treatment of edema associated with circulatory problems such as heart failure, in which impaired cardiac function may lead to a severe breathlessness caused by pulmonary edema. They are also essential in the treatment of hypertension.

Disorders of antidiuretic hormone (ADH) secretion

Diabetes insipidus and the syndrome of inappropriate secretion of ADH

Deficiency of ADH causes the condition known as diabetes insipidus, in which large amounts of dilute urine are lost. In contrast, an uncontrolled, excessive secretion of ADH may take place following major trauma or surgery. This is known as the syndrome of inappropriate antidiuretic hormone secretion (SIADH), and leads to water retention.

Please note : ADH is also known as arginine–vasopressin (AVP).

strictor molecule, angiotensin II. In the kidney, angiotensin II affects blood flow, glomerular function, and, by stimulation of aldosterone secretion which in turn affects sodium, potassium, and hydrogen ion transport. These actions of angiotensin II are mediated through two different receptors: the AT1 receptor is G-protein-linked, whereas the AT2 receptor can inhibit protein tyrosine phosphatase activity.

Apart from the classical pathway of production of angiotensin II from angiotensin I by the action of ACE, an alternative pathway leads directly from angiotensinogen to angiotensin II. This is catalyzed by a serine protease and carboxypeptidase. Drugs that inhibit ACE are now extensively used in the treatment of hypertension.

Aldosterone is produced in the adrenal cortex, and is the major mineralocorticosteroid hormone in man. Its production is stimulated by increases in the plasma concentration of potassium. Aldosterone regulates electrolyte balance in the kidney, salivary glands, sweat glands, and gastrointestinal tract. In the kidney, it increases the activity of the Na^+/K^+-ATPase that causes renal sodium retention and, in parallel, an increased excretion of potassium.

Water metabolism and sodium metabolism are closely interrelated

The actions of aldosterone and ADH complement each other, and result in a control system that is responsible for the handling of water and sodium by the kidney

Water normally enters renal tubular cells by passive diffusion; the exceptions are the ascending loop of Henle (see above) and the collecting duct, where its entry is regulated by the antidiuretic hormone (ADH). ADH is synthesized in the supraoptic and paraventricular nuclei of the hypothalamus and transported along axons to the posterior pituitary, where it is stored before being processed and released. The binding of ADH to its receptor, located on the membranes of tubule cells in the collecting duct, leads to the synthesis of a variety of proteins, including aquaporin-2, which facilitates the passage of water. The action of ADH determines the final volume and concentration of the urine (see Fig. 21.14).

Renal clearance and glomerular filtration rate

Glomerular filtration rate is an important parameter of kidney function

The renal clearance is the volume of plasma (in milliliters) that the kidney clears of a given substance per unit of time (minutes). Associated with this concept is that of glomerular filtration rate (GFR), which is the most important parameter describing kidney function. The GFR could be estimated by

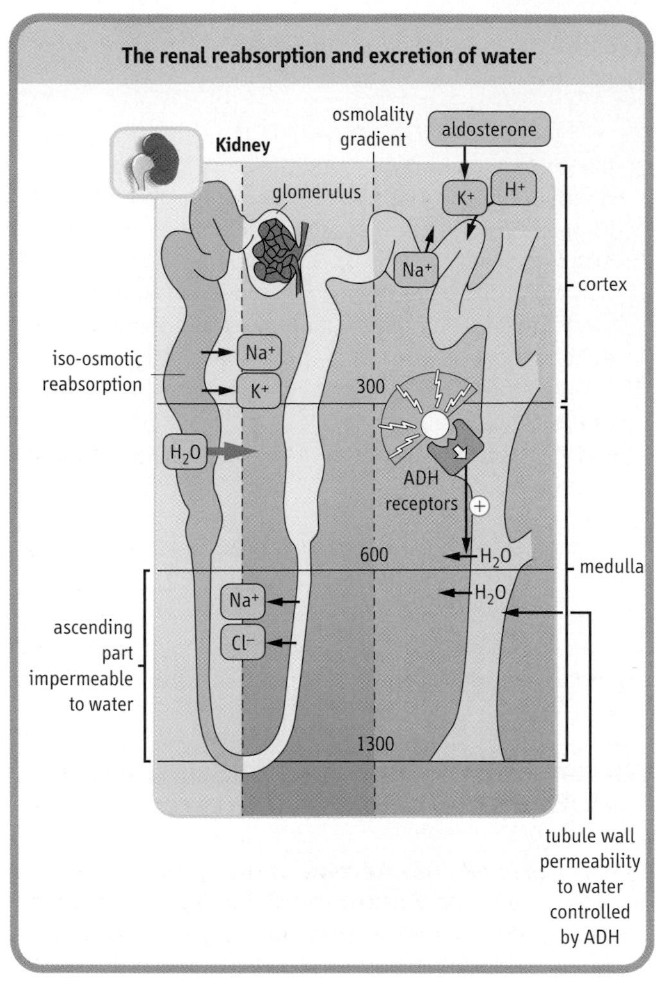

Fig. 21.14 The renal handling of water. The permeability of the tubular walls to water differs along the nephron. About 80% of filtered water is reabsorbed in the proximal tubule, by iso-osmotic reabsorption. The ascending loop of Henle is impermeable to water; sodium is reabsorbed there, creating high osmolality in the medulla. In the collecting duct, ADH controls the reabsorption of water by rendering the tubule wall permeable to water.

measuring the clearance of a substance, such as polysaccharide inulin, that is removed from the body only through the kidneys and that is neither secreted nor reabsorbed in the renal tubules. The amount of inulin filtered from plasma (plasma concentration, P_{in}, multiplied by GFR) equals the amount recovered in urine (i.e. the urinary concentration, U_{in}, times the rate of urine formation rate, V). One can present this in the form of an equation:

$$P_{in} \times \text{GFR} = U_{in} \times V$$

From this the GFR can be calculated as:

$$\text{GFR} = \frac{U_{in} \times V}{P_{in}}$$

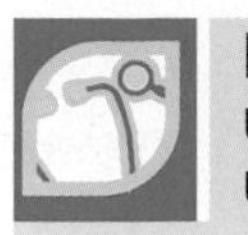

Renal failure leads to a falling urine output, and high serum urea and creatinine

A 25-year-old man is admitted to hospital unconscious after a motorcycle accident. He has a fractured skull and multiple injuries to his limbs. He receives blood transfusions. On the third day, his serum creatinine concentration is 300 μmol/L (3.9 mg/dL) and his urea concentration is 22 mmol/L (132 mg/dL). Reference values are:

- **creatinine:** 20–80 μmol/L (0.23–0.90 mg/dL)
- **urea:** 2.5–6.5 mmol/L (16.2–39 mg/dL)

Comment. This young man has developed renal failure as a consequence of hypoxia caused by blood loss. He subsequently underwent emergency renal dialysis. Renal function recovered within 2 weeks.

- **urea** mg/dL = mmol/L × 6.02
- **creatinine** mg/dL = μmol/L × 0.0113

The expression $U \times V/P$ is called the renal clearance and the renal clearance of inulin equals the GFR. The average GFR is 120 mL/min for men and 100 mL/min for women.

Creatinine clearance and serum urea and creatinine levels

Measurement of serum creatinine is a simple test of renal function that does not require the collection of urine

Inulin is a substance that would need to be administered intravenously to assess the GFR of an individual. This is quite impractical. Instead, we calculate the clearance of creatinine, a compound derived from phosphocreatine in skeletal muscles, which is normally present in plasma. The clearance of creatinine is similar to that of inulin. There is some tubular reabsorption of creatinine, but this is compensated by a roughly equivalent degree of tubular secretion.

To calculate creatinine clearance, one needs a sample of blood and urine collected over 24 hours; the concentration of creatinine in the serum and the urine is measured. Urine volume is used to calculate urine excretion per unit of time (V, above). The clearance of creatinine is then calculated according to the formula:

$$\text{creatinine clearance} = \frac{U_{\text{creatinine}} \times V}{P_{\text{creatinine}}}$$

The concentration of creatinine in serum is 20–80 μmol/L (0.28–0.90 mg/dL). The measurement of serum creatinine is the simplest test of renal function, and does not require urine collection. An increase in the plasma creatinine concentration is a marker of a decrease in GFR: plasma crea-

Severe vomiting may lead to dehydration

A 65-year-old man has been admitted to hospital after prolonged vomiting. He has dry skin and dry mucous membranes and tongue. Serum measurements reveal: sodium 150 mmol/L, potassium 2.5 mmol/L, bicarbonate 35 mmol/L, creatinine 100 mmol/L (1.13 mg/dL), and urea 15 mmol/L (90.3 mg/dL).
Reference values are:

- **sodium:** 135–145 mmol/L
- **potassium:** 3.5–5.0 mmol/L
- **bicarbonate:** 20–25 mmol/L
- **creatinine:** 20–80 μmol/L (0.28–0.90 mg/dL)
- **urea:** 2.5–6.5 mmol/L (16.2–39mg/dL)

Comment. This patient presents with dehydration, indicated by the high sodium and urea values. He also presents with a high bicarbonate concentration, which suggests metabolic alkalosis as a result of the loss of hydrogen ion from the stomach during vomiting. Alkalosis is associated with a low potassium concentration (see Chapter 22).

tinine concentration doubles when the GFR decreases to approximately 50% of its original value. Another test that is used to assess kidney function is the measurement of serum urea. However, as urea is an end product of protein catabolism, its amount in plasma is also dependent, apart from renal function, on factors such as the amount of protein in the diet and the rate of tissue breakdown (Fig. 21.15).

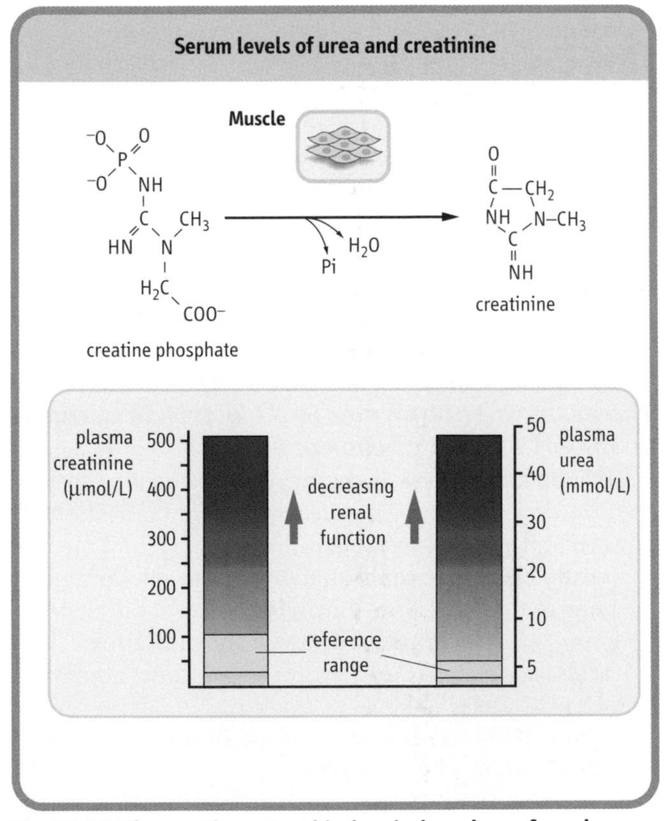

Fig. 21.15 The most important biochemical markers of renal function are serum urea and serum creatinine. The insert shows the process by which muscle phosphocreatine is converted into creatinine. Loss of 50% of nephrons results in approximate doubling of serum creatinine concentration.

The diagnostic features of renal failure are a decrease in urine formation, a decrease in creatinine clearance, and an increase in plasma concentrations of creatinine and urea. The measurements of serum creatinine and urea concentration are usually performed as first-line tests.

The importance of plasma potassium concentration

Potassium affects the contractility of the heart, and both too high a concentration (hyperkalemia) and too a low concentration (hypokalemia) can be life threatening

The measurement of serum potassium concentration is a very important test performed in clinical laboratories. The serum concentration of potassium is 3.5–5 mmol/L. Because the intracellular concentration of potassium is much greater (135–145 mmol/L), a minor shift of potassium between ECF and ICF may result in large changes in the serum potassium concentration. A potassium value less than 2.5 mmol/L or greater than 6.0 mmol/L is life threatening, and therefore maintenance of the potassium concentration is very important.

The most common cause of severe hyperkalemia is renal failure, in which potassium cannot be excreted with the urine. Low plasma potassium may result from excessive loss in urine or from the gastrointestinal tract. Changes in plasma potassium concentration may also be precipitated by acid–base disorders (see Chapter 22) (Fig. 21.16).

ADH and the renin–angiotensin system in water and electrolyte disorders

Normally, despite large variations in fluid intake, the serum osmolality is maintained within narrow limits. Osmoreceptors, which stimulate ADH release and thirst, respond to

Serum sodium concentration in disorders of fluid and electrolyte balance

Disorders of water and electrolyte metabolism
Water and electrolyte disturbances may result from an imbalance between the intake of fluids and electrolytes and their loss through either renal or extrarenal routes, a movement of water and electrolytes between body compartments, or both. A decreased sodium concentration (hyponatremia) usually indicates that the extracellular fluid is being 'diluted', whereas an increasing sodium concentration means that the extracellular fluid is being 'concentrated'. Hyponatremia may result from either excess of water in the system (common), or from the loss of sodium (rare). Conversely, hypernatremia results from either loss of water (common) or excess of sodium (rare).

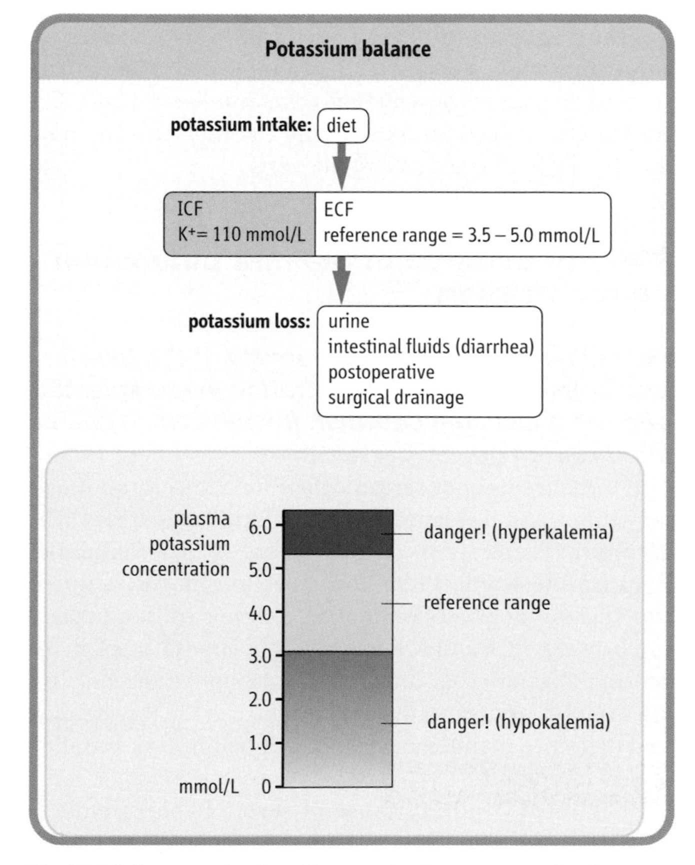

Fig. 21.16 Serum potassium. Serum potassium concentration is maintained within narrow limits of concentration. Both low (hypokalemia) and high (hyperkalemia) concentrations may be dangerous, because of the effect of potassium on the contractility of heart muscle. The figure shows some sources of potassium loss from the body.

changes in plasma osmolality. Changes in circulating volume lead to ADH secretion by triggering receptors sensitive to pressure (baroreceptors).

Water excess leads to increases in plasma volume, renal blood flow, and GFR. These suppress the production of renin and thus the whole renin–angiotensin system. The resulting decreased concentration of aldosterone leads to a decreased sodium reabsorption and to urinary sodium loss. Also, the excess of water 'dilutes' plasma and plasma osmolality decreases. In addition, through the action of hypothalamic osmoreceptors, the decrease in osmolality suppresses thirst and secretion of ADH. A decrease in ADH leads to increased excretion of water. The overall response to water excess is therefore a compensatory increase in water and sodium excretion.

Water deficit (dehydration) leads to a decrease in renal blood flow and GFR and, later, to a decrease in plasma volume. This stimulates the renin–angiotensin system. Aldosterone causes increased sodium reabsorption. Water deficit thus 'concentrates' the plasma and leads to an increase in plasma osmolality, which stimulates both thirst and ADH secretion, which leads to an increased water reabsorption in the proximal tubules. Thus the overall response to water deficit is retention of sodium and water (Fig.21.17).

The assessment of water and electrolyte status is an important part of clinical practice

In addition to the physical examination and medical history, the following measurements are required for the assessment of water and electrolyte status:

- **serum electrolyte concentrations:** the profile commonly requested by physicians includes serum sodium, potassium, chloride, bicarbonate, urea, and creatinine;
- **acid–base status:** blood hydrogen ion concentration or pH, pCO_2, pO_2;
- **serum albumin concentration and serum osmolality;**
- **urine volume and osmolality.**

For patients with water and electrolyte disorders who stay on a hospital ward, a fluid chart – that is, a daily record of fluid intake and loss – is also essential.

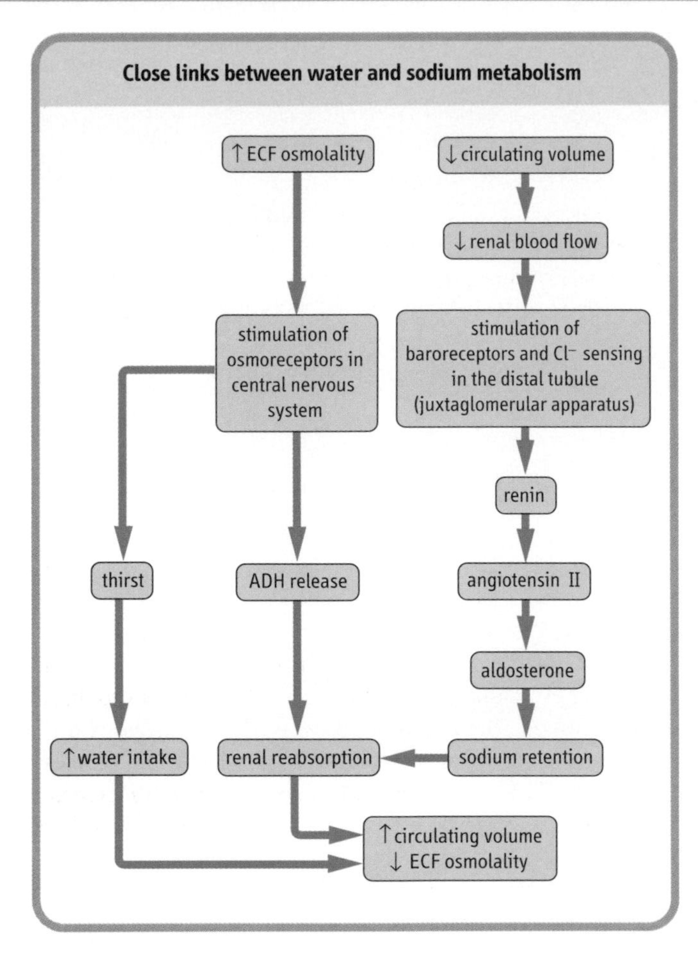

Fig. 21.17 The metabolism of water and that of sodium are closely interrelated. An increase in ECF osmolality by only as much as 2% stimulates secretion of ADH. ADH increases the reabsorption of water. This leads to 'dilution' of the ECF, and results in a decrease in osmolality. This response is reinforced by the stimulation of thirst.

Water retention can be induced, not only by an increase in osmolality, but also by a decrease in the volume of circulating plasma, through stimulation of the pressure-sensitive receptors (baroreceptors) in the juxtaglomerular apparatus.

Summary

Water is the environment of most chemical reactions in the body. Either the deficit of body water (dehydration) or its excess (overhydration) may lead to serious clinical problems. The body water balance is closely linked to the balance of dissolved ions (electrolytes), the most important of which are sodium and potassium. The state of hydration depends on the balance between intake and loss. The kidney is the main regulator of water and electrolyte status. The reabsorption of sodium, potassium and water are hormonally controlled. The renin–angiotensin system is the principal regulator of potassium excretion, and the posterior pituitary hormone, ADH, regulates water excretion and intake. The measurements of plasma electrolytes are an essential part of clinical assessment.

Further reading

Avner ED. Clinical disorders of water metabolism – hyponatremia and hypernatremia. *Pediatr Ann* 1995;**24**:23–30.

Dominiczak MH, ed. *Seminars in clinical biochemistry*, 2nd ed. Glasgow: Glasgow University Press, 1997.

Fulop M. Algorithms for diagnosing some electrolyte disorders. *Am J Emerg Med* 1998;**16**:76–84.

Okun JP. Severe hypernatremia. *Lab Med* 1995;**26**:507–509.

Soupart A, Decaux G. Therapeutic recommendations for management of severe hyponatremia. Current concepts on pathogenesis and prevention of neurologic complications. *Clin Nephrol* 1996;**46**:149–169.

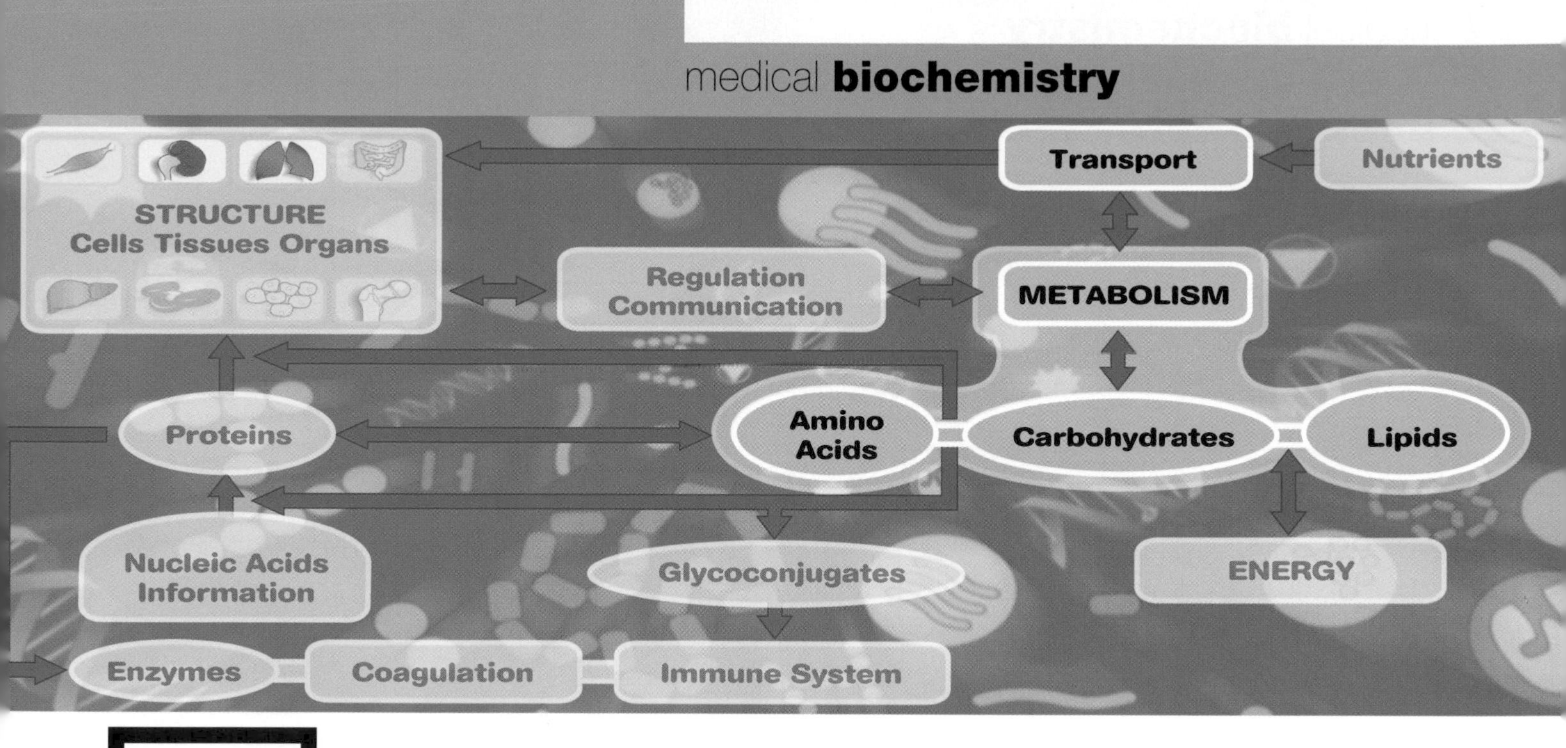

22 Lung and Kidney: The Control of Acid–Base Balance

The acid–base balance

There is a large daily flux of oxygen, carbon dioxide, and hydrogen ion through the human body. Metabolism generates CO_2, which dissolves in H_2O to form carbonic acid, which in turn dissociates to form hydrogen ion. Despite large variations in CO_2 production, for example during exercise, blood pH is surprisingly constant: the concentration of hydrogen ion in plasma remains in the nanomolar range (36–43 nmol/L; pH 7.37–7.44). Indeed, it is rare to see pH below 6.9 or above 7.6 even in severe disease. Changes in pH profoundly affect the ionization of proteins and, consequently, the activity of many enzymes. Changes in pH also affect the cardiovascular system: a decrease in pH decreases cardiac output and blood pressure, whereas an increase in pH causes constriction of small arteries and may lead to arrhythmia. Changes in pH, and in the partial pressure of carbon dioxide (pCO_2) affect tissue oxygenation by changing the shape of the hemoglobin saturation curve, and thus the ease with which hemoglobin gives up its oxygen to the tissues (see Chapter 4).

Problems with gas exchange or acid–base balance underlie a number of diseases of the respiratory system and kidney. Changes in blood pH also accompany uncontroled diabetes (see Chapter 20), trauma, shock, and many other conditions.

Maintaining the acid–base balance involves lungs, erythrocytes, and kidneys

Figure 22.1 summarizes the system of maintenance of acid–base balance. The lungs control the exchange of carbon dioxide and oxygen between the blood and the external atmosphere. Erythrocytes transport gases between lungs and tissues. Kidneys control blood bicarbonate concentration, excrete the hydrogen ion, and regulate the production of erythrocytes by secreting erythropoietin, a hormone that stimulates erythrocyte synthesis.

The body buffer systems

Metabolism produces substantial quantities of both inorganic and organic acids from two sources: the dissolving of metabolically produced CO_2 in water; the metabolism of sulfur-containing amino acids and phosphorus-containing compounds. Lactic acid and keto acids (acetoacetate and hydroxybutyrate) are also produced; they become clinically important in diabetes (see Chapter 20). Acids derived from sources other than CO_2 are known as nonvolatile acids; by definition, they cannot be removed through the lungs, and must be excreted via the kidney. The net production of nonvolatile acids is of the order of 50 mmol/24 h.

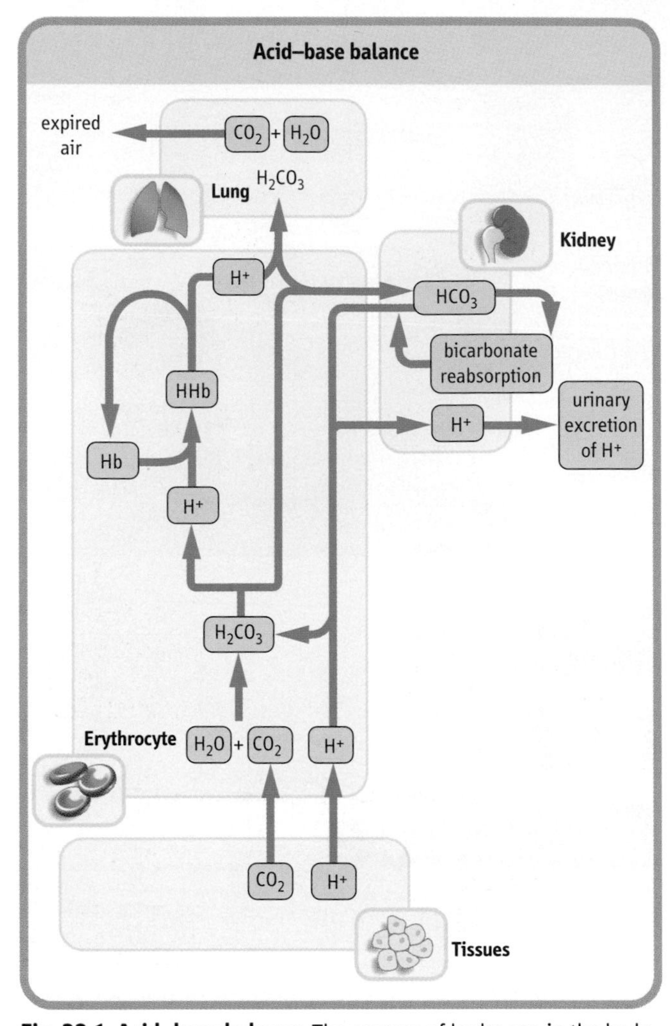

Fig. 22.1 Acid–base balance. The sources of hydrogen in the body are nonvolatile acids produced by tissues, and CO_2, which generates hydrogen ion during its conversion into bicarbonate. The lungs, kidneys, and erythrocytes have important roles in the maintenance of blood pH. The lungs control gas exchange with the atmospheric air, CO_2 is transported as bicarbonate, and hemoglobin (Hb) in erythrocytes also contributes to CO_2 transport. Hemoglobin buffers hydrogen ion derived from carbonic acid. Nonvolatile acids have to be excreted via the kidneys, which also control the reabsorption of bicarbonate.

To minimize changes in hydrogen ion concentration, blood and tissues contain several buffer systems

Hemoglobin has an important role in buffering hydrogen ion derived from the carbonic anhydrase reaction. Acids generated during metabolism and hydrogen ion generated from fixing CO_2 are also buffered by intracellular buffers, mainly proteins and phosphates. However, the main buffer that neutralizes acids emerging from metabolism is bicarbonate. These systems are summarized in Figure 22.2 (see also Chapter 2).

Buffers in the human body

Buffer	Acid	Conjugate base	Main buffering action
hemoglobin	HHb	Hb^-	erythrocytes
proteins	HProt	$Prot^-$	intracellular
phosphate buffer	$H_2PO_4^-$	HPO_4^{2-}	intracellular
bicarbonate	$CO_2 \rightarrow H_2CO_3$	HCO_3^-	extracellular

Fig. 22.2 The main buffers in the human body. (For the principles of buffering action see Chapter 2.)

The bicarbonate buffering system

This bicarbonate buffering system is unique, because it remains in equilibrium with atmospheric air, thus creating an open system with many times greater capacity than that of any 'closed' buffer system. A description of how it works follows:

Carbon dioxide produced in tissues diffuses through cell membranes and dissolves in plasma. The solubility coefficient of CO_2 in water is 0.23 if pCO_2 is measured in kPa (or 0.03 if pCO_2 is measured in mmHg). Thus, at normal pCO_2 of 5.3 kPa (40 mmHg; i.e. 1 kPa = 7.5 mmHg or 1 mmHg = 0.133 kPa), the concentration of dissolved CO_2 (dCO_2) is:

$$dCO_2\,(\text{mmol/L}) = 5.3\ \text{kPa} \times 0.23 = 1.2\ \text{mmol/L}$$

This CO_2 equilibrates with H_2CO_3, but the nonenzymatic reaction in plasma is very slow, so that there is only about 0.0017 mmol/L of H_2CO_3 present. However, because of the equilibrium between dissolved CO_2 and carbonic acid in plasma, the 'acid' component of the bicarbonate buffer is equivalent to the plasma concentration of dissolved CO_2.

The Henderson–Hasselbalch equation (see Chapter 2) provides a formula for calculating the pH of a buffer solution such as plasma. Applied to our buffer system, it reads:

$$\text{pH} = \text{p}K + \log\frac{[\text{bicarbonate}]}{pCO_2 \times 0.23}$$

pH 7.4 (hydrogen ion concentration 40 nmol/L) is the average pH of the extracellular fluid at normal concentrations of bicarbonate and at normal partial pressure of CO_2

Normally, there is approximately 24 mmol/L of bicarbonate in plasma, generated as a result of erythrocyte carbonic anhydrase 'fixing' CO_2. The pK (see Chapter 2) of the bicarbonate buffer is 6.1, and if we insert the concentrations of buffer components into the above equation:

$$pH = pK + \log\frac{[HCO_3^-]}{0.23[CO_2]}$$

$$pH = 6.1 + \log\frac{24}{1.2}$$

The bicarbonate buffer minimizes changes in hydrogen ion concentration on the addition of either acid or alkali to plasma. When an acid is added, bicarbonate reacts with it and there is release of CO_2:

$$H^+ + HCO_3^- = HHCO_3^- = H_2O + CO_2$$

Excess of CO_2 is then eliminated through the lungs. When alkali is added, carbonic acid/CO_2 neutralize it to water:

$$OH^- + H^+ + HCO_3^- = H_2O + HCO_3^-$$

Intracellular buffering

Acids generated during metabolism and hydrogen ion generated from the fixation of CO_2 are also buffered by the intracellular buffers, mainly proteins and phosphates. Hydrogen ion produced in plasma enters cells in exchange for potassium. This is important, as intracellular buffering of plasma hydrogen ion can result in an increased potassium concentration.

Conversely, excess of bicarbonate in the extracellular space can be buffered by cell-derived hydrogen ion. In this case, hydrogen ion enters the plasma in exchange for potassium, and the plasma potassium concentration decreases. Thus, an increased hydrogen ion concentration (a decrease in blood pH: acidemia) is associated with an increase in plasma potassium concentration, and a decrease in plasma hydrogen ion concentration (an increase in blood pH: alkalemia) is associated with a decrease in plasma potassium concentration (Fig. 22.3).

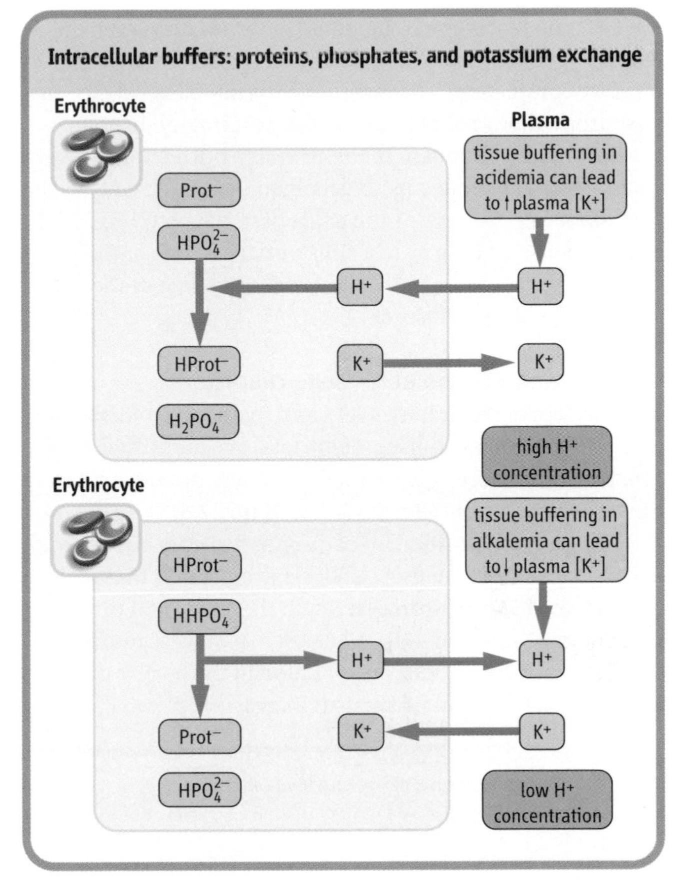

Fig. 22.3 In contrast to buffers in blood, intracellular buffers are primarily proteins and phosphates. Hydrogen ion from the plasma enters cells in exchange for potassium. Accumulation of the hydrogen ion in the plasma (acidemia) may thus lead to increased plasma potassium, and a deficit of hydrogen ion in plasma (alkalemia) to a decreased plasma potassium concentration. Prot, protein.

Respiratory and metabolic components of acid–base balance adjust to maintain optimal pH

Respiratory and metabolic compensation in acid–base disorders

When the primary disorder is respiratory, the accumulation of CO_2 leads to a compensatory increase in bicarbonate reabsorption by the kidney. Conversely, a decrease in pCO_2 decreases bicarbonate reabsorption. In the event that the primary disorder is metabolic, a decrease in bicarbonate concentration and resulting decrease in pH stimulate the respiratory center to increase the ventilation rate: CO_2 is blown off and pCO_2 in plasma decreases. This is why patients with metabolic acidosis hyperventilate. Conversely, an increase in bicarbonate, for whatever reason, leads to an increase in pH and to suppression of the ventilation rate: CO_2 is retained, and the pH returns towards normal. In this way, the interplay between lung (CO_2 transport) and kidney (HCO_3^- handling) leads to fine control of blood pH. More details will be found later in the chapter.

Respiratory and metabolic components of the bicarbonate buffer system

The bicarbonate buffer communicates with the environment through pulmonary gas exchange. The key determinant of the CO_2 concentration in blood is the ventilation rate; this is why the pCO_2 component of the buffer is referred to as the respiratory component. The disorders of acid–base balance that arise from abnormal pCO_2 are called respiratory disorders.

Because bicarbonate is the primary buffer for the nonvolatile acids generated by metabolism, it is referred to as the metabolic component of the acid–base balance (Fig. 22.4). Bicarbonate handling is, to a substantial extent, controlled by the kidney, and disturbed renal function is one of the major causes of acid–base disorders.

Classification of the acid–base disorders

The concept of respiratory (CO_2) and metabolic (bicarbonate) components of acid–base disorders, as identified in the Henderson–Hasselbalch equation above, provides a key to the classification of these disorders (Fig. 22.4).

The primary classification of disorders of acid–base balance is into acidosis or alkalosis. Acidosis is a process that leads to the accumulation of hydrogen ion in the body (and to a resultant decrease in blood pH). Alkalosis is a process leading to a decrease in hydrogen ion concentration in the body (a decrease in its concentration in plasma; an increase in plasma pH).

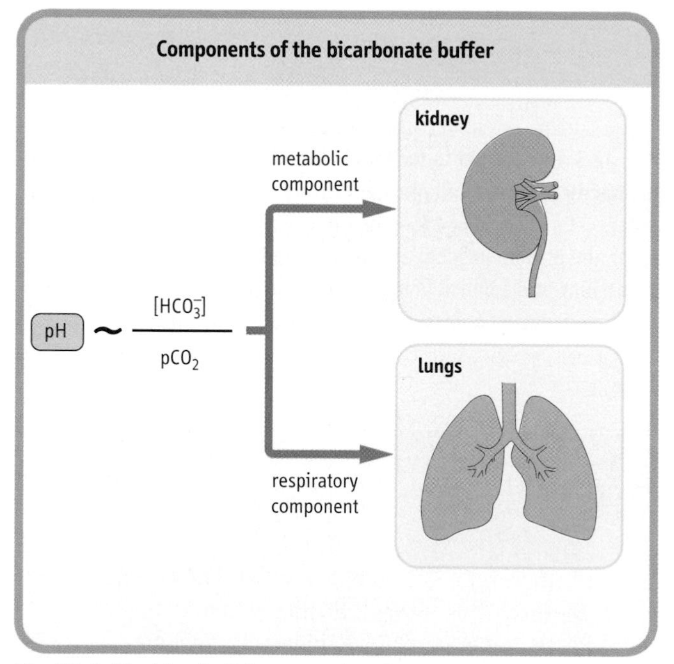

Fig. 22.4 The blood pH is proportional to the ratio of plasma bicarbonate to the partial pressure of CO_2 in blood (pCO_2). CO_2 and bicarbonate are the components of the bicarbonate buffer. Because the pCO_2 depends on the rate of respiration, it is called the respiratory component of the acid–base balance. In contrast, because the plasma concentration of bicarbonate is maintained by the kidney, and is affected by the amount of nonvolatile acids produced in tissues, it is called the metabolic component of the acid–base balance.

Depending on the primary cause, acidosis and alkalosis can each be classified further as either respiratory or metabolic. Thus there are four main disorders of acid–base balance: respiratory acidosis, metabolic acidosis, respiratory alkalosis, and metabolic alkalosis (Fig. 22.5). Mixed disorders can also develop; these are considered in more detail later in the chapter.

Blood gas measurement

Blood gas measurement is an important laboratory investigation

CO_2 and O_2 are measured whenever there is a suspicion of respiratory failure or of conditions that may be associated with acid–base disorders, such as diabetic ketoacidosis (see Chapter 20). In respiratory failure, the results of such measurements may indicate a need for oxygen treatment; in more severe conditions, artificial ventilation may be required.

To perform the measurements, a sample of arterial blood is taken, usually from the radial artery in the forearm or, less commonly, from the femoral artery in the leg. The measurements usually include pO_2, pCO_2, and pH or hydrogen ion concentration. Bicarbonate concentration is calculated using the Henderson–Hasselbalch equation, and several

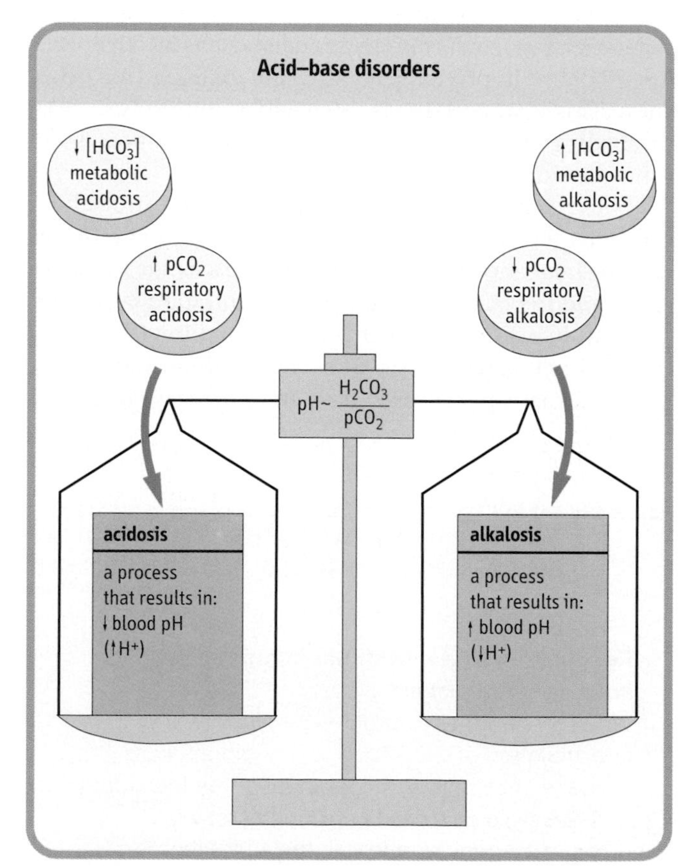

Fig. 22.5 The acid–base disorders. An increase in pCO_2 or a decrease in $[HCO_3^-]$ lead to acidosis. A decrease in pCO_2 or an increase in $[HCO_3^-]$ lead to alkalosis. If the change in pCO_2 is primary, the disorder is called respiratory. If the primary change is plasma bicarbonate, the disorder is called metabolic.

other calculated parameters can be provided. These give an impression of either the total amount of buffers in the blood (so-called buffer base) or the difference between the desired (normal) amount of buffers in the blood and the actual amount (base excess). The acid–base analysers used in the hospitals are designed to perform these calculations automatically. Normal values for pH, pCO_2, and O_2 are given in Figure 22.6.

The lung

The lung supplies oxygen necessary for tissue metabolism and removes the CO_2 produced; removal of CO_2 is essential for maintenance of blood pH

Approximately 10,000 L of air pass through the lungs of an average person each day. Anatomically, the lungs belong to the lower respiratory tract (the upper respiratory tract comprises nose, pharynx, and larynx) and lie in the thoracic cavity. They are surrounded with the pleural sac, a thin 'bag' of tissue that lines the thoracic cage at one end, and attaches to the external surface of the lungs at the other. When the thoracic cage (chest cavity) expands during inspiration, negative pressure created in the expanding pleural sac inflates the lung.

The airways are a set of tubes of progressively decreasing diameter. They consist of the trachea, large and small bronchi, and even smaller bronchioles (Fig. 22.7). At the end of the bronchioles, there are pulmonary alveoli – structures lined with endothelium and covered with a film of surfactant – the main component of which is dipalmitoylphosphatidylcholine (see Chapter 25). Surfactant decreases the surface tension of the alveoli. Bronchi and bronchioles transport and purify the air by removal of organic and particulate matter; the gas exchange occurs in the alveoli.

Reference range for blood gas results

	Arterial	Venous
[H+]	36–43 mmol/L	35–45 mmol/L
pH	7.37–7.44	7.35-7.45
pCO_2	4.6–6.0 kPa	4.8–6.7 kPa
pO_2	10.5–13.5 kPa	4.0–6.7 kPa
bicarbonate	23–30 mmol/L	

Fig. 22.6 Blood gases: normal values. pH, pCO_2, pO_2, and bicarbonate are the most important blood parameters. pH values less than 7.0 or greater than 7.7 occur only in extreme situations, and are life-threatening. Bicarbonate concentration is calculated from pH and pCO_2 values .(Adapted with permission from Hutchinson AS. In: Dominiczak MH, ed. *Seminars in clinical biochemistry*. Glasgow: Glasgow University Press, 1997.)

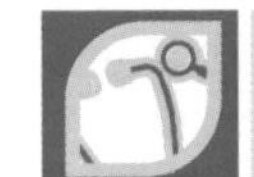

Respiratory alkalosis is caused by hyperventilation

A 25-year-old man was admitted to hospital having an asthmatic attack. His blood gas values were pO_2 9.3 kPa (70 mmHg) and pCO_2 4.0 kPa (30 mmHg), with pH 7.50 (hydrogen ion concentration 42 nmol/L). He was treated with salbutamol, a β-adrenergic stimulant (see Chapter 37) that causes dilatation of the bronchial tree, and recovered completely.

Comment. This man's blood gases show only a mild degree of respiratory alkalosis caused by hyperventilation and 'blowing off' the CO_2. Only in severe asthma does the ventilatory impairment lead to CO_2 retention and respiratory acidosis. Normal (reference) ranges are given in Fig. 22.6.

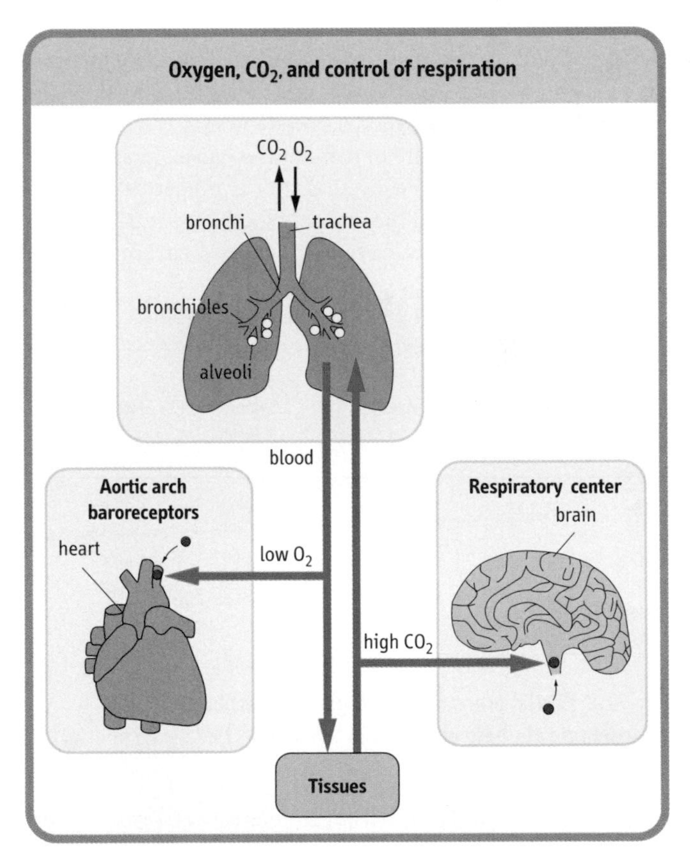

Fig. 22.7 The structure of the lungs and the regulation of respiratory rate by pCO_2 and pO_2. In the lungs, ventilation rate and blood flow (perfusion) are the main factors controling gas exchange. The partial pressure of the CO_2 affects the rate of ventilation through the central chemoreceptors located in the brainstem. When pO_2 decreases, the control switches to pO_2-sensitive peripheral receptors in the carotid bodies and in the aortic arch.

The rate of respiration is controlled by the respiratory center in the brainstem. Both pO_2 and pCO_2 affect respiratory control and thus the ventilation rate: the respiratory center located in the medulla oblongata has chemoreceptors sensitive to pCO_2 and to pH. Under normal circumstances, it is an increase in pCO_2 or a decrease in pH, but not pO_2 status, that stimulates ventilation. However, when hypoxia occurs, the pO_2 takes over the control of ventilation through another set of receptors, located in the carotid bodies in the aortic arch. When the arterial pO_2 decreases to less than 8 kPa (60 mmHg), this 'hypoxic drive' becomes the main mechanism controlling ventilation. Persons who suffer from long-term hypoxia in chronic lung diseases depend on hypoxic drive rather than on pCO_2/pH for control of ventilation (see below).

The two main factors that determine gas exchange are ventilation of the lungs and their perfusion with blood

The blood supply to the alveoli is provided by the pulmonary arteries, which carry deoxygenated blood from the periphery through the right ventricle. After passing through the lungs, the newly oxygenated blood is carried away through pulmonary veins to the left atrium. During the transit of blood through alveolar capillaries, oxygen diffuses from the inspired air to the blood, and the CO_2 diffuses from the blood into the pulmonary alveoli (see Fig. 22.7).

In the alveoli, the rate of diffusion of gases in and out of the blood is determined by the partial pressure gradient between alveolar air and the arterial blood. Figure 22.8 shows the partial pressures of oxygen (pO_2) and carbon dioxide (pCO_2) in the lungs. Compared with the atmospheric air, the alveolar air has slightly increased pCO_2 and decreased pO_2 (this is mainly because of the water vapor pressure). Carbon dioxide is much more soluble than oxygen in water, and equilibrates approximately 20 times more rapidly. This means that, when there are problems with gas exchange, one first notices a decrease in pO_2 values (hypoxia). An increase in pCO_2 (hypercapnia) occurs later, and is often an indicator of more severe disease.

Partial pressure gradients determine diffusion of gases through the alveolar/blood barrier

	Dry air	Alveoli	Systemic arteries	Tissue
pO_2	21.2 kPa	13.7 kPa	12 kPa	5.3 kPa
pCO_2	<0.13 kPa	5.3 kPa	5.3 kPa	6 kPa
water vapor		6.3 kPa		

Fig. 22.8 Partial pressures of oxygen and carbon dioxide in atmospheric air, lung alveoli, and the blood. (1 kPa = 7.5 mmHg)

Normally, the ratio of ventilation to perfusion (V_A/Q) equals 0.8: the alveolar ventilation rate is approximately 4 L/min and the pulmonary blood flow (perfusion) is 5 L/min. As shown in Figure 22.9, two conditions may occur. In the first, some parts of the lung may be well-perfused, but poorly ventilated. This occurs when there are collapsed alveoli. The pO_2 in the outflowing blood is low, because there is no diffusion of oxygen from the alveolar air. The oxygen-poor blood is shunted into the arterial circulation without gas exchange; this is known as a condition of 'shunt'. In the second condition, in which ventilation is adequate but perfusion is poor, there is also no gas exchange. In such a case, part of the lung behaves as if it had no alveoli at all; this is called the 'physiologic dead space'.

Respiratory problems may arise that are related to perfusion, ventilation, or the combination of both:

- rib-cage deformities impair ventilation by limiting the movement of lungs;
- chest trauma may decrease respiratory capacity as a result of collapse of the lung;
- the bronchial tree may be obstructed by inhaled objects, or narrowed by a growing tumor;
- constriction of the bronchi occurs in asthma, which impairs ventilation and gas exchange;
- impaired elasticity of the lung or impaired function of ventilatory muscles (diaphragm and intercostal muscles of the chest wall) reduce ventilatory efficiency;
- destruction of pulmonary alveoli occurs in emphysema;
- impairment of the blood/air barrier is caused by fluid in the alveoli (pulmonary edema);
- inadequate synthesis of surfactant leads to the collapse of lung alveoli and to the respiratory distress syndrome;
- defects in the neural control of lung movement affect its function;
- circulatory problems (inadequate blood flow through the lungs) can occur in shock and in heart failure.

Blood pCO_2 and pO_2 are affected by the perfusion and ventilation of the lungs

	Alveolar pO_2	Alveolar pCO_2	Arterial pO_2	Arterial pCO_2	Comment
poor ventilation, adequate perfusion	decreased	increased	decreased	normal	physiologic shunt
adequate ventilation, poor perfusion	increased	decreased	decreased*	increased*	physiologic dead space

Fig. 22.9 The effect of perfusion and ventilation of the lungs on blood pCO_2 and pO_2. *Depending on a degree of shunt.

The handling of carbon dioxide

The body produces CO_2 at a rate of 200–8000 mL/min. The CO_2 dissolves in water and generates carbonic acid, which in turn dissociates into hydrogen and bicarbonate ions:

$$CO_2 + H_2O \rightleftharpoons H_2CO_3 \rightleftharpoons H^+ + HCO_3^-$$

Thus, CO_2 secondarily generates large amounts of hydrogen ion.

As CO_2 diffuses from the tissues, it is subject to different treatment in the plasma and the erythrocytes

In plasma, the above reaction is nonenzymatic and proceeds slowly. This generates very small amounts of carbonic acid, which remain in equilibrium with a substantial amount of dissolved CO_2. The same, but much faster, reaction occurs in the erythrocytes. Here the reaction is catalyzed by a zinc-containing enzyme, carbonic anhydrase:

$$CO_2 + H_2O \xrightleftharpoons{\text{carbonic anhydrase}} H_2CO_3 \rightleftharpoons H^+ + HCO_3^-$$

Respiratory acidosis occurs in chronic lung diseases

A 56-year-old woman was admitted to a general ward with increasing breathlessness. She had smoked 20 cigarettes a day for the previous 25 years. Blood gas measurements revealed a pO_2 of 6 kPa (45 mmHg), pCO_2 of 8.4 kPa (53 mmHg), and pH 7.35 (hydrogen ion concentration 51 nmol/L); bicarbonate concentration was 35 mmol/L (for reference ranges, refer to Fig. 22.6).

Comment. This patient has an exacerbation of smoking-related chronic lung disease and a respiratory acidosis. Her pCO_2 is high, and her ventilation depends on the hypoxic drive. Her bicarbonate is also increased, as a result of the metabolic compensation of the respiratory acidosis. One needs to be careful when treating such patients with high concentrations of oxygen, because the increased pO_2 may remove hypoxic drive and cause respiratory arrest. Therefore, this patient was not given pure oxygen to breathe, but a 26% mixture of oxygen and air.

Carbonic anhydrase 'fixes' CO_2 as bicarbonate. The hydrogen ion resulting from the dissociation of carbonic acid is buffered by hemoglobin. Approximately 70% of all CO_2 produced is fixed as bicarbonate; approximately 20% is carried as carbamino groups on hemoglobin; and dissolved CO_2 constitutes the remaining 10% of the total.

The bicarbonate produced in erythrocytes during the above reaction is transferred to plasma in exchange for chloride (this is known as 'chloride shift') (Fig. 22.10).

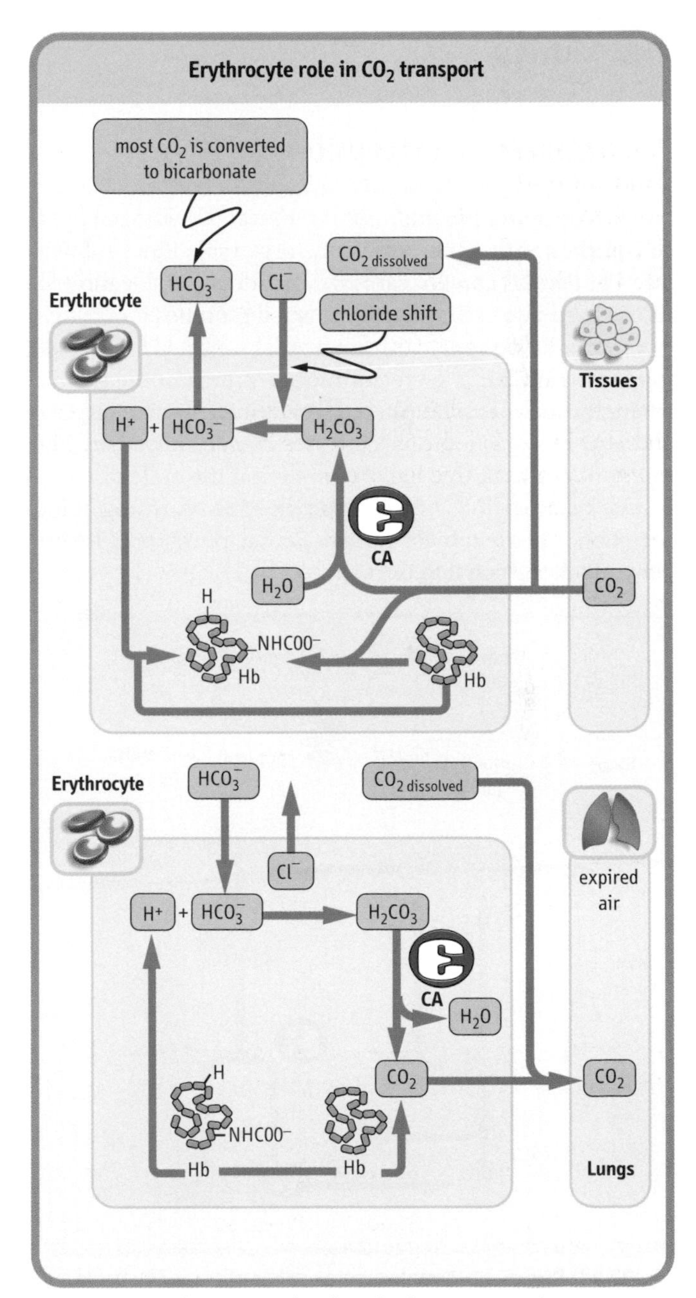

Fig. 22.10 Erythrocyte carbonic anhydrase convertions. A large amount of tissue-desired CO_2 is converted to bicarbonate, and a significant part (approximately 20% of the total) travels bound to erythrocyte hemoglobin as carbamino groups. Carbon dioxide is transported to a limited extent as a dissolved gas in plasma.

In the lungs, the dissolved CO_2 diffuses into the alveolar space

An increase in pO_2 in the lungs increases the dissociation of CO_2 from hemoglobin. This is known as the Haldane effect. In parallel, the buffering capacity of hemoglobin decreases: it releases hydrogen ion, which reacts with bicarbonate, contributing to CO_2 release:

$$H^+ + HCO_3^- \rightleftharpoons H_2CO_3 \rightleftharpoons CO_2 + H_2O$$

The kidney

Bicarbonate reabsorption

The kidney plays an essential role in the control of bicarbonate concentration and in the removal of hydrogen ions. In common with erythrocytes, the proximal kidney tubules (see Chapter 21) contain carbonic anhydrase. Inside the cell, CO_2 is fixed into carbonic acid, which dissociates into hydrogen ion and bicarbonate. Bicarbonate is returned to the plasma; hydrogen ion is secreted into the lumen of the tubule, where it combines with filtered bicarbonate, producing CO_2 and H_2O in a reaction catalyzed by carbonic anhydrase. The CO_2 diffuses back into cells, completing the cycle of bicarbonate reabsorption. At this stage, in spite of hydrogen ion secretion into the tubular lumen, no net excretion of hydrogen ion takes place (Fig. 22.11).

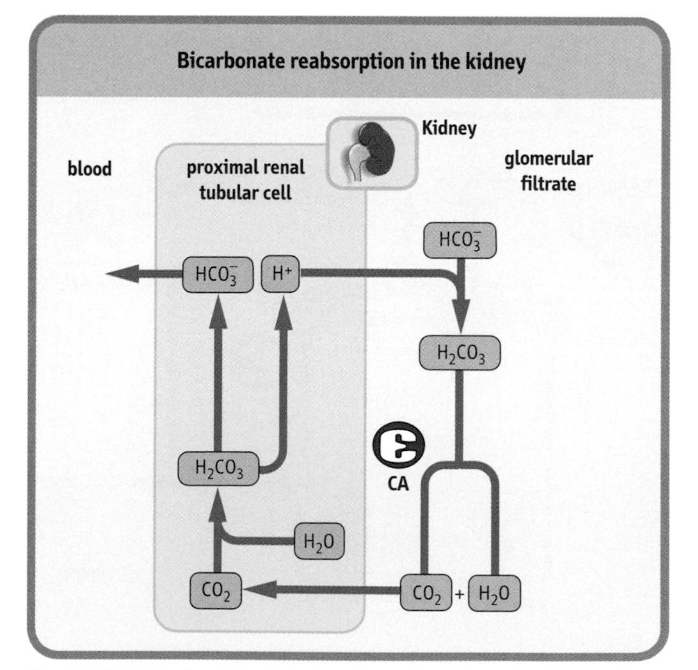

Fig. 22.11 Bicarbonate reabsorption takes place in the proximal tubule. Bicarbonate filtered through kidney glomeruli is reabsorbed in the proximal tubule. At this stage, there is no net excretion of hydrogen ion. Carbonic anhydrase (CA) is present in the luminal side of the tubule.

Hydrogen ion excretion

The kidney removes waste products of metabolism, which include nonvolatile acids, such as minute amounts of sulfuric and phosphoric acids generated from the protein metabolism, and keto acids generated from fatty acids. As in the proximal tubule, in the distal tubule cells the CO_2 is converted into carbonic acid, which immediately dissociates into hydrogen ion and bicarbonate. Bicarbonate is returned to the plasma and hydrogen ion is secreted into the tubule in exchange for sodium. Now, the fate of hydrogen ion differs from that in the proximal tubule. Normally, there is no bicarbonate available in the distal tubule lumen, because it has already been reabsorbed, and so here the hydrogen ion is buffered mainly by the phosphate ion and is excreted in the urine.

The major route of hydrogen ion excretion, however, is through the renal secretion of ammonia. Ammonia is generated from the transformation of glutamine into glutamic acid by glutaminase. Ammonia diffuses freely through the luminal membrane and the hydrogen ion is trapped inside the tubule lumen as ammonium ion (NH_4^+) to which the membranes are not permeable. The ammonium ion is then excreted in the urine (Fig. 22.12). The secretion of ammonia increases in acidosis and decreases in alkalosis.

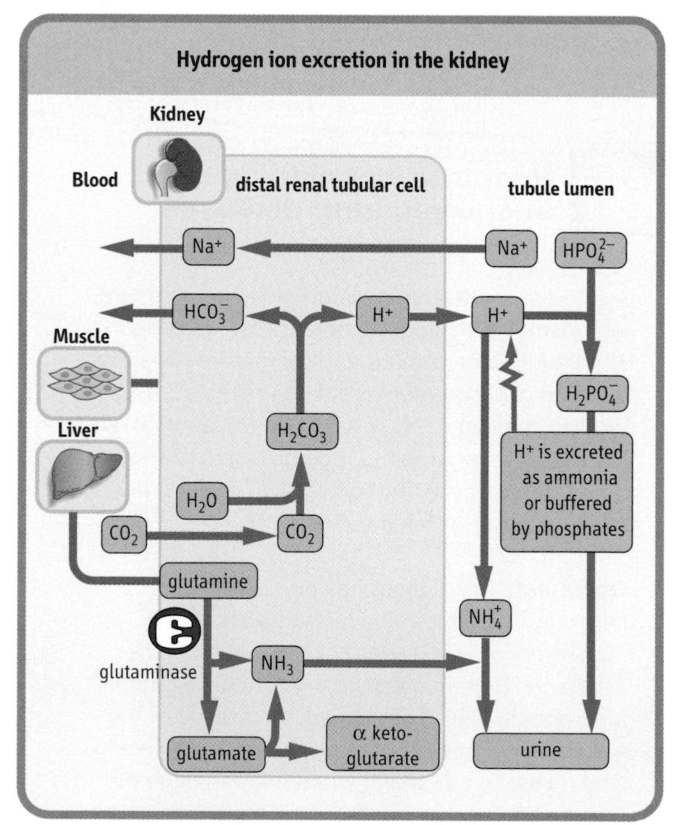

Fig. 22.12 The excretion of the hydrogen ion takes place in the distal tubules of the kidney. Hydrogen ion is excreted as an ammonium ion, or is buffered by phosphates. The daily excretion of hydrogen ion is approximately 50 mmol. The production of ammonia increases in acidemia and decreases in alkalemia.

Anemia is an important cause of shortness of breath and tiredness

A 35-year-old man presented with shortness of breath after climbing two flights of stairs. His chest radiograph was normal, as were examination of the heart and the electrocardiogram; blood gases were normal. His blood cell count revealed a hemoglobin value of 10 g/dL. (reference range for men 13–18 g/dL)

Comment. This person is anemic, and anemia is the cause of his rapid exhaustion. Anemia can present with general symptoms such as tiredness or breathlessness. It appeared that the patient had an undiagnosed gastric ulcer that caused a slow loss of blood.

The metabolic and respiratory acid–base disorders

The lung and the kidney work in a concerted way to minimize changes in plasma pH, and can compensate for each other when problems occur. Disturbances of acid–base balance can arise from either the respiratory or the metabolic component. When the primary problem occurs in one of the components, the other compensates for it, to minimize final changes in plasma hydrogen ion concentration (Fig. 22.13).

Acidosis

Acidosis is a much more common disorder than alkalosis (Fig. 22.14).

Respiratory acidosis

Respiratory acidosis is most often the result of lung disease; decreased ventilation from any cause, if severe enough, can lead to this disorder. The most common of the latter causes is chronic obstructive airways disease (COAD; or, synonymously, chronic obstructive pulmonary disease, COPD). Severe asthmatic attack can also result in respiratory acidosis because of the bronchial constriction it causes. Hypoxia (respiratory failure) is often accompanied by respiratory acidosis; in such a case, the decrease in pO_2 occurs together with an increase in pCO_2.

Metabolic acidosis

Metabolic acidosis results from the excessive production of nonvolatile acids, or from their inefficient metabolism or excretion. The classic example of metabolic acidosis is that seen in diabetes mellitus, when keto acids, acetoacetate, and hydroxybutyrate accumulate in the plasma (see Chapter 20). Physiologic acidosis occurs during extreme physical exertion, when lactate accumulates from muscle metabolism; in normal circumstances, this lactate is quickly metabolized on cessation of exercise. However, lactic acidosis becomes a life-threatening condition when large amounts of lactate are generated as a consequence of hypoxia: for instance, in shock.

Respiratory and metabolic compensation of acid–base disorders

Acid/base disorder	Primary change	Compensatory change	Timescale of compensatory change
metabolic acidosis	decrease in plasma bicarbonate concentration	decrease in pCO_2 (hyperventilation)	minutes/hours
metabolic alkalosis	increase in plasma bicarbonate concentration	increase in pCO_2 (hypoventilation)	minutes/hours
respiratory acidosis	increase in pCO_2	increase in renal bicarbonate reabsorption: increase in plasma bicarbonate concentration	days
respiratory alkalosis	decrease in pCO_2	decrease in renal bicarbonate reabsorption: decrease in plasma bicarbonate concentration	days

Fig. 22.13 Respiratory and metabolic compensation. A change in one of the components of acid–base balance stimulates a compensatory change in the other to minimize changes in pH. In respiratory disorders, there is a change in bicarbonate handling; in metabolic disorders, the ventilation rate changes. A compensatory increase or decrease in the rate of ventilation (hyper- or hypoventilation) is seen within minutes, but an adjustment of the rate of bicarbonate reabsorption by the kidney may take several days.

Excretion of nonvolatile acids is impaired also in kidney disease (renal failure), which is often accompanied by metabolic acidosis. Renal failure can occur quickly when the perfusion of the kidneys is inadequate (e.g. in trauma, shock, or dehydration, see Chapter 21) or if there is an intrinsic kidney disease caused, for instance, by the inflammation of kidney glomeruli (glomerulonephritis) (see Chapter 21).

The excessive loss of bicarbonate can also be a cause of metabolic acidosis. Bicarbonate is present in the intestinal fluid in a concentration equivalent to plasma. Thus severe diarrhea or surgical drainage after bowel surgery may result in metabolic acidosis.

Alkalosis

Alkalosis is a much more rare condition than acidosis (see Fig. 22.14). Clinically significant respiratory alkalosis is rare. Mild degrees can occur on exertion, or during anxiety, stress or fever, as a consequence of hyperventilation. Mild respiratory alkalosis also occurs in pregnancy.

The clinical causes of acid–base disorders

Metabolic acidosis	Respiratory acidosis	Metabolic alkalosis	Respiratory alkalosis
diabetes mellitus (ketoacidosis)	chronic obstructive airways disease	vomiting (loss of hydrogen ion)	hyperventilation (anxiety, fever)
lactic acidosis (lactic acid)	severe asthma	nasogastric suction (loss of hydrogen ion)	lung diseases associated with hyperventilation
renal failure (inorganic acids)	cardiac arrest	hypokalemia	anemia
severe diarrhea (loss of bicarbonate)	depression of respiratory center (drugs, e.g. opiates)	intravenous administration of bicarbonate (e.g. after cardiac arrest)	salicylate poisoning
surgical drainage of intestine (loss of bicarbonate)	weakness of respiratory muscles (e.g. poliomyelitis, multiple sclerosis)		
renal loss of bicarbonate (renal tubular acidosis type 2 – rare)	chest deformities		
impairment of renal H^+ excretion (renal tubular acidosis type 1 – rare)	airway obstruction		

Fig. 22.14 The main acid–base disorders are acidosis and alkalosis; each of them can be either respiratory or metabolic. Respiratory acidosis is common. It is caused primarily by lung diseases that affect gas exchange. Respiratory alkalosis is much more rare. It is caused by hyperventilation, which results in a decrease in blood pCO_2. Metabolic acidosis is common and results from either overproduction or retention of nonvolatile acids in the circulation. Metabolic alkalosis is more rare. The most common cause is the loss of hydrogen ion from the stomach through vomiting or gastric suction.

Respiratory and metabolic disorders of acid–base balance can occur together

During his resuscitation from a cardiorespiratory arrest, the blood gas analysis of a 60-year-old man revealed pH 7.00 (hydrogen ion concentration 100 nmol/L) and pCO_2 7.5 kPa (52 mmHg). His bicarbonate concentration was 11 mmol/L. pO_2 was 12.1 kPa (91 mmHg) while he was being treated with 48% oxygen.

Comment. This patient has a mixed disorder: a respiratory acidosis caused by lack of ventilation, and metabolic acidosis caused by the hypoxia that occurred before oxygen treatment was instituted. This most probably was caused by an accumulation of lactic acid: the measured lactate concentration was 7 mmol/L (reference range is 0.7–1.8 mmol/L [6–16 mg/dL]). The terms acidosis and alkalosis do not only describe blood pH changes: they relate to the processes that result in these changes. Therefore, in some instances, two independent processes may occur: for example a patient may be admitted to hospital with uncontroled diabetes (diabetic acidosis) and may develop breathing difficulty because of a coexisting lung disease associated with long-term smoking (respiratory acidosis). In such an instance, the final result could be a more severe change in pH than would have resulted from a simple disorder (Fig. 22.15). Any combination of disorders can occur; the skills of an experienced physician are usually required to diagnose them.

Metabolic alkalosis as a result of overproduction of bicarbonate occurs infrequently. The most common cause of metabolic alkalosis is hypokalemia (see below). A severe metabolic alkalosis may also occur as a result of the loss of hydrogen ion from the stomach through vomiting, or during nasogastric suction: for example, after surgery. Rarely, it may occur when too great a dose of bicarbonate is given intravenously: for example, in the emergency treatment of cardiac arrest.

Mixed metabolic and respiratory acidosis

Disorder	pH	pCO_2	Bicarbonate
metabolic acidosis	decrease	decrease (respiratory compensation)	decrease (primary change)
respiratory acidosis	decrease	increase (primary change)	increase (metabolic compensation)
mixed respiratory and metabolic acidosis	excessive decrease	increase (respiratory acidosis)	decrease (metabolic acidosis)

Mixed metabolic and respiratory alkalosis (rare)

Disorder	pH	pCO_2	Bicarbonate
metabolic alkalosis	increase	increase (respiratory compensation)	increase (primary change)
respiratory alkalosis	increase	decrease (primary change)	decrease (metabolic compensation)
mixed respiratory and metabolic alkalosis	excessive increase	decrease (respiratory alkalosis)	increase (metabolic acidosis)

Fig. 22.15 Mixed acid–base disorders result in a greater change in blood pH than simple disorders.

Potassium and alkalosis

An association of acid–base disorders with plasma potassium concentration

Plasma potassium concentration is one of the more important variables measured in clinical laboratories. This is because it affects the contractility of the heart. Either a too-high plasma concentration (hyperkalemia) or a too-low concentration (hypokalemia) can be life-threatening; the effects of potassium concentration on heart function can be observed on the electrocardiogram. Hypokalemia leads to an increased excretion of hydrogen ion and, consequently, to metabolic alkalosis. Conversely, metabolic alkalosis, whatever its cause, leads to an increased renal excretion of potassium, and is associated with hypokalemia. One always measures potassium concentration in plasma when a disorder of acid–base balance is suspected (see Fig. 22.3, and Chapter 21).

Vomiting can lead to metabolic alkalosis

A 47-year-old man came to the outpatient clinic with a history of vomiting after he had eaten, which he had suffered over the previous month. His blood pH was 7.55 (hydrogen ion concentration 28 nmol/L) and pCO_2 was 6.4 kPa (48 mmHg). His bicarbonate concentration was 35 mmol/L.

Comment. This patient presents with metabolic alkalosis caused by the loss of hydrogen ion through vomiting. Investigations showed an obstruction of the stomach entrance, the pylorus (pyloric stenosis), for which he subsequently underwent surgery, with a good outcome. Note his increased pCO_2 as a result of respiratory compensation of metabolic alkalosis.

Summary

The maintenance of the hydrogen ion concentration is vital for cell survival. In the human body this is regulated by a concerted action of lungs and kidney. The main buffers present in blood are hemoglobin and bicarbonate, whereas the cells main buffers are proteins and phosphate. The bicarbonate buffer system is unique in that it can communicate with atmospheric air, which enormously increases its buffering capacity. The acid–based disorders are divided into acidosis and alkalosis and each of them may be either metabolic or respiratory. The determination of the main parameters of the acid–base balance: pH, pCO_2 and bicarbonate, and also pO_2, are the first line investigations frequently required in medical and surgical emergencies.

Further reading

Androgue HJ, Madias NE. Management of life-threatening acid–base disorders (first of two parts). *N Engl J Med* 1998;**338:**26–34.

Androgue HJ, Madias NE. Management of life-threatening acid–base disorders (second of two parts). *N Engl J Med* 1998;**338:**107–111.

Dominiczak MH, ed. *Seminars in clinical biochemistry.* Glasgow: University of Glasgow; 1997.

Holmes O. *Human acid–base physiology. A student text.* London: Chapman and Hall Medical; 1993.

Thomson WST, Adams JF, Cowan RA. *Clinical acid–base balance.* Oxford: Oxford University Press; 1997.

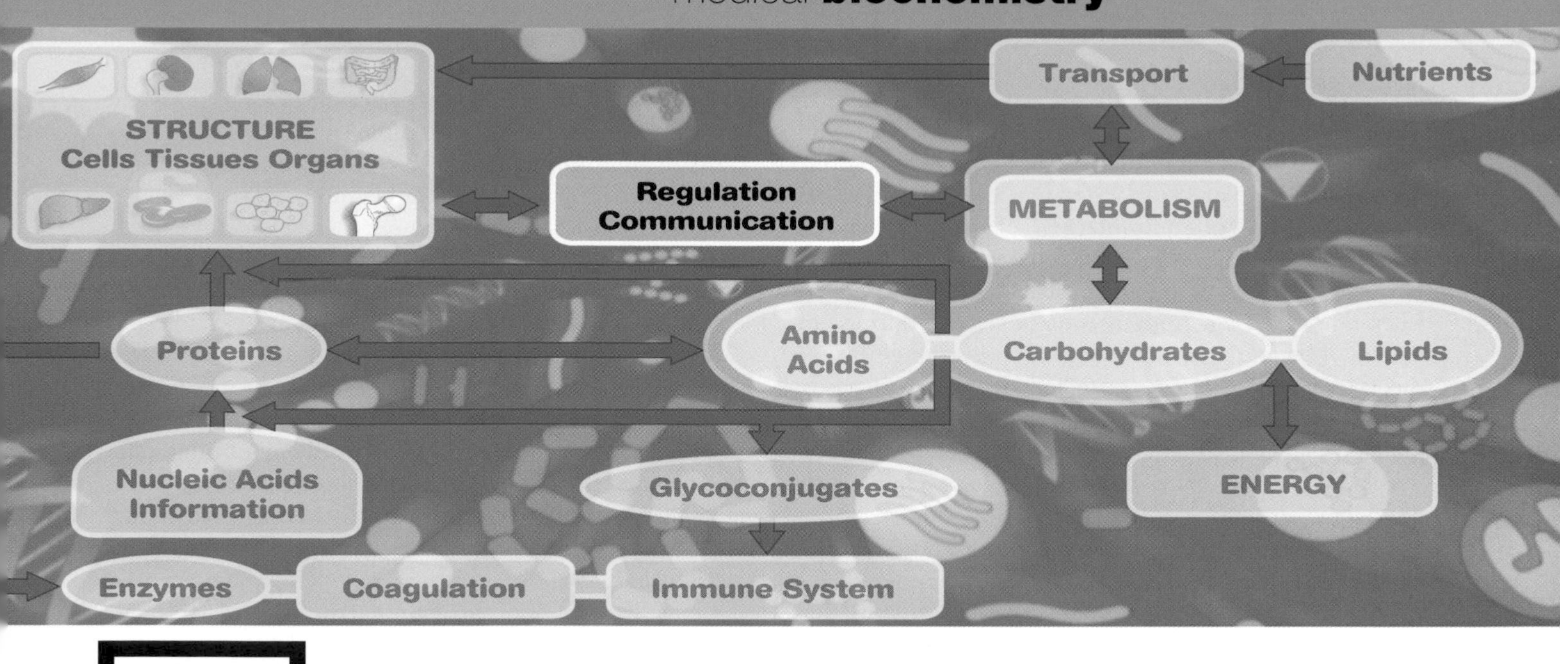

23 Calcium and Bone Metabolism

Bone is a specialized connective tissue that, along with cartilage, forms the skeletal system. In addition to serving a supportive and protective role, bone is the site of substantial metabolic activity. Two types of bone are recognized: the thick, densely calcified external bone (cortical or compact bone), and a thinner, honeycomb network of calcified tissue on the inner aspect of bone (trabecular or cancellous bone). Within the bone matrix, type 1 collagen is the major protein (90%) and calcium-rich crystals of hydroxyapatite ($3Ca[PO_4]_2 \bullet Ca[OH]_2$) are found on, within, and between the collagen fibers. The attachment of hydroxyapatite to collagen and the calcification of bone are, in part, controled by the presence of glycoproteins and proteoglycans with a high ion-binding capacity. Collagen fibers orientate so that they have the greatest density per unit volume and are packed in layers, giving the lamellar structure observed on microscopy. Post-translational modifications of collagen occur that result in the formation of intra- and intermolecular pyridinoline and pyrrole crosslinks, which have an important role in fibril strength and matrix mineralization. This microarchitecture allows bone to function as the major reservoir of calcium for the body.

The noncalcified organic matrix within bone, known as osteoid, becomes mineralized through two mechanisms

Within the bone extracellular space, plasma-membrane-derived matrix vesicles act as a focus for deposition of calcium phosphate, the lipid-rich inner membrane of these vesicles being the nidus for the formation of hydroxyapatite crystals. Crystallization proceeds rapidly, eventually obliterating the vesicle membrane and leaving a collection of clustered hydroxyapatite crystals. Within this environment, osteoblasts (bone-forming cells) can also secrete preorganized packets of matrix proteins that rapidly mineralize, and these combine with matrix-vesicle-derived crystals to form a continuous mineralized tissue within the matrix space. Molecules that inhibit this process – for example pyrophosphate – exist within the matrix environment, but the secretion of alkaline phosphatase by osteoblasts destroys pyrophosphate, allowing mineralization to occur.

Within lamellar bone, collagen fibrils are tightly packed and matrix vesicles are rarely seen. Mineralization occurs in association with matrix fibrils, especially at the spaces between collagen molecules, and is controled by noncollagenous proteins (e.g. proteoglycans and osteonectin).

Both these processes of mineralization are highly dependent on the presence of an adequate supply of calcium and phosphate. When mineral deprivation exists, there is an increase in the percentage of osteoid (the nonmineralized organic matrix) within bone, resulting in the clinical condition of osteomalacia.

Cellular activity within bone

Under normal circumstances, small amounts of calcium are exchanged daily between bone and the extracellular fluid (ECF) as a result of constant bone remodeling i.e. coupled processes of resorption by osteoclasts and formation by osteoblasts (Fig. 23.1). This exchange maintains a relative calcium balance between newly formed bone and older resorbed bone. Calcium flux takes place across the bone-lining cell layer and within bone into the ECF of the periosteocytic space.

Osteocytes are found in osteocytic lacunae and are believed to be derived from osteoblasts trapped during production of bone matrix. Cellular processes extend between osteocytes and from osteocytes to bone-lining cells, forming a network of canaliculi throughout the bone matrix. These cells may contribute to the maintenance of calcium homeostasis by resorbing bone, and they may have a role in activating bone turnover.

Osteoclasts are multinucleated giant cells found on the endosteal surface of bone, in Haversian systems and periosteal surfaces. They are in contact with a calcified surface generated by their resorptive activity. Characteristically, they contain between four and 20 nuclei, pleomorphic mitochondria, and a ruffled cell border. Osteoclasts are derived from pluripotent hematopoietic mononuclear cells in the bone marrow. Parathyroid hormone (PTH) activates osteoclasts indirectly via osteoblasts that possess PTH receptors. Calcitonin is a potent direct inhibitor of osteoclast activity, decreases proliferation of the progenitor cells, and inhibits differentiation of precursors. Local factors such as the cytokines interleukin-1 (IL-1), tumor necrosis factor (TNF), transforming growth factor-β (TGF-β) and interferon-γ (INF-γ) are important regulators of osteoclast activity. Osteoclast resorption of bone releases collagen peptides, pyridinoline crosslinks, and calcium from the bone matrix through the production, secretion, and action of lysosomal enzymes, collagenases, and cathepsins at an acidic pH. Collagen breakdown products in serum and urine (e.g. hydroxyproline) and collagen fragments can be measured as surrogate biochemical markers of bone metabolism.

Osteoblasts are metabolically active bone-forming cells. The precursor cell is the committed progenitor cell, or pre-osteoblast, which is derived from a pluripotent stromal mesenchymal cell. Mature osteoblasts synthesize type 1 collagen, osteocalcin (also known as bone Gla protein), cell attachment proteins (thrombospondin, fibronectin, bone sialoprotein, osteopontin), proteoglycans, and growth-related proteins, and they control bone mineralization. Osteoblast function and activity is now known to be altered by several hormones and growth factors. PTH binds to a specific receptor, stimulating production of cyclic adenosine monophosphate (cAMP), ion and amino acid transport, and collagen synthesis. 1,25-Dihydroxy ($1,25(OH)_2$) D_3 stimulates synthesis of alkaline phosphatase, matrix, and bone-specific proteins, and can

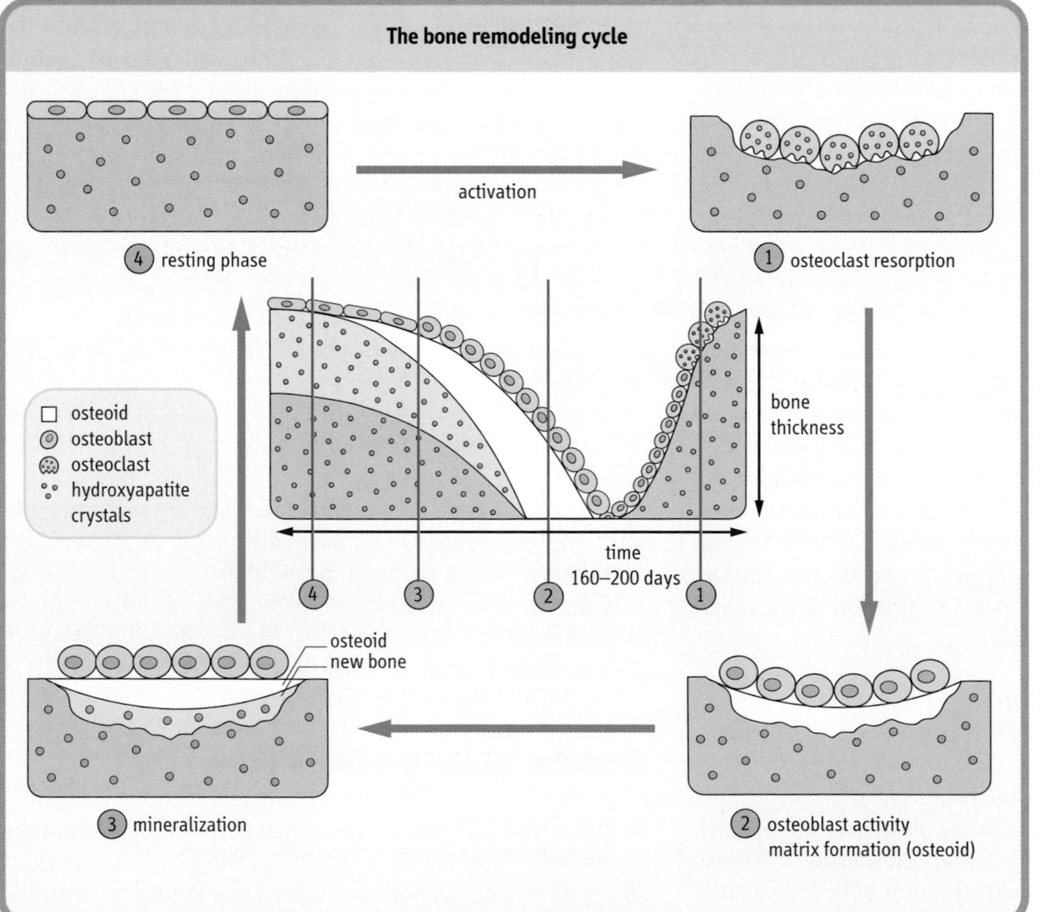

Fig. 23.1 Maintaining bone mass – the bone remodeling cycle. Resorption and formation of bone by osteoclasts and osteoblasts is coupled.

decrease osteocalcin secretion. Growth factors TGF-β, insulin-like growth factors (IGFs) -I and -II, and platelet-derived growth factor (PDGF) serve as autocrine regulators of osteoblast function. Serum biochemical markers reflecting osteoblast function are bone-specific alkaline phosphatase, osteocalcin, and markers of collagen formation such as fragments of collagen molecule, carboxy-terminal procollagen extension peptide (ICTP) and amino-terminal procollagen extension peptide (PINP).

Calcium metabolism

A complex calcium homeostatic mechanism utilizes bone as the reservoir of calcium when deficiency exists, and as a store of calcium when the body is replete

The skeleton contains 99% of the calcium present in the body, in the form of hydroxyapatite; the remainder is distributed in the soft tissues, teeth, and ECF. A multitude of cell and organ

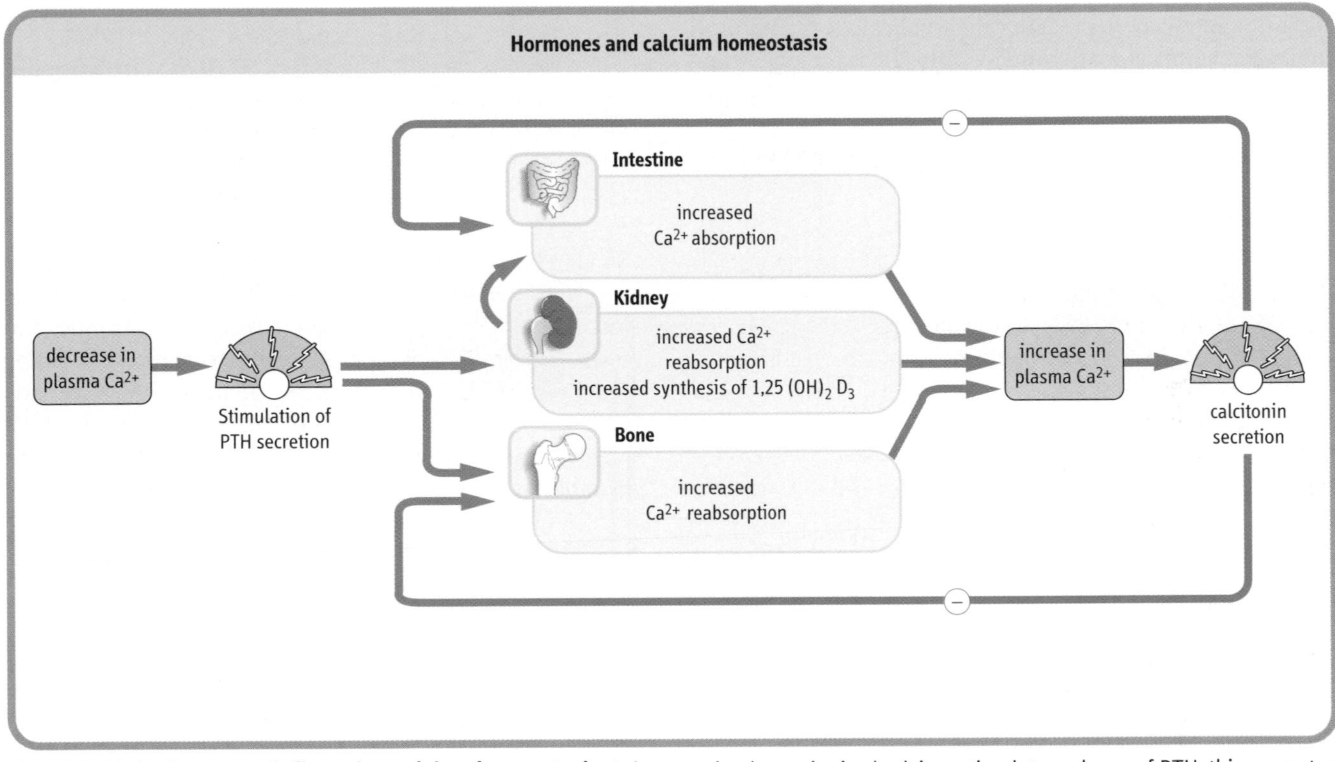

Fig. 23.2 Major hormones influencing calcium homeostasis. A decrease in plasma ionized calcium stimulates release of PTH; this promotes Ca^{2+} reabsorption from the kidney, resorption from bone, and absorption by the gut via increased production of $1,25(OH)_2$ vitamin D_3. As a result, plasma calcium increases. Conversely, an increase in plasma ionized calcium stimulates release of calcitonin, which inhibits reabsorption of calcium by the kidney and osteoclast-mediated bone resorption.

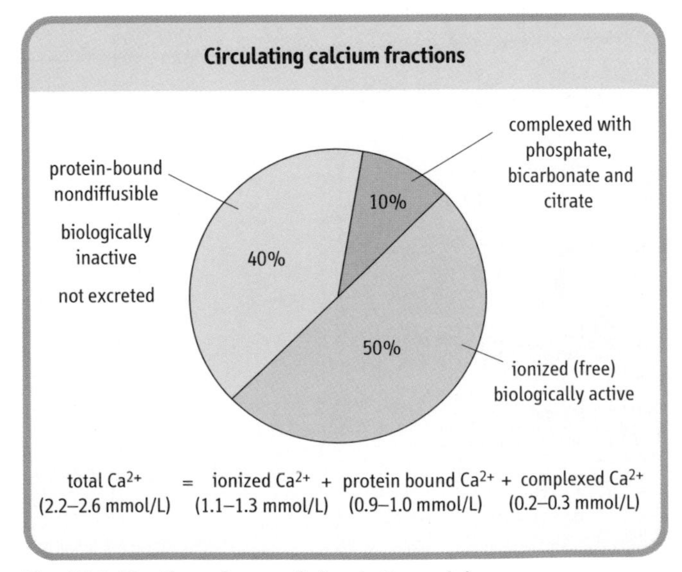

Fig. 23.3 The three forms of circulating calcium.

functions are dependent on the tight control of extracellular calcium concentration; these include neural transmission, cellular secretion, contraction of muscle cells, cell proliferation, the stability and permeability of cell membranes, blood clotting, and the mineralization of bone. Total serum calcium is maintained between 2.2 and 2.60 mmol/L (8.8–10.4 mg/dL). Hormonal control of the circulating calcium involves bone, kidneys, and the gastrointestinal tract: if plasma calcium decreases, PTH is released from the parathyroid glands, stimulating osteoclast-mediated bone resorption, reabsorption of calcium at the kidney, and absorption of calcium at the small intestine, mediated by $1,25(OH)_2D_3$; an increase in calcium results in the inverse effects and the stimulation of calcitonin release, which can inhibit osteoclast resorption of bone (Fig. 23.2).

Calcium exists in the circulation in three forms (Fig. 23.3):

- **ionized Ca^{2+}**: the most important, physiologically active form (50% of total calcium).

- **protein-bound:** the majority of the remaining calcium, mainly bound to negatively charged albumin;
- **complexed to substances such as citrate and phosphate:** a smaller fraction.

If serum protein concentration increases (as in dehydration and after prolonged venous stasis), protein-bound calcium and total serum calcium increase. In contrast, in conditions of reduced serum proteins (e.g. liver disease, nephrotic syndrome, malnutrition), the protein-bound calcium concentration is reduced, decreasing the total calcium, although ionized calcium is maintained within the reference range 1.1–1.30 mmol/L (4.4–5.2 mg/dL). Many acute and chronic illnesses decrease serum albumin concentration, which consequently decreases serum total calcium. Because of this, in clinical situations it is important to calculate the 'adjusted calcium' – total serum calcium, adjusted for the patient's prevailing albumin concentration. This is achieved by means of a formula utilizing the population mean albumin concentration 40 g/L (4g/dL):

$$\text{adjusted } Ca^{2+} = \text{measured Ca (mmol/L)} + 0.02\,(40 - \text{albumin [g/L]})$$

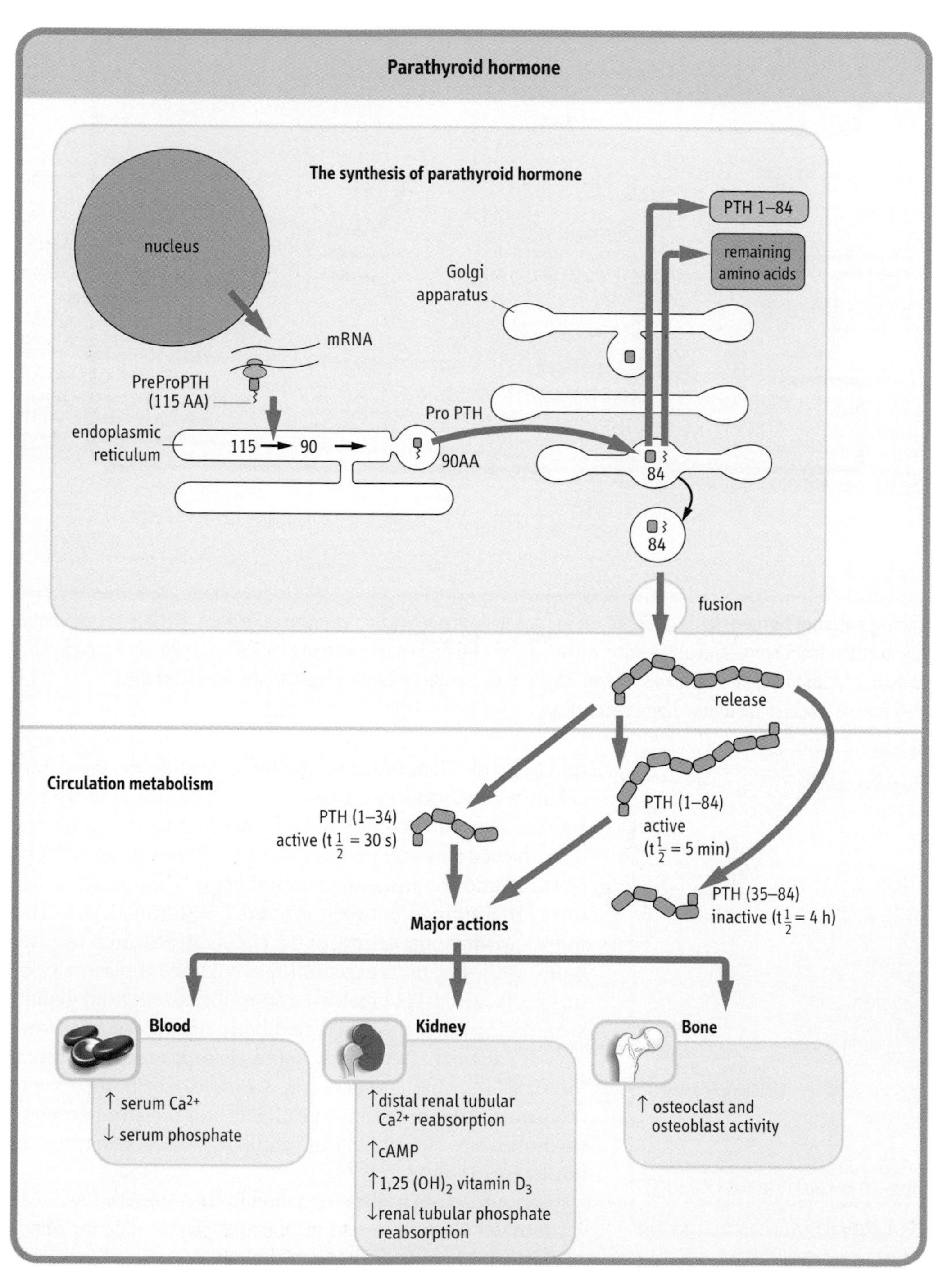

Fig. 23.4 Synthesis, metabolism and major actions of PTH.

Parathyroid hormone measurement

PTH assays

Intact $PTH_{(1-84)}$ is the most important biologically active form of PTH, and can be measured by specific immunoradiometric assays (IRMAs) without interference from its amino- or carboxy-terminal fragments. These assays utilize a double-antibody technique, with a capture antibody directed against one end of the intact molecule, and the labeled detection antibody directed against the opposite end of the molecule. Only the full-length, 1–84-amino acid molecule can bind to the two antibodies, which means that fragments of PTH produced during its metabolism (see Fig. 23.4) are not measured by these assays.

Factors influencing calcium homeostasis

Parathyroid hormone

PTH is an 84-amino-acid, single-chain peptide hormone secreted by the chief cells of the parathyroid glands. A decrease in extracellular ionized calcium or an increase in serum phosphate concentration stimulates its secretion, chronic severe magnesium deficiency can inhibit its release from secretory vesicles, and low concentrations of $1{,}25(OH)_2$ D_3 interfere with its synthesis. PTH(1–84) is metabolized into a biologically active PTH(1–34) amino-terminal fragment and an inactive carboxy-terminal fragment, PTH(35–84) (Fig. 23.4).

PTH regulates serum calcium concentrations by direct actions on bone and kidney, and indirect actions on the intestine by increasing the synthesis of $1{,}25(OH)_2$ D_3 (see Fig. 23.2). Most of the classical actions of PTH are mediated by cAMP, which is generated through G-protein-stimulated adenyl cyclase.

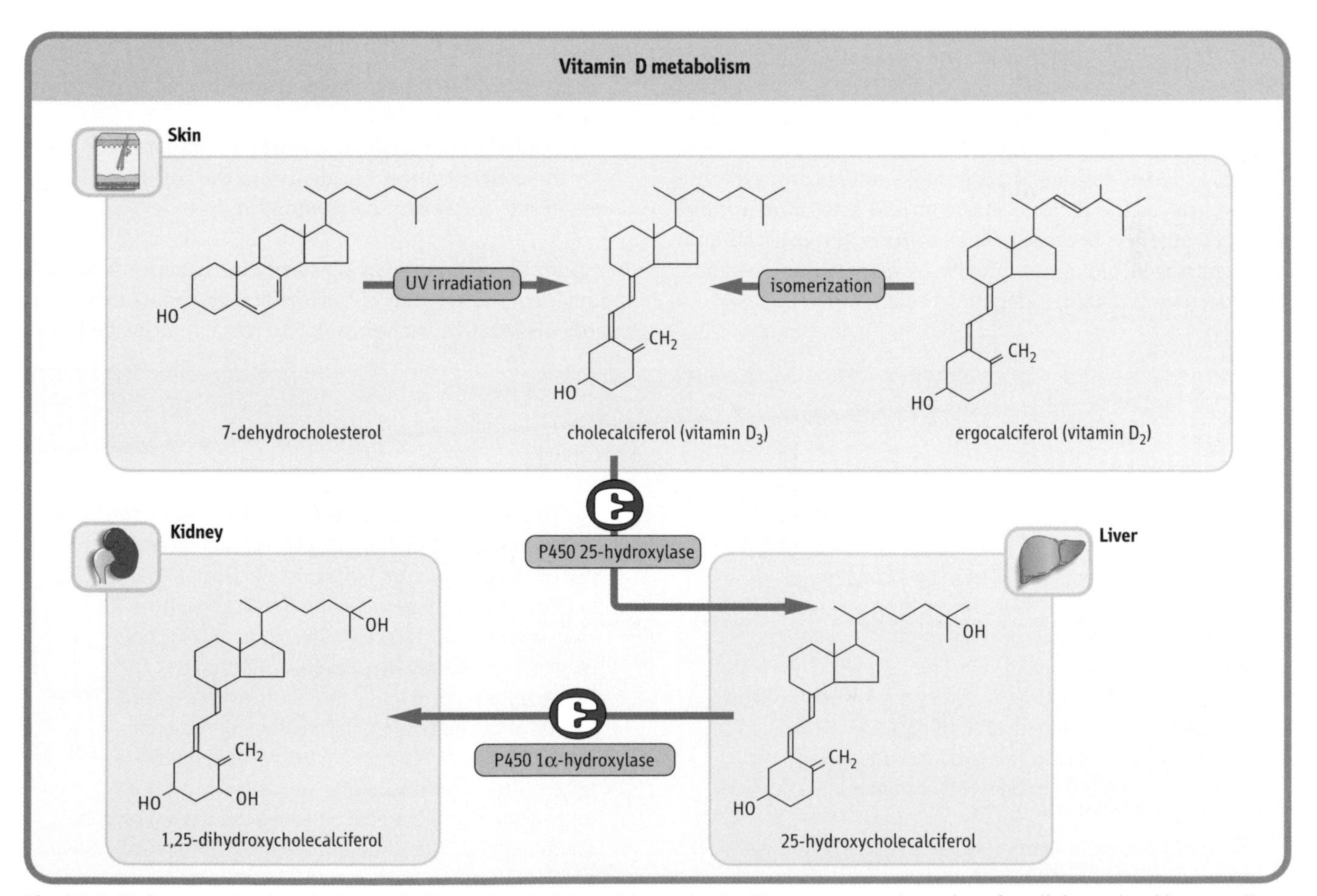

Fig. 23.5 Major steps in vitamin D metabolism. Vitamin D is mainly synthesized in response to the action of sunlight on the skin; a smaller component comes from the diet. Normal liver and kidney function are essential to the formation of the active $1{,}25(OH)_2$ D_3.

Vitamin D synthesis and metabolism

The synthesis and metabolism of vitamin D is illustrated in Figure 23.5. The final step in synthesis of $1,25(OH)_2\ D_3$ formation is catalyzed in the kidney by 1α-hydroxylase, the activity of which is stimulated by PTH, low serum concentrations of phosphate or calcium, vitamin D deficiency, calcitonin, growth hormone, prolactin, and estrogen. Conversely, the activity of 1α-hydroxylase is feedback-inhibited by $1,25(OH)_2\ D_3$, hypercalcemia, high phosphate and hypoparathyroidism.

$1,25(OH)_2\ D_3$ increases the absorption of calcium and phosphate from the gut via active transport by calcium-binding proteins. Together with PTH, $1,25(OH)_2\ D_3$ stimulates bone resorption by osteoclasts. These effects increase serum calcium and phosphate concentrations. Low $1,25(OH)_2\ D_3$ causes abnormal mineralization of newly formed osteoid as a result of low calcium and phosphate availability and reduced osteoblast function, which result in rickets (infants and children) or osteomalacia (adults).

Calcitonin

Calcitonin is a 32-amino-acid peptide synthesized and secreted primarily by the parafollicular cells of the thyroid gland (C-cells). Its secretion is regulated acutely by serum calcium: an increase in serum calcium results in a proportional increase in calcitonin, and a decrease elicits a corresponding reduction in calcitonin. Chronic stimulation results in exhaustion of the secretory reserve of the C-cells. The precise biological role of calcitonin is not known, but the main effect is inhibition of osteoclastic bone resorption. There is significant species homology for calcitonin, with a 1–7 amino-terminal disulfide bridge, a glycine residue at position 28, and a carboxy-terminal proline amide residue. Basic amino acid substitutions enhance potency; because of this, salmon and eel calcitonin have increased biological activity in mammalian systems, compared with that of endogenous calcitonin.

Steroid and peptide hormones which affect calcium homeostasis

Several hormones for which the primary action is not related to calcium regulation directly or indirectly affect calcium homeostasis and skeletal metabolism

Thyroid hormones stimulate osteoclast-mediated resorption of bone. Adrenal and gonadal steroids, particularly estrogen in women and testosterone in men, have important regulatory effects, increasing osteoblast and decreasing osteoclast function. They also decrease renal calcium and phosphate excretion, intestinal calcium excretion, and PTH function. Growth hormone has anabolic effects on bone, promoting growth of the skeleton. These effects of growth hormone on bone are believed to be mediated by insulin-like growth factors (IGF-I and IGF-II) acting on cells of the osteoblast lineage. Growth hormone increases the urinary excretion of calcium and hydroxyproline, whilst decreasing the urinary excretion of phosphate.

Calcium absorption and excretion

Calcium is absorbed predominantly in the proximal small intestine; this is regulated through the quantity of calcium ingested in the diet, and two cellular calcium transport processes:

- **active saturable transcellular absorption** which is stimulated by $1,25(OH)_2D_3$;
- **nonsaturable paracellular absorption** which is controled by the concentration of calcium in the intestinal lumen relative to the serum concentration.

In a normal adult taking a Western diet, calcium balance is maintained: the amount of calcium intake and its deposition in bone are exactly matched by the excretion in urine and feces.

Parathyroid-hormone-related protein

Parathyroid-hormone-related protein (PTHrP)
PTHrP is synthesized as three isoforms containing 139, 141 and 173 amino acids, as a result of alternative differential splicing of RNA. There is amino-terminal sequence homology with PTH: eight of the first 13 amino acids are identical in PTHrP and PTH, three are identical within residues 14–34, and a further three are identical within residues 35–84 (Fig. 23.6). Activation of the classical PTH receptor is by the amino-terminal portion of both PTH and PTHrP, and there is a common α-helical secondary structure in the binding domain of both peptides. As a result of this structural similarity, PTHrP possesses many of the biological actions of PTH.

There is, to date, little evidence to suggest that PTHrP has a role in normal adult calcium homeostasis, but evidence from animal models indicates that it is important in regulating fetal skeletal development and calcium homeostasis. Deletion of the PTHrP gene results in either severe skeletal abnormalities (heterozygote) or a lethal mutation (homozygote). PTHrP has an important role in the etiology of hypercalcemia associated with malignancy (HCM) (see below). PTHrP is subject to post-translational processing, producing fragments with biological activities that have yet to be characterized. The structural and functional relationships of PTHrP as they are currently understood are summarized in Figure 23.6.

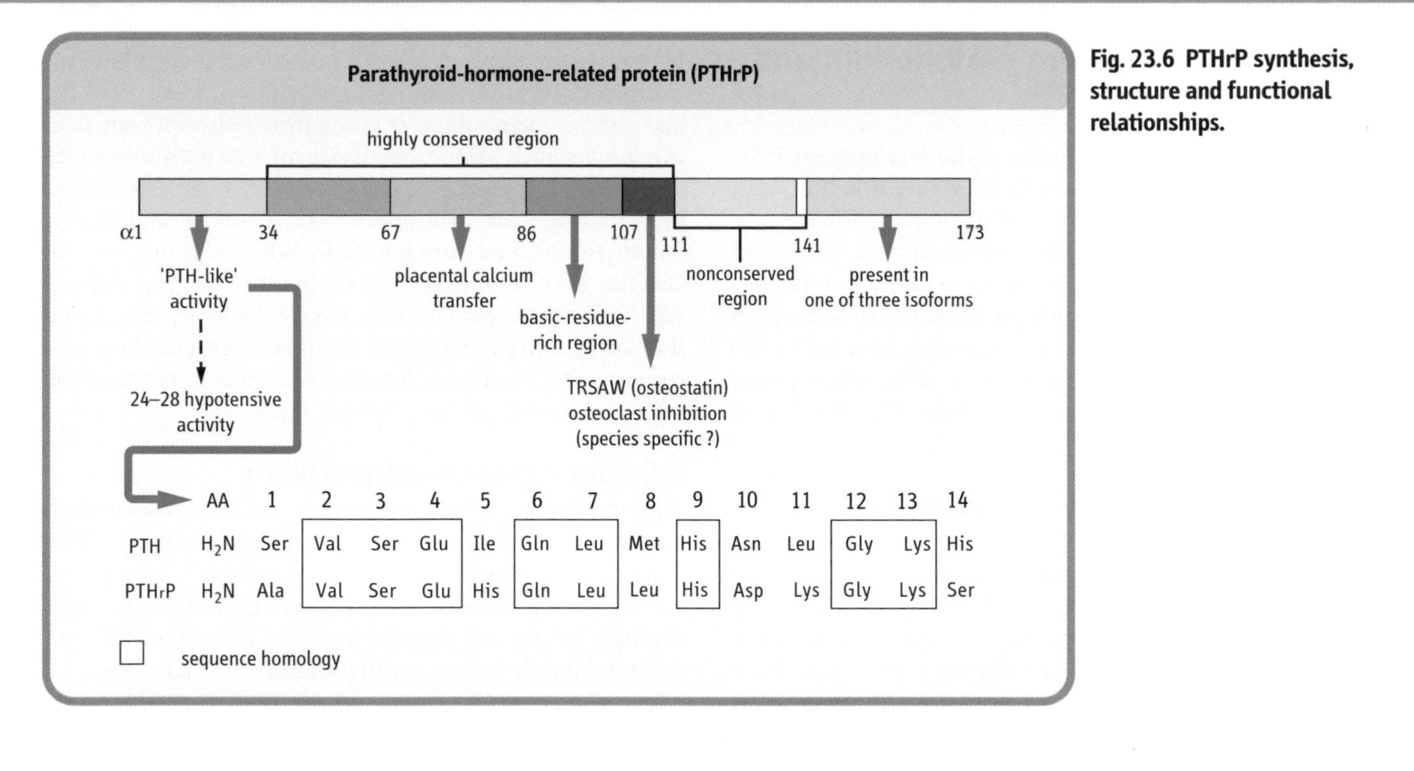

Fig. 23.6 PTHrP synthesis, structure and functional relationships.

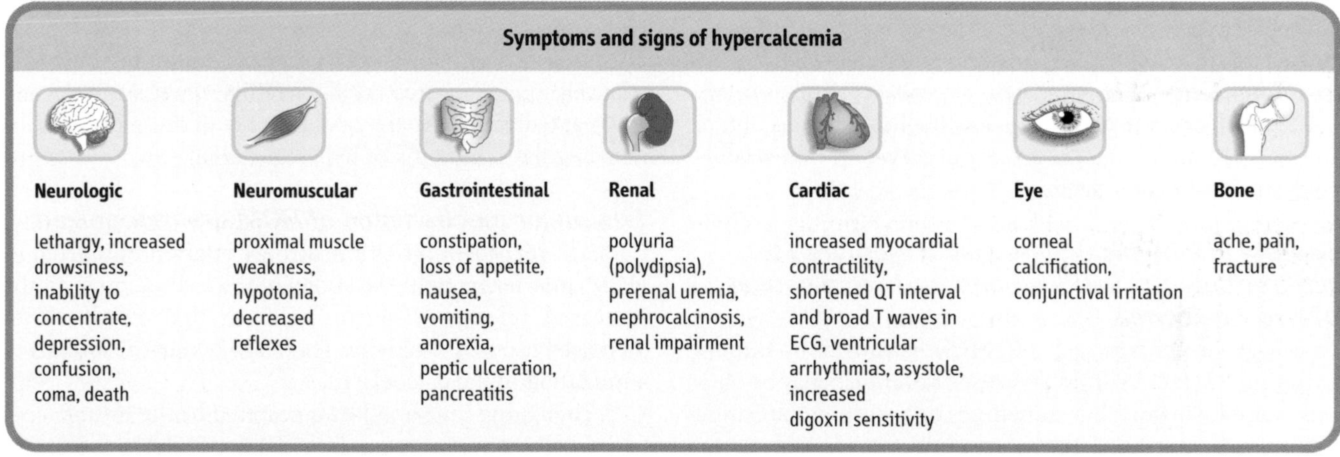

Fig. 23.7 Symptoms and signs of hypercalcemia. Symptoms are more likely as the serum concentration of calcium increases.

During growth, a child will be in positive calcium balance, whereas in the elderly and in several diseases, individuals may be in negative calcium balance. Reduced or increased calcium absorption reflect alterations in dietary calcium intake, intestinal calcium solubility, and $1{,}25(OH)_2D_3$ metabolism.

Generally, as serum calcium increases, calcium excretion increases. When hypercalcemia is caused by hyperparathyroidism (HPT), PTH will act on the renal tubule, promoting reabsorption of filtered calcium and thus diminishing the effects of the increased filtration of calcium and inhibition of renal tubular reabsorption that are caused by increased serum calcium. Decreasing serum calcium is associated with a reduction in urinary excretion of calcium, mainly as a result of decreased amounts of filtered calcium. In hypoparathyroid patients, who lack PTH secretion, renal tubular reabsorption of calcium is reduced.

Causes of hypercalcemia	
Common causes	primary HPT malignant disease iatrogenic – vitamin D or vitamin D analogues
Uncommon causes	thyrotoxicosis multiple myeloma sarcoidosis drug-induced: thiazide diuretics lithium renal failure (acute and chronic) familial hypocalciuric hypercalcemia

Fig. 23.8 Common and uncommon causes of hypercalcemia.

Disorders of calcium metabolism and bone: Hypercalcemia

There is a wide individual variation in the development of the numerous symptoms and signs of hypercalcemia (Fig. 23.7) in patients with this condition. In general, the greater the adjusted calcium value, and the more rapid the increase in calcium, the more likely are symptoms to be present. In practice, 90% of cases are due to either primary HPT or malignancy; a greater diagnostic challenge is presented to the physician when it becomes necessary to differentiate occult malignancy from the less common causes of hypercalcemia (Fig. 23.8).

Investigation of hypercalcemia

In the majority of cases, the cause of hypercalcemia will be identified by obtaining an accurate history, by clinical examination, and by appropriate biochemical tests. In some cases, however, additional valuable information on causation may be obtained from radiologic investigations and tissue biopsies. The development and ready availability of specific, sensitive, and reliable assays for intact PTH has enabled the clinician to discriminate primary HPT from nonparathyroid causes of hypercalcemia (particularly malignancy) with confidence: an increased or inappropriately detectable intact PTH in the presence of hypercalcemia is observed in primary HPT, whereas an intact PTH below the limit of detection of the assay (undetectable) is usually observed in nonparathyroid causes of hypercalcemia.

Several biochemical tests may be required to differentiate the various nonparathyroid causes of hypercalcemia

In a high percentage of cases, HCM is caused by tumors secreting PTHrP. Vitamin D excess or overdose may be obvious from the history, but sometimes only becomes apparent after measurement of the concentrations of vitamins D_3 (cholecalciferol), D_2 (ergocalciferol), and $1,25(OH)_2 D_3$. Measurement of serum electrolytes, urea, and creatinine will confirm renal failure, and estimation of protein, albumin, and immunoglobulins may indicate the presence of myeloma, which should be investigated by both serum and urine electrophoresis. Thyroid function tests (measurement of thyroid-stimulating hormone, total thyroxine, and total triiodothyronine, or estimation of free thyroid hormone, (see Chapter 35) enable diagnosis of thyrotoxic hypercalcemia. Rare causes of hypercalcemia may be diagnosed by estimation of lithium (toxicity or overdose), growth hormone (acromegaly), vitamin A (toxicity), and urine catecholamines (pheochromocytoma, see Chapter 38).

Primary hyperparathyroidism

Primary HPT is a relatively common endocrine disease, characterized by hypercalcemia associated with an increased or inappropriate intact PTH. It has an incidence ranging from 1 in 500 to 1 in 1000 of the population. In 80–85% of patients, a solitary parathyroid gland adenoma is present, and the condition is curable by successful removal of the adenoma.

Hypercalcemia associated with malignancy (HCM)

Hypercalcemia tends to occur late in the course of malignant disease, and is usually a poor prognostic sign. Effective and relatively innocuous treatment for HCM is now available, and can markedly improve the quality of life of these patients, by relieving the symptoms of hypercalcemia.

Two major mechanisms of HCM are recognized

Calcium reabsorption by the kidney is often enhanced in HCM, interfering with the ability of the kidneys to limit the increased release of serum calcium that results from increased osteoclast activity. There are two main sources of stimulation of the osteoclasts:

- a circulating humoral factor secreted by the tumor;
- a locally active secretory factor(s) produced by the tumor or metastases in bone.

Primary hyperparathyroidism (HPT)

A 52-year-old woman presented to the Accident and Emergency Department of her local hospital with severe right-sided flank pain. Blood was detected in urine (hematuria) and radiography revealed the presence of kidney stones. The pain settled with opiate analgesia. Further questioning revealed a history of recent depression, generalized weakness, recurrent indigestion, and aches in both hands. Serum adjusted calcium was 3.20 mmol/L (12.8 mg/dL; normal range 2.2–2.6 mmol/L [8.8–10.4 mg/dL]), serum phosphate 0.65 mmol/L (2.0 mg/dL; normal range 0.7–1.4 mmol/L [2.2–5.6 mg/dL]), and PTH 16.9 pmol/L (169 pg/mL); normal range 1.1-6.9 pmol/L (11-69 pg/mL).

Comment. Most patients with primary HPT are identified when asymptomatic hypercalcemia is discovered on 'routine' biochemical testing or during investigation of nonspecific symptoms. When symptomatic, primary HPT classically affects the skeleton, kidneys, and gastrointestinal tract, resulting in the well-recognized triad of complaints: 'bones, stones, and abdominal groans'. Renal stone disease is the most common single presenting complaint.

Hypercalcemia associated with malignancy (HCM)

Treatment of HCM

Treatment of hypercalcemia can greatly improve the quality of life of patients. Symptomatic hypercalcemia or serum calcium concentrations exceeding 3.00 mmol/L (12 mg/dL) would merit treatment. Dehydration results from hypercalcemia-induced polyuria, reduced fluid and food intake, associated vomiting, decreased arginine-vasopressin, AVP (also known as ADH) activity at the distal tubule, and reduced renal perfusion. Fluid replacement corrects hypovolemia and provides a moderate sodium load, which will cause a concomitant increase in calcium excretion. Bisphosphonates (Fig. 23.9) have improved the management of HCM, and are the most effective drugs available for treating hypercalcemia. This group of drugs have their major effects by inhibiting osteoclast activity immediately when infused, and then exert a more prolonged effect by being incorporated into bone matrix in a position normally occupied by pyrophosphate. Calcitonin also inhibits osteoclast activity directly, and decreases renal tubular reabsorption of calcium; downregulation of calcitonin receptors subsequently occurs, and so the calcium-decreasing effect lasts only 4–5 days.

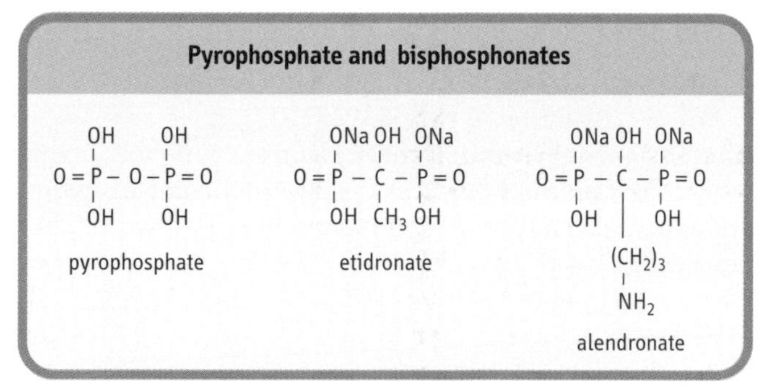

Fig. 23.9 Structural formulae of pyrophosphate and bisphosphonates. The P–C–P bonds of the biphosphonates can resist enzymic cleavage, and the potency of these drugs is determined by the sequence attached to the carbon molecule.

Causes of hypocalcemia

Hypoparathyroid	Nonparathyroid	PTH resistance
postoperative idiopathic acquired hypomagnesemia neck irradiation anticonvulsant therapy	vitamin D deficiency malabsorption liver disease renal disease vitamin D resistance hypophosphatemia	pseudohypoparathyroidism hypomagnesemia

Fig. 23.10 Causes of hypocalcemia.

Current evidence indicates that the most common cause of HCM is the production, by tumors or their metastases, of a humoral factor, PTHrP (see above), that can circulate in blood and exert its effects on the skeleton and kidneys. Production of PTHrP is common in breast, lung, kidney, or other solid tumors, but is more rare in hematologic, gastrointestinal, and head and neck malignancies. The amino-terminal portion of PTHrP possesses PTH-like activity that results in hypercalcemia, hypophosphatemia, phosphaturia, increased renal calcium reabsorption, and osteoclast activation.

The second type of hypercalcemia is the result of increased bone resorption by osteoclasts stimulated by factors produced by the primary tumor or, more usually, by metastases, in close proximity to the osteoclasts. Mediators such as PTHrP, cytokines, and growth factors (e.g. interleukin-1 [IL-1], tumor necrosis factor-α [TNF-α], lymphotoxin, and transforming growth factor-α [TGF-α]) have all been shown to possess osteoclast-stimulating activity that results in significant bone resorption. Production of prostaglandins, especially those of the E series (PGE_2), has been demonstrated in several classes of tumor, particularly breast cancer. Prostaglandins stimulate osteoclastic bone resorption, and infusion of high concentrations of prostaglandins results in hypercalcemia.

Vitamin D toxicity

Increasing therapeutic use of potent vitamin D analogues, hydroxylated at position 1 or at both positions 1 and 25, has had the consequence that vitamin D toxicity is the third most common cause of hypercalcemia.

Disorders of calcium metabolism and bone: Hypocalcemia

Many acute and chronic illnesses lead to a decrease in serum albumin, which in turn decreases serum total calcium concentration

As explained earlier, it is important that adjusted calcium is calculated in clinical situations. A serum adjusted calcium that is below the reference range (<2.20 mmol/L; 8.8 mg/dL) may occur commonly in clinical practice. Changes in ionized calcium can result from pH changes in plasma – particularly

Hypoparathyroidism

Magnesium in hypoparathyroidism
Patients with a low plasma magnesium concentration develop a state of functional hypoparathyroidism and end-organ resistance to PTH; these patients tend not to respond to treatment with calcium supplementation alone. Secretion of PTH from the parathyroid gland requires a finite amount of magnesium to allow fusion of the secretory granules with the membrane and release of PTH (see Fig. 23.4); thus, in order to restore normal parathyroid gland function and normocalcemia, magnesium supplementation is essential.

alkalemia, which increases the protein binding of calcium, causing decreased concentrations of ionized calcium.

Clinical signs of hypocalcemia are, in the main, due to neuromuscular irritability and are more obvious and severe when the onset of hypocalcemia is acute. In some cases, this irritability may be demonstrated by eliciting specific clinical signs. Chvostek's sign is the presence of twitching of the muscles around the mouth (circumoral muscles) in response to tapping the facial nerve anterior to the ear, and Trousseau's sign is the typical contraction of hand in response to reduced blood flow in the arm induced by inflation of a blood-pressure cuff. Seizures also may be stimulated by hypocalcemia.

Caues of hypocalcemia can be divided into those associated with low $PTH_{(1-84)}$, those in which the decreased serum calcium results in secondary HPT, and rare cases in which there is PTH resistance (Fig. 23.10). Development of hypocalcemia indicates that the normal physiologic compensatory mechanisms controled by PTH have failed and there are disturbed fluxes of calcium between bone, kidney, and intestine.

Hypoparathyroidism

The most common cause of hypoparathyroidism is as a complication of neck surgery, particularly that for pharyngeal or laryngeal tumors, thyroid disease, or parathyroid disease.

Pseudohypoparathyroidism

This group of syndromes are characterized by hypocalcemia, hyperphosphatemia, and increased concentrations of $PTH_{(1-84)}$ (markedly increased in some patients). The classical type of pseudohypoparathyroidism (PHP) is due to end-organ resistance to PTH, caused by a defective regulatory subunit of the G-protein of the adenylate cyclase complex. Confirmation of the diagnosis is given by a lack of increase in plasma or urinary cAMP in response to the infusion of PTH.

Abnormalities of Vitamin D metabolism

Hypocalcemia resulting from abnormalities in vitamin D metabolism can be due to vitamin D deficiency, acquired or inherited disorders of vitamin D metabolism, and vitamin D resistance. In addition, tissue insensitivity to $1,25(OH)_2 D_3$ results in the characteristic findings of hypocalcemia, secondary HPT, and increased $1,25(OH)_2 D_3$. Individually or in combination, the most common causes of vitamin D deficiency are:

- **reduced exposure to sunlight:** this is common in institutionalized elderly and immigrants to western Europe from the Middle East or Indian subcontinent who wear traditional dress;
- **poor dietary intake:** most Western diets contain sufficient vitamin D, but specialist diets (strict vegetarian, lactovegetarian) have inadequate vitamin D and, in the long term, may result in deficiency;
- **malabsorption** may be caused by celiac disease, Crohn's disease, pancreatic insufficiency, inadequate bile-salt secretion, and nontropical sprue (see Chapter 9).

Abnormalities of vitamin D metabolism are observed, among others, in patients with liver disease (deficient 25-hydroxylation, malabsorption), and renal failure (1α-hydroxylase deficiency).

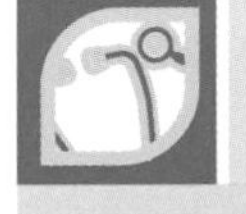

Osteoporosis

A 62-year-old woman was admitted to hospital because of sudden onset of severe back pain between the shoulder blades after a fall in her bathroom. Radiography detected two fractured thoracic vertebrae and generalized demineralization of her skeleton. She had experienced the menopause when aged 41 years, after a hysterectomy, but had been unable to tolerate hormone replacement therapy (HRT). Biochemical investigations were all within normal limits.

Comment. Symptoms of osteoporosis develop at a late stage of the disease, and are often caused by the presence of fracture(s). Hip, vertebral, and wrist fractures are common in patients with osteoporosis, and result in significant mortality and morbidity. HRT can be used to prevent the development of bone loss in postmenopausal women.

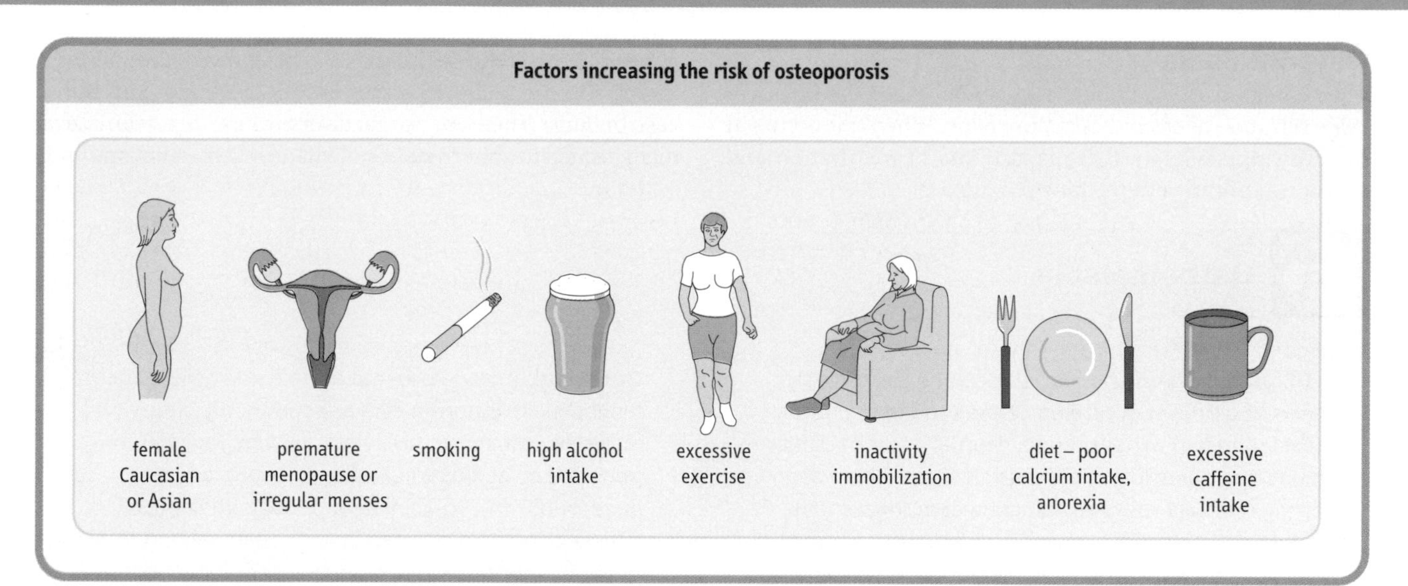

Fig. 23.11 Risk factors and secondary causes of osteoporosis.

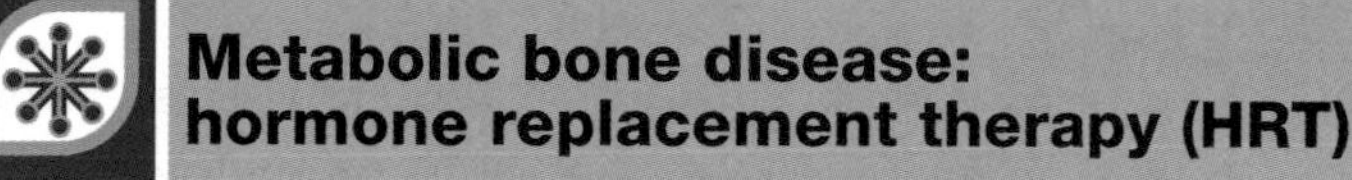

Metabolic bone disease: hormone replacement therapy (HRT)

Treatment of osteoporosis

Estrogen has a pivotal role in the maintenance of bone mass in women. An increased rate of osteoclast-mediated bone loss is observed immediately after the menopause, as a result of ovarian failure and the decrease in estrogen secretion. This can be prevented by HRT, which can be given as cyclical estrogen and progesterone in women with an intact uterus, or as unopposed estrogen in women who have had their uterus surgically removed (by hysterectomy).

Other treatments are biphosphonates, calcitonin, and, in some patients, calcium and vitamin D supplementation.

Metabolic bone disease

Osteoporosis

Osteoporosis can be defined as 'a significant reduction of bone mineral density compared with age- and sex-matched norms, with an increased susceptibility to fractures'

Osteoporosis is a disease of aging and, with increasing longevity, a greater percentage of the population will become susceptible to osteoporosis and its sequelae. Bone density decreases from a peak achieved by the age of 30 years in men and women, and the rate of bone loss is accelerated in women after loss of estrogen secretion at the menopause. The progressive loss of bone that takes place with aging is a result of uncoupling of bone turnover over a prolonged period of time, with a relative increase of bone resorption or decrease in bone formation. A number of factors have been recognized as contributing to an increased risk of osteoporosis (Fig. 23.11).

There is much debate on the best therapeutic approach to osteoporosis, and many questions are raised about the possibility of detecting 'fast bone-losers'. Treatments that are available currently have the effect of decreasing osteoclast activity, and are therefore antiresorptive.

Paget's disease of bone

Paget's disease of bone is a localized disorder characterized by accelerated bone turnover initiated by increased osteoclast-mediated bone resorption

Osteoclasts in Paget's disease are large, numerous, and multinucleate (up to 100 nuclei); their activity is coupled to increased osteoblast number and activity. A common biochemical abnormality of the disease is increased alkaline phosphatase, which is indicative of increased osteoblast activity. Increased collagen breakdown by osteoclasts results in a high urinary concentration of hydroxyproline and collagen fragments. The bisphosphonates (see Fig. 23.9) have significant antiosteoclastic activity, and are the drugs of first choice for treating Paget's disease.

Osteomalacia

The characteristic pathology in osteomalacia is the defective mineralization of osteoid in mature bone. When this occurs in the growing skeleton, there is also loss of maturation and mineralization of the cartilage cells at the growth plate; the term 'rickets' is used to describe the clinical, radiologic, and pathologic findings. There are several diverse causes of osteomalacia, many related to abnormalities of vitamin D metabolism.

Osteomalacia

A 60-year-old woman who had become increasingly infirm and housebound was referred to the Metabolic Clinic. She had experienced a gradual onset of diffuse aches and pains throughout her skeleton. She was having difficulty walking, had generalized weakness, and had recently experienced sudden onset of acute pain in her ribs and pelvis. Radiography detected fractured ribs. Adjusted serum calcium was 2.1 mmol/L (8.4 mg/dL; normal range 2.2–2.6 mmol/L [8.8–10.4 mg/dL]), serum phosphate 0.56 mmol/L (1.7 mg/dL; normal range 0.7–1.4 mmol/L [2.2–4.3 mg/dL]), alkaline phosphatase 190 IU/L (normal range 25–125 IU/L) and PTH 12.6 pmol/L (normal range 1.1–6.9 pmol/L [11-69 pg/mL]).

Comment. In severe forms of osteomalacia, biochemical abnormalities are commonly seen, including low serum adjusted calcium, low serum phosphate, increased alkaline phosphatase and increased $PTH_{(1-84)}$. Clinically, patients may have diffuse bone pain or more specific pain related to a fracture, lateral bowing of the lower limbs, and a distinctive waddling gait. Ethnic groups with dark skin are particularly at risk in countries with low-average sunlight, as the majority of vitamin D in the body comes from synthesis by the action of UV light on 7-dehydrocholesterol. This may be exacerbated by the traditional dress and a diet that is high in phytates (unleavened bread) and low in calcium and vitamin D.

Summary

Bone is a metabolically active tissue which undergoes constant remodeling. Bone metabolism is closely interrelated with the metabolism of calcium which also involves the intestine and kidney. The calcium balance is hormonally regulated by parathormone, vitamin D metabolites and calcitonin. The measurement of calcium in serum is an important test in the clinical laboratories, because both hypercalcemia and hypocalcemia lead to clinical symptoms. Osteoporosis, a decrease in bone density leading to bone fractures, is a major health problem.

Further reading

Avioli LV, Krane SM, eds. *Metabolic bone disease*, 2nd ed. Philadelphia: WB Saunders; 1990.

Favus MJ, ed. *Primer on the metabolic bone disease and disorders of mineral metabolism*, 3rd ed. Philadelphia, New York: Lippincott-Raven; 1996.

Fleisch H. Bisphosphonates in bone disease. *From the laboratory to the patient*, 3rd ed. New York, London: The Parthenon Publishing Group; 1997.

Martin TJ, Mossley JM, Williams ED. Parathyroid hormone-related protein: hormone and cytokine. *J Endocrinol* 1997;**154**(suppl):S23–S37.

Mundy GR. Bone remodelling and its disorders. London: Martin Dunitz; 1995.

Ooi CG, Fraser WD. Paget's disease of bone. *Postgrad Med J* 1997;**73**:69–74.

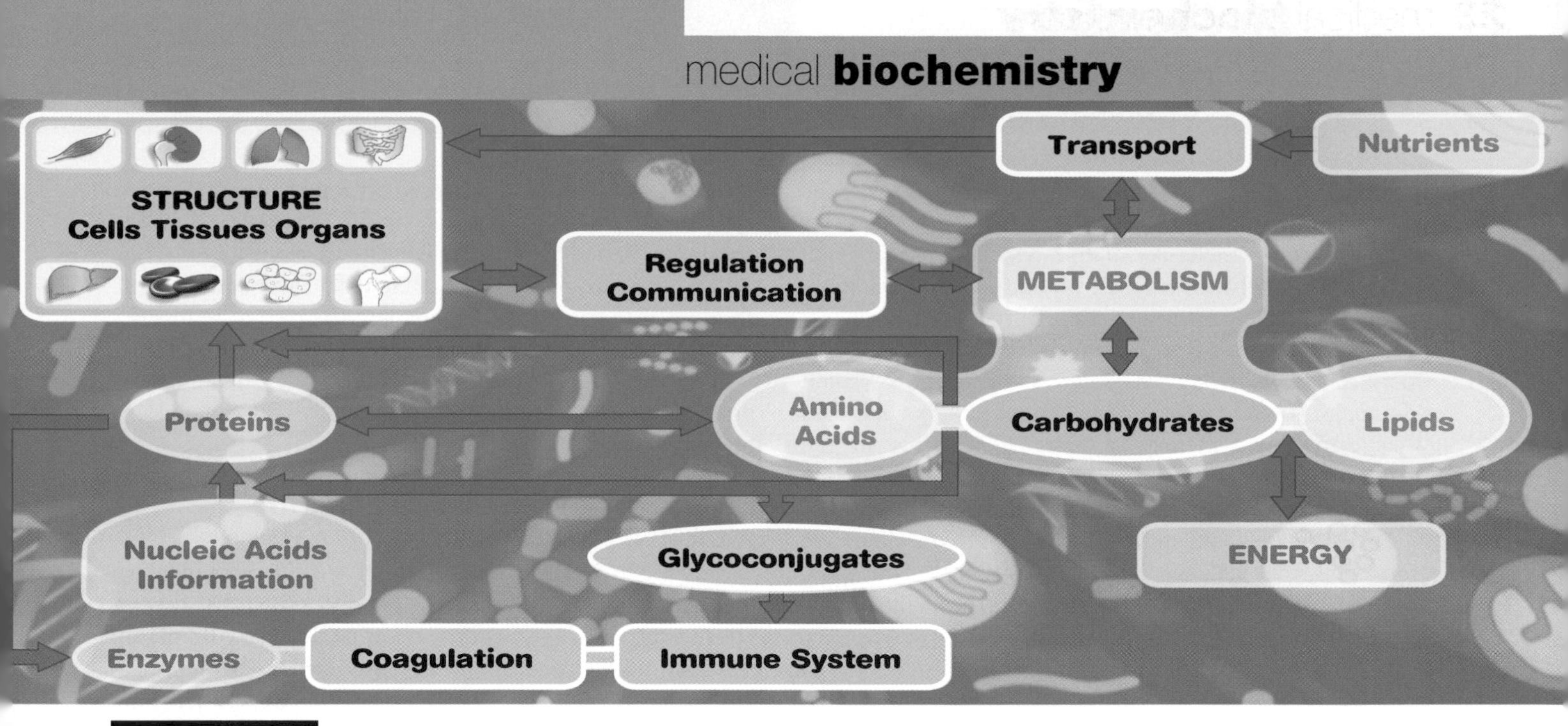

24 Complex Carbohydrates: Glycoproteins

Introduction

Most mammalian proteins contain covalently attached sugars – i.e., they are glycoproteins. There are two distinct types of sugar-containing proteins that occur in animal cells: glycoproteins and proteoglycans. Along with glycolipids, which are discussed in the next chapter, they are part of the group of sugar-containing molecules that are called glycoconjugates. Figure 24.1 presents models of the structure of glycoproteins and proteoglycans in order to demonstrate the differences between them: glycoproteins have short oligosaccharide (glycan) chains (1–20 sugars in length), which are highly branched and which generally do not have a repeating sequence. On the other hand, proteoglycans are long, linear, unbranched glycans that have a disaccharide repeating unit. This chapter will focus on the glycoproteins; Chapter 26 will consider the proteoglycans.

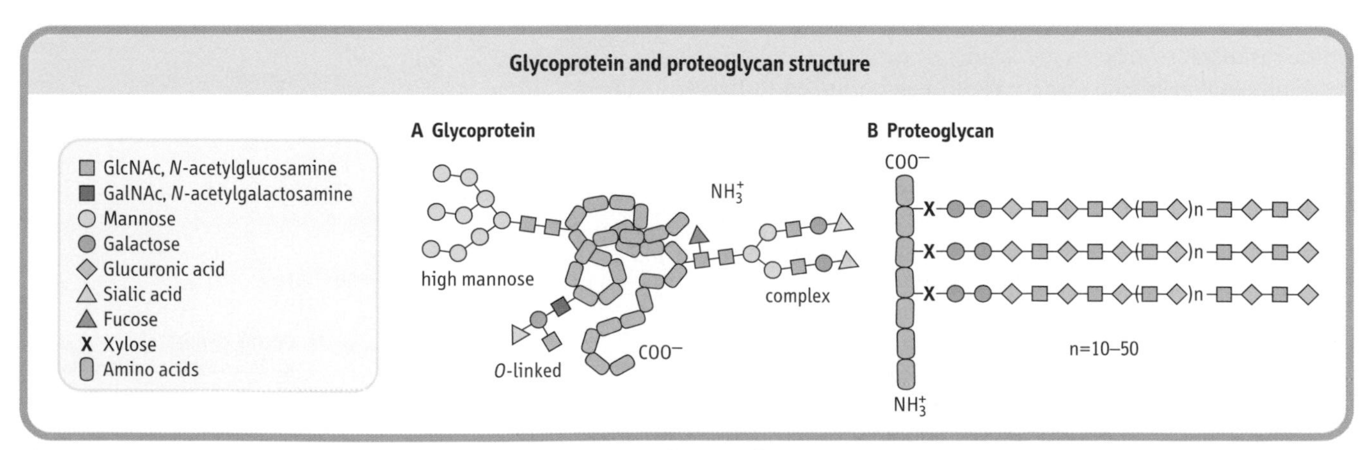

Fig. 24.1 Generalized model of the structure of glycoproteins and proteoglycans.

Most of the proteins that are integral components of the plasma membrane and that function as receptors for hormones or other molecules in the circulation, or that mediate interactions between cells, are glycoproteins. In addition, many of the proteins of the endoplasmic reticulum and Golgi apparatus, and those that are secreted by cells, including serum and mucous proteins, are glycoproteins. Indeed, glycosylation is the major postsynthetic modification of proteins; it occurs either during the course of protein synthesis in the endoplasmic reticulum or once the protein has been synthesized and transported to the Golgi apparatus. The functions of the carbohydrate chains of the resulting glycoproteins are diverse: they may stabilize the protein against denaturation, protect it from proteolytic degradation, enhance its solubility, or serve as recognition signals to facilitate cell–cell interactions.

Structures and linkages

Sugars are attached to specific amino acids in proteins

Sugars may be attached to protein in either *N*-glycosidic or *O*-glycosidic linkage. *N*-linked oligosaccharides, characteristic of plasma and membrane proteins, are always attached by a glycosylamine linkage of *N*-acetylglucosamine (GlcNAc) to the amide nitrogen of an asparagine residue (Fig. 24.2A). The asparagine at the site of glycosylation must be in an–Asn–X–Ser (Thr)–consensus sequence in order to undergo glycosylation. However, not all asparagine residues present in that consensus sequence are glycosylated, indicating that other factors, such as the conformation of the protein around the glycosylation site, are also important.

Mucins, collagens, and cytoplasmic proteins may be glycosylated

Many membrane proteins and proteins in mucous secretions (mucins) contain oligosaccharides linked by a glycosidic linkage between *N*-acetylgalactosamine (GalNAc) and the hydroxyl group of serine or threonine residues (Fig. 24.2B). There is no known consensus sequence that indicates which serine residues are to be glycosylated.

A glucosyl-galactose disaccharide may be linked to the hydroxyl group of hydroxylysine residues in collagen (Fig. 24.2C). Hydroxylysine is an unusual amino acid that occurs in mammalian collagens; it is formed by hydroxylation of lysine after this amino acid has been incorporated into the collagen chain. Collagen is first synthesized in the cell in a precursor form called procollagen. Procollagens are usually *N*-glycosylated glycoproteins, but peptide segments containing *N*-linked oligosaccharides are removed during maturation of the protein, and only the *O*-linked glycans remain in mature collagen. The less glycosylated collagens tend to form ordered, fibrous structures, such as occur in tendons; the more heavily glycosylated collagens are found in meshwork structures, such as basement membranes (chapter 26).

Finally, a recently discovered structure found in many cytoplasmic proteins involves a single GlcNAc that is *O*-glycosidically linked to serine residues in protein (Fig. 24.2D). This GlcNAc appears to be attached to the same serine residues that become phosphorylated as part of regulatory and signal transduction processes. The GlcNAc may represent a means by which the cell blocks or controls the phosphorylation of certain proteins, while allowing others to be phosphorylated.

Structures of the oligosaccharides attached to proteins

Sugars are linked to each other in glycosidic linkages. Two identical sugar residues can be linked together in many different ways (α1,2; α1,3; α1,4; α1,6; β1,2; β1,3; β1,4; β1,6), yielding distinct disaccharides; in contrast, two identical amino acids can only be combined in one way – two alanine

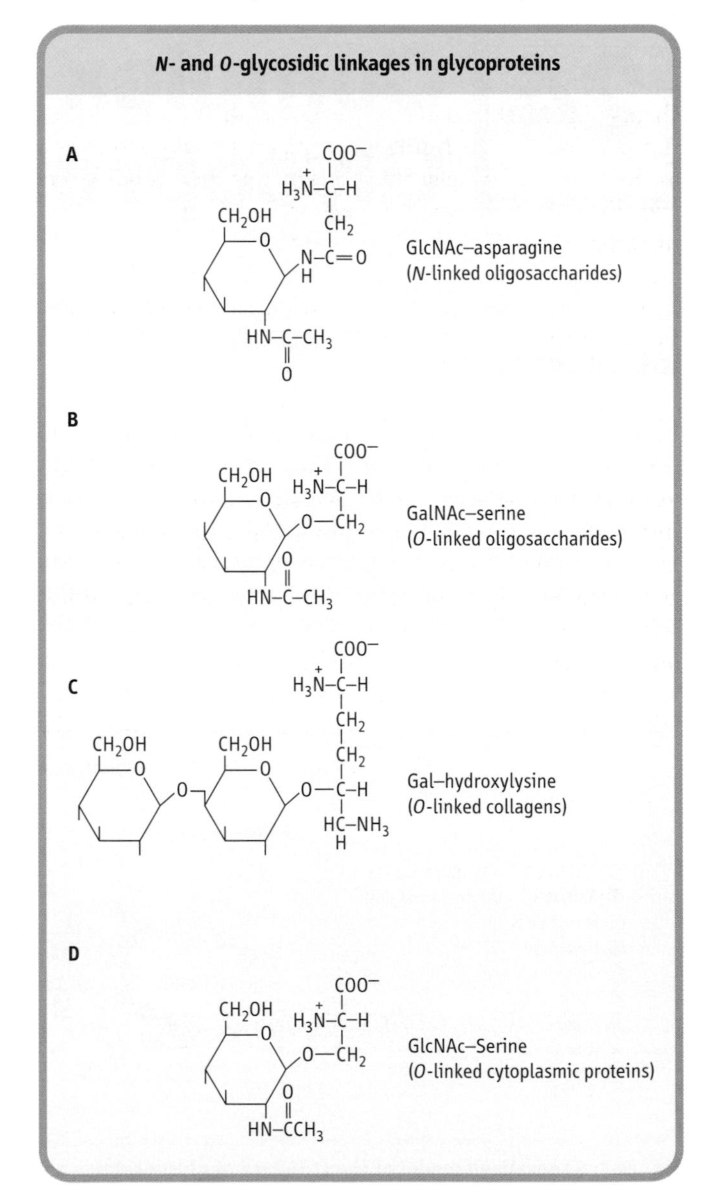

Fig. 24.2 Various linkages of sugars to amino acids in glycoproteins.

residues can form only one dipeptide, alanyl-alanine. Thus carbohydrate structures have greater structural diversity, and therefore have the potential to contain more information, than proteins of similar size. Nevertheless, although there are a large number of structures produced by living cells, most of the oligosaccharides on glycoconjugates have many sugars and linkages in common.

N-linked oligosaccharides have either 'high-mannose' or 'complex' structures built on to a common core

All the *N*-linked oligosaccharide chains that are found on proteins are, typically, branched structures having a common core structure of three mannose units and two GlcNAc residues (shown by the boxed areas in Fig. 24.3). Beyond this core region, the oligosaccharides can be very different from each other, giving rise to a vast array of structures. Thus oligosaccharides may be either 'high-mannose' structures containing only two GlcNAc residues and up to nine mannose units (Fig. 24.3A), or they may consist of 'complex' chains (Fig. 24.3B), so named because of their more complex composition. All the *N*-linked chains are initially assembled as the high-mannose structure, which is then modified to give rise to different types of complex oligosaccharides. High-mannose oligosaccharides are found to a limited extent in animal glycoproteins; they are more common in glycoproteins of lower eukaryotes and in viral envelope glycoproteins. Complex oligosaccharides are common in animals, but are generally not present in lower eukaryotes.

Complex oligosaccharides have the same core structure as the high-mannose oligosaccharides, but have terminal trisaccharide sequences composed of sialic acid–galactose–GlcNAc attached to the core mannose structure. Fucose may be found in the core (see Fig. 24.1A) or in place of sialic acid in complex oligosaccharides. In common with sialic acid, fucose is usually a terminal sugar on oligosaccharides – that is, no other sugars are attached to it. Some complex oligosaccharides have two of the terminal trisaccharide sequences (one attached to each mannose) and are called 'biantennary' complex chains, whereas others have tri- or tetra-antennary structures (Fig. 24.3B). 'Microheterogeneity' of oligosaccharide structure results from the fact that the basic structure is often found in an incomplete form on glycoproteins. Thus, a given glycoprotein may have several *N*-linked oligosaccharides and these may have the same structure or different structures. Generally, oligosaccharides located near the amino terminus are more highly processed (complex types) whereas those near the carboxy terminus are more likely to be high-mannose types. More than 100 different complex oligosaccharide structures have now been identified, giving carbohydrates great diversity as mediators of chemical signaling and recognition events.

The viscous properties of mucins derive from their content of negatively charged sialic acid

In mucins, the *O*-linked oligosaccharides are usually short, branched structures containing sialic acid (*N*-acetylneuraminic acid), galactose, and GalNAc, and sometimes other sugars such as GlcNAc and L-fucose (Fig. 24.4). Salivary mucin contains an unusually large number of serine or threonine residues, and many of these serines or threonines are glycosylated with a sialic acid–galactose–GalNAc trisaccharide. The *O*-linked oligosaccharides are negatively charged because of the presence of the sialic acid residues, and when they occur in clusters and in close proximity to each other, they repel each other and prevent the protein from folding. As a result, the protein assumes an extended state, yielding a highly viscous (mucous) solution. Mucins form a protective barrier on the surface of epithelial cells, provide lubrication between surfaces, and facilitate transport processes – for example, the movement of food through the gastrointestinal tract.

General structures of glycoproteins

A glycoprotein may contain a single *N*-linked oligosaccharide chain, or it may have several of these types of oligosaccharides. Furthermore, the *N*-linked oligosaccharides may all have identical structures, or they may be quite different in structure; such a glycoprotein may also contain *O*-linked oligosaccharide chains. Alternatively, a glycoprotein may contain only *N*-linked oligosaccharides, or only *O*-linked

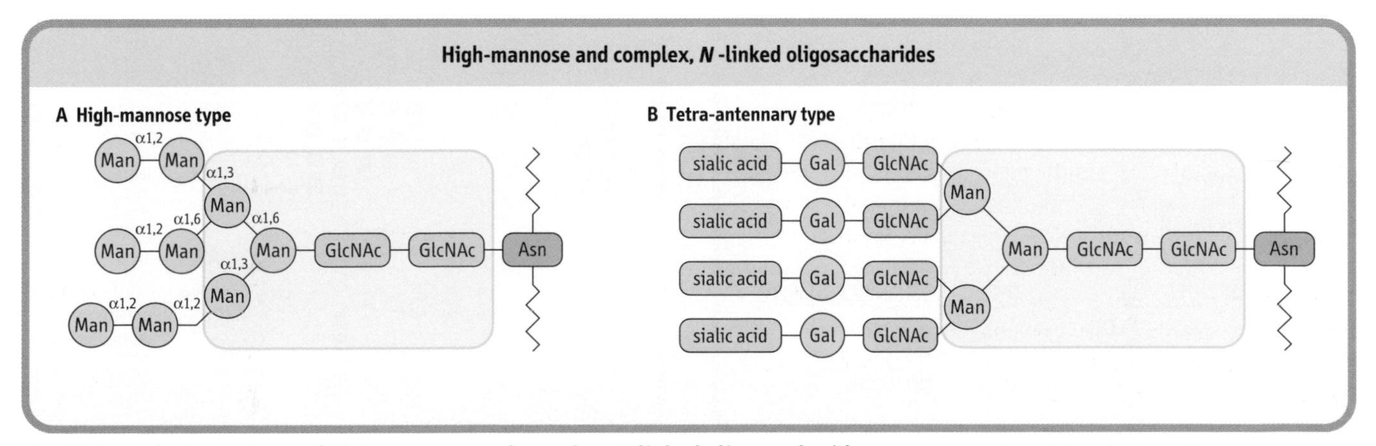

Fig. 24.3 Typical structures of high-mannose and complex, *N*-linked oligosaccharides. Asn, asparagine; Gal, galactose; Man, mannose.

oligosaccharides. The total number of oligosaccharide chains may also vary considerably, depending on the protein and its function. For example, the low density lipoprotein receptor that is found in plasma membranes of smooth muscle cells and fibroblasts contains two *N*-linked biantennary complex chains, in addition to a cluster of *O*-linked chains near the membrane-spanning region. This receptor has a membrane-spanning region of hydrophobic amino acids, an extended region on the external side of the plasma membrane that contains the cluster of *O*-linked, negatively charged oligosaccharides, and a functional domain that is involved in binding serum low-density lipoprotein (Fig. 24.5). The two *N*-linked oligosaccharides are near the functional domain. It remains unclear what role these *N*-linked chains play in the function of this receptor, but the *O*-linked oligosaccharides are believed to keep the molecule in an extended state.

Interconversions of dietary sugars

Glucose is a precursor of all sugars in the body

Humans have a requirement for some essential fatty acids and amino acids, but all the sugars in glycoconjugates can be synthesized from D-glucose. Figure 24.6 shows the reactions that occur in animal cells to convert glucose to mannose or galactose. The latter is achieved by epimerization of the nucleoside diphosphate sugar, uridine diphosphate-glucose (UDP-Glc), to UDP-galactose (UDP-Gal) by UDP-Gal 4-epimerase, providing a source of UDP-Gal for glycoconjugate biosynthesis. Glucose-6-phosphate (G-6-P), which is an early intermediate in that pathway, may also be converted to fructose-6-phosphate (F-6-P) by the glycolytic enzyme, phosphoglucose isomerase, and F-6-P can then be isomerized to mannose-6-phosphate (Man-6-P) by phosphomannose isomerase. Man-6-P is then converted by a mutase to mannose-1-phosphate (Man-1-P), which reacts with guanosine triphosphate (GTP) in a pyrophosphorylase reaction to form guanosine diphosphate-mannose (GDP-Man), the activated form of mannose. Mannose also occurs in the diet, although in small amounts; it is phosphorylated by hexokinase and enters metabolism through phosphomannose isomerase.

Galactose

Galactose is an important component of our diet, because it is one of the sugars in the milk disaccharide, lactose. The pathway of galactose metabolism and its conversion to glucose is fairly complex (see Fig. 24.6). Galactose is first phosphorylated by a specific hepatic kinase, galactokinase, to form galactose-1-phosphate (Gal-1-P). The conversion of Gal-1-P to G-1-P involves the nucleoside diphosphate sugar intermediate, UDP-Glc. In this reaction, the enzyme Gal-1-P uridyl transferase catalyzes an exchange between UDP-Glc and Gal-1-P to form UDP-Gal and G-1-P – that is, the G-1-P part of UDP-Glc is replaced with Gal-1-P, to give UDP-Gal and G-1-P. This enzyme is absent in individuals with galactosemia.

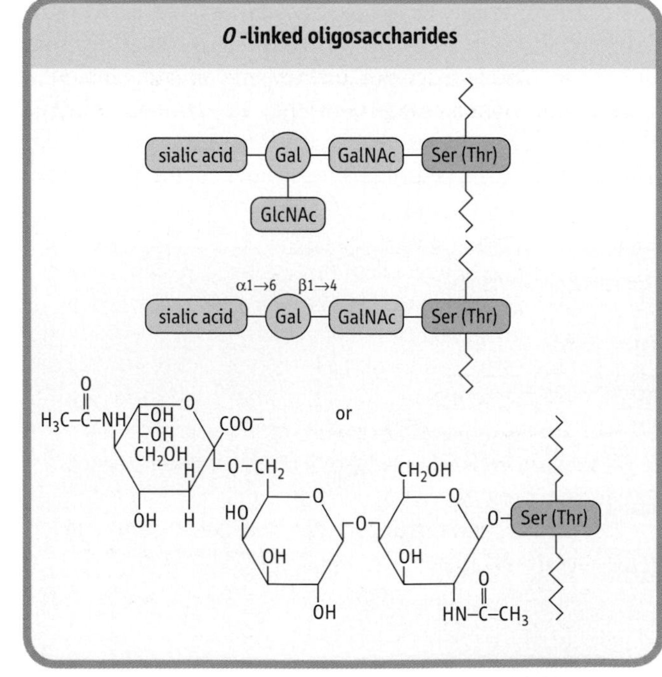

Fig. 24.4 Typical structures of *O*-linked oligosaccharides. Thr, threonine.

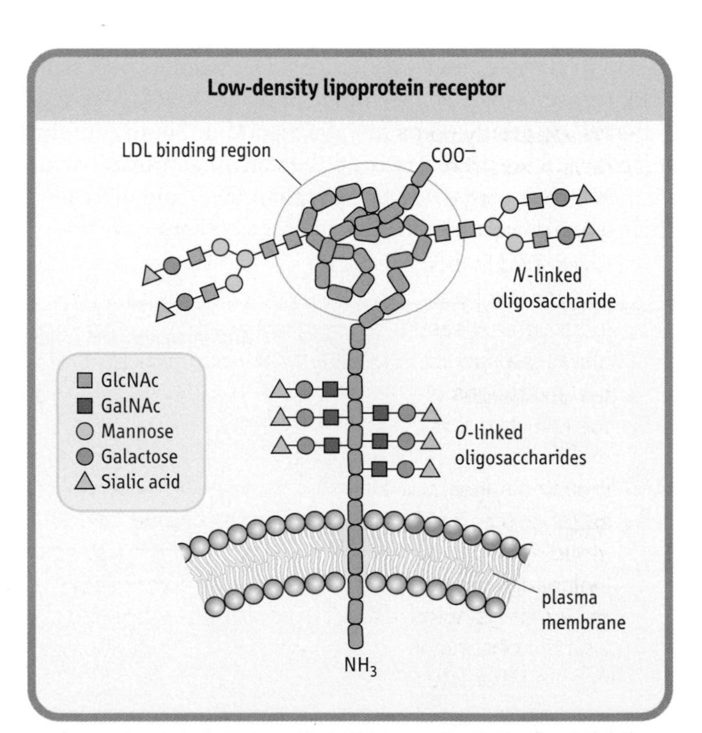

Fig. 24.5 Model of the low density lipoprotein (LDL) receptor.

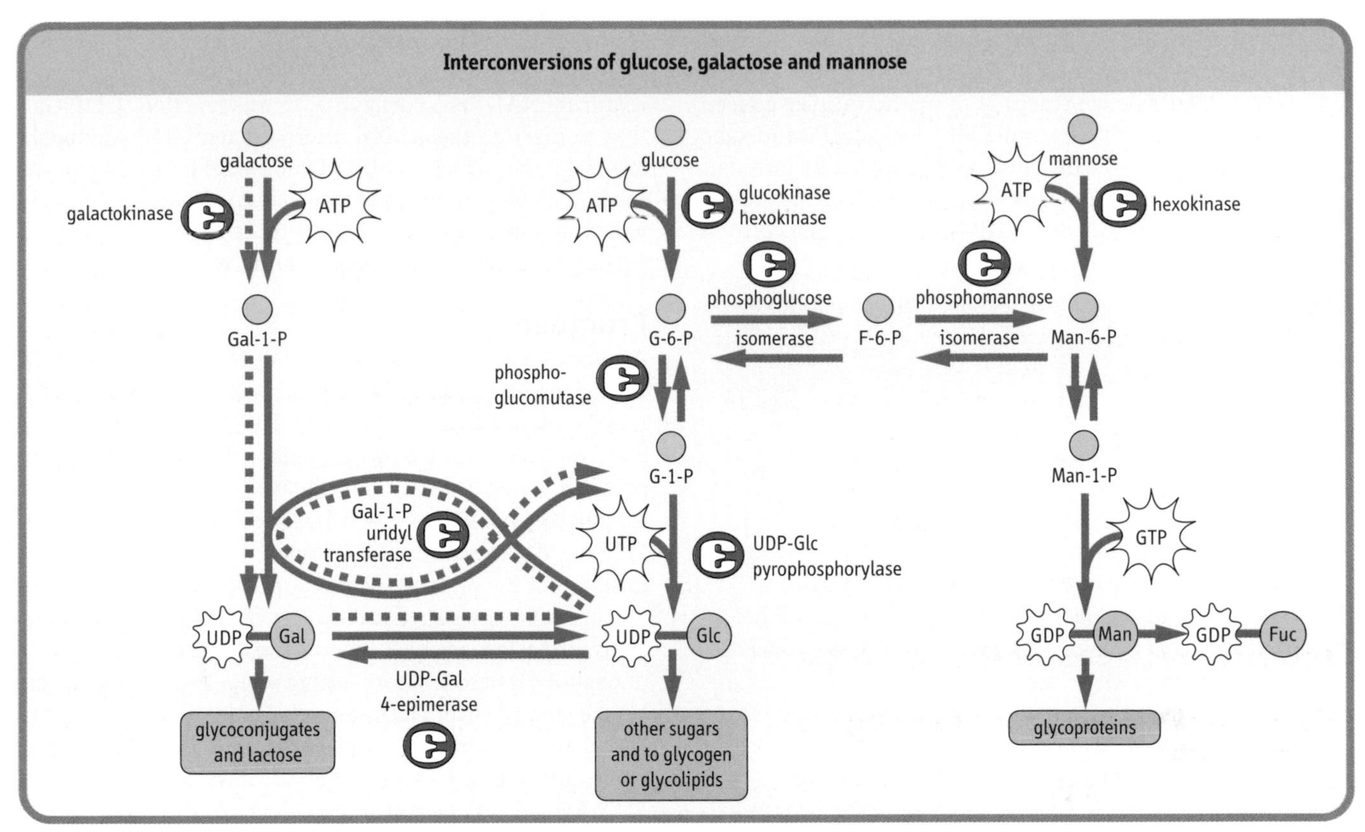

Fig. 24.6 Interconversions of glucose, mannose, galactose, and their nucleotide sugars. The dotted lines in the metabolism of galactose show the pathway by which dietary galactose is converted to G-1-P for energy metabolism. Fuc, fucose.

Galactosemia

An apparently normal baby began to vomit and develop diarrhea after breast feeding. These problems, together with dehydration, continued for several days, when the baby began to refuse food and developed jaundice, indicative of liver damage, followed by hepatomegaly, and then lens opacification (cataracts). Measurement of glucose in the blood and urine by a specific enzymatic technique indicated that concentrations of glucose were low, consistent with the failure to absorb foods. However, glucose measured by a colorimetric test that determined total reducing sugar indicated that the concentration of sugar was quite high in both blood and urine. The reducing sugar was eventually identified as galactose, indicating an abnormality in galactose metabolism known as galactosemia. This finding was consistent with the observation that, when milk was removed from the diet and replaced with an infant formula containing sucrose rather than lactose, the vomiting and diarrhea stopped, and hepatic function was gradually restored.

Comment. The accumulation of galactose in the blood is most often a result of a deficiency of Gal-1-P uridyl transferase, which prevents the conversion of galactose to glucose and leads to the accumulation of galactose and Gal-1-P in tissues. Accumulation of the latter interferes with phosphate and glucose metabolism, leading to widespread tissue damage, organ failure, and mental retardation. In addition, accumulation of galactose in tissues results in galactose conversion via the polyol pathway to galactitol, and the accumulation of galactitol in the lens results in osmotic stress and formation of cataracts. A milder form of galactosemia is caused by galactokinase deficiency.

The G-1-P arising from galactose metabolism can be converted to G-6-P by phosphoglucomutase, and thus enter glycolysis. UDP-Glc is present at only micromolar concentrations in cells, so that its availability for galactose metabolism would be quickly exhausted were it not for the presence of UDP-Gal 4-epimerase. This enzyme catalyzes the equilibrium between UDP-Glc and UDP-Gal, providing a constant source of UDP-Glc during galactose metabolism. UDP-Gal 4-epimerase is actually a dehydrogenase, a redox enzyme, requiring NAD^+ as a coenzyme. In this reaction, UDP-Gal is first oxidized to the achiral intermediate, UDP-4-ketogalactose, with the reduction of NAD^+ to NADH. The 4-keto intermediate is then reduced by the enzyme-bound NADH, but with a change in the stereochemistry of the 4-hydroxyl group, producing UDP-Glc and regenerating NAD^+.

Biosynthesis of lactose

Biosynthesis of lactose

Lactose synthase and α-lactalbumin. Lactose (galactosyl-β1,4-glucose) is synthesized from UDP-Gal and glucose in mammary glands during lactation. Lactose synthase is formed by the binding of α-lactalbumin to the galactosyl transferase that normally participates in biosynthesis of *N*-linked glycoproteins. α-Lactalbumin, which is expressed only in the mammary glands during lactation, converts galactosyl transferase to lactose synthase by reducing its K_m for glucose by about three orders of magnitude, from 1 mol/L to 1 mmol/L, leading to preferential synthesis of lactose. α-Lactalbumin is the only known example of a 'specifier' protein that alters the substrate specificity of an enzyme.

Fructose

Another common sugar in the diet is fructose, which is a component of the disaccharide sucrose, table sugar. Fructose may be metabolized by two pathways in cells (Fig. 24.7). It may be phosphorylated by hexokinase, an enzyme that is present in all cells; however, hexokinase has a strong preference for glucose, and glucose, which is present at about 5 mmol/L (100 mg/dL) concentration in blood, is a strong competitive inhibitor of the phosphorylation of fructose. The other pathway of fructose metabolism involves fructokinase and is especially important in liver after a meal. In liver, fructose is phosphorylated to fructose-1-phosphate (F-1-P) by a specific kinase, and the liver aldolase, called aldolase b, can cleave F-1-P, as well as fructose-1,6-bisphosphate (F-1,6-BP), alternatively, muscle aldolase, called aldolase a, is specific for F-1,6-BP. In this case, the products are dihydroxyacetone phosphate and glyceraldehyde (not glyceraldehyde phosphate). The glyceraldehyde must then be phosphorylated by triose kinase in order to be metabolized in glycolysis.

It should be noted that in the liver, fructose enters glycolysis at the level of triose phosphate intermediates, after the

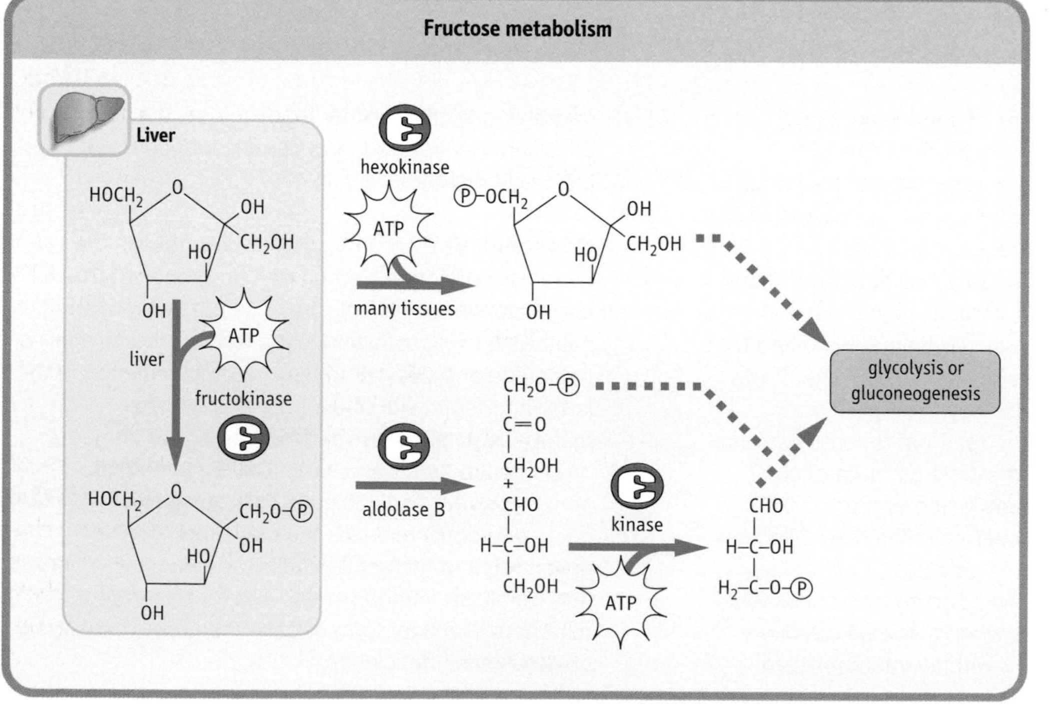

Fig. 24.7 Metabolism of fructose by fructokinase or hexokinase.

control points for the regulatory enzymes, hexokinase and phosphofructokinase-1 (PFK-1). By circumventing these two rate-limiting and regulatory enzymes, fructose provides a rapid source of energy in both aerobic and anaerobic cells. This is part of the rationale behind the development of high-fructose drinks, such as Gatorade®. The significance of the fructokinase, as opposed to the hexokinase, pathway of fructose metabolism is indicated by the pathology of hereditary fructose intolerance, which results from genetic defects in fructokinase and aldolase B. In patients with such defects, fructose or F-1-P accumulates in the body, leading to problems similar to those seen in galactosemia. Fortunately, both galactosemia and hereditary fructose intolerance can be managed by removing galactose or fructose, respectively, from the diet.

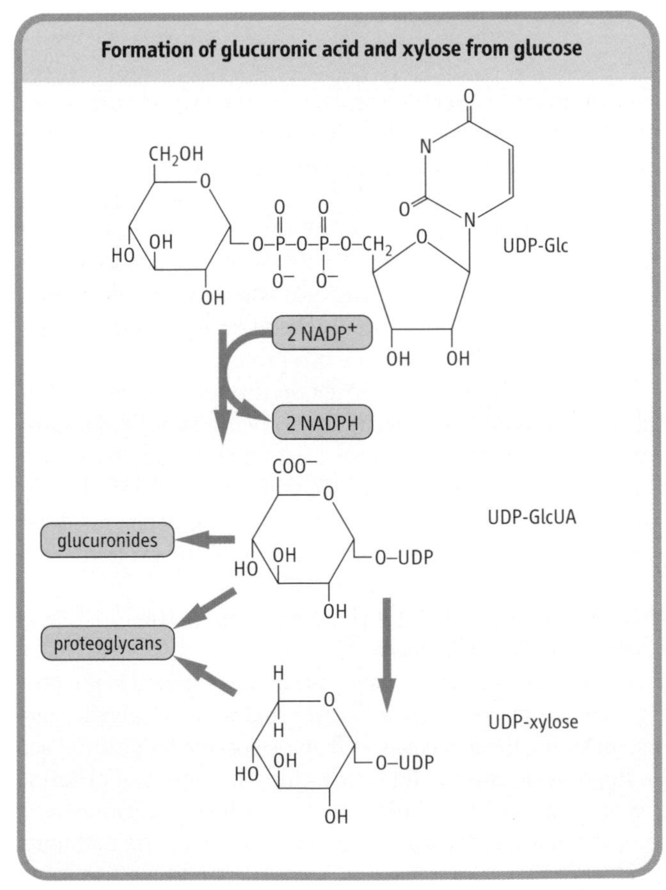

Fig. 24.8 Conversion of UDP-Glc to UDP-glucuronic acid (UDP-GlcUA) and UDP-xylose.

Other pathways of sugar nucleotide metabolism

UDP-glucose

UDP-Glc is the precursor of other essential sugars, such as glucuronic acid and xylose, which are required for proteoglycan biosynthesis. The reactions that lead to the formation of these sugars are outlined in Figure 24.8. Two-step oxidation of UDP-Glc by UDP-Glc dehydrogenase leads to the activated form of glucuronic acid, i.e. UDP-glucuronic acid (UDP-GlcUA). This nucleotide is the donor of glucuronic acid both for the formation of proteoglycans (see Chapter 26) and for conjugation and detoxification reactions that occur in the liver. In that organ, glucuronic acid is conjugated to steroid hormones, bilirubin (a degradation product of heme), and many drugs. The conjugation reaction increases the water solubility of hydrophobic compounds, facilitating their excretion in urine. UDP-GlcUA undergoes a decarboxylation reaction to form UDP-xylose, the activated form of xylose, the linkage sugar between protein and glycan in proteoglycans.

Guanosine diphosphate-mannose

GDP-Man is not only the mannose donor for glycoprotein synthesis, but it is also the precursor of L-fucose, a deoxyhexose that is an important recognition signal in many glycoproteins and glycolipids. The conversion of GDP-D-Man to GDP-L-fucose involves a complex series of oxidative and reductive steps, as well as epimerizations.

Fructose-6-phosphate (F-6-P)

F-6-P is the common precursor of amino sugars, which are found in almost all glycoconjugates. Figure 24.9 shows the pathway of formation of GlcNAc, GalNAc, and sialic acid. The initial reaction involves transfer of an amino group from glutamine to F-6-P, to form glucosamine-6-phosphate (GlcN-6-P). This is followed by transfer of acetate from acetyl Coenzyme A to the amino group to form *N*-acetylglucosamine-6-phosphate (GlcNAc-6-P). GlcNAc-6-P is converted to its activated form, UDP-GlcNAc, by mutase and pyrophosphorylase reactions. In addition to its role as a GlcNAc donor, UDP-GlcNAc can also be epimerized to UDP-GalNAc. With few exceptions, all amino sugars in glycoconjugates are acetylated; thus they are neutral and do not contribute charge to glycoconjugates.

UDP-*N*-acetylglucosamine

UDP-GlcNAc is the precursor of *N*-acetyl neuraminic acid (NANA) or sialic acids. These are 9-carbon sugar acids that are common components of *N*- and *O*-linked oligosaccharides. Sialic acid is produced by the condensation of an amino sugar with phosphoenolpyruvate. Cytidine monophosphate (CMP)-sialic acid is the sugar nucleotide donor for glycoconjugate biosynthesis. It is formed by reaction of free sialic acid with cytidine triphosphate (CTP) (see Fig. 24.9). CMP-sialic

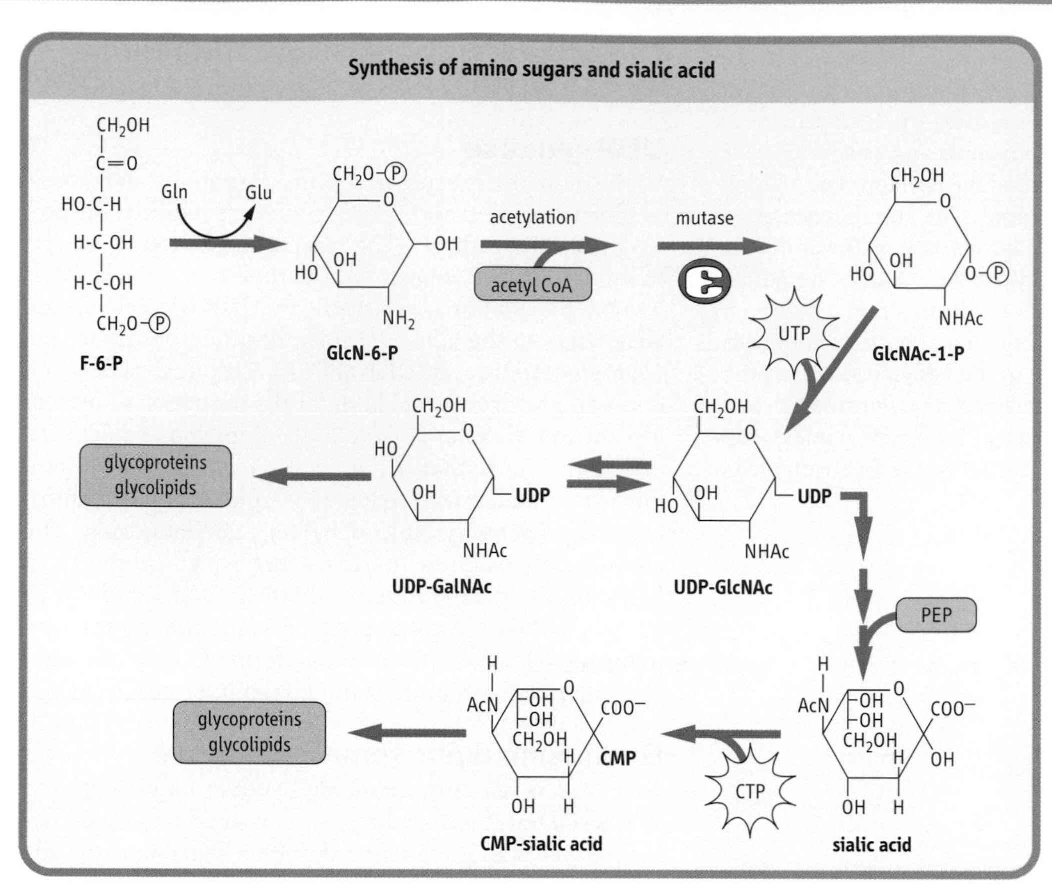

Fig. 24.9 Synthesis of amino sugars and sialic acid. Acetyl CoA, acetyl Coenzyme A; HNAc, AcHN, acetamide group; PEP, phosphoenolpyruvate.

acid is the only nucleotide-monophosphate sugar donor in glycoconjugate metabolism.

Biosynthesis of oligosaccharides

N-linked oligosaccharides

The assembly of the N-linked oligosaccharide chains begins in the endoplasmic reticulum

The pathway of assembly of *N*-linked oligosaccharides involves the participation of a lipid carrier to form an activated oligosaccharide, which is then transferred 'en bloc' to specific asparagine residues on the protein (Fig. 24.10). The lipid carrier is a long-chain polyisoprenol of about 120 carbon atoms, with a phosphate group esterified to the terminal isoprene unit. This molecule, called dolichol phosphate (Dol-P), is an integral lipid of the endoplasmic reticulum. The individual sugars – GlcNAc, mannose and glucose – are added to Dol-P, one at a time, to form $Glc_3Man_9GlcNAc_2$, which is then transferred to protein. For *N*-linked oligosaccharides, glycosylation is cotranslational, which means that it occurs while the peptide chain is still being synthesized on the membrane-bound ribosome (see Chapter 31).

The first GlcNAc residue is added with its phosphate residue, so that the core structure is dolichyl pyrophosphate-GlcNAc (Dol-PP-GlcNAc), a hydrophobic, membrane-bound analogue of a sugar nucleotide. The remaining GlcNAc and five mannose units are transferred from their sugar nucleotides (UDP-GlcNAc and GDP-Man), whereas the next four mannose and three glucose residues are donated from lipid precursors (Dol-P-mannose and Dol-P-glucose). Each of the sugars that is added to the Dol-P carrier is transferred by a specific enzyme; these enzymes are members of the class of enzymes known as glycosyl transferases. The completed oligosaccharide is finally transferred from its Dol-PP derivative to an asparagine residue in an Asn-X-Ser(Thr) sequence in a protein.

The function of the glucoses on the initial oligosaccharide is to expedite oligosaccharide transfer from lipid to protein. Oligosaccharyl transferase, the transferring enzyme, has a preference for oligosaccharides that contain three glucose units. In addition, as indicated below, the glucose residues are important for expediting the folding of the protein.

Intermediate processing continues in the endoplasmic reticulum

Once the oligosaccharide chain has been transferred to the protein, various glycosidases (enzymes that remove specific sugars, including the glucosidases that are specific for glucose) act on the protein-bound oligosaccharide. In a series of processing or 'pruning' reactions, the three glucose residues are removed in the endoplasmic reticulum, and up to six mannose residues in the Golgi apparatus. These trimming reactions give rise to a core structure of 2 GlcNAc and 3 mannose residues, and this core oligosaccharide is elongated to form complex oligosaccharides. These elongation reactions involve the addition of one or more of the sugars, GlcNAc, galactose, sialic acid, and L-fucose. The reactions involved in the modification of the oligosaccharide chains are outlined in Figure 24.11.

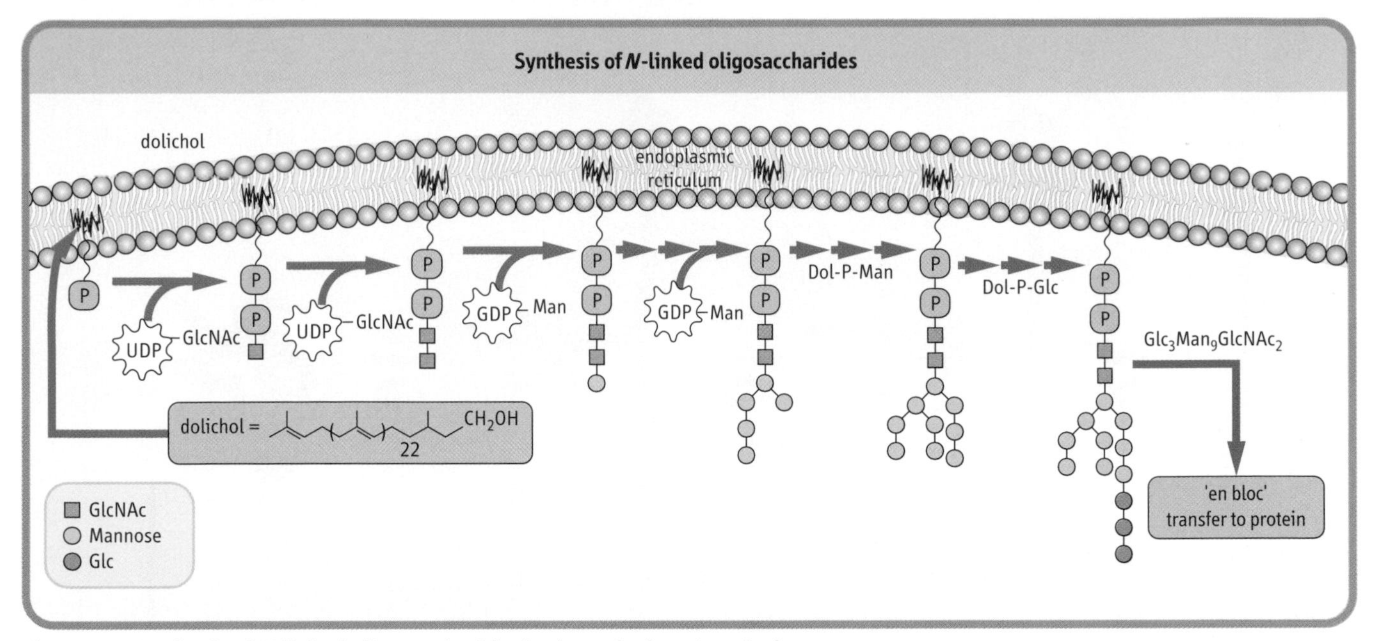

Fig. 24.10 Synthesis of *N*-linked oligosaccharides in the endoplasmic reticulum.

Biosynthesis of oligosaccharides

Inhibitors of glycosyl transferases

Several inhibitors of the biosynthesis of *N*-linked oligosaccharides have become valuable reagents for understanding of the role of specific carbohydrate structures in glycoprotein function. Tunicamycin is a glycoside antibiotic that inhibits the first step in synthesis of *N*-linked oligosaccharides – formation of Dol-PP-GlcNAc (see Fig. 24.10). It has varied effects, from profound to benign, on the synthesis and function of glycoproteins. Some proteins are insoluble without their carbohydrate, and aggregate in the cell. Others do not fold correctly, or may fail to function in recognition reactions. Tunicamycin is quite toxic to animals, as animal cells need *N*-linked oligosaccharides for essential functions.

Final modifications to the glycoprotein take place in the Golgi apparatus

When the glucose residues have been removed and the protein has folded into its correct conformation, the glycoprotein with one or more $Man_9GlcNAc_2$ oligosaccharides is transported from the endoplasmic reticulum to the Golgi apparatus, where other modification reactions can occur. Usually, in the *cis* and medial regions of the Golgi, six of the mannose residues are removed by several different α-mannosidases. The exact role of each of the mannosidases is not known, but they produce glycoproteins with oligosaccharide chains having variable numbers of mannose residues. If the oligosaccharide is trimmed to its core structure, it may then be converted to a complex structure by glycosyl transferases in the *trans*-Golgi apparatus (Fig. 24.11).

O-linked oligosaccharides

The biosynthesis of *O*-linked oligosaccharides occurs in the Golgi apparatus by the stepwise addition of sugars from their sugar nucleotide donors. No lipid intermediates are involved. Figure 24.12 describes the straightforward sequence of reactions for assembly of the oligosaccharide chains of salivary mucins. GalNAc is first transferred from UDP-GalNAc to serine or threonine residues on the protein by a GalNAc transferase in the Golgi apparatus. The GalNAc-serine-(protein) serves as the acceptor for galactose and then sialic acid, transferred from their sugar nucleotides by Golgi galactosyl and sialyl transferases, to form the final trisaccharide sequence on the mucin. Other Golgi glycosyl transferases are involved in the stepwise biosynthesis of oligosaccharides on proteoglycans and collagen. There are more than 100 glycosyl transferases involved in glycoconjugate biosynthesis in a typical cell.

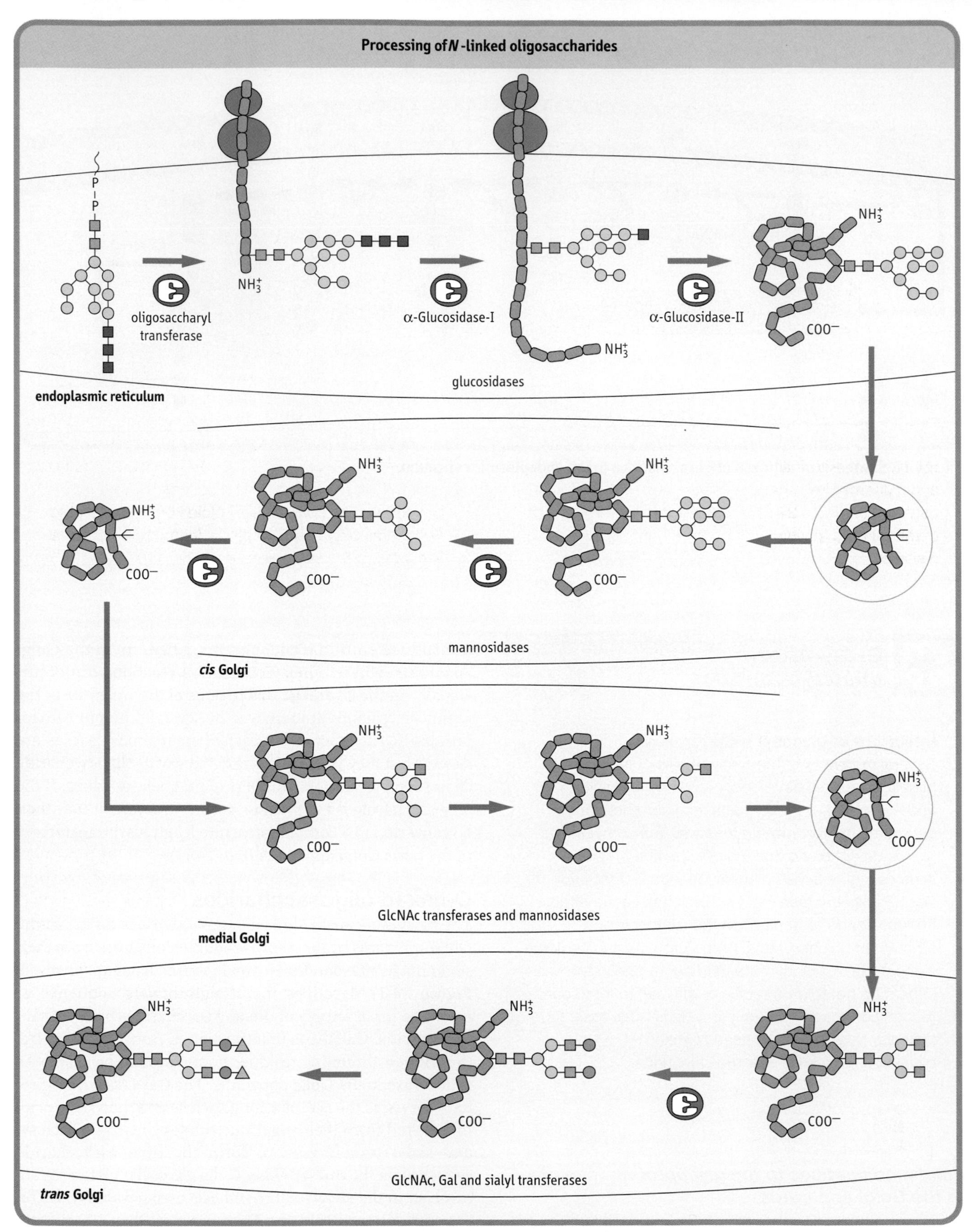

Fig. 24.11 Processing of *N*-linked oligosaccharides from high-mannose to complex forms.
Glycoproteins are transported between the endoplasmic reticulum and Golgi compartments in vesicles.

Deficiencies in glycoprotein synthesis

The carbohydrate-deficient glycoprotein syndromes (CDGSs) are a newly described group of rare genetic diseases. All patients show multisystem pathology, with severe involvement of the nervous system. Three distinct variants have been identified and are characterized by a deficiency of the carbohydrate moiety of secretory glycoproteins, lysosomal enzymes, and, probably, membrane glycoproteins. The diagnosis is made by electrophoresis of serum transferrin. In CDGS, the transferrin contains less sialic acid and migrates more slowly.

Comment. The basic defects in this group of diseases appears to be in the synthesis or processing of *N*-linked oligosaccharides. Defects in GlcNAc transferase I and mannosidase II have been identified, and it seems likely that defects in each of the enzymes depicted in Figure 24.11 will eventually be demonstrated, giving rise to a family of related diseases.

Biosynthesis of oligosaccharides

Inhibitors of glycosidases
Several plant alkaloids are potent glycosidase inhibitors, and some of these compounds inhibit the 'pruning' enzymes shown in Figure 24.11. Castanospermine inhibits glucosidases I and II, blocking the removal of the glucose residues from the $Glc_3Man_9GlcNAc_2$–protein. Many proteins, including the AIDS virus envelope glycoprotein, are assisted in folding by interacting with chaperone proteins such as calnexin. Calnexin binds to glycoproteins in which the oligosaccharide has been trimmed to contain a single glucose (that is, $Glc_1Man_9GlcNAc_2$). If removal of glucose is inhibited by castanospermine, the protein does not fold correctly. Other plant alkaloids inhibit the mannosidases required for synthesis of complex oligosaccharides (see Fig. 24.11). Swainsonine inhibits mannosidase II, yielding oligosaccharides with only a partial complex chain; this alkaloid is of considerable interest, as plants that contain it (commonly known as locoweed in the USA) are quite toxic to animals.

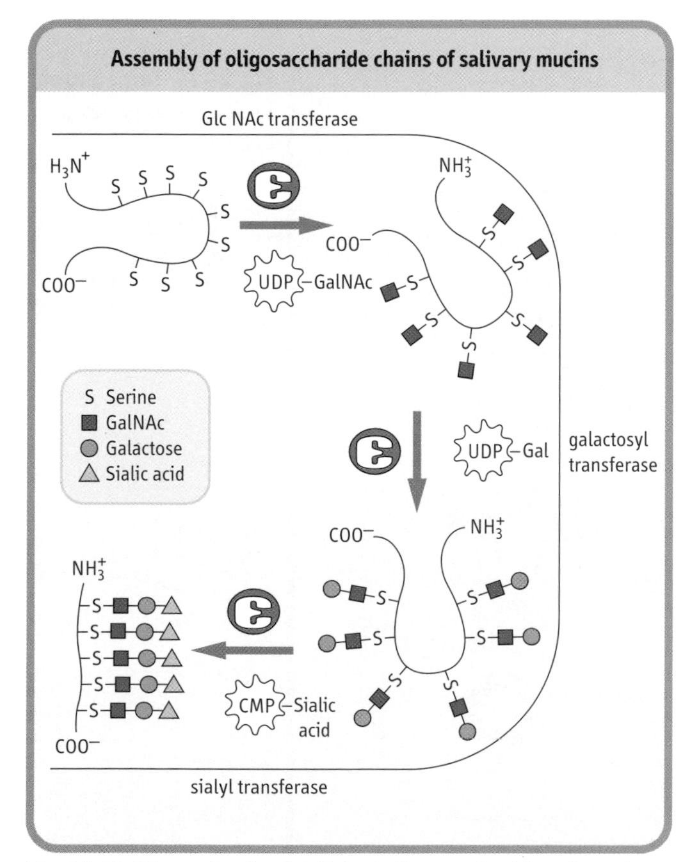

Fig. 24.12 Biosynthesis of *O*-linked oligosaccharides of mucins in the Golgi apparatus.

Functions of the oligosaccharide chains of glycoproteins

N-linked oligosaccharides

N-linked oligosaccharides have an important role in protein folding

Resident proteins in the endoplasmic reticulum, known as chaperones, assist newly synthesized membrane proteins to fold into their correct conformations. Two of these chaperones, calreticulin and calnexin, bind to unfolded glycoproteins by recognition of high-mannose oligosaccharides that have a single glucose remaining on their structure. These two chaperones are examples of a class of carbohydrate-binding proteins known as lectins – proteins that have a recognition and binding site for specific carbohydrate structures. Not all the proteins synthesized in the cell require assistance in folding, for those that do, the rate of folding is greatly increased by the chaperones. Incorrectly folded or unfolded proteins do not undergo normal transport to the Golgi apparatus, and are frequently degraded in the endoplasmic reticulum.

High-mannose oligosaccharides target some proteins to specific sites in the cell

Lysosomes are involved in the hydrolysis and turnover of many cellular components, and contain a variety of degradative hydrolytic enzymes, including proteases, lipases, and glycosidases. Most of these lysosomal enzymes are *N*-linked glycoproteins. The 'sorting' of lysosomal enzymes occurs in the *cis*-Golgi. Those enzymes destined for the lysosomes have a cluster of lysine residues formed by folding of the protein. As outlined in Figure 24.13, these lysine residues serve as a docking site for an *N*-acetylglucosamine-1-phosphate (GlcNAc-1-P) transferase that binds to the lysosomal enzyme and transfers GlcNAc-1-P from UDP-GlcNAc to the terminal mannose residue on the high-mannose chains of these proteins. An uncovering enzyme then removes the GlcNAc residues to expose Man-6-P structures, which are recognized by a Man-6-P receptor in the Golgi; the receptor binds to the modified glycoproteins and directs them to the lysosomes. The Man-6-P receptor is also present on the cell surface, so that even extracellular enzymes containing this signal are endocytosed and transferred to lysosomes.

The oligosaccharide chains of glycoproteins frequently increase the solubility of the proteins and their stability in adverse conditions

Because of the increased solubility conferred by oligosaccharide chains, most proteins secreted from cells into the environment, such as plasma proteins, or degradative enzymes released by yeast and fungi, are glycoproteins. These enzymes have high stability to heat, detergents, acids, and bases. Enzymatic removal of the carbohydrate greatly reduces their stability. Indeed, when glycoproteins are synthesized in cells in the presence of glycosylation inhibitors such as tunicamycin, which inhibits the synthesis of Dol-PP-GlcNAc, many of these proteins precipitate in the endoplasmic reticulum because of a combination of incorrect folding or processing, or decreased solubility.

I-cell disease

I-cell disease (mucolipidosis II) and pseudo-Hurler polydystrophy (mucolipidosis III) are rare inherited diseases that are due to deficiencies in the machinery that targets lysosomal enzymes to lysosomes.

Comment. These diseases result from a deficiency of the enzyme, GlcNAc-1-P transferase, such that lysosomal enzymes do not acquire the targeting signal (Man-6-P residues on their *N*-linked oligosaccharides). As a consequence, they are secreted from cells, rather than transported to lysosomes. Fibroblasts from individuals with this disease have dense inclusion bodies (hence the term 'I-cell') and are deficient in many lysosomal enzymes. The lysosomes become engorged with indigestible substrates, leading to death in infancy.

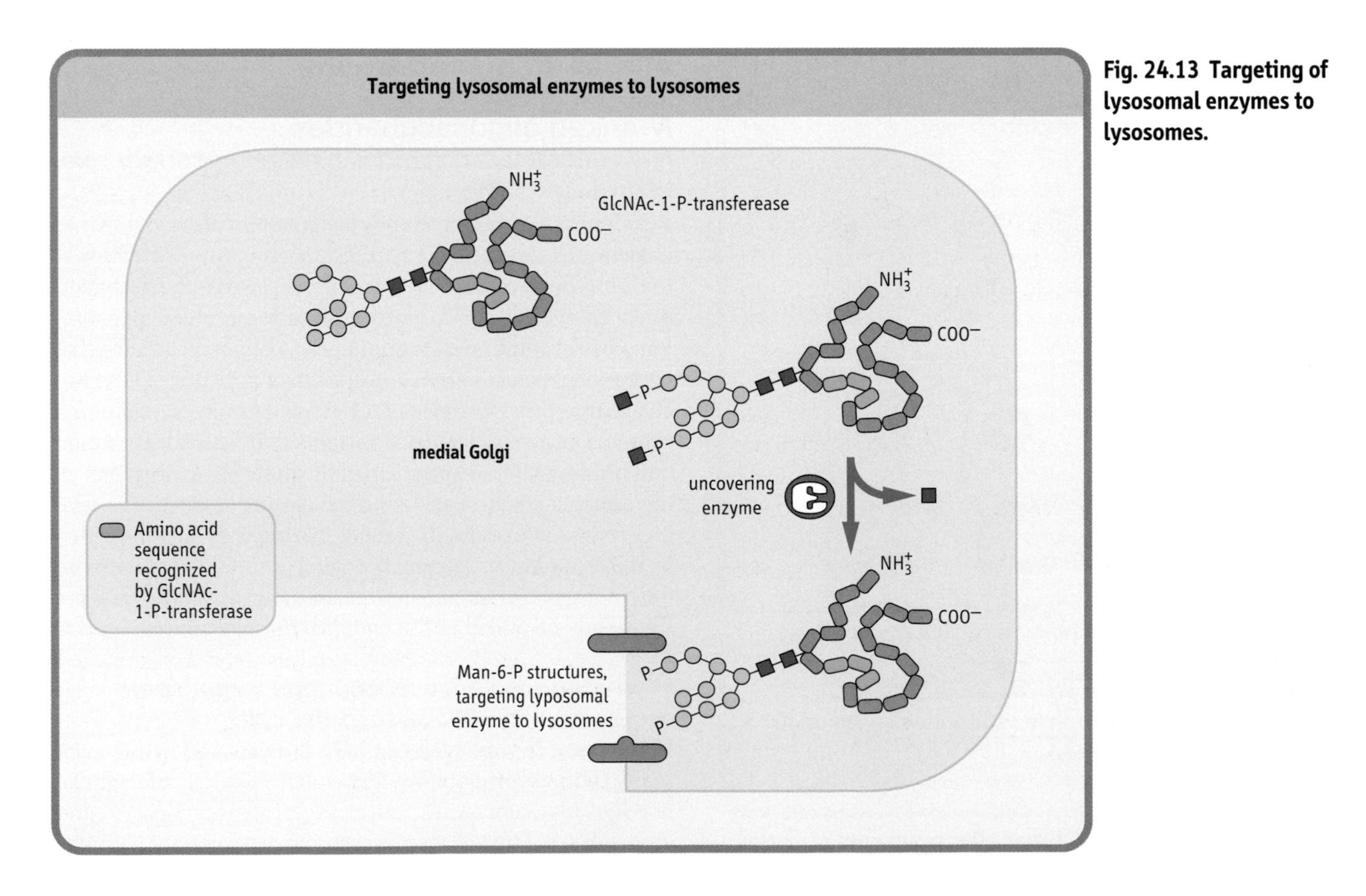

Fig. 24.13 Targeting of lysosomal enzymes to lysosomes.

Both N- and O-linked glycan structures are involved in recognition processes

N-linked glycoproteins are frequently present at the animal cell surface and have important roles in cell–cell interactions. One cell may contain on its cell surface a recognition protein – that is, a specific lectin – that binds to specific carbohydrate structures on the surface of the complementary cell. This interaction provides for specific cell recognition, and is a key factor in fertilization, inflammation development and differentiation (see Advanced Concept box below).

Variations in mucin structure appear to have a role in the specificity of fertilization, cell differentiation, development of the immune response, and virus infectivity. Glycoprotein ZP3, present on the zona pellucida of the mouse egg, functions as a receptor for sperm during fertilization. Enzymatic removal of *O*-linked oligosaccharides from ZP3 results in loss of sperm receptor activity, whereas removal of the *N*-linked oligosaccharides has no effect on sperm binding. The isolated *O*-linked oligosaccharides obtained from ZP3 also have sperm-binding activity and could be used to inhibit fertilization. Differences between the *O*-glycan structures of cytotoxic lymphocytes and helper cells involved in the immune response are also believed to be important in mediating cellular interactions during the immune response.

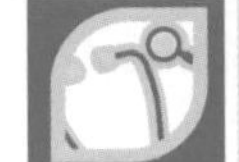

Lectin–carbohydrate interactions and infection

Some lectin–carbohydrate interactions may be harmful to animal cells. Many bacterial pathogens use this type of interaction to allow them to bind to specific host cells and gain entry into the body. One example is the influenza virus, which uses a hemagglutinin protein on its surface to bind sialic acid residues on target cells.

Comment. Pathogenic enteric bacteria, such as *Escherichia coli* and *Enterobacter cloacae*, have surface structures called pili (or fimbriae) that contain a protein subunit with lectin activity. This protein binds to high-mannose oligosaccharides that are components of *N*-linked glycoproteins on the surface of intestinal epithelial cells. This binding frequently leads to infection and, in some cases, severe disease.

Carbohydrate-dependant cell interactions

Inflammation

An important example of carbohydrate-dependent cell–cell interactions occurs during inflammation. Injury to vascular endothelial cells elicits an inflammatory response that causes the release of cytokines (proteins affecting cell migration) from the injured tissue. The cytokines attract leukocytes to the site of the injury or infection to remove the invading organisms or damaged tissue. These leukocytes must be able to exit from the blood flow and attach to the injured tissue. They are able to do this because they have a tetrasaccharide, known as sialyl Lewis-X antigen, as a component of a membrane glycoprotein or glycolipid. The sialyl Lewis-X antigen is recognized by a lectin (that is, a carbohydrate-binding protein), called E-selectin, that is present on the surface of the endothelial cells. The interaction between the selectin and the sialyl Lewis-X antigen enables the leukocytes to adhere to the vascular wall even under the shear forces of the circulation. Figure 24.14 presents a model demonstrating the chemistry of this important interaction. Selectins mediate the initial adhesive step, which is described as tethering, then a 'rolling' of leukocytes along the endothelial cell surface. In fact, leukocytes also contain a selectin, L-selectin, that probably interacts with a saccharide structure on the endothelial cells. These weak binding interactions enable leukocytes to penetrate the interstitial layer and clear up the site of injury.

While adherence of leukocytes to endothelial cells is important in fighting infection, it can be dangerous and life threatening. In myocardial infarction, the leukocytes can cause blockage of arteries, leading to ischemia. Because of the significance of this interaction, many laboratories are seeking to develop novel chemicals, known as glycomimetics, that mimic the sialyl Lewis-X structure. Administration of these drugs to patients who have suffered a heart attack should block the selectin sites, inhibiting the binding of leukocytes to the vascular wall and diminishing the probability of further ischemia.

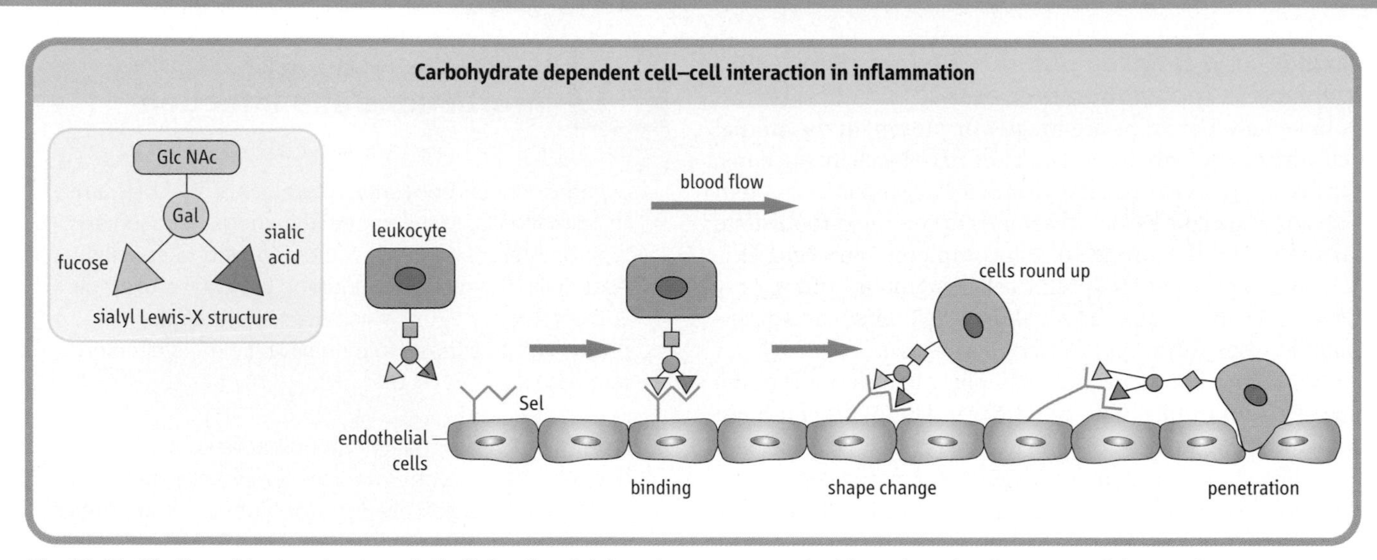

Fig. 24.14 Binding of leukocytes to endothelial cells. Sialyl Lewis-X, a tetrasaccharide antigen that forms part of the membrane structure of leukocytes, is recognized by a carbohydrate-binding protein, E-selectin (Sel), on the surface of endothelial cells. Leukocytes are first retarded by, then roll along, and eventually penetrate the endothelial monolayer. In addition, leukocytes contain L-selectins, proteins that recognize saccharide structures on endothelial cells.

Summary

Sugars are common components of proteins and can serve a number of different functions in these molecules, including modification of the physical properties of the protein (solubility, stability, viscosity), and assisting in its folding, processing, transport, and targeting. Because sugars are located on the surfaces of cells, they are in an excellent position to act as informational molecules, mediating recognition reactions between cells or with other molecules in the environment.

Further reading

Elbein AD. Glycosidase inhibitors: inhibitors of *N*-linked oligosaccharide processing. *FASEB J* 1995;**5**:3055–3063.

Kim YJ, Varki A. Perspectives on the significance of altered glycosylation of glycoproteins in cancer. *Glycoconj J* 1997;**14**:569–576.

Krasnewich D, Gahl WA. Carbohydrate-deficient glycoprotein syndrome. *Adv Pediatr* 1997;**44**:109–140.

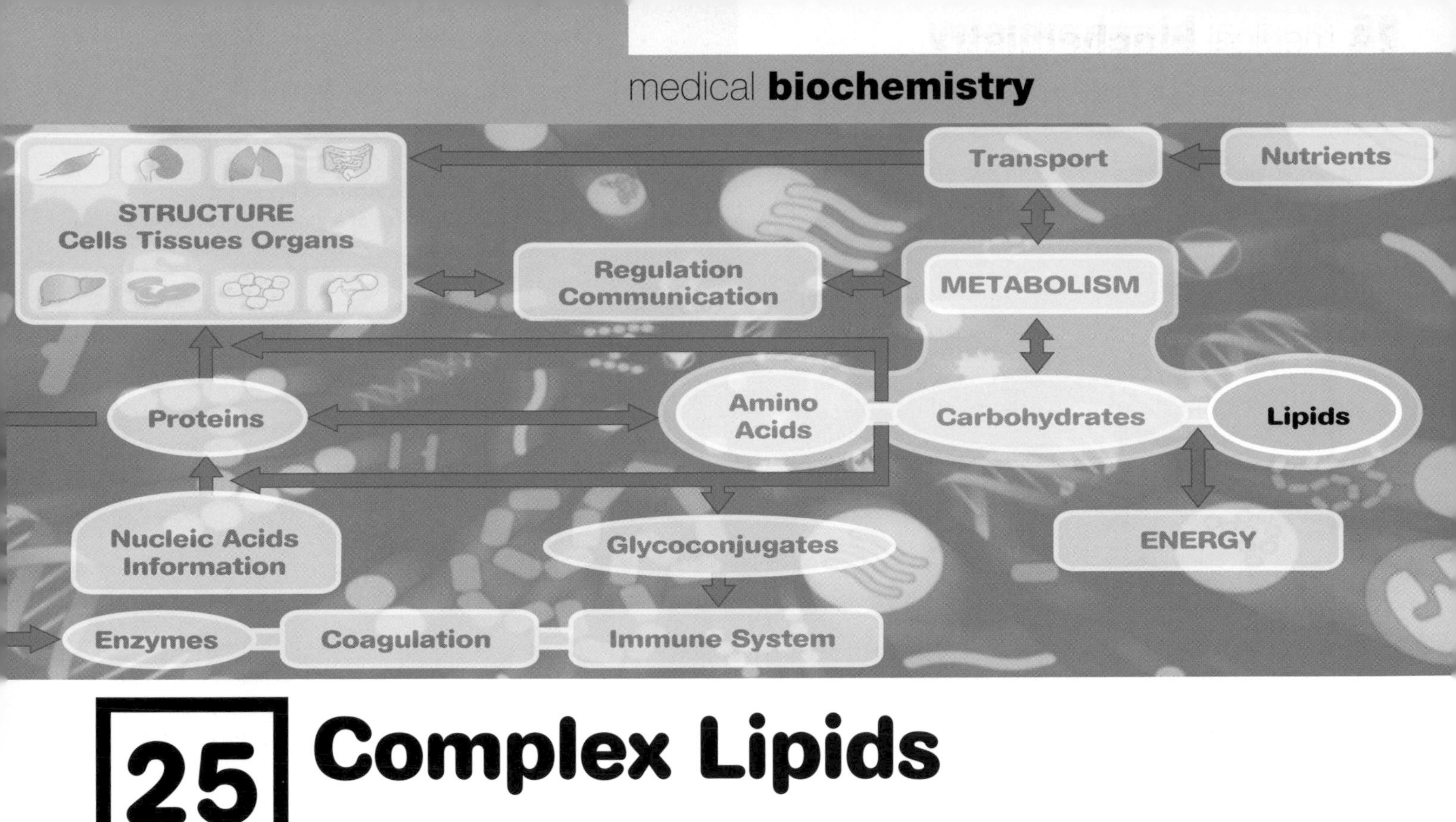

25 Complex Lipids

The term 'complex lipids' refers to a diverse group of water-insoluble compounds found in biological membranes. They are characterized by the following general properties:

- **they are generally saponifiable**: they release fatty acids when treated with strong base;
- **they are water insoluble**, but soluble in nonpolar organic solvents, such as chloroform, ether, or benzene;
- **they may be nonpolar** (triglycerides, cholesterol esters) **or they may be polar** (phospholipids, sphingolipids).

Polar lipids are amphipathic, meaning that they contain both a hydrophobic domain that interacts with the membrane environment and a hydrophilic domain that interacts with the aqueous environment. These polar lipids are major components of all biological membranes, and have diverse functions in the cell. This chapter discusses the structure, biosynthesis, and function of the two major classes of polar lipids: glycerophospholipids and sphingolipids.

Glycerophospholipids (phospholipids)

Structure of phospholipids

The term 'phospholipid', introduced in Chapter 7, commonly refers to the major class of membrane lipids, glycerophospholipids, or phosphoglycerides. They contain a diacyl glycerol (DAG) phosphate (phosphatidic acid) backbone. In addition to phosphatidylcholine, phosphatidylethanolamine, and phosphatidylserine, there are a number of more complicated structures, such as phosphatidylglycerol and diphosphatidylglycerol (cardiolipin), which are found in mitochondrial membranes, and phosphatidylinositol, which has a role in signal transduction and in anchoring proteins in the plasma membrane (Fig. 25.1).

Most phospholipids contain more than one kind of fatty acid molecule. Thus a given class of phospholipid from any source may contain a spectrum of fatty acids. For example, phosphatidylcholine usually contains palmitic acid (C-16:0) or stearic acid (C-18:0) at its carbon-1 position and an 18-carbon, unsaturated fatty acid (e.g. oleic, linoleic, or linolenic) at its carbon-2 position (C-16:0 represents a 16-carbon fatty acid with no [0] unsaturation, while C-18:1 represents an 18-carbon fatty acid with one unsaturated bond). Phosphatidylethanolamine usually has a longer-chain fatty acid at carbon-2, such as arachidonic acid (C-20:5). In addition to forming the backbone of the membrane, these complex lipids contribute charge to the membrane: phosphatidylcholine and phosphatidylethanolamine have no net charge at physiologic pH, but the other glycerophospholipids and the major species of sphingolipids are anionic in nature.

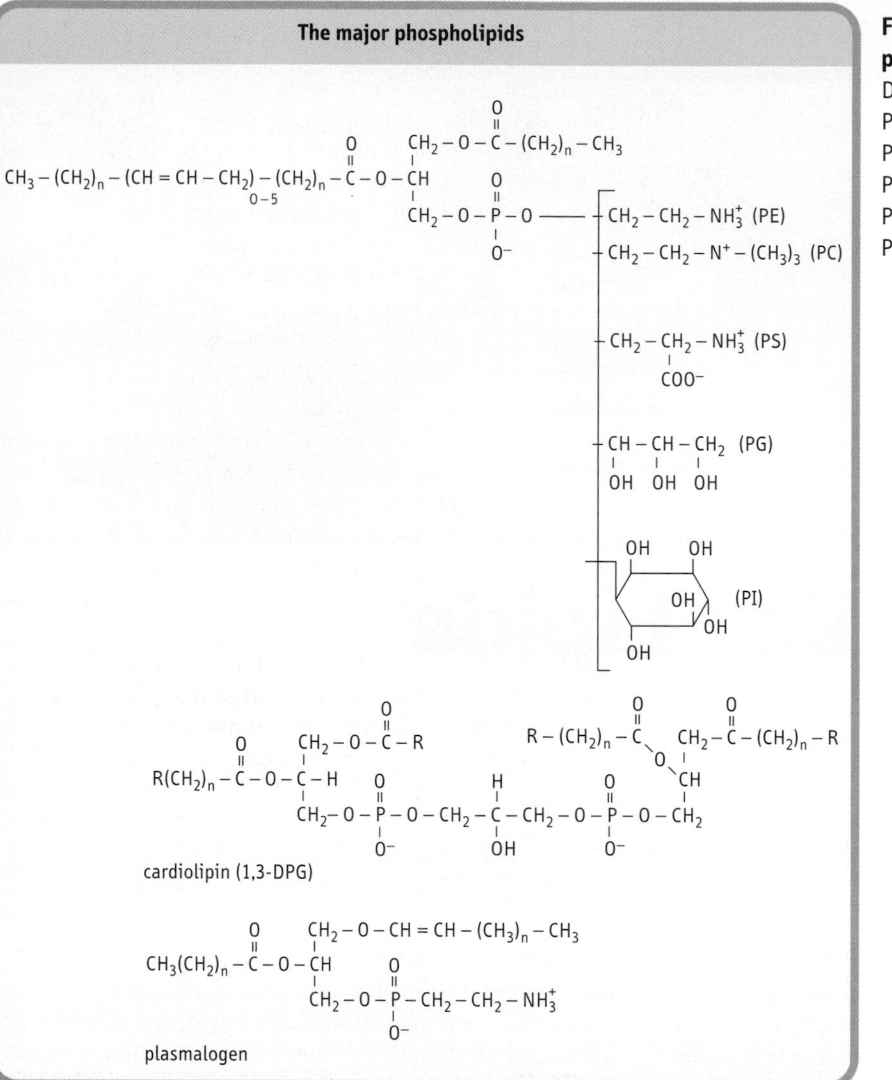

Fig. 25.1 Structure of the major phospholipids of animal cell membranes. DPG, diphosphatidylglycerol; PC, phosphatidylcholine; PE, phosphatidylethanolamine; PG, phosphatidylglycerol; PI, phosphatidylinositol; PS, phosphatidylserine.

Phosphatidylinositol

Glycosylphosphatidylinositol membrane anchors
Phosphatidylinositol is an integral component of the glycosylphosphatidylinositol (GPI) structure that anchors various proteins to the plasma membrane (Fig. 25.2). In contrast to other membrane phospholipids, including most of the membrane phosphatidylinositol, GPI has a glycan chain containing glucosamine and mannose attached to the inositol. Ethanolamine connects the GPI-glycan to the carboxyl terminus of the protein. Many membrane proteins in eukaryotic cells are anchored by a GPI structure, including alkaline phosphatase and acetylcholinesterase, and also the variable surface antigen of trypanosomes. In contrast to integral or peripheral membrane proteins, GPI-anchored proteins may be released from the cell surface in response to regulatory processes.

Biosynthesis of precursors: phosphatidic acid and DAG

All animal cells, except for erythrocytes, are able to synthesize phospholipids *de novo*, whereas triglyceride synthesis occurs mainly in liver, adipose tissue, and intestinal cells. As illustrated in Figure 25.4, phosphatidic acid and 1,2-DAG are common intermediates in the synthesis of both triglycerides and phospholipids. Glycerol-3-phosphate (glycerol-3-P) is the primary starting material for synthesis of phosphatidic acid; it is formed in most tissues by reduction of the glycolytic

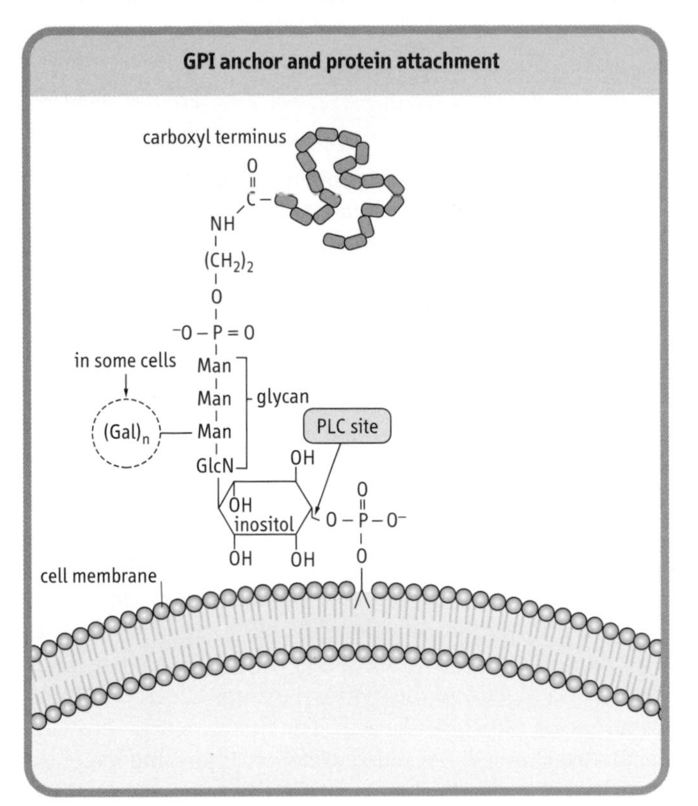

Fig. 25.2 Structure of the glycosylphosphatidylinositol (GPI) anchor and its attachment to proteins. Gal, galactose; GlcN, glucosamine; Man, mannose; PLC, phospholipase C.

Variable surface antigens of trypanosomes

The parasitic trypanosome that causes sleeping sickness, *Trypanosoma brucci*, has a protein called the variable surface antigen bound to its cell surface by a GPI anchor. This variable surface antigen elicits the formation of specific antibodies in the host, and these antibodies can attack and kill the parasite. However, some of the parasites evade immune surveillance by shedding this antigen, as if they were shedding a coat.

Comment. These organisms are able to shed the antigen because they have an enzyme, phospholipase C, that cleaves the GPI anchor at the inositol–phosphate bond, releasing the protein–glycan component into the external fluid. Surviving cells rapidly make a new coat with a different antigenic structure that will not be recognized by the antibody. Of course, this new coat will elicit the formation of new specific antibodies, but the parasite can again shed this coat, and so on.

Phosphatidylcholine

Platelet activating factor and hypersensitivity
Platelet activating factor (PAF; Fig. 25.3) contains an acetyl group at carbon-2 of glycerol and a saturated 18-carbon alkyl ether group linked to the hydroxyl group at carbon-1, rather than the usual long-chain fatty acids of phosphatidylcholine. It is a major mediator of hypersensitivity reactions, acute inflammatory reactions, and anaphylactic shock, and affects the permeability properties of membranes, increasing platelet aggregation and causing cardiovascular and pulmonary changes, edema, and hypotension. In allergic persons, cells involved in the immune response become coated with immunoglobulin E (IgE) molecules that are specific for a particular antigen or allergen, such as a pollen or insect venom. When these individuals are re-exposed to that antigen, antigen–IgE complexes form on the surface of the inflammatory cells and initiate the synthesis and release of PAF.

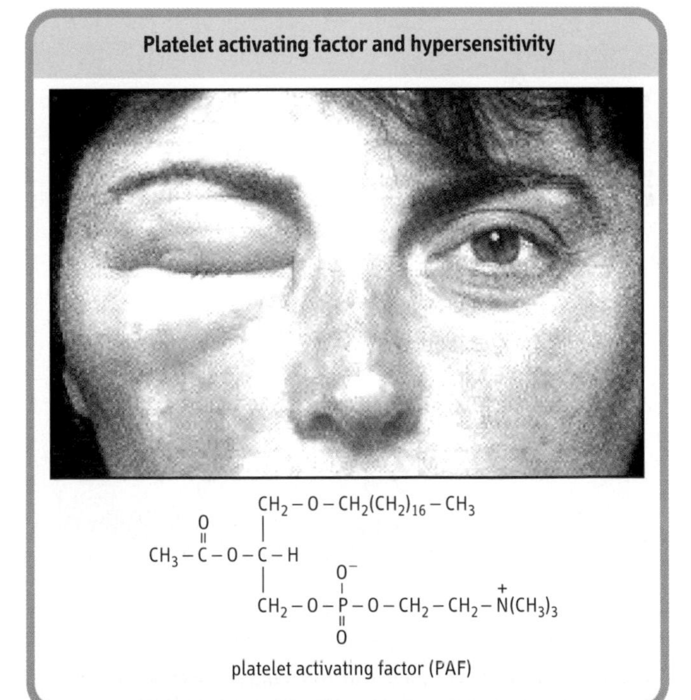

Fig. 25.3 PAF is a major mediator of acute inflammatory reactions such as this anaphylactic response to bee venom. (Courtesy of Professor Jonathan Brostoff.)

Defects in GPI anchoring associated with genetic disease

Paroxysmal nocturnal hemoglobinuria (PNH) is a complex hematologic disorder characterized by hemolytic anemia, venous thrombosis in unusual sites, and deficient hematopoiesis. The diagnosis of this disease is based on the unusual sensitivity of the red blood cells to the hemolytic action of Complement, because red cells from patients with PNH lack several proteins that are involved in regulating the activation of Complement at the cell surface.

Comment. One of these proteins is decay accelerating factor, a GPI-anchored protein that inactivates a hemolytic complex formed during complement activation; in its absence, there is increased hemolysis. There are several genetic variants of PNH. One of these involves a defect in the GlcNAc transferase that adds *N*-acetylglucosamine to the inositol moiety of phosphatidylinositol, the first step in GPI anchor formation.

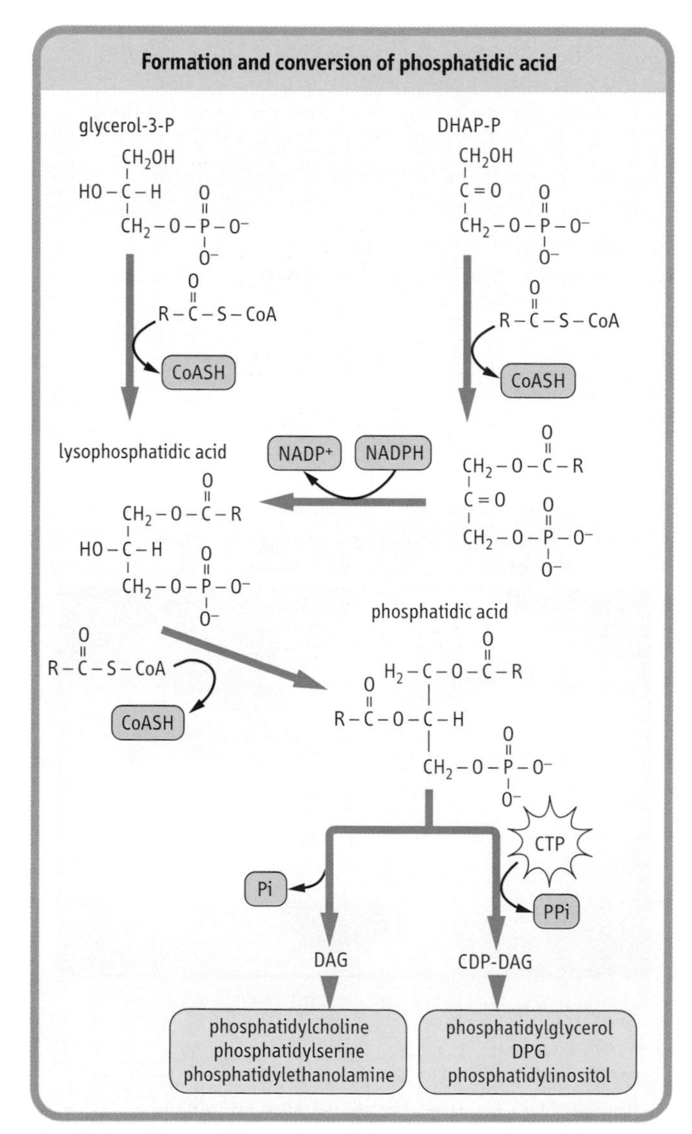

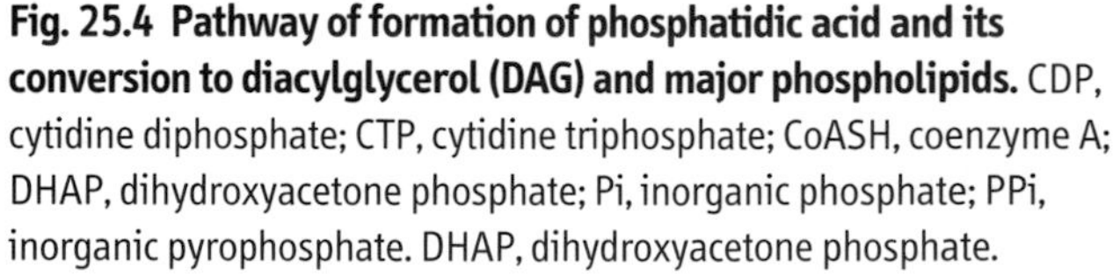

Fig. 25.4 Pathway of formation of phosphatidic acid and its conversion to diacylglycerol (DAG) and major phospholipids. CDP, cytidine diphosphate; CTP, cytidine triphosphate; CoASH, coenzyme A; DHAP, dihydroxyacetone phosphate; Pi, inorganic phosphate; PPi, inorganic pyrophosphate. DHAP, dihydroxyacetone phosphate.

intermediate, dihydroxyacetone phosphate (DHAP). In liver, kidney, and intestine, glycerol-3-P can also be formed directly via phosphorylation of glycerol by a specific kinase. Glycerol-3-P is acylated by transfer of two long-chain fatty acids from fatty acyl CoA to the hydroxyl groups at carbons-1 and -2, producing phosphatidic acid. The first fatty acid – usually a saturated fatty acid – is added to carbon-1, forming lysophosphatidic acid; the prefix 'lyso' indicates that one of the hydroxyl groups is not acylated. Then, a second fatty acid – usually an unsaturated fatty acid – is added to carbon-2 to form phosphatidic acid. DHAP may also be acylated by addition of a fatty acid to the 1-hydroxyl group, and this intermediate is then reduced and acylated to phosphatidic acid. Phosphatidic acid is converted to DAG by a specific cytosolic phosphatase.

Biosynthesis of phospholipids

The biosynthesis of lecithin (phosphatidylcholine) from DAG requires activation of choline to a cytidine diphosphate (CDP) nucleotide derivative. In this series of reactions, shown in Figure 25.5, the head-group choline is converted to phosphocholine, and then activated to CDP-choline by a pyrophosphorylase reaction. The pyrophosphate bond is cleaved, and phosphocholine is then transferred to DAG to form lecithin. This reaction is analogous to the transfer of glucose from UDP-glucose to glycogen, except that both the choline and phosphate groups are transferred to DAG. Phosphatidylethanolamine is formed by a similar pathway using cytidine triphosphate (CTP) and phosphoethanolamine, to form CDP-ethanolamine.

Both phosphatidylcholine and phosphatidylethanolamine can react with free serine by an exchange reaction to form phosphatidylserine and the free base, choline or ethanolamine (Fig. 25.6). In a secondary pathway, phosphatidylcholine can also be formed by methylation of phosphatidylethanolamine with the methyl donor, S-adenosylmethionine (SAM) (Fig. 25.7). The methylation pathway involves the sequential transfer of three activated methyl groups from three different molecules of SAM. Liver also has another route to phosphatidylethanolamine, involving decarboxylation of phosphatidylserine by a specific mitochondrial decarboxylase.

Surfactant function of phospholipids

Respiratory distress syndrome (RDS) accounts for 15–20% of neonatal mortality in Western countries. The disease affects only premature infants and its incidence is directly related to the degree of prematurity.

Comment. Immature lungs do not have enough type II epithelial cells to synthesize sufficient amounts of the phospholipid, dipalmitoylphosphatidylcholine (DPPC). This phospholipid makes up more than 80% of the total phospholipids of the extracellular lipid layer that lines the alveoli of normal lungs. DPPC decreases the surface tension of the aqueous surface layer of the lungs, facilitating opening of the alveoli during inspiration. Lack of surfactant causes the lungs to collapse during the expiration phase of breathing, leading to RDS. The maturity of the fetal lung can be determined by measuring the lecithin:sphingomyelin ratio in amniotic fluid. If there is a potential problem, a mother can be treated with a glucocorticoid to accelerate maturation of the fetal lung. RDS is also seen in adults in whom the type II epithelial cells have been destroyed as a result of the use of immunosuppressive drugs or certain chemotherapeutic agents.

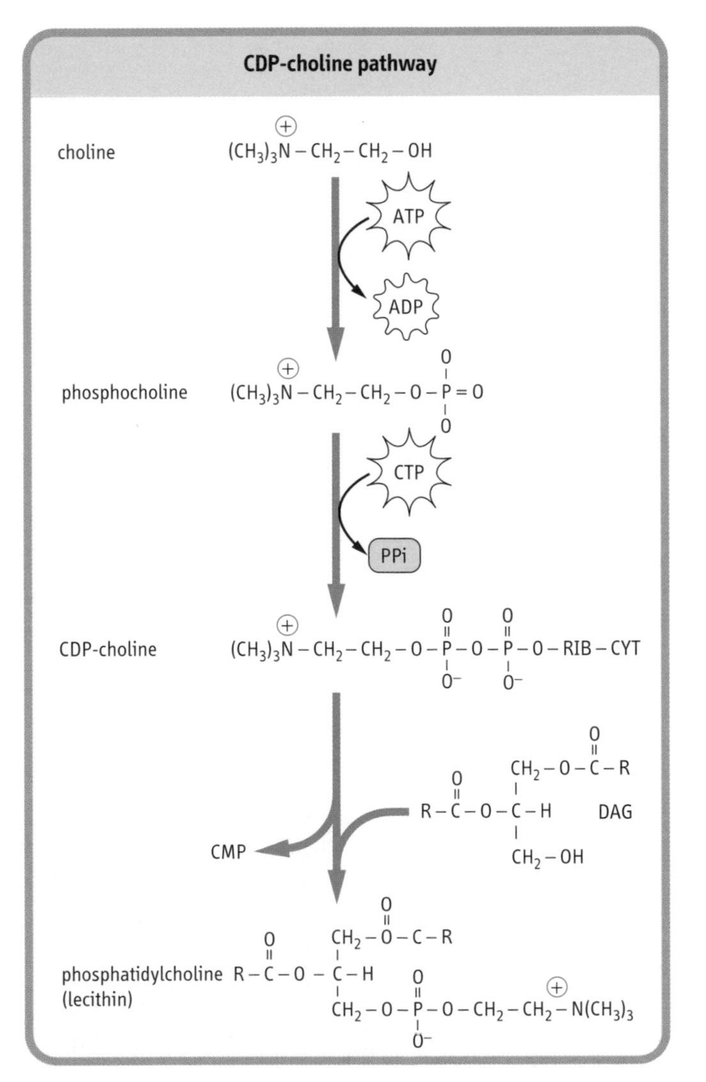

Fig. 25.5 Formation of phosphatidylcholine by the CDP-choline pathway. CYT, cytosine; CDP, cytidine diphosphate,CMP, cytidine monophosphate; DAG, diacyl glycerol; RIB, ribose. CTP, cytidine triphosphate.

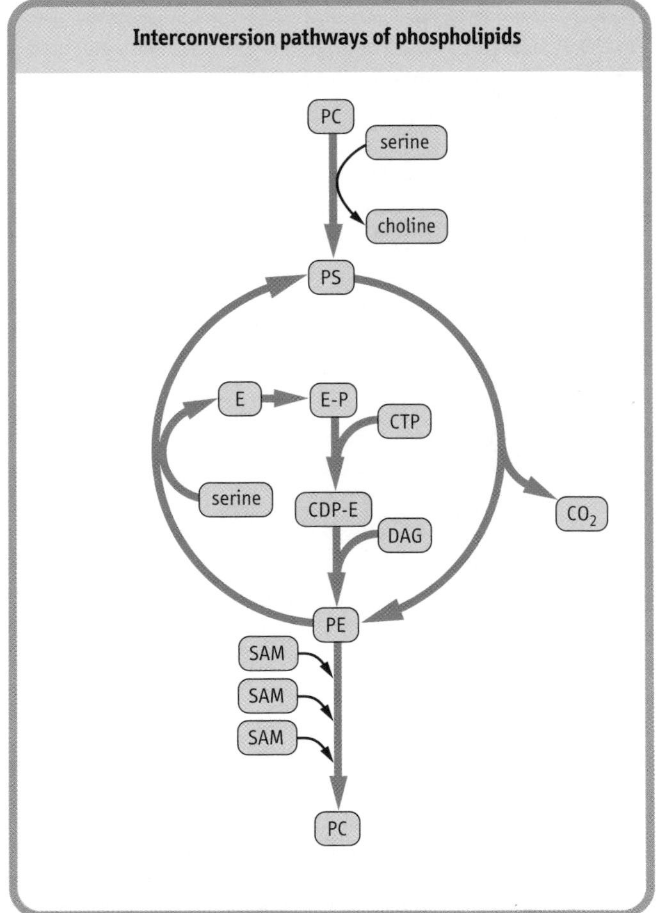

Fig. 25.6 Pathways of interconversion of phospholipids by exchange of head groups, by methylation or by decarboxylation. E, ethanolamine.

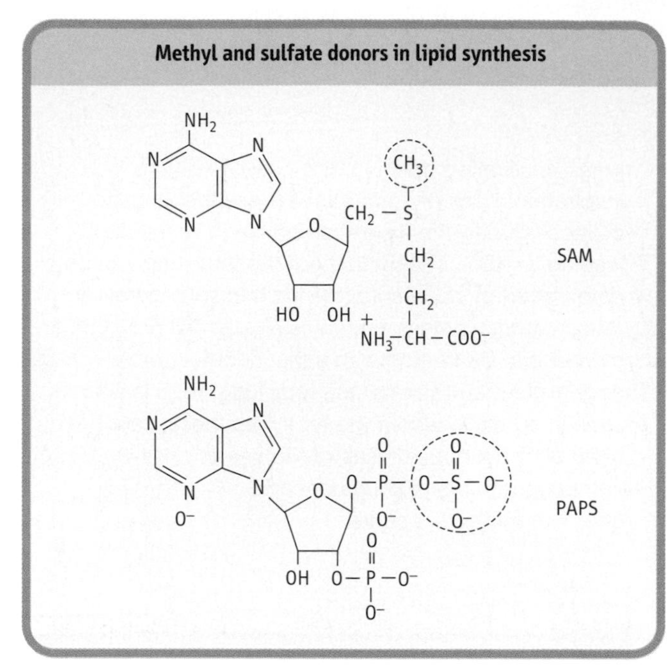

Fig. 25.7 Structures of the methyl and sulfate donors involved in the synthesis of membrane lipids. PAPS, 3´-phosphoadenosine-5´-phosphosulfate (active sulfate); SAM, S-adenosylmethionine.

Phospholipids that have an alcohol as the head group, e.g. phosphatidylglycerol and phosphatidylinositol, are synthesized by an alternative pathway, which also involves activation by cytidine nucleotides. In this case, the phosphatidic acid is activated, rather than the head group, yielding CDP-DAG (Fig. 25.8). The phosphatidic acid group is then transferred to free glycerol or inositol, to form phosphatidylglycerol or phosphatidylinositol, respectively. A second phosphatidic acid may also be added to phosphatidylglycerol to form diphosphatidylglycerol (DPG).

Plasmalogens

Plasmalogens are a subclass of glycerophospholipids that have a 1-alkenyl ether linked to the carbon-1 of glycerol (see Fig. 25.1). They are a major class of mitochondrial lipids, and are enriched in nerve and muscle tissue – in the heart they may account for nearly 50% of total phospholipids. Their function in relation to that of diacylphospholipids is not clear, but there is some evidence that they are more resistant to oxidative degradation, which may provide protection against oxidative stress in tissues with active aerobic metabolism.

Turnover of phospholipids

Phospholipids are in a continuous state of turnover in most membranes. This occurs in response to oxidative damage, during inflammation, and particularly in response to hormonal stimuli. As shown in Figure 25.9, there are number of phospholipases that act on specific bonds in the phospholipid structure. Phospholipases A_2 (PLA_2) and C (PLC) are particularly active during the inflammatory

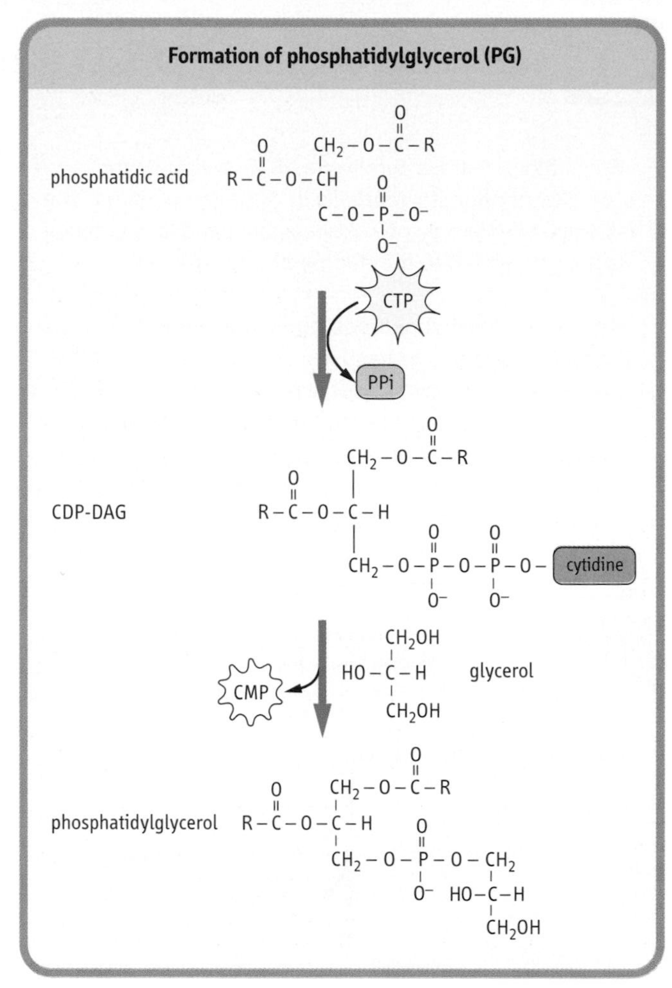

Fig. 25.8 Formation of phosphatidylglycerol by activation of phophatidic acid to form CDP-DAG, and transfer of DAG to glycerol. CMP, cytidine monophosphate; CTP, cytidine triphosphate.

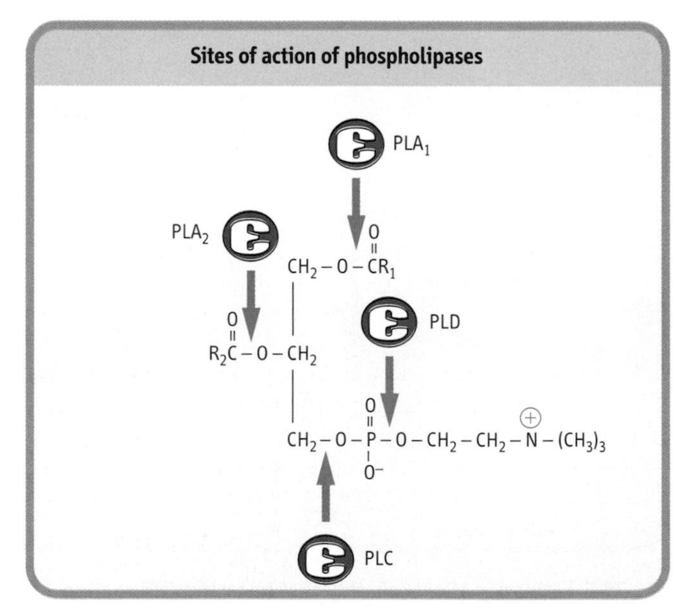

Fig. 25.9 Sites of action of phospholipases on phosphatidylcholine. PLA_1, PLA_2, PLC, PLD are phospholipases; A_1, A_2, C, and D respectively.

response and in signal transduction. Phospholipase B (not shown) is a lysophospholipase that removes the second acyl group after action of PLA_1 or PLA_2; the lysophospholipids and other products are recycled by scavenger pathways.

Sphingolipids

Structure and biosynthesis of sphingosine

Sphingolipids are a complex group of amphipathic, polar lipids. They are built on a core structure of the long-chain amino alcohol, sphingosine, which is formed by oxidative decarboxylation and condensation of palmitate with serine. In all sphingolipids, the long-chain fatty acid is attached to the amino group of the sphingosine in an amide linkage (Fig. 25.10). Because of the alkaline-stable nature of amide linkages compared with ester linkages, sphingolipids are non-saponifiable, which facilitates their separation from alkali-labile glycerophospholipids.

The synthesis of the sphingosine base of sphingolipids involves a condensation of palmitoyl CoA with serine, in which the carbon-1 of serine is lost as carbon dioxide. The product of this reaction is converted in several steps to sphingosine, which is then *N*-acylated to form ceramide (*N*-acylsphingosine). Ceramide (see Fig. 25.10) is the precursor and backbone structure of both sphingomyelin and glycosphingolipids.

Sphingomyelin

Sphingomyelin (Fig. 25.10) is found in plasma membranes, subcellular organelles, endoplasmic reticulum, and mitochondria. It comprises 5–20% of the total phospholipids in most cell types, and is mostly localized in the plasma membrane. It is the only sphingolipid that contains phosphorus, and is the major phospholipid of the myelin sheath of nerves. The phosphocholine group in sphingomyelin is transferred to the terminal hydroxyl group of sphingosine from phosphatidylcholine. The fatty acid composition varies, but long-chain fatty acids are common, including lignoceric (C-24:0), cerebronic (2-hydroxylignoceric) and nervonic acids.

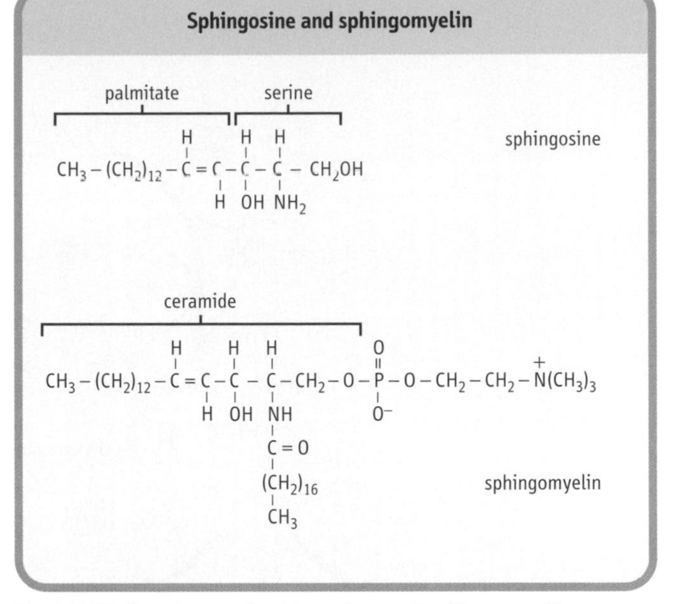

Fig. 25.10 Structures of sphingosine and sphingomyelin.

Glycolipids

Sphingolipids containing covalently bound sugars are known as glycosphingolipids, or glycolipids. The cell's complement of carbohydrate structures on glycolipids (and glycoproteins) does not depend on a template mechanism for synthesis, as do nucleic acids and proteins. Instead, it is determined by the enzymatic makeup of the cell (that is, its content of glycosyltransferases), and on how these various enzymes are expressed with respect to each other. The glycosyltransferase distribution and glycosphingolipid content of cells varies during development and in response to regulatory processes.

Glycolipids can be classified into four different groups: cerebrosides, sulfatides, globosides, and gangliosides. In all of these compounds, the polar head-group – comprising the sugars – is attached to ceramide by a glycosidic bond at

Lysosomal storage diseases resulting from defects in sphingomyelin degradation

Niemann–Pick syndrome is a lysosomal storage disease characterized by the formation of lipid-laden phagocytes, known as 'foam' cells, that are engorged with sphingomyelin. There is no treatment for this disease, but its inheritance is autosomal recessive and, if carriers are identified, prenatal diagnosis is possible by amniocentesis and assay of sphingomyelinase activity in fetal fibroblasts.

Comment. Niemann–Pick syndrome results from a deficiency in the activity of the enzyme, acid sphingomyelinase. In one of the final steps in sphingolipid turnover, this enzyme cleaves the sphingolipid to release choline phosphate and ceramide. Absence of this enzyme leads to accumulation of sphingomyelin in lysosomes.

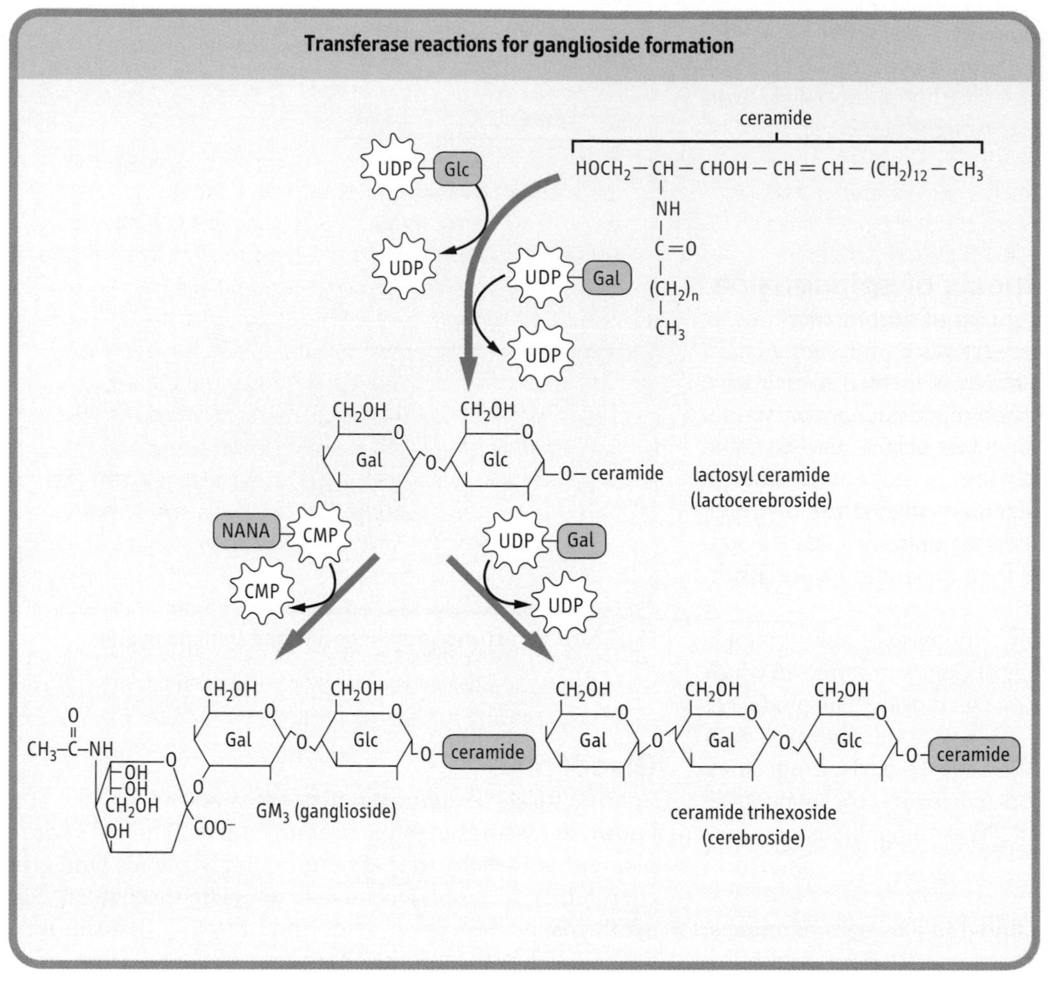

Fig. 25.11 Generalized outline of transferase reactions for elongation of glycolipids and formation of gangliosides.

the terminal hydroxyl group of sphingosine; Figure 25.11 illustrates this, together with the pathways involved in the biosynthesis of some of the more complex members of the group. The simplest of the glycosphingolipids are the (Glc) gluco- and (Gal) galactocerebrosides, which are ceramides with a single sugar attached to sphingosine. Sulfatides are formed by addition of sulfate from the sulfate donor, 3′-phosphoadenosine-5′-phosphosulfate (PAPS) (see Fig. 25.7), yielding, for example, galactocerebroside 3-sulfate. When the carbohydrate portion of the cerebroside contains two or more sugars plus an *N*-acetyl-galactosamine group (GalNAc), it is referred to as a globoside. Finally, glycolipids in which the oligosaccharide chains contain sialic acids (*N*-acetylneuraminic neuraminic acids, NANA) are called gangliosides.

Structure and nomenclature of gangliosides

The term 'ganglioside' was used because these glycolipids were originally identified in high concentrations in ganglionic cells of the central nervous system. In cells of the nervous system in general, more than 50% of the sialic acid in the cell is present in gangliosides, and the other 50% is in glycoproteins. Gangliosides are also found in the surface membranes of cells of most extraneural tissues, but, in these tissues, they account for less than 10% of the total sialic acid.

The nomenclature used to identify the various gangliosides is based on the number of sialic acid residues contained in the molecule, and on the sequence of the carbohydrates (Fig. 25.12). 'GM' means a ganglioside with a single (mono) sialic acid, whereas GD, GT and GQ would indicate two, three and four sialic acid residues in the molecule, respectively. The number after the GM – for example, GM_1 – indicates the sequence of sugars: 1 represents the sequence Gal–GalNAc–Gal–Glc–ceramide, 2 means GalNAc–Gal–Glc–ceramide, and 3 is Gal–Glc–ceramide. These numbers were derived from the relative mobility of the glycolipids on thin layer chromatographyn – the larger, GM_1, gangliosides migrating the most slowly. These complex structures are built up, one sugar residue at a time, in the Golgi apparatus, and are degraded by a series of exoglycosidases in lysosomes (Fig 25.13).

Gangliosides

Ganglioside receptor for cholera toxin

The galactose-containing cerebrosides and globosides in the plasma membranes of intestinal epithelial cells are binding sites for bacteria. The glycolipids appear to assist in retention of normal intestinal flora (symbionts) in the intestine but, conversely, binding of pathogenic bacteria to these and other glycolipids is believed to facilitate infection of the epithelial cells. The difference between symbiotic and parasitic bacteria depends, in part, on their ability to secrete toxins or to penetrate the host cell after the binding reaction.

Intestinal mucosal cells contain ganglioside GM_1 (see Fig. 25.11). This ganglioside serves as the receptor to which cholera toxin binds as the first step in its penetration of intestinal cells. Cholera toxin is a hexameric protein of 84 kDa molecular mass that is secreted by the bacterium *Vibrio cholerae*. The protein is composed of one 28 kDa A subunit and five 11 kDa B subunits. The protein binds to gangliosides through the B subunits, which enables the A subunit to enter the cell and activate adenylate cyclase on the inner surface of the membrane. The cyclic AMP that is formed then stimulates intestinal cells to export chloride ions, leading to osmotic diarrhea, electrolyte imbalances, and malnutrition. Cholera remains the number one killer of children in the world today.

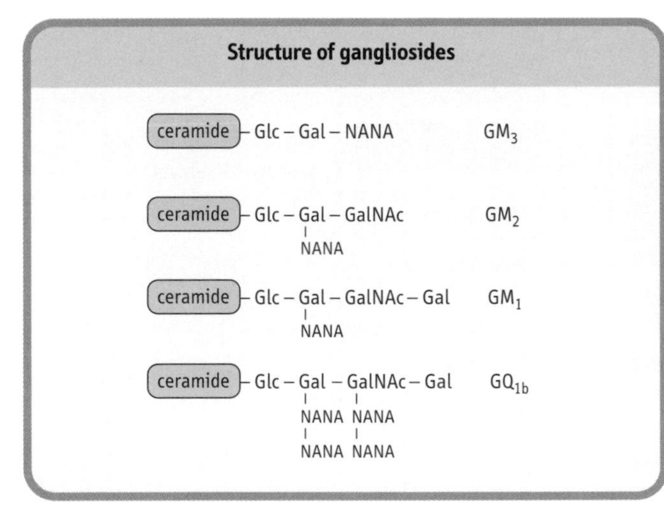

Fig. 25.12 Generalized structures of gangliosides.

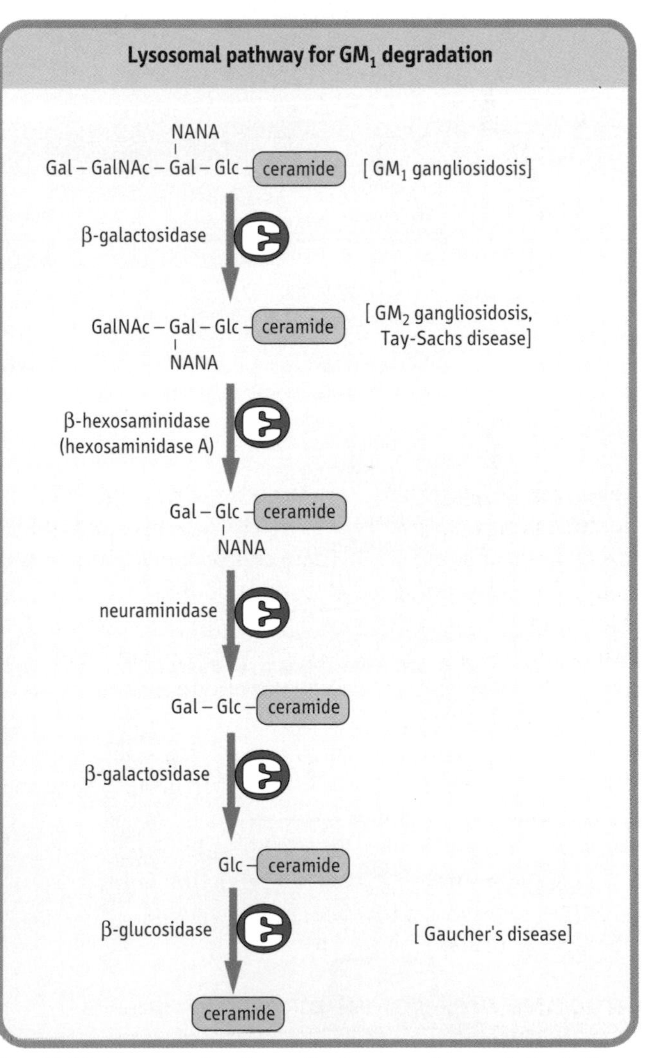

Fig. 25.13 Lysosomal pathway for turnover (degradation) of ganglioside GM_1 in human cells. Various enzymes may be missing in specific lipid storage diseases, as indicated in Figure 25.15.

Cerebrosidoses and gangliosidoses

Tay-Sachs disease is a gangliosidosis in which GM_2 accumulates as a result of an absence of hexosaminidase A. Individuals with this disease usually have mental retardation and blindness and die between 2 and 3 years of age. Fabry's disease is a cerebrosidosis resulting from deficiency of α-galactosidase and accumulation of ceramide trihexoside. The symptoms of Fabry's disease are skin rash, kidney failure, and pain in the lower extremities. Patients with this condition benefit from kidney transplants and usually live into early to mid adulthood. Most of these lysosomal storage diseases appear in several forms (variants), resulting from different mutations in the genome; some variants are more severe and debilitating than others. Although lysosomal storage diseases are relatively rare, they have had a major impact on our understanding of the function and importance of lysosomes.

Comment. When cells die, their components, including glycosphingolipids and glycoproteins, are degraded to their individual components. Figure 25.13 presents the pathway for the degradation of a ganglioside such as GM_1 in the lysosomes. These organelles contain a number of exoglycosidases and other necessary enzymes that function in a prescribed sequence to degrade complex glycolipids. A number of lysosomal diseases result from the absence of one of these essential glycosidases (Fig. 25.14). These sphingolipidoses are characterized by lysosomal accumulation of the substrate of the missing enzyme. This either causes lysosomes to lyse and release their hydrolytic enzymes into the cell, or prevents the lysosome from functioning normally.

Lipid storage diseases

Disease	Symptoms	Major storage product	Missing enzymes
Tay-Sachs	blindness, mental retardation, death between 2nd and 3rd year	GM_2 ganglioside	hexosaminidase A
Gaucher's	liver and spleen enlargement, mental retardation in infantile form	glucocerebroside	β-glucosidase
Fabry's	skin rash, kidney failure, pain in lower extremities	ceramide trihexoside	α-galactosidase
Krabbe's	liver and spleen enlargement, mental retardation	galactocerebroside	β-galactosidase

Fig. 25.14 Some lipid storage diseases.

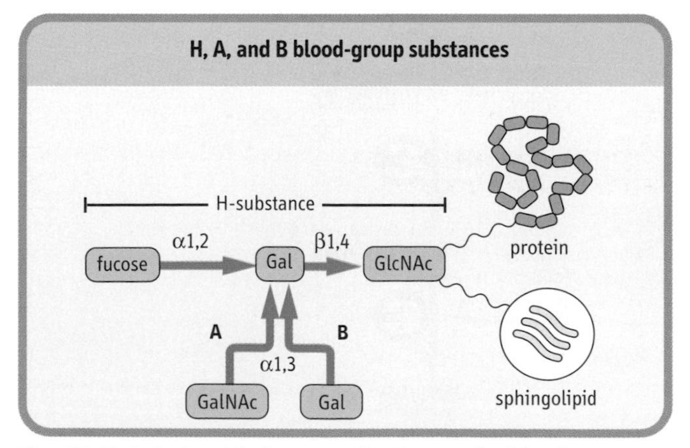

Fig. 25.15 Relationship between the H, A, and B blood-group substances. The terminal oligosaccharide is linked via other sugars to proteins and lipids of the red cell membrane.

Glycolipids

ABO blood-group antigens are glycosphingolipids
Blood transfusion replenishes the oxygen-carrying capacity of blood in persons who suffer from blood loss or anemia. The term 'blood transfusion' is something of a misnomer, because it involves only the infusion of washed and preserved red cells. The membranes of the human red blood cells contain a number of blood group antigenic determinants (as many as 100), of which the ABO blood-group system is the best understood and most widely studied. The antigens in this group are derived from a common precursor, called the H substance, and an individual may have type A, type B, type AB, or type O blood. Individuals with type A cells develop natural antibodies in their plasma that are directed against, and will agglutinate type B and type AB red blood cells; whereas those with type B red cells develop antibodies against A substance and will agglutinate type A and type AB blood. Persons with type AB blood have neither A nor B antibodies, and are called 'universal recipients', as they can be transfused with cells of either blood type. Individuals with type O blood have only H substance, not A or B substance, on their red blood cells, and are 'universal donors', as their red blood cells are not agglutinated with either A or B antibodies.

The ABO blood-group antigens are complex carbohydrates present as components of glycoproteins or glycosphingolipids of red cell membranes. The precursor of the ABO types is the H substance (Fig. 25.15); the H locus codes for a fucosyltransferase. Individuals with type A blood have, in addition to the H substance, an A gene that codes for a specific GalNAc transferase that adds GalNAc α1,3 to the galactose residue of H substance, to form the A-type glycolipid. Individuals with type B blood have a B gene that codes for a galactosyl transferase that adds galactose α1,3 to the galactose residue of H substance, to form the B-type glycolipid. Individuals with type AB blood have both the GalNAc and the galactosyl transferases, and their red blood cells contain both the A and the B substances. Those with type O blood, and only H substance on their red cell glycosphingolipids, do not make either enzyme. Enzymes such as coffee bean α-galactosidase can remove the galactose from type B red cells, approach that may be helpful for increasing the supply of type O red cells.

Summary

Complex polar lipids are essential components of all living cell membranes. Phospholipids are the major structural lipids of all membranes, but they also have important functional properties as surfactants, as cofactors for membrane enzymes, as mediators of hypersensitivity, and as components of signal transduction systems. The primary route for *de novo* biosynthesis of phospholipids involves the activation of one of the components (either DAG or the head group) with CTP to form a high-energy intermediate, such as CDP-diglyceride or CDP-choline. In addition, there are exchange and modification reactions that enable the animal cell to interconvert various phospholipids. The other major types of membrane lipids are the sphingolipids, including sphingomyelin and various glycolipids. These lipids function as receptors for cell–cell recognition and interactions, and as binding sites for symbiotic and pathogenic bacteria and for viruses. Furthermore, various carbohydrate structures on the glycosphingolipids of red cell membranes are also the antigenic determinants responsible for the ABO blood types. Glycosphingolipids are degraded in the lysosomes by a complex sequence of reactions that involve a stepwise removal of sugars from the non-reducing end of the molecule, with each step involving a specific lysosomal exoglycosidase. A number of inherited lipid-storage diseases result from defects in degradation of glycosphingolipids.

Further reading

Dowhan W. Molecular basis for membrane phospholipid diversity: why are there so many lipids? *Annu Rev Biochem* 1997;**66**:199–232.

England PT. The structure and biosynthesis of glycosyl phosphatidylinositol protein anchors. *Annu Rev Biochem* 1993;**62**:121–138.

Karlsson K. Animal glycosphingolipids as membrane attachment sites for bacteria. *Annu Rev Biochem* 1989;**58**:309–350.

Kent C. Eucaryotic phospholipid biosynthesis. *Annu Rev Biochem* 1995;**64**:315–343.

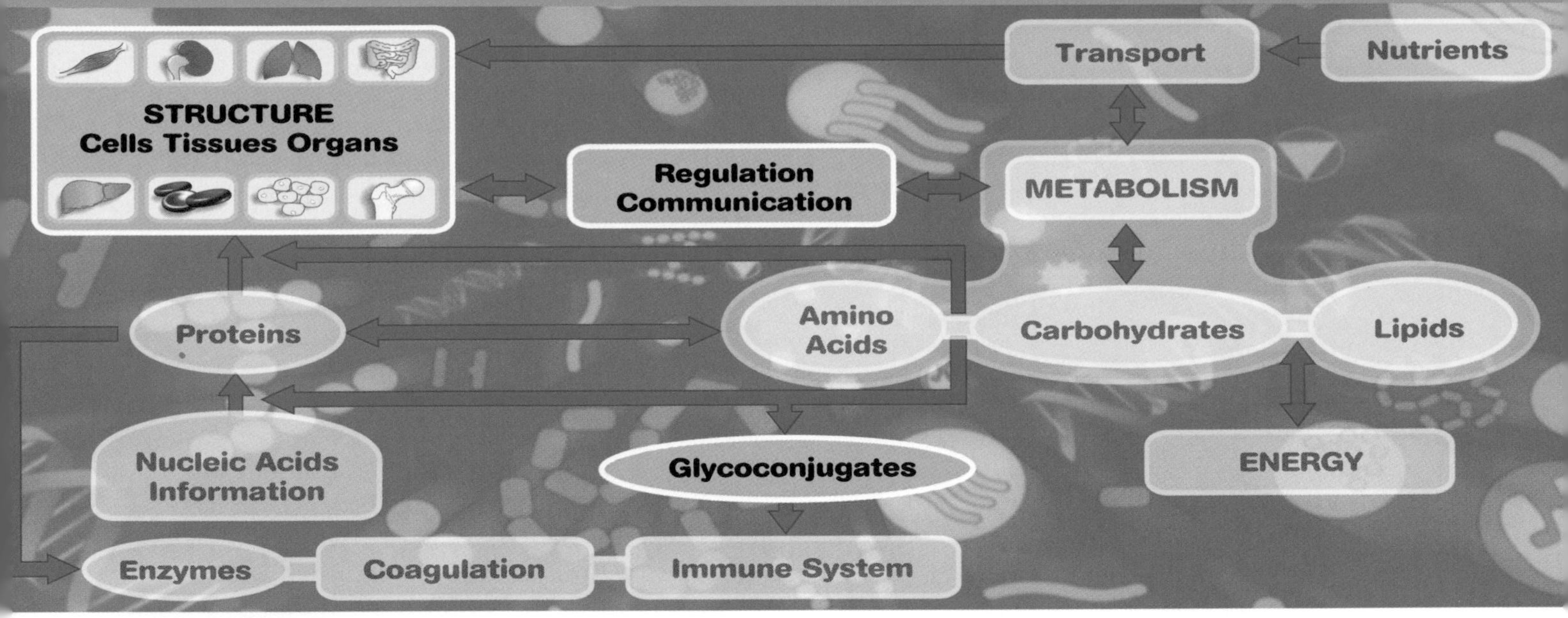

STRUCTURE
Cells Tissues Organs
Transport
Nutrients
Regulation
Communication
METABOLISM
Proteins
Amino
Acids
Carbohydrates
Lipids
Nucleic Acids
Information
Glycoconjugates
ENERGY
Enzymes
Coagulation
Immune System

26 The Extracellular Matrix

The extracellular matrix (ECM) is a complex network of secreted macromolecules located in the extracellular space. The ECM of skin and bone provides the structural framework of the body; however, in all tissues it has a central role in regulating basic cellular processes, including proliferation, differentiation, migration and cell–cell interactions. The macromolecular network of the ECM is made up of collagens, elastin, glycoproteins and proteoglycans, secreted by connective tissue cells such as fibroblasts and epithelial cells. The components of the ECM are in intimate contact with their cells of origin and form a three-dimensional gelatinous bed in which the cells thrive. Proteins in the ECM are also bound to the cell surface, so that they transmit signals resulting from stretching and compression of tissues. The relative abundance, distribution of proteins, and molecular organization of ECM components vary enormously among tissues, depending on their structure and function. Changes in the composition and turnover of the ECM are associated with chronic diseases, such as arthritis, atherosclerosis, cancer and fibrosis.

Collagen

Collagens are the major proteins in the ECM

The collagens are a family of proteins that comprise about 25% of total protein mass in the body. As the primary structural components of the ECM in connective tissues, collagens have an important role in tissue architecture, tissue strength, and a wide variety of cell–cell and cell–matrix interactions. To date, 19 different types of collagens have been identified. They are composed of 34 related, but distinct, peptide chains and vary greatly in their distribution, organization, and function in tissues.

Triple-helical structure of collagens

The structural hallmark of collagens is their triple-helical structure, formed by folding of three peptide chains. These chains vary in size from 600–3000 amino acids. X-ray diffraction analysis indicates that three left-handed helical chains are wrapped around one another in a rope-like fashion, to form a right-handed superhelix structure (Fig. 26.1). This collagen helix is more extended than the α-helix of globular proteins,

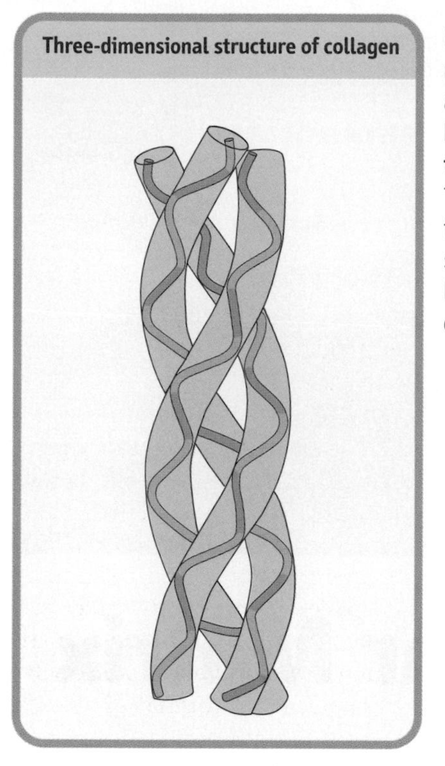
Three-dimensional structure of collagen

Fig. 26.1 Collagen monomer strands assume a left-handed, α-helical tertiary structure. They then associate to form a triple-stranded, right-handed superhelical quarternary structure.

having nearly twice the rise per turn and only three, rather than 3.6, amino acids per turn. Every third amino acid is glycine, because only this amino acid, with the smallest side chain, fits into the crowded central core. The characteristic, repeating sequence of collagen is Gly-X-Y, where X and Y can be any amino acid, but most often X is proline and Y is hydroxyproline. Because of their restricted rotation and bulk, proline and hydroxyproline confer rigidity to the helix. The intra- and interchain helices are stabilized by hydrogen bonds, largely between peptide NH and C = O groups. The side chains of the X and Y amino acids point outwards from the helix, and thus are on the surface of the protein, where they form lateral interactions with other triple helices or proteins.

Types of collagen

Some representative collagens are listed in Figure 26.2. The collagen family of proteins can be divided into two main types: the fibril-forming (fibrillar) and the nonfibrillar collagens.

Fibril-forming collagens

Collagen fibrils can be formed from a mixture of different fibrillar collagens. Dermal collagen fibrils are hybrids of type I and type III collagen, and fibrils in corneal stroma are hybrids of type I and type IV collagen. Type I is the most abundant fibrillar collagen and occurs in a wide variety of tissues; others have a more limited tissue distribution (see Fig. 26.2). Type I and related fibrillar collagens form well-organized, banded fibrils and provide high tensile strength to skin, tendons and ligaments. These collagens are composed of two different α-helical peptide chains, known as α_1(I) and α_2(I), each containing about 1000 amino acids per chain and having a triple-helical domain structure, $[\alpha_1(I)]_2\alpha_2(I)$, along almost the entire length of the molecule. The collagen fibrils are formed by lateral association of triple helices in a 'quarter-staggered' alignment in which each

Osteogenesis imperfecta

Osteogenesis imperfecta, also called brittle-bone disease, is a congenital disease caused by multiple genetic defects in the synthesis of type I collagen. The disease is characterized by fragile bones, thin skin, abnormal teeth and weak tendons. The most severe form of the disease causes death at birth or shortly thereafter. The majority of the individuals with this disease have mutations in genes encoding α_1(I) and α_2(I) chains. Many of these mutations are single-base substitutions that convert glycine in the Gly-X-Y repeat to bulky amino acids, preventing the correct folding of the collagen chains into a triple helix and their assembly to form collagen fibrils.

Members of the collagen family

Type	Class	Distribution
I	fibrillar	skin and tendon
II	fibrillar cartilage	developing cornea and vitreous humor
III	fibrillar	extensible connective tissue, e.g. skin, lung and vascular system
IV	network	basement membranes, kidney, vascular wall
VI	beaded filament	most connective tissue
IX	FACIT	cartilage, vitreous humor
XI	fibril forming	cartilage, bone, placenta
XII	FACIT	embryonic tendon and skin
XIII	transmembrane domain	widely distributed
XIV	FACIT	fetal skin and tendons

Fig. 26.2 Classification and distribution of representative collagens. FACIT, fibril-associated collagen with interrupted triple helices.

molecule is displaced by about one-quarter of its length relative to its nearest neighbor (Fig. 26.3); the quarter-staggered array is responsible for the banded appearance of collagen fibrils in connective tissues. The fibrils are stabilized by both noncovalent forces and interchain crosslinks derived from lysine residues (see below).

Nonfibrillar collagens

Nonfibrillar collagens are a heterogeneous group containing triple-helical segments of variable length, interrupted by one or more intervening nonhelical (noncollagenous) segments. This group includes basement membrane collagens (the type IV family), fibril-associated collagens with interrupted triple helices (FACITs), and collagens with multiple triple-helical domains with interruptions, known as multiplexins. Nonfibrillar collagens may associate with the fibrillar collagens, through their collagenous domains, forming microfibrils and network or mesh-like structures.

Type IV collagen, with the composition $[\alpha_1(IV)]_2\alpha_2(IV)$, is the major structural component of all basement membranes, where it assembles into a flexible network structure. It contains a long triple-helical domain interrupted by short noncollagenous sequences; these interruptions in the helical domain block continued association of two triple helices, oblige them to find another partner, and thus form the meshwork of interactions. Type VI collagen is widely distributed in connective tissues, where it assembles to form beaded filaments or microfibrillar meshwork structures. Each of its three chains contains a short triple-helical domain and large amino (N)-terminal and carboxyl (C)-terminal globular domains. Collagens of the FACIT group have short triple-helical domains interspersed by nonhelical domains. They do not assemble to form fibrils, but associate with the fibrillar collagens.

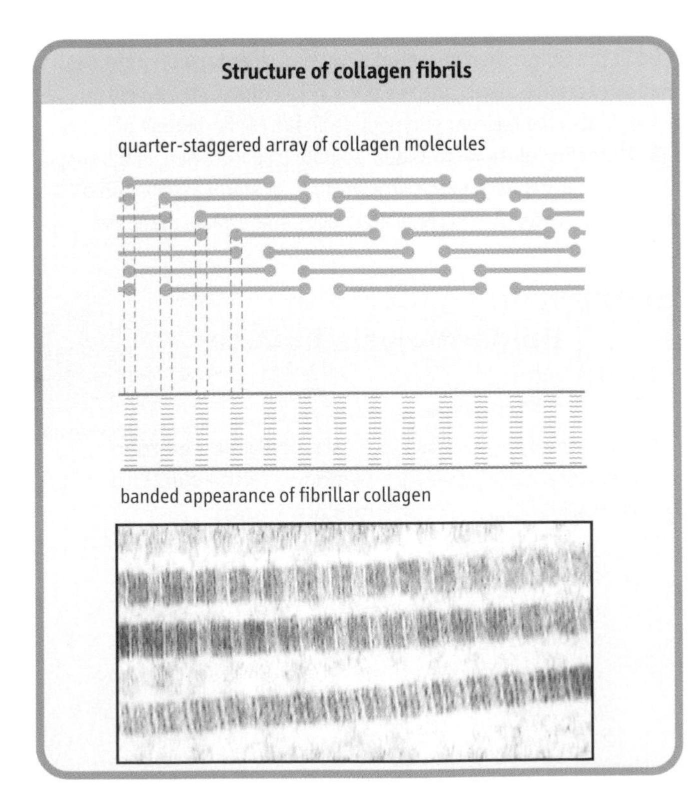

Fig. 26.3 Formation of the quarter-staggered array of collagen molecules in a fibril. The regular overlap of the short, nonhelical termini of the collagen chains yields a regular, banded pattern in the collagen fiber. (Electron micrograph courtesy of Dr Trevor Gray.)

Synthesis and post-translational modification of collagens

Collagen synthesis begins in the rough endoplasmic reticulum (RER)

After synthesis in the RER, the nascent collagen polypeptide undergoes extensive modification, first in the RER, then in the Golgi apparatus, and finally in the extracellular space, where it is modified to a mature extracellular collagen fibril (Fig. 26.4). A nascent polypeptide chain, preprocollagen, is synthesized initially with a hydrophobic signal sequence that facilitates binding of ribosomes to the endoplasmic reticulum (ER) and directs the growing polypeptide chain into the lumen of the ER. Post-translational modification of the protein begins with removal of the signal peptide in the ER, yielding procollagen. Three different hydroxylases then add hydroxyl groups to proline and lysine residues, forming 3- and 4-hydroxyprolines and δ-hydroxylysine. *O*-linked glycosylation occurs by the addition of galactosyl residues to hydroxylysine by galactosyl transferase; a disaccharide may also be formed by addition of glucose to galactosyl hydroxylysine by a glucosyl transferase (see Chapter 24). These enzymes have strict substrate specificity for hydroxylysine or galactosyl hydroxyproline, and they glycosylate only those polypeptides that are in noncollagenous domains. *N*-linked glycosylation also occurs on specific asparagine residues in nonfibrillar domains. The nonfibrillar collagens, with a greater extent of nonhelical domains, are more highly glycosylated than fibrillar collagens. Intra- and interchain disulfide bonds are formed in the C-terminal domains by a protein disulfide isomerase, facilitating the association and folding of peptide chains into a triple helix. At this stage, the procollagen is still soluble, and contains additional, nonhelical extensions at its N- and C- terminals.

Procollagen is finally modified to collagen in the Golgi apparatus

After assembly into the triple helix, the procollagen is transported from the RER to the Golgi apparatus, where it is packaged into cylindrical aggregates in secretory vesicles, then exported to the extracellular space by exocytosis. The nonhelical extensions of the procollagen are now removed in the extracellular space, by specific N- and C-terminal procollagen proteinases. The 'tropocollagen' molecules then self-assemble into insoluble collagen fibrils, which are further stabilized by the formation of aldehyde-derived intermolecular crosslinks. Lysyl oxidase – not to be confused with lysyl hydroxylase

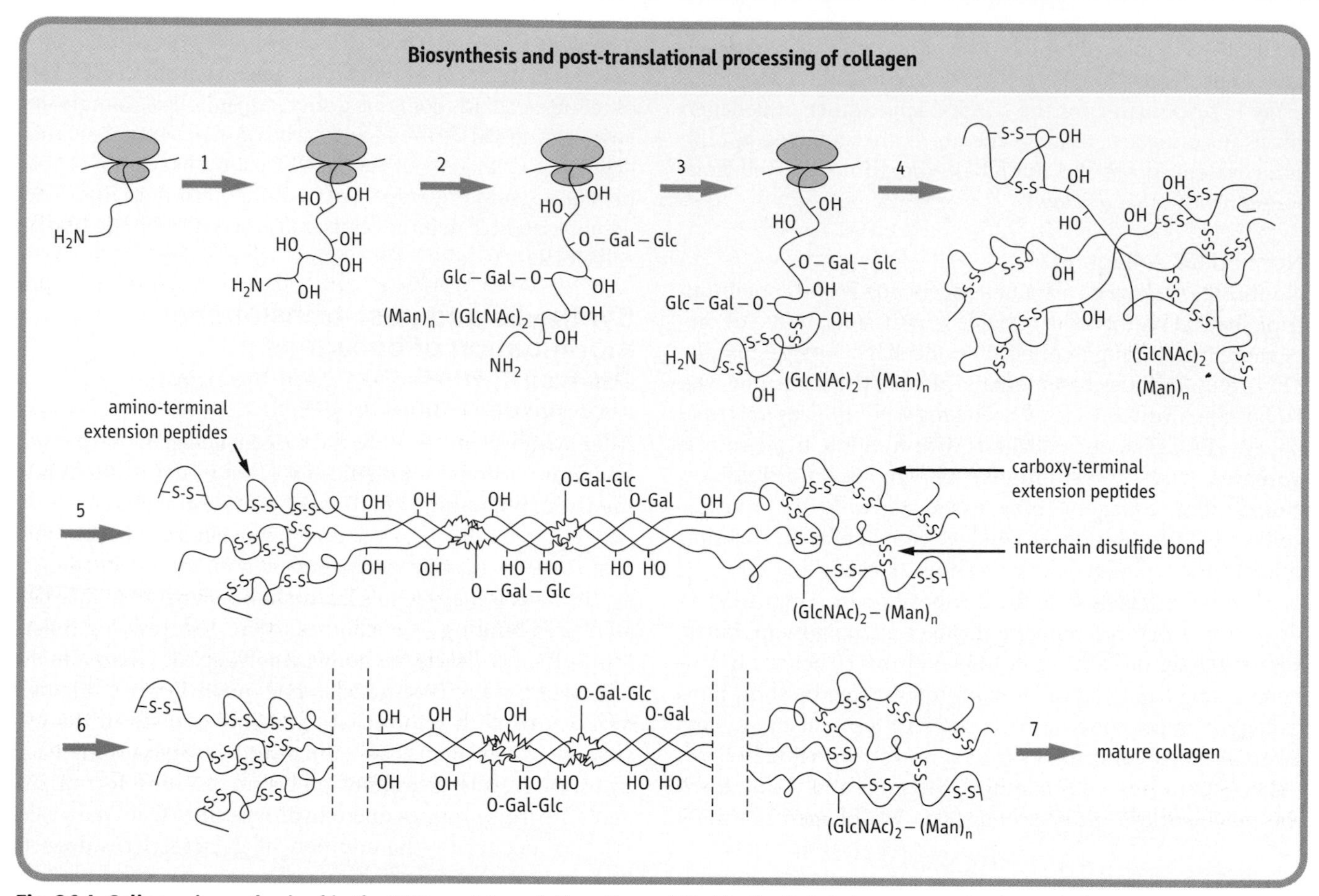

Fig. 26.4 Collagen is synthesized in the RER, post-translationally modified in the Golgi, then secreted, trimmed of extension peptides, and finally assembled into fibrils in the extracellular space. (1) Hydroxylation of proline and lysine residues. (2) Addition of *O*-linked and *N*-linked oligosaccharides. (3) Formation of intrachain disulfide bonds at the N-terminal of the nascent polypeptide chain. (4) Formation of interchain disulfides in the C-terminals, which assist in alignment of chains. (5) Formation of triple-stranded, soluble tropocollagen, and transport to Golgi vesicles. (6) Exocytosis and removal of N- and C-terminal propeptides. (7) Final stages of processing, including lateral association of triple helices, covalent crosslinking and collagen fiber formation. Gal, galactose; Glc, glucose; GlcNAc, *N*-acetylglucosamine; Man, mannose.

involved in formation of hydroxylysine – oxidatively deaminates the amino group from the side chains of some lysine and hydroxylysine residues, producing reactive aldehyde derivatives, known as allysine and hydroxyallysine. The aldehyde groups now form aldol condensation products with neighboring aldehyde groups, generating crosslinks both within and between triple-helical molecules. They may also react with the amino groups of unoxidized lysine and hydroxylysine residues to form Schiff base (imine) crosslinks (Fig. 26.5). The initial products may rearrange, or be dehydrated or reduced to form stable crosslinks, such as lysinonorleucine.

Noncollagenous proteins in the ECM

Elastin

The flexibility required for function of our blood vessels, lungs, ligaments and skin is contributed by a network of elastic fibers in the ECM of these tissues. The predominant

Epidermolysis bullosa

Epidermolysis bullosa is a rare heritable disorder characterized by severe blistering of the skin and epithelial tissue. Three kinds are known:

- **simplex:** blistering in the epidermis, caused by defects in keratin filaments;
- **junctional:** blistering in the dermal–epidermal junction, caused by defects in laminin;
- **dystrophic:** blistering in the dermis, caused by mutations in the gene encoding type VII collagen.

Epidermolysis bullosa illustrates the multifactorial nature of connective tissue diseases that have similar clinical features.

Lathyrism

Lathyrism is a dietary disease characterized by deformation of the spine, dislocation of joints, demineralization of bones, aortic aneurysms, and joint hemorrhages. These problems develop as a result of inhibition of lysyl oxidase, an enzyme required for the crosslinking of collagen chains. Lathyrism can be caused by chronic ingestion of the sweet pea, *Lathyrus odoratus*, the seeds of which contain β-aminopropionitrile, an irreversible inhibitor of lysyl oxidase. Penicillamine, a sulfhydryl agent used for chelation therapy in heavy-metal toxicity, also causes lathyrism, because of either chelation of copper required for lysyl oxidase activity or reaction with aldehyde groups of (hydroxy)allysine, inhibiting collagen crosslinking reactions.

protein of elastic fibers is elastin. Unlike the multigene collagen family, there is only one gene for elastin – a polypeptide about 750 amino acids long. In common with collagens, it is rich in glycine and proline residues, but elastin is more hydrophobic: one in seven of its amino acids is a valine residue. Unlike collagens, elastin contains little hydroxyproline and no hydroxylysine or carbohydrate chains, and does not have a regular secondary structure. Its primary structure consists of alternating hydrophilic and hydrophobic, lysine and valine- rich domains. The lysines are involved in intermolecular crosslinking, while the weak interactions between valine residues in the hydrophobic domains impart elasticity to the molecule.

Elastin has a two-way stretch

The soluble monomeric form of elastin initially synthesized on the RER is called tropoelastin. Except for some hydroxylation of proline, tropoelastin does not undergo post-translational modification. During the assembly process in the extracellular space, lysyl oxidase generates allysine in specific sequences: -Lys-Ala-Ala-Lys- and -Lys-Ala-Ala-Ala-Lys-. As with collagen, the reactive aldehyde of allysine condenses with other allysines or with unmodified lysines. Allysine and dehydrolysinonorleucine on different tropoelastin chains also condense to form pyridinium crosslinks – heterocyclic structures known as desmosine or isodesmosine (Fig. 26.6). Because of the way in which elastin monomers are crosslinked in polymers, elastin can stretch in two dimensions.

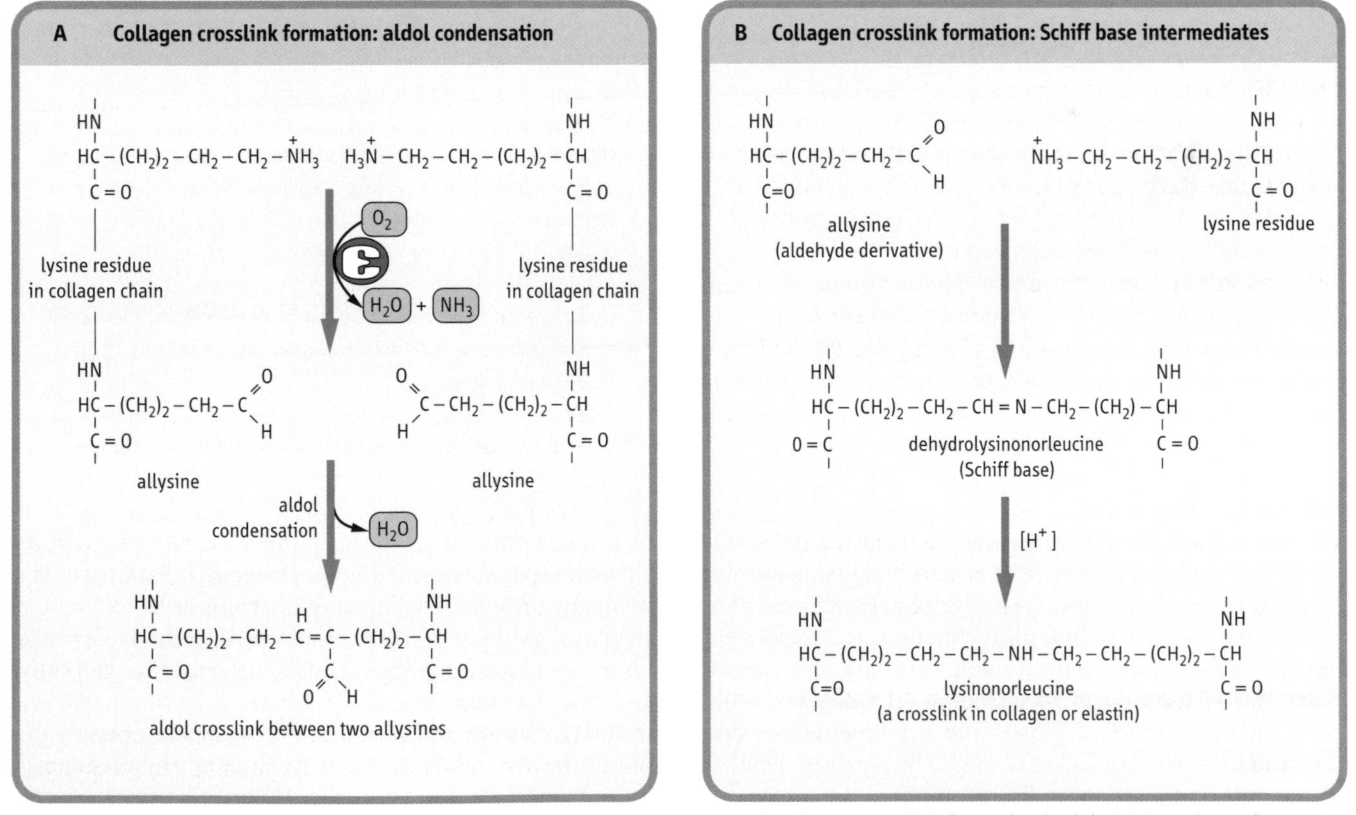

Fig. 26.5 Collagen crosslinking. Allysine and hydroxyallysine are precursors of collagen crosslink formation by (A) aldol condensation and (B) Schiff base (imine) intermediates.

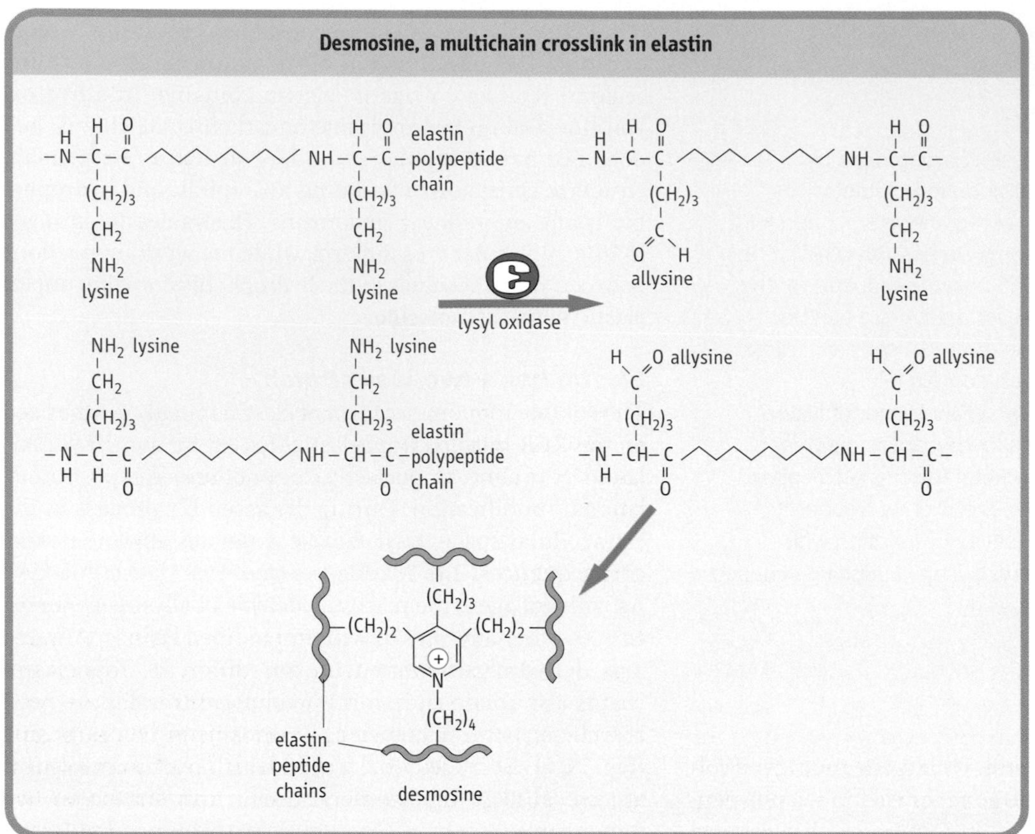

Fig. 26.6 Elastin crosslinking. Allysine and dehydrolysinonorleucine residues in adjacent elastin chains react to form the three-dimensional elastic polymer, crosslinked by desmosine.

Marfan syndrome

The ultrastructure of elastic fiber reveals elastin as an insoluble, polymeric, amorphous core covered with a sheath of microfibrils that contribute to the stability of the elastin fiber. The predominant constituent of microfibrils is the glycoprotein, fibrillin. Marfan syndrome is a relatively rare genetic disease of connective tissues caused by mutations in the fibrillin gene (frequency: 1 in 10 000 births). People with this disease have typically tall stature, long arms and legs, and arachnodactyly (long, 'spidery' fingers). The disease in a mild form causes loose joints, deformed spine, floppy mitral valves (leading to cardiac regurgitation), and eye problems such as lens dislocation. In severely affected individuals, the aorta wall is prone to rupture because of defects in elastic fiber formation.

Fibronectin

Fibronectin is a glycoprotein present in the ECM and also in plasma as a soluble protein. It is involved in a large number of biological activities, including cell adhesion, cell migration, cell morphology, embryonic differentiation, and cytoskeletal organization. It has a central role in binding together the many structures of the ECM, including collagens, proteoglycans, and the cell surface. Fibronectin is a dimer of two identical subunits, each of 230 kDa, joined by a pair of disulfide bonds at their C-terminals. Each subunit is organized into domains, known as type I, II, and III domains, and each of these has several homologous repeating units or modules in its primary structure (Fig. 26.7): there are 12 type I repeats, 2 type II repeats, and 15–17 type III repeats. Each module is independently folded, forming a 'string of beads' type of structure. At least 20 different tissue-specific isoforms of fibronectin have been identified, all produced by alternative splicing of a single precursor messenger ribonucleic acid (mRNA). The alternative splicing is regulated, not only in a tissue-specific manner, but also during embryogenesis, wound healing, and oncogenesis. Plasma fibronectin, secreted mainly by liver cells, lacks two of the type III repeats that are found in cell- and matrix-associated forms of fibronectin.

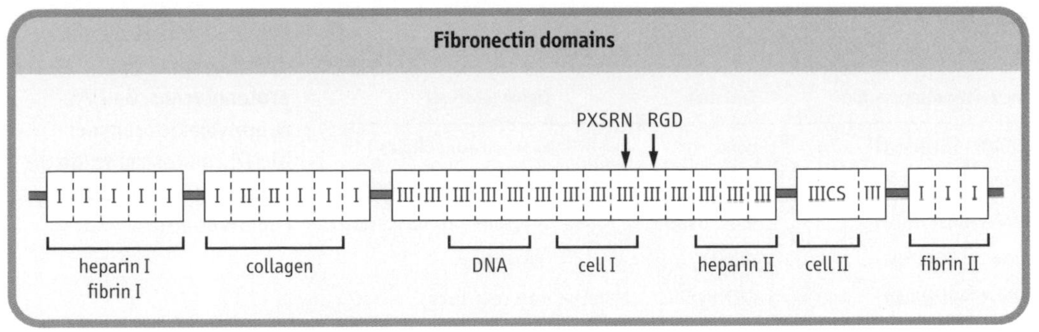

Fig. 26.7 Structural map of fibronectin. This shows various globular domains and domains involved in binding to various molecules in the cell and ECM. RGD, Arg-Gly-Asp; PXSRN, Pro-X-Ser-Arg-Asn.

Functional domains in fibronectin have been identified by their binding affinity for other ECM components, including collagen, heparin, fibrin, and the cell surface

The type I modules interact with fibrin, heparin, and collagen, type II modules have collagen-binding domains, and type III modules are involved in binding to heparin and the cell surface. The specific interactions have been further mapped to short stretches of amino acids. A short peptide containing Arg-Gly-Asp (RGD), present in the tenth type III repeat of fibronectin, binds to the integrin family of proteins present on cell surfaces; this sequence is not unique to fibronectin, but is also found in other proteins in the ECM. Another sequence, Pro-X-Ser-Arg-Asn (PXSRN), present in the ninth type III repeat, is implicated in integrin-mediated cell attachment. The integrins are a family of transmembrane proteins that bind extracellular proteins on the outside and cytoskeletal proteins, such as actin, on the inside of the cell, providing a mechanism for communication between the intracellular and extracellular environments of the cell. The loss of fibronectin from the surface of many tumor cells may contribute to their penetration through the ECM, one of the first steps in metastasis.

Laminins

Laminins are a family of noncollagenous glycoproteins in basement membranes, expressed in variant forms in different tissues. They are large (850 kDa), heterotrimeric molecules, composed of α-, β-, and γ-chains. The three chains are arranged in an asymmetric cruciform molecule, held together by disulfide linkages. Laminins undergo reversible self-assembly in the presence of calcium, to form polymers. Biochemical and electron microscopic studies indicate that all full-length short arms of laminin are required for self-assembly and that the polymer is formed by joining the ends of the short arms in a polymeric network. Like fibronectin, laminins interact with cells through multiple binding sites in several domains of the molecule. The α-chains have binding sites for integrins, dystroglycan (see Chapter 19), heparin, and heparan sulfate (see below). Laminin polymers are also connected to type IV collagen by a single-chain protein, nidogen/entactin, which has a binding site for collagen and, in common with fibronectin, also has an RGD sequence for integrin binding. Nidogen also binds to the core proteins of proteoglycans (see below). It has a central role in formation of crosslinks between laminin and type IV collagen, generating a scaffold for anchoring of cells and ECM molecules in the basement membranes.

Proteoglycans

Proteoglycans are gel-forming components of the ECM. Some proteoglycans are located on the cell surface, where they bind growth factors and other ECM components. They are composed of peptide chains containing covalently bound sugars (see Fig. 24.1). However, the peptide chains of proteoglycans are usually more rigid and extended than the protein portion of the glycoproteins, and the proteoglycans contain much larger amounts of carbohydrate – typically at least 95% carbohydrate. The sugar chains are linear, unbranched oligosaccharides that are much longer than those of the glycoproteins, and may contain 100 or more sugar residues in a chain. Furthermore, the oligosaccharide chains of proteoglycans have a repeating disaccharide unit, usually composed of an amino sugar and a uronic acid. Proteoglycan oligosaccharide chains are polyanionic because of the many negative charges of the carboxyl groups of the uronic acids, and from sulfate groups attached to some of the hydroxyl or amino groups of the sugars.

Structure of proteoglycans

The general structures of the glycosaminoglycans (GAGs), the carbohydrate part of the proteoglycans, are shown in Figure 26.8. The disaccharide repeat is different for each type of GAG, but is usually composed of a hexosamine and a uronic acid residue, except in the case of keratan sulfate, in which the uronic acid is replaced by galactose. The amino sugar in GAGs is either glucosamine ($GlcNH_2$) or galactosamine ($GalNH_2$), both of which are present mostly in their *N*-acetylated forms (GlcNAc and GalNAc), although in some of the GAGs (heparin, heparan sulfate) the amino group is sulfated, rather than acetylated. The uronic acid is usually D-glucuronic acid (GlcUA), but in some cases (dermatan sulfate, heparin) it may be L-iduronic acid (IdUA). With the exception of hyaluronic acid and keratan sulfate, all of the GAGs are attached to protein by a core trisaccharide, Gal-Gal-Xyl, attached to a serine or threonine residue of a core protein.

The proteoglycans			
Proteoglycan	Characteristic disaccharide	Sulfation	Tissue locations
hyaluronic acid	[4GlcUAβ1–3GlcNAcβ1]	none	joint and ocular fluids
chondroitin sulfates	[4GlcUAβ1–3GalNAcβ1]	GalNAc	cartilage, tendons, bone
dermatan sulfate	[4IdUAα1–3GalNAcβ1]	IdUA, GalNAc	skin, valves, blood vessels
heparan sulfate	[4IdUAα1–4GlcNAcβ1]	GlcNAc	cell surfaces
heparin	[4IdUAα1–4GlcNAcβ1]	$GlcNH_2$, IdUA	mast cells, liver
keratan sulfates	[3Galβ1–4GlcNAcβ1]	GlcNAc	cartilage, cornea

Fig. 26.8 Structure and distribution of the proteoglycans. GalNAc, *N*-acetylgalactosamine; $GlcNH_2$, glucosamine; GlcUA, D-glucuronic acid; IdUA, L-iduronic acid.

Keratan sulfate is also attached to protein, but in that case the linkage is either through an *N*-linked oligosaccharide (keratan sulfate I), or an *O*-linked oligosaccharide (keratan sulfate II). Hyaluronic acid is the only GAG that does not appear to be attached to a core protein.

Hyaluronic acid

Hyaluronic acid is composed of repeating units of GlcNAc and GlcUA. This polysaccharide chain is the longest of the GAGs, with molecular weight of $1 \times 10^5 - 1 \times 10^7$ (250–25 000 repeating disaccharide units), and is the only nonsulfated GAG.

The chondroitin sulfates

The chondroitin sulfates contain GalNAc rather than GlcNAc as the amino sugar, and their polysaccharide chains are shorter: $2–5 \times 10^5$ Da. In addition, the chondroitin chains are attached to protein via the trisaccharide linkage region (Gal-Gal-Xyl) and they contain sulfate residues linked to either the 4- or 6-hydroxyl groups of GalNAc.

Dermatan sulfate

Dermatan sulfate was originally isolated from skin, but is also found in blood vessels, tendon and heart valves. This GAG is similar in structure to chondroitin sulfate, but has a variable amount of L-IdUA, the C-5-epimer of D-GlcUA, formed in an unusual reaction by epimerization of GlcUA after it has been incorporated into the polymer. Dermatan sulfate has a higher charge density than the chondroitin sulfates, as it contains sulfate residues on the C-2 position of some of the IdUA residues, and on the 4-hydroxyl groups of GalNAc.

Heparin and heparan sulfate

Heparin and heparan sulfate consist primarily of repeating disaccharide units of $GlcNH_2$ with IdUA or GlcUA, respectively. The linkage between the amino sugar and the uronic acid is uniformly 1–4, rather than the alternating 1–4/1–3 linkages seen in other GAGs. Most of the $GlcNH_2$ units of heparin are *N*-sulfated, whereas many of the IdUA residues are sulfated at the C-2 hydroxyl group, and the $GlcNH_2$ residues at the C-6 hydroxyl group. Heparin and heparan sulfate are the most highly charged of the GAGs. Although the structures of these two polymers are closely related, their distribution in the body and their functions are quite different: heparin is found intracellularly as a proteoglycan, is released into the extracellular space, and has strong anticoagulant activity. In contrast, heparan sulfate is bound in the ECM or on the surface of cells, and has only weak anticoagulant activity.

Keratan sulfate

The final GAG structure shown in Figure 26.8 is keratan sulfate (KS). This is a rather unusual GAG because it is linked to protein either by an *N*-linked (KS I), or by an *O*-linked (KS II) oligosaccharide. Thus it has features common to both proteoglycans and glycoproteins. It is considered to be a proteoglycan, however, because the glycan portion has a repeating disaccharide unit and a long, linear chain. The repeating unit is composed of GlcNAc and galactose in place of the uronic acid. Both the GlcNAc and the galactose are generally sulfated on the C-6 hydroxyl groups.

Synthesis and degradation of proteoglycans

Proteoglycans are synthesized by a series of glycosyl transferases, epimerases and sulfotransferases, beginning with the synthesis of the core oligosaccharide while the core protein is still in the RER. Synthesis of the repeating oligosaccharide and other modifications take place in the Golgi apparatus. As with the synthesis of glycoproteins and glycolipids, separate enzymes are involved in individual steps. For example, there are separate galactosyl transferases for each of the galactose units in the core, a separate GlcUA transferase for the core and repeating disaccharides, and separate sulfotransferases for the C-4 and C-6 positions of the GalNAc residues of chondroitin sulfates. Phosphoadenosine phosphosulfate (PAPS) is the sulfate donor for the sulfotransferases. These pathways are illustrated in Figure 26.9, for chondroitin-6-sulfate.

Defects of proteoglycan degradation lead to mucopolysaccharidoses

The degradation of proteoglycans occurs in lysosomes. The protein portion is degraded by lysosomal proteases, and the GAG chains are degraded by the sequential action of a number of different lysosomal acid hydrolases. The stepwise

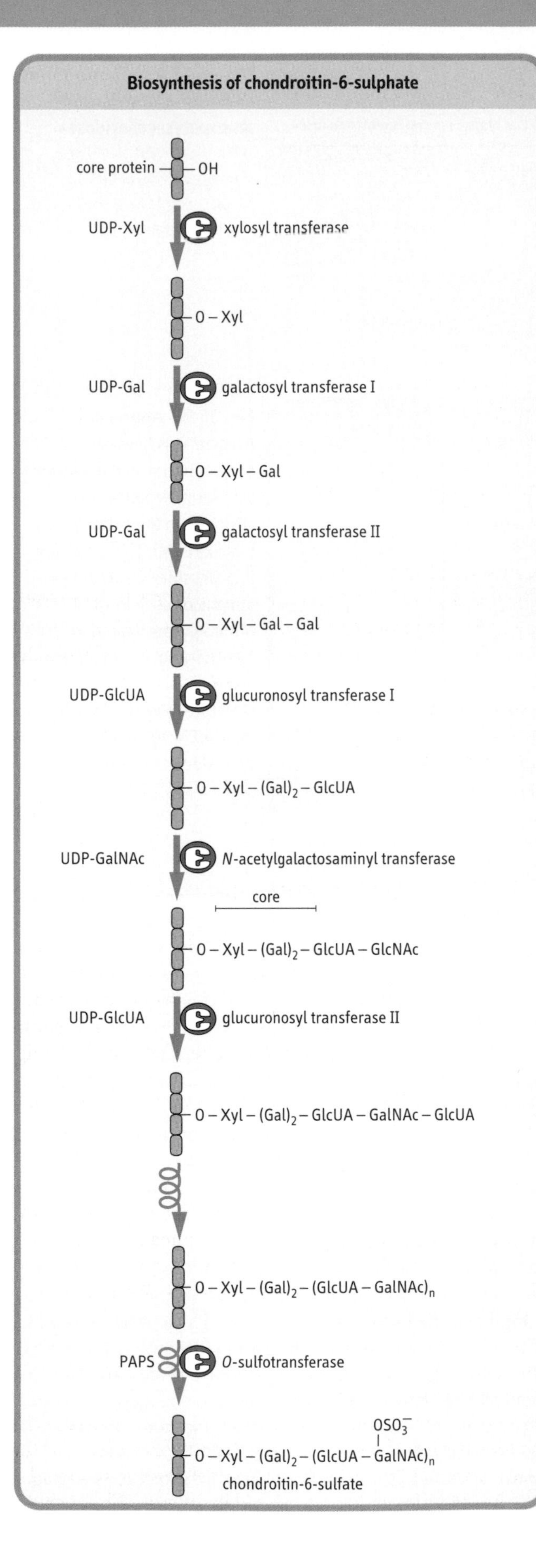

Fig. 26.9 Synthesis of the proteoglycan, chondroitin-6-sulfate. Several enzymes participate in this pathway. Xyl, xylose.

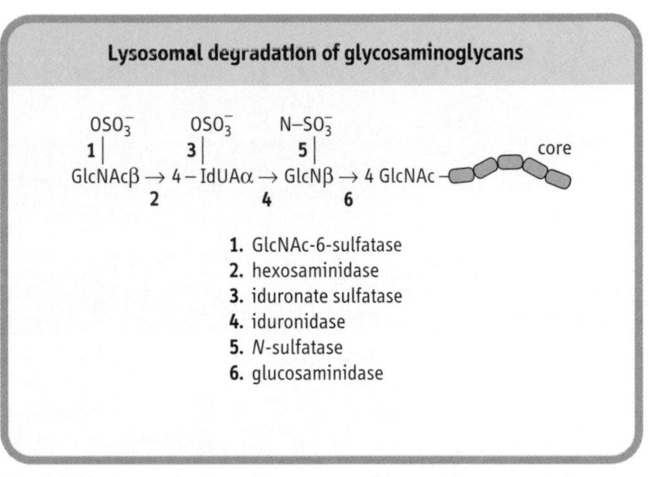

Fig. 26.10 Degradation of heparan sulfate. This proceeds by a defined sequence of lysosomal hydrolase activities.

degradation of GAGs involves exoglycosidases and sulfatases, beginning from the external end of the glycan chain. This may involve the removal of sulfate by a sulfatase, then removal of the terminal sugar by a specific glycosidase, and so on. Figure 26.10 shows the steps in the degradation of dermatan sulfate. As with degradation of glycosphingolipids, if one of the enzymes involved in the stepwise pathway is missing, the entire degradation process is halted at that point, and the undegraded molecules accumulate in the lysosome. The lysosomal storage diseases resulting from accumulation of GAGs are known as mucopolysaccharidoses (Fig. 26.11), because of the original designation of GAGs as mucopolysaccharides. There are more than a dozen such mucopolysaccharidoses, resulting from defects in degradation of GAGs. In general, these diseases can be diagnosed by the identification of specific GAG chains in the urine, followed by assay of the specific hydrolases in leukocytes or fibroblasts. Although they are rare genetic diseases, the mucopolysaccharidoses have been important in the elucidation of the role of lysosomes in health and disease, and the mechanisms involved in biosynthesis of lysosomes and targeting of lysosomal enzymes.

Functions of the proteoglycans

'Bottlebrushes and reinforced concrete'

Proteoglycans are found in association with most tissues and cells. One of their major roles is to provide structural support to tissues, especially cartilage and connective tissue. In cartilage, large aggregates, composed of chondroitin sulfate and keratan sulfate chains linked to their core proteins, are noncovalently associated with hyaluronic acid via link proteins, forming a jelly-like matrix in which the collagen fibers are embedded. This macromolecule of macromolecules, a

The mucopolysaccharidoses			
Syndrome		Deficient enzyme	Product accumulated in lysosomes and secreted in urine
Hurler's		iduronate sulfatase	heparan and dermatan sulfate
Hunter's		α-iduronidase	heparan and dermatan sulfate
Morquio's	A	galactose-6-sulfatase	keratan sulfate
	B	β-galactosidase	keratan sulfate
Sanfilippo's	A	heparan sulfamidase	heparan sulfate
	B	*N*-acetylglucosaminidase	
	C	*N*-acetylglucosamine-6-sulfatase	

Fig. 26.11 Enzymatic defects characteristic of various mucopolysaccharidoses.

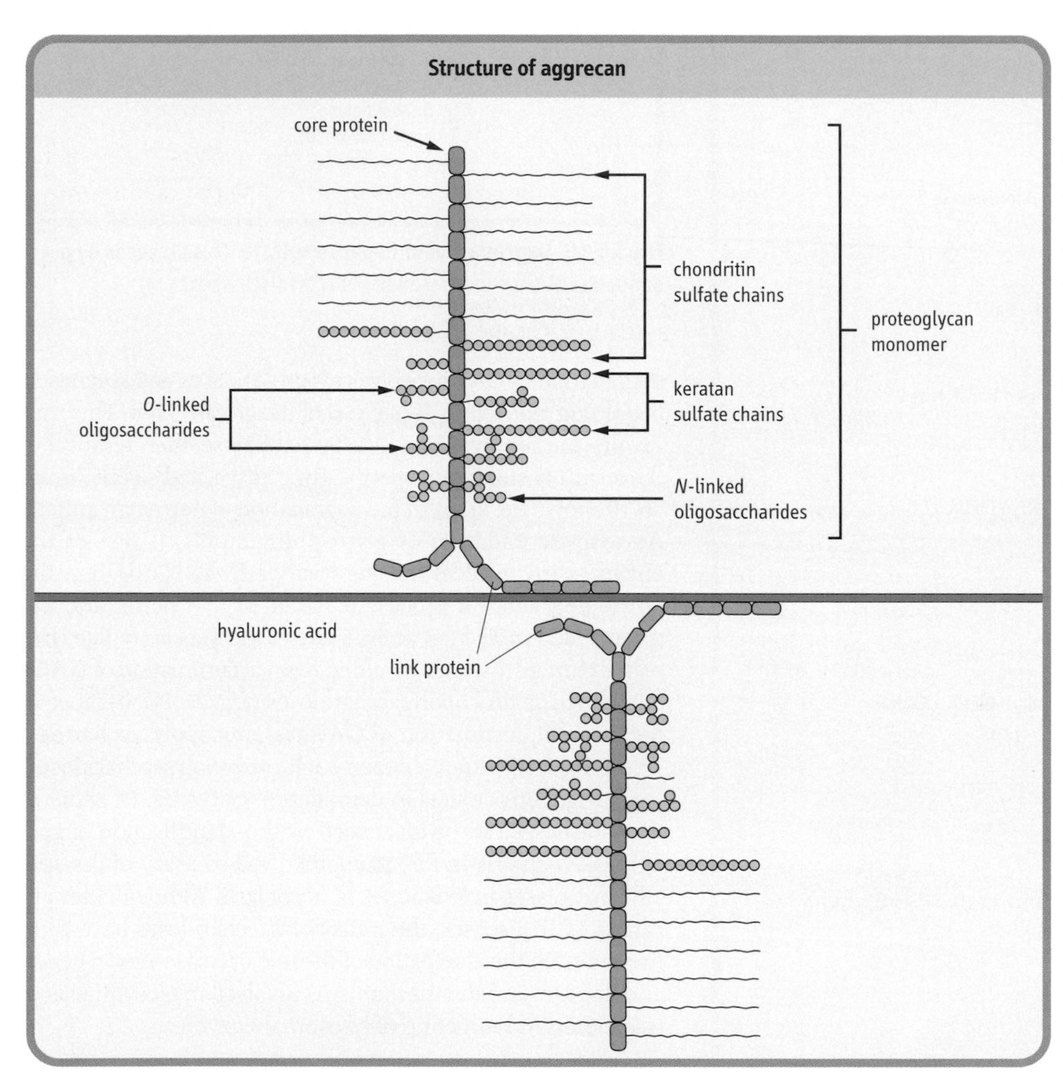

Fig. 26.12 Aggrecan. Associations between proteoglycans and hyaluronic acid form an aggrecan structure in the extracellular matrix (ECM). The extension of this structure yields a three-dimensional array of proteoglycans bound to hyaluronic acid, which creates a stiff matrix or 'bottlebrush' structure in which collagen and other ECM components are embedded.

'bottlebrush' structure known as aggrecan (Fig. 26.12), provides both rigidity and stability to the connective tissue, but also allows for a degree of flexibility and compressibility, enabling the tissue to withstand shock. Because of their negative charge, the GAG chains bind large amounts of monovalent and divalent cations: a cartilage proteoglycan molecule of 2×10^6 Da would have an aggregate charge of about 10 000. The maintenance of electrical neutrality consequently requires a high concentration of counterions. These ions draw water into the ECM, causing swelling and stiffening of the matrix, the result of tension between osmotic forces and binding interactions between proteoglycans and collagen. The overall structure of cartilage can be likened to that of the vertical reinforced-concrete slabs poured during the construction of large buildings, in which steel rods (collagen fibers) are embedded in an amorphous layer of cement (the proteoglycan aggregates). Collagen stabilizes the network of proteoglycans in cartilage in much the same way that the reinforcing rods in the concrete provide structural strength for the cement walls.

Although the amounts involved are low compared with those in skin and cartilage, organs such as the liver, brain or kidney also contain a variety of proteoglycans:

- **hepatocytes:** heparan sulfate is the principal GAG; it is present both intracellularly and on the cell surface of the hepatocyte, and the attachment of hepatocytes to their substratum in cell culture is mediated, at least in part, by this proteoglycan;
- **kidney:** changes in both the collagen and proteoglycan content of the renal basement membrane are associated with diabetic renal disease. In this case, the change in structure and charge of the proteoglycans is associated with a change in the filtration selectivity of the glomerulus;
- **cornea:** two populations of proteoglycans have been identified in the cornea, one containing keratan sulfate and the other dermatan sulfate. These molecules have a much smaller hydrodynamic size than the large cartilage proteoglycans, which may be required for interaction of the corneal proteoglycans with the tightly packed and oriented collagen fibers in this transparent tissue. Corneal clouding in macular corneal dystrophy is associated with undersulfation of keratan sulfate I proteoglycan.

Some proteoglycans or GAGs, especially heparin and heparan sulfate, probably have important physiologic roles in binding proteins or other macromolecules:

- **mast cells (granulated cells involved in the inflammatory response):** heparin is believed to function as an intracellular binding site for inactive proteinases in secretory granules;
- **the vascular wall:** proteoglycans are involved with the binding of proteins and enzymes, such as low-density lipoprotein and lipoprotein lipase, to the vascular wall. They may also inhibit clot formation on the vascular wall by surface activation of antithrombin III (see Chapter 6).

Summary

The ECM contains a complex array of fibrillar and network-forming collagens, elastin fibers, a stiff gelatinous matrix of proteoglycans, and a number of glycoproteins that mediate the interaction of these molecules with one another and with the cell surface. These molecules and their interactions afford structure, stability, and elasticity to the ECM, and provide a route for communication between the intra- and extracellular environments in tissues. The heterogeneity of both the protein and the carbohydrate components of these molecules provides for great diversity in the structure and function of the ECM in various tissues.

Further reading

Aumailley M, Gayraud B. Structure and biological activity of the extracellular matrix. *J Mol Med* 1998;**76:**253–265.

Brown JC, Timpl R. The collagen superfamily. *Int Arch Allergy Immunol* 1995;**107:**484–490.

Clark EA, Bruggse JS. Integrins and signal transduction pathways: the road taken. *Science* 1995;**268:**233–239.

Fraser JR, Laurent TC, Laurent UB. Hyaluronan: its nature, distribution, functions and turnover. *J Int Med* 1997;**242:**27–33.

Mewer UM, Engvall E. Domains of laminin. *J Cell Biochem* 1996;**61:**493–501.

Prockop DJ, Kivirikko KI. Collagens: molecular biology, diseases and potentials for therapy. *Annu Rev Biochem* 1995;**64:**403–434.

Rosenbloom J, Abrams WR, Mecham R. Extracellular matrix 4: the elastic fiber. *FASEB J* 1993;**7:**1208–1218.

Timpl R, Brown JC. Supramolecular assembly of basement membranes. *Bioessays* 1996;**18:**123–132.

Tilstra DJ, Byers PH. Molecular basis of hereditary disorders of connective tissue. *Annu Rev Med* 1994;**45:**149–163.

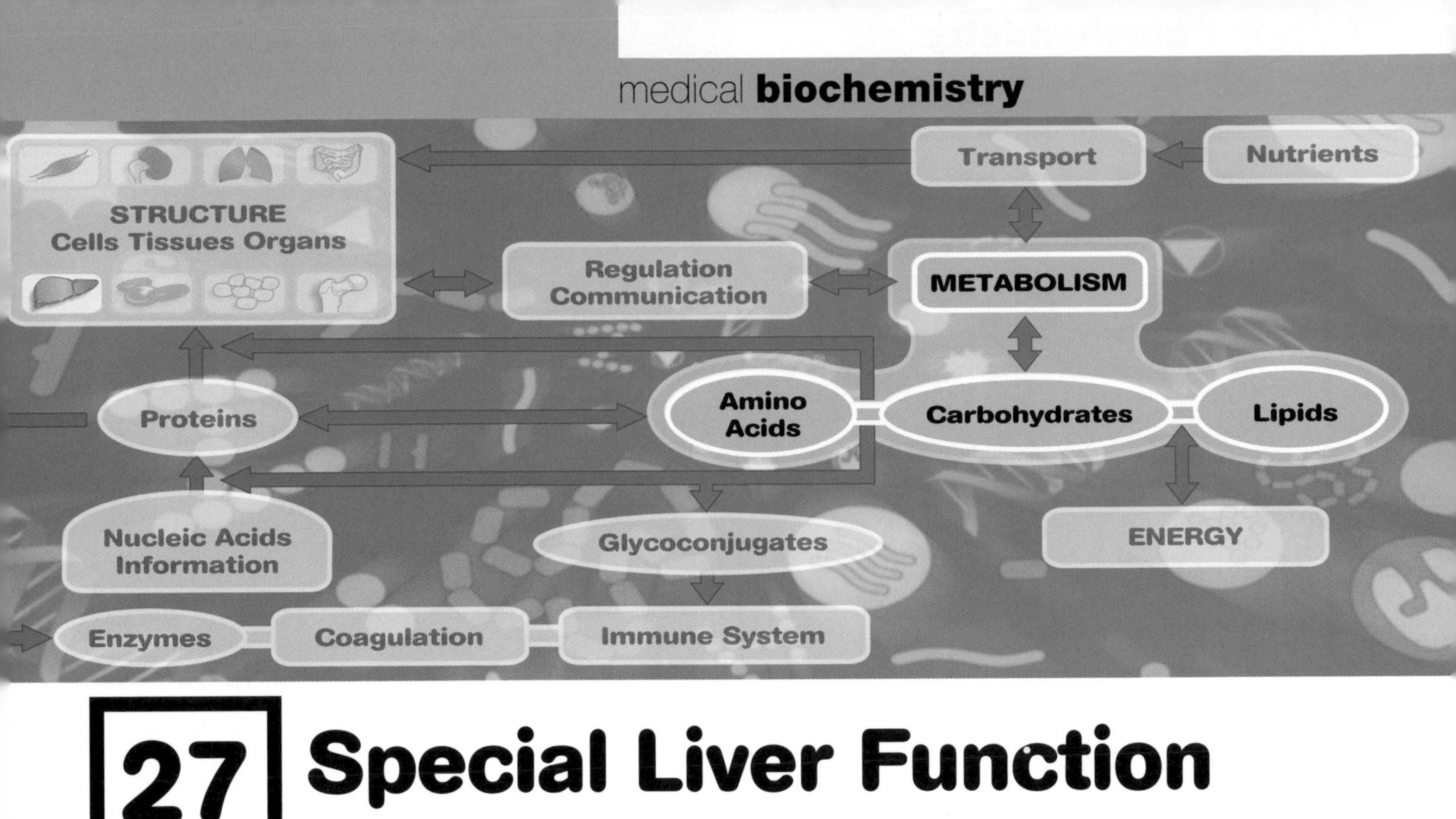

27 Special Liver Function

The liver has a central role in metabolism, because of both its anatomical connections and its many biochemical functions. It receives venous blood from the intestine, and thus all the products of digestion, in addition to ingested drugs and other xenobiotics, perfuse the liver before entering the systemic circulation. The hepatic parenchymal cells, the hepatocytes, have an immensely broad range of synthetic and catabolic functions, which are summarized in Figure 27.1. The liver has important roles in the regulation of carbohydrate and lipid metabolism, in amino acid metabolism, in the synthesis and breakdown of plasma proteins, and in the storage of vitamins and metals; it also has the ability to metabolize and so detoxify an infinitely wide range of xenobiotics. The liver also has an excretory function, in which metabolic waste products are secreted into a branching system of ducts known as the biliary tree, which in turn drains into the small intestine; the biliary constituents are then excreted fecally.

The liver has a substantial reserve metabolic capacity; mild liver disease may cause no symptoms, and be detected only as biochemical changes in the blood. However, the patient with severe liver disease has a yellow pigmentation of the skin (jaundice), bruises readily, may bleed torrentially, has an abdomen distended with fluid (ascites), and may be confused or unconscious (hepatic encephalopathy) (Fig. 27.2). This chapter will describe the specialized metabolic functions of the liver and the abnormalities that occur in liver disease.

Hepatic function

Function	Markers of impairment in plasma
heme catabolism	bilirubin ↑
carbohydrate metabolism	glucose ↓
protein synthesis	albumin ↓ prothrombin time ↑
protein catabolism	ammonia ↑ urea ↓
lipid metabolism	cholesterol ↑ triglycerides ↑
drug metabolism	drug $t_{\frac{1}{2}}$ ↑
bile acid metabolism	bile acids ↑

Fig. 27.1 Functions of hepatic parenchymal cells and their disturbances in linear disease. $t\frac{1}{2}$ = biological half-time.

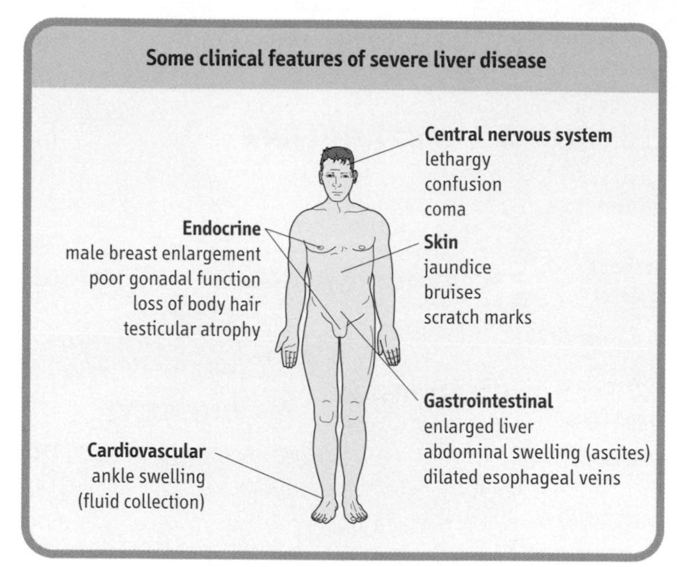

Fig. 27.2 Clinical features of severe liver disease.

The structure of the liver

The liver occupies the right upper quadrant of the abdominal cavity, lying below the diaphragm and protected by the rib cage. It is the largest solid organ in the body and, in adults, weighs about 1500 g. Approximately 75% of the blood flow to the liver is supplied by the portal vein, which arises from the intestine. Blood leaving the liver enters the venous system through the hepatic vein. The biliary component of the liver comprises the gall bladder and bile ducts.

The microscopic structure of the liver facilitates the exchange of metabolites between hepatocytes and plasma

Under the microscope, the substance of the liver is composed of a very large number of lobules, each of which is polyhedral in shape (Fig. 27.3). Blood sinusoids arise from the terminal branches of the portal vein and interconnect and

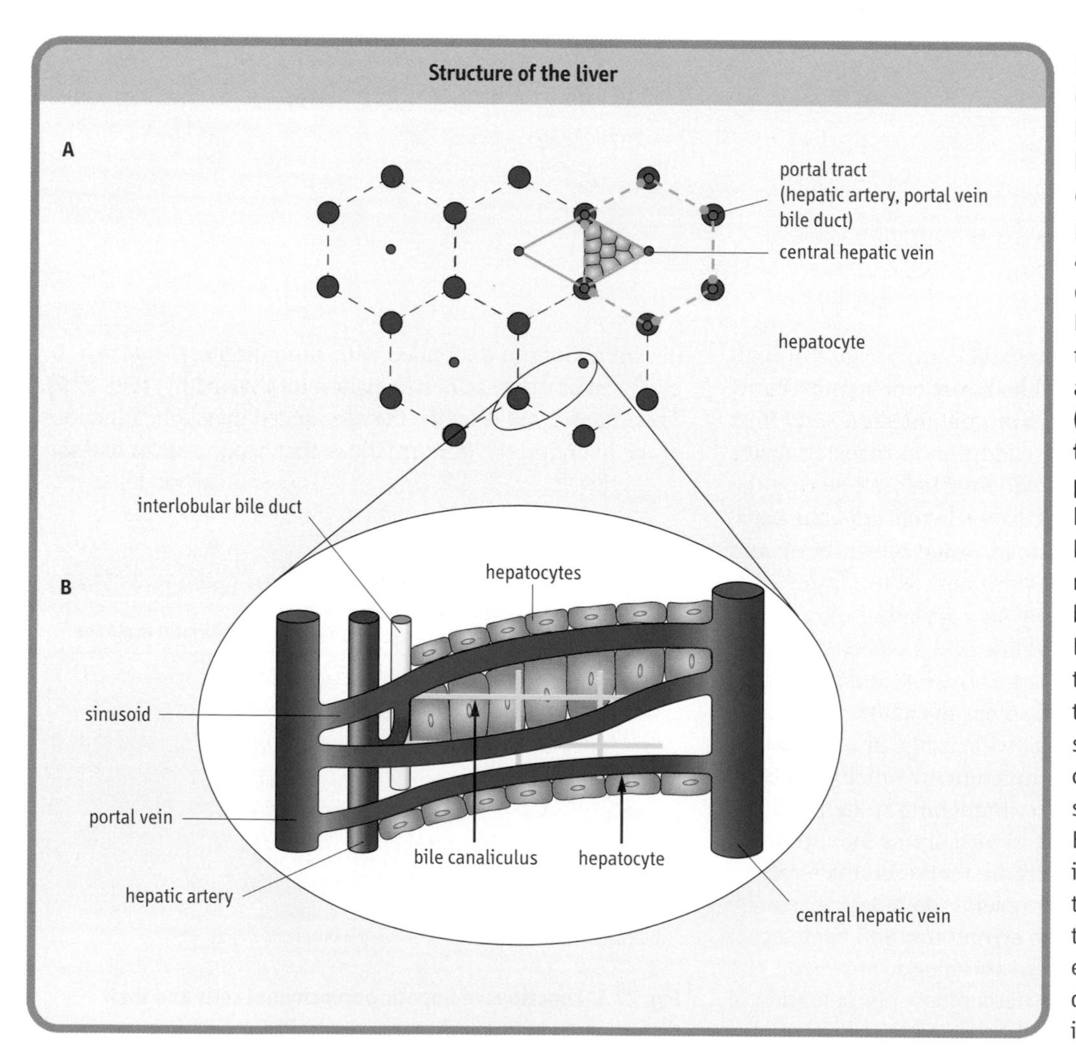

Fig. 27.3 Liver structure. (A) Cross-section through the liver, showing the hexagonal liver lobules. Portal tracts, each containing a terminal part of the portal vein and hepatic artery in addition to an interlobular bile duct, lie at each 'corner' of the liver lobule, and a terminal tributary of the hepatic vein lies at the center of each lobule. **(B) Magnified and schematic three-dimensional view of part of the hepatic lobule.** A hepatic sinusoid arises from a branch of the portal vein and receives a connection from a branch of the hepatic artery. Each sinusoid weaves inwards to the center of the lobule through sheets of hepatocytes a single cell in thickness. At the center of the lobule, the sinusoid joins the central vein. Biliary canaliculi are formed as intercellular channels between the hepatocytes. In the portal tracts, the canaliculi acquire an epithelial lining to form biliary ductules, which in turn join the interlobular bile duct.

interweave through these sheets of hepatocytes before joining the central lobular vein.

Sinusoids are lined by two cell types. The first are vascular endothelial cells, which are loosely connected one with another, leaving numerous gaps, and thus form a net-like lining to the sinusoids. Moreover, there is no basement membrane between the endothelial cells and the hepatocytes. This structural arrangement facilitates the exchange of metabolites between hepatocyte and plasma. The second type of sinusoidal cell, known as Kupffer cells, are mononuclear phagocytes; they are generally found in the gaps between adjacent endothelial cells.

Heme synthesis

Heme, a constituent of hemoglobin, myoglobin and cytochromes is synthesized in most cells of the body. The liver is the main nonerythrocyte source of its synthesis. Heme is a porphyrin, a cyclic compound which contains four pyrrole rings linked together by methenyl bridges. It is synthesized from glycine and succinyl Coenzyme A, which condense to form 5-aminolevulinate (5-ALA). This reaction is catalysed by 5-ALA synthase, located in mitochondria and is the rate limiting step in heme synthesis. Subsequently, in the cytosol, two molecules of 5-ALA condense to form a molecule containing a pyrrole ring, porphobilinogen (PBG). Then, four PBG molecules combine to form a linear tetrapyrrole compound which cyclizes to yield uroporphyrinogen III and then coproporphyrinogen III. Final stages of the pathway occur in the mitochondria where a series carboxylation and oxidation of side chains in uroporphyrinogen III yield protoporphyrin IX. At the final stage, iron (Fe^{2+}) is added by ferrochelatase to protoporphyrin IX to form heme. Heme controls the rate of its synthesis by inhibiting 5-ALA synthase (Fig. 27.4).

Bilirubin metabolism

Bilirubin is a catabolic product of heme. About 75% is derived from the hemoglobin of senescent red blood cells, which are phagocytosed by mononuclear cells of the spleen, bone marrow, and liver; in normal adults, this contributes a daily load of 250–350 mg of bilirubin. The ring structure of heme is oxidatively cleaved to biliverdin by heme oxygenase, a P450 cytochrome. Biliverdin is, in turn, enzymatically reduced to bilirubin (Fig. 27.5). The normal plasma concentration of bilirubin is less than 17 μmol/L (1.0 mg/dL); however, increased concentrations (more than 50 μmol/L or 3 mg/dL) are readily recognized clinically, because bilirubin imparts a yellow color to the skin (jaundice). Abnormalities in bilirubin metabolism are important in the diagnosis of liver disease.

Porphyrias

Defects in the heme synthetic pathway lead to rare disorders known as porphyrias. Different forms of porphyrias are caused by defects in enzymes starting from 5-ALA synthase and ending with ferrochelatase. Porphyrias are classified as hepatic or erythropoietic depending on the primary organ affected.

Three porphyrias are known as acute porphyrias and can be a cause of emergency admissions with abdominal pain (which needs to be differentiated from various surgical causes). They also cause neuropsychiatric symptoms. Acute intermittent porphyria (AIC) is caused by the deficiency of hydroxymethylbilane synthase: an enzyme converting PBG to a linear tetrapyrrole: in this disorder the concentrations of 5-ALA and PBG increase in plasma and urine.

Hereditary coproporphyria is due to the defect in the conversion of coproporphyrinogen III to protoporphyrinogen III (copro-oxidase). The third form of acute porphyria is the variegate porphyria, the clinical manifestations of which are very similar to AIC. Other porphyrias, such as porphyria cutanea tarda, present clinically as the sensitivity of skin to light (photosensitivity) which may cause disfiguration and scarring. Also, the pathway is inhibited by lead at the stage of porphobilinogen synthase.

Laboratory diagnosis of porphyrias involves both the measurements of metabolites of heme synthetic pathway and of relevant enzymes. The metabolite measurements are performed in plasma, urine, feces and erythrocytes. Also, the measurement of porphyrin intermediates is helpful in the diagnosis of lead poisoning. The most important test for the diagnosis of acute porphyrias is the measurement of PBG in urine and plasma.

Bilirubin is metabolized by the hepatocytes and excreted by the biliary system

Whereas biliverdin is water soluble, paradoxically, bilirubin is not, and so must be further metabolized before excretion (Fig. 27.6). This occurs in the liver, to which bilirubin is transported as a complex with plasma albumin; the hepatic uptake of bilirubin is mediated by a carrier, and this process may be competitively inhibited by other organic anions. The hydrophilicity of bilirubin is increased by esterification of one or both of its carboxylic acid side chains with glucuronic acid, xylose, or ribose. The glucuronide diester is the major conjugate, and its formation is catalyzed by a uridine diphosphate (UDP) glucuronyl transferase. Conjugated bilirubin is then secreted by the hepatocyte into the biliary canaliculi.

Conjugated bilirubin in the gut is catabolized by bacteria to form stercobilinogen, also known as fecal urobilinogen, which is a colorless compound. On oxidation, however, stercobilinogen forms stercobilin (otherwise known as fecal urobilin), which is colored; most stercobilin is excreted in the feces, and is responsible for the color of feces. Some stercobilin may be reabsorbed from the gut and, being water soluble, can then be re-excreted by either the liver or the kidneys.

Bile acid metabolism

Bile acids: key elements in fat metabolism

Bile acids have a detergent-like effect, solubilizing biliary lipids and emulsifying dietary fat in the gut to facilitate its digestion (Chapter 9). They are synthesized by hepatocytes, which hydroxylate cholesterol at the C-7 position to produce the primary bile acids, cholic acid and chenodeoxycholic acid. These are conjugated with glucuronic acid, or with the amino acids taurine or glycine, and secreted in bile. In the intestine, the primary bile acids are dehydroxylated by bacteria to form the secondary bile acids: cholic acid is converted to deoxycholic acid, and chenodeoxycholic acid to lithocholic acid. Both primary and secondary bile acids are also deconjugated by bacteria.

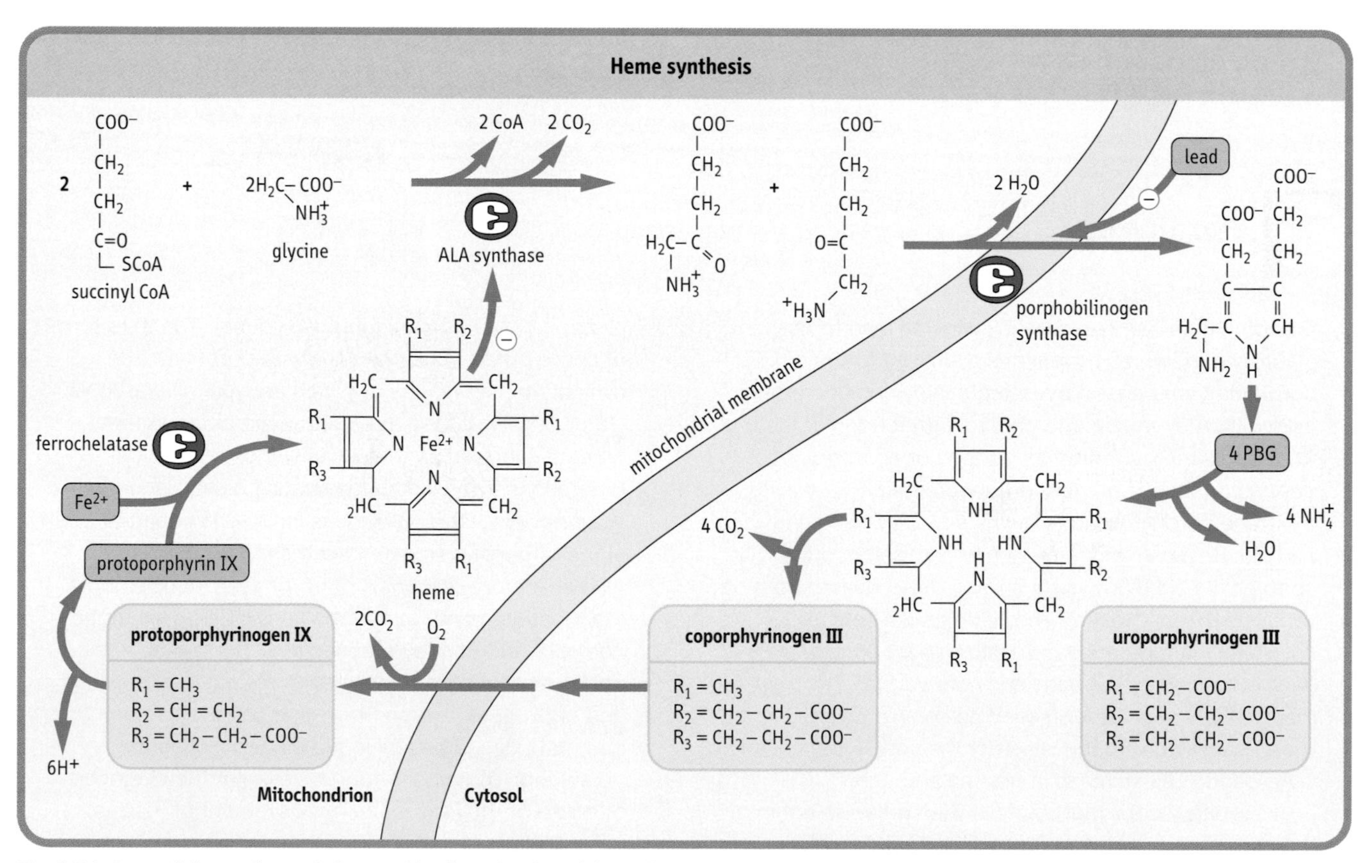

Fig. 27.4 Part of the pathway is located in the mitochondria and part in the cytosol. ALA, aminolevulinate, PBG, porphobilinogen.

Intestinal bile acids are reabsorbed and returned to the liver for resecretion – a process known as the enterohepatic circulation. Elevated plasma, bilirubin levels and levels which occur in liver disease cause itching.

The enterohepatic circulation of bile acids is a target of some lipid-lowering drugs

The liver is the major site of both synthesis and catabolism of cholesterol; indeed, the chenodeoxycholic acid/lithocholic acid pathway is the major route by which cholesterol is excreted. Consequently, the rate of production of bile acids from cholesterol can affect the plasma concentration of low-density lipoprotein (LDL) cholesterol. Increasing the rate of bile acid production lowers the intracellular concentration of cholesterol, upregulates hepatic cell membrane LDL receptors, and enhances the clearance of LDL from plasma. Interrupting the enterohepatic circulation of bile acids by the use of a nonabsorbable drug that binds them in the gut has the effect of decreasing plasma LDL, and is of therapeutic benefit in hypercholesterolemia (see Chapter 17).

The urea cycle and ammonia

Catabolism of amino acids generates ammonia (NH_3) and ammonium ions (NH_4^+). Ammonia is toxic, particularly to the central nervous system (CNS). Most ammonia is detoxified at its site of formation, by amidation of glutamate to glutamine, which is to a large extent derived from muscle and used as an energy source by enterocytes. The remaining nitrogen enters the portal vein either as ammonia or as alanine, both of which are used by the liver for the synthesis of urea.

The impaired clearance of ammonia and other nitrogenous waste products is an important cause of brain damage

The urea cycle is the major route by which waste nitrogen is excreted, and is described in Chapter 18. In neonates, inherited defects of any of the enzymes of the urea cycle lead to hyperammonemia, which impairs the function of the brain and causes a clinical condition known as encephalopathy.

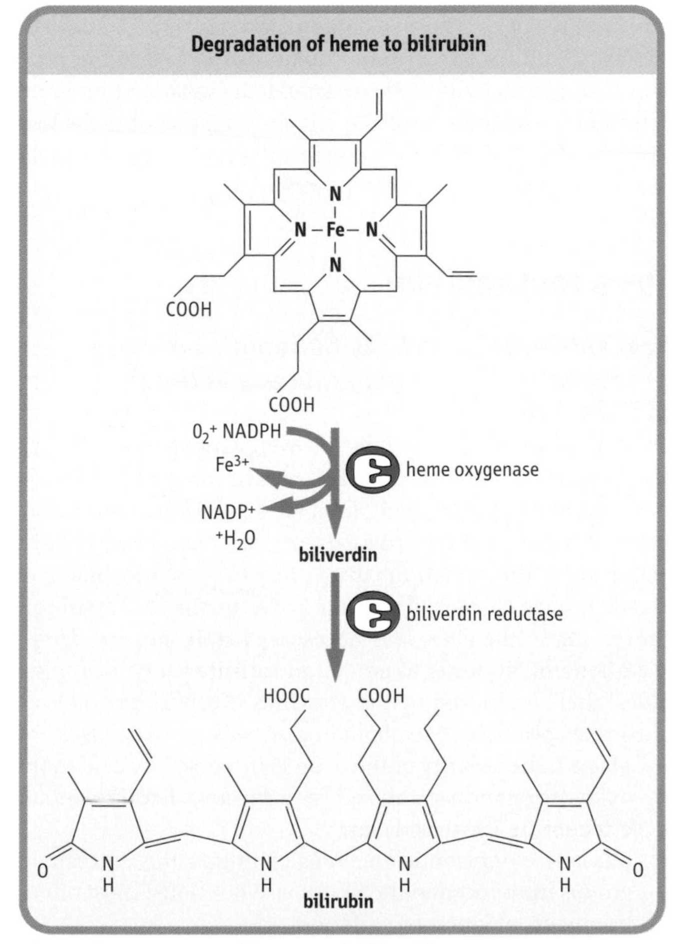

Fig. 27.5 Structures of heme and bilirubin.

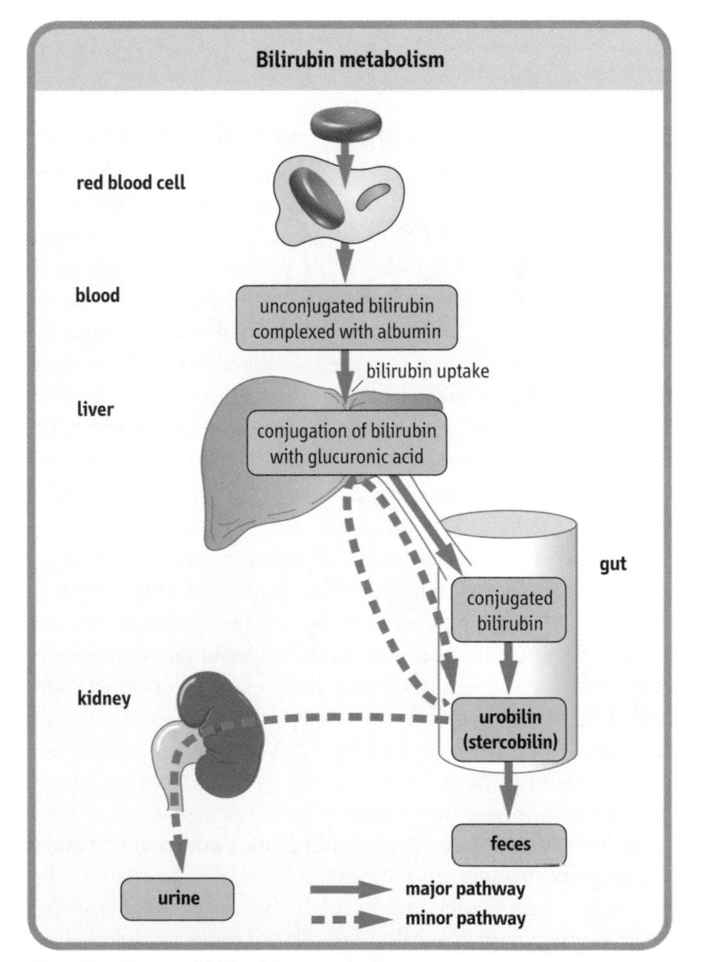

Fig. 27.6 Normal bilirubin metabolism.

Such problems arise within the first 48 hours of life, and inevitably will be made worse by protein-rich foods such as milk.

Liver turnover of proteins

Protein synthesis

Hepatic protein synthesis is important, as the majority of plasma proteins are synthesized in the liver, and hepatocellular disease may alter protein synthesis both quantitatively and qualitatively.

Albumin is the most abundant protein in blood, and is synthesized exclusively by the liver (see chapter 3). Low plasma albumin concentrations occur commonly in liver disease, but a better index of hepatocyte synthetic function is the production of the coagulation factors, II, VII, IX, and X, which all undergo post-translational γ-carboxylation of specific glutamyl residues, allowing them to bind calcium. As a group, their functional concentration can be readily assessed in the laboratory by measurement of the prothrombin time (PT) (see Chapter 6).

The liver also synthesizes most of the plasma α and β globulins. Plasma concentrations of these globulins change in hepatic disease and in systemic illness; in the latter case, the changes form part of the important 'acute phase response' to the illness.

Specific deficiencies of liver proteins may also be clinically important

Hepatic α_1 antitrypsin, belongs to serpins, one of the family of serine protease inhibitors, and has, contrary to its name, macrophage-derived elastase as its predominant target protease. Genetic deficiency of α_1-antitrypsin presents in infancy as liver disease, or in adulthood as lung disease caused by elastase-mediated tissue destruction; the severity of the liver disease associated with α_1-antitrypsin deficiency is variable. Several isoforms of α_1-antitrypsin exist as a result of allelic variation: the normal isoform is known as M, and the two common defective isoforms as S and Z; the null allele produces no α_1-antitrypsin.

Ceruloplasmin is the major copper-containing protein of the liver and plasma, and functions as an iron oxidizing (ferroxidase) enzyme: oxidation of Fe^{2+} to Fe^{3+} is necessary for the mobilization of stored iron, and nutritional copper deficiency produces anemia. Genetic deficiency of ceruloplasmin itself leads to Wilson's disease, a condition associated with damage to both the liver and the CNS. The liver also synthesizes proteins responsible for storage (ferritin) and transport (transferrin) of iron (see Chapter 3).

α-fetoprotein (AFP) and albumin have considerable sequence homology, and appear to have evolved by reduplication of a single ancestral gene. In the fetus, AFP appears to serve physiologic functions similar to those performed by albumin in the adult; furthermore, by the end of the first year of life, AFP in the plasma is entirely replaced by albumin. During hepatic regeneration and proliferation, AFP is again synthesized: particularly high plasma concentrations of AFP are associated with liver cancer.

Proteolysis

Hepatic protein turnover is highly regulated, which allows the activity of metabolic pathways to adapt to alterations in physiologic circumstances

Mammalian cells possess several proteolytic systems. Plasma proteins and membrane receptors are endocytosed, to be hydrolyzed by acid proteases within intracellular organelles known as lysosomes; intracellular proteins are degraded within structures known as proteasomes. These are large protein complexes which are present in the cytosol. Each proteasome which consists of four rings of protein subunits, the rings being stacked on top of one another to form a hollow cylindrical structure within which the intracellular proteins are hydrolyzed to small peptides of six to 12 amino acids. These peptides are released into the cytoplasm, where they are further degraded to their constituent amino acids by peptidases.

The intracellular proteins intended for proteasomal degradation are marked by conjugation with a small protein known as ubiquitin. Ubiquitin is coupled to the amino groups of protein lysyl residues, and this conjugation reaction is repeated to form chains of five or more ubiquitins linked to the protein. The specificity of the proteasomal degradation therefore relies on the specificity of the enzymes responsible for the ligation of ubiquitin to the protein substrate, without which the protein will not enter the proteasome.

Drug metabolism

Low substrate specificity of hepatic enzymes produces a wide-ranging efficacy in drug metabolism

Most drugs are metabolized by the liver. Among other effects, this hepatic metabolism usually increases the hydrophilicity of drugs, and therefore their ability to be excreted. Generally, the metabolites that are produced are less pharmacologically active than the substrate drug; however, some inactive pro-drugs are converted to their active forms as a result of processing in the liver. It is necessary for the hepatic drug-metabolizing systems to act on an infinite range of molecules; this is achieved by the enzymes involved having low substrate specificity. Metabolism proceeds in two phases:

- **phase I:** the polarity of the drug is increased by oxidation or hydroxylation catalyzed by a family of microsomal cytochrome P450 oxidases;
- **phase II:** cytoplasmic enzymes conjugate the functional groups introduced in the first phase reactions, most often by glucuronidation or sulfation.

Three of the 12 cytochrome P450 gene families share the main responsibility for drug metabolism

The cytochrome P450 enzymes are heme-containing proteins that co-localize with reduced nicotinamide adenine dinucleotide phosphate (NADPH) cytochrome P450 reductase. The reaction sequence catalyzed by these enzymes is shown in Figure 27.7. There are 12 cytochrome P450 gene families, of which three, designated *CYP1, CYP2,* and *CYP3,* are mainly responsible for drug metabolism. Each gene locus has multiple alleles, and the P450 enzyme encoded by the *CYP3A4* gene appears to be involved in the metabolism of most drugs. Liver disease is likely to impair drug metabolism, and so drugs must be prescribed carefully for such patients.

The hepatic synthesis of P450 cytochromes is increased by certain drugs and other xenobiotic agents – a process known as 'induction', which increases the rate of phase I reactions. Conversely, drugs that form a relatively stable complex with a particular cytochrome P450 inhibit the metabolism of other drugs that are normally substrates for that cytochrome P450. Allelic variation that affects the catalytic activity of a cytochrome P450 will also affect the pharmacologic activity of drugs. The best described example of such polymorphism is that of the P450 cytochrome *CYP2D6,* which was recognized initially in the 5–10% of normal individuals who were noted to be slow to hydroxylate debrisoquine, a now little-used blood-pressure-decreasing drug. *CYP2D6* also metabolizes a significant number of other commonly used drugs, so that 'debrisoquine polymorphism' remains clinically significant.

Drug hepatotoxicity

Drugs that exert their toxic effects on the liver may do so through the hepatic production of a toxic metabolite. Drug toxicity may occur in all individuals exposed to a sufficient concentration of a particular drug. A drug may even be toxic in some individuals at concentrations normally tolerated by most patients prescribed the drug. This phenomenon is known as idiosyncratic drug toxicity, and may be due to a genetic or immunologic cause.

Acetaminophen (paracetamol)

Acetaminophen is widely used as a painkiller. Taken in the usual therapeutic doses, it is conjugated with glucuronic acid or sulfate, which is then excreted by the kidneys. In overdose, the capacity of these conjugation pathways is overwhelmed, and acetaminophen is then oxidized by a liver P450 cytochrome to *N*-acetyl benzoquinoneimine (NABQI), which can cause a free-radical-mediated peroxidation of membrane lipids, and thereby hepatocellular damage. NABQI may be detoxified by conjugation with glutathione, but in acetaminophen overdose these glutathione stores also become exhausted, and hepatotoxicity ensues (Fig. 27.8). Therapeutically, the sulfhydryl compound, *N*-acetyl cysteine (NAC), which promotes detoxification of NABQI by the glutathione pathway and also scavenges free radicals, is routinely used as an antidote to acetaminophen poisoning. The risk of hepatotoxicity can be reliably predicted from measurement of the plasma concentration of acetaminophen, and then NAC can be given to those patients at risk of liver damage.

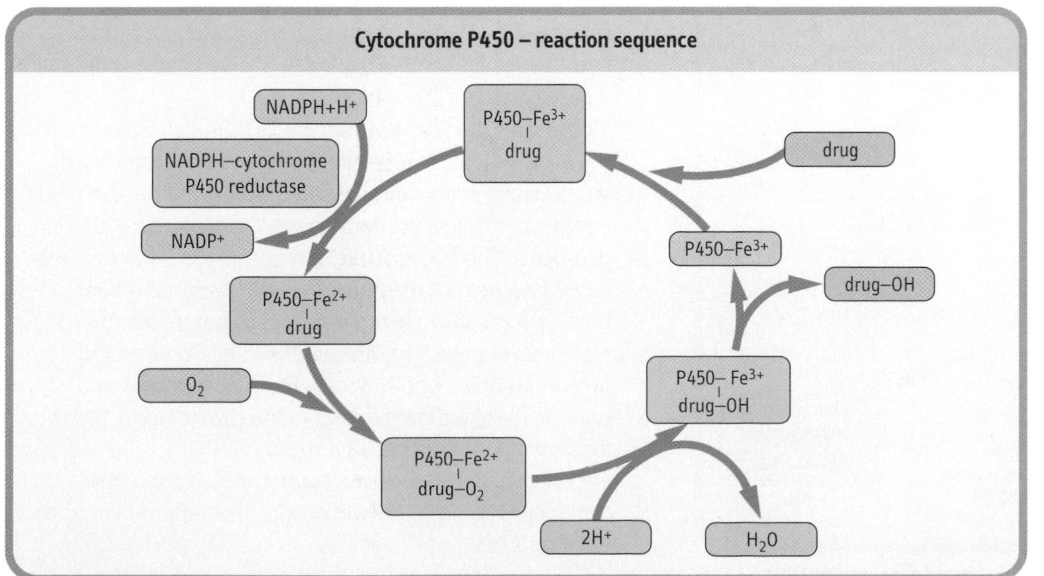

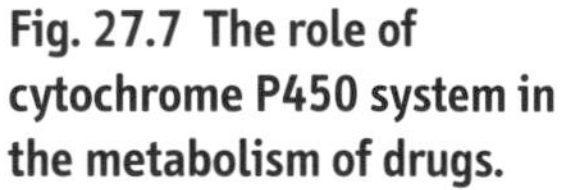
Fig. 27.7 The role of cytochrome P450 system in the metabolism of drugs.

Alcohol

Excess intake of alcohol remains the most common cause of liver disease in the Western world

Alcohol is oxidized in the liver, mainly by alcohol dehydrogenase, to form acetaldehyde, which is in turn oxidized by aldehyde dehydrogenase (ALDH) to acetyl Coenzyme A. Nicotinamide adenine dinucleotide (NAD^+) is the cofactor for both these oxidations, being reduced to NADH. A P450 cytochrome, *CYP2E1*, also contributes to ethanol oxidation, but is quantitatively less important than the alcohol dehydrogenase/ALDH pathway. Liver damage in patients who are abusers of alcohol may arise from the toxicity of acetaldehyde, which forms Schiff base adducts with other macromolecules.

The redox potential of the hepatocyte is altered by ethanol oxidation, as a result of the increased ratio of NADH to NAD^+. This inhibits the oxidation of lactate to pyruvate – a step that requires NAD^+ as a cofactor. There is the potential for lactic acidosis and, because hepatic gluconeogenesis requires pyruvate as a substrate, there is also a risk of hypoglycemia. The likelihood of hypoglycemia is also increased in alcoholics when they fast, as they often have low hepatic stores of glycogen because of poor nutrition. The shift in the $NADH/NAD^+$ ratio also inhibits β-oxidation of fatty acids and promotes triglyceride synthesis; this increases hepatic synthesis of very-low-density lipoprotein, and the excess is deposited in the liver (hepatic steatosis) and secreted into plasma.

The unpleasant symptoms of alcohol intolerance are exploited to reinforce abstinence

Both alcohol dehydrogenase and ALDH are subject to genetic polymorphisms, which have been investigated as a potential inherited basis of susceptibility to alcoholism and alcoholic liver disease. Possession of the $ALDH2^2$ allele, which encodes an enzyme with reduced catalytic activity, leads to increased plasma concentrations of acetaldehyde after the ingestion of alcohol. This causes the individual to experience unpleasant flushing and sweating, which discourages alcohol abuse. Disulfiram, a drug that inhibits ALDH, also leads to these symptoms when alcohol is taken, and may be given to reinforce abstinence from alcohol.

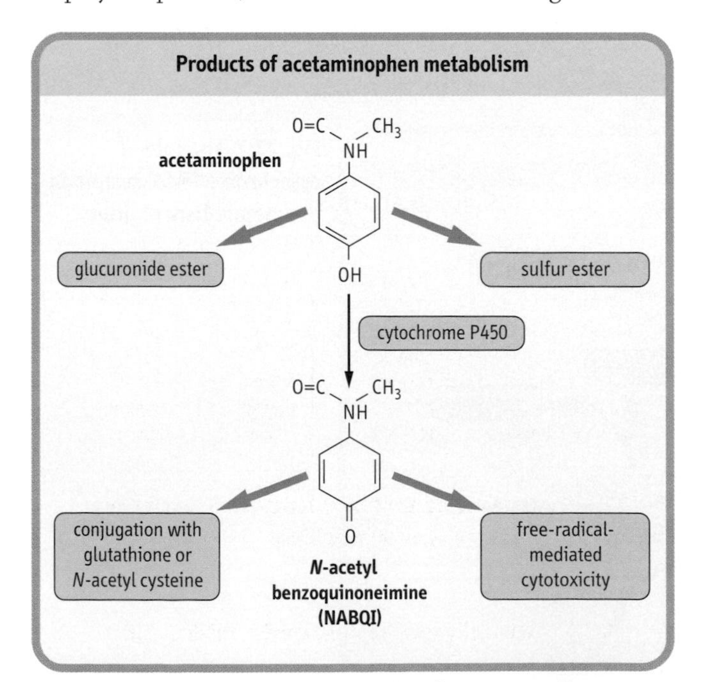

Fig. 27.8 Metabolism of acetaminophen (paracetamol).

Biochemical tests of liver function

In the clinical laboratory, a panel of biochemical measurements are routinely performed on plasma or serum specimens (Fig. 27.9). This group of tests are usually, and incorrectly, described as liver 'function' tests, which commonly include measurement of:

- **bilirubin,**
- **albumin,**
- **transaminases:** aspartate aminotransferase (AST) and alanine aminotransferase (ALT).

Acetaminophen and hepatic failure

A 22-year-old woman was admitted to hospital in a semiconscious state. She had been found with a suicide note and empty acetaminophen containers, having been perfectly well 2 days before. Tests revealed: aspartate aminotransferase (AST) 5500 IU/L, alkaline phosphatase (ALP) 125 IU/L, bilirubin 70 μmol/L (4.1 mg/dL), prothrombin time 120 s (normal value 16 s), creatinine 350 μmol/L (4.0 mg/dL) (normal range 20–80 μmol/L [0.28–0.90 mg/dL]), glucose 2.6 mmol/L (47 mg/dL) (normal range 4.0–6.0 mmol/L [72–109 mg/dL]), and blood pH 7.1. No acetaminophen was found in her plasma.

Comment. This patient had acute hepatic failure, most probably caused by acetaminophen poisoning. Blood acetaminophen is often undetectable if the patient first comes to medical attention more than 24 hours after an overdose. The hepatocellular damage worsens over the first 72 hours, but may improve spontaneously after that, as a result of regeneration of hepatocytes. However, in patients with hepatic failure caused by acetaminophen who have a metabolic acidosis or a combination of a prothrombin time greater than 100 s, creatinine >300 μmol/L (3.4 mg/dL) and encephalopathy, mortality is around 90%, and liver transplantation may be necessary. For reference ranges see Fig. 27.9.

- **alkaline phosphatase (ALP)**, sometimes in conjunction with:
- **γ-glutamyl transferase (γGT).**

(Reference ranges for these tests are given in Figure 27.9).

Transaminases

Aspartate amino transferase (AST) and alkaline phosphatase (ALP) are involved in the interconversion of amino and ketoacids, and are therefore required for hepatic metabolism of nitrogen and carbohydrate. Both these transaminases are located in the mitochondria; ALT is also found in the cytoplasm. The serum activity of ALT is a better marker of liver disease than that of AST, which is also present in muscle and red blood cells. In more severe liver disease, the synthetic functions of the hepatocytes are likely to be affected, and so the patient would be expected to have a prolonged PT and low serum albumin concentration.

Alkaline phosphatase

ALP is synthesized both by the biliary tract and by bone, but the two tissues contain different ALP isoenzymes. Thus, the origin of the ALP may be determined from the isoenzyme pattern. Alternatively, the plasma activity of another enzyme, such as γGT, which also originates in the biliary tract, may be measured.

Alcohol-related liver disease

A 45-year-old business man has a routine medical examination, at which he is found to have a slightly enlarged liver. Tests reveal: bilirubin 15 mmol/L (0.9 mg/dL), AST 434 IU/L, ALP 300 IU/L, γ-glutamyl transpeptidase (γGT) 950 IU/L, and albumin 40 g/L (4 g/dL). He seems perfectly well.

Comment. The patient has long-standing (chronic) liver disease; on the basis of the biochemical tests, this is mainly hepatocellular. The increased γGT concentration is often found in alcoholic hepatitis, and enlarged red blood cells (macrocytosis) and an increased serum uric acid concentration (hyperuricemia) are also common. Patients often deny alcohol abuse. The definitive test is a liver biopsy: the patient receives a local anesthetic, and a small fragment of liver tissue is removed with a hollow needle introduced into the liver through the skin. Microscopic examination of the tissue will then reveal the fibrosis of the hepatic lobules that characterizes alcoholic hepatitis. This may be the forerunner of cirrhosis, and the patient should abstain from alcohol. Other causes, such as chronic viral infection of the liver or an autoimmune disease producing antibodies against the liver, can be detected by blood tests. For reference ranges see Fig 27.9.

Classification of liver disorders

Hepatocellular disease

Inflammatory disease of the liver is termed hepatitis, and may be of short (acute) or long (chronic) duration. Viral infections, particularly hepatitis A, B, and C, are common infectious causes of acute hepatitis, whereas alcohol and acetaminophen are the most common toxicologic causes. Chronic hepatitis, defined as inflammation persisting for more than 6 months, may also be due to the hepatitis B and C viruses, alcohol, and immunologic diseases in which the body produces antibodies against its own tissues (autoimmune diseases). Cirrhosis is the result of chronic hepatitis, and is characterized microscopically by fibrosis of the hepatic lobules. The term 'hepatic failure' denotes a clinical condition in which the biochemical function of the liver is severely, and potentially fatally, compromised.

Cholestatic disease

Cholestasis is the clinical term for biliary obstruction, which may occur in the small bile ducts in the liver itself, or in the larger extrabiliary ducts. Biochemical tests cannot distinguish between these two possibilities, which generally have radically different causes; imaging techniques such as ultrasound are required.

Jaundice

Jaundice is clinically obvious when plasma bilirubin concentrations exceed 50 μmol/L (3 mg/dL). Hyperbilirubinemia occurs when there is an imbalance between production and excretion. The causes of jaundice (Fig. 27.10) are conventionally classified as:

- **prehepatic:** increased production of bilirubin (Fig. 27.11),
- **intrahepatic:** impaired hepatic uptake, conjugation, or secretion of bilirubin (Fig. 27.12),
- **posthepatic:** obstruction to biliary drainage (Fig. 27.13).

Differential diagnosis of jaundice

	Prehepatic	Intrahepatic	Posthepatic
conjugated bilirubin	absent	↑	↑
AST or ALT	normal	↑	normal
ALP	normal	normal	↑
urine bilirubin	absent	present	present
urine urobilinogen	present	present	absent

Fig. 27.9 Biochemical tests of liver function. The reference ranges for liver function tests are as follows:
AST 5–45 U/L,
ALT 5–40 U/L,
ALP 50–260 (ALP is physiologically elevated in children and adolescents).
bilirubin <17 μmol/L (<1.0 mg/dL),
Albumin 35–45 g/L (3.5–4.5 g/dL).

The causes of jaundice			
Type	**Cause**	**Clinical example**	**Frequency**
Prehepatic	hemolysis	autoimmune abnormal hemoglobin	uncommon depends on region
intrahepatic	infection	hepatitis A, B, C	common/very common
	chemical/drug	acetaminophen alcohol	common common
	genetic errors: bilirubin metabolism	Gilbert's syndrome Crigler–Najjar syndrome Dubin–Johnson syndrome Rotor's syndrome	1 in 20 very rare very rare very rare
	genetic errors: specific proteins	Wilson's disease α_1 antitrypsin	1 in 200 000 1 in 1000 with genotype
	autoimmune	chronic active hepatitis	uncommon/ rare
	neonatal	physiologic	very common
Posthepatic	intrahepatic bile ducts	drugs primary bilary cirrhosis cholangitis	common uncommon common
	extrahepatic bile ducts	gall stones pancreatic tumor cholangiocarcinoma	very common uncommon rare

Fig. 27.10 Causes of jaundice.

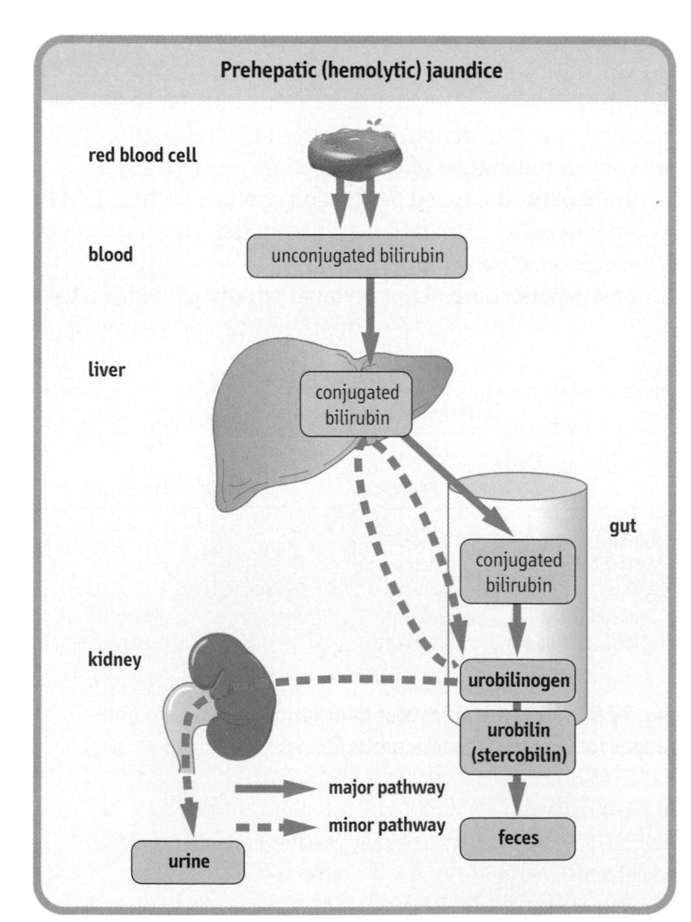

Fig. 27.11 Prehepatic jaundice.

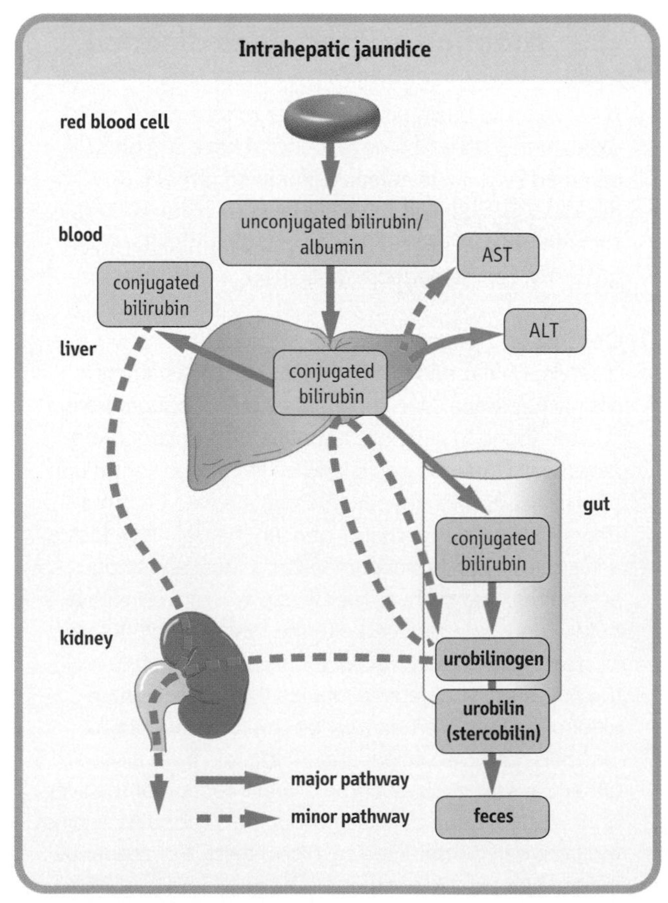

Fig. 27.12 Intrahepatic jaundice.

Prehepatic jaundice (see Fig. 27.11) results from excess production of bilirubin after hemolysis. Hemolysis is commonly the result of immune disease, structurally abnormal red cells, or breakdown of extravasated blood. Intravascular hemolysis releases hemoglobin into the plasma, where it is either oxidized to methemoglobin or complexed with haptoglobin. More commonly, red cells are hemolyzed extravascularly, within phagocytes, and hemoglobin is converted to bilirubin; such bilirubin is unconjugated. Unconjugated and conjugated bilirubin can be chemically distinguished.

Intrahepatic jaundice (see Fig. 27.12) reflects a generalized hepatocyte dysfunction. Hyperbilirubinemia is usually accompanied by other abnormalities in biochemical markers of hepatocellular function.

In neonates, transient jaundice is common, particularly in premature infants, and is due to immaturity of the enzymes involved in bilirubin conjugation. Unconjugated bilirubin is toxic to the immature brain, and causes a

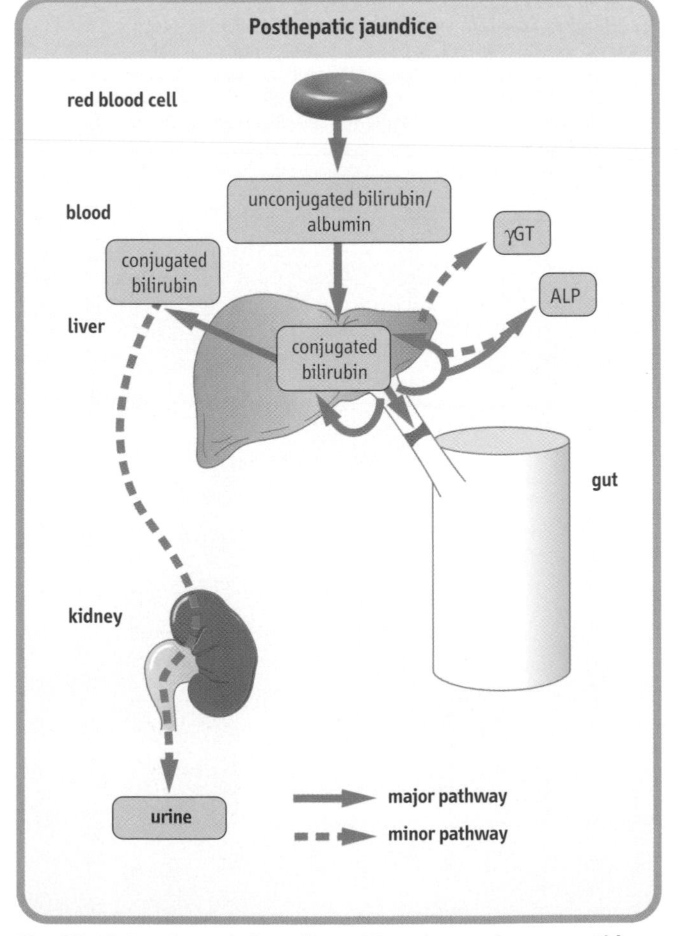

Fig. 27.13 Posthepatic jaundice. γGT-γ-glutamyl transpeptidase; ALP, alkaline phosphatase.

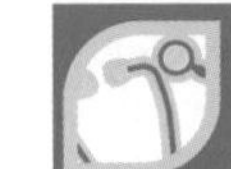

The significance of neonatal jaundice

A normal, term baby developed jaundice on the third day of life, with a bilirubin concentration of 150 μmol/L (8.8 mg/dL), predominantly of the indirect form. The baby was otherwise well.

Comment. About 50% of normal babies become jaundiced 48 hours after birth. This physiologic jaundice is caused by temporary inefficiency in bilirubin conjugation, and resolves in the first 10 days. The hyperbilirubinemia is unconjugated in nature; if severe, it may require phototherapy (ultraviolet light to photoisomerize bilirubin into a nontoxic form) or exchange blood transfusion to prevent damage to the brain (kernicterus). Bruising from delivery, infection, or poor fluid intake may exaggerate the hyperbilirubinemia. Jaundice in the first 24 hours of life is abnormal and requires investigation to exclude hemolysis, as does physiologic jaundice severe enough to require phototherapy. Jaundice that presents later – after 10 days – is always abnormal, and is likely to indicate an inborn error of metabolism or structural defects of the bile ducts.

Obstructive liver disease caused by pancreatic cancer

A 65-year-old man was admitted to hospital because of jaundice. He had no abdominal pain, but had dark urine and pale stools. Liver function tests showed: bilirubin 230 μmol/L (13.5 mg/dL), AST 32 IU/L, and ALP 550 IU/L. Urine testing revealed the presence of bilirubin, but no urobilin.

Comment. The patient had a clinical history typical of obstructive jaundice. The increased ALP and normal AST concentrations were consistent with this, and the absence of urobilin in the urine indicated that the biliary tract was totally blocked. It was important to undertake imaging of the liver, to find the site of the blockage; the absence of pain suggested that gall stones were not the cause. Ultrasound confirmed a blockage of the common bile duct, and a computed tomography scan showed an abnormal lump in the pancreas. At operation, a pancreatic cancer was found, which could not be removed.

condition known as kernicterus. If plasma bilirubin concentrations are judged to be too high, phototherapy with ultraviolet light – which oxidizes and detoxifies bilirubin – or exchange blood transfusion to remove the excess bilirubin are necessary, to avoid kernicterus.

Posthepatic jaundice (see Fig. 27.13) is caused by obstruction of the biliary tree. The plasma bilrubin is conjugated, and other biliary metabolites, such as bile acids, accumulate in the plasma. The clinical features are pale-colored stools, caused by the absence of fecal bilirubin and urobilin, and dark urine as a result of the presence of water-soluble conjugated bilirubin. In complete obstruction, urobilinogen/urobilin is absent from urine, as there can be no intestinal conversion of bilirubin to urobilinogen/urobilin, and hence no renal excretion of reabsorbed urobilinogen/urobilin.

Genetic causes of jaundice

There are a number of genetic disorders Fig 27.10 that impair bilirubin conjugation or secretion. Gilbert's syndrome, affecting up to 5% of the population, causes a mild unconjugated hyperbilirubinemia that is harmless and asymptomatic. It is due to a modest impairment in UDP glucuronyl transferase activity.

Other inherited diseases of bilirubin metabolism are rare. Crigler–Najjar syndrome, which is the result of a complete absence or marked reduction in bilirubin conjugation, causes a severe unconjugated hyperbilirubinemia that presents at birth; when the enzyme is completely absent, the condition is fatal. The Dubin–Johnson and Rotor's syndromes impair the biliary secretion of conjugated bilirubin, and therefore cause a conjugated hyperbilirubinemia, which is usually mild.

Summary

Liver is the central organ in human metabolism and is extensively involved in the metabolism of carbohydrate fat and protein. Special functions of the liver include the metabolism of blood pigment heme, the detoxification of foreign compounds by the cytochrome P450 system and synthesis of a broad range of proteins. Liver is also an organ which provides constant supply of glucose through glycogenolysis and gluconeogenesis. The most common sign of the liver and biliary system disease is jaundice. The assessment of liver status and function is an important part of the clinical investigations. Use of a panel of tests known as liver function tests (LFTs) which include AST, ALT. bilirubin, alkaline phosphatase. γGT are used in the differential diagnosis of liver disease. Prothrombin time is used as a marker of the synthetic liver function.

Further reading

Jansen PL, Bosma PL, Chowdhury JR. Molecular biology of bilirubin metabolism. *Progr Liver Dis* 1995;**13:**125–150.
Mas A, Rhodes J. Fulminant hepatic failure. *Lancet* 1997;**349:**1081–1085.
Renner EL. Liver function tests. *Baillière's Clin Gastroenterol* 1995;**9:**661–677.
Rosenthal P. Assessing liver function and hyperbilirubinaemia in the newborn. *Clin Chem* 1997;**43:**228–234.
Sherlock S, Dooley J. eds *Diseases of the liver and biliary system*, 10th ed. Oxford: Blackwell Science; 1997.
Vale JA, Proudfoot AT. Paracetamol (acetaminophen) poisoning. *Lancet* 1995;**346:**547–552.

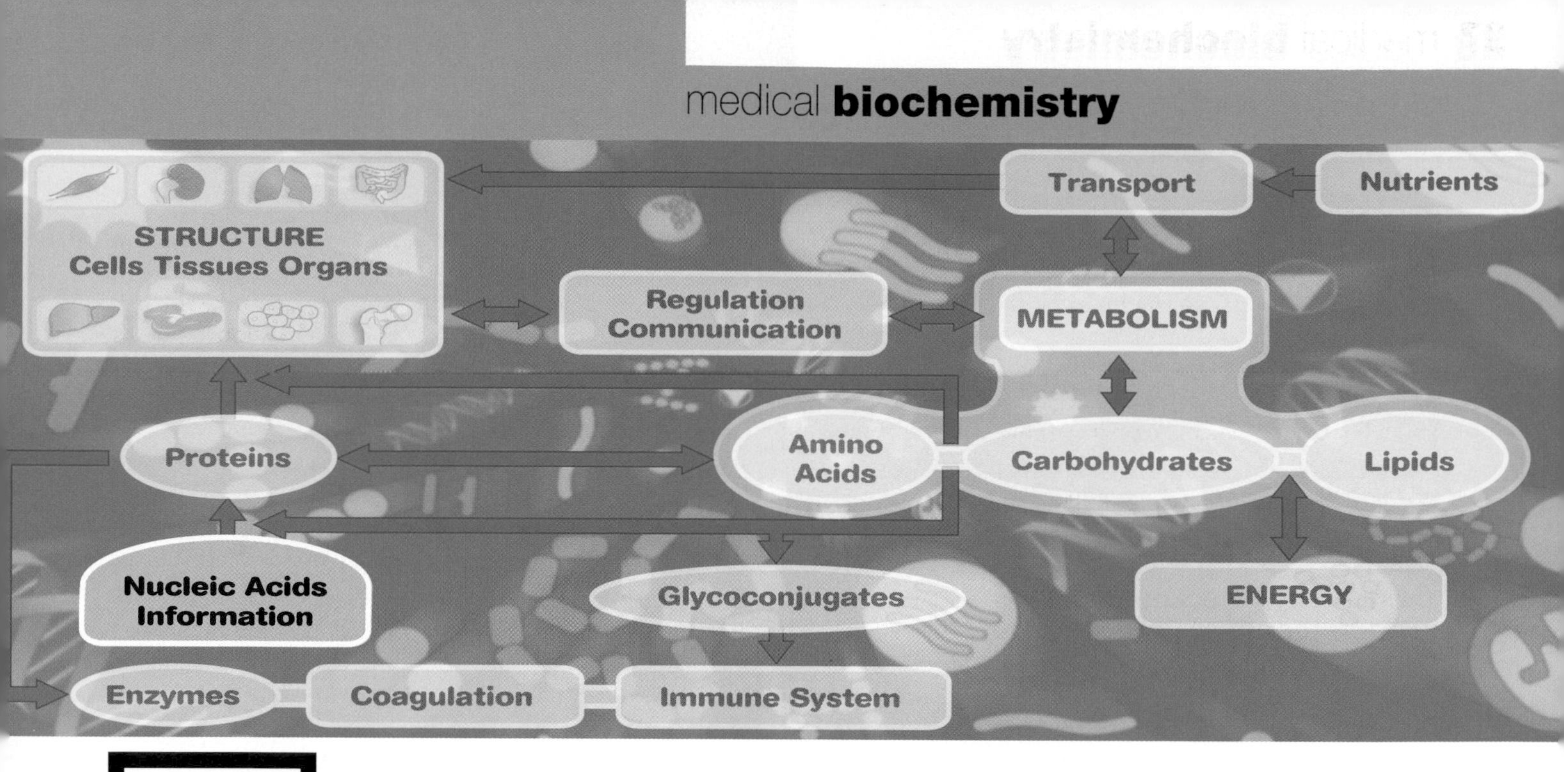

28 Biosynthesis and Degradation of Nucleotides

Overview on nucleotide metabolism

Nucleotides are intimately involved in the normal physiology of cells. In the form of nucleotide triphosphates, they constitute the currency of energy within cells. In other forms they participate at multiple levels in intermediary metabolism, from nucleotide and macromolecule biosynthesis to the biosynthesis of complex carbohydrates and the regulation of metabolism.

This chapter will discuss the structure and metabolism of nucleotides. Nucleotide metabolism is presented in three phases, the *de novo* synthesis from basic metabolites, the salvage pathways that reutilize preformed bases and nucleosides, and the metabolic breakdown leading to reutilization or excretion. *De novo* biosynthesis provides a source for the production of nucleotides from scratch. Salvage is important because cells at rest can usually meet their metabolic requirements for nucleotides by salvage. Finally, the breakdown of nucleotides is important, because the accumulation of excess nucleotides within cells leads to medically important, metabolic impairments. In addition, the enzymatic conversion of the ribonucleotides into the deoxyribonucleotides provides the precursors for deoxyribonucleic acid (DNA) synthesis, so we will examine this conversion as well.

Purines and pyrimidines

Nucleotides are formed from three components: a nitrogenous base, a five-carbon sugar, and phosphate. The nitrogenous bases found in nucleic acids belong to one of two heterocyclic groups, either purines or pyrimidines (Fig. 28.1). The major purines of both DNA and ribonucleic acid (RNA) are guanine and adenine.

In DNA the major pyrimidines are thymine and cytosine, while in RNA the major pyrimidines are uracil and cytosine. When the nitrogenous bases are combined with a five-carbon sugar, they are known as nucleosides. When the nucleosides are phosphorylated, the compounds are known as nucleotides. The phosphate can be attached either at the 5′-position or the 3′-position of ribose. Figure 28.2 gives the names and structures of the most important purines and pyrimidines.

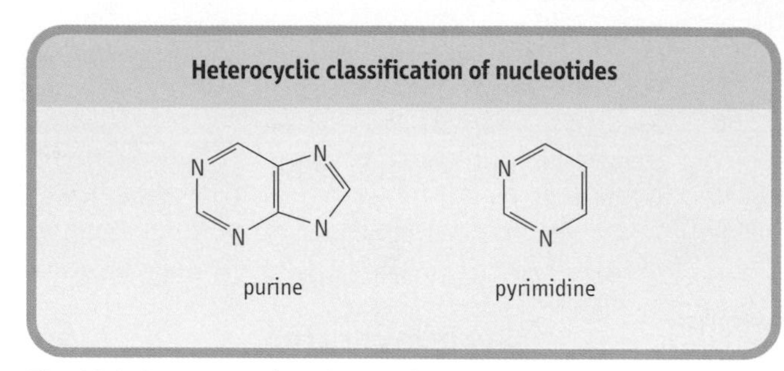

Fig. 28.1 Structure of purines and pyrimidines.

De novo synthesis of inosine monophosphate (IMP)

The *de novo* synthesis of IMP provides the cell with the capacity to construct the purine ring from scratch, thereby insuring a source of nucleotides for normal cellular processes. Figure 28.3 shows portions of the pathway of purine biosynthesis along with the origin of each of the atoms in the purine ring. Synthesis is an energy-demanding process. Adenosine 5′-triphosphate (ATP) is required at each of the kinase, synthetase, and ring-closure steps. The starting material for purine biosynthesis is ribose 5-phosphate, a product of the pentose phosphate pathway.

The first step, catalyzed by ribose phosphate pyrophosphokinase (PRPP synthetase), generates the activated form of pentose phosphate by transferring a pyrophosphate group from ATP to form 5-phosphoribosyl-pyrophosphate (PRPP). PRPP is also used in other biosynthetic reactions including nucleotide salvage, tryptophan, and histidine biosynthesis.

In the second step, catalyzed by amidophosphoribosyl transferase, an amino group from glutamine displaces the pyrophosphate group. In a reaction that is similar to the formation of a peptide bond, glycine is added to the β-5-phosphoribosylamine to yield glycinamide ribonucleotide (GAR), catalyzed by the enzyme GAR synthetase. The free amino group of GAR is then modified by addition of a formyl group by GAR transformylase to make formylglycinamide ribonucleotide (FGAR). The formyl group comes from N^{10}-formyl tetrahydrofolate.

In step 5, another amino group is added onto FGAR to form formylglycinamidine ribonucleotide (FGAM). As for other amino group additions, the donor is glutamine. The enzyme catalyzing this addition is termed FGAM synthetase. The purine imidazole ring is closed to form 5-aminoimidazole ribonucleotide (AIR) by action of AIR synthetase.

The second ring is now built by adding the final three atoms, followed by cyclization. The first atom to be added is the C-6 carbon forming carboxyaminoimidazole ribonucleotide (CAIR). In an unusual reaction, catalyzed by AIR carboxylase, this carbon atom arises directly from CO_2, without involvement of biotin. The nitrogen atom at position 1 of the mature purine ring is now added in a two-step reaction. The first step is the formation of an aspartate-CAIR covalent intermediate termed 5-aminoimidazole-4-(*N*-succinylocarboxamide) ribonucleotide (SACAIR), catalyzed by SACAIR synthetase. The four-carbon dicarboxylic acid, fumarate, is then released from SACAIR by the action of adenylosuccinate lyase to form 5-aminoimidazole-4-carboxamide ribonucleotide (AICAR).

The last purine-ring carbon is added as a formyl group from N^{10}-formyl tetrahydrofolate, by AICAR transformylase; the product is 5-formylaminoimidazole-4-carboxamide ribonucleotide (FAICAR). Finally, IMP synthase catalyzes closure of the second purine ring to form IMP.

Synthesis of purine nucleotides, ATP and guanosine triphosphate (GTP)

IMP is the precursor of other purine nucleotides; it does not accumulate significantly within the cell. It serves as the starting material for both adenosine and guanosine monophosphate (AMP and GMP), as shown in Figure 28.4. Two enzymatic reactions are required to convert the ring system of IMP into that of AMP. In reactions similar to those in steps 8 and 9 for addition of the N1 purine nitrogen to CAIR, aspartate is added at C-6 of IMP to form the adenylosuccinate intermediate by adenylosuccinate synthetase. Adenylosuccinate lyase then catalyzes hydrolysis of adenylosuccinate, yielding fumarate and AMP.

Names and structures of purines and pyrimidines

Structure	Free base	Nucleoside	Nucleotide
	adenine	adenosine	AMP ADP ATP cAMP
	guanine	guanosine	GMP GDP GTP cGMP
	hypoxanthine	inosine	IMP
	uracil	uridine	UMP UDP UTP
	cytosine	cytidine	CMP CDP CTP
	thymine	thymidine	TMP TDP TTP

Fig. 28.2 The designation NTP refers to the ribonucleotide. The prefix d, as in dATP, is used to identify deoxyribonucleotides. dTTP is usually written as TTP, with the d-prefix implied. The prefix c, as in cAMP, indicates a cyclic 3′,5′-phosphodiester.

Formation of GMP

The formation of GMP from IMP is also carried out in a two-step process. The C-2 carbon is oxidized by IMP dehydrogenase by sequential hydration and oxidation reactions catalyzed by IMP dehydrogenase. XMP is then converted into GMP by GMP synthetase, using glutamine as the nitrogen donor.

Adenylate kinase and guanylate kinase use ATP to synthesize the nucleotide diphosphates from the nucleotide monophosphates. Finally, a single enzyme, termed nucleotide diphosphokinase, converts diphosphonucleotides into triphosphonucleotides. This enzyme has activity towards all nucleotide diphosphates, regardless of which base or sugar is attached.

Preformed nucleotides can be utilized by salvage pathways

In addition to *de novo* synthesis, cells can utilize preformed nucleotides obtained from the diet or from the breakdown of endogenous nucleic acids through salvage pathways. In mammals, purines are primarily salvaged by two enzymes. Adenine phosphoribosyltransferase (APRT) converts free adenine into AMP:

$$\text{Adenine} + \text{PRPP} \rightarrow \text{AMP} + \text{PP}_i$$

Hypoxanthine-guanine phosphoribosyltransferase (HGPRT) catalyzes a similar reaction for both hypoxanthine and guanine:

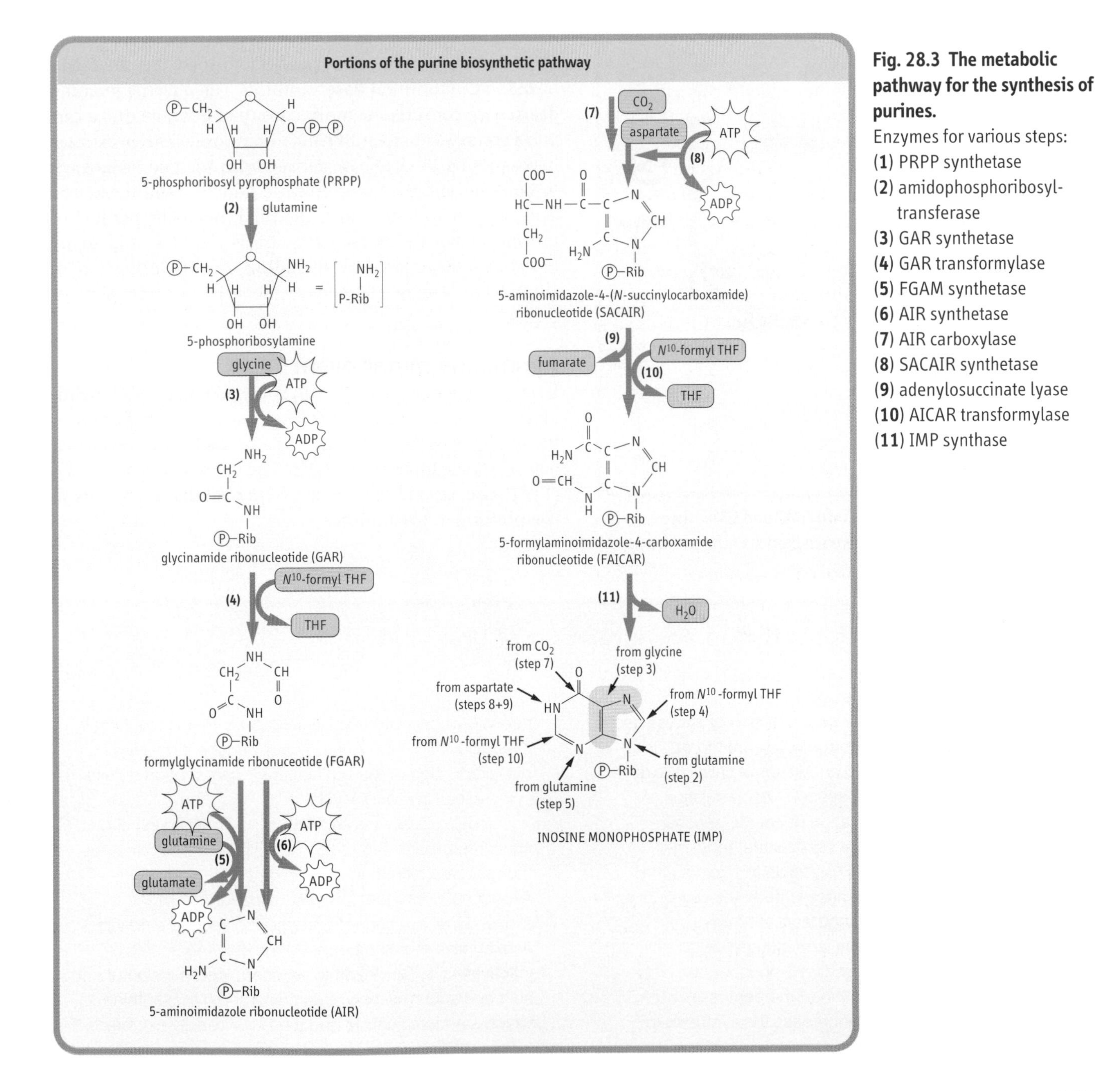

Fig. 28.3 The metabolic pathway for the synthesis of purines.
Enzymes for various steps:
(**1**) PRPP synthetase
(**2**) amidophosphoribosyltransferase
(**3**) GAR synthetase
(**4**) GAR transformylase
(**5**) FGAM synthetase
(**6**) AIR synthetase
(**7**) AIR carboxylase
(**8**) SACAIR synthetase
(**9**) adenylosuccinate lyase
(**10**) AICAR transformylase
(**11**) IMP synthase

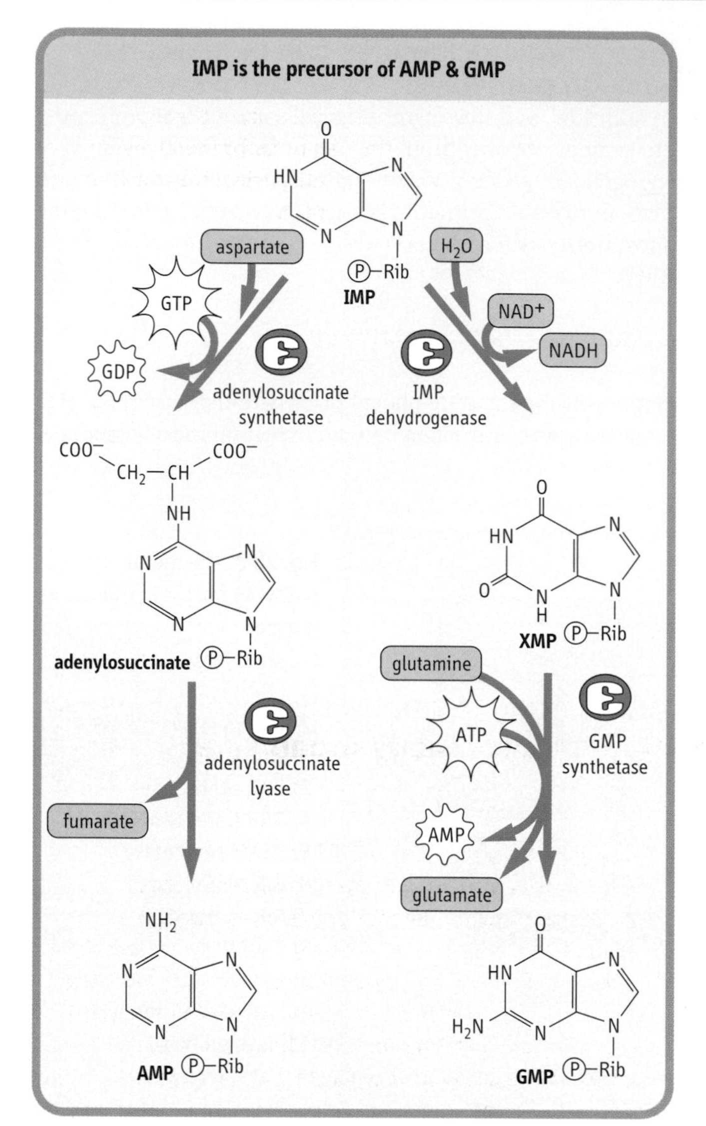

Fig. 28.4 The conversion of IMP into AMP and GMP. Two enzymatic reactions are needed for each branch of the pathway.

$$\text{Hypoxanthine} + \text{PRPP} \rightarrow \text{IMP} + \text{PP}_i$$
$$\text{Guanine} + \text{PRPP} \rightarrow \text{GMP} + \text{PP}_i$$

Degradation of purines in humans to uric acid

Each of the purine monophosphates (IMP, AMP, XMP, and GMP) can be converted into their corresponding nucleosides by 5′-nucleotidase. The enzyme purine nucleoside phosphorylase then converts the purine nucleoside (inosine, xanthosine, and guanosine) into free purine base. This enzyme, however, does not function on adenosine. Two other enzymes, AMP deaminase and adenosine deaminase, convert the amino group of AMP and adenosine into the carbonyl group of IMP and inosine respectively.

Xanthine

Once the various free bases have been formed, they are catabolized to the common base xanthine. The enzyme guanine deaminase converts the amino group of guanine into a carbonyl group, yielding xanthine. Xanthine oxidase (XO) oxidizes hypoxanthine, the free base derived from adenine via inosine, to xanthine, and this same enzyme oxidizes xanthine to uric acid (Fig. 28.5). Uric acid is the final metabolic product of purine catabolism in primates, birds, reptiles, and many insects. Other organisms, including most mammals, fish, amphibians, and invertebrates metabolize uric acid to simpler forms.

Pyrimidine metabolism

Uridine monophosphate (UMP) is the precursor of all pyrimidine nucleotides within a cell. The *de novo* pathway ends with the synthesis of UMP, and other pathways lead to the formation of cytidine triphosphate (CTP) and thymidine triphosphate (TTP). Also, several salvage pathways exist that allow cells to use preformed pyrimidines.

Gout results from tissue accumulation of uric acid

Gout is a disease caused by accumulation of excess uric acid in body fluids (normal range = 1.5–7.0 mg/dL = 0.1–0.4 mmol/L). Uric acid and its urate salts have a low solubility in water, and excessive accumulation of urates results in the precipitation of needle-shaped sodium urate crystals. These crystals are frequently deposited in the soft tissues, particularly in joints. The toxicity of uric acid can be exacerbated by a variety of biochemical defects, including decreased renal clearance, HGPRT deficiency, and glucose-6-phosphatase deficiency. Glucose-6-phosphatase deficiency causes a stimulation of the pentose phosphate pathway, which increases the synthesis of ribose 5-phosphate and consequently PRPP, resulting in overproduction of purines. Along with the increased overproduction of the purines, their degradation results in increased uric acid levels.

Gout is usually treated with allopurinol, an inhibitor of xanthine oxidase (Fig. 28.5). Xanthine oxidase catalyzes the two-step oxidation of hypoxanthine to uric acid. Allopurinol undergoes the first oxidation to yield alloxanthine, but cannot undergo the second oxidation. Alloxanthine remains bound to the enzyme, thereby inactivating it. This leads to reduced accumulation of uric acid and accumulation of xanthine and hypoxanthine, which are more soluble and thus more easily excreted.

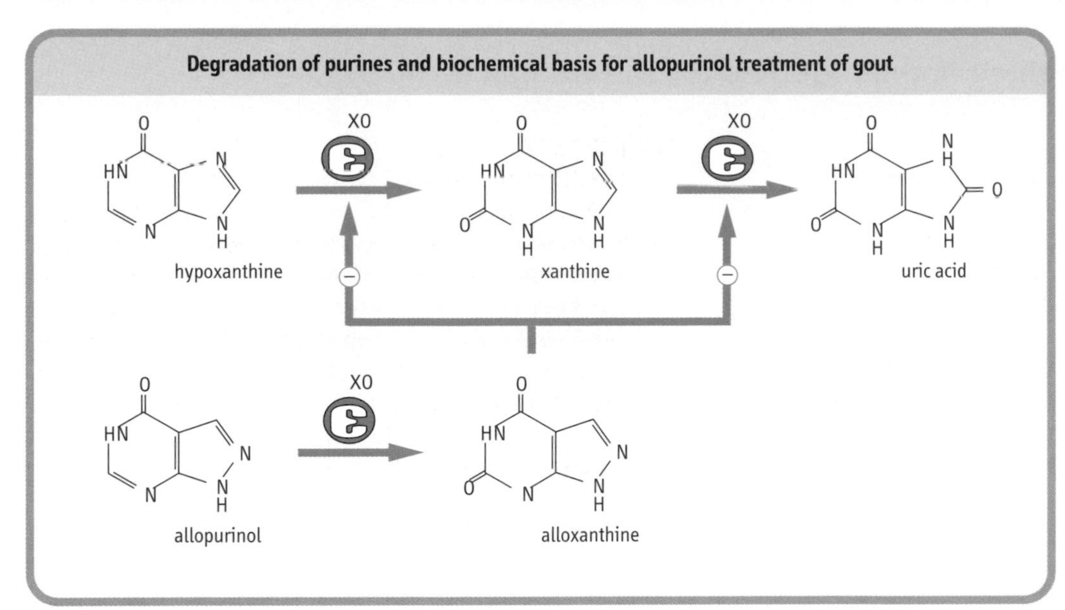

Fig. 28.5 Inhibition of xanthine oxidase (XO) by alloxanthine is the mechanism involved in allopurinol treatment of gout.

Salvage is the principal source of nucleotides for many organisms

The observation that organisms blocked in the *de novo* biosynthetic pathways can survive and grow if a source of nucleotides is available from the diet indicates the strategic importance of the nucleotide-salvage pathways. The salvage pathways are especially important for many parasites. Such organisms prey metabolically on their host, utilizing preformed metabolites, including nucleotides. Indeed, many obligate parasites, such as *Mycoplasma*, *Borrelia,* and *Chlamydia*, have lost the genes required for the *de novo* synthesis of nucleotides. They do not need them because they can obtain these important components from their host.

Even in humans, resting T lymphocytes, immune-system cells produced in the thymus, meet their metabolic requirements for nucleotides through the salvage pathway, but *de novo* synthesis is required for activation of the T-lymphocyte population. The salvage of nucleotides is especially important in HIV-infected T lymphocytes. In asymptomatic patients, resting lymphocytes show a block in *de novo* pyrimidine biosynthesis, and correspondingly reduced pyrimidine pool sizes. Following activation of the T-lymphocyte population, these cells cannot synthesize sufficient new DNA. The activation process leads to cell death, contributing to the decline in the T-lymphocyte population during the late stages of HIV infection.

Lesch–Nyhan syndrome results from hypoxanthine-guanine phosphoribosyl transferase (HGPRT) deficiency

The gene encoding HGPRT is located on the X chromosome. Its deficiency results in a rare recessive disorder termed Lesch–Nyhan syndrome. The lack of HGPRT causes an overaccumulation of PRPP, which is also the substrate for the enzyme amidophosphoribosyl transferase (see Fig. 28.3). This stimulates purine biosynthesis by up to 200-fold. Because of increased purine synthesis, the degradation product, uric acid, also accumulates to high levels. Elevated uric acid leads to a crippling gouty arthritis and severe neuropathology resulting in mental retardation, spasticity, aggressive behavior, and a compulsion towards self-mutilation by biting and scratching.

Severe combined immunodeficiency syndromes (SCIDS)

SCIDS are a group of fatal disorders resulting from defects in both cellular and humoral immune function. SCIDS patients cannot efficiently produce antibodies in response to an antigenic challenge. Approximately 50% of patients with the autosomal recessive form of SCIDS have a genetic deficiency in the purine salvage enzyme, adenosine deaminase. The pathophysiology involves lymphocytes of both thymic and bone-marrow origin (T and B lymphocytes), as well as 'self-destruction' of differentiated cells following antigen stimulation. The precise cause of cell death is not yet known, but may involve accumulation in lymphoid tissues of adenosine, deoxyadenosine, and dATP, accompanied by ATP depletion. The finding that deficiency of the next enzyme in the purine salvage pathway, nucleoside phosphorylase, is also associated with an immune-deficiency disorder suggests that integrity of the purine salvage pathway is critical for normal differentiation and function of immunocompetent cells in man.

De novo pathway

The pathways of pyrimidine biosynthesis (Fig. 28.6) are invariant in all organisms that have been examined to date. The first step, catalyzed by the enzyme carbamoyl phosphate synthetase, uses bicarbonate, glutamine, and 2 moles of ATP to form carbamoyl phosphate. Carbamoyl phosphate is also used in the synthesis of arginine. Most of the atoms that form the pyrimidine ring are added in a single step catalyzed by aspartate transcarbamoylase (ATCase). ATCase condenses carbamoyl phosphate with the amino group of aspartic acid to form carbamoyl aspartate. Carbamoyl aspartate is then cyclized to dihydroorotic acid by the action of the enzyme dihydroorotase. Dihydroorotic acid is oxidized to orotic acid by a mitochondrial enzyme, dihydroorotate dehydrogenase, which is linked to the electron-transport system through ubiquinone. The ribosyl-5′-phosphate group from PRPP is transferred onto orotic acid to form orotate monophosphate (OMP). Finally, OMP is decarboxylated to form UMP.

Metabolic channeling by CAD and UMP synthase improves metabolic efficiency

In bacteria, and some fungi, the six enzymes of pyrimidine biosynthesis exist as distinct proteins. However, during the evolution of mammals the first three enzymatic activities have been fused together into a single multifunctional polypeptide. This multifunctional protein, termed CAD for the first letter of each enzyme activity, is encoded by a single gene.

Similarly, the polypeptides having the final two enzymatic activities of pyrimidine biosynthesis, orotate phosphoribosyl transferase and orotidylate decarboxylase, have also been fused into a single polypeptide, UMP synthase, which is also encoded by a single gene.

The advantages of such multifunctional proteins are that one protein can have multiple sequential enzymatic activities. In such enzymes, the products from the first reaction remain bound to the enzyme and are directed to the second enzymatic center. As with the fatty acid synthetase complex, this avoids the diffusion of the metabolic intermediates into the intracellular milieu, thereby improving the metabolic efficiency of the individual steps.

Orotic aciduria

Orotic aciduria is a very rare genetic condition characterized by anemia resulting from defects in red cell maturation, accompanied by orotic aciduria. Orotic acid crystals form in urine on standing, but may also precipitate *in vivo*, causing obstruction of the urinary tract. It can have several causes. The most severe form of the disease (Type 1) results from a deficiency of the enzyme UMP synthase. These patients are incapable of converting orotic acid into UMP. The accumulated orotic acid is excreted in the urine. UMP synthase is a multifunctional protein containing both orotate phosphoribosyltransferase and OMP decarboxylase activity. The Type 1 form results from a complete loss of both activities. A second, rarer form of hereditary orotic aciduria (Type 2) is deficient in only OMP decarboxylase. This deficiency has been successfully treated with chronic uridine therapy.

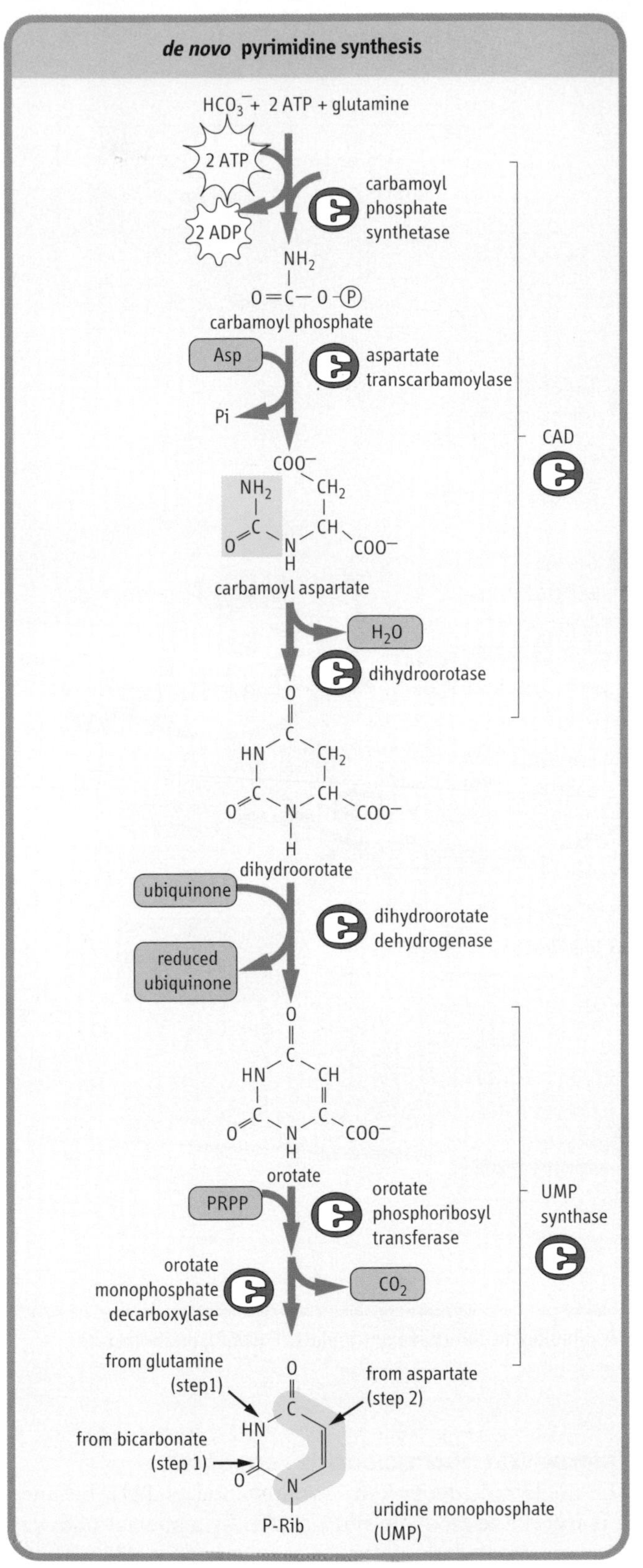

Fig. 28.6 The metabolic pathway for the synthesis of pyrimidines. This pathway has two multifunctional proteins. The first, CAD, has three enzymatic activities: **C**arbamoyl phosphate synthetase, **A**spartate transcarbamoylase, and **D**ihydroorotase. The second, UMP synthase, has two enzymatic activities: orotate phosphoribosyl transferase and OMP decarboxylase.

CTP is synthesized from UTP

UTP is synthesized in two enzymatic phosphorylation steps by the actions of UMP kinase and nucleotide diphosphokinase. UMP kinase can convert both UMP and CMP into UDP and CDP respectively. Again, as with the purines, nucleotide diphosphokinase converts the diphosphates into triphosphates. CTP synthetase converts UTP into CTP by amination of UTP. The nitrogen group is donated from glutamine in mammals. In bacteria, the nitrogen comes directly from ammonia (Fig. 28.7).

Pyrimidine-salvage pathways

As with the purines, free pyrimidine bases, available from the diet or from the breakdown of nucleic acids, can be salvaged by a series of enzyme reactions. The first of these operates on uracil and is similar to the purine-salvage systems:

$$\text{uracil} + \text{PRPP} \rightarrow \text{UMP} + \text{PPi}$$

The enzyme uracil phosphoribosyl transferase (UPRTase) is also required to activate some chemotherapeutic agents such as 5-fluorouracil (FU) or 5-fluorocytosine (FC). A second salvage reaction that is relatively specific for thymidine converts nucleosides into nucleotides by direct phosphorylation using ATP is the phosphate donor:

$$\text{thymidine} + \text{ATP} \rightarrow \text{TMP} + \text{ADP}$$

Cytosine deaminase

In bacteria and some fungi, there is an additional salvage enzyme, cytosine deaminase, that converts cytosine into uracil, which is further salvaged by UPRTase. This enzyme is not found in most higher eukaryotes, including humans. Therefore, cytosine is not salvaged in humans. Because of this, fluorocytosine is a potent antimicrobial compound. The microbial enzyme converts fluorocytosine into fluorouracil, which is toxic.

Degradation of pyrimidines is essentially the reverse of synthesis

The pyrimidine monophosphates, CMP, UMP, and TMP, are dephosphorylated to form their respective nucleosides. Cytidine is converted into uridine by action of cytidine deaminase. The ribose moiety of both uridine and thymidine (5-methyl uracil) is removed by action of an enzyme, uridine phosphorylase. The next three steps in the degradation of pyrimidines are essentially the reverse of the synthetic pathway. The double bond between carbons 5 and 6 is reduced by dihydrouracil dehydrogenase to form dihydrouracil or dihydrothymine. Next the ring is opened by the action of hydropyrimidine hydratase to form β-ureidopropionate or β-ureidoisobutyrate. The enzyme β-ureidopropionase removes ammonia and carbon dioxide to form β-alanine or β-amino-isobutyrate. Finally, following an amino transfer of the β-amino group to α-ketoglutarate, the resulting malonic semialdehyde or methylmalonic semialdehyde can be condensed with Coenzyme A for further metabolism.

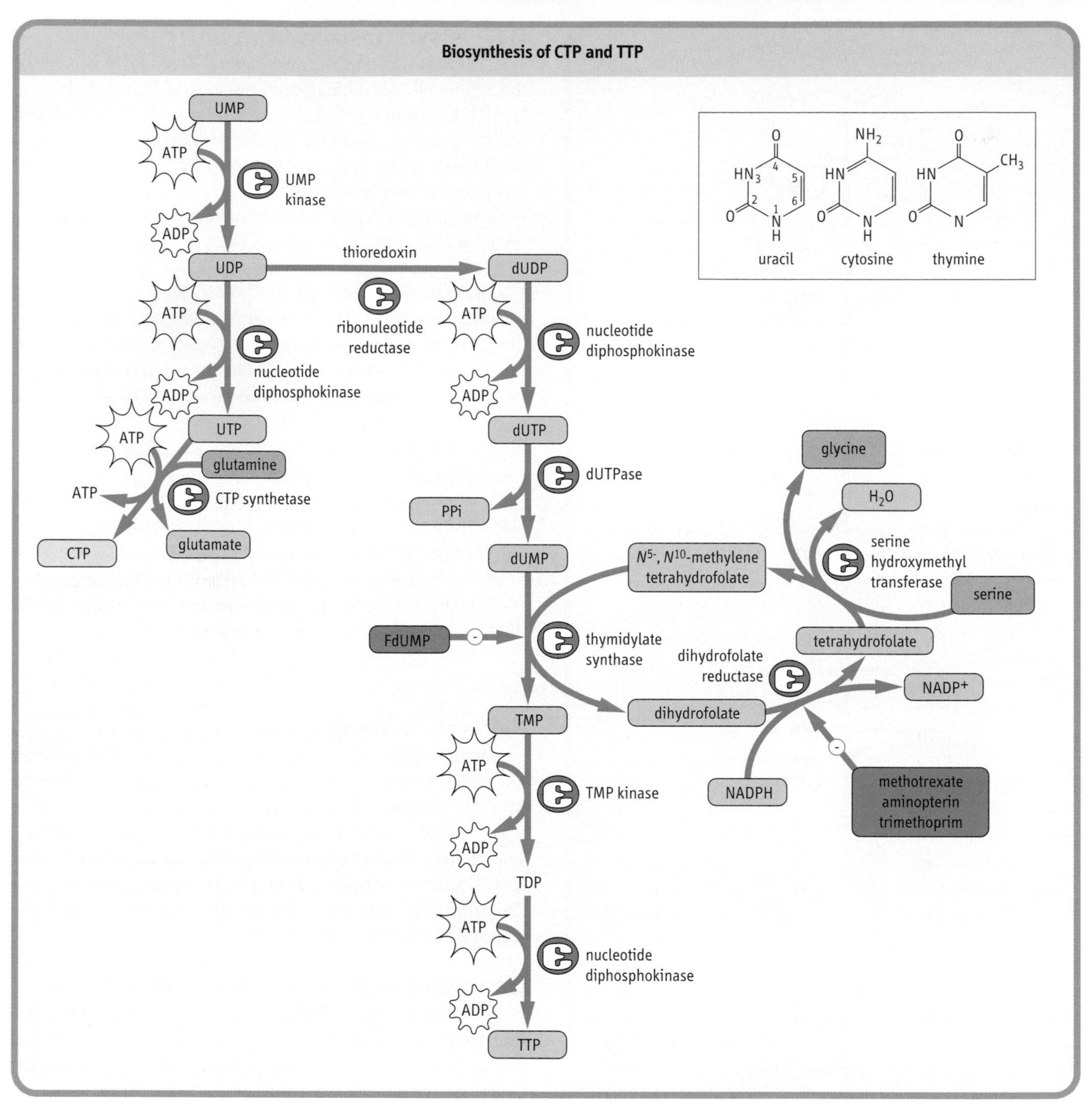

Fig. 28.7 Synthesis of pyrimidine triphosphates. Synthesis of thymidine is inhibited by fluorodeoxyuridylate (FdUMP), methotrexate, aminopterin, and trimethoprim at the indicated sites.

Formation of deoxynucleotides

Because DNA uses deoxyribonucleotides instead of the ribonucleotides found in RNA, cells require pathways to convert ribonucleotides into the deoxy forms. The adenine, guanine and uracil deoxyribonucleotides are synthesized from their corresponding ribonucleotides by direct reduction of the 2′-hydroxyl by the enzyme ribonucleotide reductase:

$$\text{NDP} \rightarrow \text{dNDP}$$

Thioredoxin ribonucleotide reductase

The nucleotide deoxy TMP, abbreviated as TMP because it is unique to DNA, is synthesized by a special pathway involving methylation of the *deoxy*ribose form of uridylate, dUMP. However, reduction of ribonucleotides to deoxyribonucleotides occurs only as diphosphates, not as monophosphates. Thus, the TMP biosynthetic pathway leads from UMP to UDP to dUDP (see Fig. 28.7). dUDP is then phosphorylated to dUTP, which creates an unexpected biochemical problem. DNA polymerase does not

Some chemotherapeutic agents block pyrimidine biosynthesis

When DNA synthesis is blocked, cells cannot divide. Because of this, several important anticancer drugs that block the synthesis of TMP are widely used as chemotheraputic agents (see Fig. 28.7). These include pyrimidine analogs such as fluorouridine and fluorocytosine, as well as the antifolates such as aminopterin, methotrexate, and trimethoprim.

One of the unique steps in DNA biosynthesis is the formation of TMP from dUMP by the enzyme thymidylate synthase. During the process of catalysis, thymidylate synthase forms a covalent bond with its substrate, dUMP. The enzyme–substrate complex then forms a complex with N^5-,N^{10}-methylene-THF. Once this ternary complex is formed, the methyl transfer occurs.

Fluorodeoxyuridylate (FdUMP) is a specific, suicide inhibitor of thymidylate synthase. In FdUMP, a highly electronegative fluorine replaces the C5 proton or uridine. This compound can begin the enzymatic conversion into dTMP by forming the enzyme–FdUMP covalent complex; however, this covalent intermediate cannot accept the donated methyl group from methylene THF, nor can it be broken down to release the active enzyme. The result is a suicide complex in which the substrate is covalently locked at the active site of thymidylate synthase. The drug is frequently administered as fluorouridine (Adrucil, Efudex, Fluracil), and the body's normal metabolism converts the fluorouridine into FdUMP. Fluorouridine is used against colorectal, gastric, and uterine cancers.

Fluorocytosine (Flucytosine, Alcobon, Ancotil) is a potent antimicrobial agent. Its mechanism of action is similar to that of FdUMP; however, it must first be converted into fluorouracil by the action of cytosine deaminase. The fluorouracil is subsequently converted into FdUMP, which blocks thymidylate synthase as above. While cytosine deaminase is present in most fungi and bacteria, it is absent in animals and plants. Therefore, in humans fluorocytosine is not converted into fluorouracil and is nontoxic, while in the microbes, metabolism of fluorocytosine results in cell death.

Aminopterin and methotrexate are folic acid analogs that bind about 1000-fold more tightly to dihydrofolate reductase (DHFR) than does dihydrofolate. See Figure 28.7 for the sites of action of these drugs. Thus, they effectively block the synthesis of tetrahydrofolate, which in turn limits the formation of N^5-,N^{10}-methylene-THF. In this manner, they block the synthesis of dTMP. Further, these compounds are competitive inhibitors of other THF-dependent enzyme reactions used in the biosynthesis of purines, histidine, and methionine. Trimethoprim is an antifolate that also binds to DHFR. It binds more tightly to bacterial DHFRs than it does to mammalian enzymes, making it an effective antibacterial agent. Folate analogs are relatively nonspecific chemotherapeutic agents. They poison rapidly dividing cells, including those in hair follicles and gut endothelia, causing the loss of hair and gastrointestinal side-effects of chemotherapy.

effectively discriminate between dUTP and TTP and can incorporate dUTP into DNA. Therefore, cells limit the concentration of dUTP by rapidly hydrolyzing dUTP to dUMP with the enzyme dUTP diphosphohydrolase. This enzyme releases pyrophosphate, which is rapidly hydrolyzed to phosphate, pulling the reaction towards the formation of dUMP. dUMP then serves as the substrate for thymidylate synthase, which transfers a methyl group from N^5-,N^{10}-methylene tetrahydrofolate to make TMP. The folate is released as dihydrofolate. Two rounds of phosphorylation yield TDP and subsequently TTP.

Regeneration of the methyl donor is accomplished in two steps. First, dihydrofolate is reduced to tetrahydrofolate. Then the hydroxymethyl side chain of serine is removed to produce formaldehyde and glycine. The formaldehyde reacts with the N^5- and N^{10}-imine groups of tetrahydrofolate to yield N^5-,N^{10}-methylene tetrahydrofolate.

Ribonucleotide reductase is regulated by a complex feedback network

The reduction of the 2′-hydroxyl of ribose uses a pair of protein-bound sulfhydryls (cystine side chains). The hydroxyl group is released as a water molecule. During the reaction, this pair of cysteines is oxidized to a cysteine disulfide. To regenerate an active enzyme, the disulfide must be reduced back to the original sulfhydryl pair by disulfide exchange with a small protein termed thioredoxin. Thioredoxin, a highly conserved Fe-S protein, is in turn reduced by the flavoprotein thioredoxin reductase. The dNTPs are produced by phosphorylation of the dNDPs. ATP serves as the phosphate donor.

Feedback regulation

Because a single enzyme is responsible for the conversion of all ribonucleotides into deoxyribonucleotides, this enzyme is subject to a complex network of feedback regulation.

The enzyme contains several allosteric sites for metabolic regulation. Levels of the different dNTPs modify the enzyme activity toward the different NDPs. By thus regulating the enzymatic activity of deoxyribonucleotide synthesis as a function of the different dNTPs, the cell insures that the proper ratios of the different deoxyribonucleotides are produced for normal growth, protein synthesis, and cell division. In addition to the posttranslational regulation of ribonucleotide reductase, there are also complex regulatory cascades that affect the *de novo* synthesis, salvage, and degradation.

Summary

Nucleotides are synthesized primarily from amino acid precursors and phosphoribosylpyrophosphate by complex, multi-step pathways. Not suprisingly, salvage pahtways therefore play a prominent role in nucleotide metabolism. With the exception of TTP, the ribonucleotides products are converted to deoxyribonucleotides by ribonucleotide reductase. TTP is synthesized from dUMP by a special pathway involving folates. The salvage pathways have proven useful for the activation of pharmaceutical agents, while the unigueness of the pathway for synthesis of TTP has provided a special target for chemotherapeutic inhibition of DNA sysnthesis and cell division in cancer cells.

Further reading

Emmerson BT. Drug therapy: the management of gout. *N Engl Med J* 1996;**334**:445–451.
Heunnekens FM. The methotrexate story: a paradigm for development of cancer chemotherapeutic agents. *Adv Enz Reg* 1994;**34**:379–419.
Jackson RC. Contributions of protein structure-based design to cancer chemotherapy. *Sem Oncol* 1997;**24**:164–172.
Kinsella AR, Smith D, Pickard M. Resistance to chemotherapeutic antimetabolites: a function of the salvage pathway involvement and cellular response to DNA damage. *Br J Cancer* 1997;**75**: 935–945.
Nocentini G. Ribonucleotide reductase inhibitors: new strategies for cancer chemotherapy. *Crit Rev Oncol Hematol* 1996;**22**:89–126.
Nyhan WL. The recognition of Lesch–Nyhan syndrome as an inborn error of purine metabolism. *J Inherited Metab Dis* 1997;**20**:171–178.
Rustum YM, Harstrick A, Cao S, Vanhoefer U, Yin M, Wilke H, Seeber S. Thymidylate synthase inhibitors in cancer therapy: direct and indirect inhibitors. *J Clin Oncol* 1997:**15**:389–400.

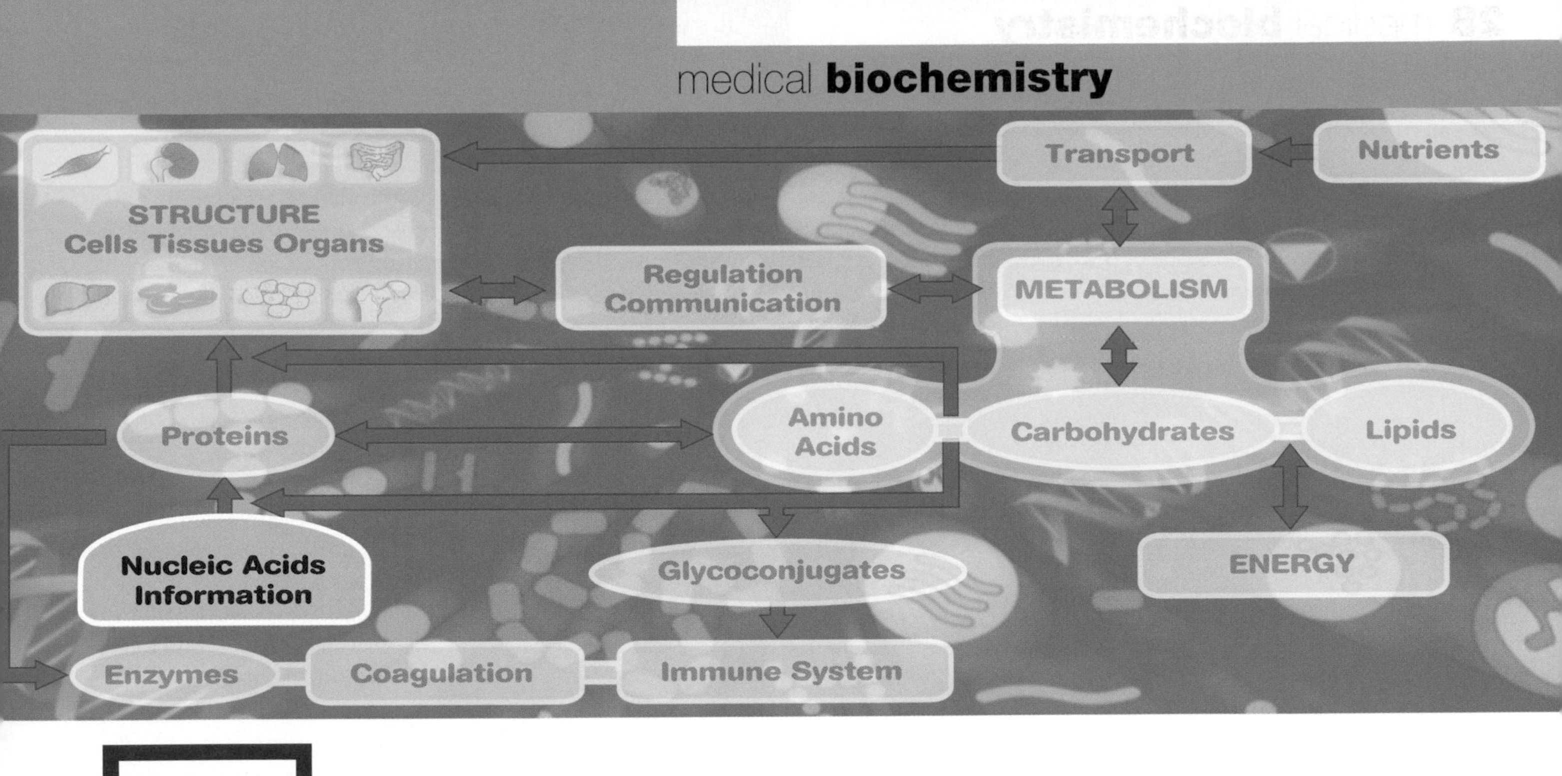

29 DNA

Structure of deoxyribonucleic acid (DNA)

Structure of cellular nucleic acids

Cellular nucleic acids exist of two forms, DNA and RNA. Approximately 90% of the nucleic acid within cells is RNA, and the remainder is DNA. DNA is the repository of genetic information within the cell. This chapter deals with the structure of DNA, the manner in which it is stored in chromosomes in the nucleus, and the mechanisms involved in its biosynthesis (semiconservative replication) and repair.

DNA is an antiparallel dimer of nucleic acid strands

DNA is composed of nucleotides containing the sugar deoxyribose. Deoxyribose is missing the hydroxyl group at the 2′-position. The chains of DNA are polymerized through a phosphodiester linkage from the 3′-hydroxyl of one subunit to the 5′-hydroxyl of the next subunit (Fig. 29.1A) Thus, DNA is a linear deoxyribose phosphate chain with purine and pyrimidine bases attached to the C-1 of the ribose subunit.

Using X-ray diffraction photographs of DNA taken by Rosalind Franklin, a structure for DNA was proposed by James Watson and Francis Crick in 1953. This model proposed that DNA was composed of two intertwined complementary strands with hydrogen bonds holding the strands together (Fig. 29.1B). The basic simplicity of this structure led to its rapid acceptance. While some of the details of the model have been modified, its essential elements have remained unchanged.

Watson and Crick model of DNA

As originally presented by Watson and Crick, DNA is composed of two strands, wound around each other in a right-handed, helical structure with the base pairs in the middle and the ribosylphosphate chains on the outside. The orientation of the DNA strands is antiparallel (i.e. the strands run in opposite directions).

The nucleotide bases on each strand interact with the nucleotide bases on the other strand to form base pairs (Fig. 29.2). The base pairs are planar and are oriented nearly perpendicular to the axis of the helix. Each base pair is formed by hydrogen bonding between a purine and a pyrimidine. Each guanine can form three hydrogen bonds with cytosine. Adenine can form two hydrogen bonds with thymine. Because of the specificity of this interaction between purines and pyrimidines on the opposite strands, the opposing strands of DNA are said to have complementary structures. The composite strength of the numerous hydrogen bonds formed between the bases of the opposite strands is responsible for the extreme stability of the DNA double helix. While the hydrogen bonds between strands are affected by temperature and ionic strength, stable complementary structures can be formed at room temperature with as few as six to eight nucleotides.

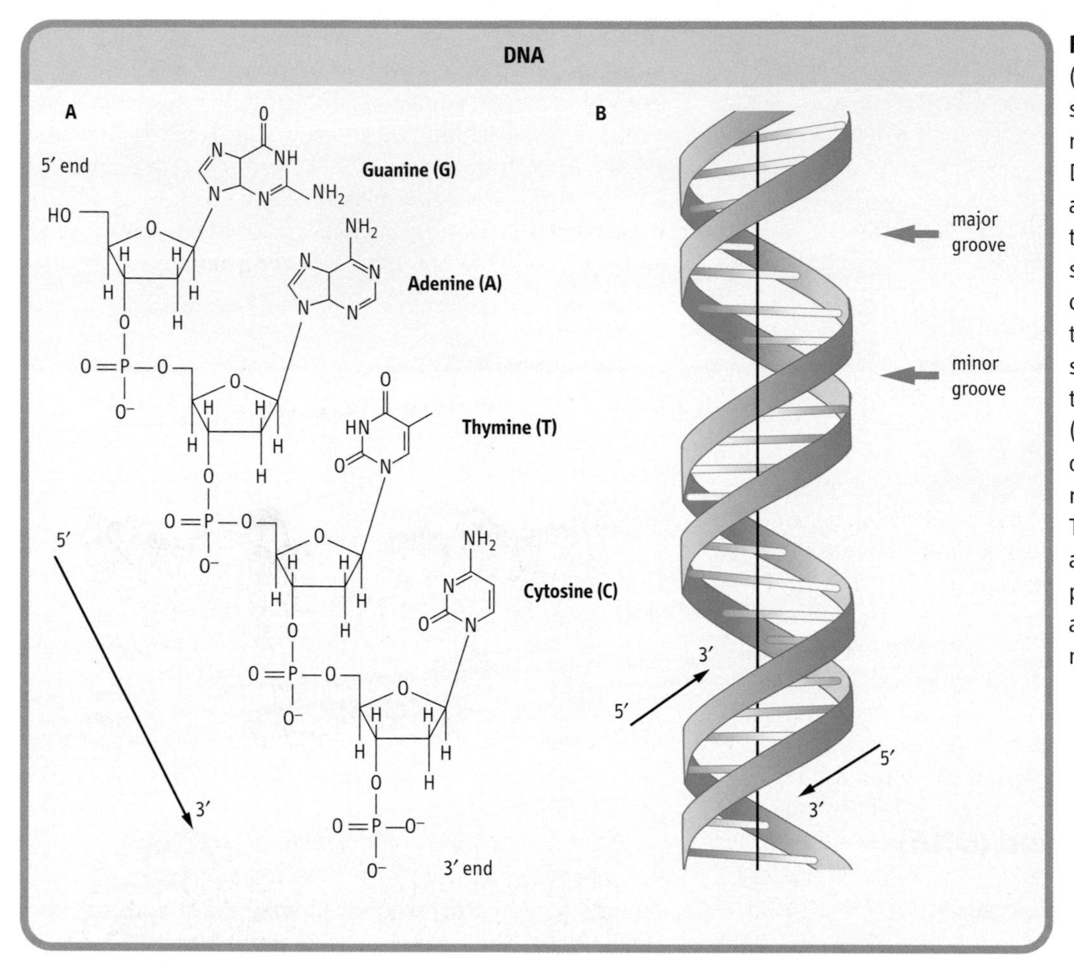

Fig. 29.1 Structure of DNA. (A) A tetranucleotide of DNA showing each of the nucleotides normally found in DNA. The deoxyribose sugars are missing the 2′-hydroxyl that is present in the ribose sugars found in RNA. By convention, DNA is read from the 5′ to 3′ end, so the sequence of this tetranucleotide is 5′-GATC-3′, (B) A graphic representation of the structure of B-DNA, the major form of DNA in the cell. The base pairs in the middle are aligned nearly perpendicular to the helical axis. The major groove and the minor groove are shown.

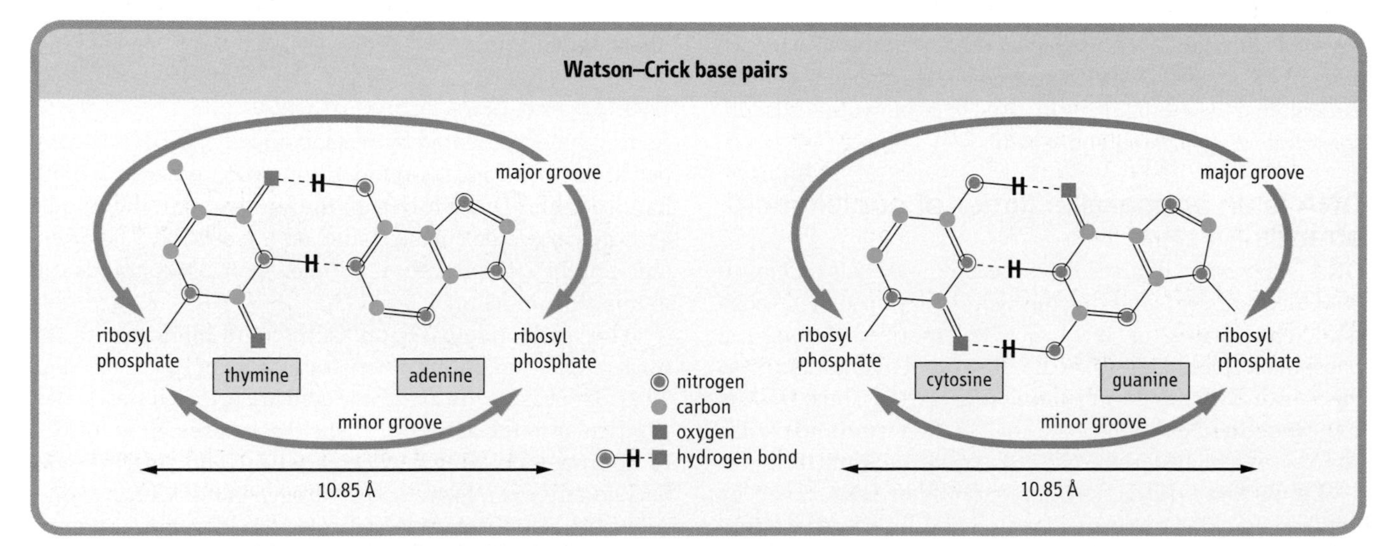

Fig. 29.2 Watson–Crick base pairing of nuclotides in DNA. The GC base pairs form three hydrogen bonds, and the AT base pairs form two hydrogen bonds. Thus, GC-rich regions are more stable than AT-rich regions.

Three-dimensional DNA

The three-dimensional structure of the DNA double helix is such that the ribosylphosphate backbones of the two strands are slightly offset from the center of the helix. Because of this, the grooves between the two strands are of different sizes. These grooves are termed the major groove and the minor groove (see Fig. 29.1B). The major groove is more open and exposes the nucleotide base pairs. The minor groove is more constricted, being partially blocked by the ribosyl moieties linking the base pairs. Binding of proteins to DNA frequently occurs in the major groove.

Alternate forms of DNA may help to regulate gene expression

Although the majority of DNA within a cell exists in the form described above (the B-form), alternative forms of DNA also exist. When the relative humidity of B-form DNA falls to less than 75%, the B-form undergoes a reversible transition into the A-form of DNA. In the A-form of DNA, the nucleotide base pairs are tilted 20° relative to the helical axis. This causes the helix to be wider than in the B form (Fig. 29.3). When the DNA strands consist of polypurine and polypyrimidine tracks, the DNA helix shows different properties. The polypurine strand becomes A-like while the polypyrimidine strand is B-like. These regions do not efficiently bind histones and are therefore unable to form nucleosomes (see below), resulting in nucleosome-free regions of DNA. Another unique form of DNA exists when the sequence of nucleotides consists of alternating purine/pyrimidine stretches. This form, termed Z-DNA, is favored at high ionic concentrations. In Z-DNA, the base pairs flip 180° around the sugar–nucleotide bond. This results in a novel conformation of the base pairs relative to sugar–phosphate backbones. The conformational change in the Z-form of DNA results in a zigzag configuration (hence the name Z-DNA) of the sugar–phosphate backbone. Surprisingly, this change in conformation leads to the formation of a left-handed DNA helix. While the Z-DNA form is favored at high ionic concentrations, it can also be induced at normal ionic concentrations by DNA methylation. The control of gene expression caused by DNA methylation may be mediated by B- to Z-DNA transition. The positioning of these alternate forms of DNA throughout the genome may participate in the regulation of gene expression.

Separated DNA strands can reassociate to form duplex DNA

Because the DNA strands are complementary and are held together only by noncovalent forces, they can be separated into individual strands. When recombined, the interactions between the complementary nucleotide sequences enable the DNA strands to reassociate into a double helix. The complementary stretches on the DNA strands can find each other and reanneal or reassociate to reform their original base pairs. To identify single-copy genes, annealing generally requires several hours. This is the basis for one of the primary methods for DNA analysis, Southern hybridization (see Chapter 33).

The human genome is very complex

The human genome contains approximately 100 000 different genes scattered over 46 chromosomes (23 pairs). These different genes represent unique DNA sequences that are present in single copies or at most only a few copies per genome. In addition, there are many types of repeated sequences within the genome. These repeated sequences have been divided into two major classes termed middle, or moderately repetitive, sequences ($< 10^6$ copies per genome) and highly repetitive ($> 10^6$ copies per genome).

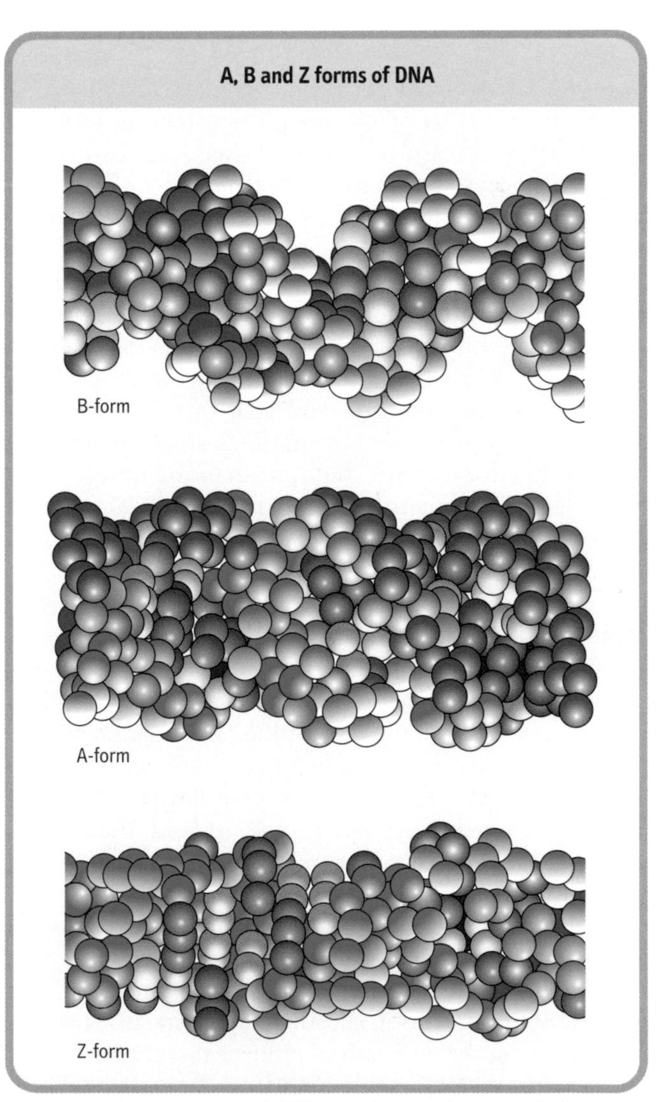

Fig. 29.3 The structures of different forms of DNA include the B, A, and Z forms. The sugar–phosphate backbone of the DNA strands are colored blue. The nucleotide bases forming the internal base pairs are yellow for pyrimidines (thymine and cytosine) and red for purines (adenine and guanine).

Middle repetitive DNA consists of a variety of types of DNA

Some middle repetitive DNA consists of genes that specify transfer RNAs, ribosomal RNAs, or histone proteins that are required in large amounts in the cell. Other middle repetitive DNA sequences have no known useful function, but as they represent a significant portion of the genome, they may participate in chromosomal rearrangements. The best characterized repetitive sequence in humans is known as the *Alu* sequence. Between 300 000 and 500 000 *Alu* I repeats of about 300 base pairs each exist in the human genome, comprising 3–6% of the total genome. Individual repeats of the *Alu* sequence may vary by 10–20% in identity. *Alu* sequences also occur in monkeys and rodents, and similar sequences occur in slime molds and other animal phyla. In addition to the *Alu* family, there are several other families of middle repetitive DNA in the human genome.

Satellite DNA

The highly repetitive DNA, which is sometimes termed satellite DNA, consists of clusters of short, species-specific, nearly identical sequences that are tandemly repeated tens of thousands of times. These clusters are found principally near the centromeres of chromosomes, suggesting that they may function to align the chromosomes during cell division to facilitate recombination.

The nucleus of eukaryotic cells contains the majority of the DNA in the cell – genomic DNA. However, DNA is found in mitochondria and also in chloroplasts of plants. The mitochondrial genome is small in size, circular, and encodes some, but not all, of the mitochondrial proteins. The majority of the mitochondrial proteins are produced from nuclear genes and are imported into the mitochondrion.

DNA is compacted into chromosomes

In eukaryotes, the DNA is arranged in linear segments termed chromosomes. Each chromosome contains between 48 million and 240 million base pairs. The B-form of DNA has a contour length of 3.4 Å per base pair. Therefore, chromosomes have contour lengths of 1.6–8.2 cm! This is much larger than a cell. Instead, the DNA is precisely condensed (> 8000-fold) into structures that fit within the nucleus.

In the native chromosome, DNA is complexed with RNA and an approximately equal mass of protein. These DNA–RNA–protein complexes are termed chromatin. The majority of the proteins in chromatin are histones. Histones are a highly conserved family of proteins that are involved in the packing and folding of DNA within the nucleus. There are five classes of histones, termed H1, H2A, H2B, H3, and H4. They are all rich (> 20%) in positively charged, basic amino acids (lysine and arginine). These positive charges interact with the negatively charged, acidic phosphate groups of the DNA strands to reduce electrostatic repulsion and permit tighter DNA packing.

Nucleosomes

The histone proteins associate into a complex termed a nucleosome (Fig. 29.4). Each of these complexes contain two molecules each of H2A, H2B, H3, and H4 and one molecule of H1. The nucleosome protein complex is encircled with about 200 base pairs of DNA that form two coils around the nucleosome core. The H1 protein associates with the outside of the nucleosome core to stabilize the complex. Because of their appearance in the electron microscope, these nucleosome–DNA complexes are sometimes termed 'beads on a string'. By forming nucleosomes, the packing density of DNA is increased by a factor of about sevenfold.

The nucleosome particles themselves are also organized into other, more tightly packed structures termed 300 Å chromatin filaments. These filaments are constructed by winding the nucleosome particles into a spring-shaped solenoid with about six nucleosomes per turn (Fig. 29.4). The solenoid is stabilized by head-to-tail associations of the H1 histones.

Finally, the chromatin filaments are compacted into the mature chromosome that is associated with a nuclear scaffold. The nuclear scaffold is about 0.4 μm in diameter and forms the core of the chromosome. The filaments are dispersed around the scaffold to form radial loops about 0.3 μm in length. The final diameter of a chromosome is about 1 μm.

Telomeres

The ends of the chromosomes are composed of unique DNA sequences called telomeres. These structures consist of tandem repeats of short, G-rich, species-specific oligonucleotides. In humans, the repeated sequence is TTAGGG. Telomeres can contain as many as 1000 copies of this sequence. In the synthesis of telomeres, the enzyme telomerase adds the preformed hexanucleotide repeats onto the 3′-end of the chromosome. There is no requirement for a DNA template in the formation of the telomere.

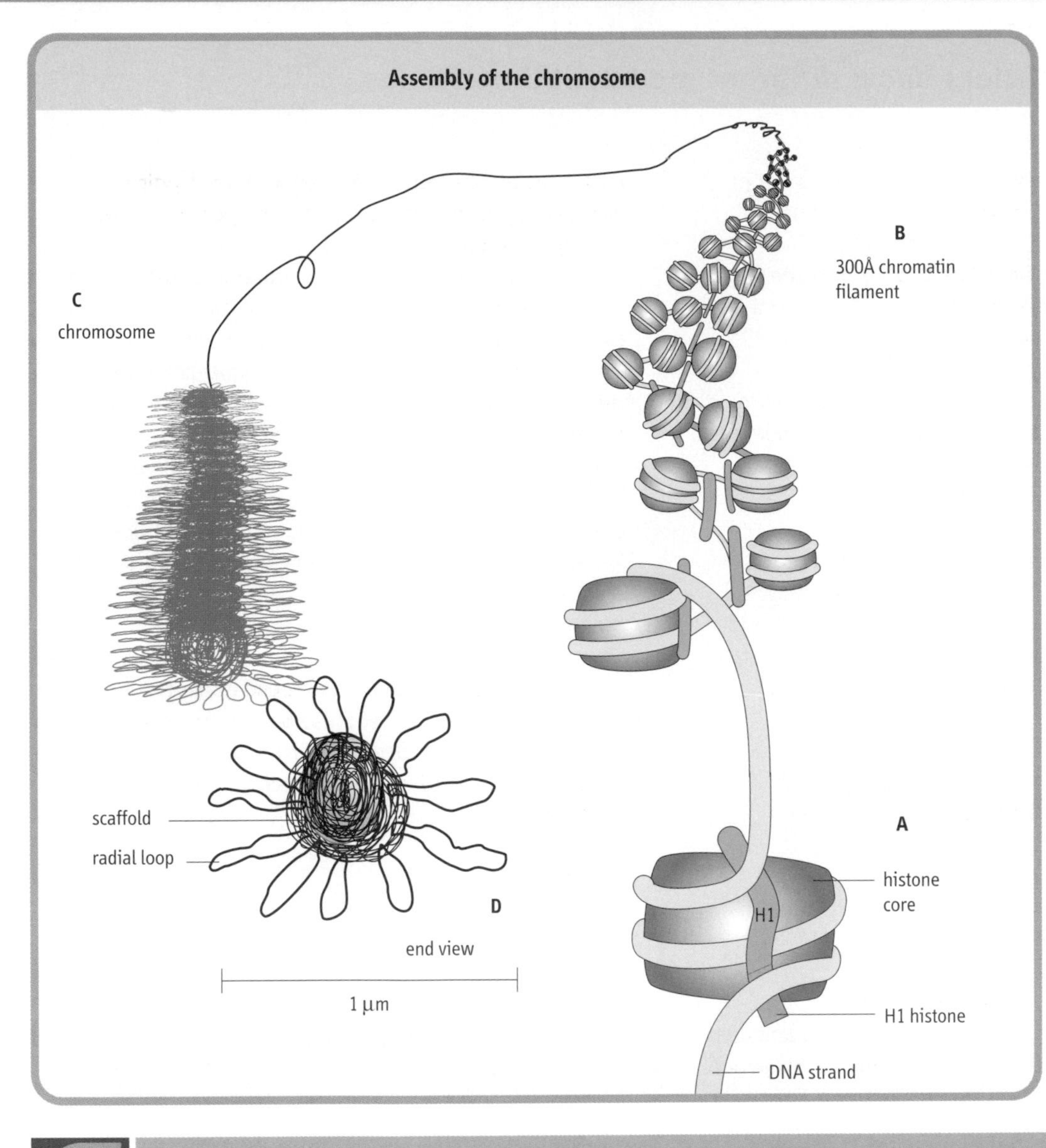

Fig. 29.4 Structures involved in chromosome packaging. (A) The nucleosome core is composed of two subunits each of H2A, H2B, H3, and H4. The core is twice wrapped with DNA, and the H1 histone binds to the completed complex. (B) The 300 Å chromatin filament is formed by wrapping the nucleosomes into a spring-shaped solenoid. (C) The chromosome is composed of the 300 Å filaments, which bind to a nuclear scaffold, forming large loops of chromatin material. (D) The end view of a chromosome shows the central nuclear scaffold surrounded by the radial loops of chromatin. The diameter of a chromosome is about 1 μm.

Telomere length correlates with age and replicative capacity

The length of the telomeres at the ends of the chromosomes are maintained by the enzyme telomerase. However, somatic cells of multicellular organisms lack telomerase activity. This leads to problems during DNA replication. DNA polymerase requires a double-stranded template for DNA biosynthesis. During DNA replication RNA primers serve this purpose and are subsequently replaced. However, at the ends of the chromosomes, these primers cannot be replaced, because there are no sequences further upstream. Therefore, each round of chromosome replication results in chromosome shortening. This has led to the hypothesis that shortening of the telomere is involved in aging of multicellular organisms. Cells isolated from younger individuals can undergo more divisions *in vitro* than cells from older donors, and the number of cell divisions is correlated with the telomere length of the parental cell. Increased expression of telomerase in human cells results in elongated telomeres and an increase in the longevity of those cells by at least 20 cell doublings. Cells from individuals with premature aging diseases (progeria) also have short telomeres. In contrast, cancer cells that are immortal have regained an active telomerase activity. All of these observations suggest that the decrease in telomere length is associated with cellular senescence and aging. 'Knockout' mice, in which the telomerase gene has been deleted, have chromosomes lacking detectable telomeres. These mice have high frequencies of aneuploidy and chromosomal abnormalities.

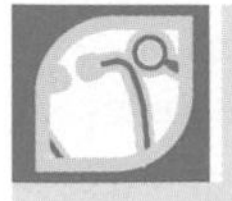

Genetic disorders show different modes of inheritance

Chromosomes are inherited according to the tenets of Mendelian genetics. Somatic cells have two copies of each chromosome, and therefore two copies of each gene. During the process of meiosis, chromosome pairs are separated into gamete cells, each having a 1*n* haploid chromosome number. The gamete cells, one from each parent, combine to form a daughter cell. Because each gamete has a 50/50 chance of inheriting either parental allele of a gene, inheritance of any particular gene can be statistically predicted.

Some genetic diseases show a dominant phenotype. For example, familial hyper-cholesterolemia is a dominant genetic defect that in the heterozygote results in reduced functional low-density-lipoprotein (LDL) receptors and plasma cholesterol levels that are twice normal. Homozygotes completely lack functional LDL receptors and have a three-to-five-fold higher plasma cholesterol level (see Chapter 17).

In contrast, the majority of genetic diseases show a recessive phenotype. Heterozygotes with these recessive diseases are frequently asymptomatic carriers of the disorder and often do not even know that they are carriers.

A special group of recessive disorders are known as X-linked genetic diseases. These genetic diseases, which include hemophilia, color blindness, Lesch–Nyhan syndrome, and glucose-6-phosphate dehydrogenase deficiency, are located on the X chromosome. Because females have two X chromosomes, these disorders behave like typical recessive disorders in females. However, with one X chromosome and one Y chromosome, they become X-linked, dominant disorders in males. With X-linked disorders females are usually asymptomatic carriers of these diseases while males are unambiguously affected.

Many diseases have a complex genetic component. Diseases such as cancer, muscular dystrophy, and xeroderma pigmentosum may be caused by any one of many different genetic defects.

The cell cycle

Figure 29.5 shows the various phases of the growth and division of cells, known as the cell cycle. The G1 phase is a period of cell growth that occurs prior to DNA replication. The phase during which DNA is synthesized or replicated is termed the S phase. A second growth phase, termed G2, occurs after DNA replication, but prior to cell division. The mitosis or M phase is the period of cell division. Following mitosis, the daughter cells either re-enter the G1 phase or enter a quiescent phase termed G0, where growth and replication ceases. The passage of cells through the cell cycle is tightly controlled by a variety of proteins termed cyclin-dependent kinases (see Chapter 39).

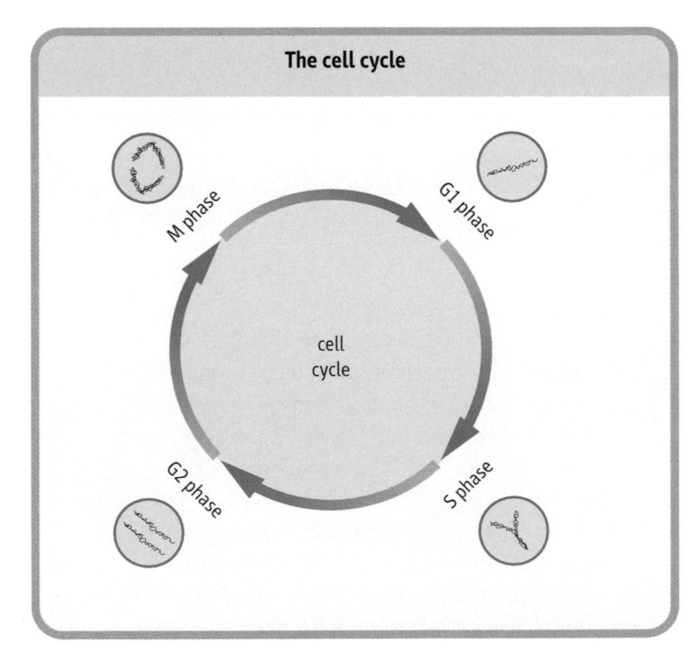

Fig. 29.5 Stages of the cell cycle. G1 and G2 are growth phases that occur before and after DNA synthesis respectively. DNA replication occurs during the S phase. Mitosis occurs during the M phase, producing new daughter cells that can re-enter the G1 phase.

DNA is replicated by separating and copying the strands

For cells to divide, their DNA must be duplicated. The very nature of the DNA double helix suggests a mechanism for DNA replication that consists of strand separation followed by strand copying. The separated strands serve as templates for the synthesis of the new daughter strands. This method of DNA replication results in a 'semiconservative' mechanism in which each replicated duplex, daughter DNA molecule contains one parental strand and one newly synthesized strand.

DNA replication site

The site at which DNA replication is initiated is termed the 'origin of replication'. In prokaryotes, a DNA-binding protein termed DnaA binds to four 9-nucleotide repeated elements located within the origin. Binding of 20–30 DnaA molecules to the origin of replication induces unwinding, resulting in separation of the strands in an AT-rich region adjacent to the DnaA-binding sites. Next the hexameric protein DnaB binds to the separated DNA strands. The DnaB protein has helicase activity that results in an ATP-mediated unwinding of the DNA helix. DNA gyrase also participates in separation of the strands. As this complex continues unwinding the DNA strands in both directions from the origin of replication, single-stranded-DNA-binding proteins coat the separated strands to inhibit their reassociation.

Once the strands are sufficiently separated, another protein, termed primase, is added, resulting in the formation of a primosome complex at each of the replication forks. The primosome synthesizes RNA primers complementary to each parental DNA strand. Once each RNA primer has been laid down, two DNA polymerase III complexes are assembled, one at each of the primed sites. Because of the antiparallel nature of the two strands, the synthesis of DNA along the two strands is different (Fig. 29.6). The two daughter strands being synthesized are termed the leading strand and the lagging strand.

DNA synthesis along the leading strand produces a single, long continuous DNA strand. However, because DNA synthesis adds new nucleotides at the 5′-end of the elongating DNA strand, DNA polymerase III cannot synthesize the lagging strand in one long continuous piece as it does for the leading strand. Instead, the lagging strand is synthesized in small fragments, 1000–5000 base pairs in length, termed Okazaki fragments (Fig. 29.6).

The primosome remains associated with the lagging strand and continues periodically to synthesize RNA primers complementary to the separated strand. As DNA polymerase III moves along the parental DNA strand, it initiates the synthesis of Okazaki fragments at the RNA primers, elongating each fragment from each primer, thereby forming the daughter strand.

When the 3′-end of the elongating Okazaki fragment reaches the 5′-end of the previously synthesized Okazaki fragment, the DNA polymerase III releases the template and finds another RNA primer further back along the lagging strand, initiating another Okazaki fragment. The Okazaki fragments are joined in a process that requires DNA polymerase I. DNA polymerase I has an exonuclease function that permits it to remove and replace a stretch of nucleotides as it proceeds along the template. With this activity, the DNA polymerase I removes the RNA primer and replaces it with DNA. Finally, DNA ligase joins the lagging-strand DNA fragments to form a continuous strand.

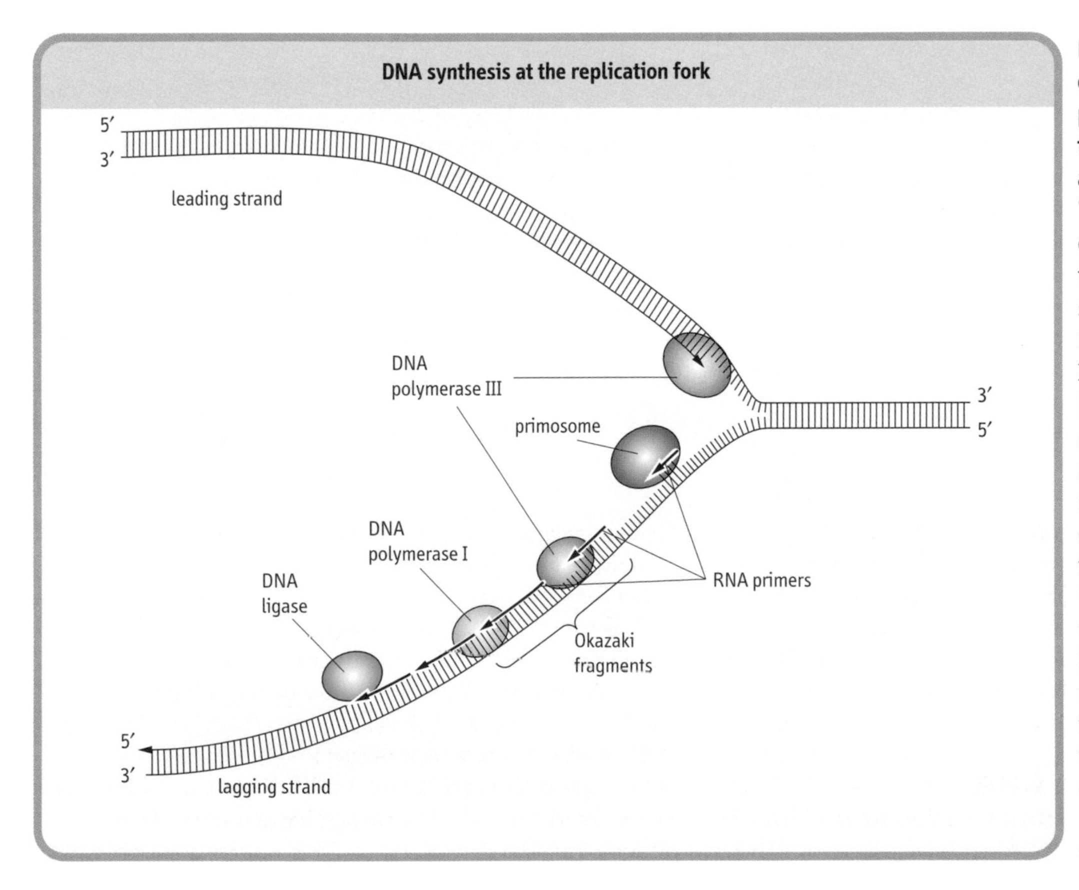

Fig. 29.6 DNA synthesis occurs at a replication fork producing new strands termed the leading strand and the lagging strand. The 'railroad tracks' represent double-stranded DNA. Some of the enzymes involved in DNA synthesis are shown (primosome, DNA polymerase III, DNA polymerase I, DNA ligase). RNA primers are periodically added by the primosome along the lagging strand. DNA polymerase III elongates these RNA primers to form Okazaki fragments. When the Okazaki fragment is complete, the DNA polymerase III on the lagging strand will shift to the next RNA primer to initiate another Okazaki fragment. DNA polymerase I replaces the RNA primers with DNA. DNA ligase seals the gaps in the DNA strands.

Mechanism of DNA polymerization

Several different DNA polymerases exist within cells. DNA polymerase III, the principal DNA-replication enzyme, has the highest rate of DNA polymerization. DNA polymerase I is involved in excision repair and in the removal of RNA primers from the Okazaki fragments.

DNA polymerase III is a complex protein, with at least 10 distinct subunits, ranging in size from 12 kDa to 130 kDa. The action of DNA polymerase III requires a single-stranded template and a primer, either DNA or RNA. After binding to the template–primer complex, the enzyme first determines which nucleotide is complementary to the next available template nucleotide and allows that nucleotide to bind to the active site. The elongation reaction occurs when the 3′-hydroxyl of the previous base attacks the α-phosphate of the incoming deoxynucleotide triphosphate (dNTP). Pyrophosphate is released. This process results in an elongation of the DNA strand by one nucleotide. Then, in a process termed proofreading, the enzyme checks whether the newly incorporated base can form an allowed Watson–Crick base pair (i.e. AT, TA, CG, or GC) with the adjacent base on the template strand. If the base pair is not allowed, the enzyme removes the last added nucleotide and repeats the process. If the base pair is allowed, the enzyme then advances one nucleotide along the template strand and repeats the process. This ability to proofread DNA sequences and replace mismatched nucleotides serves to maintain the high fidelity of DNA replication.

Azido-2′,3′-dideoxythymidine (AZT) therapy for AIDS

HIV infection results in a profound weakening of the immune system that leaves the patient vulnerable to an array of clinical complications, including opportunistic infection by bacterial and fungal pathogens, vascularized neoplasms such as Kaposi's sarcoma, and neurologic dysfunction termed AIDS dementia complex (ADC).

Effective treatments of the HIV viral infection requires detailed information about the viral life cycle. In the life cycle of the AIDS virus, an RNA viral genome is copied into a DNA form by a viral enzyme termed reverse transcriptase. Reverse transcriptase is an error-prone enzyme that does not have the proofreading capabilities of DNA polymerase. One treatment of AIDS takes advantage of this promiscuity in the enzyme's choice of substrates. A potent drug for AIDS treatment is AZT. This drug is metabolized in the body into the thymine triphosphate (TTP) analog azido-TTP. The HIV reverse transcriptase misincorporates azido-TTP into the reverse-transcribed viral genome. The incorporation of azido-TTP into DNA blocks further chain elongation, because the 3′-azido group cannot form a phosphodiester bond with subsequent nucleoside triphosphates. This results in the inability to replicate the virus (Fig. 29.7).

Eukaryotic DNA synthesis

Eukaryotic DNA synthesis is remarkably similar to prokaryotic DNA synthesis. However, because eukaryotic chromosomes contain much more DNA than does a prokaryotic chromosome, eukaryotes typically have multiple origins of replication per chromosome. These are activated during the S-phase of the cell cycle, rapidly replicating the entire chromosome.

Oxidative damage to DNA

DNA within the cell suffers many environmental insults, causing damage to both the sugar and bases of DNA. Oxidative damage to DNA is increased in aging and in age-related, degenerative diseases, including cancer. In the presence of reactive oxygen intermediates, metal ions such as iron and copper, catalyze the oxidation of DNA, which, if unrepaired, may lead to mutations in DNA. Approximately 20 000 to 60 000 modifications of DNA occur per cell per day.

8-Oxo-2′-deoxyguanosine

About 20 major oxidative modifications of DNA have been characterized. The best characterized is 8-oxo-2′-deoxyguanosine (8-oxoG) (Fig. 29.8). During the process of

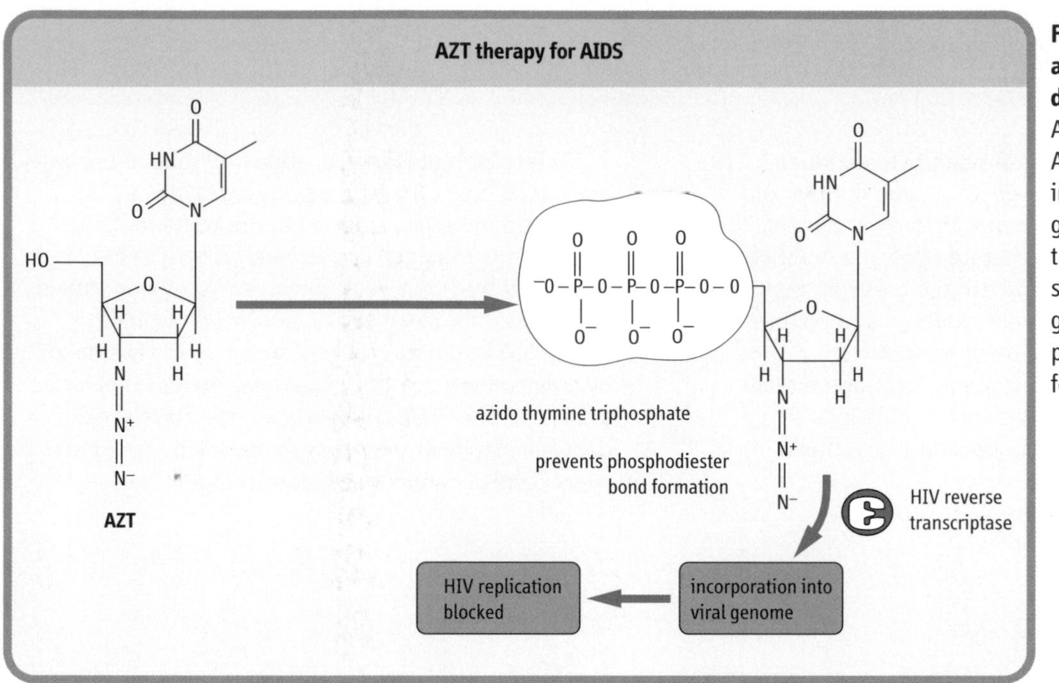

Fig. 29.7 Mechanism of action of azido-2′,3′-dideoxythymidine (AZT). AZT is phosphorylated to AZT-triphosphate, then incorporated into the viral genome by HIV reverse transcriptase. This blocks synthesis of the viral genome because the 3′-phosphate ester cannot be formed.

DNA replication, mismatches between the modified 8-oxoG nucleoside in the template strand and incoming nucleotide triphosphates result in G-to-T transversions, thereby introducing mutations into the DNA strand.

In lung cells, inhalation of some particulate materials results in an increase in 8-oxoG levels. The inflammation associated with increased 8-oxoG may play a role in asbestos-induced formation of lung tumors. Smoking also increases levels of DNA oxidation products.

Cells can repair damaged DNA

Because DNA is the reservoir of genetic information within the cell, it is extremely important to maintain the integrity of DNA. Therefore, the cell has developed highly efficient mechanisms for the repair of modified or damaged DNA.

Excision repair

There are several specific types of damage that affect DNA. These include modification of a single base, alterations involving adjacent pairs of bases, chain breaks, and crosslinkages between bases or between bases and proteins. Those nucleotides that contain amines, cytosine and adenosine, can also be spontaneously deaminated to form uracil or hypoxanthine respectively. When these bases are found in DNA, specific *N*-glycosylases remove them. This produces base-pair gaps that are recognized by specific apurinic or apyrimidinic endonucleases. These endonucleases cleave the DNA near the site of the defect. An exonuclease then removes the stretch of the DNA strand containing the defect. A repair DNA polymerase replaces the DNA, and, finally, DNA ligase rejoins the DNA strand. This repair mechanism is referred to as excision repair.

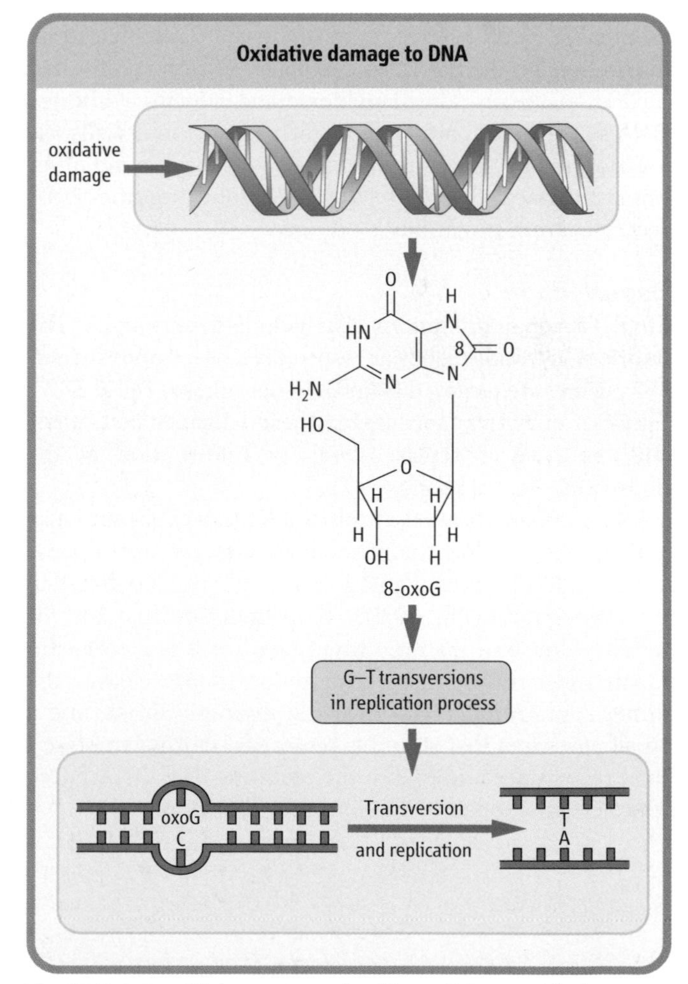

Fig. 29.8 8-Oxo-2′-deoxyguanosine (8-oxoG) is an oxidative modification of DNA that causes mutations in the DNA strand.

Ames test for mutagenesis

Mutagens are chemical compounds that induce changes in the DNA sequence. A large number of natural and man-made chemicals are mutagenic. To evaluate the potential to mutate DNA, the American biochemist Bruce Ames developed a simple test, using special *Salmonella typhimurium* strains that cannot grow in the absence of histidine (*his*⁻ phenotype). These histidine auxotroph strains contain nucleotide substitutions or deletions that prevent the production of histidine biosynthetic enzymes.

To test for mutagenesis, about 10^9 mutant bacteria are spread on a culture plate lacking histidine. The suspected mutagen is added to the bacteria. The action of the mutagen occasionally results in the reversal of the histidine mutation so that these strains can now synthesize histidine and will grow in its absence. The mutagenicity of a compound is scored by counting the number of colonies that have reverted to the *his*⁺ phenotype. There is a good correlation between results of the Ames mutagenicity test and direct tests of carcinogenic activity in animals.

Alkylation of nucleotides in DNA is a common way to induce mutations in DNA. Numerous chemical and environmental agents are known that produce specific chemical modification of the nucleotides in the DNA strand, leading to mismatches during DNA synthesis. After chromosomal replication, the resulting daughter strand contains a different DNA sequence (mutation) from the parent strand. Cells use excision repair to remove alkylated nucleotides and other unusual base analogues, thereby protecting the DNA sequence from mutations.

Depurination

Single base-pair alterations also include depurination. The purine-*N*-glycosidic bonds are especially labile. Approximately 3–7 purines are removed from DNA per min per cell at 37 °C. Specific enzymes recognize these depurinated sites, and the base is replaced without interruption of the phosphodiester backbone.

When short-wavelength ultraviolet (UV) light interacts with DNA, adjacent thymine bases undergo an unusual dimerization that results in a cyclobutylthymine dimer in the DNA strand (Fig. 29.9). The primary mechanism for repair of these intrastrand thymine dimers is an excision repair mechanism. A specific endonuclease cleaves the dimer-containing strand near the thymine dimer, and a small portion of that strand is removed. DNA polymerase I then recognizes and fills in the resulting gap. DNA ligase completes the repair by rejoining the DNA strands.

Single-stranded breaks are frequently induced by ionizing radiation. These are repaired by direct ligation or by excision repair mechanisms. Double-stranded breaks are statistically rare, but are produced by ionizing radiation. These are serious modifications of DNA that are not readily repaired *in vivo*.

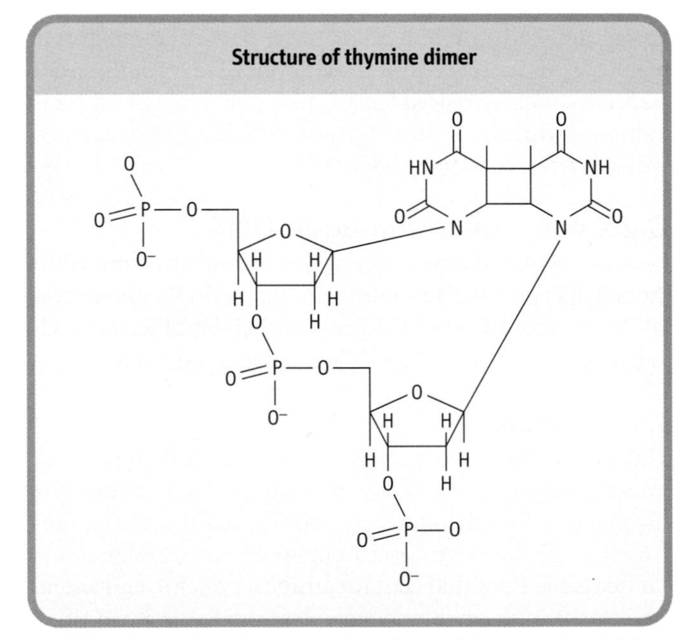

Fig. 29.9 A thymine dimer consists of a cyclobutane ring joining a pair of adjacent thymine nucleotides.

Summary

Genetic information is replicated by a semi-conservative mechanism in which parental strands are separated and both act as templates for daughter DNA. A proofreading function of DNA polymerase III ensures the high fidelity of DNA replication. DNA is the only polymer in the body that is repaired, rather than degraded, following chemical or biological modification. Repair mechanisms generally involve excision of modified bases and replacement, using the unmodified strand as a template.

Further reading

Bambara RA, Murante RS, Henricksen LA. Enzymes and reactions at the eukaryotic DNA replication fork. *J Biol Chem* 1997;**272**:4647–4650.
Fischl MA, Richman DD, Grieco MH, *et al.* The efficacy of azidothymidine (AZT) in the treatment of patients with AIDS and AIDS related complex: a double-blind, placebo-controlled trial. *N Engl J Med* 1997;**317**:185–191.
Hickey RJ, Malkas LH. Mammalian cell DNA replication. *Crit Rev Eukaryotic Gene Expression* 1997;**7**:125–157.
Lambert W, Kuo H, Lambert M. Xeroderma pigmentosum. *Dermatol Clin* 1995;**13**:169–209.
Morin GB. Telomere control of replicative lifespan. *Expl Gerontol* 1997;**32**:375–382.
Sarasin A, Stary A. Human cancer and DNA repair-deficient diseases. *Cancer Detection Prevention* 1997;**21**:406–411.
Sherratt EJ, Thomas AW, Alcolado JC. Mitochondrial DNA defects: a widening clinical spectrum of disorders. *Clin Sci* 1997;**92**:225–235.
Watson JD. The double helix: a personal account of the discovery of the structure of DNA. (Norton: New York, 1980) 298pp.

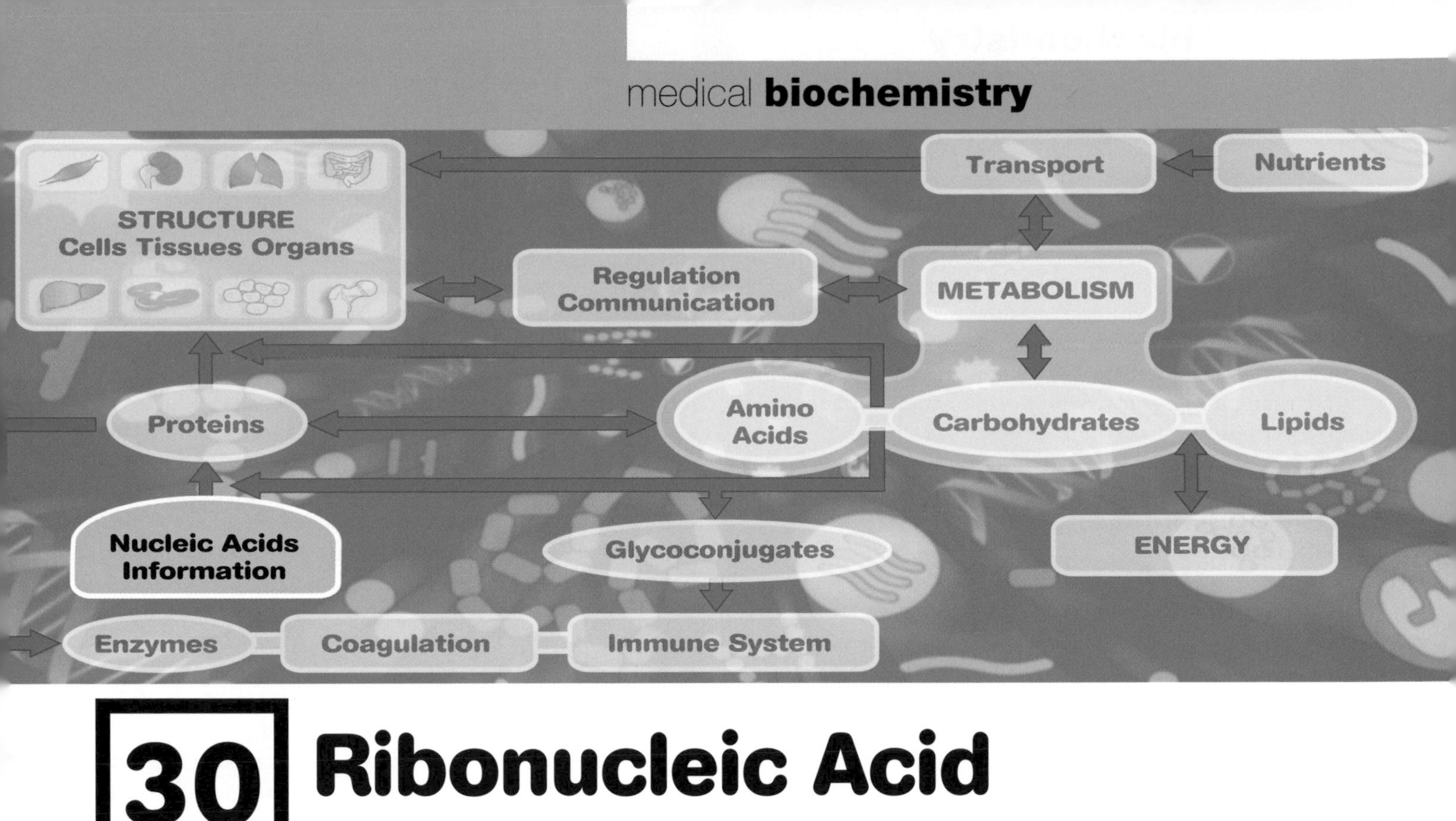

30 Ribonucleic Acid

Transcription is defined as the synthesis of a ribonucleic acid (RNA) molecule using deoxyribonucleic acid (DNA) as a template. This rather simple definition describes a series of complicated enzymatic processes that result in the transfer of the information stored in double-stranded DNA into a single-stranded RNA molecule that will be used by the cell to direct the synthesis of its proteins. There are three general classes of RNA molecules found in prokaryotic and eukaryotic cells: ribosomal RNA (rRNA), transfer RNA (tRNA), and messenger RNA (mRNA). Each class has a distinctive size and function (Fig. 30.1), described by its sedimentation rate in an ultracentrifuge (S, Svedbergs) or its number of bases (nt, nucleotides, or kb, kilobases). Prokaryotes have the same three general classes of RNA as eukaryotes, but their RNAs differ in size and in some structural features:

- **rRNA** consists of three different sizes of RNA:
  28S RNA (~5 kb)
  18S RNA (~1.9 kb)
  5.8S RNA (~120 nt)
  which interact with each other, and other proteins, to form a ribosome that provides the basic machinery on which protein synthesis takes place;
- **tRNAs** consist of one size class of RNA that are 65–110 nucleotides in length; they function as adapter molecules that translate the information stored in the mRNA nucleotide sequence to the amino acid sequence of proteins;
- **mRNAs** represent the most heterogeneous class of RNAs found in cells, ranging in size from 500 nt to >6 kb; they are carriers of genetic information, directing the synthesis of all proteins in the cell.

General classes of RNA

RNA	Size range	Percent of total cellular RNA	Function
rRNA	28S, 18S, 5.8S (26S, 16S, 5S)*	80	interact to form ribosomes
tRNA	65–110 nt	15	adapter
mRNA	0.5–6 kb	5	direct synthesis of cellular proteins

Fig. 30.1. Classes of RNA. *Size of rRNA in prokaryotic cells.

In order to understand the complex series of events that result in the production of these three classes of RNA, this chapter is divided into four parts. The first part deals with the molecular anatomy of the major types of RNA found in prokaryotic and eukaryotic cells; knowing then the chemical nature of the final products of transcription, you will be better prepared to understand the steps involved in generating these molecules. The second part describes the main enzymes involved in transcription, and their specificities. The third part describes the three steps (initiation, elongation, and termination) required to produce an RNA transcript. Finally, in the last section, the modifications that are made to the primary products of transcription (post-transcriptional processing) are described.

The molecular anatomy of RNA molecules

In general, the RNAs produced by prokaryotic and eukaryotic cells are single-stranded molecules that consist of adenine, guanine, cytosine, and uracil nucleotides joined to one another by phosphodiester linkages. The start of an RNA molecule is known as its 5′ end, and the termination of the RNA is known as its 3′ end. Even though most RNAs are single stranded, they exhibit extensive secondary structures that are important to their function. These secondary structures, one of the most common of which is called a hairpin loop (Fig. 30.2), are the product of intramolecular base pairing that occurs between complementary nucleotides within a single RNA molecule.

rRNAs: formation of the ribosome

The eukaryotic rRNAs are synthesized as a single RNA transcript with a size of 45S and about 13 kb long. This large primary transcript is processed into 28S, 18S, and 5.8S rRNAs. The 28S and 5.8S rRNAs associate with ribosomal proteins to form what is called the large ribosomal subunit. The 18S rRNA associates with other specific proteins to form the small ribosomal subunit. The large ribosomal subunit with its proteins and RNA has a characteristic size of 60S; the small ribosomal subunit has a size of 40S. These two subunits interact to form a functional ribosome that has a size of 80S (see Chapter 31). Prokaryotic rRNAs interact in a similar fashion to form these ribosomal subunits, but have a slightly smaller size, reflecting the difference in rRNA transcript size that exists between prokaryotic and eukaryotic cells (Fig. 30.3).

tRNA: the molecular cloverleaf

Prokaryotic and eukaryotic tRNAs are similar in both size and structure. They exhibit extensive secondary structure and contain several ribonucleotides that differ from the usual four by a variety of modifications. All tRNAs have a similar folded structure, with four distinctive loops, that has been described as a cloverleaf (Fig. 30.4). The D loop contains several modified bases, including methylated cytosine and dihydrouridine, for which the loop is named. The anticodon loop is the structure responsible for recognition of the complementary codon of an mRNA molecule: specific interaction of an anticodon with the appropriate codon is due to complementary base pairing between these two trinucleotide sequences. A variable loop, 3–21 bp in length, exists in most tRNAs, but its function is unknown. Finally, there is a TψC loop, which is named for the presence in this loop of the modified base, pseudouridine. Another prominent structure found in all tRNA molecules is the acceptor stem. This structure is formed by base pairing between the nucleotides found at the 5′ and 3′ ends of the tRNA. The last three bases found at the extreme 3′ end remain unpaired, and always have the same sequence: 5′-CCA-3′. The 3′ end of the acceptor stem is the point at which an amino acid is attached via an ester bond between the 3′-hydroxyl group of the adenosine and the carboxyl group of an amino acid.

mRNAs: sorting the prokaryotes from the eukaryotes

mRNAs are the most distinctive class of RNAs when prokaryotic and eukaryotic cells are compared (Fig. 30.5). This is largely because a single prokaryotic mRNA can encode multiple proteins (polycistronic), whereas a single eukaryotic

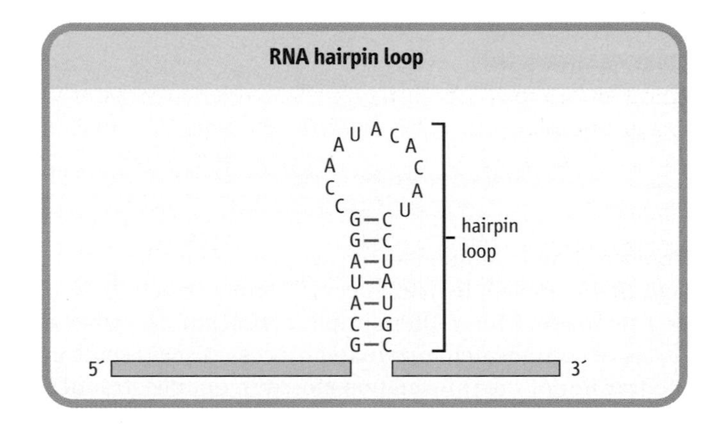

Fig. 30.2 RNA can form secondary structures called hairpin loops. These structures form when complementary bases within an individual RNA share hydrogen bonds and form base pairs. Hairpin loops are known to be important in the regulation of transcription in both eukaryotic and prokaryotic cells.

rRNAs interact to form ribosomes

Cell type	rRNA	Subunit	Size	Intact ribosome
prokaryotic	23S, 5S	large	50S	70S
	16S	small	30S	
eukaryotic	28S, 5.8S	large	60S	80S
	18S	small	40S	

Fig. 30.3 rRNAs and ribosomes.

mRNA carries the information to encode only a single protein; however, there are also a number of chemical modifications that are unique to eukaryotic mRNAs. For example, the 5′ and 3′ ends of eukaryotic mRNAs are modified, after they have been synthesized, in specific ways that protect them from exonuclease attack. All eukaryotic mRNAs contain a methylated guanine nucleotide 'cap' at their 5′ end, bound in a unique 5′–5′ triphosphate linkage. The 3′ end is even more extensively modified, by the addition of a number of adenine residues – known as a polyA tail. The number of adenine residues added to a particular transcript can vary from as few as 30 to more than 100 residues. Although the vast majority of eukaryotic mRNAs contain a polyA tail, there are notable exceptions to this rule: for example, the mRNAs that encode histone proteins and heat shock proteins do not contain such a tail. While there are many theories, there is little compelling evidence to explain why these particular mRNAs do not contain this structure at their 3′ end.

One of the most striking differences between prokaryotic and eukaryotic mRNAs is that eukaryotic mRNAs are synthesized as large precursors that have to be processed before they are functional. This processing usually involves the removal of portions of the transcript, called introns, and ligation of the remaining sequences, called exons, to one another. There is little time for this type of processing to occur in prokaryotic cells, in which the transcript may be partially translated even before transcription is completed; in contrast, simultaneous transcription/translation cannot occur in eukaryotic cells because the nuclear envelope acts as a natural barrier between the process of transcription and translation. The process of removal and ligation of sequences within a primary transcript is called splicing; it is more thoroughly discussed later in this chapter.

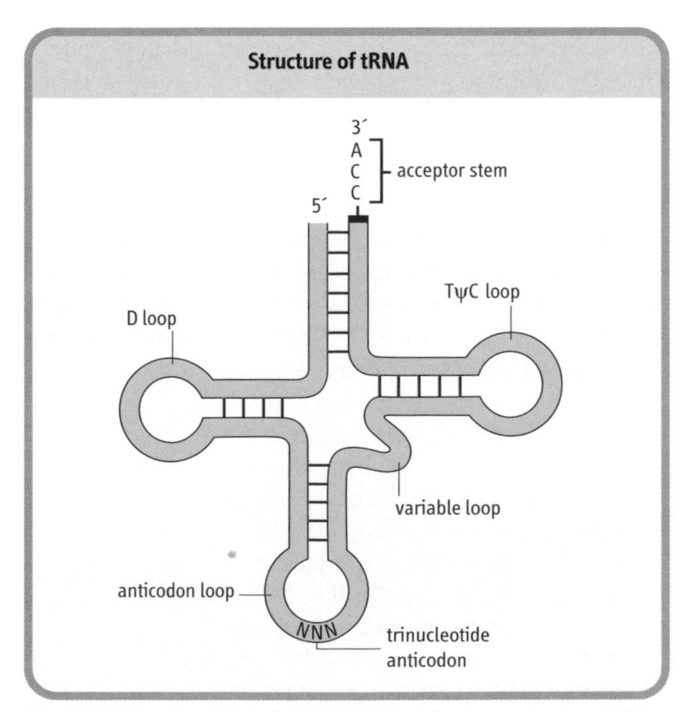

Fig. 30.4 The structure of a tRNA molecule. A prototypical tRNA molecule is shown, and the structures important to its function are indicated. The overall structure of the molecule is due to complementary base pairing between nucleotides within a single RNA. All tRNAs have this basic structure.

RNA polymerases

The enzymes responsible for the synthesis of RNA, using DNA as a template, are called RNA polymerases. All RNAs are synthesized by these enzymes, in a direction that is 5′ to 3′ with respect to their internucleotide linkage. This polarity of synthesis dictates that the DNA strand used as a template is read in the 3′ to 5′ direction (Fig. 30.6): RNA polymerase uses the DNA template strand to synthesize the complementary RNA strand, adding each new nucleotide onto the 3′ end of the growing chain. RNA polymerases have the ability to initiate the synthesis of RNA without the benefit of a free 3′-OH on which to build the new RNA strand. This activity distinguishes them from the DNA polymerases, which require RNA or DNA primers with a free 3′-OH to begin synthesis.

RNA polymerases are large multimeric enzymes that transcribe defined segments of DNA into RNA with a high degree of selectivity and specificity

The RNA polymerases generally consist of two large-molecular-weight subunits and several smaller subunits, all of which are necessary for accurate transcription to occur. In prokaryotic cells, there is only one type of RNA polymerase, which synthesizes all three of the general classes of RNA. In contrast, eukaryotic cells have three RNA polymerases (I, II, and III), which are distinguished from one another by the class of RNA for which they direct the synthesis. The function of each of the eukaryotic RNA polymerases was determined in part by using a transcription inhibitor, α-amanitine – a toxic chemical found in some mushrooms.

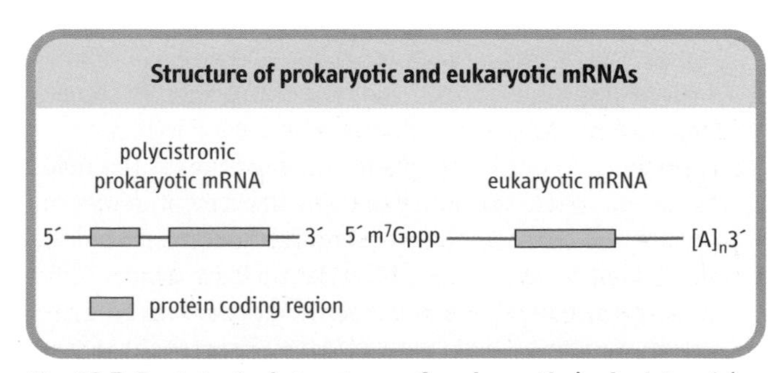

Fig. 30.5 Prototypical structures of prokaryotic (polycistronic) and eukaryotic mRNAs. The boxes indicate those portions of the mRNA that encode a protein. $[A]_n$, polyA tail of adenine residues; m^7Gppp, 7-methylguanine nucleotide cap.

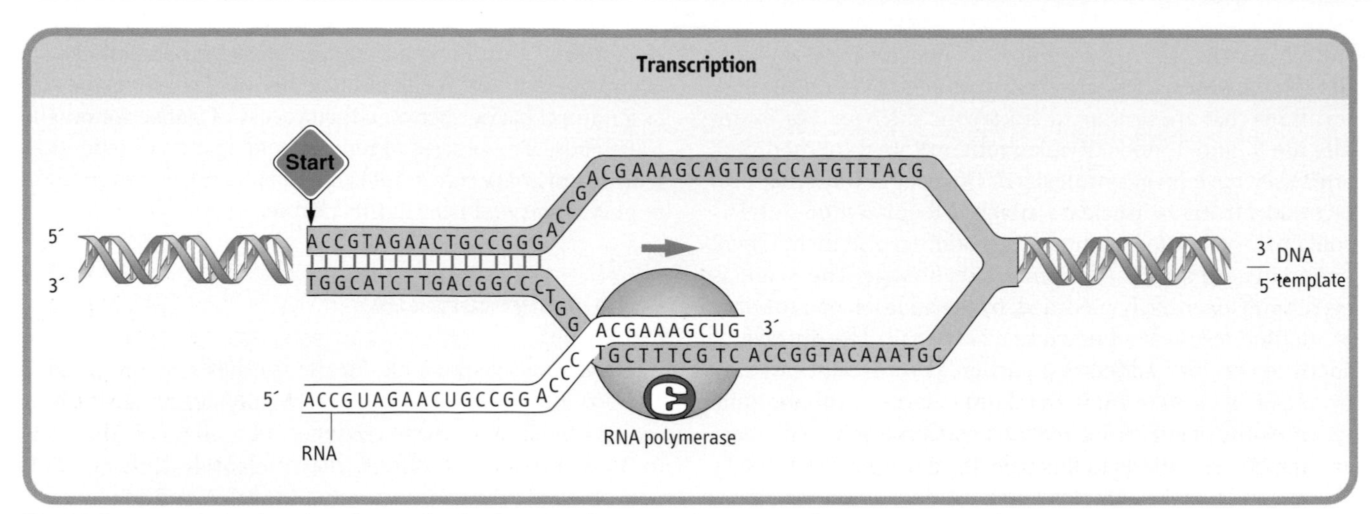

Fig. 30.6 Transcription involves the synthesis of an RNA by RNA polymerase using DNA as a template. The RNA polymerase holoenzyme uses one strand of DNA to direct the synthesis of an RNA molecule that is complementary to this strand.

- **RNA polymerase I** is responsible for synthesizing the rRNAs;
- **RNA polymerase II,** the activity of which is extremely sensitive to inhibition by α-amanitine, is responsible for mRNA synthesis;
- **RNA polymerase III** synthesizes the small RNAs, including the tRNAs.

The best-studied RNA polymerase is that of *Escherichia coli.* The core portion of this bacterial polymerase contains four proteins or subunits: one molecule each of a β (155 kDa) and a β´ (165 kDa) subunit, and two molecules of an α subunit (35 kDa). These subunits interact to form what is called the core polymerase. This polymerase is capable of synthesizing RNA, but does so in a nonspecific fashion. It gains specificity and is able to bind and initiate transcription at true initiation sites on DNA when a fifth subunit, the σ-factor, joins the complex. *E. coli* contains a number of different σ-factors that, when joined with the core polymerase, impart specificity for the transcription of certain classes of genes.

A case of mistaken identity

An otherwise healthy young man presents himself to the emergency room with severe nausea and diarrhea. After his medical history has been taken, he explains that his symptoms came on rather suddenly, about 2–3 hours after he had eaten dinner. The physician suspects some form of food poisoning, and asks the patient to recall everything he has eaten over the past 24 hours. The only suspicious food mentioned were mushrooms that the patient ate for dinner. The mushrooms become prime candidates for the cause of this patient's symptoms when he further relates that they were picked up on a recent hike through the woods. What is the biochemical basis for suspecting that the mushrooms are the cause of this man's illness?

Comment. It is likely that the patient mistakenly picked a member of the family *Amanita phalloides* and ingested them at dinner. The toxin, α-amanitine, binds preferentially to RNA polymerase II and inhibits its function; if a large quantity of the mushrooms had been ingested, even RNA polymerase III could be inhibited. The first cells that encounter the toxin are those of the digestive tract, leading to acute gastrointestinal distress: cells that are incapable of synthesizing new mRNAs and tRNAs would die, causing the diarrhea and nausea that the patient complained of when first examined.

The case of a stubborn microbe

A patient you were treating for an infected wound last week returns to your clinic, complaining that the infection is worse. You examine the wound and confirm that, indeed, the infection has become worse, even though you had prescribed a high-dose regimen of antibiotics that targeted the bacterial protein synthetic machinery. After questioning the patient to ensure that she was compliant and took the medication, you elect to prescribe rifampicin, a synthetic derivative of the naturally occurring antibiotic, rifamycin. The patient asks you why you think this antibiotic will work.

Comment. Antibiotics work by targeting specific functions in the bacterial cell. In the first round of treatment, you used antibiotics that inhibited the bacterial protein synthetic machinery. In some cases, microbes become resistant to a specific type of antibiotic and it is necessary to target other bacterial functions in order to clear up the infection. Rifampicin inhibits the transcriptional machinery of bacteria, inhibiting the β subunit of bacterial RNA polymerase. Without the ability to make RNA, the bacterial cell will die.

The process of transcription

Transcription is a dynamic process that involves the interaction of enzymes and DNA in specific ways to produce an RNA molecule. In order to understand this process better, it is convenient to divide it into three separate stages:

- **initiation**,
- **elongation**,
- **termination**.

Initiation of transcription

Initiation involves the interaction of the RNA polymerase with DNA in a site-specific fashion, so that the correct sequence of DNA can be used as a template for synthesis of an RNA molecule (Fig. 30.7). The focus on specific sequences of DNA for transcription is an important consideration, because most of the DNA in a cell does not encode proteins. This problem is solved through the interaction of RNA polymerase with specific sites on the DNA, known as promoters. Promoters are characteristic sequences of DNA, usually located in front (upstream) of the gene that is to be transcribed.

The simplest promoters are found in prokaryotic cells, in which two general types of sequence elements are found in front of most genes: one sequence element is believed to promote initial binding of the RNA polymerase, and the other element usually has a high content of adenine (A) and thymine (T). Because hydrogen bonding is weaker between A–T base pairs than between guanine–cytosine (G–C) base pairs, the increased A–T content helps in the dissociation of the two DNA strands, enabling transcription to occur. These promoter sequences, 6–8 nt in length, are generally located about 35 and 10 bp upstream from the start of transcription of prokaryotic genes. Because these sequence elements are required for any level of transcription to occur, they are called basal promoters. The promoter located at –10 base pair, i.e. 10 bp before the transcription starting site (Fig 30.7), is known commonly as the TATA or Pribnow box.

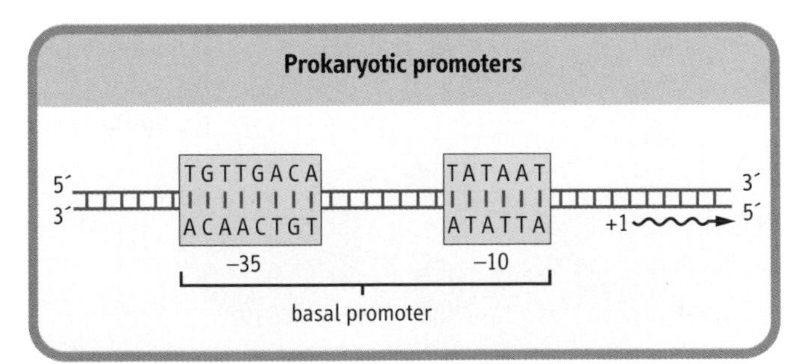

Fig. 30.7 Prokaryotic transcription promoters are located upstream of the gene. DNA sequences that act as transcription promoters are indicated at their respective positions relative to the gene. Position +1 indicates the first nucleotide that will be transcribed into RNA. TATA box, a common A–T-rich promoter element.

Promoters found in eukaryotic cells are more complicated. In addition to the sequence elements required for basal expression, they have additional sequences that are responsible for regulating the rate of initiation of transcription. These sequence elements are known as either enhancers or silencers, depending on the effect they have on transcription. They can be located at great distances either upstream or downstream of the start of transcription. Promoters with these types of sequence elements exert their effect on transcription by acting as the binding site for a variety of proteins known as trans-acting factors. The type of trans-acting factor that binds to these sequence elements will determine whether the rate of transcription is increased or decreased.

Elongation

Once RNA polymerase has bound to a promoter, it begins the process of selecting the appropriate complementary ribonucleotide and forming phosphodiester bridges between this nucleotide and the nascent chain, in a process called elongation. Elongation can be a very rapid process, occurring at the rate of 40 nt per second. For elongation to occur, the double-stranded DNA must be continually unwound, so that the template strand is accessible to the RNA polymerase; DNA topoisomerases I and II are enzymes associated with the transcription complex that have the ability to separate DNA strands so that they are accessible as templates for RNA synthesis.

Termination

In addition to knowing where to start transcription, RNA polymerase must have a defined site at which to stop RNA synthesis, so that the appropriate size of transcript is produced. This process, known as transcription termination, is probably the least understood part of RNA synthesis. In eukaryotic cells, termination is believed to involve a secondary structure formed in the newly synthesized RNA, which dislodges the RNA polymerase from the DNA template, resulting in the release of the transcript. In addition, as yet unidentified ancillary factors may be necessary for accurate termination of certain classes of transcripts in mammalian cells.

Transcription termination in bacterial cells occurs by one of two well-characterized mechanisms (Fig. 30.8). The first mechanism, rho-independent termination, is similar to that described for eukaryotic transcription, except that the secondary structure and sequences involved are much better characterized in bacterial cells. In rho-independent termination, a hairpin loop is formed just before a sequence of six to eight uridine (U) residues near the 3´ end of the newly synthesized RNA. The formation of this secondary structure dislodges the RNA polymerase from the DNA template, resulting in termination of RNA synthesis in the U stretch. The second mechanism, rho-dependent termination, requires the action of a

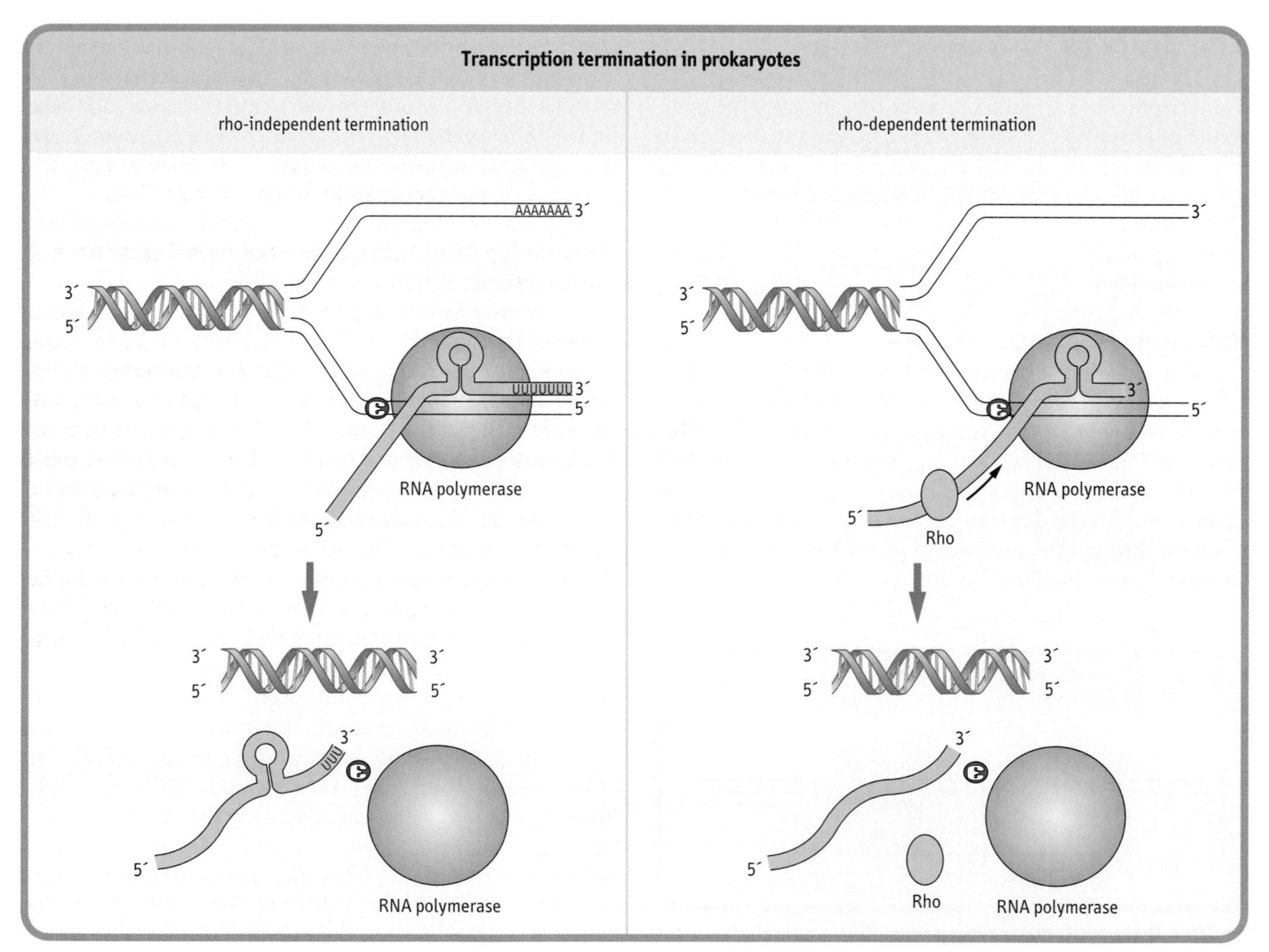

Fig. 30.8 Two mechanisms of transcription termination in bacterial cells. Rho-independent termination relies on the formation of a secondary structure in the newly transcribed RNA to dislodge the RNA polymerase from the DNA template and stop transcription. Rho-dependent termination requires the action of the rho protein. This protein will move along the newly transcribed RNA, catching up with the RNA polymerase when it pauses at the termination site, and causing the polymerase to leave the DNA template.

protein factor called rho, which has an ATP-dependent helicase activity that is required for transcription termination. The rho protein is believed to travel along the newly synthesized RNA, chasing the RNA polymerase. In this mechanism, the formation of a hairpin loop in the RNA structure causes the RNA polymerase to pause, allowing the rho protein to catch up with and displace the RNA polymerase from the template.

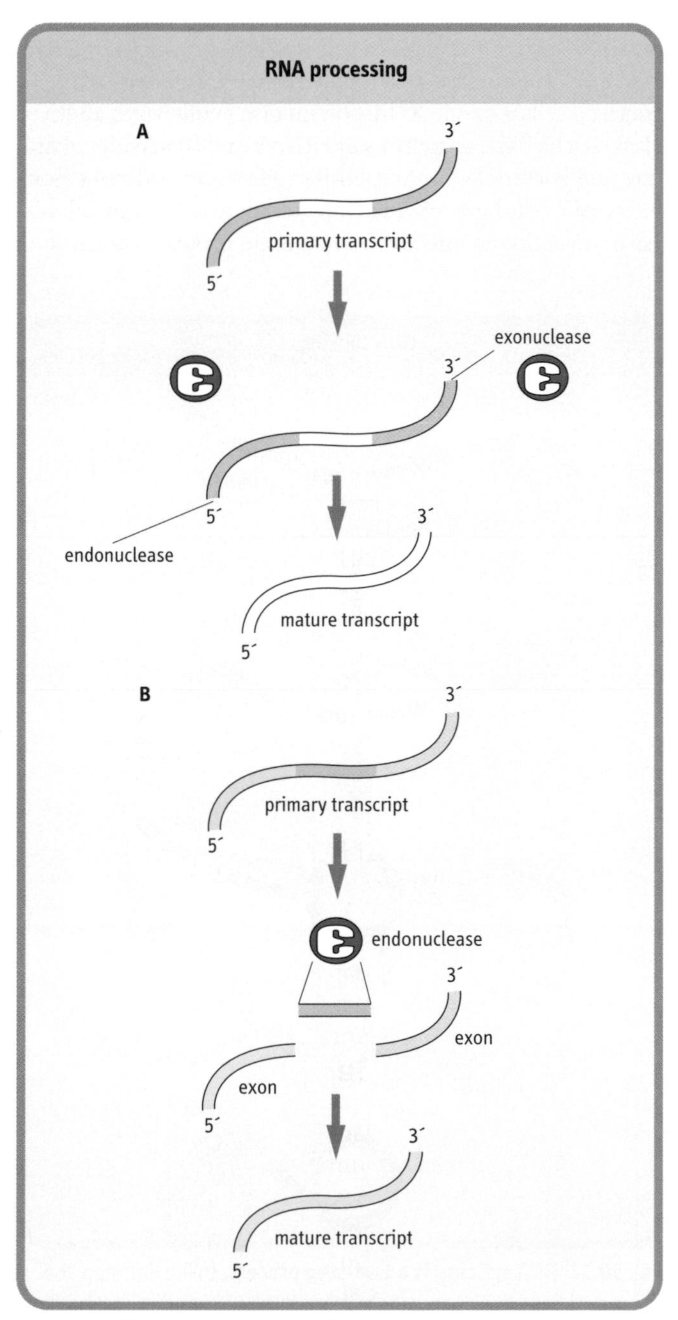

Fig. 30.9 There are two general types of RNA processing events. Processing of an RNA transcript can involve (A) the removal of excess sequences by the action of endonucleases and exonucleases, or (B) the removal of excess sequences and the rejoining of segments of the newly transcribed RNA.

Post-transcriptional processing

In eukaryotes, all three classes of RNA (tRNAs, rRNAs, and mRNAs) are synthesized as larger primary transcripts, also known as heterogeneous nuclear (heteronuclear, hn) RNAs. hnRNAs are usually much larger than the RNAs found in the cytoplasm. For any of these transcripts to be functional, it must be processed to a smaller size before leaving the nucleus. Processing may involve either removal of sequences from the primary transcript, or removal and rejoining of segments of the transcript (Fig. 30.9).

Both eukaryotic and prokaryotic tRNAs and rRNAs are subject to the simplest type of processing

As noted above, the rRNAs are initially synthesized as a large precursor RNA that contains the 28S, 18S, and 5.8S RNAs. In some cases, there are tRNAs in addition to the rRNAs contained within this primary transcript. In order to obtain RNAs of the correct size from the initial transcript, the actions of endoribonucleases and exoribonucleases are required. Endoribonucleases cleave phosphodiester bonds within the primary transcript to release individual RNAs; exoribonucleases remove excess nucleotides from the 5´ and 3´ ends of these RNAs until a molecule of the correct size is produced. After processing, specific nucleotides are modified to give the unusual complement of bases found in most tRNAs.

Processing of eukaryotic mRNAs: spliceosomes and lariats

In the more complicated post-transcriptional processing of eukaryotic mRNAs, sequences called introns (intravening sequences) are removed from the primary transcript and the remaining segments, termed exons (expressed sequences), are ligated, to form a functional RNA. This process involves a large complex of proteins and auxiliary RNAs called small nuclear RNAs (snRNAs), which interact to form a spliceosome. The function of the five snRNAs (U1, U2, U4, U5, U6) in the spliceosome is to help position reacting groups within the substrate mRNA molecule, so that the introns can be removed and the appropriate exons can be spliced together (Fig. 30.10). The snRNAs accomplish this task by binding,

snRNAs and their function in splicing mRNAs

snRNA	Size	Function
U1	165 nt	Binds the 5´ exon/intron boundary
U2	185 nt	Binds the branch site on the intron
U4	145 nt	Helps assemble the spliceosome
U5	116 nt	Binds the 3´ intron/exon boundary
U6	106 nt	Helps assemble the spliceosome

Fig. 30.10 The function of snRNAs in the splicing of mRNAs.

through base-pairing interactions, with the sites on the mRNA that represent intron/exon boundaries. Accompanying protein factors are responsible for holding the reacting components together to facilitate the reaction.

The removal of an intron and rejoining of two exons can be considered to occur in two steps (Fig. 30.11). The first step involves the breaking of the phosphodiester bond at the exon/intron boundary at the 5′ end of the intron. This is accomplished by a transesterification reaction, which occurs between the 2′-OH of an adenine nucleotide usually found about 30 nt from the 3′ end of the intron, and the phosphate in the phosphodiester bond of a guanosine residue located at the 5′ end of the intron. This reaction cleaves the nucleotide chain and produces a branched structure in which the adenine has 2′, 3′ and 5′ phosphate groups. The intron forms a looped structure similar in appearance to that of a cowboy's lariat. The second step in the reaction involves the cleavage of the phosphodiester bond at the 3′ end of the intron, which releases the lariat structure from the complex.

Splicing is completed by the joining of the 3′ end of one exon and the 5′ end of the next exon, through the formation of a regular 5′–3′ phosphodiester bond. Typically, the 3′ end of one exon will be spliced to the 5′ end of the next closest exon, producing a transcript that exhibits all the exons in the order in which they were transcribed. However, depending on cell and tissue type, differential processing of a transcript can occur. In these instances, some exons may not be represented in the final transcript, yielding an RNA that encodes a different protein. This process represents a major mechanism by which eukaryotic cells control synthesis of different proteins from the same gene transcript in a cell- or tissue-specific manner.

A new light on cellular evolution: self-splicing RNA

The vast majority of transcript processing occurs as described in the preceding parargraphs, but a small percentage of transcripts from a wide variety of organisms undergo splicing reactions without the benefit of protein cofactors. These transcripts are self-splicing and can be classified into two groups, depending on whether a cofactor is required and whether the intron forms a lariat configuration during splicing:

- **group I self-splicing introns:** guanosine acts as a cofactor in the transesterification reaction, which leads to release of the intron without formation of the lariat;
- **group II self-splicing introns:** lariats are formed, and the process is similar to that which occurs in spliceosomes, except that no proteins are required.

The discovery of self-splicing RNA – that is, RNA with an enzymatic activity – has led to new ideas about early cellular evolution, which was originally believed to start with amino acids and proteins. It is now believed that ribonucleotides and RNA may have been the most primitive biopolymers to form on earth, providing for genetic diversity, and that DNA and proteins may have developed later.

Summary

The major products of transcription are the rRNAs, tRNAs, and the mRNAs. These RNAs perform specific functions in the cell. rRNAs interact to form ribosomes, the basic cellular machinery on which protein synthesis occurs. tRNAs function as adapter molecules that translate the information stored in the mRNA nucleotide sequence to the amino acid sequence of proteins. mRNAs carry the genetic information from DNA and direct the synthesis of all proteins in the cell. In eukaryotic cells each of these classes of RNAs is produced by a specific RNA polymerase (RNA polymerase I, II, or III) while in bacterial cells a single RNA polymerase synthesizes all three classes. The basic structures of rRNAS and tRNAs in eukaryotic and bacterial cells are similar. However, mRNAs from eukaryotic cells have a 5′ (m^7Gppp) cap and a 3′ ([A]n) tail and have the ability to encode only a single protein. Prokaryotic

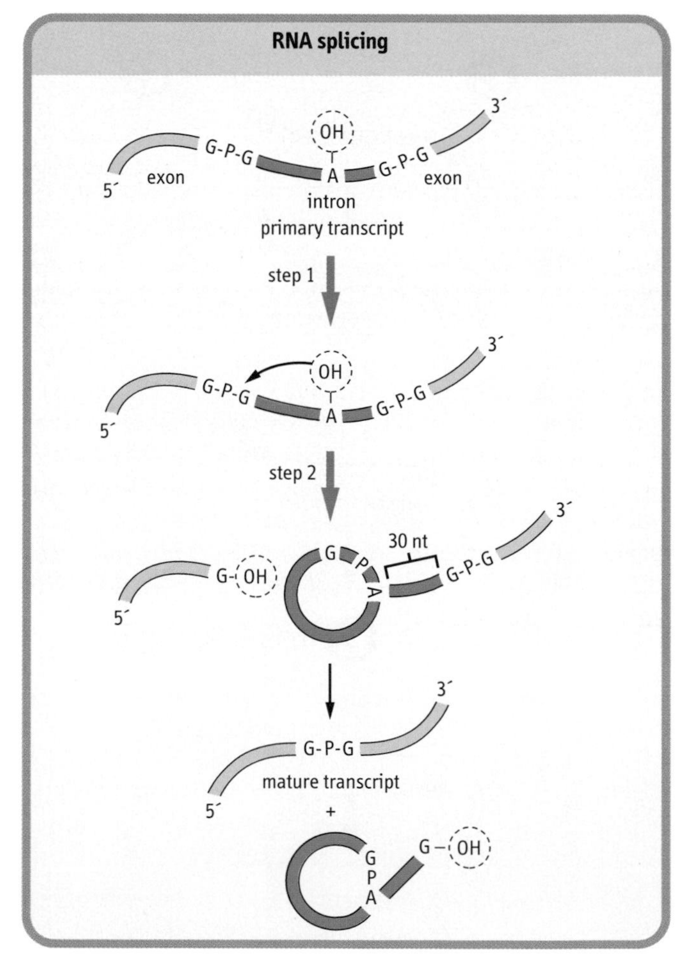

Fig. 30.11 RNA splicing is a two-step process. In the first step, the phosphate bond of a guanosine residue at the 5′ exon/intron boundary is broken and joined to the 2′-OH of an adenine residue located in the middle of the intron. In the second step of the reaction, the phosphate bond at the 3′ intron/exon boundary is first cleaved and then the two exons are spliced together by re-formation of a phosphodiester bond between the nucleotides at either end of the exons.

Ribozymes

RNAs that act like enzymes

In some instances, RNAs have a catalytic ability similar to the type of activities previously ascribed only to proteins. These special molecules, known as ribozymes, possess a catalytic activity and a substrate specificity similar to those of proteinaceous enzymes. The substrate specificity of a ribozyme is determined via nucleotide base pairing between complementary sequences contained within the enzyme and the RNA substrate. Just like enzymes that are proteins, the ribozyme will cleave its substrate RNA at a specific site and then release it, without itself being consumed in the reaction.

Ribozymes are being considered as possible therapeutic agents for diseases that are caused by the inappropriate expression of an RNA or the expression of a mutated RNA. In these cases, the development of a ribozyme that had specificity for a particular RNA could result in the selective degradation of the substrate, eliminating it from the cell and inhibiting the disease process. Many more years of research on the mechanism of ribozyme activity are required before this type of treatment will be available.

cells do not have these modifications on their 5′ and 3′ ends and can be polycistronic. In addition, most eukaryotic mRNAs must undergo a process called splicing to be functional, whereas prokaryotic mRNAs are functional as soon as they are synthesized. Splicing involves the removal of sequences called introns and the joining of other sequences called exons to each other to form a functional mRNA. The process of transcription consists of three parts; initiation, elongation, and termination. Initiation involves the recognition by RNA polymerase of the specific region of DNA that is to be transcribed. To accomplish this RNA polymerases interact with specific DNA sequences called promoters located 5′ to the start of transcription. Elongation involves the selection of the appropriate nucleotide, as determined by the DNA strand, and formation of the phosphodiester bridges that exist between each nucleotide in an RNA molecule. Finally, termination involves the dissociation of the RNA polymerase from the DNA template. This can be accomplished by either RNA secondary structure or specific protein factors.

Further reading

Barnes PJ, Karin M. Nuclear factor-κB: a pivotal transcription factor in chronic inflammatory diseases. *N Engl J Med* 1997;**336**:1066–1071.

Earnshaw DJ, Gait MJ. Progress toward the structure and therapeutic use of the hairpin ribozyme. *Antisense Nucleic Acid Drug Development* 1997;**7**:403–411.

Herman T, Westhof E. RNA as a drug target: chemical, modeling, and evolutionary tools. *Curr Opin Biotechnol* 1998;**9**:66–73.

Sassone-Corsi P. Transcriptional checkpoints determining the fate of male germ cells. *Cell* 1997;**88**:163–166.

Tichan R. Molecular machines that control genes. *Sci Am* 1995;**272**:54–61.

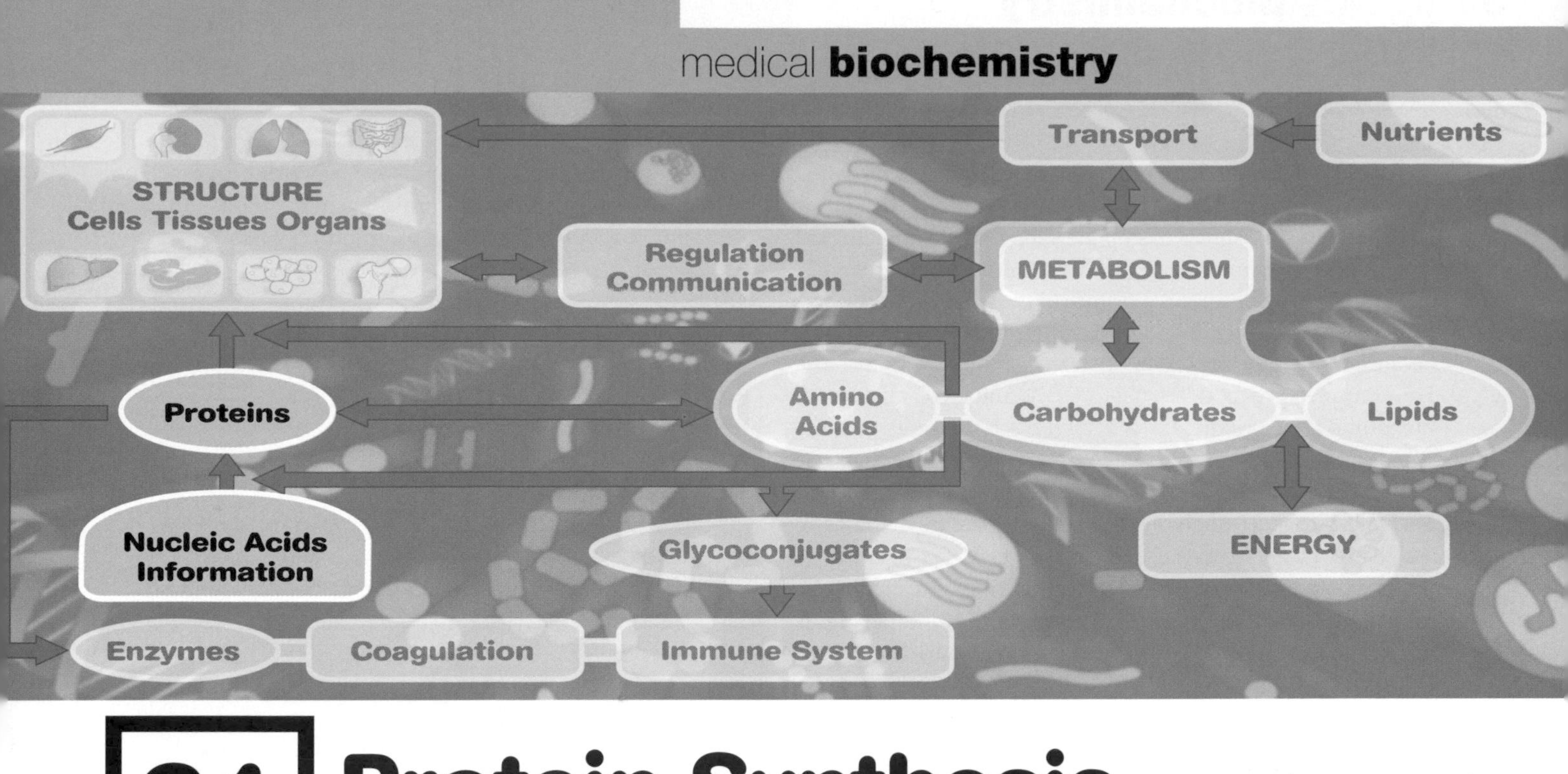

31 Protein Synthesis

Protein synthesis, also known as translation, represents the culmination of the transfer of genetic information, stored as nucleotide bases in deoxyribonucleic acid (DNA), to protein molecules that are the major structural and functional components of living cells. It is during translation that the information carried on a ribonucleic acid (RNA) molecule, expressed as a specific nucleotide sequence, is used to direct the synthesis of a protein, the three-dimensional structure of which will be defined, in large part, by its amino acid sequence. The interaction between the RNA to be translated and the protein synthetic machinery involves three main components:

- **ribosomes**,
- **messenger RNA (mRNA)**,
- **transfer RNA (tRNA)**.

The ribosome is the machine on which all proteins are synthesized. mRNA contains the information required to direct the synthesis of the primary sequence of the protein, although only a portion of that information is used to encode the protein, and tRNAs carry the amino acids that are to be incorporated into the protein; the ribosome brings together the tRNA molecule and mRNA so that the correct amino acid is incorporated. In general, the translation of mRNA begins near the 5′ end and moves towards the 3′ end, and proteins are synthesized starting with their amino-terminal ends and moving toward the carboxy-terminal end; thus the 5′ end of the RNA corresponds to the amino-terminal end of the protein, whereas the 3′ end of the RNA corresponds to the carboxy-terminal end of the protein. In this chapter, we will begin our discussion of translation by looking at the general characteristics of the genetic code. Next, we will discuss ribosomes, mRNA, and tRNA, with emphasis on the structures and interactions required to produce a protein. The process of translation (initiation, elongation, and termination of protein synthesis) will then be explained, and the mechanism by which proteins are targeted to specific locations in the cell will be discussed. Finally, the many modifications that proteins undergo after synthesis so that they can attain their full activity will be described briefly.

The genetic code

The code of life: degenerate, not quite universal, and unpunctuated!

When one considers the transfer of information from RNA containing only four different bases (adenine, A; cytosine, C; guanine, G; and uracil, U) to a protein containing 20 different amino acids, it is apparent that there is not a one-to-one correspondence between nucleotide and amino acid sequence. In fact, three nucleotides in the mRNA, known as a codon, are required to specify each amino acid. There is thus a total of 64 possible codons when all combinations of four nucleotides are taken into account three at a time (Fig. 31.1). Three of these codons (UAA, UAG, UGA) are used as signals to stop the synthesis of a protein, and do not specify an amino acid. This leaves 61 codons to specify 20 amino acids, and illustrates a feature of the genetic code known as degeneracy. The genetic code is described as degenerate because more than one codon can specify a specific amino acid. For example, codons GGU, GGC, GGA, and GGG all code for the amino acid, glycine. Indeed, all the amino acids, with the exception of methionine and tryptophan, have more than one codon. The codon AUG, which specifies only methionine, has a dual role: it encodes methionine anywhere it occurs in the RNA, and it also marks the start of protein synthesis.

The genetic code as specified by the triplet nucleotides is, for the most part, the same for bacteria and humans, and is referred to as 'universal'. However, there are notable exceptions. In bacteria, if the codons GUG and UUG occur at the beginning of protein synthesis, they can be read as a methionine codon. In addition, the protein synthesis stop-codon, UAA, can encode a tryptophan amino acid in some lower eukaryotic organisms such as *Paramecium* and *Tetrahymena*. There are also minor differences in the genetic code in mitochondria.

Another aspect of the genetic code is that it is translated without punctuation. This means that, once synthesis has started at an AUG codon, each successive triplet from that start point will be read without interruption until a termination codon is encountered. Thus the 'reading frame' of the mRNA will be dictated by the AUG codon. This means that mutations that cause the addition or deletion of single nucleotides will cause a frame shift, resulting in a protein with a different (nonsense) amino acid sequence after the mutation, or a protein that is prematurely terminated (Fig. 31.2).

The genetic code

Sickle cell anemia

Sickle cell anemia is an example of a disease in which a single nucleotide change within the coding region of the gene for the β chain of hemoglobin A, the major form of adult hemoglobin, yields an altered protein that has impaired function. The mutation that causes this disease is a single nucleotide change in a codon that normally specifies glutamate (GAG), and which produces a codon that specifies valine (GTG). Under conditions of low oxygen tension, this single amino acid change causes the protein to polymerize into rod-shaped structures, resulting in deformation and altered rheological properties of red blood cells (Fig. 31.3). This substitution of an acidic for a nonpolar, hydrophobic amino acid is known as a nonconservative mutation. Conservative mutations of one amino acid by another with similar physical and chemical properties usually have less severe consequences.

The genetic code

Codon	Amino acid	Codon	Amino acid	Codon	Amino acid	Codon	Amino acid
AAA	lysine	CAA	glutamine	GAA	glutamic acid	UAA	Stop
AAG	(Lys)	CAG	(Gln)	GAG	(Glu)	UAG	Stop
AAC	asparagine	CAC	histidine	GAC	aspartic acid	UAC	tyrosine
AAU	(Asp)	CAU	(His)	GAU	(Asp)	UAU	(Tyr)
ACA		CCA		GCA		UCA	
ACC	threonine	CCC	proline	GCC	alanine	UCC	serine
ACG	(Thr)	CCG	(Pro)	GCG	(Ala)	UCG	(Ser)
ACU		CCU		GCU		UCU	
AGA	arginine	CGA		GGA		UGA	Stop
AGG	(Arg)	CGC	arginine	GGC	glycine		
		CGG	(Arg)	GGG	(Gly)	UGG	tryptophan
AGC	serine	CGU		GGU			(Trp)
AGU	(Ser)					UGC	cysteine
						UGU	(Cys)
		CUA		GUA			
AUG	methionine	CUC	leucine	GUC	valine	UUA	leucine
	(Met)	CUG	(Leu)	GUG	(Val)	UUG	(Leu)
		CUU		GUU			
AUA						UUC	phenylalanine
AUC	isoleucine					UUU	(Phe)
AUU	(Ile)						

Fig. 31.1 Amino acids specified by each of the codons. Note that the first and second positions of codons that specify the same amino acid are generally the same.

The machinery of protein synthesis

Ribosomes consist of a small and a large subunit that, when associated with each other, possess specific sites at which tRNAs bind. These sites are known as the aminoacyl, or A site, and the peptidyl, or P site. The A site is where a tRNA molecule, carrying the appropriate amino acid on its acceptor stem, sits before that amino acid is incorporated into the protein. The P site is the location in the ribosome that contains a tRNA molecule with the amino-terminal portion of the newly synthesized protein still attached to its acceptor stem. It is within these sites that the process of peptide bond formation takes place. This process is catalyzed by peptidyl transferase, an enzyme that forms the peptide bond between the amino group of the amino acid in the A site and the carboxyl terminus of the nascent peptide attached to the tRNA in the P site.

Each amino acid has a specific synthetase that is responsible for attaching it to all the tRNAs that bind it

Each amino acid is attached to the acceptor stem of the tRNA by an enzyme called aminoacyl-tRNA synthetase; this enzyme catalyzes the formation of an ester bond linking the 3′ hydroxyl group of the adenosine nucleotide of the tRNA to the carboxyl group of the amino acid (Fig. 31.4). The attachment of an amino acid to a tRNA requires that the amino acid first be activated by reacting with adenosine triphosphate (ATP) to form an aminoacyladenylate intermediate, bound to the synthetase complex. The enzymology of activation of the carboxyl group of amino acids is similar to that for activation of fatty acids by thiokinase, but, rather than transfer of the acyl group to coenzyme A, the aminoacyl group is transferred to the tRNA, where it is now known as a charged tRNA molecule. At this point it is ready to enter the A site of the ribosome, where it will contribute its amino acid to the protein that is being synthesized. There is a different synthetase specific for each of the 20 amino acids that is responsible for attaching the appropriate amino acid to all the tRNAs that bind that amino acid. There is also a distinct tRNA molecule for each of the codons represented in Figure 31.1.

Effect of mutations on protein synthesis

Normal	AUG	GCA	UUA	CAG	GUA	UUA	CUA	CGA	GGC	ACA	CCU	GAA...	functional
gene	Met	Ala	Leu	Gln	Val	Leu	Leu	Arg	Gly	Thr	Pro	Glu	protein
insertion	AUG	GCA	UUU	ACA	GGU	AUU	ACU	ACG	AGG	CAC	ACC	UGA A...	premature
	Met	Ala	Phe	Arg	Gly	Ile	Thr	Thr	Arg	His	Thr	**Stop**	termination
deletion	AUG	GCA	UAC	AGG	UAU	UAC	UAC	GAG	GCA	CAC	CUG	AAA...	different
	Met	Ala	Tyr	Arg	Tyr	Tyr	Tyr	Glu	Ala	His	Leu	Lys	protein
altered base	AUG	GCA	UUA	CAG	GAA	UUA	CUA	CGA	GGC	ACA	CCU	GAA...	single amino
	Met	Ala	Leu	Gln	Glu	Leu	Leu	Arg	Gly	Thr	Pro	Glu	acid change
altered base	AUG	GCA	UUA	CAG	GUA	UUA	CUG	CGA	GGC	ACA	CCU	GAA...	no change
	Met	Ala	Leu	Gln	Val	Leu	Leu	Arg	Gly	Thr	Pro	Glu	

Fig. 31.2 The effect that single-base mutations in the mRNA have on the primary sequence of the encoded protein. Note that the location of the mutation will dictate the extent of change observed in the primary sequence. Insertion and deletion mutations cause frame shifts that lead to synthesis of nonsense proteins with either premature or delayed termination.

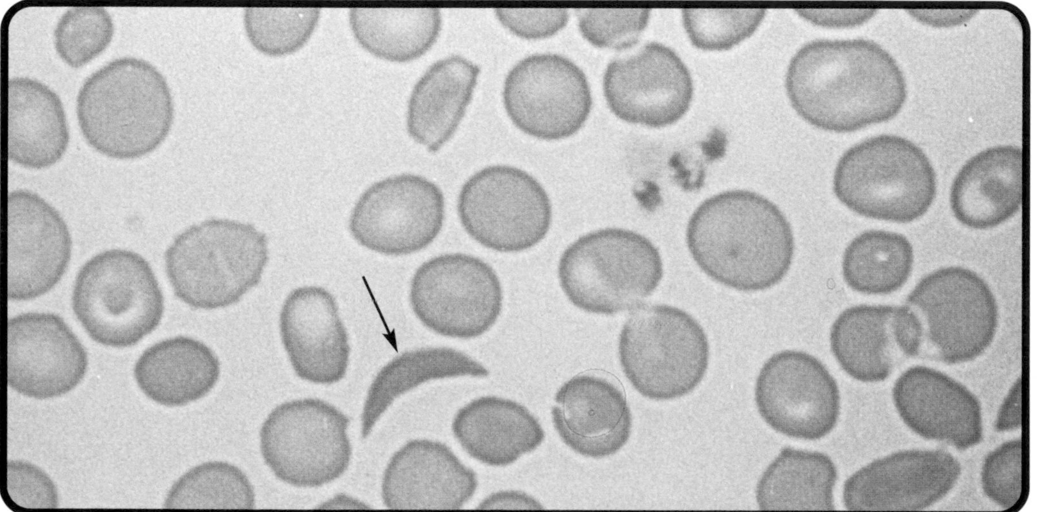

Fig. 31.3 Sickle cell anemia. Peripheral blood film, showing deeply staining sickle cell (arrowed). (Courtesy of Professor AV Hoffbrand and Dr JE Pettit.)

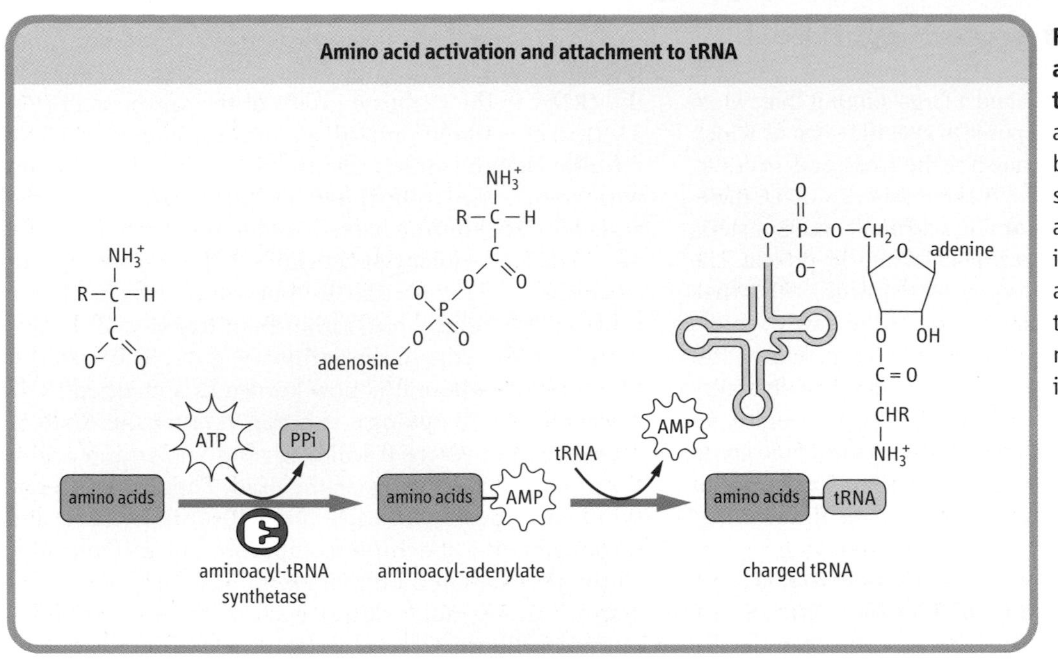

Fig. 31.4 Activation of an amino acid and attachment to its cognate tRNA. The amino acid must be activated by aminoacyl-tRNA synthetase to form an aminoacyladenylate intermediate, before its attachment to the 3 end of the tRNA. AMP, adenosine monophosphate; PPi, inorganic pyrophosphate.

Aminoacyl-tRNA synthetases

Aminoacyl-tRNA synthetases have proofreading ability

To guarantee the accuracy of protein synthesis, mechanisms have evolved to ensure selection of the correct amino acid for acylation and for proofreading of already charged tRNAs. One such mechanism is found in the enzymes responsible for attaching an amino acid to the correct tRNA. The aminoacyl-tRNA synthetases have the ability, not only to discriminate between amino acids before they are attached to the appropriate tRNA, but also to remove amino acids that are attached to the wrong tRNA. This discriminating ability exhibited by the synthetases is accomplished by a series of hydrogen bonding interactions between the enzyme and the amino acid. These two mechanisms combine to ensure accurate transfer of information from RNA to protein.

Some flexibility of base pairing occurs at the 3´ base of the mRNA codon

Interaction of the charged tRNA with its cognate codon is accomplished by association of the anticodon loop in tRNA with the codon in mRNA via hydrogen bonding of complementary base pairs (Fig. 31.5). The base-pairing rules are the same as those for DNA, except at the 3´ base of the codon (see Chapter 29). At this position, nonclassical base pairs can form between this base and the 5´ base of the anticodon. This observation led to the formulation of the *wobble* hypothesis of codon–anticodon pairing. The basis for this hypothesis lies in the fact that there appears to be less energetic constraint on the type of base pair that is formed at the 3´ position of the codon. For example, if a guanosine residue is at the 5´ position of the anticodon, it can form a base pair with either a cytosine or a uridine residue in the 3´ position of the codon. If the deaminated adenosine residue, inosine, occurs at the 5´ position of the anticodon, it can form a base pair with uracil, adenosine, or even cytosine at the 3´ position of the codon (Fig. 31.6).

How does the ribosome know where to begin protein synthesis?

The mRNA molecule carries the information that will be used to direct the synthesis of the protein. However, not all of the information carried on the mRNA is used for this purpose. Most eukaryotic mRNAs contain regions both before and after the protein coding region, called 5´ and 3´ flanking sequences, respectively. These sequences are believed to be important in regulating the rate of protein synthesis and the stability of the mRNA; however, the fact that, as a consequence of their presence, the protein coding region does not start immediately at the beginning of the mRNA raises the question of how the ribosome knows where to start synthesis. In the case of eukaryotic cells, the ribosome first binds to a 5-methylguanine 'cap' structure at the 5´ end of the mRNA, probably through recognition of a specific secondary structure, and then moves down the molecule until it encounters the first AUG codon (Fig. 31.7). This signals the ribosome to begin synthesizing the protein and to continue until it encounters one of the termination codons (UGA, UAA, or UAG). In the

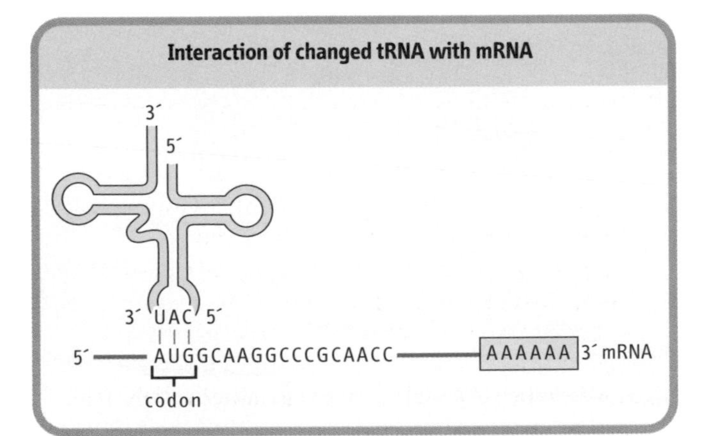

Fig. 31.5 The interaction of a charged tRNA with an mRNA occurs by base pairing of complementary nucleotides in the anticodon loop and the codon of the mRNA.

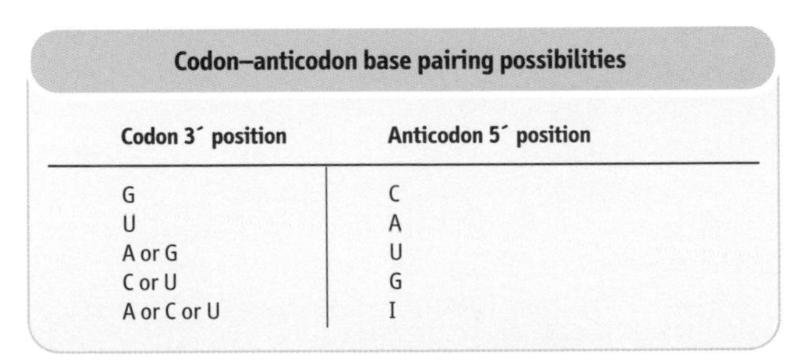

Codon–anticodon base pairing possibilities

Codon 3´ position	Anticodon 5´ position
G	C
U	A
A or G	U
C or U	G
A or C or U	I

Fig. 31.6 Base pairing possibilities between the 3´ nucleotide of the codon and the 5´ nucleotide of the anticodon.

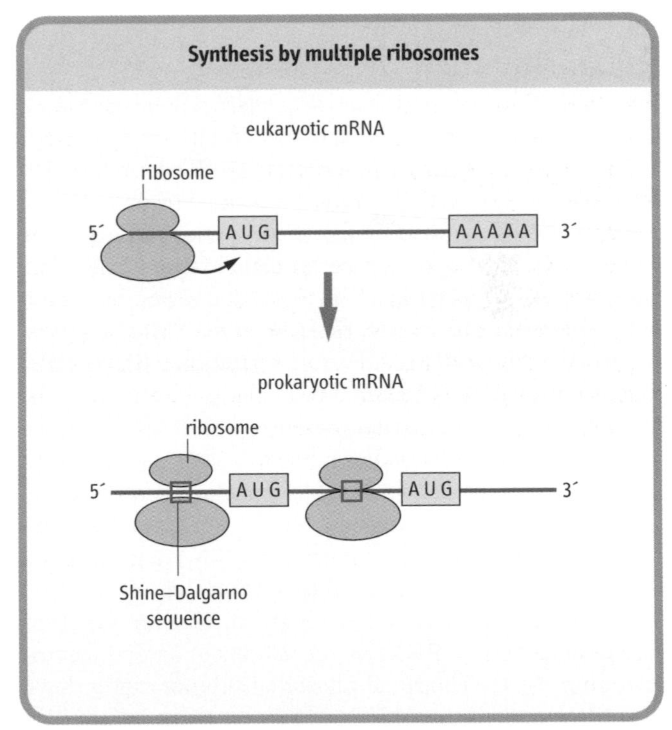

Fig. 31.7 The ribosome binds to the mRNA before locating the protein-coding region. Eukaryotic ribosomes bind to the 5´ end of mRNAs and then move down the mRNA until they encounter the first AUG codon. Bacterial ribosomes bind to complementary sequences in the mRNA – the Shine–Dalgarno sequences – that locate the protein-coding regions on the mRNA.

case of bacterial cells, knowing what portion of the mRNA is to be used to synthesize a protein is complicated by the fact that there can be several proteins encoded on a single mRNA, each out of register with one another, so that proteins of different sequence may be obtained from the same ribonucleotide sequence. This problem has been solved by the discovery of a sequence that helps to position the ribosome at the beginning of each protein coding region. This sequence, known as a Shine–Dalgarno sequence, is complementary to a portion of the 16S rRNA in the small ribosomal subunit. These sequences interact through hydrogen bonding of complementary base pairs, and this interaction helps to target the ribosome to the protein-coding regions of the mRNA.

The process of protein synthesis

Translation is a dynamic process that involves the interaction of enzymes, tRNAs, ribosomes, and mRNA in specific ways to produce a protein molecule capable of carrying out a specific cellular function. This complex process is normally divided into three steps:

- **initiation**,
- **elongation**,
- **termination**.

Initiation

Initiation of protein synthesis takes place when a ribosome (both large and small subunits) has assembled on the mRNA and the P site is occupied by a methionyl-tRNA (met-tRNA) molecule (Fig. 31.8). This complex is formed by the action of proteins known as initiation factors. In prokaryotic cells, the process (Fig. 31.8) involves three initiation factors. The initiation complex first forms just 5′ to the coding region, as a result of the interaction of the 16S rRNA with the Shine–Dalgano sequence on the mRNA. *N*-Formyl methionine (fmet) is the first amino acid in all bacterial proteins. In eukaryotic cells, there are at least 12 different initiation factors (eIFs or EFs), only a few of which have a known functon. The majority of eIFs help to promote the association of the small ribosomal subunit with the mRNA and a charged met-tRNA. For example, a complex containing the activated initiator, met-tRNA, and eIF-2 first binds to the small ribosomal subunit, which has eIF-3 bound to it. This complex then binds to the 5-methylguanine cap structure of eukaryotic mRNAs by the actions of several factors, including eIF-4F. The small ribosomal subunit moves down the mRNA until the first AUG codon is encountered, at which time the large ribosomal subunit joins the complex and protein synthesis is ready to begin. The assembly of the initiation complex is driven by the hydrolysis of guanosine triphosphate (GTP), and the movement of this complex down the mRNA is driven by the hydrolysis of ATP.

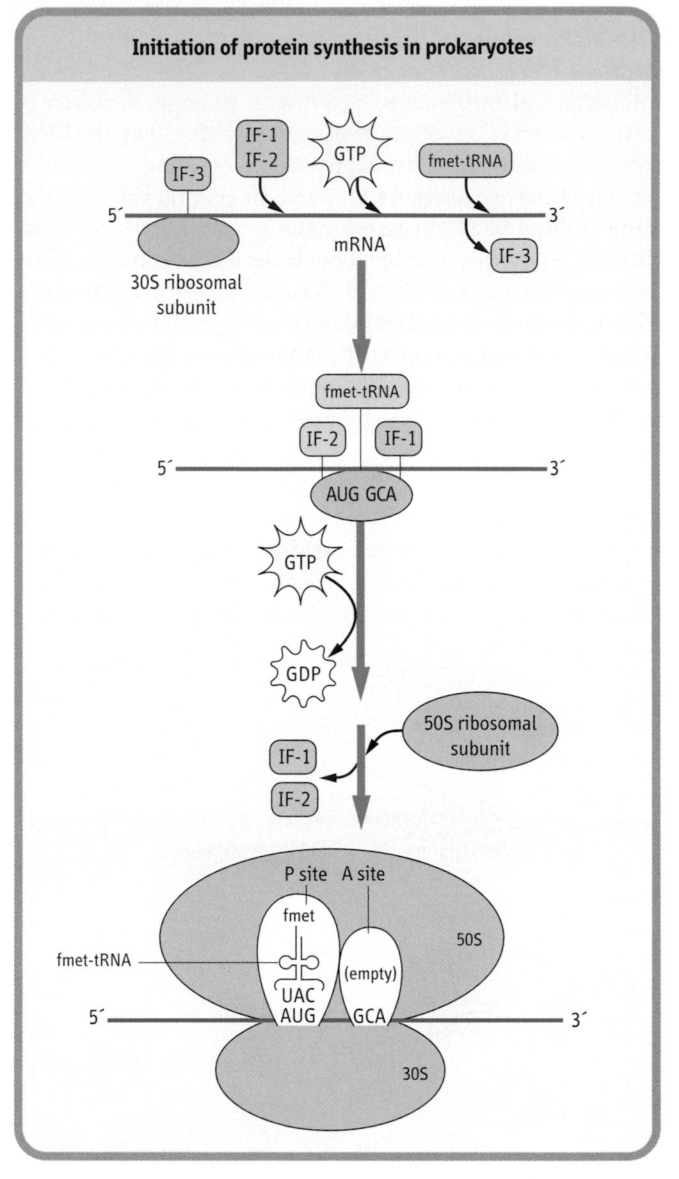

Fig. 31.8 Initiation of protein synthesis in bacterial cells. The 30S ribosomal subunit, mRNA, and formylated met-tRNA (fmet-tRNA) are brought together by the action of initiation factors. Once these components are assembled, the 50S ribosomal subunit completes the initiation complex. Note that, at initiation, the P site is occupied by the initiator tRNA, met-tRNA. GDP, guanosine diphosphate; IF, initiation factor.

Elongation

Factors involved in the elongation stage of protein synthesis are targets of some antibiotics

After initiation is complete, the process of translating the information in mRNA into a functional protein is started. Elongation begins with the binding of a charged tRNA in the A site of the ribosome. In eukaryotic cells, the charged tRNA molecule is brought to the ribosome by the action of an elongation factor called EF-1α (Fig. 31.9). For EF-1α to be active, it must have a GTP molecule associated with it. If the charged tRNA is the correct one – that is, one in which the anticodon of the tRNA forms base pairs with the codon on the mRNA – GTP is hydrolysed and EF-1α is released. For this EF-1α factor to bring another charged molecule to the ribosome, it must be regenerated by an elongation factor called EF-βγ, which will promote the association of EF-1α with GTP so that it may bind to another charged tRNA molecule. Once the correct charged tRNA molecule has been delivered to the A site of the ribosome, peptidyl transferase catalyzes the formation of a peptide bond between the amino acid in the A site and the amino acid in the P site. The ribosome is then moved one codon down the mRNA by a factor known as EF-2, and the whole process is begun again for addition of the next amino acid (Fig. 31.10). This complex process is identical in prokaryotic cells, but the factors are different; this explains the utility of antibiotics that preferentially inhibit protein synthesis in bacteria.

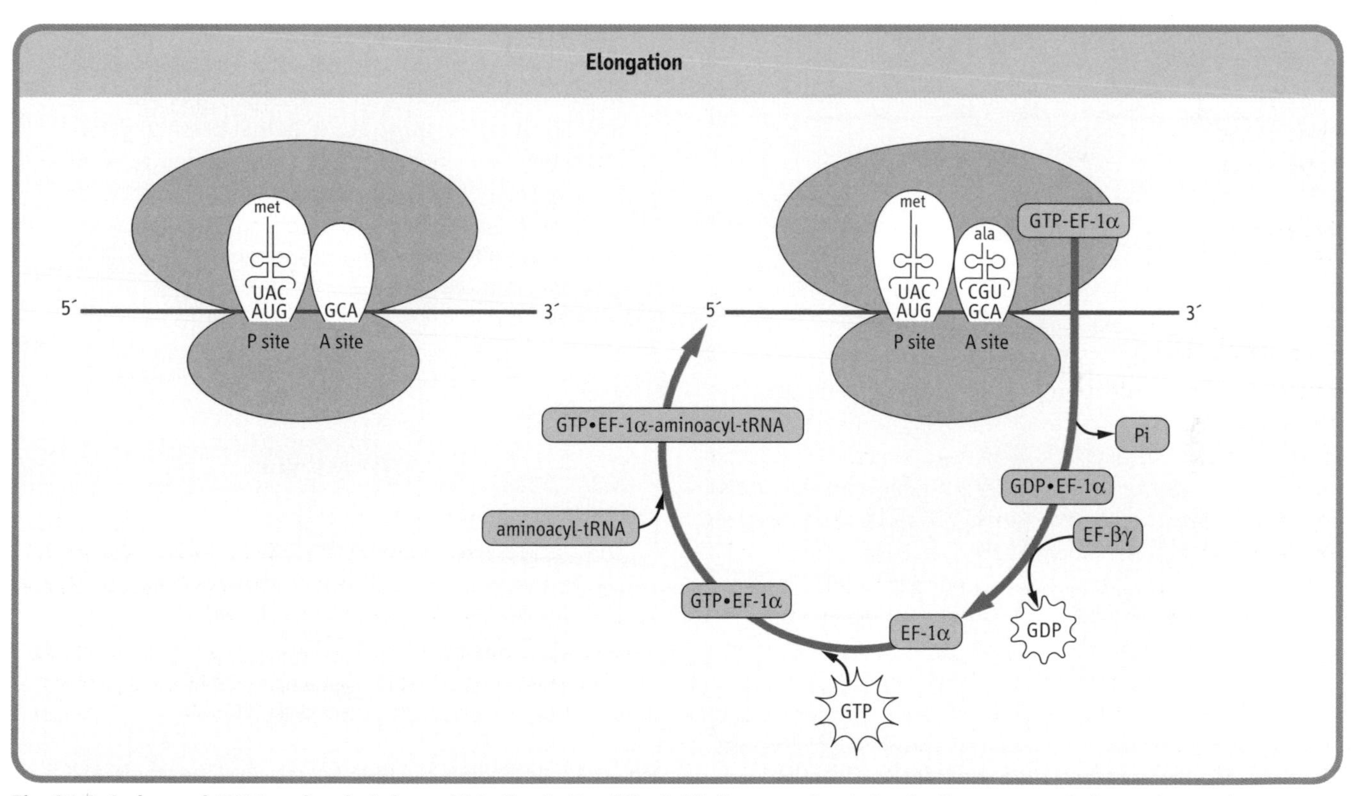

Fig. 31.9 A charged tRNA molecule is brought to the A site of the initiation complex to begin the process of elongation. Each successive amino acid addition requires that the correctly charged tRNA molecule be brought to the A site of the ribosome. ala, alanine; Pi, inorganic phosphate.

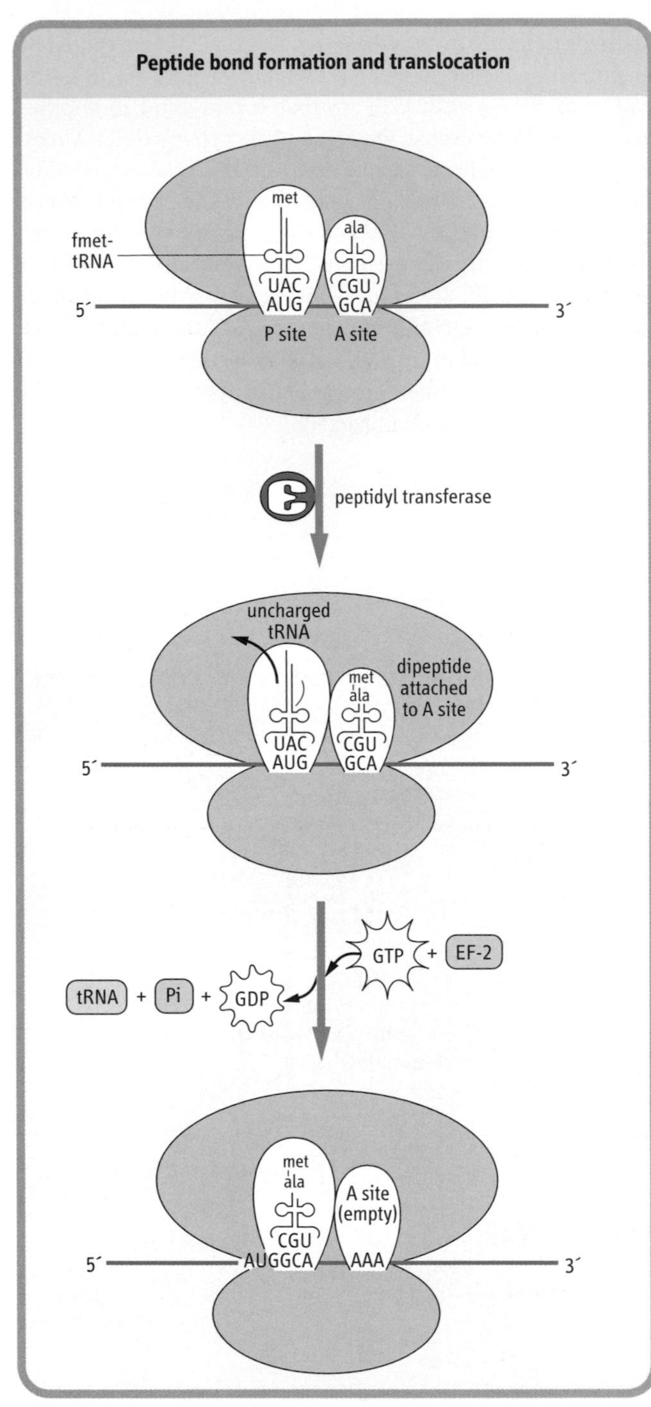

Fig. 31.10 Peptide bond formation and translocation. The formation of the peptide bond between each successive amino acid is catalyzed by peptidyl transferase. Once the peptide bond is formed, an elongation factor (EF-2) will move the ribosome down one codon on the mRNA, so that the A site is vacant and ready to receive the next charged tRNA.

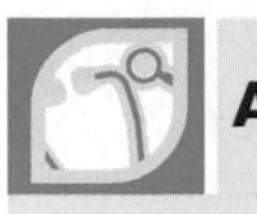

A noncompliant patient

A young man you were treating for a sinus infection returns to your clinic after 1 week, still complaining of sinus headaches and stuffiness. He explains that he began to feel better about 3 days after starting to take the antibiotic tetracycline, which you had prescribed. You inquire as to whether he continued to take the full dose of the drug, even after he began to feel better. He reluctantly admits that, as soon as he felt better, he stopped taking the drug. How do you explain to your patient that it is important that he takes the drug for as long as you prescribed it, even if he feels better after only a few days?

Comment. As a physician, you know that tetracycline inhibits the protein synthetic machinery of the bacterial cell by binding to the A site of the ribosome (Fig. 31.11). You also know that, if the drug is removed, protein synthesis can resume. If the drug is not taken for the entire period recommended, bacteria will begin to grow again, leading to the resurgence of the infection. Further, those bacteria that begin to grow after early termination of treatment are likely to be the most resistant to the drug, either because of selection of more resistant strains, or because of mutation to more resistant strains. The secondary infection is therefore likely to be more difficult to control.

Antibiotic targets

Antibiotic	Target
tetracycline	bacterial ribosome – A site
streptomycin	bacterial 30S ribosome
erythromycin	bacterial 50S ribosome
chloramphenicol	bacterial ribosome – peptidyl transferase
cycloheximide	eukaryotic 80S ribosome
ricin	eukaryotic 60S ribosome

Fig. 31.11 Antibiotics and their targets. Cycloheximide, and ricin in particular, are potent poisons to humans.

Protein synthesis: elongation

Peptidyl transferase is not your typical enzyme
Peptidyl transferase is the enzyme responsible for peptide bond formation during protein synthesis. This enzyme catalyzes the reaction between the amino group of the aminoacyl-tRNA, forming a peptide bond from an ester bond. The enzyme activity is located in the ribosome, but none of the ribosomal proteins has the capacity to catalyze this reaction. In fact, when ribosomes are stripped of all of their associated proteins, the rRNA appears to remain capable of converting an ester bond to a peptide bond. These observations have led investigators to hypothesize that peptidyl transferase activity is contained in the rRNA, rather than in the proteins that associate with ribosomes.

Termination

Termination of protein synthesis in both eukaryotic and bacterial cells is accomplished when the A site of the ribosome reaches one of the stop-codons of the mRNA. Protein factors called releasing factors recognize these codons, and cause the protein that is attached to the last tRNA molecule in the P site to be released (Fig. 31.12). This process is an energy-dependent reaction catalyzed by the hydrolysis of GTP, which transfers a water molecule to the end of the protein, thus releasing it from the tRNA. After release of the newly synthesized protein, the ribosomal subunits, tRNA, and mRNA dissociate from each other. An initiation factor, such as eIF-2, binds to the small ribosomal subunit, setting the stage for the translation of another mRNA (see Fig. 31.8).

Protein targeting and post-translational modifications

Protein targeting

More than one ribosome can translate an mRNA at the same time. An mRNA with several bound ribosomes is known as a polyribosome or polysome (Fig. 31.13). There are two general classes of polysomes found in cells: those that are free in the cytoplasm, and those that are attached to the endoplasmic reticulum. Messenger ribonucleic acids encoding membrane and secretory proteins are translated on polysomes attached to the endoplasmic reticulum, whereas those mRNAs encoding proteins destined for the cytoplasm are translated primarily on polysomes free in the cytoplasm. Translation on the endoplasmic reticulum assists in the post-translational modification of proteins and targeting of proteins to specific subcellular compartments.

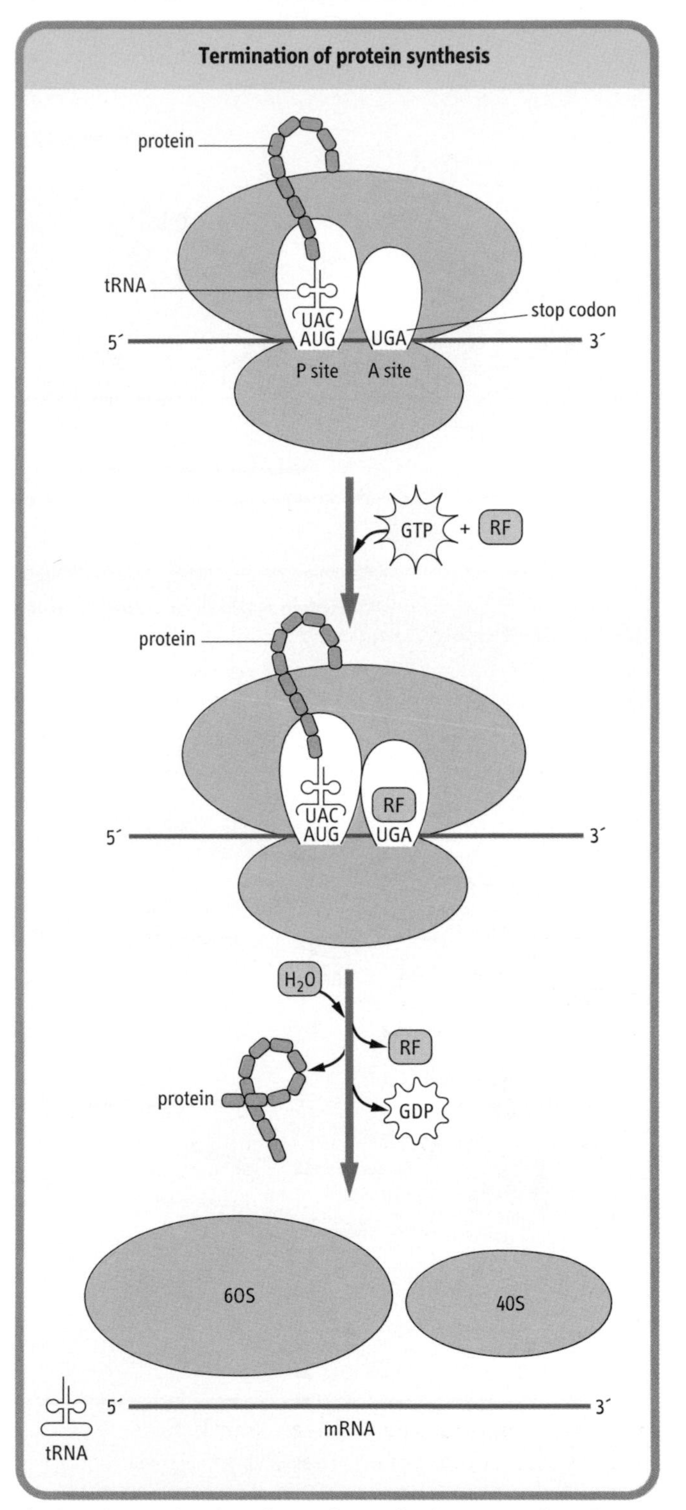

Fig. 31.12 Termination of protein synthesis occurs when the A site is over a termination codon. A releasing factor (RF) will cause the completed protein to be released and the ribosome, mRNA, and tRNA will dissociate from each other.

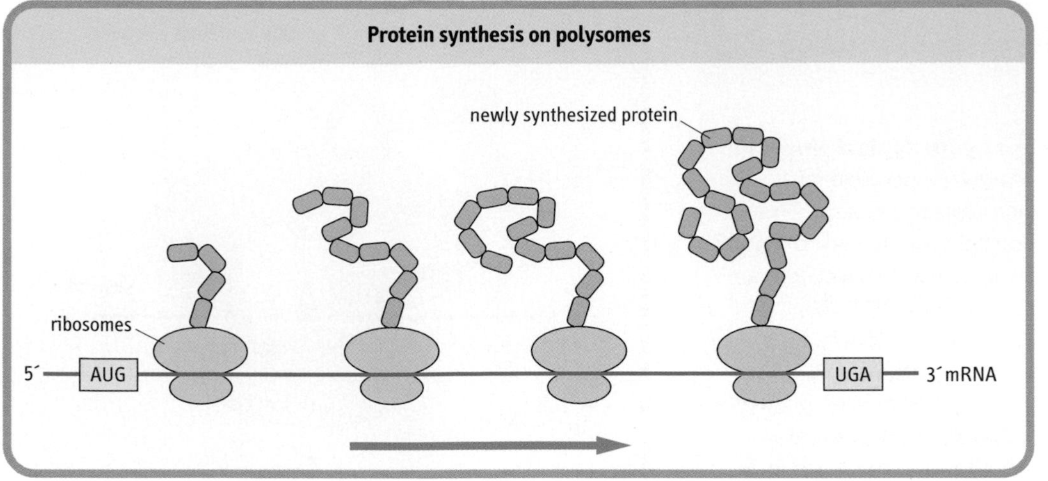

Fig. 31.13 Protein can be synthesized by several ribosomes bound to the same mRNA, forming a structure known as a polysome.

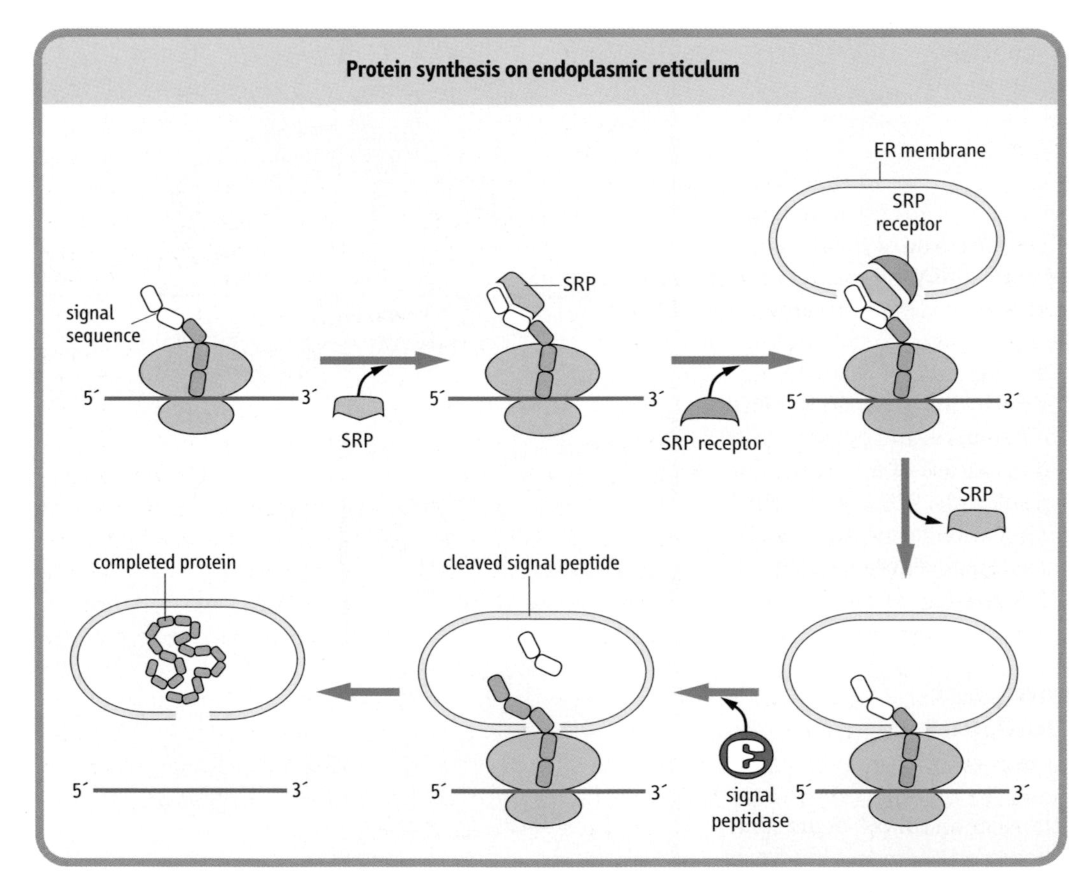

Fig. 31.14 Proteins can be targeted to specific sites in the cell by recognition of the signal sequence by an SRP. This complex is then recognized by the SRP docking protein (SRP receptor) on the endoplasmic reticulum (ER), where the signal sequence is inserted through the membrane. A signal peptidase removes the signal sequence and translation resumes. Once synthesis of the protein is complete, the protein is delivered to its cellular location.

The cellular fate of proteins is determined by signal peptide sequences

Proteins that are destined for export, for insertion into membranes, or for specific cellular organelles, must in some way be distinguished from proteins that reside in the cytoplasm. The distinguishing characteristic of all proteins targeted for these locations is that they contain a signal sequence usually comprising the first 20–30 amino acids on the amino-terminal end of the protein. In the case of secretory or membrane proteins, shortly after the signal sequence is synthesized, it is recognized by a complex that consists of proteins and RNA and is known as a signal recognition particle (SRP). The SRP binds to the signal sequence and halts translation of the remainder of the protein. This complex then binds to a receptor, known as the SRP docking protein, located on the membrane of the endoplasmic reticulum. After the SRP has delivered the ribosome-bound mRNA with its nascent protein to the endoplasmic reticulum, the signal sequence is inserted through the membrane, the SRP dissociates, and translation continues with the protein being moved across the membrane into the interstitial space of the endoplasmic reticulum (Fig. 31.14). Protein is then transferred from the endoplasmic reticulum to the Golgi apparatus, and then to its final destination.

In contrast to secretory and membrane proteins, mitochondrial and nuclear proteins are transported after their translation is complete. In common with membrane and secretory proteins, both mitochondrial and nuclear proteins have signal sequences that mark them to be transported from the endoplasmic reticulum to their respective organelles. In the case of proteins destined for the mitochondria, they may have two signal sequences on their amino-terminal ends, depending on whether they are destined for the matrix or for the intramembrane space. Mitochondrial proteins must be unfolded before they can be transported; in contrast, nuclear proteins may have nuclear localization signals throughout the entire length of the molecule and do not have to be unfolded before transport. This difference may be due to the fact that nuclear pores are very large complexes that can accommodate the transport of a protein in its native state.

Post-translational modification

Many proteins must be altered before they are biologically active; collectively, these alterations are known as post-translational modifications. It is within the endoplasmic reticulum and Golgi apparatus that two of the major post-translational modifications of proteins occur. In the endoplasmic reticulum, an enzyme called signal peptidase removes the signal sequence from the amino-terminal end of the protein, resulting in a mature protein that is 20–30 amino acids shorter than that encoded by the mRNA. In the endoplasmic reticulum and Golgi apparatus, carbohydrate side chains are added and modified at specific sites on the protein; the specific types of the carbohydrates added are important to the eventual function of the protein. A detailed description of the types of oligosaccharide side chains added to proteins, the mechanism of their addition, and the enzymes involved in this process may be found in Chapter 24.

Post-translational modifications to proteins may also include amino-terminal modification, modification of individual amino acids, proteolytic processing, and formation of disulfide bridges. One of the common amino-terminal modifications of eukaryotic cells is the removal of the methionine residue that initiates protein synthesis; in bacterial cells, the formyl group is removed from the methionine and, in some cases, the amino acid is also removed. In addition to these modifications that can occur to the amino-terminal end of the protein, amino acids within the protein can also be altered: for example, the amino acids serine, threonine, and tyrosine may have phosphate groups attached to their side chains. This type of modification is used by the cell to communicate changes in regulatory pathways resulting from environmental changes. Some amino acids, such as lysine, will be modified by the addition of a methyl group; others, such as cysteine, can have isoprenyl groups or other lipids added to their side chains, to facilitate protein binding to membranes. Cysteine residues also form specific disulfide bridges that are important to the structural integrity of the protein. Finally, many proteins are synthesized as proproteins that must be proteolytically cleaved for them to be active. The cleavage of a proprotein to its biologically active form is usually accomplished by a specific protease, and is a regulated cellular event.

Summary

Protein synthesis is the culmination of the transfer of genetic information from DNA to proteins. In this transfer, information must go from the 4-nucleotide language of DNA and RNA to the 20-amino acid language of proteins. The genetic code, in which three nucleotides (codon) specify an amino acid, represents the translation dictionary of the two languages. The tRNA molecule is the main bridge between these two languages. The tRNA accomplishes this task by virtue of its anticodon loop which interacts with specific codons on the mRNA and amino acids via its amino acid attachment site located on the 3′ end of the molecule. The process of translation consists of three parts; initiation, elongation, and termination. Initiation involves the assembly of the ribosome and charged tRNA at the initiation codon (AUG) of the mRNA. This assembly process is mediated by initiation factors and requires the expenditure of energy in the form of GTP. Elongation is the stepwise addition of individual amino acids to a growing peptide chain by the action of peptidyl transferase. The charged tRNA molecules are brought to the protein synthesis complex by elongation factors at the expense of GTP hydrolysis. Termination of protein synthesis is accomplished when the ribosome reaches one of the stop codons (UGA, UAA, UAG) of the mRNA. Releasing factors, proteins that recognize stop codons, cause the newly synthesized protein that is attached to the last tRNA molecule to be released. Many newly synthesized proteins must be modified, by a variety of chemical and structural modifications, before they are biologically active.

Further reading

Gilbert DN, Dworkin RJ, Raber SR, Leggett JE. Outpatient parenteral antimicrobial drug therapy. *N Engl J Med* 1997;**337**:829–838.
Quagliarello VJ, Scheld WM. Treatment of bacterial meningitis. *N Engl J Med* 1997;**336**:708–716.
Siegel V. A second signal recognition event required for translocation into the endoplasmic reticulum. *Cell* 1995;**82**:167–170.
Zheng N, Gierasch L. Signal sequences: the same yet different. *Cell* 1996;**86**:849–852.

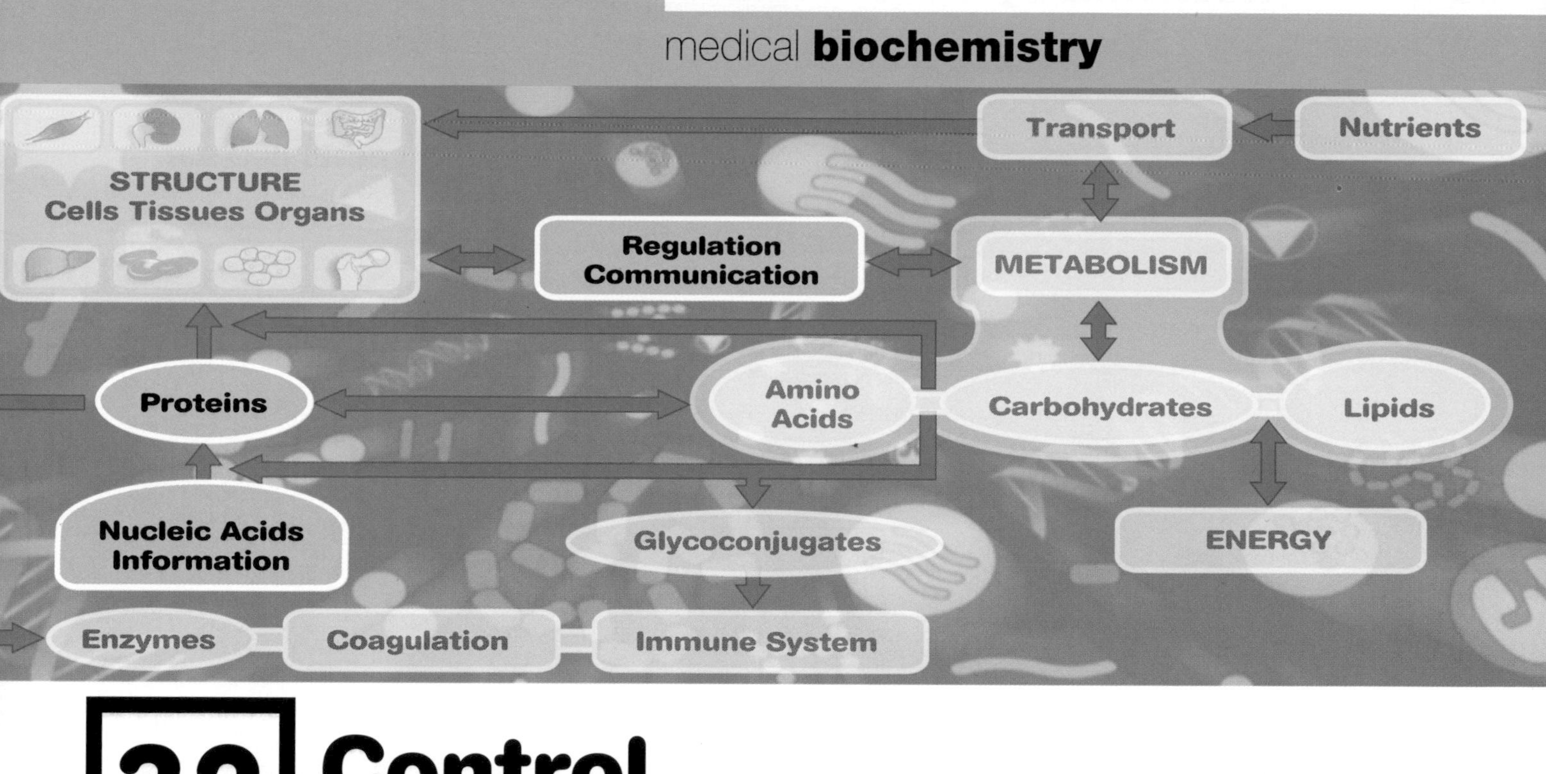

32 Control of Gene Expression

The discovery of genes and the mechanism whereby the information they hold is converted into biosynthetic enzymes, polypeptides hormones, or intracellular signaling molecules is central to the understanding of molecular biology. However, in studying human molecular biology, one of the most fascinating aspects of molecular genetics is the way in which the expression of genes is controlled, both in time and in place, and the consequences if these control mechanisms are disrupted.

The aim of this chapter is to introduce the basic concepts involved in the regulation of genes and how these processes may be involved in the causation of human disease. Initially, the basic mechanism of gene regulation will be described, followed by a discussion of a specific gene regulation system to highlight specific aspects of the basic mechanism. The chapter will end with a discussion of the specific ways in which the basic gene regulation apparatus can be adapted to suit different tissues and different situations.

The basic mechanisms of gene expression

Gene expression encompasses several different processes

The control of human gene expression occurs principally at the level of transcription. However, transcription is just one step in the conversion of the genetic information encoded by a gene into the final processed gene product, and it has become increasingly clear that post-transcription events occur that may also have an important role in regulating the expression of a gene. The sequence of events involved may be summarized as:

initiation of transcription → processing the transcript → transport to cytoplasm → translation of transcript to mRNA → post-transcription processing

Gene expression in humans can be regulated in many ways, both at specific phases in development and in differentiated tissues of mature organisms. Clearly, during the growth of a human embryo from a single fertilized ovum to a newborn infant, there must be numerous changes in the regulation of genes, to allow the differentiation of a single cell into cells that develop specific tissue characteristics (Fig. 32.1). Similarly, at puberty there are changes in the secretion of pituitary hormones that result in the cyclic secretion of ovarian and adrenal hormones in females and the production of secondary sexual characteristics. Such programed events are common in all cellular organisms, and the production of these phenotypic changes in cells – and thus the whole organism – arises as a result of changes in the expression of key genes. These key genes vary from cell to cell and also in time, but the mechanisms underlying the changes in regulation are less

Regulation of gene expression

Mechanism	Example
Transcriptional level	
transcription by tissue-specific factors	muscle-cell-specific transporter factor (MyoD): in myoblasts Ker 1=keratinocyte differentiation factor – skin cells HNF-5=hepatic nuclear factor-5 – liver cells
hormone, growth factor, or cellular messenger binding to response element	TATA box: binds transcription factor IID (TFIID) glucocorticoid response element (GRE): binds glucocorticoid receptor complex
alternative promoters	dystrophin gene
Post-transcriptional level	
tissue-specific RNA processing	alternative splicing, e.g. immunoglobulin genes alternative polyadenylation signals, e.g. calcitonin-related gene peptide
RNA editing	editing of apoB mRNA
RNA translation	ferritin and transferrin

Fig. 32.1 Mechanisms involved in the regulation of gene expression in humans.

variable. In humans and other eukaryotes, mechanisms that regulate transcription appear to be particularly frequent, as are the factors that influence them.

Gene transcription requires key elements to be present in the region of the gene

The key step in the transcription of a gene is the conversion of the message held within the gene into the nuclear template, which can then be used to form the protein product of the gene. The template is called messenger ribonucleic acid (mRNA). For expression of a gene to take place, the enzyme that catalyzes the formation of mRNA, RNA polymerase II (RNAPol II), must be able to recognize the so-called startpoint for transcription of the gene – that is, the exact position at which the gene starts and the intergenic deoxyribonucleic acid (DNA) ends. RNAPol II uses the sense strand of the DNA template to create a new polymeric nucleotide, comprising all the nucleotides transcribed from the target gene, the first nucleotide of which corresponds to the first nucleotide of the gene. However, RNA Pol II cannot initiate transcription alone. This enzyme along with other factors recognizes crucial sequences in the gene and proteins within the vicinity.

Promoters

Perhaps the most important and fundamental element required for the initiation of transcription is the promoter. Promoters are nucleotide sequences found in the DNA sequence upstream of the startpoint of the gene being transcribed, most commonly within 200 base pairs (bp) of the startpoint. The promoter sequence acts as a basic recognition unit, signaling that there is a gene nearby that can be transcribed. The structure of promoters varies from gene to gene, but they all have a number of basic key elements that can be identified within the promoter. These elements may be present in varying combinations in the vicinity of genes, some elements being present in one gene but absent in another. However, some form of promoter element is virtually always present.

The efficiency and specificity of gene expression is conferred by cis-acting elements

A promoter lies on the same strand as the gene being transcribed and is referred to as a *cis*-acting element (other *cis*-acting elements include response elements and enhancers and are discussed below).

In general, the nucleotide sequence of the startpoint of a gene varies from gene to gene. However, it has been observed that the first base in mRNA tends to be adenine (A), usually flanked by pyrimidines. This sequence is called the initiator (Inr). The Inr is the simplest form of promoter known to be recognizable by RNAPol II, and its structure varies from gene to gene. However, in general, it has the nucleotide sequence Py2CAPy5 (Py = pyrimidine base; C–cytosine, A–adenine) and is found between positions −3 to +5 in relation to the startpoint. In addition to Inr, most promoters possess a sequence known as the 'TATA box'. This element is relatively unique, because its position in relation to the startpoint is relatively fixed: approximately 25 bp from the startpoint. The TATA box has an 8 bp consensus sequence that consists entirely of adenine–thymine (A–T) base pairs, although very rarely a guanine–cytosine (G–C) pair may be present. This sequence appears to be very important in the process of transcription, as nucleotide substitutions that disrupt the TATA box result in marked reduction in the efficiency of transcription.

Consensus elements

Identifying the function and specificity of nucleotide sequences

Consensus sequences are nucleotide sequences that contain unique core elements that identify the function and specificity of the sequence, for example the TATA box. The sequence of the element may differ by a few nucleotides in different genes, but a core, or consensus, sequence is always present. In general, the differences do not influence the effectiveness of the sequence, and it is not unusual to find consensus sequences in both orientations on the same strand of DNA, both sequences being functional. Obviously, the reverse sequence of the consensus sequence will occur on the opposite strand of the DNA and, in the case of the TATA box, it may share some sequence similarity with the true TATA box on the gene being transcribed, and thus could interfere with transcription of the target gene. This problem seems to be overcome by the use of other transcription factors, which bind to the DNA strand being actively transcribed and allow for the proper orientation of the polymerase protein and the transcription apparatus on the appropriate TATA sequence.

In addition to the TATA box, other commonly found cis-acting promoters have been described. The CAAT box is often found upstream of the TATA box, most commonly approximately 80 bp from the startpoint, and may be present and functional in either orientation. As in the case of the TATA box, it does not confer specificity to the promoter – that is, it is not a tissue- or time-specific promoter – but does increase the strength of the promoter signal. Another commonly noted promoter element is the GC box. This element also can function in either orientation, and multiple copies may be found in a single promoter region. Figure 32.2 summarizes the variation in number and type of cis-acting elements seen within promoters.

Enhancers

Although the promoter is essential for the initiation of transcription, it is not necessarily alone in influencing the strength of transcription of a particular gene. Another group of elements, known as enhancers, can upregulate the level of transcription of a gene but, unlike promoters, their position may vary substantially with respect to the startpoint and their orientation has no effect on their efficiency. Enhancers often contain a number of elements that are recognized by both tissue-specific and ubiquitous transcription factors. As a result, nucleotide substitutions in enhancers often result in marked reductions in transcription efficiency compared with the effects of similar subtitutions in promoters.

Enhancers may lie upstream or downstream of the specific promoter and may be important in conferring tissue-specific transcription. For instance, a nonspecific promoter may initiate transcription only in the presence of a tissue-specific enhancer. Alternatively, a tissue-specific promoter may initiate transcription, but with a greatly increased efficiency in the presence of a nearby enhancer that is not tissue-specific. Indeed, in some genes, for example immunoglobulin genes, enhancers may actually be present downstream of the startpoint of transcription, within an intron of the gene being actively transcribed.

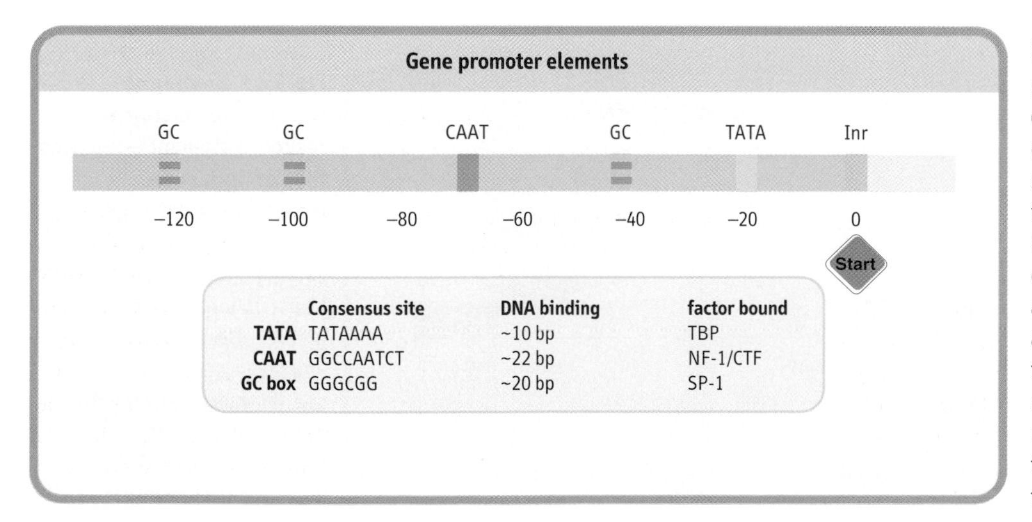

Fig. 32.2 Idealized version of a promoter comprising various different elements. Each promoter element has a specific consensus sequence that binds ubiquitous transcription-activating factors. Binding of transcription factors encompasses the consensus site and a variable number of anonymous adjacent nucleotides, depending on the promoter element. CTF, a member of a protein family whose members act as TFs; NF-1, nuclear factor-1; SP-1, ubiquitous transcription factor.

Response elements

Response elements are nucleotide sequences that allow specific stimuli, such as steroid hormones, cyclic AMP, or insulin-like growth factor-1 (IGF-1), to control gene expression. Not all genes possess response elements, and they are thus important when the role of gene expression in human development is being considered. Response elements are often found within 1 kilobase (1 kb) of the start point, and a single gene may possess any number of different response elements.

Transcription factors

The combination of a promoter, an enhancer, and a variable number of response elements linked to a gene is the basic model of a human gene, the transcription of which can be initiated and regulated by a number of different transcription factors. These factors bind to specific nucleotide sequences and bring about differential expression of the gene, not only during development, but also within tissues of the mature organism (Fig. 32.3).

Initiation of transcription requires binding of transcription factors to DNA

For transcription to occur, proteins known as transcription factors, or *trans*-acting elements, must bind to DNA. These *trans*-acting factors recognize and bind to short nucleotide sequences at the start of transcription. A protein known as TATA-binding protein (TBP) binds to the region of the TATA box. This binding of TBP to the TATA box results in positioning of the transcription apparatus (TBP in association with a variable number of other proteins) at a fixed distance from the startpoint of transcription and thus allows RNAPol II to be positioned exactly at the site of initiation of transcription. Once RNAPol II and a number of other transcription factors have bound to the region of the startpoint, transcription can occur. When transcription begins, many of the transcription factors required for binding and alignment of RNAPol II are released, and the polymerase travels along the DNA, forming the primary messenger RNA transcripts in the process, pre-mRNA.

What is the 'gene'?

Transcription unit versus gene

Exactly what a 'gene' is has become increasingly difficult to define in recent years. The initial notion that a gene was a piece of DNA that gave rise to a single gene product has been challenged. It is now clear that two or more functional products – different mRNA species or different protein products – may arise from a single region of transcribed DNA, as a result of differences either at the level of transcription or at the post-transcriptional level. Thus there is now a tendency to refer to such 'genes' as transcription units. These transcription units encapsulate, not only those parts of the gene such as the promoters, response elements, and exons and introns, classically regarded as the gene unit, but also the molecular events that modify the transcription process from the initiation of transcription to the final post-transcriptional modifications. This is a shift away from the notion of a gene being a single strand of DNA with exons and introns, to one of a gene being a complex structure of template and machinery that comprises a dynamic process giving rise to the final gene product or products at variable stages of development of an organism.

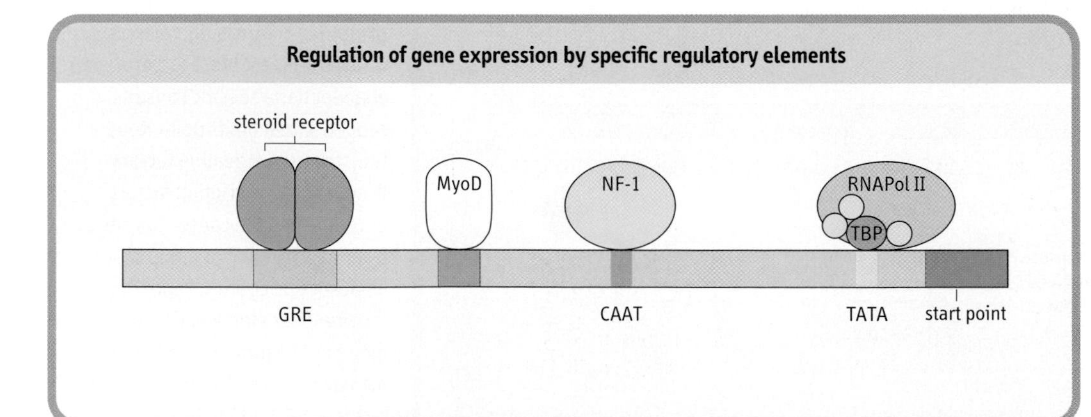

Fig. 32.3 Binding of transcription factors to response elements enhances the strength of the transcription message. Different elements have varying effects on the level of transcription, some exerting greater effects than others, and may also activate tissue-specific expression. MyoD, muscle-cell-specific transporter factor. GRE, glucocorticoid response element.

Transcription factors have common structures that permit DNA binding

The binding of transcription factors to DNA results from the ability of a relatively small area of the transcription factor protein to come into close contact with the double helix of the DNA to be transcribed. The regions of these proteins that contact the DNA are called DNA-binding regions or motifs, and are highly conserved between species. Four major classes of DNA-binding domain have been described (Fig. 32.4):

- **helix–turn–helix (HTH) motif:** mediates DNA binding by fitting into the major groove of the DNA helix, allowing precise alignment of the factor in relation to the DNA sequence recognized;
- **helix–loop–helix (HLH) motif:** promotes both DNA binding and protein dimer formation. It is believed that HLH motifs mediate mainly negative influences on gene expression;
- **zinc finger:** loops or fingers of amino acids that have a zinc ion at their core. The adjacent DNA often forms α-helices that makes contact with the DNA in the major groove;
- **leucine zipper:** forms a dimer that grips the DNA double helix like a peg, by inserting into the major groove.

These four differing DNA structures are able to bind to DNA strands as a result of the tertiary structure of the various binding motifs, which allow the regulatory proteins to come into close contact with a specific region of the DNA double helix. Although binding between amino acids and DNA is via weak hydrogens bonds, the average transcription factor may have 20 or more sites of contact, which consequently increase the strength and specificity of the contact considerably.

Steroid receptors

Steroid receptors possess many characteristics of typical transcription factors and provide a model for the role of zinc fingers in DNA binding

Steroid hormones have many functions in humans and are essential to normal life. They are derived from a single precursor, cholesterol, and thus share a similar structural backbone. However, differences in hydroxylation of certain carbon atoms and aromatization of the steroid A ring of the molecule give rise to very marked differences in biological effect. Steroids bring about their biological effects by binding to steroid hormone receptors; these receptors are found in the cell cytoplasm and, on binding steroid molecules, undergo a series of changes that result in the steroid–receptor complex binding to DNA at a specific response element, a so-called steroid response element (SRE). These SREs may be found many kilobases upstream or downstream of the startpoint, and binding of the receptor complex to the SRE results in activation of the nearby promoter and initiation of transcription (Fig. 32.5). As might be expected, because of the

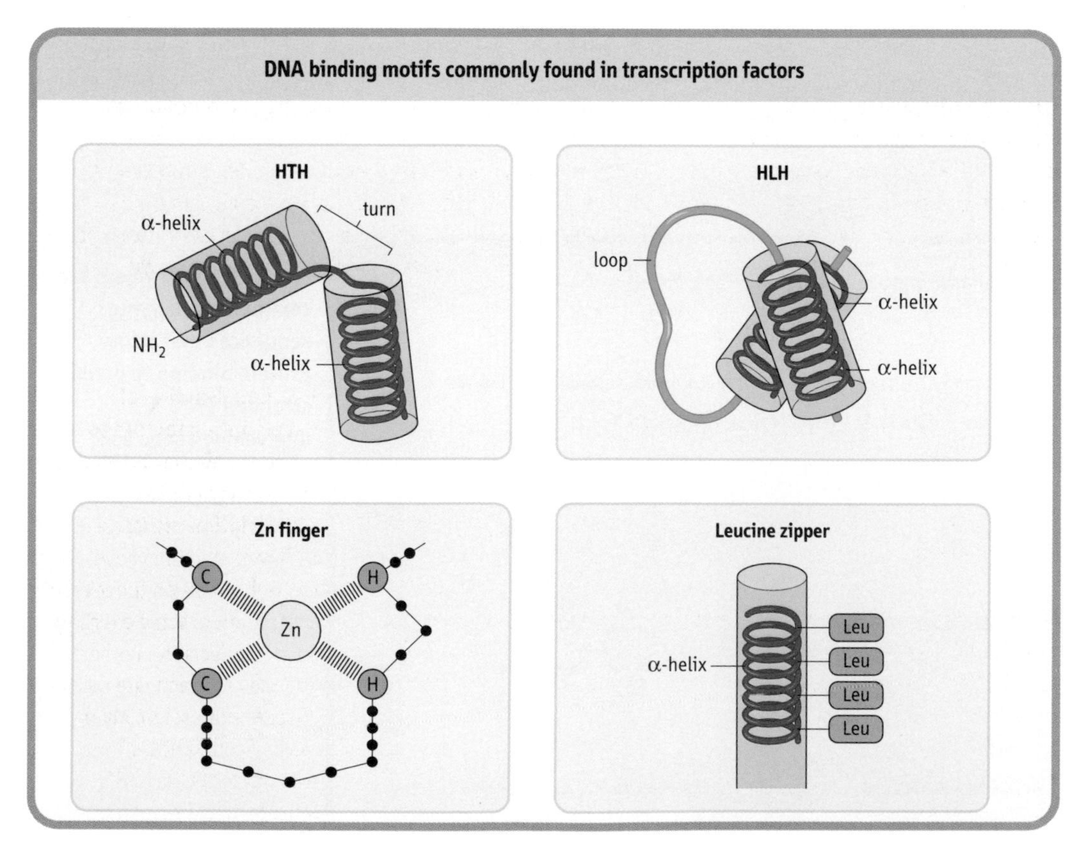

Fig. 32.4 The four main classes of DNA-binding domains. Leucine zippers have hydrophobic leucine residues consistently on one face of the helix, which allows two leucine zippers to align with their hydrophobic residues facing each other. HLH, helix–loop–helix; HTH, helix–turn–helix.

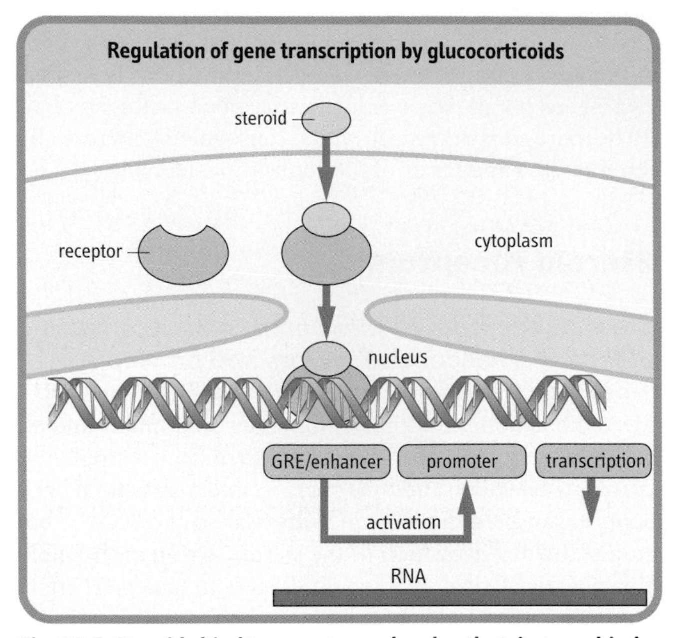

Fig. 32.5 Steroids bind to receptor molecules that, in turn, bind to an enhancer, the action of which stimulates promoter function. GRE, glucocorticoid response element.

wide number of steroids found in humans, there are a correspondingly large number of distinct steroid receptor types, and each of these recognizes a specific consensus sequence, the SRE, in the region of the promoter.

The zinc finger region

Central to the recognition of the SRE, and to the binding of the receptor to it, is the presence of the so-called zinc finger region in the DNA-binding domain of the receptor molecule. This region consists of a peptide loop with a zinc atom at the core of the loop. Each loop is identical, in that it comprises two cysteine and two histidine residues in highly conserved positions relative to each other, separated by a fixed number of intervening amino acids. The consensus sequence of a single zinc finger is:

$$\text{Cys - } X_{2-4} \text{ - Cys - } X_3 \text{ - Phe - } X_5 \text{ - Leu - } X_2 \text{ - His - } X_3 \text{ - His}$$

where X represents any intervening amino acid (Fig. 32.6).

This structure holds the zinc atom at the core of a tetrahedron and allows the receptor molecule to sit on the surface of the DNA double helix and interact with a response element, thus enhancing the efficiency of, and possibly conferring specificty to, the promoter. Zinc finger motifs are generally organized as a series of tandem repeat fingers, although the precise number varies in different transcription factors.

The precise structure of the steroid receptor zinc finger differs from the consensus shown above. Steroid receptor zinc fingers have a simpler structure (Fig. 32.6):

$$\text{Cys - } X_2 \text{ - Cys - } X_{13} \text{ - Cys - } X_2 \text{ - Cys}$$

This type of finger recognizes and binds to short palindromic DNA sequences and, in the case of steroid receptors, two fingers are found side by side, one binding DNA and the other allowing the receptor to form a larger molecule by binding with a second steroid receptor molecule, a so-called homodimer – a protein molecule comprised of two identical subunits. Studies of the genes for steroid receptors have shown that artificially created mutations in the DNA sequence of the receptor gene that disrupt the zinc finger region of the receptor result in a loss of function of the receptor.

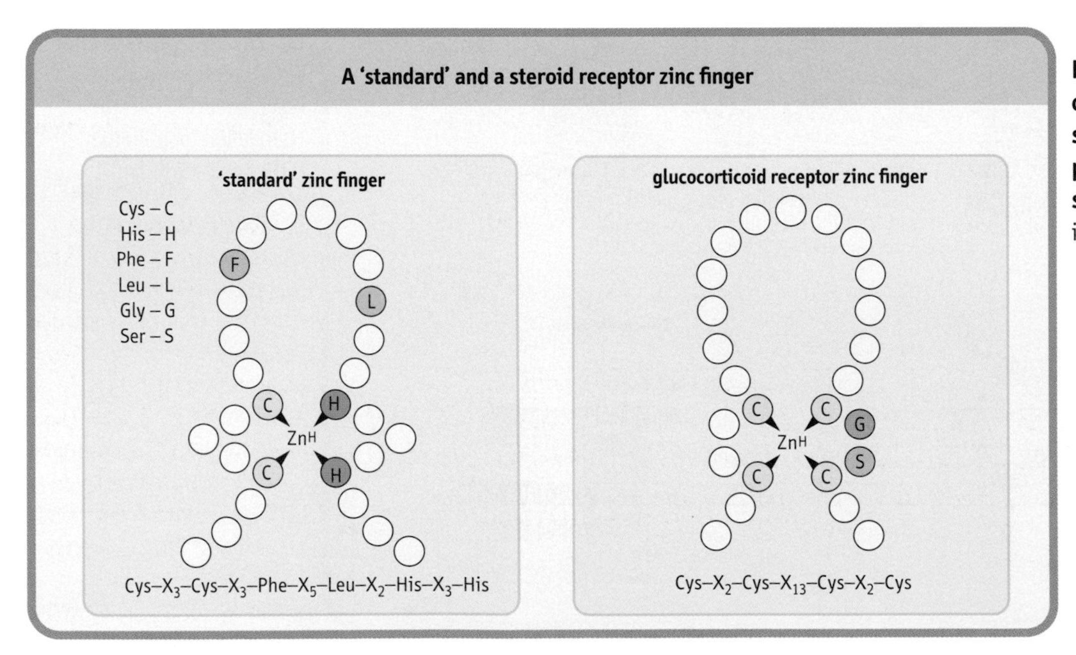

Fig. 32.6 Zinc fingers are commonly occurring sequences that allow protein binding to double-stranded DNA. X, any intervening amino acid,

Steroid receptor gene family

Gene superfamilies: steroid–thyroid–retinoic acid
The steroid receptor gene family, although large, is in fact only a subset of a much larger family of so-called nuclear hormone receptors. All members of this family have the same basic structure as the steroid hormone receptors: a hypervariable N-terminal region, a highly conserved DNA-binding region (49–85% homology), a variable hinge region, and a highly conserved ligand-binding domain (30–87% homology). They are separated into two groups. Type I receptors are a group of receptor proteins that form homodimers and bind specifically to steroid hormone response elements only in the presence of their ligand; they encompass the steroid hormone receptors (glucocorticoid, mineralocorticoid, androgen, and progesterone, but less so the estrogen receptor). Type II receptors are a group of hormone receptors that form homodimers that can bind to response elements in the absence of their ligand, and may also form heterodimers with other type II receptor subunits, to form active units. The type II receptors include thyroid hormone, vitamin D, and the retinoic acid receptors. Other gene families have been reported in both human and nonhuman organisms, although they generally lack the high degree of homology seen in the nuclear hormone receptor family.

Organization of the steroid receptor molecule

Steroid receptors are a gene family with important similarities

One central feature of all the steroid receptor proteins is the similarity of organization of their receptor molecules and the complementary DNA (cDNA) sequences of those molecules. Each receptor has a DNA-binding and transcription-activating domain, a steroid-hormone-binding and -dimerization domain, and an N-terminal region, required for initiating transcription.

There are three striking features about the organization of the steroid hormone receptors:

- The DNA-binding region always contains a zinc finger region which, if mutated, results in loss of function of the receptor;
- The DNA-binding regions of all the steroid hormone receptors show an extremely high degree of homology to one another;
- The steroid-binding regions show a high degree of homology to one another.

These common features have identified the steroid receptor proteins as a gene family. It would appear that, during the course of evolution, diversification of organisms has resulted in the need for different steroids with varied biological actions and, consequently, a single ancestral gene has undergone evolutionary change over millions of years, resulting in the presence of five different receptors (Fig. 32.7).

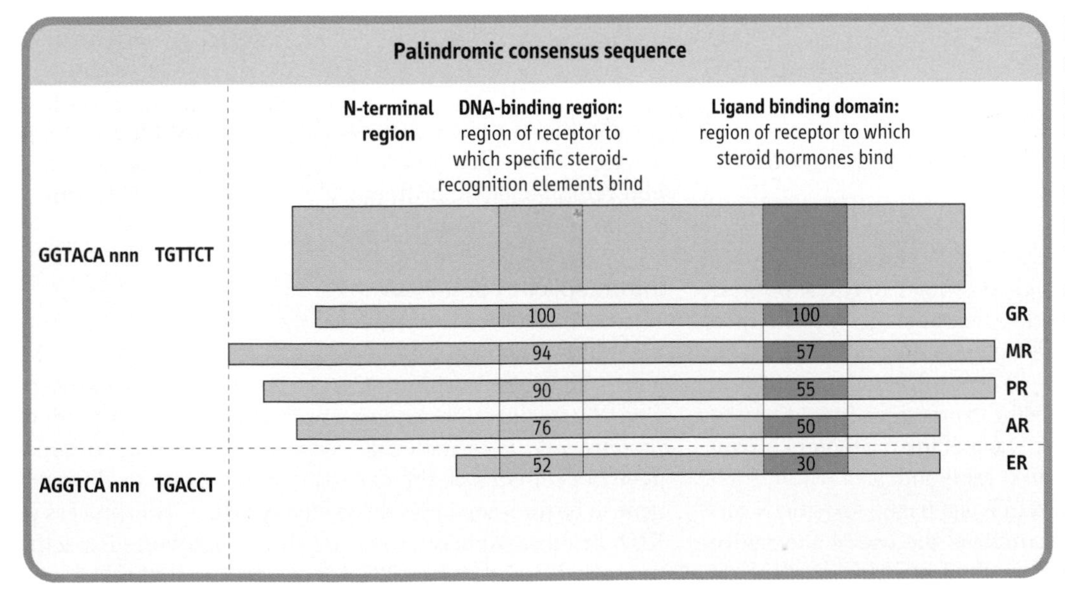

Fig. 32.7 The DNA-binding and hormone-binding regions share a high degree of homology, although the estrogen receptor is less similar to the glucocorticoid receptor than are the others. The hexameric sequences are sites of DNA binding, although the intervening nucleotides are not important. AR, androgen receptor; ER, estrogen receptor; GR, glucocorticoid receptor; MR, mineralocorticoid receptor; PR, progesterone receptor; nnn, anonymous nucleotides. Numbers denote % similarity to sequence in GR.

Alternative approaches to gene regulation in humans

How does transcription start?

In humans, two models have been proposed to describe how transcription starts and how a gene may be expressed

The first model of transcription and gene expression is the pre-emptive model. This states that the DNA, normally in a double strand surrounded by histone proteins, is constantly under 'attack' by transcription factors, and that, if a gap in the DNA occurs, then the transcription factor will bind irreversibly to that area of DNA. When the cell subsequently divides, the transcription factor remains attached to its binding site and is thus able to activate transcription at any time in the future, and also to prevent histone proteins from recombining with DNA to inactivate the transcription activity.

A second mechanism proposed is the dynamic model. The pre-emptive model assumes that there is a casual replacement of the histone protein with the transcription factor. However, evidence would suggest that an energy-dependent process also exists whereby hydrolysis of ATP is central to the process of replacing the histone protein with a transcription factor. Some examples of this process have been described in organisms such as the extensively studied fruit fly, *Drosophila*.

To complicate matters further, there are also examples of systems in which it is not necessary for the histone protein to be displaced to allow the transcription factor to bind. The best known of these is the case of the glucocorticoid steroid receptor, which can bind to apparent 'gaps' at the site of the glucocorticoid response element (GRE) and interact with the histone proteins in a process that results in transcription.

It can be seen that it is not altogether clear how transcription may occur. However, the conceptual ideas of the pre-emptive, dynamic, and glucocorticoid methods of activation of transcription give us an idea as to how the process may occur, and are the subject of intense study.

Gene expression

Methylation of DNA regulates gene expression

Certain nucleotides, principally cytosine, can undergo methylation; in mammals, up to 7% of all cytosine residues in DNA are methylated. The methylated cytosine residues are virtually always found associated with a guanidine in a CG form, and in double-stranded DNA the complementary cytosine is also methylated, giving rise to a palindromic sequence:

5′ ^{m}CpG 3′

3′ GpC^m 5′

The presence of the methylated cytosine can be exploited by examining the effects of enzymes that cut DNA at CG groups whether or not those groups are methylated. This property of the enzymes has shown that in general, DNA which has little in the way of methylated cytosine residues contains genes that are actively transcribed, whilst heavily methylated regions are not being actively transcribed. The areas at which demethylated residues exist are likely to be critical to the regulation of expression: for example, demethylation of a promoter may be required for the initiation of transcription, whereas a reduction in the level of methylation in the coding sequence of the gene may be required for transcription to proceed.

Gene expression

Not all gene expression depends on the regulation of transcription

Alternative promoters

Although it is clear that, in order for gene expression to occur, a promoter is essential, a single promoter may not possess the tissue specificity or developmental stage specificity to allow it to direct expression of a gene at the correct time and place. Therefore, some genes have evolved a series of promoters that confer tissue-specific expression. In addition to the use of different promoters, each of the promoters is usually associated with its own first exon and, as a result, each mRNA and subsequent protein has a tissue-specific N-terminal sequence. The best example of the use of alternative promoters in humans is the gene for dystrophin, the muscle protein that is deficient in Duchenne muscular dystrophy. This gene uses promoters that give rise to brain-, muscle-, and retinal-specific proteins, all with differing N-terminal amino acid structures.

Intron splicing of mRNA

When RNAPol II has completed the transcription of a gene, the initial transcript, which has the same organization as the gene – the so-called pre-mRNA – then undergoes a process whereby the intronic sequences of the gene are removed, to produce an mRNA molecule that is smaller than the original gene but contains all the necessary information to allow the gene to be translated into the protein product. This process of RNA or intron splicing relies on the recognition of specific sequences, at the start and finish of introns, which allow spe-

Tissue-specific expression may involve more than one mechanism

A 17-year-old girl notices a swelling on the left side of her neck. She is otherwise well, but her mother and maternal uncle have both had adrenal tumors removed. Blood is withdrawn and sent to the laboratory for measurement of calcitonin, which is grossly increased. Pathology of the excised thyroid mass confirms the diagnosis of medullary carcinoma of the thyroid. This suggests that this family have a genetic mutation causing the condition known as multiple endocrine neoplasia type IIA (MEN IIA).

Comment. Expression of the calcitonin gene provides an example of how different mechanisms may regulate gene expression and give rise to tissue-specific gene products. The calcitonin gene consists of five exons and uses two alternative polyadenylation signaling molecules. In the thyroid gland, the medullary C cells produce calcitonin by using one polyadenylation signaling molecule associated with exon 4 to transcribe a pre-mRNA comprising exons 1–4. The associated introns are spliced out and the mRNA is translated to give calcitonin. However, in neural tissue, a second polyadenylation signaling molecule next to exon 5 is used. This results in a pre-mRNA comprising all five exons and their intervening introns. This larger pre-mRNA is then spliced and, in addition to all the introns, exon 4 is also spliced out, leaving an mRNA comprising exons 1–3 and 5, which is then translated into the so-called calcitonin-related gene peptide (CRGP), which has important effects as a growth factor.

cific ribonuclease enzymes to cleave the DNA molecule at the exon–intron boundary and re-form the molecules into a single, intron-free, mRNA. Thus, by using processes that modify the splicing process, different tissues may splice out introns, and some exons, to give rise to tissue-specific mRNA molecules that are then translated into tissue-specific proteins. This type of alternative splicing can be seen in the genes for human leukocyte antigen (HLA) class I and II and the immunoglobulin, IgM. In addition to the alternative splicing of pre-mRNA, alternative polyadenylation signals in the transcription unit – intronic sequences that signal the site of addition of the polyA tail – may also give rise to tissue-specific gene expression.

Editing of RNA at the post-transcription level

RNA editing involves the enzyme-mediated alteration of RNA in the cell nucleus before translation. The process may involve the insertion, deletion, or substitution of nucleotides in the RNA molecule. The substitution of one nucleotide for another has been observed in humans, and can result in tissue-specific differences in transcripts. Apolipoprotein B (apoB), a lipoprotein molecule, illustrates this. In the case of the apoB

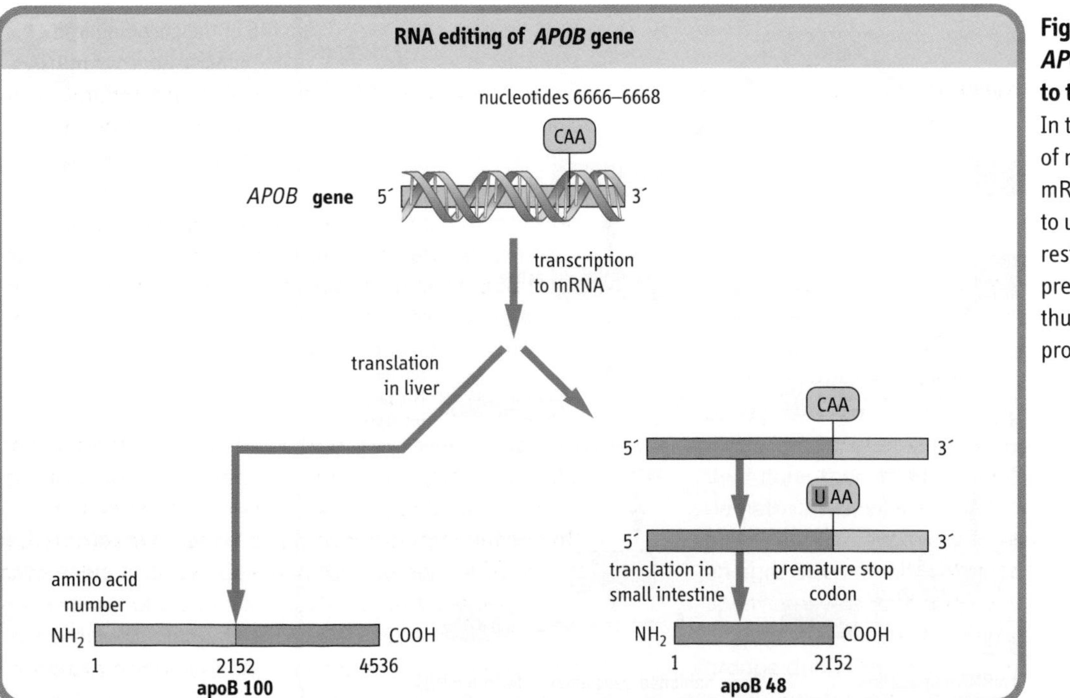

Fig. 32.8 RNA editing of the *APOB* gene in man gives rise to tissue-specific transcripts. In the small intestine, editing of nucleotide 6666 of apoB mRNA, by changing cytidine to uracil, converts a glutamine residue in apoB 100 to a premature stop codon, and thus produces the truncated product, apoB 48.

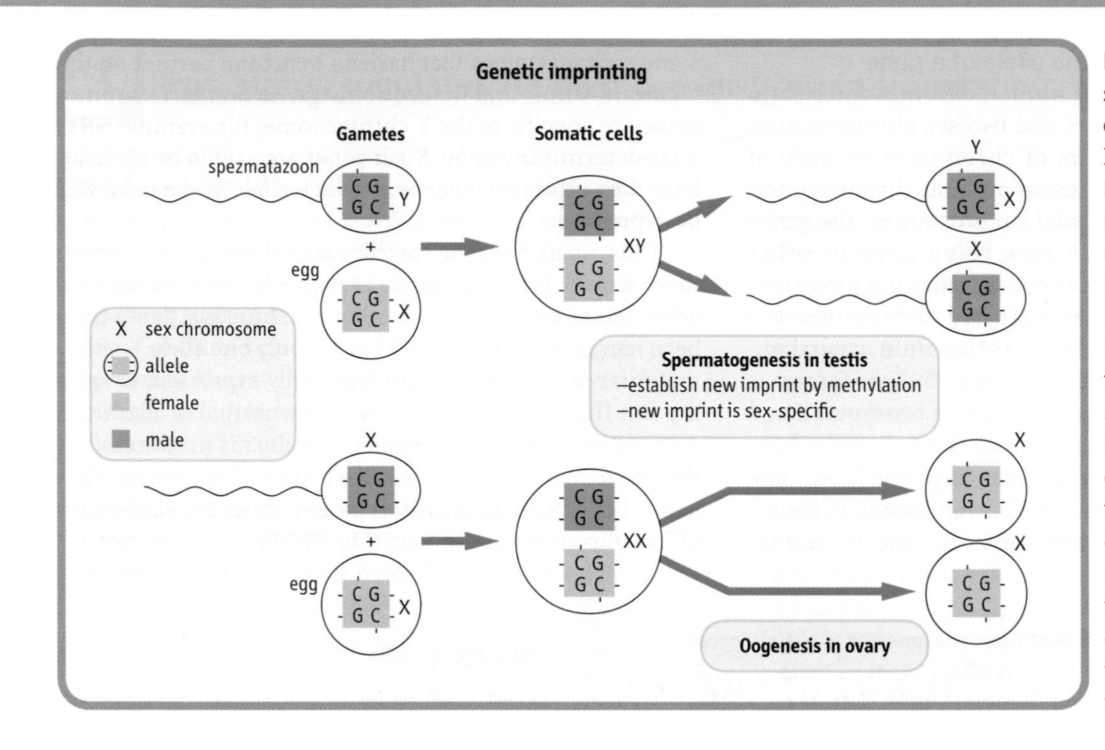

Fig. 32.11 When a gene is subject to imprinting, the sex of the donating cell is critical. In the first example shown here, the imprinted gene is paternally derived and, in the male diploid somatic cell, the paternal gene is dominant. During spermatogenesis, the maternally derived gene is imprinted and the maternal signal removed. Thus, in the next generation, paternally derived genes will be imprinted. In the second example shown here, the maternally derived gene is imprinted and, during oogenesis, the paternal imprint is erased and the new maternal imprint established.

Summary

In this chapter, the fundamental steps in the initiation of gene transcription have been discussed and an attempt has been made to highlight some of the key elements that can interact with the process. There are a number of ways in which gene expression can be regulated, the principal method being regulation of transcription. However, other important mechanisms exist, and attention has been drawn to the role of imprinting and post-transcriptional processes.

Further reading

Boyes J, Bird A. DNA methylation inhibits transcription indirectly via methyl-CpG binding protein. *Cell* 1991;**64**:1123–1134.

Latchman DS. Transcription-factor mutations and disease. *N Engl J Med* 1996;**334**:28–33.

Lewin B, Stachen T, Read AP. *Genes*, VI. Oxford: Oxford University Press; 1997.

Parker KL, Schimmer BP. Transcriptional regulation of the adrenal steroidogenic enzymes. *Trends Endocrinol Metab* 1993;**4**:46–50.

Rosenthal N. Regulation of gene expression. *N Engl J Med* 1994;**331**:931–933.

Strachan T, Read AP. *Human molecular genetics*. Oxford: BIOS Scientific Publishers Ltd; 1996.

Tsai M-J, O'Malley BW. Molecular mechanisms and action of steroid/thyroid receptor superfamily members. *Annu Rev Biochem* 1994;**63**:451–486.

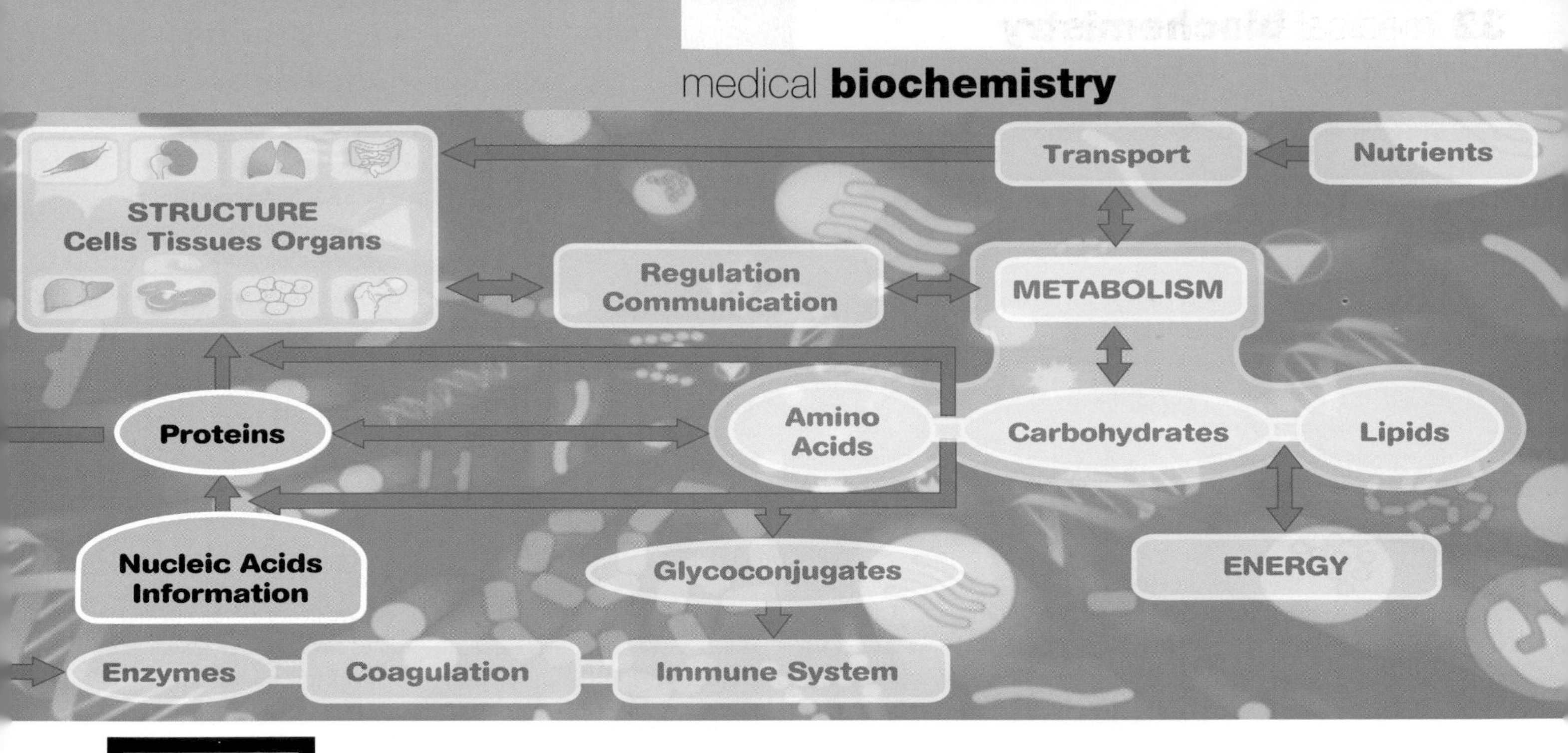

33 Recombinant DNA Technology

Introduction

There are several key features to all DNA tests

Hand in hand with the development of our understanding of deoxyribonucleic acid (DNA), genes, and their functions has been the explosion of technology for the clinical analysis of DNA and ribonucleic acid (RNA). A full description of all these processes is out of the scope of this text but an understanding of the basic principles of the methods and examples of some commonly used applications will become increasingly essential with the shift from experimental DNA analysis to diagnostic services providing genetic analysis in the clinical setting.

There are several key features central to all the DNA methods currently used and these will be discussed prior to outlining some of the commonly used techniques.

Hybridization

Hybridization is based on the annealing properties of DNA

Double-stranded DNA exists in life as a double helix, in which the two strands are held in close association by virtue of hydrogen bonds between complementary bases. Because of the structure of the four bases, it is difficult for two bases such as adenosine (A) and guanine (G), or thiamine (T) and cytosine (C), to align and form a complementary pair. Therefore, in any region of DNA where the bases of one strand are not complementary to the other strand, a 'kink' in the double helix will occur. The pairing of the two strands is thus interrupted and a continuous, stable double helix is not formed.

Hybridization is a fundamental feature of DNA technology. It is a process by which a piece of DNA or RNA of known nucleotide sequence, which can range in size from as little as 15 bp to several hundred kilobases, is used to identify another piece of DNA containing complementary sequences. Such a piece of DNA or RNA is called a probe. Probe DNA will form complementary base pairings with another strand of DNA, termed the template, if the two strands are complementary, and a sufficient number of hydrogen bonds is formed.

The principles of molecular hybridization

In molecular hybridization, it is essential that the probe and template are single-stranded

Probes can vary in both their size and their nature (Fig. 33.1). However, one essential feature of any hybridization reaction is that both the probe and the template are single-stranded.

Nucleic acid probes used for hybridization studies			
Probe type	**Origin**	**Probe characteristics**	**Labeling method**
DNA	cell-based DNA: cloning, polymerase chain reaction (PCR)	double-stranded cell-based: 0.1–100+ kb PCR-based: 0.1–10 kb	random primer nick translation
RNA	RNA transcription from phage vectors	single stranded: 1–2 kb	run-off translation
oligonucleotide	chemical synthesis	single-stranded: 15–50 nucleotides	end-labeling

Fig. 33.1 Characteristics of nucleic acid probes used for hybridization studies.

The process of separating the two strands of DNA is called DNA denaturation or melting. Probe DNA is generally denatured by heating. If the template DNA is double-stranded then it too must be denatured, either by heating or treatment with alkali. (NB: RNA is destroyed by alkali treatment.) Once both probe and template DNA are single-stranded, mixing of the two will allow complementary bases to reassociate. This process is called DNA annealing.

A single-stranded probe and template can anneal in a variety of ways:

- probe–probe complementary strands;
- template–template complementary strands, so-called homoduplexes;
- template–probe or probe–template heteroduplexes.

It is the formation of probe–template heteroduplexes that is the key to the usefulness of molecular hybridization

The conditions under which DNA hybridization occurs and the reliability and specificity, or stringency, of hybridization may be affected by several factors:

- **base composition:** GC pairs have three hydrogen bonds compared with the two in an AT pair. Double-stranded DNA with a high GC content is therefore more resistant to denaturation;
- **strand length:** the longer a strand of DNA, the greater the number of hydrogen bonds between the two strands. Longer strands require more heat or alkali treatment to denature them; this seems to be most important when the probe size is between 15–500 bp;
- **reaction mixture:** if the sodium concentration of the reaction mix is high, then the formation of double-stranded DNA is favored, whereas substances that disrupt hydrogen bonds in DNA, (e.g. urea), favor single-stranded DNA.

Thus if a probe is small, e.g. 50 bp, then when stringency is high, i.e. if the probe and template form in the presence of a high sodium concentration and at high temperature, a single base pair difference may then prevent the formation of the probe–template duplex, whereas a larger probe, e.g. 500 bp, may still form a stable duplex. The converse may also be true, i.e. if there are substantial differences between the probe and template, annealing of the probe and template may occur under conditions of low stringency (Fig. 33.2).

One means of measuring the stability of a nucleic acid duplex is assessing its melting temperature (Tm)

The **melting temperature (Tm)** is the temperature, *in vitro*, at which 50% of a double-stranded duplex has dissociated into single-strand form. There are several ways of determining Tm but one simple formula, looking at the melting temperature of oligonucleotides, is as follows:

$$\text{Tm}(^\circ\text{C}) = 2\,[\text{number of AT base pairs}] + 4\,[\text{number of GC base pairs}]$$

This formula emphasizes the importance of the number of hydrogen bonds present between DNA strands and the energy required to disrupt double-stranded DNA. The Tm for human DNA is approximately 87°C and reflects the relatively high GC component – about 40% – of human DNA.

Probes must have a label to be identified

Implicit in the use of probes to identify pieces of complementary DNA is the notion that if hybridization occurs, some useful information may be gained. This is usually provided by virtue of the fact that the probe is labeled and the probe–template duplex can be identified. There are many ways in which probes can be labeled but they fall into two categories, either **isotopic**, i.e. involving radioactive molecules, or **nonisotopic**, e.g. end-labeling probes with fluorescent tags or small ligand molecules (Fig. 33.3).

The majority of techniques involving probe hybridization and labeling still involve the use of radioisotopes such as ^{32}P, ^{35}S or ^{3}H and, as such, require a method for detecting and localizing the radioactivity. The most common method involves the process of autoradiography. **Autoradiography** allows information from a solid phase, e.g. as a gel or fixed-tissue samples, to be saved in two-dimensional form as a photographic emulsion.

Probe–template hybridization

A Hybridization characteristics using a large conventional probe (>200 bases)

Match	**Perfect**	**Single base mismatch**	**Multiple mismatch**
stringency	high	intermediate	low
example	human template + human probe 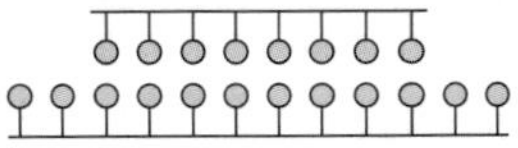	human template + human probe with mutation 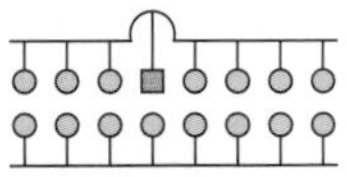	human template + mouse probe 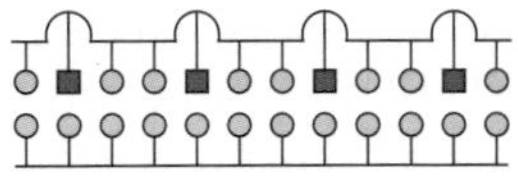
stability	stable	stable	stable

B Hybridization characteristics using a small oligonucleotide probe

Match	**Perfect**	**Single base mismatch**	
stringency	high	high	
example	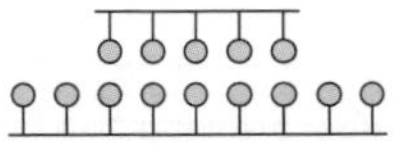	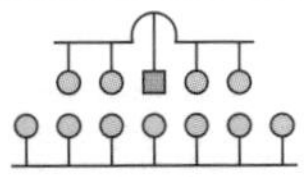	
stability	stable	unstable	

Fig. 33.2 Probe–template hybridization. (A) Large probes, e.g. 200 bases or more, can form stable heteroduplexes with the template DNA even if there are a significant number of noncomplementary bases in conditions of low stringency. (B) Oligonucleotide probes that differ by a single base from the complementary region of the template may not be able to hybridize if the conditions are sufficiently stringent.

Labels used to prepare probes used for nucleic acid hybridization

Probe label	**Characteristics**	**Examples**
radioactive	use radioisotopes emitting β-particles or γ-rays. These are bound to dNTPs and incorporated into the probe DNA	^{32}P, ^{35}S
nonradioactive	rely on the coupling of a reporter molecule to a nucleotide precursor, e.g. a dNTP. When the probe hybridizes, another protein with high affinity for the reporter group (an affinity group which has a marker group associated) binds the reporter and the bound complex can be detected	biotin and streptavidin digoxigenin

Fig. 33.3 Details of labels used to prepare probes used for nucleic acid hybridization. dNTP, deoxyribonucleoside phosphate.

Autoradiography

The most commonly used imaging technique is direct autoradiography, which involves placing the sample in direct contact with the photographic material, usually an X-ray film. The radioactivity of the bound probe produces a dark image on the X-ray film. Nuclides that emit weak β-particles, such as ^{35}S or ^{3}H, are ideally suited to this technique as the low energy of these particles determines that their energy can only be detected a short distance from their source. High-energy β-particles from ^{32}P, a commonly used radioactive label pass straight through the conventional X-ray film. In this instance, intensifying screens are placed behind the film to reflect the β-particles and convert them to light that is reflected onto the photographic emulsion.

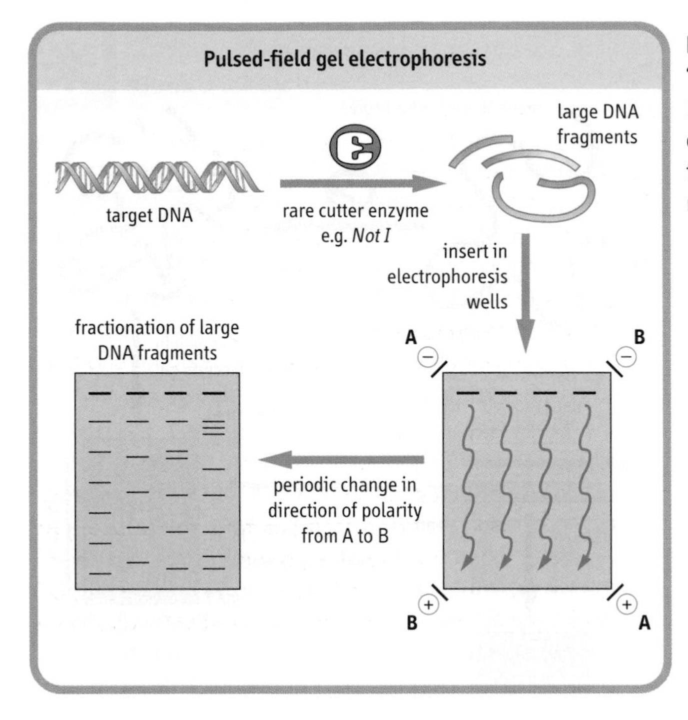

Fig. 33.11 Pulsed-field gel electrophoresis. DNA is digested with 'rare-cutting' restriction enzymes to generate large DNA fragments. During electrophoresis, the polarity of the electric field changes direction periodically. This results in the uncoiling of the large DNA fragments, which can then migrate slowly through the agarose. The end result is size-fractionated DNA of fragments over 100 kb.

- **cell-based DNA cloning:** DNA is amplified *in vivo* by a cellular host such that the number of copies of the desired DNA template increases simply due to the increase in number of replicating host cells;
- **enzyme-based DNA cloning (cell-free):** this method is represented by the polymerase chain reaction (PCR) and involves entirely *in vitro* DNA amplification.

Cell-based cloning

The majority of cell-based cloning uses recombinant DNA in replicating bacteria

Cell-based cloning is based on the ability of replicating cells, e.g. bacteria, to sustain the presence of so-called recombinant DNA within them. Recombinant DNA refers to any DNA molecule that is artificially constructed from two pieces of DNA not normally found together. One piece of DNA will be the target DNA that is to be amplified and the other will be the replicating vector or replicon (a sequence capable of initiating DNA replication in a suitable vector).

The majority of cell-based cloning is performed using bacterial cells. In addition to the bacterial chromosome, bacteria contain extrachromosomal double-stranded DNA within their cytoplasm which can undergo replication. One such example is the bacterial plasmid. Plasmids are circular double-stranded DNA molecules that undergo intracellular replication and are passed vertically from the parent cell to each daughter cell. However, unlike the bacterial chromosome, plasmids are copied many times

Bacterial plasmids contain elements that determine the properties of the plasmid

Within the circular DNA of a plasmid molecule are several elements that confer specific properties on the plasmid. Naturally occurring plasmids require several modifications to transform them into efficient DNA amplification vectors:

- **a polylinker cloning site:** this comprises a short sequence of nucleotides that contain unique recognition sites for several common restriction enzymes;
- **an antibiotic resistance gene:** the host bacterial cells used for cloning must be sensitive to an antibiotic, e.g. ampicillin or tetracycline, so that when they are transformed by the plasmid, the host cell acquires resistance to the antibiotic. This then provides a means of selecting out bacteria that contain the plasmid from those that do not. This selection step is important because of the low efficiency of forming plasmids containing recombinant DNA;
- **a recombinant screening system:** it is customary for the plasmid to have an expressible gene within it that can be used to show that the plasmid is present in the bacteria. This so-called 'marker gene' can be designed to have the polylinker sequence within its reading frame so that, if the plasmid contains the target DNA insert, the normal reading frame of the DNA is disrupted and the gene becomes nonfunctioning. The most common example is the gene for β-galactosidase. Using chromogenic substrates, bacteria containing this gene can be recognized readily on a culture plate. If there is an insert in the plasmid, the gene product is altered and there is no generation of the colored reaction product (see Fig. 33.12B).

during each cell division. Thus, plasmids represent ideal replicons for the amplification of target DNA and methods involving the use of plasmids are widespread throughout molecular biology.

Target DNA is introduced into a replicon by using restriction enzymes to cut target and replicon DNA so that the target DNA and the now open-circle plasmid DNA now have complementary sticky ends (Fig. 33.12). DNA ligase now forms a closed circular DNA plasmid with the target DNA inserted into the plasmid such that its replication efficiency is not affected.

Once the target DNA is incorporated into the plasmid vector, the next step is to introduce the plasmid into the host cell to allow replication to occur. The cell membrane of bacteria is selectively permeable and acts as a 'molecular sieve to stop the free passage of large molecules such as DNA in and out of the cell. However, the permeability of cells can be altered temporarily by factors such as electric currents (electrophoration) and high-solute concentration (osmotic stress), so that the membrane becomes permeable and will allow DNA to enter the cell. Such a process renders the cells competent, i.e. they can take up foreign DNA from the extracellular fluid. This process of DNA transformation is generally very inefficient, in so far as only 1–2% of cells may take up plasmid DNA and often only a single plasmid is introduced during transformation. However, it is this process of cellular uptake of plasmid DNA that forms the critical step in cell-based cloning and transformed cells act as gatekeepers to ensure that individual DNA molecules are sorted out into individual replication units, i.e. the bacterium they transform.

Following transformation, the cells are allowed to replicate, usually on a standard agar plate. Colonies (clones of single cells) are then 'picked', based on β-galactosidase expression, and transferred to tubes for growth in liquid culture and a second phase of exponential increase in cell number. Thus, from a single cell, an extremely large number of identical clones can be generated in a relatively small time.

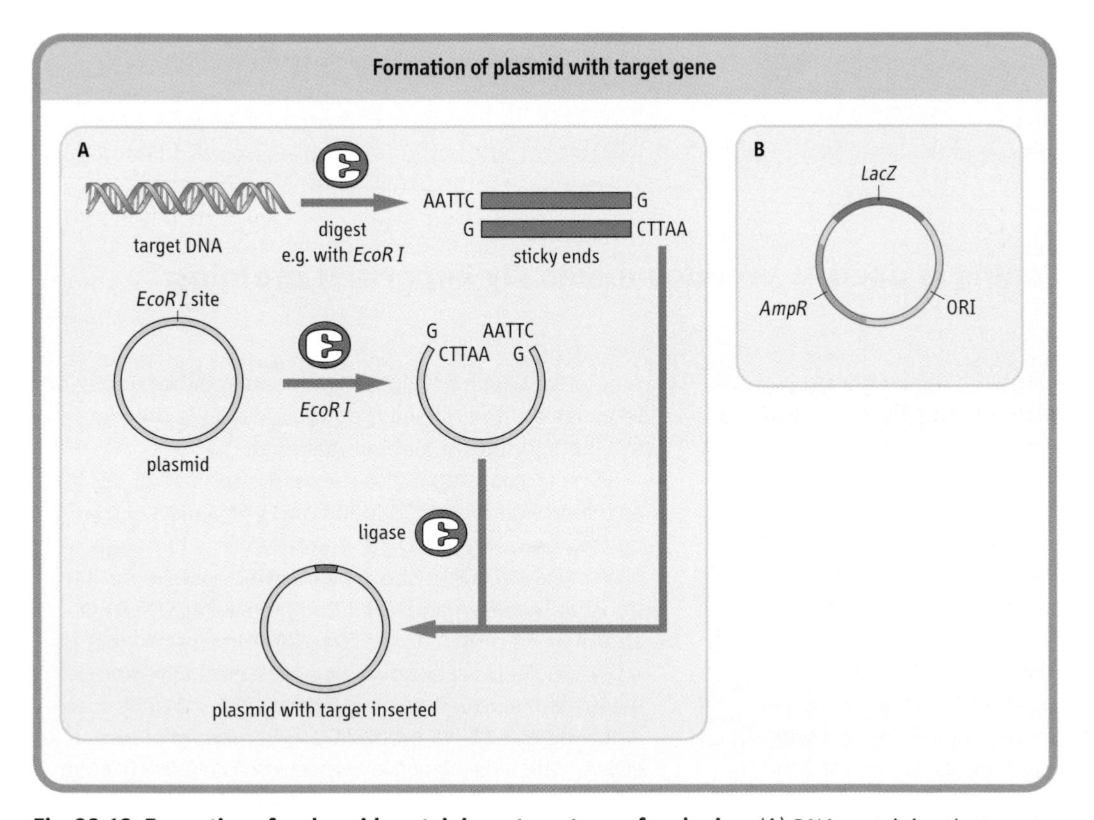

Fig. 33.12 Formation of a plasmid containing a target gene for cloning. (A) DNA containing the target gene is digested with a restriction enzyme that will produce 'sticky ends', e.g. *EcoR I*. The plasmid also has a restriction site for *EcoR I* which means that when digested with EcoR I it will become a linear DNA strand with 'sticky ends' complementary to the target, and plasmid DNA will form a new circular plasmid with the target gene inserted. (B) Structure of a typical plasmid. The plasmid contains a gene conferring resistance to the antibiotic ampicillin, *AmpR*, and the *LacZ* gene, which encodes the N-terminal region of the β-galactosidase enzyme. When introduced into host cells the N-terminal portion of the β-galactosidase associates with the remainder of the enzyme, produced by bacteria, to form a functioning enzyme. Within the *LacZ* gene is a polylinker site containing approximately 10 restriction-enzyme recognition sites, which serve as sites for insertion of target DNA. When a target DNA is ligated into this polylinker region the *LacZ* gene is disrupted and therefore abolishes the function of β-galactosidase in the host cells. The plasmid also contains ORI, the site for origin of replication.

If the transformation was successful in introducing a single plasmid, the result will be a huge amount of plasmid containing the target DNA, which can then be purified (Fig. 33.13). The recovery of the plasmid DNA is made easy because it is small, closed and circular, unlike the bacterial chromosomal DNA.

Vector systems for cloning DNA fragments

One critical factor in the ability of plasmids to replicate, and thus clone, target DNA is the size of the target DNA to be inserted. Conventional bacterial plasmids, although convenient to work with, are limited in the size of insert they can accept; 1–2 kb is the common size of the insert, with an upper limit of 5–10 kb. Some modified plasmid vectors called **cosmids** can accept larger fragments up to 20 kb. Another commonly used vector that has the ability to accept larger DNA fragments is the bacteriophage **lambda (λ)**. This viral particle contains a double-stranded DNA genome packaged within a protein coat. The λ phage can infect *E. coli* cells with high efficiency and introduce its DNA into the bacterium. The DNA is either incorporated into the bacterial chromosome and replicates during cell division, or replicates within the cell independent of the bacterial chromosome. Such independent replication leads to the synthesis of new viral particles, which can then lyze the host cell and then infect neighboring cells to replicate the process. The viral DNA is then re-isolated to obtain the recombinant DNA.

Larger inserts can be cloned by using modified chromosome from either bacteria **(bacterial artificial chromosomes, BACs)**, or yeast **(yeast artificial chromosomes, YACs)**. Such vectors can accommodate DNA fragments up to 1–2 Mb.

Cell-based cloning is used to produce medically important proteins

A 13-year-old girl is admitted with dehydration, vomiting and weight loss. Her blood glucose level is 19.1 mmol/L (344 mg/dL) and she has ketonuria. A diagnosis of type 1 diabetes mellitus is made. She is started on recombinant human insulin, is rehydrated, and makes a prompt recovery.

Comment. Prior to the advent of recombinant DNA technology, insulin therapy involved the use of animal insulins, most commonly pork or beef, which were chemically similar, but not identical, to human insulin. As a result of these differences, animal insulins often led to the development of antibodies, which reduced the efficacy of the insulin and could lead to treatment failures. Insulin was the first biologically important human molecule to be produced by means of recombinant DNA technology. Following the cloning of the human insulin gene, large-scale production of pure human insulin was performed by inserting the cloned gene into an *in vitro* amplification system. Thus, large amounts of insulin gene copies may be produced, which are then expressed in either bacteria or yeast and the resulting purified insulin is then available for use in diabetic patients.

One crucial step in the process is the use of cDNA as a template for transcription and translation of the desired gene. By using insulin mRNA as a template, human insulin cDNA can be generated by reverse transcription (formation of a complementary DNA strand using mRNA as the template) whose product is a DNA molecule that contains all the coding region of the insulin gene without any introns. The cDNA is then amplified by PCR to produce large amounts of insulin cDNA, which can then be cloned into a plasmid vector for *in vivo* amplification. This process of producing cDNA from mRNA is called **reverse transcriptase PCR or RT-PCR.** By this means, human recombinant insulin has virtually totally replaced animal insulin in the treatment of diabetes. Other important recombinant human peptides used clinically include growth hormone, erythropoietin and parathyroid hormone.

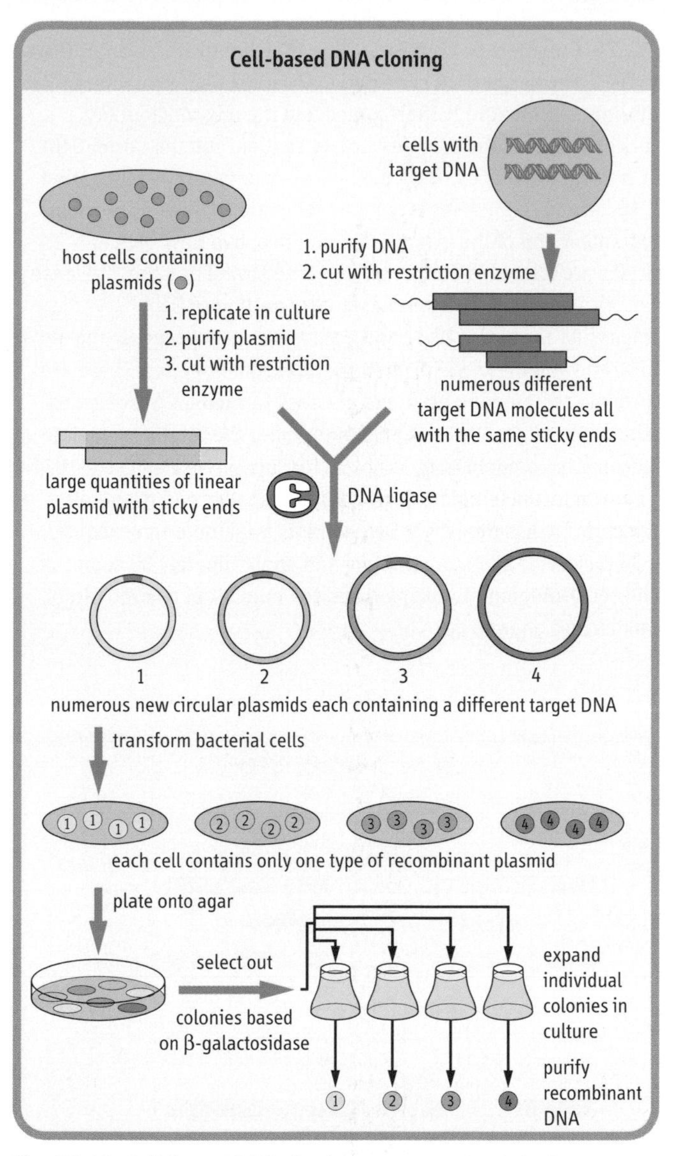

Fig. 33.13 Cell-based DNA cloning. An example of cloning genomic DNA using bacterial cells. One important feature of this type of cloning is the frequency with which transformed bacteria take up only one type of plasmid. In general each transformed bacterium will take up only one type of plasmid, even if there are more than one available at the time of transformation which in general is 100%.

Polymerase chain reaction-based DNA cloning

The use of PCR for the amplification of DNA in vitro has revolutionized molecular biology; it is a method of copying a single template DNA

PCR is a simple and quick means of generating up to 10^9 copies of a single template DNA molecule within hours. A standard PCR reaction requires the following:

- **DNA template:** the piece of DNA that is to be amplified;
- **amplimers:** small oligonucleotide primers that will hybridize with complementary DNA sequences upstream of the template and act as a starting point for the synthesis of the newly amplified DNA strands;
- **polymerase enzyme:** an enzyme that will catalyze the formation of the DNA during the amplification reaction;
- **dNTPs:** these are essential for the synthesis of the new DNA strands by the polymerase.

PCR is a cycle of reactions that relies on the completion of the previous step for the satisfactory completion of the next step (Fig. 33.14). The average PCR involves about 30–35 cycles of reactions that provide sufficient DNA, about 10^9 copies of the original template, so that it can be visualized following electrophoresis on an agarose gel.
The cycle of the PCR comprises three steps:

- **Denaturation:** heating of the reaction to approximately 95°C to ensure template and primers are single stranded;
- **Annealing:** cooling of the mixture to allow heteroduplexes of primer and template to form. The temperature for this is based on the Tm of the expected duplex (usually 3–5°C below Tm);
- **Elongation:** DNA polymerase elongates the newly synthesized DNA strand from the site of annealing of the primer along the template strand.

The process of PCR has been greatly helped by the discovery of heat-stable DNA polymerases, which can withstand the heat required during the denaturation step and still function as polymerases, e.g. *Taq* polymerase and *Pfu I*). The enzymes were isolated from organisms that were noted to be able to live and reproduce in hot springs and geysers.

The ability of PCR to amplify selectively a single template DNA relies on some prior knowledge of the nucleotide sequences in the regions flanking the target sequence

If the sequences in the regions flanking the target sequence sequences are known, the oligonucleotide primers are complementary to:

- the region on the sense strand 5′ to the target region;
- the region on the antisense strand 5′ to the target region, which corresponds to the 3′- limit of the sense strand (Fig. 33.15).

It is essential that the sequence of the primers is unique to the target DNA being amplified. If, for instance, the 3′- primer contains sequences that contain tandem repeats of bases, e.g. TGT-GTG, the specificity of the amplification may be reduced as such repetitive sequences occur widely throughout the genome, and the amplified sequence may contain many unwanted products.

Advantages and disadvantages of PCR cloning

PCR is a quick, simple and robust technique, but a prior sequence knowledge is needed, only a small target DNA amplification is achieved and DNA replication may be inaccurate

PCR has three principal advantages over cell-based cloning:

- **time:** by using PCR, a single template strand can be amplified to more than 10^5 copies within a few hours.

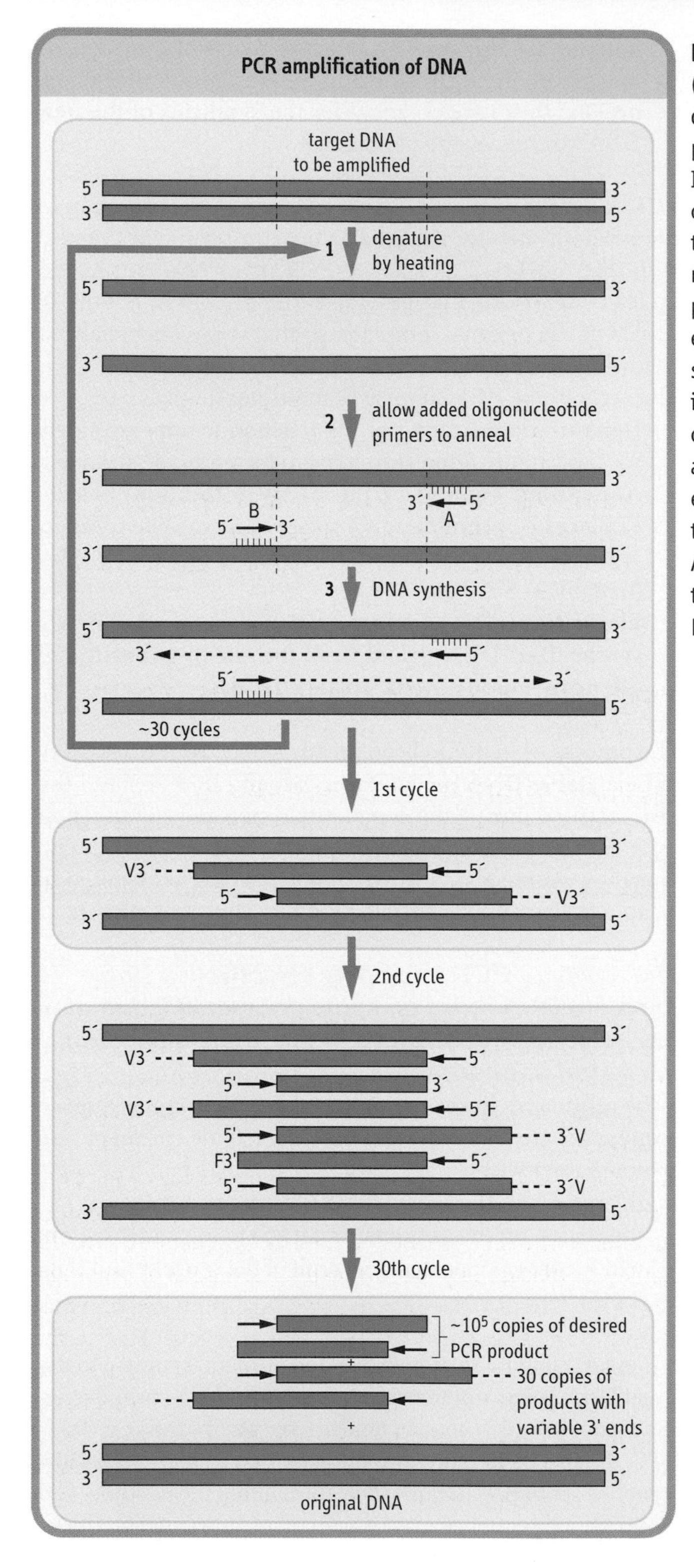

Fig. 33.14 Polymerase chain reaction (PCR) amplification of DNA. (A) Cycle 1. The target DNA is heated to 94°C to ensure complete denaturing. The mixture is then cooled and the oligonucleotide primers hybridize to the complementary sequence in the target DNA. In the presence of dNTPs, *Taq I* polymerase catalyzes the elongation of a DNA strand complementary to the template DNA with its origin at the extreme 3´end of the primer. In doing this, two new DNA molecules are formed: a duplex of the sense strand and the antisense primer, with an additional complementary portion of DNA encompassing the target DNA, and a similar duplex based on the anti-sense strand and the sense primer. (B) Subsequent cycles. These are identical but the growing numbers of newly formed DNA molecules containing the target DNA begin to outnumber the original template and become the predominant template for successive PCR cycles. With each new cycle, the template DNA becomes smaller and eventually the template DNA is the target DNA with its flanking primer region. After 30 cycles, the original template and approximately 30 copies of the shorter DNA templates remain and the number of desired target DNA molecules approaches 10^9.

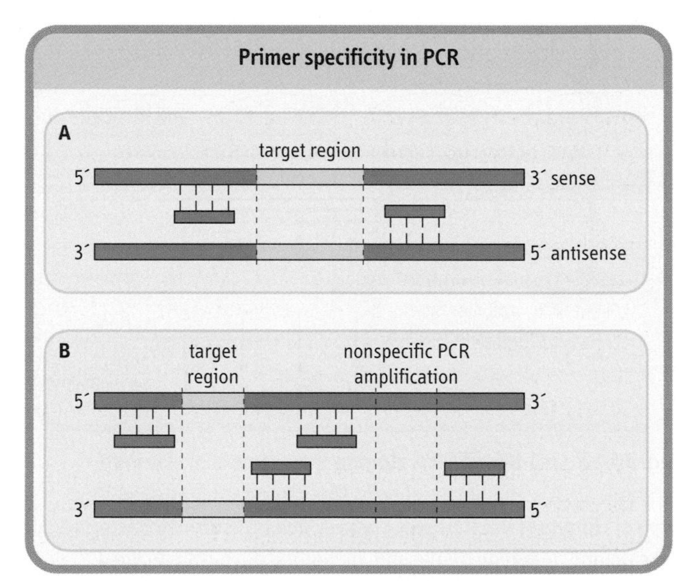

Fig. 33.15 Primer specificity in PCR. The design of PCR primers is central to the success of the reaction. (A) Well-designed, specific primers will only amplify the DNA of interest. (B) Primers which are nonspecific will amplify sequences other than the target DNA. Often this is undesirable but it may be a strategy employed to identify similar DNA sequences in similar genes of different species, or when looking for homologous (similar) genes in the same species – so-called degenerate PCR.

The denaturation step takes 1 minute, the annealing step 3–5 minutes and the elongation step 1–5 minutes. Thus 30 cycles of this PCR reaction will take 3–6 hours. Cell-based cloning is more expensive and requires days or weeks to produce a similar yield of DNA;

- **sensitivity:** PCR can amplify a DNA template to usable amounts if only a single copy of the template DNA is present. Indeed, PCR protocols have been developed to allow amplification of target DNA from a single cell. This is why PCR has found such a prominent place in forensic applications and also in paleontology;
- **robustness:** PCR is able to amplify DNA that is often badly degraded by time or the elements, or is present in previously inaccessible sites, e.g. formalin-fixed tissue.

Oligonucleotide primer design

It is the very versatility of PCR that can be its greatest weakness. If the oligonucleotide primers are not sufficiently robust then the reliability of the PCR amplification will be significantly reduced. In order to reduce the possibility of unwanted amplification, several simple rules are followed in the design of primers. Primer length should be about 20–30 nucleotides long for amplification of genomic DNA, while amplification of plasmid DNA may be performed with shorter primers. The sequence of the primer should avoid tandem repeats of nucleotides, e.g. CACACACACA, as these are common throughout the genome and will reduce the specificity of the amplification. In addition, it is important to ensure that there is no complementarity between the two primers. If the primers are complementary then they will hybridize to each other rather than the template and amplification will be inefficient.

By analyzing the DNA sequence of the target DNA and paying attention to these simple rules, a large percentage of the PCR amplification problems will be avoided. However, the use of computer software to perform the analysis of suitable sequences is increasingly common, and the software used is becoming extremely sophisticated and almost essential for many largescale PCR applications.

Despite the apparent attraction of PCR, it is not without fault and, depending on the application of the method, it may be an unacceptable means of amplifying DNA. The drawbacks of PCR are as follows:

- **prior sequence knowledge:** to perform PCR, flanking primers must be synthesized. For this to happen, the sequence of the target DNA, or at least its flanking regions, must be known. This may require prior cell-based cloning to derive the flanking nucleotide information;
- **small target DNA amplification:** PCR will reliably amplify target sequences up to about 5 kb, although products of 200–1000 bases are most readily amplified. However, larger products have been amplified recently by so-called 'long PCR' methods, allowing sequences up to 20 kb or more to be amplified;
- **DNA replication may be inaccurate:** during DNA polymerization by *Taq* polymerase in a standard PCR reaction, up to 40% of the synthesized products will have some error in the nucleotide sequence introduced by the polymerase enzyme, usually only a single nucleotide substitution. Therefore, the final reaction mixture will contain numerous copies of the target DNA, which are almost but not quite the same, and the substituted nucleotide may be different in different copies of the template. Similarly, if there is any contamination of the DNA source by other DNAs that may contain the target sequence (cf. forensic material), unwanted amplification may occur.

Thus, it is safe to say that PCR provides a simple, fast and very efficient means of amplifying DNA which is partly or fully characterized, is of relatively small size and may be present in small amounts or in a poor condition. However, it is a much weaker technique if the objective of the amplification is accurately to amplify DNA without introducing sequence errors or if the target DNA sequence is large.

Specific methods used in the analysis of DNA

Hybridization-based methods

There are several ways in which hybridization can be used in the study of DNA and the used methods exploit either the stringency of the hybridization of probe to template or rely on the effects of restriction enzymes for detecting variations in nucleotide sequences.

Restriction fragment length polymorphisms (RFLPs)

Restriction enzymes cleave DNA at specific recognition sites; if the sequence is disrupted by mutation or polymorphism, the results of a blot will differ

If that sequence is disrupted, either by a pathologic change in the DNA sequence resulting in a disease (a mutation), or a naturally occurring variation in the DNA sequence unaccompanied by disease (a polymorphism), the results of probing a Southern blot of DNA digested by a restriction enzyme may differ. Such differences in DNA sequence may lead to the creation of new restriction sites or the abolition of existing sites, and result in DNA fragments of different lengths – restriction fragment length polymorphisms (RFLPs) (Fig. 33.16). Such RFLPs can be used either to identify disease-causing mutations, because of a single point mutation creating or abolishing a restriction site, or to study variation in noncoding DNA which can be used in the study of genetic linkage.

RFLPs can also help detect larger pathologic changes in the DNA sequence, either deletions or duplications. Large deletions of a gene may abolish restriction sites; this leads to the disappearance of a fragment on a Southern blot in homozygous individuals. Alternatively, if a DNA duplication event occurs, a new gene may be formed, which has a different pattern of restriction sites which allow the presence of the new gene to be detected. This type of hybridization is performed

RFLPs can be used for detecting pathologic mutations

A 24-year-old Afro-Caribbean woman is referred for prenatal counseling. Her younger brother has sickle-cell anemia, and she has become pregnant. Her partner is known to be a carrier of the sickle-cell mutation (sickle-cell trait) and she wants to know if her child will develop sickle-cell anemia.

Since the patient is at risk of being a carrier, she opts to have chorionic villus sampling (CVS) performed to detect the presence or absence of the sickle-cell mutation in her child. Analysis of her own DNA reveals that she is a carrier, and the CVS shows that the child is also a carrier and will **not** develop sickle-cell anemia.

Comment. Occasionally, a mutation will directly abolish or create a restriction site and thus allow the use of a restriction-based method to demonstrate the presence or absence of the mutant allele. One widely examined mutation is the A→T substitution at codon 6 in the sequence for the β-globin gene responsible for causing sickle-cell disease. This results in a glutamine–valine (Glu-Va)l mutation in the amino acid sequence of the β-globin gene and also abolishes a recognition site for *Mst II* (CCTN(A→T)GG). Digestion of normal human DNA with *Mst II* and probing the Southern blot with a probe specific for the promoter of the β-globin gene yields a single band of 1.2 kb as the nearest *Mst II* site is 1.2 kb upstream in the 5′ region of the gene. The abolition of the codon 6 restriction site means that the fragment size seen when probing *Mst II* digested DNA is now 1.4 kb, as the next *Mst II* site is located 200 bases downstream in the intron after exon 1. Thus, patients with sickle-cell anemia will show only one band, 1.4 kb, while carriers will have 2 bands, one 1.4 kb and another, 1.2 kb, and unaffected individuals will have a single 1.2 kb band (Fig. 33.17).

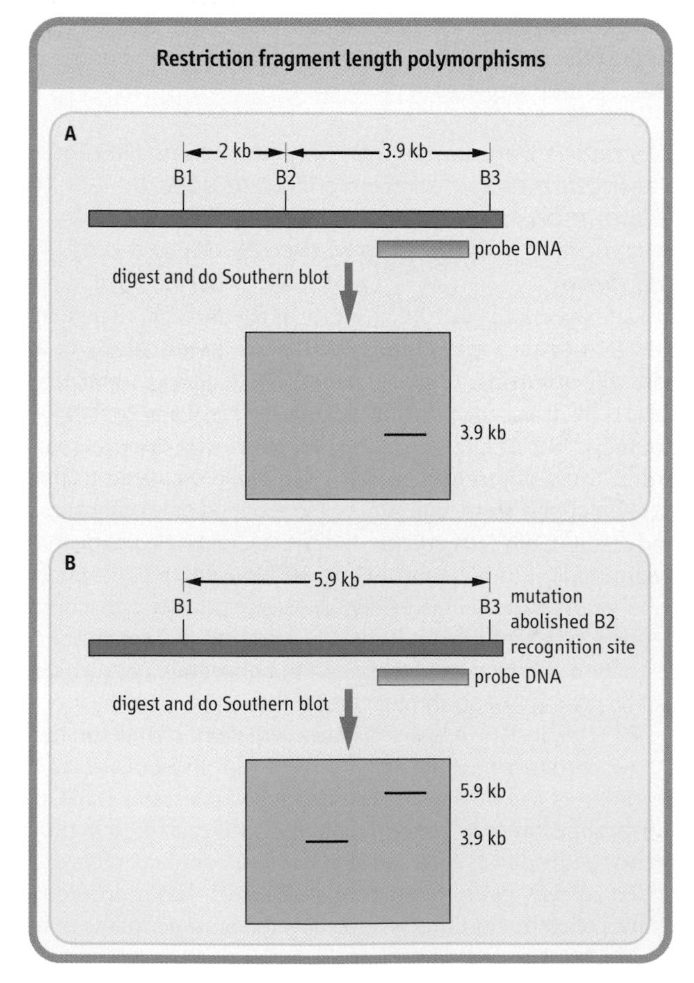

Fig. 33.16 Restriction fragment length polymorphisms (RFLP). (A) Variations in the nucleotide sequence of DNA, either due to natural variation in individuals or as a result of a DNA mutation, can abolish the recognition sites for restriction enzymes. (B) This means that when DNA is digested with the enzyme whose site is abolished, the size of the resulting fragments is altered. Southern blotting and probe hybridization can be used to detect this change. B, *Bam H1* restriction site.

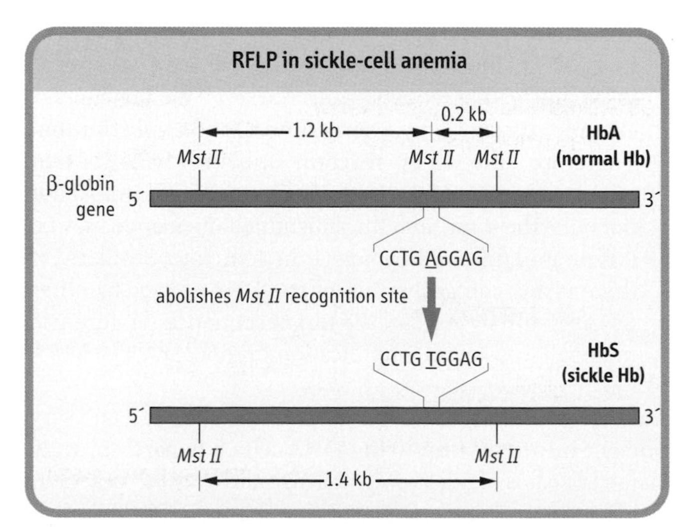

Fig. 33.17 RFLP analysis in sickle-cell anemia. An A→T substitution at codon 6 of the β-globin gene abolishes a recognition site for the enzymes, *Mst II*, which can be used to determine the presence or absence of the mutation by studying the RFLP pattern.

using large probes (0.5–5.0 kb) and is performed under moderate stringency, i.e. it is sufficiently rigid to allow hybridization of probe and target but weak enough to overlook minor differences, e.g. in introns or flanking, noncoding DNA.

Low stringency hybridization of a probe to a Southern blot of digested DNA may allow genes related to, but not identical to, the starting gene to be identified. Many genes exist in families, or have nonfunctional nearby identical copies elsewhere in the genome (pseudogenes), and thus hybridization of a probe may identify one or more restriction fragments, corresponding to related genes, which can be identified by other means. Similarly, genes in different species may be identified by using a single probe, which can hybridize to complementary sequences in blots of DNA from mouse, rat or whichever species is examined.

Allele-specific oligonucleotides and dot-blot hybridization

This method uses a small probe to probe the blotted DNA under highly stringent conditions

The other extreme of the RFLP method is to use a small oligonucleotide probe (15–30 nucleotides) and probe the blotted DNA under conditions of high stringency. This method will ensure that oligonucleotides will not hybridize to complementary sequences where there is a difference between the oligonucleotide and the target sequences, even if there is only a single nucleotide difference. Mutations that result from a single nucleotide change can be detected by using oligonucleotides that recognize either the mutant or the normal allele of the gene, so-called allele-specific oligonucleotides (ASOs). ASOs can be used to screen DNA for the presence of several mutations, particularly if the template DNA is immobilized on a membrane for dot-blot analysis. Single-stranded DNA is transferred to a membrane as in the case of Southern blotting, but rather than transferring size-fractionated digested DNA, total human genomic DNA or PCR generated fragments are blotted onto the membrane. In this way, ASOs can be applied to the membrane under high-stringency conditions and autoradiography performed to determine whether the DNA contains a specific allele. The process can then be repeated for the other allele and other genes tested (Fig. 33.18).

PCR based methods

PCR is a versatile method applied to a multitude of methods in DNA research (Fig. 33.19). The most widely used clinical application of PCR is in the identification of genetic mutations that can be detected by manipulation of PCR amplified DNA from 'at risk' individuals.

Assay of restriction site polymorphisms

Restriction site polymorphisms (RSPs) are small-scale equivalents of RFLPs

Naturally occurring variations in DNA sequence may abolish or create restriction sites that can be detected conveniently by digesting PCR amplified sequences with the appropriate restriction enzyme.

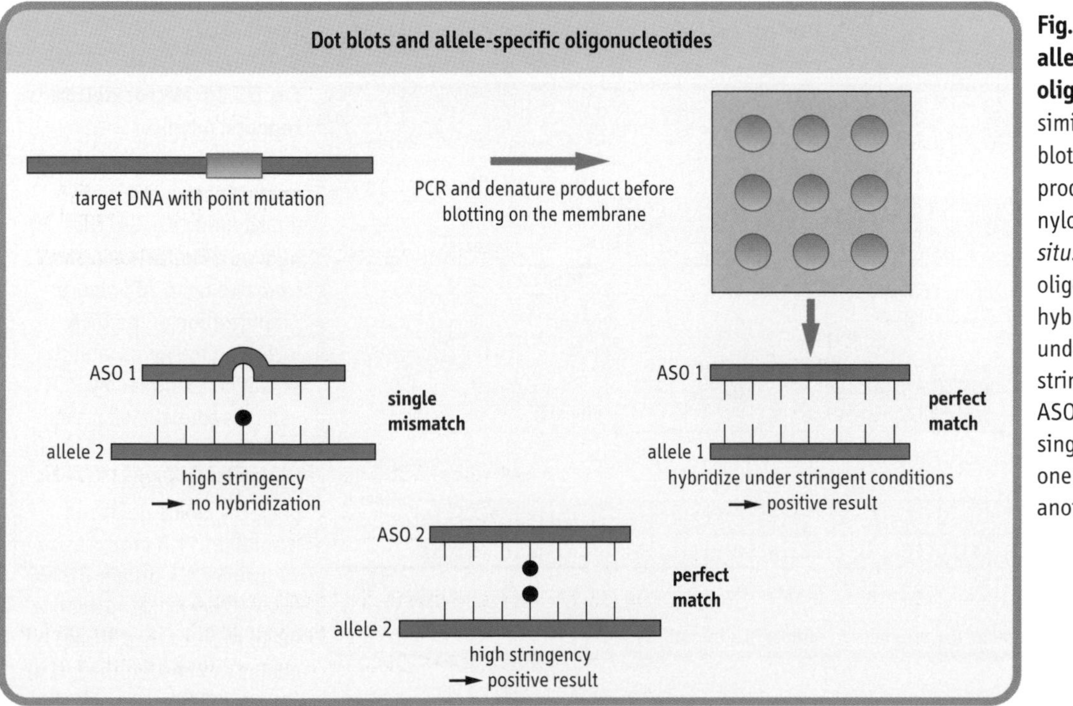

Fig. 33.18 Dot blots and allele-specific oligonucleotides. Using a similar technique to Southern blotting, single-stranded PCR products are transferred onto a nylon membrane and fixed *in situ*. Allele-specific oligonucleotides (ASOs) are hybridized to the target DNA under conditions of high stringency. In this case, the ASOs are used to identify a single base substitution, with one detecting the wild type and another the mutant allele.

Applications of PCR

Application	
genetic marker typing	restriction site polymorphisms microsatellite repeats
detection of point mutations	restriction site polymorphisms amplification refractory mutation systems
amplification of DNA templates for DNA sequencing	double-stranded DNA used in chain termination sequencing single-stranded templates by asymmetric PCR
genomic DNA cloning	PCR of gene families or genes in different species
others	genome walking, introduction of mutations *in vitro* to test their effect in biological systems

Fig. 33.19 Applications of PCR.

Detection of microsatellite repeats

This technique detects the few nucleotide repeats in tandem, in mammalian genomes

Tandem microsatellite repeats are sequences that comprise a few nucleotides – between one and four but typically two or three, e.g. CA, a tandem repeat of cytosine adrenosine nucleotides – that are repeated in tandem throughout mammalian genomes, so-called microsatellite repeats. These microsatellites are small blocks of DNA that occur in noncoding regions, either intergenic or within introns of genes and are not transcribed. The term 'satellite DNA' derives from the initial discovery that much of the DNA not involved in gene expression, so-called heterochromatin, contained large amounts of highly repetitive DNA, which was often physically localized to a few areas of the chromosome, i.e. the centromere. One important feature of these microsatellite repeats is that they are highly polymorphic, i.e. the number of tandem repeats in a particular microsatellite may vary from one individual to another. This means that the two copies of a single human microsatellite, one from each chromosome, may be of different lengths and can thus be distinguished in that individual. The number of different alleles of the microsatellite will vary depending on which microsatellite is studied, but may vary from 2–15 or more copies of the repeat per allele (Fig. 33.20). As a result of this high degree of polymorphism, microsatellite repeats are of great value in the study of genetic linkage, particularly in complex disorders such as diabetes mellitus and the forensic application of DNA fingerprinting, because they allow for the accurate identification of single-gene disorders.

Mutation detection by allele-specific PCR

Only mutant DNA will be amplified using this technique

Standard PCR uses two primers to amplify both strands of the DNA and relies on the fact that each primer will reliably amplify the target strand. However, PCR can be performed using allele-specific primers, which will only amplify one allele of a target gene when the primer and target are 100% complementary. This means that if one primer contains a sequence that recognizes a mutation, it will only hybridize with the mutant DNA. Thus, only DNA containing the mutation will amplify with this primer. Primers for this type of PCR are designed such that the 5′-primer is allele-specific while the 3′-primer is the consensus primer. The consensus primer is common to both alleles and ensures that the PCR reaction will work. The 5′-primer has its extreme 3′-nucleotides as the ones that detect the mutation, as DNA

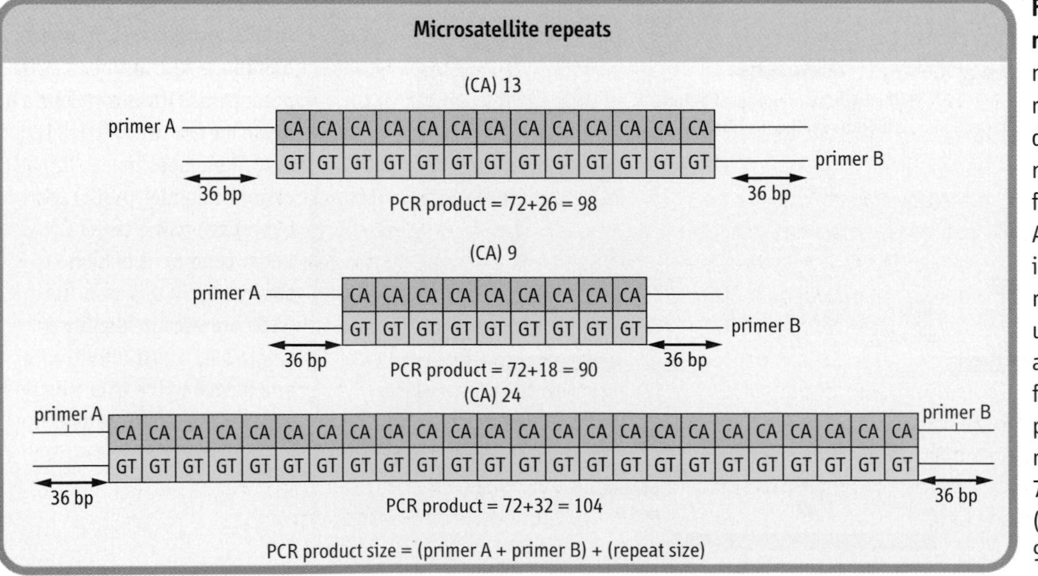

Fig. 33.20 Microsatellite repeats. A typical microsatellite contains tandem repeats of a simple dinucleotide e.g. CA. The number of repeats can vary from two up to 20 or more. Amplification of the DNA including the microsatellite repeat is performed by PCR using a radiolabeled primer and the products are then size-fractionated onto a polyacrylamide gel to aid resolution. PCR product size = 72 (primer A + primer B) +26 (13 × 2 for each dinucleotide) = 98. In this instance, three alleles can be identified: $(CA)_{13}$, $(CA)_9$ and $(CA)_{24}$.

Microsatellite repeats are ideal markers for study inheritance

A 7-year-old boy is the subject of disputed paternity. His mother claims that his father is a politician and that he is responsible for the child's financial wellbeing. All parties consent to DNA testing by means of DNA fingerprinting. Study of the DNA tests reveals that the politician is the father and is indeed responsible for the child.

Comment. The most commonly studied microsatellites contain dinucleotide repeats of (CA) or (TG) and may vary in repeat number from one or two up to 20 or more. Thus, a standard dinucleotide repeat would be denoted (CA)n, where n is the number of repeats. An ideal microsatellite marker would have an infinite number of alleles, i.e. n>100, and each allele would be equally frequent. However, most microsatellites have up to 10, or sometimes as many as 20, and there is often a preponderance of one or two particular alleles in any population. When microsatellites are typed, by PCR amplification and gel electrophoresis, both chromosomes are amplified and thus two alleles are generated from a single individual. The size of the alleles may be the same but is often different.

In forensic medicine, simple questions are often asked, e.g. 'Is this man the child's father?' or 'Does this blood stain correspond to blood of the suspect?' By selecting several microsatellite markers, typically six to nine, that are highly polymorphic and have an even spread of allele frequency, i.e. each allele is equally frequent, then a series of microsatellite markers can be generated. When an individual's markers are tested, the result is a set of alleles that are virtually unique to an individual, a so-called 'DNA fingerprint'. Thus, by comparing the pattern of microsatellites generated in the subject in question and comparing this with the pathologic sample or the DNA of offspring, the identity of a parent or suspect can be confirmed or refuted with a high degree of certainty.

Similarly, microsatellites have been used to examine the linkage of a disease trait, e.g. type 1 diabetes, to certain areas of the human genome. By studying the inheritance of microsatellite markers in families or the sharing of alleles in affected siblings, a statistical measure of the degree of inheritance of a marker can be determined. This approach has been used successfully to identify genetic loci strongly linked to the inheritance of type 1 and type 2 diabetes, and locate the genes causing Huntington's disease, myotonic dystrophy and many other single gene disorders.

Trinucleotide repeats in genes give rise to human disease

A 33-year-old woman has recently been diagnosed as having myotonic dystrophy (MD). MD is characterized by autosomal dominant inheritance, muscular weakness associated with impaired muscular relaxation, frontal balding, endocrine disorders such as diabetes and premature ovarian failure, and premature cataract formation. The patient has muscular problems and frontal balding, whereas her affected father, aged 60 years, has only just developed cataracts. She is concerned about the risk to her own children if she becomes pregnant.

Comment. While the majority of nucleotide repeats are noncoding, several trinucleotide repeats have been described, which are related to human disease. The majority of **trinucleotide repeats** are clinically silent, but 11 clinical disorders have been identified where trinucleotide repeats appear directly responsible for the disease phenotype (Fig. 33.21), and where the repeat is said to be **unstable**. Unstable repeats undergo changes in the size of the repeat during meiotic division and generally undergo expansion of the size of the repeat so that with each successive meiosis, i.e. from generation to generation, an expansion will gradually increase in size. Once trinucleotide expansions reach a critical size they begin to interfere with gene function and then result in the clinical syndrome (Fig. 33.22). Thus, in the case of myotonic dystrophy, with an expansion size of 80 repeats, the patient will exhibit the signs of the disease. Owing to the instability of the triplet repeat during meiosis, offspring of this patient might expect to have a larger expansion and also display features of the disease. Similarly, successive generations may display signs of MD at a younger age. This phenomenon of progressive worsening of the clinical phenotype with successive meioses is known as **anticipation** and is due to the progressive enlargement of the trinucleotide expansions during meiosis (Fig. 33.23). Interestingly, anticipation may be sex-limited, with the largest expansions only present in eggs, which have larger amounts of cytoplasm than spermatozoa. This observation has been suggested as the explanation for the occurrence of neonatal MD (a life-threatening form of the disease seen in newborn children) only in cases where the mother donates the triplet expansion to the offspring.

polymerization relies upon the presence of complete hybridization of primer and template at the 3′ end of the primer. This method can be applied to the detection of specific mutations in a single disease-causing gene, e.g. the gene for cystic fibrosis. This type of PCR is often called an amplification refractory mutation system (ARMS) as PCR amplification will be refractory, i.e. will not occur, in the absence of the mutant allele.

Detection of deletions in genes causing disease

This is performed using primers that generate PCR products of different sizes

In patients with Duchenne muscular dystrophy, many of the disease causing mutations are due to deletions of one of relatively few exons in this, the largest of human genes. Specific amplification of these exons using primers, which generate PCR products of different sizes for each exon, can be used to screen quickly for the presence of a disease-causing deletion in an affected individual or determine if the fetus of a carrier possesses the deletion. Such PCR is carried out in a single reaction using a single template and 5–10 different primer pairs, each generating a separate amplification product. This form of PCR is called multiplex PCR.

Other methods for the detection of variation in DNA

Single-strand conformational polymorphism (SSCP)

SSCP is useful for analysing PCR products of 200 bp and less and is moderately sensitive

Single-stranded DNA has a tendency to form complex structures by folding up. Often base pairing via hydrogen bonds occurs and this helps maintain this altered conformation. The mobility of such single-stranded DNA in an electrophoretic gel depends not only on the length of the DNA but also its

Unstable trinucleotide repeats

Disease	Repeat	Normal Length	Mutation length
Huntington's disease	(CAG)n	A:9–35	37–100
Kennedy disease	(CAG)n	17–24	40–55
Spinocerebellar ataxia I (SCA I)	(CAG)n	19–36	43–81
Dentatopallidoluysian atrophy (DRPLA)	(CAG)n	7–23	49–75+
Machado–Joseph disease (SCA III)	(CAG)n	12–36	69–79+
Fragile X site A (FRAXA)	(CGG)n	6–54	200–1000+
Fragile X site E (FRAXE)	(CCG)n	6–25	>200
Fragile X site F (FRAXF)	(GCC)n	6–29	>500
Fragile 16 site A (FRA16A)	(CCG)n	16–49	1000–2000
Myotonic dystrophy	(CTG)n	5–35	50–4000
Friedreich ataxia	(GAA)n	7–22	>200

Fig. 33.21 Unstable trinucleotide repeats. In all of the cases listed, trinucleotides repeats are present in individuals without clinical disease. However, progressive enlargement of the repeat leads to the onset of clinical signs, although a premutant phase has been described in some disorders, such as FRAXA, where expansions larger than normal are found but with minimal or no clinical features. These premutant states give rise to the finding of carriers – females who can subsequently pass on an enlarged triplet repeat to their offspring, which results in the full-blown condition.

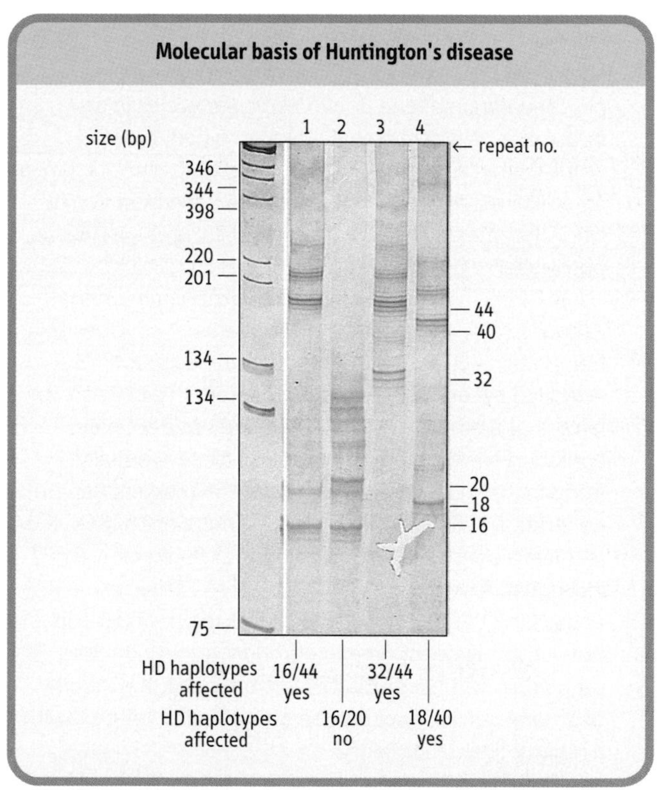

Fig. 33.22 Molecular diagnosis of Huntington's disease. This blot illustrates the use of a PCR-based method to diagnose Huntington's disease (HD). PCR amplification of the portion of the HD gene including the disease-associated triplet expansion region is performed. Each individual has two copies of the gene, one on each chromosome 4, and each PCR generates two PCR products. The PCR products are electrophoresed and compared with a size-specific ladder, which enables the size of the PCR product, and thus the number of triplet repeats in each product, to be assessed accurately. Affected patients will have triplet expansions greater than 35 repeats. HD is inherited as an autosomal-dominant trait and inheritance of a single copy of the mutant gene is required for the disease to be expressed.

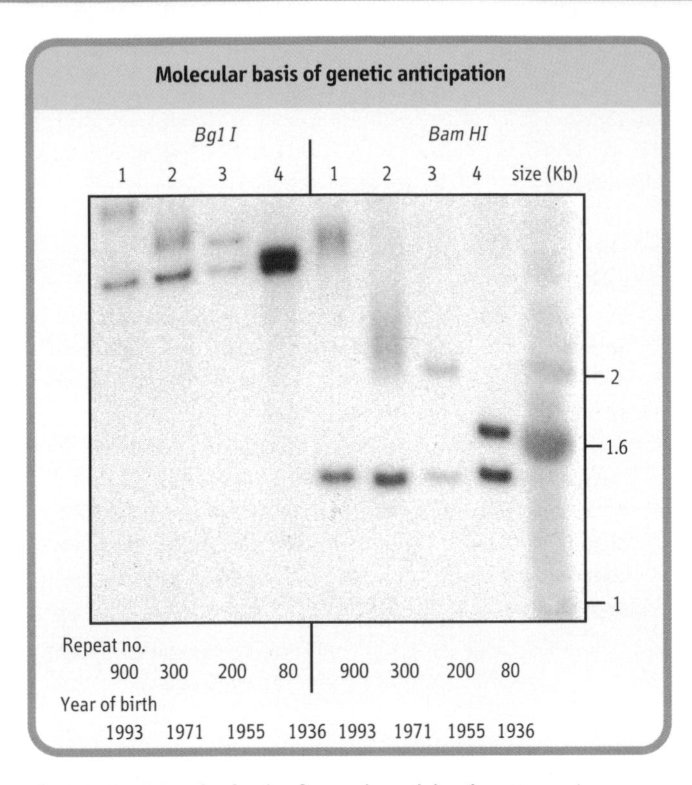

Fig. 33.23 Molecular basis of genetic anticipation. Myotonic dystrophy (MD) exhibits anticipation within families, i.e. the clinical features of the disease become progressively more severe and its age at onset of symptoms is younger with each successive generation. The underlying problem is the presence of an unstable trinucleotide expansion within the MD gene, which progressively enlarges with each meiosis. This results in the inheritance of larger mutations, which give rise to a more severe clinical picture with the tendency for the features to manifest at an earlier stage. The Southern blot illustrates the progressive increase in size of the trinucleotide expansion in successive generations of a family with MD. The normal allele size is 3.4 kb for the *Bgl I* digest and 1.4 kb for the *Bam HI* digest. It is clear that with each generation there is a progressive expansion in the size of the trinucleotide repeat using both methods. (This is a good example of an RFLP.)

conformation, which is determined by its DNA sequence and folding. Therefore, changes in DNA structure that do not alter the length of the target DNA, i.e. substitutions but not deletions or insertions, can alter the mobility of the fragment in a nondenaturing gel, so-called single-strand conformational polymorphisms (SSCP). SSCP is performed by PCR with a radioactive primer. The reaction products are denatured to render them single-stranded, and diluted to prevent intermolecular hybridization. The DNA is then electrophoresed on nondenaturing polyacrylamide gels. A control sample is also included on the gel to allow identification of variation from the wild-type allele.

Denaturant-gradient gel electrophoresis (DGGE)

DGGE uses 'melting' of double-stranded DNA

In SSCP, changes in the tertiary structure of DNA can result in differing electrophoretic mobilities of single-stranded DNA. However, this type of analysis cannot be used for double-stranded DNA. Nevertheless, if double-stranded DNA undergoes electrophoresis in a gel that is comprised of regions of increasing amounts of denaturant, e.g. urea, double-stranded DNA will then denature, or melt, to form single stands and stop migrating through the gel. A single nucleotide change between two alleles of a gene will affect the melting point of the alleles and thus their position on the denaturing gel.

To perform DGGE, PCR is performed using radiolabeled primers that have a GC clamp at their ends. This is a tail of guanine and cytosine residues added to the PCR primers such that the PCR product will have ends that have artificially created GC ends. These GC ends, or clamps, improve the sensitivity by preventing nonspecific melting of double-stranded DNA. PCR products are electrophoresed in the denaturing gel and, following autoradiography, different alleles of the suspect gene are identified (Fig. 33.24).

Protein truncation testing in breast cancer

A hereditary form of breast cancer associated with the coexistence of ovarian cancer has been shown to be due to mutations in the *BRCA1* gene. This gene, a tumor suppressor gene, was discovered following the analysis of large pedigrees with breast/ovarian cancer. Analysis of the mutations found in the genes of affected individuals showed that most of the mutations were either nonsense mutations, altered splice sites, or frameshift mutations, which gave rise to abnormally truncated protein products. Thus, following the identification of the common mutation types in breast cancer due to *BRCA1* (and the subsequently discovered *BRCA2* screening for mutations using a protein truncation test has become standard practice when examining both pathologic tissue and blood from individuals at risk of possessing a *BRCA1* or *BRCA2* mutation.

Comment. The translation of mRNA to the final protein product results in a peptide of predictable size. If a gene has a mutation that results in a premature stop codon, then the resulting peptide will be abnormally small. This difference can then be detected by electrophoresis of the protein product in a suitable gel and demonstrating the presence of a truncated protein product.

Reverse transcriptase-PCR

From whole blood, or pathologic tissue, cDNA for a gene can be prepared by the process of reverse transcription. Reverse transcription uses a reverse transcriptase enzyme to polymerize a DNA molecule complementary to the mRNA molecule, using a single poly T primer, which recognises the mRNA via its poly A tail. Following hybridization, the reverse transcriptase proceeds along the mRNA to manufacture a complementary strand, cDNA. The resulting cDNA is then used as a template for a PCR where the entire cDNA can be amplified to produce a DNA molecule containing the entire coding sequence of a gene. Clearly in some cases, gene expression is tissue-specific and one would not expect some mRNAs to be present in blood, e.g. dystrophin from muscle. However, ectopic transcription of genes occurs in white blood cells at a low level and allows analysis of transcripts of genes not normally expressed. This can be useful if the desired transcript is derived from a tissue that is not easily accessible or if study of the cDNA can give definitive information about the presence of a mutation.

The cDNA can then be incorporated into a cell-based cloning vector for further amplification or can be introduced into a cellular expression system to produce the gene product *in vitro*. Such methods are used increasingly in the study of the functional aspects of mutant gene products.

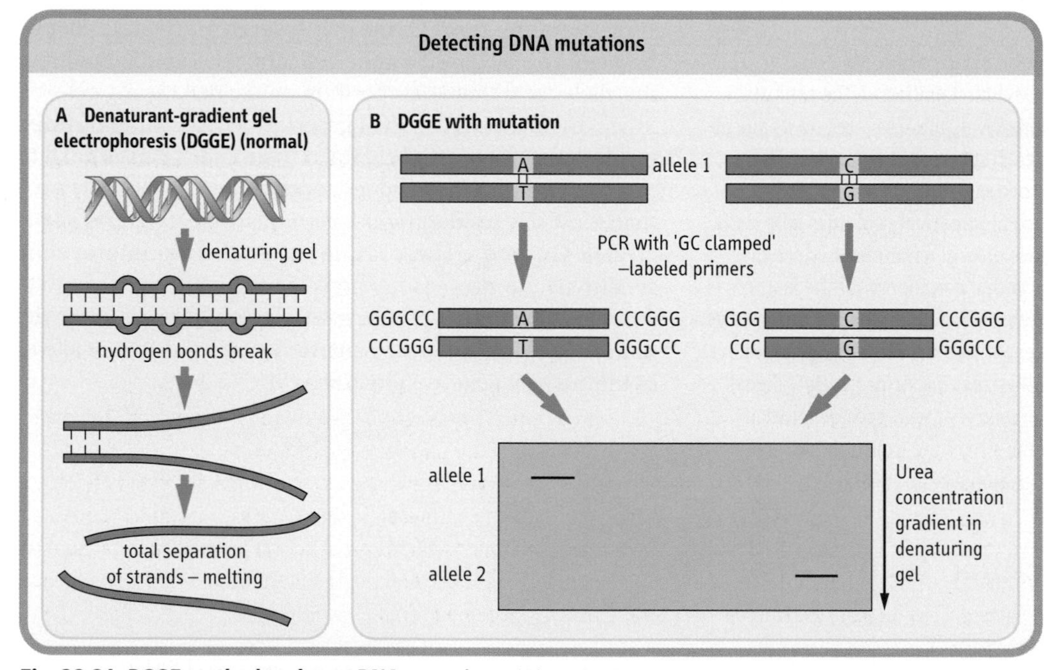

Fig. 33.24 DGGE method to detect DNA mutations. (A) As double-stranded DNA migrates through a gel comprised of increasing concentration of denaturants, e.g. urea, hydrogen bonds between the two strands are broken. At a point unique to that particular DNA, all the hydrogen bonds will be broken and the two strands separate – the so-called melting point. If the nucleotide sequence of a DNA is changed, e.g. by a mutation, the melting point of the DNA will be altered and thus the mobility of the DNA in the gel will be altered, allowing the change to be detected by autoradiography. (B) DGGE with mutation. In this case, a single A→C substitution has occurred. This increases the number of hydrogen bonds between the two strands and alters the mobility in the denaturant gradient. The difference in mobility through the gel only highlights the difference in the two strands' nucleotide sequence – it does not say what or where the difference is.

DNA sequencing

At the heart of many of the foregoing techniques is the notion that at least some of the nucleotide sequence of the DNA being studied is known, particularly in the case of PCR primers. The technologies now used in the sequencing of DNA have become extremely advanced and automated sequencing of large amounts of DNA is now standard practice, particularly in the Human Genome Project.

Chain termination DNA sequencing

Chain termination sequencing uses a DNA polymerase enzyme, a single-stranded template DNA and a sequencing primer

In this method, the sequencing primer is designed to be complementary to the region flanking the sequence of interest and acts as the starting point of chain elongation. DNA polymerization also requires dNTPs to allow the strand to be elongated but, in addition, chain terminating dideoxynucleotides (ddNTPs) are added. These are analogues of the dNTPs but differ in that they lack the 3´-hydroxyl group required for formation of a covalent bond with the 5´-phosphate group of the adjacent dNTP. Therefore, during DNA polymerization, the growing DNA chain will incorporate a ddNTP and, as a result, growth of the nucleotide chain is halted. A total of four reactions are carried out in parallel, each containing the primer, template, polymerase and dNTPs. However, to each of the four reactions, a small amount of one ddNTP is added (ddATP, ddCTP, ddTTP, ddGTP) so that four separate reactions, the A, T, G and C reactions, are conducted in parallel. As the chains elongate, ddNTPs will be incorporated into the chain in place of the corresponding dNTP on a random basis. This means that in any one reaction mixture, there will be many chains of varying lengths (partial reactions), which, when pooled together, represent the total collection of reactions for that base . The level of ddNTPs is limited so that chain termination is a random event. In practice, the number of molecules of each product declines with product length but there is a trade-off in sensitivity. Longer fragments, which are fewer, have more label incorporated, while shorter fragments are more frequent. The DNA chains of differing lengths can be separated by electrophoresis on denaturing polyacrylamide gels. These gels allow DNA fragments that differ by only one nucleotide in length to be separated and, if the reaction involves a labeled group, either a dNTP or the primer, then electrophoresis of the four reactions in parallel, with subsequent autoradiography, will allow the sequence of the DNA to be determined (Fig. 33.25). In general, this method can produce sequence data for the 300–500 bases downstream of the sequencing primer.

The chain termination method can be modified by replacing radioactive groups with fluorescence-labeled primers, which then allow the process to be automated.

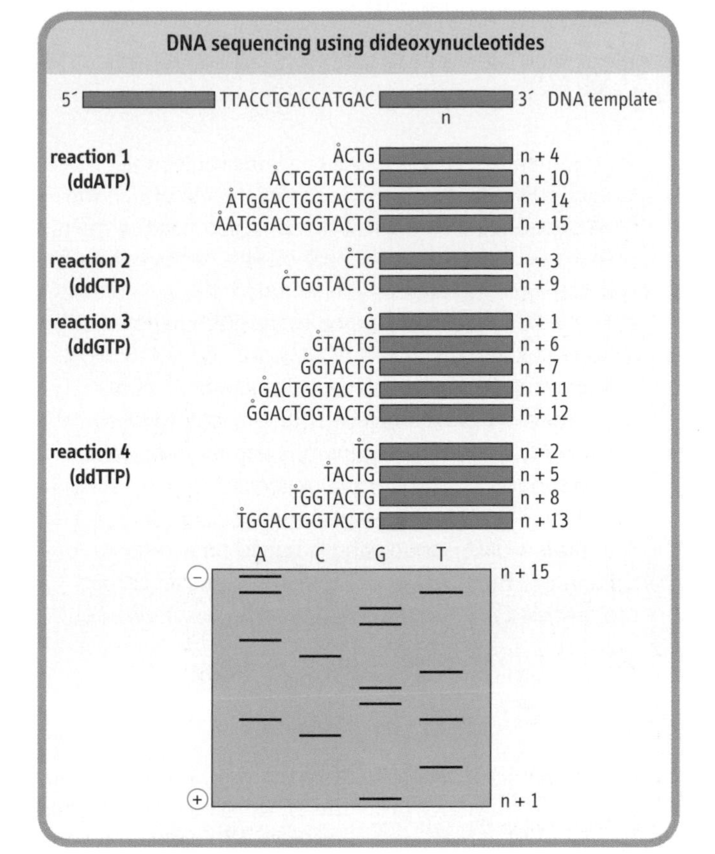

Fig. 33.25 DNA sequencing using dideoxynucleotides. A single reaction involving the DNA template and primer will proceed until a ddNTP is incorporated into the chain at random. This halts the reaction and that particular chain can grow no longer. Four reactions are performed in parallel, and each reaction is identical other than the ddNTP that is included, ddATP, ddCTP, ddGTP, and ddNTP. Each reaction generates hundreds of different reaction products with the same 5´end, the sequencing primer, but differing 3´ ends. The length of the 3´end will vary from 1 to over 300 bases depending on when a ddNTP was incorporated into the chain. In order to visualize the sequence a radiolabelled dNTP is added to all four reactions to facilitate autoradiography. Once completed, the four reactions are electrophoresed simultaneously on a polyacrylamide gel and the resultant autoradiograph is read and the sequence determined.

Summary

There are a myriad of DNA techniques available for the analysis of every aspect of DNA metabolism and function. However, regardless of the complexity of the methods, they all rely on the basic principles of hybridization and polymerization by appropriate enzymes. There is no doubt that the most significant advance in the study of DNA has been the discovery and automation of PCR. PCR is now used widely in all aspects of biomedical research, in particular, the study of human genetics. Whether in the diagnostic laboratory, or in the search for new disease genes, analysis has been totally transformed by PCR. There are now many hundreds of human genetic diseases whose diagnosis can

Automated gene sequencing and the Human Genome Project

The Human Genome Project is a large multicenter collaborative venture the aim of which is to obtain the DNA sequence of the entire human genome. The main objectives of this complex project were set out early in the history of the project and included the construction of a genetic map to facilitate genetic linkage studies, discovery of all human genes to allow further study of human genetic disease and, as a byproduct, the development of technology to allow nonexperts to use the technology and the automation of the sequencing process in general.

Automated DNA sequencing is based on a variation of the manual method but using fluorescent rather than radiolabeled products. Using four different fluorescent tags, one for each sequencing reaction, all four reactions can be run on the same lane of a sequencing gel rather than on four separate adjacent lanes. A scanner device can then read the gel and produce an automated sequence output. This enables four times as many reactions to be performed per gel and, in addition, the time required for developing autoradiographs is abolished, thus speeding up the process further. As a result, automated sequencers are highly utilized to determine sequence data and, in diagnostic genetics, are particularly useful in determining the size of trinucleotide expansions where PCR and autoradiography may yield unhelpful results, since knowing the size of the expansion may critically influence the outcome for the patient.

now be made by a single PCR-based method on a single blood sample. In many cases the PCR-based diagnosis has replaced complex biochemical tests or allowed a definitive diagnostic test to be available for the first time, e.g. in Huntington's disease.

As a result of the expansion of the use of PCR and the versatility of PCR-based techniques, the use of Southern blotting and RFLPs has become less. However, the basic principles of Southern blotting are at the center of all protocols that require radioactive labels to identify the reaction products and an understanding of the principle of probe–template hybridization is essential.

The sequencing of DNA has taken on a new dimension with the advent of the Human Genome Project. The drive to sequence the entire genome has resulted in a leap in technology and the mechanization of the previously labor-intensive task of DNA sequencing. Coupled to the replacement of radioactive labels with nonisotropic labels, the march in technology has highlighted that the primary goal of the Human Genome Project is now likely to be realized.

Recombinant DNA technology has replaced performing complex biochemical assays as the major biological scientific technology in most laboratories. The subject is continually evolving but, at the center, are several key principles that are modified to fit the application of the particular method. Understanding these key steps is the key to grasping the seemingly abstract nature of some of the newer techniques.

Further reading

Primrose SB. *Principles of Genome Analysis*. Oxford: Blackwell, 1995.
Strachan T, Read AP. *Human Molecular Genetics*. Oxford: BIOS, 1996.
Jamieson A. Dissecting hypertension: the role of the new genetics. *J Royal Coll Phys Lond* 1994; **28**: 512–518.
Davies JL, Kawaguchi Y, Bennett ST, *et al*. A genome-wide search for human type 1 diabetes susceptibility genes. *Nature* 1994; **371**:130–136.
Collins F, Galas D. A new five-year plan for the US human genome project. *Science* 1993; **262**:43–46.
Hubert C, Houot A, Corvol P, Soubrier F. Structure of the angiotensin I-converting enzyme gene: two alternate promoters correspond to evolutionary steps of a duplicated enzyme. *J Biol Chem* 1991; **266**:15377–15383.

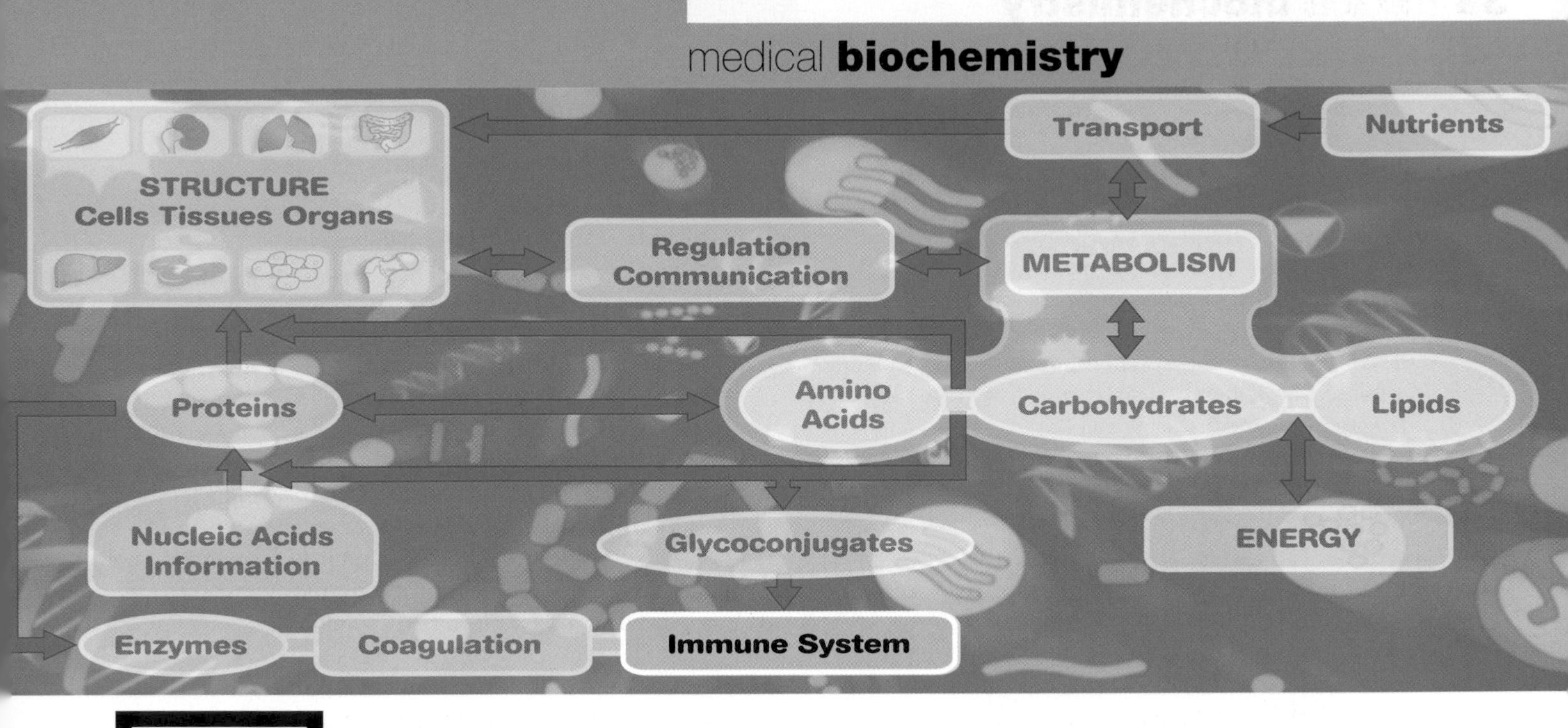

34 The Immune Response

Introduction

The immune response includes specific and nonspecific components

The immune system consists of the tissues, cells and molecules involved in the recognition, reaction with and elimination of foreign and nonself substances (antigens) that may disturb the homeostasis of the body. Immunity can be broken down and is categorized as being either non-specific and innate, or specific and adaptive.

The nonspecific immune response

Nonspecific immunity is the first line of defense

Nonspecific immunity uses physicochemical barriers, such as the skin, and their associated secreted products e.g. sweat. It also involves inflammatory response. Inflammation is the body's response to injury or tissue damage. Its purpose is to limit, and then repair, the damage brought about by the injurious agent. It involves the interaction of the microvasculature, circulating blood cells, other cell types in the tissues, and their secreted products. The endothelial cells that line the vessels play a major role in the vascular effects that include increased permeability and vasodilatation. These cellular effects are due to the activities of cells that usually circulate in the blood as well as some tissue-based cells (Fig. 34.1).

Inflammatory mediators also contribute to the immune response

All the cells involved can synthesize and secrete a wide variety of different types of soluble chemical substances termed inflammatory mediators. Since several inflammatory mediators can also be found circulating in the blood, they are also known as **humoral components** and include acute-phase reactants such as C-reactive protein (CRP) and components of the complement system.

The complement system

Complement is activated in a series of sequential steps

The complement system consists of a series of proteins that are normally present in the inactive form. The components are

Cells involved in inflammation

Cell type	Circulating cells	Tissue-based cells
polymorphonuclear leukocytes	neutrophil eosinophil basophil	
mononuclear phagocytes	monocytes lymphocytes platelets	mast cell macrophages endothelial cells

Fig. 34.1 Cells involved in inflammation.

sequentially activated, in a cascade, after contact with surfaces or particles that have a specific molecular composition or configuration. Please refer to Figure 34.2 throughout the following explanation of complement.

- The activation of the first component leads to the unlocking of serine protease activity;
- The activation of the next component in the sequence is achieved by it being cleaved into two unequally sized fragments by the first;
- The larger fragment then shows serine protease activity and can also attach to the activating surface, while the smaller fragment demonstrates its own distinctive biologic activity, which can include the facilitation of phagocytosis – a process termed opsonization, the attraction of cells – chemotaxis and stimulating degranulation of mast cells by anaphylatoxin activity;
- This series of activation events is then repeated with subsequent early components, thereby creating a cascade effect;
- By activating the next component, the large enzymatically active fragment of the previous component is inactivated. When components further down the sequence – the so-called 'late' components of the pathway – are activated, instead of demonstrating enzyme activity they also show the ability to coalesce with each other and subsequently activated components, to form a multimolecular complex that can breach the integrity of the surface containing the original activating molecular complex.

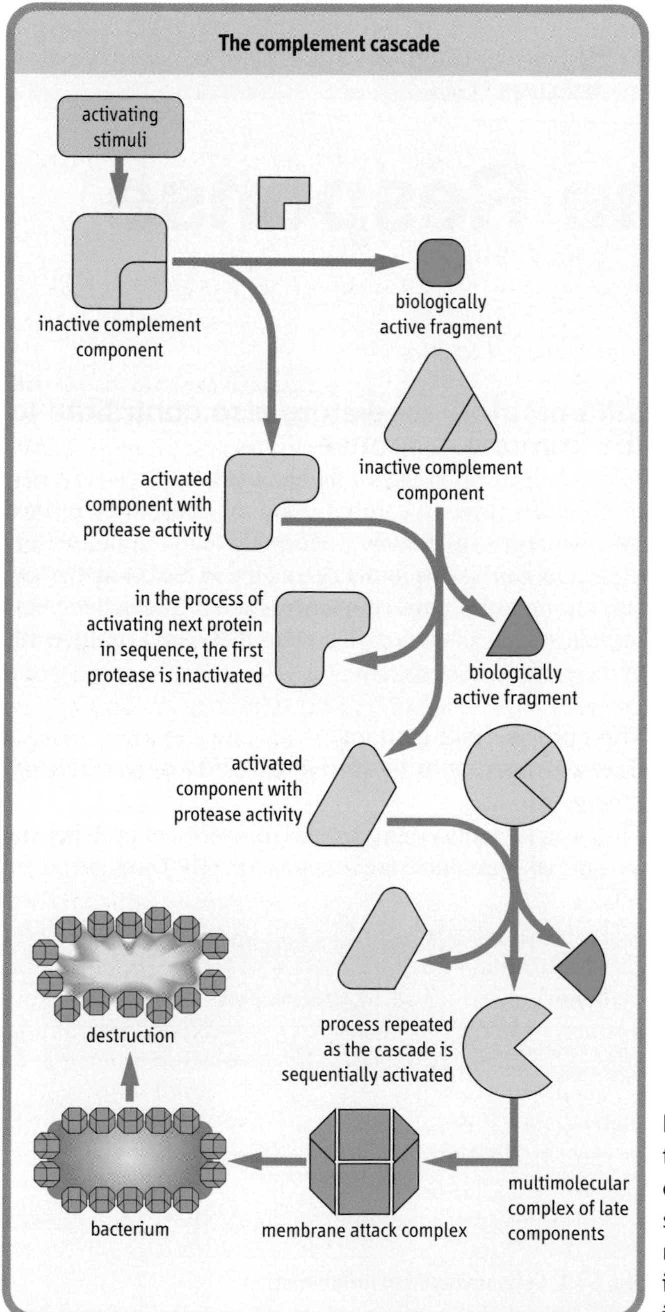

Fig. 34.2 The complement cascade. Activating stimuli bear surfaces that trigger complement activation and to which the activated component can attach itself. The late components of the cascade do not show enzymatic activity. Instead, they coalesce to form a polymeric macromolecule (the membrane attack complex), which can insert itself into the activating surface (the cell wall in the case of bacteria), breach its integrity and allow osmotic lysis.

Cytokines

Cytokines have several features and characteristics in common

Cytokines are soluble mediators of the inflammatory and immune response. They are produced by a variety of cells and tissues, are peptide or glycoprotein in nature, are active at concentrations between 10^{-9}–10^{-15} molar, and bring about their effects by interacting with receptors on the surfaces of their target cells. The majority act within short distances of the site of their production (paracrine) or on the cells that produced them (autocrine). A few, however, are capable of acting on cells distant to their site of production. They show significant overlap in their functions (redundancy), the cell types they act on may be multiple (pleiotropy) and have potential for interaction via the effects they mediate.

Cytokines may be classified into families or by their principal effect

Cytokines can be grouped into families:

- **interferons (IFNs):** while IFN-α and IFN-β have a role in protection against viral replication, IFN-γ plays a significant part in regulating the specific immune response (SIR);
- **interleukins (IL):** currently there are 18 interleukins recognized, all of which participate in regulating the cells involved with both the nonspecific immune response and the specific immune response.

- **chemokines:** these are a relatively recently described family of mediators that bring about chemokinesis – movement in response to chemical stimuli. Interest has increased dramatically in the receptors for these mediators since some appear to act as coreceptors for the infection of lymphocytes by human immunodeficiency virus.

Cytokines are, however, more practically grouped by their principal effects or roles, examples of which are shown below:

- **proinflammatory cytokines**: tumor necrosis factor (TNF)-α, IL-1, IL-6, IL-8 and other chemokines, IL-12 and IL-15;
- **antiinflammatory**: transforming growth factor (TGF)-β and IL-10;
- **immunostimulatory**: for cellular responses IL-2 and IFN-γ; for humoral, including allergic, responses, IL-4, IL-13, TGF-β and IL-10.

As will be seen later, the specific immune response is driven by direct cell interactions together with the effects of cytokines. Cells and soluble mediators of the nonspecific immune response participate both in initiating these responses and are being used by them in dealing with the antigen.

Arachidonic acid metabolites

The prostaglandins (PGs) and leukotrienes (LTs) form another group of inflammatory mediators. On each occasion, the nonspecific immune response follows the same sequence of events, which are nonspecific and stereotypical.

The specific immune response

The specificity of the immune response is achieved through antigen recognition

If the nonspecific defenses are unsuccessful, e.g. due to the persistence of the triggering agent, the specific immune response is brought into play. This system achieves specificity by using specialized receptors for any individual foreign or nonself substance antigen. The antigen and receptors can be considered to have a 'hand-in-glove' relationship, fitting together in a unique manner.

To show specificity for such an antigen therefore requires the ability to distinguish between self and nonself. This is achieved by thymic education and self-tolerance. When a specific immune response is initiated, there are likely to be relatively few components available that will be able to react specifically with any chosen foreign or nonself substance. There is a delay or lag period while these components increase to a level at which they can ensure elimination of the antigen or at least reduce it to such a level that is manageable by the nonspecific immune response. This delay can have disastrous consequences for the organism and its survival.

In addition to generating specificity, the specific immune response also employs a mechanism to remember the specific encounter so that if the same foreign or nonself substance is encountered again it can be dealt with more quickly and effectively. Thus, in comparison to the non-specific immune response, the specifc immune response shows specificity for the foreign or nonself substance.

As with nonspecific immune responses, specific immune responses are mediated by cellular and humoral elements. The cells responsible are the lymphocytes that develop and reside in the lymphoid tissues and there are two major types:

- **T cells:** responsible for cellular immunity,
- **B cells:** responsible for humoral immunity.

Lymphoid tissues and the circulation between them

The lymphoid tissues are classified as being primary or secondary.

Primary lymphoid tissues

Primary lymphoid tissue includes the fetal yolk sac, the embryonic liver in the embryo; the bone marrow and thymus

Lymphocytes originate in primary lymphoid tissue and undergo early development and differentiation. Their location alters during gestation: in the early embryo the main site of lymphocyte production is the fetal yolk sac but later in embryonic life this shifts to the liver. Later this shifts again to the sites at which it will remain throughout the rest of the individual's life – the bone marrow and the thymus. Within the bone marrow, pluripotential stem cells become commited to the lymphoid lineage. Those that remain within the bone marrow for the rest of their maturation become B cells. For T-cell development, the primitive lymphocytes have to travel to the thymus. It is still uncertain at which point in their development lymphocytes become commited to a T-cell lineage.

The thymus has epithelial and mesenchymal (supporting tissue) components in addition to the lymphoid contributions. It is a bilobed structure found in the anterior chest or mediastinum. At the microscopic level, there is an outer cortex and an inner medullary area within each lobule. Once in the thymus, T-cell development appears to progress as the immature T cells migrate from the cortex to the medulla. During this time, the early T cells interact with nonlymphoid elements, including thymic epithelia and dendritic cells. These cells are thought to be responsible for the important processes of positive and negative selection that take place as part of what is termed the 'thymic education of T cells' during which the T cells are assessed for their ability to discriminate between what is and is not self. During early development and differentiation, the T and B cells show changes in expression of surface molecules that will determine their future role and functional capabilities. The development of both early T and B cells in the primary lymphoid tissues is independent of extrinsic antigen stimulation.

Secondary lymphoid tissues

The secondary lymphoid tissues comprise lymph nodes, spleen and mucosa-associated lymphoid tissues (MALT)

These tissues are functionally organized for the interaction of lymphocytes with other cells and antigen, and it is at these sites that immune reactions are actually generated. Common to them all is a degree of compartmentalization, with specific areas for T cells and B cells and areas of overlap where they interact.

Within the lymph node, the T-cell area is the paracortex and the B-cell area the follicular areas of the medulla. Here follicular structures of two types can be found: the unstimulated primary follicle; and the stimulated secondary follicles, characterized by the presence of the germinal center. The spleen contains nonlymphoid tissue (the red pulp) as well as lymphoid areas – termed the 'white pulp', which surrounds the splenic arterioles and is called the periarteriolar lymphoid sheath. Within the white pulp, follicular B-cell areas are evident, and the T-cell areas lie between them in the interfollicular space. MALT comprises the lymphoid elements found adjacent to the mucosal surfaces lining the internal body surfaces; they are found at the entrance to the respiratory tract and gut, and include the tonsils and adenoids. Further down the digestive tract, unencapsulated aggregates of lymphoid cells are found, overlain by specialized areas of epithelium for sampling the antigenic environment as are Peyer's patches. Lymphocytes are also found singly throughout the mucosal epithelium.

Circulation between the secondary lymphoid tissues takes place via the lymphatics and blood vessels. The lymphoid tissues at which antigens localize after gaining access to the body is dependent upon the route of exposure. Entry via the circulation leads to localization in marginal zone lymphatics of the spleen. Entry via the gut and dome epithelia leads to localization in the MALT. With entry via the skin, as well as direct passage to the local lymph node via the lymphatics, antigen may be taken up by the epidermal Langerhans cells, which then pass via afferent lymphatics to the interdigitating cell area of the paracortex of the local lymph node. Entry via other sites/tissues leads to the local lymph node.

Lymphocyte circulation

The regulation of lymphocyte circulation depends on expression of markers of location and activation by the cells and tissues involved

The entry of antigen into the lymphoid tissues has several effects on lymphocyte traffic into and out of lymphoid tissues especially the lymph nodes. The effects occur in four phases:

- Within hours, increased blood flow leads to increased lymphocyte entry. In the case of some antigens this is also accompanied by decreased lymphocyte outflow;
- By 1–2 days, there is an increased outflow of small antigen-nonspecific lymphocytes, which represents the antigen-specific nonmigratory phase. Expansion of primed antigen-reactive B cells leads to the primary follicle in which the antigen-reactive B cells are located, to become a secondary follicle;
- At 4 days, activated lymphocytes that have undergone blast transformation (blast cells) emigrate from the lymph node and home with increased affinity to the high endothelial venule (HEV) at the site of their first antigen contact;
- In the final phase, blast cell output falls and memory cells are generated.

Adhesion molecules

Adhesion molecules mediate adhesion between cells

The whole process of the immune response is dependent on the expression of the molecules and ligands that mediate adhesion between cells and the connective tissue elements such as fibronectin, collagen, etc. These are termed 'adhesion molecules'. They are found on a wide variety of cell types but have important roles in interactions between the lymphocytes and the endothelia lining the small blood vessels. A major determinant of this expression is the prevailing cytokine environment.

Based on their structural relationships, adhesion molecules are grouped into the following families:

- integrins;
- immunoglobulin supergene family adhesion molecules;
- selectins.

Lymphocytes

Distinction between T and B cells is most easily made with reference to the cells' antigen receptor

The cells primarily involved in the immune response are the lymphocytes. In total, lymphocytes are present at between 1.5×10^9 and 3.5×10^9/L in the peripheral blood. Of these, approximately 50–70% are T cells and 10–20% are B cells. A third population termed 'natural killer' (NK) cells make up the remainder. NK cells are so-called because they demonstrate the ability to kill neoplastic cells without prior exposure or sensitization. Despite their different functions it is not possible to discern any morphologic features that can be used to distinguish lymphocyte populations. Instead, their identification is based on immunophenotypic or functional studies.

While each of the major populations of lymphocytes carry particular collections of markers that can be used to assist in assigning their lineage, the distinction between T and B cell is most easily made with reference to the cell's antigen receptor. In the B cell, this is a form of secretory immunoglobulin termed 'sIg' that is an integral part of the cell membrane that, on binding to its specific antigen, is responsible for bringing about the cell's activation and subsequent proliferation and differentiation. In addition to the sIg, B cells express several other markers, the best characterized and reliable of which include CD19, CD20 and the major histocompatibility complex (MHC) class II DR molecules. The antigen receptor for the T cell is termed the T-cell receptor (TCR) and it is complexed with CD3. Two other CD markers whose expression appears to be mutually exclusive are the CD4 and CD8 markers. The NK cells are currently identified by the expression of CD16 and CD56 in combinations. Use of these markers and flow cytometric technology confirms that T, B and NK cells are present in the proportions mentioned above.

Molecules involved in antigen recognition

T and B cells are involved in antigen recognition

The specificity of the immune response is made possible through the recognition of the antigen by specific receptors on T and B cells. The ability to recognize the enormous number of possible antigenic configurations is achieved by differences in amino acid sequence giving rise to differences in protein shape or conformation. As mentioned earlier, the antigen and its specific receptor have a hand-in-glove relationship. Both T- and B-cell antigen receptors show marked variability in the sequence of amino acids that actually come into contact with the antigenic element, while the rest of the molecules are relatively constant in their amino acid sequences. In terms of the actual receptors themselves, although T and B cells use a similar motif there are significant differences between them.

T-cell antigen receptors

The TCR is made up of two nonidentical polypeptide chains: these are TCR $\alpha\beta$ or TCR $\gamma\delta$.

The TCR is a heterodimer made up of two nonidentical polypeptide chains (Fig. 34.3). These chains are termed the TcR α, β, γ and δ chains, and the only functional combinations are $\alpha\beta$ and $\gamma\delta$. The structure is based on the immunoglobulin domain (see Chapter 3). Each chain comprises two domains – one constant and one variable amino acid sequence. The antigen binding site of the TCR is in the cleft formed by the adjoining single variable domains of the constituent alpha (Vα), beta (Vβ), gamma (Vγ), or delta (Vδ) chains. The effector functions of the constant domain in each of the antigen receptor chains is signal transduction. The N-terminal domains of the two chains come into close contact via the covalent bonds between them and noncovalent hydrophobic interactions between the opposing faces of the constant domains. The pocket created between the N-terminal domains forms the antigen-binding site.

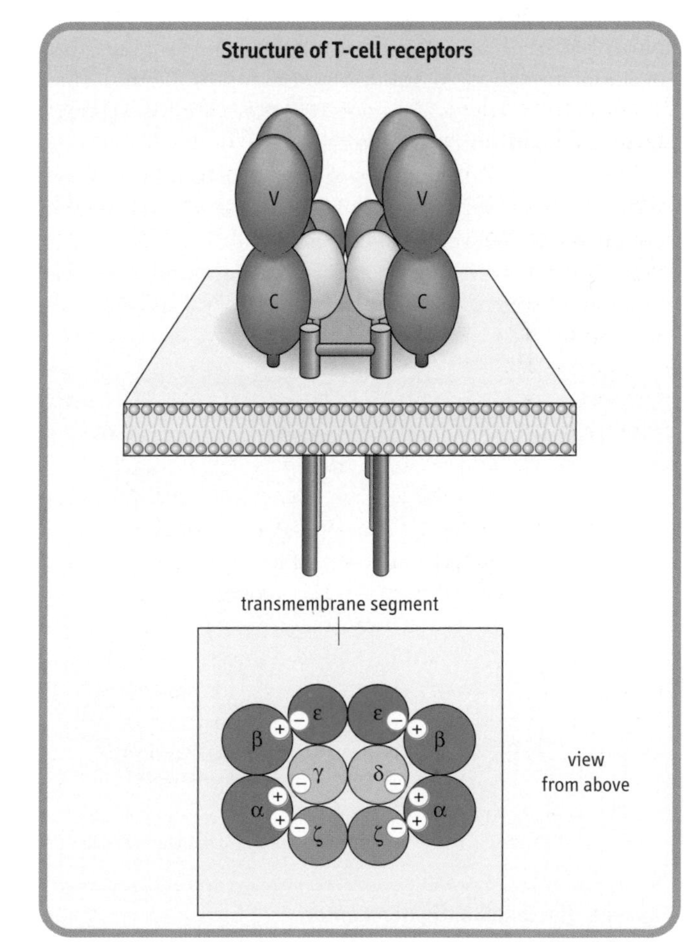

Fig. 34.3 Structure of the α/β, γ/δ T-cell receptors. V and C refer to the variable and constant sequence domains, respectively, of each of the α, β, γ and δ chains. Modified from Roitt IM *et al. Immunology*, 5th edn. London: Mosby, 1998.

The B-cell antigen receptor

The B-cell antigen receptor is merely the membrane form of the immunoglobulin found circulating in serum. Immunoglobulins are Y-shaped molecules made up of four polypeptide chains (Fig. 34.4) – a pair of heavy chains each of approximate molecular weight 50 kDa and a pair of light chains each of approximate molecular weight 23 kDa. The arms interact with antigen and have a structure based on immunoglobulin domains that demonstrate areas of constant and variable sequences of amino acids in both the heavy and the light chains. It is the variably sequenced amino (NH_2) terminal domains of both the heavy (VH – variable heavy) and the light (VL – variable light) chains that form a pocket that constitutes the antigen binding site – the fragment antigen binding (Fab) portion sites at the end of the arms. The remaining relatively constant amino acid sequence domains of the chains are termed constant heavy (CH) or constant light (CL) and form the stem that provides transduction effects.

Thus several similarities and differences are evident. While the TCR is described as a member of the immunoglobulin supergene family and has structural similarities with the B-cell antigen receptor, in terms of the immunoglobulin domain and fold, the terminology is merely the consequence of the timing of characterization of the antigen recognition molecules. Whereas there are two types of TCR there are five classes of immunoglobulin that can be found on the B cell, although three constitute consequences of antigen activation. Unlike B cells, no secreted version of the TCR is made. The most important difference between the receptors is the necessity for the TCR to interact with antigen against a background of self-molecules in the form of MHC or human leukocyte antigen (HLA) molecules.

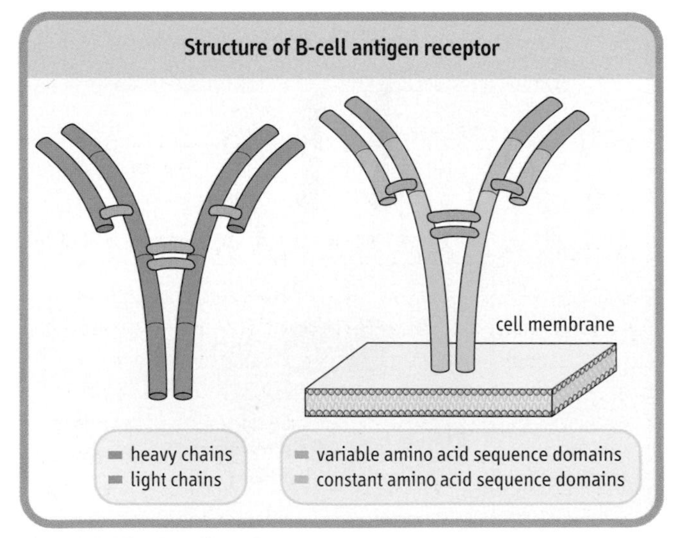

Fig. 34.4 The B-cell antigen receptor.

Immunoglobulin variable gene recombination events

The ability to generate molecules of variable amino acid sequence, using a basic template onto which some variation is superimposed provides the capability to generate as many different shapes as there are sequences. This is made possible by the organization of the genes that give rise to the T- and B-cell antigen receptors. They both undergo similar rearrangements and recombinations to generate a vast repertoire of antigen recognition units. This is illustrated by the idealized representation of the organization of a hypothetical immunoglobulin heavy-chain gene (see Fig. 34.7).

One gene from each of the variable domain gene segments combines with one gene from each of the other segments to produce a whole rearranged variable domain gene. This then associates with the gene coding for the constant domains of the particular heavy chain class being produced. In the case of a light chain, only two segments are involved in the production of a variable domain gene and this is then associated with the gene coding, whichever of the light chain types is going to be produced. The whole rearranged gene can be transcribed and subsequently translated. Thus multiple gene segments contribute to the formation of each individual variable domain gene and, together with the gene encoding, the constant domains complete light- and heavy-chain genes.

In the example shown in Figure 34.7, the number of genes available in each segment has been limited for simplicity. In reality, the actual numbers available at the three segments are 65, 10 and 6, giving rise to a huge number of possible permutations for the variable domain of just the heavy chain derived from one chromosome alone. Therefore there is an enormous potential for variation in the translated amino acid sequence and thus shape of the variable domains of both light and heavy chains, which, when combined, will be capable of fitting with the huge variety of antigenic conformations likely to be encountered. By concentrating diversity to the variable domain genes, the amount of DNA required is also conserved.

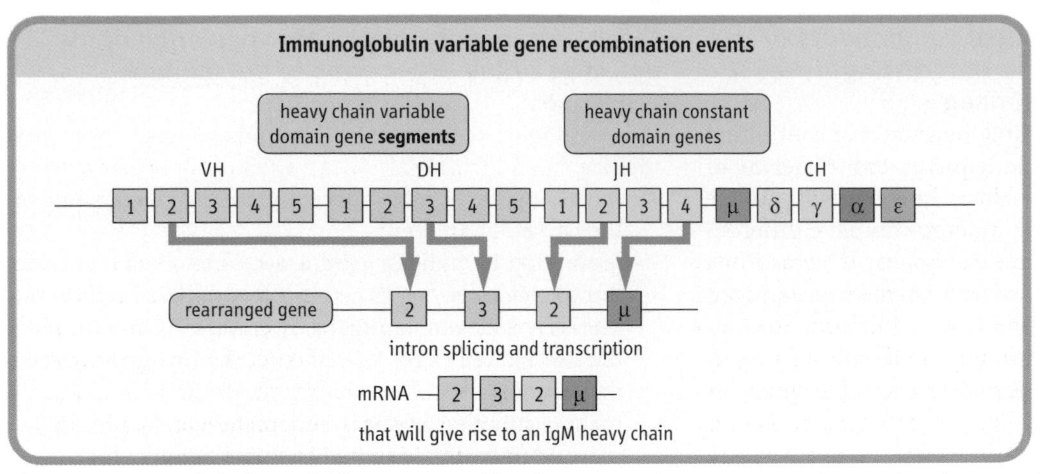

Fig. 34.5 Immunoglobulin variable gene recombination events. The apparent requirement for a huge number of genes required for antibody diversity is resolved by using a gene rearrangement strategy. The variable domains of both the heavy (H) and light (L) chains are encoded by a number of gene segments, two for the L chain (Variable L [VL] and Junctional L [JL]) and three for the H chain (Variable H [VH], Diversity H [DH], and Junctional H[JH]). The constant domains that make up the majority of the heavy chain molecules and half the light chain molecules are coded for by single gene segments, CH and CL, respectively.

The reaction with, response to, and elimination of antigen

On binding to the antigen, the cell divides and then differentiates into having an effector function or a memory function

On successful antigen binding, the cell is activated to enter the cell cycle leading to repeated cell division or proliferation, resulting in multiple generations of daughter cells or a clone. This is followed by differentiation in which the functional capabilities of the cell are altered, and this may be accompanied by a change in the cell's appearance. Differentiation can either lead to the development of an effector function or the generation of memory. The effect is to create clones of cells that have precisely the same specificity for antigen and the process is termed 'clonal selection'. Antigen therefore determines the specific lymphocyte that will undergo activation. This outcome is also assisted by the organization of the lymphoid tissues, together with the lymphoid circulation between them. This ensures that antigen is inspected by many lymphocytes and can select the cell that bears its specific and reciprocal antigen receptor for proliferation and differentiation.

Clonal selection (Fig. 34.6) ensures not only an adequate number of effector cells to deal with the threat at the time of initial stimulation but also a suitable number of part-primed memory cells that will be able to complete their activation more rapidly on subsequent exposure.

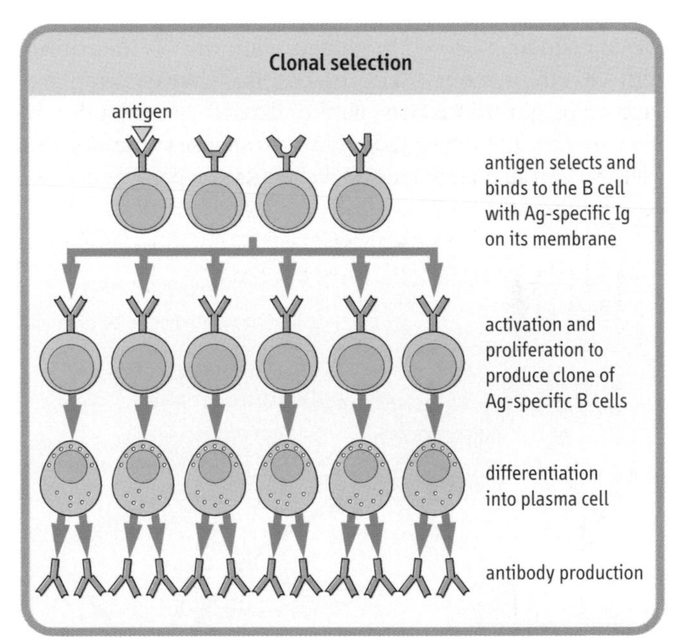

Fig. 34.6 Clonal selection in B cells. Antigen-specific (Ag-specific) secretory immunoglobulin (sIg) on the B-cell membrane has a reciprocal shape to the antigen. Antigen–immunoglobulin binding leads to activation and proliferation to produce a clone of antigen-specific B cells. Each member of the specifically activated clone then undergoes differentiation into a plasma cell, which produces large quantities of a single homogeneous immunoglobulin with identical specificity to the sIg that triggered the response in the first instance.

Memory is one of the critical components of the specific immune response that distinguishes it from the nonspecific response

How memory is generated is still the subject of continuing research but suggestions include the partial triggering of lymphocytes being an important mechanism. These partially triggered cells fail to undergo complete differentiation to the effector cell state and appear to return to a resting state. On subsequent exposure to the same antigen that partially triggered them previously, the time to completion of the cell cycle and through proliferation and activation is significantly shortened, thus leading to a quicker and more effective response.

The specific immune response is responsible for the elimination of antigen that is intracellular or integral to the cell surface

As with nonspecific immune responses, specific immune responses are mediated by cellular and humoral elements. The specific immune response has been classically described as having cellular and humoral arms, T cells being considered responsible for cellular immunity and B cells for humoral immunity. It is now established that the role of cellular, or T-cell mediated, immunity is in dealing with chronic intracellular infections, whether they are bacterial, e.g. tuberculosis, viral or caused by certain fungi and parasites, mediating the immune response to tumors, the rejection of transplanted tissue and contact hypersensitivity.

T cells are responsible for the regulation of the activities of the other arms of the immune response

The strategies to meet the demands of such antigens include:

- increasing the numbers of cells capable of responding to the particular antigen;
- preventing the antigen gaining access to unaffected cells;
- limiting damage, i.e. assisting already altered cells to rid themselves of such antigens by enhancing any capacity the affected cells have for self-cure, destroying those cells that cannot;
- making sure that any antigenic material released will be mopped up by other elements of the immune response, (both specific and nonspecific), clearing the damage and generating memory of the process.

All of these other functions can be achieved either by direct cell–cell contact or by the secretion of soluble mediators that interact with the relevant cells be they other T cells, B cells, NK cells, cells of the nonspecific immune response or cells of other tissues. The responsibilities of providing all these functions, with the exception of killing, resides in a subpopulation of T cells. Not surprisingly, as they have the role of assisting the rest of the immune response and host response, these cells have been termed **T-helper** cells (Fig. 34.7).

This population of T cells achieves its goals for the elimination of intracellular antigen by the secretion of

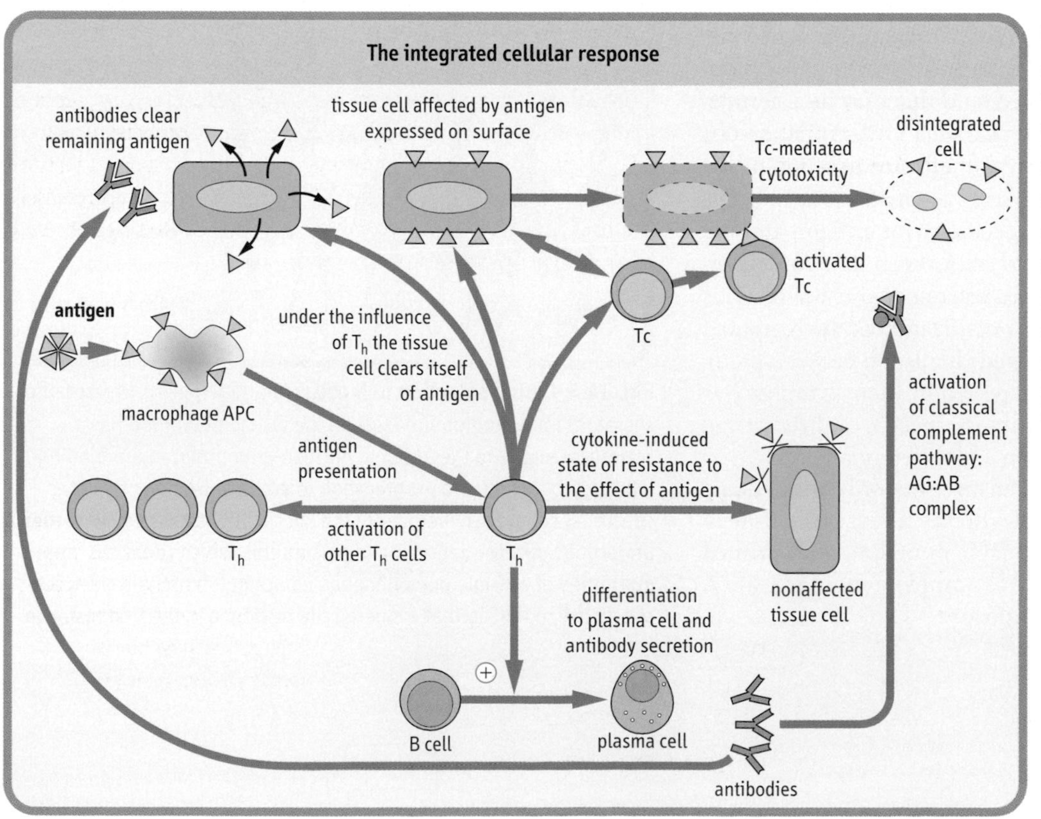

Fig. 34.7 The integrated cellular immune response. T-helper cells (T_h cells) are involved in many aspects of the cellular immune response. Cytokine and direct T_h cell-mediated activation of B cells leads to differentiation to plasma cells, which secrete antibody. AG:AB, antigen-antibody; APC, antigen presenting cell; Tc, cytotoxic T cell; T_h, T helper cell.

cytokines and direct interaction with the B cells responsible for antibody-mediated immunity, other T cells and macrophages. It can be identified phenotypically by its expression of CD4, together with its nonexpression of CD8. The term CD4 T cell is synonymous with T-helper cell.

T-helper cells can also differentiate into those that drive a proinflammatory response and those that drive an antiinflammatory response

It is now recognized that T-helper cells can be further divided into helper cells that drive the immune response towards a proinflammatory or antibody-based response and helper cells that drive the response towards an antagonistic anti-inflammatory or cellular-orientated response. This division is done on the basis of the profiles of secreted cytokines. Lymphocyte-derived cytokines:

- further augment the nonspecific immune response;
- determine the form of the specific immune response to be used;
- assist the relevant cellular components in mediating it.

Other cells are cytotoxic and kill cells that carry a 'recognizable' antigen, i.e. the antigen for which the killer cells carry a receptor

Another subpopulation within T cells are the cytotoxic T cells (Tc) responsible for specifically killing those cells that express on their surface the antigen for which the Tc carries the appropriate TCR. Like T-helper cells, cytotoxic T cells can be identified phenotypically and are $CD8^+$, $CD4^-$. These cells employ a mechanism of killing that is similar to that employed by the complement system (see earlier) but is dependent on direct cellular interaction between the Tc and the target cell. A polymeric macromolecule termed 'perforin' is inserted into the target cell wall, breaching its integrity. Unlike the complement system, reliance is not placed on this alone and enzymes (granzymes) from the Tc cell are introduced into the target cell, to hasten cell death.

Major histocompatibility complex

The MHC is how T cells see antigen against a background of self

For the TCR to interact with antigen there must be a background of self-molecules in the form of MHC or HLA molecules. For an immune response to be initiated, antigen cannot simply bind to the nearest T cell, but must be 'formally' presented to the immune system.

The MHC complex of genes is found on the short arm of chromosome 6 and is grouped into three regions termed class I, II and III with the same nomenclature being applied to the respective polypeptide products (Fig. 34.8). Class I and II molecules are directly involved with immune recognition and cellular interactions, whereas class III molecules are involved with the inflammatory response by virtue of coding for soluble mediators, including complement components and TNF.

Class I genes are organized into several loci the most important of which are those termed HLA-A, HLA-B and HLA-Cw. Alleles are transmitted and expressed in Mendelian codominant fashion. Owing to their closeness on the chromosome, they are inherited 'en-bloc' as parts of a haplotype and are expressed on the surface of all nucleated cells. The α-chains they encode have three domains, one of which is similar to those found in immunoglobulin molecules but the other two show significant differences. The α-chains combine with β2 microglobulin to give rise to a functional class I molecule (Fig. 34.9). Crystallographic studies indicate that the chains combine to produce a molecule which is designed to offer processed antigenic fragments to T cells in a cup-like structure, the sides of the molecule being formed by two α-helices and the bottom by a β-pleated sheet.

In contrast the class II subregion genes are organized into α- and β-loci, giving rise to α- and β-polypeptide chains, respectively. Both are of approximately the same molecular weight and combine to form a heterodimer with a tertiary structure similar to a class I molecule, with a peptide groove into which the processed antigenic fragment is inserted during intracellular antigen processing.

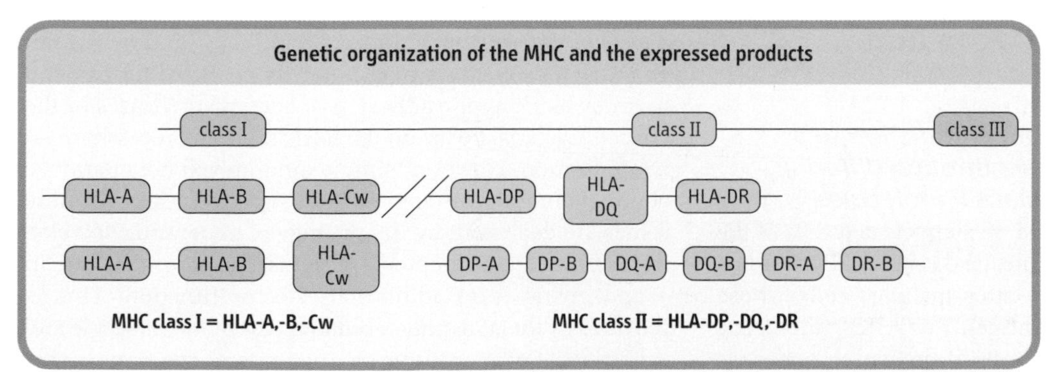

Fig. 34.8 Genetic organization of the MHC and expressed products. Genes of the MHC in humans are located on chromosome 6. The gene products are the human leukocyte antigens (HLA).

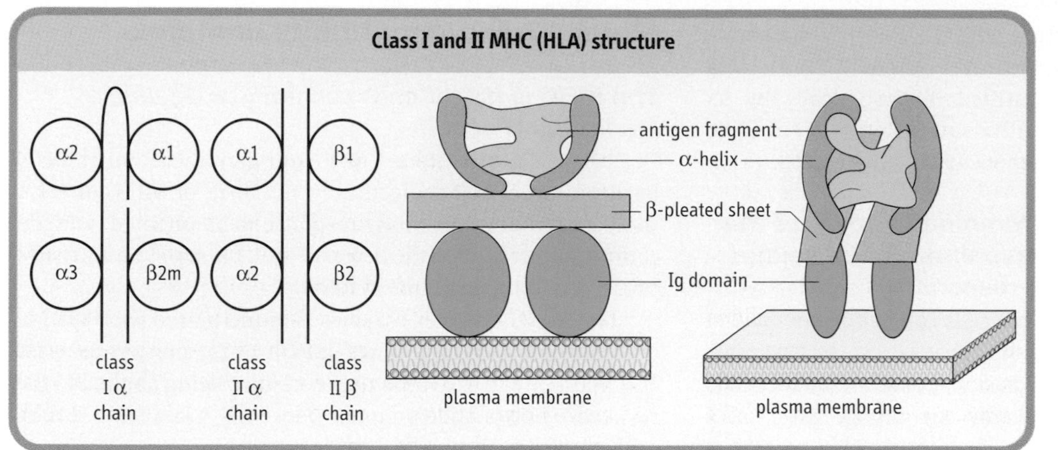

Fig. 34.9 Structures of MHC class I and II. On the left there are class I and class II MHC molecules. In class I molecules β2 microglobulin (β2m) provides the fourth domain in the class II molecule. On the right this is related to the types of protein conformation in the MHC molecules.

The HLA system is intensely polymorphic

Many allelic variants can be identified in each of the loci associated with antigen presentation. As there are six major loci each having between 10 and 60 alleles and as each parent passes on one set or haplotype on each chromosome, it is easy to appreciate that the likelihood of another individual in the same species having an identical set is remote.

The genetic polymorphism is predominantly in exon 2 of the β-chain of class II molecules and α-chain of class I molecules. These encode amino acids in the floor and sides of the peptide grooves and thus have a significant bearing on the capacity to bind differing antigenic fragments. The combination of the antigenic fragment with the edge of the peptide groove of the MHC molecule is then recognized by the relevant T cell. This is another mechanism to ensure appropriate stimulation. It limits the possibility of a single microorganism being able to subvert or circumvent the immune response of an entire species, as there is a huge potential for binding to at least some component of a microbe, together with a similarly high potential for a T cell to be able to recognize the antigen in combination with an element that says 'self + antigen'. At the same time, it prevents any single pathogen from infecting the entire population – either by using the system as a back door receptor into cells of every member of the species, or avoiding detection and elimination by mimicking one particular HLA allele.

MHC expression patterns and restriction

MHC molecules can restrict stimulation to a limited number of cell types through differing properties of class I and class II molecules

Class I molecules are expressed on all nucleated cells of the body whereas class II molecules are expressed on only a limited variety of cells. The latter includes cells whose primary role is to present antigen i.e. APCs, including macrophages and the cortical cells of the lymph nodes, as well as B cells and activated T cells. In this way, the MHC molecules can restrict stimulation to a limited number of cell types but at the same time, using similar mechanisms, permit appropriate targeting of any cell in the body. Without such a control, the immune system could be triggered repeatedly for no good reason.

As well as restricting input, the class I and II molecules also provide a differential mechanism for processing antigens that originate from within cells, e.g. viruses, and those that arise from the extracellular environment, e.g. bacterial antigens. The different class MHC molecules also lead such antigens through different pathways to interact with the immune system, in particular with the T cells, on the basis that each will be better dealt with by differing effector systems: class I leads to $CD8^+$ T-cytotoxic responses and class II leads to antibody mediated responses via $CD4^+$ T- helper pathways.

Humoral-mediated specific immune response

The humoral- or antibody-mediated specific immunity is directed at extracellular infection, especially bacteria and their products, the extracellular phase of viral infection and individual cell transplantation

Humorally mediated immune reactions require a soluble and cell-free antigen receptor to match the soluble and cell-free nature of the antigens. The act of antibody binding to antigen may go some way to counter its potential for causing harm by blocking the adhesion of bacteria or viruses or the effect of bacterial toxins on the hosts cells – a process termed neutralization. However, simple binding will not guarantee elimination of the antigen in most situations. Both of these issues are dealt with by the strategy of generating secreted forms of the B cell receptor – recognised as immunoglobulin – and giving it an additional effector function. This is provided by the nonantigen binding part of the molecule and it is capable of recruiting components of the nonspecific immune response. The secreted immunoglobulins have different effector functions in the nonantigen-binding fragment of the molecule. This is achieved by genetic recombinations of the heavy chain genes additional to those that have taken place already to generate diversity and specificity. Thus the original antigen specificity is preserved whilst effector functions can alter.

As with T cells, B-cell subsets are operative in the humoral immune responses. However, unlike the functional basis for the categorization of T-cell subsets, B-cell subsets are defined by the timing of their participation in the humoral response. The B1 subset is responsible for the production of antibody that appears during the earliest stages of development of the organism. This subset and the antibodies it produces persist as a carryover from this period. Although they possess a lesser degree of specificity than the antibodies produced later by the more mature B2 cells, they form a relatively effective first line of defense against a broad spectrum of antigens. As with T-cell subsets it is possible to distinguish these subsets on the basis of their phenotypic properties: B1 cells express CD5 in addition to CD19, and CD20 is found on the B2 population. It may be this population of B cells, when dysregulated, that is responsible for the production of autoantibodies – this is but one of many theories as to the pathogenesis of autoimmune reactions. As noted earlier, T cells, particularly Th

Recurrent infection in an immunocompromised patient

A 2-year-old child presents with a history of recurrent *Candida* and chest infections. Investigations reveal decreased neutrophils, IgG and IgA. Assessments of lymphocyte proliferative response show decreased expression of CD40 on T cells. A diagnosis of X-linked hyper IgM is made and intravenous immunoglobulin commenced.

Comment. T cell help is required for effective B-cell responses. Particular interactions are required for the switch of isotype from the IgM response that is typical of a primary antibody response, to the more mature IgG and or IgA isotypes seen in secondary antibody responses produced to subsequent challenges. CD40L on the T cell is required to interact with the CD40 on B cells to achieve this. In its absence, antibody production is limited to IgM and the affected individual is immunocompromised owing to the lack of the other important isotypes so critical to the integrity of the immune response. The problem of infection more typically associated with cellular problems suggests the T-cell defect has functional consequences for this arm of the immune response.

cells interact with B cells both directly and indirectly via cytokines. This is to such a degree that effective B cell responses are described as being T cell dependent.

Antibodies illustrate the capability of the immune system for diversity

The normal human immune system is apparently capable of producing a limitless number of antibodies with the ability of recognizing any and all nonself elements it comes into contact with. Antibodies therefore represent an excellent demonstration of the diversity of the immune response in terms of its ability to recognize antigen and of its methods for antigen elimination. The terms antibody, gammaglobulin and immunoglobulin are synonymous.

Structure of antibodies

Five classes of immunoglobulin are recognized: IgG, IgA, IgM, IgD and IgE, with subclasses being recognized for IgG (IgG1, 2, 3 and 4) and for IgA (1 and 2). When studied at the individual molecular level, no other proteins show such amino acid sequence variation between individual members of the same class or subclass. This is most evident in the NH_2-terminal domains of both heavy and light chains. As mentioned earlier, antibodies are capable of discriminating between the molecules that characterize the outer capsular coverings of differing bacterial species and may vary by a single amino acid or a monosaccharide residue. This is a consequence of the dimensions of the area recognized by the antibody molecule being 10 x 20 Å (10^{-10} m) and thus significantly influenced by the alteration in three-dimensional conformation brought about by the change of a single residue.

Antibody functions

Antibodies are also good examples of how function is intimately related to structure. They are Y-shaped molecules. The arms interact with antigen and the stem provides additional or effector functionality. This secondary or effector function endows the antibody with an ability to not only recognize the antigen but also to help eliminate it.

The activation of the complement system (component of nonspecific immune response) is one of the most important antibody effector functions of the specific immune response. This is achieved by using a collection of components termed the 'classical activation pathways', which comprise C1q, C1r, C1s, C4 and C-2. Sequential activation of these components leads to the activation of the pivotal and critically important C-3 component, which is an absolute requirement for full complement activation. Once this is achieved, the terminal membrane attack complex, which comprises the components C-5, C-6, C-7, C-8 and C-9, is activated. This complex eventually generates the polymeric ring structure that inserts into the cell membrane of bacteria and is responsible for cell lysis. This classical pathway is triggered by IgG or IgM that has been bound to its specific antigen (Fig. 34.10).

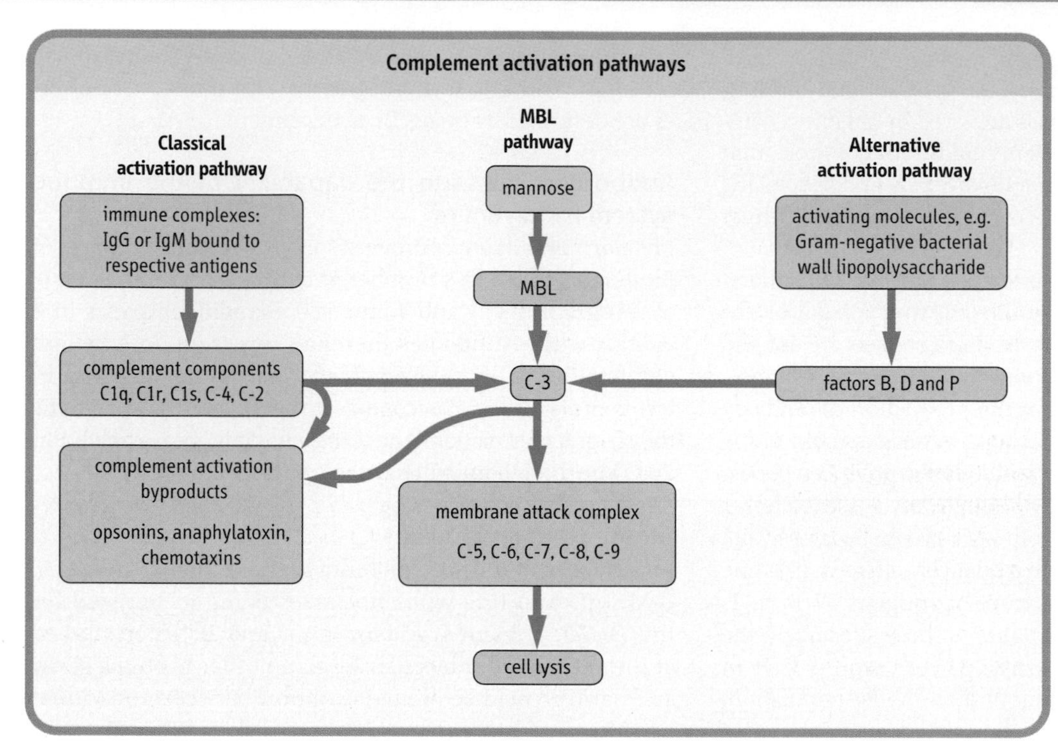

Fig. 34.10 Complement activation pathways. There are three possible pathways to complement activation. Only the classical pathway is triggered by the specific immune response. The mannose binding ligand (MBL) pathway and the alternative activation pathway are triggered directly by microbes and their products.

Two other pathways of activation exist, both constituting parts of the nonspecific immune response, and which are probably older in evolutionary terms. These are the alternative pathway, which can be activated by lipopolysaccharide such as is found in Gram-negative bacterial walls, and the mannose binding ligand (MBL) pathway, which can be activated by mannose, a carbohydrate that is found in the cell wall of fungi, bacteria and viruses. The effector functions of antibodies are summarized in Figure 34.11.

Effector functions of antibodies

Isotype	Functions	Concentration in peripheral blood
IgG	neutralization opsonization for neutrophils and macrophages Passive immunity for fetus via transplacental passage complement activation via classical pathway antibody-dependent cell-mediated cytotoxicity natural killer-cell killing of antibody bound cells achieved by FcRs (receptors for the Fc portion) major isotype used in a secondary antibody response	5.0–16.0g/L
IgA	defence of mucosal surfaces as the most predominant immunoglobulin produced by MALT neutralization	0.5–4.0 g/L
IgM	neutralization most effective classical complement pathway activator predominant isotype in primary antibody responses	0.5–2.0 g/L
IgD	possible role in signal transduction in the B cell B-cell maturation and significance of circulating IgD is undefined	0.03 g/L
IgE	major role is defence of mucosal surfaces against multicellular microorganisms	< 120 kU/L

Fig. 34.11 Effector functions of antibodies.

Bee sting allergy

A young man is brought into the emergency room in a state of shock and respiratory distress, with markedly swollen soft tissues. A companion tells the admitting medic that the patient had just been stung by a bee. Cardiorespiratory support is provided as well as epinephrine injection and rapid recovery ensues over the subsequent few hours.

Comment. While the physiologic role of the IgE response is considered to be protection against parasite infestation, this response is seen to be subverted in those who experience atopic diseases and anaphylaxis. The atopic diseases include allergic rhinitis, allergic conjunctivitis and asthma. The major fraction of IgE is bound via receptors to mast cells in the tissues. When antigen binds and crosslinks its specific IgE on the mast cells, it triggers the degranulation of the cells and release of preformed mediators (principally histamine), as well as the synthesis of other mediators, including arachidonic acid metabolites. The clinical effects depend on the location and/or extent of mast cell degranulation. When localized to one site, such as the nasal or bronchial mucosa, it usually gives rise to only localized reactions in the form of allergic rhinitis and asthma, respectively. If the degree of sensitization with the antigen-specific IgE and/or the antigenic burden is much greater, systemic degranulation can occur, with consequent anaphylactic shock. This is primarily due to the effects of the released mediators on vessels and vascular integrity. Significant vasodilation takes place, reducing the blood pressure. This is accompanied by large increases in vessel wall permeability, leading to shifts of fluid from the intravascular to the extracellular compartment and substantial swelling of any tissues with the capacity, which particularly affects the skin and other loose connective tissue such as those in the larynx. Smooth muscle spasm also occurs, leading to bronchoconstriction and contraction of the gut wall, with consequent marked respiratory difficulty and wheezing. These features are accompanied by increased secretory activity of seromucous glands in the respiratory and gastrointestinal tract as well as itching of the skin.

The cellular and molecular elements of the integrated immune response

Vaccination depends on the integrated specific immune response

Probably the single most beneficial application of the basic functioning of the immune response as we have now described it has been the effects of vaccination. This process illustrates well the combined interactions of the humoral and cellular arms of the specific immune response and the features that characterize it best – specificity and memory. On first encounter with antigen, the immune system and antigen interact to select those lymphocytes with the receptors specific for that antigen. These undergo activation, proliferation and differentiation into effector memory cells, a process that may take up to 14 days to complete (Fig. 34.12). The process of memory cell generation, however, now leaves a body of cells semiprimed for that specific antigen. On subsequent exposure, the response is more rapid in view of the partly activated state of the memory cells, and more effective as a consequence of a degree of maturation of the response, due to the differentiation of the lymphocytes that has already taken place.

With reference to antibody responses, the primary challenge elicits a predominantly IgM response. On subsequent challenge, the lymphocytes undergo further maturation and differentiation and isotype switching more rapidly, to produce a predominantly IgG response. This provides additional effector functions to that obtained with just IgM. It is this heightened and more specific response that can reduce both the severity and the duration of any damage sustained by the offending antigen; this may be critical to the survival of the organism.

Immunologic dysfunction

Autoimmunity is normally avoided by thymic education; a breakdown in the processes involved leads to autoimmune disease

While the immune system's activities are mostly beneficial, there are several situations in which they can have deleterious effects. These are best considered as aberrations of the quality, quantity or direction of the response (Fig. 34.13).

One particular aspect of these immunodysregulatory disorders, that of autoimmunity (self-reactivity), is avoided by the processes of thymic education. This is achieved via positive and negative selection and tolerance, which includes the processes of clonal deletion, clonal abortion and anergy. These terms refer to the processes whereby

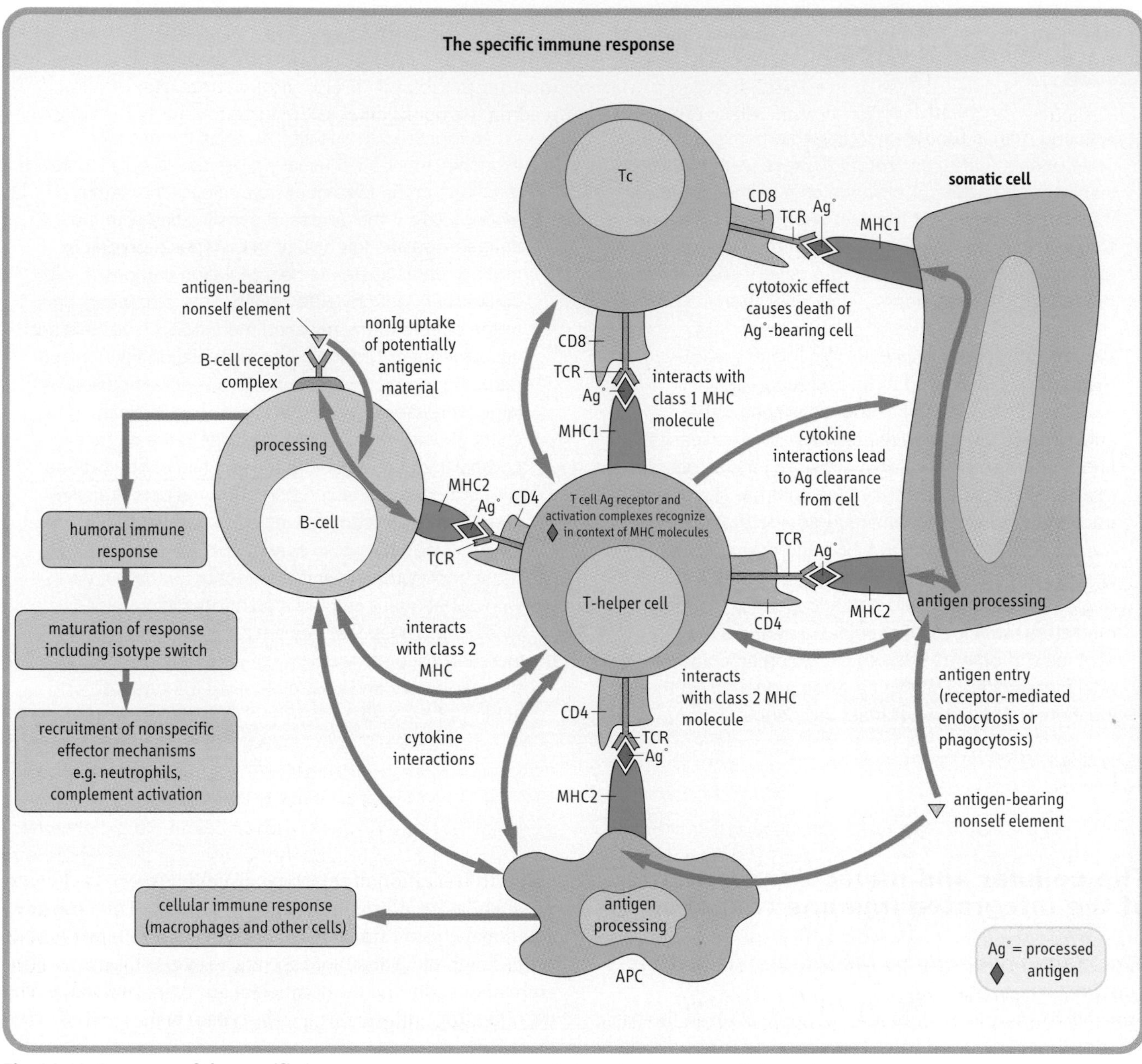

Fig. 34.12 Summary of the specific immune response. The interrelationships between cellular and humoral components of the specific immune response. APC, antigen presenting cell; Th, T-helper cell; Tc, cytotoxic T cell; MHC, major histocompatibility complex; TCR, T-cell receptor.

self-reactive clones are eliminated or rendered impotent. These mechanisms can be seen as a multilayered fail-safe strategy. Should these processes breakdown or be circumvented, the resulting state of self-reactivity and the inflammatory damage generated constitutes autoimmune disease. The form of disease is determined by the target antigen and the form of the immune response. At its simplest, those reactions against ubiquitous antigens lead to what are termed nonorgan-specific autoimmune disease, whereas those reactions to unique components of individual tissues, organs or systems are termed organ-specific diseases. The former are best exemplified by systemic lupus erythematosus (SLE), in which the apparent target antigens are components common to all nuclei. Damage is seen in several tissues, including the skin, joints, kidneys and nervous system. The latter are exemplified by autoimmune thyroiditis, where the apparent target is thyroid peroxidase. The use of the word 'apparent' should be noted here as, on deeper investigation, it becomes clear that these particular antigen systems, while acting as markers or indicators of autoimmune disease, are not pathogenically responsible for the damage being inflicted. The methods used to detect such autoantibodies include indirect and direct immunofluorescence, particle agglutination, electroimmunodiffusion and enzyme immunoassay.

These diseases and the others mentioned in Figure 34.13 are the focus of the discipline of clinical immunology and immunopathology. More information can be found by reading the books cited in the further reading list below.

Disorders of immune regulation

	Polyclonal response	**Monoclonal response**	
Antigen source	**Exogenous**	**Endogenous**	
decreased response	primary and secondary immunodeficiency		
increased response	tuberculosis, leprosy, immune complex disease		
inappropriate response	allergic disease – IgE		lymphoproliferative disease
increased and inappropriate response	allergic disease – EAA	autoimmune disease	

Fig. 34.13 Disorders of immune regulation.

Summary

Integrated immune response to non or latered-self elements, antigen(s), are made up of a number of components. Some of these show unique specificity for the particular stimulating antigen(s) and comprise the specific or adaptive immune response, whilst others do not and comprise the nonspecific or innate immune response.

The nonspecific response represents the first line response and can be considered cruder and more primitive. The cells and soluble mediators involved are primarily those associated with the processes of acute inflammation and endothelial cell activation. The specific response is more refined and usually invoked only in the face of either failure/continued stimulation of the nonspecific response.

The cells responsible for the specific immune response are the lymphocytes; T, B and NK. The specificity they show for the inciting antigen is achieved via the use of specific antigen receptors, TcRs and BcRs, expressed on their cell surface.

T cells recognise processed antigen via the TcR interacting with antigen presented by MHC bearing cells, leading to the secretion of additional cytokines and the generation of so-called effector functions such as T cell help and T cell mediated by cytotoxicity brought about by the T-helper and T-cytotoxic subsets respectively. Historically, T cell responses have been termed the Cellular Immune Response.

B cells recognise native antigen leading to the secretion of soluble forms of the individual B cell antigen receptors which we recognise as antibodies. Historically, B cells and their antibody products have been termed the Humoral Immune Response. Both T and B cells and their products are able to recruit and utilize components of the nonspecific response in a more effective and targeted manner with the aim of eliminating or eradicating the antigen.

In addition to demonstrating specificity, the specific immune reponse also demonstrates another critically important characteristic not seen with the nonspecific response – memory for its encounter with all types of antigen. The benefit being, on subsequent contact with the same antigen, it can be eliminated more quickly and effectively and with less tissue damage than on the previous occasion achieving homeostasis even more rapidly.

Further reading

Roitt IM. *Immunology*, 5th edn. London: Mosby, 1998.

Janeway C, Traver P, Hunt S, Walport M. *Immunobiology*, 3rd edn. London: Garland, 1997.

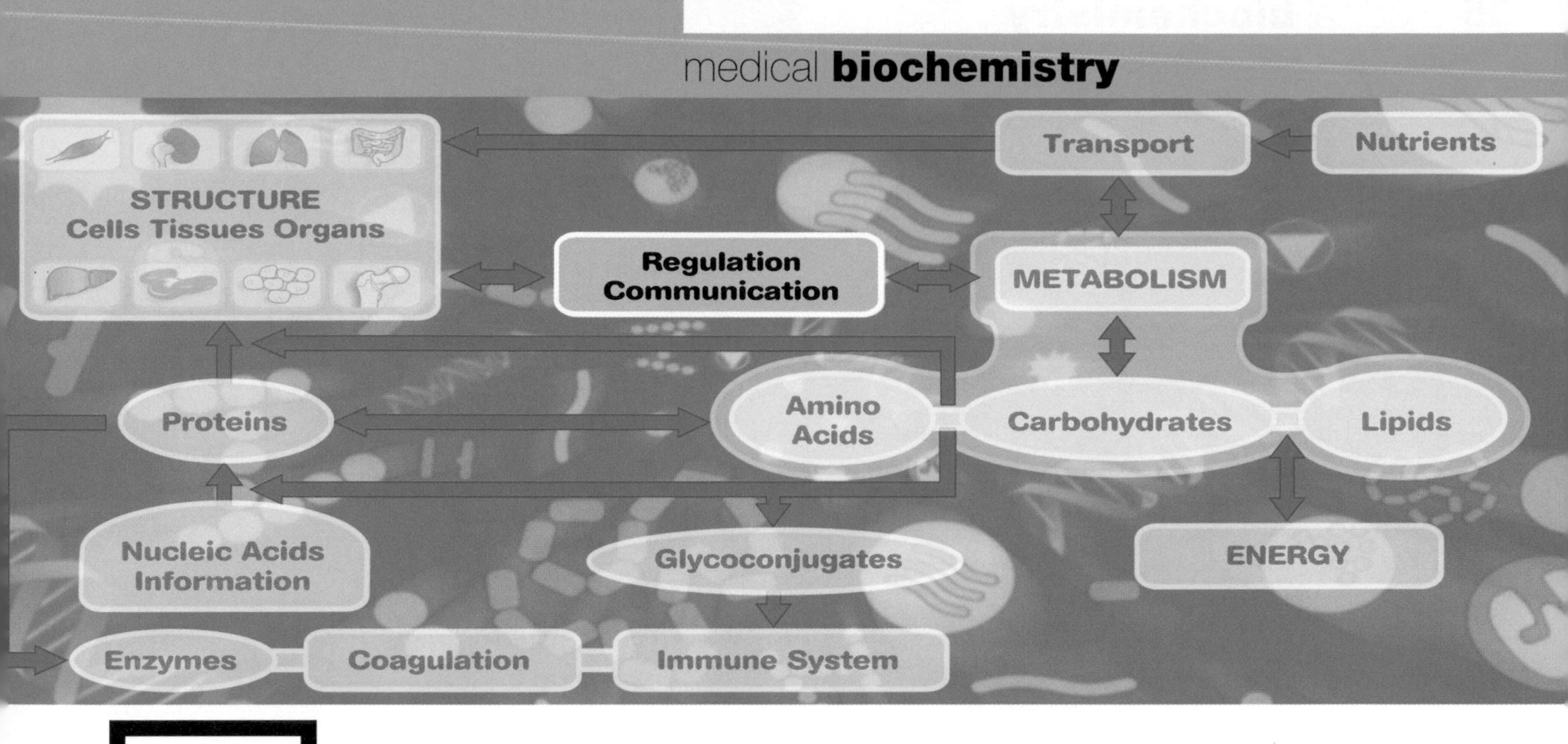

35 Biochemical Endocrinology

The endocrine system

The endocrine system provides communication between cells, tissues and organs

In higher species, such as humans, sophisticated control mechanisms are required to ensure optimal communication between cells, tissues and organs. Along with the nervous system, the endocrine system provides this communication, being responsible for the regulation of a wide range of functions including growth, development, reproduction, homeostasis and the response to external stimuli and stress. Failures in this communication channel are common and many diseases of the endocrine system exist.

Several organs secrete hormones

Several organs of the body are capable of endocrine function – the secretion of biologically active compounds called hormones, which are transported via the bloodstream to other tissues or organs where they exert a biological effect. Hormones act by binding to specific receptors, either on the cell surface or within the target cell (see Chapter 36). It is this hormone–receptor interaction that triggers one or more of a wide range of biological effects.

Negative feedback regulation is important for homeostatic control

In order to ensure tight homeostatic control, the final part of the endocrine cycle is regulation of the further secretion of the hormone or its receptor binding capacity (Fig. 35.1).

Several hormones may control one process or one hormone may control several processes

While it would be convenient to think of the endocrine system as being compartmentalized so that one hormone has control over one process, this is rarely the case. For example, at least four different hormones are involved in the regulation of plasma glucose concentration (see Chapter 20). Conversely, single hormones such as testosterone act to influence a range of metabolic processes.

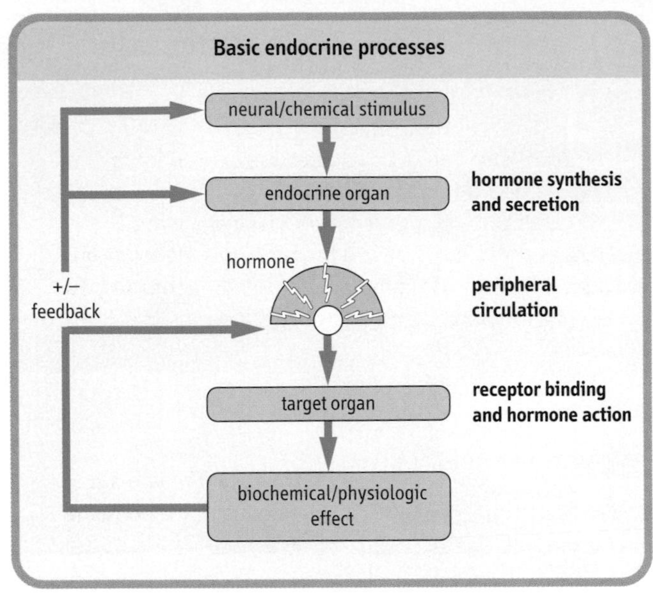

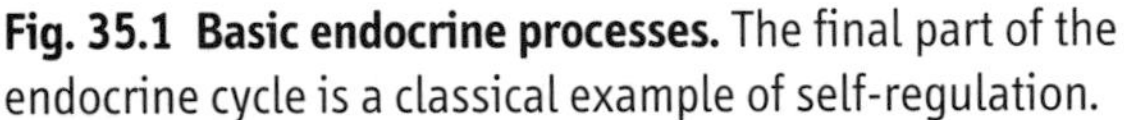

Fig. 35.1 Basic endocrine processes. The final part of the endocrine cycle is a classical example of self-regulation.

Chemical derivation of hormones

Derived from amino acids	
single amino acid derivatives	catecholamines, serotonin
dipeptides	thyroxine, triiodothyronine
small peptides	TRH, AVP, somatostatin
intermediate-size peptides	insulin, parathyroid hormone
complex polypeptides	gonadotrophins, TSH
Derived from lipid precursors	
cholesterol derivatives	cortisol, testosterone, estradiol vitamin D
fatty acid derivatives	prostaglandins

Fig. 35.2 The chemical derivation of hormones. AVP, arginine vasopressin; TRH, thyrotropin releasing hormone; TSH, thyroid stimulating hormone.

Types of hormone

Hormones may be simple or complex, and they may be derived from proteins or simple lipids

Many different types of molecule function as hormones (Fig. 35.2). At the simplest level, modified amino acids (epinephrine) may act as hormones. The assembly of amino acids during protein synthesis produces hormones varying in size from a dipeptide (e.g. thyroxine) up to complex glycoproteins (e.g. luteinizing hormone, LH). In all cases, however, these hormones are synthesized as larger peptides or proteins, known as prohormones, which require specific enzymatic cleavage to release the active hormone from the endocrine gland. Yet other hormones are derived by modification of simple lipids such as cholesterol or fatty acids. Some of these molecules require further metabolism within their target cell before they can exert their full biological effect (e.g. testosterone).

Circulation, action and inactivation of hormones

Hormones are transported to their site of action usually by carrier proteins, where they exert their action and are inactivated by further metabolism

Within the circulation many hormones are transported bound to carrier proteins. For example, thyroxine and cortisol are transported on specific plasma binding globulins (thyroid binding globulin – TBG, and cortisol binding globulins – CBG, respectively), a process that provides a further means of regulating the delivery and biological half-life of the hormone. At the cellular level, hormones exert their actions by a wide range of biochemical mechanisms, as discussed in Chapter 36. Hormone inactivation usually occurs by further metabolism followed by excretion of the metabolites. The rate of clearance of different hormones varies enormously from a few minutes (insulin), to hours (steroids), to days (thyroxine).

Hormone immunoassay – a billion dollar business

Immunoassay is the most widely applied technique for detecting and quantitating hormones in biological samples. One or more antibodies are produced, and these bind to the hormone antigen. This antibody binding may be highly specific for any one hormone and of high affinity, in this respect the hormone-antibody binding is analogous to the hormone-receptor binding during hormone action.

Antibodies are produced by utilizing the mammalian immune system. In the first immunoassays, the antibodies were produced in the serum of animal species (e.g. rabbit or sheep) that had been immunized with human hormone preparations. Such antisera contained a range of different antibodies (polyclonal) capable of binding to different sites on the hormone antigen. Today most immunoassays employ monoclonal antibodies, which are produced by fusion of spleen cells from an immunized mouse with a mouse myeloma cell line. Hybridoma cell lines result, and these may be cloned to produce a cell line that secretes a single antibody species on an indefinite basis.

Many commercial producers have designed their own particular methods of quantitating the hormone–antibody interaction in order to get into the billion-dollar worldwide market. One of the most widely used formats is the two-site immunometric assay outlined in Figure 35.3. This employs two antibodies binding to different epitopes on the hormone, one of which is modified to be capable of generating an optical signal.

Hormone measurement – a matter of timing

Since hormones are involved with homeostatic mechanisms, their secretion is not constant. Most hormones are released in bursts ranging from single spikes (e.g. growth hormone, GH) to sustained release following a specific stimulus (e.g. insulin). Hormone secretion often also conforms to strict biological rhythms which may be at intervals of 1-2 hours (e.g. LH), 24 hours (e.g. cortisol) or 28 days (e.g. progesterone). For these reasons, it is always necessary to standardize the time and status of the patient before sampling and it is often necessary to perform multiple sampling or dynamic testing.

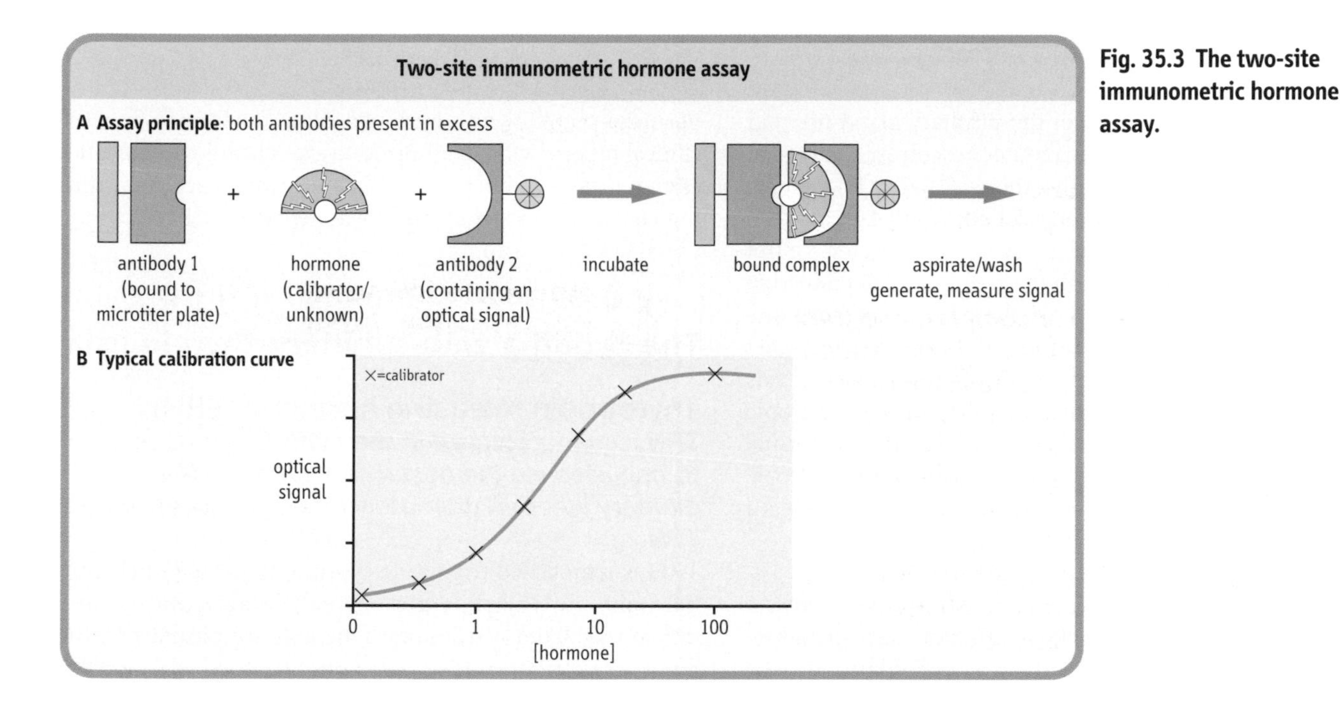

Fig. 35.3 The two-site immunometric hormone assay.

The hypothalamo–pituitary regulatory system

Hormones of the posterior pituitary gland are distinct from those of the anterior pituitary

The pituitary gland is an oval organ weighing about 0.6 g and is encased in a bony cavity of the skull (sella turcica) below the brain. It communicates with the hypothalamus of the brain via the pituitary stalk which contains a complex array of axons and the portal blood vessels (Fig. 35.4). The pituitary gland is clearly demarcated into two lobes. The anterior lobe (adenohypophysis) accounts for 80% of the gland and may be seen as a complex endocrine target organ for hormones released from hypothalamic nuclei and transported via the median eminence and the portal circulation. The hormones of the posterior pituitary (neurohypophysis) are synthesized and packaged in the supraoptic and paraventricular nuclei and transported as granules along axons.

Hormones of the posterior pituitary gland

The posterior pituitary gland secretes two hormones into the circulation. Oxytocin is a small peptide hormone with biological activities linked to parturition and lactation. Arginine vasopressin (AVP), also known as antidiuretic hormone (ADH), is a nine amino acid cyclic peptide with functions linked to the control of water metabolism, as described in Chapter 21.

Hormones of the hypothalamo–anterior pituitary regulatory system

There are five separate endocrine axes within this system (Fig. 35.5). Three of these are part of a complex three-level endocrine

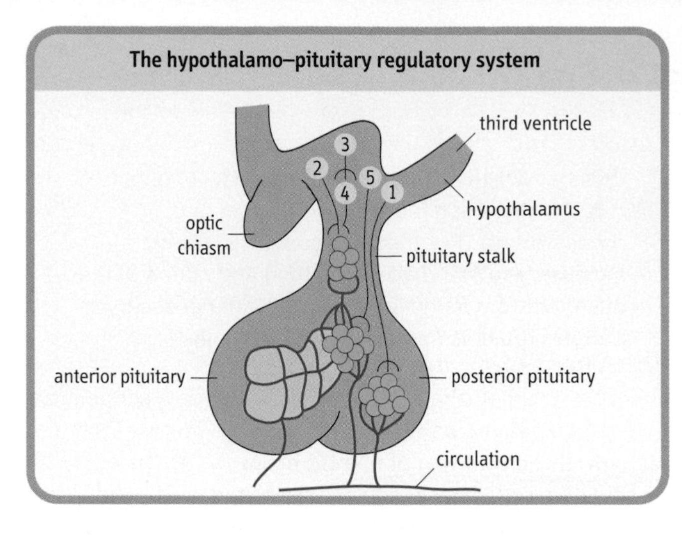

Fig. 35.4 The hypothalamo–anterior pituitary regulatory system. Hormones of the posterior pituitary are synthesized and packaged in the supraoptic and paraventricular nuclei (1) and transported along axons for storage prior to release into the circulation. The anterior pituitary releasing, or release-inhibiting hormones are synthesized in various hypothalamic nuclei (2–5) and transported to the median eminence. From there they travel to the anterior pituitary via a portal venous system.

axis in which the hormones of the pituitary gland (thyroid stimulating hormone (TSH), adrenocorticotropic hormone (ACTH), follicle-stimulating hormone (FSH)/LH) may be regarded solely as tropic hormones for other target organs (e.g. thyroid, adrenal, gonads). In these axes, sophisticated control is exercised via a cascade which ensures that each endocrine organ in the axis amplifies both the amount of hormone secreted and the biological half-life of its hormone product compared with the previous organ. The fourth endocrine axis is a hybrid, since growth hormone (GH) is both a tropic hormone and has actions in its own right. The fifth endocrine axis results in the secretion of prolactin, which is not a tropic hormone and has limited physiologic function but which is a major contributor to endocrine disease.

With one exception, there are distinct endocrine disorders associated with both deficient and excess hormone secretion by the anterior pituitary, although their incidence varies (Fig. 35.6). Tumors of the pituitary gland may be hormone secreting or nonfunctional. In the former case, the clinical picture will be of hormone excess while, in the latter case, it may be of hormone deficiency since a space-occupying lesion causes atrophy of hormone-secreting cells or the portal system of supply.

The hypothalamo–pituitary–thyroid axis

Thyrotropin releasing hormone (TRH)

TRH is manufactured in the hypothalamus and transported via the portal circulation to the pituitary where it ultimately leads to exocytosis of TSH

TRH is a modified tripeptide synthesized as a 26 kDa prohormone by peptidergic hypothalamic nuclei and transported, after activation, to the anterior pituitary by the portal circulation. It is secreted in pulsatile fashion. TRH

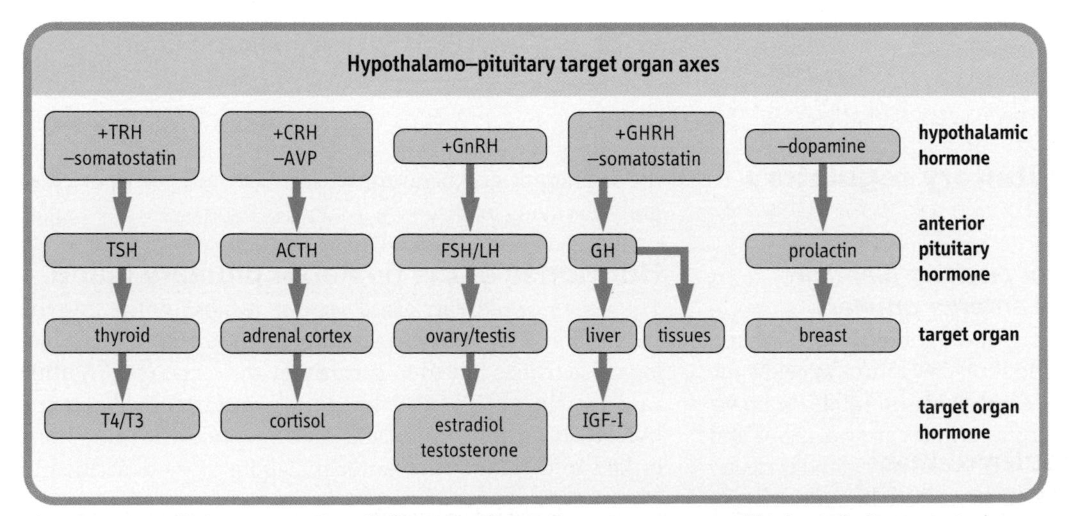

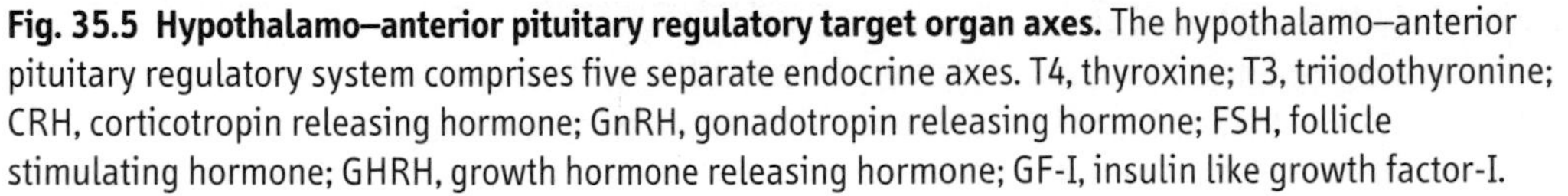
Fig. 35.5 Hypothalamo–anterior pituitary regulatory target organ axes. The hypothalamo–anterior pituitary regulatory system comprises five separate endocrine axes. T4, thyroxine; T3, triiodothyronine; CRH, corticotropin releasing hormone; GnRH, gonadotropin releasing hormone; FSH, follicle stimulating hormone; GHRH, growth hormone releasing hormone; GF-I, insulin like growth factor-I.

stimulates TSH synthesis and secretion by binding to receptors on the pituitary thyrotroph cell membranes that are linked to phospholipase C. The resulting phosphoinositides stimulate the release of calcium from intracellular storage sites and so lead to exocytosis of TSH. More chronic actions of TRH include stimulation of TSH subunit biosynthesis and TSH glycosylation. The number of TRH receptors on the thyrotrophs is downregulated both by the concentration of TRH itself and the thyroid hormones.

Thyroid stimulating hormone

TSH (also known as thyrotropin) is a 28 kDa glycoprotein synthesized by the pituitary thyrotroph. It consists of two noncovalently linked subunits and contains about 15% carbohydrate. The α-chain is identical to that found in other pituitary glycoprotein hormones and so the specificity is determined by the β-chain and the three-dimensional configuration. The synthesis of each subunit is directed by separate messenger R ribonucleic acids (mRNAs) encoded by separate genes on different chromosomes. The carbohydrate side chains are complex mixtures of acetylated sugars, sialic acid and sulfate.

TSH is secreted in a pattern that is both pulsatile and circadian. The plasma half-life of TSH is about 65 minutes. The normal plasma reference concentration is approximately 0.4–4.0 mU/L although logarithmic transformation is required to give it a Gaussian distribution (Fig. 35.7).

TSH acts on the thyroid gland and influences virtually every aspect of thyroid hormone biosynthesis and secretion

TSH acts via a specific membrane receptor on the target cell of the thyroid gland. The receptor is a 85 kDa glycoprotein with an extracellular domain of approximately 400 amino acids, seven transmembrane domains of about 250 amino acids and an intracytoplasmic domain of about 100 amino acids. The binding of TSH to the receptor activates adenylate cyclase, increasing cyclic adenosine monophosphate (cAMP) formation. Thereafter, several protein kinases are activated and these influence virtually every aspect of thyroid hormone biosynthesis and secretion. Thus, TSH affects iodide transport, iodothyronine formation, thyroglobulin proteolysis and thyroxine deiodination. In addition, TSH regulates thyroid cellular function and stimulates thyroid growth.

Negative feedback by thyroid hormones occurs at both hypothalamic and pituitary levels. At the pituitary level, thyroxine (T_4) and tri-iodothyronine (T_3) inhibit TSH secretion by decreasing both the biosynthesis and release of TSH through regulation of gene transcription and TSH glycosylation.

Thyroxine and tri-iodothyronine

T_4 is produced exclusively in the thyroid gland and is more abundant than T_3, which is the biologically active form

T_4 (also known as tetra-iodothyronine) and T_3 are structurally simple molecules, being iodinated thyronines produced by the condensation of two tyrosine molecules (Fig. 35.8). The biosynthesis of T4 and T3 occurs within thyroglobulin, a 660 kDa molecule that accounts for about 75% of the protein content of the thyroid gland. Iodination of thyroglobulin occurs late on in its synthesis and packaging, as it transfers from the Golgi apparatus to the follicular lumen.

Thyroid hormone bioactivity is regulated by controling the conversion of T_4 into T_3 by deiodination; this is mediated by at least two deiodinases enzymes

T_4 is quantitatively the most important thyroid hormone and it is produced exclusively by the thyroid gland. T_3 is the biologically active form of thyroid hormone produced by 5′-deiodination of T_4. This process may occur in the thyroid gland, in target tissues or in other peripheral tissues. Removal of an iodine from the 5- rather than the 5′ position results in reverse T_3, which is biologically inactive. Thus, control of the deiodination of T_4 is one method of controling thyroid hormone bioactivity and there are at least two distinct 5′-deiodinases (see Fig. 35.8).

The secretion of T_4 and T_3 relies on the enzymatic hydrolysis of thyroglobulin, which is presented in the form of colloid droplets. During hydrolysis and subsequent deiodination the released iodide is conserved and reutilized. About 100 µg of thyroglobulin is released unchanged from the

Clinical conditions associated with pituitary hormone disorders

Hormone	Deficiency	Excess
TSH	hypothyroidism	thyrotoxicosis
ACTH	hypoadrenalism	Cushing's disease
FSH/LH	hypogonadism	precocious puberty
GH	short stature	gigantism/acromegaly
prolactin	none	galactorrhea/infertility

Fig. 35.6 Clinical conditions associated with pituitary hormone disorders.

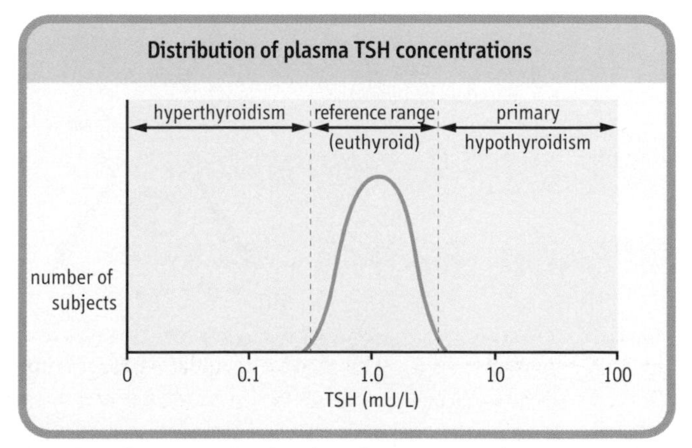

Fig. 35.7 Plasma TSH levels. The reference range is 0.4–4.0 mU/L.

thyroid gland, a tiny fraction of the 25 mg that must be hydrolyzed to meet the daily production of T_4.

Approximately 80% of T_4 is converted into equal amounts of T_3 and reverse T_3 (rT_3)

The daily production rate of T_4 is approximately 110 nmol (90 μg). The extrathyroidal pool of T_4 is about 1100 nmol, much of which is transported in the plasma bound to a variety of specific and nonspecific binding proteins. Approximately 80% of T_4 is metabolized by deiodination with about equal amounts of T_3 and rT_3 being produced. The remaining T_4 is conjugated with sulfate or glucuronide and deactivated by deamidation or decarboxylation.

The biologically active fraction of T_3 and T_4 in plasma (i.e. which does not bind to protein) represents, in each case, less than 1% of the total

Approximately 80% of the 45 nmol (35 μg) daily production rate of T_3 is by extrathyroidal deiodination. The turnover of T_3 is much greater than that of T_4. The biologically active component of T_4 and T_3 in plasma is the free fraction – that not bound to proteins. This represents a very small percentage of the total, 0.02% for T_4 and 0.3% for T_3.

Biochemical actions of thyroid hormones

The thyroid hormones may be considered as the accelerator pedal of metabolism

By acting on a wide range of tissues they influence the basal metabolic rate. Most, if not all, of these actions result from altered transcription brought about by the binding of T3 to its nuclear receptor (see Chapter 36).

Thyroid hormones augment thermogenesis by increasing mitochondrial oxidative metabolism driven by the increase in ATP utilization that occurs as a result of increased sodium–potassium (Na^+/K^+)-dependent ATPase activity. Lipolysis is stimulated by increasing the activity of hormone sensitive lipase, thus producing fatty acids that can be oxidized to generate the ATP used for thermogenesis. Increases in both glycogenolysis and gluconeogenesis occur to balance the increased use of glucose as a fuel for thermogenesis. The rate of synthesis of many structural proteins, enzymes and other hormones is thus affected as a result of thyroid hormone action.

Clinical disorders of thyroid function

Thyroid disease is common, affecting almost 3% of the population, and nine times as many women as men are affected

Thyroid disease can occur at any time from birth to old age. Greater than 95% of thyroid disease originates in the thyroid gland and much of this is autoimmune in origin. Thyroid autoantibodies can bind to the TSH receptor. If the autoantibodies bind but do not stimulate, thyroid hormone production will fall and the patient will be hypothyroid, with an increased plasma TSH and reduced free T_4. If the autoantibodies bind and stimulate, then the patient will be hyperthyroid (thyrotoxic) with increased plasma free T_4 and suppressed TSH (see Fig. 35.7).

Hypothalamic and pituitary causes of hypothyroidism occur regularly often as a result of impaired TSH secretion secondary to pressure from an adjacent tumor. Pituitary TSH-secreting tumors are known but are an extremely rare cause of hyperthyroidism.

Abnormal patterns of thyroid hormones may occur as a secondary consequence of nonthyroidal illness. It is important to distinguish this situation from a primary thyroid disorder since inappropriate treatment could be dangerous.

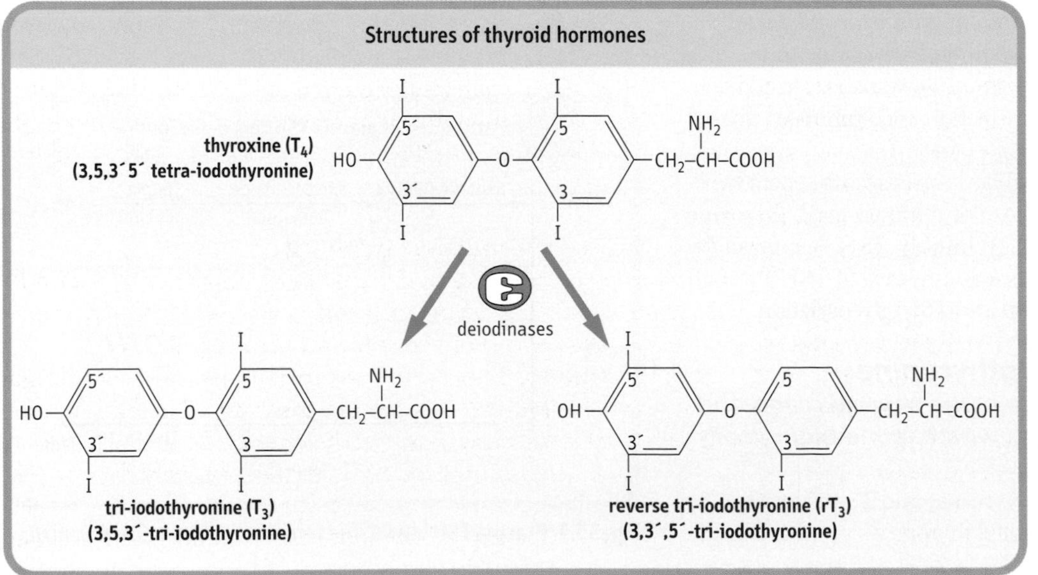

Fig. 35.8 Structures of the thyroid hormones T_4, T_3, and rT_3.

Hypothyroidism

The symptoms of hypothyroidism are nonspecific and easy to miss. A 60-year-old woman comes to the outpatient clinic and complains of weight gain, intolerance of cold and tiredness. She also says that she has recently become less alert mentally and attributes this to aging. She says that two members of her family had 'thyroid trouble'. On examination, she is moderately obese and has dry skin. The thyroid gland is not palpable. Her thyroxine is 15 nmol/L (range: 55–144 nmol/L) and TSH is 25 mU/L (range 0.4–4 mU/L) (see Fig. 35.7).

Comment. Symptoms of hypothyroidism at an early stage can be fairly non-specific, as they are in this case. The best laboratory test for the diagnosis of hypothyroidism is the blood TSH level. The elevated level of TSH suggests primary thyroid disorder. Subsequently this lady's blood was shown to be positive for the microsomal and antithyroglobulin antibodies. A diagnosis of lymphocytic thyroiditis (Hashimoto's thyroiditis) was made. She was commenced on thyroxine treatment.

Hyperthyroidism

A 35-year-old woman comes to her General Practitioner complaining of palpitations and fatigue. She also says that she has lost 4 kg of weight recently in spite of good appetite and no attempt at dieting. She reports occasional diarrhea.

Comment. On examination her skin is warm and moist and she has a fine tremor of outstretched hands. She has tachycardia 110/min. She also has a mild thyroid enlargement (goitre). Thyroid function tests show suppressed TSH level (<0.05; range 0.4–4 mU/L) (see Fig. 35.7) and increased thyroxine (T_4 = 220; range: 55–144 nmol/L) and tri-iodothyronine (T_3 = 4.0; range 0.9–2.8 nmol/L). Thyroid receptor antibodies were detected. In hyperthyroidism, the TSH level tends to be suppressed by high circulating thyroid hormones. This lady's low TSH level and high thyroid hormone concentrations suggest hyperthyroidism. The presence of thyroid receptor antibodies confirms that the cause is Graves disease, an autoimmune thyroid disorder. The antireceptor antibodies bind to the TSH receptors in the thyroid gland and mimic the effect of TSH producing thyroid oversecretion. She was commenced on treatment with an antithyroid drug, carbimazole.

The hypothalamo–pituitary–adrenal axis

Two hypothalamic peptides are the principal regulators of pituitary ACTH release

Corticotropin releasing hormone (CRH) and arginine vasopressin

Two hypothalamic peptides, CRH and AVP, are the principal but not the sole regulators of pituitary ACTH synthesis and release. They act independently and synergistically. CRH is a 41 amino acid peptide secreted by the paraventricular nucleus which acts via the cAMP second messenger system. AVP is a nine amino acid peptide synthesized by both the supraoptic and paraventricular nuclei, which acts by altering intracellular calcium ion channels. Negative feedback by cortisol reduces the actions of both CRH and AVP.

Adrenocorticotropic hormone

ACTH is part of the 241 amino acid precursor molecule pro-opiomelanocortin (POMC). POMC is unusual as a hormone precursor in that it is cleaved to release several hormonally active peptides including the endorphins and melanocyte stimulating hormones. POMC is normally only produced by the pituitary gland but it may also be produced in large quantities by certain malignancies giving rise to ectopic ACTH syndrome.

ACTH itself is comprised of 39 amino acids with the biological activity residing in the N-terminal 24 moieties. ACTH is secreted in stress-related bursts superimposed on a marked diurnal rhythm which shows a peak at 05.00 h. It is transported unbound in plasma. It has a half-life of about 10 minutes and is unstable in plasma. ACTH stimulates the synthesis and release of glucocorticoid hormones by interacting with cell-surface receptors on the adrenal cortex that stimulate the production of intracellular cAMP. Acute increases in the adrenal synthesis of cortisol occur within 3 minutes, principally by stimulating the activity of cholesterol esterase. Chronic effects of ACTH include induction of transcription of the genes that encode steroidogenic enzymes and other factors.

Negative feedback by cortisol occurs within two time frames, acting at both the hypothalamic and pituitary levels. Fast feedback alters the release of hypothalamic CRH and the CRH-mediated secretion of ACTH. Slow feedback results from reduced synthesis of CRH and AVP plus suppression of POMC gene transcription, which results in reduced ACTH synthesis.

Biosynthesis of cortisol

Cortisol is the major glucocorticoid synthesized in man in the inner two zones of the adrenal cortex under the direct control of pituitary ACTH

A simplified scheme of steroid biosynthesis is shown in Figure 35.9. Cholesterol is the precursor for all steroid hormone synthesis. Cleavage of the cholesterol side chain liberates the so-called C-21 corticosteroids, further side chain cleavage yields the C-19 androgens and aromatization of the A ring results in the C-18 estrogens. The structures of the key steroid hormones are shown in Chapter 16. Several of the steroidogenic enzymes are members of the cytochrome P450 superfamily of oxidases. The genes for many of these enzymes have been located and we now understand the details of this process.

The plasma concentration of cortisol shows a pronounced diurnal rhythm, being some 10 times higher at 08.00 h than at 24.00 h. This parallels the marked diurnal rhythm of secretion of ACTH. Approximately 95% of cortisol in plasma is bound to proteins, mainly CBG. As cortisol concentration rises, the percentage of free cortisol also rises, indicating that CBG binding is saturable. Cortisol has a half-life in plasma of about 100 minutes. Cortisol is metabolized in the liver and other organs by a combination of reduction, side chain cleavage and conjugation to produce a wide range of inactive metabolites that are excreted in urine.

Cortisol has a major influence on gluconeogenesis

As the name glucocorticoid suggests, cortisol has a major influence on increasing glucose production by enhancing virtually every step in the gluconeogenic pathway. At the same time, cortisol inhibits glucose uptake and metabolism in peripheral tissues. Glycogen synthesis and deposition are increased. Lipolysis is stimulated by cortisol. Protein and RNA synthesis are stimulated in the liver but inhibited in peripheral tissues (muscle) with the protein breakdown products acting as substrates for hepatic gluconeogenesis. In this role, cortisol is working in tandem with several other hormones including insulin, glucagon and GH.

Excess cortisol has a wide range of effects on the immune system causing overall suppression and providing a useful form of therapy. It is likely that at least some of these actions are important in the control of the immune response in normal physiology. Cortisol also influences the heart, vasculature, blood pressure, water excretion and electrolyte balance. While some of these effects are mineralocorticoid in nature others act by direct action through the cortisol receptor. Cortisol also influences bone turnover through a variety of mechanisms and osteoporosis is a recognized consequence of sustained cortisol excess. Finally, cortisol can inhibit linear growth and cell division in several individual tissues.

Clinical disorders of cortisol secretion

Hyposecretion of cortisol may occur as a result of hypothalamic, pituitary or adrenal failure

In disorders causing cortisol hyposecretion, clinical presentation plus measurement of cortisol and ACTH, and the extent of the cortisol response to synthetic ACTH (Synacthen) can help to elucidate the source of the problem.

Deficiencies of CRH and ACTH are often accompanied by deficiencies of other hypothalamic or pituitary hormones. Addison's disease is primary adrenal failure and one of the few endocrine emergencies. Biochemically, it is

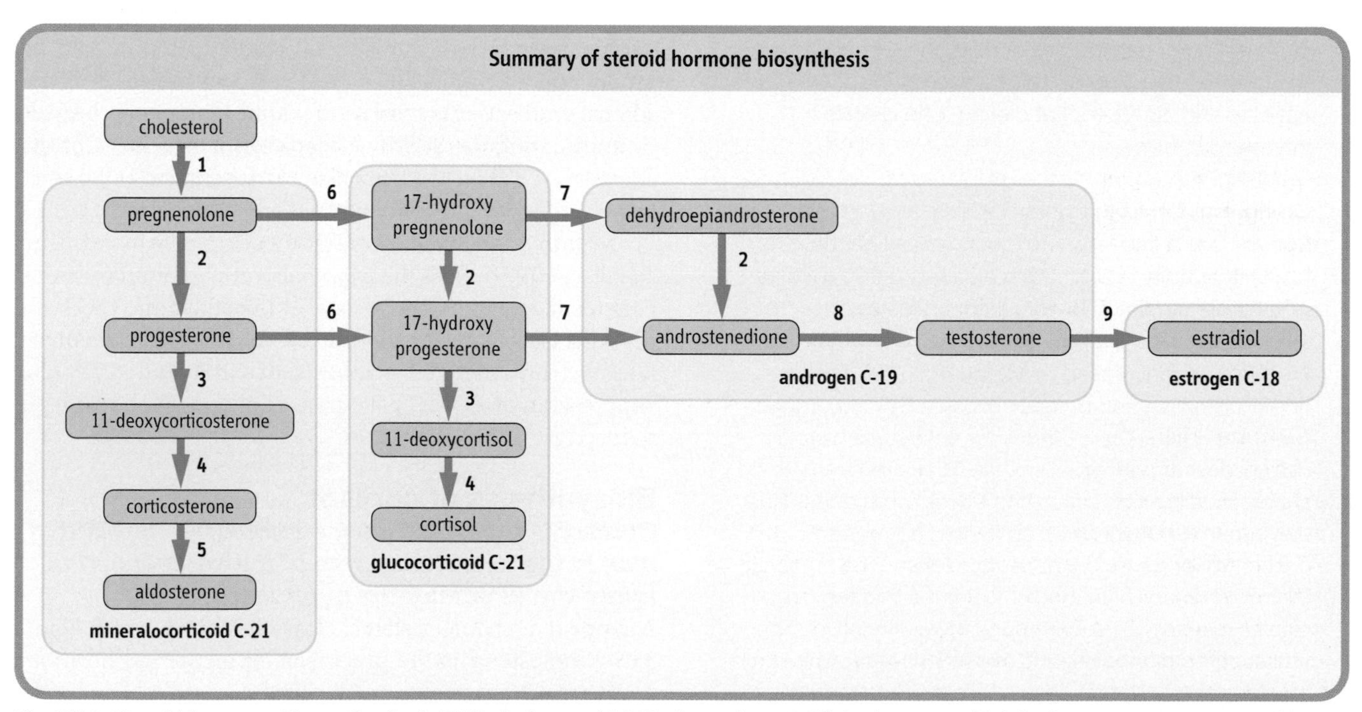

Fig. 35.9 Steroid hormone biosynthesis. 1, P450 cholesterol 20,22-desmolase; **2**,3β-hydroxysteroid dehydrogenase:Δ5-oxosteroid isomerase; **3**, P450c21 21-hydroxylase; **4**, P450c11 11-hydroxylase; **5**, corticosterone methyloxidase; **6**, P450c17 17α-hydroxylase; **7**, P450c17 17,20-lyase/desmolase; **8**, 17KR 17-ketoreductase; **9**, A aromatase.

characterized by hyponatremia, an impaired cortisol response to Synacthen and an elevated plasma ACTH. Cortisol replacement, usually together with a mineralocorticoid, is an effective treatment.

Rare causes of cortisol hyposecretion result from a genetic disorder in the steroid biosynthetic pathway. The commonest of these disorders, congenital adrenal hyperplasia, is the result of a partial or complete absence of the steroid 21-hydroxylase enzyme (see Fig. 35.9). In neonates, the presentation is of hyponatremia and may be accompanied by ambiguous genitalia resulting from the metabolism of accumulated 17-hydroxyprogesterone into androgens. The milder form presents in teenage girls where the androgen excess results in hirsutism and menstrual irregularity. A clinical example of this condition is given in Chapter 16.

Hypersecretion of cortisol results in Cushing's syndrome – easily the most challenging of all endocrine disorders

Exogenous glucocorticoids, commonly used to suppress the immune system in a range of gastrointestinal, rheumatic or dermatologic disorders, may result in clinical Cushing's syndrome. However, Cushing's syndrome may also result from a disorder of the hypothalamus, pituitary or adrenal gland and it may also be the consequence of ectopic ACTH syndrome. The symptoms are often nonspecific and cyclical. The challenge is to demonstrate cortisol excess and then to undertake the differential diagnosis by a variety of means including measuring ACTH, assessing the suppressibility to dexamethasone (a synthetic glucocorticoid) and selective catheterization of the veins draining the pituitary or the adrenal glands. Treatment is usually surgical and should be targeted at the primary cause of the condition.

Cushing's syndrome – a diagnostic dilemma

A 65-year-old man presents with a long history of central obesity, a plethoric face and abdominal bruising. He is mildly hypertensive. A urine cortisol result is 1000 nmol/24 h (<250 nmol/24 h); serum cortisol is 500 nmol/L at 24.00 h (<50 nmol/L) and his 08.00 h cortisol is 550 nmol/L after 1 mg of dexamethasone (a potent synthetic glucocorticoid) (<50 nmol/L). Plasma ACTH is 100 ng/L (<80 ng/L).

Comment. This man has hypercortisolism, known as Cushing's syndrome. The elevated urine cortisol, elevated evening serum cortisol and failure to suppress following dexamethasone support the diagnosis. The ACTH result indicates that the cause is either a pituitary tumor or ectopic ACTH secretion from a occult tumor (probably carcinoid). In this case magnetic resonance imaging of the pituitary revealed a clear tumor of 0.7 cm diameter – the differential diagnosis is not often this easy. Pituitary surgery is the appropriate treatment in this case.

The hypothalamo–pituitary–gonadal axis

Gonadotropin releasing hormone (GnRH)

GnRH is essential for the secretion of intact FSH and LH

The hypothalamus has a major role in the control of gonadal function in both males and females. The secretion of GnRH is the first stage in the onset of puberty, and the interaction between hypothalamic opiates and GnRH has much to do with regulating the pulsatile nature of GnRH secretion which, in turn, influences the relative secretion of pituitary FSH and LH.

Primary adrenal insufficiency – an endocrine emergency

A 40-year-old woman is admitted as a medical emergency after collapsing at home 2 days after contracting influenza. For 1 year she has complained of weakness, fatigue, abdominal pain, nausea, vomiting, anorexia and confusion. She is dehydrated and hypotensive with pigmentation on her face and hands. Biochemical analysis reveals Na^+ 115 mmol/L (135–145 mmol/L); K^+ 5.9 mmol/l (3.5–5.0 mmol/L); urea 12 mmol/L (2.5–6.5 mmol/L) and glucose 2.9 mmol/L (4.0–6.0 mmol/L). Baseline serum cortisol is 95 nmol/L rising to 110 nmol/L 30 minutes after administration of synthetic ACTH (Synacthen).

Comment. This is an acute presentation of Addison's disease resulting from progressive autoimmune destruction of the adrenal cortex. The nonspecific symptoms of cortisol deficiency became critical when she was unable to cope with the stress of a minor illness. The hypotension and electrolyte imbalance result from mineralocorticoid deficiency. The pigmentation is due to elevated ACTH and other POMC peptides (notably melanocyte stimulating hormone). The diagnosis is made by the impaired cortisol response to synthetic ACTH (normal increment >200 nmol/L). Treatment is urgent rehydration plus a bolus of hydrocortisone followed by daily maintenance therapy with hydrocortisone and a mineralocorticoid.

GnRH is a 10 amino acid peptide synthesized as a 92 amino acid precursor by various hypothalamic nuclei and transported to the pituitary via the portal system. GnRH has a half-life of about 3 minutes and effects the synthesis and secretion of both FSH and LH from the same gonadotroph cell type. GnRH acts through its cell surface receptor to increase intracellular calcium, hydrolyze inositol phosphates and phosphorylate protein kinase C. Estrogens increase and androgens decrease the number of GnRH receptors. Long-acting GnRH agonists can cause downregulation of GnRH receptors leading to much reduced FSH and LH secretion. Such agonists are now used to prepare infertile women for assisted-conception programs.

Follicle stimulating hormone and luteinizing hormone

Although FSH and LH have been given their names on the basis of their function in the female, it is now clear that identical hormones are secreted and function in the male

Both FSH and LH are secreted in the male and female from the same cell and appear to be under the influence of the same stimulus (GnRH). There is growing evidence that alterations in the GnRH pulse frequency and amplitude can influence the relative amounts of FSH and LH secreted by the gonadotroph. Inhibin, as a selective feedback inhibitor of FSH, also contributes to the relative output of FSH and LH from the cell.

FSH and LH are both glycoproteins with molecular weights of approximately 28 kDa. Each is comprised of an identical α-subunit (shared with TSH) and a specific β subunit. The α-subunit has 92 amino acids and two carbohydrate side chains and is encoded by a gene on chromosome 6. The β-subunits of both FSH and LH are composed of 115 amino acids and have two carbohydrate chains. The gene for FSH is on chromosome 11, while that for LH is on chromosome 19 and close to the gene for the β-subunit of human chorionic gonadotropin (HCG) with which it has considerable homology (Fig. 35.10).

Although GnRH is essential for the secretion of intact FSH and LH, feedback from estradiol and testosterone plus gonadal peptides such as inhibin have a secondary effect. Feedback by estradiol is especially interesting because it may have either negative or positive effects on gonadotropins depending on the stage of the menstrual cycle. An outline of the control of the hypothalamo–pituitary–gonadal axes, both for adult men and women, is shown in Figure 35.11.

The production rate of LH in men and ovulating women is about 200 IU/day, the corresponding figure for FSH is around 50 IU/day. These production rates are greatly increased in postmenopausal women in whom there is no negative feedback from ovarian steroids. The half-life of LH

Biochemical confirmation of pregnancy

Biochemical confirmation of pregnancy is now achieved by means of a simple and sensitive test that can give reliable results within 2 weeks of fertilization (i.e. before the next menstrual period). Pregnancy tests may be performed by doctors, nurses or even the patients themselves since they are readily purchased from pharmacies.

Pregnancy tests are immunoassays that measure the production of human chorionic gonadotropin (HCG). A specific two-site immunoassay is employed (see Fig. 35.3). One antibody recognizes the α-subunit of HCG, while the other only binds to the 32 amino acid segment of the β-subunit that is unique to HCG (see Fig. 35.10). Since nonpregnant women do not normally produce HCG, the detection of even low levels from the first stages of placental development is sufficient to confirm pregnancy.

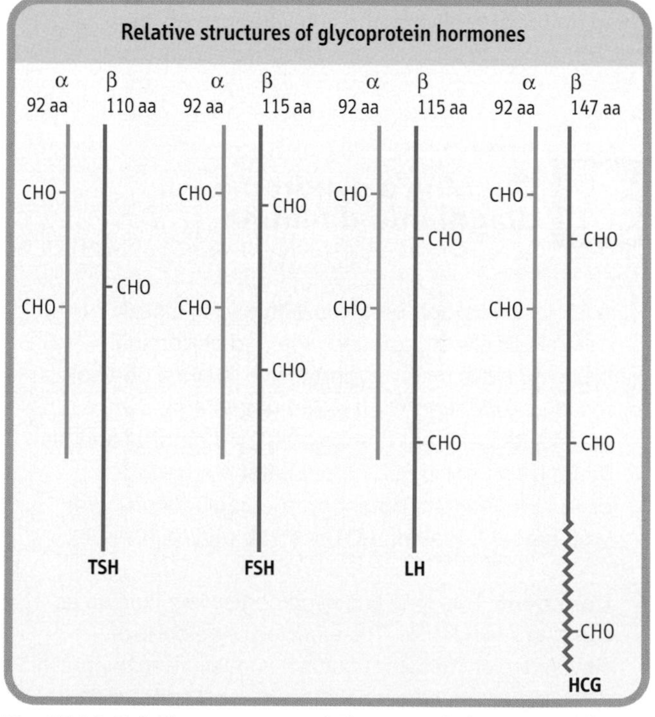

Fig. 35.10 Relative structures of glycoprotein hormones. The three pituitary glycoprotein hormones (TSH, FSH, LH) and the placental glycoprotein (HCG) share a common α-subunit. Hormone specificity is conveyed by the β-subunit and the resulting three-dimensional protein structure. CHO indicates the number of carbohydrate side chains, although no attempt has been made to position these accurately. It is worthy of note that HCG differs from LH solely by having an additional 32 amino acids (aa) in the β subunit.

in plasma is approximately 50 minutes, FSH has a longer half-life of about 4 hours. Both FSH and LH concentrations vary considerably depending on the age and sex of the subject under investigation (Fig. 35.12).

Actions of FSH and LH on the testes

FSH and LH influence spermatogenesis

In the male, testosterone biosynthesis occurs in the Leydig cells of the testes under the primary influence of LH (see Fig. 35.11). The LH receptor is a member of the G-protein coupled superfamily, which includes TSH and FSH among many others (see Chapter 36). It has an extracellular domain of 340 amino acids (which bind the β-subunit of LH) and a transmembrane domain (which contains 330 amino acids and seven transmembrane spanning elements). LH binding activity activates adenylate cyclase, leading to a cascade of events that result in the phosphorylation of proteins, which regulate the steroidogenic enzymes shown in Figure 35.9.

Testosterone has several effects outside the testes but it also facilitates the effects of FSH on the spermatic tubule, leading to spermatogenesis. FSH binds to its specific receptor on the Sertoli cell of the testes and, by a mechanism similar to LH, induces increased synthesis of several proteins, including androgen binding protein (ABP) and inhibin. ABP is secreted into the seminiferous tubular lumen where it binds testosterone (or its active form dihydrotestosterone). This ensures a high local androgen content which, together with FSH, brings about the meiotic divisions that are necessary for spermatogenesis. Inhibin may well have a role specific to FSH in the negative feedback loop (see Fig. 35.11).

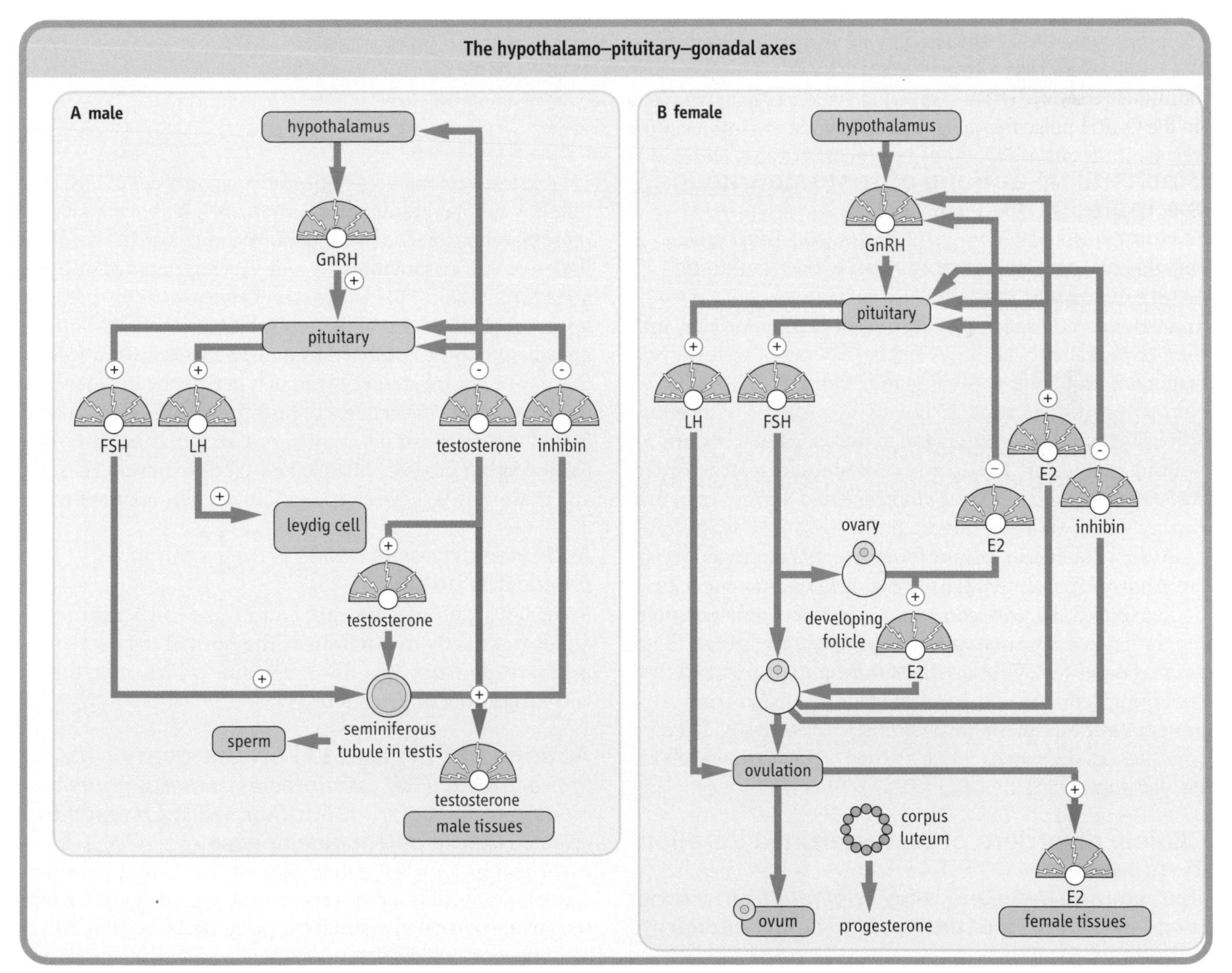

Fig. 35.11 Control of the hypothalamo-pituitary-gonadal axes. (A) In men, testosterone is produced from cholesterol n the Leydig cell under LH stimulation. Testosterone assists FSH to bring about spermatogenesis. (B) In women, estradiol (E2) is produced by the granulosa cell and the developing follicle after feedback stimulation. E2 feedback is mainly negative but, in midcycle, there is a positive E2 feedback resulting in the surge of LH that causes ovulation. Progesterone (P) is secreted by the resultant corpus luteum.

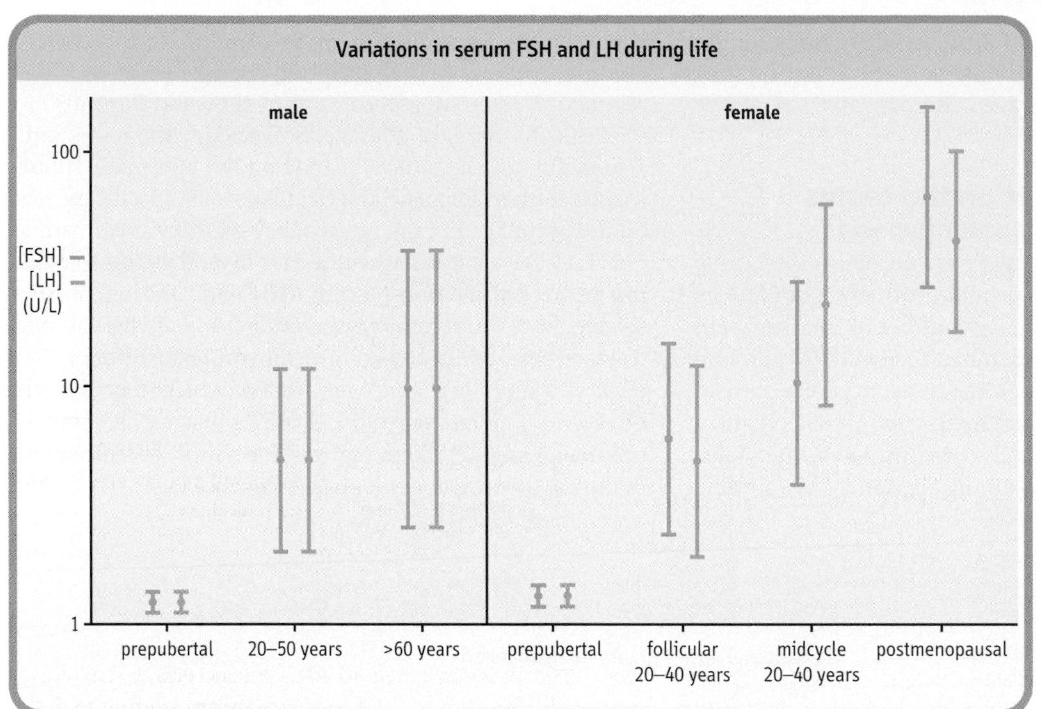

Fig. 35.12 Variations in serum FSH and LH during life. Serum FSH and LH levels vary depending on age and sex.

Biochemical actions of testosterone in the male

Testosterone not only influences gonadotropin regulation, and spermatogenesis, but is also a natural anabolic steroid

The testes secrete about 7000 μg of testosterone each day into the peripheral circulation. A further 500 μg of testosterone originates from the adrenal gland. More than 97% of the circulating testosterone is bound to protein, with equal amounts bound to albumin and to the specific sex hormone binding globulin (SHBG), which is similar in structure to ABP. Testosterone is unique among steroid hormones in that further metabolism, 5-α reduction to dihydrotestosterone, more than doubles its affinity for the nuclear androgen receptor. Androgen receptors have 919 amino acids and are widely distributed. In addition to its effects on gonadotropin regulation and spermatogenesis, testosterone brings about virilization of the Wolffian ducts during male sexual differentiation. Acting via dihydrotestosterone, it also stimulates protein synthesis and increases muscle mass in a wide range of tissues. In this capacity testosterone is the natural anabolic steroid (Fig. 35.13).

Clinical disorders of testosterone secretion in males

Testosterone deficiency may originate from a wide range of disorders of the hypothalamus, pituitary or testes

A genetic hypothalamic disorder in children (Kallmann's syndrome) results in deficient GnRH production, and affected individuals present with delayed puberty and subnormal FSH, LH and testosterone. A genetic testicular disorder of the seminiferous tubules (Klinefelter's syndrome, karyotype 47XXY) presents with gynecomastia (abnormal increase in size of the male breast), eunuchoidism and varying degrees of hypogonadism. Serum FSH is elevated, LH is usually elevated and testosterone may be subnormal. The external effects of hypogonadism may be corrected by androgen administration but these patients are sterile. Androgen deficiency in older men may have a specific cause e.g. mumps orchitis – inflammation of the testicle or pituitary tumor, or it may be part of the natural aging process. Muscle and fat distribution changes, and these may be corrected by androgen administration.

Androgen excess in males is only seen in precocious puberty

Precocious puberty is a rare condition which may result either from early maturation of the normal hypothalamo–pituitary–gonadal axis or as a result of a tumor that is secreting either androgen or HCG.

Actions of FSH and LH on the ovary

In the female, FSH promotes estradiol synthesis leading to follicular maturation, while LH leads to follicle rupture and oocyte release

In the mature female it is now clear that the GnRH pulse generator is responsible for the control of all the hormonal changes seen in the normal menstrual cycle (Fig. 35.14, see Fig. 35.11). The initiation of follicular growth begins in the last few days of the preceding menstrual cycle and terminates at the time of ovulation. FSH is the dominant hormone acting through its receptors on the ovarian granulosa cells. Rising FSH

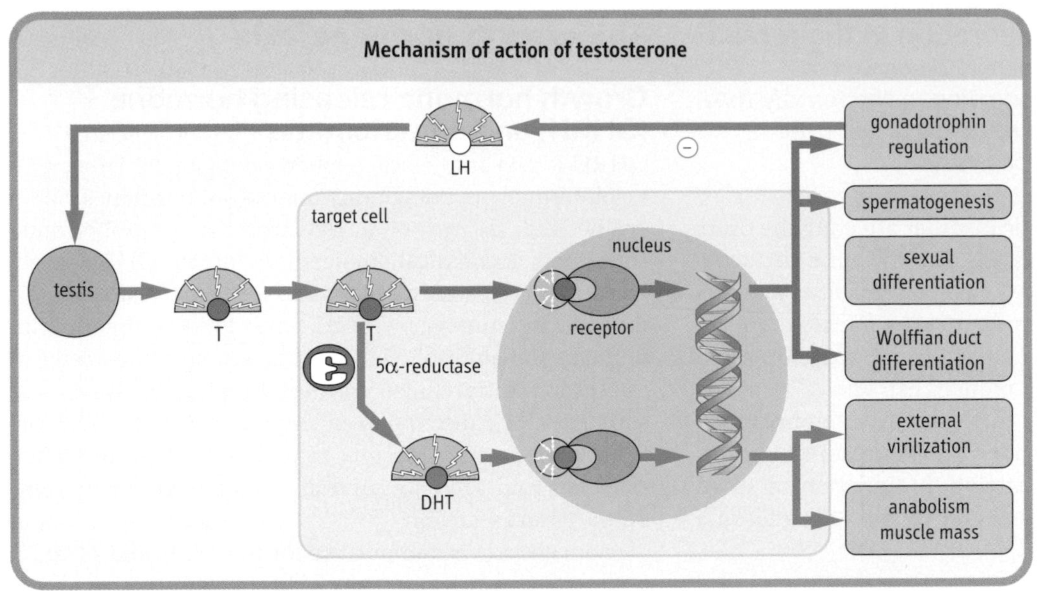

Fig. 35.13 Mechanism of action of testosterone. Testosterone (T) from the testis enters a target cell and binds to the androgen receptor, either directly or after conversion to 5α-dihydrotestosterone (DHT). DHT binds more tightly than T and the DHT-receptor complex binds more efficiently to chromatin. Actions mediated by T are shown by purple lines, those mediated by DHT are shown by blue lines.

concentrations stimulate estradiol synthesis through enzyme induction, including aromatase (see Fig. 35.9). As estradiol is secreted so FSH falls, and this combination leads to the selection of a dominant follicle for further development. Follicular maturation continues under the influence of rising estradiol concentrations. At midcycle there is a further estradiol surge, which causes positive feedback at the pituitary to initiate the LH surge. This LH binds to its receptors on the dominant follicle and, in tandem with steroid hormones and other factors such as prostaglandins, results some 36 hours later in rupture of the follicle and release of the oocyte. This is ovulation. At this time there is a sharp fall in plasma estradiol, followed by a fall in LH. The ruptured follicle transforms into the corpus luteum, which secretes progesterone and lesser amounts of estradiol both to sustain the oocyte and to prepare the estrogen primed uterine endometrium for acceptance of a fertilized ovum and the establishment of early pregnancy. In the absence of fertilization, corpus luteum function declines resulting in a decrease in progesterone and estradiol secretion. This brings about vascular changes in the endometrium leading to tissue death and menstruation. The fall in steroid secretion also stimulates FSH secretion to start the whole process in the next cycle.

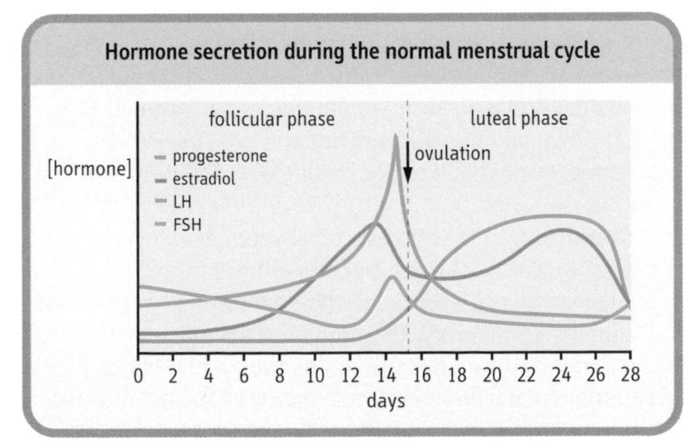

Fig. 35.14 Hormone secretion in the menstrual cycle. LH, luteinizing hormone; FSH, follicle stimulating hormone.

Biochemical actions of steroid hormones in the female

In a woman with a normal menstrual cycle, it is the progesterone level that is of diagnostic significance

It follows from the previous paragraph that the concentrations of FSH, LH, estradiol and progesterone vary considerably during the menstrual cycle. In a normally cycling woman it is only progesterone that is of diagnostic significance – a serum concentration of more than 20 nmol/L in the luteal phase being consistent with ovulation. The production rate of estradiol is 80 μg/day early in the cycle rising 10-fold during the follicular phase. In contrast the production rate of progesterone is 2 mg/day in the follicular phase and 10-fold greater in the luteal phase. Estradiol binds to SHBG in plasma, although with a lower affinity than testosterone. Progesterone binds to CBG in plasma with a lower affinity than cortisol.

Aside from their roles in the menstrual cycle and reproduction, both estradiol and progesterone have other effects, acting via their specific nuclear receptors in target cells. Estradiol, working in tandem with other hormones such as insulin-like growth factor-I (IGF-I), is responsible for linear growth, breast development and maturation of the urogenital tract and the female habitus. In adult life, both estradiol and progesterone support breast function and estradiol has an important role in influencing bone turnover. Progesterone is responsible for the rise in basal body temperature during the luteal phase of the menstrual cycle and falls in progesterone secretion may contribute to changes in mood as seen in premenstrual tension.

Disorders of steroid secretion in the female

Just as in the male, endocrine disorders of subnormal sex steroid secretion in the female may result from a wide range of disorders in the hypothalamus, pituitary or ovary

Kallmann's syndrome is seen in females as well as males, but the most common genetic disorder affecting the ovary is Turner's syndrome (karyotype 45X), which has characteristic somatic features and an elevated plasma FSH and prepubertal estradiol pattern. In the mature female there are various endocrinal causes of infertility, the scope of which is beyond this text. Although reduced estradiol secretion is a natural consequence of the menopause, hormone replacement therapy (HRT) controls vasomotor symptoms (symptoms related to contraction and dilation of small blood vessels), protects bones and reduces the risk of a cardiovascular disease.

Syndromes of excess ovarian steroid secretion lead to precocious puberty in children and infertility and/or hirsutism in the adult

Precocious puberty may arise from early maturation of the normal axis, an estradiol- or androgen-secreting cyst or tumor of the ovary or the adrenal gland, or congenital adrenal hyperplasia. In the mature female, hypersecretion of androgen in the polycystic ovary syndrome (PCOS) may result both in infertility and/or hirsutism (growth of male type and distribution of hair in women). This occurs via androgen receptors.

HRT

Women normally reach the menopause at age 45–55 years of age. The absence of ovarian follicles leads to an absence of estrogen and inhibin secretion resulting in a marked rise in pituitary FSH. There are vasomotor symptoms of estrogen deficiency (flushing) associated with the menopause. Long-term postmenopausal estrogen deficiency is known to increase the rate of bone loss, leading to osteoporosis, and also alters lipoprotein metabolism in a way that increases the risk of a cardiovascular event.

HRT with an estrogen preparation is now standard practice in most developed countries both to relieve the acute symptoms and to provide prophylaxis against the long-term risks. Considerable debate exists about the most efficacious preparation of estrogen to use (natural estradiol or synthetic estrogen +/– a progestogen) and the most appropriate route of administration (oral, transdermal, implant). There is also debate about the optimal time to commence HRT and for how long it should continue.

The growth hormone axis

Growth hormone releasing hormone (GHRH) and somatostatin

GHRH is a 44 amino acid peptide belonging to the secretin-vasointestinal peptide-glucagon family of hormones, and is synthesized as part of a 108 amino acid prohormone. Immunohistochemical studies have located GHRH in the arcuate and ventromedial nuclei of the hypothalamus and in the median eminence. GHRH binds to its receptor on the pituitary somatotroph cell and triggers both the adenylate cylase and intracellular calcium–calmodulin systems to stimulate GH transcription and secretion. Negative feedback from GH and IGF-I results in both a decrease in GHRH synthesis and secretion and an increase in somatostatin synthesis and secretion.

Somatostatin is found in two forms with 14 and 28 amino acids, both of which are produced from a 116 amino acid gene product. Somatostatin and its receptors are found throughout the brain and also in other organs, notably the gut. Binding of somatostatin to its receptor is coupled to adenylate cyclase by an inhibitory guanine nucleotide binding protein, resulting in a decrease in intracellular cAMP. In the context of growth, somatostatin inhibits the secretion of GH. GHRH and somatostatin are released in separate bursts to provide a very fine level of control of GH release. Somatostatin also inhibits basal and stimulated TSH release. A long-acting analog of somatostatin has been prepared, which is effective in the management of GH excess and tumors secreting a wide range of other hormones including TSH, insulin and glucagon.

Polycystic ovary syndrome – a common cause of infertility

A 24-year-old woman presents with a 2-year history of oligomenorrhea (infrequent menstruation) and infertility. She is obese with facial acne and mild hirsutism. Basal hormone studies show 'normal' FSH, LH prolactin and estradiol but she has a serum testosterone result of 4.5 nmol/L (0.9–3.2 nmol/L).

Comment. This woman's polycystic ovary syndrome (PCOS) was confirmed by ultrasound of the ovaries showing enlargement with the characteristic cysts that secrete testosterone (and other androgens). The androgen is the cause of the hirsutism and its interference with the normal estrogen feedback at the pituitary causes the oligomenorrhea and infertility. Antiandrogen therapy is indicated.

Growth hormone

The greatest secretion of GH occurs in children and young adults; it occurs chiefly during sleep

The gene for pituitary GH is located as part of a complex on chromosome 17. This expresses a 28 kDa prohormone, which is cleaved to release the 191 amino acid 22 kDa protein that is the principal species of GH. Variant forms of GH are found, which may be different products of the GH gene or aggregates. Nearly two-thirds of GH in the circulation is associated with a 29 kDa binding protein that is identical to the extracellular domain of the GH receptor. This binding protein prolongs the half-life of GH in plasma.

The normal human pituitary contains approximately 10 mg of GH, less than 5% of this is released each day. GH is released in bursts with a periodicity of 3–4 hours and greatest secretory activity occurs during sleep. At the peak of a secretory burst, the plasma GH concentration may be 100-fold greater than baseline; this means that no meaningful reference interval can be derived for this hormone. Secretory bursts occur most frequently in children and young adults. The plasma half-life of GH is about 20 minutes.

Control of GH secretion

GH has a wide range of actions in regulating growth and intermediary metabolism

It is not surprising, in view of the wide range of actions of GH, that several factors other than GHRH and somatostatin, can influence GH secretion. These include other hormones (estradiol) and metabolic fuels (glucose). Several pharmacologic agents are also known to influence GH secretion. These factors act at the higher centers and the hypothalamus to alter the pattern of GHRH and somatostatin secretion.

Biochemical actions of GH

Binding of GH to its receptor precipitates a complex series of intracellular events that lead to the transcription of many enzymes, hormones and growth factors, including IGF-I

The diversity of action of GH makes it difficult to understand all its functions; thus it is convenient to think in two distinct phases (Fig. 35.15). The direct actions of GH are on lipid and carbohydrate metabolism. These are synergistic with cortisol

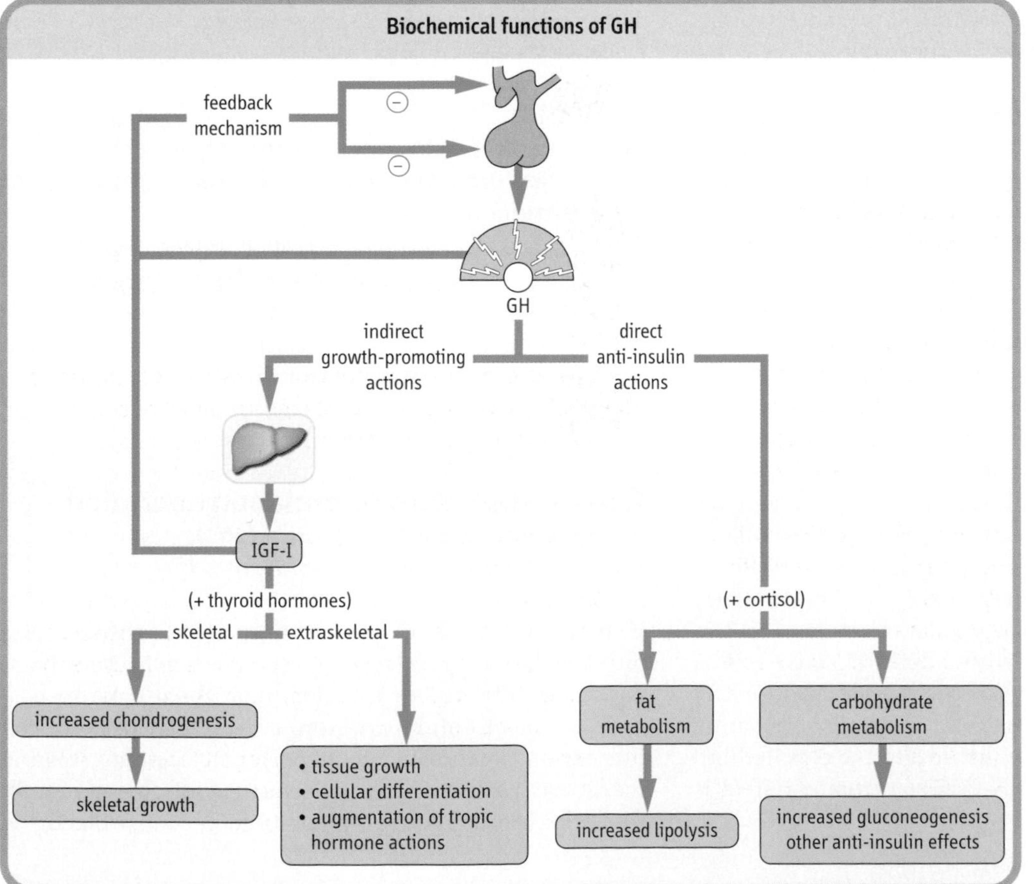

Fig. 35.15 Biochemical functions of GH. These can be divided conveniently into direct actions on lipid and carbohydrate metabolism and indirect actions on protein synthesis and cell proliferation.

and opposite to the actions of insulin and IGF-I. The indirect actions of GH are through several growth factors, known collectively as somatomedins, and illustrated by IGF-I. These actions promote protein synthesis and cell proliferation in both skeletal and nonskeletal tissues, actions which are insulin-like and opposed to cortisol. Overall, we tend to consider GH as having its major role in stimulating linear growth but it clearly has other roles in influencing the relative amount and distribution of fat and muscle. This is best seen in GH-deficient adults after they receive GH replacement therapy.

Insulin-like growth factor-I

IGF-I is the most GH-dependent of a series of growth factors

IGF-I is a 70 amino acid straight-chain basic peptide produced by the liver in response to GH action from a gene located on chromosome 12. IGF-I has considerable homology with proinsulin. In plasma and other extracellular fluids, IGF-I is complexed to a series of IGF-binding proteins (IGFBPs) of which IGFBP-3 is the most abundant and GH-dependent. IGF-I works through the type 1 IGF receptor, which is structurally similar to the insulin receptor and linked to intracellular tyrosine kinase activity. The relative affinities of insulin and IGF-I for their respective receptors mean that, in normal physiology, there is little crossbinding, although in pathologic and pharmacologic situations it is possible for insulin to have some action through the IGF-I receptor, and *vice versa*. This has some implications for the possible use of IGF-I as a therapeutic agent in type 2 diabetes mellitus.

The plasma reference interval for IGF-I in adults aged 20–60 years is fairly constant. It is much lower in young children but rises dramatically during the period of growth and progression through puberty. IGF-I concentrations fall after the sixth decade of life. Thus, IGF-I appears to be a good marker of integrated GH activity.

Clinical disorders of GH secretion

GH deficiency in children is one of several causes of short stature, while GH excess in children leads to gigantism and, in adults, acromegaly

The absence of a plasma reference range for GH means that GH deficiency may only be diagnosed by studying the dynamics of GH secretion, either during sleep or following a stimulation test. It is hoped that basal IGF-I or IGFBP-3 measurements may serve as a preliminary screening test. Treatment is by regular injection of recombinant human GH. Adults with a definite cause of GH deficiency (hypopituitarism) are also candidates for GH replacement therapy. A rare genetic cause of short stature is Laron dwarfism in which GH levels are elevated but IGF-I levels are subnormal – a GH receptor defect is responsible.

In most cases, GH excess is caused by a pituitary tumor

GH excess is almost always due to a GH-secreting pituitary tumor, although ectopic GHRH from a pancreatic tumor has been described. In children, GH excess manifests itself as gigantism. In adults the epiphyses of the long bones have closed and so further linear growth is not possible. Therefore, the adult form, acromegaly, is characterized by a thickening of tissues and a change of facial appearance. Classically, the diagnosis of GH excess has relied on showing inadequate GH suppression during a standard 75 g oral glucose tolerance test. Surgery is the preferred treatment, although long-acting somatostatin is also effective.

The prolactin axis

Dopamine

Dopamine is an inhibitor prolactin secretion

Prolactin is unique among the pituitary hormones in that it is under predominant inhibitory control from the hypothalamus (see Fig. 35.5). Furthermore, the controlling agent is the very simple molecule dopamine, which is produced by tuberoinfundibular dopamine neurons. Dopamine works by stimulating the pituitary lactotroph D2 receptor to inhibit adenylate cyclase and consequently inhibits both prolactin synthesis and secretion. Several neuropeptides, including TRH, have prolactin-releasing properties but there is little evidence for a physiologic role.

Prolactin

In association with other pregnancy-related hormones, prolactin may assist breast growth and milk formation

Prolactin is a 199 amino acid protein, molecular weight 23 kDa, which has 16% homology with GH. It is synthesized as a 28 kDa prohormone by transcription of a gene located on chromosome 6. There is only one definite role for prolactin in humans. During pregnancy prolactin binds to its receptor in mammary tissue and stimulates the synthesis of several milk proteins, including lactalbumin.

Clinical disorders of prolactin secretion

There are no known prolactin deficiency syndromes but hyperprolactinemia is very common

Hyperprolactinemia may result from a prolactin-secreting pituitary tumor (prolactinoma), a deficient supply of dopamine from the hypothalamus or the use of any of a wide range of antidopaminergic drugs. In women, the presenting features of hyperprolactinemia include menstrual irregularity and galactorrhea (discharge of milk from the breast). A grossly elevated serum prolactin is usually diagnostic of a prolactinoma. For subjects with modest hyperprolactinemia, who are not taking anti-dopaminergic drugs, the differential diagnosis is difficult; pituitary imaging and/or dynamic tests of prolactin secretion will assist the diagnosis of a microprolactinoma. Treatment options include long-acting dopamine agonist drugs or surgery. In men, there is no early indication of hyperprolactinemia, which means that prolactinomas can be very large and may present as hypopituitarism with visual field defects as the tumor expands out of the pituitary fossa and impinges on the optic chiasm. Such tumors will shrink with dopamine agonist therapy.

Prolactin exerts its effects on female reproductive function by blocking the action of FSH on estrogen secretion by the developing follicle.

Summary

The endocrine system fulfils the overall regulatory role in the organism. The body contains several endocrine systems comprising the hypothalamic releasing hormones, the pituitary tropic hormones and hormones secreted by target origin. Hormones are a multitude of molecules of different origin. Their action on cells is controled by receptors located either on cell membranes or intracellularly. The feedback mechanisms are the common way of controlling the endocrine systems. Both overactivity and underactivity of a hormone produce distinct clinical syndromes. Laboratory diagnosis of endocrine disorders relies heavily on the measurements of hormones in blood and also on function tests which test the integrity of endocrine regulatory systems.

The human body contains several other endocrine systems not considered in this chapter. Some of these are considered in other chapters as part of the physiologic function that they control. Thus the reader is referred to Chapter 20 for carbohydrate homeostasis, Chapter 23 for calcium homeostasis and Chapter 21 for water and electrolyte homeostasis and the control of blood pressure.

Further reading

Baulieu EE, Kelly PA (Eds). *Hormones: from molecules to disease.* London: Chapman and Hall, 1990.

Besser GM, Thorner MO (Eds). *Clinical endocrinology*, 2nd edn. London: Times Mirror International, 1994.

Felig P, Baxter JD, Frohman LA (Eds). *Endocrinology and metabolism*, 3rd edn. New York: McGraw-Hill, 1995.

Sheppard MC, Franklyn JA (Eds). Membrane surface receptors. *Baillière's clinical endocrinology and metabolism* 1996; **10**:1–192.

Wilson JD, Foster DW, Kronenburg HM, Larsen PR (Eds). *Williams textbook of endocrinology*, 9th edn. Philadelphia: WB Saunders, 1998.

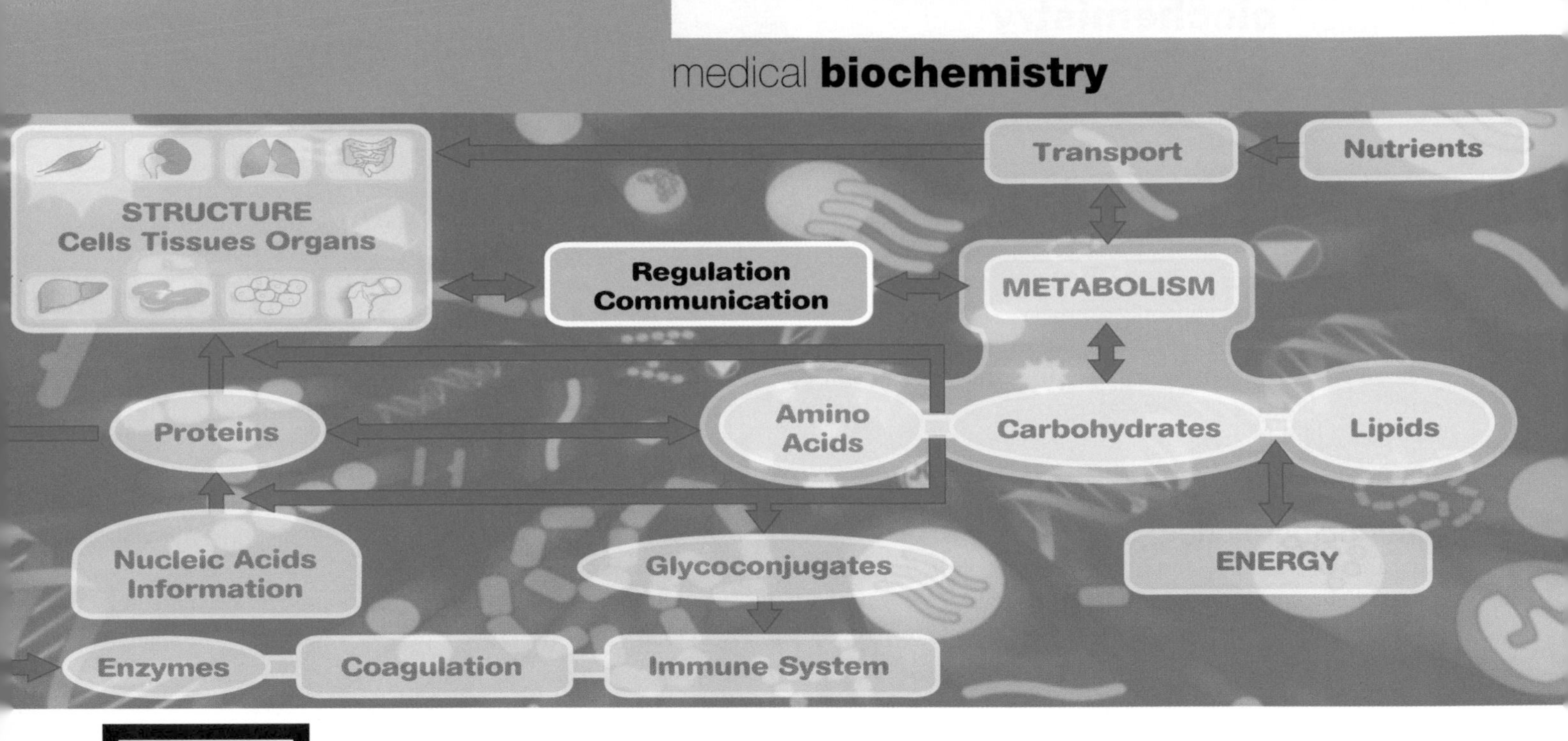

36 Membrane Receptors and Signal Transduction

Cells sense, respond to, and integrate a multiplicity of signals from their environment. Although some of these signals may be mediated by cell–cell contact, in multicellular organisms many signal molecules, such as hormones, originate in organs distant from their site of action and must be carried in the blood to their target effector cells. Likewise, immune cells such as phagocytes are recruited from the blood to sites of inflammation by migrating along chemoattractant gradients. Signals generated in these ways are sensed and processed by cellular signal transduction cassettes that comprise specific cell-surface membrane receptors, effector signaling elements, and regulatory proteins. These signaling cassettes serve to detect, amplify, and integrate diverse external signals to generate the appropriate cellular response (Fig. 36.1). In this chapter, we first discuss how cell-surface receptors sense and transduce their specific hormone signal by transmembrane coupling to effector enzyme systems, generating low-molecular-weight molecules termed second messengers. We then discuss the diversity of these second messengers and how they influence the activity of a range of key protein kinases with distinct substrates that ultimately determine the type of biological response obtained.

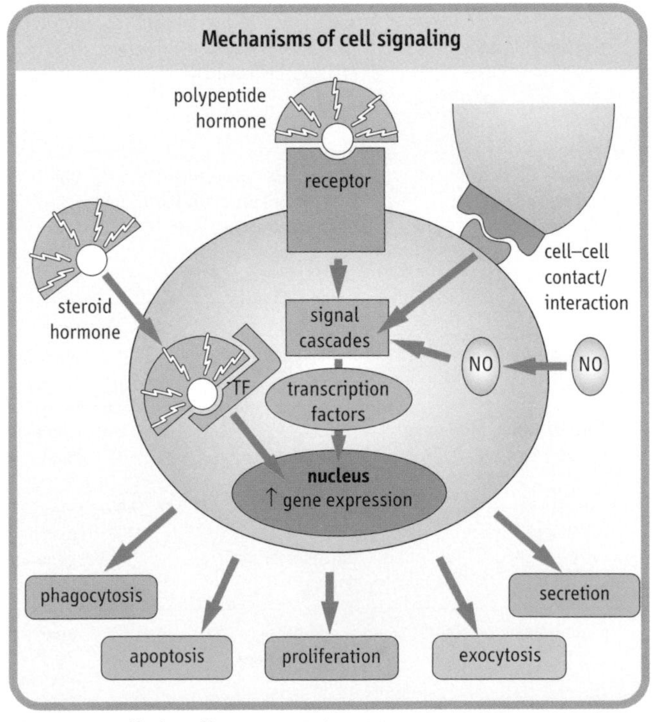

Fig. 36.1 Cell signaling. NO, nitric oxide.

Cell-surface membrane receptors

Hormones are biochemical messengers that act to orchestrate the responses of different cells within a multicellular organism. They are generally synthesized by specific tissues and secreted directly into the blood, which transports them to their target responsive organs. Hormones can broadly be subdivided into two major classes:

- **steroid hormones,**
- **polypeptide hormones.**

Steroid hormones

Because of the cholesterol-based nature of their structure, steroid hormones, such as cortisol (made in the cortex of the adrenal gland), sex hormones, and vitamin D, can directly traverse the plasma membrane of cells, to initiate their responses via cytoplasmically located receptors called steroid hormone receptors (see Fig. 36.1). These receptors belong to a superfamily of cytoplasmic receptors called the intracellular receptor superfamily, which also transduce signals from other small hydrophobic signaling molecules such as the thyroid hormones (e.g. thyroxine), which are derived from the amino acid tyrosine, and retinoids (e.g. retinoic acid), which are made from vitamin A. The intracellular receptors for these steroid and thyroid hormones and retinoids are transcription factors; they bind to regulatory regions of the DNA of genes that are responsive to the particular steroid/thyroid hormone. Such 'ligand binding' (ligation) induces a conformational change in the transcription factor that allows it to activate, or repress, gene induction.

Membrane receptors

Receptor class	Transmembrane-spanning domains	Intrinsic catalytic activity	Accessory coupling / regulatory molecules	Examples of receptor subclasses
G-protein-coupled receptors (serpentine receptors)	multipass (seven transmembrane α-helices)	none	G-proteins	β-adrenergic α-adrenergic muscarinic chemokines (IL-8) rhodopsin (vision)
ion-channel receptors (ligand-gated receptors)	multipass; and generally form multimeric complexes	none	none	neurotransmitters ions nucleotides inositol triphosphate (IP_3)
intrinsic receptor tyrosine kinases	single-pass transmembrane domain, but may be multimeric (e.g. insulin receptor)	tyrosine kinase	none	epidermal growth factor (EGF) nerve growth factor (NGF) platelet-derived growth factor (PDGF) fibroblast growth factor (FGF) insulin
tyrosine kinase-associated receptors	single-pass transmembrane domain, but generally form multimeric receptors	none	some require ITAM/ITIM-containing proteins	antigen receptors (ITAM–Src-related kinases) FcγR (ITIM–Src-related kinases) hemopoietin cytokine receptors (Janus kinases)
intrinsic tyrosine phosphatase receptors	single-pass transmembrane domain	tyrosine phosphatase	none	CD45-phosphatase receptor
intrinsic serine–threonine receptor kinases	single-pass transmembrane domain	serine–threonine kinase	none	tumor growth factor β (TGF-β)
intrinsic guanylate cyclase receptors	single-pass transmembrane domain	guanylate cyclase (generates cGMP)	none	atrial natriuretic peptide (ANP) receptors
death-domain receptors	single-pass transmembrane domain	none	death-domain accessory proteins (TRADD, FADD, RIP, TRAFS)	tumor necrosis factor (TNF-α) Fas

Fig. 36.2 Classification of membrane receptors. FADD, fas-associated death domain; FcγR, Fc-γ receptor (receptor for immunoglobin G); IL, interleukin; ITAM/ITIM, immunoreceptor tyrosine activation/inhibition motif; RIP, 'rest in peace' death domain; Src, Src-tyrosine kinase; TRADD, TNF-receptor-associated death domain; TRAFS, TNF-receptor-associated factors.

Although all the target cells have specific receptors for the individual hormones, they express distinct combinations of cell-type-specific regulatory proteins that cooperate with the intracellular hormone receptor to dictate the precise repertoire of genes that are induced. Hence the hormones induce specific differential sets of responses in different target cells.

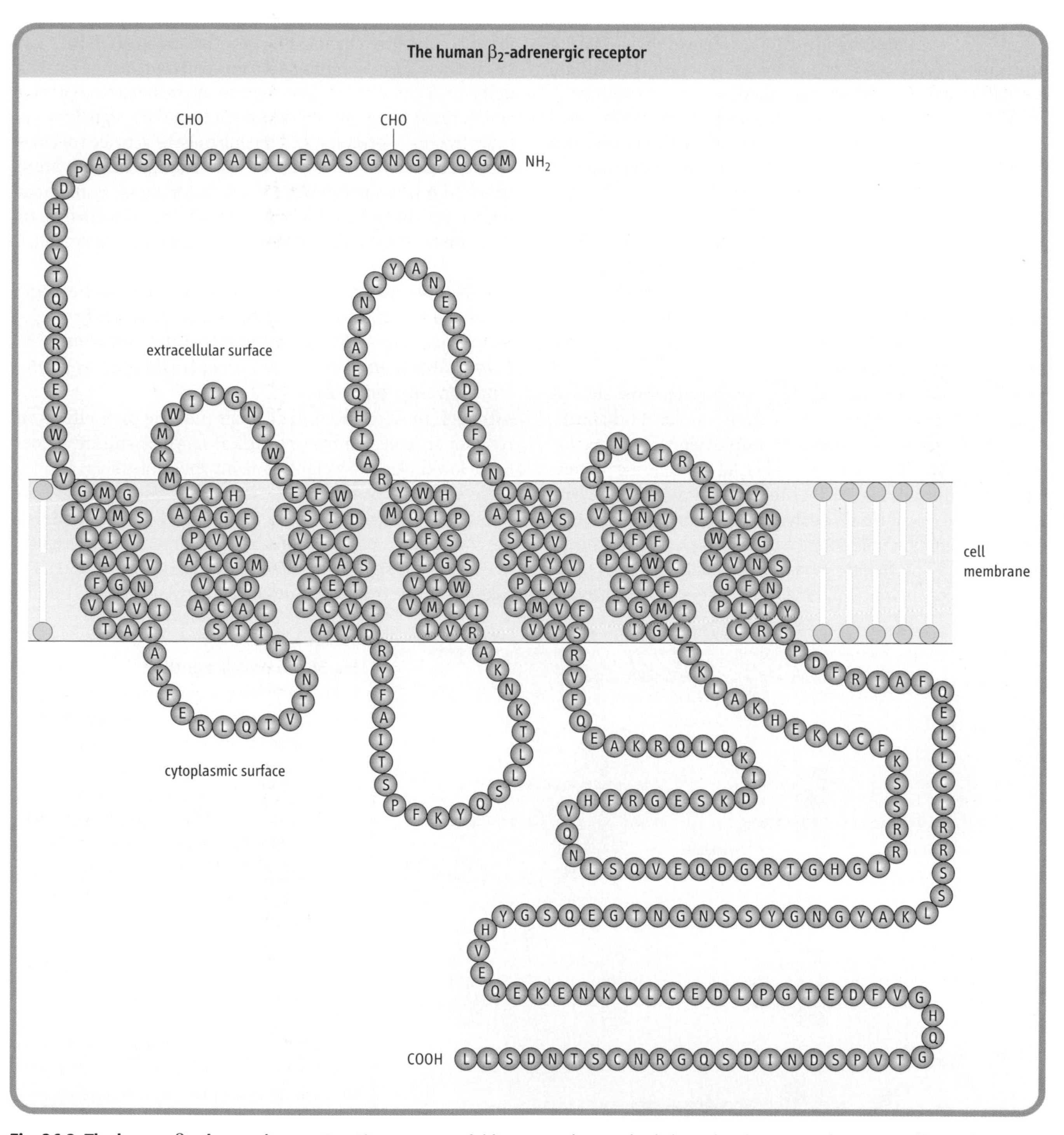

Fig. 36.3 The human β-adrenergic receptor. The sequence of this receptor is organized about the plasma membrane according to the seven membrane-spanning helices model. The ligand-binding site is deep within the plane of the membrane and is formed by amino acids from several of the transmembrane helices. Residues shown in blue in the third cytoplasmic loop and the beginning of the cytoplasmic tail represent regions that, when deleted or substituted, prevent coupling of the mutated receptor to adenylate cyclase. The blue residues in the first and second cytoplasmic loops (Leu^{64} and Pro^{138}) and the carboxyl tail (Cys^{341}) are conserved in many G-protein-coupled receptors and, when mutated, are found to affect markedly both the expression and the functionality of the receptor (see also Chapter 38).

Polypeptide hormones

In contrast to the steroid hormones, polypeptide hormones cannot cross cell membranes and must initiate their effects on their target cells via specific cell-surface receptors (see Fig. 36.1). As they do not themselves enter the target cell, they are termed 'first messengers' and their intracellular effects are mediated by low-molecular-weight signaling molecules such as cyclic adenosine monophosphate (cAMP) or calcium, which are called 'second messengers'. In fact, the term polypeptide hormones encompasses a wide range of families of hormones, growth factors, and cytokines that use transmembrane signal transduction cassettes to elicit their biological effects.

Families of specific cell-surface membrane receptors have evolved both to allow specific recognition of the target cell by a particular hormone and to initiate the appropriate signal transduction cassette to generate the appropriate response

In addition to hormone receptors, sensory systems such as vision (Chapter 37), taste, and smell use similar mechanisms of cell-surface membrane receptor-coupled signal transduction (see Fig. 36.2). Some of these receptors, for example the β-adrenergic receptors (see Fig. 36.3) or the antigen receptors on lymphocytes, have no intrinsic catalytic activity and serve simply as specific recognition units; these receptors use a variety of mechanisms, including adaptor molecules or catalytic active regulatory molecules such as G-proteins (guanosine triphosphatases, GTPases, which hydrolyze GTP), to couple them to their effector signaling elements, which are generally enzymes (often called signaling enzymes or signal transducers) or ion channels (Fig. 36.4). In contrast, other receptors – such as the intrinsic tyrosine kinase receptors for growth factors (e.g. platelet-derived growth factor, PDGF) or the intrinsic serine kinase receptors for molecules like transforming growth factor-β (TGF-β) – have extracellular ligand-binding domains and cytoplasmic catalytic domains. Thus, after receptor–ligand interactions (receptor ligation), these receptors can directly initiate their signaling cascades by phosphorylating and modulating the activities of target signal-transducing molecules (downstream signaling enzymes), which in turn propagate the growth factor signal by modulating the activity of further specific signal transducers or transcription factors, leading to gene induction (see Chapter 39).

Some low-molecular-weight signaling molecules traverse the plasma membrane and directly modulate the activity of the catalytic domain of transmembrane receptors or cytoplasmic signal-transducing enzymes

Although most extracellular signals mediate their effects via receptor ligation of either cell-surface or cytoplasmic receptors, some low-molecular-weight signaling molecules such as nitric oxide, which is involved in the processes of oxidative stress, are able to traverse the plasma membrane and directly modulate the activity of the catalytic domain of transmembrane receptors or cytoplasmic signal-transducing enzymes (see Fig. 36.1). For example, nitric oxide can stimulate guanylate cyclase, leading

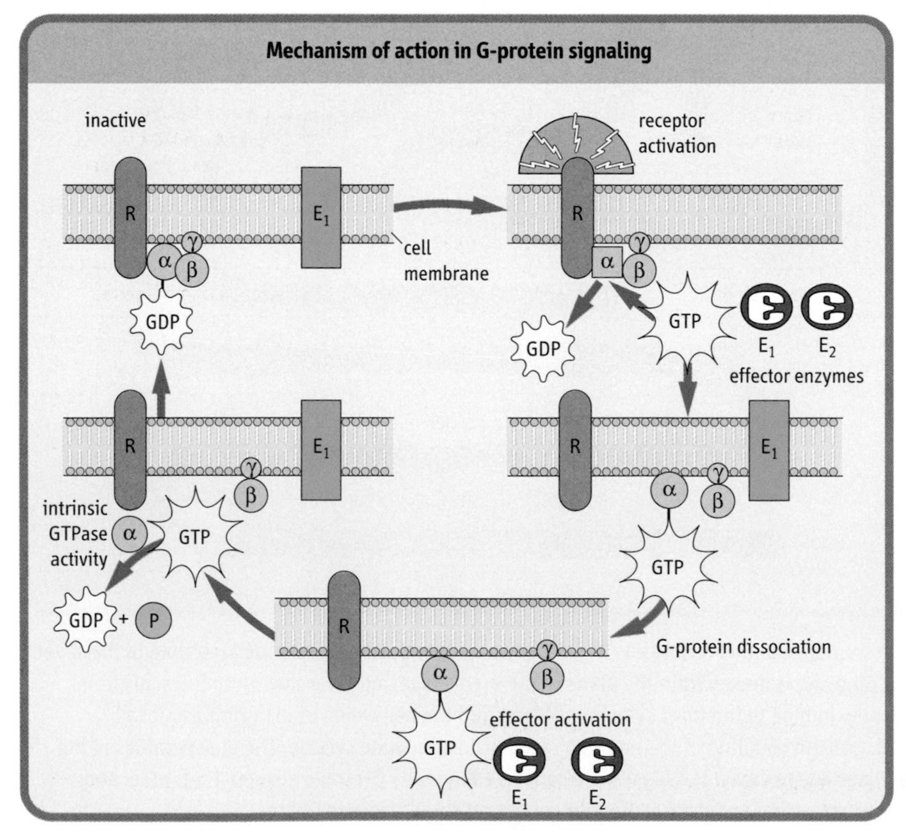

Fig. 36.4 G-protein signaling. In the inactive state, G-proteins exist as heterotrimers with GDP bound tightly to the α-subunit. None of the subunits is an integral membrane protein; however, the G-protein is anchored to the plasma membrane by lipid modification of the γ-subunits (prenylation) and some of the α-subunits (myristoylation of the Giα family). Ligation of the receptor (R) drives exchange of GDP for GTP and induces a conformational change in G_α, which results in a decrease in its affinity both for the receptor and for the βγ-subunits, leading to dissociation of the receptor–G-protein complex. The activated G_α (GTP-bound) or the released βγ-subunits, or both, can then interact with one or more effectors (E_1 or E_2), to generate intracellular second messengers that activate downstream signaling cascades. Signaling is terminated by the intrinsic GTPase activity of the α-subunit, which hydrolyzes GTP to GDP to allow reassociation of the inactive, heterotrimeric G-protein, $G_{\alpha\beta\gamma}$.

to the generation of the second messenger, cGMP. Nitric oxide has thus been postulated to have a number of signaling functions, ranging from being generated by neutrophils and macrophages in a microbicidal role, to signaling the relaxation of smooth muscle cells in blood vessels. This latter effect on smooth muscle cells provides a mechanism for the therapeutic effect of glyceryl trinitrate in patients with angina: the glyceryl trinitrate is converted to nitric oxide, resulting in relaxation of the blood vessels, and the consequent improvement in the oxygen delivery to the heart muscle results in alleviation of the pain that is caused by inadequate blood flow to the heart.

G-protein-coupled receptors: the β-adrenergic receptor

G-protein coupled receptors comprise a superfamily of structurally related receptors for hormones, neurotransmitters, imflammatory mediators, proteinases, taste and odorant molecules, and light photons. Many of these receptors have been cloned and sequenced, and a classic example of this class of receptors is the β-adrenergic receptor (for which the ligand is epinephrine), which has been extensively studied with respect to its structure–function properties in relation to its signal transduction cascade (see Fig. 36.3). G-protein-coupled receptors are integral membrane proteins characterized by the seven transmembrane-spanning helices within their structure. Typically, the receptor has an extracellular N-terminus, the seven transmembrane-spanning α-helices each comprising 20–28 hydrophobic amino acids, three extracellular and intracellular loops, and an intracellular C-terminal tail. Ligands, such as epinephrine, typically bind to the G-protein-coupled receptor by sitting in a pocket formed by the seven transmembrane helices. G-protein-coupled receptors have no intrinsic catalytic domains, and recruit guanine nucleotide-binding proteins (G-proteins) via their third cytoplasmic loop, to couple to their signal transduction elements.

G-proteins constitute a group of regulatory molecules are involved in the regulation of a diverse range of biological processes

Signal transduction, protein synthesis, and targeted delivery to the plasma membrane or intracellular organelles (intracellular trafficking); exocytosis; and cell movement, growth, proliferation, and differentiation all come under the regulatory influence of G-proteins. The G-protein superfamily predominantly comprises two major subfamilies: the small, monomeric Ras-like G-proteins (see Chapter 39), and the heterotrimeric G-proteins. Heterotrimeric G-proteins regulate the transduction of transmembrane signals from cell-surface receptors to a variety of intracellular effectors, such as adenylyl cyclase, phospholipase C (PLC), cGMP-phosphodiesterase (PDE) and ion-channel effector systems (Fig. 36.4). These G-proteins consist of three distinct classes of subunits: α (39–46 kDa),

Properties of mammalian G-proteins

G-protein subfamily	αβγ subunits	Molecular mass (kDa)	Toxin substrate	Tissue distribution	Effector
G_i	G_Z	41	none	brain, adrenal medulla, platelets	inhibits adenylate cyclase
	G_i G_0	40 40	pertussis toxin pertussis toxin	nearly ubiquitous brain, neural systems	G_i α-subunits activate PLC and PLA_2 and K^+ channels, and inhibit adenylate cyclase and Ca^{2+} channels
	G_t	40	pertussis/cholera toxin	retinal rods and cones	activates cGMP-phosphodiesterase
	G_{gust}	40	pertussis toxin	taste buds	activates phosphodiesterase
G_s	G_s	44–46	cholera toxin	ubiquitous	G_s activates adenylate cyclase and Ca^{2+} channels
	G_{olf}	45	cholera toxin	olfactory neuroepithelium	activates adenylate cyclase
G_{12}	G_{12} G_{13}	44 44	none none	ubiquitous ubiquitous	$G_{12/13}$ subunits regulate Na^+/H^+ exchange, voltage-dependent Ca^{2+} channels, and eicosanoid signaling
G_q	G_q G_{11} G_{14}	42 42 42	none none none	nearly ubiquitous nearly ubiquitous lung, kidney, liver, spleen, testis	PLC
	G_{15} G_{16}	43 44	none none	hemopoietic cells hemopoietic cells	

Fig. 36.5 Properties of the four main classes of mammalian G-protein α-subunits. Some G-proteins can be characterized using different bacterial toxin substrates. PDE, phosphodiesterase, cGMP, cyclic GMP; PLC, phospholipase C; PLA_2, phospholipase A_2.

β (37 kDa), and γ (8 kDa). In general, effector specificity is conferred by the α-subunit, which contains the GTP-binding site and an intrinsic GTPase activity; however, it is now widely accepted that βγ complexes can also directly regulate effectors such as phospholipase A_2 (PLA_2), PLC-β isoforms, adenylate cyclase, and ion channels in mammalian systems and, in addition, cellular responses such as mating-factor receptor pathways in yeast. Four major subfamilies of α-subunit genes have been identified on the basis of their cDNA homology and function: G_s, G_i, G_q, and G_{12} (Fig. 36.5). Many of these G_α subunits have been shown to exhibit a rather ubiquitous pattern of expression in mammalian systems, at least at the mRNA level, but it is also clear that certain α-subunits have a tissue-restricted profile of expression. Moreover, there is evidence of differential expression of α-subunits during cellular development.

Heterotrimeric G-proteins regulate transmembrane signals by acting as a molecular switch, coupling cell-surface G-protein-coupled receptors to one or more downstream signaling molecules (see Fig. 36.4). Ligation of the receptor initiates an interaction with the inactive, GDP-bound heterotrimeric G-protein. This interaction drives exchange of GDP for GTP, inducing a conformational change in G_α, which results in a decrease in its affinity both for the receptor and for the βγ-subunits, leading to dissociation of the receptor–G-protein complex. The activated G_α (GTP-bound) or released βγ-subunits, or both, can then interact with one or more effectors to generate intracellular second messengers, which activate downstream signaling cascades. Signaling is terminated by the intrinsic GTPase activity of the α-subunit, which hydrolyses GTP to GDP to allow reassociation of the inactive heterotrimeric G-protein ($G_{\alpha\beta\gamma}$).

Bacterial toxins which target G-proteins cause a range of diseases

A variety of bacterial toxins exert their toxic effects by covalently modifying G-proteins and hence irreversibly modulating their function. For example, cholera toxin (choleragen) from *Vibrio cholerae* contains an enzyme (subunit A) that catalyzes the transfer of ADP-ribose from intracellular NAD to the α-subunit of G_s; this modification prevents the hydrolysis of G_s-bound GTP, resulting in a constitutively (permanently) active form of the G-protein. The resulting prolonged increase in cAMP concentrations within the intestinal epithelial cells leads to phosphokinase A-mediated phosphorylation of Cl^- channels, causing a large efflux of electrolytes and water into the gut, which is responsible for the severe diarrhea that is characteristic of cholera. Enterotoxin action is initiated by specific binding of the B (binding) subunits of choleragen (AB_5) to the oligosaccharide moiety of of the monosialoganglioside, GM1, on epithelial cells. A similar molecular mechanism has been attributed to the action of the heat-labile enterotoxin, labile toxin, secreted by several strains of *Escherichia coli* responsible for 'traveler's diarrhea'.

In contrast, pertussis toxin (another AB_5 toxin) from *Bordetella pertussis*, the causative agent of whooping cough, catalyzes the ADP-ribosylation of G_i, which prevents G_i from interacting with activated receptors. Hence, the G-protein is inactivated and cannot act to inhibit adenylate cyclase, activate PLA_2 or PLC, open K^+ channels, or open and close Ca^{2+} channels, causing a generalized uncoupling of hormone receptors from their signaling cascades. For example, the α_2-adrenergic receptor mediates inhibition of insulin secretion from pancreatic islet cells by inhibiting adenylate cyclase in a G_i-dependent manner.

β-Adrenergic receptors are coupled to the generation of the second messenger, cAMP

The β-adrenergic hormone, epinephrine, induces the breakdown of glycogen to glucose in muscle and, to a lesser extent, in the liver. In the latter organ, the breakdown of glycogen is predominantly stimulated by the polypeptide hormone, glucagon, which is secreted by the pancreas when blood sugar is low (Chapter 20). One of the earliest signaling events after binding of these hormones to these receptors is the generation of cAMP, a small molecule that has a key role in the regulation of intracellular signal transduction, leading the conversion of glycogen to glucose. cAMP is derived from ATP by the catalytic action of the signaling enzyme, adenylate cyclase (Fig. 36.6). This cyclization reaction involves the intramolecular attack of the 3´-OH group of the ribose unit on the α-phosphoryl group of ATP to form a phosphodiester bond, and it is driven by the subsequent hydrolysis of the released pyrophosphate. The activity of cAMP is terminated by the hydrolysis of cAMP to 5´-AMP by specific cAMP-phosphodiesterases. The importance of cAMP in regulating glycogen breakdown was demonstrated by a series of experiments showing not only that hormones that activate adenylate cyclase activity in fat cells also stimulate glycogen breakdown, but also that cell-permeant analogs of cAMP, such as dibutyryl cAMP, can mimic the effects of these hormones in inducing glycogen breakdown.

The β-adrenergic receptor is coupled to adenylate cyclase activation by action of the α-subunit of the *stimulatory* G-protein, called G_s. Because many G_s α-subunits can be stimulated by each molecule of bound hormone, the orginal hormone signal is amplified by this form of transmembrane signaling. Although hydrolysis of GTP by the intrinsic GTPase of G_s α-subunit acts to switch off adenylate cyclase activation, the hormone–receptor complex must also be deactivated to return the cell to its resting, unstimulated state. This receptor desensitization involves phosphorylation of the C-terminal tail of the hormone-occupied β-adrenergic receptor by a kinase known as β-adrenergic receptor kinase, and it occurs after prolonged exposure to the hormone. G-protein-coupled receptors, such as α-adrenergic receptors, which act to inhibit cAMP generation, are coupled to the inhibition of adenylate cyclase via the *inhibitory*, G_i-, G-protein.

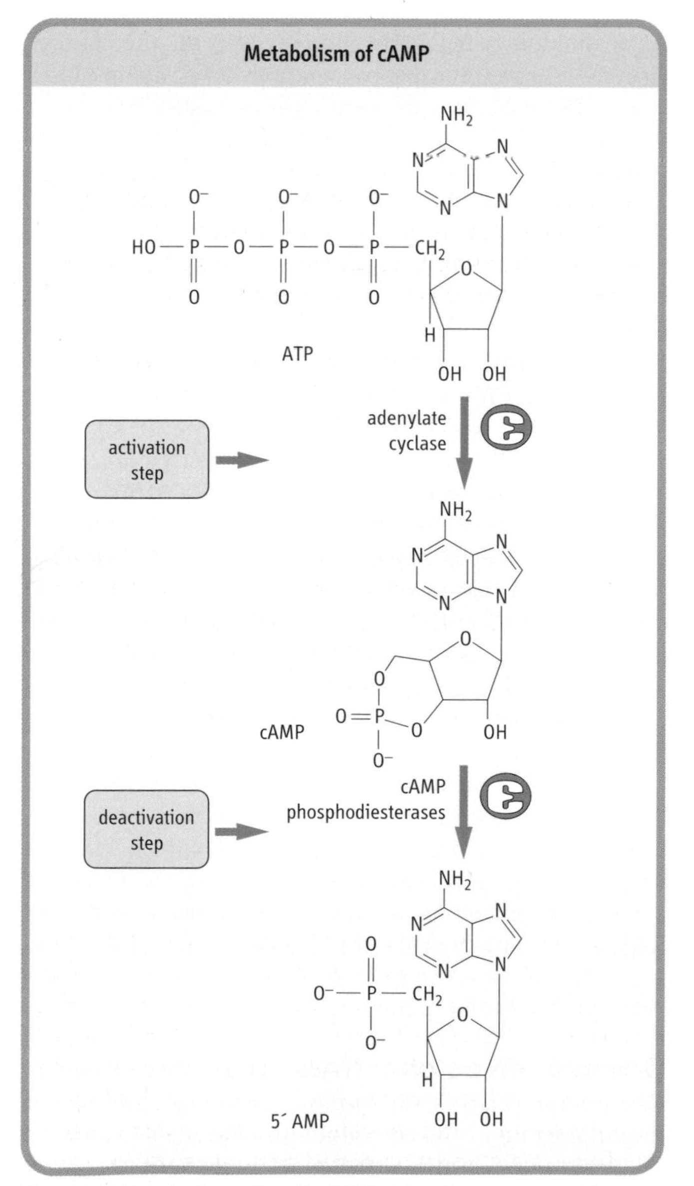

Fig. 36.6 Metabolism of cyclic AMP. Adenylate cyclase catalyzes a cyclization reaction to produce the active cAMP, which is then deactivated by cAMP-phosphodiesterases.

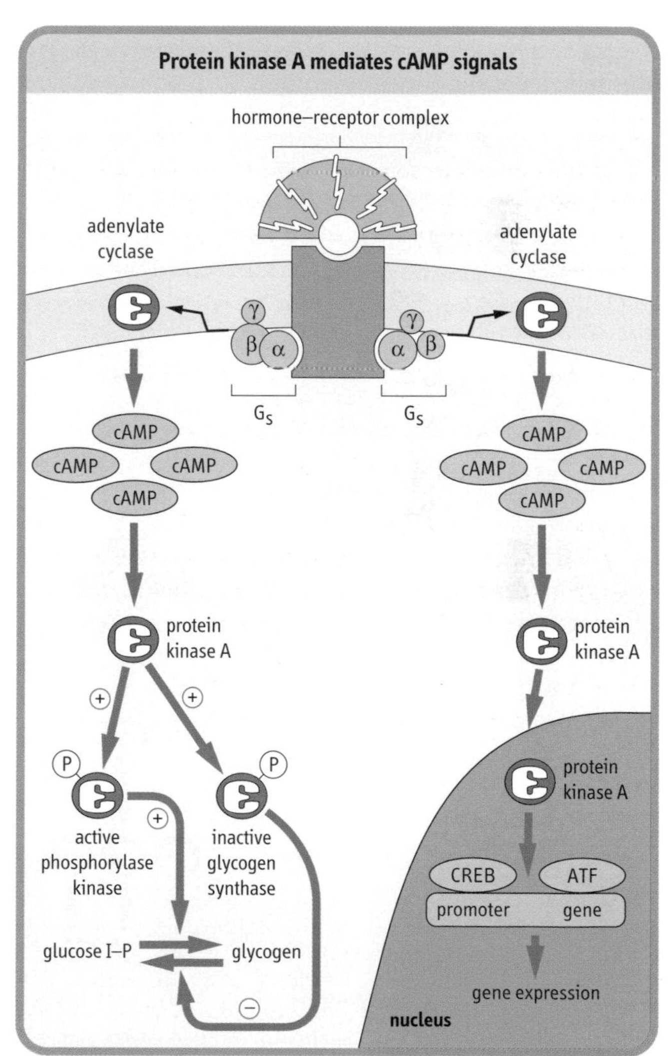

Fig. 36.7 PKA acts as a signaling enzyme for the second messenger, cAMP. Protein kinase A is activated by binding four molecules of cAMP, produced under the influence of adenylate cyclase, which in turn is activated by binding of the stimulatory G-protein (G_s) to the hormone–receptor (H–R) complex. Translocation of PKA into the nucleus modulates activity of transcription factors – for example CREB and ATF – leading to promotion or repression of gene expression.

Second messengers

Cyclic AMP

cAMP transduces its effects on glycogen–glucose interconversion by regulating a key signaling enzyme, protein kinase A (PKA), which phosphorylates target proteins on serine and threonine. PKA is a multimeric enzyme comprising two regulatory (R) subunits and two catalytic (C) subunits: the R_2C_2 tetrameric form of PKA is inactive, but binding of four molecules of cAMP to the regulatory subunits leads to the release of catalytically active C-subunits, which can then phosphorylate and modulate the activity of two key enzymes, phosphorylase kinase and glycogen synthase (see Fig. 36.7), which are involved in regulation of glycogen metabolism (see Chapter 12). Involvement of such a multilayered signal transduction cascade leads to substantial amplification of the orginal signal at each stage of the cascade (Fig. 36.8), ensuring that binding of only a few hormone molecules leads to release of a large number of sugar molecules.

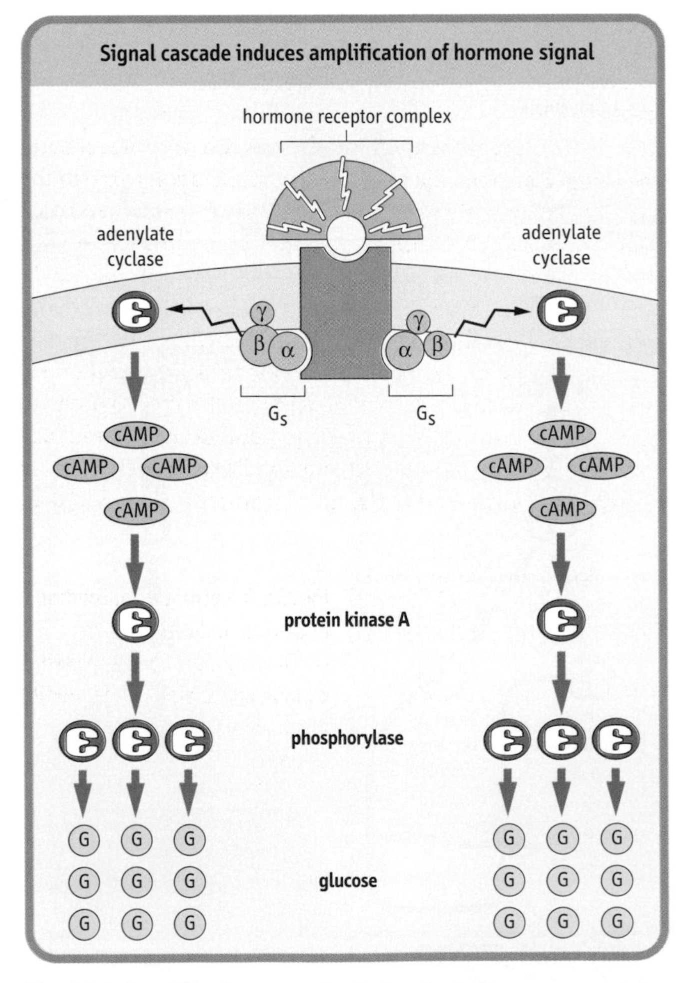

Fig. 36.8 Amplification cascade. Each activated hormone–receptor (H–R) complex can stimulate multiple G_s molecules. Each adenylate cyclase can produce many cAMP molecules, and each protein kinase A can activate many phosphorylase (P) molecules, leading to the release of a large number of glucose molecules (G) as a result of glycogen degradation.

Many other cellular responses can be mediated by the cAMP–PKA signaling cassette

PKA-mediated phosphorylation can regulate the activity of a number of ion channels such as K^+, Cl^- and Ca^{2+} channels, and that of phosphatases involved in the regulation of cell signaling. In addition, translocation of PKA into the nucleus allows modulation of the activity of transcription factors such as the cAMP-responsive-element-binding protein (CREB) or the activation transcription factor (ATF) families, leading to either the stimulation or the repression of expression of specific genes (see Fig. 36.7 and Chapter 32).

Transcription requires binding of RNA polymerase (RNAPol) to nontranscribed regions of the gene, called promoters, upstream (5′) of the start site; transcription factors are activators or repressors of gene expression that act by altering the rate of formation of the basal transcriptional complex on the promoter (Chapter 31). Transcription factors binding to regulatory sites on DNA can therefore be regarded as passwords that cooperatively open multiple locks to give RNAPol access to specific genes. It should be noted that binding of multiple transcription factors may be required for the induction of a single gene and, hence, activation of different combinations of transcription factors provides the specificity for switching on of particular genes in response to different hormones or growth factors.

Cyclic nucleotide phosphodiesterases have a key role in determining the cellular content of cAMP and, hence, PKA activation

Phosphodiesterases (PDEs) terminate the cAMP signal by converting cAMP to its 5′AMP metabolite (see Fig. 36.6); thus they have the potential to play key roles in the regulation of various physiologic responses in many different cells and tissues. Indeed, they have been postulated to regulate platelet activation, vascular relaxation, cardiac muscle contraction, and inflammation, making them attractive targets for the development of selective inhibitors as potential therapeutic agents. For example, one group of compounds, the methylxanthines – which include theophylline, papaverine, and isobutylmethylxanthine – have been used as bronchodilators in the treatment of asthma, and are also able to exert positive inotropic responses in the heart. Moreover, drugs selective for the PDE_3 class of enzymes – such as milrinone – are cardiotonic and increase the force of contraction of the heart, presumably by increasing the concentrations of cAMP and stimulating PKA, leading to phosphorylation of cardiac calcium channels and a subsequent increase in intracellular calcium concentration.

Calcium ion as an intracellular messenger

Calcium ion (Ca^{2+}) is a ubiquitous messenger and has an important role in the transduction of signals leading to cellular responses such as cell motility changes, egg fertilization, neurotransmission, protein secretion, and cell fusion, differentiation, and proliferation. Cells expend a considerable amount of energy maintaining a steep extracellular/intracellular Ca^{2+} concentration gradient: for example, the intracellular Ca^{2+} concentration in resting, unstimulated cells is of the order of 10^{-7} mol/L (100nM), whereas the extracellular Ca^{2+} concentration is typically 10^{-3} mol/L (1mM). This steep gradient allows for rapid, abrupt, transient changes in Ca^{2+} concentration: ligation of a wide range of hormone receptors, for example, leads to a rapid (seconds) and transient increase in intracellular Ca^{2+} concentration to the micromolar range. The rapid raising and lowering of Ca^{2+} concentrations is very tightly regulated and utilizes a variety of mechanisms involving cell compartmentalization. For example, intracellular Ca^{2+} concentrations can be lowered by sequestration of Ca^{2+} into the endoplasmic reticulum by Ca^{2+}-ATPases or into the mitochondria using the energy-driven electrochemical gradient. Alternatively, free Ca^{2+} can be chelated by Ca^{2+}-binding proteins such as calsequestrin.

Many downstream signaling events mediated by Ca^{2+} are modulated by a Ca^{2+}-sensing and binding protein, calmodulin

Calmodulin is a 17-kDa protein found in all animal and plant cells (it comprises up to 1% of cellular protein); it belongs to a family of proteins characterized by one or more copies of a Ca^{2+}-binding structural motif called an EF hand motif (Fig. 36.9). Calmodulin is comprised of two similar globular domains joined by a long (1.1 nm; 65Å) α-helix, each globular lobe having two EF hand motifs/Ca^{2+}-binding sites, 1.1 nm (11Å) apart. Binding of three to four calcium ions (which occurs when the intracellular calcium ion concentration is increased to about 500 nmol/L) induces a major conformational change that allows calmodulin to bind to and modify target proteins such as cAMP-PDE. Binding of several calcium ions allows cooperativity in the activation of calmodulin, such that small changes in Ca^{2+} concentration cause large changes in the concentration of an 'active Ca^{2+}–calmodulin (Ca^{2+}/CAM) complex', providing amplification of the original hormone signal.

Calmodulin has a wide range of target effectors, including Ca^{2+}/CAM-dependent protein kinases, which phosphorylate serine–threonine residues on proteins, including the broad-specificity kinase, Ca^{2+}/CAM-kinase II, which is involved in the regulation of fuel metabolism, ion permeability, neurotransmitter biology, and myosin light-chain kinase and phosphorylase kinase activity. Interestingly, calmodulin serves as permanent regulatory subunit of phosphorylase kinase and may also regulate nonkinase effectors such as certain adenylate cyclase isoforms and also cAMP-PDEs, indicating 'cross-talk' between cAMP- and Ca^{2+}-dependent signaling pathways. Moreover, mice defective in the calmodulin-dependent adenylate cyclase are deficient in spatial memory, showing that this signal transduction system is important for learning and memory in vertebrates.

The minor phospholipid species, phosphatidylinositol 4,5-biphosphate, is a second messenger signaling intracellular Ca^{2+} mobilization

The major breakthrough in the search for the second messenger responsible for calcium mobilization came in the early 1980s, when it was found that stimulation of receptors that were known to increase the intracellular free Ca^{2+} concentration could activate the prior phospholipase C-mediated hydrolysis of a minor phospholipid species, phosphatidylinositol 4,5-biphosphate (PIP_2), which typically represents about 0.4% of total phospholipids in membranes. PIP_2 is generated from phosphatidylinositol by two kinases, phosphatidylinositol-4-kinase (PI-4-K or PI kinase) and phosphatidylinositol-5-kinase (PI-5-K or PIP kinase), both of which are found in the plasma membrane (Fig. 36.10A).

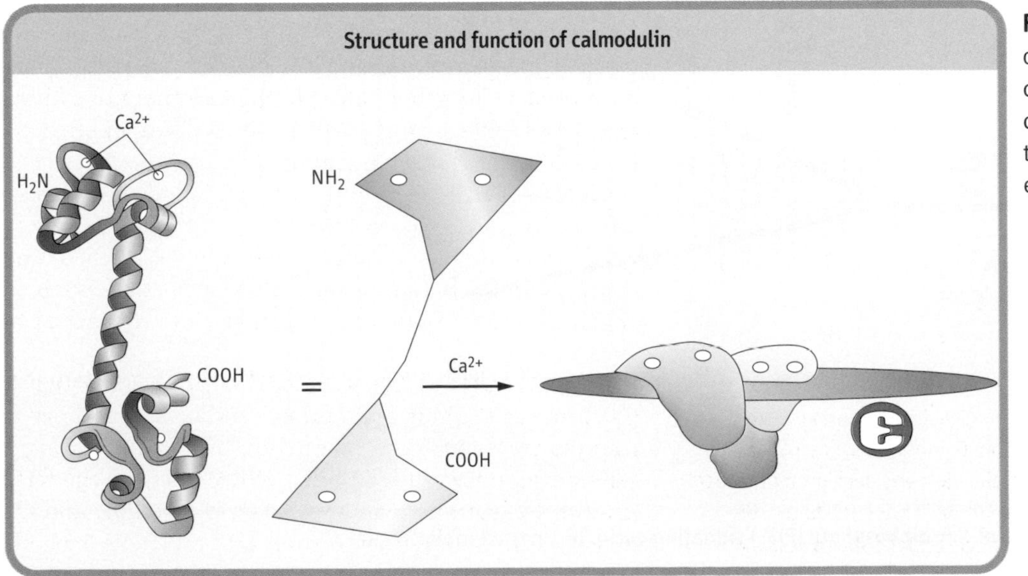

Fig. 36.9 Calmodulin. Binding of calcium induces a conformational change, allowing calmodulin to bind to and modify the activity of target signaling enzymes.

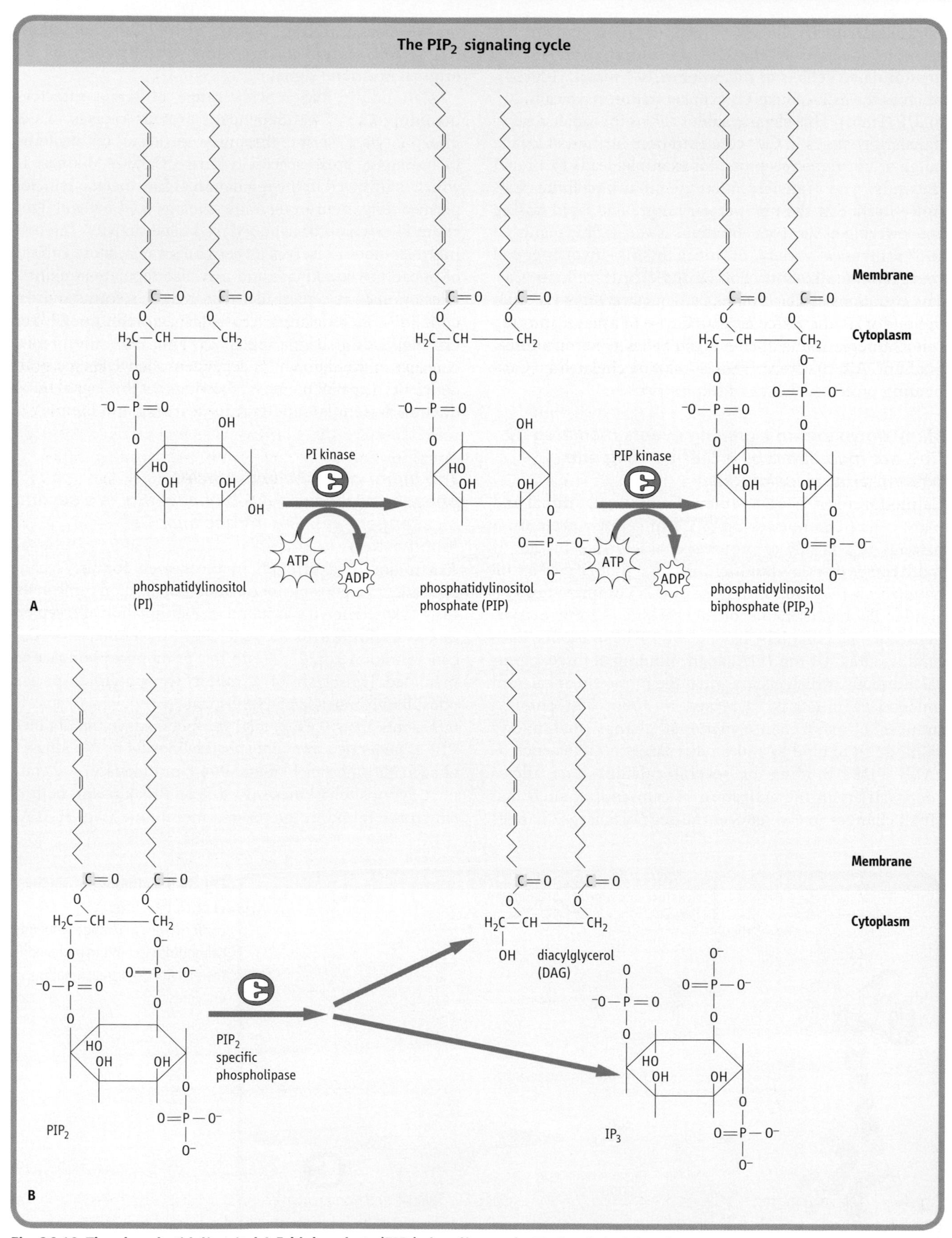

Fig. 36.10 The phosphatidylinositol 4,5-biphosphate (PIP_2) signaling cycle. IP_3, inositol triphosphate.

PIP_2 is hydrolyzed (Fig. 36. 10B) by a PIP_2-specific PLC, to generate two second messengers: inositol triphosphate (I-1,4,5-P_2 or IP_3), a water-soluble product that is released into the cytosol and which has been shown to mobilize intracellular stores of calcium, and diacyl glycerol (DAG), a membrane-anchored lipid second messenger that activates a key family of signaling enzymes known as protein kinase C (PKC). DAG is anchored in the membrane by virtue of its hydrophobic fatty-acid side chains inserted into the plasma membrane (Fig. 36.10B).

Second messenger role for IP_3

The rapid production (<1 second) and removal ($t_{1/2}$ = 4 seconds in liver) of IP_3 that precedes an increase in intracellular Ca^{2+} concentration implied that it could be a second messenger signaling calcium mobilization. That IP_3 is, indeed, a second messenger was confirmed by two types of experiment:

- **microinjection of IP_3 into cells,** or by simply including IP_3 in incubations of cells, in which detergent treatment had induced the formation of holes or pores in the plasma membrane (permeabilized cells), mobilized calcium in the absence of hormone stimulation;
- **structure–function analyses** of a variety of inositol phosphate molecules suggested that calcium mobilization was dependent on specific I-1,4,5-P_3 receptors.

IP_3 receptors that exhibit IP_3 binding and calcium release have now been cloned. These receptors have been shown to be expressed on the endoplasmic reticulum of all cells as a family of related glycoproteins (molecular mass 250 kDa) comprising six transmembrane-spanning domains. The active receptor is expressed as a multimer of four IP_3 receptor molecules; this tetrameric structure gives rise to cooperativity in Ca^{2+}-channel activity. It has been estimated that stimulated IP_3 (typically 2–3 µmol/L) releases 20–30 calcium ions. This calculation reveals the amplification inherent in this signaling cascade as, typically, only nmol/L concentrations of growth factor are required to elicit µmol/L concentrations of this second messenger. Consistent with the transient nature of the release of intracellular calcium that is observed after hormone receptor ligation, cellular concentrations of IP_3 are rapidly returned to resting values (0.1 µmol/L) by more than one route of degradation (Fig. 36.11).

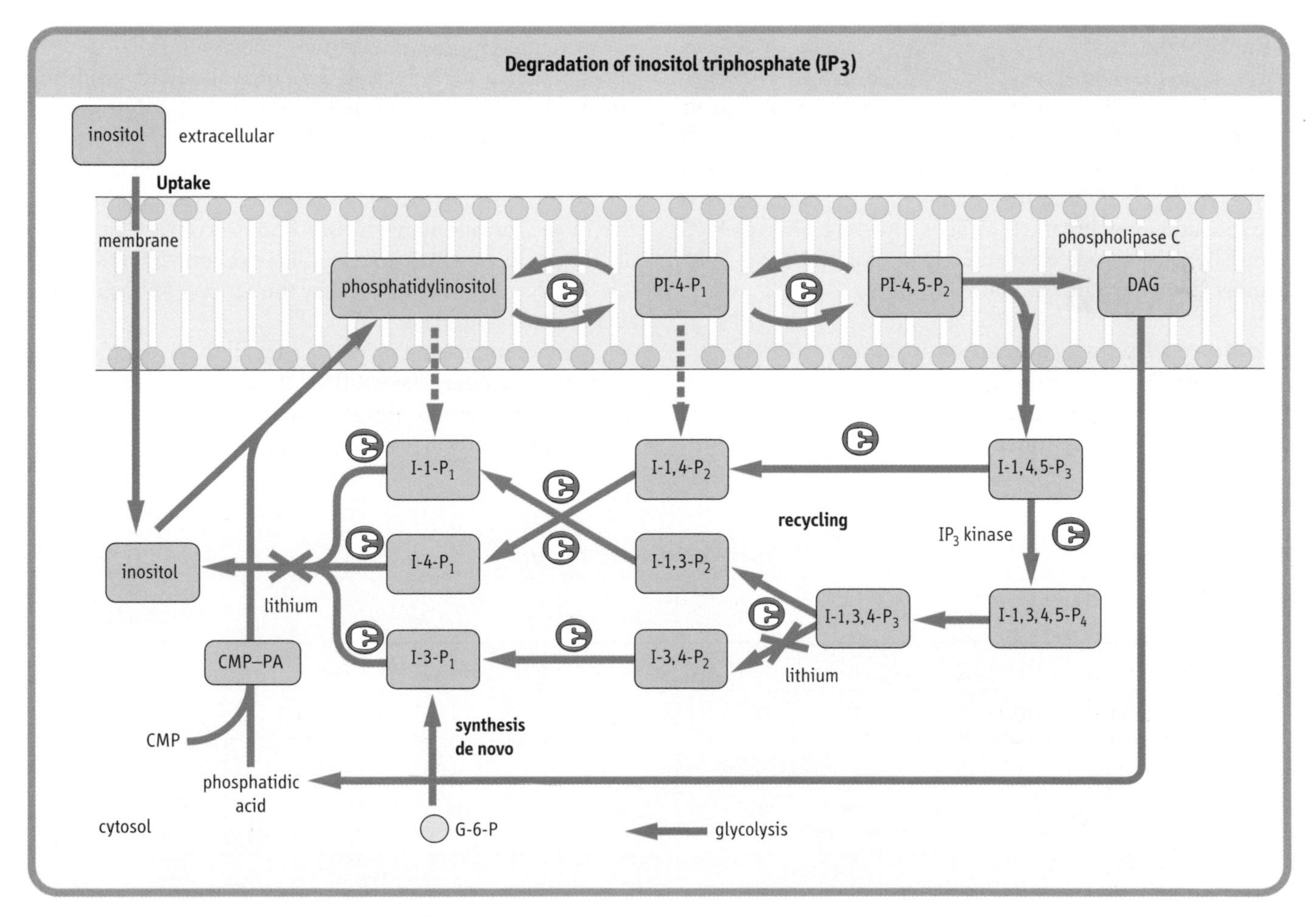

Fig. 36.11 The various routes of degradation of IP_3. The major pathways involve (i) the sequential action of phosphatases converting I-1,4,5-P_3 to inositol, and (ii) an IP_3 kinase, which generates I-1,3,4,5-P_4, which is in turn sequentially degraded to inositol by inositol phosphate-specific phosphatases. PA, phosphatidic acid; CMP–PA, cytosine monophosphate–phosphatidic acid.

DAG as second messenger

In addition to generating IP_3, PLC-mediated hydrolysis of PIP_2 generates another second messenger, DAG. DAG fulfils its second messenger role by activating the key signaling enzyme, PKC, which phosphorylates a wide range of target signal-transduction proteins on serine or threonine. PKC was originally identified as a calcium- and lipid (phosphatidylserine)-dependent kinase important in the regulation of cell proliferation. However, in recent years it has become clear that PKC is, in reality, a generic name for a superfamily of related kinases that have different tissue distribution and activation requirements (Fig. 36.12). Nevertheless, all these enzymes share some conserved features, most notably that they comprise two major domains: an N-terminal regulatory domain and a C-terminal catalytic kinase domain. The regulatory domain contains a pseudosubstrate sequence that resembles the consensus phosphorylation site in PKC substrates. In the absence of activating cofactors (Ca^{2+}, phospholipid, DAG), this pseudosubstrate sequence interacts with the substrate-binding pocket in the catalytic domain and represses PKC activity; binding of cofactors reduces the affinity of this interaction, induces a conformational change in the PKC, and allows stimulation of PKC activity. Consistent with the fact that the activator/cofactor, DAG, is anchored in the membranes, PKC activation is generally associated with translocation from the cytosol to the plasma or nuclear membranes. Although PKC is generally considered to be a serine kinase, it can phosphorylate threonine, but never tyrosine, residues. Interestingly, PKC can phosphorylate the same protein targets as does PKA but, whereas PKC generally phosphorylates the protein at serine residues, PKA usually phosphorylates threonine residues.

Are there alternative sources of DAG and PKC activators?

In many hormone systems, although receptor-stimulated hydrolysis of PIP_2 is transient and rapidly switched off (desensitized), generation of DAG is sustained. Together with the recent emergence of the multiple isoforms of PKC, some of which do not require Ca^{2+} or DAG for activation (Fig. 36.12), these findings led to the discovery of additional receptor-coupled lipid-signaling pathways involving hydrolysis of phosphatidylcholine or phosphatidylethanolamine (Fig. 36.13), which can give rise to DAG and other biologically active lipids (Fig. 36.14) in response to a wide range of growth factors and mitogens. Phosphatidylcholine comprises about 40% of the total cellular phospholipid. It can be hydrolyzed by distinct phospholipases, generating a diversity of lipid second-messengers, including arachidonic acid (generated by PLA_2) and different species of DAG (generated by PLC) and phosphatidic acid (generated by PLD). In addition, hormone-stimulated phosphatidylethanolamine-PLD activities have also been reported.

Some hormones or growth factors can stimulate only one or other of these phospholipases, but other ligands after stimulation of their specific receptors can stimulate all these pathways

Because some ligands can stimulate all these pathways after stimulation of their specific receptors, there is the potential to generate multiple distinct species of DAG and phosphatidic acid, reflecting the differential fatty-acid side chains (Fig. 36.15) expressed by the PIP_2 (predominantly stearate/arachidonate), phosphatidylcholine, and phosphatidylethanolamine substrates. Moreover, DAG species can be further metabolized to produce arachidonic acid via DAG lipase or, alternatively, DAG can be converted to phosphatidic acid via DAG kinase. Likewise, phosphatidic acid can be interconverted to DAG by the action of phosphatidic acid phosphohydrolase.

There is increasing evidence that all these distinct lipid second messengers have different targets. For example, it has recently been suggested that the saturated/monounsaturated fatty-acid-containing DAGs derived from

Protein kinase C superfamily

	Classical PKCs				Novel PKCs					Atypical PKCs	
PKC designation	α	β1	β2	γ	δ	ε/ε´	η	θ	μ	ι/λ	ζ
molecular mass (kDA)	82	80	80	80	78	90	80	79	115	74	72
activators											
$[Ca^{2+}]$	yes	yes	yes	yes	no	no	no	no		no	no
DAG	yes	yes	yes	yes	yes	yes	yes	yes		no	no
tissue distribution	all	some	many	neural	all	neural	many	muscle, skin		ovary, testis	all

Fig. 36.12 The PKC superfamily. This is a multigene family, except for β1, β2 and ε/ε´, which are alternatively spliced forms of β and ε. DAG, diacyl glycerol.

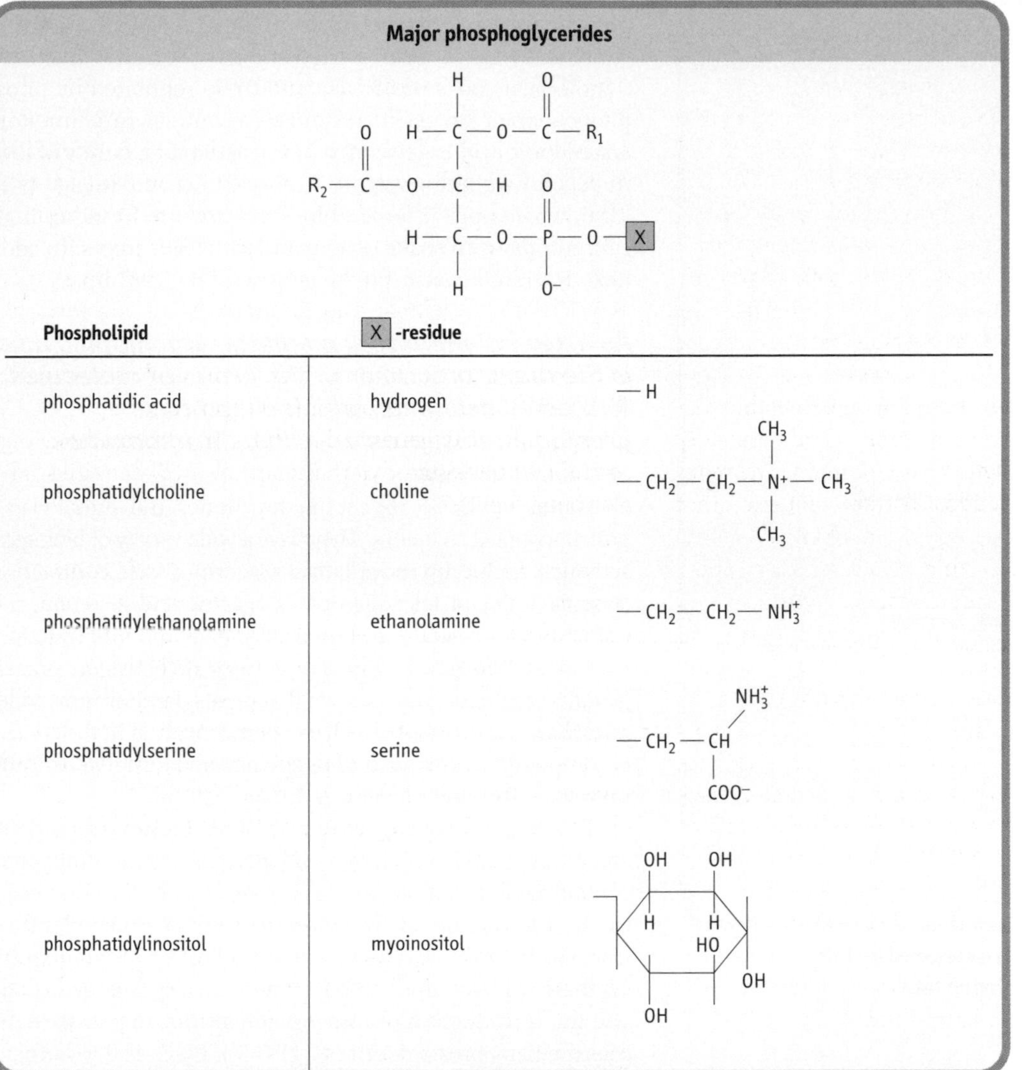

Phospholipid	X -residue
phosphatidic acid	hydrogen
phosphatidylcholine	choline
phosphatidylethanolamine	ethanolamine
phosphatidylserine	serine
phosphatidylinositol	myoinositol

Fig. 36.13 Potential lipid second messengers: phosphoglycerides. R1/R2, fatty acid side chains.

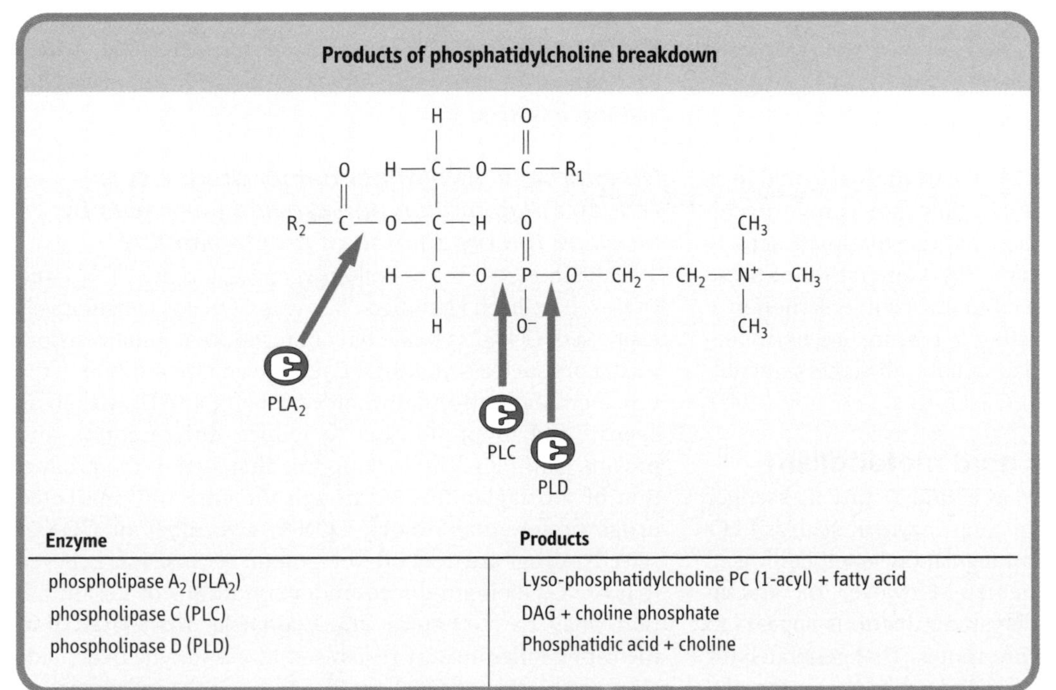

Enzyme	Products
phospholipase A_2 (PLA_2)	Lyso-phosphatidylcholine PC (1-acyl) + fatty acid
phospholipase C (PLC)	DAG + choline phosphate
phospholipase D (PLD)	Phosphatidic acid + choline

Fig. 36.14 Breakdown products of phosphatidylcholine.

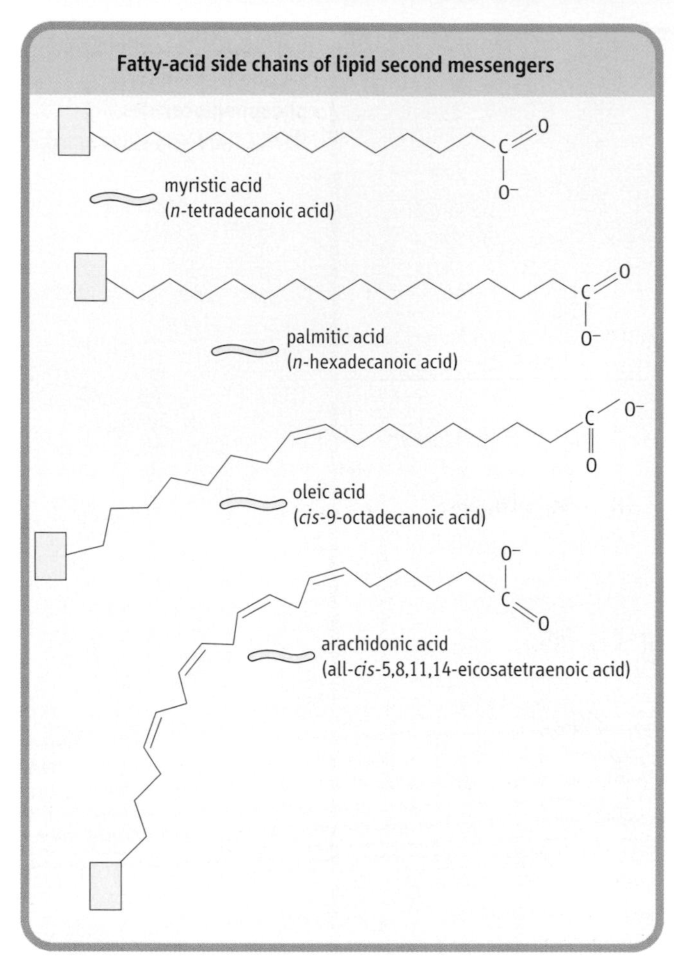

Fig. 36.15 Fatty-acid side chains expressed by PIP_2, phosphatidylcholine, and phosphatidyleth anolamine.

phosphatidylcholine-specific PLD activation are unable to activate PKC isoforms, and that it is only the stearoyl–arachidonyl–phosphatidic acid species that can modulate activity of the GTPase, Ras (see Chapter 39). Generation of these diverse, but related, lipid second messengers therefore provides a mechanism for initiating or terminating hormone-specific responses via particular signal transducers, including differential activation of PKC isoforms.

PLA_2 and arachidonic acid metabolism

In addition to being implicated as a lipid second messenger involved in the regulation of signaling enzymes, such as PLC-γ, PLC-δ, and PKC-α,β,γ isoform regulation, arachidonic acid is a key inflammatory intermediate. However, the arachidonic acid involved in these disparate functions appears to be generated by two distinct PLA_2 routes. That generated for signaling purposes appears to be derived by the action of a phosphatidylcholine-specific cytosolic PLA_2 ($cPLA_2$) activity, which has recently been cloned and has been shown to have a molecular mass of 85 kDa and to be regulated by phosphorylation of key serine residues. In contrast, inflammatory arachidonic acid is generated by the action of a family of low-molecular-weight secretory PLA_2 ($sPLA_2$) activities (14–18 kDa), which appear to be ubiquitous and are found in high concentrations in snake venom and pancreatic juices. In addition, arachidonic acid can be generated by DAG lipase.

As a key inflammatory mediator, arachidonic acid is the major precursor of the group of molecules termed eicosanoids, which encompass prostaglandins, prostacyclins, thromboxanes, and leukotrienes

Eicosanoids (Fig. 36.16) act like hormones and signal via G-protein-coupled receptors. They have a wide variety of biological activities, including modulating smooth muscle contraction (vascular tone), platelet aggregation, gastric acid secretion, and salt and water balance, and mediating pain and inflammatory responses. Moreover, 'knockout' mice defective in prostaglandin production have shown that prostaglandins have wider roles than was previously realized, being involved in the control of complex processes such as pregnancy, and tumor spread and metastasis in colon cancer (see Chapter 39).

Prostaglandins are synthesized in membranes, from arachidonic acid, which is a C20 polyunsaturated fatty acid containing four double bonds (see Fig. 36.15). The first stage in the conversion of arachidonic acid to prostaglandins involves the cyclo-oxygenase component of prostaglandin synthase, which induces the formation of a cyclopentane ring and the introduction of four oxygen atoms, to generate the intermediate, prostaglandin G_2 (PGG_2). PGG_2 is then subject to a hydroperoxidase reaction that catalyzes the two-electron reduction of the 15-hydroperoxy group to a 15-hydroxyl group and generates the highly unstable intermediate, PGH_2, which can then be converted into other prostaglandins, prostacylin, and thromboxane.

Nonsteroidal anti-inflammatory drugs act to reduce inflammation and provide pain relief by blocking the production of prostaglandins

Two distinct isoforms of cyclo-oxygenase, termed COX-1 and COX-2, have been identified. Whereas COX-1 is constitutively expressed, COX-2 is made only in response to inflammatory mediators such as cytokines. Cyclo-oxygenase inhibitors, the non-steroidal anti-inflammatory drugs (NSAIDs) such as aspirin and ibuprofen, act to reduce inflammation and provide pain relief by blocking this first step in the production of prostaglandins. Although the currently marketed drugs preferentially block COX-1, a number of COX-2-selective drugs are under development, because it is believed that COX-2, being induced under conditions of inflammation, may have a more important role than COX-1 in mediating inflammatory responses. Moreover, the major side effects of blocking COX-1 are bleeding and the inflammation of gastric mucosa, and the use of inhibitors selective for COX-2 may make it possible to target inflammatory activity and thus avoid most of the sequelae associated with the use of COX-1 inhibitors.

Arachidonic acid is also converted into other inflammatory and vasoactive mediators, called leukotrienes, by the action of a variety of structurally related lipoxygenases (see Fig. 36.16). For example, 5-lipoxygenase comprises both a dioxygenase

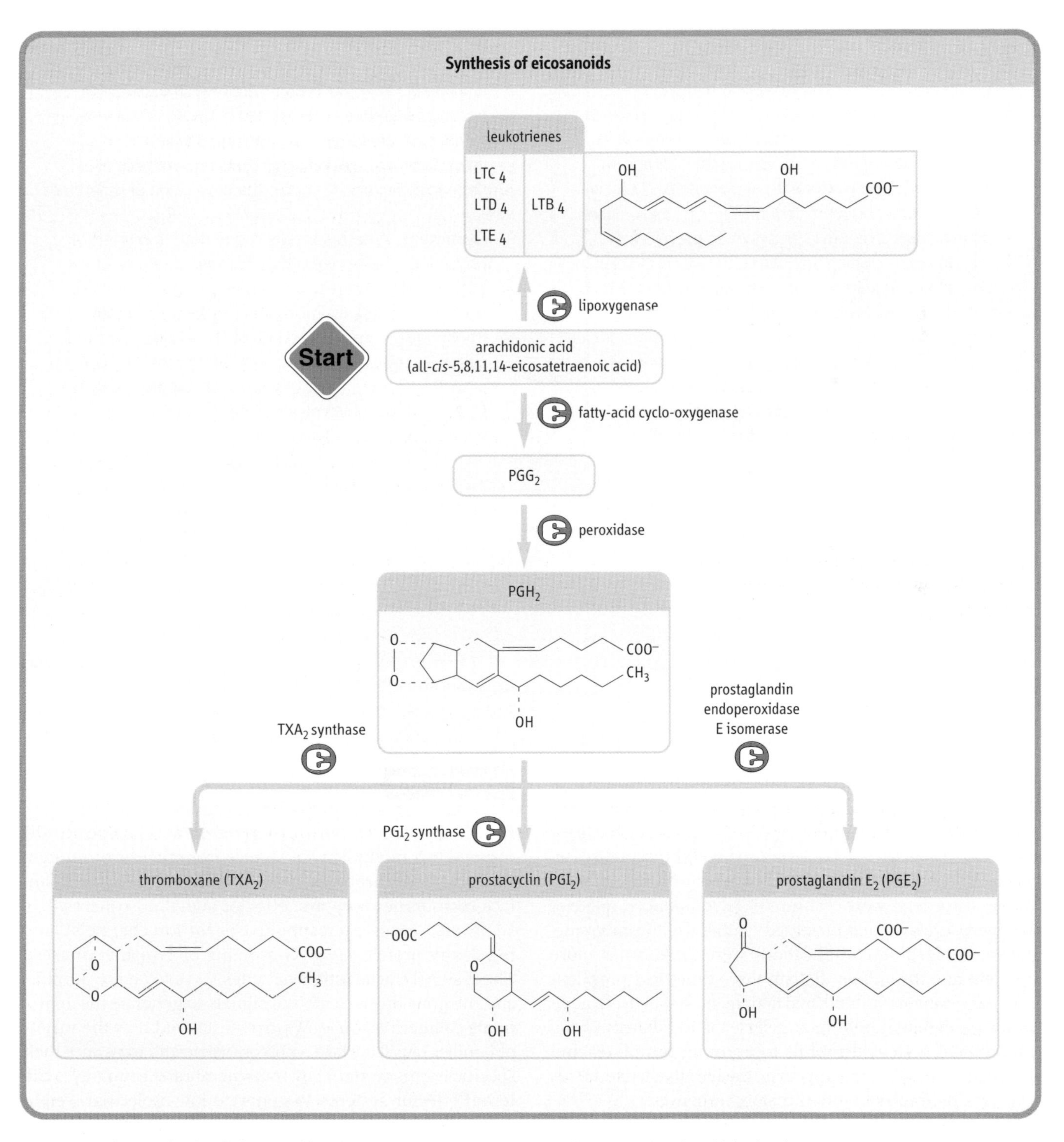

Fig. 36.16 The synthesis of eicosanoids. Eicosanoids are primarily derived from arachidonic acid, leukotrienes (LT) being derived via a lipoxygenase-dependent pathway, whereas prostaglandins (PG), prostacyclins, and thromboxanes (TX) arise from cyclo-oxygenase-dependent routes.

Anti-inflammatory drugs target prostaglandin synthesis

NSAIDs such as aspirin and ibruprofen decrease inflammation, pain, and fever through inhibition of the synthesis of prostaglandins, by blocking the first stage of the conversion of arachidonic acid to the common prostaglandin precursor, PGG_2. For aspirin (acetylsalicylate), this involves the covalent modification (acetylation) and irreversible activation of the cyclo-oxygenase activity of prostaglandin synthase. Similarly, because abrogation of cyclo-oxygenase activity will also block production of the potent vasoconstrictor and aggregator of blood platelets, thromboxane A_2 (TXA_2, also derived from the common precursor, PGG_2), aspirin can also be used as a prophylactic agent, to prevent the excessive blood clotting that can lead to heart attacks and stroke. Indeed, results from the second European Stroke Prevention Study, which assessed protection from secondary ischemic stroke, showed that the risk of stroke was reduced by 18% with aspirin alone in comparison with placebo. When aspirin was taken in conjunction with dipyridamole, which reduces platelet aggregation by increasing cAMP concentrations, the combination treatment produced a 37% reduction in risk. Corticosteroid hormones, such as cortisone, are also anti-inflammatory drugs that target prostaglandin synthesis; in this case, however, such reagents do not affect cyclo-oxygenase but, rather, appear to inhibit activation of the PLA_2 activity that generates the key intermediate, arachidonic acid. Such drugs are therefore useful for treating inflammatory responses involving leukocyte recruitment or asthma, as targeting arachidonic acid production will, in addition to blocking synthesis of prostaglandin, also abrogate the production of leukotriene.

Comment. Prostaglandins orchestrate a myriad of physiologic responses: they stimulate inflammation, regulate blood flow to particular organs, control ion transport across membranes, modulate synaptic transmission, and induce sleep. Prostaglandins are generally derived from the key inflammatory mediator, arachidonic acid, which in turn can be produced by PLA_2-mediated hydrolysis of various phospholipids such as phosphatidylcholine.

activity that converts arachidonate to 5-hydroperoxyeicosatetraenoic acid (5-HPETE), and a dehydrase activity that transforms 5-HPETE to leukotriene A_4 (LTA_4). Natural deficiencies in lipoxygenases have been associated with human disease: for example, in one study, 40% of patients with myeloproliferative disorders were found to have reduced platelet lipoxygenase activity and increased synthesis of thromboxane. Moreover, bleeding complications were three times more prevalent and thromboembolism three times less prevalent than in the patients with normal lipoxygenase activity. In addition, mice deficient in 5-lipoxygenase exhibit defects in the responses of their neutrophils to immune complexes and platelet-activating factor, supporting the idea that leukotrienes have important roles in inflammatory responses.

Summary

In this chapter we have discussed how cells specifically respond to a multiplicity of signals from their environment via signal transduction cassettes comprising specific cell-surface membrane receptors, effector signaling systems (e.g. adenylate cyclases, phospholipases, or ion channels), and regulatory proteins (e.g. G-proteins or tyrosine kinases). These signal transduction cassettes serve to detect, amplify, and integrate diverse external signals to generate the appropriate cellular response. We first discussed how the variety of families of cell-surface receptors sense and transduce their specific hormone signal by transmembrane coupling to different effector systems to generate low-molecular-weight molecules, termed second messengers, such as cAMP, IP_3, DAG, and Ca^{2+}, which mediate their signaling functions by activating key protein kinases. We then discussed how the specificity of a particular hormone response can be further heightened by the variety of phospholipase-signaling activities available (PLC, PLD, and PLA_2) that, taking into account their range of potential lipid substrates (e.g. PIP_2, phosphatidylcholine, and phosphatidylethanolamine) and products (e.g. DAG, phosphatidic acid, and arachidonic acid), can generate a diverse array of lipid second messengers to influence differentially the activity of the key protein kinases. Because, for example, many of the members of the PKC family appear to have distinct substrate repertoires of downstream signal transducers, differential activation of particular PKC isoforms can therefore ultimately determine the specific type of biological response obtained.

Further reading

Dawson AP. Calcium signalling: how do IP3 receptors work? *Curr Biol* 1997;**7**:R544–R547.
Diener HC *et al.* European Stroke Prevention Study. Dipyridamole and acetylsalicyclic acid in the secondary prevention of stroke. *J Neurol Sci* 1996;**143**:1–3.
Goetzl EJ, An S, Smith WL. Specificity of expression and effects of eicosanoid mediators in normal physiology and human diseases. *FASEB J* 1995;**9**:1051–1058.
Houslay MD, Milligan G. Tailoring cAMP-signalling responses through isoform multiplicity. *Trends Biochem Sci* 1997;**22**:217–224.
Newton AC. Regulation of protein kinase C. *Curr Opin Cell Biol* 1997;**9**:161–167.
Selbie LA, Hill SJ. G-protein-coupled-receptor cross-talk: the fine-tuning of multiple receptor-signalling pathways. *Trends Pharmacol Sci* 1998;**19**:87–93.
Toker A. The synthesis and cellular roles of phosphatidylinositol 4,5-bisphosphate. *Curr Opin Cell Biol* 1998;**10**:254–261.

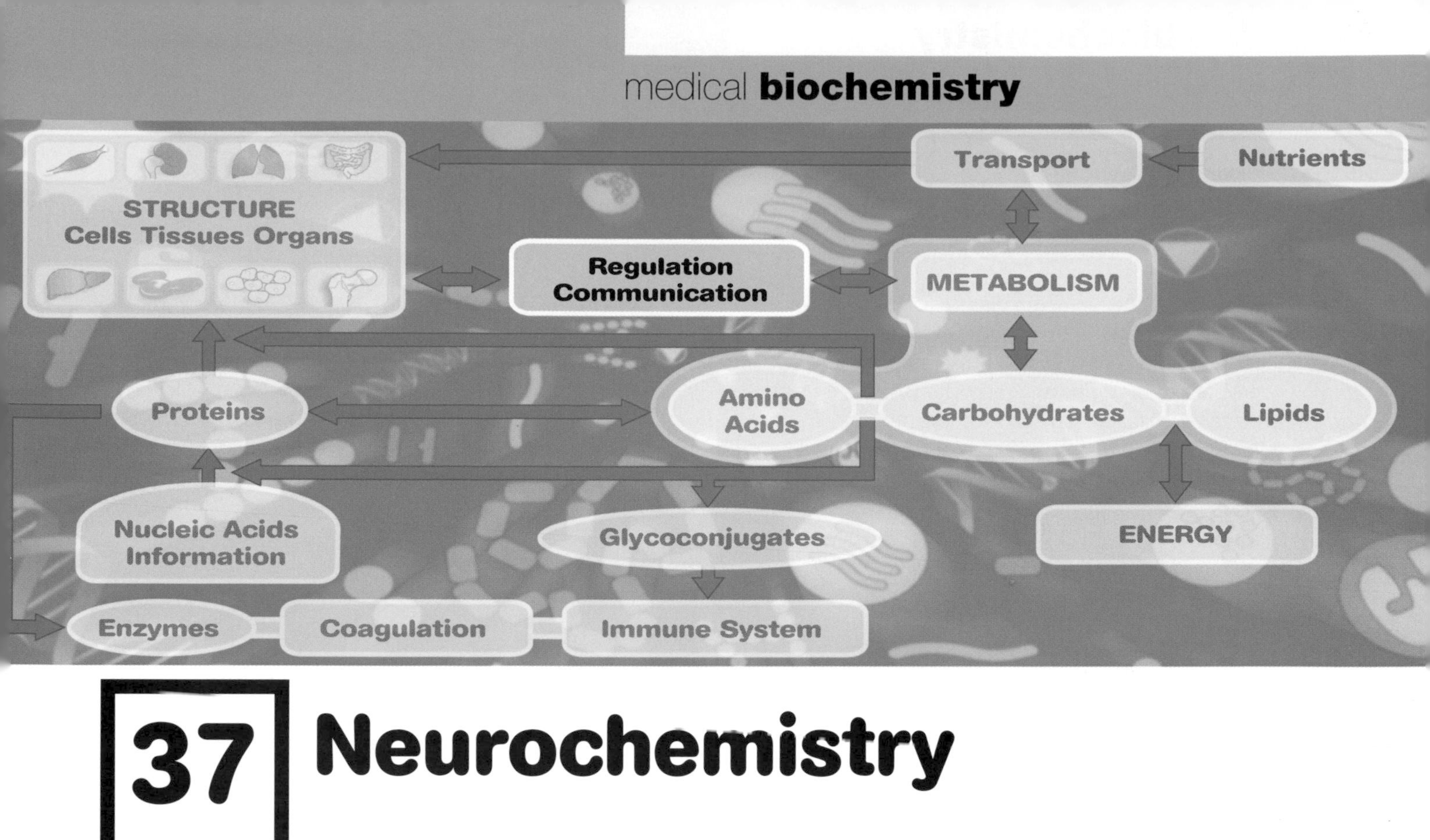

37 Neurochemistry

The brain is, in many ways, a chemist's delight. This is so because it illustrates various general principles of biology that have become highly specialized in the tissue that ultimately regulates all the other tissues of the body. This chapter will highlight the differences between the central nervous system – that is, the brain and spinal cord – and the peripheral nervous system, which is outside the dura (the thick fibrous covering that contains the cerebrospinal fluid [CSF]).

Brain and peripheral nerve

The distinction between brain and peripheral nerve essentially reflects the division between the central nervous system (CNS) and the peripheral nervous system (PNS): a convenient dividing line is afforded by the confines of the dura, within which watertight compartment is the CSF, partially produced (about one-third of the total volume) through the action of the blood/brain barrier. Myelin insulates the axons of nerves; the chemical composition of CNS myelin is quite distinct from that of PNS myelin, not least because the two forms are produced by two different cells: the oligodendrocyte within the CNS, and the Schwann cell within the PNS. The distinction between the separate functions of the CNS and those of the PNS is fundamental to differential diagnosis in neurology, and many tests exist to discriminate the two. A typical example is the difference between the demyelination of the CNS that occurs in multiple sclerosis, and the demyelination of the PNS that occurs in Guillain–Barré syndrome.

The blood/brain barrier

The term blood/brain 'barrier' is a slight misnomer, in that the barrier is not absolute, but relative: it's permeability depends on the size of a molecule

Initially, experiments based upon use of a dye (Evans' blue) bound to albumin showed that, over a period of hours, an animal progressively turned blue in all tissues, with the notable exception of the brain, which remained white. It subsequently became clear that 1 molecule in 200 of serum albumin passed normally into the CSF, which is analogous to lymph. It also became obvious that, for any given protein, the ratio of its concentrations in CSF and parallel serum was a linear function of the molecular radius of the molecules in solution.

There are a total of six 'barriers'. Under normal and pathologic conditions, proteins pass with varying degrees of filtration or local synthesis, or both, from other cellular or

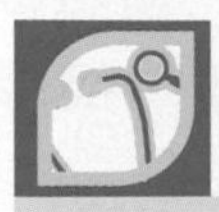

Guillain–Barré syndrome

Three weeks after an acute diarrheal illness, a 45-year-old man presented with a short history of 'pins and needles' in the feet, followed by an ascending weakness of the limbs and respiratory failure requiring assisted ventilation. On examination, he was hypotonic and areflexic, with profound global weakness. Isoelectric focusing of CSF and parallel serum samples showed an abnormal pattern of oligoclonal bands, which were being passively transferred from serum into the spinal fluid.

Comment. This is typical of Guillain–Barré syndrome, in which the patient probably has antibodies against the bacterium *Campylobacter jejuni*. This organism contains the antigen ganglioside sugar GM1, which is also found on the peripheral nerve. This is thus an example of molecular mimicry, in which antibodies against the bacteria in effect cross-react with the patient's own peripheral nerve.

tissue sources into the CSF, the total quantity of which therefore constitutes the algebraic summation of these six sources (Fig. 37.1):

- **the blood/brain barrier** (the parenchymal capillaries) gives rise to about one-third of the volume of CSF, and has been termed the interstitial fluid source;
- **the blood/CSF barrier** provides the bulk of CSF (almost all of the remaining two-thirds), termed choroidal fluid, as it is principally the choroid plexi (capillary tufts) situated in the lateral ventricles and, to a lesser degree, the plexi situated in the third and fourth ventricles;
- **the dorsal root ganglia** contain capillaries that have a much greater degree of permeability. In the animal experiments referred to above, although the brain was otherwise white, the dorsal roots took up the Evans' blue color, reflecting their greater permeability to albumin;
- **the brain parenchyma of the CNS** produces a number of brain-specific proteins. These include prostaglandin synthase (formerly called β-trace protein), which shows an 11-fold increase from the choroid plexus to the lumbar sac, and transthyretin (a protein that was formerly called prealbumin and is produced locally by the choroid plexi), for which the reverse is found: it has a much greater relative concentration in ventricular than in lumbar CSF;
- **CSF circulating cells**, mainly lymphocytes within the CNS, synthesize local antibodies: there is a strong immunosuppressive influence in the CNS. Because of this, in brain infections such as meningitis, steroids are given in addition to antibiotics, to suppress the potentially devastating effects, within this confined space, of inflammation associated with the intrathecal immune response;
- **the meninges** represent a sixth source of CSF under pathologic conditions; they can give rise to dramatic increases in the concentrations of CSF proteins.

Regional fluids within the brain

CSF leak

It is essential, on clinical grounds, to distinguish CSF rhinorrhea from local nasal secretions caused by, say, influenza infection. The ENT surgeon must know whether the fluid present is CSF, as any leak must be surgically repaired lest it remain a chronic potential source of meningitis as a result of the migration of nasal flora into the subarachnoid space. One characteristic and useful marker protein in the CSF is asialotransferrin, which is transferrin lacking sialic acid. In the systemic circulation, this absence of sialic acid gives a molecular signal for the protein to be recycled, and it is thus immediately removed from the systemic circulation by all reticuloendothelial cells. The brain has no true reticuloendothelial cells along the path of CSF flow, and hence asialotransferrin is present in quite high concentrations. The aqueous humor of the anterior chamber of the eye also produces the characteristic asialotransferrin, and the same asialotransferrin can also be found in the perilymph of the semicircular canals.

Iron in the CNS

During normal maturation of the oligodendrocytes that form the myelin sheath, iron appears to be important, as there are increased local concentrations of this cation and of the transferrin protein that is required to bind it (including the asialo form). An apparent discrepancy between the amounts of iron and those of transferrin is readily explained by the local synthesis of the latter, particularly of the asialo form; large amounts of mRNA are found in oligodendrocytes and in the choroid plexus.

There is also local synthesis of ferritin by the normal CNS. Free iron can have a particularly toxic effect on the CNS and, by binding to it, ferritin provides a store for it. This is clearly evident in the event of cerebrovascular accidents, after which there is a further dramatic increase in the local synthesis of ferritin, to bind the iron that has been released by the destruction of red cells.

CSF circulating cells

'Regional' immunology

As mentioned in the adjacent text, the brain is essentially an immunologically 'quiet' place. There is an almost 10-fold difference between the ratio of helper cells (CD4) to suppressor cells (CD8) in the CSF, where the ratio is 3.6, and that in the brain parenchyma, where the ratio is 0.4, reflecting a preponderance of suppressor cells within the CNS (see also Chapter 34). Approximately one-third of CNS cells are sessile macrophages; within the CSF, two-thirds of the cells are lymphocytes and the remaining one-third are macrophages. There are very few lymphatics within the brain itself, and CSF can be thought of as being analogous to lymph.

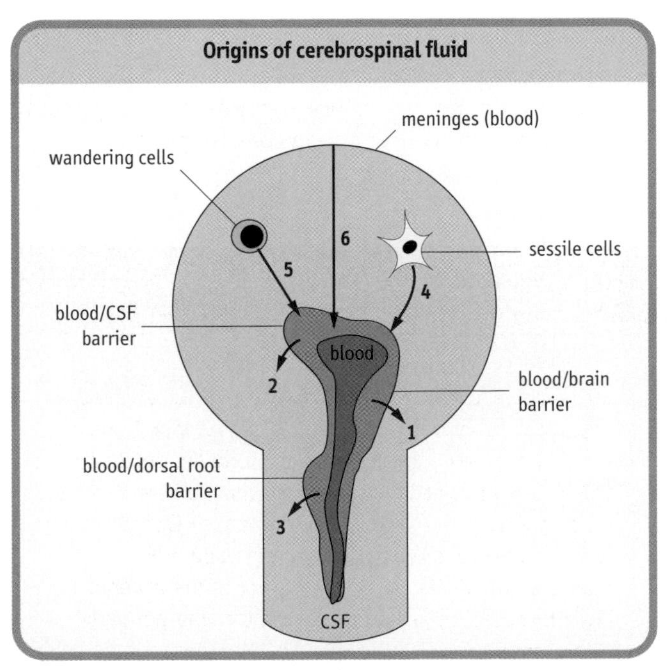

Fig. 37.1 The six main sources of CSF. These comprise passage across barriers (from the blood, i.e. 1,2,3) and direct sources of local production (CNS cells, i.e. 4,5,6).

The different cells of the nervous system

Fewer than 10% of the cells of the nervous system are large neurons. The three major cell types in the nervous system (which each constitute about 30%) are:

- **astrocytes**, which also make up part of the blood/brain barrier;
- **oligodendrocytes**, which are principally composed of fat, and serve to insulate the axons;
- **microglia**, which are essentially sessile macrophages (scavengers).

These different cell types are associated with predominant protein molecules that are of significance in various brain pathologies (Fig. 37.2). Other minor constituents of the nervous system include the ependymal cells, which are ciliated cells secreting brain-specific proteins such as prostaglandin synthase. The brain endothelial cells also merit special comment because, unlike other tissue capillaries, contiguous endothelial cells have tight junctions that bind them together; this feature is believed also to contribute to the blood/brain barrier, although the basement membrane is the major source of molecular sieving of the different sized proteins.

CNS cells and markers for brain pathology

Cell	Protein	Pathology
neuron	neuron-specific enolase	brain death
astrocyte	GFAP	plaque (or scar)
oligodendrocyte	myelin basic protein	de/remyelination
microglia	ferritin	stroke
choroid plexi	asialotransferrin	CSF leak (rhinorrhea)

Fig. 37.2 The different cells of the CNS, and their protein markers for brain pathologies. GFAP, glial fibrillary acidic protein.

Neurons

Their length, their many interconnections, and the fact that they do not divide postpartum, are significant features of neurons

There is an archetypal notion of the electrical activity of the nervous system – in particular, of the electrical activity of neurons. However, three other biological features of neurons are particularly worthy of note: their length, their prolific interconnections, and the fact that they do not divide postpartum. Further to the last of these features and the segmented nature of the embryo, many embryonic neurons are destined to die (through apoptosis), as they have no postembryonic target organ. The existence of cervical and lumbar swellings of the spinal cord, reflecting segmental neurons for which, respectively, the target organs were the hands and feet, is evidence for reduced diameter of the cord due to neuronal death. The capacity for memory and learning similarly indicates the importance of the plasticity of the synaptic system afforded by the extensive interconnections of and between neurons.

Because of their great length, neurons depend upon an efficient system of axonal transport

Neurons can typically be 1 m long; thus the nucleus, the source of information for the synthesis of neurotransmitters, is typically quite remote from the synaptic terminal, the site of release of those transmitters. Because of this extensive length of the neuron, a crucial requirement is its ability to transport material both from the nucleus towards the synapse (anterograde transport) and in the reverse direction (retrograde transport). To deal with this extreme separation of their two functional sites, and to maintain electrical activity at the nodes of Ranvier, as the remainder of the axon is electrically quiet during the saltatory process of electrical conduction, neurons have evolved special characteristics (Fig. 37.3).

The normal 'resting' movement within the axon is mediated by separate molecular 'motors' (motile proteins): kinesin in the case of anterograde transport, and dynein in retrograde transport. The materials being transported in each direction are also rather different, and the different components of axonal structure shown in Figure 37.3 possess the capacity for different speeds of transportation (Fig. 37.4). During growth, a

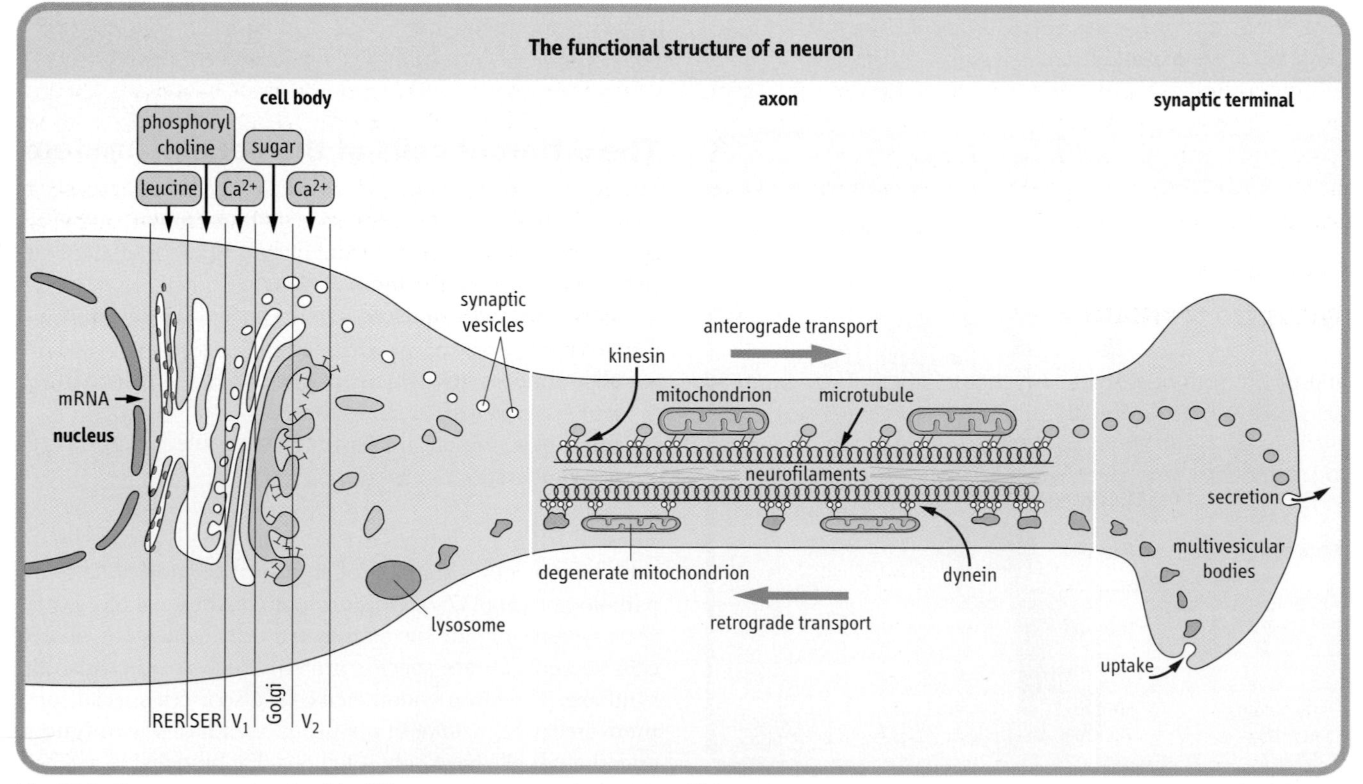

Fig. 37.3 The structures involved in different movements that occur within a neuron. Within the cell body, there is specialized movement through the Golgi stack by the components required to form synaptic vesicles (V_1, V_2). In the axon, there is fast axonal transport along microtubules via the motile proteins, kinesin (in anterograde transport) or dynein (in retrograde transport). RER, rough endoplasmic reticulum; SER, smooth endoplasmic reticulum.

separate form of transport (toward the synapse) occurs that takes place at the rate of about 1 mm/day; this flow constitutes bulk movement of the building blocks such as the filamentous proteins.

Neuroglial structures

Essentially, the astrocytes and the oligodendrocytes comprise the neuroglial structures

In the cortex, or gray matter, one typically finds a protoplasmic astrocyte with one set of processes surrounding the endothelial cells, thereby helping to 'filter' materials from the blood, and a separate set of processes surrounding the neurons, which are thereby being 'fed' selected substances that have been extracted from the blood for passage to the neurons. In the white matter, the astrocytes have a rather more fibrous appearance, and have more of a structural role. Under pathologic conditions in which there is injury to the CNS, astrocytes can play a major part in the reaction, synthesizing large amounts of the glial fibrillary acid protein (GFAP). This is the cellular equivalent of scar tissue, and is found in diseases such as multiple sclerosis, in which it is the major constituent of the characteristic plaques. Astrocytes are not found in the PNS.

The oligodendrocytes of the CNS can wrap round as many as 20 axons, forming the myelin sheath that insulates these neuronal processes from one another, and thereby stopping cross-talk between neurons. There is also intense oligodendrocyte mitochondrial activity at the nodes of Ranvier, which are parallel to the sites of depolarization within the underlying axon. In the PNS, the Schwann cells form the myelin and, typically, wrap round only a single axon. As noted previously, the chemical constituents of the myelin sheath are different when produced by Schwann cells rather than by oligodendrocytes.

Synaptic transmission

One of the unique chemical characteristics of the brain is the massively high density of synapses between different neurons, at which a locally acting neurohormone is released by one axon onto many other cell bodies. On the receiving end, a given cell body will typically receive a myriad of cellular products via its profusely branched dendritic tree: each branch can be smothered in synapses. The first chemical messenger or 'neurotransmitter' to traverse the synaptic cleft is the neurohormone, which is released by the axon of the first cell onto the dendrite of the second cell. There is usually a second messenger such as a cyclic nucleotide, which may also lead to a third messenger such as a phosphorylated protein. Typically, G-proteins are found just under the neurotransmitter receptor protein spanning the cell membrane, where they act to 'couple' the first messenger (e.g. norepinephrine) to a second messenger (e.g. cyclic AMP [cAMP]).

Neurotransmitters are normally inactivated after their postsynaptic actions on the target cell, hydrolysis being a major mechanism by which this is achieved. The best studied example is that of the enzyme, acetylcholinesterase. There can also be blockade at the level of the second messenger, such as cAMP, which is broken down by the enzyme phosphodiesterase. This enzyme is inhibited by methylxanthines and caffeine, and thereby mimics many of the effects of adrenergic neurotransmission.

Synaptic transmission also involves the recycling of membrane components

In addition to release of a specific neurohormone, there is also an extensive system for recycling of membrane constituents associated with this process. The synaptic vesicles contain a very high concentration of the relevant neurotransmitter, which is bounded by a membrane (see Chapter 38). During synaptic release of the transmitter, there is fusion of the synaptic vesicle membrane (containing the neurotransmitter) with the presynaptic membrane. This increase in total membrane mass is redressed by invagination of the lateral aspects of the nerve terminals, where an inward puckering movement of the membrane is effected by contractile movements of the protein, clathrin. There then follows a form of pinocytosis of the excess membrane, which is transported in retrograde fashion toward the nucleus, to be digested in lysosomes.

Differing speeds of axonal transport

Component	Rate (mm/day)	Structure and composition of transported substances
fast transport		
anterograde	200–400	small vesicles, neurotransmitters, membrane proteins, lipids
mitochondria	50–100	mitochondria
retrograde	200–300	lysosomal vesicles, enzymes
slow transport		
slow component a	2–8	microfilaments, metabolic enzymes, clathrin complex
slow component b	0.2–1	neurofilaments, microtubules

Fig. 37.4 Speeds of axonal transport.

Multiple myeloma

A 75-year-old woman complains of postural dizziness, dry mouth, intermittent diarrhea, and numbness in both her feet. On examination, there was a marked postural decrease in blood pressure. A chest radiogram revealed lytic lesions in the sternum. Her serum showed a monoclonal protein (gammopathy) and her urine contained Bence Jones proteins. A bone marrow examination demonstrated increased numbers of plasma cells (see also Chapter 3).

Comment. These are classical findings in amyloid neuropathy, in which the free light-chain component of myeloma globulin (Bence Jones proteins), produced by tumor of plasma cells in the bone marrow, accumulates in the peripheral nerves. The light chains adopt the configuration of a β-pleated sheet, with multiple copies that are intercalated and thereby resistant to normal proteolysis.

Types of synapse

Because of the multitude of different synaptic inputs to a given neuron, the final algebraic summation results in a 'decision' at the level of the axon hillock (the site of origin of the axon from the cell body) as to whether or not to transmit an action potential down the axon as an all-or-none phenomenon. However, even before this decision is made, the input of a particular neurotransmitter can essentially be classified as excitatory or inhibitory.

In addition to the relatively short-term decisions concerning action potentials (Chapter 38), there is a longer-term modulation of the resting membrane potential, moving it either closer to (excitation) or further from (inhibition) the critical membrane potential, which is the level at which the resting membrane potential will finally trigger an action potential at the axon hillock. Many drugs have a longer-term effect on modulation, in addition to the short-term effect, which partially explains their addictive effect; this can be seen with alcohol or the opioid drugs. There are also long-term effects during treatment with various drugs – for example those used to treat endogenous depression – such that it may be weeks before any beneficial effects are seen.

Cholinergic transmission

Acetylcholine is the neurotransmitter that has been best studied. As a model system, this transmitter can have two rather different effects, depending upon its site of origin within the nervous system (i.e. central or peripheral): those effects originally demonstrated by experiments with nicotine are characteristic of the nicotinic receptor, whereas those demonstrated with muscarine characterize the muscarinic receptor. Modern developments in pharmacology and DNA technology have produced a complex picture of the agonists and antagonists associated with the regional actions of acetylcholine (Fig. 37.5). The classical antagonist of the muscarinic effect is atropine, and the best-studied blocker for the nicotinic receptor is the poisonous snake venom, α-bungarotoxin.

In myasthenia gravis, autoantibodies are formed against the nicotinic receptor for acetylcholine. However, by blocking the hydrolysis of acetylcholine, for example by means of the drug edrophonium (which inhibits the hydrolytic enzyme, acetylcholinesterase), the concentration of acetylcholine can be effectively increased (see Chapter 38).

Adrenergic transmission

After acetylcholine, epinephrine and norepinephrine are the neurotransmitters that have received most study. Again, they have two separate receptors:

- **α-adrenergic receptor**, blocked by phentolamine;
- **β-adrenergic receptor**, blocked by propranolol.

The latter drug used to be commonly used by cardiologists (other B-blockers still are), but neurologists also use it as part of the treatment of Parkinson's disease. This disease results from a deficiency of dopamine, the precursor of epinephrine and norepinephrine. Many adrenergic effects are promoted by cAMP, but other neurotransmitters have either excitatory or inhibitory effects (Fig. 37.5) (see Chapter 38).

Ion channels

Even at rest, the neuron is working to pump ions along ionic gradients

The 'resting' neuron is, nevertheless, continually pumping sodium out of the cell and potassium in, through ion channels. During an action potential, there is a momentary reversal of these ionic movements, such that sodium enters the cell and potassium then leaves, effectively repolarizing the resting membrane potential. Mutations of sodium channels can occur at different sites and give rise to hyperkalemic periodic paralysis. The negative ion, chloride, moves through separate channels, which are implicated in specific pathologic states such as myotonia.

Calcium ions have an important role in the synchronization of neuronal activity

The movement of calcium ions within cells often provides a 'trigger' for the cells to synchronize an activity such as synaptic release of neurotransmitter; this synchronization of movement is also seen to have a prominent role in the

Ion channels

Defects of sodium channels
Different molecular lesions at various sites of the sodium channel pores can give rise to hyperkalemic periodic paralysis. As this name suggests, the patient has intermittent muscle weakness, during which time the serum potassium concentration is increased. This is caused by an imbalance of cationic movements in which sodium enters the cell and potassium leaves it. For these patients, the abnormal flux of sodium into the muscle is not correctly regulated with its counterflux of potassium ions.

Channelopathies

A 7-year-old boy presented with numerous episodes of intermittent weakness, which usually occurred 15–20 minutes after exercise. The attacks were first noted when he was aged 2 years. They are characterized by moderate limb weakness, with decreased tendon reflexes. A typical attack would last about 30–40 minutes. A provocative potassium chloride infusion test induced a typical period of paralysis.

Comment. The patient suffers from the rare yet intriguing metabolic problem of defective channels that carry sodium ions. Along with other diseases affecting channels for different ions, these diseases have been called channelopathies.

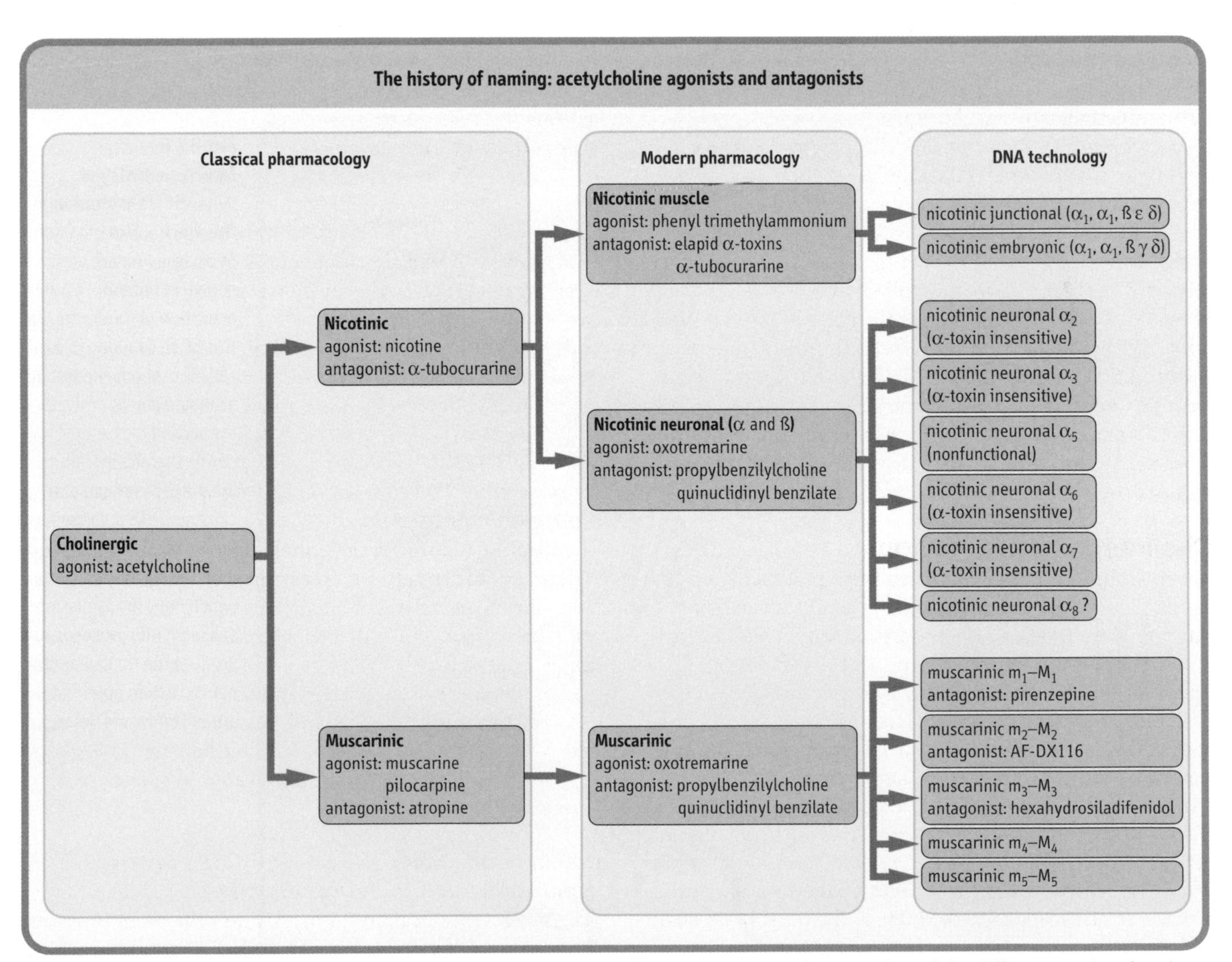

Fig. 37.5 The changes in nomenclature, from early to modern terms, for the agonists and antagonists of the different regional actions of acetylcholine e.g. central (neuronal) vs. peripheral (muscle).

sarcoplasmic reticulum of muscle (see Chapter 19). Within the central nervous system, the Lambert–Eaton syndrome is a disease that affects predominantly the P/Q subtype of calcium channels, in an example of molecular mimicry. The patient may have a primary oat-cell carcinoma of the lung; the immune system responds by making antibodies against these malignant cells. However, the malignancy and the calcium channels possess a common epitope, the effect of which is that the immune response causes the release of neurotransmitter to be blocked at the presynaptic site. This is analogous to, but nevertheless can be clearly distinguished from, the condition in myasthenia gravis, in which the block is postsynaptic.

It is also worth noting that blockade of the presynaptic release of neurotransmitter may be usefully exploited by therapeutic application of botulinus toxin (a protein derived from anerobic bacteria), which contains enzymes to hydrolyze the presynaptic proteins involved in release of neurotransmitters. This toxin is used in special cases of spasticity such as torticollis, in which the patient can be relieved of the excessive contractures of the neck muscles, which turn the head chronically to one side and thus cause pain and distraction if untreated.

The mechanism of vision

The mechanism by which the human eye can detect a single photon of light provides a marvelous example of the chemical processes underlying neuronal function. It involves both trapping of photons and the transducer effect, whereby the energy of light is converted into a chemical form, which is then ultimately transmuted into an action potential by a retinal ganglion neuron. A number of the intermediates are as yet not precisely known, but the underlying hypothesis is that the receptor protein, rhodopsin, is coupled to the G-protein. There are several sequence homologies of rhodopsin with the adrenergic β-receptor and with the muscarinic acetylcholine receptor. The main steps take place in the following order (Fig. 37.6):

- *Cis* retinal is converted to *trans* retinal,
- Rhodopsin becomes activated,
- The level of cGMP decreases,
- Na^+ entry is blocked,
- The rod cell hyperpolarizes,
- There is release of glutamate (or aspartate),
- An action potential depolarizes the adjacent bipolar cell,
- This depolarizes the associated ganglion neuron, to send an action potential out of the eye.

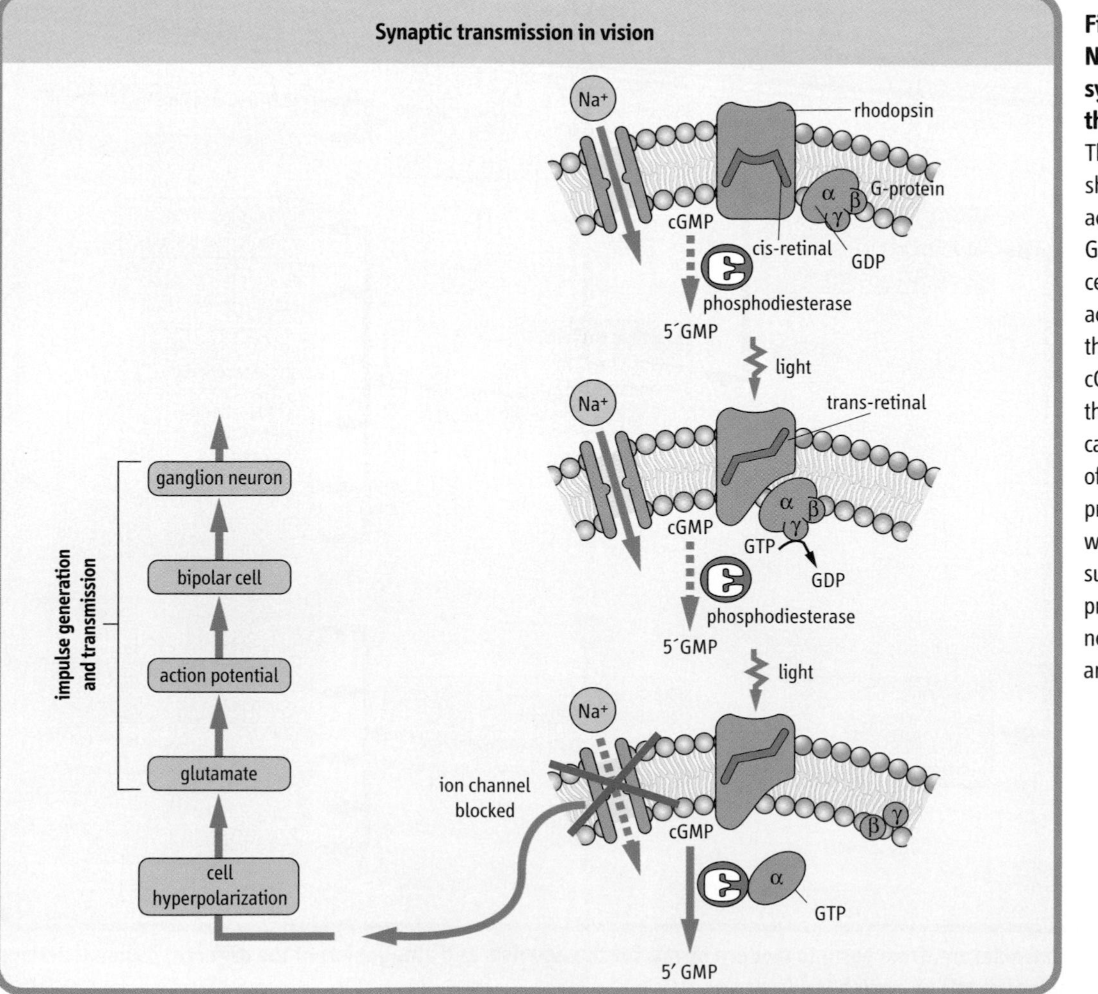

Fig. 37.6 Neurochemistry of synaptic transmission in the mechanism of vision. The consequences are shown of photon activation of rhodopsin via G-protein coupling in a rod cell. Phosphodiesterase is activated and hydrolyzes the second messenger, cGMP, thereby blocking the entry of sodium and causing hyperpolarization of the cell. Currently, the precise steps through which neurotransmission subsequently proceeds to produce the final ganglion neuron action potential are not known in detail.

Botulism

Twenty-four hours after eating home-preserved vegetables, a healthy young woman experiences blurred vision, severe vomiting, and progressive muscle weakness. Her doctor admits her to hospital, and electrophysiological studies confirm the clinical diagnosis of botulism. Trivalent antiserum, made from inactivated toxin is administered immediately and, with the help of assisted ventilation, the patient recovers within a few weeks.

Comment. The vegetables contained the exotoxin of the anerobe, *Clostridium botulinum*, which had not been destroyed during the preservation process. The toxin hydrolyzes the presynaptic proteins involved in the release of neurotransmitter, and thus the blockade is similar to the functional lesion in Lambert–Eaton myasthenic syndrome; however, in botulism the blockade can be lethal, especially at the level of the phrenic nerve which is essential for appropriate respiratory-lung movement.

Summary

The nervous system contains a number of distinct cells, each of which synthesizes its own individual proteins. The specialized functions of the nervous system mean that these proteins are effectively compartmentalized in different loci. In order to facilitate communication within the brain there are two specialized methods of moving cells, organelles and proteins, i.e. the cerebrospinal fluid and axonal transport.

Further reading

Barchi RL. Sodium channel gene defects in the periodic paralyses. *Curr Opin Neurol* 1992;**2**:631–637.

Dessauer CW, Posner BA, Gilman AG. Visualizing signal transduction: receptors, G-proteins, and adenylate cyclases. *Clin Sci* 1996;**91**:527–537.

Schiff G, Morel N. Rapid anterograde axonal transport of the SNAP-25 and VAMP (synaptobrevin) during axonal transport. *J Neurochem* 1997;**68**:1663–1667.Siegel GJ, Agranoff BW, Albers RW, Molinoff PB, eds. *Basic neurochemistry*, 5th ed. New York: Raven Press; 1993.

Taylor AW, Streilein JW. Inhibition of antigen-stimulated effector T cells by human cerebrospinal fluid. *Neuroimmunomodulation* 1996;**3**:112–118.

Thompson EJ. The CSF proteins. *A biochemical approach*. Amsterdam: Elsevier; 1988.

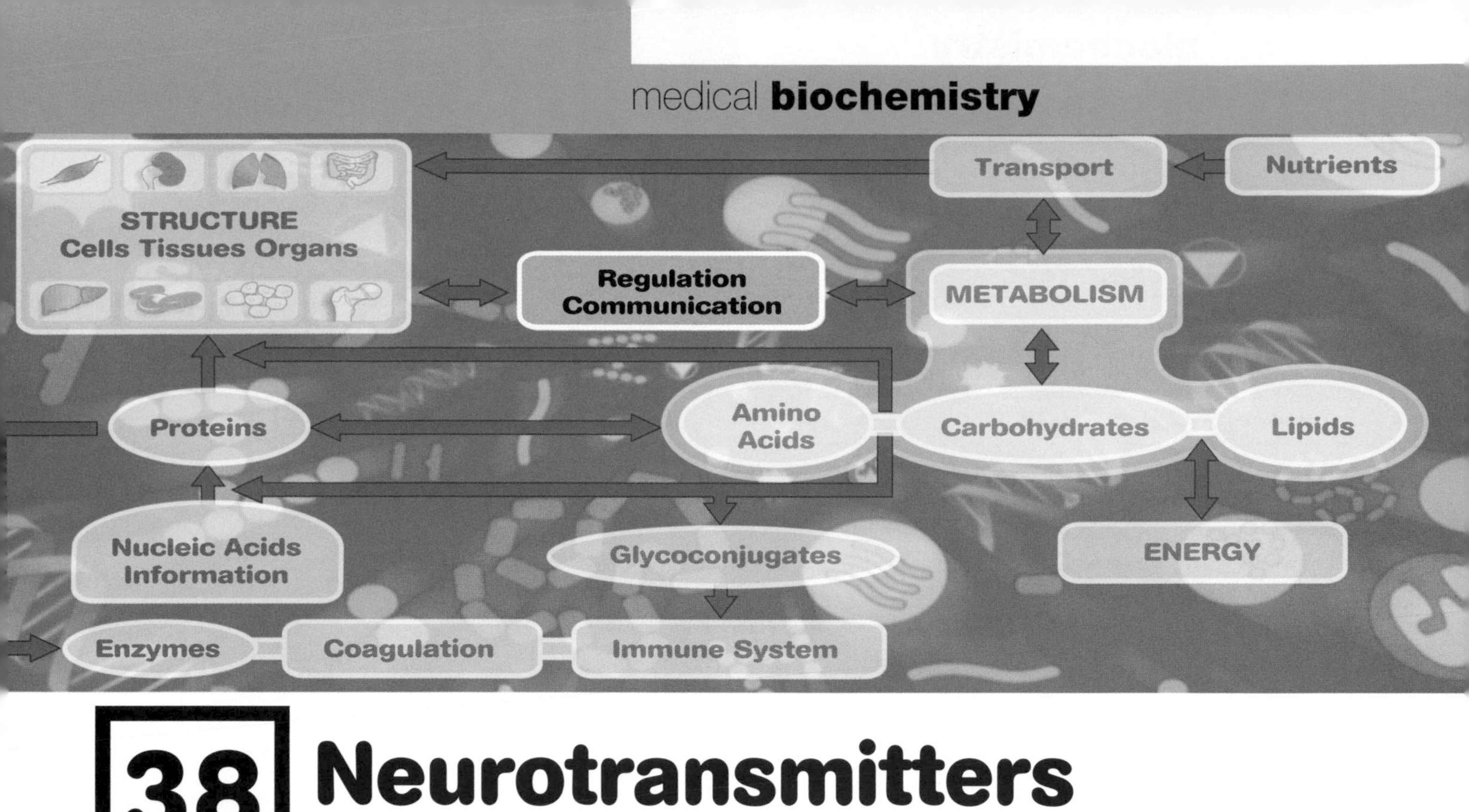

38 Neurotransmitters

Neurotransmitters

Neurotransmitters are molecules that act as chemical signals between nerve cells

Nerve cells communicate with each other and with target tissues by secreting chemical messengers, called neurotransmitters. This chapter describes the various classes of neurotransmitters and how they interact with their target cells. It will discuss their effects on the body, how alterations in their signaling may cause disease, and how pharmacologic manipulation of their concentrations may be used therapeutically.

Large numbers of potential neurotransmitters have been described. Those best studied are small molecules, such as norepinephrine and acetylcholine (ACh). They are the result of the conversion of the electrical signal that passes along a neuron, known as an action potential, to a chemical signal (neurotransmitter) that travels across the space between the nerve cells to a receptor and elicits a new action potential in the target cell.

Classification of neurotransmitters

A classification of neurotransmitters based on chemical composition is shown in Figure 38.1. Many are derived from simple compounds, such as amino acids (Fig. 38.2), but peptides are also now known to be extremely important. The principal transmitters in the peripheral nervous system are norepinephrine and ACh (Fig. 38.3).

Several transmitters may be found in one nerve

An early dogma of nerve function held that one nerve contained one transmitter. This is now known to be, at best, only a first approximation. Combinations of transmitters now appear to be the rule, although it is often difficult to establish the exact roles of the various substances that may be present. The pattern of transmitters may characterize a particular functional role, but details of this also remain unclear. A major low-molecular-weight transmitter such as an amine is often present, along with several peptides, an amino acid, and a purine. Sometimes, there may even be more than one possible transmitter in a particular vesicle, as is believed to be the case for adenosine triphosphate (ATP) and norepinephrine in sympathetic nerves. In some cases, the intensity of stimulation may control which transmitter is released, peptides often requiring greater levels of stimulus. Furthermore, different transmitters may act at different rates; sympathetic nerves are good examples of nerves for which this is the case: it is believed that ATP causes their rapid excitation, whereas norepinephrine and the neuromodulator, neuropeptide Y (NPY), cause a slower phase of action. In some tissues, NPY may be able to produce a very slow excitation on its own.

Classification of neurotransmitters

Group	Examples
	acetylocholine (ACh)
amines	norepinephrine, epinephrine, dopamine, 5-HT
amino acids	glutamate, GABA
purines	ATP, adenosine
gases	nitric oxide
peptides	endorphins, tachykinins, many others

Fig. 38.1 Neurotransmitters can be classified in several ways. The scheme shown relies on chemical similarities. All except the peptides are synthesized at the nerve ending and packaged into vesicles there; peptides are synthesized in the cell body and transported down the axon. 5-HT, 5-hydroxytryptamine; GABA, γ-amino butyric acid.

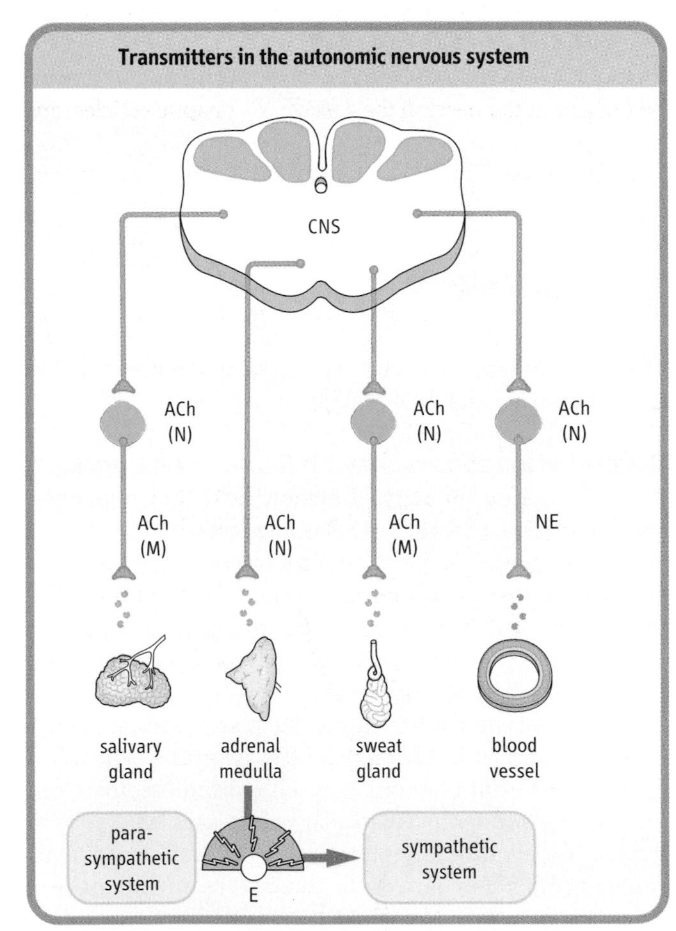

Fig. 38.3 Catecholamines and acetylcholine are transmitters in the sympathetic and parasympathetic nervous systems. Preganglionic nerves all release ACh, which binds to nicotinic (N) receptors. Most postganglionic sympathetic nerves release norepinephrine (NE), whereas postganglionic parasympathetic nerves release ACh, which acts at muscarinic (M) receptors. Motor neurons release ACh, which acts at distinct nicotinic receptors. E, epinephrine.

Neurotransmitters of low molecular weight

Compound	Source	Site of production
amino acids		
glutamate		central nervous system (CNS)
aspartate		CNS
glycine		spinal cord
amino acid derivatives		
GABA	glutamate	CNS
histamine	histidine	hypothalamus
norepinephrine	tyrosine	sympathetic nerves, CNS
epinephrine	tyrosine	adrenal medulla, a few CNS nerves
dopamine	tyrosine	CNS
5-HT	tryptophan	CNS, enterochromaffin gut cells, enteric nerves
purines		
ATP		sensory, enteric, sympathetic nerves
adenosine	ATP	CNS, peripheral nerves
gas		
nitric oxide	arginine	genitourinary tract, CNS
miscellaneous		
ACh	choline	parasympathetic nerves, CNS

Fig. 38.2 Many neurotransmitters are of low molecular weight, and are simple compounds, often derived from common amino acids.

Neurotransmission

Action potentials are caused by changes in ion flows across cell membranes

The signal carried by a nerve cell reflects an abrupt change in the voltage potential difference across the cell membrane. The normal resting potential difference is a few millivolts, with the inside of the cell being negative, and is caused by an imbalance of ions across the plasma membrane: the concentration of K^+ ions is much greater inside cells than outside, whereas the opposite is true for Na^+ ions. This difference is maintained by the action of the Na^+/ K^+-ATPase (see Chapter 21). Only those ions to which the membrane is permeable can affect the potential, as they can come to an electrochemical steady state under the combined influence of concentration and voltage differences. Because the membrane in all resting cells is comparatively permeable to K^+ as a result of the presence of voltage-independent (leakage) K^+ channels, this ion largely controls the resting potential. A change in voltage tending to drive this resting potential towards zero from the normal negative voltage is known as a depolarization, whereas a stimulus that increases the negative potential is called hyperpolarizing.

So far, this picture is common to all cells. However, nerve cells contain voltage-dependent sodium channels that open very rapidly when a depolarizing change in voltage is applied. When they open, they allow the inward passage of huge numbers of Na^+ ions from the extracellular fluid (Fig. 38.4), which swamps the resting voltage and drives the membrane potential to positive values. This reversal of voltage is the action potential. Almost immediately afterwards, the sodium channels close and so-called delayed potassium channels open. These restore the normal resting balance of ions across the membrane and, after a short refractory period, the cell can conduct another action potential. Meanwhile, the action potential has spread by electrical conductance to the next segment of nerve membrane, and the entire cycle starts again.

Neurotransmitters alter the activity of various ion channels to cause changes in the membrane potential. Excitatory neurotransmitters cause a depolarizing change in voltage, in which case an action potential is more likely to occur. In contrast, inhibitory transmitters hyperpolarize the membrane, and an action potential is then less likely to occur.

Neurotransmitters act at synapses

Neurotransmitters are released into the space between cells at a specialized area known as a synapse (Fig. 38.5). In the simplest case, they diffuse from the presynaptic membrane, across the synaptic space or cleft, and bind to receptors at the postsynaptic membrane. However, many neurons, particularly those containing amines, have several varicosities along the axon, containing transmitter. These varicosities may not be close to any neighboring cell, so transmitter released from them has the possibility of affecting many neurons. Nerves innervating smooth muscle are commonly of this kind.

When the action potential arrives at the end of the axon, the change in voltage opens calcium channels. Calcium entry is essential for mobilization of vesicles containing transmitter, and for their eventual fusion with the synaptic membrane and release through it.

Because transmitters are released from vesicles, impulses arrive at the postsynaptic cell in individual packets, or quanta. At the neuromuscular junction between nerves and skeletal muscle cells, a large number of vesicles are discharged at a time, and a single impulse may therefore be enough to stimulate contraction of the muscle cell. The number of vesicles released at synapses between neurons, however, is much smaller; consequently, the recipient cell will be stimulated only if the total algebraic sum of the various positive and negative stimuli exceeds its threshold. As each cell in the brain receives input from a huge number of neurons, this implies that there is a far greater capability for the fine control of responses in the central nervous system (CNS) than there is at the neuromuscular junction.

For a compound to be accepted as a transmitter, it must be present in the nerve at the area of the synaptic vesicles, and the nerve must be able to make it or accumulate it. It must be released on stimulation, and application of exogenous

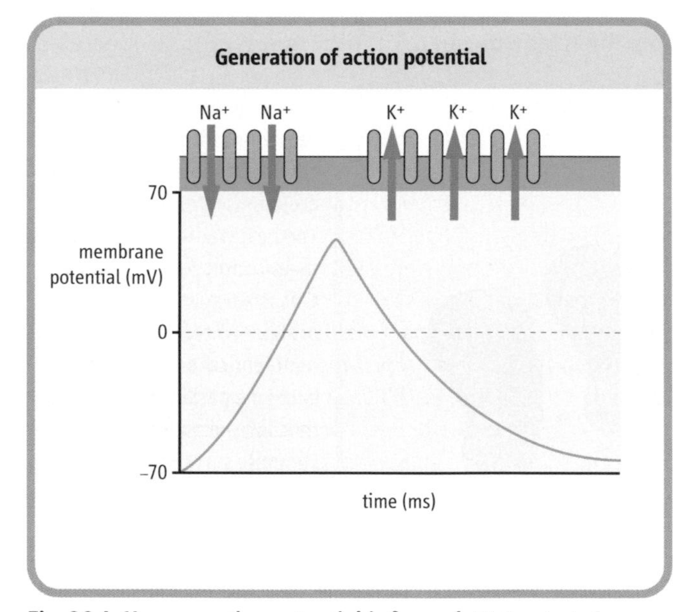

Fig. 38.4 How an action potential is formed. At the start of an action potential, the membrane is at its resting potential of about –70 mV; this is maintained by voltage-independent K^+ channels. When an impulse is initiated by a signal from a neurotransmitter, voltage-dependent Na^+ channels open. These allow inflow of Na^+ ions, which alter the membrane potential to positive values. The Na^+ channels then close, and K^+ channels, called delayed rectifier channels, open to restore the initial balance of ions and the negative membrane potential.

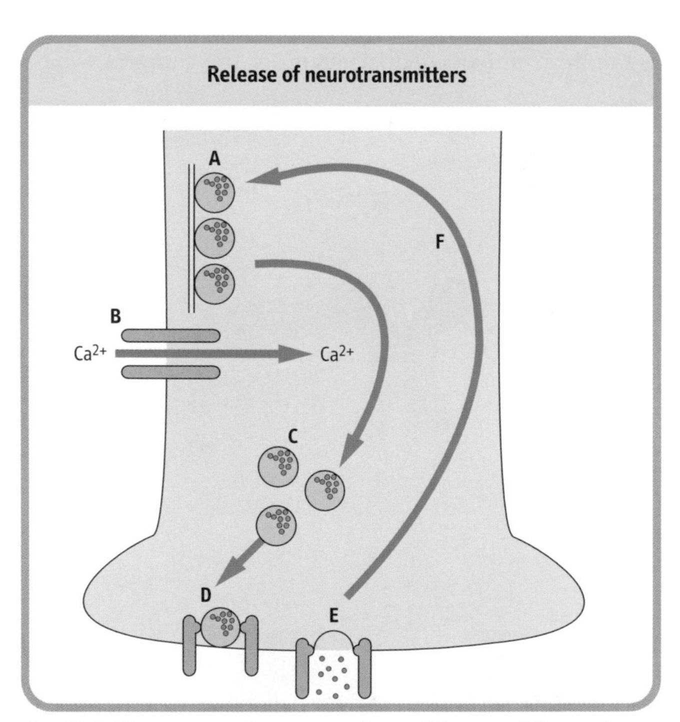

Fig. 38.5 Neurotransmitters are released from vesicles at the synaptic membrane. (A) In the resting state, vesicles are attached to microtubules. (B) When an action potential is received, calcium channels open. (C) Vesicles move to the plasma membrane, and (D) bind to a complex of docking proteins. (E) Neurotransmitter is released, and (F) vesicles are recycled.

material must mimic the effect of nerve stimulation. These criteria are difficult to apply rigorously, even in the comparatively simple environment of the autonomic ganglion, and are nearly impossible to meet in the complex network of nerve terminals in the brain.

Receptors

Neurotransmitters act by binding to specific receptors and opening or closing ion channels

There are several mechanisms by which receptors for excitatory neurotransmitters can cause the propagation of an action potential in a postsynaptic neuron. Directly or indirectly, they cause changes in ion flow across the membrane, until the potential reaches the critical point, or threshold, for initiation of an action potential. Receptors that directly control the opening of an ion channel are called ionotropic, whereas metabotropic receptors cause changes in second messenger systems, which, in turn, alter the function of channels that are separate from the receptor.

Ionotropic receptors (ion channels)

Ionotropic receptors contain an ion channel within their structure (Fig. 38.6)(see also Chapter 37). Examples include the nicotinic ACh receptor and some glutamate and γ-amino butyric acid (GABA) receptors. These are transmembrane proteins, with several subunits, usually five, surrounding a pore through the membrane. Each subunit has four transmembrane regions. When the ligand binds, there is a change in the three-dimensional structure of the complex, which allows the flow of ions through it. The effect on membrane potential depends on the particular ions that are allowed to pass: the nicotinic ACh receptor is comparatively nonspecific towards sodium and potassium, and causes depolarization, whereas the $GABA_A$ receptor is a chloride channel, and causes hyperpolarization.

Metabotropic receptors

Metabotropic receptors are coupled to second messenger pathways and act more slowly than ionotropic receptors. All known metabotropic receptors are coupled to G-proteins (see Chapter 36) and, like hormone receptors, have seven transmembrane regions. Typically, they couple either to adenylate cyclase, altering the production of cyclic adenosine monophosphate (cAMP), or to the phosphatidylinositol pathway, which alters calcium fluxes. Ion channels that are separate from the receptor are then usually modified by phosphorylation. For instance, the β-adrenergic receptor, which responds to norepinephrine and epinephrine, causes an increase in cAMP, which stimulates a kinase to phosphorylate and activate a calcium channel. Some of the muscarinic class of ACh receptors have similar effects on K^+ channels.

Regulation of neurotransmitters

The action of transmitters must be halted by their removal from the synaptic cleft

When transmitters have served their function, they must be removed from the synaptic space. Simple diffusion is probably the major mechanism of removal of neuropeptides.

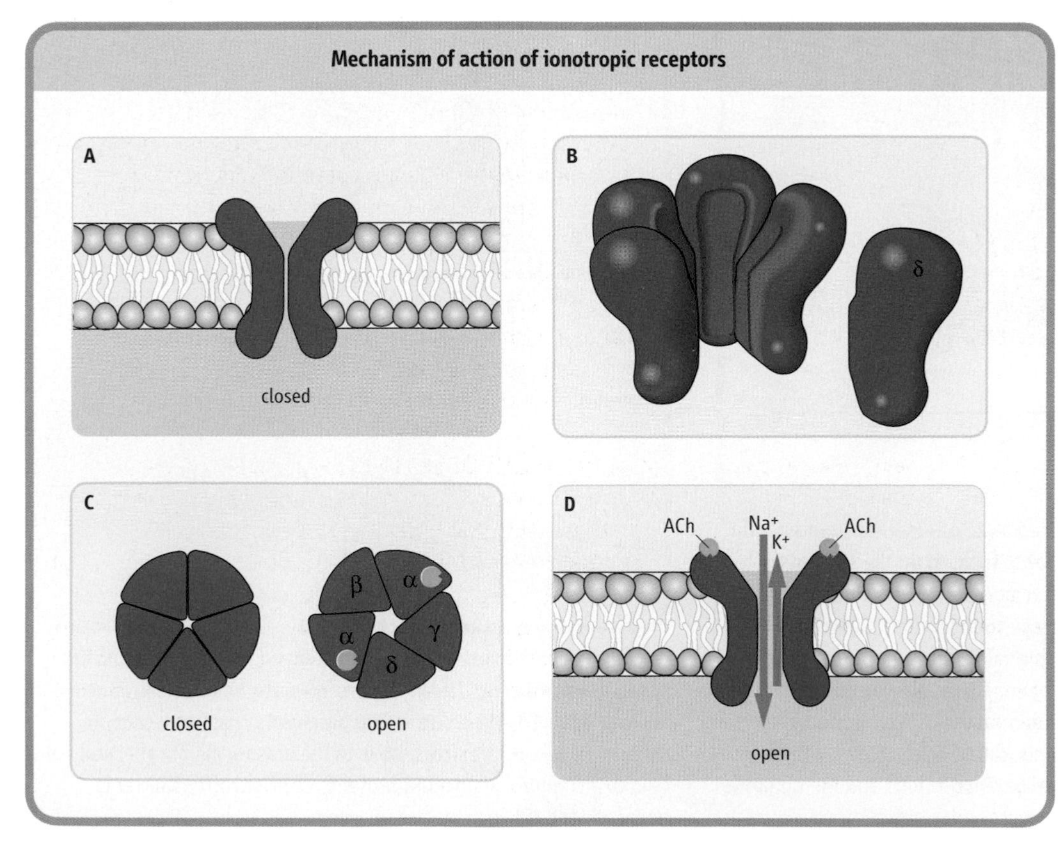

Fig. 38.6 Ionotropic receptors directly open ion channels. Ionotropic receptors are themselves ion channels. The best studied example is the nicotinic ACh receptor. This is a transmembrane protein (A) consisting of five nonidentical subunits (B), each one passing right through the membrane. The subunits surround a pore (C) that selectively allows certain ions through when it is opened by a ligand (D).

Enzymes such as acetylcholinesterase, which cleaves ACh, may destroy any remaining transmitter. Surplus transmitters may also be taken back up into the presynaptic neuron for reuse, and this is a major route of removal for catecholamines and amino acids. Interference with uptake causes an increase in the concentration of transmitter in the synaptic space; this often has useful therapeutic consequences.

Concentrations of neurotransmitters may be manipulated

The effects of neurotransmitters can be altered by changing their effective concentrations or the number of receptors. Concentrations can be altered by:

- **changing the rate of synthesis,**
- **altering the rate of release at the synapse,**
- **blocking reuptake,**
- **blocking degradation.**

Changes in the number of receptors may be involved in long-term adaptations to the administration of drugs.

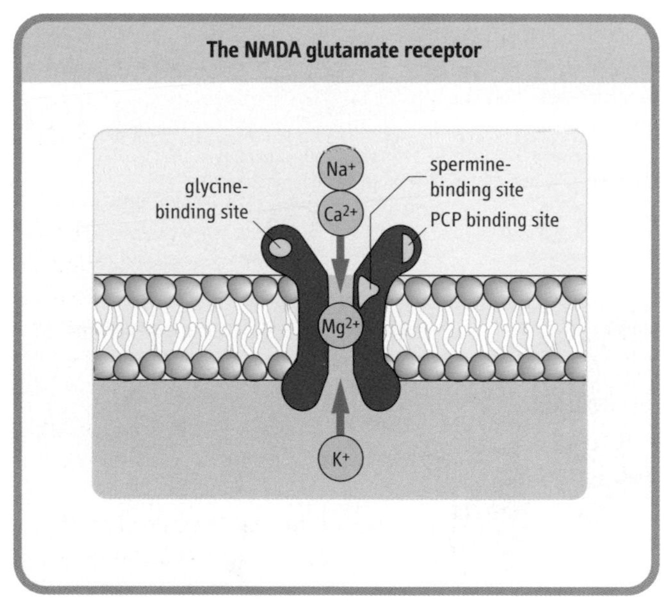

Fig. 38.7 The glutamate receptor that binds *N*-methyl-D-aspartate (NMDA) is complex. This receptor is clinically important because it may cause damage to neurons after stroke (excitotoxicity). It contains several modulatory binding sites, so it may be possible to develop drugs that could alter its function. Glycine is an obligatory cofactor, as are polyamines such as spermine. Magnesium physiologically blocks the channel at the resting potential, so the channel can open only when the cell has been partially depolarized by a separate stimulus. It therefore causes a prolongation of the excitation. This receptor also binds phencyclidine (PCP). Because this drug of abuse can cause psychotic symptoms, it is possible that dysfunction of pathways involving NMDA receptors causes some of the symptoms of schizophrenia.

Neurotransmitters: the various classes

Amino acids

It has been particularly difficult to prove that amino acids are true neurotransmitters; they are present in high concentrations because of their other metabolic roles, and therefore simple measurement of their concentrations was not conclusive. Pharmacologic studies of responses to different analogs and the cloning of specific receptors finally provided the conclusive proof.

Glutamate

Glutamate is the most important excitatory transmitter in the CNS. It acts on both ionotropic and metabotropic receptors. Clinically, the receptor characterized *in vitro* by *N*-methyl-D-aspartate (NMDA) binding is particularly important (Fig. 38.7).

Excitotoxicity

Clinical significance of NMDA receptors

Extracellular glutamate concentration is increased after trauma and strokes, during severe convulsions, and in some organic brain diseases such as Huntington's chorea, AIDS-related dementia, and Parkinson's disease, because of release of glutamate from damaged cells and damage to the glutamate uptake pathways.

Excess glutamate is toxic to nerve cells. The NMDA receptor is activated, which allows calcium entry into cells. This activates various proteases, which in turn initiate the pathway of programmed cell death, or apoptosis. There may, in addition, be changes in other ionotropic glutamate receptors that also cause aberrant calcium uptake. Uptake of sodium ions is also implicated, and causes swelling of cells. Activation of NMDA receptors also increases the production of nitric oxide, which may in itself be toxic. Cell death in some models of excitotoxicity can be prevented by inhibitors of nitric oxide production, but the mechanism of toxicity is not clear.

Attempts are being made to develop drugs to inhibit NMDA activation and suppress excitotoxicity. The hope is that damage caused by stroke can be limited or even reversed. Unfortunately, many of the drugs have side effects because they bind to the phencyclidine binding site and have unpleasant psychologic effects such as paranoia and delusions.

The hippocampus (Fig. 38.8) is an area of the limbic system of the brain that is involved in emotion and memory. Certain synaptic pathways there become more active when chronically stimulated – a phenomenon known as long-term potentiation. This represents a possible model of how memory is laid down, and it requires activation of the NMDA receptor and the consequent influx of calcium.

Glutamate is recycled by high-affinity transporters into both neurons and glial cells. The glial cell converts it into glutamine, which then diffuses back into the neuron. Mitochondrial glutaminase in the neuron regenerates glutamate for reuse.

γ-Amino butyric acid (GABA)

GABA is synthesized from glutamate by the enzyme glutamate decarboxylase (Fig. 38.9). It is the major inhibitory transmitter in the brain. There are two known GABA receptors: the $GABA_A$ receptor is ionotropic, and the $GABA_B$ receptor is metabotropic. The $GABA_A$ receptor consists of five subunits that arise from several gene families, giving an enormous number of potential receptors with different binding affinities. This receptor is the target for several useful therapeutic drugs. Benzodiazepines bind to it and cause a potentiation of the response to endogenous GABA; these drugs reduce anxiety and also cause muscle relaxation. Barbiturates also bind to the GABA receptor and stimulate it directly in the absence of GABA; because of this

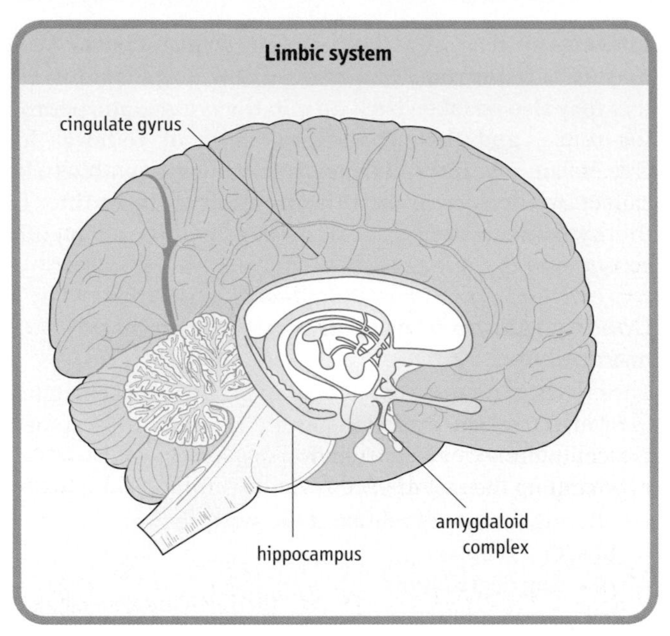

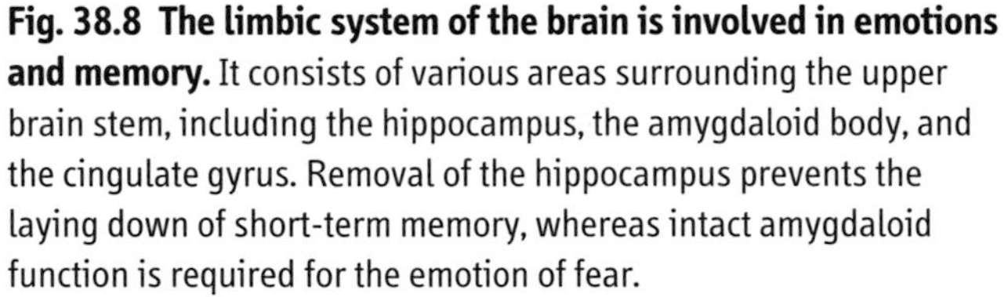
Fig. 38.8 The limbic system of the brain is involved in emotions and memory. It consists of various areas surrounding the upper brain stem, including the hippocampus, the amygdaloid body, and the cingulate gyrus. Removal of the hippocampus prevents the laying down of short-term memory, whereas intact amygdaloid function is required for the emotion of fear.

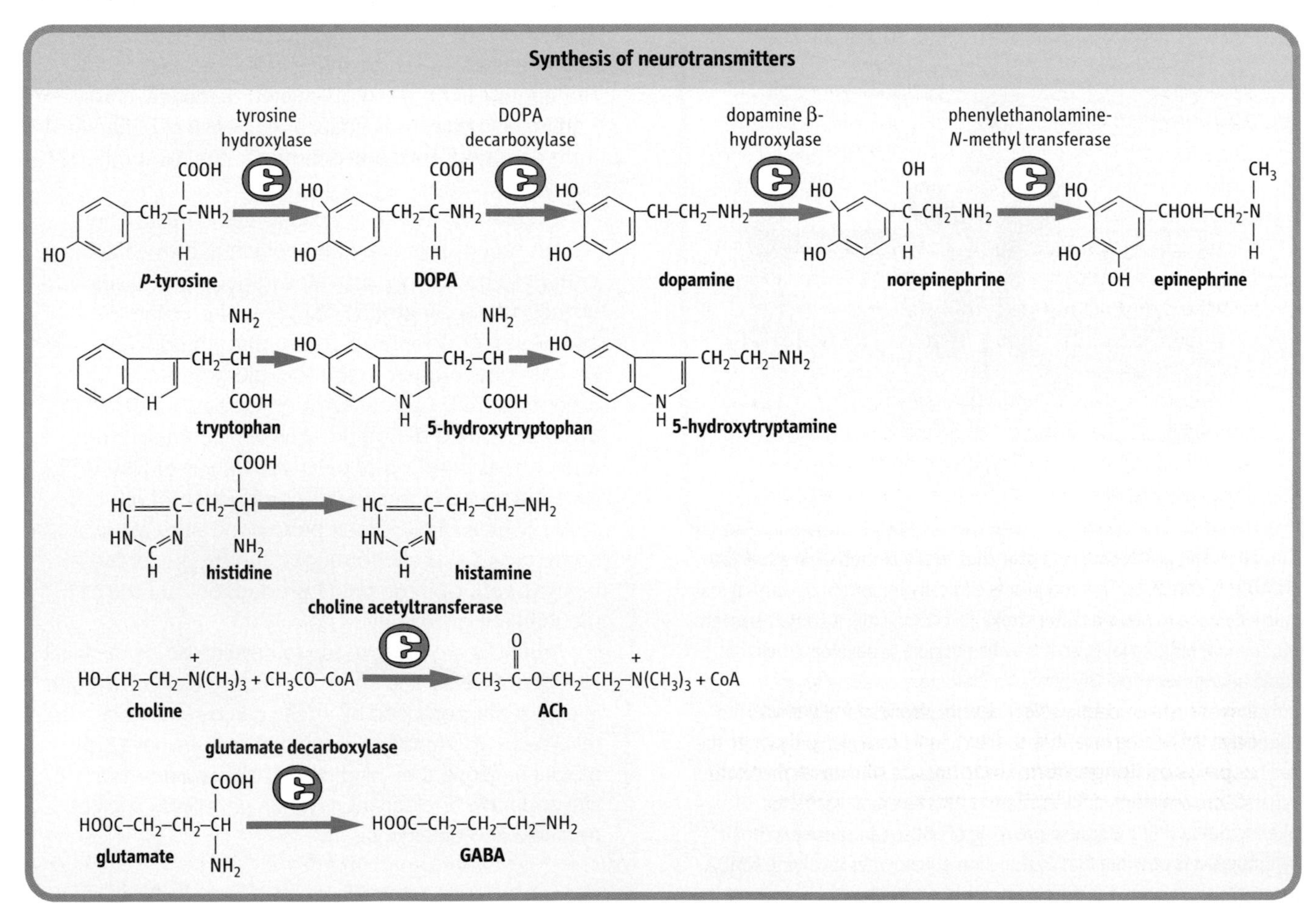

Fig. 38.9 Pathways of synthesis of neurotransmitters are simple.

lack of dependence on endogenous ligand, they are more likely to cause toxic side effects in overdose.

Glycine

Glycine is primarily found in inhibitory interneurons in the spinal cord, where it blocks impulses traveling down the cord in motor neurons to stimulate skeletal muscle. The glycine receptor on motor neurons is ionotropic and is blocked by strychnine; motor impulses can then be passed without negative control, which accounts for the rigidity and convulsions caused by this toxin.

Catecholamines

Norepinephrine, epinephrine, and dopamine, known as catecholamines, are all derived from the amino acid tyrosine (see Fig. 38.9). In common with other compounds containing amino groups, such as serotonin, they are also known as biogenic amines. Nerves that release catecholamines have varicosities along the axon, instead of a single area of release at the end. Transmitter is released from the varicosities, and diffuses through the extracellular space until it meets a receptor. This allows it to affect a wide area of tissue, and these compounds are believed to have a general modulatory effect on overall brain functions such as mood and arousal.

Norepinephrine and epinephrine

Norepinephrine (or noradrenaline) is a major transmitter in the sympathetic nervous system. Sympathetic nerves arise in the spinal cord and run to ganglia situated close to the cord, from which postganglionic nerves run to the target tissues. Norepinephrine is the transmitter for these postganglionic nerves, whereas the transmitter at the intermediate ganglia is ACh. Stimulation of these nerves is responsible for various features of the 'fight or flight' response, such as stimulation of the heart rate, sweating, vasoconstriction in the skin, and bronchodilatation.

There are also norepinephrine-containing neurons in the CNS, largely in the brain stem (Fig. 38.10). Their axons extend in a wide network throughout the cortex, and alter the overall state of alertness or attention. The stimulatory effects of amphetamines are caused by their close chemical similarity to catecholamines.

Epinephrine (adrenaline) is produced by the adrenal medulla under the influence of ACh-containing nerves analogous to the sympathetic preganglionic nerves. It is more active than norepinephrine on the heart and lungs,

Disorders of neurotransmission

Depression as a disease of amine neurotransmitters

Monoamine oxidase (MAO) inhibitors prevent the catabolism of catecholamines and serotonin. They therefore increase the concentrations of these compounds at the synapse and increase the action of the transmitters. Compounds with this property are antidepressants. Reserpine, an antihypertensive drug that depletes catecholamines, caused depression and is no longer in use. These dual findings gave rise to the 'amine theory of depression': this states that depression is caused by a relative deficiency of amine neurotransmitters at central synapses, and predicts that drugs which increase amine concentrations should improve symptoms of the condition.

In support of this theory, tricyclic antidepressants inhibit transport of both norepinephrine and serotonin into neurons, thereby increasing the concentration of amines in the synaptic cleft. Specific serotonin reuptake inhibitors (SSRIs), such as fluoxetine (Prozac), are also highly effective antidepressants.

This picture is, however, undoubtedly an oversimplification. Cocaine is also an effective reuptake inhibitor, but is not an antidepressant, and amphetamines both block reuptake and cause release of catecholamines from nerve terminals, but cause mania rather than relief of depression. In addition, symptoms of depression do not resolve for several days after treatment is started, and it is likely that, in depression, longer-term adaptations of concentrations of transmitters and their receptors are at least as important as acute changes in amine concentrations in the synaptic cleft.

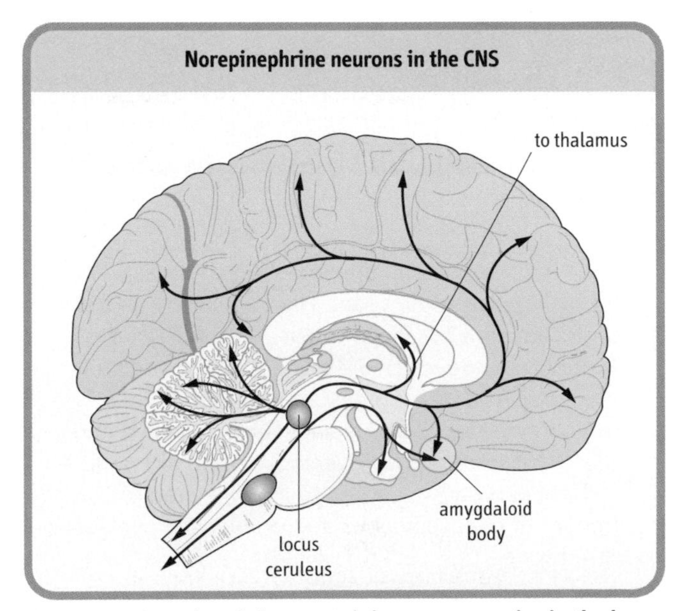

Fig. 38.10 Norepinephrine-containing neurons arise in the locus ceruleus in the brain stem and are distributed throughout the cortex.

causes redirection of blood from the skin to skeletal muscle, and has important stimulatory effects on glycogen metabolism in the liver. In response to epinephrine, a sudden extra supply of glucose is delivered to muscle, the heart and lungs work harder to pump oxygen round the circulation, and the body is then prepared to run or to defend itself (see Chapter 20). Epinephrine is not essential for life, however, as it is possible to remove the adrenal medulla without serious consequences.

The receptors for norepinephrine and epinephrine are called adrenoceptors. They are divided into α- and β-receptor classes and subclasses on the basis of their pharmacology. Epinephrine acts on all classes of the receptors, but norepinephrine is more specific for α-receptors. β-Blockers, such as atenolol, are used to treat hypertension and chest pain (angina) in ischemic heart disease because they antagonize the stimulatory effects of catecholamines on the heart. Non-specific α-blockers have limited use, although the more specific α_1-blockers such as prazosin, and α_2-blockers such as clonidine, can be used to treat hypertension. Certain subclasses of β-receptors are found in particular tissues; for instance, the β_2-receptor is present in lung, and β_2-receptor agonists such as salbutamol are therefore used to produce bronchial dilatation in asthma without stimulating the β_1-receptor in the heart.

Norepinephrine is taken up into cells by a high-affinity transporter and catabolized by the enzyme monoamine oxidase (MAO). Further oxidation and methylation by catecholamine-*O*-methyl transferase (COMT) convert the products to metanephrines and vanillyl mandelic acid (Fig. 38.11), which can be measured in the urine as indices of the function of the adrenal medulla. They are particularly increased in patients who have the tumor of the adrenal medulla known as pheochromocytoma. This tumor causes hypertension because of the vasoconstrictor action of the catecholamines it produces.

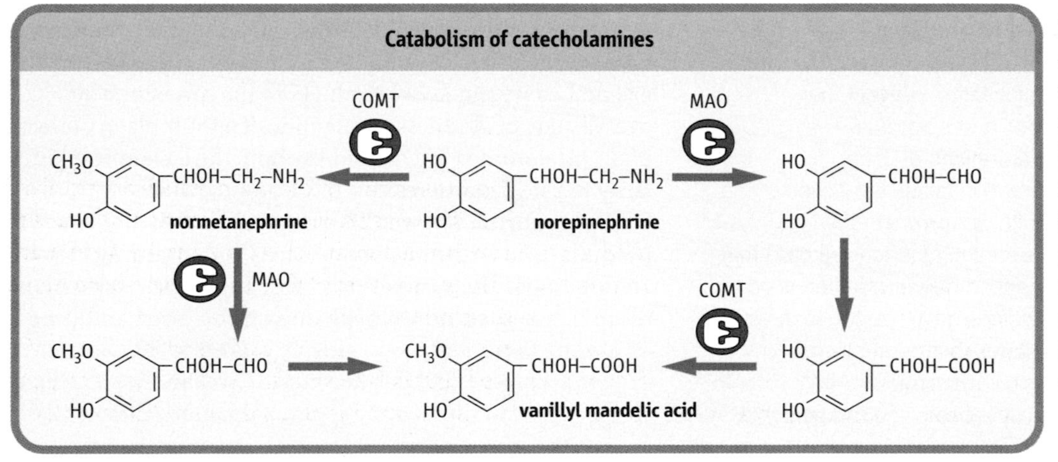

Fig. 38.11 Catecholamines are degraded by oxidation of the amino group by the enzyme monoamine oxidase (MAO), and by methylation by catecholamine-*O*-methyl transferase (COMT). The pathway shown is for norepinephrine, but the pathways for epinephrine, dopamine, and 5-HT are analogous.

An unusual reaction to cheese

A 50-year-old man had been suffering from depression for some years. His condition was well controlled with tranylcypromine, an MAO inhibitor. Suddenly, he developed a severe, throbbing headache. His blood pressure was found to be 200/110 mmHg. The only unusual occurrence had been that, the previous evening, he had attended a cocktail party at which he ate a considerable amount of cheese snacks and drank several glasses of red wine.

Comment. The patient was experiencing a hypertensive crisis caused by an interaction between the food he had eaten and his drug treatment – an MAO inhibitor. This drug inhibits the main enzyme that catabolizes catecholamines. Several foods, including cheese, pickled herring, and red wine, contain an amine called tyramine, which is similar in structure to natural amine transmitters and is also broken down by MAO. If this enzyme is not functional, the concentrations of tyramine increase and it starts to act as a neurotransmitter. This can cause a hypertensive crisis, as in this patient.

A neurotransmitter cause of hypertension

A 56-year-old woman presented with hypertension. She suffered from headaches, attacks of sweating, and palpitations. Her high blood pressure was difficult to control with conventional treatment and was episodic. The examining physician therefore decided to test for endocrine causes of hypertension, such as Cushing's syndrome and pheochromocytoma. The patient's cortisol concentration was normal, excluding a diagnosis of Cushing's syndrome, so a sample of urine was taken for measurement of catecholamines and metabolites. The rate of excretion of norepinephrine was 1500 nmol/24 h (253 µg/24 h); (reference range <900 nmol/24 h [<152 µg/24 h]), that of epinephrine 620 nmol/24 h (113 µg/24 h); (reference range <230 nmol/24 h [<42 µg/24 h]) and that of vanillyl mandelic acid 60 µmol/24 h (11.9 mg/24 h); (reference range <35.5 µmol/24 h [<7.0 mg/24 h]).

Comment. The patient had a pheochromocytoma. This is a tumor of the adrenal medulla that secretes catecholamines. Both norepinephrine and epinephrine may be secreted; they cause hypertension by acting on adrenoceptors on vascular smooth muscle, and also cause an increased heart rate by their action on β-receptors on the heart muscle. The paroxysms of hypertension may be severe and the patient is at considerable risk of stroke or other complications.

Diagnosis is made by measuring catecholamines in plasma or urine, or their metabolites, such as metanephrines and vanillyl mandelic acid, in urine.

Although this is a rare cause of hypertension, comprising only about 1% of cases, it is very important to remember it, as the condition is dangerous and usually relatively easily cured.

Dopamine

Dopamine is both an intermediate in the synthesis of norepinephrine and a neurotransmitter. It is a major transmitter in nerves that interconnect the nuclei of the basal ganglia in the brain and control voluntary movement (Fig. 38.12). Damage to these nerves causes Parkinson's disease, which is characterized by tremor and difficulties in initiating and controling movement. Dopamine is also found in pathways

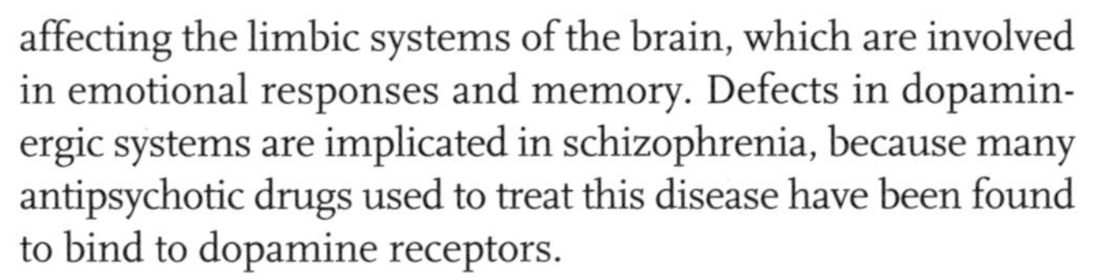

affecting the limbic systems of the brain, which are involved in emotional responses and memory. Defects in dopaminergic systems are implicated in schizophrenia, because many antipsychotic drugs used to treat this disease have been found to bind to dopamine receptors.

In the periphery, dopamine causes vasodilatation, and it is therefore used clinically to stimulate renal blood flow, and is important in the treatment of renal failure (Chapter 21).

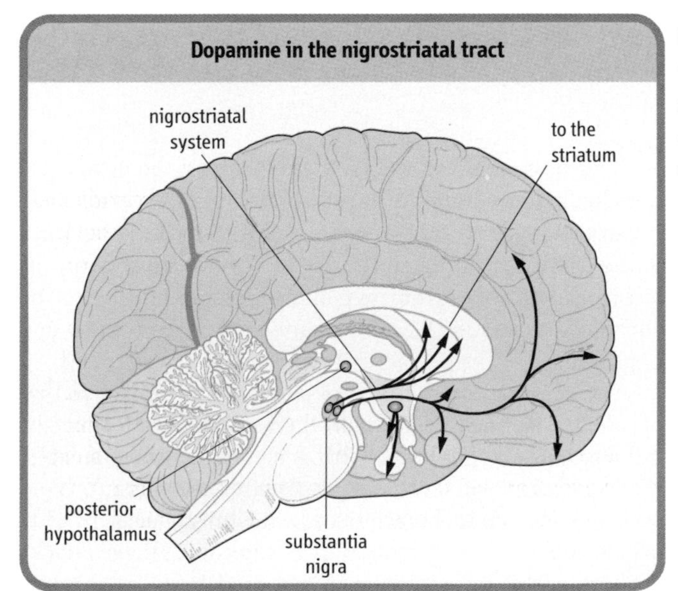

Fig. 38.12 Nerves containing dopamine run in well-defined tracts. One of the most important tracts, the nigrostriatal, connects the substantia nigra in the mid brain with the basal ganglia below the cortex. Damage to this causes Parkinson's disease, with loss of fine control of movement.

Multiple receptors for catecholamine neurotransmitters

Dopamine and serotonin (5-HT) receptors
Multiple receptors have now been isolated for dopamine and 5-HT. Although the existence of several was predicted by pharmacologic studies, the extreme complexity that has been discovered was unexpected. Not all those that have been cloned have yet been shown to be functional, but the possible relevance in terms of drug development is obvious. In some cases, specific actions on particular receptors can be exploited therapeutically.

There are five known dopamine receptors, falling into two main groups – D_1-like (D_1 and D_5) and D_2-like (D_2, D_3, and D_4) – that differ in their signaling pathways. D_1 receptors increase the production of cAMP, whereas D_2 receptors inhibit it. Antipsychotic drugs such as phenothiazines and haloperidol tend to inhibit D_2-like receptors, suggesting that excessive dopamine activity may be important in causing the symptoms of schizophrenia.

The D_2 receptor is a major receptor in the nerves that interconnect the basal ganglia. As it is known that destruction of these nerves causes Parkinson's disease, it is not surprising that antipsychotic drugs that inhibit the D_2 receptor tend to have the side effect of causing abnormal movements. Drugs, such as clozapine, that bind preferentially to the D_4 receptor appear to be free of such side effects, although that particular drug also binds to several other receptors.

More than a dozen serotonin (5-HT) receptors have been isolated using molecular biological techniques. They have been divided into classes and subclasses on the basis of their pharmacologic properties and their structures. Most are metabotropic, although the 5-HT_3 receptor is ionotropic and mediates a fast signal in the enteric nervous system. The 5-HT_{1A} receptor is found on many presynaptic neurons, where it acts as an autoreceptor to inhibit the release of 5-HT.

In general, increasing the brain concentration of 5-HT appears to increase anxiety, whereas reducing its concentration is helpful in treating the condition. The antidepressant, buspirone, acts as an agonist at 5-HT_{1A} receptors, and presumably causes a decrease in production of 5-HT. In addition to its effects on the D_4 dopamine receptor, clozapine binds strongly to the 5-HT_{2A} receptor, and it may be that a combination of a high level of 5-HT_{2a} antagonism and low D_2-binding activity is desirable for drugs that can be used to treat schizophrenia with the minimum frequency of side effects. The 5-HT_3 blocker, ondansetron, is an antiemetic, extensively used to prevent vomiting during chemotherapy. Migraine can be treated with sumatriptan, a 5-HT_{1D} agonist. SSRIs such as fluoxetine (Prozac) have been found to have activity against a wide variety of other psychiatric diseases, such as obsessive–compulsive behavior and eating disorders, in addition to depression, suggesting that 5-HT has a central role in controlling mood and behavior.

This central role of 5-HT in controling brain function and the huge number of associated receptors suggest that it may possible to tailor a large number of drugs to treat specific disorders, and that pharmacologic manipulation of the function of the nervous system is only in its infancy.

Serotonin (5-hydroxytryptamine)

Serotonin, also called 5-hydroxytryptamine (5-HT), is derived from tryptophan. Serotoninergic neurons are concentrated in the raphe nuclei in the upper brain stem (Fig. 38.13), but project up to the cerebral cortex and down to the spinal cord. They are more active when subjects are awake than when they are asleep, and serotonin may control the degree of responsiveness of motor neurons in the spinal cord. In addition, it is implicated in so-called vegetative behaviors such as feeding, sexual behavior, and temperature control.

Serotonin has effects on mood. Reuptake inhibitors, which increase its concentration at the synapse, relieve depression. An excess of serotonin may cause panic attacks; these can be controlled by 5-HT_{1A}-receptor agonists, which are believed to act on autoreceptors to reduce the production of serotonin.

Serotonin also has substantial effects on the peripheral nervous system and enteric neurons. It is a powerful vasoconstrictor, and increases motility of the gastrointestinal tract. Some of the serotonin in the gut arises not from neurons, but from enterochromaffin cells, which are similar to the chromaffin cells in the adrenal medulla that produce epinephrine.

The degradation pathway of serotonin is similar to that of catecholamines, resulting in the formation of 5-hydroxyindoleacetic acid (5-HIAA). This is a useful marker of excessive production of serotonin in diseases such as carcinoid syndrome, which is characterized by flushing attacks. In addition to the excessive production of serotonin, these attacks are caused by several other active mediators from small tumors in the gut or lungs.

Flushing attacks caused by a tumor

A 60-year-old man complained of attacks of flushing, associated with an increased heart rate. He also had troublesome diarrhea and abdominal pain. He had lost weight. The picture suggested carcinoid syndrome, which is caused by excessive secretion of serotonin and other metabolically active compounds from a tumor. To confirm this, a urine sample was taken for measurement of 5-hydroxy indoleacetic acid (5-HIAA), the major metabolite of 5-HT; the concentration was found to be 120 µmol/24 h (23 mg/24 h); (reference range 10–52 µmol/24 h [3–14 mg/24 h]).

Comment. The patient had a very rare carcinoid syndrome, which is caused by tumors of entero-chromaffin cells in the gut or lungs. These cells are related to the catecholamine-producing chromaffin cells in the adrenal medulla and convert tryptophan to serotonin (5-HT). Serotonin itself is believed to cause diarrhea, but other mediators, such as histamine and bradykinin, may be more important in the flushing attacks. The urinary concentration of 5-HIAA provides a useful diagnostic test and can also be used to follow the progress of the disease.

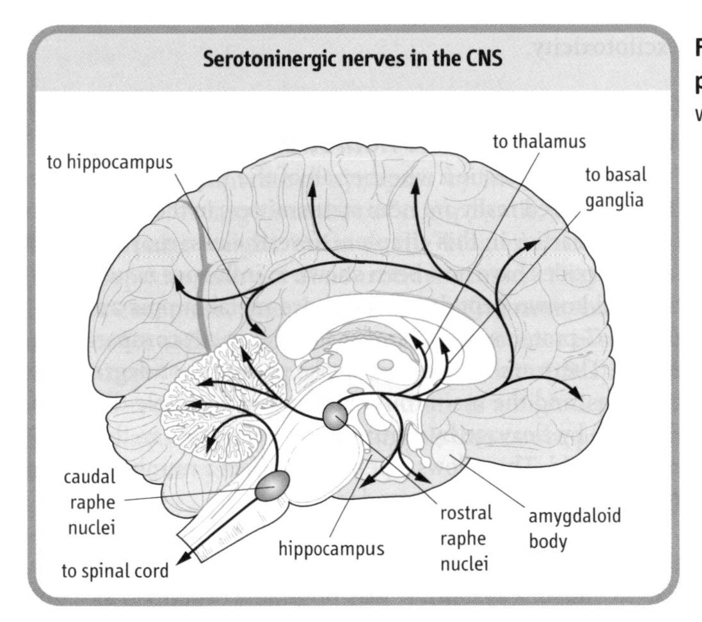

Fig. 38.13 Serotonin-containing nerves arise in the raphe nuclei, part of the reticular formation in the upper brain stem. In common with those containing norepinephrine, they are distributed widely.

Acetylcholine

Acetylcholine (ACh) is the transmitter of the parasympathetic autonomic nervous system and of the sympathetic ganglia. Stimulation of the parasympathetic system produces effects that are broadly opposite to those of the sympathetic system, such as slowing of the heart rate, bronchoconstriction, and stimulation of intestinal smooth muscle. ACh also acts at neuromuscular junctions, where motor nerves contact skeletal muscle cells and cause them to contract. Apart from these roles, ACh may be involved in learning and memory, as neurons containing this transmitter also exist in the brain.

A disease causing muscle weakness

A 30-year-old man noticed that he was steadily losing strength in his muscles. This was particularly noticeable in the facial muscles. His eyelids drooped, he had double vision when he was tired, his voice was indistinct and nasal, and he had difficulty swallowing. His physician suspected myasthenia, a disease of nerve–muscle conduction. Injection of edrophonium, an acetylcholinesterase inhibitor, dramatically improved his symptoms. A diagnostic test; the serum titre of anti-acetylcholine receptor antibodies, was performed and found to be elevated.

Comment. The patient was suffering from myasthenia gravis. This is a disease that manifests itself as weakness of voluntary muscles and is corrected by treatment with acetylcholinesterase inhibitors. It is caused by autoantibodies directed against the nicotinic acetylcholine receptor, which circulate in serum. Because of these autoantibodies, transmission of nerve impulses to muscle is much less efficient than normal.

Drugs that inhibit acetylcholinesterase increase the concentration of acetylcholine in the synaptic space, which compensates for the reduced number of receptors. Improvement in nerve–muscle conduction in response to edrophonium can be used as a diagnostic test, and long-acting acetylcholinesterase inhibitors such as pyridostigmine can be used to treat the disease.

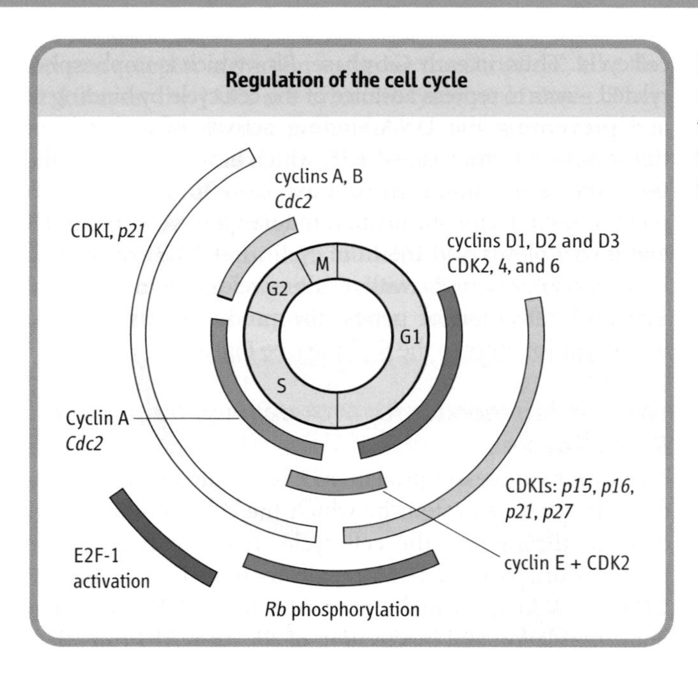

Fig. 39.5 Stimulation of growth factors leads to cyclin-dependent activation of the regulation of key steps in the cell cycle by CDKs and their inhibitors. The cyclins and their respective CDK partners acting at the different stages of cell cycle progression are shown. CDKI, CDK inhibitor; E2F-1, transcription factor E2F-1; *Rb*, retinoblastoma protein.

regulates growth in many cell types, mediates many of its effects by inhibition of the cyclin D–CDK4 and 6 complexes that promote cell cycle progression. Indeed, in some tumors, transformation appears to result from deletion or functional inactivation of TGF-β receptors or associated downstream signal-transducing elements. Second, the key signaling element, *Ras*, which has been found to be mutated to a constitutively activated form in approximately 30% of all tumors, similarly appears to exert many, if not all, its effects ultimately by upregulating concentrations of cyclin D and, hence, by stimulating cell cycle progression. This upregulation of cyclin D expression results from stimulation of the MAPK cascade by *Ras* and the induction of the transcription factor, AP-1, which regulates induction of the expression of cyclin D (Fig. 39.6).

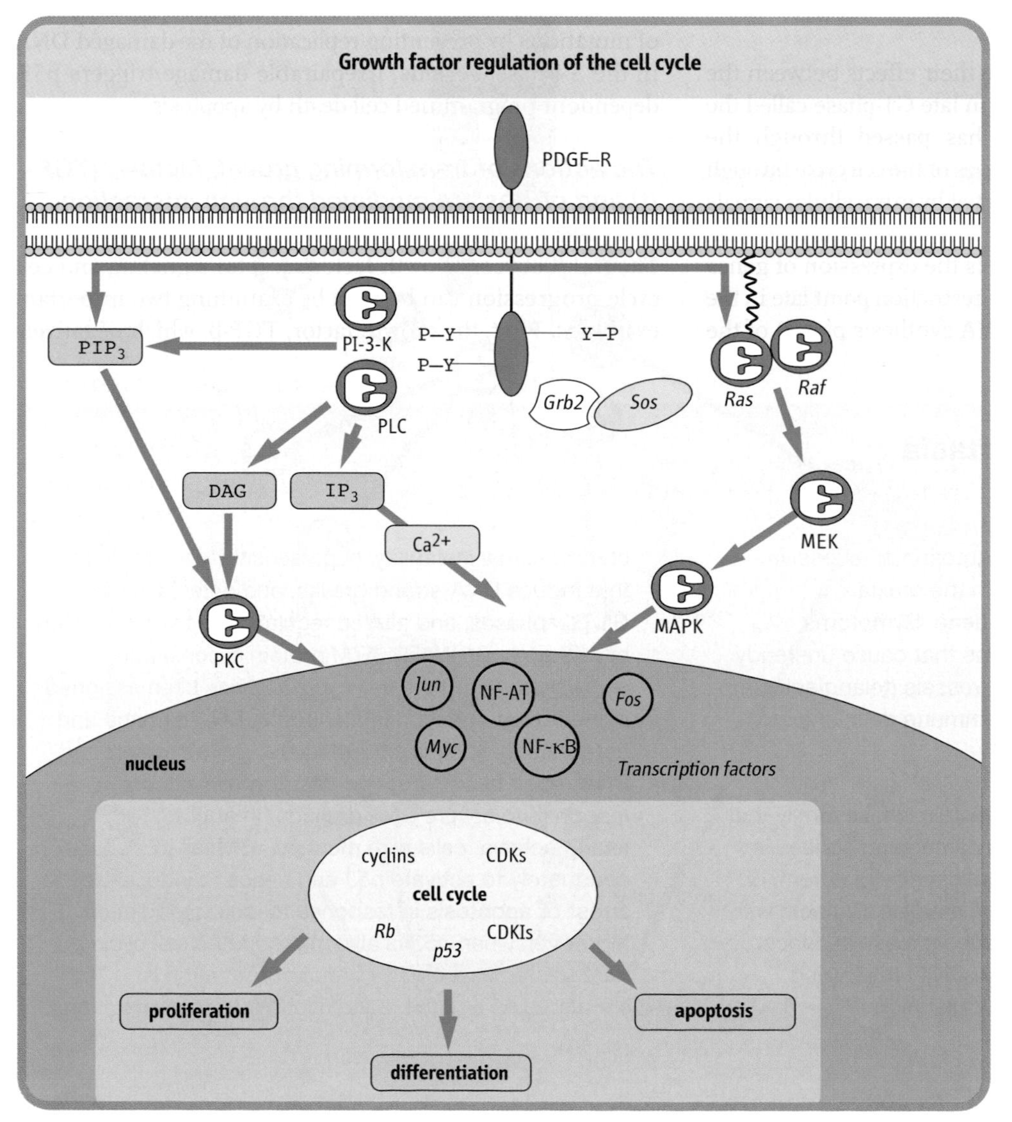

Fig. 39.6 Growth factor signals generated at the plasma membrane can transduce gene induction, cell cycle progression and proliferation differentiation, or apoptosis. Growth factor receptor signaling leads to activation of transcription factors such *Jun* and *Fos* (which dimerize to form AP-1), *Myc*, NF-AT, and NF-κB. These regulate the induction of components of the cell cycle machinery that monitor and direct cell cycle progression, cell growth arrest, cell differentiation. Induction of DNA damage that would result in the accumulation of mutations leads to cell cycle arrest and DNA repair. Alternatively if the DNA damage is too great, cell death occurs by apoptosis.

Apoptosis

Apoptosis, a form of programmed cell death, occurs during normal cellular development; it is a fundamentally important biological process that is required to maintain the integrity and homeostasis of multicellular organisms. Whereas inappropriate apoptosis can lead to degenerative conditions such as neurodegenerative disorders, subversion and disruption of the apoptosis machinery can result in cancer (Fig. 39.7) or autoimmune disease. Cells that die from damage typically swell and burst (necrosis), but apoptosis is characterized by dramatic morphologic changes in the cell, including shrinkage, chromatin condensation, cleavage and dissassembly into membrane-enclosed vesicles called apoptotic bodies that are rapidly phagocytosed by neighboring cells. It is the accumulation of sterols at the plasma membrane and translocation of phosphatidylserine to the outer leaflet of the plasma membrane that serve to flag the apoptotic cell for elimination via phagocytosis by neighboring cells or macrophages.

Apoptotic cell death is a rigorously controlled and highly ordered process

Recent research has elucidated many of the biochemical signals that regulate apoptosis, and has identified two key families of proteins that appear to be central to the regulation of mammalian cell death by apoptosis: the cysteine proteases or caspases, and the B cell lymphoma protein 2 *(Bcl-2)*-related family members, which serve to modulate cell survival by regulating caspase activity.

Caspases

The caspases act by cleaving key target proteins, resulting in the systematic disassembly of the apoptotic cell by:

- **halting cell cycle progression;**
- **disabling homeostatic and repair mechanisms;**
- **initiating the detachment of the cell** from its surrounding tissue structures;
- **dismantling structural components** such as the cytoskeleton;
- **flagging the dying cell for phagocytosis.**

Overexpression of active caspases is sufficient to cause cellular apoptosis. Caspases are expressed as inactive procaspases, and it is likely that caspase activation occurs in a cascade fashion, with the initial activation of a regulatory caspase serving to activate downstream effector caspases by proteolysis. Thus cell death receptors such as tumor necrosis factor (TNF)-R and *Fas* (also known as CD95 or APO-1), which mediate apoptotic cell death pathways in a number of cell types but particularly in cells of the immune system, initiate apoptosis by directly recruiting procaspases such as caspase 8 to their accessory 'death domain'

Cancer and the cell cycle

Cancer	Cell cycle element	Outcome
	cyclin expression increased as a result of gene amplification or chromosomal translocation	hyperphosphorylation and inactivation of retinoblastoma protein *(Rb)*, leading to deregulated malignant cell proliferation
breast cancer, leukemia	cyclin D1	
testicular carcinomas	cyclin D2	
breast tumors	cyclin E	
retinoblastomas small-cell lung carcinomas osteosarcomas	mutational inactivation of Rb genes	loss of *Rb* control of cell cycle, leading to deregulated malignant cell proliferation
cervical carcinomas	sequestration of Rb by the human papilloma virus E7 protein	loss of *Rb* control of cell cycle
variety of tumor cells	deletion of CDKI genes, *p14*, *p15*	loss of inhibition of cyclin D–CDK4/6 complexes, resulting in inappropriate hyperphosphorylation and inactivation of *Rb*
melanomas	mutant CDK4 demonstrating resistance to CDKI gene products, *p14*, *p15*	loss of inhibition of cyclin D–CDK4/6 complexes, resulting in loss of *Rb* control of cell cycle

Fig. 39.7 Relationship between various cancers and stages of the cell cycle. Some degree of disruption of the cell cycle machinery is likely to occur in every type of cancer cell (Fig. 39.9). As most proliferative decisions are made at the restriction point in late G1-phase, the key target for cellular transformation appears to be the retinoblastoma protein, *Rb*. Indeed, it is now clear that the signal transduction events leading to *Rb* phosphorylation and functional inactivation are disrupted in many cancer cells, suggesting that *Rb* inactivation may be crucial to deregulated, malignant cell proliferation.

MCQ Answers

1. (c)
2. (a) T, (b) T; (c) F; (d) F
3. (b)
4. (c)
5. (a) T, (b) T, (c) F, (d) F, (e) F
6. (d)
7. (e)
8. (b)
9. (b)
10. (b)
11. (b), (d), (e)
12. (c)
13. (a), (c), (e)
14. (a), (e)
15. (a), (c), (d), (e)
16. (c), (e)
17. (a), (b), (d), (e)
18. (a), (c), (d)
19. (a), (b), (d), (e)
20. (a), (b), (c)
21. (b), (c)
22. (b), (c), (d)
23. (a), (b), (d)
24. (a), (b), (c)
25. (a), (c), (d)
26. (a)
27. (c)
28. (d)
29. (b)
30. (b)
31. (c)
32. (d)
33. (c)
34. (a)
35. (d)
36. (b)
37. (c), (e)
38. (c)
39. (d)
40. (d)
41. (d)
42. (c)
43. (e)
44. (c)
45. (b)
46. (b)
47. (e)
48. (d)
49. (a)
50. (d)
51. (c)
52. (e)
53. (b)
54. (e)
55. (d)
56. (b)
57. (e)
58. (c)
59. (b)
60. (c)
61. (d)
62. (a)
63. (a)
64. (b)
65. (a), (b)
66. (a), (b), (c), (e)
67. (b)
68. (d)
69. (a), (b), (d), (e)
70. (b)
71. (a), (b), (c)
72. (d)
73. (d)
74. (c)
75. (b)
76. (d)
77. (b)
78. (c)
79. (a)
80. (c)
81. (e)
82. (d)
83. (d)
84. (c)
85. (e)
86. (a)
87. (b)
88. (b)
89. (a), (c), (d)
90. (b)
91. (a)
92. (c)
93. (b)
94. (a)
95. (d)
96. (a), (b)
97. (a), (c)
98. (a), (d)
99. (a)
100. (a), (d)
101. (d)
102. (a), (b), (c), (d)
103. (a)
104. (c)
105. (e)
106. (d)
107. (d)
108. (b)
109. (c)
110. (a)
111. (d)
112. (b)
113. (b)
114. (b)
115. (e)
116. (e)
117. (d)
118. (e)
119. (c)
120. (d)
121. (d)
122. (b)
123. (c)
124. (b)
125. (a)
126. (c)
127. (d)
128. (a)
129. (c)
130. (b)
131. (e)
132. (b)
133. (a), (b), (c), (d)
134. (b)
135. (b), (c)
136. (a), (c), (d)
137. (a)
138. (d)
139. (a) ,(b), (c), (d)
140. (b)
141. (d)
142. (a), (d)
143. (a), (c), (d), (e)
144. (d)
145. (b), (d)
146. (a), (b), (c), (e)
147. (a), (d)
148. (b)
149. (d), (e)
150. (a), (b), (e)
151. (b), (e)
152. (c), (d), (e)
153. (a), (b), (d)
154. (b)
155. (b)
156. (a)
157. (b)
158. (c)
159. (d)
160. (d)
161. (e)
162. (c)
163. (c)
164. (d)
165. (a), (c)
166. (b)
167. (a), (d)
168. (b), (c)
169. (b), (d)
170. (b), (d)
171. (b)
172. (a)
173. (b)

1. Increased rate of respiration due to hyperventilation causes the lungs to remove more CO_2 from the blood than normal and then this will decrease pCO_2 and increase blood pH. This is because, in plasma CO_2 is considered an acid and HCO_3^- its conjugated base. The decrease in acid relative to the level of its conjugated base causes an increase in pH, which is an example of respiratory alkalosis. Rebreathing from the paper bag in which CO_2 is accumulated will normalize the pCO_2 and decrease the blood pH to normal or near normal levels.

2. The sickle cell mutation results in the loss of two negative charges in the Hb tetramer at pH 8.6. Hb-C has an additional two positive charges because of glutamic acid to lysine and migrates more rapidly migrates to cathode (negative pole) than Hb-S and Hb-A.

3. Alanine, serine, threonine, valine, leucine, isoleucine, phenylalanine, tyrosine, tryptophan, histidine. Actually monoamine dicaboxylic amino acids such as glutamine and aspargine are also excreted.

4. (b).

5. (c), (d).

6. (b), (d) for multiple myeloma.

7. This individual cannot synthesize a sufficient level of β-globin, leading directly to the profound anemia and requirement for transfusion therapy. The homozygous β-thalaessemia genotype arises from different mutations to each of the two alleles that encode β-globin. One of her β-globin gene alleles includes an RNA translation defect that terminates prematurely; the other contains a processing defect that limits the production of normal β-globin mRNA. Together, these mutations result in the synthesis of a fraction of the amount of normal β-globin polypeptide.

8. The concentration of 2,3-BPG should shift the O_2 saturation curve further to the right. The compensatory benefit of this shift is to facilitate O_2 delivery to peripheral tissues at higher pO_2 levels. But the physiologic cost is a somewhat diminished re-oxygenation of Hb in the pulmonary alveoli.

9. (b).

10. (e).

11. Option (a) provides a definitive diagnosis, but (b) is a reliable screening test and is appropriate for the first testing. The diagnosis can be confirmed by (a).

12. To reduce the secretion of insulin from β cells, drugs that activate the K^+-ATP channel, such as diazoxide, and/or those that inhibit VDDC, such as nifedipine, would be given.

13. Consider maintenance treatment using a proton pump inihibitor, or aggressive antibiotic therapy to eradicate *Helicobacter pylori*, if that organism is detected.

14. Pentachlorophenol has been used as a wood preservative and an insecticide. Workers exposed to it and people living in log homes, where the wood is preserved with pentachlorophenol, may suffer from exposure, because it has a significant vapor pressure. One of the symptoms of pentachlorophenol exposure is chloracne, which is a type of acne caused by exposure to chlorinated hydrocarbons. Pentachlorophenol is also known to act as an uncoupler of the mitochondrial electron transport system, and it stimulates respiration and wastes metabolic energy by dissipating the proton gradient across the inner mitochondrial membrane generating larger quantities of heat, causing fever.

15. Clearly there is something metabolically wrong with the central nervous system. The high levels of pyruvate and lactate indicate that pyruvate is not being metabolized as a fuel. Defects could be in any of the enzymes leading to the TCA cycle, in the TCA cycle, in the electron transport system or ATP synthase. Such disorders fall under the category of Leigh's syndrome.

16. (c).

17. (b).

18. Distilled spirits are said to be empty calories, because they supply plenty of energy, but they are devoid of essential nutrients like vitamins and minerals, proteins and their essential amino acids, and essential fatty acids. Of the water-soluble vitamins, humans require more thiamine than any of the others, and its deficiency will be expressed first. Since thiamine is required by both pyruvate dehydrogenase and by α-ketoglutarate dehydrogenase, their metabolism will be impaired and they will accumulate to such an extent that their excess amounts will be excreted by the kidneys.

19. (c) Pyruvate carboxylase converts pyruvate to oxaloacetic acid, a key reaction during the net synthesis of glucose from alanine.

20. (b).

21. (b).

22. (a) Partial steroid 21-hydroxylase deficiency.
(b) The adrenal gland.
(c) Glucocorticoid replacement therapy (e.g. hydrocortisone).
(d) Yes, in neonates who present with hyponatremia and/or ambiguous genitalia

23. (b) Osteomalacia. 25-hydroxycholecaferol, vitamin D therapy plus a calcium enriched diet.

24. (c) Suppression of pituitary ACTH by exogenous steroid. He was unable to produce cortisol during a time of exercise-related stress.

25. (a) Chylomicrons.
(b) Very high triglyceride concentration is associated with the risk of pancreatitis.
(c) Lipoprotein lipase.

26. (a) Taking into account the level of cholesterol and the family history, it is unlikely that dyslipidaemia in this child is diet related.
(b) Fasting lipid profile including total cholesterol, triglycerides and HDL cholesterol to confirm the previous results.
(c) Familial hypercholesterolemia.

27. (a) This cardiovascular risk is high on account of a moderate increase in cholesterol level, low HDL cholesterol, smoking, hypertension, and probably diabetes. High glucose level will have to be confirmed. See chapter 20.
(b) Weight reduction, low cholesterol diet, smoking cessation, and mild regular exercise.
(c) This level of fasting glucose is consistent with diabetes mellitus. It needs to be confirmed as the diagnosis cannot be made on the basis of one measurement.

28. The key to this child's problems is likely to stem from abnormal neurotransmitter generation. This may be due to a poor supply of tyrosine (although it is presumed that the low phenylalanine diet provided to the child was supplemented with tyrosine). It is more likely that this child may suffer an abnormality in biopterin cofactor metabolism since this same cofactor is required for tyrosine hydroxylase, a prerequisite step for catecholamine neurotransmitter biosynthesis and tryptophan hydroxylase, a prerequisite step for serotonin biosynthesis.

29. The useful compound from this list would clearly be sodium benzoate. This compound will form hippuric acid after conjugation with the glycine molecule. The hippuric acid is excreted in the urine, carrying nitrogen out of the body and relieving some of the ammonia load, since glycine biosynthesis from non-nitrogen containing precursors can utilize ammonia. Phenylacetate, which conjugates with glutamine, is also used in a similar approach aimed at removing amino groups and lowering free ammonia in patients suffering from ammonia toxicity. The ultimate solution in all cases is to define the route cause of the elevated ammonia levels and when possible address the problem directly. These agents are stopgap measures.

30. In both cases (rapid growth or tissue regeneration from trauma), an increased requirement for all amino acids in new protein synthesis is seen. This will result in a higher portion of dietary essential amino acids, as well as non-essential amino acids being incorporated into new protein, rather than being channelled into energy metabolism, etc., with the loss of nitrogen. Thus, the amount of nitrogen excreted is less than that consumed and they are considered to be in positive nitrogen balance. Certain amino acids, like lysine and histidine, are required in higher amounts in the diet under these circumstances, because our ability to synthesize them is limited.

31. Diagnosis. This patient experienced severe muscle trauma during prolonged, strenuous exercise, as indicated by the release of muscle potassium, enzymes, and myoglobin into serum. Myoglobin is a low molecular weight protein which is filtered through the glomerulus and partially resorbed in renal tubules. Under acidic conditions, the heme group of myoglobin induces oxidative stress and renal tubular damage. Alkalinization of urine by administration of an antacid ($NaHCO_3$) with simultaneous dialysis therapy reduces renal injury and may lead to complete recovery. Other causes of rhabdomyolysis and subsequent renal failure include crush injuries, electrocution, severe hyperthermia, metabolic diseases (e.g. McArdle's disease), drugs.

32. (b).

33. (b), (d).

34. (a) The patient is hyperglycaemic and the increased ketones indicate increased lipolysis and ketogenesis.
(b) Hemoglobin A_{1C} level suggests that he has been hyperglycemic for the last 3–6 weeks.
(c) Urine protein level suggests diabetic nephropathy.

35. (b).

36. (b).

37. (b).

38. (c).

39. (c).

40. (b).

41. (b) T
Results consistent with overbreathing. Tetany as a result of respiratory alkalosis lowering ionized calcium.

42. This child shows symptoms of hereditary fructose intolerance, resulting from aldolase B deficiency. The disorder appears when children begin to consume fruit products and juices rich in fructose or sucrose. The liver is the primary site of fructose metabolism. Fructose 1-phosphate accumulates because of aldolase B deficiency. The resulting hypophosphatemia inhibits glycogenolysis by limiting the action of glycogen phosphorylase. Lactate accumulates in blood because fructose 1-phosphate also inhibits aldolase A, limiting gluconeogenesis.

43. (b).

44. (c).

45. (a).

46. Gout. See Fig. 28.5

47. See CE Box on page 362. Uridine provides a source of UMP by the salvage pathway, reducing the need for de novo synthesis.

48. Bacteria synthesize folic acid *de novo.* They use sulfonamide, instead of *p*-aminobenzoic acid, forming an inactive folate analog, which inhibits bacterial nucleic acid metabolism and causes cell death. Humans are protected because they cannot synthesize the vitamin, folic acid. Sulfonamides block two steps in purine synthesis (Fig. 28.3) and thymidylate synthase in pyrimidine (Fig. 28.7) biosynthesis.

49. Diagnosis of xeroderma pigmentosum can be confirmed by analysis of DNA repair activity in cultured fibroblasts. Recommendations include avoidance of sunlight, use of sunscreen, and frequent checks for skin cancer, plus genetic counseling.

50. β-Thalassemia is an autosomal recessive disease. Hemophilia is an X-linked recessive disease. Because her father had β-thalassemia, there is a 50% chance that she carries one copy of the β-thalassemia gene. Based on the frequency of β-thalassemia in the population, her husband has about 10% probability of carrying the same gene, so their son is unlikely (<5%) to suffer from β-thalassemia. Her husband does not suffer from hemophilia, so their child will be normal in this respect. Despite the genetic background of the family there is every reason to be optimistic. The advice would be the same for a female child.

51. You explain to the young man that AZT, once it is phosphorylated by the cell, acts like the nucleotide thymidine used for DNA synthesis. However, once AZT is incorporated into a newly synthesized DNA strand, no other nucleotides can be added because it lacks a 3'-OH necessary for strand elongation. This prevents the virus from replicating, helping to keep the virus in check. It is effective against the AIDS virus because the AIDS virus DNA polymerase is about 100 times more sensitive to the drug than the cellular enzyme.

52. You could start by explaining to her that diptheria is caused by a bacterium called *Corynebacterium diptheriae* and that the disease is highly contagious. You go on to explain that this bacterium produces a toxin that attacks the protein synthetic machinery of her child's cells leading to cell death. The toxin is so specific that it only targets a eukaryotic elongation factor (EF-2) and ignores the bacterial protein synthetic machinery. Since the toxin is covalently attached to the elongation factor it is impossible to rescue cells once they are affected, thus the importance of having preventive immunity against this deadly microbe.

53. Dexamethasone inhibits pituitary ACTH secretion. This results in a down regulation in the activity of ACTH-response genes. ACTH acts via generation of cAMP which binds to a cAMP response element on transcribable genes. In this case the chimeric gene is ACTH responsive and thus dexamethasone will suppress ACTH secretion which in turn switches off the chimeric gene and thus aldosterone secretion falls. This fall in aldosterone concentration leads to resolution of the clinical abnormalities.

54. In some cases of β-thalassemia these is a mutation of one or other end of an intron at the cervical splice donor or acceptor site. This means that when the ribonuclease enzyme performing the splicing reaches the mutation, it is unable to splice one exon to the next. The result is often a non-functional peptide due to the loss of a group of important amino acids. If the sequence of the exons alone is examined, the intron boundaries will be missed but the mRNA transcribed will be abnormally small.

55. The mechanism of action of steroid hormones is very complex. Although in theory the formation of a steroid-receptor-response element – complex would appear to be at the heart of their mode of action, there are many other steroid related effects which are subject to natural variation. The affinity of the steroid for the receptor, the binding of the steroid-receptor complex, and the effectiveness of transcription can all vary from one person to another and thus the effect of steroids on human disease will vary accordingly.

56. As the majority of deletions causing Duchenne muscular dystrophy occur within a limited number of exons, then PCR-based approaches to the detection of deleted exons would be the first step. The use of multiplex PCR is particularly useful in this case. If such a deletion was detected in the patient then testing of a pregnancy by multiplex PCR or exon-specifi PCR of DNA obtained by chorionic villous sampling would the be investigation of choice.

57. As the NF-II gene is large, it would be impossible to sequence the gene quickly to detect a point mutation or some other small mutation responsible for the condition. It may therefore be easier to employ methods that can detect variation between normal alleles and mutant alleles without necessarily defining the nature of the mutation. This could involve RFLP analysis of the affected man with a normal individual and then comparing his RFLP patterns with his cousin. Similar strategies employing SSCP or DGGE analysis of PCR amplified regtion of the gene are also potentially useful.

58. Fragile X syndrome (FRAXA) is due to the inheritance of an unstable trinucleotide repeat in the FMR-1 gene on the X chromosome. As the trinucleotide expansion increases in size, the ability for it to cause the FRAXA phenotype increases. Small expansions may not give rise to clinical manifestations, but can act as 'pre-mutations' which if inherited can be passed on and result in a larger disease causing expansion. FRAXA can be diagnosed by using Southern blots of digested DNA to show increasing fragment sizes in those individuals with normal, premutant and mutant alleles. In this case the boy has inherited a trinucleotide repeat sufficient to cause FRAXA from his mother. She has a smaller expansion that has caused milder learning difficulties that are also present in her sister. Both these women must have inherited the mutation from a carrier of a 'premutation', either mother or father, whereas their brother has inherited a non-FRAXA X chromosome from the parent with the FRAXA 'premutation'

59. (b).

60. (a).

61. (b).

62. (b).

63. (c).

Short Answers

1. Hydrogen bonds.

2. This is because the protein consists of two subunits with molecular mass of 30,000 Da.

3. 25 mM.

4. Digestion by trypsin yields Gly-Val-Arg, His-Met-Trp-Lys and His-Ala. These three peptides are isolated and then sequenced by Edman degradation. CNBr treatment cleaves the peptides at Met residue into two peptides. Following isolation, they are also subjected to sequencing. N-terminal amino acid sequence is obtained when the peptide is sequenced without cleavage. By these sequence data, you can determine the complete sequence.

5. HbA has a sigmoid curve with n=2.7. The isolated globin monomer has a hyperbolic curve with n =1, identical to that of Mb because of the loss of subunit cooperativity.

6. A mutational loss of a positively charged amino acid side chain in γ-globin.

7. HbS differs from HbA at residue 6 of β-globin: Glu in the normal; Val in the mutant. With the loss of a negative charge on each of the subunits, HbS migrates more slowly toward the positive electrode (anode) than HbA. Homozygotes exhibit a single band representing the $\alpha_2\beta^s_2$ tetramer. A mixture of two tetrameric structures, $\alpha_2\beta_2$ (60% of total) and $\alpha_2\beta^s_2$ (40% of total), is present in heterozygotes; the proportion of the mixed tetramer $\alpha_2\beta\beta^s$ is too slow to be detected readily.

8. Active site of enzymes is composed of catalytic site and substrate binding site. In enzyme reactions, ionizable amino acids, such as histidine and cysteine, often participate in catalytic reaction. Specific serine, histidine, and aspartate residues form a 'catalytic triad' in serine proteases. Chymotrypsin, trypsin, and elastase, three serine proteases, are similar in many respects. However, their substrate specificities are quite different because structures and charges of the binding sites are different from each other.

9. In the case of an inorganic catalyst, the reaction rate increases with the temperature of the system. The tertiary (three-dimensional) structure of a protein involves weak bonds such as hydrogen bonds, which can be disrupted by high temperature. Thus, enzymes function at body temperature 37°C, but also many of them display a temperature optimum at this range.

10. Allosteric enzyme can be converted to ideal form in response to surrounding conditions such as substrates, products, and other regulatory factors. For example, the allosteric enzyme can become a more active form in the presence of positive, allosteric effectors and produce products more efficiently than an isosteric enzyme.

11. Pepsin works in the stomach where the pH is extremely low (1.5–2.0) due to gastric juice, while trypsin exists in the intestinal tract where the pH is about neutral (7–8). Thus these enzymes have ideal characteristics for working at the maximum power where they exist.

12. In methanol poisoning, main, toxic compounds are produced by metabolic conversion of methanol. Both methanol and ethanol are metabolized by alcohol dehydrogenase. This enzyme, however, has much lower Km, i.e. higher affinity, for ethanol than for methanol. Thus, ethanol can competitively inhibit conversion of methanol to toxic metabolites at lower concentration.

13. Histidine is protonated and aspartate is ionized between pH 3 and pH 8 when the enzyme is active.

14. See Figure 7.3. Biomembranes are composed of proteins and lipids. As proposed by Singer & Nicolson, the phospholipid bilayer is a basic structure of biomembrane into which globular proteins are embedded. Polar head groups of phospholipids are located in the membrane surface and their fatty acyl chains face inside the membranes. Variations in the size and degree of unsaturation of fatty acids in the phospholipids affect the fluidity of biomembranes. While membrane lipids and proteins easily move on the membrane surface (lateral diffusion), the 'flip-flop' movement of lipids between outer and inner bilayer leaflets rarely occurs. Membrane proteins are classified into integral transmembrane proteins and peripheral membrane proteins. The former proteins are deeply embedded into lipid bilayer and some of them traverse the membrane several times. Peripheral membrane proteins are bound to the membrane lipids and proteins by electrostatic interaction.

15. See Figure 7.7. In a fashion similar to enzymic reactions, proteins participate in the facilitated diffusion. The kinetics of facilitated diffusion for transport of substrate can be described by the same equation that is used for enzyme catalysis (the Michaelis–Menten equation). Facilitated diffusion is also a saturable process, having a maximum transport rate (V_{max}). When the concentration of extracellular transport substrate becomes very high, the V_{max} value is achieved, as a result of saturation of the protein with substrate. K_m is the dissociation constant of substrate and transporter, and equal to the concentration that gives the half-maximal transport rate.

16. See Figure 7.5. Transport substrates move in the same direction during symport, whereas they move in opposite directions in antiport systems. The movement of one substrate against its concentration gradient is driven by movement of another substrate (usually cations such as Na^+ and H^+) down a gradient. Charged substrates are electrophoretically driven by membrane potential in uniport. These systems are classified as secondary active transport. Uniport of a neutral substrate is classified as facilitated diffusion (passive transport). Examples: uniporter, glucose transporter (GLUT); symporter, Na^+-coupled glucose transporter (SGLT) in kidney and intestine; antiporter, Na^+/Ca^{2+} antiporter in heart muscle and Cl^-/HCO_3^- antiporter in gastric parietal cells.

17. See Figure 7.10. Dietary glucose is transported into intestinal epithelial cells against a concentration gradient. The Na^+-coupled glucose symporter (SGLT) functions in this process. The Na^+ ions symported into the cell are pumped out by an Na^+/K^+-ATPase. The intracellular glucose concentration increases and glucose is transported across the membrane to the blood by facilitated diffusion, using the GLUT family of glucose transporters.

18. See Figure 7.11. The lumen of the stomach is highly acidic (pH-1) because of the presence of a proton pump (H^+/K^+-ATPase) that is specifically expressed in gastric parietal cells. The pump protein is localized intracellularly in vesicles in the resting state. Stimuli such as histamine induce fusion of the vesicles with the plasma membrane. The pump antiports two cytoplasmic hydrogen ions and two extracellular potassium ions, coupled with hydrolysis of a molecule of ATP. Cl^- is secreted through a Cl^- channel, producing HCl (gastric acid) in the lumen. The cytoplasmic K^+ is secreted by a K^+ channel. The CO_2 gas transported by simple diffusion into the parietal cells is hydrated immediately, producing H^+ and HCO_3^-. The HCO_3^- is exchanged with extracytoplasmic Cl^- by a Cl^-/HCO_3^- antiporter. Thus HCl is supplied in the parietal cell.

19. There are two classes of ionophores: mobile ion carriers (or caged carrier) and channel formers. Valinomycin is a typical example of a mobile ion carrier. K^+ binds to the center of the depsipeptide ring structure at the aqueous interface of membrane. The complex diffuses across the hydrophobic membrane central zone to the distal aqueous interface, where the ion is released. Gramicidin A is a linear peptide with 15 amino acid residues. A β-helical gramicidin A molecule forms a pore. The head-to-head dimer formation makes a transmembrane channel that allows movement of monovalent cations (H^+, Na^+, and K^+). (Figure 7.6)

20. Hemoglobin and myoglobin clearly require iron for oxygen transport. Mitochondrial cytochromes also require iron. Iron deficiency could affect all of these components. Both oxygen transport and energy production would be impaired, causing fatigue.

21. Ubiquinone is synthesized from the same isoprenoid intermediates that lead to cholesterol. HMG CoA reductase catalyzes a reaction that leads to these intermediates, so its inhibition will result in less synthesis of both cholesterol and ubiquinone, which functions both as a component of the respiratory chain and as an antioxidant. Its antioxidant function may by important in the prevention of atherosclerosis caused by oxidized LDL.

22. Many of the respiratory chain components have unique absorption spectra and can be identified spectrophotometrically. The absorption spectrum of each also shifts when reduced or oxidized. The point of inhibition is known as the crossover point and can be determined spectrophotometrically by monitoring the absorption spectrum of each respiratory chain component.

23. The low pH of gastric fluid denatures dietary protein. Pepsinogen is secreted by gastric chief cells and is spontaneously and autocatalytically activated to pepsin at acid pH, beginning the digestive process. The peptide digest stimulates cholecystokinin release in the duodenum, which then stimulates secretion of pancreatic enzymes.

24. Carbohydrate digestion begins with salivary amylase, continues with pancreatic amylase, yielding dextrins, which are degraded to glucose by intestinal brush border glucosidases.

25. Glucose is absorbed from the lumen primarily by an active transport process involving cotransport with Na^+. High luminal Na^+ is maintained by a Na^+,K^+-ATPase. Thus glucose uptake occurs by an indirect, active transport process. Glucose moves from the enterocyte into blood by passive transport involving GLUT-5.

26. Amino acids and di- and tri-peptides are absorbed from the gut. Amino acids are absorbed by a family of Na^+ symporters with broad specificity for similar types of amino acids (aromatic, basic, acidic, etc.). Di- and tri-peptides are cotransported with H^+ ions; digestion to amino acids is completed in the enterocyte.

27. Bile salts, with some assistance from dietary lipids and lipid digestion products (fatty acids), disperse fats into small micelles with large surface areas, facilitating degradation by pancreatic lipase. Water-soluble products (fatty acids, 2-monoacylglycerols) are absorbed by intestinal epithelial cells.

28. Fatty acids at the sn-3 position of triglycerides are released by pancreatic lipase. Very long-chain, water-insoluble fatty acids are largely unabsorbed. Long-chain fatty acids are absorbed into the enterocyte, reassembled into tryglycerides and secreted into the lymphatic circulation. Medium- and short-chain fatty acids are absorbed directly into the portal circulation.

29. Vitamin A is a precursor of the visual pigment, rhodopsin, which is essential for function of rod cells. Deficiency of vitamin A leads to poor vision at low light levels (night blindness), and may lead to complete blindness as a result of abnormalities in growth and development of corneal epithelia.

30. Vitamin D is a steroid derivative. Following activation in liver and kidney, it binds to transport proteins in target cells and affects gene expression. Its mechanism of action is identical to that of steroid hormones, in general.

31. Folate and cobalamin are required for 1-carbon transfer reactions in nucleic acid synthesis and methionine metabolism. Deficiencies in these vitamins affects cell division, particularly the production and development of red cell precursors (reticulocytes in the bone marrow) leading to anemia.

32. Vitamins A and E are fat soluble vitamins, providing antioxidant protection in membrane environments. Vitamin C is a water-soluble vitamin and provides protection in aqueous compartments.

33. Zinc status was originally evaluated by measurement of serum zinc concentration. Because many dietary, health, and environmental factors affect serum zinc concentration, measurements of red cell levels of the zinc-containing protein, metallothioneine, have proven more useful for assessing zinc status.

34. Copper is a cofactor for oxygenases and oxidases which use molecular oxygen for metabolism, and for proteins and enzymes involved in antioxidant defenses and scavenging of reactive oxygen species (ceruloplasmin and superoxide dismutase).

35. Hemolytic anemia is a common consequence of defects in glycolytic enzymes, leading to low hematocrit and inadequate oxygen transport.

36. (A) hexokinase + G-6-P dehydrogenase, yielding increase in NADPH. (B) hexokinase, pyruvate kinase, lactate dehydrogenase, yielding decrease in NADH.

37. [1-^{13}C]-glucose is used to assess the activity of the pentose phosphate pathway by measuring the rate of release of $^{13}CO_2$. Release of $^{13}CO_2$ from [6-^{13}C]-glucose provides a measure of the relative rate of other pathways of glucose metabolism.

38. Glucagon response in liver, response to neural stimulation in muscle.

39. Phosphorylase kinase provides an additional site for regulation of glycogenolysis

40. Carbohydrate loading builds up both hepatic and muscle glycogen. Storage of 500 grams of glycogen would provide sufficient energy for 4 hours of intense exercise; however, after a short duration of exercise, fatty acids become the major energy source in muscle.

41. Since one of the major precursors to acetyl-CoA is pyruvate, it would probably accumulate producing excess lactic acid, which could lead to lactic acidosis. TCA cycle defects are very rare, because when they occur, the organism probably dies before the fetal stage and aborts. Since the nervous system generally has the highest metabolic rates in the body, it would probably be severely impaired and cell death would occur, resulting in encephalopathy.

42. Heme would be synthesized at a rapid rate, as well as proteins because of the loss of blood. The TCA cycle activity would be accelerated to produce heme, because the synthesis of heme starts with succinyl CoA supplied by the TCA cycle. A source of carbons would be required to make up for the deficit created by the loss of succinyl CoA. For example, glutamate could undergo trasamination to α-ketoglutarate, an anaplerotic reaction. In addition, the cycle would supply reductive power in the form of reduced nucleotide coenzymes for the production of mitochondrial ATP, which would be needed for biosynthetic reactions.

43. Citric acid increases the net carbons in the cycle and will result in the net synthesis of oxaloacetic acid, as well as other intermediates.

44. Conversion to glycerol phosphate, then dihydroxyacetone phosphate for gluconeogenesis in liver.

45. Absence of carnitine would inhibit oxidation of long-chain fatty acids. Depending on severity, may lead to fasting hypoglycemia, which could be life-threatening. Treatment with dietary carnitine.

46. Oxidation of fatty acids in muscle mitochondria produces acetyl-CoA. Metabolism of acetyl-CoA in the TCA cycle requires oxaloacetate for condensation with acetyl-CoA in the citrate synthase reaction. Carbohydrates produce pyruvate, which is converted to oxaloacetate by pyruvate carboxylase, assuring a constant supply of oxaloacetate for acetyl-CoA metabolism.

47. Fatty acid catabolism supports gluconeogenesis and provides energy in most tissues during fasting and starvation. Because of decreased glycolysis in the liver during these conditions, oxaloacetate concentration is low and acetyl-CoA cannot be efficiently metabolized in the TCA cycle. To recover the acetyl-CoA required for continued β-oxidation of fatty acids, the liver converts acetyl-CoA to ketone bodies, exporting them from liver. The CoA is released for continued β-oxidation.

48. A ketogenic diet involves dietary restriction until ketone bodies begin to appear in urine. The presence of urinary ketone bodies signifies active fat metabolism, which will lead to a gradual weight loss.

49. One mole of glucose yields 36–38 moles ATP, or 0.2 mole ATP per g glucose. One mole of acetoacetate requires investment of 2 equivalents of ATP in the thiokinase reaction, yielding two moles of acetyl-CoA, each of which yields 24 moles ATP. Thus, the energy yield from acetoacetate is about 52 ATP/mol, or about 0.5 mole ATP per g acetoacetate.

50. Review Figures 15.2 and 15.3, discussing the role of the cysteinyl and pantetheinyl sulfhydryl groups, the charging and chain elongation reactions, and the additional enzymatic activities of the fatty acid synthase complex.

51. Acetyl-CoA carboxylase catalyzes the rate limiting step in fatty acid biosynthesis. It is activated by citrate which causes polymerization of inactive protomers into the active polymeric form of the enzyme. It is also inactivated by phosphorylation in response to glucagon, inhibiting lipogenesis during gluconeogenesis.

52. High carbohydrate diets induce a number of glycolytic and lipogenic enzymes which catalyze the conversion of carbohydrates into fatty acids and triglycerides. See AC Box on p 185.

53. Triglycerides are delivered to adipose tissue in the form of chylomicrons or VLDL. The triglycerides are hydrolyzed by lipoprotein lipase, the fatty acids are taken up into adipose tissue and esterified to glycerol 3-phosphate, derived from dihydroxyacetone phosphate. Genetic and environmental (diet, activity) factors determine total fat deposition. The hormone leptin, secreted by adipocytes, suppresses food intake by action on the hypothalamus.

54. Mammalian fatty acid desaturases cannot insert double bonds beyond the Δ^9 carbon. Linoleic acid has a double bond at $\Delta^{9,12}$ (ω-6) and linolenic acid at the $\Delta^{9,12,15}$ (ω-3 & ω-6) positions. These fatty acids are precursors of prostaglandins.

55. Cholesterol enters the cell by a receptor-mediated pathway. An increase in intracellular cholesterol levels leads to a decrease of receptor numbers and a decreased cholesterol uptake. A decrease in intracellular cholesterol (e.g. when HMG CoA reductase is inhibited) leads to an increased receptor number.

56. Cytochrome P450 mono-oxygenase enzyme regulates the two enzymes that remove the cholesterol side chain at C-17 and the aromatase enzyme that converts the A ring of testosterone into the aromatic ring of estradiol with the loss of the C-19 angular methyl group.

57. The enterohepatic circulation allows 98% of bile acids to be reabsorbed from the jejunum, transported via the portal vein and resecreted into the bile duct.

58. In normal subjects about 50% of cholesterol is synthesized *de novo*. In certain groups of subjects there is abnormal cholesterol synthesis and metabolism that can lead to a high serum cholesterol despite dietary restriction.

59. The vast majority of cholesterol in plasma is bound to hydrophilic protein molecules

60. Acetyl CoA accumulation in the mitochondrion, derived from either carbohydrate or fat metabolism, inhibits pyruvate dehydrogenase and stimulates pyruvate carboxylase, providing oxaloacetate for TCA cycle activity.

61. Pyruvate dehydrogenase is inhibited by acetyl CoA and NADH, both allosterically and by activation of pyruvate dehydrogenase kinase, which inhibits the enzyme; this limits carbohydrate utilization during periods of gluconeogenesis. Mitochondrial dehyrogenase activities are dependent on levels of NAD^+, which depends on NADH utilization to regenerate ATP; this links TCA cycle activity to energy requirements of the cell. Isocitrate dehydrogenase is especially sensitive to ATP and NADH; its inhibition during energy-rich conditions shuttles citrate to the cytoplasm for lipid biosynthesis.

62. Review data in Figure 13.9. Aerobic metabolism of glucose yields about 20 times as much ATP as glycolysis.

63. Review Figures 13.5 and 13.6.

64. Review Advanced concept box on page 165.

65. The yeast convert to more efficient oxidative metabolism of glucose, requiring less glucose and therefore evolving less CO_2.

66. Confirm increased adjusted calcium on at least two occasions. Measure PTH 1–84 and if >3pmol/L then the cause is hyperparathyroidism. Measure urea creatinine and electrolytes to exclude renal disease and urinary calcium excretion to exclude familial hypocalciuric hypocalcemia. If renal function normal and urinary calcium elevated, diagnosis is primary hyperparathyroidism. If PTH 1–84 <3pmol/L then malignancy must be excluded by a combination of history, physical examination, X-rays, CT scan, and magnetic resonance imaging where required. PTH-P can be measured if >2.6pmol/L highly suggestive of solid tumor presence. Myeloma screen should be performed (serum and urine electrophoresis for monoclonal gammopathy). Thyroid function tests, vitamin D, growth hormone, angioconverting enzymes and lithium measurement in serum would all be required when rare causes are suspected.

67. As ionized calcium decreases, this stimulates the chief cells of the parathyroid gland to produce PTH (1–84). The PTH circulates in blood and acts at receptors in the kidney promoting calcium reabsorption and phosphate excretion. It also stimulates the kidney conversion of 25-hydroxy cholecalciferol (vitamin D) to 1,25-dihydroxy cholecalciferol which promotes calcium absorption from the gut. PTH acts on osteoblasts/osteoclasts in bone stimulating osteoclast bone resorption which releases calcium and phosphate from the bone matrix. All of these effects act in combination over the short (kidney reabsorption) and longer term (gut absorption, bone resorption) to restore calcium to normal.

68. During the bone remodeling cycle resorption and formation are 'coupled' so that the amount of old bone removed by osteoclasts is equaled by the amount of new bone formed by the osteoblasts. This is achieved by hormones and cytokines released by the two cell types which regulate the cellular activity. Calcium balance is maintained. In osteoporosis osteoclast activity is increased more than osteoblast with the result that a) more bone surfaces are resorbed b) deeper 'pits' are seen in bone c) less filling in with new bone occurs. The osteoclast and osteoblast are now 'uncoupled' and with the passage of time a decrease is observed in the total amount of bone present throughout the skeleton.

69. Galactose is phosphorylated by a specific kinase to give galactose-1-P which is then converted to UDP-galactose by the Gal-1-P:uridyltransferase (the enzyme missing in severe cases of galctosemia). This UDP-galactose is epimerized to UDP-glucose which is then converted to glucose-1-P, then to glucose-6-P and then to fructose-6-P. Fructose is also phosphorylated in liver by a specific kinase to give fructose-1-P which can be cleaved by a specific liver aldolase (aldolase B) to give dihydroxyacetone-phosphose and free glyceraldehyde. The glyceraldehyde can be phosphorylated to give glyceraldehyde-3-P which can enter glycolysis. Fructose utilization is not regulated by factors (ATP, AMP levels, etc.) that control the activity of PFK.

70. N-linked oligosaccharides may be important in helping the protein to assume its proper conformation in the endoplasmic reticulum, or these oligosaccharides may increase the stability of the protein to heat or other denaturants. Specific N-linked oligosaccharides with mannose-6-P residues as part of their structure are involved in targeting hydrolytic enzymes to the lysosomes. Specific oligosaccharide structures are also involved in chemical recognition reactions that allow certain bacteria and viruses to recognize and bind to host target cells. In addition, many cell:cell interactions that occur during development involve recognition of N-linked oligosaccharides.

71. N-linked oligosaccharides are assembled on a lipid carrier whereas O-linked oligosaccharides are not. For the N-linked oligosaccharides, the individual sugars, GlcNAc, mannose, and glucose, are added sequentially to the lipid carrier, dolichol-P, in the endoplasmic reticulum by a series of glycosyltransferases. The final product is a 14 sugar oligosaccharide-PP-dolichol (i.e. $Glc^3Man^9GlcNAc^2$-PP-dolichol). This intact N-linked oligosaccharide is then transferred to specific asparagine residues on the protein. Further trimming of the protein-bound oligosaccharide, to give a variety of different oligosaccharide chains, occurs in the ER and Golgi apparatus. On the other hand, O-linked oligosaccharides are synthesized in the Golgi apparatus by the transfer of GalNAc directly to serine residues of the protein and then sequential transfer to other sugars (GlcNAc, galactose, silaic acid) to GalNAc without the participation of a lipid carrier.

72. Sugars are activated by converting them to the nucleoside diphosphate sugars (such as UDP-glucose, GDP-mannose, etc.). Polymerization, or glycoside bond formation, is an anabolic reaction and requires an input of energy, since the product if more complex than the starting material. Thus, a high energy phosphate is utilized to make the activated form of the sugar, i.e.
UTP + glucose-1-P ➜ UDP-glucose + PPi.

73. I-cell disease is an inborn error of metabolism characterized by inability of afflicted individuals properly to degrade (turnover) macromolecules such as proteoglycans, glycolipids, certain proteins, and so on. This defect results from the inability of these individuals to attach the proper targeting signal to their lysosomal proteins (which are synthesized in the ER and must have a mannose-6-P added to the N-linked oligosaccharides) in order to target these proteins to the lysosomes. Individuals with I -cell disease are missing the GlcNAc-1-P transferase that adds GlcNAc-1-P to high-mannose oligosaccharides to give GlcNAc-1-P mannose-. The GlcNAc is then removed to expose the mannose-6-P residues (possibly another form of I-cell disease will be found where people are missing the GlcNAc trimming enzyme). The mannose-6-P residues are recognized by Golgi receptors that target those proteins to the lysosomes.

74. N-linked oligosaccharides are initially synthesized in the endoplasmic reticulum and then modified in the Golgi, as the protein is transported through the various Golgi cisternae. O-linked oligosaccharides are probably mostly synthesized in the Golgi although some of the early reactions such as the xylosyltransferase (for proteoglycans) and the Gal–NAc transferase may be in the ER.

75. Phosphatidic acid is a precursor for phospholipids and for triglycerides. It is used to make DAG as well as CDP-DAG.

76. PE & PC can be made by condensation of CDP-choline or ethanolamine with DAG, or PC can be made from PE by methylation with SAM. PE can also be made by decarboxylation of PS. PS can be made from PC or PE by exchange of the headgroup (i.e., S for C or S for E).

77. Dipalmitoylphosphatidylcholine (DPPC) is a surfactant that facilitates opening of alveoli during inspiration and prevents collapse of the lungs.

78. Inositol triphosphate (IP_3) and DAG are two signalling molecules that come from PI degradation.

79. Because glycolipids can have many different sugars (and linkages) in their structure and therefore contain much more information that can be used as recognition signals.

80. Glycosphingolipidoses are a group of inborn errors where individuals are missing one of the essential enzymes in their lysosomes that are involved in degradation and turnover of membrane glycosphingolipids.

81. Glycolipids are antigens on the surfaces of red blood cells and are responsible for the ABO blood group types. People with A blood type have a certain glycolipid structure that causes formation of antibodies in individuals with type B or AB blood.

82. Review role of repeating tri-peptide, glycine, hydrogen bonding, disulfide bonds, and quarter-staggered array of collagen molecules (Figures 26.1 and 26.3).

83. Nonhelical segments interrupt fibril associations, promoting formation of nonfibrillar, network-type structures.

84. Consider three-dimensional structure, amino acid composition, post-translational modifications and crosslink structures.

85. Describe fibronectin, laminin, and nidogen, their domain structures and interactions with other components of ECM, and the role of RGD sequences in integrin binding.

86. Consider composition, linkage to protein, repeating disaccharide structure, size, and branching of saccharide chains, and polyanionic character.

87. Long, polyanionic chains form extended structure, imparting strength and compressibility to the ECM, and provide a matrix for interaction of other matrix components and cells.

88. 'Direct' bilirubin is an alternative name for conjugated bilirubin, whilst 'indirect' bilirubin refers to the unconjugated fraction of bilirubin. In patients with hyperbilirubinemia, measurement of both the direct and indirect proportions of bilirubin may help to differentiate between pre-hepatic causes, where the serum bilirubin is predominately unconjugated or indirect, and hepatic/post-hepatic causes, when the serum bilirubin is mainly conjugated or direct.

89. Patients with clinically significant disease of the hepatic parenchyma are unable to detoxify nitrogenous substances, since the functional capacity of the hepatocyte urea cycle is impaired. Amino acids and other amines arising from the gut may escape normal hepatic metabolism, and thereby gain access to the systemic circulation. Certain amines may act as neurotransmitters in the brain and alter cognitive function. Low protein diets reduce the nitrogenous load to be metabolized by the liver.

90. The liver is responsible for metabolizing most drugs, and in general the metabolic transformations increase their water-solubility and thereby their rates of excretion in urine or bile. Diseases of the hepatic parenchyma are likely to impair drug metabolism non-specifically, and increase the elimination half-life of the drug. This in turn may lead to a greater, or more prolonged, therapeutic effect in patients with liver disease. For those drugs which are subject to biliary excretion, cholestatic disease will impair the excretion of drug metabolites, which may in turn give rise to pharmacologic effects in their own right. Drug prescriptions for patients with liver disease should be carefully reviewed before any drug is administered.

91. The two essential questions to be answered in any jaundiced patient are firstly the general cause of the jaundice (pre-hepatic, hepatic or post-hepatic), and secondly of the specific disease process. Biochemical tests to delineate the general cause of jaundice.

- Serum direct/indirect bilirubin: Pre-hepatic causes of jaundice are generally associated with a rise in the serum concentration of unconjugated ('indirect') bilirubin, whilst hepatic and post-hepatic causes lead mainly to a rise in conjugated ('direct') bilirubin and to a lesser extent unconjugated bilirubin.
- Serum transaminases (AST/ALT): Hepatic causes of jaundice lead to a rise in AST and ALT.
- Serum alkaline phosphatase (ALP) and γ glutamyl transferase (GGT): Post-hepatic (obstructive/cholestatic) causes of jaundice lead to a rise in ALP/ GGT.

Biochemical tests may identify the specific disease process. In general, however, such tests may be helpful only in hepatocellular disease.
α–1 antitrypsin: Low serum concentration and phenotyping.
Ceruloplasmin: Low serum concentration (Wilson's disease).
Iron, iron-binding capacity, and ferritin: An increased ratio of iron to iron-binding capacity and/or an increased ferritin are good screening tests for hemochromatosis.
Drugs: following an acetaminophen overdose, the plasma acetaminophen concentration predicts those patients likely to suffer significant liver damage. Screening for alcohol abuse is difficult and patients do not always admit to their drinking habits. In certain circumstances a serum ethanol level may indicate whether a patient is truly abstinent.

92. Neonatal jaundice which occurs between the second and tenth day of life is common and is usually of no concern unless the degree of hyperbilirubinemia is sufficiently severe to require treatment. Jaundice in the first day of life, or which occurs after ten days, is always abnormal. Early jaundice may be caused by hemolysis, and late jaundice by inborn errors of metabolism of structural defects in the biliary tree (biliary atresia).

93. See Figures 28.3, 28.4, 28.6, and 28.7 regarding phosphoribosyltransferases.

94. See Chapter 28, pages 364–365, suicide inhibition of thymidylate synthase.

95. See Chapter 28, page 362. Substrate channeling is important in complex metabolic pathways, especially in lipid metabolism when substrates are not soluble in the aqueous phase. There is some evidence for substrate channeling in glycolysis.

96. See Chapter 28, page 361, deficiency of HGPRTase, accumulation of PRPP, leading to excessive purine biosynthesis and uric acid production.

97. Sequence element that might help to identify genes in this 450 kb sequence would be promoter and enhancer consensus sequences such as the basal promoter sequences (TATA box; TATAAT and TGTTGACA) located 5' to the start of genes. Once these areas have been located it will be possible to search the sequences downstream for protein synthesis start and stop codons and poly A addition sequences.

98. The rho protein is involved in transcription termination. It has an ATP-dependent helicase activity. This activity is required because the rho protein must travel along the newly synthesized RNA, chasing the RNA polymerase. When the rho protein catches up with the RNA polymerase it displaces it from the DNA template terminating transcription.

99. Since your goal would be to inhibit bacterial protein synthesis and have no effect on eukaryotic cells so you could use it to fight infections, the most likely candidates would be those enzymes, factors, or structures that are different between the two types of cells. For example, since bacteria typically use a formylated methionine to initiate protein synthesis while eukaryotic cells do not this would be a good target for an antibacterial drug. Another likely target would be the bacterial ribosome. Even though the ribosome in both cell types perform similar functions it is possible to target bacterial ribosomes because their constituents are different from eukaryotic cells.

100. The most likely target of the drug would be the elongation factors responsible for bringing the newly charged tRNA molecules to the protein synthetic machinery. Specific targets of the drug could be either the elongation factors which bind to the tRNA molecules or the factors that allow these molecules to be recharged with GTP.

101. The key features of a transcription unit are the gene template and the mechanism for converting this message into a peptide gene product. This machinery comprises RNA polymerase, RNA Pol II, and other proteins that allow for the correct alignment of the polymerase in relation to the start point of the gene. These are associated factors such as promoters and enhancers that confer tissue and cell phase specificity to the transcription process and determine the efficiency of the process.

102. Key features in the design of this gene would include: a tissue specific promoter to ensure expression in hair follicles, a sex-specific response element (perhaps via estrogen or androgen response elements) that would allow for male pattern baldness to occur, another time-sensitive promoter to allow graying of the hair with age, and even an enhancer that promotes red hair growth in response to bad tempers.

103. There are numerous ways to alter gene transcription - See table 32.1.

104. Yes. Genomic imprinting is a mechanism whereby the effect of a gene or one allele of a gene can be altered depending on the sex of the parent donating the gene. This means that the same gene or mutant allele can affect offspring differently depending on which parent donated the mutant. Examples of this include neonatal myotonic dystrophy (only when the mutant is donated from the mother) and Huntington's disease (where cases inherited from affected fathers have symptoms and signs at an earlier age, including childhood cases).

105. PCR amplification allows for rapid, high volume amplification of potentially very small amounts of template DNA. In addition, it can be used to amplify poor quality DNA template and DNA from other species that may share common ancestral genes. On the down side, the amplification may be inaccurate and amplification relies on knowledge of the nucleotide sequence of the neighboring DNA. In contrast, cell-based cloning is relatively slow and labor intensive. It requires the use of *in vitro* systems and the efficiency of transformation of the replicating cells is poor. However, cell-based cloning can provide large yields of pure target DNA with no sequence errors introduced. In addition, the size of the target DNA is not limited as is the case with PCR.

106. If the sequence of the gene is known then a PCR amplification reaction for each exon could be designed. Ideally the region of primer hybridization would be in the flanking intronic sequences and the resulting PCR products of different sizes. In cases where exons are of equal sizes, an attempt to design primers which allow amplification of a different sized product for each exon based on the flanking intronic sequence should be used. Such PCR products could either be sequenced directly or if of small enough size subjected to methods such as SSCP or DGGE. Alternatively, RNA could be extracted from white blood cells for instance and a reverse transcriptase reaction performed. The resulting cDNA could then be amplified by PCR and then sequenced directly.

107. The ideal PCR reaction would contain a pure template and primers that hybridized only to the regions flanking the target DNA. The polymerase enzyme would accurately and repeatedly transcribe the target DNA to provide an end-product, which consisted of large amounts of target DNA faithfully copied without any sequence errors. In addition, the reaction conditions would be constant throughout the numerous cycles and each subsequent step in the cycle would occur instantaneously. In reality, Taq I introduces a high number of sequence errors in the course of a standard PCR and the template and primers are often far from ideal. The need to change temperatures from 95°C for denaturation to 45°C for annealing and then to 72°C for extension means that each step requires time for heating and cooling which alters the efficiency of each subsequent step in the cycle.

108. Two methods are routinely used for prenatal diagnosis:
Amniocentesis – sampling of amniotic fluid at 16–18 weeks' gestation and culturing the fetally derived amniocytes. Culture of these cells takes 2–3 weeks but often gives rise to potentially large amounts of DNA for analysis.
Chorionic villous sampling (CVS) – sampling of developing chorionic villi at 9–11 weeks' gestation. Sampled tissue is then used to prepare DNA. This method produces DNA for analysis quickly, but in small amounts.
Methods that require small amounts of DNA such as PCR are therefore suited to CVS as results can be obtained very quickly.
Examples of this include detection of single gene disorders such as cystic fibrosis or sickle cell disease. However, if the method requires a large amount of DNA, e.g. a Southern blot based method, then CVS is not practical and amniocentesis is preferred. The drawback of this method is therefore the detection of mutations at 16–20 or more weeks' gestation rather than 9–12.

109. There are myriad reasons for PCR failure (and numerous books have been written to explain it!). However, simple things such as incorrect primer design, poor template preparation, incorrect timing and temperature settings for the reaction cycles, poor reaction mixture preparation, and sheer bad luck are among the most common reasons for PCR failure.

110. Essential features for microsatellite paternity testing include:
Accurate documentation and sample collection from tested parties.
Choosing microsatellites which display a large number of alleles.
Choosing microsatellites which have alleles with equal frequencies, i.e. there is no excessive bias toward one allele thus reducing its discriminatory power.
Choosing microsatellites which are easily typed and not prone to typing errors which would invalidate their usefulness.
Using a large enough battery of microsatellite markers to ensure the probability of individuals possessing similar genotypes by chance is sufficiently low.

111. GH is secreted in bursts with greatest activity during the night. The GH concentration at the peak of a burst of secretion may be 100 fold greater than the basal level. Therefore, single GH results are of no diagnostic value. Dynamic tests of GH secretory reserve are required.

112. Autoantibodies are produced which bind to the TSH receptor on the surface of the thyroid cell. However, binding by the autoantibody does not trigger the intracellular sequence of events evoked by binding of TSH. Thus, the autoantibody and TSH compete for available receptor binding sites. As the autoimmune condition progresses so more and more receptor sites are occupied by autoantibody and hypothyroidism results.

113. (a) Prolactin is under inhibitory control from hypothalamic dopamine. All other pituitary hormones are under predominantly stimulatory control from the hypothalamus.
b) Prolactin is not a trophic hormone – in other words its action does not rely on stimulating the release of another hormone
c) There is no known prolactin deficiency syndrome in normal adults

114. Steroid 21-hydroxylase stimulates the conversion of 17-hydroxy progesterone into deoxycorticosterone in the biosynthetic pathway to cortisol. If the enzyme is deficient there is a build up of 17-hydroxyprogesterone which is converted by side chain cleavage into androgens. This effect is exaggerated because the lack of cortisol feedback results in increased ACTH secretion from the pituitary which causes even more 17-hydroxyprogesterone to be synthesized.

115. FSH is a two chain polypeptide of molecular weight 28 kDa. As such it is big enough to contain many epitopes (antigen sites) and it is easy to obtain antibodies that will bind to two different epitopes on the same FSH molecule. Estradiol is a small molecule, a steroid hormone, which will only fit into one antibody binding site at any one time.

116. Plasma binding proteins provide a subtle control mechanism for the delivery of the hormone to its target cell. The bound hormone acts as a reservoir and there is sufficient excess binding capacity to prevent uncontrolled surges of hormone delivery to the target cell. The free (unbound) hormone is the biologically active form; this is in equilibrium with the bound hormone. Although plasma binding proteins most commonly exist for small molecules (e.g. thyroxine, cortisol) they are also found for polypeptide hormones (e.g. growth hormone).

A	adenine
ACE	angiotensin-converting enzyme
acetyl CoA	acetyl-coenzyme A
ACh	acetylcholine
ACTase	aspartate carbamoyl transferase
ACTH	adrenocorticotropic hormone
ADC	AIDS–dementia complex
ADH	alcohol dehydrogenase
ADH	antidiuretic hormone (same as AVP)
AFP	α fetoprotein
AGE	advanced glycation end-product
AHF	antihemophilic factor
AICAR	5-aminoimidazole-4-carboxamide ribonucleotide
AIDS	acquired immunodeficiency syndrome
AIR	5-aminoimidazole ribonucleotide
ALDH	aldehyde dehydrogenase
ALP	alkaline phosphatase
ALT	alanine aminotransferase
AML	acute myeloblastic leukemia
AMP	adenosine monophosphate
APC	adenomatous polyposis coli (gene)
APO-1	'death domain'-containing receptor
apoB	apolipoprotein B
APRT	adenosine phosphoribosyl transferase
APTT	activated partial thromboplastin time
AST	aspartate aminotransferase
ATF	activation transcription factor
ATM	ataxia telangiectasia-mutated gene
ATP	adenosine triphosphate
AVP	arginine-vasopressin (same as antidiuretic hormone)
AZT	3'-azido-3'-deoxythymidine
2,3-BPG	2,3-biphosphoglycerate
Bcl-2	B cell lymphoma protein 2
bp	base pair
BUN	blood urea nitrogen (equivalent of blood urea)
bw	body weight
C	cytosine
CA	carbonic anhydrase
CAD	caspase-dependent endonuclease
CAIR	carboxyaminoimidazole ribonucleotide
cAMP	cyclic AMP
CD	cluster designation: classiification system for cell surface molecules
CDGS	carbohydrate-deficient glycoprotein syndromes
CDK	cyclin-dependent kinase
CDKI	cyclin-dependent kinase inhibitor
CDP	cytidine diphosphate
CFTR	cystic fibrosis transmembrane conductance regulator
cGMP	cyclic GMP
CITP	carboxy-terminal procollagen extension peptide
CML	chronic myeloid leukemia
CMP	cytidine monophosphate
CNS	central nervous system
COAD	chronic obstructive airways disease (see COPD)
COMT	catecholamine-o-methyl transferase
COPD	chronic obstructive pulmonary disease (same as COAD)
CoQ_{10}	coenzyme Q_{10} (ubiquinone)
COX-1, -2	cyclo-oxygenase-1, -2
CPK	creatine phosphokinase
CPS I, II	carbamoyl phosphate synthetase I, II
CPT(-I, -II)	carnitine palmitoyl transferase(-I, -II)
CREB	cAMP-responsive-element-binding protein
CRGP	calcitonin-related gene peptide
CRH	corticotropin-releasing hormone
CRP	C-reactive protein
CSF	cerebrospinal fluid
CT	calcitonin
CTP	cytidine triphosphate
CVS	chorionic villous sampling
DAG	diacyl glycerol
DCC	'delete in colon carcinoma' gene
DEAE	diethylaminoethyl
DGGE	denaturant-gradient gel electrophoresis
DHAP	dihydroxyacetone phosphate
DIC	disseminated intravascular coagulation
DIPF	diisopropylphosphofluoride
DNA	deoxyribonucleic acid
DNP	2,4-dinitrophenol
Dol-P	dolichol phosphate
Dol-PP- GlcNAc	dolichol pyrophosphate-acetylglucosamine
DOPA	dihydroxyphenylalanine
DPPC	dipalmitoylphosphatidylcholine
EBV	Epstein–Barr virus
ECF	extracellular fluid
ECM	extracellular matrix
EDRF	endothelium-derived relaxing factor
EDTA	ethylenediamine tetra-acetic acid
EF-1	elongation factor-1
EF2	eukaryotic elongation factor
EGF	epidermal growth factor
eIF-3	eukaryotic cell initiation factor(-3)
ELK	transcription factor
ER	endoplasmic reticulum
ERK	extracellular regulated kinase
ESR	erythrocyte sedimentation rate
F-1,6-BP	fructose-1,6 biphosphate
F-2,6-BP	fructose-2,6 biphosphate
F-2,6-BPAse-1	fructose-2,6 biphosphatase-1
F-6-P	fructose-6-phosphate
F-ATPase	coupling-factor-type ATPase
FACIT	fibril-associated collagen with interrupted triple helices
FAD	flavin adenine dinucleotide
FADD	a 'death domain' accesory protein
$FADH_2$	reduced flavin adenine dinucleotide
FAICAR	5-formylaminoimidazole-4-carboxamide ribonucleotide
FAP	familial adenomatous polyposis
Fas	apoptosis signaling molecule: a 'death domain' accessory protein (CD95)
FBPase-2	fructose biphosphatase-2
FCγR	group of immunogloulin receptors
FDP	fibrin degradation product
FGAR	formylglycinamide ribonucleotide
FGF	fibroblast growth factor
FHH	familial hypocalciuric hypercalcemia
FMN	flavin mononucleotide
$FMNH_2$	reduced flavin mononucleotide
FRAXA	fragile X syndrome
FSF	fibrin-stabilizing factor
FSH	follicle stimulating hormone
G	guanine
G-1-P	glucose-1-phosphate
G-6-P	glucose-6-phosphate
G-6-Pase	glucose-6-phosphatase
G3PDH	glyceraldehyde-3-phosphate dehydrogenase
GABA	γ-amino butyric acid
GAG	glycosaminoglycan
Gal	galactose
Gal-1-P	galactose-1-phosphate
GalNAc	acetylgalactosamine
$GalNH_2$	galactosamine
GAP	guanosine triphosphatase activating protein
GAR	glycinamide ribonucleotide
GDH	glutamate dehydrogenase
GDP	guanosine diphosphate
GDP-D-Man	guanosine diphosphate-D-mannose
GDP-L-Fuc	guanosine diphosphate-L-fucose
GDP-Man	guanosine diphosphate-mannose

Abbreviations

GFAP	glial fibrillary acid protein
gGT	g-glutamyl transferase
GH	growth hormone
GHRH	growth hormone-releasing hormone
GIP	glucose-dependent insulinoptropic peptide
GIT	gastrointestinal tract
GK	glucokinase
Glc	glucose
GlcN-6-P	glucosamine-6-phosphate
GlcNAc	acetylglucosamine
GlcNAc-1P	acetylglucosamine-1-phosphate
GlcNAc-6-P	N-acetylglucosamine-6-phosphate
$GlcNH_2$	glucosamine
GlcUA	D-glucuronic acid
GLP-1	glucagon-like peptide-1
GLUT-1 etc	glucose transporters
GM1	monosialoganglioside 1
GMP	guanosine monophosphate
GnRH	gonadotropin releasing hormone
GP1b-IXa (etc)	glycoprotein receptor 1b-IXa (etc)
GRE	glucocorticoid response element
GSH	reduced glutathione
GSSG	oxidized glutathione
GTP	guanosine triphosphate
GTPase	guanosine triphosphatase
5-HIAA	5-hydroxyindoleacetic acid
5-HT	5-hydroxytryptamine
Hb	hemoglobin
HbA_{1c}	glycated hemoglobin
HbF	fetal hemoglobin
HCM	hypercalcemia associated with malignancy
Hct	hematocrit
HGF-R	hepatocyte growth factor receptor
HGPRT	hypoxanthine-guanine phosphoribosyl transferase
HIV	human immunodeficiency virus
HLA	human leukocyte antigen (system)
HLH	helix–loop–helix (motif)
HMG	hydroxymethylglutaryl
HMWK	high-molecular-weight kininogen
HNPCC	hereditary nonpolyposis colorectal cancer
hnRNA	heteronuclear ribonucleic acid
HPLC	high pressure liquid chromatography
HPT	hyperparathyroidism
HRT	hormone replacement therapy
HTH	helix–turn–helix (motif)
ICF	intracellular fluid
IDDM	insulin-dependent diabetes mellitus
IdUA	L-iduronic acid
IFN(-γ)	interferon(-γ)
Ig	immunoglobulin (etc)
IGF-I	insulin-like growth factor-I (etc)
IL	interleukin (etc)
IMP	inosine monophosphate
Inr	initiator (nucleotide sequence of a gene)
IP_1 I-1-P_1, I-4-P_1 (etc)	inositol monophosphate
IP_2, I-1, 3-P_2, I-1, 4-P_2	inositol biphosphate
IP_3, I-1,4,5-P_3	inositol triphosphate
IP_4, I-1,3,4,5-P_4	inositol tetraphosphate
IRE	iron response element
IRF-BP	IRE-binding protein
IRMA	immunoradiometric assay
ITAM	immunoreceptor tyrosine activation motif
ITIM	immunoreceptor tyrosine inhibition motif
JAK	Janus kinase
JNK	Jun N-terminal kinase
K	equilibrium constant
kb	kilobase
kbp	kilobase pair(s)
KCCT	kaolin–cephalin clotting time
KIP2	cell cycle regulatory molecule
K_m	dissociation constant
LACI	lipoprotein-associated coagulation inhibitor
LDH	lactate dehydrogenase
LDL	low-density lipoprotein
LH	luteinizing hormone
LPL	lipoprotein lipase
LPS	lipopolysaccharide
M-CSF-R	macrophage colony stimulating factor receptor
malonyl CoA	malonyl CoA
Man	mannose
Man-1-P	mannose-1-phosphate
Man-6-P	mannose-6-phosphate
MAO	monoamine oxidase
MAPK	a superfamily of signal-transducing kinase
Mb	myoglobin
MCH	mean corpuscular hemoglobin
MCHC	mean corpuscular hemoglobin concentration
MCV	mean corpuscular volume
MEK	mitogen-activated protein kinase kinase
MEN IIA	multiple endocrine neoplasia type IIA
Met	methionine
met-tRNA	methionyl-tRNA
MGUS	monoclonal gammopathy of uncertain significance
MHC	major histocompatibility complex
mRNA	messenger ribonucleic acid
MRP	multidrug-resistance-associated protein
MS	mass spectrometry
MSH	melanocyte-stimulating hormone
MSUD	maple syrup urine disease
MyoD	muscle-cell-specific transporter factor
Na^+/K^+- ATPase	sodium–potassium ATPase
NABQI	N-acetyl benzoquinoneimine
NAC	N-acetyl cysteine
NAD^+	nicotinamide adenine dinucleotide (oxidized)
NADH	nicotinamide adenine dinucleotide (reduced)
$NADP^+$	nicotinamide adenine dinucleotide phosphate (oxidized)
NADPH	nicotinamide adenine dinucleotide phosphate (reduced)
NANA	N-acetyl neuraminic acid (sialic acid)
NF	nuclear factor
NF-II	type II neurofibromatosis
NGF	nerve growth factor
NIDDM	noninsulin-dependent diabetes mellitus
NMDA	N-methyl-D-aspartate
NPY	neuropeptide Y
NSAID	nonsteroidal anti-inflammatory drug
nt	nucleotide (as measure of size/length)
$1,25(OH)_2D_3$	1,25-dihydroxy vitamin D_3
8-oxo-Gua	8-oxo-2′-deoxyguanosine
OGTT	oral glucose tolerance test
P-ATPase	phosphorylation-type ATPase
3-PG	3-phosphoglycerate
p38RK	p38-reactivating kinase
Pa	Pascal
PA	phosphatidic acid
PAF	platelet activating factor
PAGE	polyacrylamide gel electrophoresis
PAI-1	plasminogen activator inhibitor type 1
PAPS	phosphoadenosine phosphosulfate
PC	phosphatidyl choline
PC	pyruvate carboxylase
PCP	phencyclidine
PCR	polymerase chain reaction
PDE	phosphodiesterase
PDGF	platelet-derived growth factor
PDH	pyruvate dehydrogenase
PDK	phosphatidylinositol triphosphate-dependent kinase
PE	phosphatidyl ethanolamine
PEP	phosphoenolpyruvic acid
PEPCK	phosphoenolpyruvate carboxykinase
PF3	platelet factor 3
PFK-1 (-2)	phosphofructokinase-1 (-2)

PGG_2	prostaglandin G_2 (etc)
PGK	phosphoglycerate kinase
PGM	phosphoglucomutase
PHHI	persistent hyperinsulinemic hypoglycemia of infancy
PHP	pseudohypoparathyroidism
Pi	inorganic phosphate
PI-3-K	phosphoinositide-3-kinase
PICP	amino-terminal procollagen extension peptide
PIP_2/PIP_3	phosphatidylinositol biphosphate/ triphosphate
PK	pyruvate kinase
PKA/PKC	protein kinase A/C
PKU	phenylketonuria
PLA/PLC	phospholipase A/C
PLC-γ	phospholipase C-γ
PMA	phorbol myristic acetate
PNS	peripheral nervous system
PPi	inorganic pyrophosphate
PRL	prolactin
PrP	prion-protein
PRPP	5-phosphoribosyl-α-pyrophosphate
PS	phosphatidyl serine
PTA	plasma thromboplastin antecedent
PTH	parathyroid hormone
PTHrP	parathyroid hormone-related protein
PTK	protein tyrosine kinase
PTPase	phosphotyrosine phosphatase
Py	pyrimidine base (in a nucleotide sequence)
R	receptor (with qualifier, not alone)
RAIDD	a 'death domain' accessory protein
Rb	retinoblastoma protein
RBC	red blood cell
RDS	respiratory distress syndrome
RER	rough endoplasmic reticulum
RIP	a 'death domain' accessory protein
RKK	p38RK homologue of MEK
RNA	ribonucleic acid
RNAPol II	RNA polymerase II
RNR	ribonucleotide reductase
ROS	reactive oxygen species
S	Svedberg
SACAIR	5-aminoimidazole-4-(N-succinylocarboxamide) ribonucleotide
SAM	S-adenosyl methionine
SAP1	transcription factor
SAPK	a stress-activated protein kinase
SCIDS	severe combined immunodeficiency syndrome
scuPA	single-chain uPA
SDS	sodium dodecyl sulfate
SDS-PAGE	sodium dodecyl sulfate-polyacrylamide gel electrophoresis
SGLT	Na^+-coupled glucose symporter
SEK	SAPK homologue of MEK
ser-P	serine phosphate

SH (figs only)	steroid hormone
SH2	Src-homology region-2
SIADH	syndrome of inappropriate antidiuretic hormone secretion
snRNA	small nuclear RNA
SPCA	serum prothrombin conversion accelerator
Src	a protein tyrosine kinases
SRE	steroid response element
SRP	signal recognition particle
SSCP	single-strand conformational polymorphism
SSRI	selective serotonin reuptake inhibitor
STAT	signal transducer and activator of transcription
SUR	sulfonylurea receptor
T	thymine
T tubule	transverse tubule
T_3	tri-iodothyronine
T_4	thyroxine
TAG	triacylglycerol (triglyceride)
TAP	transporters associated with antigen presentation
TBG	thyroid-binding globulin
TCA	tricarboxylic acid cycle
tcuPA	two-chain uPA
TF	transcription factor (with qualifier)
TFPI	tissue factor pathway inhibitor
TGF(-β)	transforming growth factor(-β)
T_{max}	renal transport maximum
TNF	tumor necrosis factor
TNF-R	tumor necrosis factor receptor
tPA	tissue-type plasminogen activator
TRADD	a 'death domain' accessory protein
TRAFS	a 'death domain' accessory protein
TRH	thyrotrophin releasing hormone (thyotropin)
TSH	thyroid stimulating hormone
TTP	thymidine triphosphate
TXA_2	thromboxane A_2
U	uridine (base pairing etc, not in general)
UCP	uncoupling protein
UDP	uridine diphosphate
UDP-Gal	UDP-galactose
UDP-GalNAc	UDP-acetylgalactosamine
UDP-Glc	UDP-glucose
UDP-GlcNAc	UDP-acetylglucosamine
UDP-GlcUA	UDP glucuronic acid
UMP	uridine monophosphate
uPA	urinary-type plasminogen activator
UV	ultraviolet
VDCC	voltage-dependent Ca^{2+} channel
VIP	vasoactive intestinal peptide
VLDL	very-low-density lipoprotein
vWF	von Willebrand factor
WAF1	cell cycle regulator
X-SCID	X-linked severe combined immunodeficiency
XMP	xanthine monophosphate
XO	xanthine oxidase
ZP3	zona pellucida 3 glycoprotein

INDEX

INDEX

INDEX

T